Wadge Degrees and Projective Ordinals: The Cabal Seminar, Volume II

The proceedings of the Los Angeles Caltech-UCLA "Cabal Seminar" were originally published in the 1970s and 1980s. *Wadge Degrees and Projective Ordinals* is the second of a series of four books collecting the seminal papers from the original volumes together with extensive unpublished material, new papers on related topics, and discussion of research developments since the publication of the original volumes.

Focusing on the subjects of "Wadge Degrees and Pointclasses" (Part III) and "Projective Ordinals" (Part IV), each of the two sections is preceded by an introductory survey putting the papers into present context. These four volumes will be a necessary part of the book collection of every set theorist.

ALEXANDER S. KECHRIS is Professor of Mathematics at the California Institute of Technology. He is the recipient of numerous honors, including the J. S. Guggenheim Memorial Foundation Fellowship and the Carol Karp Prize of the Association for Symbolic Logic. He is also a Member of the Scientific Research Board of the American Institute of Mathematics.

BENEDIKT LÖWE is Universitair Docent in Logic in the Institute for Logic, Language and Computation at the Universiteit van Amsterdam, and Professor of Mathematics at the Universität Hamburg. He is the Vice-president of the Deutsche Vereinigung für Mathematische Logik und für Grundlagenforschung der Exakten Wissenschaften (DVMLG), and a Managing Editor of the journal *Mathematical Logic Quarterly*.

JOHN R. STEEL is Professor of Mathematics at the University of California, Berkeley. Prior to that, he was a professor in the mathematics department at UCLA. He is a recipient of the Carol Karp Prize of the Association for Symbolic Logic and of a Humboldt Prize. Steel is also a former Fellow at the Wissenschaftskolleg zu Berlin and at the Sloan Foundation.

LECTURE NOTES IN LOGIC

A Publication for The Association for Symbolic Logic

This series serves researchers, teachers, and students in the field of symbolic logic, broadly interpreted. The aim of the series is to bring publications to the logic community with the least possible delay and to provide rapid dissemination of the latest research. Scientific quality is the overriding criterion by which submissions are evaluated.

More information, including a list of the books in the series, can be found at http://www.aslonline.org/books-lnl.html.

LECTURE NOTES IN LOGIC 37

Wadge Degrees and Projective Ordinals: The Cabal Seminar, Volume II

Edited by

ALEXANDER S. KECHRIS

California Institute of Technology

BENEDIKT LÖWE

Universiteit van Amsterdam and *Universität Hamburg*

JOHN R. STEEL

University of California, Berkeley

ASSOCIATION FOR SYMBOLIC LOGIC

CAMBRIDGE UNIVERSITY PRESS

CAMBRIDGE UNIVERSITY PRESS
Cambridge, New York, Melbourne, Madrid, Cape Town,
Singapore, São Paulo, Delhi, Tokyo, Mexico City

Cambridge University Press
The Edinburgh Building, Cambridge CB2 8RU, UK

Published in the United States of America by Cambridge University Press, New York

www.cambridge.org
Information on this title: www.cambridge.org/9780521762038

Association for Symbolic Logic
Richard Shore, Publisher
Department of Mathematics, Cornell University, Ithaca, NY 14853
http://www.aslonline.org

First published 2012

Printed in the United Kingdom at the University Press, Cambridge

A catalogue record for this publication is available from the British Library

Library of Congress Cataloguing in Publication data

ISBN 978-0-521-76203-8 Hardback

CONTENTS

PART IV: PROJECTIVE ORDINALS

PREFACE

This book continues the series of volumes containing reprints of the papers in the original Cabal Seminar volumes of the Springer *Lecture Notes in Mathematics* series [CABAL i, CABAL ii, CABAL iii, CABAL iv], unpublished material, and new papers. The first volume, [CABAL I], contained papers on games, scales and Suslin cardinals. In this volume, we continue with Parts III and IV of the project: *Wadge degrees and pointclasses* and *Projective ordinals*. As in our first volume, each of the parts contains an introductory survey (written by Alessandro Andretta and Alain Louveau for Part III and by Steve Jackson for Part IV) putting the papers into a present-day context.

In addition to the reprinted papers, this volume contains papers by Steel (*More measures from* AD) and Martin (*Projective sets and cardinal numbers*) that date back to the period of the original Cabal publications but were not included in the old volumes. Jackson contributed a new paper *Regular cardinals without the weak partition property* with recent results that fit well with the topic of Part IV. The paper *Early investigations of the degrees of Borel sets* by Wadge is a historical overview of the process of the development of the basic theory of the Wadge degrees. Table 1 gives an overview of the papers in this volume with their original references.

As emphasized in our first volume, our project is not to be understood as a historical edition of old papers. In the retyping process, we uniformized and modernized notation and numbering of sections and theorems. As a consequence, references to papers in the old Cabal volumes will not always agree with references to their reprinted versions. In this volume, references to papers that already appeared in reprinted form will use the new numbering. In order to help the reader to easily cross-reference old and new numberings, we provide a list of changes after the preface.

The typing and design were partially funded by the *Marie Curie Research Training Site* GLoRiClass (MEST-CT-2005-020841) of the European Commission. Infrastructure was provided by the Institute for Logic, Language and Computation (ILLC) of the *Universiteit van Amsterdam*. Many people were involved in typing, laying out, and proofreading the papers. We

	PART III	
Andretta, Louveau	*Wadge degrees and pointclasses* *Introduction to Part III*	NEW
Van Wesep	*Wadge degrees and descriptive set theory*	[CABAL i, pp. 151–170]
Kechris	*A note on Wadge degrees*	[CABAL ii, pp. 165–168]
Louveau	*Some results in the Wadge hierarchy of Borel sets*	[CABAL iii, pp. 28–55]
Louveau, Saint-Raymond	*The strength of Borel Wadge determinacy*	[CABAL iv, pp. 1–30]
Steel	*Closure properties of pointclasses*	[CABAL ii, pp. 147–163]
Kechris, Solovay, Steel	*The axiom of determinacy and the prewellordering property*	[CABAL ii, pp. 101–125]
Jackson, Martin	*Pointclasses and wellordered unions*	[CABAL iii, pp. 56–66]
Becker	*More closure properties of pointclasses*	[CABAL iv, pp. 31–36]
Steel	*More measures from* AD	NEW
Wadge	*Early investigations of the degrees of Borel sets*	NEW
	PART IV	
Jackson	*Projective ordinals* *Introduction to Part IV*	NEW
Kechris	*Homogeneous trees and projective scales*	[CABAL ii, pp. 33–73]
Kechris	AD *and projective ordinals*	[CABAL i, pp. 91–132]
Solovay	*A* Δ^1_3 *coding of the subsets of* ω_ω	[CABAL i, pp. 133–150]
Jackson	AD *and the projective ordinals*	[CABAL iv, pp. 117–220]
Martin	*Projective sets and cardinal numbers: some questions related to the continuum problem*	NEW
Jackson	*Regular cardinals without the weak partition property*	NEW

TABLE 1.

should like to thank (in alphabetic order) Can Baskent, Hanne Berg, Pablo Cubides Kovacsics, Jined Elpitiya, Thomas Göbel, Leona Kershaw, Anston Klev, Alexandru Marcoci, Kian Mintz-Woo, Antonio Negro, Maurice Pico de los Cobos, Sudeep Regmi, Cesar Sainz de Vicuña, Stephan Schroevers, Sam van Gool, and Daniel Velkov for their important contribution as typists and diligent proofreaders. We should like to mention that the original

LaTeX stylefile for the retyping was designed by Dr. Samson de Jager. Very special thanks are due to Dr. Joel Uckelman, who took over the typesetting coordination from de Jager in 2007.

REFERENCES

ALEXANDER S. KECHRIS, BENEDIKT LÖWE, AND JOHN R. STEEL

[CABAL I] *Games, scales, and Suslin cardinals: the Cabal seminar, volume I*, Lecture Notes in Logic, vol. 31, Cambridge University Press, 2008.

ALEXANDER S. KECHRIS, DONALD A. MARTIN, AND YIANNIS N. MOSCHOVAKIS

[CABAL ii] *Cabal seminar* 77–79, Lecture Notes in Mathematics, no. 839, Berlin, Springer, 1981.

[CABAL iii] *Cabal seminar* 79–81, Lecture Notes in Mathematics, no. 1019, Berlin, Springer, 1983.

ALEXANDER S. KECHRIS, DONALD A. MARTIN, AND JOHN R. STEEL

[CABAL iv] *Cabal seminar* 81–85, Lecture Notes in Mathematics, no. 1333, Berlin, Springer, 1988.

ALEXANDER S. KECHRIS AND YIANNIS N. MOSCHOVAKIS

[CABAL i] *Cabal seminar* 76–77, Lecture Notes in Mathematics, no. 689, Berlin, Springer, 1978.

The Editors
Alexander S. Kechris, *Pasadena, CA*
Benedikt Löwe, *Amsterdam*
John R. Steel, *Berkeley, CA*

ORIGINAL NUMBERING

Numbering in the reprints may differ from the original numbering. Where numbering differs, the original designation is listed on the left, with the corresponding number in the reprint listed on the right. In rare cases where an item numbered in the reprint had neither a number nor a name in the original, we have indicated that with a '—'.

Volume I

Notes on the theory of scales, Kechris & Moschovakis, [CABAL i, pp. 1–53]

§2A	§2.1
Theorem 2A-1	Theorem 2.1
Proposition 2A-2	Proposition 2.2
§2B	§2.2
Theorem 2B-1	Theorem 2.3
Theorem 2B-2	Theorem 2.4
Corollary	Corollary 2.5
Corollary	Corollary 2.6
Theorem 2B-3	Theorem 2.7
Corollary	Corollary 2.8
Corollary	Corollary 2.9
§2C	§2.3
Theorem 2C-1	Theorem 2.10
Corollary	Corollary 2.11
1.	Claim 2.12
2.	Claim 2.13
3.	Claim 2.14
4.	Claim 2.15
§2D	§2.4
§3A	§3.1
Theorem 3A-1	Theorem 3.1
§3B	§3.2
Theorem 3B-1	Theorem 3.2
Corollary	Corollary 3.3
1.	Claim 3.4
2.	Claim 3.5
Theorem 3B-2	Theorem 3.6
Corollary	Corollary 3.7
Corollary	Corollary 3.8
1.	Claim 3.9
2.	Claim 3.10
Theorem 3B-3	Theorem 3.11
Corollary	Corollary 3.12
Corollary	Corollary 3.13
§3C	§3.3
Theorem 3C-1	Theorem 3.14
Corollary	Corollary 3.15
Corollary	Corollary 3.16
1.	Claim 3.17
2.	Claim 3.18
§3D	§3.4
§3E	§3.5
Theorem 3E-1	Theorem 3.19
§4A	§4.1
4A-1	4.1.1
4A-2	4.1.2

4A-3	4.1.3	Corollary	Corollary 8.3
Theorem 4A-4	Theorem 4.4	§8B	§8.2
Theorem 4A-5	Theorem 4.5	Theorem 8B-1	Theorem 8.4
Theorem 4A-5(a)	Theorem 4.5	Corollary	Corollary 8.5
Theorem 4A-5(b)	Theorem 4.6	§8C	§8.3
§4B	§4.2	Theorem 8C-1	Theorem 8.6
Theorem 4B-1	Theorem 4.7	Corollary	Corollary 8.7
Theorem 4B-2	Theorem 4.8	Open Problem	Open Problem 8.8
Theorem 5-1	Theorem 5.1	§9A	§9.1
Theorem 5-2	Theorem 5.2	Theorem 9A-1	Theorem 9.1
§6A	§6.1	Theorem 9A-2	Theorem 9.2
§6B	§6.2	Theorem 9A-3	Theorem 9.3
Theorem 6B-1	Theorem 6.1	§9B	§9.2
§6C	§6.3	Theorem 9B-1	Theorem 9.4
Theorem 6C-1	Theorem 6.2	—	Claim 9.5
Theorem 6C-2	Theorem 6.3	§9C	§9.3
Theorem 6C-3	Theorem 6.4	Lemma 9C-1	Lemma 9.6
Corollary	Corollary 6.5	Theorem 9C-2	Theorem 9.7
Corollary	Corollary 6.6	Corollary	Corollary 9.8
Theorem 6C-4	Theorem 6.7	—	Claim 9.9
Corollary	Corollary 6.8	§10A	§10.1
§7A	§7.1	Theorem 10A-1	Theorem 10.1
Theorem 7A-1	Theorem 7.1	Corollary	Corollary 10.2
Corollary	Corollary 7.2	§10B	§10.2
Corollary	Corollary 7.3	Theorem 10B-1	Theorem 10.3
Corollary	Corollary 7.4	Corollary	Corollary 10.4
§7B	§7.2	§11A	§11.1
7B-1	7.2.1	Theorem 11A-1	Theorem 11.1
7B-2	7.2.2	Corollary (a)	Corollary 11.2
7B-3	7.2.3	Corollary (b)	Corollary 11.3
7B-4	7.2.1	§11B	§11.2
Open Problem	Open Problem 7.5	Theorem 11B-1	Theorem 11.4
§8A	§8.1	Theorem 11B-2	Theorem 11.5
Theorem 8A-1	Theorem 8.1	1.	Claim 11.6
Theorem 8A-2	Theorem 8.2	2.	Claim 11.7

Inductive scales on inductive sets, Moschovakis,
[Cabal i, pp. 185–192]

Main Theorem	Theorem 0.1	Lemma 4	Lemma 1.5
Lemma 1	Lemma 1.1	Corollary 1	Corollary 2.1
Lemma 2	Lemma 1.2	Corollary 2	Corollary 2.2
Lemma 3	Lemma 1.3	Corollary 3	Corollary 2.3
—	Theorem 1.4	Corollary 4	Corollary 2.4

Scales on Σ^1_1 sets, Steel, [Cabal iii, pp. 72–76]

Lemma	Lemma 1.1	Theorem 2	Theorem 1.3
Theorem 1	Theorem 1.2	Corollary	Corollary 1.4

Scales on coinductive sets, Moschovakis, [Cabal iii, pp. 77–85]

Infimum Lemma	Lemma 1.1	Lemma	Lemma 2.1
Fake Supremum Lemma	Lemma 1.2	Theorem	Theorem 2.2

The extent of scales in $\mathbf{L}(\mathbb{R})$, Martin & Steel, [Cabal iii, pp. 86–96]

Corollary 1	Corollary 2	Corollary 3	Corollary 6
Lemma 2	Lemma 3	Corollary 4	Corollary 7
Theorem 1	Theorem 4	Theorem 2	Theorem 8
Corollary 2	Corollary 5		

The largest countable, this, that, and the other, Martin, [Cabal iii, pp. 97–106]

Theorem	Theorem 1.1	Theorem	Theorem 1.3
Corollary	Corollary 1.2	Sublemma 2.4	Sublemma 2.3.1

Scales in $\mathbf{L}(\mathbb{R})$, Steel, [Cabal iii, pp. 107–156]

Definition 2.2	Definition 2.4	Corollary 2.10	Corollary 2.12
Lemma 2.3	Lemma 2.5	Proposition 2.11	Corollary 2.13
Corollary 2.4	Corollary 2.6	Claim 1	Claim 3.8
Lemma 2.5	Lemma 2.7	Claim 2	Claim 3.9
Corollary 2.6	Corollary 2.8	Claim 3	Claim 3.10
Theorem 2.7	Theorem 2.9	Corollary 3.8	Corollary 3.11
Corollary 2.8	Corollary 2.10	Corollary 3.9	Corollary 3.12
Theorem 2.9	Theorem 2.11		

The real game quantifier propagates scales, Martin, [Cabal iii, pp. 157–171]

§0	§1	§2	§3
§1	§2	Lemma 2.1	Lemma 3.1
Lemma 1.1	Lemma 2.1	Lemma 2.2	Lemma 3.2
Lemma 1.2	Lemma 2.2	§3	§4

Lemma 3.1	Lemma 4.1	Lemma 4.4	Lemma 5.4
Lemma 3.2	Lemma 4.2	Theorem 4.5	Theorem 5.5
Lemma 3.3	Lemma 4.3	§5	§6
§4	§5	Theorem 5.1	Theorem 6.1
Lemma 4.1	Lemma 5.1	§6	§7
Lemma 4.2	Lemma 5.2	Theorem 6.1	Theorem 7.1
Lemma 4.3	Lemma 5.3	Corollary 6.2	Corollary 7.2

Long games, Steel, [CABAL iv, pp. 56–97]

Example	Example 1.1	Corollary 2	Corollary 2.6
Theorem 1	Theorem 1.2	Theorem 3	Theorem 3.1
Lemma 1	Lemma 1.3	Lemma 4	Lemma 3.2
Claim	Claim 1.4	Claim	Claim 3.3
Corollary	Corollary 1.5	Claim	Claim 3.4
Lemma 2	Lemma 1.6	Subclaim	Subclaim 3.5
Corollary 1	Corollary 1.7	Theorem 4	Theorem 3.6
Theorem 2	Theorem 2.1	Lemma 5	Lemma 4.1
Lemma 3	Lemma 2.2	Conjecture	Conjecture 4.2
Claim 1	Claim 2.3	Lemma 6	Lemma 4.3
Claim 2	Claim 2.4	Theorem 5	Theorem 4.4
Claim 3	Claim 2.5	Corollary 3	Corollary 4.5

The axiom of determinacy, strong partition properties and nonsingular measures, Kechris, Kleinberg, Moschovakis, & Woodin, [CABAL ii, pp. 75–100]

Lemma	Lemma 2.4	Theorem 2.5	Theorem 2.8
Lemma	Lemma 2.5	Lemma	Lemma 3.2
Open Problem	Open Problem 2.6	Lemma 1	Lemma 4.2
Theorem 2.4	Theorem 2.7	Lemma 2	Lemma 4.3

Suslin cardinals, κ-Suslin sets and the scale property in the hyperprojective hierarchy, Kechris, [CABAL ii, pp. 127–146]

Definition	Definition 1.2	Lemma 2.4	Lemma 2.6
Theorem 1.2	Theorem 1.3	Theorem	Theorem 3.1
Lemma	Lemma 1.4	Theorem 3.1	Theorem 3.2
Conjecture	Conjecture 1.5	Corollary 3.2	Corollary 3.3
Conjecture	Conjecture 1.6	Corollary 3.3	Corollary 3.4
Definition	Definition 2.3	Corollary 3.4	Corollary 3.5
Definition	Definition 2.4	Corollary 3.5	Corollary 3.6
Lemma 2.3	Lemma 2.5	§5(A)	§5.1

Conjecture	Conjecture 5.2	§5(D)	§5.4
Problem	Question 5.3	§5(E)	§5.5
§5(B)	§5.2	Conjecture	Conjecture 5.6
Question	Question 5.4	Conjecture	Conjecture 5.7
§5(C)	§5.3	Proposition	Proposition 5.8
Question	Question 5.5		

A coding theorem for measures, Kechris, [CABAL iv, pp. 103–109]

Theorem	Theorem 1.1	Corollary	Corollary 1.5
Corollary	Corollary 1.2	Corollary	Corollary 1.6
Corollary	Corollary 1.3	Lemma	Lemma 2.1
Corollary	Corollary 1.4		

Volume II

Wadge degrees and descriptive set theory, Van Wesep, [CABAL i, pp. 151–170]

Definition	Definition 1.1	Remark	Remark 4.4
Definition	Definition 1.2	Definition	Definition 4.5
Definition	Definition 1.3	Lemma 4.2	Lemma 4.6
Definition	Definition 1.4	Lemma 4.3	Lemma 4.7
Definition	Definition 1.5	Lemma 4.4	Lemma 4.8
Definition	Definition 2.2	Theorem 4.2	Theorem 4.9
Definition	Definition 2.3	Definition	Definition 4.10
Definition	Definition 2.4	Claim	Claim 4.11
Theorem 2.2	Theorem 2.5	Claim	Claim 4.12
Definition	Definition 2.6	Definition	Definition 5.1
Lemma 2.3	Lemma 2.7	Theorem 5.1	Theorem 5.2
Claim 1	Claim 2.8	Theorem 5.2	Theorem 5.3
Claim 2	Claim 2.9	Theorem 5.3	Theorem 5.4
Lemma	Lemma 3.2	Theorem 5.4	Theorem 5.5
Definition	Definition 3.3	Theorem 5.5	Theorem 5.6
Definition	Definition 3.4	Theorem 5.6	Theorem 5.7
Corollary	Corollary 3.5	Claim	Claim 5.8
Definition	Definition 4.1	Lemma	Lemma 5.9
Remark	Remark 4.2	Lemma	Lemma 5.10
Theorem 4.1	Theorem 4.3		

A note on Wadge degrees, Kechris, [CABAL ii, pp. 165-168]

Lemma 1	Lemma 2.1	Sublemma	Sublemma 2.3
Lemma 2	Lemma 2.2		

Some results in the Wadge hierarchy of Borel sets, Louveau,
[CABAL iii, pp. 28–55]

Figure a	Figure 1	Figure d	Figure 4
Figure b	Figure 2	Figure e	Figure 5
Figure c	Figure 3	Claim	Claim 2.8

The strength of Borel Wadge determinacy, Louveau & Saint-Raymond,
[CABAL iv, pp. 1–30]

Definition 1	Definition 1.1	Theorem 3	Theorem 3.3
Definition 2	Definition 1.2	Theorem 4	Theorem 3.4
Definition 3	Definition 1.3	Theorem 5	Theorem 3.5
Theorem 4	Theorem 1.4	Corollary 6	Corollary 3.6
Definition 5	Definition 1.5	Theorem	Theorem 3.7
Definition 6	Definition 1.6	Theorem	Theorem 3.8
Proposition 7	Proposition 1.7	Theorem	Theorem 3.9
Theorem 8	Theorem 1.8	Theorem 0	Theorem 4.1
Remark	Remark 1.9	Theorem 1	Theorem 4.2
Definition 9	Definition 1.10	Theorem 2	Theorem 4.3
Proposition 10	Proposition 1.11	Definition 3	Definition 4.4
Definition 1	Definition 2.1	Definition 4	Definition 4.5
Definition 2	Definition 2.2	Theorem 5	Theorem 4.6
Definition 3	Definition 2.3	Proposition 6	Proposition 4.7
Theorem 4	Theorem 2.4	Theorem 7	Theorem 4.8
Lemma 5	Lemma 2.5	Theorem 8	Theorem 4.9
Definition 6	Definition 2.6	Corollary 9	Corollary 4.10
Theorem 7	Theorem 2.7	Example 10	Example 4.11
Lemma 8	Lemma 2.8	Theorem 1	Theorem 5.1
Lemma 9	Lemma 2.9	Theorem 2	Theorem 5.2
Lemma 10	Lemma 2.10	Definition 3	Definition 5.3
Lemma 11	Lemma 2.11	Proposition 4	Proposition 5.4
Theorem 1	Theorem 3.1	Theorem 5	Theorem 5.5
Theorem 2	Theorem 3.2	Lemma 6	Lemma 5.6

Closure properties of pointclasses, Steel,
[CABAL ii, pp. 147–163]

Claim	Claim 3.2	Theorem 3.2	Theorem 3.5
Claim	Claim 3.3	Claim	Claim 3.6
Claim	Claim 3.4	Theorem 3.3	Theorem 3.7

The axiom of determinacy and the prewellordering property, Kechris, Solovay, & Steel, [CABAL ii, pp. 101–125]

Theorem	Theorem 1.1	Lemma 1	Lemma 3.5
Theorem	Theorem 1.2	Lemma 2	Lemma 3.6
§2.3	§2.1	Corollary 3.3	Corollary 3.7
Lemma 2.3.1	Lemma 2.3	Corollary 3.4	Corollary 3.8
§2.4	§2.2	Definition	Definition 4.1
Lemma 2.4.1	Lemma 2.4	Definition	Definition 4.2
§2.5	§2.3	Theorem 4.1	Theorem 4.3
Lemma 2.5.1	Lemma 2.5	Definition	Definition 5.1
Lemma 2.5.2	Lemma 2.6	Definition	Definition 5.2
Lemma 2.5.3	Lemma 2.7	Theorem 5.1	Theorem 5.3
Definition	Definition 3.1	Corollary 5.2	Corollary 5.4
Definition	Definition 3.2	Corollary 5.3	Corollary 5.5
Theorem	Theorem 3.3	Lemma	Lemma 5.6
Theorem 3.2	Theorem 3.4		

Pointclasses and wellordered unions, Jackson & Martin, [CABAL iii, pp. 56-66]

§0	§1	Lemma 2.6	Lemma 3.6
§1	§2	Lemma 2.7	Lemma 3.7
Theorem 1.1	Theorem 2.1	Lemma 2.8	Lemma 3.8
Corollary 1.1.1	Corollary 2.2	Lemma 2.9	Lemma 3.9
Corollary 1.1.2	Corollary 2.3	Lemma 2.10	Lemma 3.10
Theorem 1.2	Theorem 2.4	§3	§4
§2	§3	Theorem 3	Theorem 4.1
Theorem 2	Theorem 3.1	Lemma 3.1	Lemma 4.2
Lemma 2.1	Lemma 3.2	Lemma 3.2	Lemma 4.3
Lemma 2.2	Lemma 3.3	Lemma 3.3	Lemma 4.4
Lemma 2.3	Lemma 3.4	Lemma 3.4	Lemma 4.5
Lemma 2.4	Lemma 3.5	—	Addendum (2010)

More closure properties of pointclasses, Becker, [CABAL iv, pp. 31–36]

Definition	Definition 5	Theorem 6	Theorem 7
Lemma 5	Lemma 6	Theorem 7	Theorem 8

Homogeneous trees and projective scales, Kechris, [CABAL ii, pp 33–73]

Lemma	Lemma 2.2	Theorem	Theorem 3.2
Remark	Remark 2.3	Theorem	Theorem 5.1
Remark	Remark 3.1	Theorem	Theorem 6.1

Theorem	Theorem 7.1	Theorem	Theorem 8.1
Corollary	Corollary 7.2	Lemma A	Lemma 8.2
		Lemma B	Lemma 8.3

AD *and projective ordinals*, Kechris, [CABAL i, pp. 91–132]

Definition	Definition 1.1	Theorem 8.1	Theorem 8.2
Definition	Definition 2.1	Lemma A	Lemma 8.3
Definition	Definition 2.2	Definition	Definition 8.4
Theorem 2.1	Theorem 2.3	Lemma B	Lemma 8.5
Corollary	Corollary 2.4	Lemma C	Lemma 8.6
Theorem 2.2	Theorem 2.5	Theorem 8.2	Theorem 8.7
Definition	Definition 3.1	Theorem 8.3	Theorem 8.8
Theorem 3.1	Theorem 3.2	Theorem 8.4	Theorem 8.9
Definition	Definition 3.3	Lemma	Lemma 8.10
Lemma	Lemma 3.4	Proposition 8.5	Proposition 8.11
Theorem 3.2	Theorem 3.5	Basic Open Problem	Basic Open Problem 8.12
Definition	Definition 3.6	Basic Open Problem	Basic Open Problem 9.4
Theorem 3.3	Theorem 3.7	Definition	Definition 10.1
Claim	Claim 3.8	Remark	Remark 10.2
Definition	Definition 3.9	Definition	Definition 10.3
Definition	Definition 3.10	Theorem 10.1	Theorem 10.4
Theorem 3.4	Theorem 3.11	Definition	Definition 12.1
Definition	Definition 3.12	Theorem 12.1	Theorem 12.2
Theorem 3.5	Theorem 3.13	Definition	Definition 13.1
Theorem 3.6	Theorem 3.14	Theorem 13.1	Theorem 13.2
Theorem 3.7	Theorem 3.15	Theorem 13.2	Theorem 13.3
Claim	Claim 3.16	Corollary 13.3	Corollary 13.4
Definition	Definition 3.17	Corollary 13.4	Corollary 13.5
Theorem 3.8	Theorem 3.18	Lemma	Lemma 13.6
Theorem 3.9	Theorem 3.19	Claim	Claim 13.7
Theorem 3.10	Theorem 3.20	Theorem 13.5	Theorem 13.8
Theorem 3.11	Theorem 3.21	Theorem 13.6	Theorem 13.9
Theorem 3.12	Theorem 3.22	Definition	Definition 14.1
Theorem 3.13	Theorem 3.23	Theorem 14.1	Theorem 14.2
Theorem 3.14	Theorem 3.24	Corollary 14.2	Corollary 14.3
Lemma 1	Lemma 5.2	Theorem 14.3	Theorem 14.4
Lemma 2	Lemma 5.3	Lemma A	Lemma 14.5
Lemma 3	Lemma 5.4	Claim	Claim 14.6
Lemma 4	Lemma 5.5	Lemma B	Lemma 14.7
Lemma 5	Lemma 5.6	Claim	Claim 17.3
Claim	Claim 5.7	Theorem 17.3	Theorem 17.4
Claim	Claim 6.4	Corollary 17.4	Corollary 17.5
Theorem 6.4	Theorem 6.5		
Corollary 6.5	Corollary 6.6		
Definition	Definition 8.1		

A Δ^1_3 *coding of the subsets of* ω_ω, Solovay, [CABAL i, pp. 131–170]

Introduction	§1	§A.14	§2.14
§A	§2	Claim 1	Claim 2.17
§A.1	§2.1	Claim 2	Claim 2.18
§A.2	§2.2	Claim 3	Claim 2.19
Lemma	Lemma 2.1	Claim 4	Claim 2.20
Corollary	Corollary 2.2	Definition	Definition 2.21
§A.3	§2.3	Claim 5	Claim 2.22
Theorem	Theorem 2.3	Claim 6	Claim 2.23
§A.4	§2.4	Claim 7	Claim 2.24
Theorem	Theorem 2.4	Claim 8	Claim 2.25
§A.5	§2.5	Claim A.9	Claim 2.26
§A.6	§2.6	Fact	Fact 2.27
§A.7	§2.7	§B	§3
§A.8	§2.8	§B.1	§3.1
Definition	Definition 2.5	Theorem	Theorem 3.1
Lemma 1	Lemma 2.6	§B.2	§3.2
Lemma 2	Lemma 2.7	Lemma	Lemma 3.2
§A.9	§2.9	Lemma	Lemma 3.3
Definition A.9	Definition 2.8	—	Fact 3.4
Lemma 3	Lemma 2.9	§B.3	§3.3
Lemma 4	Lemma 2.10	Definition	Definition 3.5
§A.10	§2.10	Theorem	Theorem 3.6
§A.11	§2.11	Lemma	Lemma 3.7
Theorem A.11	Theorem 2.11	§B.4	§3.4
Lemma 1	Lemma 2.12	Theorem B.4	Theorem 3.8
Lemma 2	Lemma 2.13	§B.5	§3.5
Lemma 3	Lemma 2.14	Corollary B.5	Corollary 3.9
§A.12	§2.12	§B.6	§3.6
Lemma	Lemma 2.15	Theorem B.6	Theorem 3.10
§A.13	§2.13	Claim	Claim 3.11
Lemma	Lemma 2.16		

AD *and the projective ordinals*, Jackson, [CABAL iv, pp. 117–220]

§I	§1	Lemma 4	Lemma 2.8
§II	§2	§III	§3
Definition	Definition 2.1	Theorem	Theorem 3.1
Definition	Definition 2.2	§IV	§4
Definition	Definition 2.3	Theorem	Theorem 4.1
Definition	Definition 2.4	Definition	Definition 4.2
Lemma 1	Lemma 2.5	§V	§5
Lemma 2	Lemma 2.6	Definition	Definition 5.1
Lemma 3	Lemma 2.7	Condition D	Definition 5.2

Condition C	Definition 5.3	Lemma	Lemma 5.24
—	Definition 5.4	Cofinality Lemma	Lemma 5.26
Condition A	Definition 5.5	Main Inductive Lemma	Lemma 5.27
Lemma	Lemma 5.6	§VI	§6
Lemma	Lemma 5.7	Definition	Definition 6.1
Lemma 1	Lemma 5.9	—	Remark 6.2
Lemma 2	Lemma 5.10	—	Definition 6.3
Lemma 3	Lemma 5.11	Theorem	Theorem 6.4
Lemma 4	Lemma 5.12	§VII	§7
Lemma 5	Lemma 5.13	Definition	Definition 7.1
Lemma 6	Lemma 5.14	Definition	Definition 7.2
Definition	Definition 5.15	—	Definition 7.3
Definition	Definition 5.16	—	Definition 7.4
Lemma 7	Lemma 5.17	—	Lemma 7.5
Lemma 8	Lemma 5.18	Lemma	Lemma 7.6
Lemma 9	Lemma 5.19	§VIII	§8
Lemma 10	Lemma 5.20	Theorem	Theorem 8.1
Lemma 11	Lemma 5.21	§IX	§9
Lemma 12	Lemma 5.22	—	Definition 9.1
Definition	Definition 5.23		

PART III: WADGE DEGREES AND POINTCLASSES

WADGE DEGREES AND POINTCLASSES
INTRODUCTION TO PART III

ALESSANDRO ANDRETTA AND ALAIN LOUVEAU

§1. Introduction. One of the main objects of study in Descriptive Set Theory is that of **boldface pointclass**, that is a collection of subsets of the Baire space (or more generally: of a family of Polish spaces) closed under continuous preimages. Since in this paper we will have little use for the concept of lightface pointclass used in the effective theory, we will drop the 'boldface' and simply speak of pointclasses. Also, in order to avoid trivialities, we will always assume that a pointclass is non-empty and different from $\wp(\mathbb{R})$.

Despite the fact that the the concept of pointclass is both very simple and ubiquitous in modern Descriptive Set Theory, it is actually quite recent, at least in its modern conception. The French analysts at the turn of the twentieth century—Baire, Borel, and Lebesgue—and later Luzin, Suslin, Hausdorff, Sierpiński, Kuratowski, always worked with specific pointclasses (such as the collection of all Borel sets, or the collection of all projective sets) defined by closure under set theoretic operations, and stratified into a transfinite hierarchy, *e.g.*, the Baire classes $\boldsymbol{\Sigma}^0_\alpha$, $\boldsymbol{\Pi}^0_\alpha$, and $\boldsymbol{\Delta}^0_\alpha$ for the Borel sets, and $\boldsymbol{\Sigma}^1_n$, $\boldsymbol{\Pi}^1_n$, and $\boldsymbol{\Delta}^1_n$ for the projective sets. The fact that all these collections were closed under continuous preimages was probably considered a simple consequence of their definition, rather than a feature worth crystallizing into a mathematical definition. Even the fact that the Borel hierarchy (and similarly for the projective one) exhibited the well-known diamond-shape pattern

$$\begin{array}{ccccccccc}
\boldsymbol{\Sigma}^0_1 & & \boldsymbol{\Sigma}^0_2 & & \boldsymbol{\Sigma}^0_\alpha & & & \boldsymbol{\Sigma}^0_{\alpha+1} & \\
& \subseteq \quad \supseteq & & \supseteq & & \subseteq \quad \supseteq & & & \\
& \boldsymbol{\Delta}^0_2 & & \cdots \boldsymbol{\Delta}^0_\alpha & & & \boldsymbol{\Delta}^0_{\alpha+1} & & \cdots \\
& \supseteq \quad \subseteq & & \subseteq & & \supseteq \quad \subseteq & & & \\
\boldsymbol{\Pi}^0_1 & & \boldsymbol{\Pi}^0_2 & & \boldsymbol{\Pi}^0_\alpha & & & \boldsymbol{\Pi}^0_{\alpha+1} & \\
\longleftarrow & & & & \omega_1 & & & & \longrightarrow
\end{array}$$

apparently was not considered to be an indication of an underlying structure. Hausdorff showed that any $\boldsymbol{\Delta}^0_2$ set can be represented as a transfinite difference of open (or for that matter, closed) sets, and Kuratowski, by the trick of refining the topology, extended this to all $\boldsymbol{\Delta}^0_{\alpha+1}$ sets. Thus $\boldsymbol{\Delta}^0_{\alpha+1} = \bigcup_{\beta<\omega_1} \mathrm{D}_\beta\, \boldsymbol{\Sigma}^0_\alpha$

Wadge Degrees and Projective Ordinals: The Cabal Seminar, Volume II
Edited by A. S. Kechris, B. Löwe, J. R. Steel
Lecture Notes in Logic, 37

where $\mathrm{D}_\beta\, \mathbf{\Gamma}$ denotes the class of β-differences of sets in $\mathbf{\Gamma}$, *i.e.*, sets of the form

$$\{x \in \bigcup_{\gamma<\beta} A_\gamma \ : \ \text{the least } \gamma \text{ that } x \notin A_\gamma \text{ has parity different from } \beta\}$$

for some sequence $\langle A_\gamma \ : \ \gamma < \beta \rangle$ of sets in $\mathbf{\Gamma}$. Again we obtain a picture similar to the one for the Borel hierarchy:

$$\begin{array}{ccccccccc}
\mathbf{\Sigma}^0_\alpha = \mathrm{D}_1\,\mathbf{\Sigma}^0_\alpha & & & & \mathrm{D}_2\,\mathbf{\Sigma}^0_\alpha & & & & \mathrm{D}_3\,\mathbf{\Sigma}^0_\alpha \\
 & \rotatebox{-45}{$\subsetneq$} & & \rotatebox{45}{$\supsetneq$} & & \rotatebox{-45}{$\subsetneq$} & & \rotatebox{45}{$\supsetneq$} & \\
 & & \Delta_{\mathrm{D}_2\,\mathbf{\Sigma}^0_\alpha} & & & & \Delta_{\mathrm{D}_2\,\mathbf{\Sigma}^0_\alpha} & & \cdots \\
 & \rotatebox{45}{$\supsetneq$} & & \rotatebox{-45}{$\subsetneq$} & & \rotatebox{45}{$\supsetneq$} & & \rotatebox{-45}{$\subsetneq$} & \\
\mathbf{\Pi}^0_\alpha = \mathrm{D}_1\,\mathbf{\Pi}^0_\alpha & & & & (\mathrm{D}_2\,\mathbf{\Sigma}^0_\alpha)\breve{} & & & & (\mathrm{D}_3\,\mathbf{\Sigma}^0_\alpha)\breve{} \\
\longleftarrow & & & & \omega_1 & & & & \longrightarrow
\end{array}$$

Wadge in his Ph.D. thesis [Wad84] was the first to investigate in a systematic manner the notion of **continuous reducibility** on the Baire space ${}^\omega\omega$. The motivation for his study, and the reason as to why these matters had not been studied before is explained in [Wad84, pp. 2–3]:

> The notion of reducibility, including many-one reducibility, plays an extremely important role in recursive function theory. One would expect the same to be true in descriptive set theory; but that has not (at least till recently) been the case. Of course, there are in the literature many instances in which continuous preimage is used to derive a particular result. In Sikorski (1957), for example, this approach is used to construct for each countable ordinal μ a set in the μth but no lower level of the Borel hierarchy. Luzin and Sierpiński (1929) used preimage to show that the collection of (codes for) wellorderings of ω is not Borel; and there are a number of other examples. Yet nowhere (to our knowledge) is the relation $A = f^{-1}(B)$ for some continuous f ever explicitly defined and studied as a partial order, not even in exhaustive work such as Kuratowski (1958) or Sierpiński (1952). In the latter, Sierpiński discusses preimage in general, continuous image and homeomorphic image, but not (explicitly) continuous preimage, which is perhaps the most natural. One possible explanation is that the investigation of $\leq$ naturally involves infinite games, and it is only recently that game methods have been fully understood and appreciated.[1]

Wadge's main objective was a complete analysis of all the Borel pointclasses, *i.e.*, boldface pointclasses contained in $\mathbf{\Delta}^1_1$. Working in ZF+DC, he defined a hierarchy of Borel sets refining the usual Borel hierarchy, he proved that it is well-founded and computed its length, and, assuming the determinacy of all Borel games, he could show that every Borel pointclass fits in this classification. As explained by Wadge in [Wad11] in the present volume and

[1] The relation $\leq$ is nowadays called Wadge reducibility and it is denoted by $\leq_{\mathrm{W}}$, and the references mentioned are, in order, [Sik58], [LS29], [Kur58], and [Sie52].

in [Wad84, pp. 10–11], all these results were obtained before Martin's proof of Borel determinacy [Mar75]. The problem whether Borel determinacy is needed to prove that all Borel pointclasses fall into Wadge's analysis remained open for over a decade, until Louveau and Saint-Raymond answered in the negative, by conducting Wadge's analysis within second order arithmetic (see the paper [LSR88B] in this volume).

As we mentioned, all pointclasses considered by early descriptive set theorists were defined in terms of operation on sets, like taking complements, countable intersections, countable unions, Suslin's operation $\mathscr{A}$, *etc.* All of these operations can be thought as operations

$$\mathcal{O}\colon \wp(\mathbb{R})^\omega \to \wp(\mathbb{R})$$

assigning a new set to a countable sequence of sets, and with the property that there is a $T \subseteq \wp(\omega)$ such that for any $\langle A_n : n \in \omega\rangle$

$$\forall x \in \mathbb{R}\,(x \in \mathcal{O}\langle A_n : n \in \omega\rangle \iff \{n \in \omega : x \in A_n\} \in T)\,.$$

A function $\mathcal{O}$ as above is said to be an **ω-ary Boolean operation**, or simply a **Boolean operation**, and the set $T = T_{\mathcal{O}}$ which completely determines $\mathcal{O}$, is called the truth table of $\mathcal{O}$. We will say that such an operation is Borel, or $\boldsymbol{\Sigma}^1_1$, *etc.*, if its truth table is Borel, or $\boldsymbol{\Sigma}^1_1$, *etc.*, as a subset of ${}^{\omega}2$. For example: the operations of taking complements, countable intersections, or countable unions, as well as their compositions are all Borel, while Suslin's operation $\mathscr{A}$ is $\boldsymbol{\Sigma}^1_1$.

Wadge showed in ZFC that each non-self-dual Borel pointclass in ${}^{\omega}\omega$ is of the form

$$\{\mathcal{O}\langle A_n : n \in \omega\rangle : \forall n\,(A_n \text{ is open})\}$$

with $\mathcal{O}$ a Borel Boolean operation, and Van Wesep in [Van77], assuming AD and building on earlier results of Miller, Radin, and Steel, extended this result to all non-self-dual pointclasses, using of course arbitrary Boolean operations. Thus we have come to a full circle—non-self-dual pointclasses considered by early descriptive set theorists were defined in terms of (explicit) operations, and assuming AD every non-self-dual pointclass is defined in terms of operations on open sets.

Boolean operations are operations on the *collection* of open sets that allow us to construct all sets belonging to complicated pointclass $\boldsymbol{\Gamma}$, and they figure prominently in the work of Louveau and Saint-Raymond [Lou83, LSR87, LSR88B]. But Wadge also introduced certain specific operations on *sets* which yield *complete sets* for various $\boldsymbol{\Gamma}$. Thus a non-self-dual pointclass $\boldsymbol{\Gamma}$ can be described either as obtained via some some appropriate Boolean operation $\mathcal{O}$, $\boldsymbol{\Gamma} = \{\mathcal{O}\langle A_n : n \in \omega\rangle : A_n \in \boldsymbol{\Sigma}^0_1\}$, or else as the set of continuous preimages of a $\boldsymbol{\Gamma}$-complete set A, $\boldsymbol{\Gamma} = \{X : X \leq_{\mathrm{W}} A\}$. These operations on sets are quite useful to compute the Wadge rank of the various pointclasses, and were extensively used in [Ste81B] and [Van77]. Recently this approach to the Wadge hierarchy has been extended in the work of Duparc and others in connection with automata theory, see [Dup01, Dup03, DFR01].

In the next section we give a few basic definitions an review some basic results on the Wadge hierarchy.

§2. **Some basic facts about the Wadge hierarchy.** The relation of **Wadge reducibility**, $A \leq_{\mathrm{W}} B$, is defined as $A = f^{-1}(B)$ for some continuous function f. It can be defined for any pair of ambient topological spaces: $\mathcal{X}$ containing A and $\mathcal{Y}$ containing B, so that $f : \mathcal{X} \to \mathcal{Y}$, but the general theory becomes somewhat uninteresting if the spaces are not zero-dimensional, as there may be, in general, very few continuous maps. Following Wadge, from now on we will focus on the Baire space ${}^{\omega}\omega$, which—as customary in set theory—will be denoted by $\mathbb{R}$.

A continuous $f : \mathbb{R} \to \mathbb{R}$ is determined by a monotone $\varphi\colon {}^{<\omega}\omega \to {}^{<\omega}\omega$ such that $\lim_n \mathrm{lh}\, \varphi(x \restriction n) = +\infty$. If we require that $\mathrm{lh}\, \varphi(x \restriction n) = n$, then the resulting f is a Lipschitz function with constant ≤ 1, where we use the usual distance on ${}^{\omega}\omega$. In this case we will say that A **is Lipschitz reducible to** B. Wadge introduced the Lipschitz game $G_{\mathrm{L}}(A, B)$: it is a game on ω

player I	a_0		a_1		$\cdots$
player II		b_0		b_1	$\cdots$

where player II wins iff

$$\langle a_i : i < \omega \rangle \in A \iff \langle b_i : i < \omega \rangle \in B\,.$$

Thus player II has a winning strategy for the game $G_{\mathrm{L}}(A, B)$ if and only if $A \leq_{\mathrm{L}} B$. Conversely, if player I has a winning strategy, then there is a Lipschitz map witnessing $B \leq_{\mathrm{L}} \neg A$. Note that in this case player I's strategy yields a $\varphi\colon {}^{<\omega}\omega \to {}^{<\omega}\omega$ such that $\mathrm{lh}(\varphi(s)) = \mathrm{lh}(s) + 1$ hence the induced $f : \mathbb{R} \to \mathbb{R}$ is a Lipschitz map with constant $1/2$, and in fact the converse implication (if $B \leq_{\mathrm{L}} \neg A$ then player I wins $G_{\mathrm{L}}(A, B)$) in general does not hold.

Assuming determinacy we obtain the following simple—yet fundamental—result known as:

WADGE'S LEMMA. *Assume* AD. *Then*

$$\forall A, B \subseteq \mathbb{R}\, (A \leq_{\mathrm{L}} B \vee B \leq_{\mathrm{L}} \neg A)\,.$$

The gist of the result is that any two sets of reals are almost comparable, and that $\leq_{\mathrm{L}}$ is almost a linear order. Wadge dubbed this as the **Semi Linear Ordering** principle for **Lipschitz** reductions. As every Lipschitz reduction is, in particular, a Wadge reduction, Wadge's Lemma yields trivially the Semi Linear Ordering principle for **continuous** reductions:

$$\forall A, B \subseteq \mathbb{R}\, (A \leq_{\mathrm{W}} B \vee B \leq_{\mathrm{W}} \neg A)\,.$$

Since it turns out that these two versions are equivalent (assuming $\mathsf{DC}(\mathbb{R})$ and that all sets have the property of Baire—see [And03, And06]) we shall denote either version with SLO.

The **quasi-order**[2] $\leq_W$ induces an equivalence relation on $\wp(\mathbb{R})$ whose equivalence classes are called **Wadge degrees**. The collection of all Wadge degrees together with the induced order is called the **Wadge hierarchy**. A set of reals A or its Wadge degree $[A]_W$ is said to be self-dual if it $A \leq_W \neg A$; otherwise it is said to be non-self-dual. Wadge's Lemma implies that a self-dual degree is comparable to any other degree, and that if two degrees are incomparable, then they must be dual to each other. In other words: all antichains have size at most 2. Martin—building on previous work of Monk—showed in 1973 that AD implies that the ordering $\leq$ is well-founded. In fact these results hold *verbatim* for the **Lipschitz hierarchy**, *i.e.*, the collection of degrees $[A]_L$ obtained using Lipschitz reductions. By a result due independently to Steel [Ste77] and Van Wesep [Van77], AD implies that

$$A \leq_W \neg A \iff A \leq_L \neg A \tag{1}$$

and using this it is possible to completely determine the structure of the Lipschitz hierarchy: at the bottom of the hierarchy we have the non-self-dual pair $\{\varnothing\} = [\varnothing]_L$ and $\{\mathbb{R}\} = [\mathbb{R}]_L$, followed by an ω_1 chain of self-dual degrees formed by all clopen sets different from $\varnothing$ and $\mathbb{R}$. Above these there is the non-self-dual pair $\boldsymbol{\Sigma}^0_1 \setminus \boldsymbol{\Delta}^0_1$ and $\boldsymbol{\Pi}^0_1 \setminus \boldsymbol{\Delta}^0_1$ followed by an ω_1 chain of self-dual degrees. In general: at limit levels of uncountable cofinality we have a non-self-dual pair, while at all other levels we have a self-dual degree. The length of this hierarchy [Sol78B] is

$$\Theta \stackrel{\text{def}}{=} \sup\{\alpha : \exists f\, (f : \mathbb{R} \twoheadrightarrow \alpha)\}.$$

Thus the Lipschitz hierarchy looks like this:

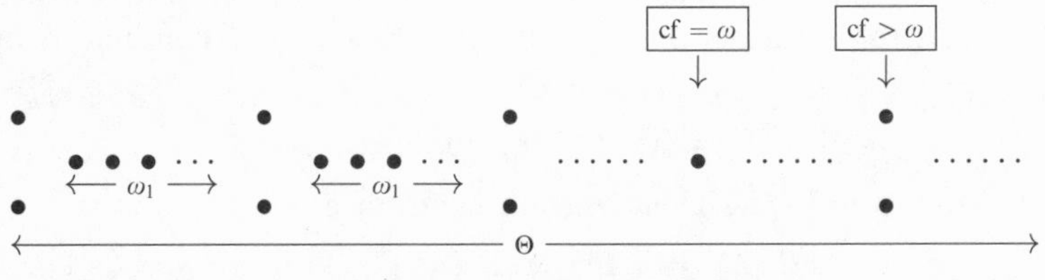

(2)

Each block of ω_1 consecutive self-dual Lipschitz degree is contained inside a single (necessarily self-dual) Wadge degree, and by the result of Steel and Van Wesep (1) nothing else is, so in the Wadge hierarchy self-dual degrees and non-self-dual pairs alternate, with the former appearing at levels of countable

[2] A quasi-order is a reflexive and transitive relation, and it is also known in the literature as a **pre-order**.

cofinality, and the latter appearing at the remaining limit levels:

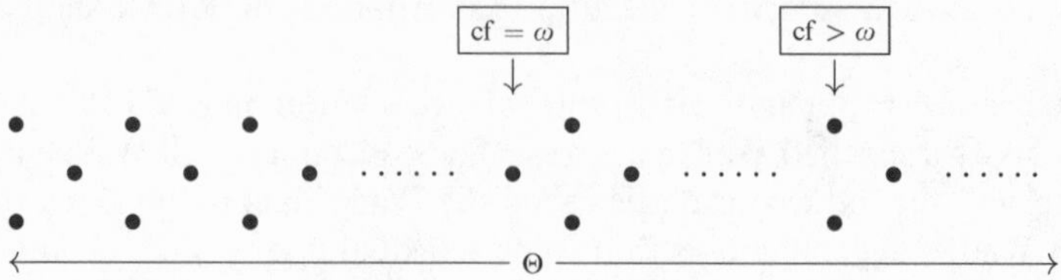

(3)

The Wadge hierarchy is the ultimate analysis of $\wp(\mathbb{R})$ in terms of topological complexity, assigning to each set $A \subseteq \mathbb{R}$ an ordinal $\|A\|_{\mathrm{W}}$, the rank of $[A]_{\mathrm{W}}$ in the hierarchy. This is somewhat surprising, since AD forbids the existence of long transfinite sequences of reals. It is not hard to check that non-self-dual pointclasses are of the form

$$\{B \subseteq \mathbb{R} : B \leq_{\mathrm{W}} A\}$$

for some non-self-dual set A, while self-dual pointclasses are all of the form

$$\{B \subseteq \mathbb{R} : B <_{\mathrm{W}} A\}$$

for some arbitrary $A \neq \mathbb{R}, \varnothing$. (Here and below, $B <_{\mathrm{W}} A$ has the obvious meaning: $B \leq_{\mathrm{W}} A$ and $A \nleq_{\mathrm{W}} B$.) Thus Wadge's Lemma yields a semi-linear ordering principle for pointclasses: for any $\boldsymbol{\Gamma}$ and $\boldsymbol{\Lambda}$,

$$\boldsymbol{\Gamma} \subseteq \boldsymbol{\Lambda} \vee \breve{\boldsymbol{\Lambda}} \subseteq \boldsymbol{\Gamma}.$$

It is a classical fact that any pointclass $\boldsymbol{\Gamma}$ of the form $\boldsymbol{\Sigma}^0_\alpha$, $\boldsymbol{\Pi}^0_\alpha$, $\boldsymbol{\Sigma}^1_n$, or $\boldsymbol{\Pi}^1_n$ has a **universal set**, *i.e.*, a set $U \subseteq \mathbb{R} \times \mathbb{R}$ that belongs to $\boldsymbol{\Gamma}$ (once it is coded as a subset of $\mathbb{R}$ via some canonical homeomorphism) and such that

$$\boldsymbol{\Gamma} = \{U_{(x)} : x \in \mathbb{R}\}$$

where $U_{(x)} = \{y \in \mathbb{R} : (x, y) \in U\}$ is the vertical section of U through x. This fact generalizes to *all* non-self-dual boldface pointclasses.

To see this, fix some canonical enumeration $\langle \ell_x : x \in \mathbb{R} \rangle$ of all Lipschitz maps $\mathbb{R} \to \mathbb{R}$ with the further property that $(x, y) \mapsto \ell_x(y)$ is continuous, and let

$$U = \{(x, y) : \ell_x(y) \in A\},$$

where A is any set in $\boldsymbol{\Gamma} \setminus \breve{\boldsymbol{\Gamma}}$. Then U is in $\boldsymbol{\Gamma}$, and since

$$\begin{aligned} B \in \boldsymbol{\Gamma} &\iff B \leq_{\mathrm{W}} A \\ &\iff B \leq_{\mathrm{L}} A \qquad \text{(by (1))} \end{aligned}$$

we obtain that U is universal for $\boldsymbol{\Gamma}$.

Another property that generalizes under AD to arbitrary pointclasses is the following: a non-self-dual pointclass $\boldsymbol{\Gamma}$ is said to have the **separation property**, in symbols $\mathsf{Sep}(\boldsymbol{\Gamma})$ if for any pair of disjoint sets $A, B \in \boldsymbol{\Gamma}$ there is a set $C \in \Delta_{\boldsymbol{\Gamma}} \stackrel{\text{def}}{=} \boldsymbol{\Gamma} \cap \breve{\boldsymbol{\Gamma}}$ that separates A from B, that is $A \subseteq C$ and $C \cap B = \varnothing$. By work of Sierpiński $\underset{\sim}{\mathbf{\Pi}}^0_\alpha$ has the separation property, while $\underset{\sim}{\mathbf{\Sigma}}^0_\alpha$ does not; assuming PD Moschovakis showed that $\underset{\sim}{\mathbf{\Sigma}}^1_{2n+1}$ and $\underset{\sim}{\mathbf{\Pi}}^1_{2n}$ have the separation, while neither $\underset{\sim}{\mathbf{\Sigma}}^1_{2n}$ nor $\underset{\sim}{\mathbf{\Pi}}^1_{2n+1}$ has it. (For $\underset{\sim}{\mathbf{\Sigma}}^1_1$ this is the classical result of Suslin, and does not require PD.) Assuming AD, given a pair of non-self-dual pointclasses $\underset{\sim}{\mathbf{\Gamma}}$ and $\check{\underset{\sim}{\mathbf{\Gamma}}}$, *at most one* of them has the separation property [Van78A], and *at least one* of them has the separation property [Ste81B], hence *exactly one* of them has the separation property.

The pointclasses $\underset{\sim}{\mathbf{\Sigma}}^0_\alpha$ can be detected inside the Wadge hierarchy by means of the rank of their complete sets. Starting from the very bottom, $\mathbb{R}$ and $\varnothing$ have least possible rank, which for technical reasons is set to be equal to 1, then the clopen set have ranks 2, and thus sets in $\underset{\sim}{\mathbf{\Sigma}}^0_1 \setminus \underset{\sim}{\mathbf{\Delta}}^0_1$ have rank 3, From this point on the $\underset{\sim}{\mathbf{\Sigma}}^0_\alpha$ are more and more spread apart. For example complete $\underset{\sim}{\mathbf{\Sigma}}^0_2$ sets have Wadge rank ω_1, complete $\underset{\sim}{\mathbf{\Sigma}}^0_3$ sets have Wadge rank $\omega_1^{\omega_1}$, and, in general, complete $\underset{\sim}{\mathbf{\Sigma}}^0_{n+1}$ sets have Wadge rank ϑ_n where $\vartheta_1 = \omega_1$ and $\vartheta_{k+1} = \omega_1^{\vartheta_k}$. The rank ϑ_ω of a complete $\underset{\sim}{\mathbf{\Sigma}}^0_\omega$ set is *not* the sup of the ϑ_ns, *i.e.*, the first fixed point of the map

$$\begin{aligned} E\colon \mathrm{Ord} &\to \mathrm{Ord} \\ \xi &\mapsto \omega_1^\xi, \end{aligned} \tag{4}$$

since this ordinal has countable cofinality, and hence it is the rank of a self-dual set. It turns out that ϑ_ω is the ω_1-st fixed point of the map E. The computation of the ranks of $\underset{\sim}{\mathbf{\Sigma}}^0_\alpha$ with $\alpha \geq \omega$ is quite technical—see [Wad11] for a summary of the results and [Wad84, Chapter V] for complete proofs. For example, the length Ξ of the Wadge hierarchy of the Borel sets, or, equivalently, the rank of a complete $\underset{\sim}{\mathbf{\Sigma}}^1_1$ or $\underset{\sim}{\mathbf{\Pi}}^1_1$ set, is computed as follows: for any cub class $C \subseteq \mathrm{Ord}$ let

$$C' = \{\xi : \xi = F_C(\xi)\}$$

be the set of fixed points of F_C, where $F_C : \mathrm{Ord} \to C$ is the enumerating function, and consider the sequence of cub classes $C = C^{(0)} \supset C^{(1)} \supset C^{(2)} \supset \dots$ given by $C^{(\alpha+1)} = \left(C^{(\alpha)}\right)'$ and $C^{(\lambda)} = \bigcap_{\alpha<\lambda} C^{(\alpha)}$ when λ is limit. Then Ξ is the least element of $C^{(\omega_1)}$ where C is taken to be the class of fixed points of the map E defined in (4). Thus the length of the Wadge degrees of Borel sets is an ordinal of cofinality ω_1 strictly smaller than ω_2. This is not just an happenstance, since under AD the length of the hierarchy of $\underset{\sim}{\mathbf{\Delta}}^1_{2n+1}$ degrees is $< \underset{\sim}{\boldsymbol{\delta}}^1_{2n+2}$. On the other hand, by a theorem due independently to Martin and Steel, the length of the hierarchy of $\underset{\sim}{\mathbf{\Delta}}^1_{2n}$ degrees is equal to $\underset{\sim}{\boldsymbol{\delta}}^1_{2n+1}$.

§3. The papers in the volume.

Early investigations of the degrees of Borel sets by W. W. Wadge.

This paper is an overview of the results of the author's Ph.D. dissertation [Wad84] and gives a glimpse on how this area of Descriptive Set Theory was uncovered. Although it contains no proofs, this article gives a quick introduction to the techniques ((α, β)-homeomorphisms, $\utilde{\mathbf{\Sigma}}^0_{1+\mu}$-separated unions, etc.) used to give a complete analysis of the Wadge degrees of the Borel sets, and a computation of its length Ξ.

Wadge degrees and descriptive set theory by R. Van Wesep, and
A note on Wadge degrees by A. S. Kechris.

Van Wesep's paper provides a good introduction to the subject, with complete (albeit terse) proofs, surveying what was known at that time (1978). The reader will find the proof of several of the results stated in the preceding section, including Martin's proof of the well-foundedness of $\leq_{\mathrm{L}}$, the result by Steel and Van Wesep on self-dual degrees stated in (1), and the proof under AD that the hierarchy of $\utilde{\mathbf{\Delta}}^1_{2n}$ degrees has length $\utilde{\boldsymbol{\delta}}^1_{2n+1}$. As this last fact is a result on projective sets, it is natural to ask for a proof assuming only PD. Such a proof is given in Kechris' paper, where $\mathsf{Det}(\utilde{\mathbf{\Delta}}^1_{2n})$ is shown to suffice.

The rest of Van Wesep's paper is devoted to the reduction and prewellordering properties. Recall that $\utilde{\mathbf{\Gamma}}$ is said to have the **reduction property**, in symbols $\mathsf{Red}(\utilde{\mathbf{\Gamma}})$, if given any two sets $A, B \in \utilde{\mathbf{\Gamma}}$ there are disjoint sets $A', B' \in \utilde{\mathbf{\Gamma}}$ such that $A' \subseteq A$, $B' \subseteq B$, and $A' \cup B' = A \cup B$; The prewellordering property $\mathsf{PWO}(\utilde{\mathbf{\Gamma}})$ means that every set in $\utilde{\mathbf{\Gamma}}$ admits a $\utilde{\mathbf{\Gamma}}$ norm [KM78B]. For any nonselfdual pointclass, $\mathsf{Red}(\utilde{\mathbf{\Gamma}}) \Longrightarrow \mathsf{Sep}(\check{\utilde{\mathbf{\Gamma}}})$ and if moreover $\utilde{\mathbf{\Gamma}}$ is closed under finite unions and intersections, then $\mathsf{PWO}(\utilde{\mathbf{\Gamma}}) \Longrightarrow \mathsf{Red}(\utilde{\mathbf{\Gamma}})$ [KM78B, Theorem 2.1]. Every $\utilde{\mathbf{\Sigma}}^0_\alpha$ has the prewellordering and hence the reduction property, and Louveau and Saint-Raymond have shown that for Borel pointclasses, the reduction property and prewellordering properties are equivalent, and have given a complete description of which Borel pointclasses possess this property—see [LSR88A]. For the sake of brevity, we say that a non-self-dual pair of pointclasses $(\utilde{\mathbf{\Gamma}}, \check{\utilde{\mathbf{\Gamma}}})$ satisfies the prewellordering property if either $\mathsf{PWO}(\utilde{\mathbf{\Gamma}})$ or else $\mathsf{PWO}(\check{\utilde{\mathbf{\Gamma}}})$, and we follow a similar convention for the reduction property. In Van Wesep's paper it is shown that there are non-self-dual pairs $(\utilde{\mathbf{\Gamma}}, \check{\utilde{\mathbf{\Gamma}}})$ that fail to have the reduction property, and since (under AD, which will be tacitly assumed from now on) the separation property holds at every level of the Wadge hierarchy, this shows that the separation property is weaker that the reduction property. Determining which non-self-dual pairs $(\utilde{\mathbf{\Gamma}}, \check{\utilde{\mathbf{\Gamma}}})$ satisfy the reduction property is a non-trivial matter. In the paper under review it is shown that

> If $\utilde{\mathbf{\Gamma}}$ is non-self-dual and closed under finite intersections then $\mathsf{Sep}(\check{\utilde{\mathbf{\Gamma}}}) \Longrightarrow \mathsf{Red}(\utilde{\mathbf{\Gamma}})$. (5)

(Notice that by the result mentioned below in (13), the hypothesis could be weakened to $\Delta_{\utilde{\mathbf{\Gamma}}}$.) A result of Steel is presented: If $\utilde{\mathbf{\Gamma}}$ is non-self-dual and closed under countable unions and intersections, then reduction holds for $(\boldsymbol{\Gamma}, \check{\boldsymbol{\Gamma}})$. This result was strengthened shortly afterwards by Steel himself in [Ste81B]:

If $\boldsymbol{\Gamma}$ is non-self-dual and $\boldsymbol{\Delta}_{\boldsymbol{\Gamma}}$ is closed under finite unions and intersections, then the reduction property holds for $(\boldsymbol{\Gamma}, \check{\boldsymbol{\Gamma}})$. (6)

Finally the proof a theorem of Kechris and Solovay is given:

Suppose $\boldsymbol{\Gamma} \subseteq \mathbf{L}(\mathbb{R})$ is non-self-dual and closed under countable unions and countable intersection. Suppose also $\exists^{\mathbb{R}}\boldsymbol{\Gamma} \subseteq \boldsymbol{\Gamma}$ *and* $\forall^{\mathbb{R}}\boldsymbol{\Gamma} \subseteq \boldsymbol{\Gamma}$. Then prewellordering holds for $(\boldsymbol{\Gamma}, \check{\boldsymbol{\Gamma}})$. (7)

The axiom of determinacy and the prewellordering property by A. S. Kechris, R. Solovay, and J. Steel.

This paper, as the title suggests, is devoted to the study of the prewellordering property under AD and, in a sense, it starts from where Van Wesep's paper ended. Firstly a criterion for PWO is established:

Suppose $\boldsymbol{\Gamma}$ is non-self-dual, closed under countable unions and intersections, and *either* $\exists^{\mathbb{R}}\boldsymbol{\Gamma} \subseteq \boldsymbol{\Gamma}$ *or else* $\forall^{\mathbb{R}}\boldsymbol{\Gamma} \subseteq \boldsymbol{\Gamma}$. Then the prewellordering property holds for the non-self-dual pair $(\boldsymbol{\Gamma}, \check{\boldsymbol{\Gamma}})$ if and only if $\boldsymbol{\Delta}_{\boldsymbol{\Gamma}}$ is *not* closed under well-ordered unions. (8)

Recall that a pointclass $\boldsymbol{\Lambda}$ is closed under well-ordered unions if $\bigcup_{\alpha<\beta} A_\alpha \in \boldsymbol{\Lambda}$ for any sequence $\langle A_\alpha : \alpha < \beta\rangle$ of sets in $\boldsymbol{\Lambda}$. Note that if $A \in \boldsymbol{\Gamma} \setminus \check{\boldsymbol{\Gamma}}$ and $\varphi\colon A \twoheadrightarrow \kappa$ is a regular $\boldsymbol{\Gamma}$-norm, then each $A_\alpha = \{x \in A : \varphi(x) < \alpha\} \in \boldsymbol{\Delta}_{\boldsymbol{\Gamma}}$, but $A = \bigcup_{\alpha<\kappa} A_\alpha \notin \boldsymbol{\Delta}_{\boldsymbol{\Gamma}}$, so one of the two directions of the equivalence is immediate. The Theorem of Kechris and Solovay stated in (7) is thus extended to the case when $\boldsymbol{\Gamma}$ is closed under only one real quantifier:

Suppose $\boldsymbol{\Gamma} \subseteq \mathbf{L}(\mathbb{R})$ is non-self-dual and closed under countable unions and countable intersection. Suppose also $\exists^{\mathbb{R}}\boldsymbol{\Gamma} \subseteq \boldsymbol{\Gamma}$ *or* $\forall^{\mathbb{R}}\boldsymbol{\Gamma} \subseteq \boldsymbol{\Gamma}$. Then prewellordering holds for $(\boldsymbol{\Gamma}, \check{\boldsymbol{\Gamma}})$. (9)

If $\boldsymbol{\Gamma}$ is $\boldsymbol{\Sigma}^1_n$ or $\boldsymbol{\Pi}^1_n$ then (9) says that exactly one among $\boldsymbol{\Gamma}$ and $\check{\boldsymbol{\Gamma}}$ has the prewellordering property—in fact by Moschovakis' First and Second Periodicity Theorems [KM78B] we can actually determine which of the two pointclasses has this property, namely PWO($\boldsymbol{\Gamma}$) iff $\boldsymbol{\Gamma} = \boldsymbol{\Pi}^1_{2n}$ or $\boldsymbol{\Gamma} = \boldsymbol{\Sigma}^1_{2n+1}$. The authors establish an analogous results for **projective-like pointclasses**, namely $\boldsymbol{\Gamma}$s which are contained in $\mathbf{L}(\mathbb{R})$, closed under countable unions and intersections, and closed under exactly one among $\exists^{\mathbb{R}}$ or $\forall^{\mathbb{R}}$. Any such pointclass can be taken to be the base of a hierarchy, obtained by taking complements and closure under $\exists^{\mathbb{R}}$ and $\forall^{\mathbb{R}}$, and if $\boldsymbol{\Gamma}$ itself is minimal, *i.e.*, it is not of the form $\exists^{\mathbb{R}}\boldsymbol{\Lambda}$ or $\forall^{\mathbb{R}}\boldsymbol{\Lambda}$ for some $\boldsymbol{\Lambda} \subset \boldsymbol{\Gamma}$, then the resulting hierarchy is maximal. Call such an object a **projective-like hierarchy**. The projective-like hierarchies are classified into four distinct types, and for each type the appropriate pattern for the prewellordering properties is established, first for the base level, and then for the higher levels by Moschovakis' periodicity. Since each projective-like pointclass is contained in a unique projective-like hierarchy, this yields a complete analysis of the prewellordering property for projective-like pointclasses.

Pointclasses and well-ordered unions by S. C. Jackson and D. A. Martin.

In this paper the general question of when a pointclass is closed under well-ordered unions is addressed. First a couple of easy facts are recalled: if $\mathsf{PWO}(\boldsymbol{\Gamma})$ holds, then $\check{\boldsymbol{\Gamma}}$ is not closed under well-ordered unions of length κ, where κ is the length of a $\boldsymbol{\Gamma}$-norm; if moreover $\boldsymbol{\Gamma}$ is closed under countable unions and intersections, and under $\exists^{\mathbb{R}}$, then $\boldsymbol{\Gamma}$ is closed under well-ordered unions of length κ. Then Jackson and Martin prove under $\mathsf{AD}+\mathsf{DC}$ that

> Suppose $\boldsymbol{\Gamma}$ is non-self-dual and closed under $\exists^{\mathbb{R}}$ and $\forall^{\mathbb{R}}$. Then either $\boldsymbol{\Gamma}$ or $\check{\boldsymbol{\Gamma}}$ is closed under well-ordered unions. (10)

Thus if $\boldsymbol{\Gamma}$ is as above and moreover $\mathsf{PWO}(\boldsymbol{\Gamma})$, then $\boldsymbol{\Gamma}$ is closed under well-ordered unions. This last result complements Lemma 2.4.1 in the preceding paper by Kechris, Solovay, and Steel, which proves the same result[3] assuming that $\boldsymbol{\Gamma}$ is closed under countable unions, countable intersections, under $\exists^{\mathbb{R}}$ but *not* under $\forall^{\mathbb{R}}$. Therefore

> If $\boldsymbol{\Gamma}$ is non-self-dual and closed under countable intersections and $\exists^{\mathbb{R}}$, and $\mathsf{PWO}(\boldsymbol{\Gamma})$ holds, then $\boldsymbol{\Gamma}$ is closed under well-ordered unions. (11)

Clearly, for a pointclass $\boldsymbol{\Gamma}$ to be closed under well-ordered unions is a meaningful property inasmuch there are well-ordered sequences of sets in $\boldsymbol{\Gamma}$ to be considered. Moreover if $\langle A_\alpha : \alpha < \nu \rangle$ is a sequence of sets in such a $\boldsymbol{\Gamma}$, then by replacing each A_α with $\bigcup_{\beta<\alpha} A_\beta$ and thinning out the sequence if needed, we may assume that the sets are strictly increasing. In this paper it is shown, assuming $\mathsf{AD}+\mathsf{DC}$, that

> If $\boldsymbol{S}(\kappa)$ has the scale property and $\mathrm{cf}(\kappa) > \omega$, then there is no strictly increasing sequence of sets in $\boldsymbol{S}(\kappa)$ of length κ^+, (12)

where $\boldsymbol{S}(\kappa)$ is the class of all κ-Suslin sets. The proof breaks down into two cases, depending whether κ is a successor or limit of uncountable cofinality.

The strength of Borel Wadge determinacy by A. Louveau and J. Saint-Raymond, and **Some results in the Wadge hierarchy of Borel sets** by A. Louveau.

Harrington proved in [Har78] that the semi-linear ordering principle restricted to the class of $\boldsymbol{\Pi}^1_1$ sets, $\mathsf{SLO}(\boldsymbol{\Pi}^1_1)$ for short, implies the existence of $x^\#$, for any

[3]Actually in that paper the assumption $\mathsf{PWO}(\boldsymbol{\Gamma})$ is replaced by the weaker $\mathsf{Red}(\boldsymbol{\Gamma})$.

real x, and therefore it implies $\mathsf{Det}(\mathbf{\Pi}^1_1)$. By work of Harrington and Martin $\mathsf{Det}(\mathbf{\Pi}^1_1)$ is equivalent to the determinacy of Boolean combinations of $\mathbf{\Pi}^1_1$ sets, hence it follows that $\mathsf{SLO}(\mathbf{\Pi}^1_1)$, the determinacy of all $G_{\mathrm{W}}(A, B)$ and $G_{\mathrm{L}}(A, B)$, with $A, B \in \mathbf{\Pi}^1_1$, and $\mathsf{Det}(\mathbf{\Pi}^1_1)$, are all equivalent. In fact the determinacy of all Wadge games G_{W}, the determinacy of all Lipschitz games G_{L}, and SLO are all equivalent [And03, And06] and the same holds true when restricted to any pointclass with sufficient closure properties, such as the $\mathbf{\Pi}^1_n$'s; for these reason we shall refer to any one of these hypotheses as **Wadge determinacy**. By [Har78] and [Ste80], $\mathsf{Det}(\mathbf{\Pi}^1_1)$ is also equivalent to the following:

$$\forall A, B \in \mathbf{\Pi}^1_1 \setminus \mathbf{\Delta}^1_1 \, \exists f \, (f : \mathbb{R} \to \mathbb{R} \text{ is a Borel isomorphism and } f(A) = B).$$

All these results lent some credibility to the conjecture that a similar pattern should occur in the Borel context, namely that Wadge determinacy for Borel sets should imply Borel determinacy, which by work of Martin [Mar75] holds in ZFC and by work of Friedman [Fri71B] is not provable in second order arithmetic. But it is not so, as proved in the first paper by Louveau and Saint-Raymond: Wadge determinacy is provable in second order arithmetic. The proof relies heavily on Wadge's analysis of the Borel classes, together with a "ramification" technique which appeared in [LSR87] for the Borel classes: one associates to each non-self-dual Borel Wadge degree, as described by Wadge, a specific game, which is somewhat of an unfolding of a Wadge game. Its determinacy implies that any set which is of this degree is strategically complete, *i.e.*, player player II wins with it the Wadge game against any other set in the class.

The second paper [Lou83] is a bit different from the other papers, as it deals with the "lightface" aspects of the Wadge hierarchy. In a previous paper [Lou80], Louveau had proved that for hyperarithmetic sets, the Borel class can be witnessed hyperarithmetically. A similar feature is proved in the paper for each Borel class in the Wadge hierarchy of Borel sets. But a great deal of work is done on introducing operations in order to build all Borel Wadge classes, and define appropriate codings of both the classes and the sets in them so that the corresponding lightface statement makes sense. It was also the first—and for quite some time the only—place where a printed account of some of Wadge's work could be found.

Closure properties of pointclasses by J. Steel, and
More closure properties of pointclasses by H. Becker.

In several of the results mentioned in the paragraphs above, in order to prove that a pointclass $\mathbf{\Gamma}$ has some structural property, like reduction or prewellordering, we must require that $\mathbf{\Gamma}$ (or perhaps $\mathbf{\Delta}_{\mathbf{\Gamma}}$) be closed under some simpler structural property, like closure under finite (or countable) unions or intersections. Notice that closure under finite union or intersections is never a problem with the Baire or projective classes, but in the realm or *arbitrary* pointclasses, closure under finite unions or intersections is a non-trivial matter. One might ask, for example: Under which assumptions on $\boldsymbol{\Gamma}$ does closure under finite unions imply closure for countable unions? Do closure properties of $\Delta_{\boldsymbol{\Gamma}}$ imply analogous properties for $\boldsymbol{\Gamma}$ or $\breve{\boldsymbol{\Gamma}}$? The paper by Steel proves several theorems under AD that address these questions. Here is just a sample of such results:

If $\Delta_{\boldsymbol{\Gamma}}$ is closed under finite (or countable) unions and $\mathsf{Sep}(\boldsymbol{\Gamma})$ holds, then $\boldsymbol{\Gamma}$ is closed under finite (or countable) unions too. (13)

If $\boldsymbol{\Gamma}$ is closed under finite unions and $\mathsf{Sep}(\breve{\boldsymbol{\Gamma}})$, then $\boldsymbol{\Gamma}$ is closed under countable unions. (14)

Suppose $\boldsymbol{\Gamma}$ is closed under finite intersections and countable unions, but *not* under countable intersections. Then $\mathsf{PWO}(\boldsymbol{\Gamma})$. (15)

Thus (14) and (15) generalize a well-known fact about the Borel hierarchy, that is: $\boldsymbol{\Sigma}^0_\alpha$ does not have the separation property but has the prewellordering property. Steel's paper contains also an interesting conjecture. Recall that Suslin's operation $\mathscr{A}$ is a Boolean operation with $\boldsymbol{\Sigma}^1_1$ truth table, and that the Boolean operations that generate the $\boldsymbol{\Sigma}^0_\alpha$s are just compositions of the operations of countable unions and countable intersections.

CONJECTURE 3.1. Assume AD and suppose $\boldsymbol{\Gamma}$ is non-self-dual and closed under both countable intersections and countable unions. Then either $\boldsymbol{\Gamma}$ or $\breve{\boldsymbol{\Gamma}}$ is closed under $\mathscr{A}$.

Becker's paper deals with closure under measure and category quantifiers. If $A \subseteq {}^\omega 2 \times {}^\omega 2$ then let

$$\forall^* y\, A = \{x : A_{(x)} \text{ is comeager}\}$$

and

$$\forall^\mu y\, A = \{x : \mu({}^\omega 2 \setminus A_{(x)}) = 0\}$$

where μ is the Lebesgue measure on ${}^\omega 2$ and $A_{(x)} = \{y : (x, y) \in A\}$ is the vertical section of A through x. In other words, $\forall^* y\, A$ is the set of all x such that $(x, y) \in A$ for comeager many y, while $\forall^\mu y\, A$ is the set of all x such that $(x, y) \in A$ for μ-almost every y; their dual quantifiers are defined by

$$\exists^* y\, A = \{x : A_{(x)} \text{ is non-meager}\}$$

and

$$\exists^\mu y\, A = \{x : \mu(A_{(x)}) > 0\}.$$

The measure and category quantifiers are very useful in many parts of Descriptive Set Theory—see for example [BK96]. In the present paper it is shown that if $\underset{\sim}{\Gamma}$ is nonselfdual and closed under countable unions and countable intersections, then it is closed under the category and measure quantifiers. In particular, $\underset{\sim}{\Delta}^1_1$ is closed under measure and category quantifiers.

More measures from AD by J. Steel. One of the early consequences of determinacy is Martin's result that ω_1 has the strong partition property, $\omega_1 \to (\omega_1)^{\omega_1}$. This in turns implies Solovay's result that ω_1 is measurable. In the following years the study of the strong partition property for cardinals $< \Theta$ became one of the main research topics of the Cabal Seminar. The construction of the normal measure from the strong partition property is usually achieved via the Boundedness Lemma together with an appropriate coding of elements of ${}^\kappa\kappa$.

In the present paper it is shown that, assuming AD, for every regular $\kappa < \Theta$ there is a measure on ${}^\kappa\kappa$. The main technical twist is the use of the Recursion Theorem instead of the Boundedness Lemma.

§4. **Recent developments.** In this last section we will try to survey some of the development that occurred after the papers in this volume were originally written.

4.1. SLO **and weaker reducibilities.** As we already mentioned, Harrington proved in [Har78] that $\mathsf{SLO}(\underset{\sim}{\Pi}^1_1)$ is equivalent to the determinacy of all $\underset{\sim}{\Pi}^1_1$ games. This was extended by Hjorth [Hjo96] to the next level, *i.e.*, $\mathsf{SLO}(\underset{\sim}{\Pi}^1_2)$ implies $\underset{\sim}{\Pi}^1_2$-determinacy—generalizations of these results to all projective levels, and beyond, have been an elusive goal, as they seem to depend on further technical advancement of core model theory. Yet the results we have now seem to lend some evidence to the following conjecture, probably due to Solovay:

CONJECTURE 4.1. Assume $\mathbf{V} = \mathbf{L}(\mathbb{R})$. Then

$$\mathsf{SLO} \Longrightarrow \mathsf{AD}.$$

Note that there is no obvious natural way to reduce a general perfect information, zero-sum game on ω into a Wadge game, so the proof—if the conjecture is true—will probably be quite indirect. Although progress on this conjecture has been essentially nil after [Hjo96], the Semi-Linear Ordering principle and some generalizations of it have been investigated in recent years. In [And03] it is shown that SLO is strong enough to prove the basic structural results on the Wadge hierarchy as embodied in diagram (3), and in [AM03], the analogue of the Wadge hierarchy using Borel functions was introduced: for any $A, B \subseteq \mathbb{R}$ let

$$A \leq_{\underset{\sim}{\Delta}^1_1} B \iff \exists f\, \left(f : \mathbb{R} \to \mathbb{R} \text{ is Borel and } f^{-1}(B) = A\right).$$

The induced equivalence relation yields the notion of $\underset{\sim}{\Delta}^1_1$ degree, and it turns out that the their structure is similar to the one of Wadge degrees, *i.e.*, it is well-founded, the self-dual degrees and non-self-dual pairs of degrees alternate, with self-dual degrees occupying the limit levels of countable cofinality, and since the length of this hierarchy is Θ, then its picture is just (3). Since all uncountable Polish spaces are Borel isomorphic, this hierarchy is independent of the underlying space, a feature sorely missing from the Wadge hierarchy. In this case there are no analogues of the games G_W or G_L, and the proofs use the principle $\mathsf{SLO}^{\mathbf{\Delta}^1_1}$, the analogue of SLO for Borel reductions,

$$\forall A, B \subseteq \mathbb{R}\big(A \leq_{\mathbf{\Delta}^1_1} B \vee \neg B \leq_{\mathbf{\Delta}^1_1} A\big). \qquad (\mathsf{SLO}^{\mathbf{\Delta}^1_1})$$

Note that $\mathsf{SLO}^{\mathbf{\Delta}^1_1}$ follows from SLO, hence from AD, and in [AM03] it is conjectured that $\mathsf{SLO}^{\mathbf{\Delta}^1_1} \implies \mathsf{SLO}$. In [And06] a similar analysis is carried out for the $\mathbf{\Delta}^0_2$ reducibility: again $\mathsf{SLO}^{\mathbf{\Delta}^0_2}$ is able to civilize this hierarchy and the familiar structure (3) is obtained, and moreover in this case it is shown that $\mathsf{SLO}^{\mathbf{\Delta}^0_2} \iff \mathsf{SLO}$. (A function is said to be $\mathbf{\Delta}^0_\alpha$ if the preimage of a $\mathbf{\Sigma}^0_\alpha$ is $\mathbf{\Sigma}^0_\alpha$.)

The results above seem to indicate that similar results should hold true of $\leq_{\mathcal{F}}$ reductions, *i.e.*,

$$A \leq_{\mathcal{F}} B \iff \exists f \in \mathcal{F}\big(A = f^{-1}(B)\big)$$

where $\mathcal{F} \subseteq {}^{\mathbb{R}}\mathbb{R}$. Obviously the class $\mathcal{F}$ must satisfy some assumptions in order for us to obtain non trivial results, *e.g.*, $\mathcal{F}$ must be closed under composition, and must contain the identity, so that $\leq_{\mathcal{F}}$ is a quasi-order, $\mathcal{F} \neq {}^{\mathbb{R}}\mathbb{R}$, *etc.* Motto Ros in [MR07] has isolated a very general class of $\mathcal{F}$ as above, with $\mathcal{F}$ a collection of Borel functions, and has shown, assuming AD+DC($\mathbb{R}$) that the structure of the $\mathcal{F}$-hierarchy can be either of Wadge-type or of Lipschitz type, *i.e.*, the ordering of the $\mathcal{F}$-degrees is as in (3) or as in (2). For example: when $\mathcal{F}$ is the collection of all $\mathbf{\Delta}^0_\xi$ functions, the resulting hierarchy is of Wadge type; when $\mathcal{F}$ is the collection of all $\mathbf{\Delta}^0_\alpha$ functions for some $\alpha < \lambda$, the resulting hierarchy is of Lipschitz type.

4.2. Connections with bqo theory. By Martin's result, Wadge reducibility $\leq_W$ is one of a few examples of "natural" quasi-orderings which are well-quasi-orderings (wqo's), *i.e.*, which admit neither infinite antichains, nor infinite strictly decreasing sequences. Other famous examples are the countable linear orders with embeddability (Laver [Lav71]) and the finite graphs with the minor ordering (Robertson and Seymour [RS04]).

As the class of wqo's lacks nice closure properties, it is usual to consider the stronger notion of a better-quasi-ordering (bqo): A quasi-ordering $(Z, \leq_Z)$ is a bqo if, for any continuous (or equivalently Borel) map $h\colon [\omega]^\omega \to Z$ there is an $X \in [\omega]^\omega$ with $h(X) \leq h(X \setminus \{\min X\})$, where $[\omega]^\omega$ is the collection of all infinite subsets of ω identified with the set of all increasing elements of the Baire space, and Z is taken with the discrete topology. (For a nice introduction to bqo theory, see Simpson's contribution in [MW85, Chapter 9].)

It is not hard to check that under AD the quasi-orders $\leq_L$ and $\leq_W$ are indeed bqo's. But one can get by similar techniques other bqo results. For example, van Engelen, Miller, and Steel prove in [vEMS87] that if $(Z, \leq_Z)$ is a bqo and one orders $\mathcal{S}_Z$, the set of all functions $h\colon {}^\omega\omega \to Z$, by

$$h_1 \preceq h_2 \iff$$
$$\exists \varphi\colon {}^\omega\omega \to {}^\omega\omega \text{ Lipschitz such that } \forall x \in {}^\omega\omega \, \big(h_1(x) \leq_Z h_2(\varphi(z))\big),$$

then $\mathcal{S}_Z$ is bqo too. ($\leq_L$ corresponds to the case $Z = \{0, 1\}$, with 0 and 1 incomparable.)

This result in turn is used to prove other bqo results, in particular in [LSR90], where Louveau and Saint-Raymond extend Laver's result about countable linear orders to Borel (or projective) linear orders embeddable in $({}^\omega\mathbb{R}, \leq_{\text{lex}})$, using AD.

4.3. Reducibility in higher dimension. An alternative way of looking at the Wadge hierarchy is to view subsets A of $\mathbb{R}$ as structures $(\mathbb{R}, A)$ in a language with a unary predicate, with the Wadge ordering being continuous homomorphisms between such structures. This of course opens the possibility of extending it to more complicated structures with domain $\mathbb{R}$ (or arbitrary Polish spaces) and, say, a n-ary relation on it. Concretely, in order to allow arbitrary Polish spaces as domains and still avoid purely topological difficulties, one prefers to consider Borel reductions rather than continuous in this context. So for $\mathcal{X}$, $\mathcal{Y}$ Polish spaces and $A \subseteq \mathcal{X}^n$, $B \subseteq \mathcal{Y}^n$, let

$$(\mathcal{X}, A) \leq_{\boldsymbol{B}} (\mathcal{Y}, B)$$

just in case

$$\exists f\colon \mathcal{X} \to \mathcal{Y} \text{ Borel, and } \forall \bar{a} \in \mathcal{X}^n \, (\bar{a} \in A \iff f(\bar{a}) \in B)$$

(where we follow the convention from model-theory and write $\bar{a}$ for the n-tuple $(a_1, \ldots, a_n)$ and $f(\bar{a})$ for $(f(a_1), \ldots, f(a_n))$; also, when the ambient spaces $\mathcal{X}$ and $\mathcal{Y}$ are understood, we simply write $A \leq_{\boldsymbol{B}} B$).

These considerations provide a natural descriptive complexity for relations. This notion was first introduced by Friedman and Stanley in [FS89], who used it to provide a classification for first order theories, by comparing their associated space of countable models with domain ω, endowed with isomorphism. It was extended soon after by Kechris and Louveau [Kec92, Lou92] to equivalence relations and even more complicated structures. It should be noted that many properties, like being an equivalence relation, a quasi-ordering, ... are downward preserved under $\leq_{\boldsymbol{B}}$, so that the subject breaks naturally into many sub-areas. And in each sub-area there is no satisfying alternative approach to descriptive complexity by using operations instead of reducibility, as in the one-dimensional case. This is because equivalence relations, for example, are not built from simpler equivalence relations, in general. And the only ways that have been proposed, like Louveau's notion of "potential Wadge class" (see [Lou94]), may be useful but are too coarse (*i.e.*, too close to the one-dimensional situation) to provide the right notion of descriptive complexity.

A lot of work has been done on linear orders, quasi-orders, even graphs, but the main part of the activity in Descriptive Set Theory over the last two decades has been to understand Polish spaces with Borel, or more generally analytic, equivalence relations. We won't try to give here an account of this theory, but refer the reader to the nice overview [HK01].

Let us just mention here that the situation for the higher dimensional theory is very different, and much more complicated that in dimension one: although some features of $\leq_B$ are nice, it is a very complicated quasi-order, ill-founded and with large antichains. And games are of little use in the new situation, so that one cannot really work by analogy with $\leq_W$.

4.3.1. *Definable cardinality.* In the context of AD and using arbitrary reductions rather than Borel ones, the classification results for equivalence relations become results on cardinality of quotients: if $\preccurlyeq$ denotes this coarser reducibility relation, any f witnessing $E \preccurlyeq F$ induces an injection $\hat{f}\colon \mathbb{R}/E \to \mathbb{R}/F$. Conversely, assuming $\mathsf{AD}_{\mathbb{R}}$, for any $g\colon \mathbb{R}/E \to \mathbb{R}/F$ we can uniformize the relation $\tilde{g} = \{(x, y) \in \mathbb{R}^2 : f([x]_E) = [y]_F\}$ by some $f\colon \mathbb{R} \to \mathbb{R}$: then f witnesses $E \preccurlyeq F$ and moreover $\hat{f} = g$. In other words, under AD the quasi-order $\preccurlyeq$ on equivalence relations yields an injection of the quotients, and under $\mathsf{AD}_{\mathbb{R}}$ any injection of the quotients lifts to a $\preccurlyeq$-reduction on $\mathbb{R}$. Many of the results on Borel or analytic equivalence relations using $\leq_B$, can be recast under AD using $\preccurlyeq$ with essentially the same proof: for example the Silver [Sil80] and Harrington-Kechris-Louveau [HKL90] dichotomies become: If $E \in \undertilde{\mathbf{\Pi}}^1_1$ is an equivalence relation on $\mathbb{R}$, then either $|\mathbb{R}/E| \leq \omega$, or else $|\mathbb{R}| \leq |\mathbb{R}/E|$; if $E \in \undertilde{\mathbf{\Delta}}^1_1$ is an equivalence relation on $\mathbb{R}$, then either $|\mathbb{R}/E| \leq |\mathbb{R}|$ or else $|\wp(\omega)/\,\mathrm{Fin}| \leq |\mathbb{R}/E|$. But in fact these dichotomies of admit a more substantial generalization:

If E an arbitrary equivalence relation on $\mathbb{R}$, then either (16)
(a) $|\mathbb{R}/E| < \Theta$, or else
(b) $|\mathbb{R}| \leq |\mathbb{R}/E|$.

If $E \in$ an arbitrary equivalence relation on $\mathbb{R}$, then either (17)
(a) $|\mathbb{R}/E| \leq |\wp(\kappa)|$, for some $\kappa < \Theta$, or else
(b) $|\wp(\omega)/\,\mathrm{Fin}| \leq |\mathbb{R}/E|$.

Dichotomies (16) and (17) were first proved under $\mathsf{AD}_{\mathbb{R}}$ by Harrington and Sami [HS79], and Ditzen [Dit92], and, independently, by Foreman and Magidor (unpublished). The consistency strength was then reduced to $\mathsf{AD} + \mathbf{V}{=}\mathbf{L}(\mathbb{R})$ by Woodin (unpublished), and Hjorth [Hjo95], respectively.

4.4. The Wadge hierarchy in set theory. Although most of the research on the notion of continuous pre-images is concerned with the general theory of pointclasses, the Wadge hierarchy has important applications in the study of models of AD^+, a generalization of AD defined by:

- every $A \subseteq \mathbb{R}$ is ∞-Borel, *i.e.*, $A = \{y \in \mathbb{R} : \mathbf{L}_\alpha[S, y] \models \varphi[S, y]\}$ for some $S \subseteq \mathrm{Ord}$ and some formula φ, and
- for every $A \subseteq \mathbb{R}$, every $\lambda < \Theta$ and every surjection $f : {}^\omega\lambda \twoheadrightarrow \mathbb{R}$, the ordinal game on λ with payoff $f^{-1}(A)$ is determined. $\qquad (\mathsf{AD}^+)$

Clearly $\mathsf{AD}^+ \implies \mathsf{AD}$, and both $\mathsf{AD} + \mathbf{V}{=}\mathbf{L}(\mathbb{R})$ and $\mathsf{AD}_\mathbb{R}{+}\mathsf{DC}$ imply AD^+; in fact every known model of AD does satisfies AD^+, and the general consensus seems to be that AD^+ is the correct axiom for the study of models of determinacy. Assuming AD^+ let

$$\Theta(A) = \sup\{\|B\|_\mathrm{W} : B \text{ is ordinal definable from reals and } A\}$$

and let

$$\begin{aligned} \Theta_0 &= \Theta(\varnothing) \\ \Theta_{\alpha+1} &= \Theta(A) \text{ for some/any } A \text{ such that } \|A\|_\mathrm{W} = \Theta_\alpha \\ \Theta_\lambda &= \sup_{\alpha<\lambda} \Theta_\alpha . \end{aligned}$$

The sequence of the Θ_α's was introduced in [Sol78B] and it is called the Solovay sequence; note that it may not be defined for all α's. For example, if $\mathbf{V} = \mathbf{L}(\mathbb{R})$ then every set is ordinal definable from a real, hence $\Theta = \Theta_0$ and the Solovay sequence is not defined for larger indexes; assuming $\mathsf{AD}_\mathbb{R}$ will ensure that the sequence is defined up to some limit ordinal λ. In general, if Θ_α is defined then $\mathbf{L}(\wp_{\Theta_\alpha}(\mathbb{R}))$ is a model for $\mathsf{AD}^+ + \Theta{=}\Theta_\alpha$, where $\wp_\nu(\mathbb{R}) = \{X \subseteq \mathbb{R} : \|X\|_\mathrm{W} < \nu\}$. The smallest model of $\mathsf{AD}_\mathbb{R}$ is $\mathbf{L}(\wp_{\Theta_\omega}(\mathbb{R}))$, and in this model Θ has cofinality ω. Even stronger theories are obtained when the model satisfies '$\Theta = \Theta_\Theta$', or '$\Theta > \Theta_0$ is regular'—see [Woo99]. Thus the Wadge hierarchy and, in particular, the Solovay sequence of the Θ_α's can be used to measure the strength of models of AD^+. Unfortunately, this method of comparing AD^+ models is not always successful since Woodin, in unpublished work, has shown that it is consistent that there are two models M and N of AD^+ having the same reals and with divergent Wadge hierarchies.

Finally we mention a fairly recent application of the Wadge hierarchy to the study, under AD, of cardinalities of pointclasses. As any pointclass $\underset{\sim}{\Gamma}$ is the surjective image of $\mathbb{R}$, *i.e.*, it is in bijection with $\mathbb{R}/E$ for some E, and as any $\mathbb{R}/E$ can be embedded into some $\underset{\sim}{\Gamma}$, it follows that the cardinalities $|\underset{\sim}{\Gamma}|$ are cofinal in the set of cardinalities of quotients of $\mathbb{R}$. The general problem is to determine which $\underset{\sim}{\Gamma}$ are **cardinality pointclasses**, *i.e.*, such that $|\underset{\sim}{\Gamma}| > |\underset{\sim}{\Delta}|$, for any $\underset{\sim}{\Delta} \subset \underset{\sim}{\Gamma}$. Examples of self-dual cardinality pointclasses are $\underset{\sim}{\Delta}^0_1$, or the pointclasses of the form $\bigcup_{\alpha<\lambda} \underset{\sim}{\Gamma}_\alpha$ with $\underset{\sim}{\Gamma}_\alpha$ increasing cardinality pointclasses and λ limit—call such a pointclass a **tower**, and say it has countable cofinality if $\mathrm{cf}(\lambda) = \omega$. In [AHN07] a complete description of the cardinality pointclasses is given, and an interesting feature of the proof is that it uses the detailed analysis of the Wadge hierarchy. Assuming $\mathsf{AD}+\mathsf{DC}(\mathbb{R})$, a non-self-dual $\utilde{\boldsymbol{\Gamma}}$ is a cardinality pointclass iff $\utilde{\boldsymbol{\Gamma}}$ is closed under pre-images of $\utilde{\boldsymbol{\Delta}}^0_2$ functions; a self-dual pointclass $\utilde{\boldsymbol{\Delta}}$ strictly larger than $\utilde{\boldsymbol{\Delta}}^0_1$ is a cardinality pointclass iff either it is a tower, or else it is the (necessarily self-dual) pointclass immediately above a tower of countable cofinality. Therefore assuming $\mathsf{AD}+\mathsf{DC}(\mathbb{R})$ the results of Hjorth [Hjo98, Hjo02]

$$\alpha < \beta \Longrightarrow |\utilde{\boldsymbol{\Sigma}}^0_\alpha| < |\utilde{\boldsymbol{\Sigma}}^0_\beta| \qquad \text{and} \qquad |\utilde{\boldsymbol{\Delta}}^1_n| < |\utilde{\boldsymbol{\Sigma}}^1_n| < |\utilde{\boldsymbol{\Delta}}^1_{n+1}|$$

are obtained as corollaries.

REFERENCES

ALESSANDRO ANDRETTA

[And03] *Equivalence between Wadge and Lipschitz determinacy*, ***Annals of Pure and Applied Logic***, vol. 123 (2003), no. 1–3, pp. 163–192.

[And06] *More on Wadge determinacy*, ***Annals of Pure and Applied Logic***, vol. 144 (2006), no. 1–3, pp. 2–32.

ALESSANDRO ANDRETTA, GREGORY HJORTH, AND ITAY NEEMAN

[AHN07] *Effective cardinals of boldface pointclasses*, ***Journal of Mathematical Logic***, vol. 7 (2007), no. 1, pp. 35–92.

ALESSANDRO ANDRETTA AND DONALD A. MARTIN

[AM03] *Borel-Wadge degrees*, ***Fundamenta Mathematicae***, vol. 177 (2003), no. 2, pp. 175–192.

HOWARD S. BECKER AND ALEXANDER S. KECHRIS

[BK96] ***The descriptive set theory of Polish group actions***, London Mathematical Society Lecture Note Series, vol. 232, Cambridge University Press, Cambridge, 1996.

ACHIM DITZEN

[Dit92] *Definable equivalence relations on Polish spaces*, ***Ph.D. thesis***, California Institute of Technology, 1992.

JACQUES DUPARC

[Dup01] *Wadge hierarchy and Veblen hierarchy. I. Borel sets of finite rank*, ***The Journal of Symbolic Logic***, vol. 66 (2001), no. 1, pp. 56–86.

[Dup03] *A hierarchy of deterministic context-free ω-languages*, ***Theoretical Computer Science***, vol. 290 (2003), no. 3, pp. 1253–1300.

JACQUES DUPARC, OLIVIER FINKEL, AND JEAN-PIERRE RESSAYRE

[DFR01] *Computer science and the fine structure of Borel sets*, ***Theoretical Computer Science***, vol. 257 (2001), no. 1–2, pp. 85–105.

HARVEY FRIEDMAN

[Fri71B] *Higher set theory and mathematical practice*, ***Annals of Mathematical Logic***, vol. 2 (1971), no. 3, pp. 325–357.

HARVEY FRIEDMAN AND LEE STANLEY

[FS89] *A Borel reducibility theory for classes of countable structures*, ***The Journal of Symbolic Logic***, vol. 54 (1989), no. 3, pp. 894–914.

LEO A. HARRINGTON

[Har78] *Analytic determinacy and* $0^{\#}$, ***The Journal of Symbolic Logic***, vol. 43 (1978), pp. 685–693.

LEO A. HARRINGTON, ALEXANDER S. KECHRIS, AND ALAIN LOUVEAU

[HKL90] *A Glimm–Effros dichotomy for Borel equivalence relations*, ***Journal of the American Mathematical Society***, vol. 3 (1990), pp. 902–928.

LEO A. HARRINGTON AND RAMEZ-LABIB SAMI

[HS79] *Equivalence relations, projective and beyond*, ***Logic Colloquium '78. Proceedings of the Colloquium held in Mons, August 24–September 1, 1978*** (Maurice Boffa, Dirk van Dalen, and Kenneth McAloon, editors), Studies in Logic and the Foundations of Mathematics, vol. 97, North-Holland, Amsterdam, 1979, pp. 247–264.

GREGORY HJORTH

[Hjo95] *A dichotomy for the definable universe*, ***The Journal of Symbolic Logic***, vol. 60 (1995), no. 4, pp. 1199–1207.

[Hjo96] Π^1_2 *Wadge degrees*, ***Annals of Pure and Applied Logic***, vol. 77 (1996), no. 1, pp. 53–74.

[Hjo98] *An absoluteness principle for Borel sets*, ***The Journal of Symbolic Logic***, vol. 63 (1998), no. 2, pp. 663–693.

[Hjo02] *Cardinalities in the projective hierarchy*, ***The Journal of Symbolic Logic***, vol. 67 (2002), no. 4, pp. 1351–1372.

GREGORY HJORTH AND ALEXANDER S. KECHRIS

[HK01] *Recent developments in the theory of Borel reducibility*, ***Fundamenta Mathematicae***, vol. 170 (2001), no. 1–2, pp. 21–52.

ALEXANDER S. KECHRIS

[Kec92] *The structure of Borel equivalence relations in Polish spaces*, ***Set theory of the continuum. Papers from the workshop held in Berkeley, California, October 16–20, 1989*** (H. Judah, W. Just, and H. Woodin, editors), Mathematical Sciences Research Institute Publications, vol. 26, Springer, New York, 1992, pp. 89–102.

ALEXANDER S. KECHRIS, BENEDIKT LÖWE, AND JOHN R. STEEL

[CABAL I] *Games, scales, and Suslin cardinals: the Cabal seminar, volume I*, Lecture Notes in Logic, vol. 31, Cambridge University Press, 2008.

ALEXANDER S. KECHRIS, DONALD A. MARTIN, AND YIANNIS N. MOSCHOVAKIS

[CABAL iii] *Cabal seminar* 79–81, Lecture Notes in Mathematics, no. 1019, Berlin, Springer, 1983.

ALEXANDER S. KECHRIS, DONALD A. MARTIN, AND JOHN R. STEEL

[CABAL iv] *Cabal seminar* 81–85, Lecture Notes in Mathematics, no. 1333, Berlin, Springer, 1988.

ALEXANDER S. KECHRIS AND YIANNIS N. MOSCHOVAKIS

[CABAL i] *Cabal seminar* 76–77, Lecture Notes in Mathematics, no. 689, Berlin, Springer, 1978.

[KM78B] *Notes on the theory of scales*, in *Cabal Seminar* 76–77 [CABAL i], pp. 1–53, reprinted in [CABAL I], p. 28–74.

CASIMIR KURATOWSKI

[Kur58] ***Topologie. Vol. I***, 4ème ed., Monografie Matematyczne, vol. 20, Państwowe Wydawnictwo Naukowe, Warsaw, 1958.

RICHARD LAVER

[Lav71] *On Fraïssé's order type conjecture*, ***Annals of Mathematics***, vol. 93 (1971), pp. 89–111.

ALAIN LOUVEAU

[Lou80] *A separation theorem for Σ^1_1 sets*, ***Transactions of the American Mathematical Society***, vol. 260 (1980), no. 2, pp. 363–378.

[Lou83] *Some results in the Wadge hierarchy of Borel sets*, this volume, originally published in Kechris et al. [CABAL iii], pp. 28–55.

[Lou92] *Classifying Borel structures*, ***Set Theory of the Continuum. Papers from the workshop held in Berkeley, California, October 16–20, 1989*** (H. Judah, W. Just, and H. Woodin, editors), Mathematical Sciences Research Institute Publications, vol. 26, Springer, New York, 1992, pp. 103–112.

[Lou94] *On the reducibility order between Borel equivalence relations*, ***Logic, Methodology and Philosophy of Science, IX. Proceedings of the Ninth International Congress held in Uppsala, August 7–14, 1991*** (Dag Prawitz, Brian Skyrms, and Dag Westerståhl, editors), Studies in Logic and the Foundations of Mathematics, vol. 134, North-Holland, Amsterdam, 1994.

ALAIN LOUVEAU AND JEAN SAINT-RAYMOND

[LSR87] *Borel classes and closed games: Wadge-type and Hurewicz-type results*, ***Transactions of the American Mathematical Society***, vol. 304 (1987), no. 2, pp. 431–467.

[LSR88A] *Les propriétés de réduction et de norme pour les classes de Boréliens*, ***Fundamenta Mathematicae***, vol. 131 (1988), no. 3, pp. 223–243.

[LSR88B] *The strength of Borel Wadge determinacy*, this volume, originally published in Kechris et al. [CABAL iv], pp. 1–30.

[LSR90] *On the quasi-ordering of Borel linear orders under embeddability*, ***The Journal of Symbolic Logic***, vol. 55 (1990), no. 2, pp. 537–560.

NIKOLAI LUZIN AND WACLAW SIERPIŃSKI

[LS29] *Sur les classes des constituantes d'un complémentaire analytique*, ***Comptes rendus hebdomadaires des séances de l'Académie des Sciences***, vol. 189 (1929), pp. 794–796.

RICHARD MANSFIELD AND GALEN WEITKAMP

[MW85] ***Recursive aspects of descriptive set theory***, Oxford Logic Guides, vol. 11, The Clarendon Press Oxford University Press, New York, 1985, With a chapter by Stephen Simpson.

DONALD A. MARTIN

[Mar75] *Borel determinacy*, ***Annals of Mathematics***, vol. 102 (1975), no. 2, pp. 363–371.

LUCA MOTTO ROS

[MR07] *General reducibilities for sets of reals*, ***Ph.D. thesis***, Politecnico di Torino, 2007.

NEIL ROBERTSON AND P. D. SEYMOUR

[RS04] *Graph minors. XX. Wagner's conjecture*, ***Journal of Combinatorial Theory. Series B***, vol. 92 (2004), no. 2, pp. 325–357.

WACLAW SIERPIŃSKI

[Sie52] ***General topology***, Mathematical Expositions, No. 7, University of Toronto Press, Toronto, 1952, Translated by C. Cecilia Krieger.

ROMAN SIKORSKI

[Sik58] *Some examples of Borel sets*, ***Colloquium Mathematicum***, vol. 5 (1958), pp. 170–171.

JACK SILVER

[Sil80] *Counting the number of equivalence classes of Borel and coanalytic equivalence relations*, ***Annals of Mathematical Logic***, vol. 18 (1980), no. 1, pp. 1–28.

ROBERT M. SOLOVAY

[Sol78B] *The independence of* DC *from* AD, in Kechris and Moschovakis [CABAL i], pp. 171–184.

JOHN R. STEEL

[Ste77] *Determinateness and subsystems of analysis*, ***Ph.D. thesis***, Berkeley, 1977.

[Ste80] *Analytic sets and Borel isomorphisms*, ***Fundamenta Mathematicae***, vol. 108 (1980), no. 2, pp. 83–88.

[Ste81B] *Determinateness and the separation property*, ***The Journal of Symbolic Logic***, vol. 46 (1981), no. 1, pp. 41–44.

FONS VAN ENGELEN, ARNOLD W. MILLER, AND JOHN R. STEEL

[vEMS87] *Rigid Borel sets and better quasi-order theory*, ***Proceedings of the AMS-IMS-SIAM joint summer research conference on applications of mathematical logic to finite combinatorics held at Humboldt State University, Arcata, Calif., August 4–10, 1985*** (Stephen G. Simpson, editor), Contemporary Mathematics, vol. 65, American Mathematical Society, Providence, RI, 1987, pp. 199–222.

ROBERT VAN WESEP

[Van77] *Subsystems of second-order arithmetic, and descriptive set theory under the axiom of determinateness*, ***Ph.D. thesis***, University of California, Berkeley, 1977.

[Van78A] *Separation principles and the axiom of determinateness*, ***The Journal of Symbolic Logic***, vol. 43 (1978), no. 1, pp. 77–81.

WILLIAM W. WADGE

[Wad84] *Reducibility and determinateness on the Baire space*, ***Ph.D. thesis***, University of California, Berkeley, 1984.

[Wad11] *Early investigations of the degrees of Borel sets*, 2011, this volume.

W. HUGH WOODIN

[Woo99] ***The axiom of determinacy, forcing axioms, and the nonstationary ideal***, De Gruyter Series in Logic and its Applications, Walter de Gruyter, Berlin, 1999.

DIPARTIMENTO DI MATEMATICA
UNIVERSITÀ DI TORINO
VIA CARLO ALBERTO 10, 10123 TORINO
ITALY
E-mail: alessandro.andretta@unito.it

EQUIPE D'ANALYSE FONCTIONNELLE
INSTITUT DE MATHÉMATIQUES DE JUSSIEU
UNIVERSITÉ PARIS VI
4, PLACE JUSSIEU
75230 PARIS, CEDEX 05
FRANCE
E-mail: louveau@math.jussieu.fr

WADGE DEGREES AND DESCRIPTIVE SET THEORY

ROBERT VAN WESEP

The work to be presented here is taken principally from the three sources Steel [Ste77], Van Wesep [Van77], and Wadge [Wad84], listed in the bibliography. There is so far nothing published on this subject except the Van Wesep paper in the *Journal of Symbolic Logic* [Van78A].

In Sections 1, 2, and 3 we provide a general picture of the Wadge degrees. In Section 4 we prove some results of Steel concerning functions from the Turing degrees to $\aleph_1$ modulo the Martin measure and apply them to a computation of the length of the Wadge ordering of $\underset{\sim}{\mathbf{\Delta}}^1_n$ sets. Then in Section 5 we prove some results about the separation, reduction, and prewellordering properties for suitable classes of sets of reals.

We work throughout in ZF+DC+AD, but the reader will be able to determine when and how the determinateness assumption may be relaxed in proving corresponding results about restricted classes of sets of reals.

§1. Definitions.

Definition 1.1. **Baire space** $:= {}^{\omega}\omega$. An element of ${}^{\omega}\omega$ is a **real**. An **interval of Baire** is a set $[s] := \{\alpha \in {}^{\omega}\omega : s \subseteq \alpha\}$ for some $s \in {}^{<\omega}\omega$.

Definition 1.2. For $A, B \subseteq {}^{\omega}\omega$, $A \leq_{\mathrm{W}} B$ iff there is a continuous $f : {}^{\omega}\omega \to {}^{\omega}\omega$ such that $A = f^{-1}[B]$

Clearly, $\leq_{\mathrm{W}}$ is reflexive and transitive.

Definition 1.3. $A \equiv_{\mathrm{W}} B$ iff $A \leq_{\mathrm{W}} B$ and $B \leq_{\mathrm{W}} A$. A **W-degree** (**Wadge degree**) is a an equivalence class $\equiv_{\mathrm{W}}$.

Consider the game $\mathrm{G_W}(A, B)$:

Player I	:	$\alpha(0)$	$\alpha(1)$	$\cdots$	$\alpha(n_0)$	$\alpha(n_0+1)$	$\cdots$	$\alpha(n_1)$	$\cdots$	α
Player II	:				$\beta(0)$			$\beta(1)$		β

Player I plays $\alpha(0)$, player II passes or plays, player I plays $\alpha(1)$, player II passes or plays, *etc.* Player II's plays in order are $\beta(0), \beta(1), \ldots$. Player II wins iff he plays infinitely often and $\alpha \in A \Leftrightarrow \beta \in B$.

It is easy to see that $A \leq_{\mathrm{W}} B \Leftrightarrow$ Player II wins $\mathrm{G_W}(A, B)$ (*i.e.*, player II has a winning strategy in $\mathrm{G_W}(A, B)$).

Wadge Degrees and Projective Ordinals: The Cabal Seminar, Volume II
Edited by A. S. Kechris, B. Löwe, J. R. Steel
Lecture Notes in Logic, 37

Definition 1.4. Let $G_L(A, B)$ be the following game:

$$\begin{array}{llllll} \text{Player I} & : \alpha(0) & & \alpha(1) & & \cdots \quad \alpha \\ \text{Player II} & : & \beta(0) & & \beta(1) & \cdots \quad \beta \end{array}$$

Player I plays $\alpha(0)$, player II plays $\beta(0)$, player I plays $\alpha(1)$, player II plays $\beta(1)$, *etc.* Player II wins iff $\alpha \in A \Leftrightarrow \beta \in B$.

Definition 1.5. $A \leq_L B$ iff player II wins $G_L(A, B)$. $A \equiv_L B$ iff $A \leq_L B$ and $B \leq_L A$. An **L-degree** (**Lipschitz-degree**) is an equivalence class $\equiv_L$.

Of course, $A \leq_L B \Rightarrow A \leq_W B$.

§2. The Lipschitz ordering.

Lemma 2.1 (Wadge's Lemma). For $A, B \subseteq {}^\omega\omega$, either $A \leq_L B$ or $B \leq_L \neg A$, *a fortiori*, either $A \leq_W B$ or $B \leq_W \neg A$.

Proof. Immediate from the determinateness of $G_L(A, B)$. ⊣

We shall generally use upper case Roman letters for sets of reals, and lower case Roman letters for their (L- or W-) degrees.

Definition 2.2. If Γ is any class of sets of reals, then $\check{\Gamma} := \{A \subseteq {}^\omega\omega : \neg A \in \Gamma\}$ is the **dual** class to Γ.

Wadge's Lemma 2.1 gives us the following information about the Lipschitz ordering:

(1) If a is a selfdual L-degree and b is any L-degree, then $b <_L a$, $b = a$, or $b >_L a$.
(2) If a is a nonselfdual L-degree, and b is any L-degree, then
  (i) $b <_L a$ and $b <_L \check{a}$, or
  (ii) $b = a$, or
  (iii) $b = \check{a}$, or
  (iv) $b >_L a$ and $b >_L \check{a}$.

Definition 2.3. Let $A_i \subseteq {}^\omega\omega$, $i \in \omega$. Then $\bigoplus_{i\in\omega} A_i := \{\langle i\rangle^\frown\alpha : i \in \omega, \alpha \in A_i\}$. For $A, B \subseteq {}^\omega\omega$, $A \oplus B := \{\langle n\rangle^\frown\alpha : n \in \omega, n \text{ even}, \alpha \in A\} \cup \{\langle n\rangle^\frown\beta : n \in \omega, n \text{ odd}, \beta \in B\}$.

Definition 2.4. For $A \subseteq {}^\omega\omega$, $s \in {}^{<\omega}\omega$, $s^\frown A := \{s^\frown\alpha : \alpha \in A\}$, $A_s = \{\alpha : s^\frown\alpha \in A\}$.

Theorem 2.5 (Martin [1973]). The relation $\leq_L$ is well-founded.

Proof. Suppose not. By DC there are sets A_n, $n \in \omega$, such that for any n, $A_{n+1} <_L A_n$. It is easy to see that player I wins each of the games $G_L(A_n, A_{n+1})$ and $G_L(A_n, \neg A_{n+1})$, say by strategies f_1^n and f_0^n, respectively.

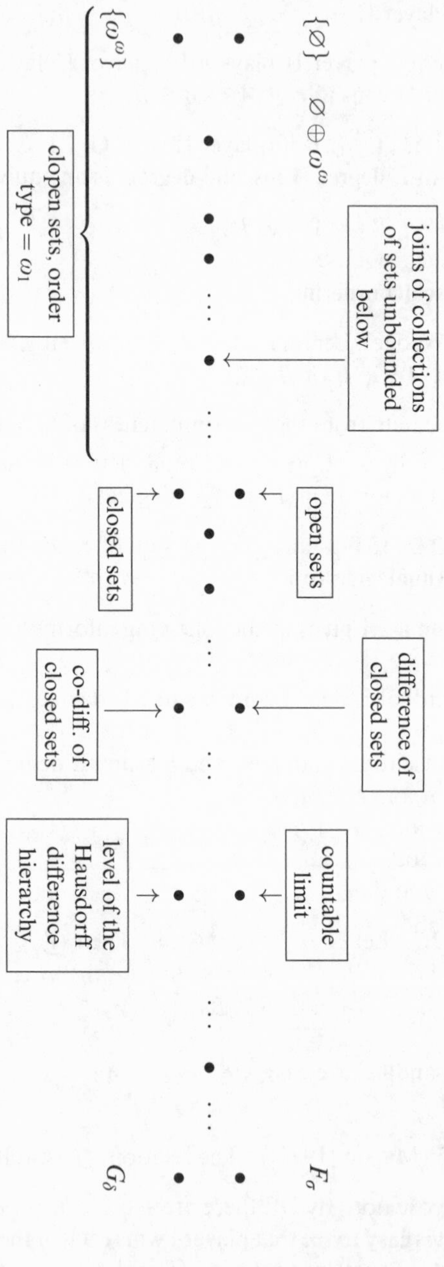

FIGURE 1. A picture of the lower L-degrees

To any $\alpha \in {}^{\omega}2$ assign the sequence $\langle f^{n}_{\alpha(n)} : n \in \omega \rangle$ of strategies and the sequence $\langle \gamma^{n}_{\alpha} : n \in \omega \rangle$ of reals, as indicated in the following diagram.

$$\begin{array}{ccccccc} & f^{n}_{\alpha(n)} & & & f^{1}_{\alpha(1)} & f^{0}_{\alpha(0)} & \\ \cdots & A_{n+1} & A_n & \cdots & A_2 & A_1 & A_0 \\ & \gamma^{n+1}_{\alpha}(0) & \gamma^{n}_{\alpha}(0) & & \gamma^{2}_{\alpha}(0) & \gamma^{1}_{\alpha}(0) & \gamma^{0}_{\alpha}(0) \\ & \gamma^{n+1}_{\alpha}(1) & \gamma^{n}_{\alpha}(1) & & \gamma^{2}_{\alpha}(1) & \gamma^{1}_{\alpha}(1) & \gamma^{0}_{\alpha}(1) \\ & \vdots & \vdots & & \vdots & \vdots & \vdots \end{array}$$

where, for each n, $\gamma^{n}_{\alpha}(0)$ is just the first move according to the strategy $f^{n}_{\alpha(n)}$, and $\gamma^{n}_{\alpha}(k)$ is the response of $f^{n}_{\alpha(n)}$ to $\gamma^{n+1}_{\alpha} \restriction k$.

Now, if α and α' are in ${}^{\omega}2$ and $n \in \omega$ is such that $\alpha(m) = \alpha'(m)$ for all $m \geq n$, then $\gamma^{n}_{\alpha} = \gamma^{n}_{\alpha'}$. For $n \in \omega$, define T^n to be the set of $\beta \in {}^{\omega}2$ such that $\gamma^{n}_{a^\frown\beta} \in A_n$ for any ${}^{n}2$. Then, for any $s \in {}^{n}2$, we have

$$T^{0}_{s} = \begin{cases} T^n & \text{if } s \text{ contains an even number of 1's} \\ \neg T^n & \text{if } s \text{ contains an odd number of 1's.} \end{cases}$$

Now no T^n is either meager or comeager, because its two details of rank one are complementary. But if T^0 is not meager, then (by the consequence of AD that all sets of reals have the Baire property) for some $s \in {}^{<\omega}2$, $T^{0}_{s} = T^{\mathrm{lh}(s)}$ is comeager, a contradiction. ⊣

Definition 2.6. We let $\mathrm{ord}_{\mathrm{L}}(a)$ be the order type of the L-degrees below a (having first coalesced each degree and its dual).

Some of the properties of L-degrees may be inferred from their ordinals.

Lemma 2.7. Suppose a is an L-degree.

(1) If $a \neq \breve{a}$, then the L-degree of joins of a set in a and a set in $\breve{a}$ is the minimum L-degree above a and $\breve{a}$.
(2) If $a = \breve{a}$, then the L-degree of $\langle 0 \rangle^\frown A$ for some $A \in a$ is the minimum L-degree above a.
(3) If $\mathrm{ord}_{\mathrm{L}}(a)$ is a limit ordinal of cofinality ω, then a is the degree of joins of ω-sequences of sets unbounded below a. So $a = \breve{a}$.
(4) If $\mathrm{ord}_{\mathrm{L}}(a)$ is a limit ordinal of uncountable cofinality, then $a \neq \breve{a}$.

Proof. (1) and (2) are easy.

(3). Let $\langle A_i : i \in \omega \rangle$ be unbounded below a. Let $A = \bigoplus_{i \in \omega} A_i$. Clearly, $A \not<_{\mathrm{L}} a$. So we need only show $A \leq_{\mathrm{L}} a$. To see this, let $B \in a$. Reduce A to B as follows:

> Suppose player I plays i. You then have to reduce A_i to B, but you must move first. You are in the position of player I attempting to show $B \leq_{\mathrm{L}} \neg A_i$. But this is true, so you have a winning strategy. Use

one. (Strategies are codable as reals, so we may use AC for countable sets of sets of reals, which follows from AD.)

(4). Suppose $a = \breve{a}$. Let $A \in a$.

CLAIM 2.8. For all n, $A_{\langle n\rangle} <_{\mathrm{L}} A$.

PROOF. That $A_{\langle n\rangle} \leq_{\mathrm{L}} A$ is immediate. Suppose $A \leq_{\mathrm{L}} A_{\langle n\rangle}$. Then $A_{\langle n\rangle} \leq_{\mathrm{L}} \neg A_{\langle n\rangle}$. Now, to show $A \not\leq_{\mathrm{L}} A_{\langle n\rangle}$, let player I play as follows in $\mathrm{G_L}(A, A_{\langle n\rangle})$:

Play n. Now play according to a winning strategy for player II in $\mathrm{G_L}(A_{\langle n\rangle}, \neg A_{\langle n\rangle})$. ⊣ (Claim 2.8)

CLAIM 2.9. $\{A_{\langle n\rangle} : n \in \omega\}$ is unbounded below A.

PROOF. Suppose not. Pick B so that for each n, $A_{\langle n\rangle} \leq_{\mathrm{L}} B <_{\mathrm{L}} A$. but in the proof of (3) we showed that $A = \bigoplus_{n\in\omega} A_{\langle n\rangle} \leq_{\mathrm{L}} B$, a contradiction. ⊣ (Claim 2.9)

Claim 2.9 violates the uncountability of the cofinality of $\mathrm{ord_L}(a)$. ⊣ (Lemma 2.7)

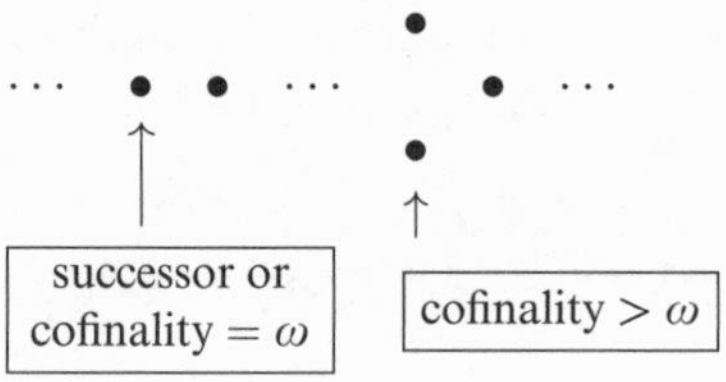

FIGURE 2. Picture of the L-degrees

§3. The Wadge Ordering. It is not hard to see that if a is a selfdual L-degree, then the next ω_1 L-degrees are $\leq_{\mathrm{W}} a$. Thus, the following Theorem 3.1 provides a complete picture of the Wadge ordering.

THEOREM 3.1 (Steel, Van Wesep). For $A \subseteq {}^{\omega}\omega$, $A \leq_{\mathrm{W}} \neg A$ implies $A \leq_{\mathrm{L}} \neg A$.

PROOF. Suppose $A \leq_{\mathrm{W}} \neg A$ and $A \not\leq_{\mathrm{L}} \neg A$. Let g_1 be a strategy for player II which witnesses the former fact, and f a strategy for player I which witnesses the latter. Let g_0 be the strategy for player II which instructs him to copy player I's moves as they are made.

Consider a sequence $\langle S_i : i \in \omega\rangle$, where, for each i, S_i is f, g_0 or g_1, and consider the following diagram.

$$\begin{array}{ccccccc} & S_2 & & S_1 & & S_0 & \\ \cdots\ A & & A & & A & & A. \end{array}$$

We imagine filling in a column of numbers below each occurrence of "A", the columns being referred to as numbered starting with zero for the rightmost column. The idea is that the entries in the $(i+1)$th and the ith columns should be a possible play of the $\left\{\begin{matrix}\text{Lipschitz}\\ \text{Wadge}\end{matrix}\right\}$ game in which $\left\{\begin{matrix}\text{player I}\\ \text{player II}\end{matrix}\right\}$ plays according to $S_i = \left\{\begin{matrix} f \\ g_0 \text{ or } g_1\end{matrix}\right\}$ when $\left\{\begin{matrix}\text{player I's}\\ \text{player II's}\end{matrix}\right\}$ moves are represented by the entries in the ith column.

A **finite filling in** of the diagram consists of a finite set of numbers arranged in columns under the A's in such a way that for each $i \in \omega$:

(1) the ith and the $(i+1)$th columns are a partial play according to S_i with the player on the left to move under the convention stated in the preceding paragraph, or

(2) for all $j \geq i$, the jth column is empty.

LEMMA 3.2. For a given sequence $\langle S_i : i \in \omega\rangle$, no two finite fillings in can clash, *i.e.*, have, for some i and n, different entries in the nth place of the ith column.

PROOF. Since all but finitely many columns are empty, we may take i_0 to be the greatest i such that the ith columns of the two finite fillings in clash. This leads easily to a contradiction. ⊣ (Lemma 3.2)

DEFINITION 3.3. The **filling in** of the diagram for $\langle S_i : i \in \omega\rangle$ is the union of all finite fillings in. The filling in is well defined by virtue of Lemma 3.2

DEFINITION 3.4. The **filling in** for a finite sequence $\langle S_i : i \leq n\rangle$ is the union of all finite fillings in for the sequence with the nth column empty.

Since the Wadge strategy g_1 has the option of passing, occurrences of g_1 in the sequence $\langle S_i : i \in \omega\rangle$ must be sufficiently rare if they are not to block the complete filling in of the diagram.

Define an increasing sequence of numbers $\langle i_k : k \in \omega\rangle$ as follows.

Set $i_0 = 0$. Suppose i_k is defined for $k \leq k_0$. For each sequence $\sigma = \langle S_i : i \leq i_{k_0}\rangle$ with the property that

$$\begin{aligned} &(1) \quad (\forall k \leq k_0)\, S_{i_k} = g_0 \text{ or } g_1 \\ \text{and}\ &(2) \quad (\forall i < i_{k_0})[(\forall k < k_0)\, i \neq i_k \Rightarrow S_i = f], \end{aligned}$$

define $i^\sigma_{k_0+1}$ to be the least number $n > i_{k_0}$ such that the filling in for $\sigma f f \cdots f$ $(n - i_{k_0} f\text{'s})$ contains at least $k_0 + 1$ entries in the zeroth column. It is easy to see that $i^\sigma_{k_0+1}$ exists for each σ. Let $i_{k_0+1} = \max\{i^\sigma_{k_0+1} : \sigma \text{ as above}\}$. Now, if $\langle S_i : i \in \omega\rangle$ is such that

$$\begin{aligned} &(1) \quad (\forall k)\, S_{i_k} = g_0 \text{ or } g_1 \\ \text{and}\ &(2) \quad \text{for all other } i,\ S_i = f, \end{aligned}$$

then by construction there are finite fillings in for the sequence $\langle S_i : i \in \omega \rangle$ with arbitrarily many entries in the zeroth column. Thus, the filling in for $\langle S_i : i \in \omega \rangle$ as above has an infinite sequence in the zeroth, *a fortiori*, in every, column.

For any $z \in {}^{\omega}2$, let $\mathcal{S}_z : \omega \to \{f, g_0, g_1\}$ be given by

$$\mathcal{S}_z(i) = \begin{cases} g_{z(k)} & \text{if } i = i_k \\ f & \text{otherwise.} \end{cases}$$

Define $h(z)$ to be the real produced in the zeroth column by the filling in for $\mathcal{S}_z$. Let $T = h^{-1}[A]$. Proceed as in the proof of Theorem 2.5. ⊣ (Theorem 3.1)

COROLLARY 3.5.

(i) $\{\varnothing\}$ and $\{{}^{\omega}\omega\}$ are the minimal W-degrees;
(ii) each non-selfdual pair of W-degrees has a selfdual successor;
(iii) each selfdual W-degree has a non-selfdual pair of successors;
(iv) each selfdual W-degree is the union of an ω_1-sequence of selfdual L-degrees;
(v) a level of cofinality ω is occupied by a selfdual W-degree;
(vi) a level of uncountable cofinality is occupied by a non-selfdual pair of W-degrees.

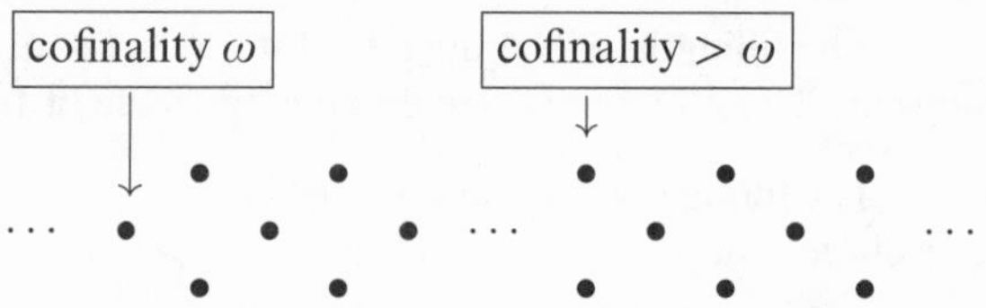

FIGURE 3. Picture of the W-degrees

§4. The Order Type of the $\underset{\sim}{\mathbf{\Delta}}^1_n$ Degrees. It is fairly easy to see that the order type of the Wadge degrees is

$$\Theta = \sup\{\xi : \xi \text{ is the length of a prewellordering of the reals}\}.$$

We seek to know the order types of the set of degrees of $\underset{\sim}{\mathbf{\Delta}}^1_n$ sets of reals; in short, the order type of the $\underset{\sim}{\mathbf{\Delta}}^1_n$ degrees.

It is easy to see that the set of degrees preceding a $\underset{\sim}{\mathbf{\Delta}}^1_n$ degree has order type less than $\underset{\sim}{\boldsymbol{\delta}}^1_{n+1}$ (just look at the prewellordering of their codes as continuous preimages of the given set). So the order type of the $\underset{\sim}{\mathbf{\Delta}}^1_n$ degrees is less than or equal to $\underset{\sim}{\boldsymbol{\delta}}^1_{n+1}$. In case n is odd, the inequality is strict. (Prewellorder the codes of $\underset{\sim}{\mathbf{\Delta}}^1_n$ sets as preimages of initial segments of a $\underset{\sim}{\mathbf{\Pi}}^1_n$-prewellordered complete

$\underset{\sim}{\Pi}^1_n$ set.) We shall show that for n even, the order type of the $\underset{\sim}{\Delta}^1_n$ degrees is $\geq \underset{\sim}{\delta}^1_{n+1}$, hence equal to $\underset{\sim}{\delta}^1_{n+1}$. So, for n odd, $\underset{\sim}{\delta}^1_n <$ the order type of the $\underset{\sim}{\Delta}^1_n$ degrees $< \underset{\sim}{\delta}^1_{n+1}$.

Our proof that the order type of the $\underset{\sim}{\Delta}^1_n$ degrees is $\underset{\sim}{\delta}^1_n$ for n even, is due to Steel and proceeds via a discussion of functions from Turing degrees to ordinals which is of interest in its own right. There is also an independent direct proof of this fact due to Martin.

Consider functions from $\mathcal{D}$ (the set of Turing degrees) into ω_1, relative to the Martin measure, μ.

Definition 4.1. Define $f_n(\boldsymbol{d}) := \delta^1_n(\boldsymbol{d})$, where $\delta^1_n(\boldsymbol{d})$ is the least ordinal which is not the order type of a wellordering of ω which is δ^1_n in $\boldsymbol{d}$.

Let $\pi\colon {}^{\mathcal{D}}\omega_1/\mu \cong \lambda$, $\lambda \in \mathrm{Ord}$, be the canonical isomorphism.

Remark 4.2. We have that $\mathsf{AD}_{\mathbb{R}}$ implies $\lambda = \Theta$, and $\mathbf{V} = \mathbf{L}(\mathbb{R})$ implies $\lambda > \Theta$.

Theorem 4.3 (Steel). For all even $n \geq 2$, we have $\pi([f_n]) = \underset{\sim}{\delta}^1_{n+1}$.

Remark 4.4. The full axiom of determinateness implies that $\underset{\sim}{\delta}^1_3 = \aleph_{\omega+1}$. Steel has shown that if $g_n(\boldsymbol{d}) = n$th $\boldsymbol{d}$-admissible beyond ω, then $\pi([g_n]) = \aleph_n$. His method also shows that the union of the first $\aleph_n$ Wadge degrees for $n \geq 2$ is just the nth level of the hierarchy based on the operation A.

Proof of Theorem 4.3. For convenience we take $n = 2$. Let W^α be a complete $\Sigma^1_2(\alpha)$ subset of ω, uniformly in α. Let $\leq^\alpha$ be the canonical prewell-ordering of W^α, and let $|m|^\alpha$ be the rank of m in $\leq^\alpha$ for $m \in W^\alpha$.

Definition 4.5. We say $\langle m, s\rangle$ is a **code** of $f : \mathcal{D} \to \omega_1$ if and only if

(1) $m \in \omega$ and s is a real coding a continuous function from ${}^\omega\omega$ to ${}^\omega\omega$,
(2) for almost all α, $m \in W^{s(\alpha)}$,
(3) for almost all $\boldsymbol{d} \in \mathcal{D}$, $f(\boldsymbol{d}) = \sup\{|m|^{s(\alpha)} : \alpha \in \boldsymbol{d}\}$,
(4) for almost all α, $s(\alpha) \equiv_{\mathrm{T}} \alpha$.

Lemma 4.6. For any $[f] < [f_2]$, there exists a code $\langle m, s\rangle$ of f.

Proof. Consider the game where player I plays m, α and player II plays β. Player I wins iff $\alpha \geq_{\mathrm{T}} \beta$, $m \in W^\alpha$, and $|m|^\alpha = f([\alpha])$.

Suppose player II wins by strategy σ. Let α be any real $\geq_{\mathrm{T}} \sigma$ and high enough that $f([\alpha]) < \delta^1_2(\alpha)$. Let $m \in W^\alpha$ be such that $|m|^\alpha = f([\alpha])$. (The length of $\leq^\alpha$ is $\delta^1_2(\alpha)$.) Then, player I can win by playing $\langle m, \alpha\rangle$; contradiction.

So player I wins, say by the strategy σ. Let m be player I's first play by σ. Let β be any real $\geq_{\mathrm{T}} \sigma$, and let player II play β. Then, if $\langle m, \alpha\rangle = \sigma * \beta$, we have $\alpha \equiv_{\mathrm{T}} \beta$, and $|m|^\alpha = f([\alpha]) = f([\beta])$. If s codes the function $\beta \mapsto \alpha$, then $\langle m, s\rangle$ codes f. ⊣ (Lemma 4.6)

Lemma 4.7. The set of codes is a complete Π^1_3 set of reals.

Proof. A pair $\langle m, s\rangle$ is a code iff

(i) s codes a continuous function,
(ii) $\forall\alpha\exists\beta \geq_T \alpha\forall\beta' \equiv_T \beta(m \in W^{s(\beta')})$.
(iii) $\exists\alpha\forall\beta \geq_T \alpha(s(\beta) \equiv_T \beta)$.

So the set of codes is Π^1_3.

To show completeness, let $P := \{\beta : \forall\alpha(m \in W^{\langle\alpha,\beta\rangle})\}$ be an arbitrary Π^1_3 set. Let k be such that $k \in W^{\langle\alpha,\beta\rangle} \iff \forall\alpha' \leq_T \alpha\; m \in W^{\langle\alpha',\beta\rangle}$. Let s_β code the continuous function $\alpha \mapsto \langle\alpha, \beta\rangle$. Then, the map $\beta \mapsto s_\beta$ is continuous, and $\beta \in P$ iff $\langle k, s_\beta\rangle$ is a code. ⊣ (Lemma 4.7)

Lemma 4.8. There are a Σ^1_2 relation R^+ and a Π^1_2 relation R^- such that if $n \in W^\beta$, then

$$R^+(m, \alpha, n, \beta) \Leftrightarrow m \in W^\alpha \,\&\, |m|^\alpha \leq |n|^\beta$$

$$R^-(m, \alpha, n, \beta) \Leftrightarrow (m \in W^\alpha \,\&\, |m|^\alpha \leq |n|^\beta) \text{ or } \delta^1_2(\alpha) \leq |n|^\beta.$$

Proof. Define

$$R^+(m, \alpha, n, \beta) \Leftrightarrow m \in W^\alpha \,\&\, \exists f\colon \{m' : m' \leq^\alpha m\} \to \{n' : n' \leq^\beta n\}$$

(f is order preserving). Define

$$\begin{aligned} R^-(m, \alpha, n, \beta) \Leftrightarrow \forall f[(&\mathrm{dom}(f) = \{n' : n' \leq^\beta n\} \,\& \\ &\mathrm{ran}(f) \subseteq W^\alpha \,\& \\ &f \text{ is order preserving} \Rightarrow \exists m' \in \mathrm{ran}(f)(m \leq^\alpha m')]. \end{aligned}$$

⊣ (Lemma 4.8)

We now define a Π^1_3-prewellordering of the codes which is very nearly the natural one.

Define $\langle m, s\rangle \leq^+ \langle n, t\rangle$ iff

(i) $\langle m, s\rangle$ and $\langle n, t\rangle$ are codes,
(ii) $\forall\alpha\exists\beta \geq_T \alpha\forall\gamma \equiv_T \beta\exists\gamma' \equiv_T \beta\; R^+(m, s(\gamma), n, t(\gamma'))$.

Define $\langle m, s\rangle \leq^- \langle n, t\rangle$ iff

(i) s, t code continuous functions,
(ii) $\exists\alpha\forall\beta \geq_T \alpha\; s(\beta) \equiv_T \beta$,
(iii) $\exists\alpha\forall\beta \geq_T \alpha\forall\gamma \equiv_T \beta\exists\gamma' \equiv_T \beta\; R^-(m, s(\gamma), n, t(\gamma'))$.

Note that if $\langle n, t\rangle$ is a code and $\langle m, s\rangle \leq^- \langle n, t\rangle$, then for almost all β, for all $\gamma, \gamma' \equiv_T \beta$, we have $s(\gamma) \equiv_T \gamma \equiv_T \gamma' \equiv_T t(\gamma')$, so $\delta^1_2(s(\gamma)) > |n|^{t(\gamma')}$, and $R^-(m, s(\gamma), n, t(\gamma')) \Leftrightarrow [m \in W^{s(\gamma)} \,\&\, |m|^{s(\gamma)} \leq |n|^{t(\gamma')}]$. Thus, if $\langle n, t\rangle$ is a code, then

(i) $\langle m, s\rangle \leq^+ \langle n, t\rangle \Rightarrow \langle m, s\rangle$ is a code,
(ii) $\langle m, s\rangle \leq^- \langle n, t\rangle \Rightarrow \langle m, s\rangle$ is a code,
(iii) $\langle m, s\rangle \leq^+ \langle n, t\rangle \Leftrightarrow \langle m, s\rangle \leq^- \langle n, t\rangle$.

The relation $\leq^\circ$ on the codes defined by $\leq^+$ (equivalently $\leq^-$) is a prelinear ordering, and clearly, if $\langle m, s\rangle \leq^\circ \langle n, t\rangle$, then the function coded by $\langle m, s\rangle$ is $\leq$ the function coded by $\langle n, t\rangle$. The converse is not quite true, as if $f(\boldsymbol{d})$ is almost always a limit ordinal the codes of $f(\boldsymbol{d})$ may occupy two consecutive levels of $\leq^\circ$. For suppose $\langle m, s\rangle$ codes f, and $f(\boldsymbol{d}) = \eta$, a limit ordinal. Then η may be the proper supremum of $\{|m|^{s(\alpha)} : \alpha \in \boldsymbol{d}\}$ or for some $\alpha \in \boldsymbol{d}$, we may find $\eta = |m|^{s(\alpha)}$. A code for which the first case almost always obtains will be $<^\circ$ a code for which the second case almost always obtains.

Nevertheless, the order type of $\leq^\circ$ is just the order type of the functions $< f_2$. Since $\leq^+$ is Π^1_3, $\leq^-$ is Σ^1_3, and the set of codes is Π^1_3, this order type is $\underset{\sim}{\boldsymbol{\delta}}^1_3$ by Moschovakis [Mos70]. Thus, the theorem is proved. ⊣ (Theorem 4.3)

THEOREM 4.9 (Martin, Steel). The order type of the $\underset{\sim}{\boldsymbol{\Delta}}^1_{2n}$ Wadge degrees is $\underset{\sim}{\boldsymbol{\delta}}^1_{2n+1}$.

PROOF. Again suppose for simplicity that $n = 1$. Let η be the order type of the $\underset{\sim}{\boldsymbol{\Delta}}^1_2$ degrees. It is easy to see that $\eta \leq \underset{\sim}{\boldsymbol{\delta}}^1_3$. (The prewellordering of codes of sets as preimages of a $\underset{\sim}{\boldsymbol{\Delta}}^1_2$ set is $\underset{\sim}{\boldsymbol{\Delta}}^1_3$).

To see that $\eta \geq \underset{\sim}{\boldsymbol{\delta}}^1_3$, we exhibit a map of η cofinal in $\{g : \mathcal{D} \to \omega_1 : g < f_2\}$. Recall that as we are assuming full AD, $\underset{\sim}{\boldsymbol{\delta}}^1_3$ is a regular cardinal. Clearly, AD($\mathbf{L}(\mathbb{R})$) would also suffice for this result.

DEFINITION 4.10. If $\langle m, s\rangle$ is a code which works above γ, *i.e.*, $\forall \beta \geq_{\mathrm{T}} \gamma (s(\beta) \equiv_{\mathrm{T}} \beta \,\&\, m \in W^{s(\beta)})$, let

$$C_{m,s,\gamma} := \{\langle p, \alpha\rangle : \alpha \geq_{\mathrm{T}} \gamma \,\&\, p \leq^{s(\alpha)} m\}.$$

Let $C^\alpha_{m,s,\gamma} = \{\langle p, e\rangle : \langle p, \{e\}^\alpha\rangle \in C_{m,s,\gamma}\}$. Note that $\alpha \equiv_{\mathrm{T}} \beta \Rightarrow C^\alpha_{m,s,\gamma} \equiv_{\mathrm{T}} C^\beta_{m,s,\gamma}$.

Let $h_{m,s,\gamma} : \mathcal{D} \to \mathcal{D}$ be given by $h_{m,s,\gamma}(\boldsymbol{d}) = \boldsymbol{d} \oplus$ degree of $C^\alpha_{m,s,\gamma}$, $\alpha \in \boldsymbol{d}$.

CLAIM 4.11. $C_{m,s,\gamma} \leq_{\mathrm{W}} C_{m',s',\gamma'} \Rightarrow h_{m,s,\gamma} \leq_{\mathrm{T}} h_{m',s',\gamma'}$ a.e.

PROOF. Let σ be a (code of a) map which reduces $C_{m,s,\gamma}$ to $C_{m',s',\gamma'}$. Let $\boldsymbol{d} \geq_{\mathrm{T}} \sigma$, and $\alpha \in \boldsymbol{d}$. Then there are e_1 and e_2 such that for all p, e

$$\begin{aligned}\langle p, e\rangle \in C^\alpha_{m,s,\gamma} &\Leftrightarrow \langle p, \{e\}^\alpha\rangle \in C_{m,s,\gamma}\\ &\Leftrightarrow \sigma(\langle p, \{e\}^\alpha\rangle) \in C_{m',s',\gamma'}\\ &\Leftrightarrow \langle \{e_1\}(\alpha, p), \{\{e_2\}(p, e)\}^\alpha\rangle \in C_{m',s',\gamma'}\\ &\Leftrightarrow \langle \{e_1\}(\alpha, p), \{e_2\}(p, e)\rangle \in C^\alpha_{m',s',\gamma'}.\end{aligned}$$

So $h_{m,s,\gamma}(\boldsymbol{d}) = \boldsymbol{d} \oplus \deg C^\alpha_{m,s,\gamma} \leq_{\mathrm{T}} \boldsymbol{d} \oplus \deg C^\alpha_{m',s',\gamma'} = h_{m',s',\gamma'}(\boldsymbol{d})$. This proves the claim. ⊣ (Claim 4.11)

Define the map $H : \eta \to \underset{\sim}{\boldsymbol{\delta}}^1_3$ as follows. Given $\xi < \eta$, look at the set of $f : \mathcal{D} \to \omega_1$ coded by $\langle m, s\rangle$ such that for some γ, $\langle m, s\rangle$ is good above γ

and $C_{m,s,\gamma}$ has ordinal $< \xi$ in the Wadge ordering of $\underset{\sim}{\mathbf{\Delta}}^1_2$. Let $H(\xi)$ be the supremum of the ordinals of these f's.

CLAIM 4.12. For any $\xi < \eta$, $H(\xi) < \underset{\sim}{\boldsymbol{\delta}}^1_3$.

PROOF. We prove this claim by showing that for any $\langle m, s\rangle$ a code good above γ, there is a $g : \mathcal{D} \to \omega_1$, $g < f_2$, such that for any $\langle n, t\rangle$ a code for g good above δ, $C_{n,t,\delta} \not\leq_{\mathrm{W}} C_{m,s,\gamma}$.

To this end let such m, s, and γ be given, let $\boldsymbol{d} \geq_{\mathrm{T}} s$ be given, and let $\alpha \in \boldsymbol{d}$. Uniformly in e such that $\{e\}^\alpha \equiv_{\mathrm{T}} \alpha$, find $n_e \in W^{\{e\}^\alpha}$ such that $\{p : p \leq^{\{e\}^\alpha} n_e\}$ has higher Turing degree than $C^\beta_{m,s,\gamma}$, $\beta \in \boldsymbol{d}$. Paste together all the ordinals $|n_e|^{\{e\}^\alpha}$ getting some $\delta^1_2(\boldsymbol{d})$ ordinal. Do this for each $\alpha \in \boldsymbol{d}$. Let $g(\boldsymbol{d})$ be the least ordinal so obtained.

Clearly, $g < f_2$. We now show that g is as desired. So let $\langle n, t\rangle$ code g above δ. Let $\boldsymbol{d} \geq_{\mathrm{T}} \langle s, \delta, \gamma\rangle$. There is $\alpha \in \boldsymbol{d}$ such that $\{p : p \leq^{t(\alpha)} n\}$ has higher Turing degree than $C^\beta_{m,s,\gamma}$, $\beta \in \boldsymbol{d}$, *i.e.*, higher Turing degree than $h_{m,s,\gamma}(\boldsymbol{d})$. But $\{p : p \leq^{t(\alpha)} n\} = \{p : \langle p, \alpha\rangle \in C_{n,t,\delta}\} = \{p : \langle p, e\rangle \in C^\alpha_{n,t,\delta}\}$ where $\{e\}$ is the identity. So $h_{n,t,\delta}(\boldsymbol{d}) \geq_{\mathrm{T}} \{p : p \leq^{t(\alpha)} n\} >_{\mathrm{T}} h_{m,s,\gamma}(\boldsymbol{d})$. So $h_{n,t,\delta} > h_{m,s,\gamma}$, a.e., whence $C_{n,t,\delta} \not\leq_{\mathrm{W}} C_{m,s,\gamma}$, and the claim is proved. ⊣ (Claim 4.12)

It remains only to show that $H : \lambda \to \underset{\sim}{\boldsymbol{\delta}}^1_3$ is cofinal. For this we need to see that if $g : \mathcal{D} \to \omega_1$, is less than f_2 then there is a $\underset{\sim}{\mathbf{\Delta}}^1_2$ set A so that any $f < g$ has a code $\langle m, s\rangle$ with the property that for some γ, $\langle m, s\rangle$ is good above γ and $C_{m,s,\gamma} \leq_{\mathrm{W}} A$. So let $\langle n, t\rangle$ be a code for g good above β, with the additional property that for all $\boldsymbol{d} \geq_{\mathrm{T}} \beta$, $f(\boldsymbol{d}) = |n|^{t(\alpha)}$, $\alpha \in \boldsymbol{d}$ (such codes are actually given by Lemma 4.6), and let $A = \{\langle m, s, \gamma, p, \alpha\rangle : \alpha \geq_{\mathrm{T}} \beta \,\&\, \alpha \geq_{\mathrm{T}} \gamma \,\&\, |m|^{s(\alpha)} \leq |n|^{t(\alpha)} \,\&\, p \leq^{s(\alpha)} m\}$. By virtue of Lemma 4.8, A is $\underset{\sim}{\mathbf{\Delta}}^1_2$. Clearly, if $f < g$, $\langle m, s\rangle$ is a code for f, and $\gamma \geq_{\mathrm{T}} \beta$ is such that $\langle m, s\rangle$ is good above γ and $f(\boldsymbol{d}) < g(\boldsymbol{d})$ for $\boldsymbol{d}$ above γ, then $\langle p, \alpha\rangle \in C_{m,s,\gamma} \Leftrightarrow \langle m, w, \gamma, p, \alpha\rangle \in A$, so $C_{m,s,\gamma} \leq_{\mathrm{W}} A$.

This concludes the proof of Theorem 4.9 ⊣ (Theorem 4.9)

§5. Separation, Reduction, and Prewellordering Properties in the Wadge Hierarchy. In this section we establish a number of descriptive set theoretic properties of nonselfdual classes closed under continuous preimage, assuming throughout ZF+DC+AD. In the following, then, Γ will always be a nonselfdual class of sets of reals closed under continuous preimages. We use **continuously closed** to mean *closed under continuous preimage*. The descriptive set theoretic properties we shall be concerned with are as follows.

DEFINITION 5.1. Γ has the **first separation property**, $\mathrm{Sep}_{\mathrm{I}}(\Gamma)$, iff

$$\forall A, B \in \Gamma(A \cap B = \varnothing \Rightarrow \exists C \in \Gamma \cap \check{\Gamma}(A \subseteq C \subseteq \neg B)).$$

Γ has the **second separation property**, $\mathsf{Sep}_{\mathrm{II}}(\Gamma)$, iff

$$\forall A, B \in \Gamma \exists A', B' \in \check{\Gamma}(A \sim B \subseteq A' \,\&\, B \sim A \subseteq B' \,\&\, A' \cap B' = \varnothing).$$

Γ has the **reduction property**, $\mathsf{Red}(\Gamma)$, iff

$$\forall A, B \in \Gamma \exists A', B' \in \Gamma(A' \subseteq A,\ B' \subseteq B,\ A' \cap B' = \varnothing \,\&\, A' \cup B' = A \cup B).$$

Γ has the **prewellordering property**, $\mathsf{PWO}(\Gamma)$, iff for every $A \in \Gamma$ there are relations $\leq^{+}$ and $\leq^{-}$ in Γ and $\check{\Gamma}$ respectively, and a prewellordering $\leq$ of A such that for any real $\beta \in A$ and any real α, $\alpha \leq^{+} \beta \Leftrightarrow \alpha \leq^{-} \beta \Leftrightarrow \alpha \leq \beta$.

Before proceeding to the statement of the results of this section we take note of some simple facts. The first is that Wadge's Lemma (Lemma 2.1) has the immediate consequence that if Γ is continuously closed and nonselfdual, and $A \in \Gamma \sim \check{\Gamma}$, then A is complete for Γ, *i.e.*, for all B in Γ, $B \leq_{\mathrm{W}} A$. It is also true that Γ has a universal set, *i.e.*, there is $A \in \Gamma$ such that for all $B \in \Gamma$ there is $\alpha \in {}^{\omega}\omega$ such that $\beta \in B \Leftrightarrow (\alpha, \beta) \in A$. One may take $A = \{(f, \beta) : f * \beta \in C\}$, where $C \in \Gamma \sim \check{\Gamma}$ and $f * \beta$ is the result of applying (the Lipschitz strategy coded by) f to β. Finally, we take note of a standard result of descriptive set theory, *viz.*, if $\mathsf{Red}(\Gamma)$ and and Γ has a universal set, then $\mathsf{Sep}_{\mathrm{I}}(\check{\Gamma})$ and $\neg\mathsf{Sep}_{\mathrm{I}}(\Gamma)$, and hence $\mathsf{Red}(\Gamma) \Rightarrow \neg\mathsf{Red}(\check{\Gamma})$. Since the proof of this fact is included in our proof of Theorem 5.2, we do not give it here.

We now state the results.

THEOREM 5.2 (Van Wesep). For any continuously closed nonselfdual class Γ, we have $\mathsf{Sep}_{\mathrm{I}}(\Gamma) \Leftrightarrow \neg\mathsf{Sep}_{\mathrm{II}}(\check{\Gamma})$.

THEOREM 5.3 (Van Wesep). For any continuously closed nonselfdual Γ, either $\mathsf{Sep}_{\mathrm{II}}(\Gamma)$ or $\mathsf{Sep}_{\mathrm{II}}(\check{\Gamma})$. Thus by Theorem 5.2, either $\neg\mathsf{Sep}_{\mathrm{I}}(\Gamma)$ or $\neg\mathsf{Sep}_{\mathrm{I}}(\check{\Gamma})$.

Using Wadge's characterizations of the continuously closed nonselfdual classes of Borel sets, Steel has shown that one of each nonselfdual pair of such classes has the first separation property. In each of the following theorems, Γ is a continuously closed nonselfdual class.

THEOREM 5.4 (Van Wesep). For some continuously closed nonselfdual Γ, we have $\neg\mathsf{Red}(\Gamma)$ and $\neg\mathsf{Red}(\check{\Gamma})$.

THEOREM 5.5 (Van Wesep). If Γ is closed under (finite) intersections and $\mathsf{Sep}_{\mathrm{I}}(\check{\Gamma})$, then $\mathsf{Red}(\Gamma)$.

THEOREM 5.6 (Steel). If Γ is closed under countable unions and countable intersections, then $\mathsf{Red}(\Gamma)$ or $\mathsf{Red}(\check{\Gamma})$.

THEOREM 5.7 (Kechris-Solovay). If Γ is closed under countable unions, countable intersections, projections, and coprojections, and $\Gamma \subseteq \mathbf{L}(\mathbb{R})$, then $\mathsf{PWO}(\Gamma)$ or $\mathsf{PWO}(\check{\Gamma})$.

We now proceed to the proofs.

Proof of Theorem 5.2. Let $A \in \Gamma \sim \check{\Gamma}$. Define

$$A_0 = \{((f_0, f_1), \alpha) : f_0 * \alpha \in A\}$$
$$A_1 = \{((f_0, f_1), \alpha) : f_1 * \alpha \in A\},$$

where $(\cdot, \cdot)$ is a reasonable pairing function for reals. Clearly, A_0 and A_1 are in Γ. Moreover, the pair $\langle A_0, A_1 \rangle$ is universal for pairs of sets in Γ, *i.e.*, for any $B_0, B_1 \in \Gamma$ there is a real γ such that $B_0 = \{\alpha : (\gamma, \alpha) \in A_0\}$, and $B_1 = \{\alpha : (\gamma, \alpha) \in A_1\}$. (To see this, note that by Wadge's Lemma 2.1, $\Gamma = \{B : B \leq_W A\}$. Then, note that this is true even if $\leq_W$ is replaced by $\leq_L$.) Likewise, the pair $\langle \sim A_0, \sim A_1 \rangle$ is universal for pairs of sets in $\check{\Gamma}$.

Now suppose $\check{\Gamma}$ has the second separation property. We shall show that Γ does not have the first separation property. Let $B_0, B_1 \in \Gamma$ be such that $A_0 \sim A_1 \subseteq B_0$, and $A_1 \sim A_0 \subseteq B_1$, with $B_0 \cap B_1 = \varnothing$. Define $C_0 = \{\alpha : (\alpha, \alpha) \in B_0\}$, $C_1 = \{\alpha : (\alpha, \alpha) \in B_1\}$. Clearly, C_0 and C_1 are disjoint sets in Γ. We claim they are not separable by a set in $\Delta = \Gamma \cap \check{\Gamma}$. To see this, suppose $C_0 \subseteq D$, and $D \cap C_1 = \varnothing$, with $D \in \Delta$. Let γ be such that $D = \{\alpha : (\gamma, \alpha) \in \neg A_0\}$ and $\neg D = \{\alpha : (\gamma, \alpha) \in \neg A_1\}$. Then, we have

$$\gamma \in D \Rightarrow (\gamma, \gamma) \in A_1 \sim A_0 \Rightarrow (\gamma, \gamma) \in B_1 \Rightarrow \gamma \in C_1 \Rightarrow \gamma \notin D,$$

and

$$\gamma \notin D \Rightarrow (\gamma, \gamma) \in A_0 \sim A_1 \Rightarrow (\gamma, \gamma) \in B_0 \Rightarrow \gamma \in C_0 \Rightarrow \gamma \in D,$$

which is a contradiction.

Now suppose Γ does not have the first separation property. Let C and D in Γ be disjoint and not separable by a set in Δ. Let A and B be in $\check{\Gamma}$. We shall find A' and B' in Γ such that $A \sim B \subseteq A'$, $B \sim A \subseteq B'$, and $A' \cap B' = \varnothing$.

Consider the following game.

Players I and II produce reals α and β respectively. Player I wins iff

$$\begin{aligned} &\beta \in C \Rightarrow \alpha \in A \sim B, \\ &\beta \in D \Rightarrow \alpha \in B \sim A, \\ \text{and} \quad &\alpha \in (A \sim B) \cup (B \sim A). \end{aligned}$$

Equivalently, player II wins iff,

$$\begin{aligned} &\alpha \in A \sim B \Rightarrow \beta \in D, \\ \text{and} \quad &\alpha \in B \sim A \Rightarrow \beta \in C. \end{aligned}$$

Now, player I can not win this game, for if he did, say by a strategy f, which we shall view as a continuous function from ${}^\omega\omega$ to ${}^\omega\omega$, then we should have $C \subseteq E$ and $E \cap D = \varnothing$, where $E = f^{-1}(A) = \neg f^{-1}(B) \in \Gamma \cap \check{\Gamma}$, which contradicts the inseparability of C and D. Thus, player II wins the game, say by the strategy f. Let $A' = f^{-1}(D)$, and $B' = f^{-1}(C)$. Then A' and B' are as desired. ⊣ (Theorem 5.2)

PROOF OF THEOREM 5.3. Let $\langle A_0, A_1\rangle$ be a complete pair of sets in Γ (see the proof of Theorem 5.2). Consider the following two games.

G_0 : Players I and II play reals α and β respectively. Player II wins iff

$$\begin{aligned} &\alpha \in A_0 \sim A_1 \Rightarrow \beta \in A_1 \sim A_0, \\ &\alpha \in A_1 \sim A_0 \Rightarrow \beta \in A_0 \sim A_1, \\ \text{and}\quad &\beta \notin A_0 \cap A_1. \end{aligned}$$

G_1 : Players I and II play reals α and β respectively. Player II wins iff

$$\begin{aligned} &\alpha \in A_0 \sim A_1 \Rightarrow \beta \in A_0 \sim A_1, \\ &\alpha \in A_1 \sim A_0 \Rightarrow \beta \in A_1 \sim A_0, \\ \text{and}\quad &\beta \in A_0 \cup A_1. \end{aligned}$$

Suppose Γ does not have the second separation property. Then player II does not win G_1. For if he did, say by the strategy f, we could let $A_0^1 = f^{-1}(\neg A_1)$ and $A_1^1 = f^{-1}(\neg A_0)$. Then $A_0 \sim A_1 \subseteq A_0^1$, $A_1 \sim A_0 \subseteq A_1^1$; $A_0^1 \cap A_1^1 = \varnothing$. Since $\langle A_0, A_1\rangle$ is complete for Γ, this contradicts our hypothesis.

Similarly, if $\check{\Gamma}$ does not have the second separation property, then player II does not win G_0.

Thus, by determinateness, we shall have proved the theorem when we have shown that player I can not win both G_0 and G_1.

So suppose player I wins G_0 and G_1 by the strategies f_0 and f_1 respectively. From the definition of G_0 it is apparent that for any β played by player II,

$$\begin{aligned} &f_0(\beta) \in A_0 \sim A_1 \,\&\, \beta \in A_1 \sim A_0, \\ \text{or}\quad &f_0(\beta) \in A_1 \sim A_0 \,\&\, \beta \notin A_0 \sim A_1, \\ \text{or}\quad &\beta \in A_0 \cap A_1. \end{aligned}$$

In other words,

$$\begin{aligned} &\beta \in A_0 \sim A_1 \Rightarrow f_0(\beta) \in A_0 \sim A_1, \\ &\beta \in A_1 \sim A_0 \Rightarrow f_0(\beta) \in A_1 \sim A_0, \\ \text{and}\quad &\beta \in \neg A_0 \cap \neg A_1 \Rightarrow f_0(\beta) \in (A_0 \sim A_1) \cup (A_1 \sim A_0). \end{aligned}$$

Similarly,

$$\begin{aligned} &\beta \in A_0 \sim A_1 \Rightarrow f_1(\beta) \in A_1 \sim A_0, \\ &\beta \in A_1 \sim A_0 \Rightarrow f_1(\beta) \in A_0 \sim A_1, \\ \text{and}\quad &\beta \in A_0 \cap A_1 \Rightarrow f_1(\beta) \in (A_0 \sim A_1) \cup (A_1 \sim A_0). \end{aligned}$$

As in the proof of Theorem 2.5, for any $\gamma \in {}^{\omega}2$, consider the sequence $\langle f_{\gamma(n)} : n \in \omega\rangle$ of strategies for player I, and consider the following diagram.

$$\begin{array}{ccccccc} & f_{\gamma(2)} & & f_{\gamma(1)} & & f_{\gamma(0)} & \\ \gamma^3(0) & & \gamma^2(0) & & \gamma^1(0) & & \gamma^0(0) \\ \gamma^3(1) & & \gamma^2(1) & & \gamma^1(1) & & \gamma^0(1) \\ \vdots & & \vdots & & \vdots & & \vdots \end{array}$$

The rule of construction in this diagram is: $\gamma^n(i)$ is the response of the strategy $f_{\gamma(n)}$ to the play $\gamma^{n+1}\lceil i$, *i.e.*, the first i in the sequence γ^{n+1}.

We consider membership of the γ^n in A_0 and A_1 for various γ. Call $(A_0 \cap A_1) \cup (\neg A_0 \cap \neg A_1)$ the **middle** and $(A_1 \sim A_0) \cup (A_0 \sim A_1)$ the **sides**.

Claim 5.8. $\{\gamma \in {}^{\omega}2 : \gamma^0 \in \text{middle}\}$ is meager.

Proof. Suppose not. Since we are assuming AD, all sets of reals have the property of Baire. So for some $s \in {}^{<\omega}2$, $\{\gamma : (s^\frown\gamma)^0 \in \text{middle}\}$ is comeager. But $(s^\frown\gamma)^0 \in$ middle implies $\gamma^0 \in$ middle, because $\beta \in \text{sides} \Rightarrow f_i(\beta) \in \text{sides}$ for $i = 0, 1$. So $\{\gamma : \gamma^0 \in \text{middle}\}$ is comeager. Without loss of generality, assume $\{\gamma : \gamma^0 \in A_0 \cap A_1\}$ is nonmeager. For any γ in this set, $(\langle 1\rangle^\frown\gamma)^0$ is in the sides. Thus, $\gamma^0 \in$ sides for a nonmeager set of γ. Contradiction. ⊣ (Claim 5.8)

Thus, $\{\gamma : \gamma^0 \in \text{sides}\}$ is nonmeager. Suppose, without loss of generality, that $\{\gamma : \gamma^0 \in A_1 \sim A_0\}$ is nonmeager. Then for some $s \in {}^{<\omega}2$, $\{\gamma : (s^\frown\gamma)^0 \in A_1 \sim A_0\}$ is comeager. There are two cases.

If s contains an odd number of 1's, then $\gamma^0 \in A_0 \sim A_1 \Rightarrow (s^\frown\gamma)^0 \in A_1 \sim A_0$, and $\gamma^0 \in A_1 \sim A_0 \Rightarrow (s^\frown\gamma)^0 \in A_0 \sim A_1$. So $\{\gamma : \gamma^0 \in A_1 \sim A_0\}$ is comeager, which contradicts our assumption.

If s contains an even number of 1's, then $\gamma^0 \in A_0 \sim A_1 \Rightarrow (s^\frown\gamma)^0 \in A_0 \sim A_1$, and $\gamma^0 \in A_1 \sim A_0 \Rightarrow (s^\frown\gamma)^0 \in A_1 \sim A_0$. So $\{\gamma : \gamma^0 \notin A_0 \sim A_1\}$ is comeager. By Claim 5.8, then, $\{\gamma : \gamma^0 \in A_1 \sim A_0\}$ is comeager. But γ in this set implies $(\langle 1\rangle^\frown\gamma)^0 \in A_0 \sim A_1$, so $\{\gamma : \gamma^0 \notin A_0 \sim A_1\}$ is nonmeager. This contradiction establishes the theorem. ⊣ (Theorem 5.3)

Proof of Theorem 5.4. We shall show that if Γ is a minimal continuously closed nonselfdual class including $F_\sigma \cup G_\delta$, then the reduction property fails for both Γ and $\check{\Gamma}$. In Section II.4.1 of Van Wesep [Van77] a more general result is proved, *viz.*, if the Wadge order type of Γ is not of the form $\omega_1^\alpha + 1$ and $\Gamma \nsubseteq F_\sigma$, then reduction fails for Γ.

Let A be G_δ but not F_σ. Let $A_0 = \{(\alpha, \beta) : \alpha \in A\}$, $A_1 = \{(\alpha, \beta) : \beta \in A\}$. It is clear that $A_0, A_1 \in G_\delta \sim F_\sigma$ and $\langle A_0, A_1\rangle$ is complete for pairs of sets in G_δ.

Now, the pair $\langle A_0, A_1\rangle$ is not reducible by sets in G_δ, for by its completeness if it were reducible by G_δ sets, then we would have $\mathsf{Red}(G_\delta)$, which is false. Moreover, $\langle A_0, A_1\rangle$ can not be reduced by sets in $F_\sigma \cup G_\delta$, for if $C \in F_\sigma$ were such that $C \subseteq A_0$ and $A_0 \sim A_1 \subseteq C$, then letting β be any real not in A, we have, for all reals α, $\alpha \in A \Leftrightarrow (\alpha, \beta) \in A_0 \Leftrightarrow (\alpha, \beta) \in C$, which is a contradiction.

We may take for A the set $\{\alpha \in {}^{\omega}\omega : \forall m \exists n > m(\alpha(n) = 0)\}$, so that for any $s \in {}^{<\omega}\omega$, $A_s \equiv_{\mathrm{W}} A$. Now, by the above considerations, for any $s \in {}^{<\omega}\omega$, $\langle (A_0)_s, (A_1)_s\rangle$ is not reducible by sets in $F_\sigma \cup G_\delta$. We shall have proved the theorem when we have shown that $\langle A_0, A_1\rangle$ is not reducible by sets in Γ, and therefore, by symmetry, in $\check{\Gamma}$.

Suppose $\langle B_0, B_1\rangle$ reduces $\langle A_0, A_1\rangle$, where B_0 and B_1 are in Γ. It is easy to show that for some $s \in {}^{<\omega}\omega$, $(B_0)_s$ and $(B_1)_s$ are in $F_\sigma \cup G_\delta$. But $\langle (B_0)_s, (B_1)_s\rangle$ reduces $\langle (A_0)_s, (A_1)_s\rangle$, a contradiction. ⊣ (Theorem 5.4)

Proof of Theorem 5.5. We use the following lemma.

Lemma 5.9. If Γ is closed under finite intersections and $\neg\mathsf{Red}(\Gamma)$, then for $C, D \in \breve{\Gamma}$, there are $C', D' \in \Gamma$, so that

$$C \sim D \subseteq C',\ C' \cap (D \sim C) = \varnothing,$$
$$D \sim C \subseteq D',\ D' \cap (C \sim D) = \varnothing,$$
$$C' \cup D' = {}^\omega\omega.$$

Proof. Let $\langle A, B\rangle$ be a complete pair for Γ, and let $C, D \in \breve{\Gamma}$. Consider the game $G(A, B, C, D)$ defined as follows:

Player I plays α, player II plays β. Player II wins iff $\beta \in A \cup B$, $\alpha \in C \sim D \Rightarrow \beta \in A \sim B$, and $\alpha \in D \sim C \Rightarrow \beta \in B \sim A$. Equivalently, player I wins iff $\beta \notin A \cup B$, or $\alpha \in C \sim D$ and $\beta \notin A \sim B$, or $\alpha \in D \sim C$ and $\beta \notin B \sim A$; in other words

$$\beta \in A \cup B \Rightarrow (\beta \in B \,\&\, \alpha \in C \sim D, \text{ or } \beta \in A \,\&\, \alpha \in D \sim C).$$

Now, player I can not win $G(A, B, C, D)$, for if he did, say by the strategy f, then letting $A'' = f^{-1}(\neg C)$, and $B'' = f^{-1}(\neg D)$, we would have $A \cup B \subseteq A'' \cup B''$, $A \sim B \subseteq A''$, $B \sim A \subseteq B''$, and $A'', B'' \in \Gamma$. Thus, letting $A' = A \cap A''$, $B' = B \cap B''$, we would have $A', B' \in \Gamma$, with $A \cup B = A' \cup B'$, $A \sim B \subseteq A'$, $B \sim A \subseteq B'$, but this contradicts $\neg\mathsf{Red}(\Gamma)$.

So player II wins $G(A, B, C, D)$, say by f. Let $C' = f^{-1}(A)$, $D' = f^{-1}(B)$. Then C' and D' are as desired, and the lemma is proved. ⊣ (Lemma 5.9)

To prove Theorem 5.5 suppose toward a contradiction that Γ is closed under intersection, $\mathsf{Sep}_{\mathrm{I}}(\breve{\Gamma})$, and $\neg\mathsf{Red}(\Gamma)$. We shall derive the absurdity $\mathsf{Red}(\breve{\Gamma})$. So let $C, D \in \breve{\Gamma}$. Let C', D' be as given by Lemma 5.9. Let $E \in \Gamma \cap \breve{\Gamma}$ separate $\neg C'$ and $\neg D'$. Let $C'' = C' \cap \neg E$, $D'' = D' \cap E$. Then $\langle C'', D''\rangle$ reduces $\langle C, D\rangle$. ⊣ (Theorem 5.5)

Proof of Theorem 5.6. Assume the hypothesis of the theorem, and suppose that $\neg\mathsf{Red}(\Gamma)$ and $\neg\mathsf{Red}(\breve{\Gamma})$. We shall show that in fact Γ has the reduction property.

Let $C, D \in \Gamma$. Define $C_{-1} = C$, $D_{-1} = D$, and for all $n \geq 0$, let C_n and D_n be C'_{n-1}, D'_{n-1} as given by Lemma 5.9. We have $C_{2n}, D_{2n} \in \Gamma$, $C_{2n+1}, D_{2n+1} \in \breve{\Gamma}$, $C_{n+1} \subseteq C_n$ and $D_{n+1} \subseteq D_n$, for each $n \geq 0$. So $C'' = \bigcap_n C_n \in \Gamma \cap \breve{\Gamma}$ and C'' separates $C \sim D$ and $D \sim C$. Let $C''' = C \cap C''$, $D''' = D \cap (\neg C'')$. Then $\langle C''', D'''\rangle$ reduces $\langle C, D\rangle$. ⊣ (Theorem 5.6)

Proof of Theorem 5.7. By Theorem 5.6 we may suppose that Γ has the reduction property. We shall show $\mathsf{PWO}(\Gamma)$. It will be useful to look at the following way of coding sets in $\Delta = \Gamma \cap \breve{\Gamma}$:

Let $\langle A, B\rangle$ be a universal pair for Γ. Let $\langle A', B'\rangle$ reduce $\langle A, B\rangle$. For any real γ, if $\forall\alpha[(\alpha,\gamma) \in A'$ or $(\alpha,\gamma) \in B']$, then let $C_\gamma = \{\alpha : (\alpha,\gamma) \in A'\}$. Then $\{\gamma : C_\gamma \text{ is defined}\} \in \Gamma$, and if C_γ is defined, $C_\gamma \in \Delta$.

Lemma 5.10 (Independently also due to Steel). Let Γ satisfy the hypothesis of the theorem, and suppose Δ is not closed under wellordered unions. Then $\mathsf{Red}(\Gamma) \Rightarrow \mathsf{PWO}(\Gamma)$.

Proof. Let λ be the least ordinal so that for some sequence $\langle A_\xi : \xi < \lambda\rangle$ of sets $A_\xi \in \Delta$, $\bigcup_{\xi<\lambda} A_\xi \notin \Delta$. Let $\Gamma^* = \{\bigcup_{\xi<\lambda} B_\xi : \forall\xi < \lambda\, B_\xi \in \Delta\}$. It is apparent from the minimality of λ, then, that $\mathsf{PWO}(\Gamma^*)$. Thus, we need only that $\Gamma^* \subseteq \Gamma$, to see that $\Gamma^* = \Gamma$, and hence $\mathsf{PWO}(\Gamma)$.

Let δ be the supremum of the lengths of prewellorderings of $\mathbb{R}$ that lie in Δ. Then $\lambda \geq \delta$ by Moschovakis [Mos70]. But $\lambda \leq \delta$. For if not, let $\bigcup_{\xi<\lambda} A_\xi \notin \Delta$. We may assume that $\xi < \eta < \lambda \Rightarrow A_\xi \subset A_\eta$, by taking cumulative unions and then a strictly increasing subsequence, noting the minimality of λ. Now, $\langle A_\xi : \xi < \delta\rangle$ provides a prewellordering in Δ of length δ, a contradiction. So $\lambda = \delta$.

Now we note that the order type of the Wadge degrees in Δ is exactly δ. Showing "$\leq$" is trivial. To see the other direction note that if $\prec$ is a prewellordering in Δ, then one may define by effective transfinite induction a $\leq_{\mathrm{W}}$-increasing sequence of Δ sets of length $|{\prec}|$. The proof of this is a routine exercise given that there are continuous functions which, acting on codes of sets in Δ, give codes for the Δ sets derived from them by the operations of union, *etc.*, which do not lead out of Δ. Indeed, the strategies which witness the corresponding closure properties of Γ provide such functions.

Now let C be the set of α for which C_α is defined. The relation $\{\langle\alpha,\beta\rangle : \alpha,\beta \in C \,\&\, (C_\alpha \leq_{\mathrm{W}} C_\beta \,\&\, C_\alpha \leq_{\mathrm{W}} \neg C_\beta)\}$ is in Γ, as is the complement of this relation relative to $C \times C$. Thus, by the Main Lemma of Moschovakis [Mos70], $\Gamma^* \subseteq \Gamma$. ⊣ (Lemma 5.10)

To finish the proof of Theorem 5.7 we must show that Δ is not closed under wellordered unions. Let $\mathfrak{J}_1, \dots, \mathfrak{J}_8$ be Gödel's operations with the property that, in the terminology of Jech [Jec71], any almost universal transitive class closed under $\mathfrak{J}_1, \dots, \mathfrak{J}_8$ is a model ZF. Set $G_i(x, y, \alpha) = \mathfrak{J}_i(x, y)$, for $i = 1, \dots, 8$, and $G_9(x, y, \alpha) = x \cap \alpha$. Let (ξ, η, ϑ) be the rank of $\langle\xi, \eta, \vartheta\rangle$ in the Gödel wellordering and let I, J, and K be such that $\xi = (I(\xi), J(\xi), K(\xi))$. Then put

$$F(\xi,\alpha) = \begin{cases} \{F(\eta,\beta) : \eta < \xi \,\&\, \beta \in \mathbb{R}\} & \text{if } I(\xi) = 0 \\ G_{I(\xi)}(F(J(\xi),(\alpha)_0), F(K(\xi),(\alpha)_1), (\alpha)_2) & \text{if } 0 < I(\xi) \leq 9 \\ \{F(J(\xi),(\alpha)_0), F(K(\xi),(\alpha)_1)\} & \text{if } I(\xi) > 9 \end{cases}$$

Then $\{F(\xi,\alpha) : \xi \in \mathrm{Ord} \,\&\, \alpha \in \mathbb{R}\} = \mathbf{L}(\mathbb{R})$.

Now suppose that Δ is closed under wellordered unions. By induction on the Gödel wellordering of pairs $\langle \xi, \eta \rangle$ one can show that for each ξ and η the relations

$$P^{=}_{\xi,\eta}(\alpha, \beta) \Leftrightarrow F(\xi, \alpha) = F(\eta, \beta)$$
$$P^{\in}_{\xi,\eta}(\alpha, \beta) \Leftrightarrow F(\xi, \alpha) \in F(\eta, \beta)$$

are in Δ. But by the preceding paragraph this means that every set of reals in $\mathbf{L}(\mathbb{R})$ is in Δ, whence $\Gamma \not\subseteq \mathbf{L}(\mathbb{R})$. This contradiction establishes Theorem 5.7.

⊣ (Theorem 5.7)

We list some other results which partake of the flavor of those presented here:

(*Steel*): Any two non-Borel analytic sets are Borel isomorphic.

(*Steel*): Jump operators on the Turing degrees are prewellordered by the Martin measure.

(*Radin, Steel, Van Wesep*): Any nonselfdual continuously closed class of sets of reals may be obtained by application of a fixed ω-ary Boolean operation to sequences of open sets.

The first two of these results appear in Steel [Ste77] and will be published elsewhere. The last appears in the union of Steel [Ste77] and Van Wesep [Van77].

§6. Conjectures and Problems. One should like to prove one of the following two competing conjectures:

(i) If Γ is a nonselfdual continuously closed class, then $\mathsf{Sep}_{\mathrm{I}}(\Gamma)$ or $\mathsf{Sep}_{\mathrm{I}}(\breve{\Gamma})$. Thus, the classical separation principles serve to distinguish each nonselfdual continuously closed class from its dual.

(ii) If S is a set of Wadge degrees, then for some nonselfdual degree a, we have $a \in S \Leftrightarrow \breve{a} \in S$.

The theory of the Wadge degrees seems to sorely need results of the following sort, of which there are now essentially no examples:

Some "closure" property of the order type of $\Gamma \cap \breve{\Gamma}$ (*e.g.*, that it is a cardinal) implies some closure property for Γ (*e.g.*, closure under intersection) or even for $\Gamma \cap \breve{\Gamma}$.

REFERENCES

THOMAS J. JECH

[Jec71] ***Lectures in set theory, with particular emphasis on the method of forcing***, Lecture Notes in Mathematics, Vol. 217, Springer-Verlag, Berlin, 1971.

YIANNIS N. MOSCHOVAKIS

[Mos70] *Determinacy and prewellorderings of the continuum*, ***Mathematical logic and foundations of set theory. Proceedings of an international colloquium held under the auspices of the Israel***

Academy of Sciences and Humanities, Jerusalem, 11–14 November 1968 (Y. Bar-Hillel, editor), Studies in Logic and the Foundations of Mathematics, North-Holland, Amsterdam-London, 1970, pp. 24–62.

John R. Steel

[Ste77] *Determinateness and subsystems of analysis*, ***Ph.D. thesis***, Berkeley, 1977.

Robert Van Wesep

[Van77] *Subsystems of second-order arithmetic, and descriptive set theory under the axiom of determinateness*, ***Ph.D. thesis***, University of California, Berkeley, 1977.

[Van78A] *Separation principles and the axiom of determinateness*, ***The Journal of Symbolic Logic***, vol. 43 (1978), no. 1, pp. 77–81.

William W. Wadge

[Wad84] *Reducibility and determinateness on the Baire space*, ***Ph.D. thesis***, University of California, Berkeley, 1984.

SINAI HOSPITAL OF BALTIMORE
2401 W. BELVEDERE AVENUE
BALTIMORE, MARYLAND, 21215
UNITED STATES OF AMERICA
E-mail: rvanwese@lifebridgehealth.org

A NOTE ON WADGE DEGREES

ALEXANDER S. KECHRIS

§**1.** It has been shown by Martin and independently Steel (see [Van78B, Theorem 4.2]), that the wellordering of Wadge degrees of $\underset{\sim}{\mathbf{\Delta}}^1_{2n}$ sets of reals has length $\underset{\sim}{\boldsymbol{\delta}}^1_{2n+1}$. Although this is a result about projective sets, their proofs require full AD as they proceed by showing that if η_{2n} is the length of the wellordering of Wadge degress of $\underset{\sim}{\mathbf{\Delta}}^1_{2n}$ sets, then cofinality$(\underset{\sim}{\boldsymbol{\delta}}^1_{2n+1}) \leq \eta_{2n}$, which by the regularity of $\underset{\sim}{\boldsymbol{\delta}}^1_{2n+1}$ (a consequence of AD) implies that $\underset{\sim}{\boldsymbol{\delta}}^1_{2n+1} \leq \eta_{2n}$. As it is easy to see that $\eta_{2n} \leq \underset{\sim}{\boldsymbol{\delta}}^1_{2n+1}$, by a direct computation, we have the desired equality. Of course, the use of full AD here can be replaced, by trivial absoluteness considerations, by Det$(\mathbf{L}(\mathbb{R}))$, *i.e.*, the hypothesis that all sets of reals in $\mathbf{L}(\mathbb{R})$ are determined. Motivated by the fact that one only needs Det$(\underset{\sim}{\mathbf{\Delta}}^1_{2n})$ to establish the fact that the Wadge degrees of $\underset{\sim}{\mathbf{\Delta}}^1_{2n}$ sets are wellordered (see [Van78B, Theorem 2.2]), Martin has asked if one can compute that also $\eta_{2n} = \underset{\sim}{\boldsymbol{\delta}}^1_{2n+1}$, using again only Det$(\underset{\sim}{\mathbf{\Delta}}^1_{2n})$. We provide such a proof below. It is based on a method of "inverting the game quantifier" which may be also useful elsewhere.

§**2.** Let us take $n = 1$ for notational simplicity. From now on we assume Det$(\underset{\sim}{\mathbf{\Delta}}^1_2)$.

2.1. For each $A \subseteq {}^\omega\omega$ let $\Gamma(A)$ be a pointclass with the following properties:

(i) $A, {}^\omega\omega \setminus A \in \Gamma(A)$,

(ii) $B, {}^\omega\omega \setminus B \in \Gamma(A) \Rightarrow \Gamma(B) \subseteq \Gamma(A)$,

(iii) $A \in \underset{\sim}{\mathbf{\Delta}}^1_m \Rightarrow \Gamma(A) \subseteq \underset{\sim}{\mathbf{\Delta}}^1_m$, $\forall m \geq 2$,

(iv) $\Gamma(A)$ is ${}^\omega\omega$-parametrized and closed under continuous substitutions,

(v) There is a map $A \mapsto C_A$, sending each A to C_A, an ${}^\omega\omega$-universal set in $\Gamma(A)$ and for each $m \geq 2$ there is a total recursive function f_m such that if $\varepsilon \in {}^\omega\omega$ is a $\underset{\sim}{\mathbf{\Delta}}^1_m$-code of A, then $f(\varepsilon)$ is a $\underset{\sim}{\mathbf{\Delta}}^1_m$-code of C_A.

For example, we can take $\Gamma(A) = {}_2\underset{\sim}{\mathbf{ENV}}({}^2\mathrm{E}, A) =$ the pointclass of all pointsets semirecursive in ^{2}E, A and a real.

Research partially supported by NSF Grant MCS-17254 A01. The author is an A. P. Sloan Foundation Fellow.

Wadge Degrees and Projective Ordinals: The Cabal Seminar, Volume II
Edited by A. S. Kechris, B. Löwe, J. R. Steel
Lecture Notes in Logic, 37

2.2. Next let us reall that if $\Game$ is the game quantifier, then $\Game\underset{\sim}{\mathbf{\Delta}}^1_2 = \underset{\sim}{\mathbf{\Delta}}^1_3$. (Here is a quick proof due to Addison:

Let P, Q be disjoint $\underset{\sim}{\mathbf{\Sigma}}^1_3$ sets. Say $P(x) \Leftrightarrow \exists\alpha P'(x,\alpha)$, $Q(x) \Leftrightarrow \exists\beta Q'(x,\beta)$, where P', $Q' \in \underset{\sim}{\mathbf{\Pi}}^1_2$. Let $P''(x,\alpha,\beta) \Leftrightarrow P'(x,\alpha)$, $Q''(x,\alpha,\beta) \Leftrightarrow Q'(x,\beta)$ and let $S(x,\alpha,\beta)$ in $\underset{\sim}{\mathbf{\Delta}}^1_2$ separate P'', Q''. Then it is easy to check that $\exists\alpha(0)\forall\beta(0)\exists\alpha(1)\forall\beta(1)\ldots S(x,\alpha,\beta)$ separates P, Q.)

Now let $W \subseteq {}^\omega\omega$ be Π^1_3 and universal for $\underset{\sim}{\mathbf{\Pi}}^1_3$ and σ a Π^1_3-norm on W. For $x \in W$, put

$$H_x = \{\langle y,z\rangle : \sigma(y) \leq \sigma(z) < \sigma(x)\}.$$

Then let f be a total recursive function such that if $x \in W$, then $f(x)$ is a $\underset{\sim}{\mathbf{\Delta}}^1_2$-code of a set, say $\underset{\sim}{\mathbf{\Delta}}_x$, such that $\Game\underset{\sim}{\mathbf{\Delta}}_x = H_x$. (The existence of such an f is clear from the proof in the preceding paragraph.) Put finally for x, $y \in W$:

$$x \lesssim y \Leftrightarrow \Gamma(\underset{\sim}{\mathbf{\Delta}}_x) \subseteq \Gamma(\underset{\sim}{\mathbf{\Delta}}_y).$$

Lemma 2.1. $\lesssim$ is a prewellordering.

Proof. $\lesssim$ is obviously reflexive and transitive.

$\lesssim$ is connected: Let x, $y \in W$. Then by Wadge, $\underset{\sim}{\mathbf{\Delta}}_x \leq_{\mathrm{W}} \underset{\sim}{\mathbf{\Delta}}_y$ or $\underset{\sim}{\mathbf{\Delta}}_y \leq_{\mathrm{W}} {}^\omega\omega \setminus \underset{\sim}{\mathbf{\Delta}}_x$, where $X \leq_{\mathrm{W}} Y$ iff X is reducible to Y via a continuous function. Say the first case occurs. Then $\underset{\sim}{\mathbf{\Delta}}_x \in \Gamma(\underset{\sim}{\mathbf{\Delta}}_y)$. But also ${}^\omega\omega \setminus \underset{\sim}{\mathbf{\Delta}}_x \leq_{\mathrm{W}} {}^\omega\omega \setminus \underset{\sim}{\mathbf{\Delta}}_y$, therefore ${}^\omega\omega \setminus \underset{\sim}{\mathbf{\Delta}}_x \in \Gamma(\underset{\sim}{\mathbf{\Delta}}_y)$, thus $\Gamma(\underset{\sim}{\mathbf{\Delta}}_x) \subseteq \Gamma(\underset{\sim}{\mathbf{\Delta}}_y)$, *i.e.*, $x \lesssim y$.

$\lesssim$ is wellfounded: Given $\varnothing \subsetneq A \subseteq W$ let $x \in A$ be such that $\underset{\sim}{\mathbf{\Delta}}_x$ has least Wadge ordinal. Then for any $y \in A$, $\underset{\sim}{\mathbf{\Delta}}_x \leq_{\mathrm{W}} \underset{\sim}{\mathbf{\Delta}}_y$ or $\underset{\sim}{\mathbf{\Delta}}_x \leq_{\mathrm{W}} {}^\omega\omega \setminus \underset{\sim}{\mathbf{\Delta}}_y$, therefore, as above $x \lesssim y$. ⊣ (Lemma 2.1)

Let $\varphi\colon W \twoheadrightarrow \lambda$ be the norm associated with $\lesssim$.

Lemma 2.2. φ is a Π^1_3-norm.

Proof. Fix $y \in W$. We want to express $x \in W \wedge x \lesssim y$ in a Δ^1_3 way uniformly in y. When both x, y are in W the condition $x \lesssim y$ is equivalent to

$$\underset{\sim}{\mathbf{\Delta}}_x \leq_{\mathrm{W}} C_{\underset{\sim}{\mathbf{\Delta}}_y} \wedge {}^\omega\omega \setminus \underset{\sim}{\mathbf{\Delta}}_x \leq_{\mathrm{W}} C_{\underset{\sim}{\mathbf{\Delta}}_y},$$

which is clearly Δ^1_3 uniformly in x, y. So it is enough to find a total recursive function g such that

(i) $y \in W \Rightarrow g(y) \in W$,
(ii) $x, y \in W \wedge x \lesssim y \Rightarrow \sigma(x) \leq \sigma(g(y))$.

Because then for $y \in W$:

$$x \in W \wedge x \lesssim y \Leftrightarrow \sigma(x) \leq \sigma(g(y)) \wedge x \lesssim y,$$

which by our preceding remarks is Δ^1_3 uniforming in y.

In order to construct g we use the following

Sublemma 2.3. There is a total recursive function h such that

(i) $y \in W \Rightarrow h(y) \in W$
(ii) $x, y \in W \wedge H_x \leq_W H_y \Rightarrow \sigma(x) \leq \sigma(h(y))$.

PROOF. Let h be a total recursive function such that if $y \in W$, then $h(y) \in W$ and

$$z \in C_{H_y} \Leftrightarrow \langle a, z\rangle \in W$$
$$\Leftrightarrow \sigma(\langle a, z\rangle) \leq \sigma(h(y)),$$

for some $a \in {}^\omega\omega$. Then if $x, y \in W \wedge H_x \leq_W H_y$, but $\sigma(x) > \sigma(h(y))$, towards a contradiction, we have $z \in C_{H_y} \Leftrightarrow \sigma(\langle a, z\rangle) \leq \sigma(h(y)) < \sigma(x)$, for some $a \in {}^\omega\omega$, therefore $C_{H_y} \leq_W H_x \leq_W H_y$, a contradiction.

⊣ (Sublemma 2.3)

To complete the proof of Lemma 2.2, we construct now g as follows: Let f^1 be total recursive such that if $y \in W$, then $f^1(y)$ is a $\underset{\sim}{\mathbf{\Delta}}^1_2$-code of $C_{\underset{\sim}{\mathbf{\Delta}}_y}$ and let f^2, f^3 be total recursive such that if $y \in W$, then $f^3(y) \in W$ and

$$\exists\alpha(\langle z, \langle t, \alpha\rangle\rangle \in C_{\underset{\sim}{\mathbf{\Delta}}_y}) \Leftrightarrow \sigma(\langle f^2(y), \langle z, t\rangle\rangle) < \sigma(f^3(y)).$$

Let $g = h \circ f_3$. Assume now $x, y \in W$ and $x \leq y$. Then $\Gamma(\underset{\sim}{\mathbf{\Delta}}_x) \subseteq \Gamma(\underset{\sim}{\mathbf{\Delta}}_y)$, so

$$\langle t, \alpha\rangle \in \underset{\sim}{\mathbf{\Delta}}_x \Leftrightarrow \langle z_0, \langle t, \alpha\rangle\rangle \in C_{\underset{\sim}{\mathbf{\Delta}}_y}, \text{ for some } z_0,$$

thus

$$t \in H_x \Leftrightarrow t \in \exists\underset{\sim}{\mathbf{\Delta}}_x \Leftrightarrow \exists\alpha(\langle t, \alpha\rangle \in \underset{\sim}{\mathbf{\Delta}}_x)$$
$$\Leftrightarrow \exists\alpha(\langle z_0, \langle t, \alpha\rangle\rangle \in C_{\underset{\sim}{\mathbf{\Delta}}_y})$$
$$\Leftrightarrow \sigma(\langle f^2(y), \langle z_0, t\rangle\rangle) < \sigma(f^3(y)).$$

So

$$H_x \leq_W H_{f^3(y)}, \text{ thus } \sigma(x) \leq h(f^3(y)) = g(y).$$ ⊣ (Lemma 2.2)

Using Lemma 2.2 we complete the proof of the result as follows: By Lemma 2.2 we have $\lambda = \underset{\sim}{\boldsymbol{\delta}}^1_3$. Since by direct computation we can easily see that $\eta_2 \leq \underset{\sim}{\boldsymbol{\delta}}^1_3$, it is enough to show $\lambda \leq \eta_2$. For that define for $\xi < \lambda$:

$$f(\xi) = \text{ Wadge ordinal of } C_{\underset{\sim}{\mathbf{\Delta}}_x},$$

where $\varphi(x) = \xi$.

Since

$$\varphi(x) = \varphi(y) \Rightarrow \Gamma(\underset{\sim}{\mathbf{\Delta}}_x) = \Gamma(\underset{\sim}{\mathbf{\Delta}}_y)$$
$$\Rightarrow C_{\underset{\sim}{\mathbf{\Delta}}_x} \leq_W C_{\underset{\sim}{\mathbf{\Delta}}_y} \wedge C_{\underset{\sim}{\mathbf{\Delta}}_y} \leq_W C_{\underset{\sim}{\mathbf{\Delta}}_x},$$

this is well-defined. Also $f : \lambda \to \eta_2$, so it is enough to show that f is order preserving. Indeed, let $\xi < \zeta < \lambda$ and $\varphi(x) = \xi$, $\varphi(y) = \zeta$. Then $x \lesssim y$ and $y \not\lesssim x$, so $\Gamma(\underset{\sim}{\mathbf{\Delta}}_x) \subsetneqq \Gamma(\underset{\sim}{\mathbf{\Delta}}_y)$. Consequently, $C_{\underset{\sim}{\mathbf{\Delta}}_x} \leq_W C_{\underset{\sim}{\mathbf{\Delta}}_y}$ but $C_{\underset{\sim}{\mathbf{\Delta}}_y} \not\leq_W C_{\underset{\sim}{\mathbf{\Delta}}_x}$, therefore, by Wadge, $C_{\underset{\sim}{\mathbf{\Delta}}_x} \leq_{\mathrm{W}} {}^{\omega}\omega \setminus C_{\underset{\sim}{\mathbf{\Delta}}_y}$, *i.e.*, the Wadge ordinal of $C_{\underset{\sim}{\mathbf{\Delta}}_x}$, which is $f(\xi)$, is smaller than the Wadge ordinal of $C_{\underset{\sim}{\mathbf{\Delta}}_y}$, which is $f(\zeta)$.

This finishes the proof of the result.

REFERENCES

Alexander S. Kechris and Yiannis N. Moschovakis

[Cabal i] *Cabal seminar* 76–77, Lecture Notes in Mathematics, no. 689, Berlin, Springer, 1978.

Robert Van Wesep

[Van78B] *Wadge degrees and descriptive set theory*, this volume, originally published in Kechris and Moschovakis [Cabal i], pp. 151–170.

DEPARTMENT OF MATHEMATICS
CALIFORNIA INSTITUTE OF TECHNOLOGY
PASADENA, CALIFORNIA 91125
UNITED STATES OF AMERICA
E-mail: kechris@caltech.edu

SOME RESULTS IN THE WADGE HIERARCHY OF BOREL SETS

ALAIN LOUVEAU

This paper has two goals: First, to provide construction principles, by means of boolean operations, of the Wadge classes of Borel sets, and in a second step, to use these construction principles to define lightface versions of the Wadge classes, and prove that the notion of Wadge class, roughly speaking, is Δ^1_1: If a Δ^1_1 set is in some (boldface) Wadge class Γ, it belongs to the corresponding Δ^1_1-recursive lightface class.

The necessary background concerning Wadge's hierarchy can be found in Van Wesep [Van78B]. Let us recall that a family $\Gamma \subseteq {}^\omega\omega$ is a **class** if it is closed under inverse images by continuous functions from ${}^\omega\omega$ into itself. If Γ is of the form $[A] = \{f^{-1}(A) : f : {}^\omega\omega \to {}^\omega\omega, \text{ continuous}\}$ for some set A, it is a **Wadge class** (the Wadge class of A). For a set A, $\check{A}$ denotes its complement, $\check{A} = {}^\omega\omega \setminus A$, and $\check{\Gamma} = \{\check{A} : A \in \Gamma\}$ is the **dual** class of the class Γ. $\Delta = \Delta(\Gamma)$ is the **ambiguous class** associated with Γ, and is defined by $\Delta = \Gamma \cap \check{\Gamma}$. A class Γ is **self dual** if $\check{\Gamma} = \Gamma$.

The Wadge hierarchy is obtained by (partially) ordering the Wadge classes by strict inclusion $\Gamma < \Gamma'$ if $\Gamma \subseteq \Gamma'$ and $\Gamma \neq \Gamma'$. We similarly define $\Gamma \leq \Gamma'$ if $\Gamma \subseteq \Gamma'$. This ordering admits a game theoretical analysis: If A, B are two subsets of ${}^\omega\omega$, let $G_{\mathrm{W}}(A, B)$ be the game where players I and II play alternatively integers, player I constructing $\alpha \in {}^\omega\omega$, and player II having the possibility of passing, as long as he constructs $\beta \in {}^\omega\omega$. player II wins this game if $\alpha \in A \iff \beta \in B$. It is easy to check that player II has a winning strategy in $G_{\mathrm{W}}(A, B) \iff [A] \leq [B]$.

Let us now consider only Wadge classes of Borel sets. Then we can use Borel Determinacy (Martin [Mar75]). This gives that $<$ is almost a linear ordering (Wadge [Wad84]): The only class not comparable to Γ is the class $\check{\Gamma}$ in case Γ is non self-dual. So by identifying twin pairs $(\Gamma, \check{\Gamma})$ of non self-dual classes, we obtain a linear ordering. Moreover, by Martin [Mar73], the ordering $<$ is well founded, so that we can associate with each Wadge class of Borel sets Γ an ordinal $o(\Gamma)$. The pattern of self-dual and non self-dual classes is as follows (see Van Wesep [Van78B]): Self-dual and non self-dual twin pairs alternate at successor stages. The hierarchy begins with the twin pair $\{\varnothing\}$, $\{{}^\omega\omega\}$; at limit

Wadge Degrees and Projective Ordinals: The Cabal Seminar, Volume II
Edited by A. S. Kechris, B. Löwe, J. R. Steel
Lecture Notes in Logic, 37

stages of cofinality ω stands a self-dual class, and at limit stages of cofinality ω_1 a non self-dual twin pair.

Of course, the Wadge hierarchy is a refinement of the classical hierarchies, the Borel hierarchy $(\underset{\sim}{\mathbf{\Sigma}}^0_\xi, \underset{\sim}{\mathbf{\Pi}}^0_\xi)_{\xi<\omega_1}$, and the hierarchy of differences of Hausdorff and Kuratowski. But it contains also a lot of more "exotic" classes. The picture presented in part 1 closely follows unpublished results of Wadge [Wad84], except that Wadge's description exhibits, for each non self-dual class Γ, a set A for which $\Gamma = [A]$, whereas we define a boolean operation which enables to construct Γ in terms of preceding classes.

I would like to thank John Steel for giving me access to Wadge's papers, and also for enlightening discussions about Wadge classes.

§1. A Description of Wadge Classes of Borel Sets. First, let us make some heuristic comments on what follows. The usual Borel hierarchy of sets, $\{\underset{\sim}{\mathbf{\Sigma}}^0_\xi : \xi < \omega_1\}$, is our starting point in analysing the Wadge classes. This obviously does not give a complete description, and we certainly must refine it by adding, between $\underset{\sim}{\mathbf{\Sigma}}^0_\xi$ and $\underset{\sim}{\mathbf{\Sigma}}^0_{\xi+1}$, the hierarchy of differences of $\underset{\sim}{\mathbf{\Sigma}}^0_\xi$ sets. This gives a first refinement, which again happens to be insufficient, at least for $\xi \geq 2$. The first expectation is that by refining again (may be a few more times), one should reach the complete picture of Wadge degrees. It is indeed what happens, and we shall define a set of levels by using successive refinements; but in order to obtain the picture between $\underset{\sim}{\mathbf{\Sigma}}^0_\xi$ and $\underset{\sim}{\mathbf{\Sigma}}^0_{\xi+1}$, ξ refinements are necessary. And because the Wadge ordering is a well-ordering, these refinements are not well-ordered by inclusion. In fact, it turns out that the reverse ordering is well-founded, and so the ordinal we shall associate with the refinements will measure the "degree of simplicity" of each level, rather than its degree of complexity: the hierarchy of differences is given level ξ, whereas the most complicated classes are those of level 1. At this last level occur all successor classes and all limit classes of cofinality ω. The "simplicity" of each class may be measured, in mathematical terms, by its closure properties. We shall see (Lemma 1.4 below) that the closure properties do increase with the level of the class.

For constructing the classes, we have selected a small set of operations, which is certainly not the least possible one, but is convenient for a nice description of the classes. Starting from the classes $\underset{\sim}{\mathbf{\Sigma}}^0_\xi$, and applying successively these operations will give the desired description. The main point here is that each operation increases the Wadge ordinal, but decreases the level: Intuitively, it means that the resulting class is more complicated than the original one. A final word of comment: By our general knowledge of the Wadge hierarchy, it is not necessary to give a construction principle from below for the self-dual classes, which are exactly the Δ-parts of successor non self-dual classes, and among the non self-dual classes, we can choose to describe only

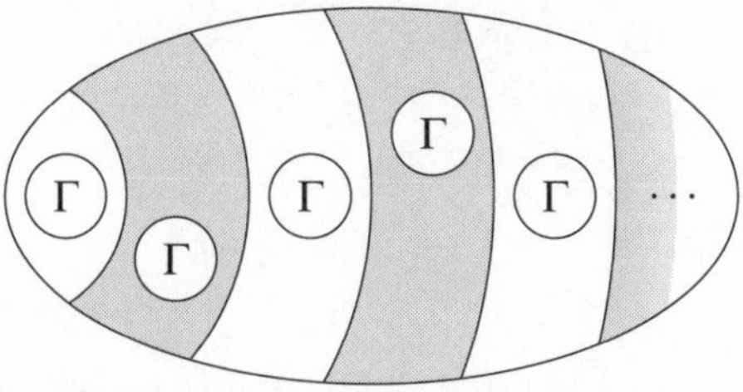

FIGURE 1. $D_\xi(\Gamma)$, for ξ even

one of the twin dual classes. One mathematically nice way for doing this choice is given by a result of Steel [Ste81B]: Among each twin pair, exactly one class has the separation property. For technical reasons, we have chosen to describe the other ones, *i.e.* the non self-dual classes which do not possess the separation property. In particular, we begin the construction with the $\mathbf{\Sigma}^0_\xi$ classes, not the $\mathbf{\Pi}^0_\xi$ ones.

Let us now begin with the definition of the operations.

DEFINITION 1.1. (a) **Differences** (Hausdorff, Kuratowski). Let $\xi \geq 1$ be a countable ordinal. If $\langle C_\eta : \eta < \xi\rangle$ is an increasing sequence of sets, the set $A = D_\xi(\langle C_\eta : \eta < \xi\rangle)$ is defined by

$$A = \begin{cases} \bigcup\{C_\eta \setminus \bigcup_{\eta'<\eta} C_{\eta'} : \eta \text{ odd}, \eta < \xi\} & \text{for } \xi \text{ even} \\ \bigcup\{C_\eta \setminus \bigcup_{\eta'<\eta} C_{\eta'} : \eta \text{ even}, \eta < \xi\} & \text{for } \xi \text{ odd} \end{cases}$$

[So $D_1(\langle C_0\rangle) = C_0, D_2(\langle C_0, C_1\rangle) = C_1 \setminus C_0, \dots$. This definition is not the usual one, as given in Kuratowski [Kur66], which deals with decreasing sequences—and is applied to $\mathbf{\Pi}^0_\xi$.]

If Γ is some class, $D_\xi(\Gamma)$ is the class of all $D_\xi(\langle C_\eta : \eta < \xi\rangle)$ for some increasing sequence of sets in Γ. $D_1(\Gamma)$ is simply written Γ.

(b) **Separated Unions** (Wadge). We define $A = \mathrm{SU}(\langle C_n : n \in \omega\rangle, \langle A_n : n \in \omega\rangle)$ in case the sets C_n are pairwise disjoint, by $A = \bigcup_n (A_n \cap C_n)$. The set $C = \bigcup_n C_n$ is the corresponding **envelope** of A.

For classes Γ, Γ', we let $\mathrm{SU}(\Gamma, \Gamma')$ be the class of all $\mathrm{SU}(\langle C_n\rangle, \langle A_n\rangle)$ with the C_n's in Γ and the A_n in Γ'. The set $\langle \Gamma, \mathrm{SU}(\Gamma, \Gamma')\rangle$ is the set of pairs $\langle C, A\rangle$ where $A = \mathrm{SU}(\langle C_n\rangle, \langle A_n\rangle)$ is in $\mathrm{SU}(\Gamma, \Gamma')$ and $C = \bigcup_n C_n$ is the corresponding envelope.

(c) **One-sided Separated Unions** (Myers, Wadge). We say that $A = \mathrm{Sep}(C, B_1, B_2)$ if $A = (C \cap B_1) \cup (B_2 \setminus C)$ [This is of course a particular case of (b)].

If Γ, Γ' are two classes, $\mathrm{Sep}(\Gamma, \Gamma')$ is the class of all $\mathrm{Sep}(C, B_1, B_2)$ where $C \in \Gamma, B_1 \in \check{\Gamma}'$ and $B_2 \in \Gamma'$ [This is not symmetric in Γ' and $\check{\Gamma}'$].

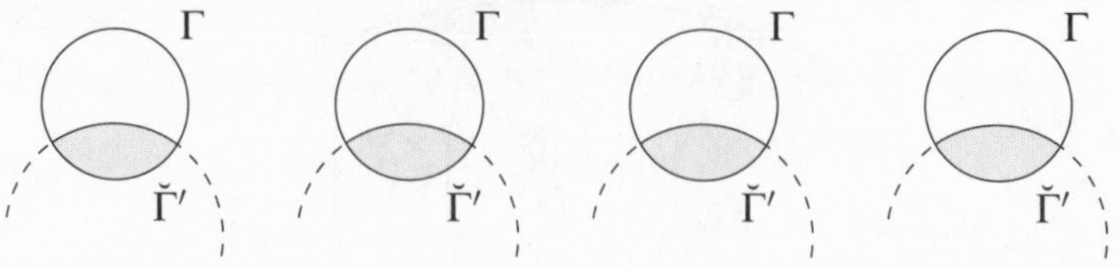

FIGURE 2. $\mathrm{SU}(\Gamma, \Gamma')$

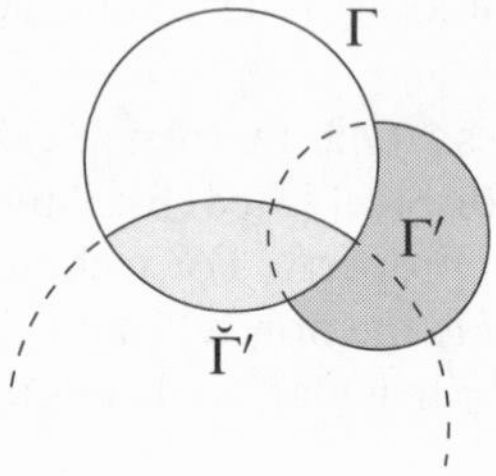

FIGURE 3. $\mathrm{Sep}(\Gamma, \Gamma')$

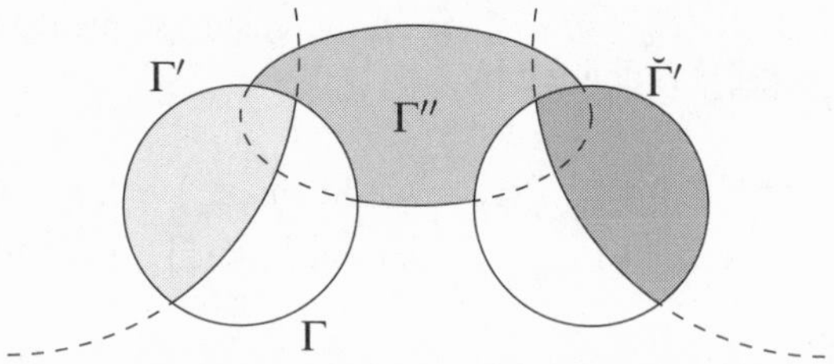

FIGURE 4. $\mathrm{Bisep}(\Gamma, \Gamma', \Gamma'')$

(d) **Two-sided Separated Unions**. [This is again a particular case of (b)]. We let $A = \mathrm{Bisep}(C_1, C_2, A_1, A_2, B)$, in case C_1 and C_2 are disjoint, be the set $A = (C_1 \cap A_1) \cup (C_2 \cap A_2) \cup (B \setminus (C_1 \cup C_2))$.

If $\Gamma, \Gamma', \Gamma''$ are classes, $\mathrm{Bisep}(C_1, C_2, A_1, A_2, B)$ with C_1, C_2 in Γ, A_1 in $\check{\Gamma}'$, A_2 in Γ' and B in Γ''. Moreover, if $\Gamma'' = \{\varnothing\}$, we just write $\mathrm{Bisep}(\Gamma, \Gamma')$ [The definition is symmetric in Γ' and $\check{\Gamma}'$].

(e) **Separated Differences**. For $\xi \geq 2$ a countable ordinal, we define $A = \mathrm{SD}_\xi(\langle C_\eta : \eta < \xi \rangle, \langle A_\eta : \eta < \xi \rangle, B)$ in case the C_η's and the A_η's are increasing, with $A_\eta \subseteq C_\eta \subseteq A_{\eta+1}$, by

$$A = \bigcup_{\eta<\xi} \Big(A_\eta \setminus \bigcup_{\eta<\eta'} C_{\eta'} \Big) \cup \Big(B \setminus \bigcup_{\eta<\xi} C_\eta \Big).$$

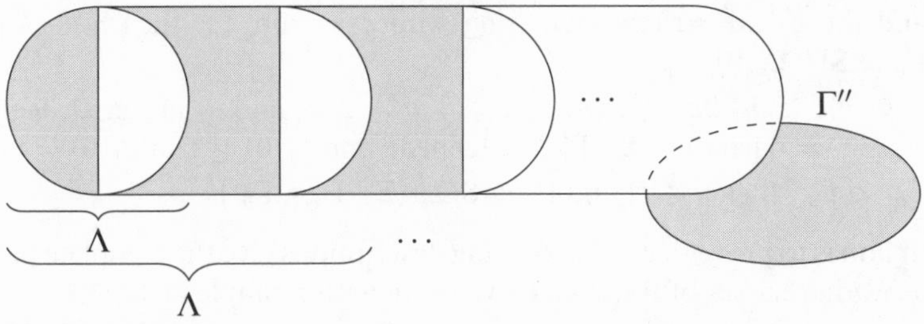

FIGURE 5. $\mathrm{SD}_\xi(\Lambda, \Gamma'')$

If $\Lambda \subseteq \Gamma \times \Gamma'$, and Γ'' is a class, $\mathrm{SD}_\xi(\Lambda, \Gamma'')$ is the set of all $\mathrm{SD}_\xi(\langle C_\eta : \eta < \xi\rangle, \langle A_\eta : \eta < \xi\rangle, B)$ for $B \in \Gamma''$ and, for each $\eta < \xi$, the pair (C_η, A_η) is in Λ. Again, for $\Gamma'' = \{\varnothing\}$ we write $\mathrm{SD}(\Lambda)$ [This operation will be used only for $\Lambda = \langle \Gamma, \mathrm{SU}(\Gamma, \Gamma')\rangle$ as defined in (b)].

We now proceed to apply these operations. Particular ways of combining these operations will be encoded by elements of ${}^\omega\omega_1$. Of course, we put restrictions on the allowed combinations. This leads to the following inductive definition of a "description" and a corresponding "described class".

Let u be some element of ${}^\omega\omega_1$. We let $u = \langle u_0, u_1\rangle$, where $u_0(n) = u(2n)$ and $u_1(n) = u(2n+1)$. Similarly we let $u = \langle (u)_n : n \in \omega\rangle$ where $(u)_n(m) = u(\langle n, m\rangle)$. $\underline{0}$ is the element of ${}^\omega\omega_1$ defined by $\underline{0}(n) = 0$. We now define inductively the relations "u is a description" and "u describes the class Γ" (written $\Gamma_u = \Gamma$). Descriptions are elements of ${}^\omega\omega_1$. The first ordinal $u(0)$ will give the level of Γ_u (as informally discussed in the introduction), $u(1)$ and sometimes $u(2)$ the operation used to obtain Γ_u, and the remaining of u the description of the classes from which Γ_u has been obtained.

DEFINITION 1.2. The relations "u is a description" (written $u \in D$), and "u describes Γ", are the least relations satisfying the following conditions:

(a) If $u(0) = 0$, u is a description, and $\Gamma_u = \{\varnothing\}$
(b) If $u(0) = \xi \geq 1, u(1) = 1$ and $u(2) = \eta \geq 1$, then $u \in D$ and $\Gamma_u = D_\eta(\boldsymbol{\Sigma}^0_\xi)$.
(c) If $u = \xi^\frown 2^\frown \eta^\frown u^*$, where $\xi \geq 1, \eta \geq 1, u^* \in D$ and $u^*(0) > \xi$, then $u \in D$ and $\Gamma_u = \mathrm{Sep}(D_\eta(\boldsymbol{\Sigma}^0_\xi), \Gamma_{u^*})$.
(d) If $u = \xi^\frown 3^\frown \eta^\frown \langle u_0, u_1\rangle$, where $\xi \geq 1, \eta \geq 1, u_0$ and u_1 are in D, $u_0(0) > \xi, u_1(0) \geq \xi$ or $u_1(0) = 0$, and $\Gamma_{u_1} < \Gamma_{u_0}$, then $u \in D$ and $\Gamma_u = \mathrm{Bisep}(D_\eta(\boldsymbol{\Sigma}^0_\xi), \Gamma_{u_0}, \Gamma_{u_1})$.
(e) If $u = \xi^\frown 4^\frown \langle u_n : n \in \omega\rangle$, where $\xi \geq 1$ and each u_n is in D, and either for all n, $u_n(0) = \xi_1 > \xi$, and the Γ_{u_n} are strictly increasing, or $u_n(0) = \xi_n$

and the ξ_n are strictly increasing with $\xi < \sup_n \xi_n$, then $u \in D$ and $\Gamma_u = \mathrm{SU}(\boldsymbol{\Sigma}^0_\xi, \bigcup_n \Gamma_{u_n})$.

(f) If $u = \xi^\frown 5^\frown \eta^\frown \langle u_0, u_1\rangle$, where $\xi \geq 1, \eta \geq 2, u_0$ and u_1 are in D, $u_0(0) = \xi, u_0(1) = 4$ [so that $\langle \boldsymbol{\Sigma}^0_\xi, \Gamma_{u_0}\rangle$ is defined], and $u_1(0) \geq \xi$ or $u_1(0) = 0$, and $\Gamma_{u_1} < \Gamma_{u_0}$, then $u \in D$ and $\Gamma_u = \mathrm{SD}_\eta(\langle \boldsymbol{\Sigma}^0_\xi, \Gamma_{u_0}\rangle, \Gamma_{u_1})$.

Our aim is to prove that the preceding descriptions give the complete picture of the Wadge classes of Borel sets. We begin with a simple fact.

Proposition 1.3. The described classes are non self-dual Borel Wadge classes.

Proof. The "Borel" part is clear. To prove that these classes are non self-dual, it is enough to exhibit a universal set, and this is easy by induction. The only fact to note here is that by using the reduction property, one can find a sequence of $D_\eta(\boldsymbol{\Sigma}^0_\xi)$ sets, in ${}^\omega\omega \times {}^\omega\omega$, which is universal for sequences of pairwise disjoint $D_\eta(\boldsymbol{\Sigma}^0_\xi)$ sets of ${}^\omega\omega$. ⊣

It is clear that for each $u \in D$, there corresponds exactly one described class Γ_u. We now show that the **level** $u(0)$ gives the closure properties of Γ_u.

Lemma 1.4. Let u be a description, with $u(0) = \xi \geq 1$. Then
(a) Γ_u is closed under union with a $\boldsymbol{\Delta}^0_\xi$ set.
(b) $\mathrm{SU}(\boldsymbol{\Sigma}^0_\xi, \Gamma_u) = \Gamma_u$ (written Γ_u is closed under $\boldsymbol{\Sigma}^0_\xi - \mathrm{SU}$).

Proof. By induction.

Case 1. $u(1) = 1$, so $\Gamma_u = D_\eta(\boldsymbol{\Sigma}^0_\xi)$.

(b) Let C_n be separating $\boldsymbol{\Sigma}^0_\xi$ sets, and $A^n = D_\eta(\langle A^n_\zeta : \zeta < \eta\rangle)$ be the $D_\eta(\boldsymbol{\Sigma}^0_\xi)$ sets. Consider $A_\zeta = \bigcup_n (A^n_\zeta \cap C_n)$. The A_ζ are clearly $\boldsymbol{\Sigma}^0_\xi$ and increasing, and moreover $\mathrm{SU}(\langle C_n : n \in \omega\rangle, \langle A^n : n \in \omega\rangle) = D_\eta(\langle A_\zeta : \zeta < \eta\rangle)$. This proves (b).

(a) Let $A = D_\eta(\langle A_\zeta : \zeta < \eta\rangle)$, with $A_\zeta \in \boldsymbol{\Sigma}^0_\xi$, and let $B \in \boldsymbol{\Delta}^0_\xi$. If η is odd, then $A \cup B = E_\eta(\langle A_\zeta \cup B, \zeta < \eta\rangle)$, and the $\langle A_\zeta \cup B : \zeta < n\rangle$ are an increasing sequence of $\boldsymbol{\Sigma}^0_\xi$ sets. If η is even, let $A'_0 = A_0 \setminus B$, and let $A'_\xi = A_\zeta \cup B$ for $\zeta \geq 1$. Then again $A \cup B = D_\eta(\langle A'_\xi : \zeta < \eta\rangle)$, and the A'_ξ are $\boldsymbol{\Sigma}^0_\xi$ and increasing.

Case 2. $u(1) = 2$, so $\Gamma_u = \mathrm{Sep}(D_\eta(\boldsymbol{\Sigma}^0_\xi), \Gamma_{u^*})$, with $u^*(0) > \xi$. By the induction hypothesis, Γ_{u^*} is closed under union with a $\boldsymbol{\Delta}^0_{\xi+1}$ set, and under $\boldsymbol{\Sigma}^0_{\xi+1} - \mathrm{SU}$. Now intersection with a $\boldsymbol{\Sigma}^0_{\xi+1}$ is a particular case of $\boldsymbol{\Sigma}^0_{\xi+1} - \mathrm{SU}$, so Γ_{u^*} and $\check{\Gamma}_{u^*}$ are closed under intersection and union with $\boldsymbol{\Delta}^0_{\xi+1}$ sets. This clearly implies that $\check{\Gamma}_{u^*}$ is closed under $\boldsymbol{\Sigma}^0_\xi - \mathrm{SU}$.

(a) Let $A_n = \mathrm{Sep}(C_n, A^n_1, A^n_2)$, with $A^n_2 \in \check{\Gamma}_{u^*}, A^n_1 \in \check{\Gamma}_{u^*}, C_n \in D_\eta(\boldsymbol{\Sigma}^0_\xi)$, and let $A = \mathrm{SU}(\langle C'_n : n \in \omega\rangle, \langle A_n : n \in \omega\rangle)$, where the C'_n are pairwise disjoint

$\mathbf{\Sigma}^0_\xi$ sets. Then clearly $A = \mathrm{Sep}(\bigcup_n(C_n \cap C'_n), \bigcup_n(A^n_1 \cap C'_n), \bigcup_n(A^n_2 \cap C'_n))$, with $\bigcup_n(C_n \cap C'_n) \in D_\eta(\mathbf{\Sigma}^0_\xi)$, $\bigcup_n(A^n_1 \cap C'_n) \in \mathrm{SU}(\langle C'_n : n \in \omega\rangle, \langle A^n_1 : n \in \omega\rangle)$ is in $\check{\Gamma}_{u^*}$ and $\bigcup_n(A^n_2 \cap C'_n) = \mathrm{SU}(\langle C'_n : n \in \omega\rangle, \langle A^n_2 : n \in \omega\rangle)$ is in Γ_{u^*}. This shows (a).

(b) Let $A = \mathrm{Sep}(C, A_1, A_2)$, $C \in D_\eta(\mathbf{\Sigma}^0_\xi)$, $A_1 \in \check{\Gamma}_{u^*}$, $A_2 \in \Gamma_{u^*}$, and let $B \in \mathbf{\Delta}^0_\xi$. Then $A \cup B = \mathrm{Sep}(C, A_1 \cup B, A_2 \cup B)$, and the induction hypothesis gives (b).

Case 3. $u(1) = 3$.

(a) If $A_n = \mathrm{Bisep}(C^n_1, C^n_2, A^n_1, A^n_2, B^n)$ where $C^n_i \in D_\eta(\mathbf{\Sigma}^0_\xi)$, $A^n_1 \in \check{\Gamma}_{u_0}$, $A^n_2 \in \Gamma_{u_0}$ and $B^n \in \Gamma_{u_1}$, with $u_0(0) > \xi$ and $u_1(0) \geq \xi$ or $u_1(0) = 0$, and $A = \mathrm{SU}(\langle C_n : n \in \omega\rangle, \langle A_n : n \in \omega\rangle)$ where the C_n are pairwise disjoint $\mathbf{\Sigma}^0_\xi$ sets, then $A = \mathrm{Bisep}(\bigcup_n(C^n_1 \cap C_n), \bigcup_n(C^n_2 \cap C_n), \bigcup_n(A^n_1 \cap C_n), \bigcup_n(A^n_2 \cap C_n), \bigcup_n(B^n \cap C_n))$ which, together with the induction hypothesis, proves (a).

(b) Let again $A = \mathrm{Bisep}(C_1, C_2, A_1, A_2, B)$ with the sets in the same classes as before, and let $D \in \mathbf{\Delta}^0_\xi$. If $u_1(0) \geq 1$, take $B' = B \cup D$, $A'_1 = A_1 \cup B$, $A'_2 = A_2 \cup B$. Then $A \cup B = \mathrm{Bisep}(C_1, C_2, A'_1, A'_2, B')$, which proves (b) in this case. If $\Gamma_{u_1} = \{\varnothing\}$, so $B = \varnothing$, consider the sets $C_1 \cup D$ and $C_2 \cup D$. These are $D_\eta(\mathbf{\Sigma}^0_\xi)$ sets, by case 1. Let C^*_1, C^*_2 reduce them. Then $A = \mathrm{Bisep}(C^*_1, C^*_2, A_1 \cup D, A_2 \cup D)$, which proves (b) in that case.

Case 4. $u(1) = 4$. In this case $\Gamma_u = \mathrm{SU}(\mathbf{\Sigma}^0_\xi, \bigcup_n \Gamma_{u_n})$, and the Γ_{u_n} are of level $> \xi$ (at least for $n \geq n_0$). (a) is almost trivial. For (b), let $A = \mathrm{SU}(\langle C_n : n \in \omega\rangle, \langle A_n : n \in \omega\rangle)$, with $A_n \in \bigcup_p \Gamma_{u_p}$, and $C_n \in \mathbf{\Sigma}^0_\xi$, and let $B \in \mathbf{\Delta}^0_\xi$. Let B^*, $\langle C^*_n : n \in \omega\rangle$ be $\mathbf{\Sigma}^0_\xi$ sets reducing the sets B, $\langle C_n : n \in \omega\rangle$. Then $A \cup B = \mathrm{SU}(\langle C'_n : n \in \omega\rangle, \langle A'_n : n \in \omega\rangle)$ where $C'_0 = B^*$, $C'_n = C^*_{n-1}$, $n \geq 1$, and $A'_0 = B$, $A'_n = A_{n-1}$, $n \geq 1$.

Case 5. $u(1) = 5$.

(a) Suppose $A_n = \mathrm{SD}_\eta(\langle C^n_\zeta : \zeta < n\rangle, \langle A^n_\zeta : \zeta < n\rangle, B^n)$, where the pairs (C^n_ζ, A^n_ζ) are in $\langle \mathbf{\Sigma}^0_\xi, \Gamma_{u_0}\rangle$, with $u_0(0) = \xi$ and $u_0(1) = 4$ and $B^n \in \Gamma_{u_1}$, and $A = \mathrm{SU}(\langle C_n : n \in \omega\rangle, \langle A_n : n \in \omega\rangle)$. Then $A = \mathrm{SD}_\eta(\langle \bigcup_n(C^n \cap C^n_\zeta), \zeta < \eta\rangle, \langle \bigcup_n(C_n \cap A^n_\zeta), \zeta < \eta\rangle, \bigcup_n(C_n \cap B^n))$, and the induction hypothesis immediately yields (a).

(b) Let $A_n = \mathrm{SD}_\eta(\langle C_\zeta : \zeta < n\rangle, \langle A_\zeta : \zeta < n\rangle)$, and let $D \in \mathbf{\Delta}^0_\xi$. Then clearly $A \cup D = \mathrm{SD}_\eta(\langle C_{\zeta \cup D} : \zeta < \eta\rangle, \langle A_\zeta \cup D : \zeta < \eta\rangle, B)$, and by the proof of case 4, (b), the $C_\zeta \cup D$ are envelopes of the $A_\zeta \cup D$. This shows (b) in this case. ⊣

Lemma 1.5. If A, B are two disjoint $D_\eta(\mathbf{\Sigma}^0_\xi)$ sets, there are sets A^*, B^* in $D_\eta(\mathbf{\Sigma}^0_\xi)$ such that

(a) A^* and B^* are disjoint

(b) $A \subset A^*, B \subset B^*$
(c) $A^* \cup B^* \in \underset{\sim}{\Sigma}^0_\xi$

Proof. Let C (resp. D) be the union of the $\underset{\sim}{\Sigma}^0_\xi$ sets in a $D_\eta(\underset{\sim}{\Sigma}^0_\xi)$ definition of A (resp. B), and let C^*, D^* reduce C, D. It is easily checked that $A^* = (A \cap C^*) \cup (D^* \setminus B)$ and $B^* = (C^* \setminus A) \cup (D^* \cap B)$ satisfy the desired properties. ⊣

Corollary 1.6 (Normal form for the Bisep operation). Suppose $A \in \text{Bisep}(D_\eta(\underset{\sim}{\Sigma}^0_\xi), \Gamma_{u_0}, \Gamma_{u_1})$, where Γ_{u_0} is of level at least $\xi + 1$, and $\Gamma_{u_1} < \Gamma_{u_0}$. Then for some set C in $\underset{\sim}{\Sigma}^0_\xi$, $A = (A \cap C) \cup (B \setminus C)$ where $A \cap C \in \text{Bisep}(D_\eta(\underset{\sim}{\Sigma}^0_\xi), \Gamma_{u_0})$, $C \setminus A \in \text{Bisep}(D_\eta(\underset{\sim}{\Sigma}^0_\xi), \Gamma_{u_0})$ and $B \in \Gamma_{u_1}$.

Proof. Extend the two $D_\eta(\underset{\sim}{\Sigma}^0_\xi)$ separating sets, using Lemma 1.5. The union C of the extended $D_\eta(\underset{\sim}{\Sigma}^0_\xi)$ sets clearly works. ⊣

Let us now define a notion of type for each description u. This type intuitively corresponds to the character (successor, limit of cofinality ω or limit of cofinality ω_1) of the class Γ_u, among the Wadge classes of level at least $u(0)$.

Definition 1.7. The type $t(u)$ is defined by the following conditions:

(a) If $u(0) = 0, t(u) = 0$ ($\{\varnothing\}$ is the first class)
(b) If $u(0) = \xi \geq 1$, and $u(1) = 1$,
 u is of type 1 if $u(2)$ is successor, and
 u is of type 2 if $u(2)$ is limit
(c) If $u(0) \geq 1$ and $u(1) = 2$, u is of type 3
(d) If $u(0) \geq 1$ and $u(1) = 3$, then

$$u \text{ is of type } \begin{cases} 1 & \text{if } u_1 \text{ is of type 0 and } u(2) \text{ is successor} \\ 2 & \text{if } u_1 \text{ is of type 0 and } u(2) \text{ is limit} \\ t(u_1) & \text{if } u_1(0) = u(0) \\ 3 & \text{if } u_1(0) > u(0) \end{cases}$$

(e) If $u(0) \geq 1$ and $u(1) = 4$, u is of type 2
(f) If $u(0) \geq 1$ and $u(1) = 5$,

$$u \text{ is of type } \begin{cases} 2 & \text{if } u_1 \text{ is of type 0} \\ t(u_1) & \text{if } u_1(0) = u(0) \\ 3 & \text{if } u_1(0) > u(0) \end{cases}$$

Definition 1.8. We define $D^0 = \{u : t(u) = 0\} = \{u : u(0) = 0\}$, $D^+ = \{u : u(0) = 1 \text{ and } t(u) = 1\}$, $D^\omega = \{u : u(0) = 1 \text{ and } t(u) = 2\}$ and $D^{\omega_1} = D \setminus (D^0 \cup D^+ \cup D^\omega) = \{u : u(0) = 1 \text{ and } t(u) = 3\} \cup \{u : u(0) > 1\}$.

We are now able to restate Wadge's result in terms of our notion of description:

THEOREM 1.9. Assume Borel Determinacy. Let $W = \{\Gamma_u : u \in D\} \cup \{\check{\Gamma}_u : u \in D\} \cup \{\Delta(\Gamma_u) : u \in D^+ \cup D^\omega\}$. Then W is exactly the set of all Borel Wadge classes.

The (long) proof of this theorem goes roughly as follows. We shall define, for each description $u \notin D^0$, a set Q_u of descriptions, satisfying the following properties:

(A) If $u \in D^+$, then $Q_u = \{\overline{u}\}, \Gamma_{\overline{u}} < \Gamma_u$ and the only Wadge class Γ such that $\Gamma_{\overline{u}} < \Gamma < \Gamma_u$ is $\Delta(\Gamma_u)$.
(B) If $u \in D^\omega$, then $Q_u = \{u_n : n \in \omega\}$, for each n $\Gamma_{u_n} < \Gamma_u$, and the only Wadge class Γ such that $\forall n\ \Gamma_{u_n} < \Gamma < \Gamma_u$ is $\Delta(\Gamma_u)$.
(C) If $u \in D^{\omega_1}$, then Q_u is a set of descriptions of cardinality ω_1, and $\Delta(\Gamma_u) = \cup\{\Gamma_{u'} : u' \in Q_u\}$.

This will finish the proof of the theorem. For suppose there is a Borel Wadge class not in W, let Γ be the $<$-least counterexample (or one of the two $<$-least counterexamples, in a non self-dual case, for both must be outside W by definition of it). As the sequence $(\mathbf{\Sigma}^0_\xi)$ is cofinal in the Borel Wadge classes, there is a $<$-least described class Γ_u such that $\Gamma < \Gamma_u$, and clearly $u \notin D^0$. Now each of the remaining cases, $u \in D^+, u \in D^\omega$ and $u \in D^{\omega_1}$ gives immediately, using (A), (B) and (C), that Γ is in W, a contradiction.

In the following proof, we shall only indicate the main steps. Some arguments are just sketched, and others are missing (they can be found either in Kuratowski [Kur66] or in Wadge [Wad84]).

We first consider the case of a description u of type 1. This will take care of (A), but as we shall see also of a part of (C).

DEFINITION 1.10. For each description u of type 1, we define a description $\overline{u}$ by the following conditions [and we indicate the corresponding classes]:

(a) $u(1) = 1, u(2) = \eta + 1$
  if $\eta = 0, \overline{u} = \underline{0}$ $[\Gamma_u = \mathbf{\Sigma}^0_\xi, \Gamma_{\overline{u}} = \{0\}]$
  if $\eta > 0, \overline{u} = u(0)^\frown 1^\frown \eta^\frown \underline{0}$ $[\Gamma_u = D_{\eta+1}(\mathbf{\Sigma}^0_\xi), \Gamma_{\overline{u}} = D_\eta(\mathbf{\Sigma}^0_\xi)]$

(b) $u(1) = 3$, $t(u_1) = 0$ and $u(2) = \eta + 1$
  if $\eta = 0, \overline{u} = u_0$ $[\Gamma_u = \text{Bisep}(\mathbf{\Sigma}^0_\xi, \Gamma_{u_0}), \Gamma_{\overline{u}} = \Gamma_{u_0}]$
  if $\eta > 0, \overline{u} = u(0)^\frown 2^\frown \eta^\frown \underline{0}$ $[\Gamma_u = \text{Bisep}(D_{\eta+1}(\mathbf{\Sigma}^0_\xi), \Gamma_{u_0}), \Gamma_{\overline{u}} = \text{Sep}(D_\eta(\mathbf{\Sigma}^0_\xi), \Gamma_{u_0})]$

(c) $u(1) = 3$ or 5, $t(u_1) = 1$ and $u_1(0) = u(0)$
  $\overline{u} = u(0)^\frown u(1)^\frown u(2)^\frown \langle u_0, \overline{u_1}\rangle$

LEMMA 1.11. For each u of type 1,

(a) $\Gamma_u = \text{Bisep}(\mathbf{\Sigma}^0_{u(0)}, \Gamma_{\overline{u}})$

(b) $\Delta(\Gamma_u) = \mathrm{Bisep}(\Delta(\mathbf{\Sigma}^0_{u(0)}), \Gamma_{\overline{u}})$

PROOF. (a)
Case 1. $u(1) = 1$, $u(2) = 0$. The equality $\mathbf{\Sigma}^0_\xi = \mathrm{Bisep}(\mathbf{\Sigma}^0_\xi, \{\varnothing\})$ is trivial.

Case 2. $u(1) = 1$, $u(2) = \eta + 1$ for $\eta \geq 1$. First $\check{\Gamma}_{\overline{u}}$ and $\Gamma_{\overline{u}}$ are contained in $\Gamma_u = D_{\eta+1}(\mathbf{\Sigma}^0_\xi)$, so by the closure properties of Γ_u, $\mathrm{Bisep}(\mathbf{\Sigma}^0_\xi, \Gamma_{\overline{u}}) \subset \Gamma_u$. On the other hand, if $A \in D_{\eta+1}(\mathbf{\Sigma}^0_\xi)$, then for some C in $\mathbf{\Sigma}^0_\xi$ and some B in $D_\eta(\mathbf{\Sigma}^0_\xi) = \Gamma_{\overline{u}}$, $A = C \setminus B$. So A is in $\mathrm{Bisep}(\mathbf{\Sigma}^0_\xi, \Gamma_{\overline{u}})$.

Case 3. $u(1) = 3$, $t(u_1) = 0$ and $u(2) = 1$. Then $\Gamma_u = \mathrm{Bisep}(\mathbf{\Sigma}^0_\xi, \Gamma_{u_0})$ and $\Gamma_{\overline{u}} = \Gamma_{u_0}$, so the equality is trivial.

Case 4. $u(1) = 3$, $t(u_1) = 0$ and $u(2) = \eta + 1$ with $\eta \geq 1$. Then $\Gamma_u = \mathrm{Bisep}(D_{\eta+1}(\mathbf{\Sigma}^0_\xi), \Gamma_{\overline{u}_0})$, and $\Gamma_{\overline{u}} = \mathrm{Bisep}(D_\eta(\mathbf{\Sigma}^0_{\xi+1}), \Gamma_{u_0})$. The inclusion $\mathrm{Bisep}(\mathbf{\Sigma}^0_\xi, \Gamma_{\overline{u}_0}) \subset \Gamma_u$ is again easy. For the converse, assume $A \in \Gamma_u$. By the normal form for the Bisep operation, let C_1, C_2 in $\mathbf{\Sigma}^0_\xi$, D_1, D_2 in $D_\eta(\mathbf{\Sigma}^0_\xi)$ be such that $C_1 \setminus D_1 \cap C_2 \setminus D_2 = \varnothing$, $D_2 \subset C_1$, $C_2 \subset D_1$, and $A \subset C_1 \cup C_2$, $A \cap C_1 \setminus D_1 \in \Gamma_{u_0}$, $A \cap C_2 \setminus D_2 \in \Gamma_{u_0}$. Let C_1^*, C_2^* be $\mathbf{\Sigma}^0_\xi$ sets reducing C_1, C_2 and remark that $A \cap C_1^* = (A \cap D_1 \cap C_1^*) \cup (A \cap (C_1 \setminus D_1) \cap C_1^*)$ is in $\check{\Gamma}_{\overline{u}}$, and $A \cap C_2^* = (A \cap D_2 \cap C_2^*) \cup (A \cap (A \cap (C_2 \setminus D_2)) \cap C_2^*)$ is in $\Gamma_{\overline{u}}$, so that, as $A \subset C_1^* \cup C_2^*$, we obtain $A \in \mathrm{Bisep}(\mathbf{\Sigma}^0_\xi, \Gamma_{\overline{u}})$.

Case 5. (Induction step) Suppose $u(1) = 3$, $t(u_1) = 1$ and $u_1(0) = u(0)$. So $\Gamma_u = \mathrm{Bisep}(D_\eta(\mathbf{\Sigma}^0_\xi), \Gamma_{u_0}, \Gamma_{u_1})$ and $\Gamma_{\overline{u}} = \mathrm{Bisep}(D_\eta(\mathbf{\Sigma}^0_\xi), \Gamma_{u_0}, \Gamma_{\overline{u}_1})$, and by the induction hypothesis, we can assume $\Gamma_{u_1} = \mathrm{Bisep}(\mathbf{\Sigma}^0_\xi, \Gamma_{\overline{u}_1})$ (for $u_1(0) = \xi$). The inclusion $\mathrm{Bisep}(\mathbf{\Sigma}^0_\xi, \Gamma_{\overline{u}}) \subset \Gamma_u$ is easy. Suppose $A \in \Gamma_u$. Then by the normal form for Bisep, we have $A = (A \cap C) \cup (B \setminus C)$, where $B \in \Gamma_{u_1}$ and $C \in \mathbf{\Sigma}^0_\xi$, $A \cap C$, and $C \setminus A$ are in $\mathrm{Bisep}(D_\eta(\mathbf{\Sigma}^0_{\xi+1}), \Gamma_{u_0})$. Now $B = (B \cap C_1) \cup (B \cap C_2)$, where C_1, C_2 are two disjoint $\mathbf{\Sigma}^0_\xi$ sets, and $B \cap C_1 \in \check{\Gamma}_{\overline{u}_1}$, $B \cap C_2 \in \Gamma_{\overline{u}_1}$. Let C_1^*, C_2^* in $\mathbf{\Sigma}^0_\xi$ reduce the pair $C \cup C_1, C \cup C_2$. Then $A = (A \cap C_1^*) \cup (A \cap C_2^*)$, and it is clear that $A \cap C_1^* \in \mathrm{Bisep}(D_\eta(\mathbf{\Sigma}^0_\xi), \Gamma_{u_0}, \check{\Gamma}_{\overline{u}_1}) = \check{\Gamma}_{\overline{u}}$, and $A \cap C_2^* \in \mathrm{Bisep}(D_\eta(\mathbf{\Sigma}^0_\xi), \Gamma_{u_0}, \Gamma_{\overline{u}_1}) = \Gamma_{\overline{u}}$, so that $A \in \mathrm{Bisep}(\mathbf{\Sigma}^0_\xi, \Gamma_{\overline{u}})$.

The second case of the induction case ($u(1) = 5, t(u_1) = 1$ and $u_1(0) = u(0)$) is entirely analogous, and we omit it.

(b) is an easy consequence of (a). The inclusion $\mathrm{Bisep}(\Delta(\mathbf{\Sigma}^0_{u(0)}), \Gamma_{\overline{u}}) \subset \Delta(\Gamma_u)$ is obvious. Suppose now $A \in \Delta(\Gamma_u)$. Then by (a), $A = (A \cap C_1) \cup (A \cap C_2)$ where C_1, C_2 are disjoint $\mathbf{\Sigma}^0_\xi$ sets, and $A \cap C_1 \in \check{\Gamma}_{\overline{u}}$, $A \cap C_2 \in \Gamma_{\overline{u}}$. Similarly $\check{A} = (\check{A} \cap D_1) \cup (\check{A} \cap D_2)$, where D_1, D_2 are disjoint $\mathbf{\Sigma}^0_\xi$ sets and $\check{A} \cap D_1 \in \check{\Gamma}_{\overline{u}}$, $\check{A} \cap D_2 \in \check{\Gamma}_{\overline{u}}$. Let C^*, D^* be $\mathbf{\Sigma}^0_\xi$ sets reducing the pair $C_1 \cup C_2, D_1 \cup D_2$. As $C^* \cup D^* = {}^\omega\omega$, C^*, D^* and the sets $C_1^* = C_1 \cap C^*$, $C_2^* = C_2 \cap C^*$, $D_1^* =$

$D_1 \cap D^*, D_2^* = D_2 \cap D^*$, are $\Delta(\boldsymbol{\Sigma}^0_\xi)$. But clearly $A = [(C_1^* \cap A) \cup (D_1^* \setminus \breve{A})] \cup [(D_2^* \cap A) \cup (C_2^* \setminus \breve{A})]$, which shows $A \in \text{Bisep}(\Delta(\boldsymbol{\Sigma}^0_\xi), \Gamma_{\overline{u}})$. ⊣

COROLLARY 1.12 (Statement (A)). Let u be some description in D^+, and let $Q_u = \{\overline{u}\}$. Then the only Wadge class Γ such that $\Gamma_{\overline{u}} < \Gamma < \Gamma_u$ is $\Delta(\Gamma(u))$.

PROOF. Using Lemma 1.11 for $u(0) = 1$, we see that $\Delta(\Gamma_u) = \text{Bisep}(\boldsymbol{\Delta}^0_1, \Gamma_{\overline{u}})$. Using the game-theoretical characterization of the Wadge ordering this immediately implies $\Delta(\Gamma_u) = \Gamma^+_{\overline{u}}$. ⊣

DEFINITION 1.13. Let u be some description in D^{ω_1}, with $u(0) = \xi + 1$, for $\xi \geq 1$, and $t(u) = 1$. We define

$$Q_u = \{\xi^\frown 3^\frown \eta \langle \overline{u}, \underline{0} \rangle \ : \ 1 \leq \eta < \omega_1\}$$

COROLLARY 1.14 (Statement (C) for $u(0) = \xi + 1$, $t(u) = 1$). Let u be a description with $u(0) = \xi + 1, \xi \geq 1$ and $t(u) = 1$. Then Q_u is a set of descriptions (of level ξ), and $\Delta(\Gamma_u) = \cup\{\Gamma_{u'} \ : \ u' \in Q_u\}$.

PROOF. (a) The case $u(0) = \xi + 1$, $u(1) = 1$, $u(2) = 1$, *i.e.* $\Gamma_u = \boldsymbol{\Sigma}^0_{\xi+1}$ is solved by the Hausdorff-Kuratowski theorem: $\boldsymbol{\Delta}^0_{\xi+1} = \cup\{D_\eta(\boldsymbol{\Sigma}^0_\xi) \ : \ 1 \leq \eta < \omega_1\}$.

(b) In the general case, we have that $\Delta(\Gamma_u) = \text{Bisep}(\Delta(\boldsymbol{\Sigma}^0_{\xi+1}), \Gamma_{\overline{u}})$ by Lemma 1.11. Using the Hausdorff-Kuratowski theorem, we obtain

$$\begin{aligned}\Delta(\Gamma_u) &= \text{Bisep}\Big(\bigcup_{\eta<\omega_1} D_\eta(\boldsymbol{\Sigma}^0_\xi), \Gamma_{\overline{u}}\Big)\\ &= \bigcup\{\text{Bisep}(D_\eta(\boldsymbol{\Sigma}^0_\xi), \Gamma_{\overline{u}}) \ : \ 1 \leq \eta < \omega_1\}\\ &= \bigcup\{\Gamma_{u'} \ : \ u' \in Q_u\} \quad \text{by the definition of } Q_u.\end{aligned}$$

⊣

We now turn to the case of $u(0)$ a limit ordinal. What we need here is the analysis, obtained by Wadge, of $\boldsymbol{\Delta}^0_\lambda$ for limit λ. This analysis is done by iterating the SU operation.

DEFINITION 1.15. Let λ be a countable limit ordinal, and $\langle \lambda_n \rangle_{n \in \omega}$ be an increasing sequence of ordinals, cofinal in λ.

(a) Let Γ be some class. We define, for each (n, η) in $\omega \times \omega_1$ a class $\text{SU}_{n,\eta}(\Gamma)$, by the following induction:
  (i) $\text{SU}_{n,0}(\Gamma) = \text{SU}(\boldsymbol{\Sigma}^0_{\lambda_n}, \Gamma)$
  (ii) $\text{SU}_{n,\eta}(\Gamma) = \text{SU}(\boldsymbol{\Sigma}^0_{\lambda_n}, \cup\{\text{SU}_{p,\eta'}(\Gamma) \ : \ p \in \omega, \eta' < \eta\})$, for $\eta > 0$.

(b) Similarly, if $s = \langle u_n \ : \ n \in \omega \rangle$ codes a sequence of descriptions, with $\sup_n u_n(0) = \lambda$, we define a family $u_{n,\eta}(s)$ by:
  (i) $u_{n,0}(s) = \lambda_n{}^\frown 4^\frown s$
  (ii) $u_{n,\eta}(s) = \lambda_n{}^\frown 4^\frown \langle u_{n,\eta}(s) \ : \ n \in \omega, \eta' < \eta \rangle$ for $\eta > 0$.

(c) If $u = \lambda^\frown 1^\frown 1^\frown u'$, we set $Q_u = \{u_{n,\eta}(s) : n \in \omega, \eta \in \omega_1\}$ where $s = \{\lambda_n{}^\frown 1^\frown 1^\frown \underline{0} : n \in \omega\}$.

LEMMA 1.16. If $s = \langle u_n : n \in \omega\rangle$ codes a sequence of descriptions with either $u_n(0) = \lambda$ and the Γ_{u_n} increasing, or $u_n(0) = \lambda_n$, then each $u_{n,\eta}(s)$ is a description, and $\Gamma_{u_{n,\eta}} = \mathrm{SU}_{n,\eta}(\Gamma_s)$, where $\Gamma_s = \bigcup_n \Gamma_{u_n}$.

PROOF. The only thing to check is that the levels of the classes on which SU is performed are acceptable, and we omit it. ⊣

The next result is a theorem of Wadge [Wad84], and gives the analysis of $\underset{\sim}{\mathbf{\Delta}}^0_\lambda$ for limit λ, in a way very similar to the Hausdorff-Kuratowski theorem.

THEOREM 1.17 (Wadge). Let λ be limit, and $\langle\lambda_n\rangle_{n\in\omega}$ be cofinal in λ. Let $\Gamma = \bigcup_{\eta<\lambda} \underset{\sim}{\mathbf{\Sigma}}^0_\eta$. Then

$$\underset{\sim}{\mathbf{\Delta}}^0_\lambda = \bigcup\{\mathrm{SU}_{n,\eta}(\Gamma) : n \in \omega, \eta < \omega_1\}$$

Hence, with our notations, if u is a description with $u(0) = \lambda$, $u(1) = u(2) = 1$, then $\Delta(\Gamma_u) = \bigcup\{\Gamma_{u'} : u' \in Q_u\}$.

DEFINITION 1.18. Let u be a description of type 1, with $u(0) = \lambda$ limit, and $\langle\lambda_n\rangle$ a cofinal sequence in λ. Define a sequence $s_u = \langle u_n : n \in \omega\rangle$ by $u_n = \lambda_n{}^\frown 3^\frown 1\langle\overline{u}, \underline{0}\rangle$. We set $Q_u = \{u_{n,\eta}(s_u) : n \in \omega, \eta < \omega_1\}$.

COROLLARY 1.19 (Statement (C) for $t(u) = 1$ and $u(0)$ limit). Let u be a description of type 1 with $u(0) = \lambda$ limit. Then

$$\Delta(\Gamma_u) = \bigcup\{\Gamma_{u'} : u' \in Q_u\}$$

PROOF. By Lemma 1.11, we know that

$$\begin{aligned}\Delta(\Gamma_u) &= \mathrm{Bisep}(\Delta(\underset{\sim}{\mathbf{\Sigma}}^0_\lambda), \Gamma_{\overline{u}})\\ &= \mathrm{Bisep}\Big(\bigcup\{\mathrm{SU}_{n,\eta}(\Gamma) : n \in \omega, \eta < \omega_1\}, \Gamma_{\overline{u}}\Big)\end{aligned}$$

with $\Gamma = \bigcup_n \underset{\sim}{\mathbf{\Sigma}}^0_{\lambda_n}$, by using Wadge's theorem. So the only thing to prove is the equality

$$\mathrm{Bisep}(\mathrm{SU}_{n,\eta}(\Gamma), \Gamma_{\overline{u}}) = \mathrm{SU}_{n,\eta}\Big(\bigcup_n \mathrm{Bisep}(\underset{\sim}{\mathbf{\Sigma}}^0_{\lambda_n}, \Gamma_{\overline{u}})\Big).$$

(a) Suppose first that $\eta = 0$. The left side of the equality is $\Gamma_\ell = \mathrm{Bisep}(\mathrm{SU}(\underset{\sim}{\mathbf{\Sigma}}^0_{\lambda_n}), \Gamma_{\overline{u}})$ and the right side is $\Gamma_r = \mathrm{SU}(\underset{\sim}{\mathbf{\Sigma}}^0_{\lambda_n}, \mathrm{Bisep}(\Gamma, \Gamma_{\overline{u}}))$. The inclusion from right to left is obvious. Let A be in Γ_ℓ. For some disjoint sets C_1, C_2 in $\mathrm{SU}(\underset{\sim}{\mathbf{\Sigma}}^0_{\lambda_n}, \Gamma)$, some A_1, A_2 in $\check{\Gamma}_{\overline{u}}, \Gamma_{\overline{u}}$ respectively, we have $A = (A_1 \cap C_1) \cup (A_2 \cap C_2)$. Now, $C_1 = \bigcup_p (H^1_p \cap C^1_p)$ with the H^1_p disjoint in $\underset{\sim}{\mathbf{\Sigma}}^0_{\lambda_n}$, and the C^1_p in Γ; and similarly we can find corresponding sets H^2_p, C^2_p for C_2. Let $K^1_p, K^2_p, p \in \omega$, be $\underset{\sim}{\mathbf{\Sigma}}^0_{\lambda_n}$ sets reducing the sets H^1_p, H^2_p. Then

$A \subset (\bigcup_p K_p^1) \cup (\bigcup_p K_p^2)$, and moreover $A \cap K_p^1 = (A \cap K_p^1 \cap C_p^1) \cup (A \cap K_p^1 \cap \check{C}_p^1)$ is in $\text{Bisep}(\Gamma, \Gamma_{\overline{u}})$, and similarly for $A \cap K_p^2$. This shows that A is in Γ_r.

(b) Suppose now $\eta > 0$. The left side is now

$$\begin{aligned} \Gamma_\ell &= \text{Bisep}\Big(\text{SU}(\mathbf{\Sigma}^0_{\lambda_n}, \bigcup\{\text{SU}_{p,\eta'}(\Gamma) : p \in \omega, \eta' < \eta\}), \Gamma_{\overline{u}}\Big) \\ &= \text{SU}\Big(\mathbf{\Sigma}^0_{\lambda_n}, \bigcup\{\text{Bisep}(\text{SU}_{p,\eta'}(\Gamma)) : p \in \omega, \eta' < \eta\}\Big) \end{aligned}$$

by the same proof as in (a), and the right hand side is

$$\Gamma_r = \text{SU}\Big(\mathbf{\Sigma}^0_{\lambda_n}, \bigcup\{\text{SU}_{p,\eta'}(\text{Bisep}(\Gamma, \Gamma_{\overline{u}})) : p \in \omega, \eta' < \eta\}\Big).$$

The induction hypothesis then immediately gives the result. ⊣

We now turn to the case of descriptions of type 2.

Definition 1.20. For each description u of type 2 we define a sequence s_u by the following conditions:

(a) If $u(1) = 1$ and $u(2) = \lambda$ is limit, with cofinal sequence $\langle \lambda_n \rangle$, let $s_u = \langle \xi^\frown 1^\frown \lambda_n{}^\frown \underline{0} : n \in \omega \rangle$.
(b) If $u(1) = 3$ and $t(u_1) = 0, u(2) = \lambda$ is limit, with cofinal sequence $\langle \lambda_n \rangle$, let $s_u = \langle \xi^\frown 2^\frown \lambda_n{}^\frown u_0 : n \in \omega \rangle$.
(c) If $u(1) = 4$, so $u = \xi^\frown 4^\frown u'$, let $s_u = u'$.
(d) If $u(1) = 5$ and $t(u_1) = 0$, then
  - if $u(2) = \eta + 1$, with $\eta > 0$, so $u_0 = \xi^\frown 4^\frown \langle u_n : n \in \omega \rangle$, let $s_u = \langle \xi^\frown 5^\frown \eta^\frown \langle u_0, u_1 \rangle : n \in \omega \rangle$.
  - if $u(2) = \lambda$ is limit with cofinal sequence $\langle \lambda_n \rangle$, let $s_u = \langle \xi^\frown 5 \lambda_n{}^\frown \langle u_0, u_1 \rangle : n \in \omega \rangle$.
(e) (induction step) If $u(1) = 3$ or 5, and $t(u_1) = 2$ and $u_1(0) = u(0)$, then writing $s_{u_1} = \langle u_n^1 : n \in \omega \rangle$, set $s_u = \langle u(0)^\frown u(1)^\frown u(2)^\frown \langle u_0, u_n^1 \rangle : n \in \omega \rangle$.

Definition 1.21. For each description u of type 2, we define a set Q_u of descriptions by the following:

(a) If $u(0) = 1$, $Q_u = \{\langle s_u \rangle_n : n \in \omega\}$.
(b) If $u(0) = \xi + 1$, with $\xi > 0$, $Q_u = \{\xi^\frown 5^\frown \eta^\frown \langle \xi^\frown 4^\frown s_u, \underline{0} \rangle : 1 < \eta < \omega_1\}$.
(c) If $u(0) = \lambda$ is limit, with cofinal sequence $\langle \lambda_n \rangle$, $Q_u = \{u_{p,\eta}(s_u) : p \in \omega, \eta < \omega_1\}$.

Definition 1.22 (Partitioned Unions (Wadge)). We say that $A = \text{PU}(\langle C_n : n \in \omega \rangle, \langle A_n : n \in \omega \rangle)$ if $A = \text{SU}(\langle C_n \rangle, \langle A_n \rangle)$, and moreover the envelope $C = \bigcup_n C_n$ is ${}^\omega\omega$ (so that $\langle C_n : n \in \omega \rangle$ is a partition of ${}^\omega\omega$). $\text{PU}(\Gamma, \Gamma')$ is the class of all $\text{PU}(\langle C_n \rangle, \langle A_n \rangle)$ with $C_n \in \Gamma$ and $A_n \in \Gamma'$, for each n.

Lemma 1.23. Let u be a description of type 2 and level $\xi \geq 1$. Then

(a) $\Gamma_u = \text{SU}(\mathbf{\Sigma}^0_\xi, \bigcup\{\Gamma_{u'} : u' \in Q_u\})$
(b) $\Delta(\Gamma_u) = \text{PU}(\mathbf{\Sigma}^0_\xi, \bigcup\{\Gamma_{u'} : u' \in Q_u\}$

(c) In particular, if $\xi = 1$, the only Wadge class Γ such that $\forall u' \in Q_u$, $\Gamma_{u'} < \Gamma < \Gamma_u$ is $\Delta(\Gamma_u)$. (Assertion (B))

PROOF. (a) is by induction.

Case 1. $u(1) = 1$ and $u(2) = \lambda$. Then $\Gamma_u = D_\lambda(\mathbf{\Sigma}^0_\xi)$, and we want to prove $\Gamma_u = \mathrm{SU}(\mathbf{\Sigma}^0_\xi, \bigcup_{\eta<\lambda} D_\eta(\mathbf{\Sigma}^0_\xi))$. From right to left the inclusion is obvious. If $A \in \Gamma_u$, let $\langle A_\eta : \eta < \lambda\rangle$ be an increasing sequence of $\mathbf{\Sigma}^0_\xi$ sets with $A = D_\lambda(\langle A_\eta\rangle)$, and let $\langle A'_\eta : \eta < \lambda\rangle$ reduce $\langle A_\eta : \eta < \lambda\rangle$. Then $A = \bigcup_{\eta<\lambda}(A \cap A'_\eta)$, and $A \cap A'_\eta = A_\eta \cap A'_\eta \cap A$ is clearly in $D_\eta(\mathbf{\Sigma}^0_\xi)$

Case 2. $u(1) = 3, t(u_1) = 0, u(2) = \lambda$. Then $\Gamma_u = \mathrm{Bisep}(D_\lambda(\mathbf{\Sigma}^0_\xi), \Gamma_{u_0})$, and we want to prove $\Gamma_u = \mathrm{SU}(\mathbf{\Sigma}^0_\xi, \bigcup_{\eta<\lambda} \mathrm{Sep}(D_\lambda(\mathbf{\Sigma}^0_\xi), \Gamma_{u_0}))$. Again the inclusion from right to left is trivial. So let $A \in \Gamma_u$, and let $C_0, C_1 \in D_\lambda(\mathbf{\Sigma}^0_\xi)$ be the biseparating sets. Using case 1, C_0 and C_1 are in $\mathrm{SU}(\mathbf{\Sigma}^0_\xi, \bigcup_{\eta<\lambda} D_\lambda(\mathbf{\Sigma}^0_\xi))$, and then it is obvious, using the closure properties of Γ_{u_0}, that $A \in \mathrm{SU}(\mathbf{\Sigma}^0_\xi, \bigcup_{\eta<\lambda} \mathrm{Sep}(D_\lambda(\mathbf{\Sigma}^0_\xi), \Gamma_{u_0}))$.

Case 3. $u(1) = 4$, so $\Gamma_u = \mathrm{SU}(\mathbf{\Sigma}^0_\xi, \bigcup_n \Gamma_{u_n})$, which is the equality we want.

Case 4. $u(1) = 5, u(2) = \eta + 1, \eta \geq 1, t(u_1) = 0$. Then, $\Gamma_u = \mathrm{SD}_{\eta+1}(\mathbf{\Sigma}^0_\xi, \Gamma_{u_0})$ where $\Gamma_{u_0} = \mathrm{SU}(\mathbf{\Sigma}^0_\xi, \bigcup_n \Gamma_{u_n})$ and we want to prove

$$\Gamma_u = \mathrm{SU}\Big(\mathbf{\Sigma}^0_\xi, \bigcup \mathrm{SD}_\eta(\mathbf{\Sigma}^0_\xi, \Gamma_{u_0}, \Gamma_{u_n})\Big)$$

It is clear that $\mathrm{SD}_\eta(\mathbf{\Sigma}^0_\xi, \Gamma_{u_0}, \Gamma_{u_n})$ is in $\Delta(\mathrm{SD}_{\eta+1}((\mathbf{\Sigma}^0_\xi, \Gamma_{u_0}))$, so the inclusion from right to left is obvious. Suppose now $A \in \Gamma_u$. Then for some increasing pairs $\langle A_\zeta, C_\zeta : \zeta \leq \eta\rangle$, with $A_\zeta = \bigcup_n (A^n_\zeta \cap C^n_\zeta)$, $\bigcup_n C^n_\zeta = C_\zeta$, where $A^n_\zeta \in \Gamma_n$ and $C^n_\zeta \in \mathbf{\Sigma}^0_\xi$, we have $A = \bigcup_\zeta (A_\zeta \setminus \bigcup_{\zeta'<\zeta} C_\zeta)$. Let $A_{<\eta} \subset C_{<\eta}$. Now let $\langle C^*_n : n \in \omega\rangle$ reduce the sequence $\langle C_{<\eta} \cup C^n_{\eta_0}\rangle$, and consider $A \cap C^*_n = (A_{<\eta} \cap C^*_n) \cup (A_\eta \setminus C_{<\eta} \cap C^*_n)$. It is clearly in $\mathrm{SD}_\eta(\mathbf{\Sigma}^0_\xi, \Gamma_{u_0}, \Gamma_{u_n})$. Moreover $A \subset \bigcup_n C^*_n = C_\eta$, so that $A \in \mathrm{SU}(\mathbf{\Sigma}^0_\xi, \bigcup \mathrm{SD}_\eta(\mathbf{\Sigma}^0_\xi, \Gamma_{u_0}, \Gamma_{u_n}))$.

Case 5. $u(1) = 5, u(2) = \lambda$ is limit and $t(u_1) = 0$. The proof is entirely analogous to case 4, and we omit it.

Case 6. (Induction step) Suppose that $u(1) = 3$ (the case $u(1) = 5$ is analogous) and $t(u_1) = 2, u_1(0) = \xi$. We have $\Gamma_u = \mathrm{Bisep}(D_\eta(\mathbf{\Sigma}^0_\xi), \Gamma_{u_0}, \Gamma_{u'})$ with $\Gamma_{u_1} = \mathrm{SU}(\mathbf{\Sigma}^0_\xi, \bigcup\{\Gamma_{u'} : u' \in Q_{u_1}\})$ and we want to prove that

$$\Gamma_u = \mathrm{SU}\Big(\mathbf{\Sigma}^0_\xi, \bigcup\{\mathrm{Bisep}(D_\eta(\mathbf{\Sigma}^0_\xi), \Gamma_{u_0}, \Gamma_{u'}) : u' \in Qu_1\}\Big)$$

Let $\Gamma^* = \bigcup\{\Gamma_{u'} : u' \in Qu_1\}$, so that the left side is
$\Gamma_u = \mathrm{Bisep}(D_\eta(\mathbf{\Sigma}^0_\xi), \Gamma_{u_0}, \mathrm{SU}(\mathbf{\Sigma}^0_\xi, \Gamma^*))$, and the right side
$\Gamma_r = \mathrm{SU}(\mathbf{\Sigma}^0_\xi, \mathrm{Bisep}(D_\eta(\mathbf{\Sigma}^0_\xi), \Gamma_{u_0}, \Gamma^*))$.

The proof that $\Gamma_r = \Gamma_u$ is entirely analogous to the one in Corollary 1.19 (a). We omit it.

(b) The proof of part (b) follows easily from part (a). For if $A, \breve{A} \in \mathrm{SU}(\mathbf{\Sigma}^0_\xi, \bigcup_n \Gamma_n)$, so that $A = \bigcup_n (A_n \cap C_n)$, $A_n \in \Gamma_n$, $C_n \in \mathbf{\Sigma}^0_\xi$, (pairwise disjoint sets) and $\breve{A} = \bigcup_n (A'_n \cap C'_n)$, $A'_n \in \Gamma_{n'}$, $C'_n \in \mathbf{\Sigma}^0_\xi$, then choosing a sequence C^*_n, C'^*_n of $\mathbf{\Sigma}^0_\xi$ sets reducing the sequence $\langle C_n : n \in \omega\rangle, \langle C'_n : n \in \omega\rangle$, we obtain $A = \bigcup_n (A_n \cap C^*_n) \cup \bigcup_n (\breve{A}'_n \cap C'^*_n)$, which, because $\bigcup_n C^*_n \cup \bigcup_n C'^*_n = {}^\omega\omega$, shows that $A \in \mathrm{PU}(\mathbf{\Sigma}^0_\xi, \bigcup_n \Gamma_n)$.

(c) Assertion (B) then follows easily from the game theoretical characterization of Wadge's ordering. ⊣

The following next two lemmas take care of assertion (C) for descriptions u of type 2 and of level, respectively, a successor and a limit ordinal.

Lemma 1.24. Assume u is a description of type 2 and level $u(0) = \xi + 1$, with $\xi \geq 1$. Then $\Delta(\Gamma_u) = \bigcup_n \{\Gamma_{u'} : u' \in Q_u\}$.

Proof. By Lemma 1.23, we know that

$$\Delta(\Gamma_u) = \mathrm{PU}\Big(\mathbf{\Sigma}^0_{\xi+1}, \bigcup_n \Gamma_{s_u(n)}\Big)$$

and we want to prove that

$$\Delta(\Gamma_u) = \bigcup\Big\{\mathrm{SD}_\eta\Big(\mathbf{\Sigma}^0_\xi, \mathrm{SU}(\mathbf{\Sigma}^0_\xi, \bigcup_n \Gamma_{s_u(n)})\Big) \;\Big|\; 2 \leq \eta < \omega_1\Big\}$$

Let $\Gamma^* = \bigcup_n \Gamma_{s_u(n)}$. By the definition of s_u, each $\Gamma_{s_u(n)}$ is of level $\geq \xi+1$. From right to left, the inclusion is easy: If $A \in \mathrm{SD}_\eta(\mathbf{\Sigma}^0_\xi, \mathrm{SU}(\mathbf{\Sigma}^0_\xi, \Gamma^*))$ for some $n_0 < \omega_1$, let $A_\eta, C_\eta, \eta < \eta_0$, be pairs witnessing this fact, $A_\eta \in \mathrm{SU}(\mathbf{\Sigma}^0_\xi, \Gamma^*)$, $C_\eta \in \mathbf{\Sigma}^0_\xi$ and $A = \bigcup_{\eta<\eta_0} (A_n \setminus \bigcup_{\eta'<\eta} C_{\eta'})$.

For each $A^* = A_\eta \setminus \bigcup_{\eta'<\eta} C_{\eta'}$ is clearly in $\mathrm{SU}(D_2(\mathbf{\Sigma}^0_\xi), \Gamma^*)$, so is in $\mathrm{PU}(\mathbf{\Sigma}^0_{\xi+1}, \Gamma^*)$. But the sets A^* are disjoint, and separated by the $C^*_\eta = C_\eta \setminus \bigcup_{\eta'<\eta} C_{\eta'}$, which are disjoint and in $D_2(\mathbf{\Sigma}^0_\xi)$. This clearly proves that $A \in \Delta(\Gamma_u) = \mathrm{PU}(\mathbf{\Sigma}^0_{\xi+1}, \Gamma^*)$.

For the other inclusion, let us look at the case $\xi = 1$. Suppose $A \in \Delta(\Gamma_u) = \mathrm{PU}(\mathbf{\Sigma}^0_2, \Gamma^*)$. Because each $\mathbf{\Sigma}^0_2$ set is the disjoint union of $\mathbf{\Pi}^0_1$ sets, we also have $A \in \mathrm{PU}(\mathbf{\Pi}^0_1, \Gamma^*)$, say $A = \bigcup_n A_n$, where $A_n \in \Gamma^*$ and for some sequence F_n of closed sets, $A_n \subset F_n$, and the F_n are a partition of ${}^\omega\omega$.

Define a transfinite sequence $O_\eta, \eta < \omega_1$, by $O_0 = \bigcup\{O \in \mathbf{\Sigma}^0_1 : A \cap O \in \Gamma^*\}$, and more generally $O_\eta = \bigcup\{O \in \mathbf{\Sigma}^0_1 : O \cap (A \setminus \bigcup_{\eta'<\eta} O_{\eta'}) \in \Gamma^*\}$. The sequence O_η is increasing, hence is stationary after $\eta_0 < \omega_1$. We claim that the sequence $\langle (A \cap O_\eta) \cup \bigcup_{\eta'<\eta} O_{\eta'}, O_\eta\rangle, \eta \leq \eta_0$ witnesses that $A \in \mathrm{SD}_{\eta_0+1}(\mathbf{\Sigma}^0_1, \mathrm{SU}(\mathbf{\Sigma}^0_1, \Gamma^*))$. We have to prove two things: First, that $A \cap$

$(O_\eta \setminus \bigcup_{\eta'<\eta} O_{\eta'})$ is in $\mathrm{SU}(\boldsymbol{\Sigma}^0_1, \Gamma^*)$ which will imply that $(A \cap O_\eta) \cup \bigcup_{\eta'<\eta} O_{\eta'}$ is also in it, and secondly that $A \subset O_{\eta_0}$. The first fact comes from the definition of O_η : O_η is the union of a disjoint family of open sets O^n_η with $O^n_\eta \cap (A \setminus \bigcup_{\eta'<\eta} O_{\eta'} O_{\eta'}) \in \Gamma^*$. This is what we wanted. The second fact is an obvious use of the Baire category theorem applied to the partition of $(\bigcup_{\eta'<\eta} O_{\eta'})$ induced by the sets F_n.

The case of arbitrary ξ can be reduced to the preceding case using Kuratowski's technique of generalized homeomorphisms. An alternative proof would use the effective topologies we introduce in Section 2. ⊣

We shall also omit the proof of the next result, which again uses an argument of transfer, as in the proof of Wadge's lemma, and in the preceding lemma.

Lemma 1.25. Assume u is a description of type 2 and level $u(0) = \lambda$ a limit ordinal. Then $\Delta(\Gamma_u) = \bigcup\{\Gamma_{u'} : u' \in Q_u\}$.

We now turn to the last step of the proof of Theorem 1.9: the case of a description of type 3.

Lemma 1.26. Let u be a description of type 3, with $u(0) = \xi$. Then Γ_u is closed under the intersection with a $\boldsymbol{\Pi}^0_\xi$ set. Moreover $\Delta(\Gamma_u)$ has the same closure property.

Proof. The second assertion is an immediate corollary of the first, and the fact that Γ_u is closed under union with a $\boldsymbol{\Sigma}^0_\xi$ set, which can easily be seen from the proof of Lemma 1.4 (b). [The only classes of level ξ not closed under union with a $\boldsymbol{\Sigma}^0_\xi$ set are the $D_\eta\boldsymbol{\Sigma}^0_\xi$ for $n \in \omega$, n even.] For if $A \in \Delta(\Gamma_u)$ and $B \in \boldsymbol{\Pi}^0_\xi$, then $A \cap B \in \Gamma_u$ by the first assertion of the lemma, and $(A\cap B)\breve{} = \breve{A} \cup \breve{B} \in \Gamma_u$ by the preceding remark.

To prove that Γ_u is closed under intersection with a $\boldsymbol{\Pi}^0_\xi$ set, argue by induction. If $u(1) = 3$ or 5, then either we have $u_1(0) > u(0)$, and then Γ_{u_1} is clearly closed under intersection with a $\boldsymbol{\Pi}^0_\xi$ set (in fact with a $\boldsymbol{\Sigma}^0_{\xi+1}$ set, by Lemma 1.4), or u_1 is itself of type 3, and we can apply the induction hypothesis. The only other case is if $u(1) = 2$, so $\Gamma_u = \mathrm{Sep}(D_\eta(\boldsymbol{\Sigma}^0_\xi), \Gamma^*_u)$. But then $u^*(0) > \xi$, so Γ_{u^*} is closed under intersection with a $\boldsymbol{\Pi}^0_\xi$, and the conclusion follows immediately. ⊣

Definition 1.27. For each u of type 3, we define a set Q_u of descriptions by the following conditions:

Case 1. $u(1) = 2$, so $\Gamma_u = \mathrm{Sep}(D_\eta(\boldsymbol{\Sigma}^0_\xi), \Gamma_{u^*})$. Define

$$Q_u = \{\xi^\frown 3^\frown \eta^\frown \langle u^*, u'\rangle : u' \in D, \Gamma_{u'} < \Gamma_{u^*} \text{ and } u'(0) \geq \xi\}.$$

Case 2. $u(1) = 3$ or 5, and $t(u_1) = 3$ (inductive step). Define

$$Q_u = \{u(0)^\frown u(1)^\frown u(2)^\frown \langle u_0, u'\rangle : u' \in Q_{u_1}\}.$$

Case 3. $u(1) = 3$ or 5, and $t(u_1) = 1$ or 2 (so $u_1(0) > u(0)$). Then Q_{u_1} has been previously defined, and define

$$Q_u = \{u(0)^\frown u(1)^\frown u(2)^\frown \langle u_0, u' \rangle \ : \ u' \in Q_{u_1} \text{ and } u'(0) \geq \xi\}.$$

LEMMA 1.28 (Assertion (C) for u of type 3). Let u be a description of type 3. Then $\Delta(\Gamma_u) = \bigcup\{\Gamma_{u'} \ : \ u' \in Q_u\}$.

PROOF.

Case 1. $u(1) = 2$, so $\Gamma_u = \text{Sep}(D_\eta(\mathbf{\Sigma}^0_\xi), \Gamma_{u^*})$, with $u^*(0) > \xi$. We want to prove that

$$\Delta(\Gamma_u) = \bigcup\{\text{Bisep}(D_\eta(\mathbf{\Sigma}^0_\xi), \Gamma_{u^*}, \Gamma_{u'}) \ : \ \Gamma_{u'} < \Gamma_{u^*}, u'(0) \geq \xi\}$$

Suppose first A is in the right hand side class. Then, by the closure properties of Γ_{u^*}, there are disjoint $D_\eta(\mathbf{\Sigma}^0_\xi)$ sets C_1 and C_2 such that $A_1 = A \cap C_1 \in \breve{\Gamma}_{u^*}$, $A_2 = A \cap C_2 \in \Gamma_{u^*}$, and $B = A \setminus (C_1 \cup C_2) \in \Delta(\Gamma_{u^*})$. Now A_2 and B are two Γ_{u^*} sets separated by the $\mathbf{\Sigma}^0_{\xi+1}$ sets C_2 and $\breve{C}_2$, so $A_2 \cup B \in \Gamma_{u^*}$, and $A \in \text{Sep}(D_\eta(\mathbf{\Sigma}^0_\xi), \Gamma_{u^*})$. Similarly A_1 and B are two $\breve{\Gamma}_{u^*}$ sets separated by the $\mathbf{\Sigma}^0_{\xi+1}$ sets C_1 and $\breve{C}_1$, so $A_1 \cup B \in \breve{\Gamma}_{u^*}$ and $\breve{A} \in \text{Sep}(D_\eta(\mathbf{\Sigma}^0_\xi), \Gamma_{u^*})$. This gives $A \in \Delta(\Gamma_u)$.

For the converse, we suppose $A \in \Delta(\Gamma_u)$, and we want to find disjoint $D_\eta(\mathbf{\Sigma}^0_\xi)$ sets C_1 and C_2 such that $A \cap C_1 \in \breve{\Gamma}_{u^*}$, $A \cap C_2 \in \Gamma_{u^*}$ and $A \setminus (C_1 \cup C_2) \in \Delta(\Gamma_{u^*})$, for then the inductive hypothesis will give the result. Let C, C' be two $D_\eta(\mathbf{\Sigma}^0_\xi)$ sets such that $A \cap C \in \breve{\Gamma}_{u^*}$, $A \setminus C \in \Gamma_{u^*}$, and $\breve{A} \cap C' \in \breve{\Gamma}_{u^*}$, $\breve{A} \setminus C' \in \Gamma_{u^*}$. Let C_1 and C_2 be $D_\eta(\mathbf{\Sigma}^0_\xi)$ sets reducing the pair (C, C'). Then $(A \cap C_1) = A \cap C \cap C_1 \in \breve{\Gamma}_{u^*}$, $A \cap C_2 = C_2 \setminus \breve{A} = C_2 \setminus (C' \cap \breve{A})$ is in Γ_{u^*}, and finally $B = A \setminus (C_1 \cup C_2) = A \setminus (C \cup C') = (A \setminus C) \setminus C'$ is in Γ_{u^*} and $\breve{B} = (C_1 \cup C_2) \cup \breve{A} = (C \cup C') \cup \breve{A} = C \cup C' \cup \breve{A} \setminus C'$ is in Γ_{u^*}. This proves case 1.

Case 2. $u(1) = 3$, $t(u_1)$ arbitrary. We have $\Gamma_u = \text{Bisep}(D_\eta(\mathbf{\Sigma}^0_\xi), \Gamma_{u_0}, \Gamma_{u_1})$, and we know that $u_0(0) > \xi = u(0)$, and either $u_1(0) > u(0)$, or $u_1(0) = u(0)$ and u_1 is of type 3. We may assume that $\bigcup\{\Gamma_{u'} \ : \ u' \in Q_{u_1} \text{ and } u'(0) \geq \xi\} = \Delta(\Gamma_{u_1})$: This is the induction hypothesis if u_1 is of type 3, and if u_1 is of type 1 or 2, a look at the definition of Q_{u_1} shows that $\bigcup\{\Gamma_{u'} \ : \ u' \in Q_{u_1} \text{ and } u'(0) \geq \xi\} = \bigcup\{\Gamma_{u'} \ : \ u' \in Q_{u_1}\}$, if $u_1(0) > \xi$. But we have already proved that this last class is $\Delta(\Gamma_{u_1})$. So, we want to prove that $\Delta(\Gamma_u) = \text{Bisep}(D_\eta(\mathbf{\Sigma}^0_\xi), \Gamma_{u_0}, \Delta(\Gamma_{u_1}))$. The inclusion from right to left is obvious. Suppose now $A \in \Delta(\Gamma_u)$. Using the normal form for the Bisep operation, plus the fact that $\Delta(\Gamma_{u_1})$ is closed under intersection with a $\mathbf{\Pi}^0_\xi$ set (Lemma 1.26), we have $A = (C_0 \cap A) \cup (C_1 \cap A) \cup B$, where C_0, C_1 are disjoint $D_\eta(\mathbf{\Sigma}^0_\xi)$ sets, with $C_0 \cup C_1 \in \mathbf{\Sigma}^0_\xi$ and $C_0 \cap A \in \breve{\Gamma}_{u_0}$, $C_1 \cap A \in \Gamma_{u_0}$ and $B \in \Gamma_{u_1}$, $B \cap (C_0 \cup C_1) = \varnothing$. Similarly we have $\breve{A} = (C'_0 \cap \breve{A}) \cup (C'_1 \cap \breve{A}) \cup B'$, with similar properties. Let C^*_0, C^*_1 be two $\mathbf{\Sigma}^0_\xi$ sets reducing the pair $C_0 \cup C_1$, $C'_0 \cup C'_1$. Then

$A \cap C_0^*$ is in $\text{Bisep}(D_\eta(\underset{\sim}{\mathbf{\Sigma}}^0_\xi), \Gamma_{u_0})$, and $\check{A} \cap C_1^*$, so $A \cap C_1^*$ is in $\text{Bisep}(D_\eta(\underset{\sim}{\mathbf{\Sigma}}^0_\xi), \Gamma_{u_0})$. So, we just have to show that $A \setminus (C_0^* \cup C_1^*) \in \Delta(\Gamma_{u_1})$. But $A \setminus C_0^* \cup C_1^* = A \setminus (C_0 \cup C_1 \cup C_0' \cup C_1') = B \setminus (C_0' \cup C_1')$. We clearly have $A \setminus (C_0^* \cup C_1^*) \in \Delta(\Gamma_{u_1})$.

Case 3, for $u(1) = 5$, is entirely similar, and we omit it. ⊣

Lemmas 1.12, 1.14, 1.16, 1.17, 1.19, 1.23, 1.24, 1.25 and 1.28 put together, give a proof of the assertions (A), (B) and (C) of page 55 and hence prove Theorem 1.9.

§2. Effective Results in the Borel Wadge hierarchy. The Wadge classes considered in the first part are boldface classes. We now are interested in their lightface counterparts, and in order to define them, we need a coding system, both for classes and for sets in each class.

For the classes, there is no problem: it is enough to code by reals sequences of countable ordinals, and this is obvious: we say that α is a **D-code**, written $\alpha \in \ulcorner D \urcorner$, if for every n $\langle \alpha \rangle_n \in \text{WO}$, and the coded sequence $u_\alpha = \langle |\langle \alpha \rangle_n| : n \in \omega \rangle$ is a description. Going back to the definition of descriptions shows immediately that $\ulcorner D \urcorner$ is a Π^1_1 set. We shall denote by Γ_α the class Γ_{u_α} (although it is a bit ambiguous, as some descriptions may be reals).

Encoding the sets in each Γ_α is also easy, but technical. Fix some recursive real $\bar{1}$ in WO, with $|\bar{1}| = 1$. Start from a pair (W, C) in Π^1_1 which is universal for $\underset{\sim}{\mathbf{\Sigma}}^0_\xi$ sets in the following sense:

1. $W \subseteq {}^\omega\omega \times {}^\omega\omega$, and $\exists \gamma (\alpha, \gamma) \in \text{W} \iff \alpha \in \text{WO}$
2. $C \subseteq {}^\omega\omega \times {}^\omega\omega \times {}^\omega\omega$ is Π^1_1, and $\forall \alpha \in \text{WO}$, $C_\alpha = \{(\gamma, \delta) : (\alpha, \gamma, \delta) \in C\}$ is universal for $\underset{\sim}{\mathbf{\Sigma}}^0_{|\alpha|}$ subsets of ${}^\omega\omega$
3. C is Δ^1_1 on W, *i.e.*, the relation $(\alpha, \gamma) \in W \wedge (\alpha, \gamma, \delta) \notin C$ is Π^1_1

It is then easy to construct a Π^1_1 pair $(W^{\text{is}}, C^{\text{is}})$ such that

1. $\exists \gamma (\alpha, \beta, \gamma) \in W^{\text{is}} \iff \alpha \in \text{WO} \wedge \beta \in \text{WO}$
2. $C^{\text{is}}_{\alpha,\beta}$ is universal for $<_\beta$-increasing sequences of $\underset{\sim}{\mathbf{\Sigma}}^0_{|\alpha|}$ sets ($C^{\text{is}} \subset {}^\omega\omega \times {}^\omega\omega \times {}^\omega\omega \times {}^\omega\omega \times {}^\omega\omega$)
3. C^{is} is Δ^1_1 on W^{is}.

[Define $(\alpha, \beta, \gamma) \in W \iff \beta \in \text{WO} \wedge \forall n (\alpha, (\gamma)_n) \in W \wedge \forall n \forall m \forall \delta ((\beta(n, m) = 0) \wedge (\alpha, (\gamma)_n, \gamma) \in C \Longrightarrow (\alpha, (\gamma)_n, \gamma) \in C)$ and $(\alpha, \beta, \gamma, n, \delta) \in C^{\text{is}} \iff (\alpha, \beta, \gamma) \in W^{\text{is}} \wedge (\alpha, (\gamma)_n, \delta) \in C$.]

Similarly, one can define a Π^1_1 pair $(W^{\text{ds}}, C^{\text{ds}})$ such that

(a) $\exists \gamma (\alpha, \gamma) \in W^{\text{ds}} \iff \alpha \in \text{WO}$

(b) For $\alpha \in \text{WO}$, C^{ds}_α is universal for pairwise disjoint sequences of $\underset{\sim}{\mathbf{\Sigma}}^0_{|\alpha|}$ sets.

(c) C^{ds} is Δ^1_1 on W^{ds}

[Let $(\alpha, \gamma) \in W^{\text{ds}} \iff \forall n (\alpha, (\gamma)_n) \in W$, and if $C_1 = \{(\alpha, \gamma, n, \delta) : (\alpha, \gamma) \in W^{\text{is}} \wedge (\alpha, (\gamma)_n, \delta) \in C\}$, let C^{ds} reduce C_1 (with respect to n), in such a way that for each α $(C)^{\text{ds}}_{\alpha,\gamma,n}$ is $\underset{\sim}{\mathbf{\Sigma}}^0_{|\alpha|}$.]

With these preliminary constructions, we can define a relation β is a Γ_α-code, together with "β codes the the set $\Gamma_{\alpha,\beta}$".

DEFINITION 2.1. For $\alpha \in \ulcorner D \urcorner$, the relations "$\beta$ is a Γ_α-code", and "The set in Γ_α coded by β is A" (written $\Gamma_{\alpha,\beta} = A$) are defined inductively by the following conditions:

(a) If $|(\alpha)_0| = 0$ and $\beta(0) = 0$, β is a Γ_α-code and $\Gamma_{\alpha,\beta} = \varnothing$

(b) Suppose $|(\alpha)_0| \geq 1$ and $|(\alpha)_1| = 1$, and $\beta = 1^\frown\beta^*$. Then if $((\alpha)_0, (\alpha)_2, \beta^*) \in W^{\text{is}}$, then β is a Γ_α-code, and

$$\Gamma_{\alpha,\beta} = D_{|(\alpha)_2|}(\langle A_\zeta : \zeta < |(\alpha)_2|\rangle),$$

where $A_\zeta = C^{\text{is}}_{(\alpha)_0,(\alpha)_2,\beta^*,n}$ for the unique n which has order type ζ in $<_{(\alpha)_2}$.

(c) Suppose $|(\alpha)_0| \geq 1$ and $|(\alpha)_1| = 2$. Let $\alpha_0 = (\alpha)_0{}^\frown 1^\frown(\alpha)_2{}^\frown\underline{0}^\frown\underline{0}\ldots$ and let $\alpha_1 = \langle(\alpha)_{n+3} : n \in \omega\rangle$. Then β is a Γ_α code if $\beta = 2^\frown\langle\beta_0, \beta_1, \beta_2\rangle$, where β_0 is a Γ_{α_0}-code, β_1 and β_2 are Γ_{α_1}-codes. Moreover $\Gamma_{\alpha,\beta} = \text{Sep}(\Gamma_{\alpha_0,\beta_0}, \check{\Gamma}_{\alpha_1,\beta_1}, \Gamma_{\alpha_2,\beta_2})$.

(d) Suppose now $|(\alpha)_0| \geq 1$ and $|(\alpha)_1| = 3$. Let $\alpha_0 = (\alpha)_0{}^\frown\bar{1}^\frown(\alpha)_2{}^\frown\underline{0}^\frown\underline{0}\ldots$ and let α_1, α_2 be such that $u_\alpha = \xi^\frown 3^\frown\eta^\frown\langle u_{\alpha_1}, u_{\alpha_2}\rangle$. (Such α_1, α_2 could be defined precisely, and are supposed to be recursive in α.) Then β is a Γ_α-code if $\beta = 3^\frown\langle\beta_0, \beta_1, \beta_2, \beta_3, \beta_4\rangle$ where β_0, β_1 are Γ_{α_0}-codes, β_1, β_2 are Γ_{α_1}-codes, β_4 is a Γ_{α_2}-code, and $\Gamma_{\alpha_0,\beta_0} \cap \Gamma_{\alpha_0,\beta_1} = \varnothing$; and then $\Gamma_{\alpha,\beta} = \text{Bisep}(\Gamma_{\alpha_0,\beta_0}, \Gamma_{\alpha_0,\beta_1}, \check{\Gamma}_{\alpha_1,\beta_2}, \Gamma_{\alpha_1,\beta_3}, \Gamma_{\alpha_2,\beta_4})$.

(e) If $|(\alpha)_1| = 4$, then let α_n be (recursively in α) a sequence such that $u_\alpha = \xi^\frown 4^\frown\langle u_{\alpha_n} : n \in \omega\rangle$; then β is a Γ_α-code if $\beta = 4^\frown\langle\beta^*, \beta^{**} : n \in \omega\rangle$, where $((\alpha)_0, \beta^*) \in W^{\text{ds}}$, $((\alpha)_0, \beta^{**}) \in W$ and codes the union of the disjoint sequence coded by β^*, and for each n β_n is a Γ_{α_n}-code. Then $\Gamma_{\alpha,\beta} = \text{SU}(\langle C^{\text{ds}}_{(\alpha)_0,\beta^*,n} : n \in \omega\rangle, \langle\Gamma_{\alpha_n,\beta_n} : n \in \omega\rangle)$.

(f) Finally, if $|(\alpha)_1| = 5$, let α_0, α_1 (recursively in α) be such that $u_\alpha = \xi^\frown 5^\frown\eta^\frown\langle u_{\alpha_0}, u_{\alpha_1}\rangle$ Then β is a Γ_α-code if $\beta = 5^\frown\langle\beta_1 : \gamma_n \in n \in \omega\rangle$, where β_1 is a Γ_{α_1}-code, for each n γ_n is a Γ_{α_0} code, say $\gamma_n = 4^\frown\langle\gamma_n^*, \gamma_n^{**}, \gamma_n^p : p \in \omega\rangle$ and the sequence of pairs $(A_\zeta, C_\zeta), \zeta < |(\alpha)_2|$ defined by $A_\zeta = \Gamma_{(\alpha)_0,\gamma_n}$, $C_\zeta = C_{(\alpha)_0,\gamma_n^{**}}$ for the only n of order type ζ in $<_{(\alpha)_2}$, is an increasing sequence with $C_\zeta \subset A_{\zeta+1}$. And then

$$\Gamma_{\alpha,\beta} = \text{SD}_{|(\alpha)_2|}(\langle C_\zeta : \zeta < |(\alpha)_2|\rangle, \langle A_\zeta : \zeta < |(\alpha)_2|\rangle, \Gamma_{\alpha_1,\beta_1}).$$

It is clear from the preceding definition, that the coding relations

$$\alpha \in \ulcorner D \urcorner$$
$$\alpha \in \ulcorner D \urcorner \wedge \beta \text{ is a } \Gamma_\alpha\text{-code}$$
$$\alpha \in \ulcorner D \urcorner \wedge \beta \text{ is a } \Gamma_\alpha\text{-code} \wedge \gamma \in \Gamma_{\alpha,\beta}$$

and

$$\alpha \in \ulcorner D \urcorner \wedge \beta \text{ is a } \Gamma_\alpha\text{-code} \wedge \gamma \notin \Gamma_{\alpha,\beta}$$

are all Π^1_1.

From the proof of the main theorem of part 1, it is also clear that some variant of the preceding coding would enable to prove "recursive" analogs of the Hausdorff-Kuratowski-type results we have quoted. Such a variant would involve coding by partial recursive functions, in the spirit of what is done for the class $\boldsymbol{\Sigma}^0_\xi$.

Anyway, we are more interested here in "Δ^1_1-recursive" results, for which the coding we defined above is good enough. From now on, Wadge classes will be written boldface, to distinguish them from their lightface counterparts.

Definition 2.2. A described Wadge class $\boldsymbol{\Gamma}$ is a Δ^1_1-class if it admits a Δ^1_1 code (*i.e.*, $\boldsymbol{\Gamma} = \boldsymbol{\Gamma}_\alpha$ for some Δ^1_1 real α in D). So in particular the Δ^1_1 classes, among the $\boldsymbol{\Sigma}^0_\xi$'s are the $\boldsymbol{\Sigma}^0_\xi$ for $\xi < \omega_1^{\mathrm{ck}}$.

Definition 2.3. Let $\boldsymbol{\Gamma}_\alpha, \alpha \in \Delta^1_1$, be a Δ^1_1-class. We define the lightface classes $\Gamma_\alpha, \Gamma_\alpha(\beta), \Gamma^1_\alpha$ by

$$\Gamma_\alpha = \{\boldsymbol{\Gamma}_{\alpha,\beta} : \beta \text{ a recursive } \boldsymbol{\Gamma}_\alpha\text{-code}\}$$
$$\Gamma_\alpha(\beta) = \{\boldsymbol{\Gamma}_{\alpha,\gamma} : \gamma \text{ a recursive in-}\beta\ \boldsymbol{\Gamma}_\alpha\text{-code}\}$$

and

$$\Gamma^1_\alpha = \bigcup\{\boldsymbol{\Gamma}_\alpha(\beta) : \beta \in \Delta^1_1\} = \{\boldsymbol{\Gamma}_{\alpha,\beta} : \beta \in \Delta^1_1\}$$

Because of the coding we chose, it is not clear that the lightface class Γ_α is really well defined, *i.e.*, does not depend on the particular code α for $\boldsymbol{\Gamma}_\alpha$, even in case α is recursive. But it can be seen that Γ^1_α does not depend on the particular choice of $\alpha \in \Delta^1_1$, but only on the class $\boldsymbol{\Gamma}_\alpha$. [This can be seen directly, but is also an immediate corollary of the main result below.]

In Louveau [Lou80], we studied $(\boldsymbol{\Sigma}^0_\xi)^1 = \bigcup_{\alpha \in \Delta^1_1} \Sigma^0_\xi(\alpha)$, for $\xi < \omega_1^{\mathrm{ck}}$, and we proved that $(\boldsymbol{\Sigma}^0_\xi)^1 = \boldsymbol{\Sigma}^0_\xi \cap \Delta^1_1$, *i.e.*, that every $\boldsymbol{\Sigma}^0_\xi$ set in Δ^1_1 admits a Δ^1_1 $\boldsymbol{\Sigma}^0_\xi$-code. The main theorem in this section is the extension of this result to all Δ^1_1 (non self-dual) Borel Wadge classes.

Theorem 2.4. *Let $\boldsymbol{\Gamma}_\alpha$ be a described Wadge class, with $\alpha \in \Delta^1_1$. Then each Δ^1_1 set in $\boldsymbol{\Gamma}_\alpha$ admits a $\boldsymbol{\Gamma}_\alpha$ code which is Δ^1_1, i.e., $\Gamma^1_\alpha = \boldsymbol{\Gamma}_\alpha \cap \Delta^1_1$.*

In order to prove this theorem, we need some tools from Louveau [Lou80]. For $\xi < \omega_1^{\mathrm{ck}}$, we define T_ξ to be the topology on ${}^\omega\omega$ generated by the Σ^1_1 sets which are in $\bigcup_{\eta<\xi} \boldsymbol{\Pi}^0_\eta$. T_∞, the Harrington topology on ${}^\omega\omega$, is the topology generated by all Σ^1_1 subsets of ${}^\omega\omega$. For this topology, ${}^\omega\omega$ is a Baire space (*i.e.*, no non-empty T_∞-open set is T_∞-meager). Say that a property of reals is true

∞-a.e. if it is false for a set of reals which is T_∞-meager. In the induction used to prove the result for $\boldsymbol{\Sigma}^0_\xi$ sets, we proved the following two results:

PROPOSITION 2.5. Let A be a $\boldsymbol{\Sigma}^0_\xi$ set. Then there is a $\boldsymbol{\Sigma}^0_\xi$ set A' which is T_ξ-open and satisfies $A = A'$ ∞-a.e.

PROPOSITION 2.6 (Separation result). Let A, B be two Σ^1_1 sets and suppose there is a $\boldsymbol{\Sigma}^0_\xi$ set C such that $A \subseteq C \subseteq B$ ∞-a.e. Then there is a $(\boldsymbol{\Sigma}^0_\xi)^1$ set D such that $A \subseteq D \subseteq B$.

Proposition 2.6 clearly implies the result for $\boldsymbol{\Sigma}^0_\xi$ classes; working analogously, we shall prove Theorem 2.4 by proving first a separation result for $\boldsymbol{\Gamma}_\alpha$ by induction on α.

THEOREM 2.7. Let $\boldsymbol{\Gamma}_\alpha$ be a described Wadge class, with $\alpha \in \Delta^1_1$. If A, B are two Σ^1_1 sets, and there is a $\boldsymbol{\Gamma}_\alpha$ set C such that $A \subseteq C \subseteq B$ ∞-a.e., then there is a Γ^1_α set D with $A \subseteq D \subseteq B$.

PROOF. It is clear that if α is Δ^1_1, the class $\boldsymbol{\Gamma}_\alpha$ has been constructed from previous classes which are also Δ^1_1, so that we can prove 2.7 by induction. There are five cases (we forget $\boldsymbol{\Gamma}_\alpha = \{\varnothing\}$!).

Case 1. $(\alpha)_1 = 1$, so $\boldsymbol{\Gamma} = D_\eta(\boldsymbol{\Sigma}^0_\xi)$, with $\eta = |(\alpha)_2|$ and $\xi = |(\alpha)_0|$ (so ξ and η are recursive ordinals). We assume η is an even ordinal, the other case being similar. We first define a sequence $\langle A_\zeta : \zeta < \eta\rangle$ by the following: If ζ is even, let A_ζ be the largest T_ξ-open set disjoint from $A \setminus \bigcup_{\zeta'<\zeta} A_{\zeta'}$. If ζ is odd, let A_ζ be the largest T_ξ-open set disjoint from $B \setminus \bigcup_{\zeta'<\zeta} A_{\zeta'}$.

Clearly the sequence $\langle A_\zeta : \zeta < \eta\rangle$ is an increasing sequence of $\boldsymbol{\Sigma}^0_\xi$ sets. Moreover we have:

(a) The relation $\beta \in A_{\zeta_n}$, where ζ_n is the order type of predecessors of n in $<_{(\alpha)_2}$, is Π^1_1 (in β and n).

(b) If $\langle C_\zeta : \zeta < \eta\rangle$ is any sequence of $\boldsymbol{\Sigma}$ sets with $A \subseteq D_\eta(\langle A_\zeta : \zeta < \eta\rangle) \subseteq B$ ∞-a.e., then for each $\zeta < \eta$ $C_\zeta \subseteq A_\zeta$ ∞-a.e., and $A \subseteq D_\eta(\langle A_\zeta : \zeta < \eta\rangle) \subseteq B$.

To prove (a). Suppose H is a Σ^1_1 set, and $\xi < \omega_1^{\mathrm{ck}}$. Then the largest T_∞-open set O disjoint from H is a Π^1_1 set. In fact

$$\begin{aligned} x \in O &\iff \exists G \in \boldsymbol{\Sigma}^0_\xi \cap \Sigma^1_1 (x \in G \wedge G \cap H = \varnothing) \\ &\iff \exists G \in (\boldsymbol{\Sigma}^0_\xi)^1 (x \in G \wedge G \cap H = \varnothing) \end{aligned}$$

the second equivalence being obtained by using Proposition 2.6. This clearly implies that the relation $\beta \in A_{\zeta_n}$ is Π^1_1.

To prove (b). Using Proposition 2.5, we may assume that C_ζ are T_∞-open. Then by induction on ζ, we prove that $C_\zeta \subseteq A_\zeta$. Suppose ζ is even. Then $C_\zeta \setminus \bigcup_{\zeta'<\zeta} C_{\zeta'}$ is disjoint from $D_\eta(\langle C_\zeta : \zeta < \eta\rangle)$, so is disjoint from A, ∞-a.e. This implies that C_ζ is ∞-a.e. disjoint from $A \setminus \bigcup_{\zeta'<\zeta} C_{\zeta'}$, and using the induction hypothesis, also from $A \setminus \bigcup_{\zeta'<\zeta} A_{\zeta'}$. But this implies, because this last set is Σ^1_1 by part (a), that $C_\zeta \cap (A \setminus \bigcup_{\zeta'<\zeta} A_\zeta = \varnothing$, using the Baire category theorem for T_∞. So by the definition of A_ζ, $C_\zeta \subseteq A_\zeta$. The odd case is similar, and we omit it. To prove the last assertion, *i.e.*, that $A \subseteq D_\eta(\langle A_\zeta : \zeta < \eta\rangle) \subseteq B$, it is enough to prove that $A \subset \bigcup_{\zeta<\eta} A_\zeta$. But $A \subseteq \bigcup_{\zeta<\eta} C_\zeta \infty$-a.e., so using the preceding result, $A \subseteq \bigcup_{\zeta<\eta} A_\zeta$ ∞-a.e.

Finally, $\bigcup_{\zeta<\eta} A_\zeta$ is Π^1_1 and A is Σ^1_1, so using again the Baire category theorem for T_∞ gives $A \subseteq \bigcup_{\zeta<\eta} A_\zeta$.

The last part of the proof consists in replacing the Π^1_1 sequence $\langle A_\zeta : \zeta < \eta\rangle$ by a Δ^1_1 sequence with the same properties. For each recursive ordinal ϑ, say that a Δ^1_1 sequence $\langle D_\zeta : \zeta < \vartheta\rangle$ of $\mathbf{\Sigma}^0_\xi$ sets is a ϑ-system for (A, B) if D_ζ is disjoint from $A \setminus \bigcup_{\zeta'<\zeta} D_\zeta$ if ζ is even, from $B \setminus \bigcup_{\zeta'<\zeta} D_\zeta$ if ζ is odd, and say that this ϑ-system covers H if $H \subseteq \bigcup_{\zeta<\vartheta} D_\zeta$. What we want to construct is an η-system covering the set A. So it is enough to prove the following claim.

Claim 2.8. Let ϑ be some even ordinal, $\vartheta \subseteq \eta$, and let H be a Σ^1_1 set, $H \subseteq \bigcup_{\zeta<\vartheta} A_\zeta$. Then there exists a ϑ-system covering H.

The proof of the claim is by induction on ϑ. Suppose first ϑ is a successor, so that $\vartheta = \vartheta' + 2$ with ϑ' even. By the hypothesis $H \subseteq A_{\vartheta'+1}$, so we can find a set $D^0 \in (\mathbf{\Sigma}^0_\xi)^1$ with $H \subseteq D^0 \subseteq A_{\vartheta'+1}$. But as $A_{\vartheta'+1}$ is disjoint from $B \setminus A_{\vartheta'}$, $D^0 \cap B \subseteq A_{\vartheta'}$, and so we can find a $(\mathbf{\Sigma}^0_\xi)^1$ set $D^1 \subseteq D^0$ with $D^0 \cap B \subseteq D^1 \subseteq A_{\vartheta'}$. By the same reasoning, we must have $A \cap D^1 \subseteq \bigcup_{\zeta<\vartheta} A_\zeta$, so by the induction hypothesis, we can find a ϑ'-system $\langle D_\zeta : \zeta < \vartheta'\rangle$ covering $A \cap D^1$. We then extend this ϑ' system by setting $D_{\vartheta'} = \bigcup_{\zeta<\vartheta'} D_\zeta \cup D^1$, and $D_{\vartheta'+1} = \bigcup_{\zeta<\vartheta'} D_\zeta \cup D^0$. This clearly gives the desired ϑ-system.

Suppose now ϑ is limit. As $H \subseteq \bigcup_{\zeta<\vartheta} A_\zeta = \bigcup_{\zeta \text{ even}} A_\zeta$, we can first choose a Δ^1_1 sequence $\langle H_\zeta : \zeta < \vartheta, \zeta \text{ even}\rangle$ with $H \subseteq \bigcup H_\zeta$, and for ζ even, $\zeta < \vartheta$, $H_{\zeta \subseteq A_\zeta}$. By the induction hypothesis, and Δ^1_1 selection, there is, for each even $\zeta < \vartheta$ a ζ-system $\langle D^\zeta_{\zeta'} : \zeta' < \zeta\rangle$, covering H_ζ, and such that the double sequence is Δ^1_1. Define then a sequence $\langle D_\zeta : \zeta < \vartheta\rangle$ by

$$D_\zeta = \bigcup_{\zeta'<\zeta} D_{\zeta'} \cup \bigcup_{\zeta<\zeta'<\vartheta} D^{\zeta'}_\zeta$$

We claim that the Δ^1_1 and increasing sequence $\langle D_\zeta : \zeta < \vartheta\rangle$ is the ϑ-system we wanted. It certainly covers H, so the only thing we have to prove is that it is a ϑ-system. Suppose $\zeta < \vartheta$ is even. Then

$$D_\zeta \cap \Big(B \setminus \bigcup_{\zeta''<\zeta} D_{\zeta''}\Big) \subseteq \bigcup_{\substack{\zeta<\zeta'<\vartheta \\ \zeta' \text{ even}}} \Big(D^{\zeta'}_\zeta \cap \Big(B \setminus \bigcup_{\zeta''<\zeta} D_{\zeta''}\Big)\Big)$$

so fix ζ' even, with $\zeta < \zeta' < \vartheta$. As $D^{\zeta'}_{\zeta''} \subseteq D_{\zeta''}$ for any $\zeta'' < \zeta$

$$D^{\zeta'}_{\zeta} \cap \Big(B \setminus \bigcup_{\zeta''<\zeta} D_{\zeta''}\Big) \subseteq D^{\zeta'}_{\zeta} \cap \Big(B \setminus \bigcup_{\zeta''<\zeta} D^{\zeta}_{\zeta''}\Big) = \varnothing$$

Similarly if ζ is odd

$$D_{\zeta} \cap \Big(A \setminus \bigcup_{\zeta''<\zeta} D_{\zeta''}\Big) \subseteq \bigcup_{\substack{\zeta<\zeta'\vartheta \\ \zeta' \text{ even}}} \Big(D^{\zeta'}_{\zeta} \cap \Big(B \setminus \bigcup_{\zeta''<\zeta} D_{\zeta''}\Big)\Big)$$

and for each ζ' even, $\zeta < \zeta'\vartheta$, as $D^{\zeta'}_{\zeta''} \subseteq D_{\zeta''}$ for any $\zeta'' < \zeta$

$$D^{\zeta'}_{\zeta} \cap \Big(A \setminus \bigcup_{\zeta''<\zeta} D_{\zeta''}\Big) \subseteq D^{\zeta'}_{\zeta} \cap \Big(A \setminus \bigcup_{\zeta''<\zeta} D^{\zeta}_{\zeta''}\Big) = \varnothing$$

This proves the claim, and so finishes the proof of case 1.

Case 2. We now suppose α codes a description u with $u(1) = 2$, so that $\underset{\sim}{\mathbf{\Gamma}}_{\alpha} = \mathrm{Sep}(D_{\eta}\underset{\sim}{\mathbf{\Sigma}}^0_{\xi}, \underset{\sim}{\mathbf{\Gamma}}_{\alpha_1})$, where $\xi = |(\alpha)_0|$ and $\eta = |(\alpha)_2|$ are recursive ordinals, and $\alpha_1 \in \Delta^1_1$, so that by the induction hypothesis we can assume the theorem is true for $\underset{\sim}{\mathbf{\Gamma}}_{\alpha_1}$. Let us suppose that η is even, the other case being similar. Given A and B, we define a sequence $\langle A_{\zeta} : \zeta < \eta\rangle$ of $\underset{\sim}{\mathbf{\Sigma}}^0_{\xi}$ sets by the following: If ζ is even, we let A_{ζ} be the union of all T_{ζ} open sets O such that for some set C in $\underset{\sim}{\mathbf{\Gamma}}_{\alpha_1}$, $(A \setminus \bigcup_{\zeta'<\zeta} A_{\zeta'}) \cap O \subseteq C \subseteq B$ ∞-a.e., and similarly if ζ is odd, A_{ζ} is the union of all T_{ζ} open sets O such that for some set C in $\underset{\sim}{\mathbf{\Gamma}}_{\alpha_1}(A \setminus \bigcup_{\zeta'<\zeta} A_{\zeta'}) \cap O \subseteq C \subseteq B$.

We claim that the sequence $\langle A_{\zeta} : \zeta < \eta\rangle$, which is clearly an increasing sequence of $\underset{\sim}{\mathbf{\Sigma}}^0_{\xi}$ sets, has the following properties:

(a) The relation $\beta \in A_{\zeta_n}$ is Π^1_1 (in β and n)
(b) If $\langle C_{\zeta} : \zeta < \eta\rangle$ is an increasing sequence of $\underset{\sim}{\mathbf{\Sigma}}^0_{\zeta}$ sets such that $C = D_{\eta}(\langle C_{\eta} : \eta < \eta\rangle)$ satisfies that $A \cap C \subseteq H \subseteq B$ ∞-a.e. for some H in $\underset{\sim}{\mathbf{\Gamma}}_{\alpha_1}$, and $A \setminus C \subseteq H' \subseteq B$ ∞-a.e. for some H' in $\underset{\sim}{\mathbf{\Gamma}}_{\alpha_1}$, then for each $\zeta < \eta$ $C_{\zeta} \subseteq A_{\zeta}$ ∞-a.e.

To prove (a), let (C, D) be the union of all T_{ξ}-open sets O such that for some set H in $\underset{\sim}{\mathbf{\Gamma}}_{\alpha_1}$, $C \cap O \subseteq H \subseteq D$ ∞-a.e. We want to prove that if C, D are Σ^1_1, (C, D) is Π^1_1. But it is Π^1_1, for

$$\begin{aligned} x \in (C, D) \iff &\exists G \in \Sigma^1_1 \cap \underset{\sim}{\mathbf{\Sigma}}^0_{\xi} \exists H (x \in G \wedge H \in \underset{\sim}{\mathbf{\Gamma}}_{\alpha_1} \wedge \\ &\qquad G \cap C \subseteq H \subseteq \check{D})\ \infty\text{-a.e.} \\ \iff &\exists G' \in (\underset{\sim}{\mathbf{\Sigma}}^0_{\xi})^1 \exists H' \in \Gamma^1_{\alpha_1} (x \in G' \wedge G' \cap C' \subseteq H' \subseteq \check{D}) \end{aligned}$$

The second equivalence is justified by the following fact: If G is Σ^1_1 and $x \in G$, and for some $H \in \underset{\sim}{\mathbf{\Gamma}}_{\alpha_1}$ $G \cap C \subseteq H \subseteq D$ ∞-a.e., then using the induction hypothesis $\exists H' \in \Gamma^1_{\alpha_1}$ with $G \cap C \subseteq H' \subseteq \check{D}$. But then G is disjoint from $C \setminus H'$, which is Σ^1_1, so we can find B' in $(\underset{\sim}{\mathbf{\Sigma}}^0_\xi)^1$ with $G \subseteq G'$ and $G \cap C \setminus H' = \varnothing$, so that $G' \cap C \subseteq H' \subseteq \check{D}$, and $x \in G'$.

To prove (b), use Proposition 2.5 to replace the $\underset{\sim}{\mathbf{\Sigma}}^0_\xi$ sets C_ζ by T_ξ-open sets. Then it is immediate, by induction on ζ, that $C_\zeta \subseteq A_\zeta$

In order to finish the proof of case 2, we argue as follows. Consider $A \setminus \bigcup_{\zeta<\eta} A_\zeta$. It is a Σ^1_1 set, by (a), and by (b) it is a subset, ∞-a.e., of $(\bigcup_{\zeta<\eta} \check{C}_\zeta)$, where $\langle C_\zeta : \zeta < \eta\rangle$ is a sequence as in (b) (we know that such a sequence exists). Now it follows that there is an $H \in \underset{\sim}{\mathbf{\Gamma}}_{\alpha_1}$ with $A \setminus \bigcup_{\zeta<\eta} A_\zeta \subseteq H \subseteq B$ ∞-a.e., so by the induction hypothesis, we can find such an H_η in $\Gamma^1_{\alpha_1}$. Now $A \cap \check{H}_\eta \subseteq \bigcup_{\zeta<\eta} A_\zeta$, and so by an argument very similar to the one we used for case 1, we can replace the sequence $\langle A_\zeta : \zeta < \eta\rangle$ by a Δ^1_1 sequence $\langle C_\zeta : \zeta < \eta\rangle$, having the same properties, namely

- it is a Δ^1_1 sequence of $\underset{\sim}{\mathbf{\Sigma}}^0_\xi$ sets, increasing, and
- if ζ is even, C_ζ is a union of T_ξ-open sets O such that for some H in $\underset{\sim}{\mathbf{\Gamma}}_{\alpha_1} (A \setminus \bigcup_{\zeta'<\zeta} C_{\zeta'}) \cap O \subseteq H \subseteq \check{B}$ ∞-a.e.
- if it is odd, C_ζ is a union of T_ζ-open sets O such that for some H in $\underset{\sim}{\mathbf{\Gamma}}_{\alpha_1}$, $(A \setminus \bigcup_{\zeta'<\zeta} C_\zeta) \cap O \subseteq H \subseteq B$ ∞-a.e.
- and finally the sequence covers $A \cap \check{H}_\eta$.

We now define a Δ^1_1 sequence $\langle H_\zeta : \zeta < \eta\rangle$ of sets: Suppose ζ is even. Then as C_ζ is a union of T_ξ sets O such that for some H in $\underset{\sim}{\mathbf{\Gamma}}_{\alpha_1}$ $A \setminus \bigcup_{\zeta'<\zeta} C_{\zeta'} \cap O \subseteq H \subseteq \check{B}$ ∞-a.e., we can, by the induction hypothesis and using Δ^1_1 selection, find two Δ^1_1 sequences $\langle C^n : n \in \omega\rangle$ and $\langle H^n : n \in \omega\rangle$ such that the C^n_ζ are pairwise disjoint $\underset{\sim}{\mathbf{\Sigma}}^0_\xi$ sets, of union C_ζ, and H^n_ζ are in $\underset{\sim}{\mathbf{\Gamma}}^1_{\alpha_1}$ with $A \setminus \bigcup_{\zeta'<\zeta} C_{\zeta'} \cap C^n_\zeta \subseteq H^n_\zeta \subset B$. Let $H_\zeta = \mathrm{SU}(\langle C^n_\zeta : n \in \omega\rangle, \langle H^n : n \in \omega\rangle)$. $H_\zeta \in \underset{\sim}{\mathbf{\Gamma}}^1_{\alpha_1}$, so that C_ζ itself satisfies $\exists H \in \underset{\sim}{\mathbf{\Gamma}}^1_{\alpha_1}$ $A \setminus \bigcup_{\zeta'<\zeta} C_{\zeta'} \cap C_\zeta \subseteq H \subseteq \check{B}$. Similarly for ζ odd, there is an $H \in \check{\underset{\sim}{\mathbf{\Gamma}}}^1_{\alpha_1}$ with $A \setminus \bigcup_{\zeta'<\zeta} C_{\zeta'} \cap C_\zeta \subseteq H \subseteq \check{B}$. Using Δ^1_1-selection, we can find such a sequence $\langle H_\zeta : \zeta < \eta\rangle$ in a Δ^1_1 way. We now put

$$H_0 = \bigcup_{\substack{\zeta \text{ odd} \\ \zeta<\eta}} \left(H_\zeta \cap \left(C_\zeta \setminus \bigcup_{\zeta'<\zeta} C_\zeta \right) \right)$$

and

$$H_1 = \bigcup_{\substack{\zeta \text{ even} \\ \zeta<\eta}} \left(H_\zeta \cap \left(C_\zeta \setminus \bigcup_{\zeta'<\zeta} C_\zeta \right) \right) \cup \left(H_\eta \setminus \bigcup_{\zeta<\eta} C_\zeta \right)$$

It is clear that $H_0 \in \check{\underset{\sim}{\mathbf{\Gamma}}}^1_{\alpha_1}$, and the set $D = \mathrm{Sep}(D_\eta(\langle C_\zeta : \zeta < \eta\rangle), H_0, H_1)$ is in Γ^1_α and satisfies $A \subseteq D \subseteq \check{B}$.

Case 3. Suppose α codes a description u with $u(1) = 3$, so that $\underset{\sim}{\mathbf{\Gamma}}_\alpha = \text{Bisep}(D_\eta(\underset{\sim}{\mathbf{\Sigma}}^0_\xi), \underset{\sim}{\mathbf{\Gamma}}_{\alpha_1}, \underset{\sim}{\mathbf{\Gamma}}_{\alpha_2})$, with η, ξ as before, and α_1, α_2 are Δ^1_1 codes for descriptions. We assume again η is even. Starting with A, B in Σ^1_1, we first construct the sequence $\langle A_\zeta : \zeta < \eta\rangle$ as in case 2, and a similar sequence $\langle B_\zeta : \zeta < \eta\rangle$, defined by exchanging the role of $\underset{\sim}{\mathbf{\Gamma}}_{\alpha_1}$ and $\check{\underset{\sim}{\mathbf{\Gamma}}}_{\alpha_1}$ in the definition. Let $A^0 = \bigcup_{\zeta<\eta} A_\zeta$, $A^1 = \bigcup_{\zeta<\eta} B_\zeta$. These sets are Π^1_1 and $\underset{\sim}{\mathbf{\Sigma}}^0_\xi$. Now using the normal forms for the Bisep operation (Lemma 1.5), we know that there are disjoint sets C^0, C^1 in $\underset{\sim}{\mathbf{\Sigma}}^0_\xi$ sets, with $\bigcup_{\zeta<\eta} C^0_\zeta = C^0$, $\bigcup_{\zeta<\eta} C^1_\zeta = C^1$, and sets $H^0_0, H^0_1, H^1_0, H^1_1, H_2$ with H^0_0, H^1_0 in $\check{\underset{\sim}{\mathbf{\Gamma}}}_{\alpha_1}$, H^0_1, H^1_1 in $\underset{\sim}{\mathbf{\Gamma}}_{\alpha_1}$ and H_2 in $\underset{\sim}{\mathbf{\Gamma}}_{\alpha_2}$ such that

$$\begin{aligned}A \subseteq\ & (D_\eta(\langle C^0_\zeta : \zeta < \eta\rangle) \cap H^0_0) \cup (C_0 \setminus D_\eta(\langle C^0_\zeta : \zeta < \eta\rangle) \cap H^0_1) \cup \\ & (D_\eta(\langle C^1_\zeta : \zeta < \eta\rangle) \cap H^1_1) \cup (C^1 \setminus D_\eta(\langle C^1_\zeta : \zeta < \eta\rangle) \cap H^1_0) \cup \\ & H_2 \setminus (C^0 \cup C^1) \\ \subseteq\ & \check{B}\ \infty\text{-a.e.}\end{aligned}$$

It easily follows that for each $\zeta < \eta$ $C^0_\zeta \subseteq A_\zeta$, $C^1_\zeta \subseteq B_\zeta$ and $C^0 \subseteq A^0$, $C^1 \subseteq A^1$ ∞-a.e. From this, it follows that $A \setminus (A^0 \cup A^1) \subseteq H_2 \subseteq \check{B}$ ∞-a.e., so using the inductive hypothesis, we can find a set $D \in \Gamma^1_{\alpha_2}$ with $A \setminus (A^0 \cup A^1) \subseteq D \subseteq \check{B}$. Now by the argument previously used, one can shrink the Π^1_1 sequences $\langle A_\zeta : \zeta < \eta\rangle$ and $\langle B_\zeta : \zeta < \eta\rangle$ into Δ^1_1 sequences $\langle D^0_\zeta : \zeta < \eta\rangle$ and $\langle D^1_\zeta : \zeta < \eta\rangle$ with the same properties, in such a way that setting $D^0 = \bigcup_{\zeta<\eta} D^0_\zeta$ and $D^1 = \bigcup_{\zeta<\eta} D^1_\zeta$, D^0, D^1 are disjoint and $D^0 \cup D^1$ covers the Σ^1_1 set $A \cap \check{D}$. Again imitating the proof of case 2, we can find sets H^0_0 and H^1_0 in $\check{\underset{\sim}{\Gamma}}^1_{\alpha_1}$ and H^0_1, H^1_1 in $\Gamma^1_{\alpha_1}$ such that

$$\begin{aligned}A \cap D_\eta(\langle D^0_\zeta : \zeta < \eta\rangle) &\subseteq H^0_0 \subseteq \check{B} \\ A \cap D^0 \setminus D_\eta(\langle D^0_\zeta : \zeta < \eta\rangle) &\subseteq H^0_1 \subseteq \check{B} \\ A \cap D_\eta(\langle D^0_\zeta : \zeta < \eta\rangle) &\subseteq H^1_1 \subseteq \check{B} \\ A \cap D^1 \setminus D_\eta(\langle D^0_\zeta : \zeta < \eta\rangle) &\subseteq H^1_0 \subseteq \check{B}\end{aligned}$$

So that the Γ^1_α set

$$\begin{aligned}H = {} & [H^0_0 \cap (D_\eta(\langle D^0_\zeta : \zeta < \eta\rangle))] \cup [H^0_1 \cap (D^0 \setminus D_\eta(\langle D^0_\zeta : \zeta < \eta\rangle))] \cup \\ & [H^1_0 \cap (D^1 \setminus D_\eta(\langle D^1_\zeta : \zeta < \eta\rangle))] \cup [H^1_1 \cap (D_\eta(\langle D^1_\zeta : \zeta < \eta\rangle))] \cup \\ & [D \setminus (D^0 \cup D^1)]\end{aligned}$$

separates A from B.

Case 4. Suppose α codes a description u with $u(1) = 4$, so that $\Gamma_\alpha = \mathrm{SU}(\underset{\sim}{\mathbf{\Sigma}}^0_\xi \bigcup_n \underset{\sim}{\mathbf{\Gamma}}_{\alpha_n})$, where the sequence $\langle \alpha_n : n \in \omega \rangle$ is a Δ^1_1 sequence of codes of descriptions. Starting from A, B in Σ^1_1, we let A_n be the union of all Y_ξ-open sets O such that for some H in $\underset{\sim}{\mathbf{\Gamma}}_{\alpha_n}$ $A \cap O \subseteq H \subseteq \breve{B}$ ∞-a.e. As before, it is not hard to see that $\langle A_n \rangle$ is a Π^1_1 increasing sequence of $\underset{\sim}{\mathbf{\Sigma}}^0_\xi$ sets, and that $A \subseteq \bigcup_n A_n$. Let $\langle C_n : n \in \omega \rangle$ be a Δ^1_1 sequence of pairwise disjoint $\underset{\sim}{\mathbf{\Sigma}}^0_\xi$ sets with $A \subseteq \bigcup_n A_n$ and $C_n \subseteq A_n$, and let $\langle H_n : n \in \omega \rangle$ be a Δ^1_1 sequence with $H_n \in \Gamma^1_{\alpha_n}$, and $A \cap C_n \subseteq H_n \subseteq \breve{B}$. This is easy to find using the induction hypothesis. Then the set $H = \mathrm{SU}(\langle C_n : n \in \omega \rangle, \langle H_n : n \in \omega \rangle)$ separates A from B and is in $\underset{\sim}{\mathbf{\Gamma}}^1_\alpha$.

Case 5. Suppose $u_\alpha(1) = 5$, so that $\underset{\sim}{\mathbf{\Gamma}}_\alpha = \mathrm{SD}_\eta(\underset{\sim}{\mathbf{\Sigma}}^0_\xi, \underset{\sim}{\mathbf{\Gamma}}_{\alpha_1, \underset{\sim}{\mathbf{\Gamma}}_{\alpha_2}})$, with $\alpha_1, \alpha_2 \in \Delta^1_1$, and $\eta, \xi < \omega_1^{\mathrm{ck}}$. We again define inductively a sequence $\langle A_\zeta : \zeta < \eta \rangle$: A_ζ is the union of all T-open sets O such that for some set H in $\underset{\sim}{\mathbf{\Gamma}}_{\alpha_1}$ with envelope O, $(A \setminus \bigcup_{\zeta' < \zeta} A_{\zeta'}) \cap O \subseteq H \subseteq \breve{B}$ ∞-a.e. Again it can be seen that the sequence $\langle A_\zeta : \zeta < \eta \rangle$ is Π^1_1 and increasing, and moreover that $A \setminus \bigcup_{\zeta < \eta} A_\zeta \subseteq H \subseteq B$ ∞-a.e. for some H in $\underset{\sim}{\mathbf{\Gamma}}_{\alpha_2}$. The rest of the proof is analogous to case 2, and we leave the details to the reader. ⊣

REMARKS.

1. All the preceding results are of effective type. But as usual, they can be translated into non-effective, uniform results concerning analytic and Borel sets in the plane, using Δ^1_1-selection.
2. The results in Section 2 do not use Borel determinacy, but without it we are unable to show that Theorem 2.7 covers all non self-dual Wadge classes of Borel sets.

In a recent paper, Thomas John [Joh86] has proved Wadge Determinacy for $\underset{\sim}{\mathbf{\Pi}}^0_4$ (*i.e.*, that all Wadge games $G_{\mathrm{W}}(A, B)$, with A, B in $\underset{\sim}{\mathbf{\Pi}}^0_4$ are determined), and the stronger statement that every $\underset{\sim}{\mathbf{\Pi}}^0_n \setminus \underset{\sim}{\mathbf{\Delta}}^0_n$ set is $\underset{\sim}{\mathbf{\Pi}}^0_n$-complete, for n in ω, in second order arithmetics. His proof uses as main tool the characterization $\underset{\sim}{\mathbf{\Pi}}^0_n \cap \Delta^1_1 = (\Pi^0_n)^1$. This fact, together with Theorem 2.7, is a bit of evidence in support of the conjecture that unlike Borel determinacy, Wadge Determinacy for $\underset{\sim}{\mathbf{\Delta}}^1_1$ could be proved in second order arithmetic.

REFERENCES

THOMAS JOHN

[Joh86] *Recursion in Kolmogorov's R-operator and the ordinal* σ_3, ***The Journal of Symbolic Logic***, vol. 51 (1986), no. 1, pp. 1–11.

ALEXANDER S. KECHRIS AND YIANNIS N. MOSCHOVAKIS

[CABAL i] *Cabal seminar* 76–77, Lecture Notes in Mathematics, no. 689, Berlin, Springer, 1978.

KAZIMIERZ KURATOWSKI
[Kur66] ***Topology***, vol. 1, Academic Press, New York and London, 1966.

ALAIN LOUVEAU
[Lou80] *A separation theorem for* Σ^1_1 *sets*, ***Transactions of the American Mathematical Society***, vol. 260 (1980), no. 2, pp. 363–378.

DONALD A. MARTIN
[Mar73] *The Wadge degrees are wellordered*, unpublished, 1973.
[Mar75] *Borel determinacy*, ***Annals of Mathematics***, vol. 102 (1975), no. 2, pp. 363–371.

JOHN R. STEEL
[Ste81B] *Determinateness and the separation property*, ***The Journal of Symbolic Logic***, vol. 46 (1981), no. 1, pp. 41–44.

ROBERT VAN WESEP
[Van78B] *Wadge degrees and descriptive set theory*, this volume, originally published in Kechris and Moschovakis [CABAL i], pp. 151–170.

WILLIAM W. WADGE
[Wad84] *Reducibility and determinateness on the Baire space*, ***Ph.D. thesis***, University of California, Berkeley, 1984.

EQUIPE D'ANALYSE FONCTIONNELLE
INSTITUT DE MATHÉMATIQUES DE JUSSIEU
UNIVERSITÉ PARIS VI
4, PLACE JUSSIEU
75230 PARIS, CEDEX 05
FRANCE
E-mail: louveau@math.jussieu.fr

THE STRENGTH OF BOREL WADGE DETERMINACY

ALAIN LOUVEAU AND JEAN SAINT-RAYMOND

One of the nice consequences of Martin's theorem that Borel games are determined is the so-called Borel Wadge Determinacy, the determinacy of all games $G(A, B)$ of the following kind: player I produces $\alpha \in {}^{\omega}\omega$, player II produces $\beta \in {}^{\omega}\omega$ and player II wins $G(A, B)$ if $\alpha \in A \iff \beta \in B$, whenever A and B are Borel subsets of ${}^{\omega}\omega$. Borel Wadge Determinacy allows to get a complete description of all classes of Borel sets, *i.e.*, of all families $\Gamma \subseteq \underset{\sim}{\mathbf{\Delta}}^1_1$ which are continuously closed.

Our main result is proved in Section 3. It is a sequel to two earlier papers: First, Louveau's paper [Lou83] which analyzes the Borel Wadge classes in a way which does not depend too heavily on Borel Wadge determinacy. We quickly review in the first section the material from [Lou83] that we need. The second source is our joint paper [LSR87], where we prove the particular instances of Borel Wadge Determinacy which correspond to the Baire classes ($\underset{\sim}{\mathbf{\Sigma}}^0_\xi$ and $\underset{\sim}{\mathbf{\Pi}}^0_\xi$) and introduce the main device for the general proof, a specific way of associating to a closed game another closed game, that we called the **ramification method**. We present the material from [LSR87] that we need in Section 2. Both papers are rather long and technical, so the information we provide in Sections 1 and 2 is a bit sketchy. In particular, the existence of ramifications will be used as a black box here, and we will also leave to the reader the verification that the results from Louveau [Lou83] we use do not depend of Borel Wadge Determinacy.

The main consequence of Borel Wadge determinacy is Wadge's lemma which asserts that any Borel set in ${}^{\omega}\omega$ which is not in a class Γ always generates by continuous preimages the dual class $\check{\Gamma}$ (of complements of sets in Γ). In Section 4, we show how Wadge's lemma extends to arbitrary Polish (even Suslin) spaces in place of ${}^{\omega}\omega$, even if no game is available in this general context. This part is much more topological in nature, and uses some transfer methods and selection results for continuous functions which might be of interest in other contexts. Finally, in Section 5, we develop a notion of Hurewicz test for a class Γ, in order to extend to all Borel Wadge classes the well known theorem of Hurewicz which characterizes among Borel sets those which are

Wadge Degrees and Projective Ordinals: The Cabal Seminar, Volume II
Edited by A. S. Kechris, B. Löwe, J. R. Steel
Lecture Notes in Logic, 37

not Polish as those which contain a relatively closed set homeomorphic to $\mathbb{Q}$. Most results in Sections 4 and 5 are again sequels of our paper [LSR87].

§1. Descriptions of Borel Wadge Classes. A family Γ of subsets of ${}^{\omega}\omega$ is a **class** if it is closed under continuous preimages, and a **Wadge class** if it is generated by one set $A \subseteq {}^{\omega}\omega$. If moreover A is Borel, it is a Borel Wadge class.

Using as main tool Borel Wadge Determinacy, Wadge analyzed in his thesis [Wad84], all Borel Wadge classes. Relying heavily on Wadge's work, Louveau proposed in [Lou83] an inductive construction of all Borel Wadge classes in terms of certain Boolean operations. Fortunately, these works do not depend too much on Borel Wadge determinacy: Although it is unclear (at this point) that the analysis in [Lou83] exhausts all Borel Wadge classes, one can still show directly that it almost does, in a precise sense (given by Theorem 1.4 below). And part of our proof of Borel Wadge Determinacy will in fact consist in showing that the analysis is exhaustive.

Let us first introduce some notations and definitions.

Definition 1.1. Let Γ be a class of subsets of ${}^{\omega}\omega$

(i) If $A \subseteq {}^{\omega}\omega$, we let $\check{A} = {}^{\omega}\omega - A$, and we let $\check{\Gamma} = \{\check{A} : A \in \Gamma\}$ be the **dual class** of Γ. We also set $\Delta(\Gamma) = \Gamma \cap \check{\Gamma}$, the **ambiguous class** of Γ. We say that Γ is **self-dual** if $\Gamma = \check{\Gamma}(= \Delta(\Gamma))$.

We also define the ordering $<$ between classes by

$$\Gamma < \Gamma' \iff \Gamma \subseteq \Delta(\Gamma').$$

(ii) $\boldsymbol{PU}(\Gamma)$ is the class of all A's $\subseteq {}^{\omega}\omega$ of form $A = \bigcup_n (A_n \cap C_n)$, where $A_n \in \Gamma$ and $(C_n)_{n\in\omega}$ is a partition of ${}^{\omega}\omega$ in clopen sets ($\boldsymbol{PU}$ stands for "partitioned union"). We will use this operation mainly in two cases: If Γ is a non self-dual class, we set $\Gamma^+ = \boldsymbol{PU}(\Gamma \cup \check{\Gamma})$. And if $\langle \Gamma_n : n \in \omega \rangle$ is a $<$-increasing sequence of classes, $\langle \Gamma_n \rangle^+ = \boldsymbol{PU}(\bigcup_n \Gamma_n)$.

Definition 1.2. Let Γ, Γ' be classes, ξ, η ordinals ≥ 1.

(a) $A \in \boldsymbol{D}_\eta(\boldsymbol{\Sigma}^0_\xi) \iff \bigcup \{A_\vartheta - \bigcup_{\vartheta' < \vartheta} A_{\vartheta'} : \vartheta < \eta, \vartheta$ of a different parity than $\eta\}$ for some increasing sequence $\langle A_\vartheta : \vartheta < \eta \rangle$ of $\boldsymbol{\Sigma}^0_\xi$ sets in ${}^{\omega}\omega$.

(b) $A \in \boldsymbol{Sep}(\boldsymbol{D}_\eta(\boldsymbol{\Sigma}^0_\xi), \Gamma) \iff A = (A_0 \cap C) \cup (A_1 \setminus C)$ for some $C \in \boldsymbol{D}_\eta(\boldsymbol{\Sigma}^0_\xi)$, $A_0 \in \check{\Gamma}, A_1 \in \Gamma$.

(c) $A \in \boldsymbol{Bisep}(\boldsymbol{D}_\eta(\boldsymbol{\Sigma}^0_\xi), \Gamma, \Gamma') \iff A = (A_0 \cap C_0) \cup (A_1 \cap C_1) \cup (B \setminus (C_0 \cup C_1))$ for some disjoint C_0, C_1 in $\boldsymbol{D}_\eta(\boldsymbol{\Sigma}^0_\xi)$, $A_0 \in \Gamma, A_1 \in \check{\Gamma}$ and $B \in \Gamma'$.

(d) $A \in \boldsymbol{SU}(\boldsymbol{\Sigma}^0_\xi, \Gamma)$ with envelope $C \iff A = \bigcup_n (A_n \cap C_n)$ for some sequence of pairwise disjoint $\boldsymbol{\Sigma}^0_\xi$ sets C_n, with $\bigcup_n C_n = C$, and $A_n \in \Gamma$.

(e) $A \in \boldsymbol{SD}_\eta(\boldsymbol{SU}(\boldsymbol{\Sigma}^0_\xi, \Gamma)\Gamma') \iff A = \bigcup_{\vartheta<\eta} (A_\vartheta \setminus \bigcup_{\vartheta'<\vartheta} C_{\vartheta'}) \cup (B \setminus \bigcup_{\vartheta<\eta} C_\vartheta)$ for some increasing sequence $\langle A_\vartheta : \vartheta < \eta \rangle$ of sets in $\boldsymbol{SU}(\boldsymbol{\Sigma}^0_\xi, \Gamma)$ with respective envelopes C_ϑ, such that $C_{\vartheta'} \subseteq A_\vartheta$ for $\vartheta' < \vartheta$, and some $B \in \Gamma'$.

In [Lou83], Louveau selects particular ways of combining the operations of Definition 1.2, encoded by what he calls "descriptions". In order to simplify later work, we will also use another notion of description. So we refer to the descriptions of [Lou83] as first type descriptions. The encoding is made by elements u in ${}^\omega\omega_1$. Such sequences are sometimes viewed as pairs $\langle u_0, u_1 \rangle$ or as sequences $\langle u_n : n \in \omega \rangle$, via some fixed bijections between ω and $\omega.2$, ${}^2\omega$, respectively. We let $\underline{0}$ be the constant function zero.

DEFINITION 1.3 ([Lou83, 1.2]). The relations "u is a first type description" and "u describes Γ" (written $u \in \boldsymbol{D}_1$ and $\Gamma_u = \Gamma$) are the least relations satisfying:

(a) If $u(0) = 0, u \in \boldsymbol{D}_1$ and $\Gamma_u = \{\varnothing\}$.
(b) If $u(0) = \xi \geq 1, u(1) = 1$ and $u(2) = \eta \geq 1, u \in \boldsymbol{D}_1$ and $\Gamma_u = \boldsymbol{D}_\eta(\boldsymbol{\Sigma}^0_\xi)$.
(c) If $u = \xi^\frown 2^\frown \eta^\frown u^*$, with $\xi \geq 1, \eta \geq 1, u^* \in \boldsymbol{D}_1$ and $u^*(0) > \xi$, then $u \in \boldsymbol{D}_1$ and $\Gamma_u = \boldsymbol{Sep}(\boldsymbol{D}_\eta(\boldsymbol{\Sigma}^0_\xi), \Gamma_{u^*})$.
(d) If $u = \xi^\frown 3^\frown \eta^\frown \langle u_0, u_1 \rangle$, with $\xi \geq 1, \eta \geq 1, u_0$ and u_1 in $\boldsymbol{D}_1$, $u_0(0) > \xi, u_1(0) \geq \xi$ or $u_1(0) = 0$ and $\Gamma_{u_1} < \Gamma_{u_0}$, then $u \in \boldsymbol{D}_1$ and $\Gamma_u = \boldsymbol{Bisep}(\boldsymbol{D}_\eta(\boldsymbol{\Sigma}^0_\xi), \Gamma_{u_0}, \Gamma_{u_1})$.
(e) If $\xi^\frown 4^\frown \langle u_n : n \in \omega \rangle$, where $\xi \geq 1, u_n \in \boldsymbol{D}_1$ for all $n, \Gamma_{u_n} < \Gamma_{u_{n+1}}$, and either for all n, $u_n(0) = \xi' > \xi$ or $\langle u_n(0) \rangle_{n \in \omega}$ is strictly increasing with $\sup \xi_n > \xi$, then $u \in \boldsymbol{D}_1$ and $\Gamma_u = \boldsymbol{SU}(\boldsymbol{\Sigma}^0_\xi, \bigcup_n \Gamma_{u_n})$.
(f) If $u = \xi^\frown 5^\frown \eta^\frown \langle u_0, u_1 \rangle$ with $\xi \geq 1, \eta \geq 1, u_0, u_1$ in $\boldsymbol{D}_1$ with $u_0(0) = \xi, u_0(1) = 4, u_1(0) \geq \xi$ or $u_1(0) = 0$, and $\Gamma_{u_1} < \Gamma_{u_0}$, then $u \in \boldsymbol{D}_1$ and $\Gamma_u = \boldsymbol{SD}_\eta(\Gamma_{u_0}, \Gamma_{u_1})$. [Note: As noted by Van Engelen, there is a slight mistake in the original definition [Lou83, 1.2, case e.]]

One easily checks that each $u \in \boldsymbol{D}_1$ codes exactly one class Γ_u. And each Γ_u is a non-self dual Borel Wadge class, as can be seen by (inductively) constructing a universal Γ_u set in ${}^\omega\omega \times {}^\omega\omega$.

The main result of [Lou83] that we will use, which does not need Borel Wadge determinacy, can be summarized as follows (it corresponds to [Lou83], Lemmas 1.11, 1.14, 1.19, 1.23, 1.24, 1.25 and 1.28).

THEOREM 1.4. Let $u \in \boldsymbol{D}_1$. The class $\Delta(\Gamma_u)$ satisfies one of the following three possibilities:

(i) There is a description $u^* \in \boldsymbol{D}_1$ with $\Delta(\Gamma_u) = (\Gamma_{u^*})^+$.
(ii) There is a sequence $\langle u_n \rangle$ in $\boldsymbol{D}_1$ with $\Delta(\Gamma_u) = (\Gamma_{u_n})^+$.
(iii) There is a family $(u_\xi)_{\xi < \omega_1}$ in $\boldsymbol{D}_1$ with $\Delta(\Gamma_u) = \bigcup_\xi \Gamma_{u_\xi}$.

In [Lou83], Theorem 1.4 is just an intermediate step towards proving that any Borel Wadge class is of form $\Gamma_u, \check{\Gamma}_u, \Gamma_u^+$ or $\langle \Gamma_{u_n} \rangle^+$ for some u, u_n in $\boldsymbol{D}_1$, as the case may be. However, this is derived from Theorem 1.4 by using Borel Wadge determinacy. As the proof is instructive, let us sketch it briefly: First, Borel Wadge Determinacy is used to show that if $A \in \Gamma_u$ but $A \notin \check{\Gamma}_u$, A generates Γ_u—and $\check{A}$ generates $\check{\Gamma}_u$. Similarly, if $A \in \Gamma_u^+$, but $A \notin \Gamma_u \cup \check{\Gamma}_u$, A generates Γ_u^+, and if $A \in \langle \Gamma_{u_n} \rangle^+$ but $A \notin \bigcup_n \Gamma_{u_n}$, A generates $\langle \Gamma_{u_n} \rangle^+$. Finally, Borel Wadge Determinacy is used again to prove that $\{\Gamma \cup \check{\Gamma} : \Gamma$ a Borel Wadge class$\}$ is well ordered by inclusion (Wadge [Wad84], Martin [Mar73]). One can argue then that if $A \subseteq {}^\omega\omega$ is Borel, there is a least (for inclusion) class Γ_u with $A \in \Gamma_u \cup \check{\Gamma}_u$. If $A \notin \Delta(\Gamma_u)$, the Wadge class of A is Γ_u or $\check{\Gamma}_u$. If $A \in \Delta(\Gamma_u)$, Theorem 1.4 applies. Case (iii) is impossible by minimality of u, and Cases (i) and (ii) are solved by the facts above. So in all cases the class of A is described.

In the next sections, we imitate the proof above, except that we will at the same time prove instances of the facts above and the corresponding instances of determinacy; so at the end we will get both that the analysis is exhaustive and that Borel Wadge Determinacy holds.

We now introduce the second type descriptions.

Definition 1.5. Let $\xi \geq 1$ be a countable ordinal, Γ, Γ' two classes. Then

$$A \in S_\xi(\Gamma, \Gamma') \iff A = \bigcup_n (A_n \cap C_n) \cup \left(B \setminus \bigcup_n C_n\right)$$

for some sequence A_n in Γ, $B \in \Gamma'$, and a sequence $\langle C_n \rangle$ of pairwise disjoint $\utilde{\mathbf{\Sigma}}^0_\xi$ sets.

Second type descriptions are also elements of ${}^\omega\omega_1$.

Definition 1.6. The relations "u is a second type description" and "u describes Γ" (written $u \in \boldsymbol{D}_2$ and $\Gamma_u = \Gamma$—ambiguously) are the least relations satisfying

(a) if $u = \underline{0}$, $u \in \boldsymbol{D}_2$ and $\Gamma_u = \varnothing$
(b) if $u = \xi^\frown 1^\frown u^*$, with $u^* \in \boldsymbol{D}_2$ and $u^*(0) = \xi$, then $u \in \boldsymbol{D}_2$ and $\Gamma_u = \check{\Gamma}_{u^*}$
(c) if $u = \xi^\frown 2^\frown \langle u_n \rangle$ with $\xi \geq 1$, $u_n \in \boldsymbol{D}_2$, $u_n(0) \geq \xi$ or $u_n(0) = 0$, then $u \in \boldsymbol{D}_2$ and $\Gamma_u = S_\xi(\bigcup_{n \geq 1} \Gamma_{u_n}, \Gamma_{u_0})$.

The $\boldsymbol{D}_2$-encoding is clearly much simpler than the first one. However, Theorem 1.4 would be hard to get using this encoding. Our next step is to show that any class admitting a $\boldsymbol{D}_1$-description also admits a $\boldsymbol{D}_2$-description.

Proposition 1.7 ([Lou83, Lemma 1.4]). Let $u \in \boldsymbol{D}_1$, with $u(0) = \xi \geq 1$. Then

(a) $\boldsymbol{SU}(\utilde{\mathbf{\Sigma}}^0_\xi, \Gamma_u) = \Gamma_u$
(b) Γ_u is closed under union with a $\utilde{\mathbf{\Delta}}^0_\xi$ set.

So in particular both Γ_u and $\check{\Gamma}_u$ are closed under unions or intersections with $\underset{\sim}{\mathbf{\Delta}}^0_\xi$ sets, and if $A = \bigcup_n (A_n \cap C_n)$ where $A_n \in \Gamma_u$ (resp. $\check{\Gamma}_u$) and (C_n) is a partition of ${}^\omega\omega$ in $\underset{\sim}{\mathbf{\Delta}}^0_\xi$ sets, then $A \in \Gamma_u$ (resp. $\check{\Gamma}_u$).

THEOREM 1.8. Every class admitting a $\boldsymbol{D}_1$-description, and every dual of such a class admit a $\boldsymbol{D}_2$-description.

PROOF. If Γ admits $u \in \boldsymbol{D}_2$ as description, $\check{\Gamma}$ admits $u(0)^\frown 1^\frown u$ as $\boldsymbol{D}_2$-description. So we prove by induction on $u \in \boldsymbol{D}_1$, that Γ_u admits a $\boldsymbol{D}_2$-description $v(u)$, with $v(u)(0) = u(0)$.

(a) Clearly if $u(0) = 0$, we can take $v(u) = \underline{0}$.

(b) Let $u(1) = 1$, *i.e.*, $\Gamma_u = \boldsymbol{D}_\eta(\underset{\sim}{\mathbf{\Sigma}}^0_\xi)$. We use induction on η. For $\eta = 1$, one has $\underset{\sim}{\mathbf{\Sigma}}^0_\xi = S_\xi(\{{}^\omega\omega\}, \{\varnothing\})$ so we can take $v(u) = \xi^\frown 2^\frown \langle v_n \rangle$ with $v_0 = 0$ and for $n \geq 1, v_n = 0^\frown 1^\frown \underline{0}$. If $\eta = \eta' + 1$, one uses similarly the equality

$$\boldsymbol{D}_\eta(\underset{\sim}{\mathbf{\Sigma}}^0_\xi) = S_\xi((\boldsymbol{D}_{\eta'}(\underset{\sim}{\mathbf{\Sigma}}^0_\xi))\check{}, \underset{\sim}{\mathbf{\Sigma}}^0_\xi)$$

and if $\lambda = \sup_n(\eta_n + 1)$ is limit

$$\boldsymbol{D}_\lambda(\underset{\sim}{\mathbf{\Sigma}}^0_\xi) = S_\xi(\bigcup_n \boldsymbol{D}_{\eta_n}(\underset{\sim}{\mathbf{\Sigma}}^0_\xi), \{\varnothing\})$$

All these equalities are easy to check.

(c) Suppose now $u = \xi^\frown 2^\frown \eta^\frown u^*$, so that $\Gamma_u = \boldsymbol{Sep}(\boldsymbol{D}_\eta(\underset{\sim}{\mathbf{\Sigma}}^0_\xi), \Gamma_{u^*})$. We again argue by induction on η, and use the induction hypothesis on Γ_{u^*} and the following equalities

$$\boldsymbol{Sep}(\underset{\sim}{\mathbf{\Sigma}}^0_\xi, \Gamma_{u^*}) = S_\xi(\check{\Gamma}_{u^*}, \Gamma_{u^*})$$

$$\boldsymbol{Sep}(\boldsymbol{D}_{\eta+1}(\underset{\sim}{\mathbf{\Sigma}}^0_\xi), \Gamma_{u^*}) = S_\xi(\boldsymbol{Sep}((\boldsymbol{D}_\eta(\underset{\sim}{\mathbf{\Sigma}}^0_\xi))\check{}, \Gamma_{u^*}), \boldsymbol{Sep}(\underset{\sim}{\mathbf{\Sigma}}^0_\xi, \Gamma_{u^*}))$$

and for limit λ

$$\boldsymbol{Sep}(\boldsymbol{D}_\lambda(\underset{\sim}{\mathbf{\Sigma}}^0_\xi), \Gamma_{u^*}) = S_\xi(\bigcup_{\eta<\lambda} \boldsymbol{Sep}(\boldsymbol{D}_\eta(\underset{\sim}{\mathbf{\Sigma}}^0_\xi), \Gamma_{u^*}), \{\varnothing\})$$

the proof of which is left to the reader.

(d) $u = \xi^\frown 3^\frown \eta^\frown \langle u_0, u_1 \rangle$, so $\Gamma_u = \boldsymbol{Bisep}(\boldsymbol{D}_\eta(\underset{\sim}{\mathbf{\Sigma}}^0_\xi), \Gamma_{u_0}, \Gamma_{u_1})$. Again by induction on η, one uses the equalities

$$\boldsymbol{Bisep}(\underset{\sim}{\mathbf{\Sigma}}^0_\xi, \Gamma_{u_0}, \Gamma_{u_1}) = S_\xi(\Gamma_{u_0} \cup \check{\Gamma}_{u_0}, \Gamma_{u_1})$$

$$\boldsymbol{Bisep}(\boldsymbol{D}_{\eta+1}(\underset{\sim}{\mathbf{\Sigma}}^0_\xi), \Gamma_{u_0}, \Gamma_{u_1}) = S_\xi(\Gamma \cup \check{\Gamma}, \Gamma_{u_1})$$

where $\Gamma = \boldsymbol{Sep}(\boldsymbol{D}_\eta(\underset{\sim}{\mathbf{\Sigma}}^0_\xi), \Gamma_{u_0})$ and

$$\boldsymbol{Bisep}(\boldsymbol{D}_\lambda(\underset{\sim}{\mathbf{\Sigma}}^0_\xi), \Gamma_{u_0}, \Gamma_{u_1}) = S_\xi(\bigcup_{\eta<\lambda} \Gamma_\eta, \Gamma_{u_1})$$

where $\Gamma_\eta = \boldsymbol{Sep}(\boldsymbol{D}_\eta(\underset{\sim}{\mathbf{\Sigma}}^0_\xi), \Gamma_{u_0})$.

The equalities do not follow immediately from Proposition 1.7, so let us sketch one of them, say the successor case (the others are similar, and a bit simpler). Denote by Γ_ℓ and Γ_r the left and right hand classes.

If $A \in \Gamma_r$, $A = \bigcup_n (A_n \cap C_n) \cup (B \setminus \bigcup_n C_n)$ with pairwise disjoint C_n's in $\boldsymbol{\Sigma}^0_\xi$, $B \in \Gamma_{u_1}$ and A_n in $\Gamma = \boldsymbol{Sep}(\boldsymbol{D}_\eta(\boldsymbol{\Sigma}^0_\xi), \Gamma_{u_0})$ or in $\check{\Gamma}$. Write $A_n = (A^0_n \cap D_n) \cup (A^1_n \setminus D_n)$ with $D_n \in \boldsymbol{D}_\eta(\boldsymbol{\Sigma}^0_\xi)$, $A^{\varepsilon(n)}_n \in \Gamma_{u_0}$ and $A^{1-\varepsilon(n)}_n \in \check{\Gamma}_{u_0}$, where $\varepsilon(n) = 0$ or 1 depending if A_n is in Γ or $\check{\Gamma}$. Let $D^0_n = D_n \cap C_n$ and $D^1_n = C_n \setminus D_n$. Both D^0_n, D^1_n are in $\boldsymbol{D}_{\eta+1}(\boldsymbol{\Sigma}^0_\xi)$, and so are $D^0 = \bigcup_n (C_n \cap D^{\varepsilon(n)}_n)$ and $D^1 = \bigcup_n (C_n \cap D^{1-\varepsilon(n)}_n)$ by an immediate computation. By Proposition 1.7 (as $u_0(0) > \xi$) $A^0 = \bigcup_n (C_n \cap A^{\varepsilon(n)}_n) \in \Gamma_{u_0}$ and $A^1 = \bigcup_n (C_n \cap A^{1-\varepsilon(n)}_n) \in \check{\Gamma}_{u_0}$. And $A = (A^0 \cap D^0) \cup (A^1 \cap D^1) \cup (B \setminus (D^0 \cup D^1))$, so that $A \in \Gamma_\ell$.

The other inclusion is a bit harder: let $A \in \Gamma_\ell$, *i.e.*, $A = (A_0 \cap C_0) \cup (A_1 \cap C_1) \cup (B \setminus (C_0 \cup C_1))$ with C_0, C_1 disjoint in $\boldsymbol{D}_{\eta+1}(\boldsymbol{\Sigma}^0_\xi)$, $A_0 \in \Gamma_{u_0}$, $A_1 \in \check{\Gamma}_{u_0}$ and $B \in \Gamma_{u_1}$. Let D_0, resp D_1, be the largest $\boldsymbol{\Sigma}^0_\xi$ sets in some constructions of C_0, resp C_1, as $\boldsymbol{D}_{\eta+1}(\boldsymbol{\Sigma}^0_\xi)$ sets, and let D^*_0, D^*_1 reduce D_0, D_1. Clearly $A \setminus (D^*_0 \cup D^*_1) = B \setminus (D^*_0 \cup D^*_1)$, so in order to show that $A \in \Gamma_r$, it is enough to prove that $A \cap D^*_0 \in \Gamma = \boldsymbol{Sep}(\boldsymbol{D}_\eta(\boldsymbol{\Sigma}^0_\xi), \Gamma_{u_0})$ and similarly $A \cap D^*_1 \in \check{\Gamma}$. Let us prove the first claim, the second one being similar. By the choice of D_0, $D_0 \setminus C_0 \in \boldsymbol{D}_\eta(\boldsymbol{\Sigma}^0_\xi)$, hence $D^*_0 \setminus C_0$ too, as $D^*_0 \subseteq D_0$. Now, $A \cap D^*_0 \cap C_0 = A_0 \cap D^*_0 \cap C_0$ is in Γ_{u_0}, and $A \cap (D^*_0 \setminus C_0) = A_1 \cap (D^*_0 \cap C_1) \cup (B \setminus (C_0 \cup C_1)) \cap D^*_0$. Both $A_1 \cap D^*_0$ and $B \setminus (C_0 \cup C_1)$ are in $\check{\Gamma}_{u_0}$, and separated by the $\boldsymbol{\Delta}^0_{\xi+1}$ set $C_0 \cup C_1$, hence by Proposition 1.7 $A \cap (D^*_0 \setminus C_0)$ is in $\check{\Gamma}_{u_0}$ and the equality is proved.

(e) One uses in case $u(1) = 4$ the equality

$$\boldsymbol{SU}(\boldsymbol{\Sigma}^0_\xi, \bigcup_n \Gamma_{u_n}) = S_\xi(\bigcup_n \Gamma_{u_n}, \{\varnothing\})$$

(f) The final case is when $\Gamma_u = \boldsymbol{SD}_\eta(\boldsymbol{SU}(\boldsymbol{\Sigma}^0_\xi, \bigcup_n \Gamma_{u_n}), \Gamma_{u^*})$ and is proved as in the $\boldsymbol{D}_\eta(\boldsymbol{\Sigma}^0_\xi)$ case, by using the following (easy) equalities, where $\Gamma = \boldsymbol{SU}(\boldsymbol{\Sigma}^0_\xi, \bigcup_n \Gamma_{u_n})$

$$\boldsymbol{SD}_{\eta+1}(\Gamma, \Gamma_{u^*}) = S_\xi\big(\boldsymbol{SD}_\eta(\Gamma, \{\varnothing\}), S_\xi(\bigcup_n \Gamma_{u_n}, \Gamma_{u^*})\big)$$

and for limit λ

$$\boldsymbol{SD}_\lambda(\Gamma, \Gamma_{u^*}) = S_\xi\big(\bigcup_{\eta<\lambda} \boldsymbol{SD}_\eta(\Gamma, \{\varnothing\}), \Gamma_{u^*}\big)$$

⊣

REMARK 1.9. There is a slight defect in the proof above: if one really wants to build a $v(u)$ for $u \in \boldsymbol{D}_1$, one needs at limit steps specific fundamental sequences below limit ordinals. This requires a form of the axiom of choice. The best way to avoid this—which would in any case be necessary for a formalization of the preceding discussion in second order arithmetics—is to replace ordinals by reals coding them, and accordingly descriptions by codes of descriptions in ${}^{\omega}\omega$. The function $v(u)$ becomes then definable in the codes. However, since working with codes would only create more notational problems to the reader, we will continue this slight kind of abuse.

Let us denote by

$$W_1 = \{\Gamma_u : u \in \boldsymbol{D}_1\} \cup \{\check{\Gamma}_u : u \in \boldsymbol{D}_1\}$$
$$W_2 = \{\Gamma_u : u \in \boldsymbol{D}_2\}, \text{ and}$$

by W the set of non self-dual Borel Wadge classes in ${}^{\omega}\omega$. The preceding theorem says that $W_1 \subseteq W_2$. It is clear that $W_2 \subseteq W$, as can be proved by inductively constructing a universal set for each $\Gamma \in W_2$ in ${}^{\omega}\omega \times {}^{\omega}\omega$. We will prove later that these inclusions are equalities.

We finish this section with the study of the effect of functions of Baire class η on classes in W_2. For ξ, η countable ordinals, with $\eta \leq \xi$, we denote by $\xi - \eta$ the unique ξ' such that $\eta + \xi' = \xi$. A function $f : {}^{\omega}\omega \to {}^{\omega}\omega$ is a Baire class η function if for each open $A \subseteq {}^{\omega}\omega$, $f^{-1}(A) \in \boldsymbol{\Sigma}^0_{1+\eta}$ (so continuous functions are Baire class 0). One easily checks by induction that if f is of Baire class η and A is $\boldsymbol{\Sigma}^0_{\xi}$, $f^{-1}(A)$ is in $\boldsymbol{\Sigma}^0_{\eta+'\xi}$, with $\eta +' \xi = 1 + \eta + (\xi - 1)$. Let also $\xi -' \eta$, for $\xi \geq 1$ and $\eta < \xi$, be defined by $\xi -' \eta = 1 + (\xi - (1 + \eta))$, so that $\eta +' (\xi -' \eta) = \xi$.

Definition 1.10. We define for each countable η and each $u \in \boldsymbol{D}_2$ a description $u^{\eta} \in \boldsymbol{D}_2$, and in case $u(0) > \eta$ or $u(0) = 0$, a description ${}^{\eta}u \in \boldsymbol{D}_2$, by the following clauses:

(a) If $u(0) = 0$, $u^{\eta} = {}^{\eta}u = u$
(b) If $u = \xi^\frown 1^\frown u^*$, with $\xi \geq 1$

$$u^{\eta} = (\eta +' \xi)^\frown 1 \cap (u^*)^{\eta}$$

and ${}^{\eta}u = (\xi -' \eta)^\frown 1^\frown {}^{\eta}(u^*)$ (for $u(0)$—hence $u^*(0)$—bigger than η)
(c) If $u = \xi^\frown 2^\frown \langle u_n \rangle$, with $\xi \geq 1$,

$$u^{\eta} = (\eta +' \xi)^\frown 2^\frown \langle u_n^{\eta} \rangle$$

and ${}^{\eta}u = (\xi -' \eta)^\frown 2^\frown \langle {}^{\eta}u_n \rangle$ (for $\xi > \eta$—note that ${}^{\eta}u_n$ is defined from some n_0 on, as $\sup u_n(0) > \xi$).

It is clear from the previous definition that $({}^{\eta}u)^{\eta} = u$, when $u(0) = 0$ or $u(0) > \eta$. And one easily gets by induction the following.

Proposition 1.11.

(i) If $f : {}^{\omega}\omega \to {}^{\omega}\omega$ is a Baire class η function, and $A \in \Gamma_u$ for some $u \in \boldsymbol{D}_2$, then $f^{-1}(A) \in \Gamma_{u^{\eta}}$.

(ii) If $u \in \boldsymbol{D}_2$ is such that $u(0) = \xi \geq 1$, there are unique $\underline{u}$ with $\underline{u}(0) = 1$, and η $(= \xi - 1)$ such that $u = (\underline{u})^\eta$.

In particular, $\boldsymbol{D}_2$ is the least subset $D \subseteq \boldsymbol{D}_2$ such that $\underline{0} \in D$, $u(0)^\frown 1^\frown u \in D$ if $u \in D$, $1^\frown 2^\frown \langle u_n \rangle \in D$ if for each n, $u_n \in D$, and for any η, $u^\eta \in D$ when $u \in D$.

Let us finally say a few words about relativization: If E is a subset of ${}^\omega\omega$, one can define for any class Γ the relativization $\Gamma(E)$ by using traces on E of sets in Γ. And clearly for u a description of any type, $\Gamma_u(E)$ is the same as the class described by u, starting from $\underset{\sim}{\boldsymbol{\Sigma}}^0_\xi$ subsets of E. And if $f : {}^\omega\omega \to E$ is continuous and $A \in \Gamma_u(E)$, $f^{-1}(A) \in \Gamma_u$. We will use these remarks in the sequel mainly for $E = {}^\omega 2$, or $E = {}^\omega 2 \times {}^\omega\omega$, viewed as a subset of ${}^\omega\omega$.

§2. **Ramifications of Closed Games.** In our first paper [LSR87] on the topic of Borel Wadge determinacy, we proved particular instances of it, namely that if Γ is one of the classes $\boldsymbol{D}_\eta(\underset{\sim}{\boldsymbol{\Sigma}}^0_\xi)$ and $A \subseteq {}^\omega 2$ is a set in $\Gamma \setminus \check{\Gamma}$, the Wadge game $G(A, B)$ is determined, for any Borel B in ${}^\omega 2$. The main technical tool we introduced to get this result is a specific way of transforming closed games that we called ramifications. We now discuss what will be needed in the sequel about this notion.

Ramifications act on the following kind of games: player I plays $\varepsilon \in {}^\omega 2$, player II plays $\beta \in {}^\omega\omega$, and the game is closed for player II, *i.e.*, specified by a tree J on $2 \times \omega$—that we will confuse with the game itself. A position $(\varepsilon \restriction k, \beta \restriction k)$ is legal in J if $(\varepsilon \restriction k, \beta \restriction k) \in J$, and a run (ε, β) is a win for player II if for all k, $(\varepsilon \restriction k, \beta \restriction k)$ is legal in J, *i.e.*, if (ε, β) is a branch through J.

We denote by $\mathcal{J} \subseteq {}^{{}^{<\omega}(2\times\omega)}\omega$ the (closed) set of all trees on $2 \times \omega$. A **strategy** for player I in games in $\mathcal{J}$ is a function $\sigma\colon {}^{<\omega}\omega \to 2$, and we denote by Σ the set ${}^{{}^{<\omega}\omega}2$.

Definition 2.1. A ramification of games is a triple (r, ρ, F) of functions, with the following properties:

(a) $r = (r_0, r_1)\colon {}^{<\omega}(2 \times \omega) \to {}^{<\omega}(2 \times \omega)$ satisfies

(i) $r_0(u, v) = r_0(u)$ depends only on $u \in {}^{<\omega}2$

(ii) If $n = \mathrm{lh}(u, v)$, $t(u, v) = \{r(u \restriction k, v \restriction k) : k \leq n\}$ is a subtree of ${}^{\leq n}(2 \times \omega)$.

We let r act on $\mathcal{J}$, by defining a function $R\colon \mathcal{J} \to \mathcal{J}$ as follows:

$$(u, v) \in R(J) \iff t(u, v) \subseteq J$$

[Intuitively when playing a position (u, v) in $R(J)$, the players are imagining a tree $t(u, v)$ of positions in J, and their position in $R(J)$ is legal if all the imagined positions are legal in J.]

(b) $\rho = (\rho_0, \rho_1)\colon {}^\omega 2 \times {}^\omega\omega \to {}^\omega 2 \times {}^\omega\omega$ satisfies

(i) $\rho_0(\varepsilon, \beta) = \rho_0(\varepsilon)$ depends only on $\varepsilon \in {}^\omega 2$.

(ii) For all ε, β, $\rho(\varepsilon, \beta)$ is a branch through the tree $T(\varepsilon, \beta) = \bigcup_k t(\varepsilon \restriction k, \beta \restriction k)$.

[Intuitively again, among the positions $T(\varepsilon, \beta)$ associated to a run in some $R(J)$, exists a complete run $\rho(\varepsilon, \beta)$. So, in particular if (ε, β) is a win for player II in some $R(J)$, $\rho(\varepsilon, \beta)$ is a win for player II in J.]

(c) $F : \mathcal{J} \times \Sigma \to \Sigma$ associates to each game J and each strategy σ (viewed as a strategy for player I in the game $R(J)$) another strategy $\sigma^* = F(J, \sigma)$, which we view as a strategy for player I in J. And F satisfies: If σ is winning in $R(J)$, σ^* is winning in J.

Note: In [LSR87], we put some more restrictions on the notion of ramification, in order to be able to inductively construct a nice family of them. But we will have no needs for these refinements.

It is easy to build ramifications—*e.g.*, the identity. But what we need are ramifications for which the function $\rho_o : {}^{\omega}2 \to {}^{\omega}2$ is as complicated a Baire class η function as possible. In order to make this idea precise, let us introduce some more definitions:

Definition 2.2. Let Γ be a class. A set $H \subseteq {}^{\omega}2$ is Γ-strategically complete if

(i) $H \in \Gamma({}^{\omega}2)$

(ii) If $A \subseteq {}^{\omega}\omega$ is a Γ set, player II wins the Wadge game $G(A, H)$ [where player I plays $\alpha \in {}^{\omega}\omega$, player II $\beta \in {}^{\omega}2$ and player II wins if $\alpha \in A \iff \beta \in H$]

Definition 2.3. Let $f : {}^{\omega}2 \to {}^{\omega}2$, and η a countable ordinal. We say that f is an independent η-function if

(i) There is a $\pi : \omega \to \omega$ such that for all ε, k, the value of $f(\varepsilon)$ at k depends only on the values of ε on $\pi^{-1}(k)$

(ii) If $\eta = \eta' + 1$ is successor, $\{\varepsilon : f(\varepsilon)(k) = 1\}$ is $\mathbf{\Pi}_{1+\eta'}$-strategically complete; and if η is limit, then for some increasing sequence $\langle \eta_n \rangle$ with supremum η, $\{\varepsilon : f(\varepsilon)(k) = 1\}$ is $\mathbf{\Pi}_{1+\eta_k}$-strategically complete.

The main result of [LSR87] about ramifications [LSR87, 3.2] can be restated as:

Theorem 2.4. For each countable η, there exists a ramification $(r_\eta, \rho^\eta, F^\eta)$ which satisfies

(a) ρ^η and F^η are Baire class η functions.

(b) If $\xi < \omega_1$ and $f : {}^{\omega}2 \to {}^{\omega}2$ is an independent ξ-function, then $\rho_0^\eta \circ f : {}^{\omega}2 \to {}^{\omega}2$ is an independent $\eta + \xi$-function.

Note: As usual, one cannot pick $(r_\eta, \rho^\eta, F^\eta)$ for each η without some choice, so we should work with a family or ramifications indexed by codes of ordinals. But we will not bother about this in the sequel.

For τ an increasing map : $\omega \to \omega$, let $\tilde{\tau}: {}^{\omega}2 \to {}^{\omega}2$ be defined by $\tilde{\tau}(\varepsilon) = \varepsilon \circ \tau$. Clearly $\tilde{\tau}$ is continuous, and in fact an independent 0-function. The next lemma is easy to check.

LEMMA 2.5. If $\tilde{\tau}: {}^{\omega}2 \to {}^{\omega}2$ is as above and $f: {}^{\omega}2 \to {}^{\omega}2$ is an independent η-function, then $\tilde{\tau} \circ f$ is an independent η-function.

We let $\mathcal{R}$ be the least set of functions: ${}^{\omega}2 \to {}^{\omega}2$ which contains the functions ρ_0^{η} associated with the ramifications of Theorem 2.4, the function $\tilde{\tau}$ for τ increasing: $\omega \to \omega$, and is closed under composition. By Theorem 2.4(b), and Lemma 2.5 each $f \in \mathcal{R}$ is an independent η-function, for some η we call the order $o(f)$ of f.

Part of our goal now is to define, for each $u \in \boldsymbol{D}_2$ a set $H_u \subset {}^{\omega}2$ which is Γ_u-strategically complete. But the inductive construction needs sets with a slightly stronger property:

DEFINITION 2.6. Let $u \in \boldsymbol{D}_2$. A set $H \subseteq {}^{\omega}2$ is u-strategically complete if
(i) $H \in \Gamma_u({}^{\omega}2)$
(ii) for each $f \in \mathcal{R}$ of order $o(f) = \eta$, the set $f^{-1}(H)$ is $\Gamma_{u^{\eta}}$-strategically complete.

THEOREM 2.7. Let $u \in \boldsymbol{D}_2$. There exists a u-strategically complete set $H_u \subseteq {}^{\omega}2$, and for each pair A_0, A_1 of disjoint $\utilde{\Sigma}^1_1$ sets in ${}^{\omega}\omega$ a closed (for player II) game $J_u(A_0, A_1)$, where player I produces $\varepsilon \in {}^{\omega}2$, player II produces $\alpha \in {}^{\omega}\omega$ and $\beta \in {}^{\omega}\omega$, and a set $C_u(A_0, A_1)$ in $\check{\Gamma}(\Sigma \times {}^{\omega}\omega)$ such that:
(i) If $(\varepsilon, \alpha, \beta)$ is a win for player II in $J_u(A_0, A_1)$, then
$$(\varepsilon \in H_u \Longrightarrow \alpha \in A_0) \text{ and } (\varepsilon \notin H_u \Longrightarrow \alpha \in A_1)$$
(ii) If for some fixed $\alpha \in {}^{\omega}\omega$, σ is a winning strategy for player I in the game $J_u(A_0, A_1)\lceil\alpha$ (where player II plays this α), then:
$$(\alpha \in A_0 \Longrightarrow (\sigma, \alpha) \in C_u(A_0, A_1)) \text{ and } (\alpha \in A_1 \Longrightarrow (\sigma, \alpha) \notin C_u(A_0, A_1))$$

This result is the main result of this section. The proof is by induction on $u \in \boldsymbol{D}_2$. Let us say that u is **nice** if it satisfies the conclusions of Theorem 2.7. Using Proposition 1.11, it is enough to prove that $\underline{0}$ is nice, that if u is nice so is $u(0)^\frown 1^\frown u$, and u^{η} for each $\eta < \omega_1$, and that if $(u_n)_{n \in \omega}$ are nice, so is $1^\frown 2^\frown \langle u_n \rangle$.

LEMMA 2.8. The description $\underline{0}$ is nice.

PROOF. We must set $H_{\underline{0}} = \varnothing$ and $C_{\underline{0}}(A_0, A_1) = \Sigma \times {}^{\omega}\omega$. For A_0, A_1 disjoint $\utilde{\Sigma}^1_1$ sets with associated trees T_0, T_1 respectively $\omega \times \omega$, let $J_{\underline{0}}(A_0, A_1)$ be the game where player I plays ε, player II plays α and β and player II wins if for all $n(\alpha\lceil n, \beta\lceil n) \in T_1$. Clearly for any $f \in \mathcal{R}$, $f^{-1}H_{\underline{0}} = \varnothing$ is strategically complete in $\Gamma_{\underline{0}}$. $J_{\underline{0}}(A_0, A_1)$ is closed for player II. If $(\varepsilon, \alpha, \beta)$ is a win for player II, $\alpha \in A_1$, so (i) is satisfied. And if for some $\alpha \in {}^{\omega}\omega$ player I has a winning strategy σ in $J_{\underline{0}}(A_0, A_1)\lceil\alpha$, $\alpha \notin A_1$ hence (ii) is satisfied too. ⊣

Lemma 2.9. Suppose u is nice. Then $\breve{u} = u(0)^\frown 1^\frown u$ is nice too.

Proof. $\Gamma_{\breve{u}} = \breve{\Gamma}_u$, and one checks that $H_{\breve{u}} = \breve{H}_u$, $J_{\breve{u}}(A_0, A_1) = J_u(A_1, A_0)$ and $C_{\breve{u}}(A_0, A_1) = \breve{C}_u(A_1, A_0)$ work. ⊣

Lemma 2.10. Suppose u is nice. For each $\eta < \omega_1$, u^η is nice.

Proof. Let $(r^\eta, \rho^\eta, F^\eta)$ be the ramification of order η, with $\rho_0^\eta \in \mathcal{R}$. Let $H_u, J_u(A_0, A_1)$ and $C_u(A_0, A_1)$ be associated to u. We define $H_{u^\eta} = (\rho_0^\eta)^{-1}(H_u)$. If now $f \in \mathcal{R}$ is of order ξ, $\rho_0^\eta \circ f \in \mathcal{R}$ is of order $\xi + \eta$, hence $(\rho_0^\eta \circ f)^{-1}(H_u) = f^{-1}(H_{u^\eta})$ is strategically complete in $\Gamma_{u^{\xi+\eta}} = \Gamma_{(u^\eta)^\xi}$, and H_{u^η} is u^η-strategically complete. We now define $J_{u^\eta}(A_0, A_1)$ by

$$J_{u^\eta}(A_0, A_1)\restriction\alpha = R^\eta(J_u(A_0, A_1)\restriction\alpha)$$

for all $\alpha \in {}^\omega\omega$. This is meaningful, for in order to check whether $(\varepsilon\restriction k, \alpha\restriction k, \beta\restriction k)$ is legal in $J_{u^\eta}(A_0, A_1)$, we need to know if $t = \{r^\eta(\varepsilon\restriction k', \beta\restriction k') : k' \leq k\}$ is contained in $J_u(A_0, A_1)\restriction\alpha$. But by the properties of ramifications, $t \subseteq {}^{\leq k}(2 \times \omega)$ so the knowledge of $\alpha\restriction k$ is enough for that. Finally we define $C_{u^\eta}(A_0, A_1)$ by

$$(\sigma, \alpha) \in C_{u^\eta}(A_0, A_1) \iff (F^\eta(J_u(A_0, A_1)\restriction\alpha, \sigma), \alpha) \in C_u(A_0, A_1).$$

As $\alpha \to J_u(A_0, A_1)$ is continuous and F^η is Baire class η, the set $C_{u^\eta}(A_0, A_1)$ is in $\breve{\Gamma}_{u^\eta}$ by Proposition 1.11. It remains to check that they satisfy (i) and (ii) of Theorem 2.7. For (i), let $(\varepsilon, \alpha, \beta)$ be a win for player II in $J_{u^\eta}(A_0, A_1)$. So $(\rho_0^\eta(\varepsilon), \alpha, \rho_1^\eta(\varepsilon, \beta))$ is a win for player II in $J_u(A_0, A_1)$. This gives

$$\varepsilon \in H_{u^\eta} \Longrightarrow \rho_0^\eta(\varepsilon) \in H_u \Longrightarrow \alpha \in A_0$$

and

$$\varepsilon \notin H_{u^\eta} \Longrightarrow \rho_0^\eta(\varepsilon) \notin H_u \Longrightarrow \alpha \in A_1$$

For (ii), let σ be winning for player I in $J_{u^\eta}(A_0, A_1)\restriction\alpha$. Then $\sigma^* = F^\eta(J_u(A_0, A_1)\restriction\alpha, \sigma)$ is winning for player I in $J_u(A_0, A_1)\restriction\alpha$. So $\alpha \in A_0 \Longrightarrow (\sigma^*, \alpha) \in C_u(A_0, A_1) \Longrightarrow (\sigma, \alpha) \in C_{u^\eta}(A_0, A_1)$ and $\alpha \in A_1 \Longrightarrow (\sigma^*, \alpha) \notin C_u(A_0, A_1) \Longrightarrow (\sigma, \alpha) \notin C_{u^\eta}(A_0, A_1)$. ⊣

The preceding proof was trivial—as everything has been embedded in the notion of ramification. The next one is on the other hand long and tedious—but more or less straightforward.

Lemma 2.11. Suppose that for all n, u_n is nice. Then so is $u = 1^\frown 2^\frown \langle u_n \rangle$.

Proof. Let $H_n, J_n = J_n(A_0, A_1)$ and $C_n = C_n(A_0, A_1)$ be associated to u_n. First we choose a bijection $\langle , \rangle$ between $(\omega \cup \{*\}) \times \omega$ and ω such that each $\tau_i = \langle i, \cdot \rangle : \omega \to \omega$ is strictly increasing, for $i \in \omega \cup \{*\}$. We view each $\varepsilon \in {}^\omega 2$ as a sequence ε^*, $\langle \varepsilon_i \rangle_{i \in \omega}$, with $\varepsilon^* = \varepsilon \circ \tau^*$, $\varepsilon_i = \varepsilon \circ \tau_i$. Recall that $\Gamma_u = S_1(\bigcup_{n \geq 1} \Gamma_{u_n}, \Gamma_{u_0})$. Intuitively, we want ε_i to correspond to u_i. As we need repetitions, we choose $\varphi\colon \omega \setminus \{0\} \to \omega \setminus \{0\}$ such that $\varphi^{-1}(i)$ is infinite for all i. Let also $\psi\colon \omega \to \omega \setminus \{0\}$ be such that $\psi^{-1}(i)$ is infinite for all i.

We set: $\varepsilon \in H_u \iff$ either $\varepsilon^* = \underline{0}$ and $\varepsilon_0 \in H_0$ or for n the least i with $\varepsilon^*(i) = 1$, $\varepsilon_{\psi(n)} \in H_{\varphi(\psi(n))}$.

We first check that H_u is u-strategically complete. Define $H_0' = \{\varepsilon : \varepsilon_0 \in H_0\}$, and for $n \geq 1$, $H_n' = \{\varepsilon : \varepsilon_n \in H_{\varphi(n)}\}$, and $C_n = \{\varepsilon : \varepsilon^* \neq \underline{0}$ and for m the least i with $\varepsilon^*(i) = 1, \psi(m) = n\}$. Clearly, the C_n are pairwise disjoint open sets, $H_0' \in \Gamma_{u_0}$ and for $n \geq 1$, $H_n' \in \Gamma_{u_{\varphi(n)}}$, so $H_u = \bigcup_{n\geq 1}(H_n' \bigcap C_n) \cup (H_0' \setminus \bigcup_n C_n)$ is in $\Gamma_u = S_1(\bigcup_{n\geq 1}\Gamma_{u_n}, \Gamma_{u_0})$. Let now $f \in \mathcal{R}$ be of order η. By Proposition 1.11, $H_u^\eta = f^{-1}(H_u)$ is in Γ_{u^η}. Let $\pi\colon \omega \to \omega$ be associated to f, and let ϑ_n be the increasing enumeration of π^{-1} (ran τ_n), and ϑ_n^* the increasing enumeration of π^{-1} $(\{i \in \operatorname{ran}\tau^* : \psi(\tau^{*-1}(i)) = n\})$. Note that the fact that $\varepsilon \in f^{-1}(H_n')$ depends only on $\varepsilon \circ \vartheta_n$. Let then $H_n^\eta = \{\varepsilon \circ \vartheta_n : f(\varepsilon) \in H_n'\}$, and also $C_n^\eta = \{\varepsilon \circ \vartheta_n^* : f(\varepsilon)^*$ takes value 1 on some i with $\psi(i) = n\}$. By the hypothesis, H_n^η is strategically complete in $\Gamma_{u^\eta_{\varphi(n)}}$ for $n \geq 1$, in $\Gamma_{u_0^\eta}$ for $n = 0$. Moreover, one easily checks that if g is an independent η-function, $\{\varepsilon : g(\varepsilon) \neq \underline{0}\}$ is $\boldsymbol{\Sigma}^0_{1+\eta}$-strategically complete. So each C_n^η is strategically complete in $\boldsymbol{\Sigma}^0_{1+\eta}$. Let then $H^* \subseteq {}^\omega\omega$ be any set in Γ_{u^η}, say $H^* = \bigcup_{n\geq 1}(H_n^* \cap C_n^*) \cup (H_0^* \setminus \bigcup_n C_n^*)$ with pairwise disjoint C_n^* in $\boldsymbol{\Sigma}^0_{1+\eta}$, H_0^* in $\Gamma_{u_0^\eta}$ and wlog H_n^* in $\Gamma_{u^\eta_{\varphi(n)}}$ (this is where repetitions are used). So player II has for each n a winning strategy σ_n in $G(H_n^*, H_n^\eta)$ and σ_n^* in $G(C_n^*, C_n^\eta)$. Let then player II play in $G(H^*, f^{-1}(H))$ against α by playing his strategies σ_n, σ_n^* at the right places—the ranges of ϑ_n and ϑ_n^* respectively—against this same α, independently. The result is some ε such that $\varepsilon \circ \vartheta_n$ wins against α in $G(H_n^*, H_n^\eta)$ and $\varepsilon \circ \vartheta_n^*$ against α in $G(C_n^*, C_n^\eta)$. This wins, for $\varepsilon \in f^{-1}(H_n')$ just in case $\alpha \in H_n^*$, and $f(\varepsilon)^*$ takes value 1 on some i with $\psi(i) = n$ just in case $\alpha \in C_n^*$. But as the C_n^* are disjoint, there is at most one n in $\{\psi(i) : f(\varepsilon)^*(i) = 1\}$, and $\varepsilon \in f^{-1}(C_n)$ just in case $\alpha \in C_n^*$. This proves that H_u is u-strategically complete.

We now define $J_u = J_u(A_0, A_1)$. We view ε as $\langle \varepsilon^*, \langle \varepsilon_i \rangle \rangle$ as before, and similarly we decompose β as $\langle \beta^*, \langle \beta_i \rangle \rangle$. In J_u, as long as player I plays 0's on his ε^*-moves, player I and player II must play the game J_0 with ε_0, α and β_0. And once player I has played 1 on his ε^*-moves, at say step k of the game, then letting $k_0 = \tau^{*-1}(k)$, $n_0 = \psi(k_0)$, $m_0 = \varphi(n_0)$, the players switch to the game J_{m_0}, played with ε_{n_0} for player I, α and the part of β_{n_0} which is played after step k for player II (*i.e.*, when switching, player I does not revise his previous moves on ε_{n_0}, when player II does for β_{n_0}). This defines a closed (for player II) game. We now check (i) of Theorem 2.7: So suppose $(\varepsilon, \alpha, \beta)$ is a win for player II in $J_u(A_0, A_1)$. There are two cases:

(a) Suppose $\varepsilon^* = \underline{0}$. Then $\varepsilon \in H_u \iff \varepsilon_0 \in H_0$. But then $\varepsilon \in H_u \Longrightarrow \varepsilon_0 \in H_0 \Longrightarrow \alpha \in A_0$ and $\varepsilon \notin H_u \Longrightarrow \varepsilon_0 \notin H_0 \Longrightarrow \alpha \in A_1$ as $(\varepsilon_0, \alpha, \beta_0)$ is a win for player II in J_0.

(b) for some least k_0, $\varepsilon^*(k_0) = 1$. Let $k = \tau^*(k_0), n_0 = \psi(k_0)$ and $m_0 = \varphi(n_0)$, and $\langle\beta_{n_0}\rangle_{\geq k}$ the sequence of β_{n_0}-moves after k. Then $(\varepsilon_{n_0}, \alpha, \langle\beta_{n_0}\rangle_{\geq k})$ is a win for player II in J_{m_0}, and again $\varepsilon \in H_u \iff \varepsilon_{n_0} \in H_{m_0}$, so that we get the same conclusion. This proves (i).

We now define $C_u = C_u(A_0, A_1)$.

Let $<$ be an ordering type of ω on ${}^{<\omega}\omega$, with $u \subseteq v \implies u < v$. Given $\alpha \in {}^{\omega}\omega$ and $\sigma \in \Sigma$, let us say that a finite sequence $w \in {}^{<\omega}\omega$ is (σ, α)-legal if the position in $J_u(A_0, A_1)$ corresponding to the play $\alpha \restriction \mathrm{lh}(w)$, w of player II and the σ-answer by player I is legal. For all (σ, α), let $w(\sigma, \alpha)$ be the least for $<$ sequence which satisfies (i) w is (σ, α)-legal; (ii) the answers by σ on the $*$-moves are 0 up to $\mathrm{lh}(w)$; and (iii) $\mathrm{lh}(w)$ corresponds to a $*$-move, and $\sigma(w) = 1$ [in other words, we are at a legal switching position in $J_u(A_0, A_1)$ where player II follows α and w, and player I answers by σ].

The function $(\sigma, \alpha) \to w(\sigma, \alpha)$ is defined on all pairs (σ, α) for which there is a legal beginning w with $\sigma(w)$ a $*$-move with value 1.

We now define, for $\alpha \in {}^{\omega}\omega$ and $\sigma \in \Sigma$, a sequence of strategies as follows:

First, to each $w_0 \in {}^{<\omega}\omega$, viewed as a play of player II in $J_0 \restriction \alpha$, associate the play w in $J_u \restriction \alpha$ consisting in playing w_0 on the 0-moves, and 0 on all other moves, with length such that the next play is the next 0-move. And define $\sigma_0(w_0) = \sigma(w)$. Note that $F_0 \colon \sigma \to \sigma_0$ is continuous: $\Sigma \to \Sigma$.

Let now $w_0 \in {}^{<\omega}\omega$. We define $F_{w_0} \colon \Sigma \to \Sigma$ as follows: We let $F_{w_0}(\sigma) = \underline{0}$ unless the answers by σ to w_0 are 0 on the $*$-moves up to $\mathrm{lh}(w_0)$, and $\mathrm{lh}(w_0)$ is a $*$-move and $\sigma(w_0) = 1$. And in this case, we associate to each $w \in {}^{<\omega}\omega$ a position w' as follows: w' is w_0 up to $\mathrm{lh}(w_0) = k$, with $\psi(k) = k_0$ say. After that, w' is 0 everywhere except on the k_0-moves, where it is w, and its length is such that the next play will be the next k_0-move. And we define $F_{w_0}(\sigma)(w) = \sigma(w')$. Again each $F_{w_0} \colon \Sigma \to \Sigma$ is continuous. We can now define $C_u = C_u(A_0, A_1) \subseteq \Sigma \times {}^{\omega}\omega$ by

$(\sigma, \alpha) \in C_u \iff$ *either* for every (σ, α)-legal sequence $w \in {}^{<\omega}\omega$ the $*$-answers by σ are 0, and $(F_0(\sigma), \alpha) \in C_0$ *or* there is a (σ, α)-legal sequence with $*$-answer 1 by σ, and if $w_0 = w(\sigma, \alpha)$ and $k_0 = \psi(\mathrm{lh}(w_0))$, $(F_{w_0}(\sigma), \alpha) \in C_{\varphi(k_0)}$.

We first check that $C_u \in \check{\Gamma}_u$: let $B_{w_0} = \{(\sigma, \alpha) : w_0 = w(\sigma, \alpha)\}$. Clearly, each B_{w_0} is clopen in $\Sigma \times {}^{\omega}\omega$, and the B_{w_0}'s are pairwise disjoint. Let $D_0 = \{(\sigma, \alpha) : (F_0(\sigma), \alpha) \in C_0\}$ and for $w_0 \in {}^{<\omega}\omega$, $D_{w_0} = \{(\sigma, \alpha) : (F_{w_0}(\sigma), \alpha) \in C_{\varphi\circ\psi(\mathrm{lh}\, w_0)}\}$. By continuity of the F_0, F_{w_0}'s, $D_{w_0} \in \Gamma_{u_0}$ and $D_{u_0} \in \Gamma_{u_k}$, $k = \varphi \circ \psi(\mathrm{lh}\, w_0)$. And $C_u = \bigcup_{w_0 \in {}^{<\omega}\omega}(B_{w_0} \cap D_{w_0}) \cup (D_0 \setminus \bigcup_{w_0 \in {}^{<\omega}\omega} B_{w_0})$, hence $C_u \in \check{\Gamma}_u$. It remains to check (ii) of Theorem 2.7. So we let $\alpha \in {}^{\omega}\omega$, and σ winning for player I in $J_u(A_0, A_1) \restriction \alpha$. There are two cases.

(a) First, if $w(\sigma, \alpha)$ is undefined, *i.e.*, for any (σ, α) legal sequence w, the $*$-answers are 0. Then $(\sigma, \alpha) \in C_u \iff (F_0(\sigma), \alpha) \in C_0$. But we claim $F_0(\sigma)$ is winning for player I in $J_0 \restriction \alpha$, for if $\beta_0 \in {}^\omega\omega$ defeats $F_0(\sigma)$ in $J_0 \restriction \alpha$, the play β corresponding to β_0 on the 0-moves and 0 everywhere else is easily seen to defeat σ in $J_u \restriction \alpha$. So we get $\alpha \in A_0 \implies (F_0(\sigma), \alpha) \in C_0 \implies (\sigma, \alpha) \in C_u$ and $\alpha \in A_1 \implies (F_0(\sigma), \alpha) \notin C_0 \implies (\sigma, \alpha) \notin C_u$.

(b) Otherwise, $w_0 = w(\sigma, \alpha)$ is defined, and by definition $(\sigma, \alpha) \in C_u \iff (F_{w_0}(\sigma), \alpha) \in C_k$ where $k = \varphi(\psi(\operatorname{lh} w_0))$. Again we claim that F_{w_0} is winning for player I in $J_k \restriction \alpha$, which as before will finish the proof. Suppose β is a play of player II which defeats $F_{w_0}(\sigma)$ in $J_k \restriction \alpha$, and let player II play in $J_u \restriction \alpha$ first w_0, then 0 on all moves except the moves corresponding to $k_0 = \psi(\operatorname{lh} w_0)$, where he plays β. One easily checks that all positions are then legal in $J_u \restriction \alpha$ against σ, a contradiction which finishes the proof. ⊣

Altogether Lemmas 2.8, 2.9, 2.10 and 2.11 prove Theorem 2.7.

§3. Proof of Borel Wadge Determinacy. As we said in the introduction, we will prove a slight generalization of Borel Wadge Determinacy. Consider, for $A \subseteq {}^\omega\omega$ and A_0, A_1 two disjoint subsets of ${}^\omega\omega$ the following extended Wadge game $G(A; A_0, A_1)$: player I plays $\alpha \in {}^\omega\omega$, player II plays $\beta \in {}^\omega\omega$, and player II wins if ($\alpha \in A \implies \beta \in A_0$ and $\alpha \notin A \implies \beta \in A_1$). The usual Wadge game $G(A, B)$ corresponds to $B = A_0 = \check{A}_1$. [We will also consider the similar game where $A \subseteq {}^\omega 2$, and player I plays $\alpha \in {}^\omega 2$, that we will denote ambiguously $G(A; A_0, A_1)$ too.] So Borel Wadge Determinacy is a particular case of

Theorem 3.1. Let $A \subseteq {}^\omega\omega$ be Borel, and A_0, A_1 two disjoint $\underset{\sim}{\mathbf{\Sigma}}^1_1$ subsets of ${}^\omega\omega$. The extended Wadge game $G(A; A_0, A_1)$ is determined.

In order to prove Theorem 3.1, we first prove particular instances of it.

Theorem 3.2. Let $u \in \boldsymbol{D}_2$, and A_0, A_1 two disjoint $\underset{\sim}{\mathbf{\Sigma}}^1_1$ sets in ${}^\omega\omega$.

(i) If $A \subseteq {}^\omega\omega$ is in Γ_u, and no set $B \in \check{\Gamma}_u$ separates A_0 from A_1 (*i.e.*, $A_0 \subseteq B \subseteq \check{A}_1$), then player II has a winning strategy in $G(A; A_0, A_1)$.

(ii) If $A \subseteq {}^\omega\omega$ is not in $\check{\Gamma}_u$, and there is a set $B \in \check{\Gamma}_u$ separating A_0 from A_1, then player I has a winning strategy in $G(A; A_0, A_1)$. In particular, if $A \subseteq {}^\omega\omega$ is in $\Gamma_u \setminus \check{\Gamma}_u$, $G(A; A_0, A_1)$ is always determined.

Proof. Let H_u, $J_u(A_0, A_1)$ and $C_u(A_0, A_1)$ be associated to u by Theorem 2.7. Being closed, the game $J_u(A_0, A_1)$ is determined. If player I has a winning strategy σ in it, let $\sigma(\alpha)$ be the corresponding winning strategy in $J_u(A_0, A_1) \restriction \alpha$ obtained by fixing the α-moves. The set $B_\sigma \subseteq {}^\omega\omega$ defined by

$\alpha \in B_\sigma \iff (\sigma(\alpha), \alpha) \in C_u(A_0, A_1)$ is then in $\check{\Gamma}_u$, and by (ii) of Theorem 2.7, separates A_0 from A_1.

(i) Assume $A \in \Gamma_u$ and no $\check{\Gamma}_u$ set separates A_0 from A_1. Then by the previous discussion, player II wins $J_u(A_0, A_1)$. And by forgetting the β-moves in this game, player II has a strategy in the game $G(H_u; A_0, A_1)$ which satisfies by (i) of Theorem 2.7, $\varepsilon \in H_u \implies \alpha \in A_0$ and $\varepsilon \notin H_u \implies \alpha \in A_1$, *i.e.*, is winning in $G(H_u; A_0, A_1)$. And as $A \in \Gamma_u$ and H_u is Γ_u-strategically complete, player II also has a winning strategy in $G(A, H_u)$. Composing his strategies gives a winning strategy for player II in $G(A; A_0, A_1)$ and (i) is proved.

(ii) Assume now $A \notin \check{\Gamma}_u$, and let for $n \in \omega$, $A(n) = \{\alpha \in {}^\omega\omega : n^\frown\alpha \in A\}$. By Proposition 1.7, one of the $A(n)$'s must satisfy $A(n) \notin \check{\Gamma}_u$. Let n_0 be the least such n. Let also $B \in \check{\Gamma}_u$ separate A_0 from A_1. Applying case (i) to $\check{B}$ and the pair $(A(n_0), \check{A}(n_0))$, we get that player II has a winning strategy τ in $G(\check{B}, A(n_0))$. Let then player I play first n_0, and then follow the strategy τ against player II's play. At the end, one gets $n_0{}^\frown\alpha$ and β, and $n_0{}^\frown\alpha \in A \iff \alpha \in A(n_0) \iff \beta \notin B$, so that $\beta \in A_0 \implies n_0{}^\frown\alpha \notin A$ and $\beta \in A_1 \implies n_0{}^\frown\alpha \in A$ and this strategy is winning for player I in $G(A; A_0, A_1)$, and (ii) is proved. And the final statement immediately follows from (i) and (ii). ⊣

Recall that we associated to each class Γ a class Γ^+ by $A \in \Gamma^+ \iff A = (B \cap D) \cup (C \setminus D)$ for some $D \in \mathbf{\Delta}^0_1$, $B \in \Gamma$ and C in $\check{\Gamma}$.

THEOREM 3.3. Let $u \in \boldsymbol{D}_2$, A_0, A_1 two disjoint $\mathbf{\Sigma}^1_1$ sets in ${}^\omega\omega$ and A a set in $\Gamma^+_u \setminus (\Gamma_u \cup \check{\Gamma}_u)$. The game $G(A; A_0, A_1)$ is determined.

PROOF. For each $s \in {}^{<\omega}\omega$, let $A(s) = \{\alpha : s^\frown\alpha \in A\}$. Let $T_A = \{s \in {}^{<\omega}\omega : A(s) \in \Gamma^+_u - (\Gamma_u \cup \check{\Gamma}_u)\}$. Clearly, T_A is a tree, $\varnothing \in T_A$, and as $A \in \Gamma^+_u$, T_A is well founded. Let $T_{A_0,A_1} = \{t \in {}^{<\omega}\omega :$ no $\Gamma_u \cup \check{\Gamma}_u$ set separates $A_0(t)$ from $A_1(t)\}$. Again T_{A_0,A_1} is a tree, which now may be empty or not well-founded. Let G^* be the game where player I and player II play integers, and player I loses if he gets off T_A before player II gets off T_{A_0,A_1}. (So if in particular T_{A_0,A_1} is empty, player I wins before the game starts). As T_A is well-founded, G^* is clopen. We claim that whoever wins G^* also wins $G(A; A_0, A_1)$:

Case (a): player I has a winning strategy in G^*. Let him play it in $G(A; A_0, A_1)$. Then a position (s, t) must be reached such that $s \in T_A$ but $t \notin T_{A_0,A_1}$ (we use $\varnothing \in T_A$ here). This means that $A_0(t)$ is separable from $A_1(t)$ by some set in $\Gamma_u \cup \check{\Gamma}_u$, say $\check{\Gamma}_u$ to be specific. As $A(s) \notin \check{\Gamma}_u$, player I has a winning strategy in $G(A(s); A_0(t), A_1(t))$ by Theorem 3.2, and switching to it is clearly winning in $G(A; A_0, A_1)$.

Case (b) is similar: By playing his winning strategy in G^*, player II reaches a position (s, t) with $t \in T_{A_0,A_1}$, but any extension of s gets off T_A. Let then player I play n_0. The set $B = \{\alpha \in A(s) : \alpha(0) = n_0\}$ is in $\Gamma_u \cup \check{\Gamma}_u$, say in

Γ_u to be specific, and $A_0(t)$ cannot be separated from $A_1(t)$ by a set in $\breve{\Gamma}_u$, hence player II has a winning strategy in $G(B; A_0(t), A_1(t))$ by Theorem 3.2, and switching to it is clearly winning in $G(A; A_0, A_1)$. ⊣

For u_n a sequence of type 2 descriptions with $\Gamma_{u_n} < \Gamma_{u_{n+1}}$, we defined $\langle \Gamma_{u_n} \rangle^+ = (\bigcup_n \Gamma_{u_n})^+$. The next result is entirely similar to the preceding one, and we omit the proof.

THEOREM 3.4. If $\langle u_n \rangle$ is a sequence in $\boldsymbol{D}_2$ with $\Gamma_{u_n} < \Gamma_{u_{n+1}}$ for all n, A_0, A_1 are two disjoint $\boldsymbol{\Sigma}^1_1$ sets in ${}^\omega\omega$ and $A \subseteq {}^\omega\omega$ is a set in $\langle \Gamma_{u_n} \rangle^+ \setminus \bigcup_n \Gamma_{u_n}$, the game $G(A; A_0, A_1)$ is determined.

The last step in the proof of Theorem 3.1 is to show that Theorems 3.2, 3.3, and 3.4 cover all possible cases. And to do this, the last ingredient is the following precise version of Martin's result on the well-foundedness of Wadge's ordering.

THEOREM 3.5 (Martin). Let $(A_n)_{n\in\omega}$ be a sequence of Borel sets in ${}^\omega\omega$. Then player I cannot at the same time have a winning strategy in all games $G(A_n, A_{n+1})$ and $G(A_n, \breve{A}_{n+1})$.

PROOF. By contradiction. Let σ_n^0 resp σ_n^1 be winning for player I in $G(A_n, A^0_{n+1})$, resp $G(A_n, A^1_{n+1})$ where by definition $A^0_{n+1} = A_{n+1}, A^1_{n+1} = \breve{A}_{n+1}$. Associate to each $\varepsilon \in {}^\omega 2$ a sequence α_n^ε by induction by $\alpha_n^\varepsilon(0) = \sigma_n^{\varepsilon(n)}(\varnothing)$, and $\alpha_n^\varepsilon(k) = \sigma_n^{\varepsilon(n)}(\alpha^\varepsilon_{n+1}\restriction k)$. This sequence is clearly obtained continuously in ε, as $\alpha^\varepsilon_n \restriction k$ depends on $\varepsilon \restriction n+k$. Moreover α_n^ε depends only on the values of ε for $m \geq n$, and for all n the pair $(\alpha_n^\varepsilon, \alpha^\varepsilon_{n+1})$ is a run in $G(A_n, A^{\varepsilon(n)}_{n+1})$ where player I follows $\sigma_n^{\varepsilon(n)}$, so $\alpha_n^\varepsilon \in A_n \iff \alpha^\varepsilon_{n+1} \notin A^{\varepsilon(n)}_{n+1}$. Let $B = \{\varepsilon : \alpha_0^\varepsilon \in A_0\}$. The set B is Borel. On the other hand, if $s \in {}^{<\omega}2$, neither B nor $\breve{B}$ is comeager on $N_s = \{\varepsilon : s \subseteq \varepsilon\}$, for if $\varepsilon \in N_s$ and $\overline{\varepsilon}$ is defined by $\overline{\varepsilon}(p) = \varepsilon(p)$ for $p \neq p_0 = \mathrm{lh}(s)$, and $\overline{\varepsilon}(p_0) = 1 - \varepsilon(p_0)$, then $\overline{\varepsilon} \in N_s$, and $\alpha_p^{\overline{\varepsilon}} = \alpha_p^\varepsilon$ for $p > p_0$, and for $p \leq p_0$, $\alpha_p^{\overline{\varepsilon}} \in A_p \iff \alpha_p^\varepsilon \notin A_p$, so that in particular $\alpha_0^{\overline{\varepsilon}} \in B \iff \alpha_0^\varepsilon \notin B$. This shows B does not possess the Baire property, a contradiction which finishes the proof. ⊣

Recall that we defined, for $u, u' \in \boldsymbol{D}_2$, $\Gamma_u < \Gamma_{u'}$ if $\Gamma_u \subseteq \Delta(\Gamma_{u'})$.

COROLLARY 3.6. The relation $<$ is well-founded on the set $W_2 = \{\Gamma_u : u \in \boldsymbol{D}_2\}$.

PROOF. If not, let $\langle u_n \rangle \in \boldsymbol{D}_2$ be such that (Γ_{u_n}) is a $<$-decreasing sequence, and A_n any set in $\Gamma_{u_n} \setminus \breve{\Gamma}_{u_n}$. By Theorem 3.2, player I wins all games $G(A_n, A_{n+1})$ and $G(A_n, \breve{A}_{n+1})$, contradicting Martin's Theorem 3.5 ⊣

PROOF OF THEOREM 3.1. We argue by contradiction. The set $W_1 = \{\Gamma_u, \breve{\Gamma}_u : u \in \boldsymbol{D}_1\}$ is a subset of W_2 (Theorem 1.8), hence is well-founded for $<$ by Corollary 3.6 above, and is cofinal in $\boldsymbol{\Delta}^1_1$. So if Theorem 3.1 fails, we can find a $<$-minimal class Γ in W_1 and sets $A \in \Gamma$, A_0 and A_1 disjoint $\boldsymbol{\Sigma}^1_1$ sets in ${}^\omega\omega$ such that $G(A; A_0, A_1)$ is not determined. Now $\Gamma = \Gamma_u$ or $\check{\Gamma}_u$ for some $u \in \boldsymbol{D}_1$, hence $\Gamma = \Gamma_v$ for some $v \in \boldsymbol{D}_2$.

Case (*a*): $A \in \Gamma \setminus \check{\Gamma}$. Then $A \in \Gamma_v \setminus \check{\Gamma}_v$, and by Theorem 1.4 $G(A; A_0, A_1)$ is determined, a contradiction. So $A \in \Delta(\Gamma)$, and we can apply Theorem 1.4. Note that Case 1.4 (iii), (*i.e.*, $\Delta(\Gamma) = \bigcup_{\xi<\omega_1}, \Gamma_{u_\xi}$ for some $u_\xi \in \boldsymbol{D}_1$) is impossible by $<$-minimality of Γ.

Case (*b*): For some $u^* \in \boldsymbol{D}_1, \Delta(\Gamma) = (\Gamma_{u^*})^+$. Then by $<$-minimality of Γ, $A \in \Gamma^+_{u^*} \setminus (\Gamma_{u^*} \cup \check{\Gamma}_{u^*})$, hence for some $v \in \boldsymbol{D}_2$, $A \in \Gamma^+_v \setminus (\Gamma_v \cup \check{\Gamma}_v)$ and $G(A; A_0, A_1)$ is determined by Theorem 3.3, a contradiction.

The last case where $\Delta(\Gamma) = \langle \Gamma_{u_n} \rangle^+$ for some (u_n) in $\boldsymbol{D}_1$ is handled similarly, using Theorem 3.4. ⊣ (Theorem 3.1)

Let us finish this section with a brief discussion of some other results which are by-products of the preceding proof.

1. We get from the preceding proof that $W = W_1 = W_2$.

2. The sets H_u, $u \in \boldsymbol{D}_2$ we constructed are in $\Gamma_u({}^\omega 2)$. If follows that (a) If Γ is a non-self dual Wadge class, so is $\Gamma({}^\omega 2)$; and (b) $\Gamma = \{f^{-1}(A) : A \subseteq {}^\omega 2, A \in \Gamma({}^\omega 2)\}$.

3. For Γ a class and η a countable ordinal, we can always define its η-expansion Γ^η as the class of $f^{-1}(A), A \in \Gamma$ and $f : {}^\omega\omega \to {}^\omega\omega$ of Baire class η. Then one gets $(\Gamma_u)^\eta = \Gamma_{u^\eta}$ for all $u \in \boldsymbol{D}_2$: inclusion $\subseteq$ is easy (Proposition 1.11), and in the other direction, the set H_{u^η} which generates Γ_{u^η} is in $(\Gamma_u)^\eta$ by its very definition. Using this and the equality $W = W_2$, one gets that the class W of non-self dual Borel Wadge classes is the least family of classes containing $\{\varnothing\}$ and closed under complementation, S_1, and η-expansion for all $\eta < \omega_1$.

4. The argument we gave for Borel Wadge classes and Borel Wadge games on ${}^\omega\omega$ easily translate to the case of ${}^\omega 2$. In particular one gets that the non-self dual Borel Wadge classes on ${}^\omega 2$ are exactly the $\Gamma_u({}^\omega 2)$ for $u \in \boldsymbol{D}_2$. One also gets the determinacy of the games $G(A; A_0, A_1)$ played on ${}^\omega 2$, by noticing that the H_u's are strategically complete in $\Gamma_u({}^\omega 2)$, and that Theorem 1.4 holds for ${}^\omega 2$ too. There is however a slight difference between the two cases: For ${}^\omega 2$, case (ii) in 1.4 trivializes as $\langle \Gamma_n \rangle^+({}^\omega 2) = \bigcup_n \Gamma_n({}^\omega 2)$ by compactness. This also shows that fact 2 above fails for self-dual classes.

Let us make now some comments on how further information can be obtained from the existence of the "unfolded" games J_u:

5. It is clear, by looking at their definition, that the games $J_u(A_0, A_1)$ are defined uniformly in A_0 and A_1. In fact if one codes the pairs of $\boldsymbol{\Sigma}^1_1$ sets reasonably—*e.g.*, by coding pairs of associated trees on $\omega \times \omega$, one easily checks that for each $u \in \boldsymbol{D}_2$ the tree of the game $J_u(A_0, A_1)$ is continuous in the codes. Using this uniformity, one easily gets the following result, where for $A \subseteq {}^\omega\omega \times {}^\omega\omega$ and $x \in {}^\omega\omega$, A_x denotes the section of A at x:

THEOREM 3.7. Let Γ be some Borel Wadge class.

(a) If A_0, A_1 are $\underset{\sim}{\Sigma}_1^1$ sets in ${}^\omega\omega \times {}^\omega\omega$

$$B_\Gamma = \{x : (A_0)_x \text{ is separable from } (A_1)_x \text{ by a } \Gamma\text{-set}\}$$

is $\underset{\sim}{\Pi}_1^1$, hence in particular for Borel $B = \{x : B_x \in \Gamma\}$ is $\underset{\sim}{\Pi}_1^1$.

(b) If A_0, A_1 are $\underset{\sim}{\Sigma}_1^1$ in ${}^\omega\omega \times {}^\omega\omega$ and for all x, $(A_0)_x$ can be separated from $(A_1)_x$ by some Γ set, there is a Borel set B with sections in Γ which separates A_0 from A_1.

To prove this, one can easily reduce the case where Γ is self dual to the non-self dual case, say $\Gamma = \Gamma_u$ for some $u \in \boldsymbol{D}_2$. One gets then a continuous map: $x \to J_{\breve{u}}((A_0)_x, (A_1)_x) = J_x$, and by Theorem 3.2, $x \in B_\Gamma \iff$ player I wins $J_x \iff$ player II does not win J_x and this last statement is $\underset{\sim}{\Pi}_1^1$. This gives (a). By general standard facts, (b) is a consequence of (a). One can also use that if $B_\Gamma = {}^\omega\omega$, one can find by the selection theorem for strategies in open games a Borel function $x \to \sigma_x$, with σ_x winning for player I in J_x. And by Theorem 3.2 again $B = \{(x, a) : (\sigma_x(\alpha), \alpha) \in C_{\breve{u}}((A_0)_x, (A_1)_x)\}$ is a set which separates A_0 from A_1, and is Borel with Γ_u sections.

6. We now discuss the uniformity of our constructions in u. This leads to lightface versions of our results.

First by encoding countable ordinals by reals, one gets a coding of descriptions. Let us say that a description u (in $\boldsymbol{D}_1$ or $\boldsymbol{D}_2$) is HYP if some code of it is Δ_1^1, and that Γ is a HYP_1 (a HYP_2) class if $\Gamma = \Gamma_u$ or $\breve{\Gamma}_u$ for some HYP u in $\boldsymbol{D}_1$ ($\Gamma = \Gamma_u$ for some HYP u in $\boldsymbol{D}_2$). The proof of Theorem 1.8 easily gives $\text{HYP}_1 \subseteq \text{HYP}_2$. Let also W_{HYP} be the family of non-self dual Wadge classes which are generated by a Δ_1^1 set. One also checks easily that $\text{HYP}_2 \subseteq W_{\text{HYP}}$.

Fix now some HYP description u (in $\boldsymbol{D}_1$ or $\boldsymbol{D}_2$). One can then define the notion of a HYP-in-Γ_u subset of ${}^\omega\omega$: Intuitively, these are the sets in Γ_u which admit a HYP construction as a Γ_u set. Formally, one has to go through codes. This is done precisely for first type descriptions in the second part of Louveau's paper [Lou83], and it is proved there that for u HYP in $\boldsymbol{D}_1$, $\Gamma_u \cap \Delta_1^1 = \text{HYP-in-}\Gamma_u$.

Our games allow to prove a similar result for $u \in \boldsymbol{D}_2$: The point is that for recursive ordinal η, one can choose the ramification $(r^\eta, \rho^\eta, F_\eta)$ so that all functions are Δ_1^1-recursive, and then it is not hard to check that for HYP $u \in \boldsymbol{D}_2$,

(a) $H_u \in \Delta_1^1$;

(b) for A_0, A_1 Σ_1^1 sets, $J_u(A_0, A_1)$ is Δ_1^1, and $C_u(A_0, A_1)$ is HYP-in-$\breve{\Gamma}_u$. Using this, one gets $\Gamma_u \cap \Delta_1^1 = \text{HYP-in-}\Gamma_u$ for u HYP in $\boldsymbol{D}_2$: For if A is Δ_1^1 in Γ_u, player I wins $J_u(A, \breve{A})$ by Theorem 3.2, and this game is open Δ_1^1 for player I, hence player I has a Δ_1^1 winning strategy σ. But then one gets $A = \{\alpha : (\sigma(\alpha), \alpha) \in C_{\breve{u}}(A, \breve{A})\}$, which proves that A is HYP-in-Γ_u.

The second step is to get a "lightface" analog of Theorem 1.4. First one easily defines the notions of HYP-in-Γ_u^+ and HYP-in-$\langle\Gamma_{u_n}\rangle^+$ for u HYP in $\boldsymbol{D}_1$, and $\langle u_n\rangle$ a Δ_1^1 sequence of HYP descriptions in $\boldsymbol{D}_1$. Then one can prove the analog of Theorem 1.4, analyzing the family $\Delta_1^1 \cap \Delta(\Gamma_u) =$ HYP-in-$\Gamma_u) \cap$ (HYP-in-$\check{\Gamma}_u$) in terms of HYP-in-$\Gamma_{u'}$ classes, for u''s HYP in $\boldsymbol{D}_1$ with $\Gamma_{u'} < \Gamma_u$. The proof essentially follows the proof in [Lou83, 1.4], and although a bit tedious, necessitates no new ideas.

By combining the two previous steps, one finally gets

THEOREM 3.8.
(i) Every W_{HYP} class Γ admits a $\boldsymbol{D}_1$ (and hence a $\boldsymbol{D}_2$) HYP description—as $\Gamma_u, \check{\Gamma}_u, \langle\Gamma_u\rangle^+$ or $\langle\Gamma_{u_n}\rangle^+$, as the case may be
(ii) if A is Δ_1^1 and in a HYP class Γ, it is HYP-in-Γ. Hence in particular the Wadge class $\Gamma(A)$ of A is a HYP class, and A is HYP-in-$\Gamma(A)$.

Finally notice that when considering a Borel Wadge game $G(A, B)$ with Δ_1^1 sets A, B, the winning strategies in this game are recursively obtained, using Theorems 3.2, 3.3, and 3.4, from winning strategies for player II in associated closed games depending on the Wadge class $\Gamma(A)$, so by the previous result which can be chosen Δ_1^1. One then finally gets

THEOREM 3.9. If A, B are Δ_1^1 in ${}^\omega\omega$, one of the two players in $G(A, B)$ has a winning strategy which is recursive in Kleene's $\mathcal{O}$.

Note that this result is the best possible along these lines, as by considering $A = {}^\omega\omega$ and B some non empty Π_1^0 set with no Δ_1^1 members, we see that there may be no HYP winning strategy.

7. A final word on the games J_u: The fact that the β-part of player II's play in J_u are in ${}^\omega\omega$ is not an essential feature, and one can define similar games with $\beta \in {}^\omega\kappa$ for κ some infinite cardinal. Such extensions were used in [LSR87] to study separation of κ-Suslin sets by $\boldsymbol{\Sigma}_\xi^0$-sets, and separation of lightface projective sets by $\boldsymbol{\Sigma}_\xi^0$ sets, using strong set theoretic hypotheses. Similar results could be obtained, along the same lines, for all Borel Wadge classes.

§4. Wadge Classes in Metric Separable Spaces.

Unless the underlying space is ${}^\omega 2$ or ${}^\omega\omega$, there is no clear notion of Wadge game available, and to extend Theorem 3.1 to more general situations, we must first rephrase (and weaken) it a bit. Let Γ be a Borel Wadge class in ${}^\omega\omega$ and A_0, A_1 a pair of disjoint sets in some metric separable space E. We say that (A_0, A_1) **reduces** Γ if for any $B \subseteq {}^\omega\omega$ in Γ, there is a continuous $f : {}^\omega\omega \to E$ with $f({}^\omega\omega) \subseteq A_0 \cup A_1$ and $f^{-1}(A_0) = B$. Note that this property is intrinsic, *i.e.*, depends only on $A_0 \cup A_1$. We say A reduces Γ if $(A, \check{A})$ does.

Let us say, for A_0, A_1 in ${}^\omega\omega$, that Γ separates (A_0, A_1) if A_0 is separable from A_1 by some B in Γ. With this terminology, Theorem 3.1 gives:

THEOREM 4.1.

(a) If Γ is a Borel Wadge class in ${}^\omega\omega$ and $A \subseteq {}^\omega\omega$ is Borel and does not reduce Γ, then $A \in \check{\Gamma}$.

(b) If Γ is a Borel Wadge class in ${}^\omega\omega$, A_0, A_1 are pairwise disjoint $\underset{\sim}{\Sigma}^1_1$ sets in ${}^\omega\omega$ and (A_0, A_1) does not reduce Γ, then $\check{\Gamma}$ separates (A_0, A_1).

Part (a) is usually called "Wadge's lemma". It is of course a consequence of part (b), but we stated the two versions, for their fate will be slightly different in the sequel.

The main point in order to extend Theorem 4.1 to other spaces is to define, for each Wadge class Γ on ${}^\omega\omega$, a corresponding class $\Gamma(E)$ of sets in E. This can be done in various ways, and most of the work will consist in showing that the various possible definitions lead to the same classes.

From now on, we will consider only classes Γ in W, *i.e.*, non-self-dual Borel Wadge Classes. In the first section, we briefly looked at the case of subsets E of ${}^\omega\omega$, where one can define $\Gamma(E)$ using traces. As any zero-dimensional metric separable space is homeomorphic to such a subset of ${}^\omega\omega$, this gives one way of defining $\Gamma(E)$ for a zero-dimensional space E. Another way is to consider continuous preimages, still another way to consider Hausdorff operations performed on open sets in E. And there is a more subtle possibility, coming from Wadge's lemma: to define the class $\Gamma(E)$ by the properties of its continuous preimages. The next result shows the equivalence of all these possible definitions, at least when E is a zero-dimensional **Suslin** space, *i.e.*, is absolute $\underset{\sim}{\Sigma}^1_1$ metrizable separable.

THEOREM 4.2. Let E be a zero-dimensional Suslin space, $\Gamma \in W$, and A a subset of E. The following are equivalent.

1. There is a 1-1 embedding $E \hookrightarrow {}^\omega\omega$ and $B \in \Gamma$, $A = j^{-1}(B)$.
2. For all 1-1 embeddings $E \hookrightarrow {}^\omega\omega$ there is a $B \in \Gamma$, $A = j^{-1}(B)$.
3. There is a continuous $f\colon E \to {}^\omega\omega$ and a $B \in \Gamma$, $A = f^{-1}(B)$.
4. There is a Hausdorff operation D and (U_n) open in E such that $\Gamma = D(\underset{\sim}{\Sigma}^0_1)$ and $A = D((U_n))$.
5. For all Hausdorff operations D with $\Gamma = D(\underset{\sim}{\Sigma}^0_1)$ there is (U_n) open in E with $A = D((U_n))$.
6. For all continuous $G\colon {}^\omega\omega \to E$, $g^{-1}(A) \in \Gamma$.

We then define $\Gamma(E)$ as the class of sets in E which satisfy one of these equivalent properties.

PROOF. $1 \Longrightarrow 3$ is trivial, and $3 \Longrightarrow 5$ and $4 \Longrightarrow 6$ come from preservation of Hausdorff operations on $\underset{\sim}{\Sigma}^0_1$ sets by continuous preimages. $2 \Longrightarrow 1$ comes from the existence of an embedding from E into ${}^\omega\omega$, *i.e.*, the fact that E is zero-dimensional. $5 \Longrightarrow 4$ comes from the existence of a Hausdorff operation generating Γ, *i.e.*, the fact that Γ is non-self dual. It remains to prove $6 \Longrightarrow 2$.

So let $j\colon E \to {}^\omega\omega$ be an embedding, and let A be such that for all continuous $g\colon {}^\omega\omega \to E$, $g^{-1}(A) \in \Gamma$. As E is Suslin, let $\pi\colon {}^\omega\omega \twoheadrightarrow E$ be a continuous surjection. Then $\pi^{-1}(A)$ is in Γ, so is Borel, and $A_0 = j(A)$ and $A_1 = j(E \setminus A)$ are both $\boldsymbol{\Sigma}^1_1$ in ${}^\omega\omega$. Applying Theorem 4.1 to $\check{\Gamma}$ and (A_0, A_1), one gets that Γ separates (A_0, A_1), else A_0, A_1 would reduce $\check{\Gamma}$ for some $f : {}^\omega\omega \to E$, $f^{-1}(A)$ would be the complete $\check{\Gamma}$-set in ${}^\omega\omega$. But if $B \in \Gamma$ separates A_0 from A_1, $A = j^{-1}(B)$ as desired. ⊣

With the previous definition, it is clear that if $f : E \to F$ is continuous, for E, F zero-dimensional Suslin spaces, and $A \in \Gamma(E)$, then $f^{-1}(A) \in \Gamma(F)$. And one immediately gets, by applying Theorem 4.2 to $A_0 \cup A_1$, the following extension of Theorem 4.1.

THEOREM 4.3. Let E be a zero-dimensional Suslin space, $\Gamma \in W$, and A_0, A_1 two disjoint $\boldsymbol{\Sigma}^1_1$ subsets of E. If the pair (A_0, A_1) does not reduce Γ, the class $\check{\Gamma}(E)$ separates (A_0, A_1).

We now study the general case, where E is still Suslin but not necessarily zero-dimensional. There are clearly difficulties: There are no more embeddings in ${}^\omega\omega$, and the only continuous functions into ${}^\omega\omega$ may be the constants, so that definitions by (1), (2), (3) of Theorem 4.2 cannot be used.

The first remaining possibility is to use descriptions and the corresponding operations, but even this approach has to be changed a bit for descriptions u with $u(0) = 1$, as there might not be enough families of pairwise disjoint open sets in E. So we adopt the following definition.

DEFINITION 4.4. Let E be metric separable. For each $u \in \boldsymbol{D}_1$ the class Γ^E_u is defined (together with $\check{\Gamma}^E_u$, where ˇ refers to the complementation inside E) by the following:

(a) If $u(0) = 0$, then $\Gamma^E_u = \{\varnothing\}$
(b) If $u = \xi^\frown 1^\frown \eta^\frown u^*$ with $u^*(0) = 0$, then $\Gamma^E_u = \boldsymbol{D}_\eta(\boldsymbol{\Sigma}^0_\xi(E))$
(c) If $u = \xi^\frown 2^\frown \eta^\frown u^*$, then $\Gamma^E_u = \boldsymbol{Sep}(\boldsymbol{D}_\eta(\boldsymbol{\Sigma}^0_\xi), \Gamma^E_{u^*})$
(d) If $u = \xi^\frown 3^\frown \eta^\frown \langle u_0, u_1\rangle$ and $\xi \geq 2$, $\Gamma^E_u = \boldsymbol{Bisep}(\boldsymbol{D}_\eta(\Sigma^E_\xi), \Gamma^E_{u_0}, \Gamma^E_{u_1})$
(d') If $u = 1^\frown 3^\frown \eta^\frown \langle v_0, v_1\rangle$ $A \subseteq E$ is in Γ^E_u if there are $\boldsymbol{D}_\eta(\boldsymbol{\Sigma}^0_\xi)$ sets C_0 and C_1 in E, and $B \in \Gamma^E_{u_1}$, such that $A \cap C_0 \in \Gamma^E_{u_0}$, $A \cap C_1 \in \check{\Gamma}^E_{u_0}$, and $A \setminus (C_0 \cup C_1) = B \setminus (C_0 \cup C_1)$
(e) If $u = \xi^\frown 4^\frown \langle u_n : n \in \omega\rangle$ and $\xi \geq 2$, $\Gamma^E_u = \boldsymbol{SU}(\boldsymbol{\Sigma}^0_\xi, (\Gamma^E_{u_n}))$
(e') If $u = 1^\frown 4^\frown \langle u_n : n \in \omega\rangle$, $A \subseteq E$ is in Γ^E_u if every point $x \in A$ admits a neighborhood V with $A \cap V \in \bigcup_n \Gamma^E_{u_n}$
(f) If $u = \xi^\frown 5^\frown \eta^\frown \langle u_0, u_1\rangle$, then $\Gamma^E_u = \boldsymbol{SD}_\eta(\boldsymbol{\Sigma}^0_\xi, \Gamma^E_{u_0}, \Gamma^E_{u_1})$

Note that in case E has dimension zero, one has the reduction property for open sets (it is actually equivalent), so that (d') and (e') above correspond then to the usual definition, *i.e.*, $\Gamma_u^E = \Gamma_u(E)$ when E is a zero-dimensional Suslin space.

Note also that although our definitions d' and e' above look natural, it is now unclear that they correspond to some Hausdorff operation on open sets—and in fact they do not—and it is also unclear that if u and v describe the same class Γ in ${}^\omega\omega$, they also describe the same class in any E. This will be true for Suslin E by Theorem 4.6 below.

Another possible way for defining Γ in a space E comes from (6) of Theorem 4.2.

DEFINITION 4.5. Let E be metric separable, and $\Gamma \in W$. We define $\Gamma(E)$ as the family of those sets $A \subseteq E$ such that every continuous $f : {}^\omega\omega \to E$, $f^{-1}(A) \in \Gamma$.

Note that in Definition 4.5, one could replace "${}^\omega\omega$" by "all zero-dimensional Suslin spaces" by using Theorem 4.2.

THEOREM 4.6. Let E be a Suslin space, and u a first type description. Then $\Gamma_u^E = \Gamma_u(E)$. Hence part (a) of Theorem 4.1 extends to arbitrary Suslin spaces, *i.e.*, if $A \subseteq E$ is Borel and A does not reduce Γ_u, $A \in \check{\Gamma}_u^E$.

The second statement follows from the first, as before. But the proof of the first statement is more involved, and uses a transfer method via open maps.

Recall that $f : E \to F$ is **open** if $f(U)$ is open in F for all U in E. It follows from a result of Hausdorff that if $f : E \twoheadrightarrow F$ is a continuous open surjection and E is Polish, then F is Polish too. Conversely, one has the following classical fact:

PROPOSITION 4.7. Let E be a (non-empty) Polish space. Then there exists a continuous open surjection $\pi : {}^\omega\omega \twoheadrightarrow E$.

PROOF. Let D be a metric for which E is complete, and construct by induction on $\mathrm{lh}(s)$, $s \in {}^{<\omega}\omega$, non-empty open sets U_s in E with $\operatorname{diam} U_s < 2^{-\mathrm{lh}(s)}$, $\overline{U_{s\frown n}} \subseteq U_s$ and $U_\varnothing = E$, $U_s = \bigcup_n U_{s\frown n}$. This is easily done. For each $\alpha \in {}^\omega\omega$, $\bigcap_{s \subseteq \alpha} U_s$ reduces to a singleton $\{\pi(\alpha)\}$, and $\pi : {}^\omega\omega \to E$ defined this way clearly works. ⊣

The next result is a variant of a selection theorem of Saint-Raymond [SR76, Section 4].

THEOREM 4.8. Let E, F be Polish spaces, π a continuous open surjection $E \twoheadrightarrow F$, and f a Baire class 1 function $E \to X$ for some Polish X. Then there exists a selection $s : F \to E$ of π (*i.e.*, $\pi \circ s = \mathrm{id}_F$) such that $f \circ s : F \to X$ is a Baire class 1 function.

PROOF. Let d_E, d_X be complete metrics on E, X respectively. Say that a function S from F into the non-empty closed subsets of E is lsc if for each open U in E $\{x \in F : S(x) \cap U \neq \varnothing\}$ is open in F. For example $x \to \pi^{-1}(x)$

is lsc, as π is open. We construct by induction a sequence S_n of applications from F into the non-empty closed subsets of E such that

(i) $S_0(x) = \pi^{-1}(x), x \in F$
(ii) $S_{n+1}(x) \subseteq S_n(x)$
(iii) For each n, there is a countable partition $(F_k^n)_{k\in\omega}$ of F in $\mathbf{\Delta}_2^0$ sets such that $S_n \upharpoonright F_k^n$ is lsc for all k, and if x, y are in F_k^n, $\alpha \in S_n(x), \beta \in S_n(y)$, then $d_E(\alpha, \beta) < \frac{1}{n}$ and $d_X(f(\alpha), f(\beta)) < \frac{1}{n}$.

Case $n = 0$ is immediate, by taking $F_k^0 = F$. Suppose $(F_k^n)_{k\in\omega}$ and S_n have been defined. Fix k, and construct by transfinite induction a sequence $\langle T_\xi \rangle$ of closed subsets of F_k^n by letting $T_0 = F_k^n$, $T_\lambda = \bigcap_{\xi<\lambda} T_\xi$ for limit λ. And if $T_\xi \neq \varnothing$, let $H_\xi = \overline{\bigcup_{x\in T_\xi} S_n(x)}$, and as f is Baire class 1, let U_ξ be open, of diameter $< \frac{1}{n+1}$, with $U_\xi \cap H_\xi \neq \varnothing$, and such that $diam_X(f(U_\xi \cap H_\xi)) < \frac{1}{n+1}$. Let then $T_{\xi+1} = T_\xi \setminus \{x : S_n(x) \cap U_\xi \neq \varnothing\}$. As F is separable, there is a countable η such that $T_\eta = \varnothing$, and the $(T_\xi \setminus T_{\xi+1})_{\xi<\eta}$ form a partition of F_k^n into $\mathbf{\Delta}_2^0$ sets. And if one defines $S_{n+1}(x)$, for $x \in T_\xi \setminus T_{\xi+1}$, by $\overline{S_n(x) \cap U_\xi}$, S_{n+1} is lsc on $T_\xi \setminus T_{\xi+1}$; so by rearranging the $T_\xi \setminus T_{\xi+1}$ for all F_k^n's, we get S_{n+1} and $(F_k^{n+1})_{k\in\omega}$ as desired. Now $\bigcap_n S_n(x)$ is a singleton $\{s(x)\}$ for each $x \in F$, and this clearly defines a selection $s\colon F \to E$ of π. It remains to show that $f \circ s$ is a first class function. Choose for all n, k a point $\alpha_{n,k}$ in the set $\bigcup\{S_n(x) : x \in F_k^n\}$, and define $g_n\colon F \to X$ by $g_n(x) = f(\alpha_{n,k})$ if $x \in F_k^n$. As the partition of F is in $\mathbf{\Delta}_2^0$ sets, g_n is a Baire class 1 function. And by condition (iii), $f \circ s$ is the uniform limit of the g_n's hence is Baire class 1 too. ⊣

We now state the transfer result from which Theorem 4.6 will follow.

THEOREM 4.9. Let E be a zero-dimensional Polish space, F a Polish space, f a continuous open surjection $E \twoheadrightarrow F$, $u \in \boldsymbol{D}_1$ and A_0, A_1 two $\mathbf{\Sigma}_1^1$ sets in F

(a) If $\Gamma_u(E)$ separates $(f^{-1}(A_0), f^{-1}(A_1))$, then the set A_0 is in $\Gamma_u^{A_0\cup A_1}$
(b) If either (i) $u(0) \geq 2$, or (ii) F is zero-dimensional, or (iii) $A_1 = \check{A}_0$, then if $\Gamma_u(E)$ separates $(f^{-1}(A_0), f^{-1}(A_1))$, Γ_u^F separates (A_0, A_1).

PROOF. Part b(iii) is of course a particular case of part (a). Assume first that $u(0) \geq 2$. One easily defines, by induction on u, a Hausdorff operation D such that $\Gamma_u, \Gamma_u(E), \Gamma_u^F$ are respectively $D(\mathbf{\Sigma}_2^0), D(\mathbf{\Sigma}_2^0(E)), D(\mathbf{\Sigma}_2^0(F))$: the only point is that for $\xi \geq 2$ the $\boldsymbol{D}_\eta(\mathbf{\Sigma}_\xi^0)$ sets do have reduction in any Polish space F. Let then C_n be a sequence of $\mathbf{\Sigma}_2^0(E)$ sets such that $C = D((C_n))$ separates $f^{-1}(A_0)$ from $f^{-1}(A_1)$, and let $g\colon E \to {}^\omega[0,1]$ be a Baire class 1 function with $C_n = \{x \in E : g_n(x) > 0\}$. Applying Theorem 4.8 to $f\colon E \to F$ and $g\colon E \to {}^\omega[0,1]$ gives a selection $s\colon F \to E$ of f with $g \circ s$ of Baire class 1. Let $B_n = \{x \in F : g_n \circ s(x) > 0\}$ and $B = D((B_n))$. The set $B \in \Gamma_u^F$ and as $B = s^{-1}(C)$, it separates A_0 from A_1. This proves both (a) and b(i). So it remains to study the case $u(0) = 1$ (We forget about $\{\varnothing\}$!). It is done by inspection of the various cases.

1. $\Gamma = \boldsymbol{D}_\eta(\boldsymbol{\Sigma}^0_1)$. If $C \in \Gamma_u$ separates $f^{-1}(A_0)$ from $f^{-1}(A_1)$, let $\langle C_\vartheta \rangle_{\vartheta<\eta}$ be an increasing sequence of open sets in E building C. The sequence $f(C_\vartheta)$ is an increasing sequence of open sets in F which build some set $B \in \boldsymbol{D}_\eta(\boldsymbol{\Sigma}^0_1(F))$, and clearly separates A_0 from A_1.

2. $\Gamma_u = \boldsymbol{Sep}(\boldsymbol{D}_\eta(\boldsymbol{\Sigma}^0_1), \Gamma_{u^*})$ with $u^*(0) > 1$. Let $C \in \boldsymbol{D}_\eta(\boldsymbol{\Sigma}^0_1)$, built say from $\langle C_\vartheta \rangle_{\vartheta<\eta}$, be such that $\check{\Gamma}_{u^*}(E)$ separates the pair $(f^{-1}(A_0) \cap C, f^{-1}(A_1) \cap C)$, and $\Gamma_{u^*}(E)$ separates the pair $(f^{-1}(A_0) \setminus C, f^{-1}(A_1) \setminus C)$. Let $D_\vartheta = f(C_\vartheta)$, and $f_\vartheta = f \restriction C_\vartheta \colon C_\vartheta \twoheadrightarrow D_\vartheta$. Clearly as C_ϑ is open, f_ϑ is a continuous open surjection. And depending on the parity of ϑ, $\Gamma_{u^*}(C_\vartheta)$ or $\check{\Gamma}_{u^*}(C_\vartheta)$ separates the pair $(f_\vartheta^{-1}(A_0 \setminus \bigcup_{\vartheta'<\vartheta} D_{\vartheta'}), f_\vartheta^{-1}(A_1 \setminus \bigcup_{\vartheta'<\vartheta} D_{\vartheta'}))$. And as $u^* \geq 2$, we can apply the first case, and the same class separates in D_ϑ $(A_0 \setminus \bigcup_{\vartheta'<\vartheta} D_{\vartheta'}, A_1 \setminus \bigcup_{\vartheta'<\vartheta} D_{\vartheta'})$. And sticking the pieces together gives, for D built from the D_ϑ's, that $\check{\Gamma}^F_{u^*}$ separates $A_0 \cap D$ from $A_1 \cap D$ and $\Gamma^F_{u^*}$ separates $(A_0 \setminus D, A_1 \setminus D)$, as desired.

3. $\Gamma_u = \boldsymbol{Bisep}(\boldsymbol{D}_\eta(\boldsymbol{\Sigma}^0_1), \Gamma_{u_0}, \Gamma_{u_1})$, with $u_0(0) \geq 2$. At first sight, this case looks very similar to the preceding case. But in fact it is where the problem arises: The same argument as before does give two sets D_0 and D_1 in $\boldsymbol{D}_\eta(\boldsymbol{\Sigma}^0_1(F))$ such that $\Gamma^F_{u_0}$ separates $(A_0 \cap D_0, A_1 \cap D_0)$, $\check{\Gamma}^F_{u_0}$ separates $(A_0 \cap D_1, A_1 \cap D_1)$ and $\Gamma^F_{u_1}$ separates $(A_0 \setminus (D_0 \cup D_1), A_1 \setminus (D_0 \cup D_1))$. But we cannot conclude that Γ^F_u separates A_0 from A_1 without sticking the pieces together, and this time D_0 and D_1 are not necessarily disjoint. So we can conclude only if there is no sticking to be done, for we work in $A_0 \cup A_1$ and the only possibility is A_0—this gives part (a) in this case, or if we can reduce D_0 and D_1, which is possible if F is zero-dimensional, and this gives b(ii). [We will see later that this obstruction cannot be avoided].

4. $\Gamma_u = \boldsymbol{SU}(\boldsymbol{\Sigma}^0_1, \langle \Gamma_{u_n} \rangle)$ with $u_n(0) \geq 2$. The proof is similar to case 3: Again one gets open sets D_n in F such that $\Gamma^F_{u_n}$ separates $A_0 \cap D_n$ from $A_1 \cap D_n$, and $A_0 \subseteq \bigcup_n D_n$. This is enough to conclude for part (a) and for b(ii). Note that in this case one can always stick the pieces together, by using locally finite refinements of the D_n's, and the closure properties of the Γ_{u_n}'s, but we won't need this.

5. $\Gamma_u = \boldsymbol{SD}_\eta(\boldsymbol{\Sigma}^0_1, \langle \Gamma_{u_n} \rangle, \Gamma_{u^*})$ creates no difficulty, the argument being as in Case 2, and is left to the reader. ⊣

Proof of Theorem 4.6. We want to show that $\Gamma^E_u = \Gamma_u(E)$ for any Suslin space E. One easily checks, using Theorem 4.2 that $\Gamma^E_u \subseteq \Gamma_u(E)$. If now $A \in \Gamma_u(E)$, let F be Polish with $E \subseteq F$, and by Proposition 4.7 let $\pi \colon {}^\omega\omega \twoheadrightarrow F$ be a continuous open surjection. By the definition of $\Gamma_u(E)$, $\pi^{-1}(A) \in \Gamma_u(\pi^{-1}(E))$, and we can apply Theorem 4.9(a) to ${}^\omega\omega$, F, π and $A_0 = A, A_1 = E \setminus A$, which gives $A \in \Gamma^E_u$. ⊣ (Theorem 4.6)

Concerning separation of analytic sets, Theorem 4.9 does not give the full strength of Theorem 4.1, part (b), even for Polish spaces. One still gets

COROLLARY 4.10. Let E be Polish, A_0, A_1 two disjoint $\boldsymbol{\Sigma}_1^1$ sets in E and Γ a class in W. The following are equivalent:

(i) $A_0 \in \Gamma(A_0 \cup A_1)$.

(ii) For any Polish zero-dimensional space F and continuous $f\colon F \to E$, $\Gamma(F)$ separates $(f^{-1}(A_0), f^{-1}(A_1))$.

(iii) There is a Polish zero-dimensional space F and an open continuous surjection $\pi\colon F \twoheadrightarrow E$ such that $\Gamma(F)$ separates $(\pi^{-1}(A_0), \pi^{-1}(A_1))$.

In most cases, (i) above can be replaced by the stronger "$\Gamma(E)$ separates (A_0, A_1)", *e.g.*, if $\dim(E) = 0$, or if Γ is Γ_u for some $u \in \boldsymbol{D}_1$ with $u(0) \geq 2$, or some specific Γ_u's with $u(0) = 1$, like the $\boldsymbol{D}_\eta(\boldsymbol{\Sigma}_1^0)$'s. But the next example shows it does not work in general (and hence that in general $\Gamma_u(E)$ is not always obtained by some Hausdorff operation performed on open sets).

EXAMPLE 4.11. Let $E = [0,1] \times {}^\omega 2$. E is a one-dimensional compact metrizable space. Let D_0, D_1 be two disjoint countable dense sets in ${}^\omega 2$, and let $\Gamma = \boldsymbol{Bisep}(\boldsymbol{\Sigma}_1^0, \boldsymbol{\Sigma}_2^0)$ and $A_0 = \{(x,\alpha) \in E : (x < 1 \text{ and } \alpha \in D_0) \vee (x = 1 \text{ and } \alpha \notin D_1)\}$, $A_1 = \{(x,\alpha) \in E : (x = 0 \text{ and } \alpha \notin D_0) \vee (x > 0 \text{ and } \alpha \in D_1)\}$. Then $A_0 \in \Gamma(A_0 \cup A_1)$, but $\Gamma(E)$ does not separate A_0 from A_1.

PROOF. On $[0,1[\times{}^\omega 2$, $\boldsymbol{\Sigma}_2^0$ separates A_0, A_1, and on $]0,1] \times {}^\omega 2$, $\boldsymbol{\Pi}_2^0$ separates (A_0, A_1), so $A_0 \in \Gamma(A_0 \cup A_1)$. Suppose, towards a contradiction, that $C \in \boldsymbol{Bisep}(\boldsymbol{\Sigma}_1^0, \boldsymbol{\Sigma}_2^0)(E)$ separates (A_0, A_1). Let $U = \{(x,\alpha) : C \text{ is locally } \boldsymbol{\Sigma}_2^0 \text{ at } (x,\alpha)\}$ and $V = \{(x,\alpha) : C \text{ is locally } \boldsymbol{\Pi}_2^0 \text{ at } (x,\alpha)\}$. By assumption, $C \subseteq U \cup V$. Now, for $x = 0$, $\langle A_0\rangle_x = \langle \check{A}_1\rangle_x = D_0$ is not $\boldsymbol{\Pi}_2^0$, hence $\{0\} \times {}^\omega 2 \cap V = \varnothing$, and similarly $\{1\}\times{}^\omega 2 \cap U = \varnothing$. Fix $\alpha \in D_0$, and let $U_\alpha = \{y : (y,\alpha) \in U\}$ and $V_\alpha = \{y : (y,\alpha) \in V\}$. U_α and V_α are open non-empty by the preceding facts, and cover $[0,1]$ as $[0,1] \times \{\alpha\} \subseteq A_0$. By connectedness, $U_\alpha \cap V_\alpha \neq \varnothing$, so $\exists x \in]0,1[$, $(x,\alpha) \in U \cap V$, hence a neighborhood W of (x,α) such that $W \cap C$ and $W \cap \check{C}$ are $\boldsymbol{\Pi}_2^0$. But both $W \cap C$ and $W \cap \check{C}$ must be dense in W by choice of D_0 and D_1, contradicting the Baire Category theorem. ⊣

§5. Hurewicz Tests and Hurewicz-Type Results. Let Γ be a Wadge class in W. A **Hurewicz-test** for Γ is a pair (K, H) consisting of a zero-dimensional metric compact space K, and a $\Gamma(K)$ subset H of K which satisfies: For every Borel set B in some (non-empty) Suslin space E, $B \notin \check{\Gamma}(E) \iff$ there is a 1-1 continuous map $\varphi\colon K \to E$ with $\varphi^{-1}(B) = H$ (in other words, E contains a homeomorphic copy of the space K on which B is the corresponding copy of H).

Our terminology comes from the well known theorem of Hurewicz on the characterization of Polish spaces, which states, with our terminology, that if D is dense countable in ${}^\omega 2$, $({}^\omega 2, D)$ is a Hurewicz test for the class $\boldsymbol{\Sigma}_2^0$.

It follows from work of Steel [Ste80] that every class Γ_u with $u(0) \geq 2$ and $\boldsymbol{D}_\omega(\boldsymbol{\Sigma}^0_2) \subseteq \Gamma_u$ admits a Hurewicz test—at least for Borel subsets of ${}^\omega\omega$. In [LSR87], we independently reproved Steel's result by a method which yields

THEOREM 5.1 ([LSR87]). Let u be a description with $u(0) \geq 2$. The class Γ_u admits a Hurewicz test (K, H), with $K = {}^\omega 2$.

We extend here this result to all Borel Wadge classes $\Gamma \in W$.

THEOREM 5.2. Every class Γ in W admits a Hurewicz test. Hence in particular, the membership of a Borel set $B \subseteq E$ Suslin in $\Gamma(E)$ depends only on the zero-dimensional compact subsets of E.

We will prove the result by induction on descriptions. For doing this, it is easier to work with type 2 descriptions, except that one has to reduce a bit the allowed combinations:

DEFINITION 5.3.

(i) Let $\langle\Gamma_n\rangle$ be a sequence of classes in W. The sequence $\langle\Gamma_n\rangle$ is admissible if it is increasing and whenever $A = (A_0 \cap C) \cup (A_1 \setminus C)$ with $C \in \boldsymbol{\Sigma}^0_1$, $A_0 \in \bigcup_{n\geq 1} \Gamma_n$ and $A_1 \in \check{\Gamma}_0$, $A \in \bigcup_{n\geq 1} \Gamma_n$.

(ii) We let $\boldsymbol{D}'_2$ be the least set of type 2 descriptions satisfying $\underline{0} \in \boldsymbol{D}'_2$, $u(0)^\frown 1^\frown u \in \boldsymbol{D}'_2$ if $u \in \boldsymbol{D}'_2$, and $\xi^\frown 2^\frown \langle u_n\rangle \in \boldsymbol{D}'_2$ if $u_n \in \boldsymbol{D}'_2$ for all n and the sequence Γ_{u_n} is admissible.

PROPOSITION 5.4. Every class Γ in W admits a $\boldsymbol{D}'_2$-description.

PROOF. It is enough to check that any Γ_u, $u \in \boldsymbol{D}_1$ admits such a description. In fact, one can check that the function $u \to v(u)$ defined in 1.8 takes values in $\boldsymbol{D}'_2$, by induction on $u \in \boldsymbol{D}_1$. ⊣

Note that the only change between $\boldsymbol{D}'_2$ and $\boldsymbol{D}_2$ descriptions occurs for $u(0) = 1$, because by the closure properties of the classes Γ_u, $u(0) \geq 2$, any sequence Γ_{u_n} with $u_n(0) \geq 2$ is automatically admissible.

It is immediate to see that if K is a singleton and $H \subseteq K$ is $\varnothing$, $\langle K, H\rangle$ is a Hurewicz-test for $\Gamma = \{\varnothing\}$, and that if $\langle K, H\rangle$ is a Hurewicz-test for Γ, $\langle K, \check{H}\rangle$ is a Hurewicz test for $\check{\Gamma}$. So the proof of Theorem 5.2 follows from Theorem 5.1 and the following:

THEOREM 5.5. Let $\langle\Gamma_n\rangle$ be an admissible sequence of classes in W which admit Hurewicz-tests. The class $\Gamma = S_1(\langle\Gamma_n\rangle)$ admits a Hurewicz-test too.

LEMMA 5.6. Let E be a Suslin space, and $\Gamma = S_1(\langle\Gamma_n\rangle)$. A Borel set $A \subseteq E$ is in $\Gamma(E)$ iff

There is a sequence $\langle C_n\rangle_{n\geq 1}$ of $\boldsymbol{\Sigma}^0_1$ sets in E such that for $n \geq 1$,

$$A \cap C_n \in \Gamma_n(C_n), \text{ and } A \setminus \bigcup_{n\geq 1} C_n \in \Gamma_0(E \setminus \bigcup_n C_n). \quad (*)$$

PROOF. If A satisfies $(*)$ and $f : {}^\omega\omega \to E$ is continuous, $f^{-1}(A)$ satisfies $(*)$ in ${}^\omega\omega$, and by reducing the $f^{-1}(C_n)$, one gets that $f^{-1}(A) \in \Gamma$ in ${}^\omega\omega$. As this is true for any such f, $A \in \Gamma(E)$. If now $A \in \Gamma(E)$, let $f : {}^\omega\omega \to F$ be a continuous open map onto some Polish space F containing E. Applying Theorem 4.9, we know that some set D in Γ separates $(f^{-1}(A), f^{-1}(E \setminus A))$ in ${}^\omega\omega$, and if $(B_n)_{n\geq 1}$ are $\mathbf{\Sigma}^0_1$ sets with $D \cap B_n \in \Gamma_n$ for $n \geq 1$, and $D \setminus \bigcup_{n\geq 1} B_n = D_0 \setminus \bigcup_{n\geq 1} B_n$ for some $D_0 \in \Gamma_{u_0}$, then by Theorem 4.9 again, the sequence $C_n = f(B_n) \cap E$ witnesses $(*)$ for A. ⊣

PROOF OF THEOREM 5.5. Let (K_n, H_n) be Hurewicz-tests for the classes Γ_n, and $\Gamma = S_1(\langle\Gamma_n\rangle)$. We define a Hurewicz-test (K, H) for Γ as follows:

First we fix a bijection $n \mapsto (n)_0, (n)_1$ between $\omega \setminus \{0\}$ and $(\omega \setminus \{0\}) \times \omega$, and a dense sequence $(x^0_p)_{p\in\omega}$ in K_0. Let K be the set defined by

$$(n, x) \in K \iff n \in \omega \wedge [(n = 0 \text{ and } x \in K_0) \text{ or } (n \neq 0 \text{ and } x \in K_{(n)_0})]$$

that we topologize as follows: the topology is generated by the sets $\{n\} \times V$ for $n \neq 0$ and V open in $K_{(n)_0}$, and by the sets $U_{q,v} = (\{0\} \times V) \cup (\bigcup\{\{n\} \times K_{(n)_0} : n \geq q, x^0_{(n)_1} \in V\})$, where $q \in \omega$ and V is open in K_0. One easily checks that K is compact metrizable of dimension zero. Note that $K^{(n)} = \{n\} \times K_{(n)_0}$ for $n \geq 1$, is clopen in K, and the sequence $K^{(\langle n,p\rangle)}$ converges, for fixed p, to the point $(0, x^0_p)$ of $K^0 = \{0\} \times K_0$.

We now define $H \subseteq K$ by $(n, x) \in H \iff (n = 0 \text{ and } x \in H_0) \vee (n \neq 0 \text{ and } x \in H_{(n)_0})$. Clearly $H \in S_1((\Gamma_n))(K)$ as witnessed by the sequence of open sets $U_n = \bigcup_{(p)_0=n} K^{(p)}, n \geq 1$.

As (K_n, H_n) is a Hurewicz-test for Γ_n, $H_n \notin \check{\Gamma}_n(K_n)$. We claim that $H \notin \check{\Gamma}(K)$: For let $V \subseteq K$ be the open set of all points $z \in K$ admitting a neighborhood V_z with $\check{H} \cap V_z \in \bigcup_{n\geq 1} \Gamma_n$. By the density of the sequence (x^0_p) in K_0 and the convergence property of the $K^{(\langle n,p\rangle)}$, any neighborhood of a $z \in K^{(0)}$ contains sets $K^{(n)}$ for arbitrary large n, hence V is disjoint from $K^{(0)}$. So if $\check{H}$ were in $\Gamma(K)$, one would get $\check{H} \cap K^{(0)}$ in $\Gamma_0(K^{(0)})$, hence H_0 in $\check{\Gamma}_0(K_0)$ a contradiction.

This proves that if E is Suslin and $B \subseteq E$ satisfies $\exists f$ 1-1 continuous: $K \to E$ with $f^{-1}(B) = H$, the set B is not in $\check{\Gamma}(E)$—else H would be in $\check{\Gamma}(K)$. It remains to prove the converse, for B Borel in E. So, we assume $B \notin \check{\Gamma}(E)$. Let for $n \geq 1$, $V_n = \{x \in E : \exists V_x \text{ open } x \in V_x \text{ and } V_x \cap \check{B} \in \Gamma_n(V_x)\}$ and let $F = E \setminus \bigcup_{n\geq 1} V_n$. By Lemma 5.6, if $\check{B} \cap F \in \Gamma_0(F)$, one gets $\check{B} \in \Gamma(E)$, a contradiction. So we can find a 1-1 continuous $f_0 \colon K_0 \to F$ with $f_0^{-1}(B) = H_0$. We now inductively construct a sequence $f_n, n \geq 1$ of 1-1 continuous functions, with $f_n \colon K_{(n)_0} \to E$, such that

(i) the images $f_0(K_0), f_n(K_{(n)_0})$ are all pairwise disjoint

(ii) $f_n^{-1}(B) = H_{(n)_0}$, for $n \geq 1$

(iii) If d is a distance on E, $d(f_0(x^0_{(n)_0}), f_n(K_{(n)_0})) \leq \frac{1}{(n)_1+1}$.

If we can do this, we are clearly done, for $f : K \to E$ defined by $f(n, x) = f_n(x)$ is 1-1 by (i), continuous by (iii) and satisfies $f^{-1}(B) = H$ by (ii).

Suppose the f_n have been constructed for $n < p$. Let $y = f_0(x_{\langle p \rangle_0}) \in f_0(K_0)$. As for $n < p$, $f_0(K_0)$ is disjoint from $f_n(K_{(n)_0})$, we can find an open U in E with $y \in U$, $U \cap f_n(K_{(n)_0}) = \varnothing$ for $n < p$ and $U \subseteq \{z : d(y, z) \leq \frac{1}{\langle p \rangle_1} + 1\}$. We claim that $\breve{B} \cap (U \setminus f_0(K_0)) \notin \Gamma_{\langle p \rangle_0}(E)$: if not, by admissibility of $\langle \Gamma_n \rangle$, the set $\breve{B} \cap U$ would be in $\bigcup_{n \geq 1} \Gamma_n$, contradicting the fact that $U \not\subset \bigcup_{n \geq 1} V_n$ (as $y \in U \setminus \bigcup_{n \geq 1} V_n$). But then we can find a 1-1 continuous $f_p \colon K_{\langle p \rangle_0} \to U \setminus f_0(K_0)$ with $f_p^{-1}(B) = H_{\langle p \rangle_0}$, and this f_p clearly works. ⊣ (Theorem 5.5)

REFERENCES

ALEXANDER S. KECHRIS, DONALD A. MARTIN, AND YIANNIS N. MOSCHOVAKIS

[CABAL iii] *Cabal seminar* 79–81, Lecture Notes in Mathematics, no. 1019, Berlin, Springer, 1983.

ALAIN LOUVEAU

[Lou83] *Some results in the Wadge hierarchy of Borel sets*, this volume, originally published in Kechris et al. [CABAL iii], pp. 28–55.

ALAIN LOUVEAU AND JEAN SAINT-RAYMOND

[LSR87] *Borel classes and closed games: Wadge-type and Hurewicz-type results*, ***Transactions of the American Mathematical Society***, vol. 304 (1987), no. 2, pp. 431–467.

DONALD A. MARTIN

[Mar73] *The Wadge degrees are wellordered*, unpublished, 1973.

JEAN SAINT-RAYMOND

[SR76] *Fonctions boréliennes sur un quotient*, ***Bulletin des Sciences Mathématiques***, vol. 100 (1976), pp. 141–147.

JOHN R. STEEL

[Ste80] *Analytic sets and Borel isomorphisms*, ***Fundamenta Mathematicae***, vol. 108 (1980), no. 2, pp. 83–88.

WILLIAM W. WADGE

[Wad84] *Reducibility and determinateness on the Baire space*, ***Ph.D. thesis***, University of California, Berkeley, 1984.

EQUIPE D'ANALYSE FONCTIONNELLE
INSTITUT DE MATHÉMATIQUES DE JUSSIEU
UNIVERSITÉ PARIS VI
4, PLACE JUSSIEU
75230 PARIS, CEDEX 05
FRANCE
E-mail: louveau@math.jussieu.fr
E-mail: raymond@math.jussieu.fr

CLOSURE PROPERTIES OF POINTCLASSES

JOHN R. STEEL

We work in ZF+AD+DC throughout this paper. Our aim is to show that certain closure and structural properties of a nonselfdual pointclass Γ follow from closure properties of the corresponding Δ together with the regularity of the Wadge ordinal of Δ. Let ${}^3\mathrm{E}$ be the type 3 object embodying quantification over the reals, and $o({}^3\mathrm{E})$ the least ordinal not the order type of a prewellorder of the reals recursive in ${}^3\mathrm{E}$. Our results imply that $o({}^3\mathrm{E})$ is the least regular limit point in the sequence of Suslin cardinals defined in [Kec81B].

The methods and most of the results of the paper fall squarely within the province of "Wadge degrees", which might more informatively be titled "the general theory of arbitrary pointclasses." (Sections 1 to 3 of [Van78B] contain the necessary background material.) Surprisingly, AD is powerful enough to yield nontrivial theorems in this generality. Now, when working in such generality, it is natural to ask: which pointclasses (identified perhaps by means of Wadge ordinals) have the closure and structural properties which make them amenable to the standard techniques of descriptive set theory. Our results bear on this question.

The author wishes to acknowledge the contribution of A. S. Kechris to this work. In a sense, the work was commissioned by him.

Some notation and terminology: We let $\mathbb{R}$ be ${}^\omega\omega$, the Baire space, and call its elements reals. If $\ell \geq 1$, then the product space $\omega^k \times ({}^\omega\omega)^\ell$ is homeomorphic to ${}^\omega\omega$, and we shall identify the two. For $A, B \subseteq {}^\omega\omega$ we say A is Wadge reducible to B, and write $A \leq_{\mathrm{W}} B$, iff $\exists f : {}^\omega\omega \to {}^\omega\omega$ (f is continuous $\wedge A = f^{-1}(B)$). The partial order $\leq_{\mathrm{W}}$ is wellfounded; $|A|_{\mathrm{W}}$ is the ordinal rank of A in $\leq_{\mathrm{W}}$. A pointclass is a class of subsets of ${}^\omega\omega$ closed downward under $\leq_{\mathrm{W}}$. The dual of a pointclass Γ, denoted $\check{\Gamma}$, is $\{-A : A \in \Gamma\}$. Here, as later, complements are taken relative to ${}^\omega\omega$. If Γ is nonselfdual, *i.e.*, $\Gamma \neq \check{\Gamma}$, then we set $\Delta = \Gamma \cap \check{\Gamma}$. If Γ is any pointclass, we set $o(\Gamma) = \sup\{|A|_{\mathrm{W}} : A \in \Gamma\}$. If Γ is nonselfdual and $o(\Gamma)$ a limit ordinal, then of course $o(\Gamma) = o(\Delta)$.

We write $\mathsf{Sep}(\Gamma)$, $\mathsf{Red}(\Gamma)$, or $\mathsf{PWO}(\Gamma)$ to mean that Γ has, respectively, the separation, reduction, or prewellordering property. The closure of Γ under existential real quantification is given by

$$\exists^{\mathbb{R}}\Gamma = \{A : \exists B \in \Gamma \forall x (x \in A \Leftrightarrow \exists y (x, y) \in B)\}.$$

Wadge Degrees and Projective Ordinals: The Cabal Seminar, Volume II
Edited by A. S. Kechris, B. Löwe, J. R. Steel
Lecture Notes in Logic, 37

We let $\forall^{\mathbb{R}}\Gamma$ be the dual of $\exists^{\mathbb{R}}\check{\Gamma}$. Similarly, the class of wellordered unions of length α of Γ sets is

$$\bigcup_{\alpha}\Gamma = \Big\{\bigcup_{\gamma<\alpha} A_{\gamma} : \forall\gamma < \alpha(A_{\gamma} \in \Gamma)\Big\}$$

and $\bigcap_{\alpha}\Gamma$ is the dual of $\bigcup_{\alpha}\check{\Gamma}$.

The selfdual pointclasses Δ which we consider will often satisfy $\exists^{\mathbb{R}}\Delta \subseteq \Delta$. In this case, Lemma 2.3.1 of [KSS81] states

$$\begin{aligned} o(\Delta) &= \sup\{\text{rank}(\prec) : \prec \text{ is a wellfounded relation in } \Delta\} \\ &= \sup\{\text{rank}(\prec) : \prec \text{ is a prewellorder in } \Delta\}. \end{aligned}$$

§1. Consequences of the separation property. The key to the transfer of closure properties from Δ to Γ is the separation property. We begin with some simple results in this vein. Notice that the hypothesis $\mathsf{Sep}(\Gamma)$ of Theorem 1.1 serves only to distinguish Γ from $\check{\Gamma}$, since by [Ste81B] and [Van78B], exactly one of $\mathsf{Sep}(\Gamma)$ and $\mathsf{Sep}(\check{\Gamma})$ holds.

THEOREM 1.1. Let Γ be nonselfdual, and suppose $\mathsf{Sep}(\Gamma)$. Then
(a) $\bigcup_2 \Delta \subseteq \Delta \Rightarrow \bigcup_2 \Gamma \subseteq \Gamma$ and $\bigcup_{\omega} \Delta \subseteq \Delta \Rightarrow \bigcup_{\omega} \Gamma \subseteq \Gamma$;
(b) $\exists^{\mathbb{R}}\Delta \subseteq \Delta \Rightarrow \exists^{\mathbb{R}}\Gamma \subseteq \Gamma$;
(c) $(\exists^{\mathbb{R}}\Delta \subseteq \Delta \wedge \alpha < \text{cf}(o(\Delta))) \Rightarrow (\bigcup_{\alpha}\Delta \subseteq \Delta \wedge \bigcup_{\alpha}\Gamma \subseteq \Gamma)$.

PROOF. (a) Suppose $\bigcup_2 \Delta \subseteq \Delta$ and $\bigcup_2 \Gamma \not\subseteq \Gamma$. Since $\bigcup_2 \Gamma$ is a pointclass, *i.e.*, closed downward under $\leq_{\mathrm{W}}$, Wadge's lemma implies $\check{\Gamma} \subseteq \bigcup_2 \Gamma$. Let $A \in \check{\Gamma} - \Gamma$, and $A = B \cup C$ where $B, C \in \Gamma$. By $\mathsf{Sep}(\Gamma)$ we have $D, E \in \Delta$ so that

$$B \subseteq D \subseteq A$$

and

$$C \subseteq E \subseteq A.$$

But then $A = D \cup E$, so $A \in \Delta$, a contradiction. The proof that $\bigcup_{\omega} \Delta \subseteq \Delta \Rightarrow \bigcup_{\omega} \Gamma \subseteq \Gamma$ is the same.

(b) It is enough to show that if A and B are disjoint sets in $\exists^{\mathbb{R}}\Gamma$, then A is separable from B by a set in Δ. We use the idea of Addison's proof of $\mathsf{Sep}(\undertilde{\Sigma}^1_3)$ for this. Let

$$\begin{aligned} A(x) &\iff \exists y P(x, y), \\ B(x) &\iff \exists y Q(x, y), \end{aligned}$$

where $P, Q \in \Gamma$. Define

$$\begin{aligned} P'(x, y, z) &\iff P(x, y), \\ Q'(x, y, z) &\iff Q(x, z), \end{aligned}$$

and by Sep(Γ) let $D \in \Delta$ and $P' \subseteq D \subseteq -Q'$. Define

$$C(x) \iff \exists y \forall z D(x, y, z).$$

Then $C \in \Delta$ since $\exists^{\mathbb{R}}\Delta \subseteq \Delta$. It is easy to check that $A \subseteq C \subseteq -B$.

(c) We extend the proof of (a). Suppose $\alpha < \mathrm{cf}(o(\Delta))$ and $\exists^{\mathbb{R}}\Delta \subseteq \Delta$, but $\bigcup_\alpha \Gamma \nsubseteq \Gamma$. Then $\check{\Gamma} \subseteq \bigcup_\alpha \Gamma$ by Wadge's lemma. Let $A \in \check{\Gamma} - \Gamma$, and $A = \bigcup_{\beta<\alpha} A_\beta$ where each $A_\beta \in \Gamma$. Since Sep(Γ) and $\alpha < \mathrm{cf}(o(\Delta))$, we can find a set $B \in \Delta$ so that

$$\beta < \alpha \Rightarrow \exists C \leq_{\mathrm{W}} B(A_\beta \subseteq C \subseteq A).$$

Let $\varphi\colon \mathbb{R} \twoheadrightarrow \alpha$ be a Δ-norm of length α. By the Coding Lemma of Moschovakis [Mos70] there is a relation R in Δ so that $\forall x \exists y R(x, y)$ and $R(x, y) \wedge \varphi(x) = \beta \Rightarrow A_\beta \subseteq B_y \subseteq A$. (Here B_y is the set $\leq_{\mathrm{W}} B$ via the strategy y.) But then

$$z \in A \iff \exists x \exists y (R(x, y) \wedge z \in B_y),$$

so $A \in \Delta$, a contradiction.

The proof that $\bigcup_\alpha \Delta \subseteq \Delta$ is the same. ⊣ (Theorem 1.1)

Part (b) is due to Kechris, and part (a) to Kechris and the author independently. Notice that the proof of (c) gives slightly more: if $\exists^{\mathbb{R}}\Delta \subseteq \Delta, \alpha < \mathrm{cf}(o(\Delta))$, and $\langle A_\beta : \beta < \alpha \rangle$ is any sequence of sets each of which is Δ-separable from a fixed set A, then $\bigcup_{\beta<\alpha} A_\beta$ is Δ-separable from A. This fact will be important in the proof of Theorem 2.1.

Theorem 1.1 leaves us the question: given that Sep(Γ) and $\exists^{\mathbb{R}}\Delta \subseteq \Delta$, must $\bigcap_2 \Gamma \subseteq \Gamma$? The next theorem provides a class of examples showing the extent to which $\bigcap_2 \Gamma \subseteq \Gamma$ can in fact fail. The theorem results from analysis of the example of a Type II hierarchy given in [Kec77B], p. 260.

If $A \subseteq \mathbb{R}$ and $\varphi\colon A \twoheadrightarrow \lambda$, we say φ is Γ-bounded just in case whenever $B \subseteq A$ and $B \in \Gamma$ there is a $\beta < \lambda$ so that $\varphi'' B \subseteq \beta$.

Theorem 1.2. Suppose Sep(Γ) and $\bigcup_2 \Delta \subseteq \Delta$. Let $A \in \Delta$ and $\varphi\colon A \twoheadrightarrow \lambda$ be $\mathbf{\Sigma}^1_1$ bounded, where $\lambda = \mathrm{cf}(o(\Delta))$. Then for some $B \in \Gamma$, $A \cap B \notin \Gamma$.

Proof. Let $\{v_\alpha : \alpha < \lambda\}$ be cofinal in $o(\Delta)$. Let $W \subseteq \mathbb{R}^2$ be a universal set in Γ. Consider the Solovay game:

Player I	x
Player II	$\langle y, z \rangle$

Player II wins iff $x \in A \Rightarrow (W_y = -W_z \wedge |W_y|_{\mathrm{W}} \geq v_{\varphi(x)})$.

Since φ is $\mathbf{\Sigma}^1_1$ bounded, player II must have a winning strategy σ. Let

$$R(x, y) \iff x \in A \wedge y \notin W_{\sigma(x)_1}.$$

Since $\check{\Gamma}$ is closed under intersection by Theorem 1.1(a), $R \in \check{\Gamma}$. But $\{|R_x|_{\mathrm{W}} : x \in A\}$ is unbounded in $o(\Delta)$, so $R \notin \Delta$, and thus $R \notin \Gamma$. On the other hand,

$$R(x, y) \iff x \in A \wedge y \in W_{\sigma(x)_0},$$

so that $R = A \cap B$ for some $B \in \Gamma$. ⊣

Theorem 1.2 implies, for example, that if $\exists^{\mathbb{R}}\Delta \subseteq \Delta$, $\mathsf{Sep}(\Gamma)$, and $\mathrm{cf}(o(\Delta)) = \boldsymbol{\delta}^1_n$, then Γ is not closed under intersection with Π^1_n sets. [Let $S \subseteq \mathbb{R}^3$ be universal Σ^1_n, and let $A = \{x : S_x \text{ is wellfounded}\}$. Then A is Π^1_n, and the map $\varphi\colon A \twoheadrightarrow \boldsymbol{\delta}^1_n$, where $\varphi(x) =$ rank of S_x, is $\boldsymbol{\Sigma}^1_1$-bounded.] On the other hand, Theorem 2.1 to follow implies that under these hypotheses on Γ and Δ, Γ is closed under intersections with $\boldsymbol{\Sigma}^1_n$ sets.

On a grosser scale, one can modify the proof of Theorem 1.2 slightly (replacing the Solovay game by the Coding Lemma) to show that if $\mathsf{Sep}(\Gamma)$, $\exists^{\mathbb{R}}\Delta \subseteq \Delta$, and $o(\Delta)$ is singular, then Γ is not closed under intersections with Δ sets.

Theorem 1.2 implies that the hypothesis "$\alpha < \mathrm{cf}(o(\Delta))$" in Theorem 1.1(c) is necessary. For consider any Γ such that $\mathsf{Sep}(\Gamma)$, $\exists^{\mathbb{R}}\Delta \subseteq \Delta$, and $\mathrm{cf}(o(\Delta)) = \omega_1$. Let $\varphi\colon A \twoheadrightarrow \omega_1$ be a Π^1_1 norm on a complete Π^1_1 set. Define R as in Theorem 1.2. Then $R = \bigcup_{\alpha<\omega_1} R_\alpha$ where

$$R_\alpha(x, y) \iff \varphi(x) \le \alpha \wedge y \in W_{\sigma(x)_1}.$$

But each R_α is in Δ by Theorem 2.1.

Kechris and Martin have located the pointclass Γ such that $o(\Delta) = \omega_2$ with the aid of Theorem 1.2. Namely, let Γ be the class of ω-$\boldsymbol{\Pi}^1_1$ sets, that is, sets of the form

$$A = \bigcup_{n<\omega} A_{2n} - A_{2n+1}$$

where $\langle A_n : n < \omega \rangle$ is a decreasing sequence of $\boldsymbol{\Pi}^1_1$ sets. Then Γ is nonselfdual, and both Γ and $\check{\Gamma}$ are closed under intersections with $\boldsymbol{\Pi}^1_1$ sets. By Theorem 1.2 we have $o(\Delta) \ge \omega_2$. By analyzing the ordinal games associated to Wadge games involving sets in Δ, Martin showed $o(\Delta) \le \omega_2$. Thus $o(\Delta) = \omega_2$.

It is unpleasant to have a natural ordinal like ω_2 assigned to an unnatural class like ω-$\boldsymbol{\Pi}^1_1$. Solovay has shown that we get a more natural assignment if we replace $\le_{\mathrm{W}}$ by the somewhat coarser $\le_\sigma$, where

$$B \le_\sigma A \iff \exists \langle A_n : n < \omega \rangle \Big(\forall n (A_n \le_{\mathrm{W}} A) \wedge B \le_{\mathrm{W}} \bigcup_{n<\omega} A_n \Big).$$

The ordinal ω_1 is assigned now to $\boldsymbol{\Pi}^1_1, \omega_2$ to $A(\boldsymbol{\Pi}^1_1), \omega_3$ to $A(A(\boldsymbol{\Pi}^1_1))$, *etc.* (Here "$A$" denotes Suslin's operation A.) The ordering $\le_\sigma$ behaves much like the order $\le_m$ of jump operators defined and studied in [Ste82]. For example, the wellfoundedness of $\le_\sigma$ can be proved by a direct diagonal argument like that of Lemma 3 of that paper.

§2. Applications of the Martin-Monk method. We return to our closure questions. The limitations established by Theorem 1.2 clearly rely on the singularity of $o(\Delta)$. Suppose then that $\mathsf{Sep}(\Gamma)$, $\exists^{\mathbb{R}}\Delta \subseteq \Delta$, and $o(\Delta)$ is regular;

does it follow that $\bigcap_2 \Gamma \subseteq \Gamma$? We believe so, but at present have a proof only for the case that every set in Δ is κ-Suslin for some $\kappa < o(\Delta)$. This partial result will be enough for our characterization of $o({}^3\mathrm{E})$, since by [Kec81B], every set A inductive over $\mathbb{R}$ is κ-Suslin for some $\kappa \leq |A|_{\mathrm{W}}$. The key to our partial result is the following theorem.

THEOREM 2.1. Let Γ be nonselfdual and suppose $\exists^{\mathbb{R}}\Delta \subseteq \Delta$. Let A be κ-Suslin, where $\kappa < \mathrm{cf}(o(\Delta))$. Then for any $B \in \Gamma$, $A \cap B \in \Gamma$.

PROOF. This follows from Theorem 1.1(a) if $\mathsf{Sep}(\check{\Gamma})$ holds, so assume $\mathsf{Sep}(\Gamma)$. Let A, B, and κ be as in the hypotheses, and suppose for a contradiction that $A \cap B \notin \Gamma$. Let σ be a winning strategy for player I in the Wadge game $G_{\mathrm{W}}(A \cap B, B)$ (*cf.* [Van78B]). Thus whenever $\sigma(x) \in A$, we have

$$x \in B \iff \sigma(x) \notin B.$$

We shall use the fact that σ flips membership in B this way to get a contradiction like that in Martin's proof that $\leq_{\mathrm{W}}$ is wellfounded.

Specifically, we define a sequence $\langle \sigma_n : n < \omega \rangle$ of winning strategies for player I in $G_{\mathrm{W}}(A \cap B, B)$. Let τ be the copying strategy for player II, *i.e.*, let $\forall x(\tau(x) = x)$. For any $x \in {}^\omega 2$, define

$$\tau_n = \begin{cases} \sigma_n, & \text{if } x(n) = 0 \\ \tau, & \text{if } x(n) = 1. \end{cases}$$

Consider the diagram of games

$$\begin{array}{cccc}
\cdots & \tau_2 & \tau_1 & \tau_0 \\
 & x_2(0) & x_1(0) & x_0(0) \\
 & . & x_1(1) & x_0(1) \\
 & . & . & x_0(2) \\
 & \vdots & \vdots & \vdots \\
\cdots & x_2 & x_1 & x_0
\end{array}$$

The rule here is: $x_n = \tau_n(x_{n+1})$. If $x(n) = 0$ for infinitely many n, then because each σ_n is a strategy for player I, there is a unique such sequence $\langle x_n : n < \omega \rangle$. We shall define the σ_n's so that for any $x \in {}^\omega 2$, if $x(n) = 0$ for infinitely many n and $\langle x_n : n < \omega \rangle$ is derived from x in this way, then $x_n \in A$ for all n.

Suppose we have done this; then the standard Martin argument leads to a contradiction. For let $I = \{x \in {}^\omega 2 : x(n) = 0 \text{ for infinitely many } n\}$. Define

$$M = \{x \in I : x_0 \in B\}.$$

Since M has the Baire property we have a basic interval [s] determined by some $s \in 2^{<\omega}$ on which M is either meager or comeager. Pick $i \notin \mathrm{dom}(s)$, and let

$$T(x)(k) = \begin{cases} x(k) & \text{if } i \neq k \\ 1 - x(k) & \text{if } i = k. \end{cases}$$

Then T is a homeomorphism and $T''[s] = [s]$. If $x \in I$, then

$$T(x)_k = x_k \text{ for } k > i,$$

and

$$T(x)_k \in B \text{ iff } x_k \notin B \text{ for } k \leq i.$$

Thus $T''(M \cap I \cap [s]) = -M \cap I \cap [s]$. Since I is comeager, this contradicts our choice of s.

We now define the σ_n's by induction on n. Let T be a tree on $\omega \times \kappa$ such that $A = p([T]) = \{x \in {}^{\omega}\omega : \exists f \in {}^{\omega}\kappa\, (x, f) \in [T])\}$. (Here [T] is the set of infinite branches of T.) As we define σ_n we shall associate to any $\langle \tau_n : i \leq n \rangle$ such that $\tau_n = \sigma_n$ and $\forall i < n(\tau_i = \sigma_i \text{ or } \tau_i = \tau)$, and any $i \leq n$ such that $\tau_i = \sigma_i$, a sequence of ordinals $\langle \xi_0, \ldots, \xi_n \rangle = \xi_{\vec{\tau}, i}$.

We arrange that for any $z \in {}^{\omega}\omega$, if the partial diagram

$$\begin{array}{ccccccccc} & \tau_n & & \tau_{n-1} & & \cdots & & \tau_0 & \\ \vdots & & \vdots & & \vdots & & \vdots & & \vdots \\ z & & x_n & & x_{n-1} & \cdots & x_1 & & x_0 \end{array}$$

is filled in as before (*i.e.*, $x_k = \tau_k(x_{k+1})$), but setting $x_{n+1} = z$, then

(i) $(x_i \restriction n + 1, \langle \xi_0, \ldots, \xi_n \rangle) \in T$, and

(ii) $z \notin B \Rightarrow \exists f (f \restriction n + 1 = \langle \xi_0, \ldots, \xi_n \rangle \wedge (x_i, f) \in [T])$.

Moreover these sequences of ordinals cohere in the natural way, that is,

(iii) $\vec{\xi}_{\vec{\tau} \restriction k+1, i} = \vec{\xi}_{\vec{\tau}, i} \restriction k + 1$, for $i \leq k < n$.

It will be enough to define $\langle \sigma_i : i < \omega \rangle \rangle$ with associated $\vec{\xi}$'s satisfying (i)–(iii). For then suppose $x \in {}^{\omega}2$ and $x(n) = 0$ for infinitely many n. Let $\tau_i = \sigma_i$ if $x(i) = 0$, and $\tau_i = \tau$ otherwise. Define x_n by: $x_n = \tau_n(x_{n+1})$. For $x(n) = 0$, let

$$f = \bigcup_{x(k)=0} \vec{\xi}_{\langle \tau_0, \ldots, \tau_k \rangle, n}.$$

Then $f \in {}^{\omega}\kappa$ by (iii), and by (i), $(x_n \restriction k, f \restriction k) \in T$ for all k. Thus $x_n \in A$. If $x(n) \neq 0$, then $x_n = x_i$ for some i such that $x(i) = 0$. Thus $x_n \in A$ for all n, and we are done.

We now define σ_n. Suppose that σ_i is defined for $i < n$, together with associates satisfying (i)–(iii) above. We define σ_n in 2^n steps, one for each $\langle \tau_i : i < n \rangle \rangle$ with $\tau_i = \sigma_i$ or $\tau_i = \tau$ for all $i < n$. After step ℓ we have a Δ-inseparable pair $C_\ell \subseteq -B$ and $D_\ell \subseteq B$ with $D_\ell \in \Gamma$. We will have $C_{\ell+1} \subseteq C_\ell$ and $D_{\ell+1} \subseteq D_\ell$ for $\ell < 2^n$.

Step 0. For each $\vec{\xi} \in \kappa^n$ set

$$A_{\vec{\xi}} = \{x : \exists f (f \restriction (n+1) = \vec{\xi} \wedge (x, f) \in [T])\}.$$

Thus $A = \bigcup_{\vec{\xi} \in \kappa^n} A_{\vec{\xi}}$. Since $\sigma''(-B) \subseteq A$ and $\kappa < \mathrm{cf}(o(\Delta))$, the remark immediately after Theorem 1.1 implies that for some $\vec{\xi}$, $-B \cap \sigma^{-1}(A_{\vec{\xi}})$ is Δ-inseparable from B. Fix such a $\vec{\xi}$, and let

$$C_0 = -B \cap \sigma^{-1}(A_{\vec{\xi}})$$

and

$$D_0 = \{x \in B : \sigma(x) \restriction n+1, \vec{\xi}) \in T\}.$$

$D_0 \in \Gamma$ since Γ is closed under intersections with clopen sets, and one easily checks that D_0 is Δ-inseparable from C_0. Finally, at this step we set $\vec{\xi}_{\vec{\tau},n} = \vec{\xi}$ for all $\vec{\tau} = \langle \tau_i : i \leq n \rangle$ such that $\tau_i = \sigma_i$ or $\tau_i = \tau$ for $i < n$, and $\tau_n = \sigma_n$. (Or more precisely, we commit ourselves to doing so once we have defined σ_n.)

Step $k+1$. We have (C_k, D_k) from the last step, and we are considering $\langle \tau_i : i < n \rangle$. Let $i < n$ be largest such that $\tau_i = \sigma_i$; if no such i exists set $C_{k+1} = C_k, D_{k+1} = D_k$, and go to step k+2 without defining any new associates. For each $j < n$ such that $\tau_j = \sigma_j$, let $\vec{\xi}_j = \vec{\xi}_{\langle \tau_\ell : \ell \leq i \rangle, j}$. Besides defining C_{k+1} and D_{k+1}, we want to extend these associates.

For each $z \in {}^\omega\omega$ consider the diagram

$$\begin{array}{ccccccc} & \sigma & \tau_{n-1} & \cdots & & \cdots & \tau_0 \\ \vdots & \vdots & \vdots & \vdots & \vdots & & \\ z & z_n & & \cdots & z_{i+1} & z_i & \cdots & z_0 \end{array}$$

filled in as before by setting $z_{n+1} = z$.

For $j < n$ so that $\tau_j = \sigma_j$, define

$$A_{\vec{\xi}_j} = \{x : \exists f (f \restriction (i+1) = \vec{\xi}_j \wedge (x, f) \in [T]\}.$$

So (ii) of our inductive hypothesis on $\langle \tau_\ell : \ell \leq i \rangle$ and its associates says:

$$z_{i+1} \notin B \Rightarrow z_j \in A_{\vec{\xi}_j}.$$

Now notice that if $z \in D_k$ (so $z \in B$) and $\sigma(z) \in A$, then $z_n = z_{i+1} \notin B$, and thus $z_j \in A_{\vec{\xi}_j}$ for all $j \leq i$. Define

$$X = \{z : \sigma(z) \in A \wedge \exists j \leq i (z_j \notin A_{\vec{\xi}_j})\}.$$

Then $X \in \Delta$, and $D_k \cap X = \varnothing$. Since (C_k, D_k) is Δ-inseparable, $(C_k - X, D_k)$ must be Δ-inseparable.

Notice that for $z \in C_k - X$ we have $z \notin B$, so $\sigma(z) \in A$, and thus $z_j \in A_{\vec{\xi}_j}$ for all $j \leq i$ with $\tau_j = \sigma_j$. This enables us to use the argument of step 0 to successively thin down $C_k - X$ and D_k, once for each $j \leq i$ so that $\tau_j = \sigma_j$, retaining at each step an inseparable pair (C'_j, D'_j) with $D_j \in \Gamma$. At the step for j we also define the associate $\vec{\xi}_{\langle\tau_i : i \leq n\rangle, j}$ extending $\vec{\xi}_{\langle\tau_\ell : \ell \leq i\rangle, j}$ so that

(i)′ $z \in D'_j \Rightarrow (z_j \restriction (n+1), \vec{\xi}_{\langle\tau_i : i \leq n\rangle, j}) \in T$,

(ii)′ $z \in C'_j \Rightarrow \exists f (f \restriction (n+1) = \vec{\xi}_{\langle\tau_i : i \leq n\rangle, j} \wedge (z_j, f) \in [T]$.

We define (C_{k+1}, D_{k+1}) to be the last pair in this process, and go to step $k+2$. Now let $(C, D) = (C_{2^n}, D_{2^n})$ be the pair we have on completion of the last step. Consider the game in which player I plays y, player II plays z, and player II wins iff

$$y \notin B \Rightarrow z \in C$$

and

$$y \in B \Rightarrow z \in D.$$

Then player I has no winning strategy in this game. For if s is such a strategy, then $C \subseteq s^{-1}(B)$ and $s^{-1}(B) \cap D = \varnothing$. Since $s^{-1}(B)$ and D are disjoint Γ sets, they can be separated by a Δ set. But such a Δ set separates C and D, a contradiction.

Fix a winning strategy s for player II, and let $\sigma_n = \sigma \circ s$. Given any $\langle \tau_j : j \leq n \rangle$ with $\tau_n = \sigma_n$, and given any $j < n$ with $\tau_j = \sigma_j$, the induction hypotheses (i) and (ii) for $\vec{\tau}$ and $\vec{\xi}_{\vec{\tau}, j}$ follow at once from (i)′ and (ii)′ for C'_j and D'_j at the step at which $\langle \tau_i : i < n \rangle$ was considered, and the fact that

$$s''(-B) \subseteq C \subseteq C'_j$$

and

$$s''(B) \subseteq D \subseteq D_j.$$

The construction of $\langle \sigma_n : n < \omega \rangle$, and hence the proof of the theorem, is complete. ⊣ (Theorem 2.1)

For Γ such that $\mathrm{Sep}(\Gamma)$ and $\exists^{\mathbb{R}} \Delta \subseteq \Delta$, it should be possible to specify exactly, as a function of $\mathrm{cf}(o(\Delta))$, those pointclasses Γ' so that Γ is closed under intersections with Γ' sets. Theorems 1.2 and 2.1 do this when $\mathrm{cf}(o(\Delta)) = \boldsymbol{\delta}^1_n$; such that Γ is closed under intersections with $\boldsymbol{\Sigma}^1_n$ sets, but not under intersections with $\boldsymbol{\Pi}^1_n$ sets.

By combining Theorems 1.2 and 2.1 we obtain the following curious fact: let A be κ-Suslin and let $\varphi \colon A \twoheadrightarrow \lambda$ be $\boldsymbol{\Sigma}^1_1$-bounded. Then $\mathrm{cf}(\lambda) \leq \kappa$. [Proof. Let Δ be such that $\exists^{\mathbb{R}} \Delta \subseteq \Delta$ and $\mathrm{cf}(o(\Delta)) = \mathrm{cf}(\lambda)$. Let $\Delta = \Gamma \cap \breve{\Gamma}$, where $\mathrm{Sep}(\Gamma)$ holds. By Theorem 1.2, Γ is not closed under intersections with A. By Theorem 2.1 then, $\mathrm{cf}(o(\Delta)) \leq \kappa$.] This fact is easy to prove for natural λ, *e.g.*, $\lambda = \boldsymbol{\delta}^1_n$, but we see no proof for arbitrary λ which does not use Theorems 1.2 and 2.1.

The basic method in the proof of Theorems 2.1 is due to D. Martin and L. Monk; as we mentioned, Martin used it to show $\leq_W$ is wellfounded. At present this method and variants on Wadge's lemma seem to be the only tools in pure Wadge theory.

We need one further preliminary closure result. Again, the Martin-Monk method is the key.

THEOREM 2.2. Suppose $\neg\mathrm{Sep}(\Gamma)$ and $\bigcup_2 \Gamma \subseteq \Gamma$. Then $\bigcup_\omega \Gamma \subseteq \Gamma$.

PROOF. Let (A_0, A_1) be a Δ-inseparable pair of Γ sets. Suppose that $\bigcup_\omega \Gamma \not\subseteq \Gamma$, so that $\check{\Gamma} \subseteq \bigcup_\omega \Gamma$ by Wadge's lemma. Since $\bigcup_2 \Gamma \subseteq \Gamma$, we have $-(A_0 \cap A_1) \in \check{\Gamma}$, and hence $-(A_0 \cup A_1) = \bigcup_n C_n$ for some sequence $\langle C_n : n < \omega\rangle$ of Γ sets.

Now $(A_0 \cup C_n, A_1)$ is a disjoint pair of Γ sets, and so the lemma of [Ste81B] gives Lipschitz continuous maps f_n, $n < \omega$, such that

$$f_{2n}(A_0 \cup C_n) \subseteq A_0 \wedge f_{2n}(A_1) \subseteq A_1,$$

and

$$f_{2n+1}(A_0 \cup C_n) \subseteq A_1 \wedge f_{2n+1}(A_1) \subseteq A_0.$$

We proceed to the usual contradiction. For any $x \in {}^\omega\omega$ let $\langle x_n : n < \omega\rangle$ be the unique sequence such that $x_n = f_n(x_{n+1})$. Suppose that $\{x : x_0 \in C_n\}$ is nonmeager; say comeager on the interval determined by $s \in \omega^{<\omega}$. Notice that

$$x_0 \in C_n \Rightarrow (s^\frown\langle 2n\rangle^\frown x)_0 \in A_0 \cup A_1,$$

so that for nonmeager many $y \supseteq s$, $y_0 \notin C_n$, a contradiction. On the other hand, suppose $\{x : x_0 \in A_i\}$ is nonmeager; say comeager on the interval determined by s. Notice that

$$x_0 \in A_i \Rightarrow (s^\frown\langle 1\rangle^\frown x)_0 \in A_{1-i},$$

so that for nonmeager many $y \supseteq s$, $y_0 \in A_{1-i}$, a contradiction. $\dashv$

The hypothesis $\neg\mathrm{Sep}(\Gamma)$ in Theorem 2.2 cannot be omitted, as witnessed by the case $\Gamma = \underset{\sim}{\Pi}^0_\alpha$. Theorem 1.2 shows that hypothesis $\bigcup_2 \Gamma \subseteq \Gamma$ cannot be omitted, even if we assume strong closure properties of Δ. However, the proof of Theorem 2.2 can be modified to show that if $\bigcup_\omega \Delta \subseteq \Delta$ and $\langle A_n : n < \omega\rangle$ is any increasing sequence of Γ sets, then $\bigcup_2 A_n \in \Gamma$.

[Van78B] shows that for any nonselfdual Γ, either $\neg\mathrm{Sep}(\Gamma)$ or $\neg\mathrm{Sep}(\check{\Gamma})$. Thus if both $\bigcup_2 \Gamma \subseteq \Gamma$ and $\bigcup_2 \check{\Gamma} \subseteq \check{\Gamma}$, then either $\bigcup_\omega \Gamma \subseteq \Gamma$ or $\bigcup_\omega \check{\Gamma} \subseteq \check{\Gamma}$. It seems quite likely that if both $\bigcup_\omega \Gamma \subseteq \Gamma$ and $\bigcup_\omega \check{\Gamma} \subseteq \check{\Gamma}$ then either $A(\Gamma) \subseteq \Gamma$ or $A(\check{\Gamma}) \subseteq \check{\Gamma}$, where "A" denotes Suslin's operation A. More vaguely, one might guess that there is always an asymmetry between the closure properties of Γ and those of $\check{\Gamma}$.

From Theorem 2.2 we obtain a prewellordering theorem for classes closed under one but not both of $\bigcup_\omega$ and $\bigcap_\omega$. The theorem is analogous to those of [KSS81].

COROLLARY 2.3. Let Γ be nonselfdual, $\bigcup_\omega \Gamma \subseteq \Gamma$ and $\bigcap_2 \Gamma \subseteq \Gamma$, but $\bigcap_\omega \Gamma \nsubseteq \Gamma$. Then $\mathsf{PWO}(\Gamma)$.

PROOF. Corollary 2.3 By Theorem 2.2, $\mathsf{Sep}(\check{\Gamma})$ holds, as otherwise $\bigcup_\omega \check{\Gamma} \subseteq \check{\Gamma}$. But then, by Theorem 1.1(a), $\bigcup_\omega \Delta \nsubseteq \Delta$, as otherwise $\bigcup_\omega \check{\Gamma} \subseteq \check{\Gamma}$. Since $\bigcup_\omega \Gamma \subseteq \Gamma$, we have $\bigcup_\omega \Delta = \Gamma$. For $A \in \Gamma$, let $A = \bigcup_n B_n$ where each $B_n \in \Delta$, and set for $x \in A$

$$\varphi(x) = \text{ least n such that } x \in B_n.$$

Since $\bigcup_2 \Delta \subseteq \Delta$, φ is a Γ norm. ⊣

The corollary generalizes the fact that $\mathbf{\Sigma}^0_\alpha$ has the prewellordering property for $\alpha < \omega_1$.

§3. Bounded unions and prewellorderings. Let Γ be a pointclass. We say that a union $\bigcup_{\alpha<\beta} A_\alpha$ is Γ-bounded iff the associated norm

$$\varphi(x) = \mu\alpha[x \in A_\alpha]$$

is Γ-bounded.

THEOREM 3.1. Let Δ be selfdual, $\exists^{\mathbb{R}}\Delta \subseteq \Delta$, and $\mathrm{cf}(o(\Delta)) > \omega$. Then the following are equivalent:

(a) $\Delta = \Gamma \cap \check{\Gamma}$ for some nonselfdual Γ such that $\mathsf{PWO}(\Gamma)$;
(b) $\bigcup_{o(\Delta)} \Delta \nsubseteq \Delta$.

PROOF. (a) $\Rightarrow$ (b) is clear. Suppose then that $\bigcup_{o(\Delta)} \Delta \nsubseteq \Delta$, and let $\vartheta \leq o(\Delta)$ be least such that $\bigcup_\vartheta \Delta \nsubseteq \Delta$. Since $\mathrm{cf}(o(\Delta)) > \omega$, Theorem 3.1 of [Van78B] gives a nonselfdual Γ such that $\Gamma \cap \check{\Gamma} = \Delta$. Assume w.l.o.g. $\mathsf{Sep}(\check{\Gamma})$. It follows that $\check{\Gamma} \neq \bigcup_\vartheta \Delta$, as otherwise $\mathsf{PWO}(\check{\Gamma})$. Thus $\Gamma \subseteq \bigcup_\vartheta \Delta$. Define now

$$\Gamma^* = \Big\{\bigcup_{\alpha<\vartheta} A_\alpha : \forall\alpha(A_\alpha \in \Delta) \wedge \bigcup_{\alpha<\vartheta} A_\alpha \text{ is } \mathbf{\Sigma}^1_1\text{-bounded}\Big\}.$$

CLAIM 3.2. $\Gamma \subseteq \Gamma^*$.

PROOF. Let $A \in \Gamma - \check{\Gamma}$, and let $S \subseteq \mathbb{R}^2$ be a universal Σ^1_1 set. Let

$$C = \{x : S_x \subseteq A\} = \{x : \forall y(y \notin S_x \text{ or } y \in A)\}.$$

Since $\mathrm{cf}(o(\Delta)) > \omega$ and every $\mathbf{\Sigma}^1_1$ set is ω-Suslin, Theorems 2.1 and 1.1(b) imply that $C \in \Gamma$. (One can show that $C \in \Gamma$ without using Theorem 2.1. For C can be defined in the form "$\forall$(open $\vee\, \Gamma$)," and it is easy to see that if Γ is nonselfdual and contains the Boolean algebra generated by the open sets, then Γ is closed under unions with open sets.) Thus $C = \bigcup_{\alpha<\vartheta} C_\alpha$ where each $C_\alpha \in \Delta$. Let

$$A_\alpha = \{y : \exists x(x \in C_\alpha \wedge y \in S_x)\}.$$

Then each $A_\alpha \in \Delta$, $A = \bigcup_{\alpha<\vartheta} A_\alpha$, and $\bigcup_{\alpha<\vartheta} A_\alpha$ is $\undertilde{\Sigma}^1_1$-bounded by construction. ⊣ (Claim 3.2)

Claim 3.3. $\Gamma \subseteq \Gamma^*$. $\Gamma^* = \Gamma$.

Proof. It is easy to check, using boundedness, that $\forall^{\mathbb{R}}\Gamma^* \subseteq \Gamma^*$. By Wadge then, if $\Gamma^* \not\subseteq \Gamma$, then $\forall^{\mathbb{R}}\Gamma \subseteq \Gamma^*$. It is enough for a contradiction to show that $\Gamma^* \subseteq \exists^{\mathbb{R}}\Gamma$. In fact, we show $\bigcup_\vartheta \Delta \subseteq \exists^{\mathbb{R}}\Gamma$. The proof is standard, granted our first claim. ⊣ (Claim 3.3)

Let $\langle A_\alpha : \alpha < \vartheta\rangle$ be a sequence of Δ sets, and let $\varphi\colon C \twoheadrightarrow \vartheta$ be a $\undertilde{\Sigma}^1_1$-bounded norm, where $C \in \Gamma$. Such a φ exists by the first claim. Let $W \subseteq \mathbb{R}^2$ be a universal set in Γ. Consider the game: Player I plays x, player II plays y. Player II wins iff

$$x \in C \Rightarrow \exists\beta\big(\varphi(x) \leq \beta \wedge \bigcup_{\alpha\leq\beta} A_\alpha = W_y\big).$$

Since φ is $\undertilde{\Sigma}^1_1$-bounded, player I has no winning strategy. Let σ be a winning strategy for player II. Then

$$x \in \bigcup_{\alpha<\vartheta} A_\alpha \iff \exists y\big(y \in C \wedge x \in W_{\sigma(y)}\big).$$

Since $\bigcap_2 \Gamma \subseteq \Gamma$, we have $\bigcup_{\alpha<\vartheta} A_\alpha \in \exists^{\mathbb{R}}\Gamma$.

Claim 3.4. $\mathsf{PWO}(\Gamma^*)$.

Proof. Let $A = \bigcup_{\alpha<\vartheta} A_\alpha$, where each A_α is in Δ and the union is $\undertilde{\Sigma}^1_1$ bounded and increasing. Define for $x \in A$

$$\varphi(x) = \mu\alpha[x \in A_\alpha].$$

To see that φ is a Γ^* norm, notice, *e.g.*, that if S is $\undertilde{\Sigma}^1_1$ and $S \subseteq \{(x, y) : x <_\varphi y\}$, then $T = \{x : \exists y\,(x, y) \in S)\}$ is $\undertilde{\Sigma}^1_1$, and $T \subseteq A$. Thus $T \subseteq A_\alpha$ for some $\alpha < \vartheta$, and so

$$S \subseteq \{(x, y) : x \in A_\alpha \wedge x <_\varphi y\} = I_\alpha.$$

Thus $<_\varphi$ is the $\undertilde{\Sigma}^1_1$-bounded union of the I_α's. Similarly, $\leq_\varphi$ is in Γ^*. ⊣ (Claim 3.4)

The three claims yield the theorem. ⊣ (Theorem 3.1)

We use "$\undertilde{\mathrm{IND}}$" to denote the class of sets definable over $\mathbb{R}$ by positive elementary induction from parameters in $\mathbb{R}$.

Theorem 3.5. Let Γ be nonselfdual and $\Gamma \subseteq \undertilde{\mathrm{IND}}$. Suppose that $\exists^{\mathbb{R}}\Delta \subseteq \Delta$ and $o(\Delta)$ is regular. Then $\bigcup_\omega \Gamma \subseteq \Gamma$.

Proof. This follows from Theorem 1.1 if $\mathsf{Sep}(\Gamma)$ holds, so assume $\mathsf{Sep}(\check{\Gamma})$. Let

$$\Gamma^* = \{ \bigcup_{\alpha<o(\Delta)} A_\alpha : \forall\alpha(A_\alpha \in \Delta) \wedge \bigcup_{\alpha<o(\Delta)} A_\alpha \text{ is } \Delta\text{-bounded}\}.$$

Clearly every set in Γ^* is a $\underset{\sim}{\Sigma}^1_1$ bounded union of Δ sets, and so the proof of Theorem 3.1 implies that $\Gamma^* \subseteq \Gamma$.

Claim 3.6. $\Gamma^* = \Gamma$.

Proof. Let $A \in \Gamma - \check{\Gamma}$, and let $A = \bigcup_{\alpha<\vartheta} A_\alpha$ where each $A_\alpha \in \Delta$ and ϑ is least such that $\bigcup_\vartheta \Delta \not\subseteq \Delta$. We may assume the A_α's are increasing. The Coding Lemma implies that $\langle |A_\alpha|_{\mathrm{W}} : \alpha < \vartheta\rangle$ is cofinal in $o(\Delta)$. For $\alpha < \vartheta$, let

$$C_\alpha = \{(x, y) : y \in A_{\alpha+1} - A_\alpha \wedge x \text{ codes a continuous function } f_x \text{ such that } f_x^{-1}(A_\alpha) \subseteq A\}.$$

Now C_α is defined in the form "$\Delta \wedge \forall z(\Delta \Rightarrow \Gamma)$." Every set in Δ is κ-Suslin for some $\kappa < o(\Delta)$ by Corollary 3.5 of Kechris [Kec81B]. Thus our Theorems 1.1 and 2.1 imply that $C_\alpha \in \Gamma$. The proof of Theorem 3.1 now shows that if $C = \bigcup_{\alpha<\vartheta} C_\alpha$, then $C \in \exists^{\mathbb{R}}\Gamma$.

Notice that $\exists^{\mathbb{R}}(\bigcup_\vartheta \Delta) = \bigcup_\vartheta \exists^{\mathbb{R}}\Delta = \bigcup_\vartheta \Delta$. So since $\Gamma \subseteq \bigcup_\vartheta \Delta$, $\exists^{\mathbb{R}}\Gamma \subseteq \bigcup_\vartheta \Delta$, and we may write

$$C = \bigcup_{\alpha<\vartheta} D_\alpha$$

where each $D_\alpha \in \Delta$, and the union is increasing. Let

$$z \in B_\alpha \iff \exists(x, y) \in D_\alpha \exists\beta \le \alpha(y \in A_{\beta+1} - A_\beta \wedge f_x(z) \in A_\beta).$$

Then each B_α is in Δ by Theorem 1.1(c) and the fact that $\exists^{\mathbb{R}}\Delta \subseteq \Delta$. It is easy to check that $\bigcup_{\alpha<\vartheta} B_\alpha = A$. Finally, the union $\bigcup_{\alpha<\vartheta} B_\alpha$ is Δ-bounded, since any Δ set is of the form $f_x^{-1}(A_\beta)$ for some $\beta < \vartheta$ and some x. This proves the claim. ⊣ (Claim 3.6)

It is enough now to show $\bigcup_2 \Gamma \subseteq \Gamma$; by Theorem 2.2 we then have $\bigcup_\omega \Gamma \subseteq \Gamma$. Let $A, B \in \Gamma$ towards showing $A \cup B \in \Gamma$. Since $\mathsf{Sep}(\check{\Gamma})$, we have $\mathsf{Red}(\Gamma)$, and so we may assume $A \cap B = \varnothing$. Let $A = \bigcup_{\alpha<\vartheta} A_\alpha$ and $B = \bigcup_{\alpha<\vartheta} B_\alpha$, where the unions are Δ bounded and increasing, and each A_α and B_α is in Δ. It is enough to show that the union $\bigcup_{\alpha<\vartheta}(A_\alpha \cup B_\alpha)$ is Δ-bounded. So let $C \in \Delta$ and $C \subseteq \bigcup_{\alpha<\vartheta}(A_\alpha \cup B_\alpha)$. Then by Theorem 1.1(a), $C \cap A \in \Gamma$. On the other hand, $C \cap A = C \cap (-B)$, and $C \cap (-B) \in \check{\Gamma}$ by Theorem 2.1 and the fact that C is κ-Suslin for some $\kappa < o(\Delta)$. Thus $C \cap A \in \Delta$, and hence $C \cap A \subseteq A_\alpha$ for some $\alpha < \vartheta$. Similarly, $C \cap B \subseteq B_\beta$ for some $\beta < \vartheta$. But then $C \subseteq A_\gamma \cup B_\gamma$, where $\gamma = \max(\alpha, \beta)$, and we are done. ⊣ (Theorem 3.5)

The hypothesis that $\Gamma \subseteq \underset{\sim}{\mathrm{IND}}$ in Theorem 3.5 was only used to conclude, via Theorem 2.1 and Corollary 3.5 of [Kec81B], that Γ is closed under intersection with Δ sets. Thus the conclusion of Theorem 3.5 holds for arbitrary Γ such that $\exists^{\mathbb{R}}\Delta \subseteq \Delta, \bigcup_{o(\Delta)} \Delta \nsubseteq \Delta$, and $o(\Delta)$ is regular, and Γ is closed under intersections with Δ sets.

One can also show that Theorem 2.1 and Corollary 3.5 of [Kec81B] apply for Γ a bit beyond $\underset{\sim}{\mathrm{IND}}$. For example, one can weaken the hypothesis "$\Gamma \subseteq \underset{\sim}{\mathrm{IND}}$" of Theorem 3.5 to "every Γ set is inductive in the complete coinductive set of reals."

We now define

$$C = \{o(\Delta) : \Delta \text{ is selfdual} \wedge \exists^{\mathbb{R}}\Delta \subseteq \Delta\}.$$

Clearly C is cub in ϑ. Theorem 3.1 of [Kec81B] implies that for $\lambda \leq o(\underset{\sim}{\mathrm{IND}})$ such that $\omega\lambda = \lambda$, the λth element of C is the λth Suslin cardinal. Thus our next theorem is actually the characterization of $o({}^3\mathrm{E})$ promised in the introduction.

THEOREM 3.7. Let $o(\Delta)$ be the least regular limit cardinal in C. Then $\Delta = \Gamma \cap \check{\Gamma}$, where Γ is the boldface 2-envelope of ${}^3\mathrm{E}$ (*i.e.*, the class of sets of reals semirecursive in ${}^3\mathrm{E}$ and a real). Thus $o(\Delta) = o({}^3\mathrm{E})$.

PROOF. Clearly C is cub in $o(\underset{\sim}{\mathrm{IND}})$, and since $o(\underset{\sim}{\mathrm{IND}})$ is regular, we have $o(\Delta) \leq o(\underset{\sim}{\mathrm{IND}})$ and $\Delta \subseteq \underset{\sim}{\mathrm{IND}}$. (Actually, Kechris has shown that $o(\underset{\sim}{\mathrm{IND}})$ is Mahlo, so $o(\Delta) < o(\underset{\sim}{\mathrm{IND}})$.) By Theorem 3.5 and its proof, we have $\Delta = \Gamma \cap \check{\Gamma}$, where Γ is the class of Δ-bounded unions of Δ sets of length $o(\Delta)$. Thus $\forall^{\mathbb{R}}\Gamma \subseteq \Gamma, \bigcup_{\omega} \Gamma \subseteq \Gamma$, and by Theorem 3.1, $\mathsf{PWO}(\Gamma)$. In order to show that Γ contains the 2-envelope of ${}^3\mathrm{E}$ it suffices to show that Δ is "uniformly closed under $\exists^{\mathbb{R}}$," in the following sense. Let $Q(x, y)$ and $R(x, y)$ be disjoint relations in Γ. Define

$$S(x) \iff \forall y(R(x, y) \vee Q(x, y)) \wedge \exists y Q(x, y)$$

and

$$T(x) \iff \forall y(R(x, y) \vee Q(x, y)) \wedge \forall y R(x, y).$$

Then we must show that S and T are in Γ.

To see this, let

$$Q = \bigcup_{\alpha<o(\Delta)} Q_\alpha \text{ and } R = \bigcup_{\alpha<o(\Delta)} R_\alpha$$

be representations of Q and R as increasing Δ-bounded unions of Δ sets.

Define

$$S_\alpha(x) \iff \forall y(R_\alpha(x, y) \vee Q_\alpha(x, y)) \wedge \exists y Q_\alpha(x, y)$$

and

$$T_\alpha(x) \iff \forall y(R_\alpha(x,y) \vee Q_\alpha(x,y)) \wedge \forall y R_\alpha(x,y).$$

Clearly $S_\alpha \subseteq S$ and $T_\alpha \subseteq T$. We show simultaneously that $S \subseteq \bigcup_\alpha S_\alpha$ and that the union is Δ-bounded. For let $D \subseteq S$ and $D \in \Delta$. Let $A = (D \times {}^\omega\omega) \cap Q$, and let $B = (D \cap {}^\omega\omega) \cap R$. Then A and B are disjoint Γ sets, and complementary on $D \times {}^\omega\omega$. Thus A and B are in Δ. Let α be such that $A \subseteq Q_\alpha$ and $B \subseteq R_\alpha$. Then $D \subseteq S_\alpha$, and we are done. An identical argument shows that $T \subseteq \bigcup_\alpha T_\alpha$ and that this union is Δ-bounded. Thus S and T are in Γ, as desired.

It is well known that the class of sets of reals is recursive in ${}^3\mathrm{E}$ and a real has the closure properties we assumed of Δ, that is, it is closed under $\exists^{\mathbb{R}}$ and its ordinal is regular. Since Δ was minimal with these properties, Δ is contained in this class. Thus Γ is contained in the 2-envelope of ${}^3\mathrm{E}$. ⊣ (Theorem 3.7)

Theorem 3.7 implies that $o({}^3\mathrm{E})$ is not Mahlo. It gives some evidence for the natural conjecture, due perhaps to Moschovakis, that $o({}^3\mathrm{E})$ is the least regular limit cardinal. Proof of this conjecture awaits further progress in computing upper bounds for the $\underset{\sim}{\delta}^1_n$'s.

We shall close with some remarks on projective-like hierarchies which extend and simplify some of the proofs of [KSS81].

For us, a projective-like hierarchy is a sequence $\langle \Gamma_i : i < \omega \rangle$ of nonselfdual pointclasses such that

(i) $\forall^{\mathbb{R}}\Gamma_i \subseteq \Gamma_i$ or $\exists^{\mathbb{R}}\Gamma_i \subseteq \Gamma_i$, but not both, for all $i < \omega$, and
(ii) $\forall^{\mathbb{R}}\Gamma_0 \subseteq \Gamma_0$, and
(iii) For all nonselfdual $\Gamma' \subseteq \Gamma_0 \cap \check{\Gamma}_0, \Gamma_0 \neq \forall^{\mathbb{R}}\Gamma'$.

Our definition is slightly more liberal then that of Kechris-Solovay-Steel, mainly because we do not require $\bigcap_\omega \Gamma_i \subseteq \Gamma_i$ and $\bigcup_\omega \Gamma_i \subseteq \Gamma_i$ for all i.

(Condition (iii) is slightly more liberal than theirs, too.) It is easy to see that if $\langle \Gamma_i : i < \omega \rangle$ is a projective-like hierarchy, then $\bigcap_\omega \Gamma_i \subseteq \Gamma_i$ and $\bigcup_\omega \Gamma_i \subseteq \Gamma_i$ for all $i \geq 1$. Thus our hierarchies differ from those of Kechris-Solovay-Steel only in that they sometimes have an extra class Γ_0 tacked on at the beginning. Consideration of this class seems to simplify some proofs.

Let Γ be a nonselfdual pointclass closed under one but not both of $\forall^{\mathbb{R}}$ and $\exists^{\mathbb{R}}$. Let $\alpha = \sup\{\beta \in C : \beta < o(\Gamma)\}$. Then $\alpha \in C$, so $\alpha = o(\Delta)$ for some Δ. It is easy to see that either Γ or $\check{\Gamma}$ is in the least projective-like hierarchy $\langle \Gamma_i : i < \omega \rangle$ such that $\Delta \subseteq \Gamma_0$. We now show that for each $i < \omega$, either $\mathsf{PWO}(\Gamma_i)$ or $\mathsf{PWO}(\check{\Gamma}_i)$, thereby reproving one of the main results of [KSS81]. We prove this by considering cases corresponding to the types I–IV of projective-like hierarchies defined in that paper. We assume throughout that $\bigcup_{o(\Delta)} \Delta \not\subseteq \Delta$.

Case 1. $\mathrm{cf}(\alpha) = \omega$. (Type I).

In this case, $\Gamma_0 = \bigcap_\omega \Delta$. It is easy to see, and in fact implied by Corollary 2.3, that $\mathsf{PWO}(\check{\Gamma}_0)$. The first periodicity theorem propagates prewellordering up to hierarchy $\langle \Gamma_i : i < \omega \rangle$.

Case 2. $\mathrm{cf}(\alpha) > \omega$.

In this case, let $\Delta = \Gamma \cap \check{\Gamma}$, where we may assume by Theorem 1.1 that $\forall^{\mathbb{R}}\Gamma \subseteq \Gamma$, and by Theorem 3.1 that $\mathsf{PWO}(\Gamma)$.

Subcase A. $\exists^{\mathbb{R}}\Gamma \not\subseteq \Gamma$.

In this case $\Gamma_0 = \Gamma$, $\mathsf{PWO}(\Gamma_0)$, and the first periodicity propagates prewellordering again. The case occurs with hierarchies of types II and III; type II in the case $\bigcup_2 \Gamma \not\subseteq \Gamma$, and type III in the case $\bigcup_2 \Gamma \subseteq \Gamma$ (and hence $\bigcup_\omega \Gamma \subseteq \Gamma$).

Subcase B. $\exists^{\mathbb{R}}\Gamma \subseteq \Gamma$. (Type IV).

In this case, $\Gamma_0 = \{A \cap B : A \in \Gamma \wedge B \in \check{\Gamma}\}$. The usual difference-hierarchy proof shows that $\mathsf{PWO}(\Gamma_0)$. Again, first periodicity propagates prewellordering.

REFERENCES

ALEXANDER S. KECHRIS

[Kec77B] *Classifying projective-like hierarchies*, ***Bulletin of the Greek Mathematical Society***, vol. 18 (1977), pp. 254–275.

[Kec81B] *Suslin cardinals, κ-Suslin sets, and the scale property in the hyperprojective hierarchy*, in Kechris et al. [CABAL ii], pp. 127–146, reprinted in [CABAL I], p. 314–332.

ALEXANDER S. KECHRIS, BENEDIKT LÖWE, AND JOHN R. STEEL

[CABAL I] *Games, scales, and Suslin cardinals: the Cabal seminar, volume I*, Lecture Notes in Logic, vol. 31, Cambridge University Press, 2008.

ALEXANDER S. KECHRIS, DONALD A. MARTIN, AND YIANNIS N. MOSCHOVAKIS

[CABAL ii] *Cabal seminar* 77–79, Lecture Notes in Mathematics, no. 839, Berlin, Springer, 1981.

ALEXANDER S. KECHRIS AND YIANNIS N. MOSCHOVAKIS

[CABAL i] *Cabal seminar* 76–77, Lecture Notes in Mathematics, no. 689, Berlin, Springer, 1978.

ALEXANDER S. KECHRIS, ROBERT M. SOLOVAY, AND JOHN R. STEEL

[KSS81] *The axiom of determinacy and the prewellordering property*, this volume, originally published in Kechris et al. [CABAL ii], pp. 101–125.

YIANNIS N. MOSCHOVAKIS

[Mos70] *Determinacy and prewellorderings of the continuum*, ***Mathematical logic and foundations of set theory. Proceedings of an international colloquium held under the auspices of the Israel Academy of Sciences and Humanities, Jerusalem, 11–14 November 1968*** (Y. Bar-Hillel, editor), Studies in Logic and the Foundations of Mathematics, North-Holland, Amsterdam-London, 1970, pp. 24–62.

JOHN R. STEEL

[Ste81B] *Determinateness and the separation property*, ***The Journal of Symbolic Logic***, vol. 46 (1981), no. 1, pp. 41–44.

[Ste82] *A classification of jump operators*, ***The Journal of Symbolic Logic***, vol. 47 (1982), no. 2, pp. 347–358.

Robert Van Wesep
[Van78B] *Wadge degrees and descriptive set theory*, this volume, originally published in Kechris and Moschovakis [Cabal i], pp. 151–170.

DEPARTMENT OF MATHEMATICS
UNIVERSITY OF CALIFORNIA
BERKELEY, CALIFORNIA 94720-3840
UNITED STATES OF AMERICA
E-mail: steel@math.berkeley.edu

THE AXIOM OF DETERMINACY AND THE PREWELLORDERING PROPERTY

ALEXANDER S. KECHRIS[1], ROBERT M. SOLOVAY[2], AND JOHN R. STEEL[3]

§1. Introduction. Let $\omega = \{0, 1, 2, \dots\}$ be the set of natural numbers and $\mathbb{R} = {}^{\omega}\omega$ the set of all functions from ω into ω, or for simplicity **reals**. A **product space** is of the form

$$\mathcal{X} = X_1 \times X_2 \times \cdots \times X_k,$$

where $X_i = \omega$ or $\mathbb{R}$. Subsets of these product spaces are called **pointsets**. A **boldface pointclass** is a class of pointsets closed under continuous preimages and containing all clopen pointsets (in all product spaces).

The following results have been proved in Steel [Ste81B]:

If Γ is a boldface pointclass, $\check{\Gamma}$ its **dual**, *i.e.*, $\check{\Gamma} = \{\neg A : A \in \Gamma\}$, and Δ its **ambiguous** part, *i.e.*, $\Delta = \Gamma \cap \check{\Gamma}$, then assuming ZF+DC+AD,

(1) Either Γ or $\check{\Gamma}$ has the separation property.
(2) If Δ is closed under (finite) intersections and unions, then either Γ or $\check{\Gamma}$ has the reduction property.

Note that by a result of van Wesep [Van78A], if Γ is not closed under complements, then it is impossible for both Γ and $\check{\Gamma}$ to have the separation property (assuming again ZF+DC+AD).

Our purpose here is to investigate the situation concerning a stronger structural property of pointclasses, namely the prewellordering property. We establish in §2 the following criterion:

THEOREM 1.1 (ZF+DC+AD). Let Γ be a boldface pointclass, closed under countable intersections and unions and *either* existential *or* universal quantification over $\mathbb{R}$, but not complements. Then the following are equivalent,

(1) Γ or $\check{\Gamma}$ has the prewellordering property.
(2) Δ is not closed under wellordered unions (of arbitrary length).

[1]Research partially supported by NSF Grant MCS79-20465. The author is an A. P. Sloan Foundation Fellow.

[2]Partially supported by NSF Grant MCS77-01640.

[3]Partially supported by NSF Grant MCS78-02989.

Wadge Degrees and Projective Ordinals: The Cabal Seminar, Volume II
Edited by A. S. Kechris, B. Löwe, J. R. Steel
Lecture Notes in Logic, 37

By an application of this criterion and some further analysis done in §§3, 4, 5 we obtain for example the following,

THEOREM 1.2 (ZF+DC+AD). Let Γ be a boldface pointclass, closed under countable intersections and unions and either existential or universal quantification over $\mathbb{R}$, but not complements. Then

$$\Gamma \subseteq \mathbf{L}(\mathbb{R}) \Rightarrow \Gamma \text{ or } \breve{\Gamma} \text{ has the prewellordering property.}$$

Here $\mathbf{L}(\mathbb{R})$ is the smallest inner model of ZF containing all the reals.

Finally we address ourselves to the following problem: If a pointclass Γ as above is closed under only one kind of quantification over $\mathbb{R}$, *i.e.*, either only existential or only universal (in which case it is reasonable to call such a pointclass *projective-like*), then in view of the inherent asymmetry in the closure properties between Γ and $\breve{\Gamma}$, is it possible to determine in which side of the pair $(\Gamma, \breve{\Gamma})$ we have the prewellordering property, as it is done by the work of Martin [Mar68] and Moschovakis (see Addison-Moschovakis [AM68]) for the projective pointclasses $\utilde{\mathbf{\Sigma}}^1_n, \utilde{\mathbf{\Pi}}^1_n$? We provide an affirmative answer in §§4, 5: We first embed each projective-like pointclass Γ in a uniquely determined projective-like hierarchy. We then classify all possible projective-like hierarchies into four types: I–IV, and demonstrate that (at least within $\mathbf{L}(\mathbb{R})$) each of these types exhibits a unique prewellordering pattern, identical to that of the classical projective hierarchy for the types I and III, but dual to that for the other two types II and IV. Of course ZF+DC+AD is assumed throughout.

The concept of and some results about Wadge degrees will be of the essence in this paper. The papers [Van78B, Van78A] and [Ste81B] provide all the necessary information. For general results in descriptive set theory needed below we refer to [Mos80]. Finally, it is convenient to assume ZF+DC throughout this paper and explicitly indicate only any further hypotheses as they are needed.

§2. A criterion for PWO(Γ)**.** Let Λ be a pointclass (*i.e.*, an arbitrary collection of pointsets). We say that Λ is **closed under wellordered unions** if for any sequence $\{A_\eta\}_{\eta<\xi}$ of members of Λ, $\bigcup_{\eta<\xi} A_\eta \in \Lambda$. (Here ξ is an arbitrary ordinal, and it is understood that all A_η are subsets of some arbitrary space $\mathcal{X}$.) Recall also that a pointclass Γ has the *prewellordering property* if for any $A \in \Gamma$, there is a norm $\varphi\colon A \to \kappa$ such that the associated relations

$$x \leq^*_\varphi y \Leftrightarrow x \in A \wedge [y \notin A \vee \varphi(x) \leq \varphi(y)],$$
$$x <^*_\varphi y \Leftrightarrow x \in A \wedge [y \notin A \vee \varphi(x) < \varphi(y)]$$

are in Γ. Such a norm Γ is called a **Γ-norm.**

We now have

THEOREM 2.1 (AD). Let Γ be a boldface pointclass, closed under countable unions and intersections and either $\exists^{\mathbb{R}}$ or $\forall^{\mathbb{R}}$, but not complements. Then the following are equivalent:

(i) $\mathsf{PWO}(\Gamma)$,

(ii) $\mathsf{Red}(\Gamma)$ and Δ is not closed under wellordered unions.

In view of the result of Steel mentioned in the introduction, we have

COROLLARY 2.2 (AD). Let Γ be a boldface pointclass closed under countable unions and intersections, and either $\exists^{\mathbb{R}}$ or $\forall^{\mathbb{R}}$, but not complements. Then the following are equivalent:

(i) $\mathsf{PWO}(\Gamma) \vee \mathsf{PWO}(\check{\Gamma})$,

(ii) Δ is not closed under wellordered unions.

PROOF. If Γ has the prewellordering property, let $A \in \Gamma \setminus \Delta$ and let φ be a regular Γ-norm on A, of length κ. **Regular** means that $\varphi\colon A \twoheadrightarrow \kappa$.) For $\xi < \kappa$, let $A_\xi = \varphi^{-1}[\{\xi\}]$. Then each $A_\xi \in \Delta$ but $\bigcup_{\xi<\kappa} A_\xi = A \notin \Delta$, so Δ is not closed under wellordered unions. Finally we clearly have $\mathsf{Red}(\Gamma)$. ⊣ (Theorem 2.1)

So assume now that Δ is not closed under wellordered unions, and $\mathsf{Red}(\Gamma)$ holds. We shall distinguish three cases, depending on the closure properties of Γ.

2.1. Γ is closed under both $\exists^{\mathbb{R}}$ and $\forall^{\mathbb{R}}$. We shall start in this case with a lemma which will be also useful later on.

Call a boldface pointclass Λ **strongly closed** if it is closed under (finite) unions and intersections, complements and quantification (of both types) over $\mathbb{R}$. For example $\Lambda = \bigcup_n \utilde{\mathbf{\Delta}}^1_n$ is strongly closed. Note that Λ need not be closed under countable unions.

LEMMA 2.3. Let Λ be strongly closed. Then the following three ordinals associated with Λ are equal:

(i) $\sup\{\xi : \xi \text{ is the length of a } \Lambda \text{ prewellordering of } \mathbb{R}\}$,

(ii) $\sup\{\xi : \xi \text{ is the rank of a } \Lambda \text{ wellfounded relation on } \mathbb{R}\}$,

(iii) $\sup\{|A|_{\mathrm{W}} : A \in \Lambda\}$.

Here $|A|_{\mathrm{W}}$ is the Wadge ordinal of $A \subseteq \mathbb{R}$. (Note that $|A|_{\mathrm{W}} = |\neg A|_{\mathrm{W}}$.)

PROOF. Clearly (i) $\leq$ (ii) and (iii) $\leq$ (i). So it is enough to show that (ii) $\leq$ (iii).

For each list of pointsets $A_1, A_2, \ldots, A_n$, let $\utilde{\mathbf{\Sigma}}^1_1(A_1, A_2, \ldots, A_n)$ be the smallest boldface pointclass closed under unions and intersections, $\exists^\omega$, $\forall^\omega$ and $\exists^{\mathbb{R}}$, which contains $A_1, \ldots, A_n$. By Moschovakis [Mos70] §3, $\utilde{\mathbf{\Sigma}}^1_1(A_1, A_2, \ldots, A_n)$ is $\mathbb{R}$-parametrized *i.e.*, has universal sets. Let $\utilde{\mathbf{\Pi}}^1_1(A_1, \ldots, A_n)$ be the dual of $\utilde{\mathbf{\Sigma}}^1_1(A_1, \ldots, A_n)$. Note that $\utilde{\mathbf{\Pi}}^1_1(A, \neg A)$ cannot be closed under $\exists^{\mathbb{R}}$, so that we can build up $\utilde{\mathbf{\Sigma}}^1_2(A, \neg A)$, $\utilde{\mathbf{\Pi}}^1_2(A, \neg A), \ldots$ in the usual way.

Recall now the 0th Periodicity Theorem (see Kechris [Kec77B]), which asserts that if Γ is a boldface pointclass closed under countable intersections and unions, then if $\forall^{\mathbb{R}}\Gamma = \{\forall\alpha(x,\alpha) \in A : A \in \Gamma\}$ and similarly for $\exists^{\mathbb{R}}\Gamma$, we have:

(i) $\exists^{\mathbb{R}}\Gamma \subseteq \Gamma \wedge \mathsf{Red}(\Gamma) \Rightarrow \mathsf{Red}(\forall^{\mathbb{R}}\Gamma)$,
(ii) $\forall^{\mathbb{R}}\Gamma \subseteq \Gamma \wedge \mathsf{Red}(\Gamma) \Rightarrow \mathsf{Red}(\exists^{\mathbb{R}}\Gamma)$.

(For the convenience of the reader we repeat this proof in an appendix.) From this and Steel's Theorem (mentioned in the introduction), we can find for each pointset A an integer $N > 1$, so that $\undertilde{\mathbf{\Pi}}^1_N(A, \neg A)$ has the reduction property. ($N = 2$ or $N = 3$ will of course suffice.)

From these remarks it follows that, if $\prec$ is a wellfounded relation in Λ, there is a boldface pointclass $\Gamma \subsetneqq \Lambda$ such that Γ is $\mathbb{R}$-parametrized, closed under countable intersections and unions and $\forall^{\mathbb{R}}$, such that Γ has the reduction property and moreover $\Delta \supseteq \undertilde{\mathbf{\Sigma}}^1_1(\prec)$. (Note that if Λ is strongly closed and $A_1, \dots, A_n \in \Lambda$, $\undertilde{\mathbf{\Pi}}^1_n(A_1, \dots, A_n) \subsetneqq \Lambda$.)

For a Γ with these structural properties we can introduce the following Γ coding of the Δ sets of reals:

Let W^0, W^1 be a universal pair for Γ (*i.e.*, for each A, $B \subseteq \mathbb{R}$ in Γ there is $\varepsilon \in \mathbb{R}$ with $A = W^0_\varepsilon = \{\alpha : (\varepsilon, \alpha) \in W^0\}$, $B = W^1_\varepsilon$). Let $\bar{W}^{\bar{0}}$, $\bar{W}^1$ in Γ reduce W^0, W^1. Put

$$\varepsilon \in C \Leftrightarrow \bar{W}^{\bar{0}}_\varepsilon \cup \bar{W}^1_\varepsilon = \mathbb{R}$$

and for $\varepsilon \in C$

$$H_\varepsilon = \bar{W}^0_\varepsilon (= \neg \bar{W}^1_\varepsilon).$$

Then $C \in \Gamma$ and

$$\{H_\varepsilon : \varepsilon \in C\} = \{A \subseteq \mathbb{R} : A \in \Delta\}.$$

We can now finish the proof as follows: By the Recursion Theorem, we can find a partial continuous function f such that

(1) $\alpha \in \mathrm{Field}(\prec) \Rightarrow f(\alpha) \in C$,
(2) $\alpha \prec \beta \Rightarrow |H_{f(\alpha)}|_{\mathrm{W}} < |H_{f(\beta)}|_{\mathrm{W}}$.

We use here the fact that $\undertilde{\mathbf{\Sigma}}^1_1(\prec) \subseteq \Delta$. Thus

$$\mathrm{rank}(\prec) \leq \sup\{|A|_{\mathrm{W}} : A \in \Delta\} \leq \text{(iii)}. \qquad \dashv \text{(Lemma 2.3)}$$

We proceed now to complete the proof of case 2.1. Let Γ satisfy its hypotheses. Let δ be the ordinal associated to $\Lambda \equiv \Delta$ by the preceding lemma. Let also θ be the least ordinal such that there is a θ-sequence $\{A_\xi\}_{\xi<\theta}$ of members of Δ with $\bigcup_{\xi<\theta} A_\xi \notin \Delta$. We claim that $\theta = \delta$.

To see that $\delta \leq \theta$, notice first that we can code Δ sets using $\mathsf{Red}(\Gamma)$ as in the proof of Lemma 2.3. (That Γ is $\mathbb{R}$-parametrized follows from the nonclosure of Γ under complements—see Van Wesep [Van78A].) Let C, $\varepsilon \mapsto H_\varepsilon$ denote the set of codes and the coding map respectively. If $\xi < \delta$, let $\varphi\colon \mathbb{R} \to \xi$ be a Δ-norm and then for each ξ-sequence $\{B_\eta\}_{\eta<\xi}$ of Δ sets let, by Moschovakis [Mos70], $R(w, \varepsilon)$ be a Δ relation such that

(i) $\varphi(w) = \varphi(v) \Rightarrow [R(w, \varepsilon) \Leftrightarrow R(v, \varepsilon)]$
(ii) $R(w, \varepsilon) \Rightarrow \varepsilon \in C$
(iii) $\forall w \exists \varepsilon [R(w, \varepsilon) \wedge H_\varepsilon = B_{\varphi(w)}]$.

Then

$$\alpha \in \bigcup_{\eta<\xi} B_\eta \Leftrightarrow \exists w \exists \varepsilon [R(w, \varepsilon) \wedge \alpha \in H_\varepsilon].$$

Thus $\bigcup_{\eta<\xi} B_\eta \in \Delta$.

To prove that conversely $\theta \leq \delta$, we use the minimality of θ. It easily implies that there is a θ-sequence of Δ sets $\{A_\xi\}_{\xi<\theta}$ such that $A_\xi \subsetneqq A_\eta$, if $\xi < \eta < \theta$, $A_\lambda = \bigcup_{\xi<\lambda} A_\xi$, if $\lambda < \theta$ is limit, and $A = \bigcup_{\xi<\theta} A_\xi \notin \Delta$. For each $x \in A$, let

$$\varphi(x) = \text{ least } \xi, x \in A_{\xi+1} \setminus A_\xi.$$

Then φ is a regular norm on A of length θ. But if $x \in A$ and $\varphi(x) = \xi < \theta$, then the prewellordering

$$\{(y, z) : \varphi(y) \leq \varphi(z) < \varphi(x)\},$$

of length ξ, is in Δ, since it is equal to

$$\bigcup_{\eta<\xi} \Big(A_\eta \times \bigcap_{\eta'<\eta} \neg A_{\eta'} \Big),$$

while $\xi < \theta$. So $\xi < \delta$, thus $\theta \leq \delta$.

Let now

$$\Gamma^* = \Big\{ \bigcup_{\xi<\theta} A_\xi \; : \; \forall \xi < \theta (A_\xi \in \Delta) \Big\}.$$

First notice that, by an argument of Martin [Mar71B], Γ^* has the prewellordering property. Indeed, if $A = \bigcup_{\xi<\theta} A_\xi$ is in Γ^* and we let

$$\psi(x) = \text{ least } \xi, x \in A_\xi,$$

be the norm associated to this wellordered union, then ψ is a Γ^*-norm. This is because if $\leq^*_\psi$, $<^*_\psi$ are its two associated relations, then

$$x \leq^*_\psi y \Leftrightarrow \exists \xi < \theta [x \in A_\xi \wedge \forall \xi' < \xi (y \notin A_{\xi'})],$$
$$x <^*_\psi y \Leftrightarrow \exists \xi < \theta [x \in A_\xi \wedge \forall \xi' \leq \xi (y \notin A_{\xi'})],$$

so these are in Γ^*, by the minimality of θ again.

So it is enough to show that $\Gamma^* = \Gamma$. Since $\Gamma^* \supsetneqq \Delta$, by Wadge's Lemma it is enough to show that $\Gamma^* \subseteq \Gamma$. For that let

$$\varepsilon \in W \Leftrightarrow \varepsilon \in C \wedge H_\varepsilon \text{ is wellfounded}$$

(view H_ε as a subset of $\mathbb{R}^2$ here), and for $\varepsilon \in W$, let

$$|\varepsilon| = \text{rank}(H_\varepsilon).$$

Then $W \in \Gamma$ and $\{|\varepsilon| : \varepsilon \in C\} = \delta = \theta$. Given now $\{B_\xi\}_{\xi<\theta}$, a θ-sequence of Δ sets, consider the following Solovay-type game:

$$\begin{array}{ccl} \text{I} & \text{II} & \text{Player II wins iff } \varepsilon \in W \Rightarrow \\ \varepsilon & \alpha & \qquad \alpha \in C \wedge \exists\eta[|\varepsilon| < \eta < \delta \wedge H_\alpha = \bigcup_{\xi<\eta} B_\xi]. \end{array}$$

If player I has a winning strategy f, then the relation

$$(\varepsilon, x) \prec (\varepsilon', y) \Leftrightarrow \varepsilon = \varepsilon' \in f[\mathbb{R}] \wedge (x, y) \in H_\varepsilon$$

is wellfounded and in Δ, so if player II plays any $\alpha \in C$ with $H_\alpha = \bigcup_{\xi<\eta} B_\xi$ for any $\delta > \eta > \text{rank}(\prec)$, he beats f. So player II must have a winning strategy g. Then

$$x \in \bigcup_{\xi<\theta} B_\xi \Leftrightarrow \exists\varepsilon[\varepsilon \in W \wedge x \in H_{g(\varepsilon)}],$$

so $\bigcup_{\xi<\theta} B_\xi \in \Gamma$, thus $\Gamma^* \subseteq \Gamma$, and we are done.

2.2. Γ is closed under $\exists^{\mathbb{R}}$ but not $\forall^{\mathbb{R}}$. In this case the result follows from the following lemma which will be also useful later on.

Lemma 2.4. Assume Γ is a boldface pointclass, closed under countable unions and intersections and $\exists^{\mathbb{R}}$ but not $\forall^{\mathbb{R}}$ (thus Γ is not closed under complements). Then if $\mathsf{Red}(\Gamma)$ holds, Γ is closed under wellordered unions.

Proof. Let θ be least such that for some $\{A_\xi\}_{\xi<\theta}$ with all A_ξ in Γ, $\bigcup_{\xi<\theta} A_\xi \notin \Gamma$, towards a contradiction. It is easy to check that

$$\theta \text{ is a regular uncountable cardinal.}$$

Put

$$\Gamma' = \Big\{\bigcup_{\xi<\theta} A_\xi : \forall\xi < \theta(A_\xi \in \Gamma)\Big\}.$$

Since $\Gamma' \supsetneqq \Gamma$, by Wadge's Lemma $\Gamma' \supseteq \check{\Gamma}$, and since clearly Γ' is closed under $\exists^{\mathbb{R}}$, $\Gamma' \supseteq \exists^{\mathbb{R}}\check{\Gamma}$. Put

$$\Gamma_+ = \forall^{\mathbb{R}}\Gamma, \text{ so that } \check{\Gamma}_+ = \exists^{\mathbb{R}}\check{\Gamma},$$

therefore $\Gamma' \supseteq \check{\Gamma}_+$. Let also

$$\delta_+ = \sup\{\xi : \xi \text{ is the rank of a } \check{\Gamma}_+ \text{ wellfounded relation}\}.$$

Since $\Gamma' \supsetneqq \Delta_+$ $(= \Gamma_+ \cap \check{\Gamma}_+)$ and $\Gamma \subseteq \Delta_+$ (because $\forall^{\mathbb{R}}\Gamma = \Gamma^+ \not\subseteq \Gamma$, thus $\Gamma \subseteq \check{\Gamma}_+$ by Wadge), Δ_+ is not closed under wellordered unions, so let θ_+ be the least ordinal such that some union of a θ_+-sequence of Δ_+ sets is not in Δ_+. Clearly $\theta_+ \leq \theta$.

We have now that $\theta_+ \geq \delta_+$, thus

$$\theta \geq \delta_+.$$

(This is essentially an argument of Martin [Mar71B]: If $\theta_+ < \delta_+$, then as in the proof in case 2.1 an application of the Moschovakis Coding Lemma shows that

$$\Big\{\bigcup_{\xi<\theta_+} A_\xi : \forall \xi < \theta_+(A_\xi \in \Delta_+)\Big\} = \check{\Gamma}_+$$

and thus again as in case 2.1, $\check{\Gamma}_+$ has the prewellordering property, contradicting the fact that Γ_+ has the reduction property by the 0th Periodicity Theorem.)

Let

$$\delta = \sup\{\xi : \xi \text{ is the rank of a } \Gamma \text{ wellfounded relation}\}.$$

By the standard argument (see for example Kechris [Kec74]) $\delta^+ > \delta$, thus

$$\theta > \delta.$$

Let now $\prec$ be a $\check{\Gamma}_+$ wellfounded relation. Then, since $\check{\Gamma}_+ \subseteq \Gamma'$,

$$\prec = \bigcup_{\xi<\theta} \prec_\xi,$$

where each $\prec_\xi$ is a Γ wellfounded relation, and by the minimality of θ we can assume that $\xi \leq \eta < \theta \Rightarrow \prec_\xi \subseteq \prec_\eta$. For each $x \in \text{Field}(\prec)$, put

$$f_x(\xi) = \begin{cases} 0 & \text{if } x \notin \text{Field}(\prec_\xi) \\ |x|_{\prec_\xi} \equiv \text{ rank of } x \text{ in } \prec_\xi & \text{otherwise} \end{cases}$$

Then $f_x : \theta \to \delta$ is nondecreasing, so since $\delta < \theta$ and θ is regular, we have

$$f_x(\xi) = \text{ constant } \equiv \xi(x) < \delta,$$

for all large enough $\xi < \theta$. then

$$x \prec y \Rightarrow \xi(x) < \xi(y),$$

thus $\text{rank}(\prec) \leq \delta$, so $\delta_+ \leq \delta$, a contradiction. ⊣ (Lemma 2.4)

2.3. Γ is closed under $\forall^{\mathbb{R}}$ but not $\exists^{\mathbb{R}}$. Let again θ be least such that some union of a θ-sequence of Δ sets is not in Δ. Put

$$\Gamma' = \Big\{\bigcup_{\xi<\theta} A_\xi : \forall \xi < \theta(A_\xi \in \Delta)\Big\}.$$

Since Γ' has the prewellordering property, $\Gamma' \neq \check{\Gamma}$ so, since $\Gamma' \supsetneq \Delta$, $\Gamma' \supseteq \Gamma$.

Call now a sequence $\{A_\xi\}_{\xi<\theta}$ Λ-*bounded* (for any pointclass Λ), if

$$\forall X \in \Lambda\Big[X \subseteq \bigcup_{\xi<\theta} A_\xi \Rightarrow \exists \xi < \theta(X \subseteq A_\xi)\Big].$$

Put

$$\Gamma^* = \Big\{\bigcup_{\xi<\theta} A_\xi \,:\, \forall \xi < \theta (A_\xi \in \Delta) \wedge \{A_\xi\}_{\xi<\theta} \text{ is } \check{\Gamma}\text{-bounded}\Big\}.$$

The result will follow from the following three lemmas.

LEMMA 2.5. Γ^* is a boldface pointclass with the prewellordering property.

LEMMA 2.6. If $\Gamma^* \supsetneqq \Delta$, then $\Gamma^* = \Gamma$.

LEMMA 2.7. $\Gamma^* \supsetneqq \Delta$.

PROOF OF LEMMA 2.5. Since $\check{\Gamma}$ is closed under $\exists^{\mathbb{R}}$ it is trivial to check that Γ^* is closed under continuous preimages. Let now $\{A_\xi\}_{\xi<\theta}$ be a $\check{\Gamma}$-bounded sequence of Δ sets. Let $A = \bigcup_{\xi<\theta} A_\xi$ and let

$$\varphi(x) = \text{least } \xi,\ x \in A_\xi$$

the associated norm on A. Then

$$x \leq^*_\varphi y \Leftrightarrow \exists \xi < \theta \big[\overbrace{\exists \xi' \leq \xi \big(x \in A_{\xi'} \wedge \forall \eta' < \xi' (y \notin A_{\eta'})\big)}^{B_\xi(x,y)}\big],$$

$$x <^*_\varphi y \Leftrightarrow \exists \xi < \theta \big[\overbrace{\exists \xi' \leq \xi \big(x \in A_{\xi'} \wedge \forall \eta' \leq \xi' (y \notin A_{\eta'})\big)}^{C_\xi(x,y)}\big].$$

Let $B_\xi(x, y)$, $C_\xi(x, y)$ be the two pointsets indicated above. It is enough to show that $\{B_\xi\}_{\xi<\theta}$, $\{C_\xi\}_{\xi<\theta}$ are $\check{\Gamma}$-bounded. Take for example $\{B_\xi\}_{\xi<\theta}$. Let $Y \in \check{\Gamma}$, $Y \subseteq \bigcup_{\xi<\theta} B_\xi$. If $X = \{x : \exists y (x, y) \in Y\}$, then $X \in \check{\Gamma}$ and $X \subseteq \bigcup_{\xi<\theta} A_\xi$ so for some $\xi < \theta$, $X \subseteq A_\xi$. Then $Y \subseteq B_\xi$. ⊣ (Lemma 2.5)

PROOF OF LEMMA 2.6. Assume $\Gamma^* \supsetneqq \Delta$. If $\Gamma^* \neq \Gamma$, then by Wadge's Lemma, $\Gamma^* \supseteq \check{\Gamma}$. Since, by the 0th Periodicity Theorem, $\mathsf{Red}(\Gamma) \Rightarrow \mathsf{Red}(\exists^{\mathbb{R}}\Gamma)$, we have by Lemma 2.4 that $\Gamma^* \subseteq \exists^{\mathbb{R}}\Gamma$, so

$$\check{\Gamma} \subseteq \Gamma^* \subseteq \exists^{\mathbb{R}}\Gamma.$$

Let now $A \in \check{\Gamma}$ be such that

$$x \in B \Leftrightarrow \forall \alpha (x, \alpha) \in A$$

is not in $\exists^{\mathbb{R}}\Gamma$. Write

$$A = \bigcup_{\xi<\theta} A_\xi,$$

where each $A_\xi \in \Delta$ and $\{A_\xi\}_{\xi<\theta}$ is $\check{\Gamma}$-bounded. Then

$$x \in B \Leftrightarrow \forall \alpha \exists \xi < \theta (x, \alpha) \in A_\xi.$$

So if $x \in B$, then $\{x\} \times \mathbb{R} \subseteq \bigcup_{\xi<\theta} A_\xi$, thus by $\check{\Gamma}$-boundedness, $\{x\} \times \mathbb{R} \subseteq A_\xi$ for some $\xi < \theta$. Thus

$$x \in B \Leftrightarrow \exists \xi < \theta \forall \alpha (x, \alpha) \in A_\xi.$$

If

$$x \in B_\xi \Leftrightarrow \forall \alpha (x, \alpha) \in A_\xi,$$

then $B_\xi \in \Gamma \subseteq \exists^{\mathbb{R}}\Gamma$, so by Lemma 2.4 again $B = \bigcup_{\xi<\theta} B_\xi \in \exists^{\mathbb{R}}\Gamma$, a contradiction. ⊣ (Lemma 2.6)

PROOF OF LEMMA 2.7. Let $A \in \Gamma \setminus \Delta$. Let $\mathcal{S}$ be universal for $\check{\Gamma}$, and let

$$B = \{\varepsilon : \mathcal{S}_\varepsilon \subseteq A\}.$$

Thus $B \in \Gamma$. As $\Gamma' \supseteq \Gamma$, we can write

$$B = \bigcup_{\xi<\theta} B_\xi,$$

where each B_ξ is in Δ. Put

$$A_\xi = \bigcup \{\mathcal{S}_\varepsilon : \varepsilon \in B_\xi\}.$$

Then each $A_\xi \in \check{\Gamma}$, $\bigcup_{\xi<\theta} A_\xi = A$ and $\{A_\xi\}_{\xi<\theta}$ is $\check{\Gamma}$-bounded. Since, by the minimality of θ, we can choose the sequence $\{B_\xi\}_{\xi<\theta}$ to be increasing and continuous, the same can be assumed to be true for the sequence $\{A_\xi\}_{\xi<\theta}$. We claim now that $E = \{\xi < \theta : A_\xi \in \Delta\}$ is ω-closed unbounded in θ, which of course completes the proof since if $\{\rho_\xi\}_{\xi<\theta}$ is its increasing enumeration and $A'_\xi = A_{\rho_\xi}$, then $\{A_{\xi'}\}_{\xi<\theta}$ is a $\check{\Gamma}$-bounded sequence of Δ sets with union $A \notin \Delta$.

Clearly E is ω-closed. To see that it is unbounded, let $\xi < \theta$ be given. As $A_\xi \in \check{\Gamma}$, $A \in \Gamma$ and $A_\xi \subseteq A$, we can find, using $\mathsf{Sep}(\check{\Gamma})$, a set $X_0 \in \Delta$, with $A_\xi \subseteq X_0 \subseteq A$. By $\check{\Gamma}$-boundedness, let $\xi_0 > \xi$ be such that $X_0 \subseteq A_{\xi_0}$. Repeat now this process with ξ_0 to define X_1, ξ_1, etc. If $\xi' = \lim_{i<\omega} \xi_i < \theta$ (recall that θ must be a regular uncountable cardinal), then $A_{\xi'} = \bigcup_{\eta<\xi'} A_\eta = \bigcup_{i<\omega} A_{\xi_i} = \bigcup_{i<\omega} X_i \in \Delta$. ⊣ (Lemma 2.7)

§3. Inductive-like pointclasses and projective algebras. We shall examine now the extent to which the condition (ii) of the criterion 2.2 is satisfied. We consider first the case when Γ is closed under both types of real quantification. Let us give the following definitions.

DEFINITION 3.1. A boldface pointclass Γ is called **inductive-like** if it is closed under countable intersections and unions, $\exists^{\mathbb{R}}$ and $\forall^{\mathbb{R}}$, but not complements.

The typical inductive-like pointclasses are $\underset{\sim}{\mathrm{IND}}(\mathbb{R})$, the class of all inductive over the structure of analysis $\mathbb{R}$ pointsets, and its dual $\underset{\sim}{\mathrm{IND}}(\mathbb{R})\check{}$.

DEFINITION 3.2. A boldface pointclass Λ is called a **projective algebra** iff it is closed under complements, wellordered unions, $\exists^{\mathbb{R}}$ and $\forall^{\mathbb{R}}$.

Then Corollary 2.2 implies immediately

THEOREM 3.3 (AD). Let Γ be an inductive-like pointclass. Then the following are equivalent

(i) $\mathsf{PWO}(\Gamma) \vee \mathsf{PWO}(\check{\Gamma})$,
(ii) Δ is not a projective algebra.

The following result now provides a good first impression about the concept of projective algebra.

THEOREM 3.4 (AD). Let Λ be a projective algebra and let $\mathbf{L}[\Lambda]$ be the smallest inner model of ZF containing $\mathbb{R} \cup \Lambda$. Then $\wp(\mathbb{R}) \cap \mathbf{L}[\Lambda] = \Lambda$.

PROOF. Let Λ be a projective algebra. let $\langle \xi, \eta \rangle$ be the bijection between $\mathrm{Ord} \times \mathrm{Ord}$ and Ord corresponding to the Gödel wellordering of pairs of ordinals. Let $\langle \xi, \eta, \zeta \rangle = \langle \xi, \langle \eta, \zeta \rangle \rangle$ and $\xi = \langle I_\xi, J_\xi, K_\xi \rangle$. Let $\mathbb{F}_1, \mathbb{F}_2, \ldots, \mathbb{F}_9$ be the (binary) Gödel operations as given in Shoenfield [Sho67]. Put for $\alpha \in \mathbb{R}$, $A \in \Lambda \cap \wp(\mathbb{R})$:

$$\begin{aligned} G_i(x, y; \alpha, A) &= \mathbb{F}_i(x, y), \quad 1 \leq i \leq 9, \\ G_{10}(x, y; \alpha, A) &= x \cap \alpha, \\ G_{11}(x, y; \alpha, A) &= x \cap A. \end{aligned}$$

Then define $F\colon \mathrm{Ord} \times \mathbb{R} \times \Lambda \to \mathbf{V}$ by

$$F(\xi; \alpha, A) = \begin{cases} \{F(\eta; \beta, B) : \eta < \xi, \beta \in \mathbb{R}, B \in \Lambda\} & \text{if } I_\xi = 0 \\ G_{I_\xi}(F(J_\xi; (\alpha)_0, (A)_0), F(K_\xi; (\alpha)_1, (A)_1); (\alpha)_2, (A)_2) & \\ & \text{if } 0 < I_\xi \leq 11, \\ \{F(J_\xi; (\alpha)_0, (A)_0), F(K_\xi; (\alpha)_1, (A)_1)\} & \text{if } I_\xi \geq 12, \end{cases}$$

where $(\alpha)_i(n) = \alpha(2^i \cdot 3^n)$, $(A)_i = \{\alpha : [i]^\frown\alpha \in A\}$. We have then

$$\mathbf{L}(\Lambda) = \{F(\xi; \alpha, A) : \xi \in \mathrm{Ord}, \alpha \in \mathbb{R}, A \in \Lambda\}.$$

(Notice here that if M is an inner model of ZF containing $\mathbb{R}$ and $\Lambda \subseteq M$ then actually $\Lambda \in M$, since either $\Lambda = M \cap \wp(\mathbb{R})$ or else if $A \notin \Lambda$ has least Wadge ordinal, $A \in M$, so $\Lambda \in M$.)

Let us say now that a relation $\Phi(x, A_1, \ldots, A_n)$, where $x \in \mathcal{X}$, $A_i \subseteq \mathbb{R}$ and $A_i \in \Lambda$, is in the class $\bar{\Lambda}$, if for each *fixed* $P_1, \ldots, P_n \in \Lambda$ the pointset $\Phi^* \equiv \Phi^*_{P_1,\ldots,P_n}$ given by

$$\Phi^*(x, \varepsilon_1, \ldots, \varepsilon_n) \Leftrightarrow \varepsilon_1, \ldots, \varepsilon_n \text{ code continuous functions } f_{\varepsilon_1}, \ldots, f_{\varepsilon_n} \\ \text{and } \Phi(x, f_{\varepsilon_1}^{-1}[P_1], \ldots, f_{\varepsilon_n}^{-1}[P_n]),$$

is in Λ.

LEMMA 3.5. $\bar{\Lambda}$ contains $\Phi(\alpha, A) \Leftrightarrow \alpha \in A$, and is closed under continuous substitutions, complements, $\exists^{\mathbb{R}}$, $\forall^{\mathbb{R}}$, $\exists^{\Lambda}$, $\forall^{\Lambda}$ and wellordered unions.

PROOF. Everything is obvious except perhaps $\exists^{\Lambda}$. Let $\Phi(x, A, B)$ be in $\bar{\Lambda}$ and consider $\Psi(x, A) \Leftrightarrow \exists B \in \Lambda\Phi(x, A, B)$. Fix $P \in \Lambda$. Then

$$\begin{aligned}\Psi^*_P(x, \varepsilon) &\Leftrightarrow \varepsilon \text{ codes a continuous function } f_\varepsilon \wedge \Psi(x, f_\varepsilon^{-1}[P])\\ &\Leftrightarrow \varepsilon \text{ codes a continuous function } f_\varepsilon \wedge \exists B \in \Lambda\Phi(x, f_\varepsilon^{-1}[P], B)\\ &\Leftrightarrow \varepsilon \text{ codes a continuous function } f_\varepsilon \wedge\\ &\qquad \exists \xi < \lambda \exists B[|B|_{\mathrm{W}} \leq \xi \wedge \Phi(x, f_\varepsilon^{-1}[P], B)],\end{aligned}$$

where $\lambda = \sup\{|B|_{\mathrm{W}} : B \in \Lambda\}$. So it is enough to show that if

$$\begin{aligned}S_\xi(x, \varepsilon) \Leftrightarrow{} & \varepsilon \text{ codes a continuous function } f_\varepsilon \wedge\\ & \exists B[|B|_{\mathrm{W}} \leq \xi \wedge \Phi(x, f_\varepsilon^{-1}[P], B)],\end{aligned}$$

then $S_\xi \in \Lambda$, as $\Psi^*_P = \bigcup_{\xi<\lambda} S_\xi$. For that assume, without loss of generality, that ξ is such that $|Q|_{\mathrm{W}} = \xi \Rightarrow Q$ is self-dual, and pick such a Q. Then $Q \in \Lambda$ and

$$\begin{aligned}S_\xi(x, \varepsilon) &\Leftrightarrow \varepsilon \text{ codes a continuous function } f_\varepsilon \wedge \exists B \leq_w Q\Phi(x, f_\varepsilon^{-1}[P], B)\\ &\Leftrightarrow \varepsilon \text{ codes a continuous function } f_\varepsilon \wedge \exists \varepsilon'[\varepsilon' \text{ codes a continuous}\\ &\qquad \text{function } f_{\varepsilon'} \wedge \Phi(x, f_\varepsilon^{-1}[P], f_{\varepsilon'}^{-1}[Q])]\\ &\Leftrightarrow \exists \varepsilon' \Phi^*_{P,Q}(x, \varepsilon, \varepsilon'),\end{aligned}$$

so $S_\xi \in \Lambda$. ⊣ (Lemma 3.5)

LEMMA 3.6. For each fixed ξ, η the following relations are in $\bar{\Lambda}$:

(i) $\Phi^{\in}_{\xi,\eta}(\alpha, A; \beta, B) \Leftrightarrow F(\xi; \alpha, A) \in F(\eta, \beta, B)$,

(ii) $\Phi^{=}_{\xi,\eta}(\alpha, A; \beta, B) \Leftrightarrow F(\xi; \alpha, A) = F(\eta; \beta, B)$,

(iii) $\Psi^1_{\xi,\eta}(\alpha; \beta, B) \Leftrightarrow \alpha \in F(\max\{\xi, \eta\}; \beta, B)$,
$\Psi^2_{\xi,\eta}(n; \beta, B) \Leftrightarrow n \in \omega \wedge n \in F(\max\{\xi, \eta\}; \beta, B)$.

PROOF. Routine induction on $\langle \xi, \eta \rangle$. ⊣ (Lemma 3.6)

To complete the proof, let now $S \subseteq \mathbb{R}$ be in $\mathbf{L}[\Lambda]$ and find $\xi_0 \in \mathrm{Ord}$, $\alpha_0 \in \mathbb{R}$, $A_0 \in \Lambda$ with $S = F(\xi_0; \alpha_0, A_0)$. Then

$$\begin{aligned}\alpha \in S &\Leftrightarrow \alpha \in F(\xi_0; \alpha_0, A_0)\\ &\Leftrightarrow \Psi^1_{\xi_0,\xi_0}(\alpha; \alpha_0, A_0).\end{aligned}$$

So if ε_0 codes the identity function and $(\Psi^1_{\xi_0,\xi_0})^*_{A_0} \equiv \mathbb{R}$, we have

$$\alpha \in S \Leftrightarrow (\alpha, \alpha_0, \varepsilon_0) \in \mathbb{R}$$

thus $S \in \Lambda$. ⊣ (Theorem 3.4)

In particular for any projective algebra Λ,

$$\wp(\mathbb{R}) \cap \mathbf{L}(\mathbb{R}) \subseteq \Lambda,$$

and if $A \in \Lambda$ and $\exists B \subseteq \mathbb{R}(B \notin \mathbf{L}[A])$, or equivalently, by Steel-Van Wesep [SVW82], $A^{\#}$ exists, then $A^{\#} \in \Lambda$. Thus any projective algebra is quite big. As a simple consequence we have

COROLLARY 3.7 (AD). Let Γ be any inductive-like pointclass contained in $\mathbf{L}(\mathbb{R})$. Then $\mathsf{PWO}(\Gamma) \vee \mathsf{PWO}(\check{\Gamma})$.

Also we immediately obtain,

COROLLARY 3.8 (AD). If Γ is any inductive-like pointclass, then $\Gamma \supseteq \underset{\sim}{\mathrm{IND}}(\mathbb{R})$ or $\Gamma \supseteq \underset{\sim}{\mathrm{IND}}(\mathbb{R})\check{}$. In particular, $\underset{\sim}{\mathrm{IND}}(\mathbb{R})$ is the smallest inductive-like pointclass satisfying reduction (or even not satisfying separation).

PROOF. By Moschovakis [Mos74], $\underset{\sim}{\mathrm{IND}}(\mathbb{R})$ is the smallest inductive-like pointclass satisfying prewellordering. ⊣ (Corollary 3.8)

We conclude this section by offering some speculations on the extent of projective algebras.

Let $\boldsymbol{P}_\infty$ denote the smallest projective algebra, $\boldsymbol{B}_\infty$ the smallest boldface pointclass closed under complements and wellordered unions (the class of ∞-Borel sets) and $\boldsymbol{S}(\infty) = \bigcup_\kappa \boldsymbol{S}(\kappa)$, where $\boldsymbol{S}(\kappa)$ is the class of κ-Suslin sets. Clearly

$$\boldsymbol{S}(\infty) \subseteq \boldsymbol{B}_\infty \subseteq \boldsymbol{P}_\infty.$$

By results of Kechris, Solovay, Martin, and Steel (see [MS83]),

$$\mathsf{AD} + \mathbf{V}{=}\mathbf{L}(\mathbb{R}) \Rightarrow \boldsymbol{S}(\infty) = \underset{\sim}{\Sigma}^2_1.$$

Using this it can be shown that

$$\mathsf{AD} + \mathbf{V}{=}\mathbf{L}(\mathbb{R}) \Rightarrow \boldsymbol{B}_\infty = \wp(\mathbb{R}).$$

A proof of this is given in an appendix. On the other hand, if $\mathsf{AD}_\mathbb{R}$ denotes the Axiom of Determinacy for games on reals and Θ is the sup of the lengths of prewellorderings on $\mathbb{R}$, then the following question has been raised in Solovay [Sol78B]:

$$\mathsf{AD}_\mathbb{R} + \Theta \text{ is regular } \Rightarrow \boldsymbol{S}(\infty) = \wp(\mathbb{R})?$$

An affirmative answer would also imply that $\boldsymbol{P}_\infty = \wp(\mathbb{R})$ and thus extend Corollary 3.7 and further results of the present paper to all appropriate Γ's. Even if the above question admits a negative answer, it is still conceivable that some reasonable hypothesis extending AD (not necessarily properly) might still imply that $\boldsymbol{P}_\infty = \wp(\mathbb{R})$.[1]

[1] In the 1980's, W. H. Woodin showed that AD+ "Every binary relation on the reals can be uniformized" implies every set of reals is Suslin. His unpublished proof made use of results in this direction of H. Becker [Bec85].

§4. Projective-like pointclasses and hierarchies. We proceed now to discuss the case of pointclasses closed under only one kind of real quantification.

DEFINITION 4.1. A boldface pointclass is **projective-like** if it is closed under countable intersections and unions, and either $\exists^{\mathbb{R}}$ or $\forall^{\mathbb{R}}$ *but not both*, and is not closed under complements.

The prototypes of such pointclasses are of course $\underset{\sim}{\mathbf{\Sigma}}^1_n$, $\underset{\sim}{\mathbf{\Pi}}^1_n$.

It is convenient for the work below to embed every projective-like pointclass in a (unique) projective-like hierarchy, where this concept is defined as follows:

DEFINITION 4.2. A **projective-like hierarchy** is a sequence $\Gamma_1, \Gamma_2, \Gamma_3, \dots$ of projective-like pointclasses such that:

(i) Γ_1 is closed under $\forall^{\mathbb{R}}$,

(ii) $\Gamma_{i+1} = \exists^{\mathbb{R}}\Gamma_i$, if Γ_i is closed under $\forall^{\mathbb{R}}$, and $\Gamma_{i+1} = \forall^{\mathbb{R}}\Gamma_i$, if Γ_i is closed under $\exists^R$,

(iii) $\{\Gamma_i\}$ is maximal, *i.e.*, there is no projective-like pointclass Γ_0 closed under $\exists^{\mathbb{R}}$ such that $\Gamma_1 = \forall^{\mathbb{R}}\Gamma_0$.

We say that a projective-like pointclass Γ **belongs to the hierarchy** $\{\Gamma_i\}$ if $\Gamma = \Gamma_i$ or $\Gamma = \breve{\Gamma}_i$, for some i.

Again $\underset{\sim}{\mathbf{\Pi}}^1_1, \underset{\sim}{\mathbf{\Sigma}}^1_2, \underset{\sim}{\mathbf{\Pi}}^1_3, \underset{\sim}{\mathbf{\Sigma}}^1_4, \dots$ is the prototype of this notion.

Note first that each projective-like Γ belongs to exactly one projective-like hierarchy $\{\Gamma_i\}$. This is easy to see from the following two facts, where we note that $\Game\Gamma = \exists^{\mathbb{R}}\Gamma$, if $\forall^{\mathbb{R}}\Gamma \subseteq \Gamma$, and $\Game\Gamma = \forall^{\mathbb{R}}\Gamma$, if $\exists^{\mathbb{R}}\Gamma \subseteq \Gamma$ ($\Game$ is the game quantifier) and we define $|\Gamma|_{\mathrm{W}} \equiv |A|_{\mathrm{W}}$, for each $A \in \Gamma \setminus \Delta$:

(i) $|\Gamma|_{\mathrm{W}} < |\Game\Gamma|_{\mathrm{W}}$,

(ii) If Γ', Γ'' are projective-like and $\Game\Gamma' = \Game\Gamma''$, then $\Gamma' = \Gamma''$ (otherwise, by Wadge, $\Gamma' \subseteq \Delta''$ or $\Gamma'' \subseteq \Delta'$, so, let us say in the first case, $\Game\Gamma'' = \Game\Gamma' \subseteq \Game\breve{\Gamma}'' =$ (by AD) $(\Game\Gamma'')\breve{}$, a contradiction).

We shall classify now all the projective-like hierarchies into four types I–IV. Let us recall a basic fact about Wadge degrees first.

Call $A \subseteq \mathbb{R}$ **self-dual** iff $A \leq_{\mathrm{W}} \neg A$. (The Wadge reducibility $\leq_{\mathrm{W}}$ is defined by $A \leq_{\mathrm{W}} B \Leftrightarrow \exists f : \mathbb{R} \to \mathbb{R}$ (f continuous $\wedge f^{-1}[B] = A$).) Then Steel-Van Wesep (see Van Wesep [Van78A]) prove: If $|A|_{\mathrm{W}}$ is limit, then

$$A \text{ is self-dual iff } |A|_{\mathrm{W}} \text{ has cofinality } \omega.$$

We describe now the classification. (AD is assumed throughout.)

Type I. $\{\Gamma_i\}$ is such that $\Gamma_1 = \underset{\sim}{\mathbf{\Pi}}^1_1(A)$, for some A with $\mathrm{cf}(|A|_{\mathrm{W}}) = \omega$ and with the property that $\Lambda = \{B : |B|_{\mathrm{W}} < |A|_{\mathrm{W}}\}$ is strongly closed (recall the beginning of 2.1 here).

We visualize this by Figure 1 in the Wadge hierarchy.

By convention we accept the case $|A|_{\mathrm{W}} = 1$ as being within this type, so that the classical projective hierarchy is of type I. The next example is generated by

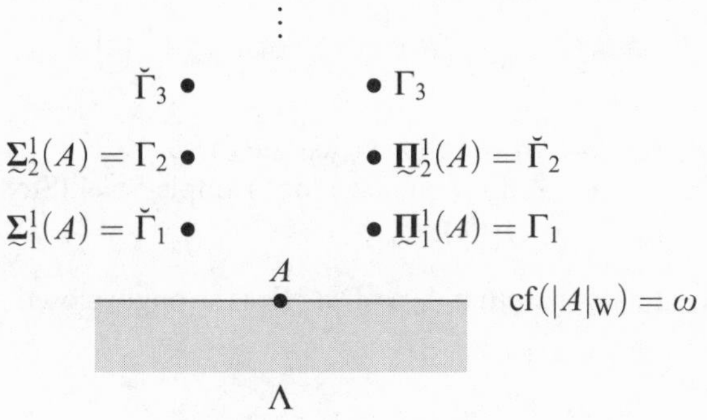

FIGURE 1.

taking A to be a set of least Wadge ordinal above the projective sets, so that $\Lambda = \{B : |B|_{\mathrm{W}} < |A|_{\mathrm{W}}\} = \bigcup_n \boldsymbol{\Delta}^1_n$.

Type II. $\{\Gamma_i\}$ is such that $\Gamma_1 = \boldsymbol{\Pi}^1_1(A)$ for some A with $\mathrm{cf}(|A|_{\mathrm{W}}) > \omega$, with $\neg A \in \boldsymbol{\Pi}^1_1(A)$, and $\Lambda = \{B : |B|_{\mathrm{W}} < |A|_{\mathrm{W}}\}$ strongly closed.

The relevant picture is in Figure 2.

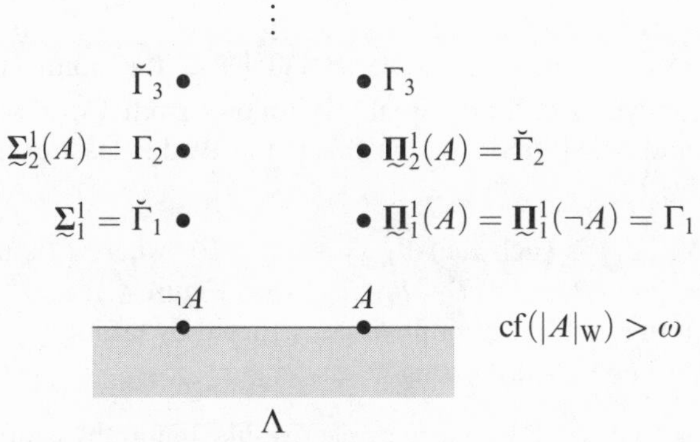

FIGURE 2.

The smallest example of such a hierarchy is constructed as follows: Let Γ_η, $1 \leq \eta < \omega_1$ be defined by

$$\Gamma_1 = \boldsymbol{\Pi}^1_1,$$
$$\Gamma_{\eta+1} = \Game\Gamma_\eta,$$
$$\Gamma_\lambda = \text{all countable unions of sets in } \bigcup_{\eta<\lambda} \Gamma_\eta, \text{ if } \lambda = \bigcup \lambda > 0.$$

Let then A be such that

$$\Lambda = \{B : |B|_W < |A|_W\} = \bigcup_{\eta<\omega_1} \Gamma_\eta.$$

Since $\mathrm{cf}(|A|_W) = \omega_1$ it follows that $\{B : B \leq_W A\}$ is not closed under both intersections and unions (see for example Steel [Ste81B]), so $\underset{\sim}{\boldsymbol{\Pi}}{}^1_1(A) = \underset{\sim}{\boldsymbol{\Pi}}{}^1_1(\neg A)$, thus $\neg A \in \underset{\sim}{\boldsymbol{\Pi}}{}^1_1(A)$.

Type III. $\{\Gamma_i\}$ is such that $\Delta_1 = \Gamma_1 \cap \breve{\Gamma}_1$ is strongly closed. (See Figure 3.)

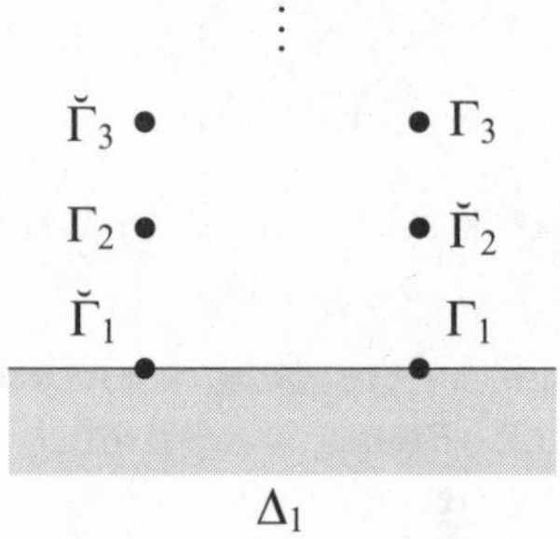

FIGURE 3.

A typical example is $\Gamma_1 =_2 \underset{\sim}{\mathrm{ENV}}({}^3\mathrm{E}) \equiv$ the pointclass of sets Kleene semirecursive in ${}^3\mathrm{E}$ and a real. It follows from Corollary 5.4(i) (see also Theorem 3.3 of [Ste81A]) that this is the Wadge-least example of a type III hierarchy.

Type IV. $\{\Gamma_i\}$ is such that $\Gamma_1 = \forall^{\mathbb{R}}(\Gamma \vee \breve{\Gamma})$, where Γ is inductive-like, and $\Gamma \vee \breve{\Gamma} = \{A \cup B : A \in \Gamma \wedge B \in \breve{\Gamma}\}$. (See Figure 4.)

Again the smallest example is constructed by taking $\Gamma = \underset{\sim}{\mathrm{IND}}(\mathbb{R})$.

We have now the following

THEOREM 4.3 (AD). Every projective-like hierarchy is of exactly one of the types I–IV.

PROOF. As no projective-like hierarchy can be of two different types, it is enough to prove that every projective-like hierarchy $\{\Gamma_i\}$ is of one of the types I–IV.

For the classical projective hierarchy, no strongly closed pointclass is included in Δ_1. We include this hierarchy under alternative I. Let Λ be the largest strongly closed pointclass contained in Δ_1. Let $\lambda = \sup\{|A|_W : A \in \Lambda\}$. Then exactly one of the four possibilities below must hold:

I. $\mathrm{cf}(\lambda) = \omega$

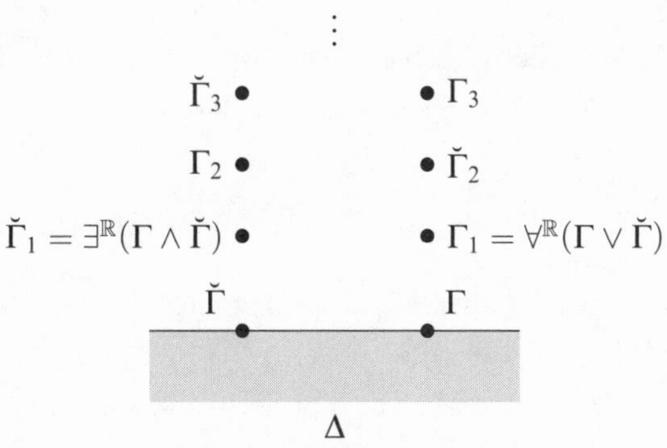

FIGURE 4.

II. $\mathrm{cf}(\lambda) > \omega$ and if A is such that $|A|_W = |\neg A|_W = \lambda$, then $\{B : B \leq_W A\}$ (and thus $\{B : B \leq_W \neg A\}$) is neither projective-like nor inductive-like.

III. $\mathrm{cf}(\lambda) > \omega$ and if A, $\neg A$ are as above, then $\{B : B \leq_W A\}$ (and thus $\{B : B \leq_W \neg A\}$) is projective-like.

IV. $\mathrm{cf}(\lambda) > \omega$ and if A, $\neg A$ are as above, then $\{B : B \leq_W A\}$ (and thus $\{B : B \leq_W \neg A\}$) is inductive-like.

If I holds, let A be such that $|A|_W = \lambda$. By the maximality of Λ we must have $\underset{\sim}{\mathbf{\Pi}}^1_1(A) = \Gamma_1$, so that $\{\Gamma_i\}$ is of type I. If II holds, then (by Wadge) $\neg A \in \underset{\sim}{\mathbf{\Pi}}^1_1(A)$, and $\Gamma_1 = \underset{\sim}{\mathbf{\Pi}}^1_1(A)$, so that $\{\Gamma_i\}$ is of type II. If III holds and, without loss of generality, $\{B : B \leq_W A\}$ is closed under $\forall^{\mathbb{R}}$ but not $\exists^{\mathbb{R}}$, then $\Gamma_1 = \{B : B \leq_W A\}$ and $\Delta_1 = \Lambda$, so that $\{\Gamma_i\}$ is of type III. Finally, if IV holds and $\Gamma = \{B : B \leq_W A\}$, we have $\Gamma_1 = \forall^{\mathbb{R}}(\Gamma \vee \check{\Gamma})$, so that $\{\Gamma_i\}$ is of type IV. $\dashv$ (Theorem 4.3)

§5. The prewellordering pattern in projective-like hierarchies. We conclude now our analysis by establishing (under certain assumptions) that each type of projective-like hierarchy has only one prewellordering pattern. For convenience let us introduce the following terminology.

DEFINITION 5.1. A projective-like hierarchy $\{\Gamma_i\}$ is of **character** Π iff $\mathsf{PWO}(\Gamma_1)$ holds (iff $\mathsf{PWO}(\Gamma_i)$ holds for each $i \geq 1$, by the First Periodicity Theorem of Martin and Moschovakis). A projective-like hierarchy is of **character** Σ iff $\mathsf{PWO}(\check{\Gamma}_1)$ holds (iff $\mathsf{PWO}(\check{\Gamma}_i)$ holds for all $i \geq 1$).

DEFINITION 5.2. The **ground** of a projective-like hierarchy $\{\Gamma_i\}$ is defined to be the largest strongly closed Λ such that $\Lambda \subseteq \Delta_1$. Thus the ground Λ coincides with the strongly closed Λ occurring in the definition of $\{\Gamma_i\}$ being of type I or II, with Δ_1 if $\{\Gamma_i\}$ is of type III, and with $\Delta = \Gamma \cap \check{\Gamma}$ if $\{\Gamma_i\}$ is of type IV and Γ is given in the definition of this type.

We now have

THEOREM 5.3 (AD). Let $\{\Gamma_i\}$ be a projective-like hierarchy. Then

(i) If $\{\Gamma_i\}$ is of type I, then $\{\Gamma_i\}$ has character Π.
(ii) If $\{\Gamma_i\}$ is of type II, then $\{\Gamma_i\}$ has character Σ, provided its ground is not a projective algebra.
(iii) If $\{\Gamma_i\}$ is of type III, then $\{\Gamma_i\}$ has character Π iff its ground is not a projective algebra.
(iv) If $\{\Gamma_i\}$ is of type IV, then $\{\Gamma_i\}$ has character Σ, provided its ground is not a projective algebra.

COROLLARY 5.4 (AD). Let $\{\Gamma_i\}$ be a projective-like hierarchy contained in $\boldsymbol{P}_\infty$. Then

(i) If $\{\Gamma_i\}$ is of type I, III, then $\{\Gamma_i\}$ has character Π.
(ii) If $\{\Gamma_i\}$ is of type II, IV, then $\{\Gamma_i\}$ has character Σ.

In particular, this holds if $\{\Gamma_i\}$ is contained in $\mathbf{L}(\mathbb{R})$.

COROLLARY 5.5 (AD). Let Γ be a boldface pointclass closed under countable intersections and unions and $\exists^{\mathbb{R}}$ or $\forall^{\mathbb{R}}$, but not complements. If $\Gamma \subseteq \boldsymbol{P}_\infty$, in particular if $\Gamma \subseteq \mathbf{L}(\mathbb{R})$, then $\mathsf{PWO}(\Gamma)$ or $\mathsf{PWO}(\check{\Gamma})$.

PROOF OF THEOREM 5.3. (i) Let A be as in the definition of a type I hierarchy, so that in particular $\Gamma_1 = \underset{\sim}{\Pi}^1_1(A)$. Let $\Lambda = \{B : |B|_{\mathrm{W}} < |A|_{\mathrm{W}}\}$. Since $\mathrm{cf}(|A|_{\mathrm{W}}) = \omega$, Λ is not closed under countable unions, so if

$$\Gamma_0 = \Big\{\bigcup_n A_n \,:\, \forall n(A_n \in \Lambda)\Big\},$$

then Γ_0 is a boldface pointclass, closed under intersections and unions, countable unions and $\exists^{\mathbb{R}}$. Moreover we have $\mathsf{PWO}(\Gamma_0)$. Since $\Gamma_1 = \forall^{\mathbb{R}}\Gamma_0$, we have $\mathsf{PWO}(\Gamma_1)$ by the First Periodicity Theorem.

(iii) By Theorem 2.1 and the fact that Δ_1 is not a projective algebra, it is enough to show $\mathsf{Red}(\Gamma_1)$ and by Steel's Theorem it is enough to show $\neg\mathsf{Sep}(\Gamma_1)$. This is immediate from the following:

LEMMA 5.6. If Γ is a boldface pointclass not closed under complements, and Δ is closed under $\exists^{\mathbb{R}}, \forall^{\mathbb{R}}$ then

$$\mathsf{Sep}(\Gamma) \Rightarrow \exists^{\mathbb{R}}\Gamma \subseteq \Gamma.$$

PROOF. Let P, Q be disjoint sets in $\exists^{\mathbb{R}}\Gamma$. Let

$$P(x) \Leftrightarrow \exists\alpha(x,\alpha) \in A, Q(x) \Leftrightarrow \exists\beta(x,\beta) \in B,$$

where $A, B \in \Gamma$ are also disjoint. Put

$$A'(x,\alpha,\beta) \Leftrightarrow A(x,\alpha)$$
$$B'(x,\alpha,\beta) \Leftrightarrow B(x,\beta).$$

Then $A', B' \in \Gamma$ are disjoint, so find $C \in \Delta$ separating them. Then

$$S(x) \Leftrightarrow \exists\alpha\forall\beta(x, \alpha, \beta) \in C$$

is in Δ and separates P, Q (this is a variant of an argument of Addison).

So if $\exists^{\mathbb{R}}\Gamma \not\subseteq \Gamma$, then by Wadge $\check{\Gamma} \subseteq \exists^{\mathbb{R}}\Gamma$, thus if $A \in \Gamma$ then $P = A$, $Q = \neg A$ are disjoint in $\exists^{\mathbb{R}}\Gamma$ and thus can be separated by $S \in \Delta$. So $A = S \in \Delta$, a contradiction. ⊣ (Lemma 5.6)

(iv) Let Γ be the inductive-like pointclass appearing in the definition of a projective-like hierarchy of type IV, so that $\check{\Gamma}_1 = \exists^{\mathbb{R}}(\Gamma \wedge \check{\Gamma})$. Assume without loss of generality that $\mathsf{Red}(\Gamma)$ holds. Then by Theorem 2.1 $\mathsf{PWO}(\Gamma)$ holds. Then $\Pi = \Gamma \wedge \check{\Gamma}$ is a boldface pointclass closed under intersections and $\forall^{\mathbb{R}}$ and has the prewellordering property. (If $A \in \Gamma$, $B \in \check{\Gamma}$ and φ is a Γ-norm on A, then $\varphi{\restriction}A \cap B$ is a $\Gamma \wedge \check{\Gamma}$-norm on $A \cap B$.) So, by the usual proof, $\check{\Gamma}_1 = \exists^{\mathbb{R}}(\Gamma \wedge \check{\Gamma})$ has the prewellordering property.

(ii) Let A be as in the definition of type II hierarchy so that $\Gamma_1 = \utilde{\Pi}^1_1(A)$ and A is chosen (from $A, \neg A$) so that

$$\Pi = \{B : B \leq_{\mathrm{W}} A\}$$

has the reduction property. Put $\Sigma = \check{\Pi}$. Then, by Lemma 5.6, Π is closed under $\forall^{\mathbb{R}}$ and hence, since Π is $\mathbb{R}$-parametrized, under countable intersections. Thus $\utilde{\Sigma}^1_1(A) = \exists^{\mathbb{R}}\Pi$.

Since $\Lambda = \{B : |B|_{\mathrm{W}} < |A|_{\mathrm{W}}\}$ is not a projective algebra, it is not closed under wellordered unions. Let θ be least such that some θ-sequence of elements of Λ has union outside Λ. (Again θ is a regular cardinal $> \omega$, since Λ is closed under countable unions as $|A|_{\mathrm{W}}$ has cofinality $> \omega$.) Let $\lambda = |A|_{\mathrm{W}} = \sup\{|B|_{\mathrm{W}} : B \in \Lambda\}$. As in 2.1, $\theta \leq \lambda$. Put

$$\Gamma' = \Big\{\bigcup_{\xi<\theta} A_\xi : \forall\xi < \theta(A_\xi \in \Lambda)\Big\}.$$

Then Γ' is a boldface pointclass closed under countable intersections and unions and $\exists^{\mathbb{R}}$. So $\Gamma' \supseteq \utilde{\Sigma}^1_1(A)$. If $\Gamma' = \utilde{\Sigma}^1_1(A) = \check{\Gamma}_1$ we are done, since clearly $\mathsf{PWO}(\Gamma')$.

Otherwise, $\Gamma' \supsetneq \utilde{\Sigma}^1_1(A)$, so by Wadge, $\Gamma' \supseteq \utilde{\Pi}^1_1(A) = \Gamma_1$, thus $\Gamma' \supseteq \Gamma_2 = \exists^{\mathbb{R}}\Gamma_1$. Now notice that there is a wellfounded relation $\prec$ of rank $\geq \lambda (\geq \theta)$ in Γ_2. Indeed, let

$$(\sigma, \alpha) \prec (\sigma', \beta) \Leftrightarrow \sigma = \sigma' \text{ codes a continuous function } f_\sigma \ \& \\ f_\sigma^{-1}[A] \text{ is wellfounded } \& \ (\alpha, \beta) \in f_\sigma^{-1}[A].$$

So by Moschovakis [Mos80], Ch. 7, Γ_2 is closed under wellordered unions of length $\mathrm{rank}(\prec)$, thus $\Gamma' \subseteq \Gamma_2$. So

$$\Gamma' = \Gamma_2.$$

Put now

$$\Gamma^* = \Big\{ \bigcup_{\xi<\theta} A_\xi \;:\; \forall \xi < \theta (A_\xi \in \Lambda) \wedge \{A_\xi\}_{\xi<\theta} \text{ is } \utilde{\Sigma}^1_1\text{-bounded} \Big\}.$$

Note first that Γ^* is closed under $\forall^{\mathbb{R}}$. The argument is similar to that in Lemma 2.6. Indeed let $(x, \alpha) \in B \Leftrightarrow \exists \xi < \theta (x, \alpha) \in B_\xi$, where $\{B_\xi\}_{\xi<\theta}$ is $\utilde{\Sigma}^1_1$-bounded. Let $x \in C \Leftrightarrow \forall \alpha (x, \alpha) \in B$. Then as in Lemma 2.6, $x \in C \Leftrightarrow \exists \xi < \theta \forall \alpha (x, \alpha) \in B_\xi$. Let $C_\xi = \{x : \forall \alpha (x, \alpha) \in B_\xi\}$. It is enough to show that $\{C_\xi\}_{\xi<\theta}$ is $\utilde{\Sigma}^1_1$-bounded. Let $X \in \utilde{\Sigma}^1_1$, $X \subseteq \bigcup_{\xi<\theta} C_\xi$. Then $\forall x \in X \exists \xi \forall \alpha (x, \alpha) \in B_\xi$, thus $\forall x \in X \forall \alpha \exists \xi (x, \alpha) \in B_\xi$, so $X \times \mathbb{R} \subseteq B$, thus for some $\xi < \theta$, $X \times \mathbb{R} \subseteq B_\xi$, therefore $\forall x \in X \forall \alpha (x, \alpha) \in B_\xi$ *i.e.*, $X \subseteq C_\xi$, and we are done.

So if $\Gamma^* \not\subseteq \Lambda$, then either $\Gamma^* = \Pi$ in which case we have $\mathsf{PWO}(\Pi)$ (as in Lemma 2.5), thus $\mathsf{PWO}(\breve{\Gamma}_1)$ since $\breve{\Gamma}_1 = \exists^{\mathbb{R}}\Pi$, and we are done, or else $\Gamma^* \supseteq \Sigma$, so $\Gamma^* \supseteq \forall^{\mathbb{R}}\Sigma = \Gamma_1$. Then we must have $\Gamma^* = \Gamma_1$ (since otherwise $\Gamma^* \supseteq \breve{\Gamma}_1$, so $\Gamma^* \supseteq \forall^{\mathbb{R}}\breve{\Gamma}_1 = \breve{\Gamma}_2$, contradicting the fact that $\Gamma^* \subseteq \Gamma' = \Gamma_2$). So if $W \in \Gamma_1 \setminus \Delta_1$, we can write $W = \bigcup_{\xi<\theta} W_\xi$, with $W_\xi \in \Lambda$ and $\{W_\xi\}_{\xi<\theta}$ a $\utilde{\Sigma}^1_1$-bounded sequence. Let

$$\varphi(x) = \text{least } \xi,\ x \in W_\xi$$

be the associated norm, which by the argument in Lemma 2.5 is a $(\Gamma^* =)\Gamma_1$-norm. But $A \in \Delta_1$, so let f continuous be such that $x \in A \Leftrightarrow f(x) \in W$. By the usual boundedness argument (recall that Γ_1 is closed under $\forall^{\mathbb{R}}$), there is $\xi < \theta$, with $x \in A \Leftrightarrow f(x) \in W_\xi$, so $A \in \Lambda$, a contradiction.

So we can complete the proof by showing that $\Gamma^* \not\subseteq \Lambda$. For that we use the same argument as in Lemma 2.7. Let $\mathcal{S}$ be universal $\utilde{\Sigma}^1_1$ and put

$$C = \{\varepsilon : \mathcal{S}_\varepsilon \subseteq A\}.$$

Then $C \in \utilde{\Pi}^1_1(A) = \Gamma_1 \subseteq \Gamma'$, so $C = \bigcup_{\xi<\theta} C_\xi$, where each $C_\xi \in \Lambda$. Put

$$A_\xi = \bigcup\{\mathcal{S}_\varepsilon : \varepsilon \in C_\xi\}.$$

Then $A_\xi \in \Lambda$, $\{A_\xi\}_{\xi<\theta}$ is $\utilde{\Sigma}^1_1$-bounded, and $A = \bigcup_{\xi<\theta} A_\xi \notin \Lambda$, so we are done. ⊣ (Theorem 5.3)

An alternative approach to the proof of Theorem 5.3 is given in §3 of Steel [Ste81A].

§6. Problems and conjectures. Assume AD in this section. Let $\Lambda \subsetneq \boldsymbol{P}_\infty$ be a strongly closed boldface pointclass and $\{\Gamma_i\}$ the first projective-like hierarchy not contained in Λ or in other words the projective-like hierarchy with ground Λ. Let λ be the ordinal associated with Λ as in Lemma 2.3, *i.e.*, the supremum of the ordinals $|B|_{\mathrm{W}}$ for $B \in \Lambda$. The question is whether λ determines the type of $\{\Gamma_i\}$ (λ is always a limit cardinal).

Clearly if $\mathrm{cf}(\lambda) = \omega$, $\{\Gamma_1\}$ is of type I. It is easy also to see that if $\mathrm{cf}(\lambda) > \omega$ and λ is singular, then $\{\Gamma_i\}$ is of type II (otherwise $\Lambda = \Delta_1$ and $\mathsf{PWO}(\Gamma_1)$ for a type III $\{\Gamma_i\}$, so, since $\forall^{\mathbb{R}}\Gamma_1 = \Gamma_1$, we have by Moschovakis [Mos70] that $\lambda = \delta_1 \equiv \sup\{\xi : \xi \text{ is the rank of a } \Delta_1 \text{ prewellordering}\}$, thus λ is regular). We offer now the following *conjecture*:

If λ is regular, then $\{\Gamma_i\}$ is of type III or IV.

(As pointed out above the converse is true.) As positive evidence we consider the fact that this conjecture is true for at least $\Lambda \subseteq \underset{\sim}{\mathrm{IND}}(\mathbb{R})$, as shown by Steel [Ste81A].

It is not clear what property of a regular λ would distinguish between types III and IV. However the results in §5(A) of Kechris [Kec81B] suggest that some combination of indescribability and Mahlo properties of λ could guarantee that $\{\Gamma_i\}$ is of type IV.

And we conclude with one more conjecture and two problems concerning closure of classes under wellordered unions:

(a) Conjecture. If Γ is inductive-like and has the prewellordering property, then Γ is closed under wellordered unions (compare with Lemma 2.4).
(b) If Γ is projective-like, can Δ be closed under wellordered unions?
(c) Can the ground of type II or III hierarchies be a projective algebra?

Appendix A. We give here a proof of the 0th Periodicity Theorem. Assume below AD. Let Γ be a boldface pointclass closed under countable unions and intersections, which has the reduction property. We show that

(i) $\exists^{\mathbb{R}}\Gamma \subseteq \Gamma = \mathsf{Red}(\forall^{\mathbb{R}}\Gamma)$,
(ii) $\forall^{\mathbb{R}}\Gamma \subseteq \Gamma = \mathsf{Red}(\exists^{\mathbb{R}}\Gamma)$.

Take first (i). Assume $\exists^{\mathbb{R}}\Gamma \subseteq \Gamma$. Then note that $\forall^{\mathbb{R}}\Gamma = \{\text{Ↄ}\alpha A(x,\alpha) : A \in \Gamma \text{ and } A \text{ is Turing invariant on } \alpha\}$, where $A(x,\alpha)$ is **Turing invariant** on α iff $\alpha \equiv_{\mathrm{T}} \beta \wedge A(x,\alpha) \Rightarrow A(x,\beta)$. This is because $\forall\alpha B(x,\alpha) \Leftrightarrow \text{Ↄ}\alpha\forall\beta \leq_{\mathrm{T}} \alpha B(x,\beta)$. So let P, Q be in $\forall^{\mathbb{R}}\Gamma$ and say $x \in P \Leftrightarrow \text{Ↄ}\alpha(x,\alpha) \in A$, $x \in Q \Leftrightarrow \text{Ↄ}\alpha(x,\alpha) \in B$, with $A, B \in \Gamma$ and Turing invariant on α. By Burgess and Miller [BM75], we can find $A_1, B_1 \in \Gamma$ Turing invariant reducing A, B. [Indeed, let $A', B' \in \Gamma$ reduce A, B and then put

$$A_1(x,\alpha) \Leftrightarrow \exists\alpha' \equiv_{\mathrm{T}} \alpha A'(x,\alpha')$$
$$B_1(x,\alpha) \Leftrightarrow \forall\alpha' \equiv_{\mathrm{T}} \alpha B'(x,\alpha).]$$

Now let $x \in P_1 \Leftrightarrow \text{Ↄ}\alpha(x,\alpha) \in A_1$ and $x \in Q_1 \Leftrightarrow \text{Ↄ}\alpha(x,\alpha) \in B_1$. Then $P_1, Q_1 \in \forall^{\mathbb{R}}\Gamma$ and they reduce P, Q.

The proof of (ii) is similar, utilizing the equivalence $\exists\alpha B(x,\alpha) \Leftrightarrow \text{Ↄ}\alpha\exists\beta \leq_{\mathrm{T}} \alpha B(x,\beta)$.

Appendix B. We prove here the fact that

$$\mathsf{AD} + \mathbf{V{=}L}(\mathbb{R}) \Rightarrow \boldsymbol{B}_\infty = \wp(\mathbb{R}).$$

Assume $\mathsf{AD} + \mathbf{V{=}L}(\mathbb{R})$. Assume also that $\boldsymbol{B}_\infty \neq \wp(\mathbb{R})$, towards a contradiction. Pick then $A \subseteq \mathbb{R}$ such that $A \notin \boldsymbol{B}_\infty$. As every set of reals is ordinal definable from a real, let $\varphi(x, \xi, \alpha)$ be a formula and $\alpha \in \mathbb{R}$, $\xi_0 \in \mathrm{Ord}$ be such that

$$x \in A \Leftrightarrow \varphi(x, \xi_0, \alpha_0).$$

Fix now α_0 and pick the least ordinal ξ such that $\{x \in \mathbb{R} : \varphi(x, \xi, \alpha_0)\} \notin \boldsymbol{B}_\infty$. Clearly this ξ is definable from α_0, so we conclude that there is a set of reals B definable from α_0, say

$$x \in B \Leftrightarrow \psi(x, \alpha_0),$$

such that $B \notin \boldsymbol{B}_\infty$. By Skolem-Lowenheim, let $\lambda < \Theta$ be least such that

$$\mathbf{L}_\lambda(\mathbb{R}) \models \mathsf{ZF}_N{+}\mathsf{DC}{+}\mathsf{AD} + \{x : \psi(x, \alpha_0)\} \notin \boldsymbol{B}_\infty.$$

Here ZF_N is a large enough finite fragment of ZF. Let

$$C = \{x \in \mathbb{R} : \mathbf{L}_\lambda(\mathbb{R}) \models \psi(x, \alpha_0)\}.$$

We have that

$$\mathbf{L}_\lambda(\mathbb{R}) \models C \notin \boldsymbol{B}_\infty.$$

But then we claim that actually $C \notin \boldsymbol{B}_\infty$. Indeed, if $\Delta = \{D \subseteq \mathbb{R} : \mathbf{L}_\lambda(\mathbb{R}) \models D \in \boldsymbol{B}_\infty\}$ and

$$\delta = \sup\{\eta : \eta \text{ is the length of a } \Delta \text{ prewellordering of } \mathbb{R}\},$$

then Δ is closed under wellordered unions of length δ. To see this, let $\{A_\xi\}_{\xi<\delta}$ be a sequence of Δ sets. Let $S \in \mathbf{L}_\lambda(\mathbb{R})$ be such that all $D \in \Delta$ are Wadge reducible to S (S exists since $\mathbf{L}_\lambda(\mathbb{R}) \models \boldsymbol{B}_\infty \neq \wp(\mathbb{R})$). Let $C_\xi = \{\varepsilon : \varepsilon$ codes a continuous function f_ε and $f_\varepsilon^{-1}[S] = A_\xi\}$. Notice now that there is a norm $\chi\colon \mathbb{R} \twoheadrightarrow \delta$ in $\mathbf{L}_\lambda(\mathbb{R})$, thus by the Moschovakis Coding Lemma there is a function h in $\mathbf{L}_\lambda(\mathbb{R})$ such that $\forall \xi < \delta (h(\xi) \neq \varnothing \wedge h(\xi) \subseteq C_\xi)$. Consequently, $\{A_\xi\}_{\xi<\delta} = \{\bigcap_{\varepsilon \in h(\xi)} f_\varepsilon^{-1}[S]\}_{\xi<\delta} \in \mathbf{L}_\lambda(\mathbb{R})$, so $\bigcup_{\xi<\delta} A_\xi \in \Delta$.

Since Δ is closed under δ unions, clearly Δ is closed under arbitrary wellordered unions, thus $\Delta \supseteq \boldsymbol{B}_\infty$, so $C \notin \boldsymbol{B}_\infty$.

We shall complete the proof by showing that $C \in \underset{\sim}{\Delta}{}^2_1$. This leads immediately to a contradiction, since $\underset{\sim}{\Sigma}{}^2_1 = \boldsymbol{S}(\infty)$ and so $C \in \boldsymbol{B}_\infty$. To see that $C \in \underset{\sim}{\Delta}{}^2_1$ notice that

$$x \in C \Leftrightarrow \exists \lambda < \Theta[\mathbf{L}_\lambda(\mathbb{R}) \models (\mathsf{ZF}_N{+}\mathsf{DC}{+}\mathsf{AD} + \{x : \psi(x, \alpha_0)\} \notin \boldsymbol{B}_\infty) \wedge \\ \lambda \text{ is least with that property } \wedge \mathbf{L}_\lambda(\mathbb{R}) \models \psi(x, \alpha_0)].$$

As structures $\mathbf{L}_\lambda(\mathbb{R})$, for $\lambda < \Theta$, can be coded in a straightforward fashion by sets of reals, this shows that $C \in \utilde{\mathbf{\Sigma}}^2_1$ and a similar computation shows that $C \in \utilde{\mathbf{\Pi}}^2_1$, so we are done.

REFERENCES

JOHN W. ADDISON AND YIANNIS N. MOSCHOVAKIS

[AM68] *Some consequences of the axiom of definable determinateness*, ***Proceedings of the National Academy of Sciences of the United States of America***, no. 59, 1968, pp. 708–712.

HOWARD S. BECKER

[Bec85] *A property equivalent to the existence of scales*, ***Transactions of the American Mathematical Society***, vol. 287 (1985), pp. 591–612.

J. BURGESS AND D. MILLER

[BM75] *Remarks on invariant descriptive set theory*, ***Fundamenta Mathematicae***, vol. 90 (1975), pp. 53–75.

ALEXANDER S. KECHRIS

[Kec74] *On projective ordinals*, ***The Journal of Symbolic Logic***, vol. 39 (1974), pp. 269–282.

[Kec77B] *Classifying projective-like hierarchies*, ***Bulletin of the Greek Mathematical Society***, vol. 18 (1977), pp. 254–275.

[Kec81B] *Suslin cardinals, κ-Suslin sets, and the scale property in the hyperprojective hierarchy*, in Kechris et al. [CABAL ii], pp. 127–146, reprinted in [CABAL I], p. 314–332.

ALEXANDER S. KECHRIS, BENEDIKT LÖWE, AND JOHN R. STEEL

[CABAL I] *Games, scales, and Suslin cardinals: the Cabal seminar, volume I*, Lecture Notes in Logic, vol. 31, Cambridge University Press, 2008.

ALEXANDER S. KECHRIS, DONALD A. MARTIN, AND YIANNIS N. MOSCHOVAKIS

[CABAL ii] *Cabal seminar* 77–79, Lecture Notes in Mathematics, no. 839, Berlin, Springer, 1981.

[CABAL iii] *Cabal seminar* 79–81, Lecture Notes in Mathematics, no. 1019, Berlin, Springer, 1983.

ALEXANDER S. KECHRIS AND YIANNIS N. MOSCHOVAKIS

[CABAL i] *Cabal seminar* 76–77, Lecture Notes in Mathematics, no. 689, Berlin, Springer, 1978.

DONALD A. MARTIN

[Mar68] *The axiom of determinateness and reduction principles in the analytical hierarchy*, ***Bulletin of the American Mathematical Society***, vol. 74 (1968), pp. 687–689.

[Mar71B] *Projective sets and cardinal numbers: some questions related to the continuum problem*, this volume, originally a preprint, 1971.

DONALD A. MARTIN AND JOHN R. STEEL

[MS83] *The extent of scales in* L(ℝ), in Kechris et al. [CABAL iii], pp. 86–96, reprinted in [CABAL I], p. 110–120.

YIANNIS N. MOSCHOVAKIS

[Mos70] *Determinacy and prewellorderings of the continuum*, ***Mathematical logic and foundations of set theory. Proceedings of an international colloquium held under the auspices of the Israel Academy of Sciences and Humanities, Jerusalem, 11–14 November 1968*** (Y. Bar-Hillel, editor), Studies in Logic and the Foundations of Mathematics, North-Holland, Amsterdam-London, 1970, pp. 24–62.

[Mos74] ***Elementary induction on abstract structures***, North-Holland, 1974.

[Mos80] ***Descriptive set theory***, Studies in Logic and the Foundations of Mathematics, no. 100, North-Holland, Amsterdam, 1980.

JOSEPH R. SHOENFIELD

[Sho67] ***Mathematical logic***, Addison-Wesley, 1967.

ROBERT M. SOLOVAY

[Sol78B] *The independence of* DC *from* AD, in Kechris and Moschovakis [CABAL i], pp. 171–184.

JOHN R. STEEL

[Ste81A] *Closure properties of pointclasses*, this volume, originally published in Kechris et al. [CABAL ii], pp. 147–163.

[Ste81B] *Determinateness and the separation property*, ***The Journal of Symbolic Logic***, vol. 46 (1981), no. 1, pp. 41–44.

JOHN R. STEEL AND ROBERT VAN WESEP

[SVW82] *Two consequences of determinacy consistent with choice*, ***Transactions of the American Mathematical Society***, (1982), no. 272, pp. 67–85.

ROBERT VAN WESEP

[Van78A] *Separation principles and the axiom of determinateness*, ***The Journal of Symbolic Logic***, vol. 43 (1978), no. 1, pp. 77–81.

[Van78B] *Wadge degrees and descriptive set theory*, this volume, originally published in Kechris and Moschovakis [CABAL i], pp. 151–170.

DEPARTMENT OF MATHEMATICS
CALIFORNIA INSTITUTE OF TECHNOLOGY
PASADENA, CALIFORNIA 91125
UNITED STATES OF AMERICA
E-mail: kechris@caltech.edu

DEPARTMENT OF MATHEMATICS
UNIVERSITY OF CALIFORNIA
BERKELEY, CALIFORNIA 94720-3840
UNITED STATES OF AMERICA
E-mail: solovay@gmail.com
E-mail: steel@math.berkeley.edu

POINTCLASSES AND WELLORDERED UNIONS

STEVE JACKSON AND DONALD A. MARTIN

§1. **Introduction.** Our basic notation and terminology is that of [KSS81]. In particular, recall that a **boldface pointclass** is a pointclass containing all clopen sets and closed under continuous preimages.

A. S. Kechris [KSS81] proved the following theorem, assuming the Axiom of Determinacy (AD) and the Axiom of Dependent Choice (DC):

Let $\utilde{\boldsymbol{\Gamma}}$ be a boldface pointclass closed under countable unions, countable intersections, and $\exists^{\mathbb{R}}$ but not under $\forall^{\mathbb{R}}$. If $\mathsf{Red}(\utilde{\boldsymbol{\Gamma}})$, then $\utilde{\boldsymbol{\Gamma}}$ is closed under wellordered unions.

Kechris' theorem raises two sorts of questions:

Q1. Does Kechris' theorem remain true when we replace "but not under $\forall^{\mathbb{R}}$" by "and under $\forall^{\mathbb{R}}$ but not under complements"?

It seems strange that a failure of closure is needed to prove another kind of closure.

The second question is, roughly: Are there indeed any interesting well-ordered unions for $\utilde{\boldsymbol{\Gamma}}$ to be closed under? If $\utilde{\boldsymbol{\Gamma}}$ is closed under countable unions, countable intersections, and $\exists^{\mathbb{R}}$ but not under complements, and $\utilde{\boldsymbol{\Gamma}}$ has the prewellordering property, then it follows easily from AD+DC and the Moschovakis Coding Lemma [Mos80, 7D.5] that $\utilde{\boldsymbol{\Gamma}}$ is closed under wellordered unions of length $\leq \kappa$, where κ is the length of any $\utilde{\boldsymbol{\Gamma}}$ prewellordering.

Q2. Can one show (assuming AD+DC) that, whenever $\langle A_\beta : \beta < \gamma \rangle$ is a strictly increasing sequence of $\utilde{\boldsymbol{\Gamma}}$ sets and $\utilde{\boldsymbol{\Gamma}}$ is as above, then there is a $\utilde{\boldsymbol{\Gamma}}$ prewellordering of length γ?

The second question is especially important when $\utilde{\boldsymbol{\Gamma}}$ is the class of κ-Suslin sets, for then it bears on the problem of reliable cardinals. A set A is **κ-Suslin** if there is a scale on A of length $\leq \kappa$. κ is a **Suslin cardinal** if some set is κ-Suslin but not λ-Suslin for any $\lambda < \kappa$. A cardinal κ is **reliable** if there is a scale on some set A of length exactly κ.

The second author was supported in part by NSF Grant #MCS 78-02989.

Wadge Degrees and Projective Ordinals: The Cabal Seminar, Volume II
Edited by A. S. Kechris, B. Löwe, J. R. Steel
Lecture Notes in Logic, 37

To see how Q2 relates to the problem of whether all reliable cardinals are Suslin, assume DC and suppose that there are no non-trivial unions of κ-Suslin sets of length κ^+, *i.e.*, suppose that there is no strictly increasing sequence $\langle A_\beta : \beta < \kappa^+\rangle$ of κ-Suslin sets. Let γ be the least reliable cardinal greater than κ. We show that γ is Suslin. If not, let $\langle \varphi_i : i \in \omega\rangle$ be a scale of length γ on a set A. By DC, κ^+ does not have cofinality ω. Hence some φ_i has length $\geq \kappa^+$. We may then assume that $\kappa^+ \subseteq \operatorname{ran} \varphi_i$. Let $A_\beta = \{x : x \in A \;\&\; \varphi_i(x) \leq \beta\}$ for $\beta < \kappa^+$. Since each A_β is γ-Suslin, we may suppose that each A_β is κ-Suslin. But the sequence $\langle A_\beta : \beta < \kappa^+\rangle$ is strictly increasing, contradicting our hypothesis.

Following Kechris [Kec81B], we are led to the following special case of Q2:

Q2′. If the class of κ-Suslin sets has the scale property, does it follow from AD+DC that there is no strictly increasing sequence of κ-Suslin sets of length κ^+?

The genesis of this paper was as follows:

Jackson answered Q1 positively (with the minor change of strengthening the assumption $\mathsf{Red}(\underset{\sim}{\boldsymbol{\Gamma}})$ to $\mathsf{PWO}(\underset{\sim}{\boldsymbol{\Gamma}})$). Jackson's method was completely different from Kechris'.

By a minor variant of his proof, Jackson also answered Q2′ positively when the κ-Suslin sets are closed under $\forall^{\mathbb{R}}$ as well as $\exists^{\mathbb{R}}$. Martin next mixed Jackson's methods with other ideas to settle Q2′ positively for a large class of successor κ. J. Steel remarked that Martin's proof could be modified to deal with many limit cardinals, assuming a certain technical lemma. Kechris then proved this technical lemma. These results, together with Steel's analysis [Ste83] of the scale property in $\mathbf{L}(\mathbb{R})$, were sufficient to answer Q2′ in $\mathbf{L}(\mathbb{R})$ and also to show that in $\mathbf{L}(\mathbb{R})$ all reliable cardinals are Suslin. (See [Ste83].) Chuang [Chu82] used Jackson's methods to produce a full positive answer to Q1. He proved several other results about wellordered unions and Suslin cardinals. We used these further results to give a full positive answer to Q2′ for κ of cofinality greater than ω. (The case $\kappa = \omega$ probably does not arise, but this is not yet proved.)

In §2 we present the results on classes closed under both $\exists^{\mathbb{R}}$ and $\forall^{\mathbb{R}}$. In §3 we deal with the case κ a successor cardinal. In §4, we indicate how to modify the proof of §3 when κ is a limit cardinal. (We state, but do not prove, Kechris' lemma. See [Ste83] for a proof.)

In §3 and §4, we make use of the results of [Chu82], [Kec81B], and [KSS81]. This is purely to make our results general. The reader not familiar with these papers may think of a concrete case (*e.g.*, $\underset{\sim}{\boldsymbol{\Gamma}} = \underset{\sim}{\boldsymbol{\Sigma}}^1_2$ in §3), whereupon the lemmas using results from these papers will become obvious from older standard results.

§2. Inductive-like pointclasses.

THEOREM 2.1 (AD+DC). Let $\underset{\sim}{\Gamma}$ be a boldface pointclass closed under $\exists^{\mathbb{R}}$ and $\forall^{\mathbb{R}}$ but not under complements. Then either $\underset{\sim}{\Gamma}$ or $\check{\underset{\sim}{\Gamma}}$ is closed under wellordered unions.

PROOF. We let $\underset{\sim}{\Gamma}$ be as in the hypothesis of the theorem and assume that $\underset{\sim}{\Gamma}$ and $\check{\underset{\sim}{\Gamma}}$ are not closed under wellordered unions. We let ϑ_1 denote the least ordinal such that, for some ϑ_1-sequence of sets in $\underset{\sim}{\Gamma}$, A_ξ, $\xi < \vartheta_1$, we have $A = (\bigcup_{\xi<\vartheta_1} A_\xi) \notin \underset{\sim}{\Gamma}$. We similarly let ϑ_2 be the least ordinal such that $\check{\underset{\sim}{\Gamma}}$ is not closed under wellordered unions of length ϑ_2. We let $\bigcup \underset{\sim}{\Gamma}$ denote the pointclass obtained by taking ϑ_1-unions of sets in $\underset{\sim}{\Gamma}$, that is $B \in \bigcup \underset{\sim}{\Gamma}$ if there exists a sequence B_ξ, $\xi < \vartheta_1$, such that $B_\xi \in \underset{\sim}{\Gamma}$ for $\xi < \vartheta_1$ and $B = \bigcup_{\xi<\vartheta_1} B_\xi$. We are then assuming that $\underset{\sim}{\Gamma} \subset \bigcup \underset{\sim}{\Gamma}$. Hence it follows (from Wadge's lemma) that $\check{\underset{\sim}{\Gamma}} \subseteq \bigcup \underset{\sim}{\Gamma}$. Hence we may write a $\check{\underset{\sim}{\Gamma}}$ complete set, which we also denote by A as $A = \bigcup_{\xi<\vartheta_1} A_\xi$ where $A_\xi \in \underset{\sim}{\Gamma}$ for $\xi < \vartheta_1$.

Since $\underset{\sim}{\Gamma}$ is not closed under complements, we get a coding of $\underset{\sim}{\Gamma}$ sets by reals using a universal $\underset{\sim}{\Gamma}$ set. We let $B = \{x : x \text{ codes a } \underset{\sim}{\Gamma} \text{ subset of } A\}$. It is easy to see that B is $\check{\underset{\sim}{\Gamma}}$ using the closure hypotheses (which imply that $\underset{\sim}{\Gamma}$ is closed under countable unions and intersections). Hence we may write $B = \bigcup_{\xi<\vartheta_1} B_\xi$ where $B_\xi \in \underset{\sim}{\Gamma}$. We now let

$$A_\xi^{(2)} = \{x : \exists y (y \in B_\xi \text{ and } x \text{ belongs to the } \underset{\sim}{\Gamma} \text{ set coded by } y)\}.$$

We then have that $A_\xi^{(2)} \in \underset{\sim}{\Gamma}$ and $A = \bigcup_{\xi<\vartheta_1} A_\xi^{(2)}$ since each $A_\xi^{(2)} \subseteq A$ and $\bigcup_{\xi<\vartheta_1} A_\xi^{(2)} \supseteq A_\xi$ for each $\xi < \vartheta_1$ since each A_ξ is a $\underset{\sim}{\Gamma}$ subset of A. We may further assume that the sequence $A_\xi^{(2)}$ is strictly increasing in the sense that $A_\xi^{(2)} \supset \bigcup_{\xi'<\xi} A_{\xi'}^{(2)}$ for all $\xi < \vartheta_1$.

Now the union $\bigcup_{\xi<\vartheta_1} A_\xi^{(2)}$ has the property that it is $\underset{\sim}{\Gamma}$-bounded, that is, if $B \subseteq \bigcup_{\xi<\vartheta_1} A_\xi^{(2)}$ and $B \in \underset{\sim}{\Gamma}$, then for some $\eta < \vartheta_1$ we have $B \subseteq A_\eta^{(2)}$. This follows from the definition of the $A_\xi^{(2)}$.

We now play the game where player I plays a real x and player II plays reals y, z. We say that player II wins provided that if x codes a $\underset{\sim}{\Gamma}$ subset of $A = \bigcup_{\xi<\vartheta_1} A_\xi^{(2)}$ then y is a $\underset{\sim}{\Gamma}$-code for some $A_\eta^{(2)}$, $\eta < \vartheta_1$, where η is larger than the least ordinal ξ such that $A_\xi^{(2)} \supseteq$ the set coded by x, and $z \in A_\eta^{(2)} - \bigcup_{\xi<\eta} A_\xi^{(2)}$. We then have that player II has a winning strategy for the game above, for, if not, then there would be a Σ_1^1 set C of codes of $\underset{\sim}{\Gamma}$ subsets of A with the property that for any $\eta < \vartheta_1$ there is an $x \in C$ such that some member of the set coded by x does not belong to $A_\eta^{(2)}$. Since $\Sigma_1^1 \subseteq \underset{\sim}{\Gamma}$ we would then have a

$\underset{\sim}{\Gamma}$ subset of A unbounded in $A = \bigcup_{\xi<\vartheta_1} A^{(2)}_\xi$. This contradicts the fact that $A = \bigcup_{\xi<\vartheta_1} A^{(2)}_\xi$ is $\underset{\sim}{\Gamma}$-bounded.

Hence we may assume that player II has a winning strategy s. We let $B = \{x : x \text{ codes a } \underset{\sim}{\Gamma} \text{ subset of } A\}$, and so $B \in \check{\underset{\sim}{\Gamma}}$ and is $\check{\underset{\sim}{\Gamma}}$-complete (otherwise A would be in $\underset{\sim}{\Gamma}$). We define an ordering on B by

$$x_1 < x_2 \Leftrightarrow x_1 \in B \;\&\; x_2 \in B \;\&\; s(x_2)_2 \notin \text{the set coded by } s(x_1)_1,$$

where $s(x)_1$ and $s(x)_2$ are the y and z respectively played against x according to s. The ordering is in $\check{\underset{\sim}{\Gamma}}$ and is easily seen to be a prewellordering. ($x_1 < x_2 \Leftrightarrow \xi_1 < \xi_2$, where $s(x_1)_1$ codes $A^{(2)}_{\xi_1}$ and $s(x_2)_1$ codes $A^{(2)}_{\xi_2}$.) It follows from the regularity of ϑ_1 that the rank of the prewellordering $<$ is ϑ_1.

It now follows from the Coding Lemma and the fact that $\check{\underset{\sim}{\Gamma}}$ is closed under $\exists^{\mathbb{R}}$ that $\check{\underset{\sim}{\Gamma}}$ is closed under wellordered unions of length ϑ_1. Hence $\vartheta_1 < \vartheta_2$.

The argument above, however, is symmetric in $\underset{\sim}{\Gamma}$ and $\check{\underset{\sim}{\Gamma}}$, and hence we also get $\vartheta_2 < \vartheta_1$. This contradiction establishes that either $\underset{\sim}{\Gamma}$ or $\check{\underset{\sim}{\Gamma}}$ is closed under wellordered unions. ⊣ (Theorem 2.1)

Corollary 2.2 (AD+DC). If $\underset{\sim}{\Gamma}$ is a boldface pointclass closed under $\exists^{\mathbb{R}}$ and $\forall^{\mathbb{R}}$ but not under complements and has the Prewellordering property, then $\underset{\sim}{\Gamma}$ is closed under wellordered unions.

Proof. This follows from the fact that, since $\underset{\sim}{\Gamma}$ has the Prewellordering property, $\check{\underset{\sim}{\Gamma}}$ cannot be closed under wellordered unions. ⊣ (Corollary 2.2)

Corollary 2.3 (AD+DC). The class of inductive sets is closed under wellordered unions.

The class of inductive sets also has the scale property. We now turn to the question Q2 for such pointclasses.

Theorem 2.4 (AD+DC). Let $\underset{\sim}{\Gamma}$ be a boldface pointclass closed under $\exists^{\mathbb{R}}$ and $\forall^{\mathbb{R}}$ but not under complements. Let κ be a cardinal and suppose that some complete $\underset{\sim}{\Gamma}$ set admits a $\underset{\sim}{\Gamma}$ norm of length κ. Suppose also that every set in $\underset{\sim}{\Gamma}$ is κ-Suslin. There is no strictly increasing sequence of $\underset{\sim}{\Gamma}$ sets of length κ^+.

Our last two hypotheses hold if $\underset{\sim}{\Gamma}$ has the scale property via a scale of length κ. They also hold by [Chu82] if the other hypotheses hold and κ is a Suslin cardinal and $\underset{\sim}{\Gamma} = \boldsymbol{S}(\kappa)$.

Proof. We let $\underset{\sim}{\Gamma}$ satisfy the hypotheses of the theorem and assume that A_ξ for $\xi < \kappa^+$ is a strictly increasing sequence of $\underset{\sim}{\Gamma}$ sets. We may assume $A_\xi \supset \bigcup_{\xi'<\xi} A_{\xi'}$ for all $\xi < \kappa^+$. We let $A = \bigcup_{\xi<\kappa^+} A_\xi$. We have that $A \in \underset{\sim}{\Gamma}$ from Corollary 2.2. Since there is a $\underset{\sim}{\Gamma}$ norm of length κ on a complete $\underset{\sim}{\Gamma}$ set, we have a $\underset{\sim}{\Gamma}$ coding for ordinals $< \kappa$, which we denote by $|x|$ for $x \in P$, the complete $\underset{\sim}{\Gamma}$ set. By the Coding Lemma, using the closure properties of $\underset{\sim}{\Gamma}$, we have that every subset X of κ is $\underset{\sim}{\Gamma}$ in the codes, that is $\{x : |x| \in X\} \in \underset{\sim}{\Gamma}$. Thus there is a $\underset{\sim}{\Gamma}$ prewellordering of length η for each $\eta < \kappa^+$.

We choose a $\underset{\sim}{\Gamma}$ coding of $\underset{\sim}{\Gamma}$ relations and let

$$C = \{x : x \text{ codes a wellfounded relation}\}.$$

We have that $C \in \breve{\underset{\sim}{\Gamma}}$. We then play the game where player I plays x and player II plays y, z and player II wins provided

$$x \in C \Rightarrow y \text{ is a } \underset{\sim}{\Gamma} \text{ code for some } A_\eta,\ \eta < \kappa^+$$

where $\eta >$ the rank of the wellfounded relation coded by x, and $z \in A_\eta - \bigcup_{\eta' < \eta} A_{\eta'}$. We have that player II has a winning strategy, for, if player I had a winning strategy, then there would be a Σ^1_1 set of codes for wellfounded relations unbounded in κ^+, and, since $\underset{\sim}{\Sigma}^1_1 \subseteq \underset{\sim}{\Gamma}$, there would be a $\underset{\sim}{\Gamma}$ wellfounded relation of height κ^+. This, however, contradicts the fact that $\underset{\sim}{\Gamma}$ is κ-Suslin, since κ-Suslin wellfounded relations have height $< \kappa^+$ by [Mos80, 2G.2].

Hence player II has a winning strategy s. We then consider the relation defined by

$$x_1 < x_2 \Leftrightarrow x_1 \in C \ \&\ x_2 \in C \ \&\ s(x_2)_2 \notin \text{the set coded by } s(x_1)_1.$$

We have, as before, that $<$ is a $\breve{\underset{\sim}{\Gamma}}$ prewellordering. It follows from the regularity of κ^+ (which follows from the Coding Lemma and the fact that $\kappa^+ = \sup\{\beta : \beta \text{ is the height of a } \underset{\sim}{\Gamma} \text{ wellfounded relation}\}$) that $<$ has length κ^+. It then follows from the Coding Lemma that $\breve{\underset{\sim}{\Gamma}}$ is closed under wellordered unions of length κ^+, and hence κ, which contradicts the existence of a $\underset{\sim}{\Gamma}$ norm of length κ on a complete $\underset{\sim}{\Gamma}$ set. ⊣ (Theorem 2.4)

§3. κ-Suslin sets for κ a successor cardinal. For each cardinal κ, let $\boldsymbol{S}(\kappa)$ be the class of all κ-Suslin sets.

Theorem 3.1 (AD+DC). *Let κ be a Suslin cardinal such that $\kappa = \lambda^+$ for some cardinal λ. There is no strictly increasing sequence $\langle A_\beta : \beta < \kappa^+\rangle$ such that each $A_\beta \in \boldsymbol{S}(\kappa)$.*

Proof. We begin by cataloguing some useful facts about κ and $\boldsymbol{S}(\kappa)$.

Lemma 3.2. *Let γ be the supremum of the Suslin cardinals $< \kappa$. γ is a Suslin cardinal.*

Proof of Lemma 3.2. Since κ is a successor cardinal, κ has cofinality greater than ω. By [Kec81B], $\boldsymbol{S}(\kappa)$ is not closed under complements.

Let $\langle \varphi_i : i \in \omega\rangle$ be a scale of length κ on a set $A \in \boldsymbol{S}(\kappa) - \breve{\boldsymbol{S}}(\kappa)$. Since κ has cofinality $> \omega$, one of the φ_i, say φ_n, has length κ. For $\beta < \alpha < \kappa$ let $A_{\alpha,\beta} = \{x : \sup_i \varphi_i(x) + 1 < \alpha \text{ or } (\sup_i \varphi_i(x) + 1 = \alpha \ \&\ \varphi_n(x) \leq \beta)\}$. It is easily seen that $A_{\alpha,\beta}$ is a $|\alpha|$-Suslin. Since κ has cofinality $> \omega$, $A = \bigcup_{\beta < \alpha < \kappa} A_{\alpha,\beta}$. If $\varphi_n(x) = \beta$, there is an α, $\beta < \alpha < \kappa$, such that $x \in A_{\alpha,\beta}$ and $x \notin A_{\alpha,\beta'}$

for any $\beta' < \beta$ and $x \notin A_{\alpha',\beta'}$ for any $\alpha' < \alpha$ and any β'. Thus there are κ distinct sets $A_{\alpha,\beta}$. If we order $\{(\alpha, \beta) : \kappa > \alpha > \beta\}$ lexicographically, we arrange the $A_{\alpha,\beta}$ in an increasing sequence. Thus we have shown that there is a strictly increasing sequence $\langle B_\alpha : \alpha < \kappa \rangle$ of sets such that each B_α is δ-Suslin for $\delta < \kappa$.

Fix, for the moment, a Suslin cardinal $\delta < \kappa$, by [KSS81], either $\forall^{\mathbb{R}}\boldsymbol{S}(\delta)$ or $\forall^{\mathbb{R}}\exists^{\mathbb{R}}\check{\boldsymbol{S}}(\delta)$ contains $\boldsymbol{S}(\delta)$, is closed under $\forall^{\mathbb{R}}$ and countable unions and intersections, and has the prewellordering property. Let Γ be whichever of these classes satisfies these conditions. If δ_1 and δ_2 are the next two Suslin cardinals after δ, $\exists^{\mathbb{R}}\check{\boldsymbol{S}}(\delta) \subseteq \boldsymbol{S}(\delta_1)$ and $\exists^{\mathbb{R}}\forall^{\mathbb{R}}\boldsymbol{S}(\delta) \subseteq \boldsymbol{S}(\delta_2)$. Thus every $\check{\Gamma}$ prewellordering has length $< {\delta_2}^+$, by [Mos80, 2G.1]. By [Chu82], there is no strictly increasing sequence of Γ sets of length ${\delta_2}^{++}$. Hence either there is a greatest Suslin cardinal $< \gamma$, and we are done, or there is no strictly increasing sequence of length γ of sets in $\boldsymbol{S}(\delta)$.

We may then assume that, for each Suslin cardinal $\delta < \kappa$, there are fewer than γ of the B_α which belong to $\boldsymbol{S}(\delta)$. Since there are at most γ Suslin cardinals $< \kappa$, we get the contradiction that there are no more than γ of the B_α. ⊣ (Lemma 3.2)

Lemma 3.3. The cardinal λ is a Suslin cardinal. $\check{\boldsymbol{S}}(\lambda)$ has the prewellordering property via an $\check{\boldsymbol{S}}(\lambda)$ norm of length κ. $\boldsymbol{S}(\kappa) = \exists^{\mathbb{R}}\check{\boldsymbol{S}}(\lambda)$. $\boldsymbol{S}(\kappa)$ has the scale property.

Proof of Lemma 3.3. Since κ has uncountable cofinality and γ is the greatest Suslin cardinal $< \kappa$, the lemma follows from [Chu82, Theorem 5] and its proof. ⊣ (Lemma 3.3)

Lemma 3.4. $\boldsymbol{S}(\lambda) \cap \check{\boldsymbol{S}}(\lambda) \supseteq \mathbb{B}_\kappa$, the closure of the open sets under wellordered unions of length $< \kappa$, wellordered intersections of length $< \kappa$, and complements.

Proof of Lemma 3.4. The proof of [Mos80, 7D.9] essentially proves the lemma. ⊣ (Lemma 3.4)

Lemma 3.5. Let μ be the ω-closed unbounded filter on κ. μ is a normal, κ-complete ultrafilter. If j is the embedding associated with the ultrapower by μ, $j(\kappa) = \kappa^+$.

Proof of Lemma 3.5. The proofs are like those of [KSS81, Theorems 11.2, 14.3]. ⊣ (Lemma 3.5)

Our next aim is to introduce the **Kunen trees** associated with $\boldsymbol{S}(\kappa)$ wellfounded relations. These trees provide us with our method for representing ordinals $< \kappa^+$.

Let $R \subseteq ({}^\omega\omega)^3$ belong to $\boldsymbol{S}(\kappa)$ and be universal for $\boldsymbol{S}(\kappa)$ subsets of $({}^\omega\omega)^2$. Let T be a tree on $\omega \times \omega \times \omega \times \kappa$ witnessing that R is κ-Suslin.

Let $(i,j) \mapsto n_{ij}$ be a one-one surjection of $\omega \times \omega$ onto ω. If $\tau \in {}^\omega\omega$ let $\tau_i(j) = \tau(n_{ij})$ if $n_{ij} < \operatorname{lh}\tau$. Similarly define $\rho_i(j)$ for $\rho \in {}^{<\omega}\kappa$.

We define our Kunen tree $\mathcal{T}$ by

$$\mathcal{T} = \{(\sigma,\tau,\rho) : \operatorname{lh}\sigma = \operatorname{lh}\tau = \operatorname{lh}\rho \;\&\; \forall i,k \le \operatorname{lh}\sigma(\text{if } \tau_i\restriction k, \tau_{i+1}\restriction k, \text{ and } \rho_i\restriction k \text{ are all defined, then } (\sigma\restriction k, \tau_{i+1}\restriction k, \tau_i\restriction k, \rho_i\restriction k) \in T)\}.$$

For $x \in {}^\omega\omega$ let $\mathcal{T}_x = \{(\tau,\rho) : (x\restriction \operatorname{lh}\tau, \tau, \rho) \in \mathcal{T}\}$. For $\alpha < \kappa$, let $\mathcal{T}\restriction\alpha = \{(\sigma,\tau,\rho) : (\sigma,\tau,\rho) \in \mathcal{T} \;\&\; \rho \in {}^{<\omega}\alpha\}$. Similarly define $\mathcal{T}_x\restriction\alpha\ (= (\mathcal{T}\restriction\alpha)_x)$.

Lemma 3.6 (Kunen). The relation R_x is wellfounded if and only if $\mathcal{T}_x$ is wellfounded. If R_x is wellfounded, $|R_x| \le |\mathcal{T}_x|$, where $|\;|$ is the height function for wellfounded relations.

Proof of Lemma 3.6. A branch through $\mathcal{T}_x$ is essentially an infinite descending chain in R_x with witnesses. For the second assertion, let $R_x^* = \{((y,f),(z,g)) : \forall i(x\restriction i, y\restriction i, z\restriction i, f\restriction i) \in T\}$. Clearly the tree of all finite descending chains in R_x^* can be embedded in $\mathcal{T}_x$. Hence $|R_x^*| \le |\mathcal{T}_x|$. It is not hard to see that $|R_x| = |R_x^*|$. ⊣ (Lemma 3.6)

Choose a universal $\boldsymbol{S}(\kappa)$ set and so a coding of $\boldsymbol{S}(\kappa)$ sets by elements of ${}^\omega\omega$.

Assume, contrary to the theorem, that $\langle A_\beta : \beta < \kappa^+\rangle$ is a sequence of $\boldsymbol{S}(\kappa)$ sets with $\bigcup_{\beta'<\beta} A_{\beta'} \subset A_\beta$ for each $\beta < \kappa^+$.

We play a game as follows:

Player I chooses $x \in {}^\omega\omega$.

Player II chooses $y \in {}^\omega\omega$ and $z \in {}^\omega\omega$.

If R_x and R_y are both wellfounded, player II wins just in case $|\mathcal{T}_x| < |\mathcal{T}_y|$ and z is a code for $A_{|\mathcal{T}_y|}$.

If R_x and R_y are not both wellfounded, let $\alpha < \kappa$ be the least ordinal such that $\mathcal{T}_x\restriction\alpha$ or $\mathcal{T}_y\restriction\alpha$ is not wellfounded. Player I wins $\Leftrightarrow$ $\mathcal{T}_x\restriction\alpha$ is wellfounded.

Lemma 3.7. Player I has no winning strategy.

Proof of Lemma 3.7. Suppose s is a winning strategy for player I. Let $\alpha < \kappa$. Let $B_\alpha = \{y : \forall\beta < \alpha(\mathcal{T}_y\restriction\beta \text{ is wellfounded and } |\mathcal{T}_y\restriction\beta| < \alpha)\}$. $B_\alpha \in \mathbb{B}_\kappa$ and so $B_\alpha \in \boldsymbol{S}(\lambda) \cap \check{\boldsymbol{S}}(\lambda)$. Let S be a tree on $\omega \times \lambda$ witnessing that B_α is λ-Suslin. Let $\mathcal{T}^*$ be the tree defined as follows:

$$\mathcal{T}^* = \{(\sigma_0,\tau_0,\rho_0,\sigma,\tau,\rho) : (\sigma_0,\rho_0) \in S \;\&\; (\sigma,\tau,\rho) \in \mathcal{T}\restriction\alpha \;\&\; \sigma \text{ agrees with the reply to } (\sigma_0,\tau_0) \text{ according to } s\}.$$

Since s is a winning strategy $\mathcal{T}_x\restriction\alpha$ is wellfounded whenever $x = s(y,z)$ and $y \in B_\alpha$. Thus $\mathcal{T}^*$ is wellfounded. Since $\mathcal{T}^*$ is a tree on $\max(\alpha,\lambda)$, $|\mathcal{T}^*| < \kappa$. It is easy to see that if $y \in B_\alpha$ and $x = s(y,z)$, then $|\mathcal{T}_x\restriction\alpha| \le |\mathcal{T}^*|$.

We have thus shown that there is an $f(\alpha) < \kappa$ (namely, $|\mathcal{T}^*|$) such that if player II plays a $y \in B_\alpha$ and player I plays according to s, then $|\mathcal{T}_x\restriction\alpha| < f(\alpha)$. Let $C \subseteq \kappa$ be closed, unbounded and satisfy $(\eta \in C \;\&\; \alpha < \eta) \Rightarrow f(\alpha) < \eta$.

Now let $\beta \in j(C)$, $\beta > \kappa$. Let $h\colon \kappa \leftrightarrow \beta$ be a bijection. Let φ be a $\boldsymbol{S}(\kappa)$ norm of length κ. Let $(z, w) \in E \Leftrightarrow h(\varphi(z)) < h(\varphi(w))$. $E \in \boldsymbol{S}(\kappa)$. Let $E = R_y$. R_y is wellfounded and $|R_y| = \beta$. Thus $|\mathcal{T}_y| \geq \beta$. Let player II play y and z coding $A_{|\mathcal{T}_y|}$. Let player I play x according to s. There is a closed, unbounded set of $\alpha < \kappa$ such that $y \in B_\alpha$. For each α in this set, $|\mathcal{T}_x{\restriction}\alpha|$ is less than the next member of C after α. By the normality of μ, $|\mathcal{T}_x|$ is less than the next element of $j(C)$ after κ. Thus $|\mathcal{T}_x| < |\mathcal{T}_y|$. It follows that the play is a win for player II, a contradiction. ⊣ (Lemma 3.7)

Let t be a winning strategy for player II.

LEMMA 3.8. There is a closed, unbounded $C^* \subseteq \kappa^+$ such that, if $\beta \in C^*$, player I plays x with $|\mathcal{T}_x| < \beta$, and player II plays according to t, then $|\mathcal{T}_y| < \beta$.

PROOF OF LEMMA 3.8. If $\alpha, \gamma < \kappa$, let

$$B_\alpha^\gamma = \{x : \mathcal{T}_x{\restriction}\alpha \text{ is wellfounded and } |\mathcal{T}_x{\restriction}\alpha| < \gamma\}.$$

$B_\alpha^\gamma \in \mathbb{B}_\kappa$. As before we get an $f(\alpha, \gamma) < \kappa$ such that if $x \in B_\alpha^\gamma$ and player II plays y according to t, then $|\mathcal{T}_y{\restriction}\alpha| < f(\alpha, \gamma)$. Let $C \subseteq \kappa$ be closed and unbounded such that

$$(\eta \in C \;\&\; \alpha < \eta \;\&\; \gamma < \eta) \Rightarrow f(\alpha, \gamma) < \eta.$$

Now let $C^* = j(C) - (\kappa + 1)$. ⊣ (Lemma 3.8)

LEMMA 3.9. For each $\beta \in C^*$, $\{x : \mathcal{T}_x \text{ is wellfounded and } |\mathcal{T}_x| < \beta\} \in \breve{\boldsymbol{S}}(\kappa)$.

PROOF OF LEMMA 3.9. For each x, let $(y(x), z(x))$ be player II's play against x according to t. Let $\beta \in C^*$ and $w \in A_\beta - \bigcup_{\beta' < \beta} A_{\beta'}$. Suppose $\mathcal{T}_x$ is wellfounded. Then

$$|\mathcal{T}_x| < \beta \Leftrightarrow |\mathcal{T}_{y(x)}| < \beta \Leftrightarrow w \text{ does not belong to the set coded by } z(x).$$

This last condition is $\breve{\boldsymbol{S}}(\kappa)$. Since $\breve{\boldsymbol{S}}(\kappa)$ is closed under $\forall^{\mathbb{R}}$,

$$\{x : \mathcal{T}_x \text{ is wellfounded}\} = \{x : R_x \text{ is wellfounded}\} \in \breve{\boldsymbol{S}}(\kappa).$$

Since $\breve{\boldsymbol{S}}(\kappa)$ is closed under intersections, the lemma is proved. ⊣ (Lemma 3.9)

LEMMA 3.10. For every sufficiently large $\beta < \kappa^+$, every $\boldsymbol{S}(\kappa)$ set is Wadge reducible to $\{x : \mathcal{T}_x \text{ is wellfounded and } |\mathcal{T}_x| < \beta\}$.

NOT A PROOF.[1] Let φ be an $\boldsymbol{S}(\kappa)$ norm of length κ on a set $H \in \boldsymbol{S}(\kappa) - \breve{\boldsymbol{S}}(\kappa)$. Let

$$\begin{aligned} R^* = \{(x, y, z)\colon\ & y \in H \;\&\; z \in H \;\&\; \\ & \exists y' \exists z' (\varphi(y') = \varphi(y) \;\&\; \varphi(z') = \varphi(z) \;\&\; (x', y', z') \in R)\}. \end{aligned}$$

[1]This proof, as published in the original paper in 1983, is faulty. A corrected version may be found in the newly added §5.

Define R_x^* in the obvious way. R_x^* is thus just R_x fixed up to be well defined on ordinals as coded by φ.

Let f be continuous such that, for each (x, y),

$$R_{f(x,y)} = \{(z, w) : z \in H \,\&\, w \in H \,\& \\ (x \notin H \vee (\varphi(x) > \varphi(z) \,\&\, \varphi(x) > \varphi(w))) \,\&\, (z, w) \in R_y^*\}.$$

Let

$$V = \{(x, y) : x \in H \,\& \\ R_y^* \cap \{(z, w) : \varphi(z) < \varphi(x) \,\&\, \varphi(w) < \varphi(x)\} \text{ is wellfounded}\}.$$

To see that $V \in \boldsymbol{S}(\kappa)$, note that

$$(x, y) \in V \Leftrightarrow \exists \alpha < \kappa(\varphi(x) = \alpha \,\&\, R_y^* \cap \\ \{(z, w) : \varphi(z) < \alpha \,\&\, \varphi(w) < \alpha\} \text{ is wellfounded}).$$

Thus V is a union of κ sets, so it suffices by Lemma 3.3 and the Coding Lemma to show that each of these sets belongs to $\boldsymbol{S}(\kappa)$. $\{x : \varphi(x) = \alpha\} \in \boldsymbol{S}(\kappa) \cap \check{\boldsymbol{S}}(\kappa)$, so it is enough to show that the second conjunct defines a set in $\boldsymbol{S}(\kappa)$. $\{(x, y) : R_y^* \cap \{(z, w) : \varphi(z) < \alpha \,\&\, \varphi(w) < \alpha\}\}$ is induced by a relation on α. If it is wellfounded, it has height $< \kappa$. Thus the second conjunct defines a union of κ sets of the form

$$\{(x, y) : R_y^* \cap \{(z, w) : \varphi(z) < \alpha \,\&\, \varphi(w) < \alpha\} \text{ has height } < \gamma\}.$$

These sets all belong to $\mathbb{B}_\kappa$ and so to $\boldsymbol{S}(\lambda) \cap \check{\boldsymbol{S}}(\lambda)$.

For each $(x, y) \in V$, $R_{f(x,y)}$ is wellfounded. If $\{|\mathcal{T}_{f(x,y)}| : (x, y) \in V\}$ were unbounded in κ^+, we could, using the fact that V is κ-Suslin, put together all the $\mathcal{T}_{f(x,y)}$ for $(x, y) \in V$ to get a wellfounded tree of height κ^+.

Thus let $\kappa^+ > \beta > |\mathcal{T}_{f(x,y)}|$ for all $(x, y) \in V$. As in the proof of Lemma 3.7, let y be such that $R_y = R_y^*$ and is a wellordering of order type $\geq \beta$.

$$x \in H \Rightarrow (x, y) \in V \Rightarrow \mathcal{T}_{f(x,y)} \text{ wellfounded } \& \ |\mathcal{T}_{f(x,y)}| < \beta.$$

Also

$$x \notin B \Rightarrow R_{f(x,y)} = R_y^* \Rightarrow |R_{f(x,y)}| \geq \beta \Rightarrow |\mathcal{T}_{f(x,y)}| \geq \beta.$$

We have thus shown that $H \in \boldsymbol{S}(\kappa) - \check{\boldsymbol{S}}(\kappa)$ is Wadge reducible to the required set.

⊣ (Lemma 3.10)

Lemmas 3.9 and 3.10 give the contradiction which proves the theorem.

⊣ (Theorem 3.1)

§4. κ-Suslin sets for κ a limit cardinal of uncountable cofinality.

THEOREM 4.1. Let κ be a Suslin cardinal, and a limit cardinal of uncountable cofinality. Assume that $\boldsymbol{S}(\kappa)$ has the reduction property. There is no strictly increasing sequence $\langle A_\beta : \beta < \kappa^+ \rangle$ such that $A_\beta \in \boldsymbol{S}(\kappa)$.

NOTE. The hypothesis that $\boldsymbol{S}(\kappa)$ has the reduction property can be eliminated.

PROOF. We prove the theorem from a lemma of Kechris proved in [Ste83].

LEMMA 4.2. The cardinal κ is a limit of Suslin cardinals.

PROOF OF LEMMA 4.2. The proof Lemma 3.2 shows that either there is a greatest Suslin cardinal $\gamma < \kappa$ or else κ is a limit of Suslin cardinals. In the former case, [Chu82, Theorem 5] implies that $\kappa = \gamma^+$. ⊣ (Lemma 4.2)

LEMMA 4.3. $\boldsymbol{S}(\kappa)$ has the scale property.

PROOF OF LEMMA 4.3. This follows from [Chu82] and [KSS81]. ⊣ (Lemma 4.3)

LEMMA 4.4. There is an ω-closed, unbounded set of $\alpha < \kappa$ such that $\mathbb{B}_{\alpha^+} \subseteq \boldsymbol{S}(\alpha) \cap \check{\boldsymbol{S}}(\alpha)$.

PROOF OF LEMMA 4.4. If δ is a Suslin cardinal $< \kappa$, there is a $\delta' < \kappa$ such that every set in $\boldsymbol{S}(\delta)$ admits a scale of length δ which belongs to $\boldsymbol{S}(\delta')$. To see this, let $A \in \boldsymbol{S}(\delta)$ and let $\langle \varphi_i : i \in \omega \rangle$ be a scale on A of length δ. Since $\{x : x \in A \,\&\, \varphi_i(x) < \alpha\}$ and $\{x : x \in A \,\&\, \varphi_i(x) \leq \alpha\}$ are both δ-Suslin for each $i \in \omega$ and $\alpha < \delta$, $\{x : x \in A \,\&\, \varphi_i(x) = \alpha\}$ is a difference of δ-Suslin sets. If δ_1 is the next Suslin cardinal after δ, $\{x : x \in A \,\&\, \varphi_i(x) = \alpha\} \in \boldsymbol{S}(\delta_1)$ for each $i \in \omega$ and $\alpha < \delta$. Either $\boldsymbol{S}(\delta_1)$ or $\exists^{\mathbb{R}} \check{\boldsymbol{S}}(\delta_1)$ is closed under $\exists^{\mathbb{R}}$, under countable unions and intersections, and has the prewellordering property, by [KSS81]. Thus one of these classes is closed under wellordered unions, by Theorem 2.1 and the Kechris theorem quoted in §1. Hence we may take δ' as the least Suslin cardinal $> \delta_1$.

There is an ω-closed unbounded set of $\delta < \kappa$ such that δ is a limit of Suslin cardinals and, if $\delta_1 < \delta$, there is a $\delta' < \delta$ such that every set in $\boldsymbol{S}(\delta_1)$ admits a scale in $\boldsymbol{S}(\delta')$. For δ in this ω-cub set, let $\Lambda_\delta = \bigcup_{\delta' < \delta} \boldsymbol{S}(\delta')$. Let Γ_δ be the collection of countable unions of members of Λ_δ. It is easily seen that every set in Γ_δ admits a Γ_δ scale of length $\leq \delta$. By first periodicity, $\forall^{\mathbb{R}} \Gamma_\delta$ has the prewellordering property. Thus, as in Lemma 3.4, $\mathbb{B}_{\delta^+} \subseteq \forall^{\mathbb{R}} \Gamma_\delta \cap \exists^{\mathbb{R}} \check{\Gamma}_\delta$. But $\exists^{\mathbb{R}} \check{\Gamma}_\delta \subseteq \boldsymbol{S}(\delta)$. By [Mos80, Theorem 2E.2], $\boldsymbol{S}(\delta) \cap \check{\boldsymbol{S}}(\delta) \subseteq \mathbb{B}_{\delta^+}$, so $\exists^{\mathbb{R}} \check{\Gamma}_\delta = \boldsymbol{S}(\delta)$, and the lemma is proved. ⊣ (Lemma 4.4)

LEMMA 4.5 (Kechris). There is a measure μ on κ such that μ extends the ω-closed, unbounded filter and, in the ultrapower by μ, the function $f(\alpha) = \alpha^+$ represents κ^+.

For a proof, see [Ste83].

We define the Kunen tree $\mathcal{T}$ exactly as in §3.

The rest of the proof is like that in §3. Lemma 3.6 goes through as before. Our game is as before. The analogue of Lemma 3.7 is proved as before, except that $f(\alpha) < \alpha^+$ and we choose β such that $\beta < \kappa^+$ and $\beta >$ the ordinal represented by f. The analogue of Lemma 3.8 is proved as before except that $f(\alpha, \gamma)$ is defined for α in the set given by Lemma 4.4 and $\gamma < \alpha^+$, and $f(\alpha, \gamma) < \alpha^+$. We let $C_\alpha \subseteq \alpha^+$ be closed and unbounded such that

$$(\eta \in C_\alpha \mathbin{\&} \gamma < \eta) \Rightarrow f(\alpha, \gamma) < \eta.$$

We let C^* be the subset of κ represented by $g(\alpha) = C_\alpha$. The analogue of Lemma 3.9 is proved as before, as is the analogue of Lemma 3.10.

⊣ (Theorem 4.1)

§5. **Addendum (2010).** The proof of Lemma 3.10 contains an error which we fix here. A correction for this was pointed out in [Jac90B], and Theorem 3.1 was there also extended a ways past $\boldsymbol{S}(\kappa)$ (in particular, the statement of Theorem 3.1 holds when the A_β are Boolean combinations of $\boldsymbol{S}(\kappa)$ sets). Hjorth [Hjo01], using techniques of inner-model theory, later obtained an optimal result showing that there are no ω_2 increasing sequences of sets A_β with each A_β in the pointclass $\Game\omega\text{–}\boldsymbol{\Pi}^1_1$. The error in Lemma 3.10 occurs in the statement that the κ sequence of sets defined towards the end of the proof of the lemma are in $\boldsymbol{B}_\kappa$. In fact, there doesn't seem to be any obvious reason why this should be the case. The correction given in [Jac90B] uses the scale property for the pointclass $\check{\boldsymbol{S}}(\lambda)$ of Lemma 3.3. While this can be shown from just AD, the arguments are more involved than those for just getting the prewellordering property for $\check{\boldsymbol{S}}(\lambda)$. Because of this, we present here a correction which only needs the prewellordering property as stated in Lemma 3.3.

Let $\mathcal{T}$ be the Kunen tree of Lemma 3.6, which we now view as a tree on $\omega \times \kappa$. So, $\{|\mathcal{T}_y| : \mathcal{T}_y \text{ is wellfounded}\}$ is unbounded in κ^+. We define an auxiliary tree $\mathcal{T}'$ as follows. As in the incorrect proof of Lemma 3.10, let φ be an $\boldsymbol{S}(\kappa)$ norm of length κ on an $\boldsymbol{S}(\kappa)$-complete set H. For $\alpha < \kappa$, let $H_\alpha = \{x \in H : \varphi(x) = \alpha\}$, so $H_\alpha \in \boldsymbol{S}(\kappa) \cap \check{\boldsymbol{S}}(\kappa)$. From the coding lemma, the tree $\mathcal{T}$ is $\boldsymbol{S}(\kappa)$ in the codes with respect to the norm φ. That is, the relation

$$\begin{aligned} B(a, z) \Leftrightarrow a = \langle a_0, \dots, a_k \rangle \mathbin{\&} z = \langle z_0, \dots, z_k \rangle \mathbin{\&} z_0, \dots, z_k \in H \\ \mathbin{\&} ((a_0, \dots, a_k), (\varphi(z_0), \dots, \varphi(z_k))) \in \mathcal{T} \end{aligned}$$

is in $\boldsymbol{S}(\kappa)$. Let U be a tree on $\omega \times \omega \times \kappa$ with $B = \mathrm{p}[U]$ (we make the slight abuse of allowing the integer a to be regarded as a real). Let V be a tree on $\omega \times \omega \times \kappa$ with $\mathrm{p}[V] = \{(z, x) : z \in H \mathbin{\&} (x \notin H \vee (x \in H \mathbin{\&} \varphi(z) < \varphi(x)))\}$.

$\mathcal{T}'$ will be a tree on $\omega \times \omega \times \omega \times \kappa \times \kappa$ such that a branch (x, y, z, f, g) through $\mathcal{T}'$ will be such that f witnesses that for all i that $(y \restriction k, \langle z_0, \dots, z_i \rangle) \in \mathrm{p}[U]$ (here z codes the reals $z_0, z_1, \dots$) and g witnesses that $(z_i, x) \in \mathrm{p}[V]$ for each i. More precisely,

$$\begin{aligned}(s, t, u, \rho, \sigma) \in \mathcal{T}' \Leftrightarrow\ & \mathrm{lh}(s) = \mathrm{lh}(t) = \cdots = \mathrm{lh}(\sigma)\\ & \&\ \exists x, y, z, f, g \text{ extending } s, t, u, \rho, \sigma\\ & [\forall i, k\ (y \restriction k, \langle z_0, \dots, z_i \rangle \restriction k, f_i \restriction k) \in U)\\ & \&\ \forall i, k\ (z_i \restriction k, x \restriction k, g_i \restriction k) \in V]\end{aligned}$$

where $f_i(j) = f(\langle i, j \rangle)$ and likewise for g. Clearly $\mathcal{T}'_{x,y}$ is wellfounded iff $\mathcal{T}_y \restriction \varphi(x)$ is wellfounded (where $\varphi(x) = \infty$ if $x \notin H$). Also, an easy argument as in the proof of the Kunen-Martin theorem shows that $|\mathcal{T}'_{x,y}| \geq |\mathcal{T}_y \restriction \varphi(x)|$. For $x \notin H$, this becomes: $|\mathcal{T}'_{x,y}| \geq |\mathcal{T}_y|$.

Define

$$\begin{aligned}W(x, y) &\Leftrightarrow x \in H \ \&\ \mathcal{T}_y \restriction \varphi(x) \text{ is wellfounded}\\ &\Leftrightarrow \exists \alpha, \beta < \kappa\ (x \in H_\alpha \ \&\ |\mathcal{T}_y \restriction \alpha| < \beta)\end{aligned}$$

Now, $\{y : |\mathcal{T}_y \restriction \alpha| < \beta\} \in \boldsymbol{B}_\kappa \subseteq \underset{\sim}{\boldsymbol{\Delta}} = \boldsymbol{S}(\kappa) \cap \check{\boldsymbol{S}}(\kappa)$ by the closure of $\underset{\sim}{\boldsymbol{\Delta}}$ under $< \kappa$ unions and intersections. Since $\boldsymbol{S}(\kappa)$ is closed under wellordered unions (we actually only need κ unions here, which follows from the coding lemma), we have $W \in \boldsymbol{S}(\kappa)$. We can therefore put together all the trees $\mathcal{T}'_{x,y}$ for $(x, y) \in W$ to obtain a single wellfounded tree on κ which therefore has rank less than κ^+. So, let $\beta_0 < \kappa^+$ be such that $\beta_0 > |\mathcal{T}'_{x,y}|$ for all $(x, y) \in W$. Let y be such that $\mathcal{T}_y$ is wellfounded and $|\mathcal{T}_y| > \beta_0$. If $x \in H$, then $(x, y) \in W$ and $|\mathcal{T}'_{x,y}| < \beta_0$. If $x \notin H$ then $\mathcal{T}'_{x,y}$ is still wellfounded (since $\mathcal{T}_y$ is) and $|\mathcal{T}'_{x,y}| \geq |\mathcal{T}_y| > \beta_0$.

Thus we have shown that the tree $\mathcal{T}'$ has the property that every $\boldsymbol{S}(\kappa)$ set is Wadge reducible to $\{(x, y) : \mathcal{T}'_{x,y} \text{ is wellfounded } \&\ |\mathcal{T}'_{x,y}| < \beta\}$ for all sufficiently large β below κ^+. This corrects Lemma 3.10 of the paper. Also, $\mathcal{T}'$ has the "Kunen tree property" that $\{|\mathcal{T}'_{x,y}| : \mathcal{T}'_{x,y} \text{ is wellfounded}\}$ is unbounded in κ^+. Thus, the previous lemmas of the paper go through without change using $\mathcal{T}'$ in place of the tree $\mathcal{T}$ of the paper.

REFERENCES

Chen-Lian Chuang

[Chu82] *The propagation of scales by game quantifiers*, **Ph.D. thesis**, UCLA, 1982.

Gregory Hjorth

[Hjo01] *A boundedness lemma for iterations*, **The Journal of Symbolic Logic**, vol. 66 (2001), no. 3, pp. 1058–1072.

STEPHEN JACKSON

[Jac90B] *Partition properties and well-ordered sequences*, ***Annals of Pure and Applied Logic***, vol. 48 (1990), no. 1, pp. 81–101.

ALEXANDER S. KECHRIS

[Kec81B] *Suslin cardinals, κ-Suslin sets, and the scale property in the hyperprojective hierarchy*, in Kechris et al. [CABAL ii], pp. 127–146, reprinted in [CABAL I], p. 314–332.

ALEXANDER S. KECHRIS, BENEDIKT LÖWE, AND JOHN R. STEEL

[CABAL I] *Games, scales, and Suslin cardinals: the Cabal seminar, volume I*, Lecture Notes in Logic, vol. 31, Cambridge University Press, 2008.

ALEXANDER S. KECHRIS, DONALD A. MARTIN, AND YIANNIS N. MOSCHOVAKIS

[CABAL ii] *Cabal seminar* 77–79, Lecture Notes in Mathematics, no. 839, Berlin, Springer, 1981.

[CABAL iii] *Cabal seminar* 79–81, Lecture Notes in Mathematics, no. 1019, Berlin, Springer, 1983.

ALEXANDER S. KECHRIS, ROBERT M. SOLOVAY, AND JOHN R. STEEL

[KSS81] *The axiom of determinacy and the prewellordering property*, this volume, originally published in Kechris et al. [CABAL ii], pp. 101–125.

YIANNIS N. MOSCHOVAKIS

[Mos80] ***Descriptive set theory***, Studies in Logic and the Foundations of Mathematics, no. 100, North-Holland, Amsterdam, 1980.

JOHN R. STEEL

[Ste83] *Scales in* $L(\mathbb{R})$, in Kechris et al. [CABAL iii], pp. 107–156, reprinted in [CABAL I], p. 130–175.

DEPARTMENT OF MATHEMATICS
UNIVERSITY OF NORTH TEXAS
P.O. BOX 311430
DENTON, TEXAS 76203-1430
UNITED STATES OF AMERICA
E-mail: jackson@unt.edu

DEPARTMENT OF MATHEMATICS
UNIVERSITY OF CALIFORNIA
LOS ANGELES, CALIFORNIA 90024
UNITED STATES OF AMERICA
E-mail: dam@math.ucla.edu

MORE CLOSURE PROPERTIES OF POINTCLASSES

HOWARD S. BECKER

We work in ZF+DC+AD. The theory of arbitrary boldface pointclasses, also known as the theory of Wadge degrees, is one of the topics in descriptive set theory (under AD) which has been a major area of study in recent years. This paper is a contribution to that topic. Specifically, it is about closure properties of arbitrary pointclasses. It will be shown that pointclasses closed under countable union and intersection are also closed under quantification by various types of measure and category quantifiers. The canonical reference for descriptive set theory and AD is Moschovakis [Mos80], whose notation and terminology we will generally follow. Oxtoby [Oxt71] is a good reference for the subject of measure and category. Van Wesep [Van78B] is an introduction to Wadge degrees, Kechris [Kec73] is about measure and category in descriptive set theory, and Steel [Ste81A] is a paper which is also about closure properties of arbitrary pointclasses.

In this paper we work with the Cantor space, ${}^{\omega}2$, which we also denote by $\mathcal{C}$; all of our results are also valid for ${}^{\omega}\omega$. Let μ denote the product measure on ${}^{\omega}2$, where the measure on each factor space is the usual probability measure m on 2, defined by setting $m(\{0\}) := m(\{1\}) := \frac{1}{2}$. To begin with, we work solely with the measure μ; at the end of the paper we will discuss more general measures. For any $B \subseteq \mathcal{C}^2$ and any $x \in \mathcal{C}$, let $B_x := \{y \in \mathcal{C} : B(x, y)\}$. To say that a pointclass $\underset{\sim}{\Gamma}$ is **closed under quantification of the form "for a comeager set of y's"** means that if $B \subseteq \mathcal{C}^2$ is in $\underset{\sim}{\Gamma}$, then the set $\{x \in {}^{\omega}2 : B_x \text{ is comeager}\}$ is also in $\underset{\sim}{\Gamma}$. A similar interpretation holds for other quantifiers. A pointclass $\underset{\sim}{\Gamma}$ is said to be **nice** if it is $\mathcal{C}$-parametrized, contains all open sets, and is closed under continuous preimages, countable unions, and countable intersections.

THEOREM 1. Let $\underset{\sim}{\Gamma}$ be a nice pointclass. Then $\underset{\sim}{\Gamma}$ is closed under quantification of the forms:

(a) For a comeager set of y's.
(b) For μ-a.e. y.

This is the main theorem of this paper. Before proving it we will point out some of its corollaries, which follow very easily.

Research partially supported by NSF Grant MCS 82-11328.

Wadge Degrees and Projective Ordinals: The Cabal Seminar, Volume II
Edited by A. S. Kechris, B. Löwe, J. R. Steel
Lecture Notes in Logic, 37

Corollary 2. Let $\undertilde{\Gamma}$ be a nice pointclass and let N_i be the ith basic open set in $\mathcal{C}$. Then $\undertilde{\Gamma}$ is closed under quantification of the following forms:
(a) For a non-meager set of y's.
(b) For a comeager-in-N_i set of y's.
(c) For a non-meager-in-N_i set of y's.
(d) For a set of y's of positive μ-measure.

Corollary 3. Let $\undertilde{\Gamma}$ be a nice pointclass and let r be a real number. If $\undertilde{\Gamma}$ is closed under either $\exists^{\mathbb{R}}$ or $\forall^{\mathbb{R}}$, then $\undertilde{\Gamma}$ is closed under quantification of the following forms (uniformly in r):
(a) For a set of y's of μ-measure $> r$.
(b) For a set of y's of μ-measure $\geq r$.

Proof. We prove (a), where $\undertilde{\Gamma}$ is closed under $\forall^{\mathbb{R}}$.

$$\mu(A) > r \Leftrightarrow \forall z[\text{if } (z \text{ encodes a } G_\delta\text{-set } S_z \text{ and } \mu(S_z) \leq r) \text{ then } \mu(A \setminus S_z) > 0].$$

By Kechris [Kec73], the set $\{z : \mu(S_z) \leq r\}$ is Borel. ⊣ (Corollary 3)

Theorem 1 also has a different sort of closure property as a corollary.

Corollary 4. Let $\undertilde{\Gamma}$ be a nice pointclass. If $\undertilde{\Gamma}$ is closed under $\exists^{\mathbb{R}}$, then $\undertilde{\Gamma}$ is closed under quantification of the form:

For uncountably many y.

Proof.

$$(\text{For uncountably many } y)P(y) \Leftrightarrow \exists z\,[z \text{ encodes a perfect tree } T_z \subseteq {}^{<\omega}2 \text{ and (for a comeager set of branches } y \in [T_z])P(y)].$$

⊣ (Corollary 4)

In Corollary 4, closure under $\exists^{\mathbb{R}}$ is necessary. If $\undertilde{\Gamma}$ is closed under "for uncountably many", then it is also closed under $\exists^{\mathbb{R}}$, because if $Q(x, y) \Leftrightarrow P(x)$, then $\exists x P(x)$ is equivalent to

$$(\text{For uncountably many } \langle x, y\rangle)Q(x, y).$$

I do not know whether the hypothesis of closure under $\exists^{\mathbb{R}}$ or $\forall^{\mathbb{R}}$ in Corollary 3 is necessary.

Theorem 1 (and its corollaries) were already known for many specific examples of nice pointclasses. Call $\undertilde{\Gamma}$ a **Kleene-type** pointclass if $\undertilde{\Gamma}$ is nice, $\undertilde{\Gamma}$ is closed under $\forall^{\mathbb{R}}$ but not under $\exists^{\mathbb{R}}$ and $\undertilde{\Delta}$ is closed under $\forall^{\mathbb{R}}$. The smallest Kleene-type pointclass is $\undertilde{\text{Env}}({}^3\text{E})$, the class of sets Kleene semirecursive in ${}^3\text{E}$ and a real (see Moschovakis [Mos67]). To use the terminology of Kechris-Solovay-Steel [KSS81], Theorem 1 was known to hold for all inductive-like pointclasses, and for all projective-like pointclasses *except* Kleene-type classes and their duals (that is, *except* the first class in a projective-like hierarchy of type III). For these classes, Theorem 1 was essentially proved in Kechris [Kec73] (and Corollary 4 was essentially proved in Kechris [Kec75]). It was the open question for Kleene-type classes which motivated this work, but it turns out that Theorem 1 holds in much greater generality than that, as there are many nice pointclasses which are not projective-like, *e.g.*, $\underset{\sim}{\mathrm{Env}}(A, {}^2\mathrm{E})$, the smallest boldface Spector pointclass containing A and $\neg A$, where $A \subseteq \mathcal{C}$ is not Borel (see Moschovakis [Mos67]).

Many of the properties of measure and category that were proved in Kechris [Kec73] for the projective pointclasses can be generalized to other classes using Theorem 1 (together with the techniques of Kechris [Kec73]). For example, if $\underset{\sim}{\Gamma} = \underset{\sim}{\mathrm{Env}}({}^3\mathrm{E})$, then every $\underset{\sim}{\Gamma}$ relation with non-meager sections has a $\underset{\sim}{\Delta}$-uniformization.

As usual in the theory of Wadge degrees, the results here are all boldface. I do not know whether Theorem 1 holds for lightface pointclasses, or whether it holds for the particular pointclass $\mathrm{Env}({}^3\mathrm{E})$. But it does follow from the S-m-n Theorem that for nice $\underset{\sim}{\Gamma}$, Theorem 1 and its corollaries hold for the lightface classes $\Gamma(x)$, for a cone of x's.

We now prove Theorem 1.

DEFINITION 5. For any $A \subseteq \mathcal{C}$, let

$$\tilde{A} := \{y : \exists y'(y \text{ and } y' \text{ agree on all but finitely many coordinates and } y' \in A)\}.$$

For any $B \subseteq \mathcal{C}^2$, let

$$\hat{B} := \{(x, y) : y \in (\widetilde{B_x})\}.$$

LEMMA 6.

(a) For any $A \subseteq \mathcal{C}$, $\tilde{A}$ is either meager or comeager. Moreover, $\tilde{A}$ is meager iff A is meager.

(b) For any $A \subseteq \mathcal{C}$, $\mu(\tilde{A})$ is either 0 or 1. Moreover, $\mu(\tilde{A}) = 0$ iff $\mu(A) = 0$.

PROOF. For the measure case, use Kolmogorov's zero-one law (Oxtoby [Oxt71]). ⊣ (Lemma 6)

PROOF OF THEOREM 1. (a) Let Q be the quantifier "for a comeager set of y's". (Thus, QB denotes the pointset $\{x : B_x \text{ is comeager}\}$.) Consider the pointclass $Q\underset{\sim}{\Gamma} = \{QB : B \in \underset{\sim}{\Gamma}\}$. To prove (a) we must show that $Q\underset{\sim}{\Gamma} \subseteq \underset{\sim}{\Gamma}$. Since $\underset{\sim}{\Gamma}$ is parametrized, $Q\underset{\sim}{\Gamma}$ is also parametrized, hence not self-dual. Let $C \subseteq \mathcal{C}$ be a set in $Q\underset{\sim}{\Gamma}$ such that $\neg C$ is not in $Q\underset{\sim}{\Gamma}$. By definition of $Q\underset{\sim}{\Gamma}$, there is a $D \subseteq \mathcal{C}^2$, D in $\underset{\sim}{\Gamma}$, such that $C(x) \Leftrightarrow D_x$ is comeager. Let B be the complement of D. Then $B \in \check{\underset{\sim}{\Gamma}}$.

$$\begin{aligned} \neg C(x) &\iff D_x \text{ is not comeager} \\ &\iff B_x \text{ is not meager} \\ &\iff (\widetilde{B_x}) \text{ is comeager (by Lemma 6)} \\ &\iff (\hat{B})_x \text{ is comeager} \\ &\iff x \in Q\hat{B} \end{aligned}$$

The pointclass $Q\underset{\sim}{\Gamma}$ is closed under Q – this is the Kuratowski-Ulam Theorem (Oxtoby [Oxt71]):

$$Qx_1 Qx_2 P(x_1, x_2) \iff Q\langle x_1, x_2\rangle P(x_1, x_2).$$

Since $\neg C$ is not in $Q\underset{\sim}{\Gamma}$ and $\neg C$ is $Q\hat{B}$, clearly $\hat{B}$ is not in $Q\underset{\sim}{\Gamma}$. Since B is in $\check{\underset{\sim}{\Gamma}}$, by definition of $\hat{B}$ and the closure properties of $\underset{\sim}{\Gamma}$, $\hat{B}$ is also in $\check{\underset{\sim}{\Gamma}}$. Therefore $\check{\underset{\sim}{\Gamma}}$ is not a subclass of $Q\underset{\sim}{\Gamma}$. So by Wadge's Lemma, $Q\underset{\sim}{\Gamma}$ is a subclass of $\underset{\sim}{\Gamma}$, which completes the proof.

(b) The proof of (b) is just like that of (a), using Fubini's Theorem rather than the Kuratowski-Ulam Theorem. ⊣ (Theorem 1)

We now consider more general measures. For any σ-finite Borel measure on $\mathcal{C}$, Corollaries 2 and 3 can be deduced from Theorem 1 by the same proof as for μ. So the only question remaining is whether arbitrary measures satisfy Theorem 1. (Note that Lemma 6 is false for arbitrary measures.) Theorem 7 below gives an affirmative answer to this question for all nice pointclasses which are closed under preimages by Borel-measurable functions.

In the original version of this paper, I asked the following question: Is *every* nice pointclass closed under preimages by Borel-measurable functions? About a year and a half later, John Steel proved that this is indeed the case. The proof of Steel's theorem (which will not be given here) is similar to the proof of Theorem 2.1 of Steel [Ste81A]. Hence the extra hypothesis on $\underset{\sim}{\Gamma}$ can be removed from Theorem 7.

THEOREM 7. Let ν be an arbitrary σ-finite Borel measure on $\mathcal{C}$, and let $\underset{\sim}{\Gamma}$ be a nice pointclass which is closed under preimages by Borel-measurable functions. Then $\underset{\sim}{\Gamma}$ is closed under quantification of the form:

$$\text{For } \nu\text{-a.e. } y.$$

By a **Borel measure** we mean that every Borel set is measurable, hence by AD, every set is measurable (in the completed measure). Call a measure ν **regular** if $\nu(\mathcal{C}) = 1$ and for any $x \in \mathcal{C}$, $\nu(\{x\}) = 0$. Call ν **principal** if there is an $x \in \mathcal{C}$ such that for any $A \subseteq \mathcal{C}$,

$$\nu(A) = \begin{cases} 1 & \text{if } x \in A \\ 0 & \text{if } x \notin A. \end{cases}$$

Call a measure **good** if it satisfies Theorem 7; we must show that all measures are good. To prove Theorem 7, we need a result from measure theory (which is a theorem of ZF).

THEOREM 8. Let ν_1 and ν_2 be arbitrary regular Borel measures on $\mathcal{C}$. There is a bijection $I : \mathcal{C} \leftrightarrow \mathcal{C}$ such that:

(a) I is a Borel isomorphism, that is, I and I^{-1} are both Borel-measurable functions.

(b) For any (measurable) set $A \subseteq \mathcal{C}$, $\nu_1(A) = \nu_2(I[A])$.

Theorem 8 is a combination of Lemma 6.2 of Aumann-Shapley [AS74] and Theorem 1G.4 of Moschovakis [Mos80].

PROOF OF THEOREM 7. Since $\underset{\sim}{\Gamma}$ is closed under preimages by Borel-measurable functions, it follows from Theorem 8 that if one regular measure is good, then all regular measures are. So by Theorem 1 (b), all regular measures are good. It is trivial that all principal measures are good. Now let ν be an arbitrary σ-finite Borel measure on $\mathcal{C}$ such that $\nu(\mathcal{C}) > 0$. There is a sequence $\nu_0, \nu_1, \nu_2, \ldots$ of measures on $\mathcal{C}$ and a sequence $r_0, r_1, r_2, \ldots$ of positive real numbers such that for any $A \subseteq \mathcal{C}$,

$$\nu(A) = \sum_{i=0}^{\infty} (r_i \cdot \nu_i(A)),$$

and such that each measure ν_i is either regular or principal. Then

$$(\text{For } \nu\text{-a.e. } y)P(y) \iff (\forall i \in \omega)(\text{For } \nu_i\text{-a.e. } y)P(y),$$

and since each measure ν_i is good, ν is also good. ⊣ (Theorem 7)

REFERENCES

ROBERT J. AUMANN AND LLOYD S. SHAPLEY

[AS74] ***Values of non-atomic games***, Princeton University Press, 1974.

ALEXANDER S. KECHRIS

[Kec73] *Measure and category in effective descriptive set theory*, ***Annals of Mathematical Logic***, vol. 5 (1973), no. 4, pp. 337–384.

[Kec75] *The theory of countable analytical sets*, ***Transactions of the American Mathematical Society***, vol. 202 (1975), pp. 259–297.

ALEXANDER S. KECHRIS, DONALD A. MARTIN, AND YIANNIS N. MOSCHOVAKIS

[CABAL ii] *Cabal seminar* 77–79, Lecture Notes in Mathematics, no. 839, Berlin, Springer, 1981.

ALEXANDER S. KECHRIS AND YIANNIS N. MOSCHOVAKIS

[CABAL i] *Cabal seminar* 76–77, Lecture Notes in Mathematics, no. 689, Berlin, Springer, 1978.

ALEXANDER S. KECHRIS, ROBERT M. SOLOVAY, AND JOHN R. STEEL

[KSS81] *The axiom of determinacy and the prewellordering property*, this volume, originally published in Kechris et al. [CABAL ii], pp. 101–125.

YIANNIS N. MOSCHOVAKIS

[Mos67] *Hyperanalytic predicates*, ***Transactions of the American Mathematical Society***, vol. 129 (1967), pp. 249–282.

[Mos80] ***Descriptive set theory***, Studies in Logic and the Foundations of Mathematics, no. 100, North-Holland, Amsterdam, 1980.

JOHN C. OXTOBY

[Oxt71] ***Measure and category***, Springer, 1971.

JOHN R. STEEL

[Ste81A] *Closure properties of pointclasses*, this volume, originally published in Kechris et al. [CABAL ii], pp. 147–163.

ROBERT VAN WESEP

[Van78B] *Wadge degrees and descriptive set theory*, this volume, originally published in Kechris and Moschovakis [CABAL i], pp. 151–170.

4840 FOREST DR., STE. 6-B
COLUMBIA, SOUTH CAROLINA 29206
E-mail: hsbecker@hotmail.com

MORE MEASURES FROM AD

JOHN R. STEEL

We assume ZF+AD+DC throughout this paper. Let $\kappa < \vartheta$ be regular. We shall define a measure on ${}^\kappa\kappa$ using some ideas of Kunen. If κ has the strong partition property $\kappa \to (\kappa)^\kappa$, then our measure is just the standard strong partition measure. But while the usual proof that $\kappa \to (\kappa)^\kappa$ requires a coding of elements of ${}^\kappa\kappa$ with certain boundedness properties, our measure can be obtained without appeal to such properties.

We show our measures have the properties of the strong partition measures that allow the Martin-Solovay construction of homogeneous trees to go through. As a corollary, we have that PD is equivalent to the existence of measures of the sort we construct defined on all the projectively coded subsets of any projective ordinal. This answers a question of A. S. Kechris. (Kechris' question inspired our work here.) After we had done this work (in 1980), Steve Jackson showed that PD implies that if n is odd, then $\underset{\sim}{\delta}^1_n \to (\underset{\sim}{\delta}^1_n)^{\underset{\sim}{\delta}^1_n}$ holds with respect to projectively coded partitions. (See [Jac11].) The converse is true by Martin-Solovay, so Jackson's work also gives a combinatorial equivalent of PD.

Fix a norm $\varphi\colon \mathbb{R} \twoheadrightarrow \kappa$, and let $\Gamma = \mathrm{IND}(\leq \varphi, \neg \leq \varphi)$ be the pointclass of relations on $\mathbb{R}$ which are inductive in the associated prewellorder and its complement. So Γ is closed under real quantification, and ω-parametrized. For $e \in \omega$ and $x \in \mathbb{R}$, let $[e]^x$ be the eth relation on reals Γ-recursive in x.

Definition 1. For $\alpha \leq \kappa$, we say (e, x) is good up to α iff there is a function $f^\alpha_{e,x}\colon \alpha \to \kappa$ such that $(y, z) \in [e]^x$ iff $f^\alpha_{e,x}(\varphi(y)) = \varphi(z)$.

Definition 2. For any e, x,

$$f_{e,x} = \bigcup \{ f^\alpha_{e,x} \; : \; (e, x) \text{ is good up to } \alpha \}.$$

Thus $f_{e,x}$ is a partial function defined on a perhaps improper initial segment of κ. We write $f_{e,x}(\alpha)\downarrow$ if $\alpha \in \mathrm{dom}(f_{e,x})$, and $f_{e,x}(\alpha)\uparrow$ otherwise.

For any $f : \kappa \to \kappa$, we define $\hat{f} : \kappa \to \kappa$ by

$$\hat{f}(\xi) = \sup_{n<\omega} f_{e,x}(\omega\xi + n),$$

Wadge Degrees and Projective Ordinals: The Cabal Seminar, Volume II
Edited by A. S. Kechris, B. Löwe, J. R. Steel
Lecture Notes in Logic, 37

for all $\xi < \kappa$. Now, for $A \subseteq {}^\kappa\kappa$, consider the following game G_A:

$$\begin{array}{ll} \text{Player I} & e, x \\ \text{Player II} & i, y \end{array}$$

Here $e, i \in \omega$, and $x, y \in \mathbb{R}$. The order of play is digit-by-digit: first e, then i, then alternating the digits of x and y. Player II wins G_A iff

(a) $\exists \alpha < \kappa$
  (i) $f_{e,x}(\alpha)\uparrow$, or
  (ii) $f_{i,y}(\alpha)\uparrow$, or
  (iii) $f_{i,x}(\alpha) < f_{i,y}(\beta)$ for some $\beta < \alpha$, or
  (iv) $f_{i,y}(\alpha) < f_{e,x}(\alpha)$,

and if α_0 is the least such α, then either (i) or (ii) holds for α_0, or

(b) Case (a) fails, and $\hat{f}_{e,x} \in A$.

Note that in case (b), we have $\hat{f}_{e,x} = \hat{f}_{i,y}$.

Of course, G_A is just the usual strong partition game, but relative to a crude coding for ${}^\kappa\kappa$. The usual proof of the strong partition property uses boundedness properties of more subtle codings to simulate a game of length κ in which the ordinals themselves, rather than codes for them, are played. In our arguments, the Recursion Theorem will do the work that was done by boundedness.

Definition 3. For $A \subseteq {}^\kappa\kappa$, $\mu(A) = 1$ iff player II has a winning strategy in G_A.

Lemma 1. $\mu({}^\kappa\kappa) = 1$.

Proof. Player II copies player I's play. ⊣

Lemma 2. $\mu(A) = 1 \wedge A \subseteq B \Rightarrow \mu(B) = 1$.

For $C \subseteq \kappa$, we put $C^{\curlywedge} = \{\hat{f} : f : \kappa \to C\}$.

Lemma 3. $\mu(\varnothing) = 0$. In fact, if $C \subseteq \kappa$ is club, then $\mu({}^\kappa\kappa \setminus C^{\curlywedge}) = 0$.

Proof. Let $A = {}^\kappa\kappa \setminus C^{\curlywedge}$, and suppose σ were a winning strategy for player II in G_A. Pick $t \in \mathbb{R}$ so that $\{W : \varphi(W) \in C\}$ is Δ in t. Consider now plays of G_A of the form

$$\begin{array}{ll} \text{Player I} & e, \langle \sigma, t \rangle \\ \text{Player II} & i_e, y_e \end{array}$$

where player II plays by σ. We can recursively in $\langle \sigma, t \rangle$ find an e' such that

$$f_{e', \langle \sigma, t \rangle}(0)\downarrow$$

and is the least element of C, and

$$f_{i_e, y_e}(\alpha)\downarrow \Rightarrow f_{e', \langle \sigma, t \rangle}(\alpha + 1)\downarrow$$

and is the least element of $C > f_{i_e,y_e}(\alpha)$, and if λ is a limit ordinal,

$$\forall\beta < \lambda(f_{i_e,y_e}(\beta)\downarrow) \Rightarrow \Big(f_{e',\langle\sigma,t\rangle}(\lambda)\downarrow \wedge f_{e',\langle\sigma,t\rangle}(\lambda) = \sup_{\beta<\lambda} f_{i_e,y_e}(\beta)\Big).$$

(e' is found using the $S-m-n$ theorem and the closure properties of Γ.)

By the Recursion Theorem, fix e such that $f_{e,\langle\sigma,t\rangle} = f_{e',\langle\sigma,t\rangle}$. It is easy to see that if player I plays $e, \langle\sigma, t\rangle$, then he defeats σ. This is a contradiction. ⊣

LEMMA 4. Either $\mu(A) = 1$ or $\mu(-A) = 1$.

PROOF. Let σ be a winning strategy for player I in G_A and τ a winning strategy for I in G_{-A}. Fix $y = \langle\sigma, \tau\rangle$. For $i \in \omega$ consider the plays

Player I	e_i, x_i
Player II	i, y

of G_A by σ, and

Player I	k_i, z_i
Player II	i, y

of G_{-A} by τ.

Given i, we can find effectively an i' so that whenever $\forall\alpha < \kappa$ $(f_{e_i,x_i}(\alpha)\downarrow$ and $f_{k_i,z_i}(\alpha)\downarrow)$, then $f_{i',y}(\alpha)\downarrow$ and $f_{i',y}(\alpha) = \max(f_{e_i,x_i}(\alpha), f_{k_i,z_i}(\alpha))$.

By the Recursion Theorem, we can fix i so that $[i]^y = [i']^y$. Consider the plays of G_A, G_{-A} by σ, τ when player II plays i, y for this i.

By induction on α

$$f_{i,y}(\alpha) = \max(f_{e_i,x_i}(\alpha), f_{k_i,z_i}(\alpha)),$$

and all are defined. Thus condition (a) operates in neither game, and $\hat{f}_{i,y}$ is the function determined in both games. So $\hat{f}_{i,y} \in A$ as τ won for player I in G_{-A}, and $\hat{f}_{i,y} \notin A$ as σ won for player I in G_A. This is a contradiction. ⊣

LEMMA 5. $\mu(A) = 1 \wedge \mu(B) = 1 \Rightarrow \mu(A \cap B) = 1$.

PROOF. Let σ_A and σ_B be winning strategies for player II in G_A and G_B, but τ a winning strategy for player I in $G_{A\cap B}$. Set $y = \langle\tau, \sigma_A, \sigma_B\rangle$. For any (i_0, i_1, i_2) consider the plays

Player I	e_{i_0}, x_{i_0}
Player II	i_0, y

of $G_{A\cap B}$ by τ,

Player I	i_1, y
Player II	e_{i_1}, x_{i_1}

of G_A by σ_A, and

Player I	i_2, y
Player II	e_{i_2}, x_{i_2}

of G_B by σ_B. Given (i_0, i_1, i_2), we can find effectively on (i'_0, i'_1, i'_2) so that

$$f_{i'_0,y} = f_{e_{i_2},x_{i_2}}$$

and

$$f_{i'_1,y} = f_{e_{i_0},x_{i_0}}$$

and

$$f_{i'_2,y} = f_{e_{i_1},x_{i_1}}.$$

By the simultaneous recursion theorem, we can fix (i_0, i_1, i_2) so that

$$f_{i_0,y} = f_{i'_0,y}$$

and

$$f_{i_1,y} = f_{i'_1,y}$$

and

$$f_{i_2,y} = f_{i'_2,y}.$$

Consider the associated plays of $G_{A\cap B}$, G_A, and G_B for this (i_0, i_1, i_2). Since the strategies in question were winning, we see by induction on α that

$$f_{e_{i_0},x_{i_0}}(\alpha) = f_{i_1,y}(\alpha) \leq f_{e_{i_1},x_{i_1}}(\alpha) = f_{i_2,y}(\alpha) \leq f_{e_{i_2},x_{i_0}}(\alpha) = f_{i_0,y}(\alpha),$$

and all are defined and $\leq f_{e_{i_0},x_{i_0}}(\alpha + 1)$.

Thus all f's above determine the same $\hat{f}$, which must then be in A and B but not $A \cap B$, a contradiction. ⊣

Notice that Lemma 5 implies, by AD, that μ is countably additive.

Now let W be a well order of order type $\leq \kappa$; we consider only order type $= \kappa$ for simplicity. Using the natural map from ${}^W\kappa \leftrightarrow {}^\kappa\kappa$, we can transfer μ to a measure μ_W on ${}^W\kappa$. These measures are compatible with one another, in the following sense:

Lemma 6. Let $j\colon W \to V$ be order-preserving and $j^*\colon {}^W\kappa \leftrightarrow {}^{j\text{"}W}\kappa$ the associated map. Suppose $\mu_V(B) = 1$. Then $\mu_W(\{j^*\text{"}(f \restriction j\text{"}W) : f \in B\}) = 1$.

Proof. Clearly we may assume $V = \kappa$, $W \subseteq \kappa$, and j is inclusion. So we want to see $\mu_W(\{f \restriction W : f \in B\}) = 1$. Let $h\colon W \leftrightarrow \kappa$ be order preserving, and let $h^*\colon {}^W\kappa \leftrightarrow {}^\kappa\kappa$ be associated. Let

$$A = h^*\text{"}\{f \restriction W : f \in B\}$$

and suppose for a contradiction that τ is a winning strategy for player I in G_A. Fix a winning strategy σ for player I in G_B. Fix also a real t so that

$$\{(x, y) : h(\varphi(x)) = \varphi(y)\}$$

and

$$\{x : \varphi(x) \in W\}$$

are Δ in t. Fix $y = \langle \sigma, \tau, t \rangle$.

For any (i_0, i_1) consider the plays

Player I	i_0, y
Player II	e_{i_0}, x_{i_0}

of G_B by σ, and

Player I	e_{i_1}, x_{i_1}
Player II	i_1, y

of G_A by τ.

Given (i_0, i_1), we can find effectively an (i_0', i_1') so that $\forall \xi < \kappa$:

(1) if $\xi \notin W$, then

$$f_{i_0',y}(\omega\xi) = \sup_{\beta<\omega\xi}(f_{e_{i_0},x_{i_1}}(\beta))$$

and

$$f_{i_0',y}(\omega\xi + n + 1) = f_{e_{i_0},x_{i_0}}(\omega\xi + n),$$

(2) if $\xi \in W$, then

$$f_{i_0',y}(\omega\xi) = \max(f_{e_{i_1},x_{i_1}}(\omega \cdot h(\xi)), \sup_{\beta<\omega\xi}(f_{e_{i_0},x_{i_0}}(\beta))),$$

and

$$f_{i_0',y}(\omega\xi + n + 1) = f_{e_{i_1},x_{i_1}}(\omega \cdot h(\xi) + n + 1),$$

and

(3) $f_{i_1,y}(\omega\xi + n) = f_{e_{i_0},x_{i_0}}(\omega \cdot h^{-1}(\xi) + n)$.

By the simultaneous recursion theorem, we get (i_0, i_1) such that $f_{i_0,y} = f_{i_0',y}$ and $f_{i_1,y} = f_{i_1',y}$. Consider the associated plays of G_A and G_B.

We now see by induction on $\alpha < \kappa$ that

$$\begin{aligned} f_{e_{i_1},x_{i_1}}(\omega\alpha + n) &\leq f_{i_0,y}(\omega \cdot h^{-1}(\alpha) + n) \\ &\leq f_{e_{i_0},x_{i_0}}(\omega \cdot h^{-1}(\alpha) + n) \\ &= f_{i_1,y}(\omega\alpha + n) \end{aligned}$$

and that all are defined and $\leq f_{e_{i_1},x_{i_1}}(\omega\alpha + n + 1)$.

Clearly,

$$\hat{f}_{i_0,y}(h^{-1}(\alpha)) = \hat{f}_{i_1,y}(\alpha),$$

for all $\alpha < \kappa$. Since τ was winning for player I in G_A, $\hat{f}_{i_1,y} \notin A$. But $\hat{f}_{i_0,y} \in B$ since σ was winning, and $h^*(\hat{f}_{i_0,y} \restriction W) = \hat{f}_{i_1,y}$. This is a contradiction. ⊣

It is not hard now to show that PD is equivalent to the existence of measures on ${}^{\kappa}\kappa$, defined on all the projectively coded subsets of any projective ordinal κ, which satisfy Lemmas 1–6. To obtain PD from the measures, one proves by induction on n that every $\underset{\sim}{\Sigma}^1_{2n}$ set is the projection of a projectively-coded, projectively-weakly-homogeneous tree. This is because the properties of the measures μ_W given by Lemmas 1–6 allow one to carry out the proof of Theorem 4.14 of [Jac08].

Martin's proof of the strong partition property for ω_1 builds on Solovay's original use of boundedness arguments to simulate games in which ordinals are played. Solovay showed this way that Δ^1_2 determinacy implies there is an inner model of ZFC + "there is a measurable cardinal". Martin, Simms, and others extended Solovay's arguments so as to obtain inner models with many measurable cardinals. (*Cf.* [Sim79].) It is also possible to replace the use of boundedness in these arguments by the Recursion Theorem, in the style of the arguments above. In this way, one can re-prove the main results of Martin and Simms in this area.

REFERENCES

Stephen Jackson

[Jac08] *Suslin cardinals, partition properties, homogeneity. Introduction to Part II*, in Kechris et al. [Cabal I], pp. 273–313.

[Jac11] *Projective ordinals. Introduction to Part IV*, 2011, this volume.

Alexander S. Kechris, Benedikt Löwe, and John R. Steel

[Cabal I] *Games, scales, and Suslin cardinals: the Cabal seminar, volume I*, Lecture Notes in Logic, vol. 31, Cambridge University Press, 2008.

John Simms

[Sim79] *Semihypermeasurables and* $\Pi^0_1(\Pi^1_1)$ *games*, ***Ph.D. thesis***, Rockefeller University, 1979.

DEPARTMENT OF MATHEMATICS
UNIVERSITY OF CALIFORNIA
BERKELEY, CALIFORNIA 94720-3840
UNITED STATES OF AMERICA
E-mail: steel@math.berkeley.edu

EARLY INVESTIGATIONS OF THE DEGREES OF BOREL SETS

WILLIAM W. WADGE

In this paper, I give an overview/summary of the techniques used, and the results derived, in my 1984 PhD dissertation, *Reducibility and Determinateness on the Baire Space*. In particular, I focus on the calculation of the order type (and structure) of the collection of degrees of Borel sets.

§1. Introduction. I would like in this article to present a overview of the main results of my PhD dissertation, and of the game and other techniques used to derive them.

My first thought was to print the entire dissertation but I quickly realized that it was too long—about ten times too long! Hopefully, this condensed version will still be useful. In producing such a drastically shortened account, I have omitted detailed proofs, and many less important or intermediate results. Also, the remaining definitions and results are for the most part given informally.

In writing this I have in mind, first, colleagues (whether in Mathematics or Computing) who are not familiar with descriptive set theory but nevertheless would like to learn about "Wadge Degrees". To make the material accessible to these readers I have included some basic information about, say, Borel sets that will be very familiar to Cabal insiders. However, my hope is that even experts in descriptive set theory may learn something, if not about my results, at least about the manner in which they were discovered. In particular, I would like to give some 'classic' notions, such as Boolean set operations, the attention they deserve.

As already indicated, the approach will be technical but fairly informal. I will skip many precise definitions and statements of results; firstly, because the details can take up precious space and obscure the important issues; and secondly, because these detailed formulations can be found elsewhere.

I would like to acknowledge above all the expert guidance of my PhD advisor, John W. Addison, Jr. He not only suggested the right questions to ask, but also time and again he introduced me to the techniques that in the end allowed me to answer these questions. I am also very grateful for the financial support I received, as a graduate student, from the Woodrow Wilson National Fellowship Foundation, from the UC Berkeley Science Division, and from the Canada Council.

Wadge Degrees and Projective Ordinals: The Cabal Seminar, Volume II
Edited by A. S. Kechris, B. Löwe, J. R. Steel
Lecture Notes in Logic, 37

Readers who need more precise formulations can find them in the dissertation itself [Wad84], which, if all goes to plan, will soon be published as a book.

§2. **Definability.** My research grew out of a seminar Prof. J. W. Addison, Jr. gave in the theory of definability in the Fall of 1967, at UC Berkeley. The theory of definability (founded, according to Addison [Add04], by Tarski) studies the relationship between the grammatical complexity of definitions and the semantic complexity of the objects (typically sets) that they define. A perfect example is the theorem that formulas whose sets of models are closed under ordinary extension are exactly those equivalent to existential formulas.

In definability, it is usually easy to show that an object has a definition of a certain degree of complexity—just come up with it. However, proving the contrary—that no such definition exists—can be extremely difficult. Many of the most important results of definability theory help with this problem by reducing proving nonexistence of a definition to proving existence (of something else). For example, we can prove that there does not exist an existential equivalent of a formula by proving that there exists a model of the formula that can be extended to a model of its negation.

§3. **Descriptive set theory.** In the 1967 seminar we learned of intriguing analogies (due largely to Addison himself) between apparently distinct results in predicate logic, recursive function theory and descriptive set theory.

Descriptive set theory is the oldest of these topics, and it grew out of classical analysis. The study of continuity, differentiation and integration, and the limit process revealed the existence of sets (of real numbers) and functions (over the reals) that failed to possess some highly desirable properties. Measure theory, for example, extended the Riemann integral to a much wider collection of functions. Analysts discovered, however, that the axiom of choice implied the existence of functions and sets that are not measurable. On the other hand, every set/function that actually arose in practice was measurable. Similary, it seems that every set actually encountered has the perfect set property—it is either countable or has a perfect subset (and hence has the power of the continuum). Nevertheless, the Axiom of Choice implies the existence of sets without the perfect set property.

The conviction grew that any set that was somehow constructible or definable was much better behaved than the mysterious sets that the axiom of choice allows us to produce like rabbits out of a magician's hat. This led to a systematic study of the ways in which sets of reals can be constructed, and to a comparative study of the power of different ways of defining these sets.

For example, the open sets are almost the simplest, and are easily seen to have the two properties just mentioned. So do closed sets (complements of open sets) and in general any finite Boolean combination of open sets.

§4. The Borel sets. In the 1967 seminar Addison suggested I work on the problem of providing constructive examples to verify the nontriviality of the famous Borel hierarchy.

The class of *Borel* sets is the least class containing all the open sets and closed under complement and countable union. Borel sets can be very complex but they are all measurable and have the perfect set property (as well as many others).

The Borel hierarchy is determined directly by the inductive definiton just given. The Borel hierarchy classifies Borel sets according to how many alternations of negation and countable union are required to construct them. The simplest are the open sets and their complements, the closed sets. Traditionally $\mathcal{G}$ denotes the class of open sets and $\mathcal{F}$ the class of closed sets.

At the next level in the Borel hierarchy we find the collection $\mathcal{F}_\sigma$ of countable unions of closed sets and its dual, the collection $\mathcal{G}_\delta$ of countable intersections of open sets (again, traditional notation). (The dual of a collection of sets is the collection of complements.) Continuing in this way we form the collection $\mathcal{G}_{\delta\sigma}$ of countable unions of countable intersections of open sets, and its dual $\mathcal{F}_{\sigma\delta}$; the class $\mathcal{F}_{\sigma\delta\sigma}$ and its dual $\mathcal{G}_{\delta\sigma\delta}$; and so on.

The finite levels by no means exhaust the Borel sets. The simplest sets not necessarily in any finite level are those that are the union of a sequence of sets each at some finite level. If the individual sets are from higher and higher finite levels, their union will not in general be at any finite level. It is not hard to see that we cannot close out under countable union until we have a level for every countable ordinal.

Nor does the class of Borel sets include all sets that can be (somehow) defined. If B is a Borel set and f a continuous function, the *image* under f of B ($\{f(\beta) \mid \beta \in B\}$) will not in general be a Borel set. These sets (called *analytic* sets) are nevertheless all measurable, and all have the perfect set property. Closing the Borel sets out under complementation and continuous image gives us the *projective* sets, and once again they form a hierarchy.

The bottom level of the projective hierarchy consists of the class of analytic sets and its dual. On the next level we find that class of continuous images of complements of analytic sets, together with its dual, and so on as before (the hierarchy has only ω levels). Almost all examples of sets explicitly defined in analysis are in the lower levels of the projective hierarchy.

§5. The analogy with recursive function theory. Recursive function theory is much younger than descriptive set theory. It began with the work of Church and Turing, who formalized the notion of "effective" and showed that (in modern terminology) there are recursively enumerable (r.e.) sets that are not recursive. Recursive enumerability is clearly a form of definability—an r.e. set is 'defined' by the Turing machine that enumerates it.

In 1946 Tarski [Tar00] pointed out an analogy between the analytic sets and the r.e. sets. In this analogy recursive sets correspond to the Borel sets: every set that is r.e. and co-r.e. is recursive, just as (a classical result) every set that is analytic and coanalytic is Borel. However in 1950 Kleene [Kle50] found a pair of r.e. sets that were not separable by a recursive set. This upset the analogy, because (by a stronger form of the classical result) every disjoint pair of analytic sets is separable by a Borel set. Kleene asked his new graduate student Addison to look into the anomaly.

Addison discovered that Tarski had, arguably, got the analogy wrong, and that it made much more sense to pair the r.e. sets with the open sets. What Addison [Add54] called the "fundamental principle" of the analogies is the fact that the continuous functions are those that are, in a very natural sense, computable. Addison's principle is indeed of fundamental importance both for the study of infinite games and of infinite (nonterminating) computations. To explain it, however, I must first introduce the Baire Space.

§6. The Baire Space. Originally descriptive set theory, which grew out of analysis, studied sets of real numbers (subsets of the continuum). However the set of real numbers, considered as a topological or metric space, was far from convenient. All the spaces formed by taking the product of the reals with itself a number of times were distinct (not isomorphic), although in terms of foundational issues, these differences were of no importance. For example, it is just as easy to prove that some subset of the real line is nonmeasurable as to prove that some subset of the plane is. Furthermore, the decimal expansion of real numbers is annoyingly irregular; for example, 1.0 and 0.99999999... represent the same number. Also, there are no nontrivial sets of reals that are both open and closed.

The first descriptive set theorists soon found that by considering only sets of *irrational* numbers, little was lost and much was gained. The space of irrationals, with the induced topology, (soon called the Baire space) has many clopen sets and is isomorphic to all its finite or countable powers. Furthermore, continued fraction expansion (necessarily infinite) is much better behaved than the decimal expansion. In fact the continued fraction expansion was used so much they eventually worked directly in the space ${}^{\omega}\omega$ of all ω-sequences of natural numbers. This is what contemporary mathematicians mean by the Baire space.

It is easy to describe the topology of the Baire space directly in terms of ω-sequences. Given any finite sequence s, the interval of Baire $[s]$ is the set of all infinite sequences that extend s. For example, $[\langle 7,3,2\rangle]$ is the set of all ω-sequences whose 0-index element is 7, whose 1-index element is 3, and whose 2-index element is 2. The intervals of Baire form a basis for the Baire topology: a set A is open iff every element of A is in an interval included in A.

§7. Clopen sets as recursive sets. To see the connection between the Baire topology and computability, let us start at the other end and ask, which subsets of the Baire space are recursive? In conventional recursive function theory, a subset of ω is recursive iff there is a computer M that implements the membership test. This means that given any n, if we give n as the input to M, M will compute for a while, then eventually output either 1 (meaning n is a member) or 0 (meaning it is not). By a computer we mean a deterministic and purely mechanical device. A computer must be finite but may have an unbounded memory. There are several ways of formalizing these ideas, all equivalent; for example, we can take our computers to be Turing machines with infinite (or extendable) tapes, initially blank. We also need an input-output convention/protocol; we can agree that the machine will be started with nothing but its input (in binary) on the tape, and that when it halts nothing but the output will be left.

There is no real difficulty in carrying this definition over to the Baire space, especially if we use the Turing machine model. An element of the Baire space is infinite but so are the Turing machine tapes. So we can say that a subset A of the Baire space is recursive iff there is a Turing machine that, when started with an arbitrary α (in ${}^{\omega}\omega$) on its tape, eventually halts with either a 0 or 1 on its tape, indicating that α either is (1) or is not (0) an element of A. We should be a bit more precise about the input-output convention. For example, we can specify that the components of the sequence α are written on the right half of the tape, and that when the machine halts the 0 or 1 is under the read head. (We obviously cannot require the machine to erase all its input).

The really important point is that the machine must give an answer after only a finite amount of computation, during which it can have examined only a finite number of α's values (let k be the largest number for which α_k is so examined). That means it must give the same answer for any α' that agrees with α on at least the first k values. Suppose the machine concludes that α is in A, and let $\alpha|k$ be the sequence $\langle \alpha_0, \alpha_1, \alpha_2, \ldots, \alpha_{k-1} \rangle$. Then not only α, but every element of the interval $[\alpha|k]$ is in A. Similarly, if the machine had concluded that α was not in A, then $[\alpha|k]$ would have to be a subset of $-A$. Putting it all together, we see that if A is Turing decidable then (1) every element of A is in an interval included in A; and (2) every element of $-A$ is in an interval included in $-A$. Bearing in mind the fact that the intervals are a basis for the Baire topology, (1) says that A is *open* and (2) says that $-A$ is open, i.e., that A is *closed*. This means that decidable subsets of the Baire space are both closed and open—they are *clopen* sets.

Are all clopen sets decidable? No, and it is simple to find a counterexample. Let K be any nonrecursive set of natural numbers and let A be the set of ω-sequences whose first element is in K. Set A is clearly clopen but not machine-decidable. However there is still a sense in which it is 'easy' to decide whether or not α is in A: we 'just' look at α_0 and ask whether it is in K. In other words, A is machine-decidable modulo a countable amount of information (the membership list of K). We can make this precise by allowing our Turing machines to have an extra, read-only tape with a single sequence δ written on it. If membership in A can be decided by such a machine, we say that A is recursive in δ. It is not hard to see that a set is clopen iff it is recursive in some δ. (Alternatively, we could allow our machines to have a countably infinite number of states, and impose no constraint on their transition functions.)

§8. **Open sets as r.e. sets.** If the recursive sets correspond to the clopen sets, which sets correspond to the r.e. sets? Since a set is recursive iff both it and its complement are r.e., it is likely that the r.e. sets correspond to either the closed or open sets. We know that a set is r.e. if there is a machine that enumerates its elements; but it is not clear how a machine could enumerate an uncountably infinite set of ω-sequences. There is another definition, however, that does carry over. A set K is r.e. iff it is half decidable, in the following sense: there is a machine M that, given n as input, eventually halts with 1 on its tape iff n is in K. (The machine is not required to halt if n is not in K.) Moving to the Baire space, a set A is analog-r.e. iff there is a machine M that, given α as input, eventually halts with a 1 on its tape iff α in A.

Given this definition, it is not hard to see that *analog-r.e.* is *open*. Suppose first that A is analog-r.e. and α in A. If we start M with $\alpha_0, \alpha_1, \alpha_2, \dots$ on its tape, M will eventually halt with 1 on its tape. Before halting, M will have had a chance to examine only finitely many components of α, none (for some k) of index k or greater. This means M will do likewise for any α' that agrees with α on at least the first k components; in other words, any α' in $[\alpha|k]$. Thus α is in an interval (namely $[\alpha|k]$) included in A, so A must be open. Conversely, if A is open, we program a machine M (with an extra read-only tape, as above) to examine one by one the components of α until, for some k, $[\alpha|k]$ is a subset of A.

§9. **Continuous functions as computable functions.** Finally, given that r.e. sets correspond to open sets, it should come as no surprise that recursive functions correspond to continuous functions. To see this, consider what it might mean to say that a function over the Baire space is Turing computable. In the case of the natural numbers, we say that a function f from ω to ω is computable by M if, for any n, if we give n as input to M, M eventually halts with $f(n)$ as output. To carry this over to the Baire space, we need a protocol that tells us how to give an α to M as input, and how M presents β $(= f(\alpha))$ as output. For input, we can simply write the components of α on half of the tape, as above. Output is not so simple, because it is infinite.

There are actually (at least) two ways to do this. One is to present β using what computer scientists now call a "demand-driven" protocol. This allows us to compute any particular component of $f(\alpha)$ for any particular α. More precisely, M accepts an arbitrary α in ${}^{\omega}\omega$ and an arbitrary n in ω as input, computes, and eventually halts with $f(\alpha)_n$ as output.

The other approach uses what is now called a "data-driven" approach. We present M with α as input, and M computes $\beta_0, \beta_1, \beta_2, \ldots$, in turn, writing them in order. In this protocol M never stops; it is a continuously operating device.

Fortunately, it does not matter which protocol we use. In either case it is easy to verify that a function is machine computable (in some δ) iff it is continuous. To see this, suppose first that f is continuous and that $\beta = f(\alpha)$. For any k, $[\beta|k]$ is open, so $f^{-1}([\beta|k])$ is also open; and since α is in $f^{-1}([\beta|k])$, α must be in some interval (necessarily of the form $[\alpha|j]$) in $f^{-1}([\beta|k])$. In other words, the fact that $f(\alpha)$ begins with $\beta|k$ follows from the fact that α begins with $\alpha|j$. Thus an arbitrary finite amount of information about $f(\alpha)$ follows from a finite amount of information about α.

Conversely, if f has this finitary property, it follows immediately that the inverse image of an interval containing β includes an interval containing α. Since the intervals are a basis for the Baire topology, f must be continuous.

§10. Luzin's examples. As my contribution to the seminar, Addison set me the problem of proving that each set in a particularly simple series of sets is exactly as complex as its obvious definition. The first set S_1 is that of all sequences in which 0 occurs at least once. This is easily seen to be open; could I prove that it is not closed? The next is the set P_2 of sequences in which 0 occurs infinitely often. It is a $\mathcal{G}_\delta$; prove it's not a $\mathcal{F}_\sigma$. The third, S_3, is the set of sequences in which some number (not necessarily 0) occurs infinitely often (a $\mathcal{G}_{\delta\sigma}$), and the fourth, P_4, is that of all sequences in which infinitely many numbers occur infinitely often.

The examples are from the famous 1930 book [Luz30] by the Russian mathematician Nikolai Luzin, one of the founders of descriptive set theory (I have used modern terminology). Luzin and his colleagues had found topological proofs of the 'properness' of these sets. Addison wanted to know if I could do better (than a founder of descriptive set theory!).

The first set was relatively easy to deal with. Since any initial segment of any arbitrary sequence has an extension in the set (with a 0 in it), it follows that any such arbitrary sequence is a limit point of the set—the set is dense. This means it cannot be closed—the only closed dense set is the entire space. Thus the set of sequences with at least one 0 is a 'proper' or 'true' open set.

§11. The difference hierarchy. With the second set—that of all sequences with infinitely many 0's—things already get much more complicated. Here we need an important result of Hausdorff [Hau57], namely if a set is both $\mathcal{G}_\delta$ and F_σ, then it must be a difference of a number of open sets. Hausdorff's difference hierarchy orders these sets (in modern notation, the $\utilde{\mathbf{\Delta}}^0_2$ sets) into a hierarchy of ω_1 levels. At the third level, for example, are those sets that are of the form $(G_0 - (G_1 - G_2))$ for open sets G_0, G_1, and G_2.

Hausdorff showed that one can identify the level a set appears in the difference hierarchy by taking what he calls *adjoins* and *residues*. This involves seeing how far back and forth one can take limits between the set and its complement. For example, the first adjoin of a set A is the set of limit points of A in $-A$; the first residue is the set of limit points of the first adjoin that are members of A; and the second adjoin is the set of limit points of the first residue that lie in $-A$. The point at which the residues and adjoins become empty determines the set's position in the difference hierarchy.

The set P_2, however, is dense and codense, so the adjoin/residue process can never terminate. It therefore cannot be a $\utilde{\mathbf{\Delta}}^0_2$ set, and so must be a proper or true $\mathcal{G}_\delta$.

§12. Many-one reducibility. The proofs for S_3 and P_4 were much more complex, so I started looking for another approach. Addison always impressed on me the importance of analogies, and he pointed me to a paper by Hartley Rogers [Rog59] in which Rogers considers sets of natural numbers that were strikingly similar to those in Luzin's book. Rogers assumed a fixed system for indexing r.e. sets, and considered the following sets of natural numbers: (1) the set of all indices of nonempty sets; (2) the set of indices of infinite sets; (3) the set R_3 of indices of sets that contain the index of an infinite set; and the set of indices of sets that contain infinitely many indices of infinite sets.

Rogers proved analogous results about these sets, but not using topology; instead, he used reducibility by recursive function (called "many-one reducibility"). A set K of natural numbers is (many-one) reducible to a set L of natural numbers iff there is a recursive function f such that for any n, n is in K iff $f(n)$ is in L. In other words, the function f allows us to 'reduce' the question of n's membership in K to the membership of another number $(f(n))$ in L. Since recursive functions are computable, this means that the computability complexity of K is no greater than that of L. Another way to put it is that K is the inverse image of L under a recursive function (namely f).

For example, it is not hard to see that set R_3 can be defined by an $\exists\forall\exists$ formula—in modern notation, it is Σ^0_3. Rogers showed that the set in question is Σ^0_3-complete: any other Σ^0_3 set is the inverse image of Roger's third set under a recursive function. It follows that if this set were also Π^0_3, then *every* Σ^0_3 would also be Π^0_3, and the arithmetic hierarchy would collapse.

§13. Continuous reducibility. It took no great insight to see that an analogy to many-one reducibility by recursive function might do the trick. Furthermore, given the algorithmic description of the Baire topology discussed above, it was obvious to me that the analogous notion must be reducibility by continuous function: $A \leq B$ iff there is a continuous function f such that for any α, $\alpha \in A$ iff $f(\alpha) \in B$ (or, more concisely, $A = f^{-1}(B)$). However, if I had not learned to think in terms of infinite games, I probably would not have taken it any farther.

To see where games come in, suppose I claim that a set A is reducible to a set B, and you doubt me. I produce a machine that, I claim, computes the reducing function incrementally. You are still skeptical and decide to call my bluff. You start to enumerate the values $\alpha_0, \alpha_1, \alpha_2, \dots$, of some sequence and demand that I turn on my machine and start enumerating the values $\beta_0, \beta_1, \beta_2, \dots$, of the corresponding sequence β. What follows, to put it melodramatically, is an epic contest between a human and a machine. You will try to discredit the machine, by choosing an α that the machine fails to reduce correctly. You can do this by enumerating an α in A that causes the machine to produce a β that is not in B; or by enumerating an α in $-A$ that causes the machine to enumerate a β in B.

Obviously, in the course of the enumerations my machine has access to your values of α, because they are its input. Conversely, I lose nothing if I allow you, my machine's opponent, to see the values of β as they are produced—because I believe my machine will cope with any α no matter what its origin.

§14. The game $G(A, B)$. In that case the real nature of the struggle between you and my machine becomes apparent: it is an infinite two-player game of perfect information. More precisely, given any subsets A and B of the Baire space, the rules of the game $G(A, B)$ are as follows:

Players I (you) and II (my machine) play alternately, player I moving first. On each move player I plays a single natural number. On each move player II plays a natural number, *or passes* (plays nothing). Let α be the sequence of all player I's moves (necessarily infinite) and β be the sequence of all player II's moves. Player II wins if β is infinite and either β in B and α in A, or else β in $-B$ and α in $-A$. Otherwise player I wins.

And now the moral of the story is apparent: if A is reducible to B by a continuous function, then there is a machine as described above; and this machine can be used to win the game $G(A, B)$. Arguably, the converse is true; if player II has a winning strategy, we can build a machine that implements the strategy, and that machine must compute a continuous function that reduces A to B.

To formalize this result, we have to formalize the game. The melodramatic account just given is of course not formal mathematics, but neither is the set of rules just given; for example, it refers to the sequence of all of player I's moves, presumably collected at the 'end' of an endless game.

In fact, formalizing a game amounts to formalizing what constitutes a strategy for player I, what constitutes a strategy for player II, and what is the result of playing a given strategy for player I against a given strategy for player II. (Addison calls this result the *clash* of the two strategies.) A winning strategy for one player is one that defeats all strategies for the opponent. Fortunately, this is fairly easy for $G(A, B)$—one way is to transform it to an equivalent Gale–Stewart game by having the players take turns choosing the components of a single sequence.

When we do this (I will omit the details) we find that a strategy for player II is specified by (and we can take it to be) a monotonic function on finite sequences whose result is never longer than its argument. To play using such a function, player II ensures that after each of his moves, the sequence of moves that he made up to that point is the result of applying the function to the 'history' of player I's moves. Since the function is monotonic, player II never has to retract moves (which of course he is not allowed to do).

Now suppose that player II has a winning strategy σ for $G(A, B)$. Given any element α of the Baire space, let $\sigma^*(\alpha)$ be the union (limit) of the sequence $\sigma(\langle\rangle), \sigma(\langle\alpha_0\rangle), \sigma(\langle\alpha_0, \alpha_1\rangle), \sigma(\langle\alpha_0, \alpha_1, \alpha_2\rangle), \ldots$. The limit exists because σ is monotonic, and is infinite because σ is a winning strategy. It is then easy to show that σ^* is continuous and reduces A to B.

Conversely, suppose that there is a continuous function f that reduces A to B. For any finite sequence s of length k, let $\sigma(s)$ be the longest sequence of length at most k that is an initial segment of $f(\alpha)$ whenever s is an initial segment of α. (One can think of t as the output obtained when the machine that computes is given s and allowed to run until (i) more input is needed; or (ii) k output values are produced.)

What we just proved (informally) is the following game characterization of $\leq$; for any subsets A and B of the Baire space: $A \leq B$ iff player II has a winning strategy for $G(A, B)$.

(It is possible to give a concise formal definition of $G(A, B)$ and a short proof of this theorem; the informal approach above was used for expository purposes. In particular, the formalism does not need machines, although I find them a useful heuristic guide.) Then σ is a winning strategy for player II for $G(A, B)$.

§15. Completeness of Luzin's sets. Once the game was perfected, I was able to make short work of Luzin's examples. In each case it was relatively straightforward to show that the set in question was complete for its 'natural' complexity class, and thus a 'proper' member of that class.

Consider first the set S_1 of all sequences in which 0 occurs at least once. To show that it is complete for the open sets, let A be an arbitrary open set; here is the winning strategy for $G(A, S_1)$. Player II simply plays 1's until the interval corresponding to player I's moves is included in A. In other words, until player I has committed himself to A because all infinite extensions of player I's current (finite) sequence of moves are in A. When (if) player I decisively 'enters' A in this sense, player II switches to playing 0's.

If at some point player I enters A in this sense, player II's final sequence will have 0's past a certain point and be in S_1. And since player I entered A, his sequence must be in A, and so player II wins. Conversely, if player I's final sequence is in A, player I must enter A at some point, because A is open. So if player I never enters A, his final sequence will be in $-A$; and if player I never enters A, player II will never play a 0, and thus player II's final sequence will be in $-S_1$. Player II wins in this case as well.

Next, we show that P_2 (the set of sequences with infinitely many 0's) is complete for $\mathcal{G}_\delta$ sets ($\utilde{\mathbf{\Pi}}^0_2$ sets). Let A be an arbitrary $\mathcal{G}_\delta$ set, the intersection of open sets $G_0, G_1, G_2, \ldots$. Without loss of generality we can assume that each G_{i+1} is a subset of G_i. Here is player II's strategy for $G(A, P_2)$.

Player II plays 1's by default, as above, but takes into account every time player I 'enters' one of the G_i, in the sense described above. (To avoid pathological cases, assume player II pays no attention to G_i before the ith move.) Every time player I enters at least one new G_i, player II plays a 0.

If player I's sequence ends up in A, it must be in all the G_i's, and player II will act on infinitely many entries so that his sequence has infinitely many 0's and ends up in P_2. On the other hand, if player I's sequence is not in A player I can enter only finitely many G_i's, and player II will play only finitely many 0's. Either way player II wins.

It should be clear now how to proceed with the others; in particular $G(A, S_3)$ is just countably many copies of $G(A, P_2)$ played in parallel. Furthermore, it is not hard to continue the Luzin examples through at least the finite levels of the Borel hierarchy. Naturally I was happy at having bested the father of descriptive set theory; but it was all thanks to infinite games, which were unknown in Luzin's day.

§16. The $\utilde{\mathbf{\Delta}}^0_2$ degrees. I was so pleased at having done an end run around highly technical topological arguments that I had not really thought ahead. Addison urged me to investigate the structure of the degrees. In particular, I should find out if it was just an accident that Luzin's examples of proper sets were in fact complete for their classes.

In the case of S_1, the complete open set, the question reduced to the analog of Post's problem. Turing's original proof yielded a set that was r.e. but not recursive, and complete for the r.e. sets: membership in any r.e. set can be decided algorithmically by a machine that has access to an oracle for the halting problem. Post's problem was to decide whether or not all nonrecursive r.e. sets have the same property (and are therefore Turing equivalent to the halting problem).

The reducibility involved here, Turing reducibility, is much coarser than many-one reducibility. If a set K is Turing reducible to a set L, that means in deciding whether or not n is in K we can ask any membership questions we want about L, and do whatever we want with the answers. In many-one reducibility we can ask about only $f(n)$, and must simply repeat this answer.

Nevertheless, Muchnik and Friedberg (using the priority method) proved that there are many different Turing degrees of r.e. sets; depressingly many; for example, every finite distributive lattice can be embedded in the r.e. degrees. It was only logical to expect that game arguments (very similar to priority arguments) would show that the class of open sets is similarly distressingly complex.

In fact, nothing of the kind emerged. Recall the strategy for $G(A, S_1)$ with A open; it relied on the fact that the sequence $\langle 1, 1, 1, \dots \rangle$ is not in A, but that every finite initial segment has an extension that is in A. If we refer to discussion of the Baire topology given above, we see that $\langle 1, 1, 1, \dots \rangle$ is a limit point of S_1. Our strategy for $G(A, S_1)$ is based on the fact that S_1 has a limit point that is not in S_1; that is, on the fact that S_1 is not closed.

In other words, any set that is not closed is complete for the open sets; so that in particular, any 'proper' open set is complete and of the same degree as S_1.

Encouraged by this unexpectedly pleasant result, I began looking at the degrees just beyond those of the degree of the true open sets and the degree of the true closed sets. In terms of the game, what counts is the number of times a player can 'switch' between entering a set and entering its complement. For example, suppose that, like S_1, there is a sequence β in $-B$ but that every $\beta|k$ can be extended to a sequence δ_k in B. But suppose each δ_k has the property that any initial segment can in turn be extended to something back in $-B$. Then in the game $G(A, B)$ player II can 'feint' towards $-B$, then if necessary 'feint' towards B, and finally if necessary enter $-B$ (the last time is not a feint). Player II can win the game if A is either open or closed, because open or closed sets allow only one feint.

A sequence with the property described above is easily seen to be an element of $-B$ that is a limit point of elements of B each of which is a limit point of elements of $-B$. In other words, an element of the *second adjoin* of B (in Hausdorff's terminology). A set has such a point iff its complement is not a difference of two open sets. Furthermore, an easy game strategy shows that such a set is complete for those that are differences of open sets.

Continuing in this way, I was able to show that the degree of a $\utilde{\boldsymbol{\Delta}}^0_2$ set is determined by how many back and forth feints are possible, and this in turn is entirely determined once one knows which of the residues and adjoins are nonempty. The structure of the $\underset{\sim}{\mathbf{\Delta}}_2^0$ degrees coincides exactly with that of the Hausdorff hierarchy.

The contrast to the situation in recursive function theory is striking; the degrees are almost linearly ordered (the exception being incomparable dual degrees). In fact they are almost (in the same sense) well-ordered, the order type being Ω $(= \omega_1)$, the first uncountable ordinal. Furthermore, the nonselfdual pairs of degrees alternate with selfdual degrees. At the very bottom we have the degree of the empty set and its dual. Right above we have the degree of a true clopen set, which is selfdual. This degree has two incomparable successors, the degrees of a true closed set and the degree of a true open set. There is a selfdual degree above them (not discussed), then another dual pair, namely the degree of a proper difference of open sets and its dual. Above them, another selfdual degree, then the degree of a proper ternary difference of open sets and its dual. This is the pattern through the $\underset{\sim}{\mathbf{\Delta}}_2^0$ sets, with selfdual degrees at (countable) limit ordinals.

§17. The determinacy of $G(A, B)$**.** Once the $\underset{\sim}{\mathbf{\Delta}}_2^0$ sets were mapped out, I faced the problem of solving the 'Post' problem for the $\mathcal{G}_\delta$ sets. Is there a topological criterion for $\mathcal{G}_\delta$-completeness? As it turns out, yes—a set is $\mathcal{G}_\delta$-complete if it is comeager on a perfect set (these are classical topological notions). Using this, it is possible to show that any proper $\mathcal{G}_\delta$ is complete. But then, like Luzin before me, I faced extending this result to the $\mathcal{F}_{\sigma\delta}$ sets, to the $\mathcal{G}_{\delta\sigma\delta}$ sets, and so on.

Fortunately, games allowed me to make another end run. It just involved examining a little more closely the relationship between the players in $G(A, B)$.

We saw that if player II has a winning strategy, then the complexity of A is comparable to that of B: it is either of the same complexity, or strictly less. But what if player I is the one with a winning strategy?

The game is not symmetric; the rules are looser for player II than for player I, because player II is allowed to pass. As a result, a strategy for player I cannot be specified by a simple monotonic function on finite sequences. We can avoid this minor complication by considering a 'fairer' game $G_L(A, B)$ in which player II cannot pass. Then winning strategies for both players are determined by monotonic finite sequence functions (strategies for player I are length preserving, strategies for player II must increase the length by exactly 1).

Now suppose again that player I has a winning strategy for $G(A, B)$. Since $G_L(A, B)$ is harder for player II, it follows that player I must also have a winning strategy for $G_L(A, B)$. This strategy corresponds to a monotonic function τ for which it is *not* the case that $\beta \in B \Leftrightarrow \tau^*(\beta) \in A$. Simple logic allows us to conclude that $\beta \in B \Leftrightarrow \tau^*(\beta) \in -A$. In other words, τ is a winning strategy for $G(B, -A)$.

Putting this all together, we see that if $G(A, B)$ is determinate (one of the players has a winning strategy), then either $A \leq B$ or $B \leq -A$. And if all such games are determinate, then $\leq$ is almost a linear order—given any A and B, A is comparable with either B or its complement.

It is not hard to see that this result—called the Semi Linear Ordering principle (SLO)—solves the 'Post' problem in its most general form. Suppose that $\mathcal{B}$ is an *initial class*—a collection of sets closed downwards under continuous preimage (so that B in $\mathcal{B}$ and $A \leq B$ implies A in $\mathcal{B}$). It follows directly that any proper element of $\mathcal{B}$—any set in $\mathcal{B}$ whose complement is not also in $\mathcal{B}$—is $\mathcal{B}$-complete. In particular, proper $\mathcal{F}_{\sigma\delta}$ sets are $\mathcal{F}_{\sigma\delta}$-complete and proper $\mathcal{G}_{\delta\sigma\delta}$ sets are $\mathcal{G}_{\delta\sigma\delta}$-complete. Once again I was spared intricate topological proofs.

Actually, I have simplified the narrative; soon after discovering the game (in late 1967, while Addison's seminar was still in progress) I realized that its determinateness had the dramatic consequences just described. In fact it took several days for me to convince myself that what is now known as "Wadge's Lemma" was correct. Initially I expected the degrees of sets of reals to be just as complicated as those of sets of natural numbers.

The picture that was emerging seemed too good to be true. But it was true—assuming, of course, that the appropriate $G(A, B)$ games are determinate.

§18. The Axiom of Determinacy. How could a game not be determinate? There are no ties, so must not someone win? In fact, the determinacy of infinite games does not follow from *a priori* reasoning. It is conceivable that given any strategy for player II, no matter how good, there is always a strategy for player I that beats it; and that in turn, given any strategy for player I, no matter how sophisticated, there is always a strategy for player II that outplays it. *A priori* reasoning guarantees only that at least one player has a collection S of strategies with the following property: given any strategy τ for the opponent, there is a strategy σ in S that defeats τ. There is no reason to think that this player can combine all the strategies in S into a single master strategy that uniformly defeats all comers.

This problem was recognized long ago, and Zermelo [Zer13] first proved that all finite games are determined. On the other hand, it is fairly easy, using the axiom of choice, to show that there must be a (Gale–Stewart) game that is not determined. This is hardly a satisfactory state of affairs.

One approach is to take the failure of determinacy to be yet one more implausible consequence of the Axiom of Choice (along with, for example, the Banach–Tarski paradox or the existence of nonmeasurable sets). In 1962 Mycielski and Steinhaus suggested [MS62] we drop (or weaken) AC and replace it with the Axiom of Determinacy (AD): every infinite (Gale–Stewart) game is determined. AD certainly has the intuitive plausibility required of an axiom; perhaps its consequences are more palatable. We have one very pleasant consequence at hand: AD implies that $\leq$ is (almost) a linear order.

The study of determinacy began in earnest in 1953 when Gale and Stewart formalized infinite games and showed that those with an open winning condition are determinate. By 1964, Morton Davis had shown [Dav64] that all $\mathcal{G}_{\delta\sigma}$ games are determinate. However, at the time I was working on my dissertation, no-one knew how far determinacy held (without assuming extra axioms). I would not have been surprised had it fizzled out at the third or fourth level of the Borel hierarchy. In 1968, just as I was beginning my research, encouraging news arrived: Martin showed [Mar70] that the existence of a measurable cardinal implied that all analytic (and hence all Borel) sets are determined. I therefore decided to assume Borel determinateness and embark on a detailed analysis of the degrees of Borel sets. My choice was vindicated in 1975, three years after finishing the research, when Martin [Mar75] proved Borel determinacy. Once again, too good to be true, but true nevertheless.

§19. Degree arithmetic. The structure of the $\mathbf{\Delta}^0_2$ sets suggested (correctly, as it turned out) that the degrees are semi-wellordered. That means that in principle, at least, we can define operations on the degrees that correspond to ordinal operations like successor, limit, sum, and product. Do these operations have game characterizations? I was able to show that many of them do—in fact enough of them to allow me to find the exact order type of the degrees of the Borel sets.

The simplest operator acts as a (countable) least upper bound. Given B_0, $B_1, B_2, \ldots$, we define $B_0 \sqcup B_1 \sqcup B_2 \sqcup \cdots$ to be

$$\{i\beta\}_{i\in\omega,\beta\in B_i}$$

We can think of the game $G(A, B_0 \sqcup B_1 \sqcup B_2 \sqcup \cdots)$ as follows: Player I plays as usual, but player II's first move i is not part of his sequence. Instead, it is used to single out B_i, so that after this first move the game is like $G(A, B_i)$, with player II's first move a pass. It was not hard to verify that $\sqcup$ induces a *least upper bound* operation on degrees.

This operation had in fact been studied by one of my predecessors, John Barnes [Bar65]. He and Addison called it the *Kalmar union*. Barnes showed that the clopen sets are generated by closing $\{\varnothing, {}^{\omega}\omega\}$ out under the Kalmar union, forming in the process a hierarchy with Ω levels.

Probably the most surprising result is that there is a simple binary addition operation on sets that induces an addition operation on degrees, one that corresponds to ordinal addition. Given two subsets B and C of the Baire space, let $B + C$ be the set

$$\{\gamma + 1\}_{\gamma\in C} \cup \{(s+1)0\beta\}_{s\in\mathrm{Sq},\beta\in B}$$

(where $\gamma+1$ and $s+1$ are the sequences formed by adding 1 to each component of γ and s respectively) . In other words, an element of $B+C$ is either an element of C with 1 added to each component, or a finite sequence of nonzero numbers followed by a 0 followed by an element of B.

Addition of sets induces a corresponding operation on degrees, which does indeed act like addition, provided the first argument is a *lub degree*: a least upper bound of strictly simpler degrees. If b is such a degree, we can show that for any degree d strictly greater than b, $d = b + c$ for some degree c (assuming determinateness, unless we restrict the result to Borel degrees). In particular (as we shall soon see), the ordinal associated with d is the sum of the ordinals associated with b and c.

In the game definition of addition 0 is used as a coding device to indicate a switch from C to B. Given any A, we can describe the game $G(A, B+C)$ as follows: it is like $G(A, C)$ except that player II has the option, at any point, of taking all his moves back, at which point the rules switch to those of $G(A, B)$. Player II does not have to exercise this option, and can do it at most once. Player I is not allowed to take his own moves back.

In particular, $G(A, B+B)$ is like $G(A, B)$ except that player II has the option of taking all his moves back once. Similarly, $B+B+B$ is B with two take-back coupons, $B+B+B+B$ offers three, and so on. The most powerful operator I found is $\sharp$—set $B^\sharp$ can be thought of as $B+B+B+\cdots$. With this set, player II can take all his moves back as often as he wants. He can even do so infinitely often, although in that case his sequence is considered as lying outside $B+B+B+\cdots$. I was able to show that this operation corresponded to multiplying (on the left) by Ω (not by ω, basically because the Baire space is not compact). In other words, for suitable degrees b, if degree a is less than both $b^\sharp$ and its dual, then a is less than $b \cdot \mu$ for some countable ordinal μ.

If we begin with the two minimal degrees, namely $\varnothing$ and its dual, we can, using the degree operations just described, generate an initial segment of the degrees of length Ω^Ω. In this initial segment, every dual pair is followed by a single selfdual degree and vice versa. At limit ordinals of cofinality ω we find a single selfdual degee, while at limit ordinals of cofinality Ω we have a dual pair.

§20. (α, β)-homeomorphisms. I have characterized the operations as "powerful" but in fact they are very weak. The inital segment of order type Ω^Ω just described takes us through the $\mathbf{\Delta}^0_3$ sets. Conceivably we could go further with more powerful operators, but it was never clear to me what these operators could be.

A few years earlier (in the 1950s, most likely) Kuratowski faced a similar problem extending Hausdorff's result that the difference hierarchy over the open sets exhausts the $\mathbf{\Delta}^0_2$ sets. He wanted to extend the result to all levels of the

Borel hierarchy—to show that for any positive μ, the difference hierarchy over the $\underset{\sim}{\mathbf{\Sigma}}^0_\mu$ sets exhausts the $\underset{\sim}{\mathbf{\Delta}}^0_{\mu+1}$ sets. Hausdorff's original result used adjoins and residues which, as we have seen, are closely related to game characterizations. However it is not at all clear that there are notions analogous to those of residue and adjoin at higher levels of the Borel hierarchy.

Instead, Kuratowski developed [Kur58] a very general technique for 'lifting' results from lower to higher levels of the Baire hierarchy: (α, β)-homeomorphisms. He used in particular $(\mu, 0)$-homeomorphisms, which are continuous in one direction but limits of limits of limits of ... of continuous functions in the other. More precisely, a function is class 0 iff it is continuous; class 1 if it is the (pointwise) limit of an ω-sequence of continuous functions; class 2 iff it is the limit of an ω-sequence of class-1 functions; and in general, of class μ iff it is the limit of an ω-sequence of functions each of which is of class ν for some $\nu < \mu$.

These homeomorphisms are a kind of point set microscope that blows up the complexity of a set. My plan was to measure the effect of this kind of magnification operation by discovering the corresponding effect on degrees. In other words, suppose R consists of all sets whose degrees are among the first κ degrees. When we enlarge R under the microscope, the result is (hopefully) all sets of degree less than λ for some ordinal λ possibly much larger than κ. If we can determine how λ depends on κ, we can lift our picture of the degrees of the $\underset{\sim}{\mathbf{\Delta}}^0_3$ sets to give us a corresponding picture of the degrees of the Borel sets.

§21. The expansion operations. Kuratowski's basic result was that every $\underset{\sim}{\mathbf{\Sigma}}^0_{1+\mu}$ open set is $(\mu, 0)$-homeomorphic to a $\underset{\sim}{\mathbf{\Sigma}}^0_1$ set on a closed set. More precisely, given any $\underset{\sim}{\mathbf{\Sigma}}^0_{1+\mu}$ set H, there is a $\underset{\sim}{\mathbf{\Sigma}}^0_1$ set G, a closed set E, and a one-one class-μ map f from ${}^\omega\omega$ onto E such that f^{-1} is continuous and $H = f^{-1}(G)$. (Unfortunately we cannot always take E to be ${}^\omega\omega$.) Let us adopt the following notation from my dissertation: given any class $\mathcal{H}$ of subsets of ${}^\omega\omega$, the μth *expansion* $\mathcal{H}^\mu$ of $\mathcal{H}$ is the collection of all sets that are, as above, $(\mu, 0)$-homeomorphic to an element of $\mathcal{H}$ modulo a closed set. Kuratowski's basic result, then, is that $\mathcal{G}^\mu$ is $\underset{\sim}{\mathbf{\Sigma}}^0_{1+\mu}$.

Kuratowski was able, with the help of $(\mu, 0)$-homeomorphisms, to extend the Hausdorff difference hierarchy result to all levels of the Borel hierarchy. In fact he proved a stronger result, namely that the μth expansion of a particular level of the difference hierarchy over $\underset{\sim}{\mathbf{\Sigma}}^0_1$ is the corresponding level of the difference hierarchy over $\underset{\sim}{\mathbf{\Sigma}}^0_{1+\mu}$.

We can express this more concisely by letting ∂_ν denote the νth set difference operation, and by extending the classical notation so that $\mathcal{G}_{\partial_\nu}$ is the set of all ν-ary differences of open sets. Then Kuratowski's result is that

$$(\mathcal{G}_{\partial_\nu})^\mu = (\mathcal{G}^\mu)_{\partial_\nu}.$$

This result is plausible but by no means obviously true. To see the problem, consider the simplest case: showing that $(\mathcal{G}_{\partial_2})^1$ is $(\mathcal{G}^1)_{\partial_2}$. This amounts to showing that any difference of $\underset{\sim}{\mathbf{\Sigma}}^0_2$ sets is $(\mu, 0)$-homeomorphic to the difference of two $\underset{\sim}{\mathbf{\Sigma}}^0_1$ sets (modulo a closed set). Let H_0 and H_1 be the $\underset{\sim}{\mathbf{\Sigma}}^0_2$ sets. We know that H_0 is $(\mu, 0)$-homeomorphic to an some open set G_0, and H_1 is $(\mu, 0)$-homeomorphic to some open set G_1. It does *not*, however, follow that $H_1 - H_0$ is $(\mu, 0)$-homeomorphic to $G_1 - G_0$. The problem is that the homeomorphism reducing H_0 to G_0 is not necessarily the same as the one reducing H_1 to G_1. To make the argument work, we need to be able to reduce H_0 and H_1 to G_0 and G_1 simultaneously (uniformly) using a *single* $(\mu, 0)$-homeomorphism.

This is in fact possible, and follows from a result of Kuratowski much stronger than the one cited above; namely, that any ω-sequence of $\underset{\sim}{\mathbf{\Sigma}}^0_{1+\mu}$ sets can be uniformly reduced to some ω-sequence of $\underset{\sim}{\mathbf{\Sigma}}^0_1$ sets by a (single) $(\mu, 0)$-homeomorphism (as usual, modulo some closed set).

§22. The ordinal jump functions. Our basic strategy is to measure the power of the expansion operators in terms of ordinal functions. The selfdual Borel degrees are well ordered, and so can be enumerated by a sequence. Let Ξ be the domain of this sequence. Of course this sequence omits the nonselfdual pairs, but since they alternate with selfdual degrees, their omission does not (at least at limit ordinals) affect the order type. Thus Ξ cleary deserves to be called the "order type" of the Borel degrees, and by 1971 or so my main objective was to define or at least characterize Ξ.

Let r_λ $(\lambda \in \Xi)$ be the λth selfdual Borel degree. For technical reasons (to simplify the statements of the results) we begin the enumeration at $\lambda = 1$, so that r_1 is the collection of all clopen sets that are neither empty nor coempty. It follows easily from the results on degree arithmetic that $r_{\kappa+\lambda} = r_\kappa + r_\lambda$ for any positive κ and λ in Ξ.

Next, we define a corresponding Ξ-sequence of initial classes and study the result of expanding these classes. For any λ in Ξ, $\mathcal{R}_\lambda$ is the collection of Borel sets whose degrees are strictly *less* than $r_{1+\lambda}$.

Thus $\mathcal{R}_0$ is $\{0, {}^\omega\omega\}$, $\mathcal{R}_1$ is the class of sets that are open or closed; and $\mathcal{R}_2$ consists of sets that are the difference of two open sets, or whose complement is. It follows from what was said earlier that $\mathcal{R}_\Omega$ is $\mathcal{F}_\sigma \cup \mathcal{G}_\delta$, $\mathcal{R}_{\Omega^2}$ consists of sets that are the difference of two $\mathcal{G}_\delta$, or whose complements are; and that R_{Ω^Ω} is $\mathcal{F}_{\sigma\delta} \cup \mathcal{G}_{\delta\sigma}$.

The plan is to show that the $\mathcal{R}$ sequence is closed under expansion; to show that for each μ in Ω there is an ordinal 'jump' function ϑ_μ with domain Ξ such that for any λ in Ξ,

$$(\mathcal{R}_\lambda)^\mu = \mathcal{R}_{\vartheta_\mu(\lambda)}$$

This is rather difficult, mainly because the vital properties of the expansion operation (as just described) do not (obviously) extend from the $\underset{\sim}{\Sigma}^0_{1+\eta}$ classes to arbitrary collections of subsets of the Baire space. For example, if $\mathcal{H}$ is $\underset{\sim}{\Sigma}^0_{1+\eta}$ for some η, it follows from what we have said that

$$\mathcal{H}^{\nu+\mu} = (\mathcal{H}^{\mu})^{\nu}$$

This equation would allow us to prove that in general

$$\vartheta_{\nu+\mu}(\lambda) = \vartheta_{\nu}(\vartheta_{\mu}(\lambda))$$

Unfortunately, there was (and as far as I know, still is) no reason to think that the second last equation holds for arbitrary subsets $\mathcal{H}$ of $\wp({}^{\omega}\omega)$, even if we add the assumption that $\mathcal{H}$ is an initial class (closed under continuous preimage). It is not hard to see that the inclusion

$$\mathcal{H}^{\nu+\mu} \subseteq (\mathcal{H}^{\mu})^{\nu}$$

is true—it follows directly from the fact that the composition of a class μ function with a class ν function is class $\nu+\mu$. However, the opposite inclusion,

$$(\mathcal{H}^{\mu})^{\nu} \subseteq \mathcal{H}^{\nu+\mu}$$

is far from obvious. It suggests that a function of class $\nu + \mu$ can always be factored as the composition of two functions of class μ and ν respectively.

§23. Boolean set operations. Fortunately, there is a large collection of classes $\mathcal{H}$ for which the cited property (and others) do hold—those that can be defined as the range of values of a *Boolean set operation* applied to an arbitrary ω-sequence of open sets.

A Boolean set operation is an operation on subsets of the Baire space that is, roughly speaking, purely set-theoretical; it makes no use of the underlying topology or other structure. Countable union and intersection are Boolean, as are the difference operations ∂_{μ}. On the other hand, the closure and interior operations are not.

Our result about expanding differences of open sets can be understood as follows: the 1-expansion operation commutes with the binary difference operation ∂_2 on open sets.

Our result immediately extends to μ-expansion for any μ. It should be clear that, in the place of binary difference, we could put any other Boolean (purely set-theoretic) operation of finite or countable arity. For example, we could prove that a set is a countable union of 8-ary differences of $\underset{\sim}{\Sigma}^0_2$ sets iff it can be 1-reduced to a countable union of 8-ary differences of open sets.

This notion, of a generalized "set-theoretic operation", was first formalized by Kantorovich and Liveson in 1932 [KL32]. (They called these operations "analytical", but that term is overused and we prefer "Boolean" as more appropriate.)

We say that an I-ary set operation Γ over the Baire space (a map from ${}^I\wp({}^\omega\omega)$ to $\wp({}^\omega\omega)$) is *Boolean* iff, roughly speaking, membership of any α in $\Gamma(H)$ is determined once we know, for each i, whether or not α is in H_i. More precisely, Γ is Boolean iff there is a subset $\mathcal{K}$ of $\wp(I)$ such that for any α,

$$\alpha \in \Gamma(H) \Leftrightarrow \{i : \alpha \in H_i\}_{i \in I} \in \mathcal{K}$$

We can think of $\mathcal{K}$ as being the 'truth table' of Γ.

For example, complementation is a 1-ary Boolean operation, difference is a 2-ary operation, and countable union is an ω-ary Boolean operation. If I is finite, the I-ary Boolean operations are exactly those that can be constructed using union, intersection and complementation. The first Boolean operation to be studied (other than finite ones and countable union and intersection) was Suslin's operation $\mathcal{A}$ [Sus17]; in our notation,

$$\mathcal{A}(F) = \bigcup_{\alpha \in {}^\omega\omega} \bigcap_{k \in \omega} F_{\alpha|k}$$

for any $F \in {}^{\mathrm{Sq}}({}^\omega\omega)$. (In our terminology, operation $\mathcal{A}$ is Sq-ary.)

Kuratowski's uniform reduction theorem allows us to conclude, then, that expansion commutes with any countable Boolean set operation over the open sets. To express this principle as an equation, suppose that Γ is an I-ary Boolean set operation and $\mathcal{H}$ a collection of subsets of the Baire space; we define $\mathcal{H}_\Gamma$ to be the collection of all sets produced by applying Γ to some I-ary family of H sets. More precisely,

$$\mathcal{H}_\Gamma = \{\Gamma(H)\}_{H \in {}^I\mathcal{H}}$$

Then Kuratowski's uniform reduction result allows us to prove that

$$(\mathcal{G}_\Gamma)^\mu = (\mathcal{G}^\mu)_\Gamma$$

(Notice that even Kuratowski's uniform reduction theorem is a commutativity result; it says that expansions commute with countable product.)

§24. ${}^\omega\mathcal{G}$-Boolean classes. The ${}^\omega\mathcal{G}$-Boolean classes are those that are defined or generated by a Boolean set operation in the way that the class of $\mathcal{G}_\delta$ sets is defined/generated by the countable-union operation. Their importance lies in the fact that they possess the properties we need.

A class is ${}^\omega\mathcal{G}$-Boolean iff it is the range of an ω-ary Boolean set operation applied to sequences of open sets; in other words, iff it is of the form $\mathcal{G}_\Gamma$ for some ω-ary Γ. For example, the class of $\underset{\sim}{\mathbf{\Pi}}^0_2$ sets is the range of the countable-intersection operation δ applied to ω-sequences of open sets. In our new notation, the class $\underset{\sim}{\mathbf{\Pi}}^0_2$ is $\mathcal{G}_\delta$, which (happily) coincides with the classical notation.

Suslin proved that any analytic set can be produced by applying the operation $\mathcal{A}$ to an Sq-family of closed sets; in our notation, that the class of $\boldsymbol{\Sigma}^1_1$ sets is $\mathcal{F}_{\mathcal{A}}$.

If $\mathcal{H}$ is ${}^{\omega}\mathcal{G}$-Boolean, our additive expansion result then follows directly from the fact that it holds for the open sets, and that expansions commute with Boolean operations and countable product: if $\mathcal{H} = \mathcal{G}_\Gamma$, then

$$\mathcal{H}^{\nu+\mu} = (\mathcal{G}_\Gamma)^{\nu+\mu} = (\mathcal{G}^{\nu+\mu})_\Gamma = \left((\mathcal{G}^\mu)^\nu\right)_\Gamma = \left((\mathcal{G}^\mu)_\Gamma\right)^\nu = \left((\mathcal{G}_\Gamma)^\mu\right)^\nu = (\mathcal{H}^\mu)^\nu$$

Incidentally, our notational shortcuts obscure the fact we are using Kuratowski's uniform reduction result. For example, spelling out the omitted steps, we have

$$(\mathcal{G}_\Gamma)^\mu = \left(\{\Gamma(G)\}_{G\in{}^\omega\mathcal{G}}\right)^\mu = \{\Gamma(H)\}_{H\in({}^\omega\mathcal{G})^\mu} = \{\Gamma(H)\}_{H\in{}^\omega(\mathcal{G}^\mu)} = (\mathcal{G}^\mu)_\Gamma$$

and the second last step uses uniform reduction.

The ${}^{\omega}\mathcal{G}$-Boolean classes have many other pleasant (and easily verified) properties. They are all nonselfdual initial classes with complete elements. They are closed under countable product and expansion, and furthermore, if $\mathcal{H}$ is ${}^{\omega}\mathcal{G}$-Boolean, then so is $\mathcal{H}_\Gamma$. (Since the class $\mathcal{F}$ of closed sets is ${}^{\omega}\mathcal{G}$-Boolean, it follows that the class of analytic sets, which is $\mathcal{F}_{\mathcal{A}}$, is also ${}^{\omega}\mathcal{G}$-Boolean.)

The collection of ${}^{\omega}\mathcal{G}$-Boolean classes is also very large; both $\boldsymbol{\Sigma}^0_{1+\mu}$ and $\boldsymbol{\Pi}^0_{1+\mu}$ are ${}^{\omega}\mathcal{G}$-Boolean for any countable μ, as are $\boldsymbol{\Sigma}^1_1$ and $\boldsymbol{\Pi}^1_1$ and all the levels of the corresponding difference hierarchy. In fact, it is hard to think of any nonselfdual initial class of Borel sets that is not ${}^{\omega}\mathcal{G}$-Boolean; for good reason, because, as we shall soon see, there are none.

§25. Separated and partitioned unions. Recall that our strategy is to show that in general $(\mathcal{R}_\kappa)^\mu$ is of the form $\mathcal{R}_\lambda$, and to calculate λ as a function of κ and μ. To carry out these calculations, we need to know that $\mathcal{R}_\kappa$ is well behaved in a strong sense, and to prove that $\mathcal{R}_\lambda$ is therefore well behaved as well. The notion of "well behaved" (called "regular" in the dissertation) includes the requirement that $\mathcal{R}_\kappa$ be a union of ${}^{\omega}\mathcal{G}$-Boolean classes, that the degrees of sets in $\mathcal{R}_\kappa$ are semi-wellordered, and that the initial class of any nonselfdual set in $\mathcal{R}_\kappa$ be ${}^{\omega}\mathcal{G}$-Boolean.

This last requirement is very strong, because it implies that, in particular, for any nonselfdual degree a of a set in $\mathcal{R}_\kappa$, the initial class $\mathrm{In}(a)$ is of the form $\mathcal{G}_\Gamma$ for some ω-ary Boolean operation Γ. In other words, it implies that every such nonselfdual degree is in a sense defined by some Boolean set operation.

We prove this by induction, and to do this we show that degrees that are definable in this way are closed under degree operations and expansion. Closure under expansion follows from our commutativity results, but proving closure under degree operations involves finding classical, non-game characterizations of the initial classes of arithmetic combinations of degrees.

Fortunately, the classical descriptive set theorists had already identified the required notions—namely, *separated* and *partitioned* unions. In general, the union of a sequence of sets, or even just two sets, can be much more complex than any of the sets involved—there is no way to predict this. The problem, it seems, is that the two sets may be very close together, which complicates determining membership in their union. However, if they are far enough apart, the problem is manageable. And we can limit the 'distance' between them by requiring that there exist a simple set that includes one and whose complement includes the other. We can extend this notion to countable unions by requiring a partition of the space into simple sets, with each union set included in its own partition.

The simplest case is that in which the separating sets are clopen; a $\underset{\sim}{\mathbf{\Delta}}^0_1$ *partitioned union* of sets from a class $\mathcal{C}$ is a set of the form

$$C_0 \cup C_1 \cup C_2 \cup \cdots$$

where each C_i is in $\mathcal{C}$, and there is a sequence $D_0, D_1, D_2, \ldots$ of disjoint clopen sets such that each C_i is a subset of the corresponding D_i, and the union of the D sequence is ${}^\omega\omega$.

In our notation, the collection of all sets of this form is $\mathrm{Pt}_0(\mathcal{C})$ (in general $\mathrm{Pt}_\mu(\mathcal{C})$ is defined similarly, with the sets forming the partition required only to be $\underset{\sim}{\mathbf{\Delta}}^0_{1+\mu}$). In a sense, we have already met the operation Pt_0; it is the *Kalmar closure* of $\mathcal{C}$. In other words, $\mathrm{Pt}_0(\mathcal{C})$ is (given simple constraints on $\mathcal{C}$) the least class containing $\mathcal{C}$ that is closed under Kalmar union. The Kalmar union, as we have already seen, is closely connected to the degree least upper bound operation, so we already have a link between the classical and game-based notions.

If we want to go beyond the lub operations, we need two operators that take us just beyond Pt_0. In general, a $\underset{\sim}{\mathbf{\Sigma}}^0_{1+\mu}$*-separated union* of sets in $\mathcal{C}$ is a set of the form

$$C_0 \cup C_1 \cup C_2 \cup \cdots$$

where each C_i is in $\mathcal{C}$, and there is a sequence $G_0, G_1, G_2, \ldots$ of disjoint $\underset{\sim}{\mathbf{\Sigma}}^0_{1+\mu}$ sets with each C_i a subset of the corresponding G_i. In our notation, this class is $\mathrm{Sp}^+_\mu(\mathcal{C})$. We also define a dual operator, Sp^-_μ, defined like Sp^+_μ except that the complement of the union of the separating sets is added to the union of the A_i's. It is easy to see that

$$\mathrm{Sp}^-_\mu(\mathcal{C}) = \mathrm{Sp}^+_\mu(\mathcal{C}^-)^-$$

There is a simple and direct connection between the operators Pt_0, Pt_1, Sp^+_0 and Sp^+_1 on the one hand, and the lub, addition, $\sharp$, and ordinal multiplication operators on the other hand.

Suppose first that b is a selfdual degree, the lub of the nonselfdual degrees $a_0, a_1, a_2, \ldots$, and let $\mathcal{C}$ be $\bigcup_{i\in\omega} \mathrm{In}(a_i)$. As we already indicated, $\mathrm{In}(b)$ is the Kalmar closure of $\mathcal{C}$; which, in turn, is $\mathrm{Pt}_0(\mathcal{C})$. Using similar (but more complex) techniques, we can show that the $\mathrm{In}(b+1)$ is $\mathrm{Sp}_0^+(\mathcal{C})$, that $\mathrm{In}(b^\sharp)$ is $\mathrm{Sp}_1^+(\mathcal{C})$, and that $\mathrm{Pt}_1(\mathcal{C})$ is $\bigcup_{\mu\in\Omega}\mathrm{In}(b\cdot\mu)$. These proofs make use of game arguments and a vital lemma that to the effect that $\mathrm{Sp}_1^+(\mathcal{C})\cap\mathrm{Sp}_1^-(\mathcal{C})$ is $\mathrm{Pt}_1(\mathcal{C})$, which is in turn the result of closing $\mathcal{C}$ out under Sp_0^+ and Sp_0^-.

We can express this concisely by defining $\mathrm{Sp}_0(\mathcal{C})$ to be $\mathrm{Sp}_0^+(\mathcal{C})\cup\mathrm{Sp}_0^-(\mathcal{C})$, and letting $\mathrm{Sp}_0^\mu(\mathcal{C})$ denote the μth stage in closing $\mathcal{C}$ out under Sp_0. Then

$$\bigcup_{\mu\in\Omega}\mathrm{In}(b\cdot\mu)=\mathrm{Sp}_1^+(\mathcal{C})\cap\mathrm{Sp}_1^-(\mathcal{C})=\mathrm{Pt}_1(\mathcal{C})=\bigcup_{\mu\in\Omega}\mathrm{Sp}_0^\mu(\mathcal{C})$$

§26. Determining ϑ_1. We are finally in a position to calculate the ordinal jump functions, starting with ϑ_1. As with everything else, we do it by induction. Since $(\mathcal{R}_0)^1=\{\varnothing,{}^\omega\omega\}^1=\{\varnothing,{}^\omega\omega\}=\mathcal{R}_0$, we see that $\vartheta_1(0)=0$.

Now suppose that $\vartheta_1(\kappa)=\lambda$. What is $\vartheta_1(\kappa+1)$? By the definition of ϑ_1, we know $(\mathcal{R}_\kappa)^1=\mathcal{R}_\lambda$. We need to calculate $(\mathcal{R}_{\kappa+1})^1$. The class $\mathcal{R}_{\kappa+1}$ is the collection of all sets of degree less than $r_{\kappa+1}$ (to simplify things, we assume κ is infinite). Since $r_{\kappa+1}$ is clearly the lub of $r_\kappa+1$ and $r_\kappa+1^-$, $\mathcal{R}_{\kappa+1}$ must be the union of the initial classes of these two degrees, and this in turn (as we have already seen) is $\mathrm{Sp}_0(\mathrm{In}(r_\kappa))$, and since $\mathrm{In}(r_\kappa)$ is $\mathrm{Pt}_0(\mathcal{R}_\kappa)$, we see that

$$\mathcal{R}_{\kappa+1}=\mathrm{Sp}_0(\mathcal{R}_\kappa)$$

(This is true even when κ is finite.)

Taking the 1-expansion of both sides,

$$\begin{aligned}(\mathcal{R}_{\kappa+1})^1=\left(\mathrm{Sp}_0(\mathcal{R}_\kappa)\right)^1&=\mathrm{Sp}_1\left((\mathcal{R}_\kappa)^1\right)=\mathrm{Sp}_1(\mathcal{R}_\lambda)\\&=\mathrm{Sp}_1^+(\mathcal{R}_\lambda)\cup\mathrm{Sp}_1^-(\mathcal{R}_\lambda)=\mathrm{In}(r_\lambda^\sharp)\cup\mathrm{In}\left((r_\lambda^\sharp)^-\right)\end{aligned}$$

However, $r_\lambda^\sharp$ and its dual lie (by previous results) just below $r_{\lambda\cdot\Omega}$. Putting it all together,

$$(\mathcal{R}_{\kappa+1})^1=\mathrm{Sp}_1\left((\mathcal{R}_\kappa)^1\right)=\mathrm{Sp}_1(\mathcal{R}_\lambda)=\mathcal{R}_{\lambda\cdot\Omega}$$

and this in turn implies that $\vartheta_1(\kappa+1)=\vartheta(\kappa)\cdot\Omega$. Once we verify that ϑ_1 is continuous (which we do) the conclusion is that, for positive κ,

$$\vartheta_1(\kappa)=\Omega^\kappa$$

It should be clear that we are omitting many important details. In particular, as part of our induction we have to show that the nonselfdual degrees of sets in $(\mathcal{R}_{\kappa+1})^1$ are ${}^\omega\mathcal{G}$-Boolean. This follows from our 'classical' characterizations of the initial classes of the results of the degree operations, but not directly.

§27. The arithmetic degrees. Once I determined ϑ_1, I could calculate ϑ_n for finite n using the additive property $\vartheta_{\nu+\mu}(\kappa) = \vartheta_\nu(\vartheta_\mu(\kappa))$. For example, $\vartheta_2(\kappa) = \vartheta_1(\vartheta_1(\kappa)) = \Omega^{(\vartheta_1(\kappa))} = \Omega^{\Omega^\kappa}$, $\vartheta_3(\kappa) = \vartheta_2(\vartheta_1(\kappa)) = \Omega^{\Omega^{\Omega^\kappa}}$, and in general

$$\vartheta_n(\kappa) = \Omega^{\Omega^{\Omega^{\cdot^{\cdot^{\cdot^\kappa}}}}}$$

there being n Ω's in the ladder of exponents.

Now that we know ϑ_n for every finite n, we have a nearly complete picture of the arithmetic degrees—the degrees of sets appearing at some finite level of the Baire hierarchy.

As we have already seen, $\mathcal{R}_0$ is $\{\varnothing, {}^\omega\omega\}$, $\mathcal{R}_1$ is $\mathcal{F}_\sigma \cup \mathcal{G}_\delta$, and $\mathcal{R}_2$ is the class of sets that are differences of open sets, or duals of such differences. Let us define $\mathcal{A}^\pm$ to be $\mathcal{A} \cup \mathcal{A}^-$. Then the first components of the $\mathcal{R}$ sequence are as follows

$$\begin{aligned} \mathcal{R}_0 &= 0^\pm \\ \mathcal{R}_1 &= \mathcal{G}^\pm \\ \mathcal{R}_2 &= (\mathcal{G}_{\partial_2})^\pm \\ \mathcal{R}_3 &= (\mathcal{G}_{\partial_3})^\pm \end{aligned}$$

and in general for any countable μ, $\mathcal{R}_\mu$ is the union of both sides of level μ of the difference hierarchy over the open sets.

If we take the 1-expansion of both sides of these equations, we see that $\mathcal{R}_\Omega$ is $(\mathcal{F}_\sigma)^\pm$ and in general $\mathcal{R}_{\Omega^\mu}$ is the union of both sides of the μth level of the difference hierarchy over the $\mathcal{F}_\sigma$ sets. Continuing up the Borel hierarchy, for any finite n, $(\underset{\sim}{\mathbf{\Sigma}}^0_{1+n})^\pm$ is

$$\mathcal{R}\left(\Omega^{\Omega^{\cdot^{\cdot^{\cdot^\Omega}}}}\right)$$

where there are n Ω's in the ladder of exponents; and the the union of both sides of the κth level of the difference hierarchy over $\underset{\sim}{\mathbf{\Sigma}}^0_{1+n}$ is

$$\mathcal{R}\left(\Omega^{\Omega^{\cdot^{\cdot^{\cdot^{\Omega^\kappa}}}}}\right)$$

where again there are n Ω's in the ladder of exponents. (Note that for infinite λ, the order type of the collection of nonselfdual degrees of sets in $\mathcal{R}_\lambda$ is λ itself.)

Throwing together all sets at finite levels of the Borel hierarchy gives us an ordinal that could be written

$$\Omega^{\Omega^{\cdot^{\cdot^{\cdot}}}}$$

To define this and subsequent ordinals more precisely, we must use the sequence of so-called "epsilon numbers". The epsilon numbers are the fixed points of exponentiation by ω. The first, ε_0, is the limit of the sequence $\omega, \omega^\omega, \omega^{\omega^\omega}, \ldots$. There are uncountably many countable fixed points of exponentiation-by-ω so that Ω is the Ωth such fixed point; in other words, $\Omega = \varepsilon_\Omega$. The collection of degrees of nonselfdual sets at finite levels of the Borel hierarchy therefore has order type $\varepsilon_{\Omega+1}$, and

$$\underset{\sim}{\mathbf{\Delta}}^0_{(\omega)} = \bigcup_{n\in\omega} \underset{\sim}{\mathbf{\Delta}}^0_{1+n} = \mathcal{R}_{\varepsilon_{\Omega+1}}$$

§28. Luzin's problem. There are, however, many sets in $\underset{\sim}{\mathbf{\Delta}}^0_\omega$ that are not in $\underset{\sim}{\mathbf{\Delta}}^0_{(\omega)}$ (not in $\underset{\sim}{\mathbf{\Delta}}^0_{1+n}$ for any n). At this point I was stumped, because the classical descriptive set theorists never found a hierarchy for the $\underset{\sim}{\mathbf{\Delta}}^0_\omega$ sets—the difference hierarchy over finite level sets collapses. Luzin himself noticed this gap and declared the problem worth studying; although as far as I could tell, no one had done so (successfully) before me.

Fortunately my study of the degree operations had allowed me to make a good guess, using partitioned and separated unions.

It is not hard to check that the $\underset{\sim}{\mathbf{\Delta}}^0_\omega$ sets are closed under $\underset{\sim}{\mathbf{\Delta}}^0_{1+n}$ partitioned unions for all n. Does closing under these unions exhaust the $\underset{\sim}{\mathbf{\Delta}}^0_\omega$ sets? I thought so. It would make a great story if I could say I discovered an elegant game proof of this, but that is not what happened. Instead, I looked very carefully at the classical results concerning (α, β)-homeomorphisms and discovered that the result for the $\underset{\sim}{\mathbf{\Delta}}^0_\omega$ sets could be obtained by 'lifting' the result that the clopen sets are generated by closing under Kalmar union.

For any class $\mathcal{K}$, define $\mathrm{Sp}_{(\omega)}(\mathcal{K})$ to be

$$\mathrm{Sp}_0(\mathcal{K}) \cup \mathrm{Sp}_1(\mathcal{K}) \cup \mathrm{Sp}_2(\mathcal{K}) \cup \cdots$$

and let $\mathcal{A}$ be the class $\underset{\sim}{\mathbf{\Delta}}^0_{(\omega)}$ (also called the class of *arithmetic* sets). My conjecture, which I verified, was that the class of $\underset{\sim}{\mathbf{\Delta}}^0_\omega$ sets is the union of the Ω-chain

$$\mathrm{Sp}_{(\omega)}(\mathcal{A}) \subset \mathrm{Sp}_{(\omega)}(\mathrm{Sp}_{(\omega)}(\mathcal{A})) \subset \mathrm{Sp}^3_{(\omega)}(\mathcal{A}) \subset \cdots \subset \mathrm{Sp}^\mu_{(\omega)}(\mathcal{A}) \subset \cdots$$

To calculate the ordinal to which this class corresponds, we need to calculate the power of $\mathrm{Sp}_{(\omega)}$ in terms of ordinals. To simplify the notation, let υ be $\varepsilon_{\Omega+1}$, so that, as we just saw, $\mathcal{A} = \mathcal{R}_\upsilon$. Our original result, proved using degrees, is that Sp_0 increases the $\mathcal{R}$ index by one; thus

$$\mathrm{Sp}_0(\mathcal{A}) = \mathcal{R}_{\upsilon+1}$$

Now take the n expansion of both sides (n finite); on the left we have

$$\mathrm{Sp}_0(\mathcal{A})^n = \mathrm{Sp}_n(\mathcal{A}^n) = \mathrm{Sp}_n(\mathcal{A})$$

because the arithmetic sets are clearly closed under n-expansion. On the other side, we have

$$(\mathcal{R}_{\upsilon+1})^n = \mathcal{R}_{\vartheta_n(\upsilon+1)} = \mathcal{R}\Big(\Omega^{\Omega^{\cdot^{\cdot^{\cdot^{\Omega^{\upsilon+1}}}}}}\Big)$$

with n Ω's in the ladder of exponents. Thus $\mathrm{Sp}_{(\omega)}(\mathcal{A})$ is $\mathcal{R}_{\upsilon'}$ where υ' is the limit of the sequence

$$\upsilon + 1, \Omega^{\upsilon+1}, \Omega^{\Omega^{\upsilon+1}}, \dots$$

and this is easily seen to be the next epsilon number (fixed point of exponentiation-by-Ω) after υ. Since υ is $\varepsilon_{\Omega+1}$, υ' must be $\varepsilon_{\Omega+2}$. What we have just described is only the first step in closing the arithmetic sets out under $\mathrm{Sp}_{(\omega)}$; the second step corresponds to the ordinal $\varepsilon_{\Omega+3}$, the third to the ordinal $\varepsilon_{\Omega+4}$, and in general the $\underset{\sim}{\mathbf{\Delta}}^0_\omega$ sets are exhausted by the hierarchy

$$\mathcal{A} = \mathcal{R}_{\varepsilon_{\Omega+1}} \subset \mathcal{R}_{\varepsilon_{\Omega+2}} \subset \mathcal{R}_{\varepsilon_{\Omega+3}} \subset \cdots \subset \mathcal{R}_{\varepsilon_{\Omega+\mu}} \subset \cdots$$

The limit of the ordinal sequence

$$\varepsilon_{\Omega+1}, \varepsilon_{\Omega+2}, \varepsilon_{\Omega+3}, \dots, \varepsilon_{\Omega+\mu}, \dots$$

is $\varepsilon_{\Omega+\Omega}$. This last ordinal is therefore the order type of the (nonselfdual) degrees of $\underset{\sim}{\mathbf{\Delta}}^0_\omega$ sets, and

$$\underset{\sim}{\mathbf{\Sigma}}^0_\omega \cup \underset{\sim}{\mathbf{\Pi}}^0_\omega = \mathcal{R}_{\varepsilon_{\Omega+\Omega}}$$

§29. Determining Ξ. Once Luzin's problem was solved, I was quickly able to generalize it. First, I found a general rule for ϑ_ω:

$$\vartheta_\omega(\lambda) = \varepsilon\big(\Omega \cdot (1+\lambda)\big)$$

for positive λ ($\vartheta_\omega(0)$ is of course 0).

To generalize this result to, say, ϑ_{ω^2}, we need (in Veblen's terminology [Veb08]) the higher *derivatives* of the epsilon series. The first derivative ε' (= $\varepsilon^{(1)}$) enumerates the fixed points of ε in the same way that ε itself enumerates the fixed points of exponentiation-by-ω. The ordinal $\varepsilon^{(1)}_0$ is the limit of the sequence

$$\varepsilon_0, \varepsilon_{\varepsilon_0}, \varepsilon_{\varepsilon_{\varepsilon_0}}, \dots$$

and can be thought of as

$$\varepsilon_{\varepsilon_{\varepsilon_{\cdot_{\cdot_{\cdot}}}}}$$

The function ϑ_{ω^2} is given by

$$\vartheta_{\omega^2}(\lambda) = \varepsilon^{(1)}\big(\Omega \cdot (1+\lambda)\big)$$

(for any positive λ).

I generalized this to arbitrary positive (countable) powers of ω:

$$\vartheta_{\omega^{1+\mu}}(\lambda) = \varepsilon^{(\mu)}(\Omega \cdot (1+\lambda))$$

(for any positive λ).

Since every countable ordinal is a finite sum of powers of ω, this last result, together with the formula for ϑ_1 and the additive rule, allow us to calculate ϑ_μ for any particular μ.

After that, determining the structure of the Borel sets required only tedious calculation. For example, the order type of the collection of degrees of sets in level six of the difference hierarchy over the class of $\underset{\sim}{\mathbf{\Sigma}}^0_{\omega^\omega+\omega^{12}+\omega^8+4}$ sets is (after some simplification)

$$\varepsilon^{(\omega)}\left(\varepsilon^{(11)}\left(\varepsilon^{(7)}\left(\Omega^{\Omega^{\Omega^6}}\right)\right)\right)$$

because the class in question is

$$((\mathcal{G}^{\omega^\omega+\omega^{12}+\omega^8+4})_{\partial_5})^{\pm}$$

Once we pass the arithmetic sets the ordinals get bigger even faster (if "big" and "fast" have any meaning in this context ...). In general, to describe the order type of the $\underset{\sim}{\mathbf{\Delta}}^0_{\omega^{1+\mu}}$ sets, we need the μth derivative of the epsilon series. This suggests that Ξ is a fixed point of every countable ordinal derivative of the epsilon series; in otherwords, Ξ is in $\varepsilon^{(\Omega)}$, the Ωth derivative of ε, the sequence of all ordinals that are fixed points of $\varepsilon^{(\mu)}$ for all countable μ.

At this stage the countable ordinals are all out of the running, so that Ω is $\varepsilon_0^{(\Omega)}$. The next ordinal in the series is $\varepsilon_1^{(\Omega)}$; it is the least ordinal greater, for every countable μ, than the order type $\varepsilon^{(\mu)}_{(\Omega+1)}$ of the $\underset{\sim}{\mathbf{\Delta}}^0_{(\omega^{1+\mu})}$ sets. This, finally, is Ξ.

If we assume, just for the moment, analytic determinacy, we can summarize this result with the equation

$$\underset{\sim}{\mathbf{\Sigma}}^1_1 \cup \underset{\sim}{\mathbf{\Pi}}^1_1 = \mathcal{R}_{\varepsilon_1^{(\Omega)}}$$

§30. The Borel degrees. Once the induction takes us through the Borel hierarchy and exhausts the degrees of Borel sets, a very orderly picture emerges (the final pieces of which fell into place in the summer of 1972, when I solved Luzin's problem). (Of course, assuming Borel determinateness.)

To begin with, the degrees are semiwellordered with selfdual degrees at successor and cofinality ω limit ordinals, dual pairs elsewhere.

Furthermore, every nonselfdual degree ${}^\omega\mathcal{G}$-Boolean. In other words (returning to the basic definitions), if a Borel set A is incomparable with $-A$, then $\{B : B \leq A\}_{B\in\wp({}^\omega\omega)}$ is of the form $\{\Gamma(G)\}_{G\in{}^\omega\mathcal{G}}$. The surprising aspect to this result is that the single set A is somehow associated with a set operation Γ that in turn generates the whole class of sets that are no more complex than A.

As for the selfdual degrees, they are clearly all least upper bounds of countable collections of nonselfdual degrees, so that even selfdual sets are associated with set operations.

This in turn implies that every initial class of Borel sets is a union of ${}^{\omega}\mathcal{G}$-Boolean classes. One surprising consequence is that the expansion $\mathcal{B}^{\mu}$ of a class $\mathcal{B}$ of Borel sets can be defined directly and simply as

$$\mathcal{B}^{\mu} = \left\{ f^{-1}(B) \,:\, f \text{ of class } \mu \right\}_{B \in \mathcal{B}, f \,:\, {}^{\omega}\omega \to {}^{\omega}\omega}.$$

We can also conclude that there are only two kinds of $\mathcal{R}_{\lambda}$ classes of Borel sets. If λ is a successor ordinal, or a limit ordinal of cofinality greater than ω, $\mathcal{R}_{\lambda}$ is of the form $\text{In}(a) \cup \text{In}(a)^{-}$ for some nonselfdual degree a. And if λ is a limit ordinal of cofinality ω, $\mathcal{R}_{\lambda}$ is a union of $\mathcal{R}_{\kappa}$ sets of the first kind, for countably many κ less than λ.

Even more remarkable is the fact that, on the Borel sets at least, there is a definition of $\leq$ that makes no reference to games or even continuous functions. The idea, natural enough, is that we can measure the complexity of a Borel set by looking at how powerful a Boolean set operation must be in order to generate the set in question, from an omega sequence of open sets. More precisely, we say that $A \preceq B$ iff A is in every ${}^{\omega}\mathcal{G}$-Boolean class containing B; that is, iff

$$B \in \mathcal{G}_{\Gamma} \Rightarrow A \in \mathcal{G}_{\Gamma}$$

for every ω-ary Boolean set operation Γ. Then it follows from the big induction argument that $A \leq B$ iff $A \preceq B$, for all Borel sets A and B.

This definition would have made perfect sense to the descriptive set theorists of Luzin's generation. Imagine their surprise had they learned that (on the Borel sets) $\preceq$ is well founded, and that for any Borel sets A and B, either $A \preceq B$ or $B \preceq -A$!

Finally, the obvious question to ask is, does this simple picture generalize beyond the Borel degrees (assuming enough determinateness)? We now know that it does, thanks to the work of Martin, van Wesep, Steel and others, beginning in 1973 with Martin's proof that AD implies $\leq$ is well founded. Van Wesep and Steel even spotted an important phenomenon that I had missed (for the Borel degrees): that in every dual pair exactly one of the initial classes has the first separation property [Ste81B].

Of course, by now a great deal more has been discovered about $\leq$, but I will leave that story for others to tell.

REFERENCES

JOHN W. ADDISON

[Add54] *On Certain Points of the Theory of Recursive Functions*, ***Ph.D. thesis***, University of Wisconsin–Madison, 1954.

[Add04] *Tarski's theory of definability: common themes in descriptive set theory, recursive function theory, classical pure logic, and finite-universe logic*, ***Annals of Pure and Applied Logic***, vol. 126 (2004), no. 1-3, pp. 77–92.

JOHN F. BARNES

[Bar65] *The classification of the closed-open and the recursive sets of number theoretic functions*, ***Ph.D. thesis***, UC Berkeley, 1965.

MORTON DAVIS

[Dav64] *Infinite games of perfect information*, ***Advances in game theory*** (Melvin Dresher, Lloyd S. Shapley, and Alan W. Tucker, editors), Annals of Mathematical Studies, vol. 52, 1964, pp. 85–101.

FELIX HAUSDORFF

[Hau57] ***Set theory***, Chelsea, New York, 1957, translated by J. R. Aumann.

L. KANTOROVICH AND E. LIVENSON

[KL32] *Memoir on the analytical operations and projective sets I*, ***Fundamenta Mathematicae***, vol. 18 (1932), pp. 214–279.

STEPHEN C. KLEENE

[Kle50] *A symmetric form of Gödel's theorem*, ***Indagationes Mathematicae***, vol. 12 (1950), pp. 244–246.

CASIMIR KURATOWSKI

[Kur58] ***Topologie. Vol. I***, 4ème ed., Monografie Matematyczne, vol. 20, Państwowe Wydawnictwo Naukowe, Warsaw, 1958.

NIKOLAI LUZIN

[Luz30] ***Leçons sur les ensembles analytiques et leurs applications***, Collection de monographies sur la théorie des fonctions, Gauthier-Villars, Paris, 1930.

DONALD A. MARTIN

[Mar70] *Measurable cardinals and analytic games*, ***Fundamenta Mathematicae***, vol. 66 (1970), pp. 287–291.

[Mar75] *Borel determinacy*, ***Annals of Mathematics***, vol. 102 (1975), no. 2, pp. 363–371.

JAN MYCIELSKI AND HUGO STEINHAUS

[MS62] *A mathematical axiom contradicting the axiom of choice*, ***Bulletin de l'Académie Polonaise des Sciences***, vol. 10 (1962), pp. 1–3.

HARTLEY ROGERS

[Rog59] *Computing degrees of unsolvability*, ***Mathematische Annalen***, vol. 138 (1959), pp. 125–140.

JOHN R. STEEL

[Ste81B] *Determinateness and the separation property*, ***The Journal of Symbolic Logic***, vol. 46 (1981), no. 1, pp. 41–44.

MIKHAIL YA. SUSLIN

[Sus17] *Sur une définition des ensembles mesurables* B *sans nombres transfinis*, ***Comptes Rendus Hebdomadaires des Séances de l'Académie des Sciences***, vol. 164 (1917), pp. 88–91.

ALFRED TARSKI

[Tar00] *Address at the Princeton University Bicentennial Conference on Problems of Mathematics* (*December* 17–19, 1946), ***The Bulletin of Symbolic Logic***, vol. 6 (2000), no. 1, pp. 1–44.

OSWALD VEBLEN

[Veb08] *Continuous increasing functions of finite and transfinite ordinals*, ***Transactions of the American Mathematical Society***, vol. 9 (1908), no. 3, pp. 280–292.

WILLIAM W. WADGE

[Wad84] *Reducibility and determinateness on the Baire space*, ***Ph.D. thesis***, University of California, Berkeley, 1984.

ERNST ZERMELO

[Zer13] *Über eine Anwendung der Mengenlehre auf die Theorie des Schachspiels*, ***Proceedings of the Fifth International Congress of Mathematicians*** (E. W. Hobson and A. E. H. Love, editors), vol. 2, 1913, pp. 501–504.

COMPUTER SCIENCE DEPARTMENT
UNIVERSITY OF VICTORIA
VICTORIA, CANADA
E-mail: wwadge@cs.uvic.ca

PART IV: PROJECTIVE ORDINALS

PROJECTIVE ORDINALS
INTRODUCTION TO PART IV

STEVE JACKSON

§1. **Introduction.** In this paper we introduce and survey the theory of the projective ordinals, the $\underset{\sim}{\delta}^1_n$, and the theory of descriptions. We work throughout in the theory ZF+AD+DC. Recall AD, the axiom of determinacy, is the axiom that every two player integer game is determined. The projective ordinals, introduced by Moschovakis, are defined by (see the next section for more background):

$$\underset{\sim}{\delta}^1_n = \sup\{|\preceq| : \preceq \text{ is a } \underset{\sim}{\mathbf{\Delta}}^1_n \text{ prewellordering of } {}^\omega\omega\}.$$

The projective ordinals are important since the theory of the projective sets (assuming projective determinacy), the $\underset{\sim}{\mathbf{\Sigma}}^1_n$, $\underset{\sim}{\mathbf{\Pi}}^1_n$ sets, is developed in terms of them. Recall a set of reals $A \subseteq {}^\omega\omega$ is said to be λ-Suslin if there is a tree T on $\omega \times \lambda$ such that

$$\begin{aligned} A &= \mathrm{p}[T] \\ &= \{x \in {}^\omega\omega : \exists f \in \lambda^\omega\ \forall n\ (x\restriction n, f\restriction n) \in T\}. \end{aligned}$$

Suslin representations are essentially the same thing as *scales* (see Fact 2.1 of §2) and this forms a central notion in descriptive set theory. The *scale property*, isolated by Moschovakis, combines a Suslin representation for the set together with a notion of definability for such a representation (we give the definition in §2). A classical result (phrased in modern terminology) is that the pointclasses $\underset{\sim}{\mathbf{\Pi}}^1_1$ and $\underset{\sim}{\mathbf{\Sigma}}^1_2$ have the scale property. In some sense, an early prototype of this notion was present in the Novikov–Kondo proof of the uniformization property for $\underset{\sim}{\mathbf{\Pi}}^1_1$ sets. The *periodicity theorems* of Moschovakis give that, assuming projective determinacy, the pointclasses $\underset{\sim}{\mathbf{\Sigma}}^1_{2n}$, $\underset{\sim}{\mathbf{\Pi}}^1_{2n+1}$ also have the scale property (these results cannot be proved in ZF). A consequence is that the $\underset{\sim}{\mathbf{\Pi}}^1_{2n+1}$ sets are $\underset{\sim}{\delta}^1_{2n+1}$-Suslin, and in fact the $\underset{\sim}{\delta}^1_{2n+1}$-Suslin sets are exactly the $\underset{\sim}{\mathbf{\Sigma}}^1_{2n+2}$ sets (we need full AD for this direction). So, the $\underset{\sim}{\delta}^1_{2n+1}$ are *Suslin cardinals*, that is, places where new Suslin representations appear. In fact, general scale type arguments along these lines show, assuming AD, that below the supremum of the projective ordinals the Suslin cardinals are

Wadge Degrees and Projective Ordinals: The Cabal Seminar, Volume II
Edited by A. S. Kechris, B. Löwe, J. R. Steel
Lecture Notes in Logic, 37

precisely the $\underset{\sim}{\delta}^1_{2n+1}$ and their cardinal predecessors λ_{2n+1}. The λ_{2n+1}-Suslin sets are the $\underset{\sim}{\Sigma}^1_{2n+1}$ sets (*cf.* Theorem 2.9 below). Thus, the scale property gives a structural representation for the projective sets, and this representation involves the projective ordinals. We refer the reader to [Mos80], [Kec78] for more details and history along these lines. We also give a quick overview in §2.

We emphasize that in the modern theory the projective ordinals (particularly the "odd" projective ordinals $\underset{\sim}{\delta}^1_{2n+1}$) play a much more extensive role than just providing the sizes of the Suslin representations. In fact, the entire modern theory hinges on the properties of the $\underset{\sim}{\delta}^1_{2n+1}$, and in particular on their partition properties. These properties are used not only to inductively propagate the theory, but to analyze the cardinal structure between the projective ordinals.

Descriptions are the combinatorial objects which are central to the modern theory of the projective ordinals $\underset{\sim}{\delta}^1_n$ (and beyond) assuming AD. For example, they are used to compute the values of the $\underset{\sim}{\delta}^1_n$, establish their partition properties, and provide the framework for analyzing the cardinal structure between them (in particular to compute the cofinalities of the cardinals). Through the projective ordinals at least, descriptions will be hereditarily finite objects which code how to build ordinals with respect to iterated ultrapowers by certain canonical measures. These measures are defined using the partition properties of the $\underset{\sim}{\delta}^1_{2n+1}$. The proofs of the partition properties for the $\underset{\sim}{\delta}^1_{2n+1}$ in turn need the theory of descriptions, but at a lower level. Thus, the entire theory is developed inductively. This has the disadvantage that the theory (at least in its current form) breaks down completely past the least point where we are able to complete the full cycle of inductive arguments. The theory as outlined here and presented in detail in [Jac99] and [Jac88] extends with only trivial modifications to the $\underset{\sim}{\delta}^1_\alpha$ where $\alpha < \omega_1$, and this analyzes the cardinal structure up to $\aleph_{\omega_1}$. The description analysis works past this point, but description cease to be purely finitary objects. These arguments, which have not yet appeared, are believed to work through the level of the first inaccessible cardinal. On the other hand, results of [Jac91] show that serious problems arise by the time one reaches $\kappa^{\mathbb{R}}$, the first $\mathbb{R}$-admissible ordinal (which is the ordinal of the *inductive sets* and corresponds to the least non-selfdual pointclass closed under real quantification). Finding ways to propagate the theory further remains an important program.

This paper serves in part as an introduction to the papers [Mar71B], [Kec78], [Sol78A], [Kec81A], and [Jac88] of this volume. It also serves as a self-contained survey of and introduction to the theory of the projective ordinals. The papers [Mar71B], [Kec78] and [Kec81A] are directly concerned with the theory of the projective ordinals. These papers, written before the theory of descriptions, give the earlier development of the subject. Martin's paper [Mar71B] represented a fundamental advance in the development of this theory. In fact, many of the ideas of the current program can be found in some form in that paper. In this earlier work the values of $\underset{\sim}{\delta}^1_1 = \omega_1$ (classical), $\underset{\sim}{\delta}^1_2 = \omega_2$ (Kunen, Martin independently), and $\underset{\sim}{\delta}^1_3 = \omega_{\omega+1}$ (Martin) were computed. Also, the strong partition relation on $\underset{\sim}{\delta}^1_1$ was established (Martin), as was the weak partition relation on $\underset{\sim}{\delta}^1_3$ (Kunen). [Mar71B] and [Kec78] present all of these results except the last, which is the subject of [Sol78A]. It was also shown that the even projective ordinals are obtained simply from the odd by $\underset{\sim}{\delta}^1_{2n+2} = (\underset{\sim}{\delta}^1_{2n+1})^+$ (Kunen, Martin; see [Kec78] and the discussion in the next section). At this time, the value of the next odd projective ordinal, $\underset{\sim}{\delta}^1_5$, was unknown as was the strong partition property on $\underset{\sim}{\delta}^1_3$. To develop the theory further requires the notion of a description.

In the early 80's Martin proved a key result which in this area we refer to simply as Martin's Theorem (Theorem 6.10 of this paper). This theorem gives the existence of the *Martin tree* which provides an analysis of functions $f : \underset{\sim}{\delta}^1_3 \to \underset{\sim}{\delta}^1_3$ with respect to one of the normal measures on $\underset{\sim}{\delta}^1_3$ (there are three normal measures on $\underset{\sim}{\delta}^1_3$ corresponding to the three regular cardinals ω, ω_1, ω_2 below $\underset{\sim}{\delta}^1_3$). Versions of this tree play a similar role for the general $\underset{\sim}{\delta}^1_{2n+1}$. This extended Kunen's earlier theorem giving the *Kunen tree* (see Theorem 2.11) which had a similar role for the cofinality ω normal measure on $\underset{\sim}{\delta}^1_{2n+1}$. Building on this and some joint work with Martin, the author developed descriptions and used them to compute $\underset{\sim}{\delta}^1_5$. This can be viewed as bringing to fruition a plan developed by Martin and Kunen for computing the projective ordinals via certain ultrapowers by homogeneity measures (again, some of these important ideas can be found Martin's paper [Mar71B]). We will outline this program in more detail in §3. The notion of a *homogeneous tree* (see Definition 3.2) is thus important to the overall analysis. This notion arose independently in the work of Kunen and Martin, and the precise formulation was given independently by Kechris and Martin. The concept is explained in detail in [Kec81A].

In hindsight, the earlier theory of the projective ordinals can be viewed as an instance of the more general theory. When these earlier results are recast into the theory of descriptions, the underlying descriptions are rather trivial objects, basically just integers. Nevertheless, one can use these "trivial descriptions" to redo these results. The reader can consult [Jac10] for details on how the trivial descriptions can be used, starting from scratch, to compute $\underset{\sim}{\delta}^1_3$, establish the strong partition relation on $\underset{\sim}{\delta}^1_1$, and prove the weak partition relation on $\underset{\sim}{\delta}^1_3$ (technically one starts from the fact that $\underset{\sim}{\delta}^1_1 = \omega_1$ and the weak partition relation on $\underset{\sim}{\delta}^1_1$, both of which are easily shown). Although we discuss the trivial descriptions here as well, our focus here is showing how descriptions allow these earlier results to be unified and generalized into a theory which gives an analysis of all the projective ordinals and beyond.

Whether one presents the arguments of [Kec78] or [Sol78A] in their original form or as arguments with (trivial) descriptions is mainly a matter of taste, the underlying mathematics is essentially the same. However, there are some points which are important when extending these arguments. The earlier arguments are presented using the theory of indiscernibles for $\mathbf{L}[x]$ and the fact that every subset of ω_1 is (assuming AD) in $\mathbf{L}[x]$ for some real x. For example, both Martin's proof of the strong partition relation on ω_1 (see [Kec78]) and Kunen's proof of the weak partition relation on $\underset{\sim}{\delta}^1_3$ (see [Sol78A]) rely on this method. These arguments are not known to generalize to higher levels. Instead, the description theory relies heavily on partition properties. Partition arguments replace indiscernibility arguments throughout. Descriptions and proofs involving them are thus closely connected to partition properties, and this gives the description theory a more combinatorial nature. In particular, establishing the strong partition relation on the $\underset{\sim}{\delta}^1_{2n+1}$ becomes a central part of the inductive analysis.

This paper also aims to help bridge the gap between the earlier pre-description theory and the full analysis of [Jac88]. In that rather long and technical paper the general theory of descriptions needed to analyze the general projective ordinal is laid out. Since then, other papers have appeared which present various aspects of this theory. In [Jac99] the complete theory at the first level past the previous results is given. Namely, $\underset{\sim}{\delta}^1_5$ is computed, the strong partition relation on $\underset{\sim}{\delta}^1_3$ is proved, and then the weak partition relation on $\underset{\sim}{\delta}^1_5$ is proved. This constitutes the complete first step of the general inductive analysis. In [Jac88] the framework for the general projective case is presented. The corresponding proofs of the partition properties are not given in [Jac88] as they use the same machinery presented, and proceed as in the corresponding proofs of [Jac99] (using the more general descriptions of [Jac88]).

In §2 we present some background material and fix some notation. The material in this section in only a brief overview, and the reader wishing to see further details could consult [Mos80], [Mar71B], and [Kec78].

In §3 we give a outline of how the inductive analysis of the projective ordinals goes. The precise definitions and proofs are not given in this section, rather it is an attempt to give the reader the overall picture. In the remaining sections we attempt to systematically fill in some of the details.

In §4 we present a sketch of the "first level theory" using the "trivial descriptions". This involves calculating $\underset{\sim}{\delta}^1_3$, proving the strong partition relation on $\underset{\sim}{\delta}^1_1 = \omega_1$, and proving the weak partition relation on $\underset{\sim}{\delta}^1_3$. Again, all of these were known results from the earlier theory, but it is perhaps instructive to see them redone using methods that will generalize to the higher levels. Some of the proofs are given completely and others just sketched or illustrated. In [Jac10] the complete proofs can be found.

In §5 and §6 we discuss the second level of the theory, which computes $\underset{\sim}{\delta}^1_5$, proves the strong partition relation on $\underset{\sim}{\delta}^1_3$, and prove the weak partition relation on $\underset{\sim}{\delta}^1_5$. Now the descriptions will be less trivial objects, although they will still be hereditarily finite objects. To ease the transition from the trivial to the non-trivial descriptions, we present this analysis in two sections. In §5 we introduce the "level-2" descriptions (the trivial descriptions of §4 are the level-1 descriptions), and in §6 these are extended in a rather trivial way to get the set of level-3 descriptions. The distinction is small enough that in [Jac99] it is not even explicitly made. Although the formal difference between the level-2 and level-3 descriptions is minor, they are used in somewhat different contexts. Here the distinction is useful as it allows to introduce all of the combinatorial machinery of descriptions quickly in §5 without the burden of the extra framework needed for full next step of the induction (*e.g.*, defining the Martin tree, analyzing the measures on $\underset{\sim}{\delta}^1_3$, proving the measure domination theorems). To illustrate the combinatorics and some of the later ideas we use the level-2 descriptions to solve two ad hoc problems: (1) analyze the cardinal structure of the iterated ultrapower of ω_n by measures from the canonical families of measures W^m_1, S^m_1 (these measures, defined in §5, are measures on $(\omega_1)^m \approx \omega_1$ and ω_{m+1} respectively) and (2) compute the number of descriptions defined with respect to a fixed sequence of measures. In §6 we say how these descriptions (technically the level-3 descriptions) are used to carry out the next level of the actual inductive analysis. We don't give complete proofs here, but we present all the necessary ingredients. We give an example of what a general measure on $\underset{\sim}{\delta}^1_3$ looks like.

In §7 we make some comments concerning how the extension to the higher levels goes which should ease the transition to [Jac88]. In §8 we make some concluding remarks including an alternate way to describe the cardinal structure below the projective ordinals which does not use descriptions for the presentation. However, the theory of descriptions is needed to show this alternate formulation works. Nevertheless, this alternate formulation may have applications of its own, as well as serve to make the theory more accessible and applicable.

§2. Background and Preliminaries. For general background in descriptive set theory we refer the reader to [Mos80] and [Kec94]. [Kec78] also gives more background related to the projective ordinals. We recall here some of the more important definitions and results, and fix some notation.

We let $(n, m) \mapsto \langle n, m \rangle$ be a recursive bijection from $\omega \times \omega$ to ω. We let $n \mapsto ((n)_0, (n)_1)$ denote the inverse (decoding) map. When there is no danger of confusion we frequently drop the parentheses and just write $n \mapsto (n_0, n_1)$. We also use this notation for variations of these coding maps.

Let $\mathrm{WO} \subseteq {}^\omega\omega$ be the standard set of codes of countable ordinals, that is, $x \in \mathrm{WO}$ iff $<_x \doteq \{(n, m) : x(\langle n, m \rangle) = 1\}$ is a wellordering. For $x \in \mathrm{WO}$, let $|x|$ be the rank of $<_x$. Let WO_α be the set of those $x \in \mathrm{WO}$ such that $|x| < \alpha$.

By a **tree** on a set X we mean a set $T \subseteq {}^{<\omega}X$ closed under initial segment. We identify trees on $X \times Y$ with subsets of ${}^{<\omega}X \times {}^{<\omega}Y$, that is, we may view an element of such a tree as a pair $(s,t) \in {}^{<\omega}X \times {}^{<\omega}Y$ with $\mathrm{lh}(s) = \mathrm{lh}(t)$ ($\mathrm{lh}(s)$ denotes the length of the sequence s). We extend this convention to longer products as well. If T is a tree on X, we write $[T] = \{x \in {}^{\omega}X : \forall n\, x{\restriction}n \in T\}$ for the **body** of T. If T is a tree on $X \times Y$, we write $\mathrm{p}[T]$ for the projection of T, so $\mathrm{p}[T] = \{x \in {}^{\omega}X : \exists y \in {}^{\omega}Y\, \forall n\, (x{\restriction}n,\, y{\restriction}n) \in T\}$. If T is a tree on a higher product, say on $X_1 \times \cdots \times X_m$, then in writing $\mathrm{p}[T]$ we should specify which coordinate we are projecting on. If T is a tree on $\omega \times Y$, then for $x \in {}^{\omega}\omega$ we let $T_x = \{s \in {}^{<\omega}Y : (x{\restriction}\mathrm{lh}(s), s) \in T\}$ be the **section** of T at x. If T is wellfounded (*i.e.*, $[T] = \varnothing$) then $|T|$ denotes the rank of T. A frequently occurring case is when T is a tree on $\omega \times \lambda$ for some $\lambda \in \mathrm{Ord}$ (or T a tree on $\omega \times \lambda_1 \times \lambda_2$, *etc.*). In this case, for $x \notin \mathrm{p}[T]$, $|T_x|$ makes sense and will be an ordinal $< \lambda^+$. We let $T_x{\restriction}\alpha = T_x \cap {}^{<\omega}\alpha$ be the restriction of T_x to ordinals less than α. If $T_x{\restriction}\alpha$ is wellfounded and $s \in T_x{\restriction}\alpha$, we let $|T_x{\restriction}\alpha(s)|$ be the rank of s in the tree $T_x{\restriction}\alpha$. For notational ease we usually identify finite sequences of ordinals s with ordinals β, and so write $|T_x{\restriction}\alpha(\beta)|$. This is only for convenience and this assumption cause no harm (see the remark after Theorem 2.11). In fact, using the Brouwer-Kleene ordering on T_x we can assume that each T_x is actually a linear ordering. Recall this ordering is defined by: $s <_{\mathrm{BK}} t$ iff s extends t or there is a least i such that $s(i) \neq t(i)$ and we have $s(i) < t(i)$.

We adopt the convention in this paper that "lexicographic ordering" on tuples actually refers to the Brouwer-Kleene ordering, that is, longer sequences correspond to smaller elements of the ordering.

Recall $A \subseteq {}^{\omega}\omega$ is said to be α-Suslin if $A = \mathrm{p}[T]$ for some tree T on $\omega \times \alpha$. Let $\boldsymbol{S}(\kappa)$ denote the collection of κ-Suslin sets, and $\boldsymbol{S}(<\kappa) = \bigcup_{\lambda<\kappa} \boldsymbol{S}(\lambda)$. We say κ is a **Suslin cardinal** if $\boldsymbol{S}(\kappa) - \boldsymbol{S}(<\kappa) \neq \varnothing$. By a **norm** on a set A we mean a map $\varphi\colon A \to \mathrm{Ord}$. We say φ is **regular** if it is onto an ordinal. We identify norms with **prewellorderings** $\preceq$ of A (reflexive, transitive, connected binary relations). We let $\prec$ denote the strict part of the prewellordering $\preceq$ (*i.e.*, $x \prec y$ iff $x \preceq y$ and $\neg\, y \preceq x$). If $\prec$ is a relation, we let $\mathrm{fld}(\prec)$ be the field of $\prec$, that is, $\mathrm{fld}(\prec) = \{z : (\exists x\; x \prec z) \vee \exists y\; z \prec y)\}$. For φ a norm we let $|\varphi|$ denote the length of φ (*i.e.*, the rank of the associated prewellordering). A **semi-scale** on $A \subseteq {}^{\omega}\omega$ (or more generally $A \subseteq \lambda^{\omega}$ for some $\lambda \in \mathrm{Ord}$) is a sequence of norms $\{\varphi_n\}_{n\in\omega}$ on A such that if $\{x_m\} \subseteq A$, x_m converges to x, and for each n, the values $\varphi_n(x_m)$ are eventually constant, then $x \in A$. $\{\varphi_n\}$ is a **scale** if in addition it satisfies the **lower-semicontinuity** property: $\varphi_n(x) \leq \lambda_n \doteq \lim_m \varphi_n(x_m)$. The scale is **good** if the convergence of norms property also implies that $\{x_m\}$ converges. The scale is **very-good** if for all $x, y \in A$, if $\varphi_n(x) \leq \varphi_n(y)$, then for all $k < n$ we have $\varphi_k(x) \leq \varphi_k(y)$. We say a norm (or semi-scale, *etc.*) is a κ-norm if all the norms map into κ. The notions of scales and Suslin representations are essentially equivalent by the following fact (see [Mos80] or [Jac10]).

Fact 2.1. For every cardinal κ, $A \subseteq {}^\omega\omega$ is κ-Suslin iff A admits a κ semiscale iff A admits a κ very-good scale.

A **pointclass** $\mathbf{\Gamma}$ is a collection of subsets of ${}^\omega\omega$ (or $({}^\omega\omega)^n$) closed under continuous preimages. A norm φ (or prewellordering) is a $\mathbf{\Gamma}$-norm if the norm relations $<^*_\varphi$, $\leq^*_\varphi$ are in $\mathbf{\Gamma}$ where:

$$x <^*_\varphi y \leftrightarrow x \in A \wedge (y \notin A \vee (y \in A \wedge \varphi(x) < \varphi(y)))$$
$$x \leq^*_\varphi y \leftrightarrow x \in A \wedge (y \notin A \vee (y \in A \wedge \varphi(x) \leq \varphi(y)))$$

Note that the initial segments of a $\mathbf{\Gamma}$-prewellordering are in $\mathbf{\Delta} = \mathbf{\Gamma} \cap \check{\mathbf{\Gamma}}$. A semiscale $\{\varphi_n\}$ (or scale) is a $\mathbf{\Gamma}$-semiscale if all the norm relations $<^*_n$, $\leq^*_n$ corresponding to the φ_n are in the pointclass $\mathbf{\Gamma}$. $\mathbf{\Gamma}$ has the **prewellordering property**, PWO($\mathbf{\Gamma}$), if every $\mathbf{\Gamma}$ set admits a $\mathbf{\Gamma}$-prewellordering, and $\mathbf{\Gamma}$ has the **scale property**, Scale($\mathbf{\Gamma}$), if every $\mathbf{\Gamma}$ set admits a $\mathbf{\Gamma}$-scale. We note that if $\mathbf{\Gamma}$ is closed under $\wedge$, $\vee$, and Scale($\mathbf{\Gamma}$), then every $\mathbf{\Gamma}$ set admits a $\mathbf{\Gamma}$-very-good scale.

Assuming projective determinacy, the *periodicity theorems* of Moschovakis (see [Mos80]) propagate the scale property through the projective hierarchy is the following "periodic" manner:

$$\text{Scale}(\mathbf{\Pi}^1_1),\ \text{Scale}(\mathbf{\Sigma}^1_2),\ \text{Scale}(\mathbf{\Pi}^1_3),\ \text{Scale}(\mathbf{\Sigma}^1_4), \dots, \text{Scale}(\mathbf{\Pi}^1_{2n+1}),\ \text{Scale}(\mathbf{\Sigma}^1_{2n+2}), \dots$$

The scale property is the basic structural property in descriptive set theory. It gives Suslin representations of the projective sets in terms of the projective ordinals, which we define next.

Definition 2.2 (Moschovakis). The ordinal $\boldsymbol{\delta}^1_n$ is the supremum of the lengths of the $\mathbf{\Delta}^1_n$ prewellorderings of ${}^\omega\omega$.

The next basic fact, due to Moschovakis, says that all the projective ordinals are actually cardinals. We give the quick proof which uses the Moschovakis *coding lemma*, an important tool in determinacy theory. The coding lemma has several versions, including one for prewellorderings, a more general one for wellfounded relations, and a "uniform" version. We refer the reader to [Mos80, 7D.5, 7D.6] as well as [Jac10, Theorem 2.12] for precise statements and proofs.

Lemma 2.3 (Moschovakis). All the $\boldsymbol{\delta}^1_n$ are cardinals.

Proof. Suppose $\lambda < \boldsymbol{\delta}^1_n$ and $|\lambda| = |\boldsymbol{\delta}^1_n|$. Fix a bijection $\pi\colon \lambda \to \boldsymbol{\delta}^1_n$. There is a $\mathbf{\Delta}^1_n$ prewellordering $\preceq$ of length λ, and we get one of length λ^2 by defining:

$$z \preceq' w \leftrightarrow ((z)_0 \prec (w)_0) \vee (((z)_0 \preceq (w)_0) \wedge ((w)_0 \preceq (z)_0) \wedge (z)_1 \preceq (w)_1)).$$

Let A be the field of $\preceq$, so $A \in \mathbf{\Delta}^1_n$. So, $\preceq'$ has field $A \times A$. By the coding lemma applied to $\preceq'$, the relation $B(x,y) \leftrightarrow (x, y \in A \wedge \pi(|x|_\preceq) < \pi(|y|_\preceq))$ is $\mathbf{\Delta}^1_n$. This B gives a $\mathbf{\Delta}^1_n$ prewellordering of length $\boldsymbol{\delta}^1_n$, a contradiction. ⊣

If $\{\varphi_n\}$ is a $\underset{\sim}{\boldsymbol{\Pi}}^1_{2n+1}$-scale on a $\underset{\sim}{\boldsymbol{\Pi}}^1_{2n+1}$ set P, then all the initial segments of each φ_n norm are $\underset{\sim}{\boldsymbol{\Delta}}^1_{2n+1}$ and so have length $< \underset{\sim}{\boldsymbol{\delta}}^1_{2n+1}$ (all of the $\underset{\sim}{\boldsymbol{\delta}}^1_n$ are easily limit ordinals). Thus, each φ_n has length $\leq \underset{\sim}{\boldsymbol{\delta}}^1_{2n+1}$ and so P is $\underset{\sim}{\boldsymbol{\delta}}^1_{2n+1}$-Suslin. On the other hand, if P is $\underset{\sim}{\boldsymbol{\Pi}}^1_{2n+1}$-complete, then P cannot be λ-Suslin for any $\lambda < \underset{\sim}{\boldsymbol{\delta}}^1_{2n+1}$. For if so, then since there is a $\underset{\sim}{\boldsymbol{\Delta}}^1_{2n+1}$ prewellordering of length λ, the coding lemma gives that we may code any tree on $\omega \times \lambda$ within the pointclass $\underset{\sim}{\boldsymbol{\Sigma}}^1_{2n+1}$. It would follow then that $P \in \underset{\sim}{\boldsymbol{\Sigma}}^1_{2n+1}$, a contradiction. So, $\underset{\sim}{\boldsymbol{\delta}}^1_{2n+1}$ is a Suslin cardinal.

Moreover, $\underset{\sim}{\boldsymbol{\Gamma}}$ is a non-selfdual pointclass closed under $\forall^{\omega\omega}$, $\vee$, and φ is a $\underset{\sim}{\boldsymbol{\Gamma}}$-norm on a $\underset{\sim}{\boldsymbol{\Gamma}}$-complete set, then a general argument using the recursion theorem (see [Mos80, 4C.14]) shows that $|\varphi|$ is at least the supremum of the lengths of the $\check{\underset{\sim}{\boldsymbol{\Gamma}}}$ wellfounded relations. It follows that any $\underset{\sim}{\boldsymbol{\Pi}}^1_{2n+1}$-norm on a $\underset{\sim}{\boldsymbol{\Pi}}^1_{2n+1}$-complete set has length exactly $\underset{\sim}{\boldsymbol{\delta}}^1_{2n+1}$. This also shows the following fact for odd n.

FACT 2.4. The ordinal $\underset{\sim}{\boldsymbol{\delta}}^1_n$ is the supremum of the lengths of the $\underset{\sim}{\boldsymbol{\Sigma}}^1_n$ wellfounded relations.

Alternatively, one can prove the fact using the *Kunen-Martin Theorem*, which will also work for even n (this proof uses scales, though, while the proof above used only the prewellordering property and so is more general). We recall the Kunen-Martin Theorem. This result was proved independently by Kunen and Martin. Martin's original proof appears in [Mar71B] (see also [Mos80, 2G.2] for another proof).

THEOREM 2.5 (Kunen-Martin). If $\prec$ is a κ-Suslin, wellfounded relation on ${}^\omega\omega$, then $|\prec| < \kappa^+$.

For the even case of Fact 2.4, recall $\underset{\sim}{\boldsymbol{\Sigma}}^1_{2n}$ has the scale property. If $A \in \underset{\sim}{\boldsymbol{\Delta}}^1_{2n}$, then the norm relations of a $\underset{\sim}{\boldsymbol{\Sigma}}^1_{2n}$ scale on A are actually $\underset{\sim}{\boldsymbol{\Delta}}^1_{2n}$, and thus each has length less than $\underset{\sim}{\boldsymbol{\delta}}^1_{2n}$. Easily all of the $\underset{\sim}{\boldsymbol{\delta}}^1_n$ have uncountable cofinality, and it follows that A is λ-Suslin for some $\lambda < \underset{\sim}{\boldsymbol{\delta}}^1_{2n}$. So, every $\underset{\sim}{\boldsymbol{\Sigma}}^1_{2n}$ set is also λ-Suslin. From the Kunen-Martin Theorem, it now follows that a $\underset{\sim}{\boldsymbol{\Sigma}}^1_{2n}$ wellfounded relation $\prec$ has rank $< \lambda^+$ for some $\lambda < \underset{\sim}{\boldsymbol{\delta}}^1_{2n}$. Since $\underset{\sim}{\boldsymbol{\delta}}^1_{2n}$ is a cardinal, it follows that $\prec$ has rank less than $\underset{\sim}{\boldsymbol{\delta}}^1_{2n}$.

For the odd case of the fact, note that a $\underset{\sim}{\boldsymbol{\Pi}}^1_{2n+1}$ scale on a $\underset{\sim}{\boldsymbol{\Delta}}^1_{2n+1}$ set is actually a $\underset{\sim}{\boldsymbol{\Delta}}^1_{2n+1}$ scale, so all the norms have length less than $\underset{\sim}{\boldsymbol{\delta}}^1_{2n+1}$. As before, this shows every $\underset{\sim}{\boldsymbol{\Delta}}^1_{2n+1}$, and hence every $\underset{\sim}{\boldsymbol{\Sigma}}^1_{2n+1}$ set is λ-Suslin for some $\lambda < \underset{\sim}{\boldsymbol{\delta}}^1_{2n+1}$. Using the Kunen-Martin Theorem we now finish as in the even case.

As a corollary of Fact 2.4 we have the following which was shown by Martin for the odd n and Kunen for the even n.

LEMMA 2.6 (Martin, Kunen). All of the $\underset{\sim}{\boldsymbol{\delta}}^1_n$ are regular cardinals.

Lemma 2.6 follows immediately from Fact 2.4 and the following Lemma 2.8. We first make a definition which generalizes the definition of projective ordinal to a general pointclass.

DEFINITION 2.7. Let $\utilde{\Gamma}$ be a pointclass. Then $\delta(\utilde{\Gamma})$ is the supremum of the lengths of the $\utilde{\Delta} = \utilde{\Gamma} \cap \check{\utilde{\Gamma}}$ prewellorderings of the reals.

LEMMA 2.8. Let $\utilde{\Gamma}$ be a non-selfdual pointclass closed under $\exists^{\mathbb{R}}$ and $\wedge$. Then the supremum $\delta'(\utilde{\Gamma})$ of the lengths of the $\utilde{\Gamma}$ wellfounded relations on ${}^{\omega}\omega$ is a regular cardinal.

PROOF. Suppose $f : \lambda \to \delta'(\utilde{\Gamma})$ is cofinal, where $\lambda < \delta'(\utilde{\Gamma})$. Fix a wellfounded relation $\prec$ of rank λ in the pointclass $\utilde{\Gamma}$. From the coding lemma, there is a $A \subseteq \text{fld}(\prec) \times {}^{\omega}\omega$ in the pointclass $\utilde{\Gamma}$ such that:

1. If $A(x, y)$ then y is a code (via a $\utilde{\Gamma}$ universal set U, which exists as $\utilde{\Gamma}$ in non-selfdual) of a $\utilde{\Gamma}$ wellfounded relation U_y of length $|U_y| = f(|x|_{\prec})$.
2. For all $\alpha < \lambda$ there is an $x \in \text{dom}(\prec)$ with $|x|_{\prec} = \alpha$ with $x \in \text{dom}(A)$.

Define $(x_1, y_1, z_1) \prec' (x_2, y_2, z_2)$ iff $x_1 = x_2$, $y_1 = y_2$, $A(x_1, y_1)$, and $U_{y_1}(z_1, z_2)$. Then easily $\prec'$ is a wellfounded relation of length $\delta'(\utilde{\Gamma})$, a contradiction. ⊣

Clearly $\delta(\utilde{\Gamma}) \leq \delta'(\utilde{\Gamma})$ for any $\utilde{\Gamma}$ closed under $\exists^{\mathbb{R}}$ and $\wedge$ (actually this holds for all $\utilde{\Gamma}$ with $\utilde{\Delta} \wedge \utilde{\Delta} \subseteq \utilde{\Gamma}$ if we use the definition of $\delta'(\utilde{\Gamma})$ given in Lemma 2.8 for arbitrary $\utilde{\Gamma}$; we need the closure hypothesis to get that the strict part of a $\utilde{\Delta}$ prewellordering lies in $\utilde{\Gamma}$). If $\utilde{\Gamma}$ is closed under $\exists^{\mathbb{R}}$, $\wedge$, and PWO($\check{\utilde{\Gamma}}$), then the result of Moschovakis mentioned above [Mos80, 4C.14] shows that $\delta(\utilde{\Gamma}) = \delta'(\utilde{\Gamma})$.

Using just general scale and pointclass arguments as the one just given, we can in fact say more. The next result is proved in [Kec78] and exactly places the Suslin cardinals below the projective ordinals. In the statement of the theorem, the cardinals λ_{2n+1} are defined. We will use this notation elsewhere as well.

THEOREM 2.9. For all n, $\utilde{\delta}^1_{2n+2} = (\utilde{\delta}^1_{2n+1})^+$ (Kunen-Martin). Also, $\utilde{\delta}^1_{2n+1} = \lambda^+_{2n+1}$, where λ_{2n+1} is a cardinal of cofinality ω (Kechris). All of the $\utilde{\delta}^1_n$ are regular cardinals (Martin for odd n, Kunen for even n). The Suslin cardinals below the projective ordinals are exactly $\lambda_1 = \omega$, $\utilde{\delta}^1_1 = \omega_1$, $\lambda_3 = \omega_\omega$, $\utilde{\delta}^1_3 = \omega_{\omega+1}$, $\lambda_5, \utilde{\delta}^1_5, \ldots, \lambda_{2n+1}, \utilde{\delta}^1_{2n+1}, \ldots$. The corresponding Suslin classes are $\boldsymbol{S}(\lambda_{2n+1}) = \utilde{\Sigma}^1_{2n+1}$, and $\boldsymbol{S}(\utilde{\delta}^1_{2n+1}) = \utilde{\Sigma}^1_{2n+2}$ (Martin, Moschovakis).

The scale property and Suslin cardinal analysis of Theorem 2.9 can be extended much further. Martin and Steel [MS83] and Steel [Ste83], assuming $\mathsf{AD} + \mathbf{V}{=}\mathbf{L}(\mathbb{R})$, determine the scaled Levy pointclasses and classify the Suslin cardinals throughout the Wadge hierarchy (a Levy pointclass is a non-selfdual pointclass closed under $\exists^{{}^{\omega}\omega}$ or $\forall^{{}^{\omega}\omega}$). In fact, assuming just AD one can get a classification of the Suslin cardinals (see [Jac10] for the arguments; the main new ingredient is due to Martin).

Another result which is used frequently in the theory of the projective ordinals, and which uses only general pointclass arguments is the following result of Martin which appears in [Mar71B].

THEOREM 2.10 (Martin). Let $\underset{\sim}{\Gamma}$ be non-selfdual, closed under $\forall^{\omega\omega}$, $\vee$ and assume PWO($\underset{\sim}{\Gamma}$). Then $\underset{\sim}{\Delta}$ is closed under wellordered unions of length $< \delta(\underset{\sim}{\Gamma})$.

An immediate consequence of this result is that $\mathcal{B}_{<\underset{\sim}{\delta}^1_{2n+1}} \subseteq \underset{\sim}{\Delta}^1_{2n+1}$, where $\mathcal{B}_{<\kappa}$ is the smallest collection of sets containing the open and closed sets and closed under complements and wellordered unions of length $< \kappa$. If T is a tree on $\omega \times \underset{\sim}{\delta}^1_{2n+1}$ and $\alpha, \beta < \underset{\sim}{\delta}^1_{2n+1}$, then $A_{\alpha,\beta} = \{x : T_x \restriction \alpha$ is wellfounded of rank $< \beta\}$ is in $\mathcal{B}_{<\underset{\sim}{\delta}^1_{2n+1}}$ by a standard tree computation (see [Kec78]), and so is in $\underset{\sim}{\Delta}^1_{2n+1}$. This computation is important in many of the arguments involving the projective ordinals. We note that the other direction is true also, namely $\underset{\sim}{\Delta}^1_{2n+1} \subseteq \mathcal{B}_{<\underset{\sim}{\delta}^1_{2n+1}}$. This follows from the general Suslin-Kleene Theorem which says that if A and ${}^{\omega}\omega \setminus A$ are both κ-Suslin then $A \in \mathcal{B}_{\kappa^+}$, using here $\kappa = \lambda_{2n+1}$ (see [Mos80, 3E.2]).

As we mentioned before, the Kunen tree is an important tool at the bottom level of the projective analysis. We will give some examples of how this is used in the next section. For the sake of completeness, and because of the importance of the Kunen tree and its generalization the Martin tree in the theory of the projective ordinals, we give (following [Jac10]; see Lemma 4.1 and Theorem 4.2) the Kunen tree construction. The result may be stated as follows.

THEOREM 2.11 (Kunen). There is a tree T on $\omega \times \omega_1$ such that for any $f : \omega_1 \to \omega_1$ there is an $x \in {}^{\omega}\omega$ such that T_x is wellfounded and such that for all $\alpha \geq \omega$ we have $|T_x \restriction \alpha| > f(\alpha)$.

PROOF. Let S be a tree on $\omega \times \omega$ with $A = \mathrm{p}[S]$ a $\underset{\sim}{\Sigma}^1_1$-complete set. Note that $\sup\{|S_y| : S_y$ is wellfounded$\} = \omega_1$ as otherwise A would be Borel. Let U be the tree on $\omega \times \omega_1$ defined by:

$$((a_0, \ldots, a_{n-1}), (\alpha_0, \ldots, \alpha_{n-1})) \in U \leftrightarrow \forall i, j < n\, (\langle i, j\rangle < n \wedge a_{\langle i,j\rangle} = 1 \to \alpha_i < \alpha_j)$$

Note that $\mathrm{p}[U] = \mathrm{WF}$, where WF is the set of $x \in {}^{\omega}\omega$ such that the binary relation $\{(i, j) : x(\langle i, j\rangle) = 1\}$ coded by x is wellfounded. View every real $\tau \in {}^{\omega}\omega$ as coding a strategy for player II in a game on ω in some standard manner (*e.g.*, the response of τ to $s \in \omega^{<\omega}$ is given by $\tau(\langle s\rangle)$ where here $s \mapsto \langle s\rangle$ denote a bijection between $\omega^{<\omega}$ and ω). Abusing notation slightly, for $s = (a_0, \ldots, a_{n-1})$ we write $\tau(s)$ for the play $(\tau(a_0), \tau(a_0, a_1), \ldots, \tau(a_0, \ldots, a_{n-1}))$

by player II following the strategy τ when player I plays s. If $a \in {}^{\omega}\omega$, we let $\tau(a) \in {}^{\omega}\omega$ be the real extending all of the $\tau(a \restriction n)$.

Let V be the tree on $\omega \times \omega \times \omega_1 \times \omega \times \omega$ given by:

$$(\vec{s}, \vec{a}, \vec{\alpha}, \vec{b}, \vec{c}) \in V \leftrightarrow (\vec{a}, \vec{\alpha}) \in U \wedge (\vec{b}, \vec{c}) \in S \wedge \exists \tau \text{ extending } \vec{s}\ (\tau(\vec{a}) = \vec{b}).$$

Suppose $f : \omega_1 \to \omega_1$. Consider the game where player I, player II play out $x, y \in {}^{\omega}\omega$ respectively, and player II wins iff

$$x \in \mathrm{WF} \to S_y \text{ is wellfounded} \wedge |S_y| > \sup\{f(\beta) : \beta \leq |x|\}.$$

By boundedness, player I cannot have a winning strategy for this game [If player I had a winning strategy σ, then $\sigma[{}^{\omega}\omega]$ would be a $\undertilde{\Sigma}^1_1$ subset of WF and then easily we have $\alpha = \sup\{|x| : x \in \sigma'' {}^{\omega}\omega\} < \omega_1$. Player II could then defeat σ by playing a y with S_y wellfounded of rank greater than $\sup\{f(\beta) : \beta \leq \alpha\}$]. So, let τ be a winning strategy for player II (more precisely, a real coding such a strategy as above). Then V_τ is wellfounded and for all $\alpha \geq \omega$ we have $|V_\tau \restriction \alpha| > f(\alpha)$. To see this, let $x \in \mathrm{WF}$ of rank α. Fix $\vec{\alpha} = (\alpha_0, \alpha_1, \dots) \in \alpha^{\omega}$ such that $(x, \vec{\alpha}) \in [U]$. Let $y = \tau(x)$, so S_y is wellfounded of rank $|S_y| > \sup\{f(\beta) : \beta \leq |x|\} \geq f(|x|) = f(\alpha)$. Thus, $V_{\tau,x,\vec{\alpha},y} = S_y$ has rank greater that $f(\alpha)$. So certainly $V_\tau \restriction \alpha$ has rank greater than $f(\alpha)$.

So, V is our Kunen tree except for the minor fact that it is not quite a tree on $\omega \times \omega_1$. To get the actual Kunen tree T, simply weave the last four coordinates of V into the second coordinate of T. This does not decrease rank, and so T has the desired property. ⊣

Finally, we recall some terminology and facts concerning partition relations. As we said before, partition properties play a central role in the theory of the projective ordinals, and in description theory in particular. The classical Erdős-Rado partition property is stated in the next definition. Recall $(\kappa)^\lambda$ denotes the set of increasing functions from λ to κ.

Definition 2.12. $\kappa \longrightarrow (\kappa)^\lambda$ if for every partition $\mathcal{P} : (\kappa)^\lambda \to \{0, 1\}$, there is an $H \subseteq \kappa$ of size κ and an $i \in \{0, 1\}$ such that for all $f \in H^\lambda$ we have $\mathcal{P}(f) = i$. We say H is **homogeneous** for the partition $\mathcal{P}$. We say κ has the **weak partition property** if $\kappa \longrightarrow (\kappa)^\lambda$ for all $\lambda < \kappa$. We say κ has the **strong partition property** if $\kappa \longrightarrow (\kappa)^\kappa$.

Any non-trivial (*i.e.*, $\lambda > 1$) exponent partition relation on κ implies κ is regular, so we henceforth assume that. Also, any infinite exponent partition property is inconsistent with AC, so it is important that we are in the full AD context when discussing such relations.

A simple but important reformulation of the partition properties is used heavily. In this reformulation, the homogeneous sets are required to be c.u.b. subsets of κ. In order to be able to do this, we must specify the *type* of the functions $f : \lambda \to \kappa$ being considered. The simplest type is what in [Jac88] is called of *the correct type.* This notion arises naturally in Martin's proof of the strong partition property on ω_1. In [Kec78] the notation $C^{\not\uparrow\lambda}$ is used to describe the set of functions from λ to C of the correct type.

DEFINITION 2.13. We say $f\colon \lambda \to \mathrm{Ord}$ is of **uniform cofinality** ω if there is an $f'\colon \lambda \times \omega \to \mathrm{Ord}$ which is increasing in the second argument such that for all $\alpha < \lambda$, $f(\alpha) = \sup_n f'(\alpha, n)$. We say f is of the **correct type** if f is increasing, everywhere discontinuous (*i.e.*, $f(\alpha) > \sup_{\beta<\alpha} f(\beta)$), and of uniform cofinality ω. We say f has uniform cofinality ω almost everywhere with respect to a measure μ on λ if there is an $f'\colon \lambda \times \omega \to \mathrm{Ord}$ which is increasing in the second argument such that for all α is a μ measure one set we have $f(\alpha) = \sup_n f'(\alpha, n)$. We say f is of the correct type almost everywhere if f is of uniform cofinality ω almost everywhere, and there is a μ measure one set A such that $f \restriction A$ is strictly increasing and discontinuous.

Saying $f\colon \lambda \to \mathrm{Ord}$ is of the correct type is equivalent to saying that there is a $g\colon \lambda \times \omega \to \mathrm{Ord}$ which is increasing with respect to lexicographic ordering on $\lambda \times \omega$ which induces f in the sense that $f(\alpha) = \sup_n g(\alpha, n)$. This is also equivalent to saying that there is an increasing $h\colon \omega \cdot \lambda \to \mathrm{Ord}$ such that $f(\alpha) = \sup_{\beta<\omega\cdot(\alpha+1)} h(\beta)$. Both of these terminologies are used frequently. However, as one moves further into the theory of the projective ordinals it becomes important to have the more general notions of type and uniform cofinality. We give the general definition next.

DEFINITION 2.14. Let $g\colon \lambda \to \mathrm{Ord}$. We say $f\colon \lambda \to \mathrm{Ord}$ is of uniform cofinality g if there is a function f' with domain $\{(\alpha, \beta) : \alpha < \lambda, \beta < g(\alpha)\}$ which is increasing in the second argument and such that $f(\alpha) = \sup_{\beta<g(\alpha)} f'(\alpha, \beta)$.

If g is the constant function δ, then we simply say f has uniform cofinality δ. We say f is of **type** g if it is increasing, everywhere discontinuous, and of uniform cofinality g. In the same manner as above, we define the almost everywhere versions of these notions with respect to a measure μ on λ.

DEFINITION 2.15. We say $\kappa \overset{\text{c.u.b.}}{\longrightarrow} (\kappa)^\lambda$ if for every partition $\mathcal{P}$ of the functions $f\colon \lambda \to \kappa$ of the correct type, there is a c.u.b. $C \subseteq \kappa$ which is homogeneous for $\mathcal{P}$.

In [Sol78A] an easy argument is given which shows the essential equivalence of these two forms of the partition property. Specifically we have the following.

FACT 2.16. For all λ, κ, we have $\kappa \overset{\text{c.u.b.}}{\longrightarrow} (\kappa)^\lambda$ implies $\kappa \longrightarrow (\kappa)^\lambda$. Also, $\kappa \longrightarrow (\kappa)^{\omega\cdot\lambda}$ implies $\kappa \overset{\text{c.u.b.}}{\longrightarrow} (\kappa)^\lambda$.

Henceforth, in writing $\kappa \longrightarrow (\kappa)^\lambda$ we will be referring to the c.u.b. version of the partition property.

In particular, the definitions coincide for κ having the weak or strong partition property. More generally, we get a partition property for function of general types. For example, if κ has the strong partition property then for any $g: \kappa \to \kappa$ we have the c.u.b. version of the partition property for functions of type g.

In proving the strong partition property on ω_1, Martin established a general result which all subsequent proofs of partition properties from AD have used. It says, roughly, that to show $\kappa \longrightarrow (\kappa)^\lambda$ it suffices to have a coding of the functions $f : \lambda \to \kappa$ by reals with a certain boundedness property. A general version of this principle (the following general version is proved in [Jac10]; see also [Kec78]) can be stated as follows.

THEOREM 2.17 (Martin). Let $\undertilde{\Gamma}$ be non-selfdual, closed under $\exists^{\omega\omega}$, $\wedge$, and $\mathrm{PWO}(\check{\undertilde{\Gamma}})$. Assume there is a map $\varphi : {}^\omega\omega \to \wp(\lambda \times \kappa)$ satisfying the following:

1. $\forall f : \lambda \to \kappa\, \exists x \in {}^\omega\omega\, (\varphi(x) = f)$.
2. $\forall \alpha < \lambda\, \forall \beta < \kappa\; R_{\alpha,\beta} \in \undertilde{\Delta} = \undertilde{\Gamma} \cap \check{\undertilde{\Gamma}}$, where
$$x \in R_{\alpha,\beta} \leftrightarrow \varphi(x)(\alpha, \beta) \wedge \forall \beta' \, (\varphi(x)(\alpha, \beta') \to \beta' = \beta)$$
3. Suppose $\alpha < \lambda$, $A \in \exists^{\omega\omega}\undertilde{\Delta}$, and $A \subseteq R_\alpha \doteq \{x : \exists \beta < \kappa\; x \in R_{\alpha,\beta}\}$. Then there is a $\beta_0 < \kappa$ such that $\forall x \in A\, \exists \beta < \beta_0\, \varphi(x)(\alpha, \beta)$.

Then $\kappa \longrightarrow (\kappa)^\lambda$.

REMARK 2.18. With some additional pointclass arguments one can see that the hypotheses (1)–(3) of Theorem 2.17 and the closure of $\undertilde{\Gamma}$ under $\exists^{\omega\omega}$ actually imply $\undertilde{\Gamma}$ is closed under $\wedge$ and $\mathrm{PWO}(\check{\undertilde{\Gamma}})$ if we assume κ is regular (see [Jac10]; if $\undertilde{\Gamma}$ is closed under quantifiers, the statement $\mathrm{PWO}(\check{\undertilde{\Gamma}})$ just defines the $\check{\undertilde{\Gamma}}$ side).

We say a coding map $\varphi : {}^\omega\omega \to \wp(\lambda \times \kappa)$ is **λ-reasonable** if it satisfies the hypotheses of Theorem 2.17 (for some $\undertilde{\Gamma}$).

The countable exponent relations $\undertilde{\delta}^1_{2n+1} \longrightarrow (\undertilde{\delta}^1_{2n+1})^\lambda$, $\lambda < \omega_1$, follow from Theorem 2.17 using a straightforward coding map φ. Namely, let $\pi : \lambda \to \omega$ be a bijection. Fix a (regular) $\undertilde{\Pi}^1_{2n+1}$ norm ψ on a $\undertilde{\Pi}^1_{2n+1}$-complete set $P \subseteq {}^\omega\omega$. So, ψ is onto $\undertilde{\delta}^1_{2n+1}$. Define φ by $\varphi(x)(\alpha, \beta)$ iff $(x)_{\pi(\alpha)} \in P$ and $\psi((x)_{\pi(\alpha)}) = \beta$. Using $\undertilde{\Gamma} = \undertilde{\Sigma}^1_{2n+1}$, it is straightforward to verify (1)–(3) of Theorem 2.17. In particular, $\undertilde{\delta}^1_1 = \omega_1$ has the weak partition property, and we may take this as the start of our inductive analysis of the projective ordinals.

In [Kec78] the countable exponent relation for the even $\undertilde{\delta}^1_{2n+2}$ is also shown, a result due to Kunen (this does not use Theorem 2.17 directly, but uses a partition argument on the odd $\undertilde{\delta}^1_{2n+1}$). Kechris [Kec77A] has also shown the strong partition relation on ω_1 by an argument using generic codes for countable ordinals. Kechris and Woodin [KW80] have developed a theory of generic codes for uncountable ordinals which allowed them to show directly the relation $\undertilde{\delta}^1_{2n+1} \longrightarrow (\undertilde{\delta}^1_{2n+1})^{\undertilde{\delta}^1_{2n-1}}$. Both of these results also use Theorem 2.17. The

Kechris-Woodin theory of generic codes also has numerous other applications in the theory of the projective ordinals and beyond.

A notational convention. We describe a certain notational convention which is used throughout in the theory of descriptions and here as well. It is a convention for making statements about iterated ultrapowers. Suppose $\mu_1, \ldots, \mu_n$ is a sequence of measures. Suppose $\vartheta \in \mathrm{Ord}$ and $P \subseteq \mathrm{Ord}$ (we view P as a property or statement about an ordinal). We write $\forall_{\mu_1}\alpha_1 \cdots \forall_{\mu_n}\alpha_n\, P(\vartheta(\alpha_1, \ldots, \alpha_n))$ to abbreviate the following statement: if we fix a function $\alpha_1 \to \vartheta(\alpha_1)$ representing ϑ with respect to μ_1 (that is, ϑ is represented by the equivalence class of $\alpha_1 \mapsto \vartheta(\alpha_1)$ in the ultrapower by μ_1), then for μ_1 almost all α_1 we have that if $\alpha_2 \mapsto \vartheta(\alpha_1, \alpha_2)$ represents $\vartheta(\alpha_1)$ with respect to μ_2, then for μ_2 almost all α_2 we have that ..., for μ_{n-1} almost all α_{n-1} if $\alpha_n \mapsto \vartheta(\alpha_1, \ldots, \alpha_n)$ represents $\vartheta(\alpha_1, \ldots, \alpha_{n-1})$ with respect to μ_n, then for μ_n almost all α_n we have $P(\vartheta(\alpha_1, \ldots, \alpha_n))$.

Thus, the statement $\forall_{\mu_1}\alpha_1 \cdots \forall_{\mu_n}\alpha_n\, P(\vartheta(\alpha_1, \ldots, \alpha_n))$ is equivalent to saying $\vartheta \in j_{\mu_1} \circ \cdots \circ j_{\mu_n}(P)$. However, in practice the convention is always used in the form as stated.

§3. Outline of the Arguments. We give in this section an overview of the arguments used in the projective ordinal analysis. We present the overall plan and leave it to the remaining sections to fill in more details. As we said before, the arguments are inductive in nature. At stage n in the induction our main inductive assumptions will be the following.

We let $\omega(0) = 1$ and $\omega(n+1) = \omega^{\omega(n)}$ (ordinal exponentiation).

Inductive Hypotheses (stage *n*):

1. For $m \leq n$, $\underset{\sim}{\boldsymbol{\delta}}^1_{2m+1} = \aleph_{\omega(2m-1)+1}$ (for $m = 0$, $\underset{\sim}{\boldsymbol{\delta}}^1_1 = \omega_1$).
2. $\underset{\sim}{\boldsymbol{\delta}}^1_{2m+1}$ for $m < n$ has the strong partition property.
3. $\underset{\sim}{\boldsymbol{\delta}}^1_{2n+1}$ has the weak partition property.

In addition to these main inductive hypotheses at stage n we also carry along several more technical induction hypotheses. Specifically, we also assume:

Auxiliary Inductive Hypotheses:

(a) For $m \leq n$, $\underset{\sim}{\boldsymbol{\delta}}^1_{2m+1}$ is closed under ultrapowers. That is, if μ is a measure on $\vartheta < \underset{\sim}{\boldsymbol{\delta}}^1_{2m+1}$ and $\alpha < \underset{\sim}{\boldsymbol{\delta}}^1_{2m+1}$, then $j_\mu(\alpha) < \underset{\sim}{\boldsymbol{\delta}}^1_{2m+1}$.

(b) For $m \leq n$ there is a $\underset{\sim}{\boldsymbol{\Delta}}^1_{2m+1}$ coding of the subsets of λ_{2m+1}. That is, there is a map π from ${}^\omega\omega$ onto $\wp(\lambda_{2m+1})$ such that for all $\alpha < \lambda_{2m+1}$ we have that $\{x \in {}^\omega\omega : \alpha \in \pi(x)\} \in \underset{\sim}{\boldsymbol{\Delta}}^1_{2m+1}$.

(c) For all regular cardinals $\kappa < \lambda_{2n+1}$ there is a set $P \subseteq {}^\omega\omega$, and a homogeneous tree T on $\omega \times \lambda$, $\lambda \leq \lambda_{2n+1}$, with $P = \mathrm{p}[T]$ and such that if $\{\psi_i\}_{i \in \omega}$ is the semiscale on P from T, then $\{\psi_0(x) : x \in P\}$ is an unbounded subset of κ (recall $\psi_i(x)$ is the ith coordinate of the leftmost branch of T_x, for $x \in P$).

(d) Every measure μ on λ_{2n+1} is $\utilde{\mathbf{\Delta}}^1_{2n+1}$ in the codes given by (b). That is, $\{x : \mu(\pi(x)) = 1\} \in \utilde{\mathbf{\Delta}}^1_{2n+1}$.

REMARK 3.1. The hypotheses (b) and (d) above actually both follow from a single more technical statement which gives the analysis of measures on λ_{2n+1}. We will illustrate this analysis in the following sections. Also, (a) follows easily from (b) and (d). Thus, we could replace (a), (b), and (d) with the statement that we have a reasonable analysis of the measures on λ_{2n+1}. Statement (c) is necessary to propagate the existence of the Martin tree, which we discuss later.

To complete the induction we must compute $\utilde{\boldsymbol{\delta}}^1_{2n+3} = \aleph_{\omega(2n+1)+1}$, prove the strong partition relation of $\utilde{\boldsymbol{\delta}}^1_{2n+1}$, and prove the weak partition relation at $\utilde{\boldsymbol{\delta}}^1_{2n+3}$ an in addition establish (a), (b), (c), and (d) at $n+1$.

Consider first the problem of computing $\utilde{\boldsymbol{\delta}}^1_{2n+3}$. The upper and lower bounds are obtained by different methods. We consider first the upper-bound. Historically, obtaining the upper-bound for $\utilde{\boldsymbol{\delta}}^1_5$ was the first problem for which the previous pre-description arguments proved inadequate. The upper-bound for $\utilde{\boldsymbol{\delta}}^1_{2n+3}$ involves centrally the notion of a homogeneous tree, and this was the original idea of the Kunen-Martin plan.

It turns out that the notion of homogeneous tree plays, somewhat indirectly, another important role in the theory. Namely, when we analyze an arbitrary measure μ on $\utilde{\boldsymbol{\delta}}^1_{2n+1}$ or λ_{2n+3} (which we will do at this stage of the induction) that they are closely related to the measures appearing in the homogeneous tree construction on $\utilde{\mathbf{\Pi}}^1_{2n+1}$, $\utilde{\mathbf{\Pi}}^1_{2n+2}$ sets respectively. With a sufficiently general definition of homogeneous tree, an arbitrary measure is equivalent to a product of such homogeneity measures. More precisely, the general measure μ will be a "lift-up" of a product of homogeneity measures. The lift-up process will involve descriptions and the Martin tree. In §4 we show arbitrary measures are generated on $\utilde{\boldsymbol{\delta}}^1_1 = \omega_1$ and $\lambda_3 = \omega_\omega$ using homogeneity measures on trees for $\utilde{\mathbf{\Pi}}^1_1$ and $\utilde{\mathbf{\Pi}}^1_2$ sets. The lifting-up here will only use the trivial descriptions and the Kunen tree. In §5, 6 we show how the next level (non-trivial) descriptions generate arbitrary measures on $\utilde{\boldsymbol{\delta}}^1_3$, using these description to lift-up what are essentially homogeneity measures on $\utilde{\mathbf{\Pi}}^1_3$ sets.

Because of the importance of homogeneous trees in the inductive analysis, we recall the definition and the important Martin-Solovay construction for propagating Suslin representations using homogeneous trees. If μ is a measure on a set X and $f : X \to Y$, let $f(\mu)$ denote the measure on Y given by $f(\mu)(B) = \mu(f^{-1}(B))$.

DEFINITION 3.2. A tree T on $\omega \times \lambda$ is **homogeneous** if there are measures μ_s, for $s \in \omega^{<\omega}$ such that $T_s \neq \varnothing$, satisfying:

1. $\mu_s(T_s) = 1$ (recall $T_s = \{\vec{\alpha} \in \lambda^{\mathrm{lh}(s)} : (s, \vec{\alpha}) \in T\}$.
2. If t extends s, then $\pi_{s,t}(\mu_t) = \mu_s$. Here $\pi_{s,t} : \lambda^{\mathrm{lh}(t)} \to \lambda^{\mathrm{lh}(s)}$ is the natural projection map: $\pi_{s,t}(\alpha_0, \ldots, \alpha_{\mathrm{lh}(t)-1}) = (\alpha_0, \ldots, \alpha_{\mathrm{lh}(s)-1})$.
3. If $x \in {}^{\omega}\omega$ and $[T_x] \neq \varnothing$ (*i.e.*, T_x is illfounded), and if $\{A_n\}_{n\in\omega}$ are given with $\mu_{x\restriction n}(A_n) = 1$, then $\exists f \in \lambda^{\omega}\, \forall n\, (x\restriction n, f\restriction n) \in A_n$.

Clause (3) is the key homogeneity condition. It says that T is homogeneous in the sense that whenever T_x has a branch, then a branch can be found in any sequence of measure one sets for the $\mu_{x\restriction n}$. We write $(T, \vec{\mu})$ for the homogeneous tree T along with the measures $\vec{\mu}$ witnessing homogeneity.

We extend Definition 3.2 in an obvious way to trees on $\omega \times \omega \times \lambda$, *etc.*, (in this case, the measures are indexed as $\mu_{s,t}$, where $\mathrm{lh}(s) = \mathrm{lh}(t)$).

We say $A \subseteq {}^{\omega}\omega$ is **homogeneously Suslin** if there is a homogeneous tree T with $A = \mathrm{p}[T]$. We say A is **weakly homogeneously Suslin** if A is the existential quantification of a set $B \subseteq {}^{\omega}\omega \times {}^{\omega}\omega$ (*i.e.*, $A(x) \leftrightarrow \exists y\ B(x, y)$) and B is homogeneously Suslin. So, A is weakly homogeneously Suslin if there is a homogeneous tree T on $\omega \times \omega \times \lambda$ such that $A = \mathrm{p}[T]$, that is, $A(x) \leftrightarrow \exists y \in {}^{\omega}\omega\ \exists f \in \lambda^{\omega}\ (x, y, f) \in [T]$. We can code the last two coordinates of T into a single ordinal coordinate, and this leads to the notion of a weakly homogeneous tree: a tree which is isomorphic to such a coded tree. In practice, we work directly with the homogeneous tree on $\omega \times \omega \times \lambda$. One can also give an "abstract" definition of a tree T on $\omega \times \lambda$ being weakly homogeneous. We will not need this here, and refer the reader to Definition 4.4 of [Jac08] for the precise statement.

We recall next the Martin-Solovay construction. Let $\{t_i\}$ be an enumeration of all $t \in \omega^{<\omega} \times \omega^{<\omega}$ with all sequences preceding any proper extension in the enumeration.

Definition 3.3 (Martin-Solovay Tree). Let $(T, \vec{\mu})$ be a homogeneous tree on $\omega \times \omega \times \lambda$ (so $A = \mathrm{p}[T]$ is weakly homogeneous). Let $\delta \geq \lambda^+$. Then the Martin-Solovay tree $\mathrm{ms}(T, \vec{\mu}, \delta)$ is the tree defined as follows. We define $(s, \vec{\alpha}) \in \mathrm{ms}(T, \vec{\mu}, \delta)$ if there is an $f : T_s \to \delta$ which is order-preserving with respect to the Brouwer-Kleene order on T_s such that $\forall i < \mathrm{lh}(s)\ (\alpha_i = [f^i]_{\mu_{s\restriction \mathrm{lh}(t_i)), t_i}})$. Here $f^i(\vec{\gamma}) = f(t_i, \vec{\gamma})$ (if $T_{s\restriction \mathrm{lh}(t_i), t_i} = \varnothing$, we set $\alpha_i = 0$).

Remark 3.4. In specifying the Brouwer-Kleene ordering on T_s, we must say how $\omega \times \lambda$ is identified with an ordinal. Usually we do this by ordering by reverse lexicographic ordering (*i.e.*, order by the ordinal coordinate first, then the integer coordinate). Also, we may restrict the type of the function f, for example we may require that f be of the correct type. We officially adopt the correct type restriction for our Martin-Solovay trees.

The basic property of the Martin-Solovay tree is given in the next theorem. We refer the reader to [Jac08, Theorem 4.10] for a proof (we give the proof in a special case after Corollary 4.4, the proof in the general case is similar).

THEOREM 3.5. Let $(T, \vec{\mu})$ be a homogeneous tree on $\omega \times \omega \times \lambda$, $\delta \geq \lambda^+$, and $S = \mathrm{ms}(T, \vec{\mu}, \delta)$. Then for all $x \in {}^\omega\omega$, T_x is illfounded iff S_x is wellfounded (*i.e.*, $\mathrm{p}[S] = {}^\omega\omega \setminus \mathrm{p}[T]$).

To propagate the Martin-Solovay construction through the projective hierarchy requires showing that the Martin-Solovay tree is itself homogeneous. This requires the use of partition properties. The exact manner in which this is done varies depending on whether we are at an odd or an even level of the projective hierarchy. Let us consider the case of propagating from an even to an odd level. Suppose we assume inductively that every $\utilde{\mathbf{\Pi}}^1_{2n}$ set is the projection of a homogeneous tree on $\omega \times \lambda_{2n+1}$. Thus, every $\utilde{\mathbf{\Sigma}}^1_{2n+1}$ set A is the projection of a homogeneous tree T on $\omega \times \omega \times \lambda_{2n+1}$. We apply Definition 3.3 with $\delta = \lambda^+_{2n+1} = \utilde{\boldsymbol{\delta}}^1_{2n+1}$. Assume (according to our hypotheses at stage n of the analysis) that $\utilde{\boldsymbol{\delta}}^1_{2n+1}$ has the weak partition property. Then $S = \mathrm{ms}(T, \delta)$ projects to ${}^\omega\omega \setminus A$, a $\utilde{\mathbf{\Pi}}^1_{2n+1}$ set. It will be homogeneous according to the following fact.

LEMMA 3.6. Let T, $\vec{\mu}$, λ, δ be as in Definition 3.3, and assume $\delta \to (\delta)^{<\delta}$. Then $S = \mathrm{ms}(T, \vec{\nu}, \delta)$ is a homogeneous tree om $\omega \times \utilde{\boldsymbol{\delta}}^1_{2n+1}$ with $\utilde{\boldsymbol{\delta}}^1_{2n+1}$-complete measures.

PROOF. For $s \in \omega^{<\omega}$, the measure ν_s on S_s is the one induced by the weak partition relation on $\utilde{\boldsymbol{\delta}}^1_{2n+1}$ and functions $f \colon T_s \to \utilde{\boldsymbol{\delta}}^1_{2n+1}$ of the correct type. That is, $E \subseteq (\utilde{\boldsymbol{\delta}}^1_{2n+1})^{\mathrm{lh}(s)}$ has ν_s measure one if there is a c.u.b. $C \subseteq \utilde{\boldsymbol{\delta}}^1_{2n+1}$ such that for all $f \colon T_s \to C$ of the correct type we have (here $j = \mathrm{lh}(s)$):

$$\vec{\alpha} = ([f^0]_{\mu_{s \restriction \mathrm{lh}(t_0), t_0}}, \ldots, [f^{j-1}]_{\mu_{s \restriction \mathrm{lh}(t_{j-1}), t_{j-1}}}) \in E.$$

Here the subfunctions f^i of f are as in Definition 3.3. The weak partition relation on $\utilde{\boldsymbol{\delta}}^1_{2n+1}$ easily gives that ν_s is a $\utilde{\boldsymbol{\delta}}^1_{2n+1}$-complete measure, and the closure of $\utilde{\boldsymbol{\delta}}^1_{2n+1}$ under ultrapowers (one of our inductive hypotheses) gives that ν_s is a measure on $\utilde{\boldsymbol{\delta}}^1_{2n+1}$ (note that the $\utilde{\boldsymbol{\delta}}^1_{2n+1}$-completeness of the ω-c.u.b. filter on $\utilde{\boldsymbol{\delta}}^1_{2n+1}$ follows by an easy argument from the c.u.b. version of the partition relation $\utilde{\boldsymbol{\delta}}^1_{2n+1} \to (\utilde{\boldsymbol{\delta}}^1_{2n+1})^2$).

To show S is homogeneous, suppose $x \in {}^\omega\omega$ and S_x is illfounded, which by Theorem 3.5 is equivalent to saying T_x is wellfounded. Let $A_i \subseteq (\utilde{\boldsymbol{\delta}}^1_{2n+1})^i$ have $\nu_{x \restriction i}$ measure one. Let $C_i \subseteq \utilde{\boldsymbol{\delta}}^1_{2n+1}$ be c.u.b. and witness that A_i has $\nu_{x \restriction i}$ measure one. Let $C = \bigcap_i C_i$, so C is c.u.b. in $\utilde{\boldsymbol{\delta}}^1_{2n+1}$. Since T_x is a wellfounded tree on λ, it has rank less than $\lambda^+ \leq \delta$. Moreover, we an get an order-preserving map $f \colon T_x \to C$ of the correct type. Let $F^i \colon T_{x \restriction i} \to C$ be the subfunctions induced by f as in Definition 3.3. Let $\alpha_i = [f^i]_{\mu_{x \restriction \mathrm{lh}(t_i), t_i}} < \utilde{\boldsymbol{\delta}}^1_{2n+1}$. Then by definition of S we have $(x, \vec{\alpha}) \in [S]$. ⊣

The propagation from the odd to the even levels is similar, but with a slight difference. Now we start with a $\utilde{\mathbf{\Sigma}}^1_{2n+2}$ set, which is the projection of a homogeneous tree on $\omega \times \omega \times \underset{\sim}{\boldsymbol{\delta}}^1_{2n+1}$. We employ now a slight variation of the Martin-Solovay construction in which we use $\delta = \underset{\sim}{\boldsymbol{\delta}}^1_{2n+1}$ (so $\delta = \lambda$ now instead of $\delta \geq \lambda^+$). That we can do this follows from a property of the Martin-Solovay tree T constructed above on a $\underset{\sim}{\boldsymbol{\Pi}}^1_{2n+1}$ set. Namely, the tree T has the property that if $(s, \vec{\alpha}) \in T$, where $\vec{\alpha} = (\alpha_0, \ldots, \alpha_{\mathrm{lh}(s)-1})$, then for all i we have $\alpha_i \leq \sup_\mu j_\mu(\alpha_0)$, where the supremum ranges over the measures μ in the homogeneous tree used to construct T (so the μ are measures on λ_{2n+1}). Because of this boundedness property it follows that for any $\alpha < \underset{\sim}{\boldsymbol{\delta}}^1_{2n+1}$ that $T^\alpha_x = \{\vec{\alpha} \in T_x : \alpha_0 \leq \alpha\}$ is a tree on an ordinal less than $\underset{\sim}{\boldsymbol{\delta}}^1_{2n+1}$. Hence, if T_x is wellfounded then for all α, T^α_x has rank less than $\underset{\sim}{\boldsymbol{\delta}}^1_{2n+1}$. In particular, if T_x is wellfounded than the rank of any $\vec{\alpha} \in T_x$ with respect to the Brouwer-Kleene ordering is less than $\underset{\sim}{\boldsymbol{\delta}}^1_{2n+1}$. Thus, we may use $\delta = \underset{\sim}{\boldsymbol{\delta}}^1_{2n+1}$ in Definition 3.3. The proof of the homogeneity of S now follows as in even-to-odd case, except that to get the measures ν_s on the equivalence classes of functions $f : T_x \to \underset{\sim}{\boldsymbol{\delta}}^1_{2n+1}$ of the correct type we must now use the strong partition relation on $\underset{\sim}{\boldsymbol{\delta}}^1_{2n+1}$ (since T_x is a tree on $\underset{\sim}{\boldsymbol{\delta}}^1_{2n+1}$ now).

Part of our job at stage n is to prove the strong partition relation on $\underset{\sim}{\boldsymbol{\delta}}^1_{2n+1}$, so assuming we do this, this will propagate the existence of homogeneous trees from $\underset{\sim}{\boldsymbol{\Pi}}^1_{2n}$ to $\underset{\sim}{\boldsymbol{\Pi}}^1_{2n+1}$ and $\underset{\sim}{\boldsymbol{\Pi}}^1_{2n+2}$. Note that the propagation of homogeneous trees uses only the partition properties of the $\underset{\sim}{\boldsymbol{\delta}}^1_{2n+1}$, and makes no direct reference to descriptions.

As an immediate corollary we get a certain computation for the upper-bound for $\underset{\sim}{\boldsymbol{\delta}}^1_{2n+3}$. This expresses the idea of the Kunen-Martin program.

Corollary 3.7. Assuming the inductive hypotheses at stage n,

$$\underset{\sim}{\boldsymbol{\delta}}^1_{2n+3} \leq \left(\sup_\mu j_\mu(\underset{\sim}{\boldsymbol{\delta}}^1_{2n+1})\right)^+,$$

where the supremum ranges over the measures μ in the homogeneous tree on a $\underset{\sim}{\boldsymbol{\Pi}}^1_{2n+1}$-complete set.

Proof. Assuming the stage n hypotheses we have shown that every $\underset{\sim}{\boldsymbol{\Pi}}^1_{2n+2}$ is λ-Suslin where $\lambda = \sup_\mu j_\mu(\underset{\sim}{\boldsymbol{\delta}}^1_{2n+1})$ where the measures μ are from the homogeneous tree of a $\underset{\sim}{\boldsymbol{\Pi}}^1_{2n+1}$ set. Thus, $\underset{\sim}{\boldsymbol{\Pi}}^1_{2n+2}$ and hence $\underset{\sim}{\boldsymbol{\Sigma}}^1_{2n+3}$ is λ-Suslin. If the $\underset{\sim}{\boldsymbol{\Pi}}^1_{2n+2}$ set is $\underset{\sim}{\boldsymbol{\Pi}}^1_{2n+2}$-complete, then every $\underset{\sim}{\boldsymbol{\Sigma}}^1_{2n+3}$ will be λ-Suslin. This shows $\lambda^1_{2n+3} \leq \lambda$, and so $\underset{\sim}{\boldsymbol{\delta}}^1_{2n+3} \leq \lambda^+$. Note that the construction of the Martin-Solovay tree for the $\underset{\sim}{\boldsymbol{\Pi}}^1_{2n+2}$ set uses only the weak partition relation at $\underset{\sim}{\boldsymbol{\delta}}^1_{2n+1}$ (the strong partition relation at $\underset{\sim}{\boldsymbol{\delta}}^1_{2n+1}$ was only needed to get the homogeneity of this tree). ⊣

The computation of the lower-bound $\underset{\sim}{\boldsymbol{\delta}}^1_{2n+3} \geq \aleph_{\omega(2n+1)+1}$ is done using the strong partition relation on $\underset{\sim}{\boldsymbol{\delta}}^1_{2n+1}$ (which must be proved at stage n of the analysis). The lower-bound argument does not use descriptions, but proceeds independently from that analysis just using the partition properties of the $\underset{\sim}{\boldsymbol{\delta}}^1_{2n+1}$. In this sense it is similar to the propagation of homogeneous trees. The main tool used for the lower-bound is the following result of Martin (see [Jac10, Theorem 4.17] for a proof).

THEOREM 3.8 (Martin). Suppose $\kappa \to (\kappa)^\kappa$. Then for any measure μ on κ, $j_\mu(\kappa)$ is a cardinal.

Assuming the weak partition relation on $\underset{\sim}{\boldsymbol{\delta}}^1_{2n+1}$, one defines a collection of measures on $\underset{\sim}{\boldsymbol{\delta}}^1_{2n+1}$ corresponding to the Cantor normal forms for the ordinals below $\omega(2n+1)$. One then proves embedding results which show that if μ_1 corresponds to a smaller ordinal than μ_2 then $j_{\mu_1}(\underset{\sim}{\boldsymbol{\delta}}^1_{2n+1}) < j_{\mu_2}(\underset{\sim}{\boldsymbol{\delta}}^1_{2n+1})$. From Theorem 3.8 the lower-bound $\lambda_{2n+3} \geq \aleph_{\omega(2n+1)}$ follows (one computes directly that the ultrapowers $j_\mu(\underset{\sim}{\boldsymbol{\delta}}^1_{2n+1})$ are $\underset{\sim}{\boldsymbol{\Delta}}^1_{2n+3}$ in the codes and thus are below $\underset{\sim}{\boldsymbol{\delta}}^1_{2n+3}$, and being cardinals of uncountable cofinality are therefore below λ_{2n+3}). The reader can see [Jac99] for the full details of these arguments below $\underset{\sim}{\boldsymbol{\delta}}^1_5$. Also, see [JL] for the definitions of these measures in the general projective case.

To compute the right-hand side of the inequality in Corollary 3.7 and to carry out the remaining steps of the analysis at stage n of the induction requires the theory of descriptions. We show how the $n = 0$ arguments are done in §4. As we said before, this will use only "trivial descriptions," but will show how the previous theory fits into the modern point of view. In §§5, 6 we show how descriptions are used to do the $n = 1$ stage arguments. In §7 we make some comments about the general stage n. In the remainder of this section we make some general comments about what descriptions are and how they are used to do the inductive analysis.

Using the partition properties of the $\underset{\sim}{\boldsymbol{\delta}}^1_{2n+1}$ we will define certain families of *canonical measures*. The first canonical family, used in §4, will consist of the measures W^m_1, where W^m_1 is just the m-fold product of the normal measure on ω_1. The "W" in the notation stands for "weak," denoting that the measures are defined using the weak partition relation on $\underset{\sim}{\boldsymbol{\delta}}^1_1$. The second canonical family will be the measures S^m_1, used in §5. Each S^m_1 will be a measure on ω_{m+1}. The notation S^m_1 represents the fact that the measures are defined using the strong partition relation on $\underset{\sim}{\boldsymbol{\delta}}^1_1$. The third family will be the measures W^m_3, which are measures on $\underset{\sim}{\boldsymbol{\delta}}^1_3$ defined using the weak partition relation on $\underset{\sim}{\boldsymbol{\delta}}^1_3$. In the general case, defined in §7, there will be a family W^m_{2n+1} of measures on $\underset{\sim}{\boldsymbol{\delta}}^1_{2n+1}$ for each odd projective ordinal $\underset{\sim}{\boldsymbol{\delta}}^1_{2n+1}$, and finitely many families $S^{\ell,m}_{2n+1}$, $1 \leq \ell \leq 2^{n+1} - 1$, of measures on λ_{2n+3}. One can think of the measures W^m_{2n+1} as simplified versions of the measures occurring in the homogeneous tree on a $\underset{\sim}{\boldsymbol{\Pi}}^1_{2n+1}$ set, and the $S^{\ell,m}_{2n+1}$ as simplified versions of those occurring for a $\underset{\sim}{\boldsymbol{\Pi}}^1_{2n+2}$ set. It will be important along the way to show that these canonical measures dominate the more general homogeneity measures. This is made precise in two theorems called the local and global embedding theorems. We state these for the families W_1^m, S_1^m, and W_3^m in Theorems 6.11 and 6.15. Their complete proofs can be found in [Jac99], although we give an example of each in §7. The statements of the embedding theorems in the general case as well as their proofs can be found in [Jac88].

Descriptions are hereditarily finite objects and are defined with respect to a finite sequence of canonical measures $K_1, \dots, K_t$. They "describe" how to build an ordinal in the iterated ultrapower by these measures (we note that the iterated ultrapower is not the same as the ultrapower by the product measure $K_1 \times \cdots \times K_t$ under AD). The level-1 descriptions will be the trivial descriptions, which are just the positive integers. They describe how to build an ordinal in the ultrapower by a single measure of the form $K_1 = W_1^m$. The description $d = i$ describes the function $f(\alpha_1, \dots, \alpha_m) = \alpha_i$. This function represents the cardinal ω_i in the ultrapower by W_1^m (see §4 for details). Thus, the cardinals below $\underset{\sim}{\delta}_3^1$ are exactly the ordinals represented by the level-1 descriptions with respect to the measures W_1^m. The trivial descriptions are also used to prove the strong partition on $\underset{\sim}{\delta}_1^1$ and the weak partition relation on $\underset{\sim}{\delta}_3^1$ and thereby complete the $n = 0$ stage of the analysis.

The level-2 descriptions describe how to generate ordinals in the iterated ultrapower by a sequence of measures $K_1, \dots, K_t$ where each K_i is one of the canonical measures $K_i = W_1^{m_i}$ or $K_1 = S_1^{m_i}$. The level-3 descriptions will be just minor variations of the level-2 descriptions. They will be defined relative to a sequence $K_1, \dots, K_t$ where $K_1 = W_3^m$ and the K_i for $i \geq 2$ are of the form $K_i = W_1^{m_i}$ or $K_1 = S_1^{m_i}$. This pattern continues throughout the projective hierarchy. We give the exact definition of the level-2 and level-3 descriptions in §§5, 6. Again it will be the case that the cardinals below λ_5 exactly correspond to the set of level-3 descriptions. We note that in [Jac99] and [Jac88] the level-2 and level-3 descriptions were grouped together and likewise at the higher levels. For expository purposes we find it convenient to separate them here. The general level descriptions are defined in [Jac88].

Aside from describing the cardinal structure, the level-2 and level-3 descriptions (and similarly at the higher levels) are used to analyze arbitrary ordinals below λ_{2n+3} as well as analyze arbitrary measures on λ_{2n+3}. Arbitrary ordinals below λ_{2n+3} will be generated as the "lift" of a finite set of descriptions via the Martin tree. Roughly speaking, the main theorem on descriptions says that if an ordinal ϑ is less than the ordinal (actually cardinal) represented by a description d defined relative to the measure sequence $K_1, \dots, K_t$, then ϑ is less than the lift of a smaller description $\mathcal{L}(d)$ by some function $g \colon \underset{\sim}{\delta}_{2n+1}^1 \to \underset{\sim}{\delta}_{2n+1}^1$. The function g is in turn dominated by the ranking function on a wellfounded section of the Martin tree. However, unlike the Kunen tree where the relevant ranking function is given by $\alpha \mapsto |T_x \restriction \alpha|$ (see Theorem 2.11), for the Martin tree M the relevant function is

$$\alpha \mapsto \left| M_x \restriction \sup_{\mu} j_\mu(\alpha) \right|,$$

where the supremum ranges over the measures in a homogeneous tree for a $\underset{\sim}{\mathbf{\Pi}}^1_{2n}$-complete set. From the embedding theorems, we can replace the measures in this supremum by those in the canonical families. This forces us, however, to lengthen the sequence of measures. That is, the lift of $\mathcal{L}(d)$ by g will be bounded by the cardinal successor of the supremum of the cardinals represented by the smaller description $\mathcal{L}(d)$ with respect to sequences $K_1, \dots, K_t, K_{t+1}$ as K_{t+1} ranges over the canonical measures. Putting this together, this gives gives an upper bound for λ_{2n+3} in terms of the rank of a certain lowering operator $\mathcal{L}$ on the descriptions and measure sequences. Computing this rank is then a purely combinatorial problem, with the result being $\omega(2n+3)$. So, for example, this gives $\lambda_5 \leq \aleph_{\omega^{\omega^\omega}}$.

A variation of the analysis of ordinals of the previous paragraph allows us to analyze arbitrary measures on $\underset{\sim}{\boldsymbol{\delta}}^1_{2n+1}$ and then on λ_{2n+3}. In §4 we show how the trivial descriptions allow to analyze the measure on ω_1 and $\lambda_3 = \omega_\omega$. This was done originally using the theory of indiscernibles for $L[x]$ by Kunen (see [Sol78A]). In §6 we show how a typical measure on $\underset{\sim}{\boldsymbol{\delta}}^1_3$ is generated. The full details for this analysis, as well as the analysis of measure of λ_5 are given in [Jac99].

Finally, the analysis of measures on $\underset{\sim}{\boldsymbol{\delta}}^1_{2n+1}$ and λ_{2n+3} can be converted, via a clever argument of Kunen, into an analysis of arbitrary subsets of $\underset{\sim}{\boldsymbol{\delta}}^1_{2n+1}$ and λ_{2n+3}. This produces a coding of these subsets good enough to satisfy the hypotheses of Theorem 2.17 and get the strong partition relation on $\underset{\sim}{\boldsymbol{\delta}}^1_{2n+1}$ and the weak relation on $\underset{\sim}{\boldsymbol{\delta}}^1_{2n+3}$. The argument of Kunen referred to above appears in [Sol78A]. We present this argument explicitly in Lemma 4.8. We note one technical point here. In proving the strong partition relation on $\underset{\sim}{\boldsymbol{\delta}}^1_{2n+1}$, Martin's method (Theorem 2.17) requires us have a good coding of the functions $f : \underset{\sim}{\boldsymbol{\delta}}^1_{2n+1} \to \underset{\sim}{\boldsymbol{\delta}}^1_{2n+1}$. If we just view such a function as a subset of $\underset{\sim}{\boldsymbol{\delta}}^1_{2n+1} \times \underset{\sim}{\boldsymbol{\delta}}^1_{2n+1}$ and use the coding of subsets of $\underset{\sim}{\boldsymbol{\delta}}^1_{2n+1}$ (actually $\underset{\sim}{\boldsymbol{\delta}}^1_{2n+1} \times \underset{\sim}{\boldsymbol{\delta}}^1_{2n+1}$) directly given from the analysis of subsets of $\underset{\sim}{\boldsymbol{\delta}}^1_{2n+1}$, then the coding is not good enough to satisfy the requirement of Theorem 2.17. Instead we must modify the analysis of subsets to directly work with functions. The technical changes, however, end up being minor. In showing the weak partition relation on $\underset{\sim}{\boldsymbol{\delta}}^1_{2n+3}$ this technical problem does not arise.

Aside from the descriptions, the other main ingredients that go into the analysis of measures are the notions of a *tree of uniform cofinalities* and a *complex*. A tree of uniform cofinalities is a code for building a basic type of measure, roughly speaking a measure which occurs in the homogeneous tree construction. A level-n tree of uniform cofinalities will correspond to a measure in the homogeneous tree on a $\underset{\sim}{\mathbf{\Pi}}^1_n$ set. A level-n complex will be a level-n tree together with a way of "lifting" the basic measure to generate a more general measure. This lifting process will involve descriptions, though at the bottom level the descriptions are trivial. These more general measures will capture all measures.

In the next sections we fill in more details starting with the $n = 0$ stage of the induction in the next section. Again, there we will only use trivial descriptions. The reader wishing to see non-trivial descriptions as quickly as possible (and skip the measure analysis at the $n = 0$ stage) can skip directly to §5. In the following sections we will not always give complete proofs or even complete definitions. Rather, we attempt to illustrate the main concepts and arguments that are used in this theory. We will typically consider some illustrative examples rather than the general definitions or proofs. Hopefully this will give the reader a general understanding of the modern theory of the projective ordinals and how it generalizes and unifies the previous results. It should also give the reader a good background for [Jac99] or [Jac88]. We will reference the papers of this volume, as well as others, for many details.

§4. The First Level Theory. As we previously mentioned, the first level (stage $n = 0$) theory of the projective ordinals involves establishing the strong partition relation on $\underset{\sim}{\delta}^1_1 = \omega_1$, computing $\underset{\sim}{\delta}^1_3$, and showing the weak partition relation on $\underset{\sim}{\delta}^1_3$. Our starting hypothesis is the weak partition relation on $\underset{\sim}{\delta}^1_1 = \omega_1$ (which we observed earlier). Martin's original proof of the strong partition relation on $\underset{\sim}{\delta}^1_1$ can be found in [Kec78], and Kunen's original proof of the weak relation on $\underset{\sim}{\delta}^1_3$ can be found in [Sol78A]. Again, those proofs both used the theory of indiscernibles for $L[x]$. We will show how these arguments can be viewed as "trivial description" arguments.

If κ has the weak partition property and $\delta < \kappa$ is regular then there is a unique normal measure on κ concentrating on points of cofinality δ which we call the δ-cofinal normal measure. It is generated by sets of the form $C \cap S_\delta$, where $C \subseteq \kappa$ is c.u.b. and S_δ is the points of cofinality δ (to see this is a measure, consider the partition of $f : \delta \to \kappa$ of the correct type according to whether $\sup(f)$ lies in a given set).

In particular, the c.u.b. filter on ω_1 is a normal measure. Similarly, the m-fold product of the normal measure is generated by sets of the form C^m for $C \subseteq \omega_1$ a c.u.b. set. We may describe this measure as being induced by the weak partition relation on ω_1 and function $f : m \to \omega_1$. We make this into the following definition.

Definition 4.1. W^1_1 is the normal measure on ω_1. W^m_1 is the m-fold product of the normal measure on ω_1. We let $\mathcal{W}_1 = \{W^m_1\}_{m\in\omega}$ be the family of these measures.

Consider functions $F : \omega_1^m \to \omega_1$. There are m canonical functions given by $F_i(\alpha_1, \dots, \alpha_m) = \alpha_i$. We can view this as an instance of a description evaluation by viewing the descriptions as the set of positive integers, and the description $d = i$ corresponds to the function F_i. Although this is just a trivial notational change, it nevertheless allows us to introduce a notational framework which will generalize to the higher levels. We make this into the following definition.

DEFINITION 4.2. A level-1 description is a positive integer. We let $\mathcal{D}^1$ be the set of level-1 descriptions. We say $d \in \mathcal{D}^1$ is defined with respect to W_1^m if $d \leq m$. Given $\vec{\alpha} = (\alpha_1, \dots, \alpha_m) \in \omega_1^m$ and d defined with respect to W_1^m, let $(\vec{\alpha}; d) = \alpha_d$, which we call the **interpretation** of the description. We define $(W_1^m; d)$ to be the ordinal represented with with respect to the measure W_1^m by the function $\vec{\alpha} \mapsto (\vec{\alpha}; d)$. Given $g\colon \omega_1 \to \omega_1$ we define the **lift by** $\boldsymbol{g}$ by: $(g; \vec{\alpha}; d) = g(\vec{\alpha}; d) = g(\alpha_d)$, and define $(g; W_1^m; d)$ to be the ordinal represented by $\vec{\alpha} \mapsto (g; \vec{\alpha}; d)$.

Of course $(W_1^m; d)$ is just the ordinal represented by the canonical function F_d mentioned above. If we identify functions and the ordinals they represent, we may identify the canonical functions with the $(W_1^m; d)$. We let id stand for the identity function, so $(\text{id}; W_1^m; d) = (W_1^m; d)$.

Suppose now $F\colon \omega_1^m \to \omega_1$ is given and $F(\vec{\alpha}) < \alpha_i$ for W_1^m almost all $\vec{\alpha}$. In the language of descriptions this reads $[F] < (W_1^m; d)$, where $d = i$. An easy partition argument shows that there is a function $g\colon \omega_1 \to \omega_1$ such that $F(\vec{\alpha}) < g(\alpha_{i-1})$ almost everywhere (unless $i = 1$ in which case F is constant almost everywhere). To see this, consider the partition of tuples $(\alpha_1, \dots, \alpha_{i-1}, \beta, \alpha_i, \dots, \alpha_m)$ according to whether $\beta > F(\alpha_1, \dots, \alpha_m)$. Easily on the homogeneous side this property must hold. Let $C \subseteq \omega_1$ be c.u.b. and homogeneous for the partition. Let $g(\alpha) = N_C(\alpha) \doteq$ the least element of C greater than α. Then for almost all $\vec{\alpha}$ we have by homogeneity of C that $F(\vec{\alpha}) < g(\alpha_{i-1})$. Putting this back into the language of descriptions this becomes $[F] < (g; W_1^m; d - 1)$.

This suggests introducing a lowering operator $\mathcal{L}$ on $\mathcal{D}^1$ defined by $\mathcal{L}(d) = d-1$, unless $d = 1$ in which case d will be declared minimal. We can then state our observations in the following "main theorem" for level-1 descriptions.

THEOREM 4.3. If $[F] < (\text{id}; W_1^m; d)$ and d is not $\mathcal{L}$-minimal, then there is a $g\colon \omega_1 \to \omega_1$ such that $[F] < (g; W_1^m; \mathcal{L}(d))$. If d is $\mathcal{L}$-minimal, then $[F] < \omega_1$.

Continuing the analysis, fix $g\colon \omega_1 \to \omega_1$ such that $[F] < (g; W_1^m; \mathcal{L}(d))$. From Theorem 2.11, fix $x \in {}^\omega\omega$ such that the section of the Kunen tree T_x is wellfounded and $\forall^*_{W_1^1}\alpha\ g(\alpha) < |T_x{\restriction}\alpha|$ (actually this holds for all infinite α, but we don't need this). This now gives a map $\vartheta \mapsto \vartheta'$ from $(\text{id}; W_1^m; \mathcal{L}(d))$ onto $(\ell; W_1^m; \mathcal{L}(d)) > [F]$, where ℓ is the function $\ell(\alpha) = |T_x{\restriction}\alpha|$. Namely, If $\vartheta < (\text{id}; W_1^m; \mathcal{L}(d))$, let ϑ' be represented with respect to W_1^m by $\vartheta'(\vec{\alpha}) = |T_x{\restriction}\beta(\vartheta(\vec{\alpha}))|$, where $\beta = (\vec{\alpha}; \mathcal{L}(d))$. This shows that $(W_1^m; d) \leq (W_1^m; \mathcal{L}(d))^+$. As an immediate corollary we get the computation for the ultrapowers $j_{W_1^m}(\omega_1)$ of ω_1 by the measures W_1^m (for the lower-bound of the corollary we are using the strong partition relation on ω_1 which is discussed later in this section).

Corollary 4.4. $j_{W_1^m}(\omega_1) = \omega_{m+1}$.

Proof. The upper bound was shown above. From Theorem 3.8, all of the $j_{W_1^m}(\omega_1)$ are cardinals and the lower-bound follows. ⊣

That this gives an upper bound for $\underset{\sim}{\delta}^1_3$ follows from the homogeneous tree construction of Corollary 3.7 and the fact that the homogeneity measures for the bottom level homogeneous tree—the Shoenfield tree—on a $\underset{\sim}{\Pi}^1_1$ set are isomorphic to measures of the form W_1^m. For the sake of completeness, and to illustrate the proof of Theorem 3.5, we construct this tree and the homogeneous tree on a $\underset{\sim}{\Pi}^1_2$ set. This will also help motivate the definition of a level-2 tree of uniform cofinalities (Definition 4.23). Again, the reader can consult [Kec81A] for more details on the homogeneous tree construction.

If $A \subseteq {}^\omega\omega \times {}^\omega\omega$ is $\underset{\sim}{\Sigma}^1_1$, then $A = \mathrm{p}[T]$ where T is a tree on $\omega \times \omega \times \omega$. The Shoenfield tree S is a tree on $\omega \times \omega \times \omega_1$ such that $p_{1,2}[S] = B \doteq {}^\omega\omega \setminus A$ (here $p_{1,2}[S]$ means the projection to the first two coordinates). Thus, the projection $\mathrm{p}[S] \subseteq {}^\omega\omega$ of S to the first coordinate is the $\underset{\sim}{\Sigma}^1_2$ set $A_2 = \exists^{\mathbb{R}} B$, and S witnesses A_2 is weakly homogeneous. Let $\{u_i\}$ enumerate $\omega^{<\omega}$ such that any sequence precedes any of its proper extensions in the enumeration. We may define S by:

$$(s, t, \vec{\alpha}) \in S \leftrightarrow \alpha_0 > \max\{\alpha_i\} \wedge \forall i, j < \mathrm{lh}(\vec{\alpha})\, [(\alpha_i < \alpha_j) \\ \leftrightarrow (u_i, u_j \in T_{s,t} \wedge u_i <_{\mathrm{BK}} u_j) \vee (u_j \in T_{s,t} \wedge u_i \notin T_{s,t}) \\ \vee (u_i, u_j \notin T_{s,t} \wedge i < j)].$$

In this definition, the ordinals in $S_{s,t}$ are ranking the sequences in $T_{s,t}$, using the Brouwer-Kleene ordering on these sequences, and we declare the sequences not in $T_{s,t}$ to be below any that are. It is straightforward to check that $\mathrm{p}[S] = {}^\omega\omega \setminus \mathrm{p}[T]$. Furthermore, for any $(s, \vec{\alpha}) \in S$, $\alpha_0 > \max\{\alpha_i\}$. For any s, t, there is a unique permutation $\pi_{s,t}$ of length $\mathrm{lh}(s)$ such that $\vec{\alpha} \in S_{s,t}$ iff $\vec{\alpha}$ is order-isomorphic to $\pi_{s,t}$ (that is, $\alpha_i < \alpha_j$ iff $\pi_{s,t}(i) < \pi_{s,t}(j)$, viewing $\pi_{s,t}$ as a bijection from $\mathrm{lh}(s)$ to $\mathrm{lh}(s)$). This last property is the homogeneity property of S. For s, t of length n, let $\nu_{s,t}$ be the measure on n-tuples of countable ordinals order isomorphic to $\pi_{s,t}$ which is induced by W_1^n and the permutation $\pi_{s,t}$. Clearly if s', t' extends s, t, then $\nu_{s',t'}$ projects to $\nu_{s,t}$ under the restriction map. The measures $\nu_{s,t}$, which are isomorphic to $W_1^{\mathrm{lh}(s)}$, witness that the tree S is homogeneous.

We now carry out explicitly, starting from S, the Martin-Solovay construction of the homogeneous tree U for the $\underset{\sim}{\Pi}^1_2$ set $B_2 = {}^\omega\omega \setminus A_2$. Order $\omega \times \omega_1$ by reverse lexicographic order (*i.e.*, order by the second coordinate first). Let $<_s$ be the Brouwer-Kleene ordering on S_s, using this ordering on pairs. For notational clarity, let also $t_0, t_i, \ldots$ also enumerate $\omega^{<\omega}$ with each sequence preceding its proper extensions in the enumeration (so $\mathrm{lh}(t_i) \leq i$).

Define the tree U by $(s, \vec{\beta}) \in U$ iff there is an $f : S_s \to \omega_1$ which is order-preserving with respect to $<_s$ and of the correct type and such that for all $i < \text{lh}(s)$ $(\beta_i = [f^{s_i,t_i}]_{\nu_{s_i,t_i}})$, where $s_i = s \upharpoonright \text{lh}(t_i)$, and f^{s_i,t_i} is the subfunction of f obtained by restricting f to S_{s_i,t_i}. By Corollary 4.4, U is a tree on $\omega \times \omega_\omega$.

We show that $\text{p}[U] = B_2 = {}^\omega\omega \setminus A_2 = {}^\omega\omega \setminus \text{p}[S]$, as in Theorem 3.5. So, we must show that for all $x \in {}^\omega\omega$ that U_x is illfounded iff S_x is wellfounded. If S_x is wellfounded, let $f_x : S_x \to \omega_1$ be order-preserving of the correct type. This is possible since the rank of any $(t, \vec{\alpha}) \in S_x$ is countable. Here we use the fact that $\alpha_0 > \max_{i<\text{lh}(t)} \alpha_i$ and the fact that we ordered $\omega \times \omega_1$ in reverse lexicographic ordering (so that if $(j, \beta) < (i, \alpha)$ then we must have $\beta \leq \alpha$). For each i let $\beta_i = [f_x^{s_i,t_i}]_{\nu_{s_i,t_i}}$, where $s_i = x \upharpoonright \text{lh}(t_i)$ and $f_x^{s_i,t_i}$ is the restriction of f_x to S_{s_i,t_i}. Clearly $(x, \vec{\beta}) \in [U]$. Conversely, suppose U_x is illfounded, say $(x, \vec{\beta}) \in [U]$. For each n, let $f_n : S_{x \upharpoonright n} \to \omega_1$ be order-preserving with respect to $<_{x \upharpoonright n}$ and of the correct type such that for all $i < n$, $[f_n^{s_i,t_i}]_{\nu_{s_i,t_i}} = \beta_i$, where $s_i = x \upharpoonright \text{lh}(t_i)$. For each i and all $n, m \geq i$, there is a c.u.b. set C defining a ν_{s_i,t_i} measure one set on which $f_n^{s_i,t_i}$ and $f_m^{s_i,t_i}$ agree. Intersecting countably many c.u.b. sets gives a c.u.b. set $C \subseteq \omega_1$ restricted to which $f_n^{s_i,t_i} = f_m^{s_i,t_i}$ for all $n, m, \geq i$. Restricted to C, the f_n define a single function f from $S_x \upharpoonright C$ to ω_1 which is order-preserving of the correct type. So, $S_x \upharpoonright C$ is wellfounded, and this implies S_x must also be wellfounded since if S_x were illfounded, we could find a branch through S_x with ordinals in the set C (this is the homogeneity property of S).

We have now constructed homogeneously Suslin representations for $\boldsymbol{\Pi}^1_1$ and $\boldsymbol{\Pi}^1_2$ sets. The former uses measure on ω_1 while the latter uses measures on ω_ω. In particular, every $\boldsymbol{\Pi}^1_2$ set, and so also every $\boldsymbol{\Sigma}^1_3$ set, is ω_ω-Suslin. Thus, $\lambda_3 \leq \omega_\omega$ and hence $\boldsymbol{\delta}^1_3 \leq \omega_{\omega+1}$.

The lower bound for $\boldsymbol{\delta}^1_3$ can be obtained in several ways. As $\text{cf}(\lambda_3) = \omega$ (*cf.* Theorem 2.9), and all the ω_n have uncountable cofinality (by countable choice), it follows that $\lambda_3 \geq \omega_\omega$. Thus, $\boldsymbol{\delta}^1_3 \geq \omega_{\omega+1}$. Alternatively, one can get the lower bound for $\boldsymbol{\delta}^1_3$ by computing directly that the ultrapower relations corresponding to the W_1^m are $\boldsymbol{\Delta}^1_3$. From Corollary 4.4 the lower bound $\boldsymbol{\delta}^1_3 \geq \omega_\omega$ follows. Since $\boldsymbol{\delta}^1_3$ is a cardinal of uncountable cofinality, we in fact have $\boldsymbol{\delta}^1_3 \geq \omega_{\omega+1}$.

Corollary 4.5. $\boldsymbol{\delta}^1_3 = \omega_{\omega+1}$

To get the strong partition relation on ω_1, Theorem 2.17 says that we must get a sufficiently good coding of the functions from $\omega_1 \to \omega_1$. Martin's original proof used indiscernibles for the $\mathbf{L}[x]$ to do this. This proof is given in [Kec78]. This coding is not known to generalize to the higher levels of the projective hierarchy. In [Jac90A] a proof of the strong partition relation on ω_1 is given using a coding coming from an analysis of measures on ω_1. This proof makes use of a result of Kunen (see [Sol78A]) who gave a general argument which shows how convert an analysis of measures on κ into an analysis of the subsets of κ. Kunen's argument plays an important role in the proofs of the partition relations. We give Kunen's result in Theorem 4.8. We use the following general lemma. Recall that Θ is the supremum of the lengths of the prewellorderings of ${}^\omega\omega$ (Θ is much larger than all the projective ordinals).

LEMMA 4.6. Let $\kappa < \Theta$ and let $\mathcal{F}$ be a countably additive filter on κ. Then there is a measure μ on κ extending $\mathcal{F}$ (*i.e.*, $\mathcal{F} \subseteq \mu$).

REMARK 4.7. From AD (which we are assuming) it follows that every ultrafilter on a set is necessarily countably additive. That is, every ultrafilter is a measure. So, Lemma 4.6 could be phrased as saying that every countably additive filter on $\kappa < \Theta$ can be extended to an ultrafilter.

PROOF. Since $\kappa < \Theta$, there is a prewellordering $\preceq$ of length κ. The coding lemma then gives a map $\pi \colon {}^\omega\omega \to \wp(\kappa)$ which is onto. To see this, let $\undertilde{\Gamma}$ be a non-selfdual pointclass containing $\preceq$ which is closed under $\exists^{\mathbb{R}}$ and $\wedge$. Let $U \subseteq {}^\omega\omega \times {}^\omega\omega$ be a universal $\undertilde{\Gamma}$ set (From Wadge's Lemma it follows that every non-selfdual pointclass has a universal set). The coding lemma implies that every $A \subseteq \kappa$ is $\undertilde{\Gamma}$ in the codes given by $\preceq$. That is, the code set $\{x \in \text{fld}(\preceq) : |x|_\preceq \in A\}$ is a $\undertilde{\Gamma}$ set. Thus, we can take $\pi(x) = \{\alpha < \kappa : \exists y \in \text{fld}(\preceq) \colon |y|_\preceq = \alpha \wedge U(x, y)\}$.

Recall from AD that there is a measure $\mathcal{M}$, the Martin measure, on the set $\mathcal{D}$ of Turing degrees (the countable set of reals Turing equivalent to fixed real). A set $A \subseteq \mathcal{D}$ has $\mathcal{M}$ measure one iff it contains a **cone** of degrees. That is, it contains a set of the form $\{d \in \mathcal{D} : d \geq_{\text{T}} x\}$ for some $x \in {}^\omega\omega$, where $\geq_{\text{T}}$ denotes Turing reduction (there is a slight abuse of notation here as d is a set of reals; what we mean is $y \geq_{\text{T}} x$ for any $y \in d$).

For d a degree, let $\alpha(d) < \kappa$ be the least ordinal in $\cap\{\pi(x) : x \in d \wedge \pi(x) \in \mathcal{F}\}$. This is well-defined by the countable additivity of $\mathcal{F}$. Let $\mu = \alpha(\mathcal{M})$ be the push-forward by α of the Martin measure (*i.e.*, $E \subseteq \kappa$ has μ measure one iff $\{d \in \mathcal{D} : \alpha(d) \in E\}$ has $\mathcal{M}$ measure one). Then μ is a measure on κ giving all elements of $\mathcal{F}$ measure one. To see this, suppose $E \in \mathcal{F}$. Let $\pi(y) = E$. Then if $d \geq_{\text{T}} y$, that is y is in the degree d, we have that $\cap\{\pi(x) : x \in d \wedge \pi(x) \in \mathcal{F}\} \subseteq E$ and so $\alpha(d) \in E$. ⊣

THEOREM 4.8 (Kunen). Let $\kappa < \Theta$, and suppose $\mathcal{S} \subseteq \wp(\kappa)$ is a base for the measures on κ. That is, suppose that for every measure μ on κ and every μ measure one set B, there is an $S \in \mathcal{S}$ with $S \subseteq B$ and $\mu(S) = 1$. Then every $A \subseteq \kappa$ is a countable union $A = \bigcup_i S_i$ of sets $S_i \in \mathcal{S}$.

PROOF. Suppose A cannot be written as a countable union of sets in $\mathcal{S}$. Let $\mathcal{I}$ be the σ-ideal on κ generated by $\kappa \setminus A$ and $\{S \in \mathcal{S} : S \subseteq A\}$. By assumption, $A \notin \mathcal{I}$. Let $\mathcal{F}$ be the corresponding filter, that is, $\mathcal{F} = \{F \subseteq \kappa : \kappa \setminus F \in \mathcal{I}\}$. So, $\mathcal{F}$ is a countably additive filter concentrating on A.

From Lemma 4.6, let μ be a measure on κ extending $\mathcal{F}$. So, $\mu(A) = 1$. By assumption, there is an $S \in \mathcal{S}$ with $S \subseteq A$ and $\mu(S) = 1$. However, $S \in \mathcal{I}$ by definition of $\mathcal{I}$. This is a contradiction since $\kappa \setminus S \in \mathcal{F}$ yet $\mu(\kappa \setminus S) = 0$. ⊣

The strong partition relation on ω_1. To prove partition relations we must analyze sets (actually functions), and to analyze sets, in view of Theorem 4.8 we must analyze measures. So, getting the strong partition relation on ω_1 reduces to analyzing the measures on ω_1. The weak partition relation on $\utilde{\delta}^1_3$ will reduce to analyzing measures on $\lambda_3 = \omega_\omega$, or equivalently the measures on the ω_n (at the next level of the theory where we need to get the strong partition property on $\utilde{\delta}^1_3$, we will need to analyze the measures on $\utilde{\delta}^1_3$).

First we give a rough outline of how analyzing measures in general goes. Given a measure μ on an ordinal κ, we analyze μ through a sequence of "pressing down" arguments. At stage n we will have two functions $g_n, r_n : \kappa \to \kappa$, that is, we have a map $\alpha \mapsto (g_n(\alpha), r_n(\alpha))$. Roughly speaking, the function g_n will constitute the known part of the measure, and the function r_n the part yet to be analyzed (the "remainder"). More precisely, $g_n(\mu)$ will be a known canonical measure. The two values $g_n(\alpha)$, $r_n(\alpha)$ will together determine α (for μ almost all α). A pressing down argument will give at the next stage functions g_{n+1}, r_{n+1}. Some sort of monotonicity of the functions is important in the pressing down argument. The measure $g_{n+1}(\mu)$ will be a more complicated extension of $g_n(\mu)$ (more precisely, both of these measures will be naturally measures on tuples of ordinals, and $g_{n+1}(\mu)$ will project to $g_n(\mu)$). On the other hand we will have $r_{n+1}(\alpha) < r_n(\alpha)$ for μ almost all α. Thus, after finitely many steps the process must stop (it will stop when the r function becomes constant almost everywhere), and at this point the measure is analyzed. The exact manner in which $g_n(\alpha)$ and $r_n(\alpha)$ determine α will involve non-trivial descriptions at the higher levels, but for the strong partition relation on ω_1 and the weak on $\utilde{\delta}^1_3$ will only involve trivial descriptions.

To illustrate, let us consider in more detail the case of a measure μ on ω_1. We prove the following result of [Jac90A].

THEOREM 4.9. *Every non-principal measure μ on ω_1 is equivalent a measure W^n_1. That is, there is a $h : (\omega_1)^n \to \omega_1$ which is one-to-one on a W^n_1 measure one set such that $\mu = h(W^n_1)$ (i.e., for all $A \subseteq \omega_1$, $A \in \mu$ iff $h^{-1}(A) \in W^n_1$).*

PROOF. To begin, let $f_1 : \omega_1 \to \omega_1$ be such that:

1. There is a μ measure one set A on which f_1 is monotonically increasing. (*i.e.*, if $\alpha \leq \beta$ are both in A then $f_1(\alpha) \leq f_1(\beta)$).
2. f_1 is not constant almost everywhere (*i.e.*, there does not exist a μ measure one set B such that $f_1 \restriction B$ is constant).

3. $[f_1]_\mu$ is minimal with respect to (1) and (2) (*i.e.*, if $[f']_\mu < [f_1]_\mu$ then f' does not satisfy (1) and (2)).

Note that the identity function satisfies (1) and (2), and so f_1 is well-defined. Also, $f_1(\alpha) \leq \alpha$ for μ almost all α. Fix A of μ measure one such that $f_1 \restriction A$ is monotonically increasing. By (1) and (2), for all $\alpha < \omega_1$ we have $h_1(\alpha) \doteq \sup\{\beta \in A : f_1(\beta) \leq \alpha\} < \omega_1$. Let $g_1 = f_1$. Note that $g_1(\mu) = W_1^1$ as otherwise there would be a c.u.b. $C \subseteq \omega_1$ such that $g_1(\mu)$ gives C measure zero, that is, $\forall^*_\mu \alpha\ (g_1(\alpha) \notin C)$. However, in that case we could let $f'(\alpha) = \ell_C \circ f_1(\alpha)$ for α in A, where $\ell_C(\beta)$ is the largest element of C less than or equal to β. We would then have a μ measure one set $A' \subseteq A$ and an f' which is strictly less than f_1 on A' and which also satisfies (1) and (2). This violates (3). So, $g_1(\mu)$ is the canonical measure W_1^1. Fix a real x such that the section T_x of the Kunen tree is wellfounded and $h_1(\beta) < |T_x \restriction \beta|$ for W_1^1 almost all β, say for all $\beta \in C$ (actually for all infinite β, but we don't have this at the higher levels so we use only the weaker statement here). By thinning out A we may assume $g_1(\alpha) \in C$ for all $\alpha \in A$. For $\alpha \in A$ let $r_1(\alpha) < g_1(\alpha) \leq \alpha$ be least such that $\alpha = |(T_x \restriction g_1(\alpha))(r_1(\alpha))|$. So, $g_1(\alpha)$ and $r_1(\alpha)$ determine α by this equation. This completes the first step of the analysis.

If there is a μ measure one set B such that for all $\alpha, \beta \in B$ if $g_1(\alpha) = g_1(\beta)$ then $r_1(\alpha) = r_1(\beta)$, then μ is equivalent to W_1^1 and we are done. To see this, suppose B of μ measure one is as stated. We may also assume that $r_1 < g_1$ on B as this holds μ almost everywhere. There is a c.u.b. $C \subseteq \omega_1$ such that $C \subseteq g_1(B)$ [Otherwise there would be a c.u.b. C with $C \cap g_1(B) = \varnothing$. But $B' \doteq g_1^{-1}(C)$ must have μ measure one, and so $B \cap B'$ has μ measure one and so is non-empty. But if $\alpha \in B \cap B'$ then $g_1(\alpha) \in C$, a contradiction to $\alpha \in B$ and the definition of C.] For $\delta \in C$, let $\eta(\delta)$ be the unique value of $r_1(\alpha)$ for any $\alpha \in B$ with $g_1(\alpha) = \delta$. Since $r_1 < g_1$ on B, we have that that $\eta(\delta) < \delta$ for all $\delta \in C$. Hence there is a $C' \subseteq C$ such that $\delta \mapsto \eta(\delta)$ is constant on C', say with constant value η_0. Let $B' = \{\alpha \in B : g_1(\alpha) \in B'\}$, so $\mu(B') = 1$. For $\delta \in C'$, let $h(\delta) = |T_x \restriction \delta(\eta_0)|$. Then h is a bijection between C' and B' (the functions $h \colon C' \to B'$ and $g_1 \colon B' \to C'$ are inverses). Also, if B'' has μ measure one, then by the argument above there is a c.u.b. C'' with $C'' \subseteq g_1(B'')$, and this shows $h(W_1^1) = \mu$.

For the second step, let $f_2 \colon \omega_1 \to \omega_1$ satisfy:

1. There is a μ measure one set A such that if $\alpha, \beta \in A$, $g_1(\alpha) = g_1(\beta)$, and $r_1(\alpha) \leq r_1(\beta)$, then $f_2(\alpha) \leq f_2(\beta)$.
2. There does not exists a μ measure one set B such that if $\alpha, \beta \in B$ and $g_1(\alpha) = g_1(\beta)$ then $f_2(\alpha) = f_2(\beta)$.

3. $[f_2]_\mu$ is minimal with respect to (1) and (2).

Note that f_2 exists since r_1 satisfies (1) and (2). Also, $f_2(\alpha) \le r_1(\alpha) < f_1(\alpha)$ for μ almost all α. Next observe that $f_2(\mu) = W_1^1$. For suppose C were c.u.b. and $\forall^*_\mu \alpha\ (f_2(\alpha) \notin C)$. Then $f_2' \doteq \ell_C \circ f_2$ is strictly less than f_2 almost everywhere with respect to μ. Also, f_2' still satisfies (1) and (2), a contradiction (for (2) we use the fact that for any μ measure one set B, for W_1^1 almost all δ we have that $\{r_1(\alpha) : \alpha \in B \wedge g_1(\alpha) = \delta\}$ is cofinal in δ as otherwise we would have a μ measure one set on which r_1 is constant, which contradicts the assumption that r_1 satisfies (2)). Let now $g_2(\alpha) = (f_2(\alpha), f_1(\alpha))$. So, $g_2(\mu) = W_1^2$. Fix a μ measure one set A for which (1) holds. Let $h_2(\delta_1, \delta_2) = \sup\{r_1(\alpha) : \alpha \in A \wedge g_2(\alpha) = (\delta_1, \delta_2)\}$. It follows from (1) and (2) that $h_2(\delta_1, \delta_2) < \delta_2$ for W_1^2 almost all (δ_1, δ_2). [Suppose C were c.u.b. and for all $\delta_1 < \delta_2$ in C we have $h_2(\delta_1, \delta_2) = \delta_2$. Thinning A we may assume that $g_2(\alpha) \in C^2$ for all $\alpha \in A$. From (2), let $\alpha, \beta \in A$ with $g_1(\alpha) = g_1(\beta)$ and $f_2(\alpha) < f_2(\beta)$. Let $(\delta_1, \delta_2) = g_2(\alpha)$. By (1), if $\gamma \in A$ with $g_1(\gamma) = \delta_2$ and $r_1(\gamma) \ge r_1(\beta)$ then $f_2(\gamma) \ge f_2(\beta) > f_2(\alpha) = \delta_1$. Thus, $h_2(\delta_1, \delta_2) \le r_1(\beta) < \delta_2$, a contradiction.] Let x_2 be such that T_{x_2} is wellfounded and for W_1^2 almost all (δ_1, δ_2) we have $h_2(\delta_1, \delta_2) < |T_{x_2} \restriction \delta_1|$. For μ almost all α we may then define $r_2(\alpha)$ to be the unique ordinal such that $r_1(\alpha) = |(T_{x_2} \restriction f_2(\alpha))(r_2(\alpha))|$. So, for μ almost all α, if $g_2(\alpha) = (\delta_1, \delta_2)$, we have $\alpha = |(T_{x_1} \restriction (\delta_2))(|(T_{x_2} \restriction (\delta_1))(r_2(\alpha))|)|$, and thus $g_2(\alpha)$, $r_2(\alpha)$ determine α. Note that $r_2(\alpha) < \delta_1 = f_2(\alpha) \le r_1(\alpha)$ for μ almost all α. This completes the second step.

The remaining steps are essentially identical to the second step. At the end of step n (if the argument has gone on this far), we have a g_n with $g_n(\mu) = W_1^n$ and an r_n with $[r_n]_\mu < [r_{n-1}]_\mu < \cdots < [r_2]_\mu < [r_1]_\mu$. Also, for μ almost all α, if $g_n(\alpha) = (\delta_1, \dots, \delta_n)$ we have $r_n(\alpha) < \delta_1 < \cdots < \delta_n \le \alpha$ and:

$$\alpha = |(T_{x_1} \restriction \delta_n)(\alpha_1)|$$

where

$$\begin{aligned} \alpha_1 &= |(T_{x_2} \restriction \delta_{n-1})(\alpha_2)|, \\ \alpha_2 &= |(T_{x_3} \restriction \delta_{n-2})(\alpha_3)|, \\ &\vdots \qquad \vdots \\ \alpha_{n-1} &= |(T_{x_n}(\delta_1))(\alpha_n)|, \text{ and} \\ \alpha_n &= r_n(\alpha) \end{aligned}$$

By wellfoundedness, after finitely many steps, say n steps, it must be that the analog of (2) fails. As we argued above, μ is then equivalent to

W_1^n. In fact, similarly to how we argued above, the r_n function is constant μ almost everywhere, and μ is equivalent to W_1^n in the following manner.

Let η_0 be the constant value of r_n almost everywhere. Recall $x_1, \dots, x_n$ are reals with the Kunen tree sections $T_{x_1}, \dots, T_{x_n}$ all wellfounded. For convenience, we change notation and call these same reals $x_n, x_{n-1}, \dots, x_1$ (*i.e.*, we enumerate in the reverse order). Define $h(\delta_1, \dots, \delta_n) = |(T_{x_n} \restriction (\delta_n))(\alpha_{n-1})|$, $\alpha_{n-1} = |(T_{x_{n-1}} \restriction \delta_{n-1})(\alpha_{n-2})|, \dots, \alpha_1 = |(T_{x_1}(\delta_1))(\eta_0)|$. In writing this equation, we assume that $\eta_0 < \delta_1$, each $|T_{x_i} \restriction \delta_i(\alpha_{i-1})|$ is less than δ_{i+1} and that $|T_{x_n} \restriction \delta_n(\alpha_{n-1})| \geq \delta_n$, otherwise we leave $h(\vec{\delta})$ undefined. Then $\mu(A) = 1$ iff $\exists$ c.u.b. $C\ \forall \vec{\delta} \in C^n\ h(\vec{\delta}) \in A$. ⊣

We call $h(\delta_1, \dots, \delta_n)$ the **lift** of $(\delta_1, \dots, \delta_n)$ via $x_1, \dots, x_n$ and η_0. So, h demonstrates the equivalence of W_1^n and μ. We make this into the following definition.

DEFINITION 4.10. We say a level-1 complex $\mathcal{C}$ is a sequence of reals $\langle x_1, \dots, x_n \rangle$ such that for each i, the section of the Kunen tree T_{x_i} is wellfounded.

Given a level-1 complex $\mathcal{C}$ and an ordinal $\eta_0 < \omega_1$, let $h = h^{\mathcal{C}}_{\eta_0} : (\omega_1)^n \to \omega_1$ be the lifting function as defined above. Let $\nu(\eta_0, \mathcal{C})$ be the measure on ω_1 equal to $h(W_1^n)$. That is, $A \subseteq \omega_1$ has $\nu(\eta_o, \mathcal{C})$ measure one iff there is a c.u.b. $C \subseteq \omega_1$ such that for all $\vec{\delta} \in C^n$ we have $h^{\mathcal{C}}_{\eta_0}(\vec{\delta}) \in A$.

We have thus shown the following.

THEOREM 4.11. Let μ be a measure on ω_1. Then there is a level-1 complex $\mathcal{C}$ and an ordinal $\eta_0 < \omega_1$ such that $\mu = \nu(\eta_0, \mathcal{C})$.

The above analysis of measures on ω_1 together with Theorem 4.8 gives a coding for the subsets of ω_1. Following the terminology of [Sol78A] we make the following definition.

DEFINITION 4.12. $A \subseteq \omega_1$ is **simple** if either $A = \{\alpha\}$ is a singleton or there is a c.u.b. $C \subseteq \omega_1$, a level-1 complex $\mathcal{C}$, and an $\eta_0 < \omega_1$ such that $A = \{h^{\mathcal{C}}_{\eta_0}(\vec{\delta}) : \vec{\delta} \in C^n\}$.

Theorems 4.8 and 4.11 then give the following.

THEOREM 4.13. Every subset of ω_1 is a countable union of simple sets.

This in turn gives a coding $\pi : {}^{\omega}\omega \to \wp(\omega_1)$ of the subsets of ω_1 as follows. If $\sigma \in {}^{\omega}\omega$, let C_σ be the closed (not necessarily unbounded) subset of ω_1 defined by $C_\sigma = \{\alpha : \forall \beta < \alpha\ |T_\sigma \restriction \beta| < \alpha\}$. From Theorem 2.11 applied to the function N_C (recall $N_C(\alpha)$ is the least element of C greater than α), we get that for any c.u.b. $C \subseteq \omega_1$ there is a σ with C_σ a c.u.b. subset of C. View every real z as coding a real σ, reals $x_1, \dots x_n$ for some n, and a real w (we allow n to be 0, which we specify by some syntactic condition on z).

Then z codes the set A_z as follows. If $w \notin \mathrm{WO}$, then $A_z = \varnothing$. Otherwise, if $n = 0$ then $A_z = \{|w|\}$. If $n > 0$ then A_z is the set of $h^{\mathcal{C}}_{|w|}(\delta_1, \dots, \delta_n)$ such that $\vec{\delta} \in (C_\sigma)^n$ and where $\mathcal{C} = \langle x_1, \dots, x_n \rangle$. In writing $h^{\mathcal{C}}_{|w|}(\vec{\delta})$ here we mean the value as defined above assuming the necessary wellfoundedness. For example, in writing $h(\delta_1, \dots, \delta_n) = |(T_{x_n} \restriction \delta_n)(\alpha_{n-1})|$, we assume here that α_{n-1} is in the wellfounded part of the tree $T_{x_n} \restriction \delta_n$, *etc.*, otherwise we say $h^{\mathcal{C}}_{|w|}(\vec{\delta})$ is undefined. Let then $\pi(z) = \bigcup_i A_{z_i}$, where z codes the sequence $z_0, z_1, \dots$.

The map π is a $\underset{\sim}{\mathbf{\Delta}}^1_1$ coding of the subsets of ω_1 in the following sense.

THEOREM 4.14. *The map* $\pi\colon {}^\omega\omega \to \wp(\omega_1)$ *is onto. For every* $\alpha < \omega_1$, $\{z : \alpha \in \pi(z)\} \in \underset{\sim}{\mathbf{\Delta}}^1_1$.

PROOF. Theorem 4.13 gives that π is onto. For the second part, note that if $\alpha \in A_z$, where z codes $\sigma, x_1, \dots, x_n, w$ as above, then $w \in \mathrm{WO}$ and $|w| \le \alpha$. This is a $\underset{\sim}{\mathbf{\Delta}}^1_1$ condition on w as $\mathrm{WO}_{\alpha+1} \in \underset{\sim}{\mathbf{\Delta}}^1_1$. Similarly, if $h^{\mathcal{C}}_{|w|}(\delta_1, \dots, \delta_n) = \alpha$ then then $|w| < \delta_1 < \dots < \delta_n \le \alpha$ and $\alpha_{i+1} = |T_{x_i} \restriction \delta_i(\alpha_{i-1})| < \delta_{i+1}$ by our convention (that is, α_{i-1} is in the wellfounded part of $T_{x_i} \restriction \delta_i$ and has rank less than δ_{i+1}). For any fixed $\alpha, \beta, \gamma < \omega_1$, the set $\{x : |T_x \restriction \alpha(\beta)| < \gamma\}$ is $\underset{\sim}{\mathbf{\Delta}}^1_1$. It follows that for fixed $\eta_0 < \delta_1 < \dots < \delta_n \le \alpha$ that $\{(x_1, \dots, x_n) : h^{\langle \vec{x} \rangle}_{\eta_0}(\vec{\delta}) = \alpha\}$ is a $\underset{\sim}{\mathbf{\Delta}}^1_1$ set. Also, $\{\sigma : \delta \in C_\sigma\} \in \underset{\sim}{\mathbf{\Delta}}^1_1$ for any $\delta < \omega_1$, and from these observations the result follows. ⊣

To prove the strong partition relation on ω_1 we need a suitable coding of the functions from ω_1 to ω_1. If we simply view functions as being subsets of $\omega_1 \times \omega_1 \approx \omega_1$, the resulting coding given by Theorem 4.14 is not quite good enough to satisfy Theorem 2.17. Instead, we must redo the analysis of measures above to work directly with functions. The changes are minor. Given a function $f\colon \omega_1 \to \omega_1$, instead of ω_1 we work on the space $X = \{(\alpha, f(\alpha)) : \alpha < \omega_1\}$ which of course is isomorphic to ω_1 by identifying $(\alpha, f(\alpha))$ with α. In the fist step of the measure analysis, instead of the h_1 used there we use $h'_1(\alpha) = \sup\{\max(\beta, f(\beta)) : (\beta, f(\beta)) \in A \wedge g_1(\beta) \le \alpha\}$. The resulting function r_1 returns pairs of ordinals as values. The argument proceeds as before, with the pressing down arguments (*i.e.*, the f_i) done with respect to the first component of the pair and the functions h_i taken to dominate both components.

We define the lift for functions similarly to that for sets. Given a pair (η_0, η_1), reals $\mathcal{C} = \langle x_1, \dots, x_n \rangle$ with $T_{x_1}, \dots, T_{x_n}$ sufficiently wellfounded and $\vec{\delta} \in (\omega_1)^n$, let $h^{\mathcal{C}}_{\eta_0, \eta_1}(\vec{\delta}) = (\alpha, \beta)$ where $\alpha = |(T_{x_n} \restriction (\delta_n))(\alpha_{n-1})|$, $\beta = |(T_{x_n} \restriction (\delta_n))(\beta_{n-1})|$, *etc.*, and $\alpha_1 = |(T_{x_1} \restriction \delta_1)(\eta_0)|$, $\beta_1 = |(T_{x_1} \restriction \delta_1)(\eta_1)|$. We define the notion of a **simple** partial function f from ω_1 to ω_1 as before, that is $f = \{h^{\mathcal{C}}_{\eta_0, \eta_1}(\vec{\delta}) : \vec{\delta} \in C^n\}$ for some c.u.b. C (we allow also f to be a single pair). Theorem 4.13 now becomes: every function $f\colon \omega_1 \to \omega_1$ is a countable union of simple subfunctions. This gives a coding for the functions from ω_1 to

ω_1 exactly as in Theorem 4.14. It is straightforward to check that this coding satisfies the requirements of Theorem 2.17. The point of this modification is that when $h^C_{\eta_0,\eta_1}(\delta_1,\dots,\delta_n) = (\alpha,\beta)$, then $\delta_n \le \alpha$ and not just $\delta_n \le \beta$. This is necessary in verifying (3) of Theorem 2.17.

The weak partition relation on $\underset{\sim}{\delta}^1_3$. We next briefly indicate how the analysis of measures on ω_ω of [Sol78A] can be given from the current point of view. We first need to introduce some notation concerning functions from $(\omega_1)^n$ to ω_1.

Let $\pi = (i_1, i_2, \dots, i_n)$ be a permutation of $\{1, 2, \dots, n\}$ beginning with n (*i.e.*, $i_1 = n$).

DEFINITION 4.15. Let $\pi = (n, i_2, \dots, i_n)$ be a permutation beginning with n. We say $f : (\omega_1)^n \to \mathrm{Ord}$ is **ordered by π** if $f(\alpha_1,\dots,\alpha_n) \le f(\beta_1,\dots,\beta_n)$ iff $(\alpha_{i_1},\dots,\alpha_{i_n}) \le_{\text{lex}} (\beta_{i_1},\dots,\beta_{i_n})$. for all $\vec{\alpha},\vec{\beta} \in (\omega_1)^n$. We say f is ordered by π almost everywhere if there is a c.u.b. $C \subseteq \omega_1$ such that $f \restriction C$ is ordered by π.

REMARK 4.16. We adopt the convention that we may write the arguments to a function $f : (\omega_1)^n \to \mathrm{Ord}$ in any order. This causes no ambiguity since $(\omega_1)^n$ consists of increasing tuples. For example, if f is ordered by $\pi = (3, 1, 2)$, then instead of $f(\alpha_1, \alpha_2, \alpha_3)$ we may write $f(\alpha_3, \alpha_1, \alpha_2)$ (the arguments are now listed in their order of significance in determining the size of $f(\vec{\alpha})$).

We say $f : (\omega_1)^n \to \mathrm{Ord}$ depends on all of its arguments (almost everywhere) if there does not exist a c.u.b. $C \subseteq \omega_1$ such that $f \restriction C^n$ only depends on a proper subset its arguments. An easy partition argument shows that if $f : (\omega_1)^n \to \omega_1$ and f depends on all its arguments, then there is a unique π starting with n such that f is ordered by π almost everywhere. If we remove the assumption that f depends on all its arguments, then there is a **partial permutation** $\pi = (i_1,\dots,i_j)$, $j \le n$, with $i_1 \ge \max\{i_2,\dots,i_j\}$ such that f is ordered by π, that is, $f(\alpha_1,\dots,\alpha_n) \le f(\beta_1,\dots,\beta_n)$ iff $(\alpha_{i_1},\dots,\alpha_{i_j}) \le_{\text{lex}} (\beta_{i_1},\dots,\beta_{i_j})$. For $f : (\omega_1)^n \to \mathrm{Ord}$, the same results hold except we remove the restriction that i_1 be maximal.

The following definition is used frequently.

DEFINITION 4.17. Suppose $f : (\omega_1)^n \to \omega_1$ is ordered by $\pi = (n, i_2, \dots, i_n)$. For $1 \le j \le n$ we define the ***j*th invariant** of f, $f(j)$, to be function from $(\omega_1)^j$ to ω_1 defined by:

$$f(j)(\alpha_n, \alpha_{i_2}, \dots, \alpha_{i_j}) = \sup\{f(\alpha_n,\dots,\alpha_{i_j},\alpha_{i_{j+1}},\dots,\alpha_{i_n}) : (\alpha_1,\dots,\alpha_n) \in (\omega_1)^n\}.$$

We also define

$$f^s(j)(\alpha_n, \alpha_{i_2}, \dots, \alpha_{i_j}) = \sup\{f(\alpha_n,\dots,\beta,\alpha_{i_{j+1}},\dots,\alpha_{i_n}) : \beta < \alpha_{i_j}, \alpha_{i_{j+1}},\dots,\alpha_{i_n}\};$$

where the supremum is over $\alpha_{i_{j+1}}, \ldots, \alpha_{i_n}$ such that $(\alpha_n, \alpha_{i_2} \ldots, \beta, \alpha_{i_{j+1}}, \ldots, \alpha_{i_n})$ is order-isomorphic to π.

For a general f, we may get a c.u.b. C so that f is ordered by π on C^n (perhaps using a smaller value of n if f doesn't depend on all its arguments), and then apply the above definition. It is easy to see that the equivalence class of $f(j)$ with respect to W_1^j is well-defined.

EXAMPLE 4.18. If $f : (\omega_1)^4 \to \omega_1$ is ordered by $\pi = (4, 1, 3, 2)$, then for all $\alpha_1 < \alpha_2 < \alpha_3$ we have $f(3)(\alpha_1, \alpha_2, \alpha_3) = \sup_{\alpha_1 < \beta < \alpha_2} f(\alpha_1, \beta, \alpha_2, \alpha_3)$. For $\alpha_1 < \alpha_2$ we have $f(2)(\alpha_1, \alpha_2) = \sup_{\alpha_1 < \beta < \gamma < \alpha_2} f(\alpha_1, \beta, \gamma, \alpha_2)$. Note that we also have $f(2)(\alpha_1, \alpha_2) = \sup_{\alpha_1 < \delta < \alpha_2} f(3)(\alpha_1, \delta, \alpha_2)$. That is, $f(2)$ is also the second invariant of $f(3)$. Also. $f^s(2)(\alpha_1, \alpha_2) = \sup_{\beta < \alpha_1, \beta < \gamma < \delta < \alpha_2} f(\beta, \gamma, \delta, \alpha_2) = \sup_{\beta < \alpha_1} f(2)(\beta, \alpha_2)$.

It remains to describe the possible uniform cofinality of an $f : (\omega_1)^n \to \omega_1$. This is done in the next lemma, whose proof uses the Kunen tree and easy partition arguments (see [Jac10] for details).

LEMMA 4.19. Let $f : (\omega_1)^n \to \omega_1$. Then almost everywhere $f(\alpha_1, \ldots, \alpha_n)$ has either uniform cofinality ω, α_1, …, or α_n. These $n + 1$ possibilities are all distinct (*i.e.*, f can only have one of these uniform cofinalities).

REMARK 4.20. A generic coding argument shows that for $f : (\omega_1)^n \to \lambda < \Theta$, the possible uniform cofinalities are those listed in the lemma and the constant functions $g(\vec{\alpha}) = \kappa$ for κ a regular cardinal.

We say a permutation $\pi' = (i_1, \ldots, i_{n+1})$ of length $n + 1$ **extends** the permutation π of length n if $(i_1, \ldots, i_n)$ is order-isomorphic to π. For example, $(3, 1, 2)$ is extended by $(4, 2, 3, 1)$, $(4, 1, 2, 3)$, and $(4, 1, 3, 2)$.

We may assume $f : (\omega_1)^n \to \omega_1$ depends on all its arguments, as otherwise we may consider f as a function from $(\omega_1)^m \to \omega_1$ for some $m < n$. Say f is ordered by $\pi = (n, i_2, \ldots, i_n)$. One possibility is that f is continuous almost everywhere, that is, there is a c.u.b. C such that for all $\vec{\alpha} \in C^n$ we have $f(\alpha_{i_1}, \ldots, \alpha_{i_n}) = \sup_{\beta < \alpha_{i_n}} f(\alpha_{i_1}, \ldots, \beta)$ (in this case $f(\vec{\alpha})$ must have uniform cofinality α_{i_n}). Suppose f is discontinuous almost everywhere. Then f must have one of the uniform cofinalities listed in Lemma 4.19. Suppose $f(\vec{\alpha})$ has uniform cofinality α_j for some $1 \leq j \leq n$. Then there is a unique permutation π' of $n + 1$ extending π and a $f' : (\omega_1)^{n+1} \to \omega_1$ ordered by π' such that $f = f'(n)$. For example, if f is ordered by $\pi = (4, 1, 3, 2)$ and $f(\vec{\alpha})$ has uniform cofinality α_3, then $\pi' = (5, 1, 4, 2, 3)$.

Summarizing, we have the following cases.

DEFINITION 4.21. Let $f : (\omega_1)^n \to \omega_1$. We say f is of **type** π if f is ordered by π, discontinuous, and of uniform cofinality ω. We say f is of type π^s is f is ordered by π and is continuous (and of uniform cofinality ω at points of successor rank). We say f is of type (π, π') if π' extends π and there is an f' ordered by π' with $f'(n) = f$.

We extend these definitions to their "almost everywhere" versions in the usual manner. We then have:

LEMMA 4.22. Let $f : (\omega_1)^n \to \omega_1$ depend on all its arguments almost everywhere. Then f is almost everywhere of type π, of type π^s, or of type (π, π') for some permutation(s) π and π'.

The analysis of measures on ω_ω involves putting together finitely many functions of various types from $(\omega_1)^n$ to ω_1 for some n. The object which describes how these are put together we call a (level-2) **tree of uniform cofinalities**. It is basically a finite tree (with root node $\varnothing$) which at each node assigns a possible uniform cofinality for the measure associated to the previous node. The node is terminal if the uniform cofinality is ω. The precise definition follows.

DEFINITION 4.23. A level-2 tree of uniform cofinalities is a function $\mathcal{R}$ with domain a finite tree satisfying the following.

1. $\mathcal{R}(\varnothing) = (1)$, the unique permutation of length 1.
2. For each $(i_1) \in \text{dom}(\mathcal{R})$, $\mathcal{R}(i_1)$ is either the symbol ω (to denote uniform cofinality ω), or is the unique permutation $(2, 1)$ of length 2 extending $\mathcal{R}(\varnothing)$. If $\mathcal{R}(i_1) = \omega$, then (i_1) is a terminal node in $\text{dom}(\mathcal{R})$.
3. In general, if $(i_1, \ldots, i_k) \in \text{dom}(\mathcal{R})$, then $\mathcal{R}(i_1, \ldots, i_{k-1})$ is a permutation of length k (beginning with k) and $\mathcal{R}(i_1, \ldots, i_k)$ is either a permutation of length $k+1$ (beginning with $k+1$) which extends $\mathcal{R}(i_1, \ldots, i_{k-1})$, or else is the symbol ω. In the latter case, $(i_1, \ldots, i_k)$ is a terminal node in $\text{dom}(\mathcal{R})$.

REMARK 4.24. We didn't make the definition of a level-1 tree of uniform cofinalities earlier, but we could have. A level-1 tree $\mathcal{Q}$ could be taken to be a finite set $\{1, 2, \ldots, n\}$. The ordering $<_{\mathcal{Q}}$ is the usual ordering on this set. The measure $\mathcal{M}^{\mathcal{Q}}$ is the measure induced by the weak partition relation on ω_1, and order-preserving functions $f\colon \text{dom}(<_{\mathcal{Q}}) \to \omega_1$. This is, of course, just the measure W_1^n. With this terminology, the above definition of a level-2 tree can be viewed as an instance of an inductive definition starting from the level-1 trees. We could view $\mathcal{R}$ as assigning to nodes s of the level-2 tree values $\mathcal{R}(s)$ which are level-1 trees. If t immediately extends s, then $\mathcal{R}(t)$ must immediately extend $\mathcal{R}(s)$ which means that $\mathcal{R}(s) = \{1, \ldots, n\}$, $\mathcal{R}(t) = \{1, \ldots, n, n+1\}$ and we have a one-to-one map from the first set into the second (which gives a way of identifying elements of first set with the second). A sequence of such extensions going down a branch corresponds to a permutation in a natural way (we require that n be identified with $n + 1$ so that the permutations begin with the largest integer). For example, if $\mathcal{Q}_1 = \{1\}$, $\mathcal{Q}_2$ adds the new element

1 before the old element 1 (which is now identified with the new element 2 of $\mathcal{Q}_2$), and $\mathcal{Q}_3$ adds the new element 2 between the old elements 1 and 2 (the old 2 becomes the new 3 of $\mathcal{Q}_3$), then $\pi = (3, 1, 2)$.

We frequently implicitly assume that $\{i_k : (i_1, \ldots, i_{k-1})^\frown i_k \in \mathrm{dom}(\mathcal{R})\}$ is an integer.

EXAMPLE 4.25. A simple level-2 tree of uniform cofinalities is the following (the nodes s of the tree are labeled with the values $\mathcal{R}(s)$):

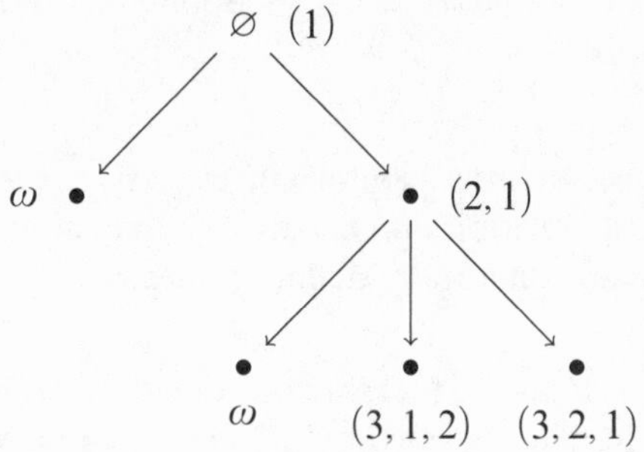

We view the domain of $\mathcal{R}$ in this example as being $\{\varnothing, (0), (1), (1, 0), (1, 1), (1, 2)\}$.

Given a tree of uniform cofinalities $\mathcal{R}$, we let $<_{\mathcal{R}}$ be lexicographic ordering on the set of tuples $(\alpha_1, i_1, \ldots, \alpha_k, i_k)$ such that $(i_1, \ldots, i_k) \in \mathrm{dom}(\mathcal{R})$ and $(\alpha_1, \alpha_2, \ldots, \alpha_k)$ is order-isomorphic to $\mathcal{R}(i_1, \ldots, i_{k-1})$. Note that $<_{\mathcal{R}}$ has order-type ω_1 as all permutations begin with their largest element (so $\alpha_1 > \max\{\alpha_2, \ldots, \alpha_k\}$).

For Example 4.25, the domain of $<_{\mathcal{R}}$ consists of tuples of the form $(\alpha, 0)$, $(\alpha, 1)$, $(\alpha, 1, \beta, 0)$, $(\alpha, 1, \beta, 1)$, and $(\alpha, 1, \beta, 2)$, where $\beta < \alpha < \omega_1$.

DEFINITION 4.26. We say a function $f : \mathrm{dom}(<_{\mathcal{R}}) \to \omega_1$ is **of type $\mathcal{R}$** if satisfies the following.

1. f is order-preserving from $\mathrm{dom}(<_{\mathcal{R}})$ to ω_1.
2. If $(i_1, \ldots, i_k)$ is not a terminal node of $\mathrm{dom}(\mathcal{R})$, then $f(\alpha_1, i_1, \ldots, \alpha_k, i_k)$ is the supremum of the values $f(\alpha_1, i_1, \ldots, \alpha_k, i_k, \beta, 0)$ where $(\alpha_1, \ldots, \alpha_k, \beta)$ is order-isomorphic to $\mathcal{R}(i_1, \ldots, i_k)$ (*i.e.*, $(\alpha_1, i_1, \ldots, \alpha_k, i_k, \beta, 0)$ is in the domain of $<_{\mathcal{R}}$).
3. If $(i_1, \ldots, i_k)$ is terminal in $\mathrm{dom}(\mathcal{R})$, then $f(\alpha_1, i_1, \ldots, \alpha_k, i_k)$ is greater than $\sup\{f(\alpha_1, i_1, \ldots, i_{k-1}, \beta, j) : \beta < \alpha_k, (i_1, \ldots, i_{k-1}, j) \in \mathrm{dom}(\mathcal{R})\}$ (note that this is automatically true if $i_k > 0$).
4. $f(\alpha_1, i_1, \ldots, \alpha_k, i_k)$ has uniform cofinality specified by $\mathcal{R}(i_1, \ldots, i_k)$. That is:
   (a) If $\mathcal{R}(i_1, \ldots, i_k) = \omega$, then $f(\alpha_1, i_1, \ldots, \alpha_k, i_k)$ has uniform cofinality ω.

(b) Otherwise, $f(\alpha_1, i_1, \dots, \alpha_k, i_k)$ has uniform cofinality

$\{\beta : (\alpha_1, \dots, \alpha_k, \beta)$ is order-isomorphic to $\mathcal{R}(i_1, \dots, i_k)\}$.

(when this set has successor order-type, we interpret this as requiring uniform cofinality ω).

Note that the requirement in (4b) is equivalent to saying the map $(\alpha_1, \dots, \alpha_k) \mapsto f(\alpha_1, i_1, \dots, \alpha_k, i_k)$ is of type (π, π') where $\pi = \mathcal{R}(i_1, \dots, i_{k-1})$ and $\pi' = \mathcal{R}(i_1, \dots, i_k)$.

If f is of type $\mathcal{R}$, then for all $(i_1, \dots, i_k) \in \mathrm{dom}(\mathcal{R})$, f induces a subfunction $f^{i_1,\dots,i_k} : (\omega_1)^k \to \omega_1$ by

$$f^{i_1,\dots,i_k}(\alpha_1, \dots, \alpha_k) = f(\alpha_1, i_1, \alpha_2, i_2, \dots, \alpha_k, i_k)$$

where here $(\alpha_1, \dots, \alpha_k)$ is order-isomorphic to $\mathcal{R}(i_1, \dots, i_{k-1})$.

The strong partition relation on ω_1 induces a partition property for functions of type $\mathcal{R}$, and this naturally defines a measure $\mathcal{M}^{\mathcal{R}}$ associated to $\mathcal{R}$. Namely:

Definition 4.27. $\mathcal{M}^{\mathcal{R}}$ is the measure on tuples $(\dots, \alpha^{i_1,\dots,i_k}, \dots\}$ (indexed by nodes in the tree $\mathrm{dom}(\mathcal{R})$) given by: $\mathcal{M}^{\mathcal{R}}(A) = 1$ iff there is a c.u.b. $C \subseteq \omega_1$ such that for all $f : \mathrm{dom}(<_{\mathcal{R}}) \to C$ of type $\mathcal{R}$, $(\dots, \alpha^{i_1,\dots,i_k}, \dots\} \in A$, where $\alpha^{i_1,\dots,i_k} = [f^{i_1,\dots,i_k}]_{W_1^k}$.

Example 4.28. For the level-2 tree $\mathcal{R}$ of Example 4.25, $\mathcal{M}^{\mathcal{R}}$ is a measure on tuples $(\alpha^0, \alpha^1, \alpha^{1,0}, \alpha^{1,1}, \alpha^{1,2})$ where $\alpha^0 < \alpha^1 < \omega_2$ and $\alpha^{1,0} < \alpha^{1,1} < \alpha^{1,2} < \omega_3$. The measure concentrates on tuples satisfying $\alpha^{1,0}(1) = \alpha^{1,1}(1) = \alpha^{1,2}(1) = \alpha^1$, where $\alpha^{1,0}(1)$ denotes the ordinal represented by the first invariant of a function of the appropriate type representing $\alpha^{1,0}$. It is not difficult to see that the top level splitting of the tree corresponds to product measures, so $\mathcal{M}^{\mathcal{R}}$ in this case is equal to $S_1^1 \times \mu$ where μ is a measure on tuples $(\alpha^1, \alpha^{1,0}, \alpha^{1,1}, \alpha^{1,2})$. Here S_1^1 is the ω-cofinal normal measure on ω_2, which is the measure induced by the strong partition property on ω_1 and functions $f : \omega_1 \to \omega_1$ of the correct type (and the measure W_1^1 on ω_1). The general definition of the measure S_1^m is given in Definition 5.1.

Theorem 4.29. Every measure on ω_ω is equivalent to a measure of the form $W_1^m \times \mathcal{M}(\mathcal{R})$ for some m and some (level-2) tree of uniform cofinalities $\mathcal{R}$.

To see how this equivalence takes place, we extend the trivial descriptions (which are just integers) slightly to **extended trivial descriptions**, defined relative to a fixed tree of uniform cofinalities $\mathcal{R}$.

Definition 4.30. An extended trivial description, defined relative to $\mathcal{R}$ and W_1^m, is a sequence of the form $\vec{d} = (d_1, i_1, \dots, d_k, i_k)$ or $\vec{d} = (d_1, i_1, \dots, d_k, i_k)^s$ where the s is a formal symbol (standing for "sup"). The d_i are trivial descriptions with $d_i \leq m$, $(i_1, \dots, i_k) \in \mathrm{dom}(\mathcal{R})$, and $(d_1, \dots, d_k)$ is order-isomorphic to $\mathcal{R}(i_1, \dots, i_{k-1})$.

We write $(d_1, i_1, \dots, d_k, i_k)^{(s)}$ to denote that the symbol s may or may not appear. We order the extended descriptions lexicographically, and where

$$(d_1, i_1, \dots, d_k, i_k)^s < (d_1, i_1, \dots, d_k, i_k),$$

$(d_1, i_1, \dots, d_k, i_k)^{(s)} < (d_1, i_1, \dots, d_\ell, i_\ell)^{(s)}$ if $k < \ell$ and s does appear on the left side, or $k > \ell$ and s does not appear on the right side.

We interpret the extended descriptions as follows.

DEFINITION 4.31. Suppose $\vec{d} = (d_1, i_1, \dots, d_k, i_k)^{(s)}$ is defined relative to $\mathcal{R}$ and W_1^m. If $f\colon \operatorname{dom}(<_{\mathcal{R}}) \to \omega_1$ is of type $\mathcal{R}$ and $\vec{\alpha} = (\alpha_1, \dots, \alpha_m) \in (\omega_1)^m$, then $(f; \vec{\alpha}; \vec{d}) = f((\vec{\alpha}; d_1), i_1, \dots, (\vec{\alpha}; d_k), i_k)$ if $\vec{d}$ does not have the symbol s and if s appears then

$$(f; \vec{\alpha}; \vec{d}) = \sup\{f((\vec{\alpha}; d_1), i_1, \dots, i_{k-1}, \beta, j) : \beta < (\vec{\alpha}; d_k), \\ (i_1, \dots, j) \in \operatorname{dom}(\mathcal{R})\}.$$

We let $(f; W_1^m; \vec{d}) < \omega_{m+1}$ be the ordinal represented with respect to W_1^m by the map $\vec{\alpha} \mapsto (f; \vec{\alpha}; \vec{d})$.

To generate an an arbitrary measure on ω_ω we must "lift-up" a measure of the form $\mathcal{M}^{\mathcal{R}}$ using sections of the Kunen tree and finitely many extended descriptions. The precise ingredients are given in the following definition.

DEFINITION 4.32. A (level-2) **complex** is a sequence

$$\mathcal{C} = \langle \mathcal{R}; W_1^m; x_1, \dots, x_n; \vec{d_1}, \dots, \vec{d_n} \rangle$$

where $\mathcal{R}$ is a (level-2) tree of uniform cofinalities, the Kunen tree sections $T_{x_1}, \dots, T_{x_n}$ are wellfounded, and the $\vec{d_i}$ are extended (trivial) descriptions defined with respect to $\mathcal{R}$ and W_1^m and satisfy $\vec{d_1} < \cdots < \vec{d_n}$.

If μ is a measure on ω_1, and $\mathcal{C}$ is a complex (as above), then μ and $\mathcal{C}$ define a measure $\mathcal{V}^{\mu, \mathcal{C}}$ as follows.

DEFINITION 4.33. $\mathcal{V}^{\mu, \mathcal{C}}(A) = 1$ iff for μ almost all γ there is a c.u.b. $C \subseteq \omega_1$ such that for all $f\colon \operatorname{dom}(<_{\mathcal{R}}) \to C$ of type $\mathcal{R}$, $(f; \mathcal{C}; W_1^m; \gamma) \in A$. Here, $(f; \mathcal{C}; W_1^m; \gamma)$ is the ordinal represented with respect to W_1^m by the function $(\beta_1, \dots, \beta_m) \mapsto (f; \mathcal{C}; \vec{\beta}; \gamma)$. Also, $(f; \mathcal{C}; \vec{\beta}; \gamma) = |(T_{x_1} \restriction (f; \vec{\beta}; \vec{d_1}))(\alpha_2)|$, where $\alpha_2 = |(T_{x_2} \restriction (f; \vec{\beta}; \vec{d_2}))(\alpha_3)|, \dots,$ and $\alpha_{n+1} = \gamma$.

We abbreviate this definition by writing

$$\mathcal{V}^{\mu, \mathcal{C}}(A) = 1 \leftrightarrow \forall^*_\mu \gamma \, \forall^* f\colon \operatorname{dom}(<_{\mathcal{R}}) \to \omega_1 \, [(f; \mathcal{C}; W_1^m; \gamma) \in A].$$

The precise theorem is then:

THEOREM 4.34. Let ν be a measure on ω_{m+1}. Then there is a measure μ on ω_1 and a complex $\mathcal{C}$ such that $\nu = \mathcal{V}^{\mu, \mathcal{C}}$.

The proof of Theorem 4.34 is very similar to that of Theorem 4.9. [Sol78A] proves essentially the same theorem using different terminology and a somewhat different argument. In [Jac10] the details of the current proof can be found.

Theorem 4.34 and Theorem 4.8 give a coding for the subsets of ω_ω, exactly as the proof of Theorem 4.14 for subsets of ω_1. This results in a $\utilde{\mathbf{\Delta}}^1_3$ coding of the subsets of ω_ω. That is:

Corollary 4.35. There is a map π from ${}^\omega\omega$ onto $\wp(\omega_\omega)$ such that for any $\alpha < \omega_\omega$, $\{x : \alpha \in \pi(x)\} \in \utilde{\mathbf{\Delta}}^1_3$.

Corollary 4.35 is enough to get the weak partition relation on $\utilde{\boldsymbol{\delta}}^1_3$, as shown in [Sol78A]. Namely, we code functions from ω_ω to $\utilde{\boldsymbol{\delta}}^1_3$ via relations $B \subseteq \omega_\omega \times \omega_\omega \times \omega_\omega$ such that for all $\alpha < \omega_\omega$, $B_\alpha = \{(\beta,\gamma) : R(\alpha,\beta,\gamma)\}$ is wellfounded (we are using Corollary 4.35 for subsets of $(\omega_\omega)^3$, which follows easily from the version for ω_ω by taking a bijection between $(\omega_\omega)^3$ and ω_ω). For example, to show $\utilde{\boldsymbol{\delta}}^1_3 \to (\utilde{\boldsymbol{\delta}}^1_3)^\lambda$ for $\lambda = \omega_\omega$ we set, in the notation of Theorem 2.17, $\varphi(x)(\alpha,\beta)$ iff B_α is wellfounded of rank β, where $B = \pi(x) \subseteq (\omega_\omega)^3$. Using Corollary 4.35 and Theorem 2.10 applied to $\utilde{\mathbf{\Delta}} = \utilde{\mathbf{\Delta}}^1_3$ it is straightforward to verify (2) of Theorem 2.17. To see (3) of 2.17, suppose $A \subseteq R_\alpha$ is $\utilde{\mathbf{\Sigma}}^1_3$, where R_α is as in Theorem 2.17. Then $\sup_{x\in A} \varphi(x)(\alpha)$ is bounded by the length β_0 of the following wellfounded relation $\prec$:

$$(x,y) \prec (x',y') \leftrightarrow$$
$$(x = x' \in A) \wedge \exists \eta,\eta' < \omega_\omega\, (\psi(y) = \eta \wedge \psi(y') = \eta' \wedge \pi(x)(\alpha,\eta,\eta')$$

where ψ is a $\utilde{\mathbf{\Delta}}^1_3$ norm of length ω_ω. From Theorem 2.10 it is easy to see that $\prec$ is $\utilde{\mathbf{\Sigma}}^1_3$ and so has length less that $\utilde{\boldsymbol{\delta}}^1_3$.

The reader can consult [Sol78A] for further details. We note that here the minor annoyance of having to slightly modify the argument to work for functions instead of sets (as we had to do for ω_1) does not arise, although it does when when have to prove the strong partition relation on $\utilde{\boldsymbol{\delta}}^1_3$ in the next level of the induction.

§5. The Second Level of the Induction. At the second level (stage $n = 1$ of the inductive analysis) we must compute $\utilde{\boldsymbol{\delta}}^1_5$, prove the strong partition relation on $\utilde{\boldsymbol{\delta}}^1_3$, and prove the weak partition relation on $\utilde{\boldsymbol{\delta}}^1_5$. The arguments follow in outline those of the previous section, but we begin to use non-trivial descriptions. Before saying what these are, we introduce two new families of canonical measures, besides the family W^m_1 we already have. Their definitions will use the strong partition property on ω_1 and the weak partition property on $\utilde{\boldsymbol{\delta}}^1_3$ respectively.

Let π_n be the permutation $(n, 1, 2, \dots, n-1)$. We abbreviate the order $<_{\pi_n}$ on $(\omega_1)^n$ by $<_n$. Recall this means:

$$(\alpha_1, \dots, \alpha_n) <_n (\beta_1, \dots, \beta_n) \leftrightarrow (\alpha_n, \alpha_1, \dots, \alpha_{n-1}) <_{\text{lex}} (\beta_n, \beta_1, \dots, \beta_{n-1}).$$

DEFINITION 5.1. By S_1^m we denote the measure on ω_{m+1} induced by the strong partition relation on ω_1, functions $f\colon \text{dom}(<_m) \to \omega_1$ which are order-preserving, discontinuous, and of uniform cofinality ω (which we also call the correct type), and the measure W_1^m on $(\omega_1)^m$. That is, $A \subseteq \omega_{m+1}$ has S_1^m measure one iff there is a c.u.b. $C \subseteq \omega_1$ such that for all $f\colon \text{dom}(<_m) \to C$ of the correct type, $[f]_{W_1^m} \in A$.

DEFINITION 5.2. By W_3^m we denote the measure on $\underset{\sim}{\boldsymbol{\delta}}_3^1$ induced by the weak partition relation on $\underset{\sim}{\boldsymbol{\delta}}_3^1$, functions $f\colon \omega_{m+1} \to \underset{\sim}{\boldsymbol{\delta}}_3^1$ of the correct type, and the measure S_1^m on ω_{m+1}.

We let $\mathcal{S}_1 = \{S_1^m\}_{m\in\omega}$ and $\mathcal{W}_3 = \{W_3^m\}_{m\in\omega}$ be the corresponding families of measures.

The fact that W_3^m is a measure on $\underset{\sim}{\boldsymbol{\delta}}_3^1$ requires the observation that for any $\alpha < \underset{\sim}{\boldsymbol{\delta}}_3^1$ we have $j_{S_1^m}(\alpha) < \underset{\sim}{\boldsymbol{\delta}}_3^1$. This is actually part of the stage-1 induction hypotheses, but can be seen in several different ways. One way is note that this is equivalent (since $\underset{\sim}{\boldsymbol{\delta}}_3^1 = (\omega_\omega)^+$) to saying $j_{S_1^m}(\omega_\omega) < \underset{\sim}{\boldsymbol{\delta}}_3^1$, and since $\text{cf}(\omega_\omega) = \omega$ to saying $j_{S_1^m}(\omega_n) < \underset{\sim}{\boldsymbol{\delta}}_3^1$ for all m, n. In fact, a description computation (see below) shows that $j_{S_1^m}(\omega_n) = \omega_k$ for some k. Another way is to use Corollary 4.35 and directly compute that the ultrapower $j_{S_1^m}(\alpha)$ is $\underset{\sim}{\boldsymbol{\Delta}}_3^1$ in the codes, and so has length less than $\underset{\sim}{\boldsymbol{\delta}}_3^1$.

The level-2 descriptions will describe how, given a sequence of functions $h_1, \dots, h_t$, to build an ordinal below $\lambda_3 = \omega_\omega$. Each h_i will either be a function of the form $h_i\colon r \to \omega_1$ for some $r \in \omega$ (*i.e.*, $h_i \in (\omega_1)^r$), or a function $h_i\colon \text{dom}(<_r) \to \omega_1$ of the correct type. We think of the h_i as representing functions for the measures $K_i = W_1^r$ or $K_i = S_1^r$. So, in the case $K_i = W_1^r$, h_1 represents (in this case actually is) an element of $(\omega_1)^r$. In the case $K_i = S_1^r$, h_i represents an ordinal $[h_i]_{W_1^r} < \omega_{r+1}$. The descriptions will in fact generate well-defined ordinals ϑ in the iterated ultrapowers by the sequence of measures $K_1, \dots, K_t$. For a level-2 description, this ordinal ϑ will also be less than λ_3. A level-3 description will build an ordinal below $\underset{\sim}{\boldsymbol{\delta}}_3^1$ given a sequence of functions $f, h_1, \dots, h_t$ where the h_i are as before and $f\colon \omega_{m+1} \to \underset{\sim}{\boldsymbol{\delta}}_3^1$ is of the correct type (and $m \in \omega$). The level-3 descriptions will generate ordinals in the iterated ultrapower by the measures $W_3^m, K_1, \dots, K_t$. This will describe an ordinal ϑ below $j_{W_3^m}(\underset{\sim}{\boldsymbol{\delta}}_3^1) < \lambda_5$. As we said before, the level-3 descriptions will be just minor variations of the level-2 descriptions. For the rest of this section we consider level-2 descriptions (level-3 descriptions will be considered in §6).

A level-2 description will have an *index* associated to it which will be written as a superscript. In the notation for the general level description this would be written as $d^{W_1^m}$, where the index W_1^m is a formal symbol (which, of course, suggests the measure W_1^m). In practice, to avoid overly cumbersome notation, we will write d^m as an abbreviation. The set of level-2 descriptions with index m will be denoted $\mathcal{D}_2^m$. A description d^m will be defined, as we said above, relative to a sequence of measures $K_1, \dots, K_t$. We will let $\mathcal{D}_2^m(K_1, \dots, K_t)$ denote the level-2 descriptions defined relative to $K_1, \dots, K_t$ with index m. As we said above, the measures $\vec{K}$ will define the iterated ultrapower in which the description is building an ordinal. The index m (or W_1^m) tells us that the ordinal that the description produces, given fixed functions $h_1, \dots, h_t$, will be defined by first generating an ordinal in the ultrapower by W_1^m. That is, it will produce an ordinal $\eta < \omega_{m+1}$. In the case of level-2 descriptions, the final ordinal the description produces can be viewed as a single iterated ultrapower by the measures $K_1, \dots, K_t, W_1^m$. For level-3 descriptions, however, the ordinal η will used indirectly and we cannot view the final ordinal produced as a iterated ultrapower by $K_1, \dots, K_t, W_1^m$. This is why the measures in the index are separated from the $K_1, \dots, K_t$. For the rest of this section we will suppress writing the subscript 2 and just write, for example $\mathcal{D}^m(K_1, \dots, K_t)$, which we frequently abbreviate as $\mathcal{D}^m(\vec{K})$.

We fix some notation to make the above discussion more precise. Given the sequence of measures $\vec{K} = K_1, \dots, K_t$ and a description $d \in \mathcal{D}^m(\vec{K})$, the ordinal in the iterated ultrapower by the measures $\vec{K}$ that d defines will be denoted $(d; K_1, \dots, K_t)$, or just $(d; \vec{K})$. The ordinal $(d; \vec{K})$ will be represented with respect to the measure K_1 by the function which assigns to $[h_1]$ the ordinal $(d; h_1, K_2, \dots, K_t)$. Here $h_1 \colon \operatorname{dom}(<_{r_1}) \to \omega_1$ is of the correct type if $K_1 = S_1^{r_1}$ (and $[h_1]$ means $[h_1]_{W_1^{r_1}}$), and if $K_1 = W_1^{r_1}$ then $h_1 \colon r_1 \to \omega_1$ (and $[h_1]$ just means h_1). Similarly, $(d; h_1, K_2, \dots, K_t)$ is represented with respect to K_2 by the function $[h_2] \mapsto (d; h_1, h_2, K_3, \dots, K_t)$. Finally, the ordinal $(d; h_1, \dots, h_t) < \omega_{m+1}$ will be defined. This will be the ordinal represented with respect to W_1^m by the function $\vec{\alpha} = (\alpha_1, \dots, \alpha_m) \mapsto (d; h_1, \dots, h_t; \vec{\alpha}) < \omega_1$. The ordinal $(d; \vec{h}; \vec{\alpha})$ will be defined directly from the functions $\vec{h}$ and the $\vec{\alpha}$ according to d. Given a function $g \colon \omega_1 \to \omega_1$, we will also define the "lift" by g, which we denote $(g; d; \vec{K})$.This is defined as above except at the bottom we define $(g; d; h_1, \dots, h_t, \vec{\alpha}) = g((d; \vec{K}; \vec{\alpha}))$. It will be important to observe that the descriptions are really generating well-defined ordinals in this manner. We make this notion more precise.

DEFINITION 5.3. Let $\mathcal{S}_1^m$ be the measure on functions $h \colon \operatorname{dom}(<_m) \to \omega_1$ of the correct type induced by the strong partition relation on ω_1. For consistency, let also $\mathcal{W}_1^m = W_1^m$.

So, $\mathcal{S}_1^m$ induces the ordinal measure S_1^m under the map $h \mapsto [h]_{W_1^m}$. If K_i denotes a measure of the form $S_1^{r_i}$ or $W_1^{r_1}$, then $\mathcal{K}_i$ denotes the corresponding function space measure $\mathcal{S}_1^{r_i}$ or $\mathcal{W}_1^{r_1}$. Note that if $K_i = W_1^{r_i}$, then $h_i : r_i \to \omega_1$, and in this case if we write $[h_i] = [h_i']$ we simply mean $h_i = h_i'$.

DEFINITION 5.4. Let $K_1, \dots, K_t$ be measures with $K_i = S_1^{r_i}$ or $K_i = W_1^{r_i}$, and let $\mathcal{K}_1, \dots, \mathcal{K}_t$ be the corresponding function space measures. Suppose $F : \mathcal{K}_1 \times \cdots \times \mathcal{K}_t \to \text{Ord}$. We say F is well-defined in the iterated ultrapower sense if: there is a c.u.b. $C_1 \subseteq \omega_1$ such that for all $h_1 : \text{dom}(<_{r_1}) \to C_1$ of the correct type (or $h_1 : r_1 \to C$ if $K_1 = W_1^{r_1}$) if $[h_1'] = [h_1]$, then there is a c.u.b. $C_2 \subseteq \omega_1$ such that for all $h_2 : \text{dom}(<_{r_2}) \to C_2$ of the correct type if $[h_2'] = [h_2]$, …, there is a c.u.b. $C_t \subseteq \omega_1$ such that for all $h_t : \text{dom}(<_{r_t}) \to \omega_1$ if $[h_t'] = [h_t]$, then $F(h_1, \dots, h_t) = F(h_1', \dots, h_t')$.

If $F : \mathcal{K}_1 \times \cdots \times \mathcal{K}_t \to \text{Ord}$ is well-defined in the iterated ultrapower sense, then F does in fact define an ordinal in the iterated ultrapowers by the measures $K_1, \dots, K_t$ (again, this is not the same as saying that F gives a well-defined function on the product of the ordinal measures spaces $K_1 \times \cdots \times K_t$; this is usually not true). So, when F is well-defined in the iterated ultrapower sense it makes sense to write $\forall^*_{K_1}[h_1] \cdots \forall^*_{K_t}[h_t]\, P(F([h_1], \dots, [h_t])$ for any $P \subseteq \text{Ord}$.

REMARK 5.5. We use the notation $\forall^* h_1, \dots, h_t$ to denote quantification by the iterated function space measures $\mathcal{K}_i$. So, "$\forall^* h_1, \dots, h_t \cdots$" abbreviates: "there is a c.u.b. $C_1 \subseteq \omega_1$ such that for all functions $h_1 : \text{dom}(<_{r_1}) \to \omega_1$ of the correct type, there is a c.u.b. $C_2 \subseteq \omega_1$ such that for all functions $h_2 : \text{dom}(<_{r_2}) \to C_2$ of the correct type, $\cdots$". Recalling our notational convention from the end of §2, if $P \subseteq \text{Ord}$ and $\vartheta \in \text{Ord}$, it also makes sense to write $\forall^*_{K_1}[h_1] \dots \forall^*_{K_t}[h_t]\, P(\vartheta([h_1], \dots, [h_t])$. Suppose now that $F : \mathcal{K}_1 \times \cdots \times \mathcal{K}_t \to \text{Ord}$ is well-defined in the iterated ultrapower sense (Definition 5.4), $P \subseteq \text{Ord} \times \text{Ord}$ and $\vartheta \in \text{Ord}$. Then it makes sense to write $\forall^*_{K_1}[h_1] \cdots \forall^*_{K_t}[h_t]\ P(\vartheta([h_1], \dots, [h_t]), F([h_1], \dots, [h_t]))$. This type of statement will occur frequently in the theory of descriptions. When it is understood that F is well-defined in the iterated ultrapower sense (and this will be the case whenever F arises from a description), then we simplify the notation and just write $\forall^* h_1, \dots, h_t\ P(\vartheta(h_1, \dots, h_t), F(h_1, \dots, h_t))$.

Before giving the formal definition of a level-2 description, we consider a simple case to see how a description d might generate an ordinal. Suppose $K_1 = K_2 = S_1^1$, and $m = 1$. Given $h_1, h_2 : \omega_1 \to \omega_1$ and $\alpha < \omega_1$, we could output α itself, $h_1(\alpha)$, $h_2(\alpha)$, or $h_1(h_2(\alpha))$. All of these are well-defined in the iterated ultrapower sense. That is, for S_1^1 almost all $[h_1]$, if $[h_1'] = [h_1]$, then for S_1^1 almost all $[h_2]$, if $[h_2'] = [h_2]$ then (for example) for W_1^1 almost all α we have $h_1(h_2(\alpha)) = h_1'(h_2'(\alpha))$. Note though that the composition $h_2(h_1(\alpha))$ is not well-defined in this sense. So, roughly speaking. a description describes a pattern of "backwards composition" among the h_i.

EXAMPLE 5.6. Consider one more example which is close to (but not quite) the general form. Say $K_1 = K_2 = S_1^3$, $K_3 = W_1^3$, $K_4 = S_1^3$, and $m = 4$. We might then let

$$(d;\vec{h};\vec{\alpha}) = h_1(\alpha_1, h_2(\beta_3, h_4(\beta_2, \alpha_2, \alpha_3), h_4(\alpha_1, \alpha_2, \alpha_4)), h_4(\alpha_1, \alpha_3, \alpha_4)),$$

where $h_3 = (\beta_1, \beta_2, \beta_3)$. The reader can check that this definition is well-defined in the previous iterated ultrapower sense.

In defining the level-2 descriptions, there will be two kinds of "basic" descriptions and then non-basic description which, roughly speaking, will be compositions of the basic ones. One of the basis descriptions, which we denote $\cdot_i$, describes the function $(d;\vec{h};\vec{\alpha}) = \alpha_i$. The other is a pair of integers $(k;i)$ where $K_k = W_1^{r_k}$ and $i \leq r_k$. This describes the value $(d;\vec{h};\vec{\alpha}) = h_k(i)$ (recall $h_k\colon r_k \to \omega_1$ is an r_k-tuple of countable ordinals). As we define $\mathcal{D}^m(K_1, \dots, K_t)$, we also define a function $d \mapsto k(d) \in \{1, \dots, t\} \cup \{\infty\}$ on $\mathcal{D}^m(\vec{K})$ which records the outermost function used in the composition described by d (with value ∞ if none are used).

The actual definitions follow. Although it was not done in [Jac99] or [Jac88], we find it convenient (following a suggestion of a previous referee) to first give the definition of the set $\mathcal{D}'(\vec{K})$ of *pre-descriptions*. The set $\mathcal{D}(\vec{K})$ of descriptions will be the pre-descriptions that satisfy an extra condition which guarantees that they produce well-defined ordinals in the iterated ultrapower. We will use pre-descriptions only as a step in defining descriptions. In the following definition the symbol "s" appears. This is regarded as a formal syntactical symbol. Its intended meaning is to stand for "sup," as will become apparent when we interpret the descriptions. The formal definition is actually by reverse induction on the value $k(d) \in \{1, \dots, t\} \cup \{\infty\}$ (that is, we assume the set of pre-descriptions having $k(d) > k$ has been defined, and proceed to define those with $k(d) = k$).

DEFINITION 5.7 (Pre-descriptions). Fix $m > 0$ and $K_1, \dots, K_t$ where $K_i = S_1^{r_i}$ or $K_i = W_1^{r_i}$. The set of pre-descriptions $\mathcal{D}'^m(\vec{K})$ and the function k are defined through the following cases (the first two cases define the *basic* descriptions).

1. $d = \cdot_i$ where $1 \leq i \leq m$. In this case $k(d) = \infty$.
2. $d = (k;i)$ where $K_k = W_1^{r_k}$ and $1 \leq i \leq r_k$. We set $k(d) = k$.
3. $d = (k; d_0, d_1, \dots, d_\ell)$ where $1 \leq k \leq t$, $K_k = S_1^r$, $0 \leq \ell \leq r-1$, and $k(d_0), k(d_1), \dots, k(d_\ell) > k$ (if $\ell = 0$ we have $d = (k; d_0)$). We set $k(d) = k$.
4. $d = (k; d_0, d_1, \dots, d_\ell)^s$ where $1 \leq k \leq t$, $K_k = S_1^r$, $r \geq 2$, $1 \leq \ell \leq r-1$, and $k(d_0), k(d_1), \dots, k(d_\ell) > k$. We set $k(d) = k$.

We write $(k; d_r, d_1, \dots, d_\ell)^{(s)}$ to indicate that the formal symbol s (which again stands for "sup") may or may not appear. We require $r \geq 2$ and $\ell \geq 1$ in the last case so as to avoid $(k;d)^s$. Allowing this would create a redundancy, as this will evaluate to the same as d.

REMARK 5.8. In [Jac99] and [Jac88] the non-basic descriptions were denote $(k;d_r,d_1,\dots,d_\ell)^{(s)}$. That is, the first description was denoted d_r (where $K_k = S_1^r$) instead of d_0.

The following definition gives the evaluation $(d;\vec{h};\vec{\alpha})$ of the description. Recall that $h(j)$ denotes the jth invariant of h (Definition 4.17). The notation $h^s(j)$ was also defined in Definition 4.17. In the following definition, the functions h_i are of the form $h_i\colon \mathrm{dom}(<_{r_i}) \to \omega_1$ and of the correct type if $K_i = S_1^{r_i}$, and of the form $h_i\colon r_i \to \omega_1$ if $K_i = W_1^i$.

DEFINITION 5.9 (Evaluation of Pre-descriptions). With notation as in Definition 5.7, and fixed functions $h_1,\dots,h_t$ and $\vec{\alpha} = (\alpha_1,\dots,\alpha_m) \in (\omega_1)^m$, we define $(d;\vec{h};\vec{\alpha})$ as follows.

1. If $d = \cdot_i$, then $(d;\vec{h};\vec{\alpha}) = \alpha_i$.
2. If $d = (k;i)$, then $(d;\vec{h};\vec{\alpha}) = h_k(i)$.
3. If $d = (k;d_0,d_1,\dots,d_\ell)^{(s)}$, then

$$(d;\vec{h};\vec{\alpha}) = h_k^{(s)}(\ell+1)((d_1;\vec{h};\vec{\alpha}),\dots,(d_\ell;\vec{h};\vec{\alpha}),(d_0;\vec{h};\vec{\alpha})).$$

We also let $(d;\vec{h})$ be the ordinal represented by the function $[\vec{\alpha} \mapsto (d;\vec{h};\vec{\alpha})]_{W_1^m}$.

Note that for $d \in \mathcal{D}^m(\vec{K})$, we have $(d;\vec{h};\vec{\alpha}) < \omega_1$ for any $\vec{h}$ and $\vec{\alpha}$, whereas $(d;\vec{h}) < \omega_{m+1}$.

EXAMPLE 5.10. For the function described in Example 5.6 the description is given by $d = (1;(4;\cdot_4,\cdot_1,\cdot_3),\cdot_1,d')$ where $d' = (2;(4;\cdot_4,\cdot_1,\cdot_2),(3;3),(4;\cdot_3,(3;2),\cdot_2))$.

REMARK 5.11. In [JK] a slightly different notation for the (pre) descriptions was introduced called the *functional representation*. There the notation is more closely tied with the way the description is evaluated. The above example would have functional representation

$$d = h_1(\cdot_1,d',h_4(\cdot_1,\cdot_3,\cdot_4)),$$

where

$$d' = h_2(\gamma_{3,3},h_4(\gamma_{3,2},\cdot_2,\cdot_3),h_4(\cdot_1,\cdot_2,\cdot_4)).$$

Here we staying with the notation of [Jac88].

The "backward composition" requirement is built into the definition of a pre-description in Definition 5.7. To have well-defined interpretations (in the sense described before), another requirement is necessary. Namely, in the last two cases, the subdescriptions $d_0,d_1,\dots d_\ell$ must give values in the correct order. We list these subdescriptions in their order of significance in determining the size of the output value. Recalling the definition of the order $<_r$, this means we want to assume that $d_1 < d_2 < \cdots < d_\ell < d_0$. More precisely, we order the descriptions by $d' < d$ iff $\forall^* h_1, \dots, h_t\ \forall^* \vec{\alpha}\ (d'; \vec{h}; \vec{\alpha}) < (d; \vec{h}; \vec{\alpha})$. This ordering is easily described combinatorially, which we give in the following (inductive) definition. In the following main case (III), where both descriptions are non-basic, the ordering is more or less lexicographic ordering.

DEFINITION 5.12 (Ordering on Pre-descriptions). Fix $m > 0$ and $K_1, \dots, K_t \in \mathcal{W}_1 \cup \mathcal{S}_1$. The ordering $d' < d$ on the pre-descriptions $\mathcal{D}'^m(K_1, \dots, K_t)$ is defined by reverse induction on $\min\{k(d'), k(d)\}$ through the following cases:

I. Suppose $k' = k(d') < k(d) = k$.
  1. $K_{k'} = W_1^{r'}$. In this case we set $d' < d$.
  2. $K_{k'} = S_1^{r'}$. In this case $d' = (k'; d_0', d_1', \dots, d_\ell')^{(s)}$. We define $d' < d$ to hold iff $d_0' < d$.

II. Suppose $k' = k(d') > k(d) = k$.
  1. $K_k = W_1^r$. In this case we do not set $d' < d$.
  2. $K_k = S_1^r$. In this case $d = (k; d_0, d_1, \dots, d_\ell)^{(s)}$. We define $d' < d$ to hold iff $d' < d_0$ or $d' = d_0$.

III. Suppose $k(d') = k(d) = k < \infty$.
  1. $K_k = W_1^r$. In this case $d' = (k; i')$, $d = (k; i)$. We set $d' < d$ iff $i' < i$.
  2. $K_k = S_1^r$. In this case $d = (k; d_0, d_1, \dots, d_\ell)^{(s)}$ and $d' = (k; d_0', d_1', \dots, d_{\ell'}')^{(s)}$.
    a. Suppose there is a least j with $0 \leq j \leq \ell$ such that $d_j' \neq d_j$. Then we define $d' < d$ iff $d_j' < d_j$. In the remaining cases assume there is no such j.
    b. If $\ell' < \ell$, then $d' < d$ iff d' has the symbol s.
    c. If $\ell' > \ell$ then $d' < d$ iff d does not have the symbol s.
    d. If $\ell' = \ell$, then $d' < d$ iff d' has the symbol s and d does not.

IV. $k(d') = k(d) = \infty$. In this case $d' = \cdot_{r'}$ and $d = \cdot_r$. We set $d' < d$ iff $r' < r$.

We can now state the well-definedness condition, which for historical reasons we call "condition C." The definition is also inductive, by reverse induction on $k(d)$.

DEFINITION 5.13. A pre-description d satisfies condition C if either d is basic, or $d = (k; d_0, d_1, \dots, d_\ell)^{(s)}$ is non-basic, $d_0, d_1, \dots, d_\ell$ satisfy C, and $d_1 < \cdots < d_\ell < d_0$.

DEFINITION 5.14. Let $m > 0$ and $K_1, \dots, K_t \in \mathcal{W}_1 \cup \mathcal{S}_1$. A **description** (with index m defined relative to $K_1, \dots, K_t$) is a pre-description $d \in \mathcal{D}'^m(\vec{K})$ which satisfies condition C. We let $\mathcal{D}^m(K_1, \dots, K_t)$ denote this set of descriptions.

The reader can check that if d is a description then the evaluation of d is well-defined in the sense that $\forall^* h_1$ if $[h_1] = [h_1']$, then $\forall^* h_2$ if $[h_2] = [h_2']$, $\dots$, $\forall^* h_t$ if $[h_t] = [h_t']$ then $\forall^* \vec{\alpha}\ (d;\vec{h};\vec{\alpha}) = (d;\vec{h}';\vec{\alpha})$. This is equivalent to saying that the function space map $(h_1,\dots,h_t) \mapsto (d;h_1,\dots,h_t) = [\vec{\alpha} \mapsto (d;h_1,\dots,h_t;\vec{\alpha})]_{W_1^m}$ is well-defined in the iterated ultrapower sense (Definition 5.4). Thus, the evaluation of the description is well-defined in the iterated ultrapower sense.

We summarize this discussion in the following definition.

Definition 5.15. Let $d \in \mathcal{D}^m(K_1,\dots,K_t)$. Then $(d;K_1,\dots,K_t)$ is the ordinal represented in the ultrapower with respect to the measure K_1 by the function which assign to $[h_1]$ the ordinal $(d;h_1,K_2,\dots,K_t)$. This, in turn, is represented with respect to the measure K_2 by the function which assign to $[h_2]$ the ordinal $(d;h_1,h_2,K_3,\dots,K_t)$. Finally, $(d;h_1,\dots,h_t)$ is the ordinal less than ω_{m+1} which is represented with respect to W_1^m by the function which assigns to $\vec{\alpha} \in (\omega_1)^m$ the ordinal $(d;\vec{h};\vec{\alpha})$ defined in Definition 5.9. If $g\colon \omega_1 \to \omega_1$, then $(g;d;\vec{K})$ is defined similarly except $(g;d;\vec{h};\vec{\alpha}) = g((d;\vec{h};\vec{\alpha}))$.

We next introduce the lowering operator $\mathcal{L}$, as we did for the trivial descriptions. $\mathcal{L}(d)$ will be the immediate predecessor of d in the above ordering on descriptions, except if d is minimal in this ordering in which case $\mathcal{L}(d)$ is not defined. We will actually define approximations $\mathcal{L}^k$ to $\mathcal{L}$, where $\mathcal{L}^k(d)$ is defined for those d with $k(d) \geq k$. We will then take $\mathcal{L} = \mathcal{L}^1$. These approximations to $\mathcal{L}$ also correspond to the steps in the proof of the main theorem 5.17 below.

Definition 5.16. Let $d \in \mathcal{D}^m(K_1,\dots,K_t)$. We define $\mathcal{L}^k(d)$, where $k \in \{1,\dots,t\} \cup \{\infty\}$ and $k(d) \geq k$, by reverse induction on $k(d)$ through the following cases.

1. $k = \infty$. In this case $d = \cdot_i$. If $i > 1$ we set $\mathcal{L}^\infty(d) = \cdot_{i-1}$, and if $i = 1$ then d is minimal with respect to $\mathcal{L}^\infty$.

In the remaining cases we assume $1 \leq k \leq t$.

2. $k = k(d)$. We have the following sub-cases.
   (a) d is basic, so $d = (k;i)$. Then $\mathcal{L}^k(d) = (k;i-1)$ if $i > 1$ and if $i = 1$ then d is minimal with respect to $\mathcal{L}^k$.
   (b) $d = (k;d_0,d_1,\dots,d_\ell)$ where $\ell = r-1$. Then $\mathcal{L}^k(d) = (d_r,d_1,\dots,d_\ell)^s$ if $\ell \geq 1$ and if $\ell = 0$ then $\mathcal{L}^k(d) = d_0$.
   (c) $d = (k;d_0,d_1,\dots,d_\ell)$ where $\ell < r-1$. If $\mathcal{L}^{k+1}(d_0)$ is defined and satisfies $\mathcal{L}^{k+1}(d_0) > d_\ell$ if $\ell \geq 1$, then $\mathcal{L}^k(d) = (k;d_0,d_1,\dots,d_\ell,\mathcal{L}^{k+1}(d_0))$. Otherwise $\mathcal{L}^k(d) = (k;d_0,d_1,\dots,d_\ell)^s$ if $\ell \geq 1$ and if $\ell = 0$ then $\mathcal{L}^k(d) = d_0$.

(d) $d = (k; d_0, d_1, \ldots, d_\ell)^s$. If $\mathcal{L}^{k+1}(d_\ell)$ is defined and satisfies $\mathcal{L}^{k+1}(d_\ell) > d_{\ell-1}$ if $\ell > 1$, then $\mathcal{L}^k(d) = (k; d_0, d_1, \ldots, d_{\ell-1}, \mathcal{L}^{k+1}(d_\ell))$. Otherwise, $\mathcal{L}^k(d) = (k; d_0, d_1, \ldots, d_{\ell-1})^s$ if $\ell > 1$ and $\mathcal{L}^k(d) = d_0$ if $\ell = 1$.

3. $k < k(d)$ and $K_k = W_1^r$.
   (a) d is not minimal with respect to $\mathcal{L}^{k+1}$. Then $\mathcal{L}^k(d) = \mathcal{L}^{k+1}(d)$.
   (b) d is minimal with respect to $\mathcal{L}^{k+1}(d)$. Then $\mathcal{L}^{k+1}(d) = (k; r)$.
4. $k < k(d)$ and $K_k = S_1^r$.
   (a) d is not minimal with respect to $\mathcal{L}^{k+1}(d)$. Then $\mathcal{L}^k(d) = (k; \mathcal{L}^{k+1}(d))$.
   (b) d is minimal with respect to $\mathcal{L}^{k+1}(d)$. Then d is minimal with respect to $\mathcal{L}^k$.

We formulate the "main theorem" for level-2 descriptions as we did in Theorem 4.3 for trivial descriptions.

THEOREM 5.17 (Main theorem for level-2 descriptions). Suppose $d \in \mathcal{D}^m(K_1, \ldots, K_t)$ and $\vartheta < (d; \vec{K})$. If d is non-minimal with respect to $\mathcal{L}$, then there is a $g : \omega_1 \to \omega_1$ such that $\vartheta < (g; \mathcal{L}(d); \vec{K})$. If d is $\mathcal{L}$-minimal then $\vartheta < \omega_1$.

We will give a brief sketch of how Theorem 5.17 is proved below.

If $g : \omega_1 \to \omega_1$, then there is a real x such that the section of the Kunen tree T_x is wellfounded and $g(\alpha) < |T_x \restriction \alpha|$ for all infinite α. This easily shows that $(g; \mathcal{L}(d); \vec{K}) < (\mathcal{L}(d); \vec{K})^+$. Namely, we define a map π from $(\mathcal{L}(d); \vec{K})$ onto $(g; \mathcal{L}(d); \vec{K})$ as follows. If $\vartheta < (\mathcal{L}(d); \vec{K})$, then $\pi(\vartheta)$ is defined by

$$\forall [h_1], \ldots, [h_t]\, \forall \vec{\alpha}\; \pi(\vartheta)([\vec{h}]; \vec{\alpha}) = |(T_x \restriction (\mathcal{L}(d); \vec{h}; \vec{\alpha})(\vartheta(\vec{h}; \vec{\alpha}))|.$$

So, we have the following immediate consequence of Theorem 5.17

COROLLARY 5.18. $(d; \vec{K}) \leq (\mathcal{L}(d); \vec{K})^+$.

In fact, we will see shortly that equality holds in Corollary 5.18. Granting this for the moment, we then have the following exact computation of the ordinal represented by a level-2 description.

COROLLARY 5.19. Let $d \in \mathcal{D}^m(\vec{K})$. Then $(d; \vec{K}) = \omega_{r+1}$ where $r = |d|_{\mathcal{L}}$ is the rank of d in the ordering of the descriptions in $\mathcal{D}^m(\vec{K})$ given in Definition 5.16.

We pause to give some examples and consequences. In the first example we compute the ultrapower $j_{S_1^m}(\omega_n)$ of ω_n by the measure S_1^m.

EXAMPLE 5.20. Consider the descriptions $d \in \mathcal{D}^{n-1}(K_1)$, where $K_1 = S_1^m$. For each such d, $(d; K_1)$ is by definition an ordinal in the ultrapower $j_{S_1^m}(\omega_n)$. A small variation of Theorem 5.17 shows that if $\vartheta < j_{S_1^m}(\omega_n)$, then there is a

$g\colon \omega_1 \to \omega_1$ such that $\vartheta < (g; d_{\max})$, where $d_{\max}$ is the maximal description in $\mathcal{D}_{n-1}(K_1)$ which is easily seen to be $d_{\max} = (1; \cdot_n)$. From Theorem 5.17 it follows that $j_{S_1^m}(\omega_n) = \omega_{\ell+1}$, where ℓ is the number of descriptions in $\mathcal{D}^{n-1}(K_1)$. These descriptions are all of the form $(1; \cdot_{i_0}, \cdot_{i_1}, \dots, \cdot_{i_{p-1}})^{(s)}$, where $1 \le i_1 < \cdots < i_{p-1} < i_0 \le n$ and $p \le m$ (recall (s) means that s may or may not appear). For convenience here we are deviating slightly from the official definition of $\mathcal{D}^{n-1}(K_1)$ in that we are allowing descriptions of the form $(1; \cdot_i)^s$ (these "extra" descriptions are just equivalent to the official descriptions $\cdot_i$, and make the counting more obvious).

Clearly there are $2\left[\binom{n-1}{1} + \cdots + \binom{n-1}{m}\right]$ such descriptions, where we regard $\binom{n-1}{k} = 0$ if $k > n-1$. Thus we have:

$$j_{S_1^m}(\omega_n) = \omega_k \text{ where } k = 1 + 2\left[\binom{n-1}{1} + \cdots + \binom{n-1}{m}\right].$$

The formula of Example 5.20 when iterated allows us to compute any iterated ultrapower of the form $j_{S_1^{m_1}} \circ \cdots \circ j_{S_1^{m_t}}(\omega_n)$. For the next example, we do this by directly counting the corresponding descriptions. This example appeared in [Jac10].

Example 5.21. We compute $j_{S_1^2} \circ j_{S_1^2}(\omega_3)$. From Example 5.20, we know the answer should be $j_{S_1^2} \circ j_{S_1^2}(\omega_3) = j_{S_1^2}(\omega_7) = \omega_{43}$. The cardinals below the iterated ultrapower $j_{S_1^2} \circ j_{S_1^2}(\omega_3)$ correspond now to the descriptions in $\mathcal{D}^2(K_1, K_2)$, where $K_1 = K_2 = S_1^2$. The maximal description is now $d_{\max} = (1; (2; \cdot_2))$. The remaining descriptions can be obtained by iterating the $\mathcal{L}$ operation, which will list the descriptions in decreasing order. Figure 1 lists these descriptions. Thus, there are 42 descriptions in $\mathcal{D}_2(K_1, K_2)$, which agrees with $j_{S_1^2} \circ j_{S_1^2}(\omega_3) = \omega_{43}$.

Remark 5.22. One can use the formula of Example 5.20 and the fact that $j_{W_1^m}(\omega_n) = \omega_{m+n}$ to compute the number of descriptions in $\mathcal{D}^m(K_1, \dots, K_t)$, in a manner similar to Example 5.21. The number of such descriptions is k, where $j_{K_1} \circ \cdots \circ j_{K_t}(\omega_{m+1}) = \omega_{k+1}$. For example, consider $\mathcal{D}^4(S_1^3, S_1^3, W_1^3, S_1^3)$ of Example 5.20. We have $j_{S_1^3} \circ j_{S_1^3} \circ j_{W_1^3} \circ j_{S_1^3}(\omega_5) = j_{S_1^3} \circ j_{S_1^3} \circ j_{W_1^3}(\omega_{29}) = j_{S_1^3} \circ j_{S_1^3}(\omega_{32}) = j_{S_1^3}(\omega_{9983}) = \omega_{331536189167}$. Thus, there are 331 536 189 166 descriptions in $\mathcal{D}^4(S_1^3, S_1^3, W_1^3, S_1^3)$.

We give sketches of two arguments. First we show why the lower bound holds in Corollary 5.18, and then we give a brief outline of how the proof of Theorem 5.17 goes.

To show that $(d; \vec{K}) = (\mathcal{L}(d); \vec{K})^+$ it suffices to show that the conclusion of Corollary 5.19 holds. Let $d \in \mathcal{D}^m(\vec{K})$ be non-minimal, and let $r = |d|_{\mathcal{L}}$ be its rank in the ordering of descriptions in $\mathcal{D}^m(\vec{K})$. We must show that $(d; \vec{K}) \ge$

$$d = d_{\max} = (1; (2; \cdot_2)) \qquad \mathcal{L}(d) = (1; (2; \cdot_2); (2; \cdot_2, \cdot_1))$$

$$\mathcal{L}^2(d) = (1; (2; \cdot_2); (2; \cdot_2, \cdot_1))^s \qquad \mathcal{L}^3(d) = (1; (2; \cdot_2); (2; \cdot_2, \cdot_1)^s)$$

$$\mathcal{L}^4(d) = (1; (2; \cdot_2); (2; \cdot_2, \cdot_1)^s)^s \qquad \mathcal{L}^5(d) = (1; (2; \cdot_2); \cdot_2)$$

$$\mathcal{L}^6(d) = (1; (2; \cdot_2); \cdot_2)^s \qquad \mathcal{L}^7(d) = (1; (2; \cdot_2); (2; \cdot_1))$$

$$\mathcal{L}^8(d) = (1; (2; \cdot_2); (2; \cdot_1))^s \qquad \mathcal{L}^9(d) = (1; (2; \cdot_2); \cdot_1)$$

$$\mathcal{L}^{10}(d) = (1; (2; \cdot_2); \cdot_1)^s \qquad \mathcal{L}^{11}(d) = (2; \cdot_2)$$

$$\mathcal{L}^{12}(d) = (1; (2; \cdot_2, \cdot_1)) \qquad \mathcal{L}^{13}(d) = (1; (2; \cdot_2, \cdot_1); (2, \cdot_2, \cdot_1)^s)$$

$$\mathcal{L}^{14}(d) = (1; (2; \cdot_2, \cdot_1); (2, \cdot_2, \cdot_1)^s)^s \qquad \mathcal{L}^{15}(d) = (1; (2; \cdot_2, \cdot_1); \cdot_2)$$

$$\mathcal{L}^{16}(d) = (1; (2; \cdot_2, \cdot_1); \cdot_2)^s \qquad \mathcal{L}^{17}(d) = (1; (2; \cdot_2, \cdot_1); (2; \cdot_1))$$

$$\mathcal{L}^{18}(d) = (1; (2; \cdot_2, \cdot_1); (2; \cdot_1))^s \qquad \mathcal{L}^{19}(d) = (1; (2; \cdot_2, \cdot_1); \cdot_1)$$

$$\mathcal{L}^{20}(d) = (1; (2; \cdot_2, \cdot_1); \cdot_1)^s \qquad \mathcal{L}^{21}(d) = (2; \cdot_2, \cdot_1)$$

$$\mathcal{L}^{22}(d) = (1; (2; \cdot_2, \cdot_1)^s) \qquad \mathcal{L}^{23}(d) = (1; (2; \cdot_2, \cdot_1)^s, \cdot_2)$$

$$\mathcal{L}^{24}(d) = (1; (2; \cdot_2, \cdot_1)^s, \cdot_2)^s \qquad \mathcal{L}^{25}(d) = (1; (2; \cdot_2, \cdot_1)^s, (2; \cdot_1))$$

$$\mathcal{L}^{26}(d) = (1; (2; \cdot_2, \cdot_1)^s, (2; \cdot_1))^s \qquad \mathcal{L}^{27}(d) = (1; (2; \cdot_2, \cdot_1)^s, \cdot_1)$$

$$\mathcal{L}^{28}(d) = (1; (2; \cdot_2, \cdot_1)^s, \cdot_1)^s \qquad \mathcal{L}^{29}(d) = (2; \cdot_2, \cdot_1)^s$$

$$\mathcal{L}^{30}(d) = (1; \cdot_2) \qquad \mathcal{L}^{31}(d) = (1; \cdot_2, (2; \cdot_1))$$

$$\mathcal{L}^{32}(d) = (1; \cdot_2, (2; \cdot_1))^s \qquad \mathcal{L}^{33}(d) = (1; \cdot_2, \cdot_1)$$

$$\mathcal{L}^{34}(d) = (1; \cdot_2, \cdot_1)^s \qquad \mathcal{L}^{35}(d) = \cdot_2$$

$$\mathcal{L}^{36}(d) = (1; (2; \cdot_1)) \qquad \mathcal{L}^{37}(d) = (1; (2; \cdot_1), \cdot_1)$$

$$\mathcal{L}^{38}(d) = (1; (2; \cdot_1), \cdot_1)^s \qquad \mathcal{L}^{39}(d) = (2; \cdot_1)$$

$$\mathcal{L}^{40}(d) = (1; \cdot_1) \qquad \mathcal{L}^{41}(d) = \cdot_1$$

FIGURE 1. The descriptions in $\mathcal{D}^2(S_1^2, S_1^2)$.

ω_{r+1}. Note that there are r descriptions below d, say $d_1 < d_2 < \cdots < d_r < d$. We define an embedding π from the ultrapower $j_{W_1^r}(\omega_1)$ into $(d; \vec{K})$. Since $j_{W_1^r}(\omega_1) = \omega_{r+1}$, this suffices. For $f : (\omega_1)^r \to \omega_1$ we define $\pi([f]_{W_1^r})$ by:

$$\forall^* h_1, \ldots, h_t \, \forall^* \vec{\alpha} \in (\omega_1)^m \; \pi([f])(\vec{h}; \vec{\alpha}) = f((d_1; \vec{h}; \vec{\alpha}), \ldots, (d_r; \vec{h}; \vec{\alpha})).$$

It is straightforward to check that π is well-defined and gives an embedding. For example, to see that π is well-defined, suppose $[f] = [f']$. Let $C \subseteq \omega_1$ be c.u.b. such that $f(\vec{\beta}) = f'(\vec{\beta})$ for all $\vec{\beta} \in C^r$. Since for almost all $h_1, \dots, h_t$ we have that all the h_i have range in C, it follows that all the $(d_i; \vec{h}; \vec{\alpha})$ are in C. From the ordering of descriptions we also have $(d_1; \vec{h}; \vec{\alpha}) < \cdots < (d_r; \vec{h}; \vec{\alpha})$ for almost all $\vec{h}$ and almost all $\vec{\alpha}$. This shows that $\forall^* \vec{h}\ \forall^* \vec{\alpha}\ \pi(f)(\vec{h}; \vec{\alpha}) = \pi(f')(\vec{h}; \vec{\alpha})$.

We next outline the proof of Theorem 5.17. In general, the proof proceeds by a sequence of partition arguments, starting with t (where $\vec{K} = K_1, \dots, K_t$) and working back to 1. At each step k we prove the analog of the statement of Theorem 5.17 for the approximation $\mathcal{L}^k$ of the $\mathcal{L}$ operation (recall that $\mathcal{L} = \mathcal{L}^1$). After the last step, we have proved the theorem for the true $\mathcal{L}$ operation.

We illustrate this proof by considering an example. Let $d = d_{\max} = (1; (2; \cdot_2)) \in \mathcal{D}^2(S_1^2, S_1^2)$ (see Figure 1). So, $\mathcal{L}(d) = (1; (2; \cdot_2), (2, \cdot_2, \cdot_1))$. Let $\vartheta \in \mathrm{Ord}$ be such that $\forall^* \vec{h}\ \forall^* \vec{\alpha} = (\alpha_1, \alpha_2)\ \vartheta(\vec{h}; \vec{\alpha}) < (d; \vec{h}; \vec{\alpha}) = h_1(1)(h_2(1)(\alpha_2))$, where we recall that $h_1(1)(\beta) = \sup_{\gamma<\beta}(h_1(\gamma, \beta))$ and similarly for $h_2(1)$. So,

$$\forall^* \vec{h}\ \forall^* \vec{\alpha}\ \exists \beta < h_2(1)(\alpha_2)\ \vartheta(\vec{h}; \vec{\alpha}) < h_1(\beta, h_2(1)(\alpha_2)),$$

and so

$$\forall^* \vec{h}\ \forall^* \vec{\alpha}\ \exists \gamma < \alpha_2\ \vartheta(\vec{h}; \vec{\alpha}) < h_1(h_2(\gamma, \alpha_2), h_2(1)(\alpha_2)).$$

It follows that

$$\forall^* h_1, h_2\ \exists g : \omega_1 \to \omega_1\ \forall^* \vec{\alpha}\ \vartheta(\vec{h}; \vec{\alpha}) < h_1(h_2(g(\alpha_1), \alpha_2), h_2(1)(\alpha_2)).$$

We now consider the first of the partition arguments. For every ordinal $\vartheta(h_1)$ (this is just an ordinal, though we write it this way to be suggestive) and function h_1 such that $\forall^* h_2\ \exists g : \omega_1 \to \omega_1\ \forall^* \vec{\alpha}\ \vartheta(h_1)(h_2; \vec{\alpha}) < h_1(h_2(g(\alpha_1), \alpha_2), h_2(1)(\alpha_2))$, we consider the following partition. Fix a function $[h_2] \mapsto \vartheta(h_1)(h_2)$ representing $\vartheta(h_1)$ with respect to $K_2 = S_1^2$.

$\mathcal{P}(h_1)$: We partition functions h_2: $\mathrm{dom}(<_2) \to \omega_1$ of the correct type according to whether $\forall^* \vec{\alpha}\ \vartheta(h_1)(h_2; \vec{\alpha}) < h_1(h_2(\alpha_1 + 1, \alpha_2), h_2(1)(\alpha_2))$.

We claim that on the homogeneous side of the partition the stated property must hold. For suppose $C \subseteq \omega_1$ were c.u.b. and homogeneous for the contrary side. Fix h_2: $\mathrm{dom}(<_2) \to C$ of the correct type such that $\exists g : \omega_1 \to \omega_1\ \forall^* \vec{\alpha}\ \vartheta(h_1)(h_2; \vec{\alpha}) < h_1(h_2(g(\alpha_1), \alpha_2), h_2(1)(\alpha_2))$ (the last condition holds for almost all h_2). Fix a function $g : \omega_1 \to \omega_1$ witnessing this statement, that is,

$$\forall^* \vec{\alpha}\ \vartheta(h_1)(h_2; \vec{\alpha}) < h_1(h_2(g(\alpha_1), \alpha_2), h_2(1)(\alpha_2)). \tag{1}$$

Let $D \subseteq C$ be c.u.b. and closed under g. Define h_2': $\mathrm{dom}(<_2) \to \omega_1$ by "sliding up" h_2 along D. That is, define $h_2'(\alpha, \beta) = h_2(N_D(\alpha), N_D(\beta))$, where we recall $N_D(\delta) =$ the least element of D which is $\geq \delta$. Clearly $[h_2'] = [h_2]$ (they agree on D^2). Since h_2 has the correct type, so does h_2'. Finally,

$\operatorname{ran}(h_2') \subseteq \operatorname{ran}(h_2) \subseteq C$. Since $[h_2'] = [h_2]$, we have $\vartheta(h_2)(h_2) = \vartheta(h_1)(h_2')$. Since h_2' is of the correct type with range in C, it follows from the assumed homogeneity of C for the contrary side that:

$$\forall^* \vec{\alpha}\, \vartheta(h_1)(h_2; \vec{\alpha}) \geq h_1(h_2'(\alpha_1 + 1, \alpha_2), h_2(1)(\alpha_2)). \tag{2}$$

Note here that $h_2(1)(\alpha_2) = h_2'(1)(\alpha_2)$ for almost all α_2. Equations (1) and (2) contradict each other as $h_2'(\alpha_1 + 1, \alpha_2) \geq h_2(g(\alpha_1), \alpha_2)$ from the definition of D and h_2'.

So, on the homogeneous side of the partition $\mathcal{P}(h_1)$ the stated property holds. Fix $C \subseteq \omega_1$ c.u.b. and homogeneous for $\mathcal{P}(h_1)$. If we let $g(\beta) =$ the least element of C greater than β, then we have:

$$\exists g\colon \omega_1 \to \omega_1\, \forall^* h_2\, \forall^* \vec{\alpha}\, \vartheta(h_1)(h_2; \vec{\alpha}) < h_1(g(h_2(\alpha_1, \alpha_2)), h_2(1)(\alpha_2)).$$

Since for almost all $[h_1]$ the ordinal $\vartheta(h_1)$ satisfies our hypotheses, we therefore have:

$$\forall^* h_1\, \exists g\colon \omega_1 \to \omega_1\, \forall^* h_2\, \forall^* \vec{\alpha}\, \vartheta(h_1)(h_2; \vec{\alpha}) < h_1(g(h_2(\alpha_1, \alpha_2)), h_2(1)(\alpha_2)).$$

The net effect of the partition argument is that we have moved the $\exists g\colon \omega_1 \to \omega_1$ quantifier to the left past the rightmost $\forall^* h_2$ quantifier.

The second partition argument, which involves partitioning the functions h_1 now, is essentially identical to the one just given, and likewise allows us to move the $\exists g$ quantifier past the $\forall^* h_1$ quantifier. Thus we get:

$$\exists g\colon \omega_1 \to \omega_1\, \forall^* h_1, h_2\ \forall^* \vec{\alpha}\, \vartheta(h_1)(h_2; \vec{\alpha}) < g(h_1(h_2(\alpha_1, \alpha_2), h_2(1)(\alpha_2))).$$

This verifies the conclusion of Theorem 5.17 in this case.

§6. Level-3 descriptions. To compute $\underset{\sim}{\boldsymbol{\delta}}^1_5$, show the strong partition property on $\underset{\sim}{\boldsymbol{\delta}}^1_3$, and the weak partition property on $\underset{\sim}{\boldsymbol{\delta}}^1_5$, we need the level-3 descriptions. As we said before, these are only fairly trivial variations of the level-2 descriptions we explored in the last section. Thus, in some sense the combinatorics we need is already present in the formalism of the last section.

We next introduce the level-3 descriptions and show how to interpret them. Recall the definition of the canonical measure W_3^m from Definition 5.2. So, $A \subseteq \underset{\sim}{\boldsymbol{\delta}}^1_3$ has W_3^m measure one if there is a c.u.b. $C \subseteq \underset{\sim}{\boldsymbol{\delta}}^1_3$ such that for all $f\colon \omega_{m+1} \to C$ of the correct type we have $[f]_{S_1^m} \in A$.

Definition 6.1 (Level-3 pre-descriptions). Given a measure W_3^m and a sequence $K_1, \dots, K_t$ (each $K_i = S_1^{m_i}$ or $K_i = W_1^{m_i}$), a level-3 pre-description defined relative to this sequence is an object of the form (d) or $(d)^s$, where $d \in \mathcal{D}^m(K_1, \dots, K_t)$ is a level-2 description. We let $\mathcal{D}'(W_3^m, \vec{K})$ denote the set of level-3 pre-descriptions defined with respect to this sequence. We also add the one extra pre-description () which we declare to be defined relative to this sequence.

REMARK 6.2. We again write $(d)^{(s)}$ to denote that the symbol s may or may not appear. This occurrence of s should not be confused with those that occur within d itself.

The next definition gives the interpretation of the level-3 pre-descriptions.

DEFINITION 6.3. Given $f : \omega_{m+1} \to \underset{\sim}{\delta}^1_3$ of the correct type and $h_1, \dots, h_t$, and given $(d)^{(s)} \in \mathcal{D}(W^n_3; \vec{K})$, we define $((d)^{(s)}; f; h_1, \dots, h_t) = f((d; \vec{h}))$ if the symbol s does not appear, and $((d)^s; f; h_1, \dots, h_t) = \sup\{f(\beta) : \beta < (d; \vec{h})\}$. We also define $((); f; \vec{h}) = \sup(f)$. If $g : \underset{\sim}{\delta}^1_3 \to \underset{\sim}{\delta}^1_3$ we also define $(g; (d)^{(s)}; f; h_1, \dots, h_t) = g(((d)^{(s)}; f; h_1, \dots, h_t))$.

Note that $((d)^{(s)}; f; h_1, \dots, h_t)$ is now an ordinal below $\underset{\sim}{\delta}^1_3$.

As before, in going from the pre-descriptions to the descriptions we need a well-definedness condition.

DEFINITION 6.4. We say $(d) \in \mathcal{D}'(W^m_3; \vec{K})$ satisfies condition D if for almost all $h_1, \dots, h_t$, $(d; \vec{h})$ is the equivalence class of a function $g : (\omega_1)^m \to \omega_1$ of the correct type. We say $(d)^{(s)}$ satisfies condition D if for almost all $h_1, \dots, h_t$, $(d; \vec{h})$ is a supremum of ordinals represented by g of the correct type.

REMARK 6.5. This definition can be given in purely combinatorial terms, as we did for the ordering $<$ on descriptions. Also, it is easy to see that if $g : (\omega_1)^m \to \omega_1$, then $[g]_{W^m_1}$ is the supremum of ordinals of the correct type iff there is a c.u.b. $C \subseteq \omega_1$ such that restricted to C the following two properties hold: (1) $g : \operatorname{dom}(<_m) \to \omega_1$ is monotonic. That is, if $\vec{\alpha} <_m \vec{\beta}$ then $g(\vec{\alpha}) \leq g(\vec{\beta})$. This is equivalent to saying that for some $j < m$, g depends almost everywhere only on $(\alpha_1, \dots, \alpha_j, \alpha_m)$ and is order-preserving with respect to $<_{j+1}$. Also (2) $g(\alpha_1, \dots, \alpha_j, \alpha_m)$ has uniform cofinality almost everywhere equal to α_m.

DEFINITION 6.6 (Level-3 descriptions). A level-3 pre-description $(d)^{(s)}$ is a description if it satisfies condition D or if it is the distinguished description $()$. We let $\mathcal{D}(W^m_3, K_1, \dots, K_t)$ denote the set of level-3 descriptions.

As before, the descriptions give well-defined ordinals in the iterated ultrapowers as stated in the next definition.

DEFINITION 6.7. Suppose $(d)^{(s)} \in \mathcal{D}(W^m_3; K_1, \dots, K_3)$ is a level-3 description. Then $((d)^{(s)}; W^m_3; K_1, \dots, K_t)$ is the ordinal represented in the ultrapower by W^m_3 by the function $[f]_{S^m_1} \mapsto ((d)^{(s)}; f; K_1, \dots, K_t)$, *etc.*, and where $((d)^{(s)}; f; h_1, \dots, h_t)$ is given in Definition 6.3. We similarly define $(g; (d)^{(s)}; W^m_3; K_1, \dots, K_t)$.

It is easy to check that if $(d)^{(s)}$ is a description then $((d)^{(s)}; W_3^m; \vec{K})$ is well-defined in the iterated ultrapower sense (the obvious extension of Definition 5.4). Note that by definition, $((d)^{(s)}; W_3^m; \vec{K}) < j_{W_3^m}(\underset{\sim}{\delta}_3^1)$. We see below that $j_{W_3^m}(\underset{\sim}{\delta}_3^1) = \omega_{\omega^{\omega^m}+1}$.

We next extend the $\mathcal{L}$ operation to level-3 descriptions.

Definition 6.8 ($\mathcal{L}$ on level-3 descriptions). Let $(d)^{(s)} \in \mathcal{D}(W_3^m; K_1, \dots, K_3)$. If s does not appear we set $\mathcal{L}((d)) = (d)^s$. When s appears we set $\mathcal{L}((d)^s)$ to be the first pre-description in the sequence $(\mathcal{L}(d))$, $(\mathcal{L}(d))^s$, $(\mathcal{L}^{(2)}(d))$, $(\mathcal{L}^{(2)}(d))^s, \dots$ which satisfies condition D (*i.e.*, is a description), where $\mathcal{L}^{(i)}$ denote the ith iterate of the $\mathcal{L}$ operation (on level-2 descriptions).

The main theorem for level-3 descriptions is now entirely analogous to that for level-1 (trivial) descriptions (Theorem 4.3) and for level-2 descriptions (Theorem 5.17). We state it next.

Theorem 6.9 (Main theorem for level-3 descriptions). Let $d \in \mathcal{D}(W_3^m; K_1, \dots, K_t)$ and suppose $\vartheta < (d; W_3^m; \vec{K})$. If d is non-minimal with respect to $\mathcal{L}$ then there is a $g : \underset{\sim}{\delta}_3^1 \to \underset{\sim}{\delta}_3^1$ such that $\vartheta < (g; \mathcal{L}(d); W_3^m; \vec{K})$. If d is $\mathcal{L}$-minimal, then $\vartheta < \underset{\sim}{\delta}_3^1$.

Sketch of proof. The proof of Theorem 6.9 is a small addendum to the proof of Theorem 5.17. The non-trivial case is when d is of the form $(d)^s$ for d a level-2 description in $\mathcal{D}^m(K_1, \dots, K_t)$. Given $\vartheta < ((d)^s; W_3^m; \vec{K})$ we have that

$$\forall^* f : \omega_{m+1} \to \underset{\sim}{\delta}_3^1 \; \forall^* h_1, \dots, h_t \; \exists \beta < (d; \vec{h}) \; [\vartheta(f; \vec{h}) < f(\beta)].$$

From Theorem 5.17 we have that

$$\forall^* f : \omega_{m+1} \to \underset{\sim}{\delta}_3^1 \; \exists g : \omega_1 \to \omega_1 \; \forall^* h_1, \dots, h_t \; [\vartheta(f; \vec{h}) < f((g; \mathcal{L}(d); \vec{h}))].$$

We assume here that the level-3 description $(\mathcal{L}(d))$ satisfies condition D (otherwise we will repeat the following argument until it does). We then consider the following partition $\mathcal{P}$, using the weak partition property on $\underset{\sim}{\delta}_3^1$.

$\mathcal{P}$: We partition functions $f : \omega_{m+1} \to \underset{\sim}{\delta}_3^1$ of the correct type according to whether $\forall^* \vec{h} \; [\vartheta(f; \vec{h}) < f((\mathcal{L}(d); \vec{h}) + 1)]$.

A "sliding argument" as in the proof of Theorem 5.17 (for the example considered) shows that on the homogeneous side of the partition $\mathcal{P}$ the stated property holds. Fix $C \subseteq \underset{\sim}{\delta}_3^1$ homogeneous for $\mathcal{P}$. Let now $g : \underset{\sim}{\delta}_3^1 \to \underset{\sim}{\delta}_3^1$ be given by $g(\alpha)$ = the least element of C greater than α. The homogeneity of C and the definition of $\mathcal{P}$ easily show that $\forall^* f \; \forall^* \vec{h} \; [\vartheta(f; \vec{h}) < g(f((\mathcal{L}(d)); \vec{h}))$, which is the conclusion of Theorem 6.9. ⊣

Theorem 6.9 leads to an upper bound for the ordinal $(d; W_3^m; K_1, \dots, K_t)$ that the level-3 description represents, as in the case of level-1 and level-2 descriptions considered earlier. There is one crucial new difference, however.

For the level-1 and 2 descriptions, the analysis of the "g function" required only the Kunen tree. In those cases, g was a function from ω_1 to ω_1, and the only possible cofinality below ω_1 was ω (that is, what mattered was $[g]_{W_1^1}$). For level-3 descriptions, g is a now a function from $\underset{\sim}{\delta}_3^1$ to $\underset{\sim}{\delta}_3^1$, and what matters now is the equivalence class of g with respect to one of the three possible normal measures on $\underset{\sim}{\delta}_3^1$ (corresponding to the three regular cardinals ω, ω_1, ω_2 below $\underset{\sim}{\delta}_3^1$). The analysis of a function g with respect to the ω-cofinal normal measure uses the analog of the Kunen tree as before, but for the other cofinalities (and in general) requires a new construction, the *Martin tree*.

The next theorem states the property of the Martin tree. The reader can refer to [Jac99] for a proof, or to [Jac10] where two somewhat different arguments are given.

Theorem 6.10. There is a tree T, called the Martin tree, on $\omega \times \underset{\sim}{\delta}_3^1$ such that for any $g: \underset{\sim}{\delta}_3^1 \to \underset{\sim}{\delta}_3^1$ there is an $x \in {}^\omega\omega$ with T_x wellfounded and a c.u.b. $C \subseteq \underset{\sim}{\delta}_3^1$ such that for all $\alpha \in C$ we have $g(\alpha) < |T_x \restriction \sup_\nu j_\nu(\alpha)|$ where the supremum ranges over the ultrapower embeddings j_ν corresponding to measures ν in the homogeneous tree construction on $\underset{\sim}{\Pi}_1^1$ and $\underset{\sim}{\Pi}_2^1$ sets.

We can be a little more specific in the statement of Theorem 6.10. Namely, for $\alpha \in C$ of cofinality ω we need no measures at all (*i.e.*, $g(\alpha) < |T_x \restriction \alpha|$), for $\mathrm{cf}(\alpha) = \omega_1$ we need the measures ν in the homogeneous tree for a $\underset{\sim}{\Pi}_1^1$ set (which are the measures W_1^m), and for $\mathrm{cf}(\alpha) = \omega_2$ we need the measures ν in the homogeneous tree on a $\underset{\sim}{\Pi}_2^1$ set.

In the $\mathrm{cf}(\alpha) = \omega_2$ case, it is desirable to simplify the conclusion of Theorem 6.10 by replacing the more general measures ν in the homogeneous tree on a $\underset{\sim}{\Pi}_2^1$ set by the measures in the canonical family S_1^m.

We state two embedding theorems which state, roughly speaking, that the canonical measures S_1^m dominate all the other measures (for points of cofinality ω_2 in one theorem). We call one the "local" embedding theorem and the other the "global" embedding theorem. We give an example of a proof of each of these results in §7.

The local embedding theorem, which we state next, is the one that simplifies the conclusion of Theorem 6.10.

Theorem 6.11 (Local embedding theorem). Let ν be a measure occurring in the homogeneous tree on a $\underset{\sim}{\Pi}_1^1$ or a $\underset{\sim}{\Pi}_2^1$ set. Then there is an $m \in \omega$ and a c.u.b. $C \subseteq \underset{\sim}{\delta}_3^1$ such that for all $\alpha \in C$ with $\mathrm{cf}(\alpha) = \omega_2$ we have $j_\nu(\alpha) \leq j_{S_1^m}(\alpha)$.

T will henceforth denote the Martin tree.

From Theorem 6.10 it follows (similarly to an earlier argument, see the argument just before Corollary 5.18)) that $(g; \mathcal{L}(d); W_3^m; \vec{K}) < (\ell; \mathcal{L}(d); \vec{K})^+$ where $\ell: \underset{\sim}{\delta}_3^1 \to \underset{\sim}{\delta}_3^1$ is the function $\ell(\alpha) = \sup_m j_{W_1^m}(\alpha)$ if $\mathrm{cf}(\alpha) = \omega_1$ and

$\ell(\alpha) = \sup_m j_{S_1^m}(\alpha)$ if $\mathrm{cf}(\alpha) = \omega_2$ (and $\ell(\alpha) = \alpha$ if $\mathrm{cf}(\alpha) = \omega$). So, $(d; W_3^m; \vec{K}) \leq (\ell; \mathcal{L}(d); W_3^m; \vec{K})^+$. By countable additivity, $(\ell; \mathcal{L}(d); W_3^m; \vec{K}) = \sup_\nu (j_\nu; \mathcal{L}(d); W_3^m; \vec{K})$, where the supremum ranges over the embeddings j_ν for the measure $\nu = W_1^m$ or $\nu = S_1^m$. By definition of these ordinals we have that $(j_\nu; \mathcal{L}(d); W_3^m; \vec{K}) = (\mathcal{L}(d); W_3^m; \vec{K}, \nu)$.

So, we have $(d; W_3^m; K_1, \dots, K_t) \leq [\sup_{K_{t+1}} (\mathcal{L}(d); W_3^m; K_1, \dots, K_t, K_{t+1})]^+$. The new feature is now apparent; as we lower the description by the lowering operator $\mathcal{L}$ we must now also add a new measure K_{t+1} to the sequence. More precisely, we make the following definition.

DEFINITION 6.12. We define an ordering on the sequences $p = (d; W_3^m; K_1, \dots, K_t)$ where $d \in \mathcal{D}(W_3^m; K_1, \dots, K_t)$ (here m is fixed, and t can vary) by taking the transitive closure of the relation

$$(\mathcal{L}(d); W_3^m; K_1, \dots, K_t, K_{t+1}) < (d; W_3^m; K_1, \dots, K_t).$$

We define the rank $|p|$ of p in the (slightly non-standard) manner $|p| = (\sup_{q<p} |q|) + 1$.

REMARK 6.13. The non-standard method of computing rank simply increases the rank by 1 at points of (usual) limit rank. So, the rank of all sequences p is now a successor ordinal.

Putting this together we have now shown the following.

THEOREM 6.14. *Let $d \in \mathcal{D}(W_3^m; K_1, \dots, K_t)$ be a level-3 description. Then we have $(d; W_3^m; \vec{K}) \leq \omega_{\omega+1+\alpha}$, where $\alpha = |(d; W_3^m; \vec{K})|$ is the rank of the sequence $(d; W_3^m; K_1, \dots, K_t)$ as in Definition 6.12.*

The homogeneous tree construction as discussed in §3 (*cf.* Corollary 3.7) shows that $\underset{\sim}{\boldsymbol{\delta}}_5^1 = \lambda_5^+$, where $\lambda_5 \leq \sup_j j(\underset{\sim}{\boldsymbol{\delta}}_3^1)$, where the supremum ranges over the ultrapower embeddings corresponding to the measures occurring in the homogeneous trees for $\underset{\sim}{\boldsymbol{\Pi}}_3^1$ sets. To use Theorem 6.14 to get an upper bound for $\underset{\sim}{\boldsymbol{\delta}}_5^1$ we need the second embedding theorem (see [Jac99] for a proof and [Jac88] for the general case of the local and global embedding theorems).

THEOREM 6.15 (Global embedding theorem). *Let ν be a measure in the homogeneous tree for a $\underset{\sim}{\boldsymbol{\Pi}}_3^1$ set. Them for some $m \in \omega$ we have $j_\nu(\underset{\sim}{\boldsymbol{\delta}}_3^1) \leq j_{W_3^m}(\underset{\sim}{\boldsymbol{\delta}}_3^1)$.*

As before, a minor modification of Theorem 6.9 and the argument following shows that $j_{W_3^m}(\underset{\sim}{\boldsymbol{\delta}}_3^1) \leq (\ell; d; W_3^m)^+ = (\sup_{K_1} (d_{\max}; W_3^m; K_1))^+$, where here $d_{\max} = ()$ is the distinguished maximal level-3 description and and $\ell(\alpha) = \sup_\nu j_\nu(\alpha)$ (the supremum corresponding to ultrapower embeddings from measures ν in the families W_1^i, S_1^i). So, $j_{W_3^m}(\underset{\sim}{\boldsymbol{\delta}}_3^1) \leq \omega_{\alpha+1}$, where $\alpha = \sup_{K_1} |((); W_3^m; K_1)|$.

Computing the ranks $|((); W_3^m; K_1)|$ is fairly straightforward. The answer turns out to be $\omega^{\omega^{m-1} \cdot r_1} + 1$, where $K_1 = S_1^{r_1}$ (details are given in [Jac99]). This gives the upper bound $j_{W_3^m}(\underset{\sim}{\delta}_3^1) \leq \omega_{\omega^{\omega^m}+1}$. In particular we have:

Corollary 6.16. $\underset{\sim}{\delta}_5^1 \leq \omega_{\omega^{\omega^\omega}+1}$.

The lower-bound for $\underset{\sim}{\delta}_5^1$ follows from some embedding arguments which we briefly indicate in the next section. In fact, further arguments (see [JK]) show that all of the ordinals $((d)^{(s)}; W_3; \vec{K})$ are actually cardinals, and that equality actually holds in the statement of Theorem 6.14.

The computation of the cardinal $((d)^{(s)}; W_3^m; \vec{K})$ represented by the description $(d)^{(s)} \in \mathcal{D}(W_3^m; \vec{K})$ is thus reduced to a purely combinatorial rank computation. This rank can be computed in several ways. The analysis of [JK] implicitly gives a method of computing this rank. Basically, one lists all the descriptions in $\mathcal{D}(W_3^m; \vec{K})$ below $(d)^{(s)}$, and for each of them an ordinal is written down, and these ordinals are then added. If $d_1 < d_2$ are consecutive descriptions in this list, then the ordinal written down represents the supremum over measure sequences $\vec{K}'$ of the rank of $(d_2; W_3^m; \vec{K}, \vec{K}')$ in the ordering as in Definition 6.12 except any sequence $(d_1; W_3^m; \vec{K}, \vec{K}')$ is declared minimal (*i.e.*, rank 0). Another method is to use the formulas of [Jac99] or [Jac88] for the general projective case. It was shown there that these formulas give upper-bounds for the ranks of these sequences in the ordering of Definition 6.12. In fact, with a little care these formulas will give will give the exact ranks. Rather than sketch these arguments, we illustrate with a specific example, and leave it to the reader to check the details. This example was mentioned in [Jac10].

Example 6.17. Consider $(d) \in \mathcal{D}(W_3^2; S_1^2, S_1^2)$, where $d = (1; (2; \cdot_2, \cdot_1)^s, (2, \cdot_1))^s$. This d is listed in Example 5.21 as $\mathcal{L}^{26}(d_M)$, where $d_M = (1; (2; \cdot_2))$ is the maximal description defined relative to S_1^2, S_1^2. Following the method of [JK] we can give a calculation of $|((d); W_3^2, S_1^2, S_1^2)|$ as seen in Figure 2 and obtain $((d); W_3^2; S_1^2, S_1^2) = \aleph_{\omega^{\omega+1}+\omega \cdot 2+1}$.

The alternate way to compute the rank of this description, following [Jac99] and [Jac88], is to use formulas which are given inductively, similar to the inductive definition of the $\mathcal{L}$ operation (that is, by a "reverse induction" on $k(d)$). We again start with $|((d); W_3^2, S_1^2, S_1^2)| = \sup_{K_3} |(\mathcal{L}^{26}(d_M))^s; W_3^2, S_1^2, S_1^2, K_3)| + 1$. After this initial step, we inductively assign to each subdescription d' of d with $k(d') = i$, say, an ordinal $f^i(d')$ which represents the supremum over measure sequences $\vec{K}'$ of the $\mathcal{L}^i$-rank of $(d'; \vec{K}; \vec{K}')$. For the example currently being considered, since $\mathcal{L}^{26}(d) = (1; (2; \cdot_2, \cdot_1)^s, (2; \cdot_1))^s$, keeping in mind the definition of the $\mathcal{L} = \mathcal{L}^1$ operation in terms of the $\mathcal{L}^2$ operation, we

$$
\begin{aligned}
|((d);W_3^2,S_1^2,S_1^2)| &= \sup_{K_3}|(\mathcal{L}^{26}(d_M))^s;W_3^2,S_1^2,S_1^2,K_3)|+1\\
&= \sup_{\vec{K}}|((\mathcal{L}^{27}(d_M));W_3^2,S_1^2,S_1^2,\vec{K})|+\omega+1\\
&= \sup_{\vec{K}}|((\mathcal{L}^{29}(d_M));W_3^2,S_1^2,S_1^2,\vec{K})|+\omega+\omega+1\\
&= \sup_{\vec{K}}|((\mathcal{L}^{30}(d_M));W_3^2,S_1^2,S_1^2,\vec{K})|+\omega^\omega\cdot\omega+\omega+\omega+1\\
&= \sup_{\vec{K}}|((\mathcal{L}^{31}(d_M));W_3^2,S_1^2,S_1^2,\vec{K})|+\omega^\omega+\omega^{\omega+1}+\omega+\omega+1\\
&= \sup_{\vec{K}}|((\mathcal{L}^{33}(d_M));W_3^2,S_1^2,S_1^2,\vec{K})|+\omega+\omega^\omega+\omega^{\omega+1}+\omega+\omega+1\\
&= \sup_{\vec{K}}|((\mathcal{L}^{35}(d_M));W_3^2,S_1^2,S_1^2,\vec{K})|+\omega+\omega+\omega^\omega+\omega^{\omega+1}+\omega+\omega+1\\
&= \sup_{\vec{K}}|((\mathcal{L}^{36}(d_M));W_3^2,S_1^2,S_1^2,\vec{K})|+\omega^\omega+\omega+\omega+\omega^\omega+\omega^{\omega+1}+\omega+\omega+1\\
&= \sup_{\vec{K}}|((\mathcal{L}^{37}(d_M));W_3^2,S_1^2,S_1^2,\vec{K})|+\omega+\omega^\omega+\omega+\omega+\omega^\omega+\omega^{\omega+1}+\omega+\omega+1\\
&= \sup_{\vec{K}}|((\mathcal{L}^{39}(d_M));W_3^2,S_1^2,S_1^2,\vec{K})|+\omega\cdot 2+\omega^\omega+\omega+\omega+\omega^\omega+\omega^{\omega+1}+\omega+\omega+1\\
&= \sup_{\vec{K}}|((\mathcal{L}^{40}(d_M));W_3^2,S_1^2,S_1^2,\vec{K})|+\omega\cdot 3+\omega^\omega+\omega+\omega+\omega^\omega+\omega^{\omega+1}+\omega+\omega+1\\
&= \sup_{\vec{K}}|((\mathcal{L}^{41}(d_M));W_3^2,S_1^2,S_1^2,\vec{K})|+\omega\cdot 4++\omega^\omega+\omega+\omega+\omega^\omega+\omega^{\omega+1}+\omega+\omega+1\\
&= \omega^{\omega+1}+\omega\cdot 2+1
\end{aligned}
$$

FIGURE 2.

first write

$$f^1(\mathcal{L}^{26}(d)) = \sum_{\beta < f^2((2;\cdot_2,\cdot_1))} \beta + f^2((2;\cdot_1)).$$

We then have $f^2((2;\cdot_1)) = f^2((\cdot_1)) + f^2((\cdot_1))$. By inspection, $f^2((\cdot_1)) = \omega$, and so $f^2((2;\cdot_1)) = \omega \cdot 2$. Next we write $f^2((2;\cdot_2,\cdot_1)) = \sum_{\beta < f^2((\cdot_2))} \beta + f^2((\cdot_1))$. We easily have $f^2((\cdot_2)) = \omega^\omega$, and so $f^2((2;\cdot_2,\cdot_1)) = \sum_{\beta<\omega^\omega} \beta + \omega = \omega^\omega + \omega$. So,

$$f^1(\mathcal{L}^{26}(d)) = \sum_{\beta<\omega^\omega+\omega} \beta + \omega \cdot 2 = \omega^{\omega+1} + \omega \cdot 2.$$

Thus, $|((d); W_3^2; S_1^2, S_1^2)| = \omega^{\omega+1} + \omega \cdot 2 + 1$, which agrees with the first computation.

Measures on $\underset{\sim}{\delta}_3^1$. The proof of the strong partition relation on $\underset{\sim}{\delta}_3^1$ is similar both in method and in notation to that of the argument for ω_1, that is, the analysis of measures we gave in the proof of Theorem 4.13, and also the analysis of measures on the ω_m we stated in Theorem 4.29.The main differences are that we must now use (generalized) level-3 descriptions and level-3 trees of uniform cofinalities, which we next introduce. Recall the notion of level-2 tree of uniform cofinalities from Definition 4.23. We call them simply level-2 trees here.

We say a level-2 tree $\mathcal{R}'$ is an **immediate extension** of the tree $\mathcal{R}$ if $\mathrm{dom}(\mathcal{R})$ is a subtree of $\mathrm{dom}(\mathcal{R}')$, $\mathrm{dom}(\mathcal{R}')$ has one extra node in it, and $\mathcal{R}' \restriction \mathrm{dom}(\mathcal{R}) = \mathcal{R}$. We frequently assume that $\mathrm{dom}(\mathcal{R})$ is a finite subtree of $\omega^{<\omega}$. In this case, when we speak of immediate extensions we implicitly assume that we have an embedding π from $\mathrm{dom}(\mathcal{R})$ to $\mathrm{dom}(\mathcal{R}')$ allowing us to identify nodes of $\mathrm{dom}(\mathcal{R})$ with nodes of $\mathrm{dom}(\mathcal{R}')$. If $s = (i_1, \dots, i_k) \in \mathrm{dom}(\mathcal{R}') - \pi(\mathrm{dom}(\mathcal{R}))$, then we say that s is the new or extra node of $\mathcal{R}'$.

A **partial** level-2 tree $\mathcal{R}^-$ is a level-2 tree except that there is a single maximal node s in $\mathrm{dom}(\mathcal{R})$ such that $\mathcal{R}(s)$ is not defined. A level-2 tree **completes** the partial tree $\mathcal{R}^-$ if $\mathrm{dom}(\mathcal{R}) = \mathrm{dom}(\mathcal{R}^-)$ and $\mathcal{R}(s)$ is defined. We say the partial tree $\mathcal{R}^-$ immediately extends the level-2 tree whose domain is $\mathrm{dom}(\mathcal{R}^-) \setminus \{s\}$, where again s is the distinguished node.

If $\mathcal{R}$ is a partial level-2 tree of uniform cofinalities, then notice that the ordering $<_{\mathcal{R}}$ is still defined (this definition didn't make use of the values $\mathcal{R}(s)$ for s a terminal node in $\mathrm{dom}(\mathcal{R})$). We say a function $f\colon \mathrm{dom}(<_{\mathcal{R}}) \to \mathrm{Ord}$ is of **partial type $\mathcal{R}$** is it satisfies the requirements for being of type $\mathcal{R}$ (including being order-preserving), except that we now have no requirement on the uniform cofinality of $f(\alpha_1, i_1, \dots, \alpha_k, i_k)$ when $s = (i_1, \dots, i_k)$ is the distinguished node in $\mathrm{dom}(\mathcal{R})$.

Recall that for convenience we said that a level-1 tree of uniform cofinalities $\mathcal{Q}$ is a finite set, which we generally identify with $\{1, 2, \dots, n\}$. We say $\mathcal{Q}'$

immediately extends $\mathcal{Q}$ if $|\mathcal{Q}'| = |\mathcal{Q}| + 1$. In this case, we implicitly assume that we have an injection π from $\mathcal{Q}$ to $\mathcal{Q}'$ allowing us to identify points of $\mathcal{Q}$ with points of $\mathcal{Q}'$. The element of $\mathcal{Q}' \setminus \pi[\mathcal{Q}]$ will be called the new or extra element of $\mathcal{Q}'$. Intuitively, we think of $\mathcal{Q}$ as a code for the measure W_1^n. If $\mathcal{Q}$ is a level-1 tree, we let $<_{\mathcal{Q}}$ be the finite linear ordering on $\{1, 2, \dots, n\}$ given by the ordering on integers.

A level ≤ 2 tree of uniform cofinalities is a pair $(\mathcal{Q}, \mathcal{R})$ where $\mathcal{Q}$ is a level-1 tree and $\mathcal{R}$ is a level-2 tree. We say $(\mathcal{Q}', \mathcal{R}')$ immediately extends $(\mathcal{Q}, \mathcal{R})$ if $\mathcal{Q} = \mathcal{Q}'$ and $\mathcal{R}'$ immediately extends $\mathcal{R}$, or $\mathcal{R} = \mathcal{R}'$ and $\mathcal{Q}'$ immediately extends $\mathcal{Q}$. A partial level ≤ 2 tree is a pair $(\mathcal{Q}, \mathcal{R})$ where $\mathcal{R}$ is a partial level-2 tree. A partial extension of $(\mathcal{Q}, \mathcal{R})$ is a $(\mathcal{Q}', \mathcal{R}')$ where either $\mathcal{Q}'$ immediately extends $\mathcal{Q}$ and $\mathcal{R}' = \mathcal{R}$ or else $\mathcal{Q}' = \mathcal{Q}$ and $\mathcal{R}'$ is a partial extension of $\mathcal{R}$. If $(\mathcal{Q}, \mathcal{R})$ is a level ≤ 2 tree or a partial level ≤ 2 tree, the ordering $<_{\mathcal{Q},\mathcal{R}}$ is the (disjoint) union of $<_{\mathcal{Q}}$ and $<_{\mathcal{R}}$, where every element of $\operatorname{dom}(<_{\mathcal{Q}})$ precedes every element of $\operatorname{dom}(<_{\mathcal{R}})$. We say $f : \operatorname{dom}(<_{\mathcal{Q},\mathcal{R}}) \to \mathrm{Ord}$ is of type $(\mathcal{Q}, \mathcal{R})$ (or partial type $(\mathcal{Q}, \mathcal{R})$) if f is order-preserving and $f \restriction \operatorname{dom}(<_{\mathcal{R}})$ is of (partial) type $\mathcal{R}$.

Definition 6.18. A level-3 tree of uniform cofinalities is a function $\mathcal{S}$ with domain a finite tree satisfying the following.

1. $\mathcal{S}(\varnothing) = \varnothing$
2. For each $(i_1) \in \operatorname{dom}(\mathcal{S})$, $\mathcal{S}(i_1)$ is either the symbol ω, or $\mathcal{S}(i_1) = (\mathcal{Q}, \mathcal{R})$ is a partial ≤ 2 tree immediately extending the empty tree $(\varnothing, \varnothing)$. This means either $\mathcal{S}(i_1) = (\mathcal{Q}, \varnothing)$ where $\mathcal{Q} = \{1\}$, or $\mathcal{S}(i_1) = (\varnothing, \mathcal{R})$ where $\mathcal{R}$ is the unique partial level-2 tree of uniform cofinalities immediately extending the empty tree ($\mathcal{R}$ has one node s other than $\varnothing$ in its domain, $\mathcal{R}(\varnothing) = (1)$, and $\mathcal{R}(s)$ is undefined).
3. In general, if $(i_1, \dots, i_k) \in \operatorname{dom}(\mathcal{S})$, then either (a) $\mathcal{S}(i_1, \dots, i_k)$ is the symbol ω, in which case $(i_1, \dots, i_k)$ must be a terminal node in $\operatorname{dom}(\mathcal{S})$, or (b) $\mathcal{S}(i_1, \dots, i_k)$ is a partial level ≤ 2 tree of uniform cofinalities immediately extending a completion of $\mathcal{S}(i_1, \dots, i_{k-1})$.

In (3) above, we mean that for each completion of $\mathcal{S}(i_1, \dots, i_{k-1})$ a certain set of i_k is allowed as an extension and for different completions these are disjoint sets. We say that i_k is associated to this particular completion.

Remark 6.19. There is a slight asymmetry in the definitions of level-n trees $\mathcal{T}$ for n odd and even. For odd n we have $\mathcal{T}(\varnothing) = \varnothing$ while for even n we have that $\mathcal{T}(\varnothing) \neq \varnothing$ ($\mathcal{T}(\varnothing)$ is a partial level $\leq n - 1$ tree in this case). This is a reflection of the slight asymmetry in the definitions of the homogeneous trees on $\boldsymbol{\Pi}^1_n$ sets for n odd and even.

Recall that the level-2 trees gave directly measures on the ω_n, which were more or less the measures in the homogeneous tree on $\boldsymbol{\Pi}^1_2$ sets. These measures were then lifted up using generalized trivial descriptions (and sections of the Kunen tree) to generate arbitrary measures on the ω_n. Similarly, the level-3 trees $\mathcal{S}$ will directly define measures $\mathcal{M}^{\mathcal{S}}$ on $\underset{\sim}{\delta}^1_3$ which are essentially the measures occurring in the homogeneous tree on a $\underset{\sim}{\Pi}^1_3$ set. We generate arbitrary measures on $\underset{\sim}{\delta}^1_3$ by lifting these using generalized level-3 descriptions (and sections of the Martin tree).

To define $\mathcal{M}^{\mathcal{S}}$ we first define an ordering $<_{\mathcal{S}}$ corresponding to $\mathcal{S}$. This is lexicographic ordering on sequences $(i_1, \beta_1, \dots, i_{k-1}, \beta_{k-1}, i_k)$ satisfying:

1. $(i_1, \dots, i_k) \in \operatorname{dom}(\mathcal{S})$.
2. If $\mathcal{S}(i_1) = \omega$, then $k = 1$ (*i.e.*, there are no β's). If $\mathcal{S}(i_1) = (\mathcal{Q}, \varnothing)$ (where $\mathcal{Q} = \{1\}$), then $\beta_1 < \omega_1$. If $\mathcal{S}(i_1) = (\varnothing, \mathcal{R})$ (where $\mathcal{R}$ is the minimal partial level-2 tree), then $\beta_1 < \omega_2$.
3. We inductively require that $(i_1, \beta_1, \dots, i_{k-1}) \in \operatorname{dom}(<_{\mathcal{S}})$ and also that $\mathcal{S}(i_1, \dots, i_{k-1}) \neq \omega$. If $\mathcal{S}(i_1, \dots, i_{k-1}) = (\mathcal{Q}_{k-1}, \mathcal{R}_{k-1})$, a partial level ≤ 2 tree, then there is a function $f\colon \operatorname{dom}(<_{\mathcal{Q}_{k-1}, \mathcal{R}_{k-1}}) \to \omega_1$ of partial type $(\mathcal{Q}_{k-1}, \mathcal{R}_{k-1})$ which represents $(\beta_1, \dots, \beta_{k-1})$. Also, i_k is associated to the completion of $(\mathcal{Q}_{k-1}, \mathcal{R}_{k-1})$ giving the type of β_{k-1}.

To say f represents $(\beta_1, \dots, \beta_{k-1})$ in (3) above means precisely the following. For $\ell \leq k-1$ let s be the new node in $\mathcal{S}(i_1, \dots, i_\ell)$ (viewed as an immediate extension of $\mathcal{S}(i_1, \dots, i_{\ell-1})$). If this node is a new node of $\mathcal{Q}_\ell$, say corresponding to the integer a, then $\beta_\ell = f(a) < \omega_1$ (here a is viewed as an element of the domain of $<_{\mathcal{Q}_{k-1}}$). If $s = (j_1, \dots, j_p)$ is a new node of $\mathcal{R}_\ell$, then $\beta_\ell = [f^{j_1, \dots, j_p}]_{W_1^p}$, where $f^{j_1, \dots, j_p}$ is the subfunction of f defined by

$$f^{j_1, \dots, j_p}(\alpha_1, \dots, \alpha_p) = f(\alpha_1, j_1, \dots, \alpha_p, j_p).$$

We say $F\colon \operatorname{dom}(<_{\mathcal{S}}) \to \mathrm{Ord}$ is of **type** $\mathcal{S}$ is F is order-preserving and satisfies the following. If $(i_1, \dots, i_k)$ is not maximal in the domain of $\mathcal{S}$, then $F(i_1, \alpha_1, \dots, \alpha_{k-1}, i_k)$ is the supremum of F at points of lower $<_{\mathcal{S}}$ rank (unless this rank is a successor ordinal in which case we require uniform cofinality ω as usual). If $\mathcal{S}(i_1, \dots, i_k) = \omega$, then $F(i_1, \beta_1, \dots, \beta_{k-1}, i_k)$ has uniform cofinality ω. If $(i_1, \dots, i_k)$ is maximal in the domain of $\mathcal{S}$ and $\mathcal{S}(i_1, \dots, i_k) = (\mathcal{Q}_k, \mathcal{R}_k)$, then $f(i_1, \beta_1, \dots, \beta_{k-1}, i_k)$ has uniform cofinality $\{\beta : \exists f \text{ of partial type } (\mathcal{Q}_k, \mathcal{R}_k) \text{ representing } (\beta_1, \dots, \beta_{k-1}, \beta)\}$.

For $(i_1, \dots, i_k) \in \operatorname{dom}(\mathcal{S})$, let $\mu^{i_1, \dots, i_k}$ be the measure on k-tuples induced by the strong partition relation on ω_1, functions $f\colon \operatorname{dom}(<_{\mathcal{Q}, \mathcal{R}}) \to \omega_1$ of type $(\mathcal{Q}, \mathcal{R})$ = the level ≤ 2 tree associated with i_k which extends the partial ≤ 2 tree $\mathcal{S}(i_1, \dots, i_{k-1})$, and the appropriate W_1^p.

Definition 6.20. $\mathcal{M}^{\mathcal{S}}$ is the measure on tuples $(\dots, \gamma^{i_1, \dots, i_k}, \dots)$, $(i_1, \dots, i_k) \in \operatorname{dom}(\mathcal{S})$, of ordinals from $\underset{\sim}{\delta}^1_3$ induced by the weak partition relation on $\underset{\sim}{\delta}^1_3$, functions $F\colon \operatorname{dom}(<_{\mathcal{S}}) \to \underset{\sim}{\delta}^1_3$ of type $\mathcal{S}$, and the measures $\mu^{i_1, \dots, i_k}$.

We illustrate these definitions with an example.

Example 6.21. Let $\mathcal{S}$ be the level-3 tree determined as follows.

1. The domain of $\mathcal{S}$ is the tree $\mathrm{dom}(\mathcal{S}) = \{\varnothing, (0), (0,0), (0,1)\}$.
2. $\mathcal{S}(0)$ is the the partial level ≤ 2 tree $(\varnothing, \mathcal{R}_0)$, where $\mathcal{R}_0$ is partial level-2 tree with $\mathrm{dom}(\mathcal{R}_0) = \{\varnothing, (0)\}$, $\mathcal{R}_0(\varnothing) = (1)$ (the permutation of length 1), and $\mathcal{R}_0(0)$ is undefined.
3. $\mathcal{S}(0,0)$ is the partial level ≤ 2 tree $(\mathcal{Q}_0, \mathcal{R}_1)$ where $\mathcal{Q}_0 = \{1\}$, and $\mathcal{R}_1$ is the level-1 tree which completes $\mathcal{R}_0$ by defining $\mathcal{R}_1(0,0) = \omega$.
4. $\mathcal{S}(0,1) = (\varnothing, \mathcal{R}_2)$, where $\mathcal{R}_2$ is the partial level-2 tree extending $\mathcal{R}_0$ with extra node $(0,0)$ in its domain and with $\mathcal{R}_2(0) = (2,1)$, $\mathcal{R}_2(0,0)$ is undefined.

The reader can check that for this $\mathcal{S}$, the ordering $<_{\mathcal{S}}$ has domain consisting of (0), which is maximal in the ordering, sequences $(0,\beta,0)$ where $\beta < \omega_2$ and $\mathrm{cf}(\beta) = \omega$, and $(0,\beta,1)$ where $\beta < \omega_2$ and $\mathrm{cf}(\beta) = \omega_1$ (the two different choices for $\mathrm{cf}(\beta)$ correspond to the different ways of completing $\mathcal{R}_0$ to a level-2 tree). If $F\colon \mathrm{dom}(<_{\mathcal{S}}) \to \boldsymbol{\delta}^1_3$ has type $\mathcal{S}$, then $F(0,\beta,0)$ has uniform cofinality ω_1 and $F(0,\beta,1)$ has uniform cofinality $\{\gamma < \omega_3 : \gamma(1) = \beta\}$, where $\gamma(1)$ denotes the first invariant of γ (this is equivalent to saying $F(0,\beta,1)$ has uniform cofinality $j_{W^1_1}(\beta)$).

The measure $\mathcal{M}^{\mathcal{S}}$ is a measure on triples $(\gamma^0, \gamma^{0,0}, \gamma^{0,1})$ from $\boldsymbol{\delta}^1_3$. One can see that this measure concentrates on triples with $\gamma^0 < \gamma^{0,0} < j_{S^1_1}(\gamma^0) < \gamma^{0,1} < j_{\bar{S}^1_1}(\gamma^0)$, where $\bar{S}^1_1$ is the ω_1-cofinal normal measure on ω_2.

Similarly to Definition 4.30 we now generalize slightly the level-3 descriptions to extended level-3 descriptions which we use to lift up the measures $\mathcal{M}^{\mathcal{S}}$.

Definition 6.22. Let $\mathcal{S}$ be a level-3 tree of uniform cofinalities and $K_1, \dots, K_t$ a sequence of measures each in $W_1 \cup S_1$. An extended level-3 description defined relative to $\mathcal{S}$ and $\vec{K}$ is a sequence $d = \langle i_1, d_1, \dots, d_{k-1}, i_k\rangle$ where $(i_1, \dots, i_k) \in \mathrm{dom}(\mathcal{S})$, each d_i is a level-2 description in $\mathcal{D}^m(\vec{K})$ (for some m), and for almost all $h_1, \dots, h_t$, if $\beta_i = (d_i; \vec{h})$, then $\langle i_1, \beta_1, \dots, \beta_{k-1}, i_k\rangle \in \mathrm{dom}(<_{\mathcal{S}})$. We also allow d^s if for almost all $h_1, \dots, h_t$ we have β_{k-1} is the supremum of β' such that $\langle i_1, \beta_1, \dots, \beta', i_k\rangle \in \mathrm{dom}(<_{\mathcal{S}})$.

If $F\colon \mathrm{dom}(<_{\mathcal{S}}) \to \boldsymbol{\delta}^1_3$ is of type $\mathcal{S}$, then we define $(F; d; \vec{K})$ in the usual manner via the iterated ultrapowers by the measures $K_1, \dots, K_t$, where

$$(F; d; h_1, \dots, h_t) = F(\langle i_1, \beta_1, \dots, \beta_{k-1}, i_k\rangle),$$

with the $\beta_i = (d_i; \vec{h})$ as above. Similarly,

$$(F; d^s; h_1, \dots, h_t) = \sup_{\beta' < \beta_{k-1}} F(\langle i_1, \beta_1, \dots, \beta', i_k\rangle).$$

The extended descriptions are ordered in the natural manner (lexicographically, using the ordering on level-2 descriptions).

The necessary ingredients to generate arbitrary measures on $\boldsymbol{\delta}^1_3$ are given in the next definition, which is the analog of Definition 4.32.

DEFINITION 6.23. A level-3 complex is a sequence of the form

$$\mathcal{C} = \langle \mathcal{S}; x_0, \ldots, x_t; d_0, \ldots, d_t; K_1, \ldots, K_t \rangle,$$

where $\mathcal{S}$ is a level-3 tree of uniform cofinalities, $x_i \in {}^\omega\omega$ are reals with the section T_{x_i} of the Martin tree T wellfounded, $d_0, \ldots, d_t$ are extended level-3 descriptions with d_i defined relative to $\mathcal{S}$ and the measure sequence $K_1, \ldots, K_i$. Also, $d_0 > d_1 > \cdots > d_t$.

If $\mathcal{C}$ is a level-3 complex as above, $F\colon \operatorname{dom}(<_\mathcal{S}) \to \underset{\sim}{\boldsymbol{\delta}}^1_3$ is of type $\mathcal{S}$, and $\gamma < \underset{\sim}{\boldsymbol{\delta}}^1_3$, then we define the ordinal $(F; \mathcal{C}; \gamma)$ as follows. We set

$$(F; \mathcal{C}; \gamma) = |(T_{x_0} \restriction \sup_j j(F; d_0))(\alpha_1)|,$$

where α_1 is represented with respect to K_1 by the function which assign to h_1 the value $\alpha_1(h_1) = |(T_{x_1} \restriction \sup_j j(F; d_1; h_1))(\alpha_2(h_1))|$, and $\alpha_2(h_1)$ is represented with respect to the measure K_2 by the function that assign to h_2 the value $\alpha_2(h_1, h_2)$, *etc.*, and where $\alpha_t(h_1, \ldots, h_t) = |(T_{x_t} \restriction (\sup_j j(F; d_t; h_1, \ldots, h_t))$ $(\alpha_{t+1}(h_1, \ldots, h_t))|$, and where $\alpha_{t+1}(h_1, \ldots, h_t) = \gamma$. In these formulas, the $\sup_j$ refers to the supremum over the ultrapower embeddings j for the measures in $\mathcal{W}_1 \cup \mathcal{S}_1$.

DEFINITION 6.24. Let μ be a measure on $\vartheta < \underset{\sim}{\boldsymbol{\delta}}^1_3$, and $\mathcal{C}$ a level-3 complex. Then $\mathcal{V}^{\mu,\mathcal{C}}$ is the measure on $\underset{\sim}{\boldsymbol{\delta}}^1_3$ defined as follows. $A \subseteq \underset{\sim}{\boldsymbol{\delta}}^1_3$ has $\mathcal{V}^{\mu,\mathcal{C}}$ measure one if for μ almost all $\gamma < \vartheta$ there is a c.u.b. $C \subseteq \underset{\sim}{\boldsymbol{\delta}}^1_3$ such that for all $F\colon \operatorname{dom}(<_\mathcal{S}) \to C$ of type $\mathcal{S}$ we have $(F; \mathcal{C}; \gamma) \in A$.

The next theorem is the analysis of measures on $\underset{\sim}{\boldsymbol{\delta}}^1_3$.

THEOREM 6.25. Let $\mathcal{V}$ be a measure on $\underset{\sim}{\boldsymbol{\delta}}^1_3$. Then there is a measure μ on an ordinal $\vartheta < \underset{\sim}{\boldsymbol{\delta}}^1_3$ and a level-3 complex $\mathcal{C}$ such that $\mathcal{V} = \mathcal{V}^{\mu,\mathcal{C}}$.

The proof of Theorem 6.25 is similar to that for the measures on ω_1 (Theorem 4.9), but uses level-2 descriptions and Theorem 5.17. The proof is given in [Jac99].

We illustrate with an example which shows how we generate a typical measure on $\underset{\sim}{\boldsymbol{\delta}}^1_3$.

EXAMPLE 6.26. Let $\mathcal{S}$ be the level-3 tree of uniform cofinalities from Example 6.21. We lift this to a measure on $\underset{\sim}{\boldsymbol{\delta}}^1_3$. Let μ be the principal measure on 0. Let $K_1 = S^2_1$, $K_2 = W^1_1$, and let d_0 be the extended level-3 description $d_0 = \langle i_1 \rangle = \langle 0 \rangle$. Let $d_1 = \langle 0, d^1_1; 1 \rangle^s$ where $d^1_1 \in \mathcal{D}^1(K_1)$ is the level-2 description $d^1_1 = (1; \cdot_1)$. Let $d_2 = \langle 0, d^1_2; 0 \rangle$, where $d^2_2 \in \mathcal{D}^1(K_1, K_2)$ is the level-2 description $d^2_2 = (1; \cdot_1; (2; 1))$. Finally, let x_0, x_1, x_2 be reals with the sections of the Martin tree T_{x_i} well-founded. $\mathcal{S}$ together with the x_i, the measures K_1, K_2, and the descriptions d_1, d_2, d_3 define a complex $\mathcal{C}$.

Suppose $F\colon \operatorname{dom}(<_{\mathcal{S}}) \to \underset{\sim}{\delta}^1_3$ if of type $\mathcal{S}$. We can view F in this case as a function from ω_2 to $\underset{\sim}{\delta}^1_3$ such that $F(\alpha)$ has uniform cofinality ω_1 for $\alpha < \omega_2$ of cofinality ω, and $F(\alpha)$ has uniform cofinality $j_{W^1_1}(\alpha)$ for $\alpha < \omega_2$ of cofinality ω_1. F represents a triple of ordinals $[F] = (\gamma^0, \gamma^{0,0}, \gamma^{0,1})$ as in Example 6.21. Note that $\gamma^0 = \sup(F)$.

Given F (actually its equivalence class $[F]$), the ordinal $\alpha(F) = (F;\mathcal{C};0)$ in this case is given as follows. $\alpha(F) = |(T_{x_0}\restriction \sup_j j(\sup(F))(\alpha_1)|$. α_1 is represented with respect to the measure K_1 by the function which assigns to $h_1\colon \operatorname{dom}(<_2) \to \omega_1$ of the correct type the value $\alpha_1(h_1) = |(T_{x_1}\restriction \sup_j j(\beta))(\alpha_2(h_1))|$, where $\beta = (F;d_1;h_1) = \sup\{F(\gamma) : \gamma < [h_1(1)]\}$, where we recall $h_1(1)\colon \omega_1 \to \omega_1$ is the first invariant of h_1. $\alpha_2(h_1)$ is represented with respect to $K_2 = W^1_1$ by the function which assigns to $h_2 < \omega_1$ the value $\alpha(h_1, h_2) = |(T_{x_2}\restriction \beta')(0)|$, where $\beta' = F([h])$ and $h(\eta) = h_1(h_2(0), \eta)$ (note that $h_2\colon 1 \to \omega_1$ and $h_2(0) < \omega_1$).

The measure $\nu^{\mathcal{C},\mu}$ is the measure on $\underset{\sim}{\delta}^1_3$ described by: A has measure one if there is a c.u.b. $C \subseteq \underset{\sim}{\delta}^1_3$ such that for all $F : \omega_2 \to \underset{\sim}{\delta}^1_3$ of type $\mathcal{S}$, $\alpha(F) \in A$.

To complete the second stage of the inductive analysis we must prove the weak partition relation on $\underset{\sim}{\delta}^1_5$. The main new ingredient is that of a level-4 tree of uniform cofinalities. This is defined similarly to how level-3 trees were defined from level-2 trees. So, a level-4 tree will be a function with domain a finite tree, and to each node is assigned a partial level ≤ 3 tree. For the precise definition and proofs the reader can consult [Jac99]. The measures on $(\underset{\sim}{\delta}^1_5)^-$ are described by level-4 complexes, which lift-up the measures given by the level-4 trees of uniform cofinalities in a manner similar to that described above for level-3 trees for measures on $\underset{\sim}{\delta}^1_3$. This gives a $\underset{\sim}{\Delta}^1_5$ coding of the subsets of $(\underset{\sim}{\delta}^1_5)^-$ which then gives the weak partition relation on $\underset{\sim}{\delta}^1_5$,

§7. Higher Levels. We have outlined so far the first two stages of the inductive analysis of the projective ordinals. In the first stage we used trivial descriptions to compute $\underset{\sim}{\delta}^1_3$, prove the strong partition relation on $\underset{\sim}{\delta}^1_1$, and prove the weak partition relation on $\underset{\sim}{\delta}^1_3$. In the second stage, we used the next level (non-trivial) descriptions to compute $\underset{\sim}{\delta}^1_5$, prove the strong partition relation on $\underset{\sim}{\delta}^1_3$, and prove the weak partition relation on $\underset{\sim}{\delta}^1_5$. Recall that in §3 we presented the general stage-n inductive hypotheses. Our stage-n inductive hypotheses were called I_{2n-1} in [Jac88] and our stage-n auxiliary hypotheses were called K_{2n+1} in [Jac88] (actually we have included a little more in our auxiliary hypotheses for clarity). So, at stage-n we must compute $\underset{\sim}{\delta}^1_{2n+3}$, prove the strong partition relation on $\underset{\sim}{\delta}^1_{2n+1}$ and the weak partition relation on $\underset{\sim}{\delta}^1_{2n+3}$, as well as prove the auxiliary hypotheses.

The partition properties on $\underset{\sim}{\boldsymbol{\delta}}^1_{2n+1}$ and $\underset{\sim}{\boldsymbol{\delta}}^1_{2n+3}$ follow from the existence of sufficiently good codings of the subsets of $\underset{\sim}{\boldsymbol{\delta}}^1_{2n+1}$ and λ_{2n+3} (again, there is there is the small point that one must work directly with functions, rather than subsets of $\underset{\sim}{\boldsymbol{\delta}}^1_{2n+1}$ in proving the strong partition relation on $\underset{\sim}{\boldsymbol{\delta}}^1_{2n+1}$). As we mentioned in §3, these codings, as well as (a), (b), (d) of the auxiliary inductive hypotheses, follow from the analysis of measures on $\underset{\sim}{\boldsymbol{\delta}}^1_{2n+1}$ and λ_{2n+3}. The arguments for analyzing these measures are essentially identical to those for the analysis of measures on ω_1 (Theorem 4.9), the measures on λ_3 (Theorem 4.29) and the measures on $\underset{\sim}{\boldsymbol{\delta}}^1_3$ (Theorem 6.25) sketched in this paper (the details for the measure analysis on λ_3 can be found in [Jac10], and for the measures on $\underset{\sim}{\boldsymbol{\delta}}^1_3$ and λ_5 in [Jac99]). In particular, one defines the notions of level-$2n+1$ and level-$2n+2$ trees of uniform cofinalities which generate more or less the measures in the homogeneous tree on $\underset{\sim}{\boldsymbol{\Pi}}^1_{2n+1}$ and $\underset{\sim}{\boldsymbol{\Pi}}^1_{2n+2}$ sets respectively. These define basic measures (which are essentially the homogeneity measures in the homogeneous tree for the $\underset{\sim}{\boldsymbol{\Pi}}^1_{2n+1}$, $\underset{\sim}{\boldsymbol{\Pi}}^1_{2n+2}$ sets) which are then lifted up by descriptions to generate arbitrary measures on $\underset{\sim}{\boldsymbol{\delta}}^1_{2n+1}$ and λ_{2n+3}. The process is entirely similar to that for level-2 and level-3 trees discussed in this paper, given the higher analogs of the descriptions (*cf.* Definitions 4.27, 4.33 and Definitions 6.20, 6.24). Because the general measure analysis arguments are very similar to these, they are not given in [Jac88]. Instead, [Jac88] focuses on the part of the arguments which are different (or at least need to be generalized).

The main points are to (1) isolate the correct families of canonical measures (the higher level analogs of the W_1^m, S_1^m), (2) prove the corresponding embedding theorems (which say that it is enough to define descriptions with respect to sequences of these measures), (3) give the correct next level definition of description, and (4) prove the analog of the "main theorem" for descriptions (the analog of Theorem 5.17). These points, along with a rank computation which gives the upper-bound for $\underset{\sim}{\boldsymbol{\delta}}^1_{2n+3}$ are what is proved in [Jac88]. The lower-bound for $\underset{\sim}{\boldsymbol{\delta}}^1_{2n+3}$ follows from an independent, easier argument just assuming the weak partition property on $\underset{\sim}{\boldsymbol{\delta}}^1_{2n+3}$ (this argument is self-contained and does not use descriptions; the reader can see this argument for $\underset{\sim}{\boldsymbol{\delta}}^1_5$ in [Jac99] or [JL]). We give a very brief overview of how these generalizations take place.

The general families of canonical measures are denoted W^m_{2n+1} and $S^{\ell,m}_{2n+1}$ (for $1 \leq \ell \leq 2^{n+1}-1$). They are defined using the weak and strong partition relations on $\underset{\sim}{\boldsymbol{\delta}}^1_{2n+1}$ respectively. We will let $S_1^{1,m} = S_1^m$ for consistency. From the stage-n inductive hypotheses we will have that $S^{\ell,m}_{2n-1}$ and W^m_{2n+1} are defined. For simplicity we assume below the strong partition property on all the $\underset{\sim}{\boldsymbol{\delta}}^1_{2n+1}$ and define all of the canonical measures. Each of these families of measures will be associated to a regular cardinal below the projective ordinals and conversely. Each measure μ in one of these families will be a measure on $\mathrm{dom}(\mu) \in \mathrm{Ord}$ except for the measures W_1^m which are measures on $(\omega_1)^m$.

Recall the order $<_m$ on $(\omega_1)^m$ was defined just before Definition 5.1. We define in a similar way the ordering $<_{2n+1}^m$ on $(\underset{\sim}{\boldsymbol{\delta}}_{2n+1}^1)^m$. That is,

$$(\alpha_1, \dots, \alpha_m) <_{2n+1}^m (\beta_1, \dots, \beta_m) \text{ iff }$$
$$(\alpha_m, \alpha_1, \dots, \alpha_{m-1}) <_{\mathrm{lex}} (\beta_m, \beta_1, \dots, \beta_{m-1}).$$

We assume inductively that the measures $W_{2n'+1}^m$, $S_{2n'+1}^{\ell,m}$ have been defined for $n' < n$ and $1 \leq \ell \leq 2^{n'+1} - 1$ (and all m). The total number of families so far defined is $\sum_{1 \leq n' < n} 2^{n'+1} = 2^{n+1} - 2$. We enumerate these families in the order $W_1^m, S_1^m, W_3^m, S_3^{1,m}, S_3^{2,m}, S_3^{3,m}, W_5^m, S_5^{1,m}, \dots, S_5^{7,m}, \dots, S_{2n-1}^{2^n-1,m}$ (which is the order in which they are defined). We then define W_{2n+1}^m and $S_{2n+1}^{\ell,m}$ as follows.

Definition 7.1. W_{2n+1}^m is the measure on $\underset{\sim}{\boldsymbol{\delta}}_{2n+1}^1$ induced by the weak partition relation on $\underset{\sim}{\boldsymbol{\delta}}_{2n+1}^1$, functions $f\colon \mathrm{dom}(S_{2n-1}^{2^n-1,m}) \to \underset{\sim}{\boldsymbol{\delta}}_{2n+1}^1$ of the correct type and the measure $S_{2n-1}^{2^n-1,m}$.

$S_{2n+1}^{1,m}$ is the measure induced by the strong partition relation on $\underset{\sim}{\boldsymbol{\delta}}_{2n+1}^1$, functions $F\colon \mathrm{dom}(<_{2n+1}^m) \to \underset{\sim}{\boldsymbol{\delta}}_{2n+1}^1$ of the correct type and the m-fold product of the ω-cofinal normal measure on $\underset{\sim}{\boldsymbol{\delta}}_{2n+1}^1$.

For $\ell \geq 2$, $S_{2n+1}^{\ell,m}$ is the measure induced by the strong partition relation on $\underset{\sim}{\boldsymbol{\delta}}_{2n+1}^1$, functions $F\colon \underset{\sim}{\boldsymbol{\delta}}_{2n+1}^1 \to \underset{\sim}{\boldsymbol{\delta}}_{2n+1}^1$ of the correct type, and the measure μ on $\underset{\sim}{\boldsymbol{\delta}}_{2n+1}^1$. Here μ is the measure induced by the weak partition relation on $\underset{\sim}{\boldsymbol{\delta}}_{2n+1}^1$, functions $f\colon \mathrm{dom}(\nu^m) \to \underset{\sim}{\boldsymbol{\delta}}_{2n+1}^1$ of the correct type and the the measure ν^m. Finally, ν^m is the $(\ell - 1)$st measure in the list $W_1^m, S_1^m, W_3^m, \dots, S_{2n-1}^{2^n-1,m}$ (for $\ell = 2$, we identify $\mathrm{dom}(W_1^m) = (\omega_1)^m$ with the ordinal ω_1^m by lexicographic ordering on the m-tuples).

The canonical measures W_1^1, $S_1^1, \dots,$ W_{2n+1}^m, $S_{2n+1}^{\ell,m}$ dominate all of the measures arising in the homogeneous tree construction on $\underset{\sim}{\boldsymbol{\Pi}}_1^1, \dots, \underset{\sim}{\boldsymbol{\Pi}}_{2n+2}^1$ sets in a precise sense given by two embedding theorems which are proved in [Jac88]. These are called there the "local" and "global" embedding theorems. The two embedding theorems follow from two more technical results which are proved in a simultaneous induction. Rather than state precisely these results here, we illustrate by considering a case of each of these results at the first level. Aside from illustrating the method of proof, this shows an essential difference between the natures of the two embedding results.

Recall S_1^2 is a measure on ω_3 corresponding to the permutation $(2, 1)$. Let $\bar{S}_1^3$ be the measure on ω_4 corresponding to the permutation $\sigma = (3, 2, 1)$. In the first example we consider an instance of the local embedding theorem.

EXAMPLE 7.2. We show that there is a c.u.b. $C \subseteq \underset{\sim}{\delta}^1_3$ such that for all $\delta \in C$ with $\mathrm{cf}(\delta) = \omega_2$ we have $j_{\bar{S}^3_1}(\delta) \leq j_{S^2_1}(\delta)$ (it actually then follows that $j_{\bar{S}^3_1}(\delta) = j_{S^2_1}(\delta)$). For δ sufficiently closed we define an embedding $\pi\colon j_{\bar{S}^3_1}(\delta) \to j_{S^2_1}(\delta)$. Fix a function $f : \omega_4 \to \delta$, and we define a function $g\colon \omega_3 \to \delta$ such that $\pi([f]_{\bar{S}^3_1}) = [g]_{S^2_1}$. To do this, we use the auxiliary measure $\nu = S^2_1$. Suppose $h\colon \mathrm{dom}(<_2) \to \omega_1$ of the correct type represents $[h]_{W^2_1} \in \mathrm{dom}(S^2_1)$. We represent $g([h])$ with respect to the measure ν by the function which assigns to $[h']_{W^2_1}$, for $h'\colon \mathrm{dom}(<_2) \to \omega_1$ of the correct type, the value $g([h],[h']) \doteq f([k]_{W^3_1})$ where k is defined by:

$$k(\alpha_1, \alpha_2, \alpha_3) = h(h'(\alpha_1, \alpha_2), \alpha_3).$$

For fixed functions h and h', $k(\alpha_1, \alpha_2, \alpha_3)$ is well-defined for almost all $\vec{\alpha}$, and furthermore on a c.u.b. set we have that k is order-preserving from $<_\sigma$ to ω_1. Also, k has uniform cofinality ω. This shows that if f and f' agree almost everywhere with respect to $\bar{S}^3_1$, then for almost all h, for almost all h' we have $f([k]) = f'([k])$. Next observe that for almost all h that for almost all h', if $[h''] = [h']$ then $[k(h,h')]_{W^3_1} = [k(h,h'')]_{W^3_1}$, where $k(h,h')$ is the k function defined using h, h' and likewise for $k(h,h'')$. Next observe that if $[h]_{W^2_1} = [\tilde{h}]_{W^2_1}$, then for almost all h' we have $[k(h,h')] = [k(\tilde{h},h')]$ (this is because almost all h' are into a c.u.b. set on which h and $\tilde{h}$ agree). Finally, if $D \subseteq \omega_1$ is c.u.b., then for almost all h and almost all h' we have that $h(h,h')$ takes values in D for almost all $\vec{\alpha}$. Putting these observations together shows that π is well-defined. If δ is closed under ultrapowers by ν, then it is easy to see now that π is an embedding from $j_{\bar{S}^3_1}(\delta)$ to $j_{S^2_1}(\delta)$. Note here that for almost all $[h]$ that for almost all $[h']$ that the equivalence class $[k(1)]$ of the first invariant of the function $k = k(h,h')$ is less than or equal to $[h(1)]$. We use also the fact that any $f\colon \mathrm{dom}(\bar{S}^3_1) \to \mathrm{Ord}$ must almost everywhere have the property that if $[k_1(1)] < [k_2(1)]$, then $f([k_1]) \leq f([k_2])$.

For the second example we consider an instance of the global embedding theorem.

EXAMPLE 7.3. Let μ be the measure on $\underset{\sim}{\delta}^1_3$ induced by the weak partition relation on $\underset{\sim}{\delta}^1_3$, function $f : \omega_3 \to \underset{\sim}{\delta}^1_3$ of the correct type, and the measure S^2_1. Let μ_σ be the measure on $\underset{\sim}{\delta}^1_3$ induced by the weak partition relation on $\underset{\sim}{\delta}^1_3$, functions $f : \omega_4 \to \underset{\sim}{\delta}^1_3$ and the measure $\bar{S}^3_1$ on ω_4.

We show that $j_{\mu\times\mu}(\underset{\sim}{\delta}^1_3) \leq j_{\mu_\sigma}(\underset{\sim}{\delta}^1_3)$. Note that $\mu \times \mu$ is the measure induced by the weak partition relation on $\underset{\sim}{\delta}^1_3$ and functions $f = f_1 \oplus f_2\colon \omega_3 \cdot 2 \to \underset{\sim}{\delta}^1_3$ of the correct type (and the measure S^2_1 on ω_3).

We again define an embedding π from the first ultrapower into the second. Fix $F : \underset{\sim}{\delta}^1_3 \to \underset{\sim}{\delta}^1_3$ representing $[F]_{\mu\times\mu}$, and we define $G : \underset{\sim}{\delta}^1_3 \to \underset{\sim}{\delta}^1_3$ with $[G]_{\mu_\sigma} =$

$\pi([F]_{\mu\times\mu})$. Fix $g\colon \omega_4 \to \underset{\sim}{\delta}^1_3$, and we define $G([g]_{\bar{S}^3_1})$. We again use an auxiliary measure, this time the measure $\nu = S^2_1 \times S^2_1$. Note that ν is induced by the strong partition relation on ω_1 and functions from $\mathrm{dom}(<_\nu) \to \omega_1$ of the correct type, where $<_\nu$ is lexicographic ordering on tuples (α, i, β) with $\beta < \alpha < \omega_1$ and $i \in \{0, 1\}$. For $(\eta_1, \eta_2) = [h] \in \mathrm{dom}(\nu)$ (where $h\colon \mathrm{dom}(<_\nu) \to \omega_1$ is of the correct type), we assign to (η_1, η_2) an ordinal $\vartheta(\eta_1, \eta_2)$ which is the equivalence class of an order-preserving function $f = f_1 \oplus f_2\colon \omega_3 \cdot 2 \to \omega_4$. For such a function f we will let $[f]$ denote the pair $([f_1]_{S^2_1}, [f_2]_{S^2_1})$. We will have that the following property is satisfied:

> For any $A \subseteq \omega_4$ of $\bar{S}^3_1$ measure one, for ν almost all (η_1, η_2) we have that f is almost everywhere into A. $(*)$

(That is, f_1 and f_2 are both almost everywhere into A.)

Granting $(*)$, we then represent $G([g])$ with respect to the measure ν by the function which assigns to $(\eta_1, \eta_2) = [h]$ the value $G([g], h) \doteq F([g \circ f])$, where $[f] = \vartheta(\eta_1, \eta_2)$. From $(*)$ it is immediate that π is a well-defined embedding from $j_{\mu\times\mu}(\underset{\sim}{\delta}^1_3)$ to $j_{\mu_\sigma}(\underset{\sim}{\delta}^1_3)$.

It remains to define $\vartheta(\eta_1, \eta_2)$ and show $(*)$. Fix $h\colon \mathrm{dom}(<_\nu) \to \omega_1$ of the correct type representing (η_1, η_2). Let $\gamma = [k] < \omega_3$, where $k\colon \mathrm{dom}(<_2) \to \omega_1$ is of the correct type, and let $i \in \{0, 1\}$, and we define $f(i, \gamma)$ (we are identifying $\omega_3 \cdot 2$ with lexicographic ordering on $2 \times \omega_3$). We set $f(i, \gamma) = [\ell]_{W^3_1}$ where ℓ is defined by

$$\ell(\alpha_1, \alpha_2, \alpha_3) = h(\alpha_3, i, k(\alpha_1, \alpha_2)).$$

For fixed h, $[\ell]$ only depends on the equivalence class of k, and so $f(i, \gamma)$ is well-defined. Also, ℓ is order-preserving with respect to $<_\sigma$ on a c.u.b. set. Since h is order-preserving with respect to $<_\nu$, it follows that if $(i_1, \gamma_1) <_{\mathrm{lex}} (i_2, \gamma_2)$ then $f(i_1, \gamma_1) < f(i_2, \gamma_2)$, so f is order-preserving from $\omega_3 \cdot 2$ to ω_4. If $[h] = [h']$, then almost all γ will be represented by a function k having range in a c.u.b. set C on which h, h' agree. Thus, the equivalence class of f depends only on $[h]$. Finally, if $C \subseteq \omega_1$ is c.u.b. and defines a measure one set A with respect to $\bar{S}^3_1$, then almost all h have range in C, and for such h the function f will have range in A. This shows $(*)$.

Remark 7.4. The previous two examples show an essential difference between the local and global embedding theorems. For the global theorem, the fact that $\bar{S}^3_1$ has domain ω_4, which is bigger that $\omega_3 \cdot 2$, was the relevant point. For the local theorem, it was not the order-type of the domain which was relevant, but the fact that $\sigma = (3, 2, 1)$ was ordered this particular way. Even if we had used $\sigma = (10, 9, 8, 7, \dots, 1)$ to generate a measure μ on ω_{11}, we would still have that $j_\mu(\delta) \leq j_{S^2_1}(\delta)$ on a c.u.b. set. So, for the local result the particular permutation matters, but not for the global result.

We will not give here the precise definition of a higher level description, nor prove the main theorem. These results are given in [Jac88]. Instead, we will make a few comments about how descriptions are defined at the next level (level-4 descriptions). This should give the reader a feel of how the general inductive step takes place.

A level-4 description is defined with respect to a finite sequence $K_1, \dots, K_t$ of canonical measures each of the form W_1^m, S_1^m, W_3^m, or $S_3^{\ell,m}$ (where $1 \leq \ell \leq 3$). A description will have an index associated to it, which will be a sequence $\mathcal{I} = (\bar{K}_1, \dots, \bar{K}_u)$ where $\bar{K}_1 = W_3^m$, and $\bar{K}_i \in \mathcal{W}_1 \cup \mathcal{S}_1$ (there are actually two other cases, but they are similar). We let $\mathcal{D}^{\mathcal{I}}(K_1, \dots, K_t)$ denote the descriptions defined relative to $K_1, \dots, K_t$ with index $\mathcal{I}$.

As before, there are basic and non-basic descriptions in $\mathcal{D}^{\mathcal{I}}(\vec{K})$. The rough idea is that the general level-3 descriptions give the basic level-4 descriptions. As before, the non-basic level-4 descriptions are "compositions" of these. Also as before, a description will represent an ordinal with respect to the iterated ultrapower by the measures $K_1, \dots, K_t$. One type of basic description (there are other types) $d \in \mathcal{D}^{\mathcal{I}}(\vec{K})$ is a level-3 description d defined relative to $\mathcal{I} = (\bar{K}_1, \dots, \bar{K}_u)$. For $h_1, \dots, h_t$ representing ordinals in the domains of $K_1, \dots, K_t$, this basic description gives the ordinal $(d; h_1, \dots, h_t)$ which is in turn represented with respect to the measures $W_3^m, \bar{K}_1, \dots, \bar{K}_u$ by the function which assign to $(f; \bar{h}_1, \dots, \bar{h}_u)$ the value $(d; h_1, \dots, h_t; f, \bar{h}_1, \dots, \bar{h}_u)$ which we define to be $(d; f, \bar{h}_1, \dots, \bar{h}_u)$ (the evaluation of the level-3 description d). This does not depend on the functions $h_1, \dots, h_t$. This type of basic description is analogous to one type of basic level-2 description. There we had $\cdot_r$ as a basic level-2 description, and $\cdot_r$ is essentially a lower-level (*i.e.*, trivial) description.

An example of a non-basic description in $\mathcal{D}^{\mathcal{I}}(\vec{K})$ is one of the form $(k; d')$, where $d' \in \mathcal{D}^{\mathcal{I}'}(\vec{K})$, and $\mathcal{I}' = \mathcal{I}^\frown \bar{K}_{u+1}$ where $K_k = S_3^{\ell,m}$ and the measure $S_3^{\ell,m}$ is defined using the strong partition on $\underset{\sim}{\delta}_3^1$, and functions $f \colon \operatorname{dom}(\bar{K}_{u+1}) \to \underset{\sim}{\delta}_3^1$. For fixed $h_1, \dots, h_t$, and fixed $f, \bar{h}_2, \dots, \bar{h}_u$, this description produces the ordinal $(d; \vec{h}; f; \vec{\bar{h}}) = h_k([g])$ where $g \colon \operatorname{dom}(\bar{K}_{u+1}) \to \underset{\sim}{\delta}_3^1$ is given by: $g([\bar{h}_{u+1}]) = (d'; \vec{h}; f; \vec{\bar{h}}, \bar{h}_{u+1})$.

When K_k is of the form $K_k = S_3^{1,m}$ we allow descriptions of the form $d = (k; d_0, d_1, \dots, d_\ell)^{(s)}$ as before, with a similar interpretation as in the level-2 case.

The proof of the main theorem (the analog of 5.17) and the rank computations are similar to the level-2 case. We refer the reader to [Jac88] for further details.

§8. Concluding Remarks. The descriptions, defined relative to canonical measure sequences, describe the cardinal structure below the projective ordinals. For example, it follows from Theorem 6.9 that every cardinal $\kappa <$

λ_5 is of the form $\kappa = (\mathrm{id}; d; W_3^m; \vec{K})$ for some level-3 description d and measure sequence $W_3^m, \vec{K}$. We have also seen how descriptions are used to lift-up certain basic measures (more or less the measures in the homogeneous tree construction) to generate arbitrary measures on the ordinals below the projective ordinals. However. while general embedding arguments suffice to give the lower bounds for the projective ordinals (these arguments don't use descriptions), the arguments sketched in this paper don't show that the ordinals represented by descriptions are cardinals. For example, below λ_5 we haven't shown that the ordinals $(\mathrm{id}; d; W_3^m; \vec{K})$ are cardinals.

To show that all the descriptions represent cardinals, another argument is needed which also has independent interest. This involves producing another representation for the cardinal structure below the projective ordinals which does not involve descriptions. This allows for a simpler, self-contained presentation of the cardinal structure. The description analysis, however, is needed to show that this alternate formulation works. This alternate method is described and proved for the cardinal structure below $\underset{\sim}{\boldsymbol{\delta}}_5^1$ in [JK], and described below the projective ordinals in general in [JL]. We do not give the full details here, but sketch the method and indicate how it gives the cardinal structure below $\underset{\sim}{\boldsymbol{\delta}}_5^1$.

Following the terminology of [JL], we say an **ordinal algebra** is a free associative, left-distributive algebra with operations $\oplus, \otimes$ over a set of generators $\{\mathsf{V}_\beta\}_{\beta<\alpha}$, for some $\alpha \in \mathrm{Ord}$. We let $\mathfrak{A}_\alpha$ be the algebra with generators $\{\mathsf{V}_\beta\}_{\beta<\alpha}$. We inductively define a function o from the algebra to the ordinals as follows. We set $o(\mathsf{V}_0) = 0$. We set $o(s \oplus t) = o(s) + o(u)$ and $o(s \otimes u) = o(s) \cdot o(u)$ (ordinal addition and multiplication). We let $\mathrm{ht}(\mathfrak{A}_\alpha) = \sup\{o(s) : s \in \mathfrak{A}_\alpha\}$. We then set $o(\mathsf{V}_\alpha) = \mathrm{ht}(\mathfrak{A}_\alpha)$.

For example, for the first ω many generators we have $o(\mathsf{V}_0) = 0$ and $o(\mathsf{V}_1) = 1$. Since

$$o(\mathsf{V}_1 \underbrace{\oplus \cdots \oplus}_{n} \mathsf{V}_1) = n,$$

we have $o(\mathsf{V}_2) = \mathrm{ht}(\mathfrak{A}_2) = \omega$. Since

$$o(\mathsf{V}_2 \underbrace{\otimes \cdots \otimes}_{n} \mathsf{V}_2) = \omega^n,$$

we have $o(\mathsf{V}_3) = \omega^\omega$. Similarly, $o(\mathsf{V}_\beta) = \omega^{\omega^{\beta-2}}$ for all $\beta < \omega$, and $o(\mathsf{V}_\beta) = \omega^{\omega^\beta}$ for all $\beta \geq \omega$.

We then inductively assign measures $\mathfrak{m}(\mathsf{V}_\beta)$ to each of the generators. We do not give the general definition here (it is given in [JL]), but simply give the result for the first ω generators. We also let $\boldsymbol{ot}(\mathsf{V}_\beta)$ be the ordinal on which $\mathfrak{m}(\mathsf{V}_\beta)$ is a measure. We let $\mathfrak{m}(\mathsf{V}_0)$ be the empty measure, with $\boldsymbol{ot}(\mathsf{V}_0) = 0$

(this is for notational convenience). Let $\mathfrak{m}(\mathsf{V}_1)$ = the principal measure on $\{0\}$, and $\boldsymbol{ot}(\mathsf{V}_1) = 1$. Let $\mathfrak{m}(\mathsf{V}_2) = W_1^1$, and let $\boldsymbol{ot}(\mathsf{V}_1) = \omega_1$. For $3 \leq n \in \omega$, let $\mathfrak{m}(\mathsf{V}_n) = S_1^{m-2}$, and $\boldsymbol{ot}(\mathsf{V}_m) = \omega_{m-1}$.

We extend the assignment of measures and order-types from the generators to general terms s in the algebra as follows. Let $\boldsymbol{ot}(s)$ be the order-type obtained by interpreting $\oplus$ by $+$ and $\otimes$ by $\times$ (ordinal addition and multiplication). For example, if $s = \mathsf{V}_4 \oplus (\mathsf{V}_3 \otimes \mathsf{V}_3)$, then $\boldsymbol{ot}(s) = \omega_3 + \omega_2 \cdot \omega_2$. We may identify terms s in the algebra with finite trees T_s whose nodes (except for the root node) are labeled with generators, in the same standard way as for ordinal expressions involving sums and products. So, $\oplus$ corresponds to splittings of a node, and $\otimes$ corresponds to descending in the tree. More precisely, the tree associated to $s \oplus t$ is the tree consisting of side-by-side copies of the trees for s and t, and the tree for $s \otimes t$ is the tree obtained by replacing every terminal node of the tree for t with a copy of the tree for s. For example, for the term $s = \mathsf{V}_4 \oplus (\mathsf{V}_3 \otimes \mathsf{V}_3)$ above we have the tree:

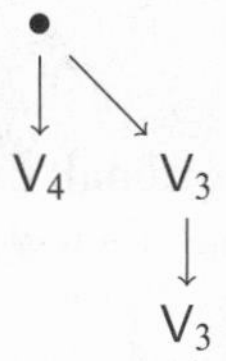

To every terminal node of the tree is associated a product measure obtained by taking the product of the measures corresponding to the generators as we descend along the branch to the terminal node. In fact, to each node of the tree we assign the product of the measures along the (non-maximal) branch ending with that node. Let $\nu(p)$ denote this product measure, for p a node of T_s. We identify the tree T_s with a subtree of $\omega^{<\omega}$ in the obvious manner. We then associate to s an ordering $<_s$ of order-type $\boldsymbol{ot}(s)$. This is lexicographic ordering on the set of tuples $(i_1, \alpha_1, \ldots, i_k, \alpha_k)$ where $(i_1, \ldots, i_k) \in T_s$ and α_j is in the domain of the measure associated to the node $(i_1, \ldots, i_j)$ for all $j \leq k$.

For the example $s = \mathsf{V}_4 \oplus (\mathsf{V}_3 \otimes \mathsf{V}_3)$ considered above, $<_s$ has domain tuples of the form $(0, \alpha)$ where $\alpha < \omega_3$ together with $(1, \beta)$ where $\beta < \omega_2$, and $(1, \beta, 0, \gamma)$ where $\beta, \gamma < \omega_2$. Note that $\operatorname{dom}(<_s)$ has order-type $\boldsymbol{ot}(s)$.

To each term s in the algebra $\mathfrak{A}_\omega$ we assign a measure $\mu(s)$ on $\underset{\sim}{\boldsymbol{\delta}}{}_3^1$ as follows.

Definition 8.1. $\mu(s)$ is the measure on $\underset{\sim}{\boldsymbol{\delta}}{}_3^1$ induced by the weak partition relation on $\underset{\sim}{\boldsymbol{\delta}}{}_3^1$, functions $f\colon \operatorname{dom}(<_s) \to \underset{\sim}{\boldsymbol{\delta}}{}_3^1$ which are order-preserving and continuous (and of uniform cofinality ω at points of successor rank), and the product measures $\nu(p)$ associated to nodes p of T_s.

For the example s we are considering, a function $f\colon \operatorname{dom}(<_s) \to \underset{\sim}{\boldsymbol{\delta}}{}_3^1$ represents a triple of ordinals $[f] = (\eta_1, \eta_2, \eta_3)$. Here η_1 is represented with respect to S_1^2 by the map $\alpha \mapsto f(0, \alpha)$. η_2 is represented with respect to S_1^1 by $\beta \mapsto f(1, \beta)$, and η_3 is represented with respect to $S_1^1 \times S_1^1$ by $(\beta, \gamma) \mapsto f(1, \beta, 0, \gamma)$.

Martin's Theorem 3.8 gives that for any term $s \in \mathfrak{A}_\omega$, $j_{\mu(s)}(\underset{\sim}{\boldsymbol{\delta}}^1_3)$ is a cardinal. The next theorem gives our alternate representation of the cardinal structure.

THEOREM 8.2. The cardinals below λ_5 are precisely the ordinals $j_{\mu(s)}(\underset{\sim}{\boldsymbol{\delta}}^1_3)$ for $s \in \mathfrak{A}_\omega$. Moreover, $j_{\mu(s)}(\underset{\sim}{\boldsymbol{\delta}}^1_3) = \omega_{\omega+o(s)+1}$.

For the example $s = \mathsf{V}_4 \oplus (\mathsf{V}_3 \otimes \mathsf{V}_3)$, we have $j_{\mu(s)}(\underset{\sim}{\boldsymbol{\delta}}^1_3) = \omega_{\omega^{\omega^2}+\omega^\omega\cdot 2+1}$.

This alternate representation has applications, for example, it makes reading off the cofinalities of the cardinals below the projective ordinals easy. The following result from [JK] computes this below $\underset{\sim}{\boldsymbol{\delta}}^1_5$.

THEOREM 8.3. Suppose $\kappa = \omega_{\alpha+1}$ is a successor cardinal with $\underset{\sim}{\boldsymbol{\delta}}^1_3 < \kappa < \underset{\sim}{\boldsymbol{\delta}}^1_5$. Let $\alpha = \omega^{\beta_1} + \cdots + \omega^{\beta_n}$, where $\omega^\omega > \beta_1 \geq \cdots \geq \beta_n$, be the normal form for α. Then:

- If $\beta_n = 0$, then $\mathrm{cf}(\kappa) = \underset{\sim}{\boldsymbol{\delta}}^1_4 = \omega_{\omega+2}$.
- If $\beta_n > 0$, and is a successor ordinal, then $\mathrm{cf}(\kappa) = \omega_{\omega\cdot 2+1}$.
- If $\beta_n > 0$ and is a limit ordinal, then $\mathrm{cf}(\kappa) = \omega_{\omega^\omega+1}$.

REFERENCES

STEPHEN JACKSON

[Jac88] AD *and the projective ordinals*, this volume, originally published in Kechris et al. [CABAL iv], pp. 117–220.

[Jac90A] *A new proof of the strong partition relation on* ω_1, ***Transactions of the American Mathematical Society***, vol. 320 (1990), no. 2, pp. 737–745.

[Jac91] *Admissible Suslin cardinals in* L($\mathbb{R}$), ***The Journal of Symbolic Logic***, vol. 56 (1991), no. 1, pp. 260–275.

[Jac99] ***A computation of*** $\underset{\sim}{\boldsymbol{\delta}}^1_5$, vol. 140, Memoirs of the AMS, no. 670, American Mathematical Society, July 1999.

[Jac08] *Suslin cardinals, partition properties, homogeneity. Introduction to Part II*, in Kechris et al. [CABAL I], pp. 273–313.

[Jac10] *Structural consequences of* AD, in Kanamori and Foreman [KF10], pp. 1753–1876.

STEPHEN JACKSON AND FARID KHAFIZOV

[JK] *Descriptions and cardinals below* $\underset{\sim}{\boldsymbol{\delta}}^1_5$, in submission.

STEPHEN JACKSON AND BENEDIKT LÖWE

[JL] *Canonical measure assignments*, in submission.

AKIHIRO KANAMORI AND MATTHEW FOREMAN

[KF10] ***Handbook of Set Theory***, Springer, 2010.

ALEXANDER S. KECHRIS

[Kec77A] AD *and infinite exponent partition relations*, circulated manuscript, 1977.

[Kec78] AD *and projective ordinals*, this volume, originally published in Kechris and Moschovakis [CABAL i], pp. 91–132.

[Kec81A] *Homogeneous trees and projective scales*, this volume, originally published in Kechris et al. [CABAL ii], pp. 33–74.

[Kec94] ***Classical descriptive set theory***, Graduate Texts in Mathematics, vol. 156, Springer, 1994.

ALEXANDER S. KECHRIS, BENEDIKT LÖWE, AND JOHN R. STEEL

[CABAL I] *Games, scales, and Suslin cardinals: the Cabal seminar, volume I*, Lecture Notes in Logic, vol. 31, Cambridge University Press, 2008.

ALEXANDER S. KECHRIS, DONALD A. MARTIN, AND YIANNIS N. MOSCHOVAKIS

[CABAL ii] *Cabal seminar* 77–79, Lecture Notes in Mathematics, no. 839, Berlin, Springer, 1981.

[CABAL iii] *Cabal seminar* 79–81, Lecture Notes in Mathematics, no. 1019, Berlin, Springer, 1983.

ALEXANDER S. KECHRIS, DONALD A. MARTIN, AND JOHN R. STEEL

[CABAL iv] *Cabal seminar* 81–85, Lecture Notes in Mathematics, no. 1333, Berlin, Springer, 1988.

ALEXANDER S. KECHRIS AND YIANNIS N. MOSCHOVAKIS

[CABAL i] *Cabal seminar* 76–77, Lecture Notes in Mathematics, no. 689, Berlin, Springer, 1978.

ALEXANDER S. KECHRIS AND W. HUGH WOODIN

[KW80] *Generic codes for uncountable ordinals, partition properties, and elementary embeddings*, circulated manuscript, 1980, reprinted in [CABAL I], p. 379–397.

DONALD A. MARTIN

[Mar71B] *Projective sets and cardinal numbers: some questions related to the continuum problem*, this volume, originally a preprint, 1971.

DONALD A. MARTIN AND JOHN R. STEEL

[MS83] *The extent of scales in* $\mathbf{L}(\mathbb{R})$, in Kechris et al. [CABAL iii], pp. 86–96, reprinted in [CABAL I], p. 110–120.

YIANNIS N. MOSCHOVAKIS

[Mos80] ***Descriptive set theory***, Studies in Logic and the Foundations of Mathematics, no. 100, North-Holland, Amsterdam, 1980.

ROBERT M. SOLOVAY

[Sol78A] *A* Δ^1_3 *coding of the subsets of* ω_ω, this volume, originally published in Kechris and Moschovakis [CABAL i], pp. 133–150.

JOHN R. STEEL

[Ste83] *Scales in* $\mathbf{L}(\mathbb{R})$, in Kechris et al. [CABAL iii], pp. 107–156, reprinted in [CABAL I], p. 130–175.

DEPARTMENT OF MATHEMATICS
UNIVERSITY OF NORTH TEXAS
P.O. BOX 311430
DENTON, TEXAS 76203-1430
UNITED STATES OF AMERICA
E-mail: jackson@unt.edu

HOMOGENEOUS TREES AND PROJECTIVE SCALES

ALEXANDER S. KECHRIS

This exposition is a sequel to [Kec78]. Its main purpose is to show how set theoretical techniques, among them infinite exponent partition relations can be used to produce homogeneous trees for projective sets. The work here is again understood as being carried completely with $\mathbf{L}(\mathbb{R})$, with the hypothesis that AD+DC holds this model. As applications, one has Kunen's important reduction of the problem of computing $\underset{\sim}{\boldsymbol{\delta}}^1_5$ to the problem of computing certain ultrapowers of $\underset{\sim}{\boldsymbol{\delta}}^1_3 = \omega_{\omega+1}$ and also a result of Martin on constructibility relative to subsets of $\omega_{\omega+1}$. An observation on the Victoria Delfino 3rd problem concludes the present paper.

Most of the results and constructions of trees presented in §§3, 4, 5 below are due to Kunen, and go back to his [Kun71B]. On the other hand, some of the effective calculations, in §§3, 4, for the scales resulting from these trees need the recent results of Kechris and Martin [KM78] and Harrington and Kechris [HK81].

§0. Introduction. In this introductory section we collect various notational conventions and some prerequisites needed to follow this paper.

0.1. Trees. If X is a set, ${}^{\omega}X$ is the set of all infinite sequences from X and ${}^{<\omega}X$ the set of all finite sequences from X. If $s, t \in {}^{<\omega}X$ and $f \in {}^{\omega}X$, then $s \subseteq t$ and $s \subseteq f$ denote the extension relation in each case. If $s = (x_0, \dots, x_{n-1})$, we let $s(i) \equiv s_i = x_i$ for $i < n$ and we put $\operatorname{lh} s = n$. Sometimes we also write $(x_1, \dots, n_n)$ for $(y_0, \dots, y_{n-1})$, where $y_i = x_{i+1}$, $0 \leq i < n$. If $s = (x_0, \dots, x_{n-1})$ and $m \leq n$, then $s \restriction m = (x_0, \dots, x_{m-1})$.

We reserve usually letters $\sigma, \tau, \dots$ for members of ${}^{<\omega}\omega$ and u, v, $w, \dots$ for members of ${}^{<\omega}\mathrm{Ord}$. As usual α, β, $\gamma, \dots$ denote **reals**, *i.e.*, elements of $\mathbb{R} = {}^{\omega}\omega$. We fix a recursive 1-1 correspondence $\tau_0, \tau_1, \tau_2, \dots$ between ω and ${}^{<\omega}\omega$, such that $\tau_j \supsetneqq \tau_i \Rightarrow j > i$ and if $\ell_i = \operatorname{lh} \tau_i$, then $\ell_i \leq i$. Moreover we agree to take $\tau_0 = \varnothing$, $\tau_1 = (0)$.

The preparation of this paper was partially supported by NSF Grant MCS 76-17254 A01. The author is an A. P. Sloan Foundation Fellow. We would like to thank Y. N. Moschovakis for making a number of valuable suggestions for improving the presentation of this paper.

Wadge Degrees and Projective Ordinals: The Cabal Seminar, Volume II
Edited by A. S. Kechris, B. Löwe, J. R. Steel
Lecture Notes in Logic, 37

By a tree on ${}^{k}\omega \times \lambda$, where $k \in \omega$ and $\lambda \in \mathrm{Ord}$, we mean a set T of $(k+1)$-tuples of the form $(\sigma_1, \dots, \sigma_k, u) \in {}^{k}({}^{<\omega}\omega) \times {}^{<\omega}\lambda$, where each σ_i and u have all the same length and such that if $(\sigma_1, \dots, \sigma_k, u) \in T$ and $n \leq \mathrm{lh}\,\sigma_1$, then $(\sigma_1{\restriction}n, \dots, \sigma_k{\restriction}n, u{\restriction}n) \in T$. For such a tree T and for $\ell \leq k$, we let $T(\sigma_1, \dots, \sigma_\ell) = \{(\sigma_{\ell+1}, \dots, \sigma_k, u) : (\sigma_1, \dots, \sigma_k, u) \in T\}$, $T(\alpha_1, \dots, \alpha_\ell) = \bigcup_{n \in \omega} T(\alpha_1{\restriction}n, \dots, \alpha_\ell{\restriction}n)$ and finally $T^{\subsetneq}(\sigma_1, \dots, \sigma_\ell) = \bigcup_{n \leq \mathrm{lh}\,\sigma_1} T(\sigma_1{\restriction}n, \dots, \sigma_\ell{\restriction}n)$. By $[T]$ we denote the set of all infinite branches through T, *i.e.*, $[T] = \{(\alpha_1, \dots, \alpha_k, f) : \forall n (\alpha_1{\restriction}n, \dots, \alpha_k{\restriction}n, f{\restriction}n) \in T\}$. Let also $\mathrm{p}[T] = \{(\alpha_1, \dots, \alpha_k) : \exists f (\alpha_1, \dots, \alpha_k, f) \in [T]\}$. For $X \subseteq \lambda$, $T{\restriction}X = \{(\sigma_1, \dots, \sigma_k, u) \in T : u \in {}^{<\omega}X\}$.

A tree T on ${}^{k}\omega \times \lambda$ is **wellfounded** iff $[T] = \varnothing$. Equivalently T is wellfounded if the **Brouwer-Kleene ordering** $<_{\mathrm{BK}}$ on $\bigcup_n {}^{k}(\omega^n) \times {}^{n}\lambda$ is a wellordering when restricted to T. Here $<_{\mathrm{BK}}$ is defined as follows: Take the case $k = 1$ for notational simplicity:

$$((a_0, \dots, a_{n-1}), (\xi_0, \dots, \xi_{n-1})) <_{\mathrm{BK}} ((b_0, \dots, b_{m-1}), (\eta_0, \dots, \eta_{m-1}))$$

$$\Longleftrightarrow$$

$$[(a_0, \dots, a_{n-1}) \supsetneq (b_0, \dots, b_{m-1}) \wedge (\xi_0, \dots, \xi_{n-1}) \supsetneq (\eta_0, \dots, \eta_{m-1})] \vee$$

[if ℓ is least such that $(a_\ell, \xi_\ell) \neq (b_\ell, \eta_\ell)$,

then $\xi_\ell < \eta_\ell$ or $(\xi_\ell = \eta_\ell$ and $a_\ell < b_\ell)$].

(The use of the anti-lexicographical ordering of pairs $(a, \xi) \in \omega \times \mathrm{Ord}$ will be convenient later on.)

The usual rank function for a wellordering or wellfounded relation W will be denoted by rank_W and in the case of a wellfounded tree T by rank_T.

0.2. Scales. If $P \subseteq {}^{k}\omega \times {}^{m}\mathbb{R}$ is a pointset, a **scale** on P is a sequence $\bar{\varphi} = \{\varphi_n\}$ of **norms** on P, *i.e.*, mappings into the ordinals, with the following property:

If $x_0, x_1, \dots \in P$ and $x_i \to x$ and for all n, $\varphi_n(x_i)$ is eventually constant (as $i \to \infty$), say equal to λ_n, then $x \in P$ and $\varphi_n(x) \leq \lambda_n$. If each φ_n maps P into λ, we call $\bar{\varphi}$ a λ-scale. A scale $\bar{\varphi} = \{\varphi_n\}$ is **regular** if each norm φ_n maps P onto an ordinal.

Finally, if Γ is a pointclass and $\bar{\varphi}$ a scale on P, we call $\bar{\varphi}$ a **Γ-scale** if the following two relations are in Γ:

$$R(n, x, y) \Leftrightarrow x \in P \wedge [y \notin P \vee \varphi_n(x) \leq \varphi_n(y)],$$

$$Q(n, x, y) \Leftrightarrow x \in P \wedge [y \notin P \vee \varphi_n(x) < \varphi_n(y)].$$

0.3. Indiscernibles. Assuming that $\forall \alpha (\alpha^{\#}$ exists), let for each real α, $\mathcal{I}_\alpha$ be the class of Silver indiscernibles for $\mathbf{L}[\alpha]$ and

$$\mathcal{U} = \bigcap_\alpha \mathcal{I}_\alpha = \{\boldsymbol{u}_1, \boldsymbol{u}_2, \dots, \boldsymbol{u}_\xi, \dots\}_{\xi \in \mathrm{Ord}}$$

the class of **uniform** indiscernibles. Under AD, $\boldsymbol{u}_n = \omega_n$ for $n \leq \omega$. By a result of Solovay each ordinal $< \boldsymbol{u}_{n+1}$ can be written in the form $t^{\mathbf{L}[\alpha]}(\boldsymbol{u}_1, \ldots, \boldsymbol{u}_n)$, for some term t and some real α. For each $f : {}^n[\omega_1] \to \omega_1$ in $\tilde{\mathbf{L}} = \bigcup_\alpha \mathbf{L}[\alpha]$, where ${}^n[\omega_1] = \{(\xi_0, \ldots, \xi_{n-1}) \in {}^n\omega_1 : \xi_0 < \xi_1 < \cdots < \xi_{n-1}\}$, define $\tilde{f}(\boldsymbol{u}_1, \ldots, \boldsymbol{u}_n) = t^{\mathbf{L}[\alpha]}(\boldsymbol{u}_1, \ldots, \boldsymbol{u}_n)$, where $f(\xi_1, \ldots, \xi_n) = t^{\mathbf{L}[\alpha]}(\xi_1, \ldots, \xi_n)$ for $\xi_1 < \cdots < \xi_n < \omega_1$. Thus every ordinal $< \boldsymbol{u}_{n+1}$ has the form $\tilde{f}(\boldsymbol{u}_1, \ldots, \boldsymbol{u}_n)$ for some $f : {}^n[\omega_1] \to \omega_1$ in $\tilde{\mathbf{L}}$. If now $H : \omega_1 \to \omega_1$ is in $\tilde{\mathbf{L}}$, define $\tilde{H} : \boldsymbol{u}_\omega \to \boldsymbol{u}_\omega$ by $\tilde{H}(\tilde{f}(\boldsymbol{u}_1, \ldots, \boldsymbol{u}_n)) = \widetilde{H \circ f}(\boldsymbol{u}_1, \ldots, \boldsymbol{u}_n)$. For $X \subseteq \omega_1$ let also $\tilde{X} \subseteq \boldsymbol{u}_\omega$ be defined by:

$$\tilde{f}(\boldsymbol{u}_1, \ldots, \boldsymbol{u}_n) \in \tilde{X} \Leftrightarrow \exists C \subseteq \omega_1\ (C \text{ is closed unbounded(cub) in } \omega_1 \wedge \text{ for all } \xi_1 < \cdots < \xi_n \text{ in } C, f(\xi_1, \ldots, \xi_n) \in C).$$

0.4. The work in this paper takes place in ZF+DC until otherwise specified.

§1. Π^1_1 sets; the tree S_1.

1.1. Definition of S_1.

(a) Let $A \subseteq \mathbb{R}$ be a Π^1_1 set of reals. Then there is a recursive tree T on $\omega \times \omega$ such that

$$\alpha \in A \Leftrightarrow T(\alpha) \text{ is wellfounded.}$$

If $\sigma \in {}^{<\omega}\omega$ and $\operatorname{lh}\sigma = n$ define the following ordering $<_\sigma$ on $\{0, 1, \ldots, n-1\} = n$:

$$i <_\sigma j \overset{\text{def}}{\Leftrightarrow} \begin{aligned} &1.\ (\tau_i, \tau_j \notin T^{\subseteq}(\sigma) \wedge i < j) \text{ or} \\ &2.\ (\tau_i \notin T^{\subseteq}(\sigma) \wedge \tau_j \in T^{\subseteq}(\sigma)) \text{ or} \\ &3.\ (\tau_i, \tau_j \in T^{\subseteq}(\sigma) \wedge \tau_i <_{\text{BK}} \tau_j). \end{aligned}$$

Thus identifying τ_i with i here, we can visualize $<_\sigma$ as being the Brouwer-Kleene ordering of $T^{\subseteq}(\sigma) \cap n$, with the rest of n thrown in at the bottom with its natural ordering. Note now the following:

(i) If $\sigma \neq \varnothing$, 0 is the top element of $<_\sigma$.
(ii) $\sigma \subseteq \sigma' \Rightarrow <_\sigma \subseteq <_{\sigma'}$ (*i.e.*, $<_\sigma$ is a subordering of $<_{\sigma'}$).

That (i) holds comes from the fact that $\varnothing = \tau_0$ is the top element of $<_{\text{BK}}$. That (ii) holds is an immediate consequence of the remark that for $i < \operatorname{lh}\sigma$, $\sigma \subseteq \sigma'$

$$\tau_i \in T^{\subseteq}(\sigma) \Leftrightarrow \tau_i \in T^{\subseteq}(\sigma').$$

This is of course because always $\ell_i = \operatorname{lh}\tau_i \leq i$, thus $\tau_i \in T^{\subseteq}(\sigma) \Leftrightarrow (\sigma\restriction\ell_i, \tau_i) \in T \Leftrightarrow (\sigma'\restriction\ell_i, \tau_i) \in T \Leftrightarrow \tau_i \in T^{\subseteq}(\sigma')$.

Put now for each $\alpha \in \mathbb{R}$:

$$<_\alpha \overset{\text{def}}{=} \bigcup_n <_{\alpha\restriction n},$$

so that $<_\alpha$ is a linear ordering of (all of) ω, with top element 0 again. Note that $<_\alpha$ is just the Brouwer-Kleene ordering on $T(\alpha)$ with the rest of ω thrown in at the bottom with its natural ordering. Thus

$$\begin{aligned} \alpha \in A &\Leftrightarrow T(\alpha) \text{ is wellfounded} \\ &\Leftrightarrow \langle T(\alpha), <_{\mathrm{BK}} \rangle \text{ is wellordered} \\ &\Leftrightarrow <_\alpha \text{ is a wellordering.} \end{aligned}$$

(b) Define now:

$$\begin{aligned} S_1(A;T) = S_1 \stackrel{\text{def}}{=} \{(\sigma, u) :\ & \sigma \in {}^{<\omega}\omega, u \in {}^{<\omega}\omega_1 \wedge \mathrm{lh}\,\sigma = \mathrm{lh}\,u \ (= n, \text{say}) \ \wedge \\ & u \colon n \to \omega_1 \text{ is order preserving relative to } <_\sigma, \\ & \textit{i.e.}, \text{for } 0 \le i, j < n \colon i <_\sigma j \Leftrightarrow u_i < u_j \}. \end{aligned}$$

Then we obviously have that:

$$\begin{aligned} \alpha \in A &\Leftrightarrow \exists f (\alpha, f) \in [S_1] \\ &\Leftrightarrow S_1(\alpha) \text{ is not wellfounded.} \end{aligned}$$

1.2. Scales for Π^1_1 sets.

(a) If J is a tree on λ we shall say that J has an **honest leftmost branch** if there is a branch $f \in [J]$ such that for all branches $g \in [J]$:

$$\forall i, f(i) \le g(i).$$

Every non-wellfounded tree J has a leftmost branch $h \in [J]$, which is by definition characterized by the property that for $g \in [J]$:

$$h \le_{\mathrm{lex}} g, \textit{ i.e.}, h = g \vee \exists i [h(i) < g(i) \wedge \forall j < i (h(j) = g(j))].$$

But only special trees J have honest leftmost branches. We shall show that those of the form $S_1(\alpha)$ are among them.

Indeed, let for $\alpha \in A$:

$$f_\alpha(i) = \mathrm{rank}_{<_\alpha}(i).$$

Clearly $f_\alpha \in [S_1(\alpha)]$. On the other hand, if $g \in [S_1(\alpha)]$, then $g \colon \omega \to \omega_1$ is such that

$$i <_\alpha j \Leftrightarrow g(i) < g(j),$$

thus

$$f_\alpha(i) = \mathrm{rank}_{<_\alpha}(i) \le g(i),$$

and we are done.

(b) Define now for $\alpha \in A$:

$$\varphi_i(\alpha) = f_\alpha(i).$$

Then $\{\varphi_i\}$ is a scale on A (and thus an ω_1-scale). Indeed, letting $\bar{\varphi}(\alpha) = (\varphi_0(\alpha), \varphi_1(\alpha), \dots)$ and assuming that for all n, $\alpha_n \in A$, while $\alpha_n \to \alpha$ and

$\bar{\varphi}(\alpha_n) \to g$ (*i.e.*, $\varphi_i(\alpha_n) = g(i)$, for all large enough n) we conclude that $(\alpha, g) = \lim(\alpha_n, \bar{\varphi}(\alpha_n)) \in [S_1]$ and moreover $f_\alpha(i) = \varphi_i(\alpha) \leq g(i)$.

Now by the usual argument (see for example Kechris and Moschovakis [KM78B]) one can show that if for $\alpha \in A$ we put

$$\psi_i(\alpha) = \langle \varphi_0(\alpha), \varphi_i(\alpha) \rangle,$$

where $\langle \xi, \eta \rangle \stackrel{\text{def}}{=}$ the ordinal attached to the pair (ξ, η) in the lexicographical ordering of $\omega_1 \times \omega_1$, then $\{\psi_i\}$ is a Π^1_1-scale on A. Thus we have shown that Π^1_1 has the scale property.

1.3. Homogeneity properties of S_1.

(a) Fix now $\sigma \in {}^{<\omega}\omega$, $\sigma \neq \varnothing$. Let $\operatorname{lh} \sigma = n$ and define

$$\pi_\sigma \colon n \to n$$

to be the unique permutation of n defined by:

$$i <_\sigma j \Leftrightarrow \pi_\sigma(i) < \pi_\sigma(j).$$

In particular,

$$\pi_\sigma(0) = n - 1.$$

Then note that if for any $A \subseteq \text{Ord}$ and any $\eta \in \text{Ord}$, we let

$${}^\eta[A] \stackrel{\text{def}}{=} \text{the set of all increasing maps from } \eta \text{ into } A,$$

we have

$$(\sigma, v) \in S_1 \Leftrightarrow \operatorname{lh} v = n \wedge \exists u \in {}^n[\omega_1](v = u \circ \pi_\sigma, \textit{i.e.}, \quad v = (u_{\pi_\sigma(0)}, u_{\pi_\sigma(1)}, \dots, u_{\pi_\sigma(n-1)})).$$

Thus

$$S_1(\sigma) = ({}^{\operatorname{lh}\sigma}[\omega_1])\pi_\sigma \stackrel{\text{def}}{=} \{(\eta_{\pi_\sigma(0)}, \dots, \eta_{\pi_\sigma(n-1)}) : \eta_0 < \eta_1 < \cdots < \eta_{n-1} < \omega_1\},$$

so that $S_1(\sigma)$ is just a "permutation" of ${}^{\operatorname{lh}\sigma}[\omega_1]$.

Since $\pi_\sigma(0) = n - 1$ we have the following important "boundedness" property:

$$(\xi_0, \dots, \xi_{n-1}) \in S_1(\sigma) \Leftrightarrow \xi_0 > \xi_1, \xi_2, \dots, \xi_{n-1}.$$

This will be quite useful in §2.

(b) Assume now $H \colon \omega_1 \to \omega_1$ is an increasing map, *i.e.*, $H \in {}^{\omega_1}[\omega_1]$. For any $u = (\xi_0, \dots, \xi_{n-1}) \in {}^n\omega_1$, we put also $H(u) = (H(\xi_0), \dots, H(\xi_{n-1}))$ and for any $(\sigma, u) \in {}^n\omega \times {}^n\omega_1$ we let again

$$H(\sigma, u) = (\sigma, H(u)).$$

The claim now is that

$$H \colon S_1 \to S_1,$$

i.e., S_1 is invariant under H, so that in particular for any α:

$$H\colon S_1(\alpha) \to S_1(\alpha).$$

This is of course because if $(\sigma, v) \in S_1$, then for some $u \in {}^{\mathrm{lh}\,\sigma}[\omega_1]$, $v = u \circ \pi_\sigma$, so that $H(v) = H(u) \circ \pi_\sigma$, where, since H is increasing, $H(u) \in {}^{\mathrm{lh}\,\sigma}[\omega_1]$. Thus $(\sigma, H(v)) \in S_1$. Since $(\sigma, u) \mapsto H(\sigma, u)$ clearly preserves the relation of proper extension between sequences, we have, letting $D = \mathrm{ran}(H)$:

$$S_1(\alpha) \text{ is not wellfounded } \Rightarrow S_1 \restriction D(\alpha) \text{ is not wellfounded.}$$

So we have shown that for each $D \subseteq \omega_1$, D uncountable, we have for all α:

$$\begin{aligned} \alpha \in A &\Leftrightarrow S_1(\alpha) \text{ is not wellfounded} \\ &\Leftrightarrow S_1 \restriction D(\alpha) \text{ is not wellfounded,} \end{aligned}$$

so that the non-wellfoundedness of $S_1(\alpha)$ depends only on its restriction to any uncountable subset of ω_1. This too will be useful in §2.

NOTE. The construction of S_1 is due to Shoenfield [Sho61]. The homogeneity properties of S_1 have been studied and used by Solovay [Sol78A], Mansfield [Man71] and Martin [Mar71B].

§2. Π^1_2 sets; the tree S_2. *Assume from now on and for the rest of this paper that for all α, $\alpha^\#$ exists.*

2.1. Definition of S_2.

(a) Let $A \subseteq \mathbb{R}$ be Π^1_2. Then for some Π^1_1 set $B \subseteq \mathbb{R} \times \mathbb{R}$ we have

$$\begin{aligned} \alpha \in A &\Leftrightarrow \neg\exists\beta B(\alpha, \beta) \\ &\Leftrightarrow \neg\exists\beta\exists f (\alpha, \beta, f) \in [S_1] \\ &\Leftrightarrow S_1(\alpha) \text{ is wellfounded,} \end{aligned}$$

where S_1 is the tree associated with B as in §1. Strictly speaking we have talked in §1 only about subsets of $\mathbb{R}$ but it is obvious how to modify this discussion so that it applies to Π^1_1 subsets of any $\mathbb{R} \times \mathbb{R} \times \cdots \times \mathbb{R}$. Thus the tree S_1 associated with B will be a tree on $\omega \times \omega \times \omega_1$, so that $S_1(\alpha)$ is a tree on $\omega \times \omega_1$. A typical element of S_1 will be a triple (σ, τ, u), where $\mathrm{lh}\,\sigma = \mathrm{lh}\,\tau = \mathrm{lh}\,u = n$ and $u\colon n \to \omega_1$ is order preserving relative to $<_{\sigma,\tau}$, which is defined as in §1.1 by replacing σ by σ, τ everywhere.

Let now for $\sigma \neq \varnothing$, $\mathrm{lh}\,\sigma = n$:

$$S_1^*(\sigma) \overset{\text{def}}{=} \{(\tau_i, u) : (\sigma \restriction \ell_i, \tau_i, u) \in S_1 \wedge 1 \leq i \leq n\}.$$

Then $(S_1^*(\sigma), <_{\mathrm{BK}})$ is clearly a wellordering as $S_1^*(\sigma)$ contains only sequences of bounded length. We shall denote it by W_σ and its order type by ρ_σ. Note that

$$\varnothing \neq \sigma \subseteq \sigma' \Rightarrow W_\sigma \subseteq W_{\sigma'}.$$

If $S_1(\alpha)$ is wellfounded, let again W_α denote the wellordering $\langle S_1(\alpha), <_{\mathrm{BK}}\rangle$ and ρ_α its order type. As $S_1(\alpha) = \bigcup_{n\geq 1} S_1^*(\alpha\restriction n)$, we have $W_\alpha = \bigcup_{n\geq 1} W_{\alpha\restriction n}$.

One should note now that for each $\sigma \neq \varnothing$,

$$\rho_\sigma = \omega_1.$$

Clearly $\rho_\sigma \geq \omega_1$, since $S_1^*(\sigma)$ is uncountable, as

$$S_1^*(\sigma) \supseteq \{((0), u) : (\sigma\restriction 1, (0), u) \in S_1\} = \{((0), (\xi)) : \xi < \omega_1\},$$

since $\tau_1 = (0)$, $\ell_1 = 1$. On the other hand, if $(\tau_i, u), (\tau_j, v) \in S_1^*(\sigma)$, say with $v = (\xi_0, \dots, \xi_{\ell_i - 1})$, $u = (\eta_0, \dots, \eta_{\ell_j - 1})$, then, since $\ell_j \geq 1$ (as we took $j \geq 1$), we must have $\eta_0 > \eta_1, \dots, \eta_{\ell_j - 1}$, so that if $(\tau_j, u) <_{\mathrm{BK}} (\tau_i, v)$, then $\eta_0 \leq \xi_0$ (recall that use of antilexicographical ordering of pairs in $<_{\mathrm{BK}}$ here), so that $\eta_0, \eta_1, \dots, \eta_{\ell_j - 1} \leq \xi_0$, *i.e.*, there are only countably many predecessors of (τ_i, v). Thus $\rho_\sigma \leq \omega_1$. (The reason we have excluded $(\varnothing, \varnothing)$ from $S_1^*(\sigma)$ was precisely so that $\rho_\sigma = \omega_1$; otherwise it would have been $\omega_1 + 1$, which would be technically awkward.) Similarly if $S_1(\alpha)$ is wellfounded, $\rho_\alpha = \omega_1$.

Let as usual

$$\tilde{\mathbf{L}} = \bigcup_{\alpha \in \mathbb{R}} \mathbf{L}[\alpha].$$

Let also for each wellordering W and set $A \subseteq \mathrm{Ord}$, ${}^W[A]$ be the set of all increasing mappings from (the domain of) W into A. As in this notation, we sometimes do not distinguish between W and its domain, when convenient. To each $h \in {}^{W_\sigma}[\omega_1] \cap \tilde{\mathbf{L}}$, where $\sigma \neq \varnothing$, we will assign a tuple of ordinals $p_\sigma(h) = (\xi_1, \dots, \xi_{\mathrm{lh}\,\sigma})$ as follows:

Since h maps $S_1^*(\sigma)$ into ω_1, it splits into $\mathrm{lh}\,\sigma = n$ many maps $h_1, \dots, h_n$, where $\mathrm{dom}(h_i) = S_1(\sigma\restriction\ell_i, \tau_i)$ and

$$h_i(u) = h(\tau_i, u).$$

Now h_i can be identified with the map $h_i' \colon {}^{\ell_i}[\omega_1] \to \omega_1$ given by

$$h_i'(v) = h_i(v \circ \pi_{\sigma\restriction\ell_i, \tau_i}),$$

since $S_1(\sigma\restriction\ell_i, \tau_i) = {}^{\ell_i}[\omega_1]\pi_{\sigma\restriction\ell_i, \tau_i}$, as we saw in 1.3(a). Put now

$$p_\sigma(h) = (\tilde{h}_i'(\boldsymbol{u}_1, \dots \boldsymbol{u}_{\ell_i}))_{1\leq i\leq n}.$$

We are now ready to define:

$$(\sigma, u) \in S_2 \overset{\text{def}}{\Leftrightarrow} \exists h \in {}^{W_\sigma}[\omega_1] \cap \tilde{\mathbf{L}}(p_\sigma(h) = u) \vee (\sigma = u = \varnothing).$$

(Again S_2 depends on A, B, S_1 but we won't indicate this explicitly.)

We first verify that this is indeed a tree: Let $(\sigma, u) \in S_2$ and $1 \leq n \leq \mathrm{lh}\,\sigma = \mathrm{lh}\,u$. Put $\sigma\restriction n = \sigma'$. We can find $h \in {}^{W_\sigma}[\omega_1] \cap \tilde{\mathbf{L}}$ such that $p_\sigma(h) = u$. Let $h' \in {}^{W_{\sigma'}}[\omega_1] \cap \tilde{\mathbf{L}}$ be defined by $h' = h\restriction W_{\sigma'}$. Note here that S_1, W_σ, $W_{\sigma'}$ are in $\mathbf{L}$, so that $h' \in \tilde{\mathbf{L}}$. Then clearly $p_{\sigma'}(h') = u\restriction n$, therefore $(\sigma\restriction n, u\restriction n) \in S_2$.

(b) We shall verify now that

$$\alpha \in A \Leftrightarrow S_2(\alpha) \text{ is not wellfounded.}$$

Indeed, if $\alpha \in A$, $S_1(\alpha)$ is wellfounded, so let $h\colon W_\alpha \to \omega_1$ be the rank function of W_α, *i.e.*, the unique isomorphism between W_α and ω_1. Clearly $h \in \mathbf{L}[\alpha]$. Then if for each $n \geq 1$, $h^n = h \restriction W_{\alpha\restriction n}$ we have that $p_{\alpha\restriction n}(h^n) \in S_2(\alpha\restriction n)$ and also $p_{\alpha\restriction 1}(h^1) \subseteq p_{\alpha\restriction 2}(h^2) \subseteq \dots$, so $S_2(\alpha)$ is not wellfounded. Conversely, assume $(\eta_1, \eta_2, \dots)$ is an infinite branch through $S_2(\alpha)$. Then for each $n \geq 1$, let ${}^n h \in {}^{W_\alpha\restriction n}[\omega_1] \cap \tilde{\mathbf{L}}$ be such that $p_{\alpha\restriction n}({}^n h) = (\eta_1, \dots, \eta_n)$. Then clearly for all $i \geq 1$ and all $n, m \geq i$, $\widetilde{{}^n h_i'}(\boldsymbol{u}_1, \dots, \boldsymbol{u}_{\ell_i}) = \widetilde{{}^m h_i'}(\boldsymbol{u}_1, \dots, \boldsymbol{u}_{\ell_i})$. So find $C^i_{n,m}$ a cub subset of ω_1 such that for all $v \in {}^{\ell_i}[C^i_{n,m}]$:

$${}^n h_i'(v) = {}^m h_i'(v).$$

This allows us to define the following map q from $S_1 \restriction C(\alpha)$ into the ordinals, where

$$C = \bigcap_{1 \leq i \leq n,m} C^i_{n,m}:$$

$$q(\tau_i, u) = {}^n h_i'(v) = {}^n h(\tau_i, u),$$

where $n \geq i$ and $u = v \circ \pi_{\alpha\restriction \ell_i, \tau_i}$. We now claim that this is an order preserving map from $\langle S_1 \restriction C(\alpha), <_{\mathrm{BK}}\rangle$ into the ordinals. Indeed, if $(\tau_i, u), (\tau_j, w) \in S_1 \restriction C(\alpha)$ and

$$(\tau_i, u) <_{\mathrm{BK}} (\tau_j, w),$$

then $(\tau_i, u), (\tau_j, w) \in S_1^*(\alpha\restriction n)$, for $n \geq i, j$, thus

$${}^n h(\tau_i, u) < {}^n h(\tau_j, w),$$

therefore

$$q(\tau_i, u) < q(\tau_j, w).$$

So $S_1 \restriction C(\alpha)$ is wellfounded and therefore $S_1(\alpha)$ is wellfounded by 1.3(b), *i.e.*, $\alpha \in A$.

2.2. Scales for Π^1_2 sets. We claim now that for each $\alpha \in A$, $S_2(\alpha)$ has an honest leftmost branch. Indeed in the notation of 2.1(b), if $h\colon W_\alpha \to \omega_1$ is as defined there and we let for each $i > 0$, $h_i(u) = h(\tau_i, u)$, for $u \in S_1(\alpha\restriction \ell_i, \tau_i)$ and $\xi_i = \widetilde{h_i'}(\boldsymbol{u}_1, \dots, \boldsymbol{u}_{\ell_i})$, then clearly $(\xi_1, \xi_2, \dots) \in [S_2(\alpha)]$. Moreover if $(\eta_1, \eta_2, \dots)$ is any branch of $[S_2(\alpha)]$, then again in the notation of 2.1(b), q is an order preserving map from $S_1 \restriction C(\alpha)$ into ω_1. Let $H\colon \omega_1 \to C$ be the normal function enumerating C. Then by 1.3 $H\colon S_1(\alpha) \to S_1 \restriction C(\alpha)$ and of course H preserves $<_{\mathrm{BK}}$, so $h(\tau_i, u) \leq q(\tau_i, H(u))$. Now let $D \subseteq C$ be cub such that $H(\xi) = \xi$ for $\xi \in D$. Then $h(\tau_i, u) \leq q(\tau_i, u)$, for $u \in {}^{\ell_i}D$, thus $h_i'(v) \leq {}^n h_i'(v)$, for $v \in {}^{\ell_i}[D]$, so $\xi_i = \widetilde{h_i'}(\boldsymbol{u}_1, \dots, \boldsymbol{u}_{\ell_i}) \leq \widetilde{{}^n h_i'}(\boldsymbol{u}_1, \dots, \boldsymbol{u}_{\ell_i}) = \eta_i$ and we are done.

We can now define for each $\alpha \in A$ and each $i \geq 1$,

$$\varphi_i(\alpha) = f_\alpha(i),$$

where f_α is the leftmost branch of $S_2(\alpha)$ as above. Clearly

$$\bar{\varphi}(\alpha) = (\varphi_1(\alpha), \varphi_2(\alpha), \dots)$$

is a scale on A. We want to show actually that it is a Δ^1_3-scale. For this note that

$$\varphi_i(\alpha) = \widetilde{h'_i}(\boldsymbol{u}_1, \dots, \boldsymbol{u}_{\ell_i}),$$

where $h'_i(v) = \text{rank}_{W_\alpha}(v \circ \pi_{\alpha \restriction \ell_i, \tau_i})$, for $v \in {}^{\ell_i}[\omega_1]$. Thus for $\alpha, \beta \in A$

$$\begin{aligned}\varphi_i(\alpha) \leq \varphi_i(\beta) \Leftrightarrow\ & \text{rank}_{W_\alpha}(v \circ \pi_{\alpha\restriction \ell_i,\tau_i}) \leq \text{rank}_{W_\beta}(v \circ \pi_{\beta\restriction \ell_i,\tau_i}),\\ & \text{for all } v \in {}^{\ell_i}[C], \text{ where } C \text{ is some cub subset of } \omega_1\\ \Leftrightarrow\ & \mathbf{L}[\alpha,\beta] \models \theta_1(\alpha,\beta,v), \text{ for all } v \in {}^{\ell_i}[C], \text{ where } C \subseteq \omega_1\\ & \text{is some cub set,}\end{aligned}$$

(here θ_i is some formula recursively determined from i),

$$\begin{aligned}&\Leftrightarrow \mathbf{L}[\alpha,\beta] \models \theta_i(\alpha,\beta,\boldsymbol{u}_1,\dots,\boldsymbol{u}_{\ell_i})\\ &\Leftrightarrow \langle\alpha,\beta\rangle^\#(\ulcorner\theta_i\urcorner) = 0,\end{aligned}$$

which is obviously Δ^1_3.

Remark 2.1. Note that we can also describe $\varphi_i(\alpha)$ as follows: Let

$$\begin{aligned}\hat{S}_i = \{(\sigma,\tau,u)\colon\ & \text{lh}\,\sigma = \text{lh}\,\tau = \text{lh}\,u \ (= \text{say}, n) \wedge\\ & \sigma,\tau \in {}^{<\omega}\omega \wedge u \in {}^{<\omega}\text{Ord} \wedge\\ & u\colon n \to \text{Ord is order preserving relative to } <_{\sigma,\tau}\}.\end{aligned}$$

Thus $\hat{S}_1$ is the "liftup" of S_1 to all ordinals. Clearly $\hat{S}_1 \restriction \omega_1 = S_1$ and $\hat{S}_1$ is a definable class in $\mathbf{L}$. Now by an easy indiscernibility argument we have for all $i \geq 1$:

$$\varphi_i(\alpha) = f_\alpha(i) = \text{rank}_{\langle \hat{S}_1(\alpha), <_{\text{BK}}\rangle}(\tau_i, (\boldsymbol{u}_1,\dots,\boldsymbol{u}_{\ell_i}) \circ \pi_{\alpha\restriction \ell_i,\tau_i}).$$

2.3. Homogeneity properties of S_2. The tree S_2^-. *From now on and for the rest of this paper (except for a portion of §2.5) assume full* AD. (What we have in mind of course is that we work completely inside $\mathbf{L}[\mathbb{R}]$ granting that $\mathbf{L}[\mathbb{R}] \models$ AD, so that every one of our results below which is sufficiently absolute holds also in the real world.)

(a) For each $A \subseteq \text{Ord}$ and W a wellordering, let ${}^W A\!\uparrow$ denote the subset of ${}^W[A]$ defined as follows (where $W = \langle S, < \rangle$):

$$h \in {}^W A\!\uparrow \Leftrightarrow h \in {}^W[A] \text{ and}$$

(i) $\forall x \in S(h(x) > \sup\{h(y) : y < x\})$
(ii) There is $\{\xi_n^x\}_{x \in S}$ with $\xi_0^x < \xi_1^x < \cdots \to h(x)$, for all $x \in S$.

According to a result of Martin (see [Kec78] for example) we have for any W of order type ω_1 and any $X \subseteq {}^W[\omega_1]$, that there is $C \subseteq \omega_1$ cub such that

$$^W C\uparrow \subseteq X \text{ or } {}^W C\uparrow \subseteq\sim X.$$

This clearly defines a measure (*i.e.*, countably additive ultrafilter) on ${}^W\omega_1\uparrow$.

Since W_σ has order type ω_1 and since $p_\sigma \colon {}^{W_\sigma}[\omega_1] \twoheadrightarrow S_2(\sigma)$, this induces a measure on $S_2(\sigma)$, when $\sigma \neq \varnothing$. For the purpose of making this measure more explicit we shall actually consider the subtree S_2^- of S_2 defined by

$$(\sigma, u) \in S_2^- \Leftrightarrow \exists h \in {}^{W_\sigma}\omega_1\uparrow(p_\sigma(h) = u) \vee (\sigma = u = \varnothing).$$

It is trivial of course to check that also,

$$\alpha \in A \Leftrightarrow S_2^-(\alpha) \text{ is not wellfounded.}$$

Let $\bar{\mathcal{U}}_\sigma$ be the above mentioned measure on ${}^{W_\sigma}\omega_1\uparrow$, *i.e.*,

$$X \in \bar{\mathcal{U}}_\sigma \Leftrightarrow \exists C \subseteq \omega_1(C \text{ cub} \wedge {}^{W_\sigma}C\uparrow \subseteq X),$$

and let $\mathcal{U}_\sigma$ be the measure on $S_2^-(\sigma)$ induced by p_σ, *i.e.*, $\mathcal{U}_\sigma = (p_\sigma)_* \bar{\mathcal{U}}_\sigma$ or explicitly, for $A \subseteq S_2^-(\sigma)$:

$$A \in \mathcal{U}_\sigma \Leftrightarrow \exists C \subseteq \omega_1[C \text{ cub} \wedge p_\sigma[{}^{W_\sigma}C\uparrow] \subseteq A].$$

We shall actually show that $\mathcal{U}_\sigma$ is generated by the sets of the form

$$S_2^-(\sigma) \cap {}^{\mathrm{lh}\,\sigma}\tilde{C},$$

where $C \subseteq \omega_1$ is cub. In other words, we claim that for $A \subseteq S_2^-(\sigma)$:

$$A \in \mathcal{U}_\sigma \Leftrightarrow \exists C \subseteq \omega_1[C \text{ cub} \wedge S_2^-(\sigma) \cap {}^{\mathrm{lh}\,\sigma}\tilde{C} \subseteq A].$$

For that is clearly enough to prove the following:

Lemma 2.2. For any uncountable $C \subseteq \omega_1$,

$$p_\sigma[{}^{W_\sigma}C\uparrow] = S_2^-(\sigma) \cap {}^{\mathrm{lh}\,\sigma}\tilde{C}.$$

Proof. ($\subseteq$): If $h \colon W_\sigma \to C$ is order preserving, then

$$p_\sigma(h) = (\tilde{h}_i'(\boldsymbol{u}_i, \ldots, \boldsymbol{u}_{\ell_i}))_{1 \le i \le \mathrm{lh}\,\sigma},$$

where $h_i'(v) = h(\tau_i, v \circ \pi_{\sigma\restriction\ell_i,\tau_i}) \in C$, so $\tilde{h}_i'(\boldsymbol{u}_1, \ldots, \boldsymbol{u}_{\ell_i}) \in \tilde{C}$, thus $p_\sigma(h) \in S_2^-(\sigma) \cap {}^{\mathrm{lh}\,\sigma}\tilde{C}$.

($\supseteq$) Let $(\xi_1, \ldots, \xi_n) \in S_2^-(\sigma) \cap {}^{\mathrm{lh}\,\sigma}\tilde{C}$. Then for some $h \in {}^{W_\sigma}\omega_1\uparrow$, $p_\sigma(h) = (\xi_1, \ldots, \xi_n)$, *i.e.*, $\xi_i = \tilde{h}_i'(\boldsymbol{u}_1, \ldots, \boldsymbol{u}_{\ell_i})$, where $h_i'(v) = h(\tau_i, v \circ \pi_{\sigma\restriction\ell_i,\tau_i})$, for $v \in {}^{\ell_i}[\omega_1]$. Moreover, $\tilde{h}_i'(\boldsymbol{u}_1, \ldots, \boldsymbol{u}_{\ell_i}) \in \tilde{C}$, so there is $D \subseteq C$, D cub such that

$$h_i'(v) = h(\tau_i, v \circ \pi_{\sigma\restriction\ell_i,\tau_i}) \in C,$$

for all $v \in {}^{\ell_i}[D]$. Let $H : \omega_1 \to D$ be the normal enumeration of D. Put

$$g(\tau_i, u) = h(\tau_i, H(u)).$$

Then $g \in {}^{W_\sigma}[C]$. Also if $g_i(u) = g(\tau_i, u)$ and $g_i'(v) = g_i(v \circ \pi_{\sigma\restriction\ell_i,\tau_i})$, for $v \in {}^{\ell_i}[\omega_1]$, then $g_i'(v) = h_i'(v)$ for $v \in {}^{\ell_i}[E]$, where E is cub and $\xi \in E \Rightarrow H(\xi) = \xi$. Thus $\xi_i = \tilde{h}_i'(\boldsymbol{u}_1, \dots, \boldsymbol{u}_{\ell_i}) = \tilde{g}_i'(\boldsymbol{u}_1, \dots, \boldsymbol{u}_{\ell_i}) \in C$.

So it only remains to show that actually $g \in {}^{W_\sigma}C\uparrow$. Recalling the definition of ${}^{W_\sigma}C\uparrow$ before, it is clear that condition (ii) holds for g. So it is enough to verify (i), *i.e.*, that for $x \in S_1^*(\sigma)$ we have

$$g(x) > \sup\{g(y) : y \in S_1^*(\sigma) \wedge y <_{\mathrm{BK}} x\}.$$

If $x = (\tau_i, u)$, then

$$\begin{aligned} g(\tau_i, u) &= h(\tau_i, H(u)) \\ &> \sup\{h(\tau_j, w) : (\tau_j, w) \in S_1^*(\sigma) \wedge (\tau_j, w) <_{\mathrm{BK}} (\tau_i, H(u))\} \\ &\geq \sup\{h(\tau_j, H(v)) : (\tau_j, v) \in S_1^*(\sigma) \wedge (\tau_j, v) <_{\mathrm{BK}} (\tau_i, u)\} \\ &= \sup\{g(\tau_j, v) : (\tau_j, v) \in S_1^*(\sigma) \wedge (\tau_j, v) <_{\mathrm{BK}} (\tau_i, u)\} \end{aligned}$$

and we are done. ⊣ (Lemma 2.2)

(b) We examine now a further homogeneity property of S_2^-. Let $H : \omega_1 \to \omega_1$ be normal. Then if $h \in {}^{W_\sigma}\omega_1\uparrow$ clearly $H \circ h \in {}^{W_\sigma}\omega_1\uparrow$ too (this does not happen in general if H is just increasing). Now it is easy to check that

$$p_\sigma(H \circ h) = \tilde{H}(p_\sigma(h)).$$

Indeed, if $p_\sigma(h) = (\xi_1, \dots, \xi_n)$, where $\xi_i = \tilde{h}_i'(\boldsymbol{u}_1, \dots, \boldsymbol{u}_{\ell_i})$, then

$$\tilde{H}(\xi_i) = \widetilde{H \circ h_i'}(\boldsymbol{u}_1, \dots, \boldsymbol{u}_{\ell_i}) = \widetilde{(H \circ h)_i'}(\boldsymbol{u}_1, \dots, \boldsymbol{u}_{\ell_i}).$$

Thus $\tilde{H}$ maps S_2^- into S_2^- (in the usual sense that if $(\tau, u) \in S_2^-$, then $(\tau, \tilde{H}(u)) \in S_2^-$), where

$$\tilde{H}(\xi_1, \dots, \xi_n) = (\tilde{H}(\xi_1), \dots, \tilde{H}(\xi_n)).$$

As $\tilde{H}$ obviously preserves the proper extension relation among sequences, we have for any cub $C \subseteq \omega_1$:

$$\begin{aligned} \alpha \in A &\Leftrightarrow S_2^-(\alpha) \text{ is not wellfounded} \\ &\Leftrightarrow S_2^- \restriction \tilde{C}(\alpha) \text{ is not wellfounded.} \end{aligned}$$

Remark 2.3. It is easy also to check that S_2 is preserved under any $\tilde{H}$, where $H : \omega_1 \to \omega_1$ is just order preserving, and so for any unbounded $C \subseteq \omega_1$: $\alpha \in A \Leftrightarrow S_2(\alpha)$ is not wellfounded $\Leftrightarrow S_2 \restriction \tilde{C}(\alpha)$ is not wellfounded.

2.4. Some definability estimates for S_2^-. Consider the structure

$$\mathcal{Q}_3 = \langle \boldsymbol{u}_\omega, <, \{\boldsymbol{u}_n\}_{n<\omega}\rangle,$$

as in Kechris-Martin [KM78]. (Recall that since we are working with AD, $\boldsymbol{u}_n = \omega_n, \forall n \leq \omega$.) We want to show that:

(i) S_2^- is Δ_1^1 on $\mathcal{Q}_3$.
(ii) The second order relation

$$A \subseteq S_2^-(\sigma) \wedge A \in \mathcal{U}_\sigma$$

is also Δ_1^1 on $\mathcal{Q}_3$.

Now (ii) follows immediately from (i) as for $A \subseteq S_2^-(\sigma)$:

$$\begin{aligned} A \in \mathcal{U}_\sigma &\Leftrightarrow \exists C(C \subseteq \omega_1 \text{ is cub } \wedge {}^{\mathrm{lh}\,\sigma}\tilde{C} \cap S_2^-(\sigma) \subseteq A) \\ &\Leftrightarrow \forall C(C \subseteq \omega_1 \text{ is cub } \Rightarrow {}^{\mathrm{lh}\,\sigma}\tilde{C} \cap A \neq \varnothing). \end{aligned}$$

It is needed of course to verify here that the map

$$X \mapsto \tilde{X}$$

for $X \subseteq \omega_1$, is also Δ_1^1, or equivalently that the second order relation

$$X \subseteq \omega_1 \wedge \xi \in \tilde{X}$$

is Δ_1^1 on $\mathcal{Q}_3$. This follows from the fact that for $X \subseteq \omega_1$:

$$\begin{aligned} \xi \in \tilde{X} \Leftrightarrow \exists\alpha\exists \text{ term } t \text{ such that} & \\ & [t^{\mathbf{L}[\alpha]}(\boldsymbol{u}_1, \dots, \boldsymbol{u}_{r(t)}) = \xi \\ & \wedge \exists C \subseteq \omega_1(C \text{ cub} \\ & \wedge \forall v \in {}^{r(t)}[C](t^{\mathbf{L}[\alpha]}(v) \in X))] \\ \Leftrightarrow & \forall\alpha\forall \text{ term } t[t^{\mathbf{L}[\alpha]}(\boldsymbol{u}_1, \dots, \boldsymbol{u}_{r(t)}) = \xi \\ \Rightarrow & \forall C \subseteq \omega_1(C \text{ cub} \Rightarrow \exists v \in {}^{r(t)}[C](t^{\mathbf{L}[\alpha]}(v) \in X))]. \end{aligned}$$

So it is enough to prove (i). For that let us say that a sequence $(\xi_1, \dots, \xi_n)$, where $\xi_i < \boldsymbol{u}_{\ell_i+1}$, has the **gap property** if there are f_i with $\tilde{f}_i(\boldsymbol{u}_1, \dots, \boldsymbol{u}_{\ell_i}) = \xi_i$ and there is cub $D \subseteq \omega_1$ such that for all $1 \leq i \leq n$ and all $v \in {}^{\ell_i}[D]$:

$$f_i(v) > \sup\{f_j(w) : f_j(w) < f_i(v) \wedge w \in {}^{\ell_j}[D]\}.$$

Note that that is independent of the particular choice of f_i's which represent the ξ_i's. Now we have the equivalence (for $\sigma \neq \varnothing$)

$$(\sigma, (\xi_1, \dots, \xi_n)) \in S_2^- \Leftrightarrow \mathrm{lh}\,\sigma = n$$
$$\& \text{ (a) } \xi_i < \boldsymbol{u}_{\ell_i+1} \wedge \mathrm{cf}(\xi_i) = \omega$$
$$\& \text{ (b) } (\xi_1, \dots, \xi_n) \text{ has the gap property}$$
$$\& \text{ (c) for any (all) } f_1, \dots, f_n \text{ such that } \tilde{f}_i(\boldsymbol{u}_1, \dots, \boldsymbol{u}_{\ell_1}) = \xi_i, \text{ if we let } h(\tau_i, v \circ \pi_{\sigma\restriction\ell_i,\tau_i}) = f_i(v), \text{ then for some cub set } C \subseteq \omega_1, h\restriction(S_1^*\restriction C(\sigma)) \text{ is order preserving relative to } <_{\mathrm{BK}}.$$

The proof of this equivalence relation is similar to the proof of Lemma 2.2.

Note now that (a) is easily Δ_1^1 on $\mathcal{Q}_3$, while in (c) C can be taken to be equal to I_α (= set of Silver indiscernibles of $\mathbf{L}[\alpha]$ below ω_1), for any α such that $f_1, \dots, f_n \in \mathbf{L}[\alpha]$, thus (c) is also Δ_1^1 in $\mathcal{Q}_3$. It only remains to check that (b) is Δ_1^1 in $\mathcal{Q}_3$. For that is clearly enough to show that in the definition of gap property one can take D to be I'_α = all limit points of I_α, for any real α such that $f_1, \dots, f_n \in \mathbf{L}[\alpha]$. This is exactly what is proved by an indiscernibility argument in Part A8, Lemma 1 of Solovay [Sol78A] and we won't repeat the argument.

2.5. An alternative tree S_2^+. We shall now describe an alternative version of a tree for a given Π_2^1 set A. This version is relevant to Martin's recent proof of Determinacy($\boldsymbol{\Pi}_2^1$) from very large cardinals. We assume only $\forall\alpha(\alpha^\#$ exists) until further notice.

In the notation of §2.1 define (for $\sigma \neq \varnothing$)

$$(\sigma, (\xi_1, \dots, \xi_n)) \in S_2^+ \Leftrightarrow$$
$$\mathrm{lh}\,\sigma = n \text{ and}$$
$$\text{(i) } \xi_i < \boldsymbol{u}_{\ell_i+1} \text{ and}$$
$$\text{(ii) there are } f_i \colon {}^{\ell_i}[\omega_1] \to \omega_1 \text{ in } \tilde{\mathbf{L}} \text{ such that } \tilde{f}_i(\boldsymbol{u}_1, \dots, \boldsymbol{u}_{\ell_i}) = \xi_i \text{ and if we let } h(\tau_i, v \circ \pi_{\sigma\restriction\ell_i,\tau_i}) = f_i(v), \text{ then for some cub } C \subseteq \omega_1 \text{ and all } u \in {}^{\ell_j}C,\ w \in {}^{\ell_i}C, \text{ if } (\tau_j, u), (\tau_i, w) \in S_1^*(\sigma), \text{ then: } (\tau_j, u) \prec^1 (\tau_i, w) \Rightarrow h(\tau_j, u) < h(\tau_i, w),$$

where

$$(\tau, u) \prec^1 (\tau', u') \Leftrightarrow$$
$$\tau \text{ is a 1-point extension of } \tau' \text{ and } u \text{ is a 1-point extension of } u'.$$

Clearly, S_2 is a subtree of S_2^+.

Now it is not hard to check as in §2.1 that

$$\alpha \in A \Leftrightarrow S_2^+(\alpha) \text{ is not wellfounded,}$$

and that for each $\alpha \in A$, $S_2^+(\alpha)$ has an honest leftmost branch, say f_α, and that if $\bar{\varphi}(\alpha) = f_\alpha$, then $\bar{\varphi}$ is a Δ_3^1-scale on A.

Our next goal will be to get an explicit description of the condition $(\xi_1, \dots, \xi_n) \in S_2^+(\sigma)$ in terms of the ordinals themselves instead of their representing functions.

For that recall that for $(\sigma, (\xi_1, \dots, \xi_n)) \in S_2^+$ there must be functions $f_i \in \tilde{\mathbf{L}}$, $f_i \colon {}^{\ell_i}[\omega_1] \to \omega_1$ such that $\tilde{f}_i(\boldsymbol{u}_1, \dots, \boldsymbol{u}_{\ell_i}) = \xi_i$ and for some cub $C \subseteq \omega_1$ and all $v \in {}^{\ell_j}[C]$, $v' \in {}^{\ell_i}[C]$ we have

$$(\tau_j, v \circ \pi_{\sigma \restriction \ell_j, \tau_j}) \prec^1 (\tau_i, v' \circ \pi_{\sigma \restriction \ell_i, \tau_i}) \Rightarrow f_j(v) < f_i(v'). \qquad (*)$$

Now the hypothesis of $(*)$ implies that τ_j is a 1-point extension of τ_i, so that in particular $\ell_j = \ell_i + 1$ and moreover if we put for simplicity $\pi_j = \pi_{\sigma \restriction \ell_j, \tau_j}$ and similarly for π_i, we must also have that

$$v_{\pi_j(k)} = v'_{\pi_i(k)}, \forall k < \ell_i$$

or equivalently

$$v_{\pi_j \circ \pi_i^{-1}(k)} = v'_k, \forall k < \ell_i.$$

Note now that from the following commutative diagram of order preserving maps:

$$\begin{array}{ccc} \langle \ell_j, <_{\sigma \restriction \ell_j, \tau_j} \rangle & \xrightarrow{\ \pi_j\ } & \langle \ell_j, < \rangle \\ {\scriptstyle \text{inclusion}} \uparrow & & \uparrow {\scriptstyle \pi_j \circ \pi_i^{-1}} \\ \langle \ell_i, <_{\sigma \restriction \ell_i, \tau_i} \rangle & \xrightarrow[\ \pi_i\ ]{} & \langle \ell_i, < \rangle \end{array}$$

we must have that

$$\pi_j \circ \pi_i^{-1} \colon \ell_i \to \ell_j$$

is order preserving, *i.e.*, for some

$$m = m(\sigma, i, j) \leq \ell_i$$

$$\pi_j \circ \pi_i^{-1}(k) = \begin{cases} k & \text{if } 0 \leq k < m \\ k+1 & \text{if } m \leq k < \ell_i - 1. \end{cases}$$

Thus $(v'_0, \dots, v'_{\ell_i - 1}) = (v_0, \dots, \hat{v}_m, \dots, v_{\ell_j - 1})$, where $\hat{v}_m$ signifies the fact that v_m is omitted.

Recall now that for each $m \geq 1$ we have the following embedding

$$j_m \colon \boldsymbol{u}_\omega \to \boldsymbol{u}_\omega,$$

where

$$j_m(\boldsymbol{u}_n) = \begin{cases} \boldsymbol{u}_n & \text{if } n < m, \\ \boldsymbol{u}_{n+1} & \text{if } n \geq m, \end{cases}$$

and $j_m(\tilde{f}(\boldsymbol{u}_1,\dots,\boldsymbol{u}_t)) = \tilde{f}(j_m(\boldsymbol{u}_1),\dots,j_m(\boldsymbol{u}_t))$. Then an easy indiscernibility argument plus the above analysis easily yields that (for $\sigma \neq \varnothing$)

$$\begin{aligned}(\sigma,(\xi_1,\dots,\xi_n)) \in S_2^+ \Leftrightarrow\ & \mathrm{lh}\,\sigma = n \\ & \&\ \text{(i) } \xi_i < \boldsymbol{u}_{\ell_i+1} \\ & \&\ \text{(ii) For all } 1 \leq i,j \leq n\text{:} \\ & \tau_j \prec^1 \tau_i \Rightarrow \xi_j < j_{m(\sigma,i,j)+1}(\xi_i).\end{aligned} \qquad (**)$$

This is the particular form that is relevant in Martin's proof.

From this explicit form of S_2^+ and *using again full* AD one can easily check that each $S_2^+(\sigma)$ is a finite union of Kunen sets $A^t_{m,n}$, where $m = \max_{1\leq i\leq n} \ell_i$ and $n = \mathrm{lh}\,\sigma$. The notion and the notation involved here are as in Solovay [Sol78A]. Now each of these $A^t_{m,n}$ carries a canonical measure generated by the sets of the form ${}^n\tilde{C} \cap A^t_{m,n}$, with $C \subseteq \omega_1$ cub; see again Solovay [Sol78A]. This establishes a homogeneity property of S_2^+. It is relevant to notice here that $S_2^-(\sigma)$ is exactly *one* of the sets $A^t_{m,n}$ that get into $S_2^+(\sigma)$, so that the passage from S_2^+ to S_2^- has the effect of canonically choosing from the finitely many Kunen sets $A^t_{m,n}$ involved in each $S_2^+(\sigma)$, exactly one which is then equal to $S_2^-(\sigma)$. Although one could write a description of each $S_2^-(\sigma)$ using the embeddings j_m as in $(**)$ it would be a bit messy and not as elegant or useful as $(**)$ itself.

As a final comment we mention that it would be easy to show again that S_2^+ has also the following homogeneity property: For all unbounded $C \subseteq \omega_1$:

$$\begin{aligned}\alpha \in A &\Leftrightarrow S_2^+(\alpha) \text{ is not wellfounded} \\ &\Leftrightarrow S_2^+ \restriction \tilde{C}(\alpha) \text{ is not wellfounded.}\end{aligned}$$

NOTE. The construction of S_2 is due to Mansfield [Man71], Martin [Mar71B] following work of Martin-Solovay [MS69]. The homogeneity properties of these trees have been studied and used by Kunen [Kun71B] and Martin.

§3. Π^1_3 sets; the tree S_3.

3.1. Definition of S_3.

(a) Let now $A \subseteq \mathbb{R}$ be Π^1_3. Then for some $B \in \Pi^1_2$,

$$\begin{aligned}\alpha \in A &\Leftrightarrow \neg\exists\beta B(\alpha,\beta) \\ &\Leftrightarrow \neg\exists\beta\exists f(\alpha,\beta,f) \in [S_2^-] \\ &\Leftrightarrow S_2^-(\alpha) \text{ is wellfounded.}\end{aligned}$$

Let again for $\sigma \neq \varnothing$, $\mathrm{lh}\,\sigma = n$:

$$S_2^*(\sigma) = \{(\tau_i, u) : (\sigma\restriction\ell_i, \tau_i, u) \in S_2^- \wedge i < n\}.$$

Note that we allow here $(\varnothing, \varnothing) \in S_2^*(\sigma)$, while we have excluded it from $S_1^*(\sigma)$: Again for each such σ, let W_σ denote the wellordering $\langle S_2^*(\sigma), <_{\mathrm{BK}}\rangle$ and ρ_σ its order type. Then ρ_σ is a successor ordinal (as $(\varnothing, \varnothing)$ is the top element of it) and $\rho_\sigma < \boldsymbol{u}_{k+1}$, where $k = \max\{\ell_i + 1 : i < n\}$. If $\alpha \in A$ let W_α be the wellordering $\langle S_2^-(\alpha), <_{\mathrm{BK}}\rangle$ and ρ_α its order type. Again $W_\alpha = \bigcup_{n\geq 1} W_{\alpha\restriction n}$. Note also that $\rho_\alpha < (\boldsymbol{u}_\omega)^+ = \omega_{\omega+1}$.

Remark 3.1. We have been using the same notation W_σ, ρ_σ in both §2 and in the present §3, although it would be more accurate to distinguish them by superscripts: $W_\sigma^1, \rho_\sigma^1, W_\sigma^2, \rho_\sigma^2$. However, in §3 we will be only using the present $W_\sigma = W_\sigma^2$, $\rho_\sigma = \rho_\sigma^2$, so there will be no danger of confusion. Similar remarks will apply also to W_α, ρ_α, p_σ (to be defined below) and in the subsequent sections.

Given now $h \in {}^{W_\sigma}[\omega_{\omega+1}]$, let h_i, for $i < n = \mathrm{lh}\,\sigma$, be defined as follows: The domain of h_i is $S_2^-(\sigma\restriction\ell_i, \tau_i)$ and for $v \in S_2^-(\sigma\restriction\ell_i, \tau_i)$:

$$h_i(v) = h(\tau_i, v).$$

Thus for $i = 0$, h_0 is the single ordinal $h(\varnothing, \varnothing)$. Recall now from §2 that $S_2^-(\sigma\restriction\ell_i, \tau_i)$ carries the measure

$$\mathcal{U}_{\sigma\restriction\ell_i,\tau_i} \equiv \mathcal{U}_{\sigma,i} \text{ (for simplicity)}$$

and let

$$\xi_0 = h_0,$$
$$\xi_i = [h_i]_{\mathcal{U}_{\sigma,i}}.$$

Finally put

$$p_\sigma(h) = (\xi_0, \xi_1, \dots, \xi_{n-1}).$$

Then define the tree S_3 by:

$$(\sigma, u) \in S_3 \Leftrightarrow \exists h \in {}^{W_\sigma}[\omega_{\omega+1}](p_\sigma(h) = u) \vee (\sigma = u = \varnothing).$$

(b) We now show that

$$\alpha \in A \Leftrightarrow S_3(\alpha) \text{ is not wellfounded.}$$

First, if $\alpha \in A$, then $S_2^-(\alpha)$ is wellfounded, so let $h\colon W_\alpha \to \omega_{\omega+1}$ be the unique isomorphism between W_α and an initial segment of $\omega_{\omega+1}$, which is of course equal to ρ_α. Let $h_i(v) = h(\tau_i, v)$, for $i > 0$ and $v \in S_2(\alpha\restriction\ell_i, \tau_i)$ and let $\xi_i = [h_i]_{\mathcal{U}_{\alpha\restriction n,i}}$ for any $n > i$. Let also $\xi_0 = h(\varnothing, \varnothing)$. Then $(\xi_0, \dots, \xi_{n-1}) = p_{\alpha\restriction n}(h^n)$, where $h^n = h\restriction W_{\alpha\restriction n} \in {}^{W_{\alpha\restriction n}}[\omega_{\omega+1}]$, so that $(\xi_0, \dots, \xi_{n-1}) \in S_3(\alpha\restriction n)$, *i.e.*, $(\xi_0, \xi_1, \dots) \in S_3(\alpha)$. Conversely, assume $(\eta_0, \eta_1, \dots) \in [S_3(\alpha)]$. For each $n > 0$, let ${}^n h \in {}^{W_{\alpha\restriction n}}[\omega_{\omega+1}]$ be such that $p_{\alpha\restriction n}({}^n h) = (\eta_0, \eta_1, \dots, \eta_{n-1})$. Then for all $n > 0$, ${}^n h(\varnothing, \varnothing) = \eta_0$ and for all $i > 0$ and all $m, n > i$

$$[{}^m h_i]_{\mathcal{U}_{\alpha\restriction m,i}} = [{}^n h_i]_{\mathcal{U}_{\alpha\restriction n,i}};$$

where $\mathcal{U}_{\alpha\restriction m,i} = \mathcal{U}_{\alpha\restriction n,i}$ of course. By the results in §2.3, we can now find $C^i_{n,m} \subseteq \omega_1$ cub such that

$$v \in {}^{\ell_i}\widetilde{C^i_{n,m}} \cap S_2^-(\alpha\restriction\ell_i, \tau_i) \Rightarrow {}^m h_i(v) = {}^n h_i(v).$$

Let $C = \bigcap_{0<i<n,m} C^i_{n,m}$. Then $C \subseteq \omega_1$ is cub and the conclusion of $(*)$ in §2.5 holds for all $0 < i < n, m$ and all $v \in {}^{\ell_i}\tilde{C} \cap S_2^-(\alpha\restriction\ell_i, \tau_i)$. This allows us to define the following map q from $S_2^-\restriction\tilde{C}(\alpha)$ into the ordinals:

$$q(\tau_i, v) = {}^n h_i(v),$$

where $n > i$. We now claim that this is an order preserving map from $\langle S_2^-\restriction\tilde{C}(\alpha), <_{\mathrm{BK}}\rangle$ into the ordinals. Indeed, if $(\tau_i, u), (\tau_j, w) \in S_2^-\restriction\tilde{C}(\alpha)$ and

$$(\tau_i, u) <_{\mathrm{BK}} (\tau_j, w),$$

then $(\tau_i, u), (\tau_j, w) \in S_2^*(\alpha\restriction n)$ for $n > i, j$, thus

$${}^n h(\tau_i, u) < {}^n h(\tau_j, w),$$

i.e.,

$${}^n h_i(u) < {}^n h_j(w),$$

thus

$$q(\tau_i, u) < q(\tau_j, w).$$

So $S_2^-\restriction\tilde{C}(\alpha)$ is wellfounded, thus $\alpha \in A$ by 2.3(b).

(c) We note also the following two basic properties of S_3:

(i) S_3 is a tree on $\omega \times \omega_{\omega+1}$, *i.e.*, all the ordinals occurring in it are $< \omega_{\omega+1}$. This is because by a result of Kunen (see [Kun71B]) if $\mathcal{U}$ is a measure on any set I of cardinality $< \omega_{\omega+1}$, then for any $f : I \to \omega_{\omega+1}$, $[f]_{\mathcal{U}} < \omega_{\omega+1}$.

(ii) Let for any measure $\mathcal{U}$ on a set I, $i^{\mathcal{U}} : \mathrm{Ord} \to \mathrm{Ord}$ be the embedding it generates, *i.e.*,

$$\begin{aligned} i^{\mathcal{U}}(\xi) &= \sup\{[f]_{\mathcal{U}} : f : I \to \xi\}. \\ &= [C_\xi]_{\mathcal{U}}, \text{ for } C_\xi \text{ the constant } \xi \text{ function.} \end{aligned}$$

Then we claim that if $\mathrm{lh}\,\sigma = n \neq 0$:

$$(\sigma, (\xi_0, \dots, \xi_{n-1}) \in S_3 \Rightarrow \xi_i < i^{\mathcal{U}_{\sigma,i}}(\xi_0), \text{ for } i > 0.$$

This is because for $i > 0$, $\xi_i = [h_i]_{\mathcal{U}_{\sigma,i}}$, where $h\colon W_\sigma \to \omega_{\omega+1}$ is order preserving and $h_i(v) = h(\tau_i, v)$, so that

$$\xi_0 = h(\varnothing, \varnothing) > h_i(v), \text{ for all } i > 0,$$

as $(\tau_i, v) <_{\mathrm{BK}} (\varnothing, \varnothing)$. Thus

$$i^{\mathcal{U}_{\sigma,i}}(\xi_0) > [h_i]_{\mathcal{U}_{\sigma,i}}.$$

This is analogous to a property of S_1 which we established in §1, and it will be useful in §4.

3.2. Scales for Π^1_3 sets. As usual we verify now that if $\alpha \in A$, then $S_3(\alpha)$ has an honest leftmost branch. For that, in the notation of 3.1(b), if $h\colon W_\alpha \to \omega_{\omega+1}$ is as defined there and we let h_i and ξ_i be again as defined there, we have $(\xi_0, \xi_1, \dots) \in [S_3^-(\alpha)]$. Moreover, if $(\eta_0, \eta_1, \dots)$ is any branch of $S_3^-(\alpha)$ then, again in the notation of 3.1(b), q is an order preserving map from $S_2^- \restriction \tilde{C}(\alpha)$ into $\omega_{\omega+1}$. Let $H\colon \omega_1 \to C$ be the normal function enumerating C. Then by 2.3, $\tilde{H}\colon S_2^-(\alpha) \to S_2^- \restriction \tilde{C}(\alpha)$, and of course $\tilde{H}$ preserves $<_{\mathrm{BK}}$, so $h(\tau_i, u) \leq q(\tau_i, \tilde{H}(u))$. Now let $D \subseteq C$ be cub such that $H(\xi) = \xi$ for $\xi \in D$. Then $\tilde{H}(u) = u$ for $u \in {}^{<\omega}\tilde{D}$, so $h(\tau_i, u) \leq q(\tau_i, u)$ for $u \in {}^{\ell_i}\tilde{D}$, therefore $h_i(u) \leq {}^n h_i(u)$ for $u \in {}^{\ell_i}\tilde{D} \cap S_2^-(\alpha \restriction n, \tau_i)$, if $n > i$, thus $[h_i]_{\mathcal{U}_{\alpha\restriction n,i}} = \xi_i \leq [{}^n h_i]_{\mathcal{U}_{\alpha\restriction n,i}} = \eta_i$ for $i > 0$. Also $\xi_0 = h(\varnothing, \varnothing) \leq q(\varnothing, \varnothing) = {}^n h(\varnothing, \varnothing) = \eta_0$, *i.e.*, $\xi_i \leq \eta_i$, $\forall i \geq 0$ and we are done.

This implies now that if for each $\alpha \in A$ we put

$$\varphi_i(\alpha) = f_\alpha(i),$$

where f_α is the leftmost branch of $S_3(\alpha)$, then $\bar{\varphi} = \{\varphi_i\}$ is an $\omega_{\omega+1}$-scale on A. By modifying this slightly (for reasons that will become apparent in a moment), we will obtain a Π^1_3-scale on A. Indeed put for $\alpha \in A$:

$$\psi_i(\alpha) = \langle \varphi_0(\alpha), \bar{\alpha}(i), \varphi_i(\alpha) \rangle,$$

where $\langle \xi, \eta, \theta \rangle$ refers to the ordinal associated to the triple (ξ, η, θ) in the lexicographical ordering of ${}^3(\omega_{\omega+1})$. Now we claim that $\bar{\psi} = \{\psi_i\}$ is a Π^1_3-scale on A. For that just note that for $\alpha, \beta \in A$:

$$\begin{aligned}\psi_i(\alpha) \leq \psi_i(\beta) \Leftrightarrow \varphi_0(\alpha) < \varphi_0(\beta) &\vee [\varphi_0(\alpha) = \varphi_0(\beta) \wedge \bar{\alpha}(i) < \bar{\beta}(i)] \\ &\vee [\varphi_0(\alpha) = \varphi_0(\beta) \wedge \bar{\alpha}(i) = \bar{\beta}(i) \\ &\wedge [\varphi_i(\alpha) \leq \varphi_i(\beta)].\end{aligned}$$

So if $\beta \in A$:

$$\begin{aligned}\alpha \in A \wedge \psi_i(\alpha) \leq \psi_i(\beta) \Leftrightarrow {} & [\alpha \in A \wedge \varphi_0(\alpha) < \varphi_0(\beta)] \vee \\ & [(\alpha \in A \wedge \varphi_0(\alpha) = \varphi_0(\beta)) \wedge \bar{\alpha}(i) < \bar{\beta}(i)] \vee \\ & [(\alpha \in A \wedge \varphi_0(\alpha) = \varphi_0(\beta)) \wedge \bar{\alpha}(i) = \bar{\beta}(i) \\ & \wedge \varphi_i(\alpha) \leq \varphi_i(\beta)],\end{aligned}$$

for which we have to calculate that it is Δ^1_3 uniformly in β. But by the results of Kechris-Martin [KM78], it is enough to show that it is Δ^1_1 over $\mathcal{Q}_3$, uniformly in β. To check this notice that

$$\begin{aligned}\alpha \in A \wedge \varphi_0(\alpha) \leq \varphi_0(\beta) &\Leftrightarrow \text{There is an embedding form } \langle S_2^-(\alpha), <_{\mathrm{BK}}\rangle \text{ into } \langle S_2^-(\beta), <_{\mathrm{BK}}\rangle \\ &\Leftrightarrow \langle S_2^-(\alpha), <_{\mathrm{BK}}\rangle \text{ is a wellordering and there is no embedding of } \langle S_2^-(\beta), <_{\mathrm{BK}}\rangle \text{ into a proper initial segment of } \langle S_2^-(\alpha), <_{\mathrm{BK}}\rangle.\end{aligned}$$

Since in §2.4 we have shown that S_2^- is Δ_1^1 in $\mathcal{Q}_3$ the above equivalences show that "$\alpha \in A \wedge \varphi_0(\alpha) \leq \varphi_0(\beta)$" is Δ_1^1 in $\mathcal{Q}_3$, uniformly in β. A similar calculation applies to the predicate "$\alpha \in A \wedge \varphi_0(\alpha) = \varphi_0(\beta)$." So to complete this proof, it is sufficient to check that in case $\alpha \in A$, $\beta \in A$ and $\bar{\alpha}(i) = \bar{\beta}(i)$, the predicate

$$\varphi_i(\alpha) \leq \varphi_i(\beta), \text{ for } i > 0$$

is (uniformly in i, α, β) Δ_1^1 in $\mathcal{Q}_3$. Recall that for $i > 0$

$$\varphi_i(\alpha) = f_\alpha(i) = [\lambda u.\, \mathrm{rank}_{W_\alpha}(\tau_i, u)]_{\mathcal{U}_{\alpha\restriction \ell_i, i}};$$

here the u varies over $S_2^-(\alpha\restriction\ell_i, \tau_i)$. Since $\alpha\restriction i = \beta\restriction i$ and $i \geq \ell_i$, we clearly have that $S_2^-(\alpha\restriction\ell_i, \tau_i) = S_2^-(\beta\restriction\ell_i, \tau_i)$ and $\mathcal{U}_{\alpha\restriction\ell_i, i} = \mathcal{U}_{\beta\restriction\ell_i, i}$. Then

$$\varphi_i(\alpha) \leq \varphi_i(\beta) \Leftrightarrow \{u \in S_2^-(\alpha\restriction\ell_i, \tau_i) : \mathrm{rank}_{W_\alpha}(\tau_i, u) \leq \mathrm{rank}_{W_\beta}(\tau_i, u)\} \in \mathcal{U}_{\alpha\restriction\ell_i, i}.$$

Since as above the relation

$$\mathrm{rank}_{W_\alpha}(\tau_i, u) \leq \mathrm{rank}_{W_\beta}(\tau_i, u)$$

is Δ_1^1 in $\mathcal{Q}_3$, uniformly in all the parameters involved, the results in §2.4 imply that "$\varphi_i(\alpha) \leq \varphi_i(\beta)$" is Δ_1^1 in $\mathcal{Q}_3$, uniformly in α, β as above.

3.3. Homogeneity properties of S_3. The tree S_3^-.

(a) By analogy with the work in §2.3, we define a subtree S_3^- of S_3 as follows:

$$(\sigma, u) \in S_3^- \Leftrightarrow \exists h \in {}^{W_\sigma}\omega_{\omega+1}\!\uparrow (p_\sigma(h) = u) \vee \sigma = u = \emptyset.$$

Again we have: $\alpha \in A \Leftrightarrow S_3^-(\alpha)$ is not wellfounded.

By a result of Kunen [Kun71B], we have

$$\omega_{\omega+1} \to (\omega_{\omega+1})^\rho_\nu, \forall \rho, \nu < \omega_{\omega+1}.$$

This implies that we have the following $\omega_{\omega+1}$-additive measure $\bar{\mathcal{V}}_\sigma$ on ${}^{W_\sigma}\omega_{\omega+1}\!\uparrow$:

$$X \in \bar{\mathcal{V}}_\sigma \Leftrightarrow \exists C \subseteq \omega_{\omega+1}(C \text{ cub} \wedge {}^{W_\sigma}C\!\uparrow \subseteq X).$$

Since

$$p_\sigma \colon {}^{W_\sigma}\omega_{\omega+1}\!\uparrow \twoheadrightarrow S_3^-(\sigma),$$

this induces the $\omega_{\omega+1}$-additive measure

$$(p_\sigma)_* \bar{\mathcal{V}}_\sigma = \mathcal{V}_\sigma$$

on $S_3^-(\sigma)$, for $\sigma \neq \varnothing$. Thus for $A \subseteq S_3^-(\sigma)$:

$$A \in \mathcal{V}_\sigma \Leftrightarrow \exists C \subseteq \omega_{\omega+1}(C \text{ cub } \wedge p_\sigma[{}^{W_\sigma}C\uparrow] \subseteq A).$$

We shall now try to get a more explicit form of this measure $\mathcal{V}_\sigma$. This will be based on the analog of Lemma 2.2.

Let $\mathcal{U}$ be a measure on a set I of cardinality $< \omega_{\omega+1}$. Then, as we mentioned before, $i^{\mathcal{U}}(\omega_{\omega+1}) = \omega_{\omega+1}$, where $i^{\mathcal{U}}$ is the associated embedding generated by $\mathcal{U}$. For each $X \subseteq \omega_{\omega+1}$, let $i^{\mathcal{U}}(X)$ be the image of X under this embedding, *i.e.*,

$$i^{\mathcal{U}}(X) = \{[f]_{\mathcal{U}} : \{t : f(t) \in X\} \in \mathcal{U}\}.$$

Then since $i^{\mathcal{U}}(\omega_{\omega+1}) = \omega_{\omega+1}$, clearly $i^{\mathcal{U}}(X) \subseteq \omega_{\omega+1}$. (*Caution*: In general $i^{\mathcal{U}}[X] = \{i^{\mathcal{U}}(\xi) : \xi \in X\} \subsetneq i^{\mathcal{U}}(X)$.) Now we have

LEMMA 3.2. For any unbounded $C \subseteq \omega_{\omega+1}$:

$$p_\sigma[{}^{W_\sigma}C\uparrow] = S_3^-(\sigma) \cap (C \times i^{\mathcal{U}_{\sigma,1}}(C) \times i^{\mathcal{U}_{\sigma,2}}(C) \times \cdots \times i^{\mathcal{U}_{\sigma,n-1}}(C)),$$

where $\operatorname{lh} \sigma = n > 0$.

PROOF. First let $h \in {}^{W_\sigma}C\uparrow$. Then, if $p_\sigma(h) = (\xi_0, \ldots, \xi_{n-1})$, $\xi_0 = h(\varnothing, \varnothing) \in C$, and if $h_i(u) = h(\tau_i, u)$ for $i > 0$, then $\xi_i = [h_i]_{\mathcal{U}_{\sigma,i}}$ so that (since $h_i(u) \in C$) $\xi_i \in i^{\mathcal{U}_{\sigma,i}}(C)$. Thus $(\xi_0, \ldots, \xi_{n-1}) \in S_3^-(\sigma) \cap (C \times i^{\mathcal{U}_{\sigma,1}}(C) \times \cdots \times i^{\mathcal{U}_{\sigma,n-1}}(C))$. The proof of the converse is very similar to the proof of Lemma 2.2 and we omit the details. ⊣ (Lemma 3.2)

Thus the measure $\mathcal{V}_\sigma$ on each $S_3^-(\sigma)$ (for $\sigma \neq \varnothing$) is generated by the sets of the form

$$S_3^-(\sigma) \cap \left(C \times \prod_{0<i<\operatorname{lh}\sigma} i^{\mathcal{U}_{\sigma,i}}(C) \right)$$

for C cub, $C \subseteq \omega_{\omega+1}$, *i.e.*,

$$A \in \mathcal{V}_\sigma \Leftrightarrow \exists C \subseteq \omega_{\omega+1} \left[C \text{ cub } \wedge A \supseteq S_3^-(\sigma) \cap \left(C \times \prod_{0<i<\operatorname{lh}\sigma} i^{\mathcal{U}_{\sigma,i}}(C) \right) \right].$$

This bears some resemblance to the corresponding result about the generation of the measures $\mathcal{U}_\sigma$ on $S_2^-(\sigma)$.

(b) Finally we establish the usual further homogeneity of S_3^-. Let $H : \omega_{\omega+1} \to \omega_{\omega+1}$ be a function. If $\mathcal{U}$ is a measure on I, where I has cardinality $< \omega_{\omega+1}$, we let $i^{\mathcal{U}}(H)$ be the image of H under $i^{\mathcal{U}}$, *i.e.*, $i^{\mathcal{U}}(H) \colon \omega_{\omega+1} \to \omega_{\omega+1}$ and

$$i^{\mathcal{U}}(H)([f]_{\mathcal{U}}) = [H \circ f]_{\mathcal{U}}.$$

Then if $\operatorname{ran}(H) = C$, $i^{\mathcal{U}}(H)\colon \omega_{\omega+1} \twoheadrightarrow i^{\mathcal{U}}(C)$.

Assume now $H\colon \omega_{\omega+1} \to \omega_{\omega+1}$ is order preserving and for any $\sigma \neq \varnothing$, if $\operatorname{lh}\sigma = n$, define for $u = (\xi_0, \dots, \xi_{n-1})$:

$$H^{\sigma}(u) = (H(\xi_0), i^{\mathcal{U}_{\sigma,1}}(H)(\xi_1), \dots, i^{\mathcal{U}_{\sigma,n-1}}(H)(\xi_{n-1})).$$

Then it is easy to check that

$$(\sigma, u) \in S_3 \Rightarrow (\sigma, H^{\sigma}(u)) \in S_3,$$

while if H is also normal,

$$(\sigma, u) \in S_3^- \Rightarrow (\sigma, H^{\sigma}(u)) \in S_3^-.$$

In particular, if $\alpha \in \mathbb{R}$ and we let $H^{\alpha} = \bigcup_{n\geq 1} H^{\alpha\restriction n}$, so that $H^{\alpha}(u) = H^{\alpha\restriction n}(u)$, where $n > \operatorname{lh} u$, then H^{α} maps $S_3^-(\alpha)$ into $S_3^-(\alpha)$. Thus if for each $X \subseteq \omega_{\omega+1}$, we let

$$S_3^- \restriction X^{\alpha} = \{(\sigma, (\xi_0, \dots, \xi_{n-1})) \in S_3^- : \xi_0 \in X \wedge \text{ for } 0 < i < n, \\ \xi_i \in i^{\mathcal{U}_{\alpha\restriction n,i}}(X)\},$$

then for any cub $C \subseteq \omega_{\omega+1}$ we have

$$\begin{aligned} \alpha \in A &\Leftrightarrow S_3^-(\alpha) \text{ is not wellfounded} \\ &\Leftrightarrow S_3^- \restriction C^{\alpha}(\alpha) \text{ is not wellfounded.} \end{aligned}$$

(For S_3 we have this for any unbounded C.)

Note. The construction of S_3 (and S_3^-) is due to Kunen [Kun71B]. The calculation of a Π^1_3 scale from this tree is orginally due to Martin by a different argument than the one we gave in §3.2.

§4. Π^1_4 sets; the tree S_4.

4.1. Definition of S_4.

(a) Let $A \subseteq \mathbb{R}$ be a Π^1_4 set of reals. Then for some $B \in \Pi^1_3$,

$$\begin{aligned} \alpha \in A &\Leftrightarrow \neg\exists\beta B(\alpha, \beta) \\ &\Leftrightarrow \neg\exists\beta\exists f(\alpha, \beta, f) \in [S_3^-] \\ &\Leftrightarrow S_3^-(\alpha) \text{ is wellfounded.} \end{aligned}$$

Again for each $\sigma \neq \varnothing$, $\operatorname{lh}\sigma = n$, we let

$$S_3^*(\sigma) = \{(\tau_i, u) : (\sigma\restriction \ell_i, \tau_i, u) \in S_3^- \wedge 1 \leq i \leq n\},$$

and we define W_σ to be $\langle S_3^*(\sigma), <_{\text{BK}}\rangle$ and ρ_σ, W_α (for $\alpha \in A$), as in §2.1. Note that $\rho_\sigma = \omega_{\omega+1}$ and for $\alpha \in A$, $\rho_\alpha = \omega_{\omega+1}$. This is because of 3.1(c), (ii).

Now for each $h \in {}^{W_\sigma}[\omega_{\omega+1}]$, let h_i, for $1 \leq i \leq n = \operatorname{lh}\sigma$, be defined as follows:

$$\operatorname{dom}(h_i) = S_3^-(\sigma\restriction \ell_i, \tau_i)$$

and for $v \in S_3^-(\sigma\restriction\ell_i, \tau_i)$

$$h_i(v) = h(\tau_i, v).$$

Recall that $S_3^-(\sigma\restriction\ell_i, \tau_i)$ carries the measure

$$\mathcal{V}_{\sigma\restriction\ell_i,\tau_i} \equiv \mathcal{V}_{\sigma,i} \text{ (for simplicity)}$$

and put

$$\xi_i = [h_i]_{\mathcal{V}_{\sigma,i}}.$$

Finally, let

$$p_\sigma(h) = (\xi_1, \dots, \xi_n) = ([h_i]_{\mathcal{V}_{\sigma,i}})_{1\le i\le n}$$

and define

$$(\sigma, u) \in S_4 \Leftrightarrow \exists h \in {}^{W_\sigma}[\omega_{\omega+1}](p_\sigma(h) = u) \vee (\sigma = u = \varnothing).$$

It is easy now to verify as in §2.1 that

$$\alpha \in A \Leftrightarrow S_4(\alpha) \text{ is not wellfounded.}$$

Put

$$\lambda_5 = \sup\{i^{\mathcal{V}_{\sigma,i}}(\omega_{\omega+1}) : \sigma \neq \varnothing, 1 \le i \le \operatorname{lh}\sigma\}.$$

Then clearly S_4 is a tree on $\omega \times \lambda_5$.

4.2. Scales for Π^1_4 sets. Again as in §2.2 we shall verify that for each $\alpha \in A$, $S_4(\alpha)$ has an honest leftmost branch. For that let $h\colon W_\alpha \to \omega_{\omega+1}$ be the rank function of W_α. For each $i \ge 1$, let h_i be the function on $S_3^-(\alpha\restriction\ell_i, \tau_i)$ given by $h_i(v) = h(\tau_i, v)$ and let $\xi_i = [h_i]_{\mathcal{V}_{\alpha\restriction\ell_i,i}}$. Clearly $(\xi_1, \xi_2, \dots)$ is a branch through $S_4(\alpha)$. Now let $(\eta_1, \eta_2, \dots)$ be a branch through $S_4(\alpha)$. We want to show that $\xi_i \le \eta_i$, $\forall i \ge 1$. Since $(\eta_1, \eta_2, \dots, \eta_n) \in S_4(\alpha\restriction n)$, let ${}^nh \in {}^{W_{\alpha\restriction n}}[\omega_{\omega+1}]$ be such that $p_{\alpha\restriction n}[{}^nh] = (\eta_1, \dots, \eta_n)$. Let ${}^nh_i(v) = {}^nh(\tau_i, v)$. Then for $n, m \ge i$, $[{}^nh_i]_{\mathcal{V}_{\alpha\restriction\ell_i,i}} = [{}^mh_i]_{\mathcal{V}_{\alpha\restriction\ell_i,i}}$, so there is a cub $C^i_{n,m} \subseteq \omega_{\omega+1}$ such that ${}^nh_i(v) = {}^mh_i(v)$ for $v \in (C^i_{n,m} \times \prod_{0<j<\ell_i} i^{\mathcal{U}_{\alpha\restriction\ell_i,\tau_i,j}}(C^i_{n,m})) \cap S_3^-(\alpha\restriction\ell_i, \tau_i)$. Let $C = \bigcap_{1<i<n,m} C^i_{n,m}$. Then $C \subseteq \omega_{\omega+1}$ is cub and ${}^nh_i(v) = {}^mh_i(v)$ for all $v \in (C \times \prod_{0<j<\ell_i} i^{\mathcal{U}_{\alpha\restriction\ell_i,\tau_i,j}}(C)) \cap S_3^-(\alpha\restriction\ell_i, \tau_i)$ and all $n, m \ge i \ge 1$. Then we can define the following function q from

$$\begin{aligned} S_3^-\restriction C^\alpha(\alpha) = \{&(\tau_i, (\xi_0, \dots, \xi_{\ell_i-1})) : (\alpha\restriction\ell_i, \tau_i, (\xi_0, \dots, \xi_{\ell_i-1})) \in S_3^- \\ &\wedge \xi_0 \in C \wedge \xi_j \in i^{\mathcal{U}_{\alpha\restriction\ell_i,\tau_i,j}}(C) \text{ for } 0 < j < \ell_i\}, \end{aligned}$$

into the ordinals:

$$q(\tau_i, v) = {}^nh_i(v), \text{ for any } n \ge i.$$

As usual q is order preserving from $\langle S_3^- \restriction C^\alpha(\alpha), <_{\mathrm{BK}}\rangle$ into the ordinals. Let now H be the normal function enumerating C. Then if $i \geq 1$ and we let

$$H^{\alpha\restriction\ell_i,\tau_i}(\xi_0,\ldots,\xi_{\ell_i-1}) = (H(\xi_0), i^{\mathcal{U}_{\alpha\restriction\ell_i,\tau_i,1}}(H)(\xi_1),\ldots, i^{\mathcal{U}_{\alpha\restriction\ell_i,\tau_i,\ell_i-1}}(H)(\xi_{\ell_i-1})),$$

the map

$$(\tau_i, v) \mapsto (\tau_i, H^{\alpha\restriction\ell_i,\tau_i}(v))$$

maps $\langle S_3^-(\alpha), <_{\mathrm{BK}}\rangle$ in an order preserving way into $\langle S_3^- \restriction C^\alpha(\alpha), <_{\mathrm{BK}}\rangle$ so that if $h\colon \rho_\alpha \to \omega_{\omega+1}$ is as before then $h(\tau_i, v) \leq q(\tau_i, H^{\alpha\restriction\ell_i,\tau_i}(v)) = {}^n h_i(H^{\alpha\restriction\ell_i,\tau_i}(v))$, for $n \geq i$. So $h_i(v) \leq {}^n h_i(H^{\alpha\restriction\ell_i,\tau_i}(v))$, for $n \geq i$. Now we can find a cub $D \subseteq \omega_{\omega+1}$ such that $H\restriction D = \mathrm{id}\restriction D$. Then for $v = (\xi_0,\ldots,\xi_{\ell_i-1}) \in D \times \prod_{0<j<\ell_i} i^{\mathcal{U}_{\alpha\restriction\ell_i,\tau_i,j}}(D)$, $h_i(v) \leq {}^n h_i(v)$. Indeed if $j > 0$ and $\xi_j \in i^{\mathcal{U}_{\alpha\restriction\ell_i,\tau_i,j}}(D)$ we have $\xi_j = [f]_{\mathcal{U}_{\alpha\restriction\ell_i,\tau_i,j}}$, where $f(x) \in D$ a.e. (mod $\mathcal{U}_{\alpha\restriction\ell_i,\tau_i,j}$). But then $H(f(x)) = f(x)$ a.e. (mod $\mathcal{U}_{\alpha\restriction\ell_i,\tau_i,j}$), so $\xi_j = [f]_{\mathcal{U}_{\alpha\restriction\ell_i,\tau_i,j}} = [H\circ f]_{\mathcal{U}_{\alpha\restriction\ell_i,\tau_i,j}} = i^{\mathcal{U}_{\alpha\restriction\ell_i,\tau_i,j}}(H)([f]_{\mathcal{U}_{\alpha\restriction\ell_i,\tau_i,j}}) = i^{\mathcal{U}_{\alpha\restriction\ell_i,\tau_i,j}}(H)(\xi_j)$. Similarly for $j = 0$.

Thus $\xi_i = [h_i]_{\mathcal{V}_{\alpha\restriction\ell_i,i}} \leq [{}^n h_i]_{\mathcal{V}_{\alpha\restriction\ell_i,i}} = \eta_i$.

If now for each $\alpha \in A$, we let f_α be the leftmost branch of $S_4(\alpha)$ and

$$\varphi_i(\alpha) = f_\alpha(i) = \xi_i = [h_i]_{\mathcal{V}_{\alpha\restriction\ell_i,i}}, \text{ (in the preceding notation)},$$

then we can verify again that

$$\psi_i(\alpha) = \langle \bar{\alpha}(i), \varphi_i(\alpha)\rangle$$

is a Δ^1_5-scale on A. Indeed, for $\alpha, \beta \in A$:

$$\begin{aligned}\psi_i(\alpha) \leq \psi_i(\beta) &\Leftrightarrow \bar{\alpha}(i) < \bar{\beta}(i) \vee (\bar{\alpha}(i) = \bar{\beta}(i) \wedge [\lambda v.\,\mathrm{rank}_{W_\alpha}(\tau_i, v)]_{\mathcal{V}_{\alpha\restriction\ell_i,i}} \\ &\qquad\qquad \leq [\lambda v.\,\mathrm{rank}_{W_\beta}(\tau_i, v)]_{\mathcal{V}_{\alpha\restriction\ell_i,i}}) \\ &\Leftrightarrow \bar{\alpha}(i) < \bar{\beta}(i) \vee (\bar{\alpha}(i) = \bar{\beta}(i) \wedge \exists C \subseteq \omega_{\omega+1}[C \text{ cub} \wedge \\ &\qquad \forall(\xi_0,\ldots,\xi_{\ell_i-1})[[(\alpha\restriction\ell_i, (\xi_0,\ldots,\xi_{\ell_i-1})) \in S_3^- \wedge \\ &\qquad \xi_0 \in C \wedge \forall j[0 < j < \ell_i \Rightarrow \xi_i \in i^{\mathcal{U}_{\alpha\restriction\ell_i,\tau_i,j}}(C)]] \\ &\qquad \Rightarrow \mathrm{rank}_{W_\alpha}(\tau_i, (\xi_0,\ldots,\xi_{\ell_i-1})) \leq \\ &\qquad\qquad \mathrm{rank}_{W_\beta}(\tau_i, (\xi_0,\ldots,\xi_{\ell_i-1}))]]).\end{aligned}$$

This relation can be verified to be Σ^1_1 over the structure $\langle \omega_{\omega+1}, <, S_2^-, S_3^-\rangle$, by using the results of Kechris-Martin [KM78]. By the Moschovakis Coding Lemma (see [Mos70] or Kechris [Kec78]) and the techniques of Harrington-Kechris [HK81] one can verify then that every relation on reals which is Σ^1_1 on $\langle \omega_{\omega+1}, <, S_2^-, S_3^-\rangle$ is Σ^1_5 (and conversely), so "$\psi_i(\alpha) \leq \psi_i(\beta)$" is also Σ^1_5. Similarly "$\psi_i(\alpha) < \psi_i(\beta)$" is Σ^1_5 and we are done.

4.3. Homogeneity properties of S_4. Unfortunately not much can be said at this time about the homogeneity properties of S_4 as the combinatorial property

$$\omega_{\omega+1} \to (\omega_{\omega+1})^{\omega_{\omega+1}}$$

is still an open question (recall that the fact that $\omega_1 \to (\omega_1)^{\omega_1}$ is the key to establishing the homogeneity properties of S_2).

NOTE. The construction of S_4 is due to Kunen [Kun71B].

§5. On $\underset{\sim}{\delta}^1_5$. Recall that we have defined in §4:

$$\lambda_5 = \sup\{i^{\mathcal{V}_{\sigma,i}}(\omega_{\omega+1}) : \sigma \neq \varnothing, 1 \leq i \leq \operatorname{lh}\sigma\}.$$

The following result reduces the problem of computing $\underset{\sim}{\delta}^1_5$ to the problem of computing these $i^{\mathcal{V}_{\sigma,i}}(\omega_{\omega+1})$.

THEOREM 5.1 ([Kun71B]). $\underset{\sim}{\delta}^1_5 = (\lambda_5)^+ =$ smallest cardinal $> \lambda_5$.

PROOF. By the results in §4, every Π^1_4, and thus every Σ^1_5 set, is λ_5- Suslin, *i.e.*, it can be written in the form

$$\exists f (\alpha, f) \in S,$$

where S is a tree on $\omega \times \lambda_5$. Thus by the Kunen-Martin theorem (see Martin [Mar71B]) $\underset{\sim}{\delta}^1_5 \leq (\lambda_5)^+$. So it is enough to prove that $\lambda_5 < \underset{\sim}{\delta}^1_5$, *i.e.*, that for each σ, i as above $i^{\mathcal{V}_{\sigma,i}}(\omega_{\omega+1}) < \underset{\sim}{\delta}^1_5$. Put $\mathcal{V}_{\sigma,i} \equiv \mathcal{V}$. Then $\mathcal{V}$ is a measure on $S_3^-(\sigma \restriction \ell_i, \tau_i) \equiv I$. Now the relation

$$f \prec g \Leftrightarrow f, g \colon I \to \omega_{\omega+1} \wedge [f]_{\mathcal{V}} < [g]_{\mathcal{V}}$$

can be easily seen to be Σ^1_1 on the structure $\langle \omega_{\omega+1}, <, S_3^- \rangle$. One can now use the Moschovakis Coding Lemma to code functions $f : I \to \omega_{\omega+1}$ by reals. Say ε codes f_ε. Then as in §4.2 one can verify that $\prec$ is Σ^1_5 in the codes, *i.e.*, the following relation is Σ^1_5:

$$\varepsilon \prec^* \delta \Leftrightarrow \varepsilon, \delta \text{ codes functions } f_\varepsilon, f_\delta \text{ (resp.) from } I \text{ into } \omega_{\omega+1} \wedge f_\varepsilon \prec f_\delta.$$

As $\prec^*$ is a wellfounded relation of rank $i^{\mathcal{V}_{\sigma,i}}(\omega_{\omega+1})$ (since $\operatorname{rank}_{\prec^*}(\varepsilon) = \operatorname{rank}_{\prec}(f_\varepsilon) = [f_\varepsilon]_{\mathcal{V}}$) we have that $i^{\mathcal{V}_{\sigma,i}}(\omega_{\omega+1}) < \underset{\sim}{\delta}^1_5$. ⊣ (Theorem 5.1)

§6. Homogeneous trees in general. We shall discuss now a general notion of homogeneity shared by the trees constructed before. We shall also formulate the type of tree construction utilized in §§2–4 as a general transfer theorem for homogeneous trees.

A tree T on $\omega \times \lambda$ is **homogeneous** if for each $\sigma \neq \varnothing$ in ${}^{<\omega}\omega$ there is a measure μ_σ on $T(\sigma)$ with the following two properties:

(i) Let for $\sigma' \supseteq \sigma$, $\pi_{\sigma'\sigma} \colon T(\sigma') \to T(\sigma)$ be the restriction map:

$$\pi_{\sigma'\sigma}(u) = u \restriction \operatorname{lh}\sigma.$$

Then $(\pi_{\sigma'\sigma})_* \mu_{\sigma'} = \mu_\sigma$ (*i.e.*, for $X \subseteq T(\sigma)$, $X \in \mu_\sigma \Leftrightarrow \pi^{-1}_{\sigma'\sigma}[X] \in \mu_{\sigma'}$).

(ii) If $T(\alpha)$ is not wellfounded and for each $n \geq 1$, $X_n \subseteq T(\alpha\restriction n)$ and $X_n \in \mu_{\alpha\restriction n}$, then there is $f \in \lambda^\omega$ with $f\restriction n \in X_n$, for all n.

The basic way in which homogeneous trees have been obtained in this paper is as follows:

Suppose T is a tree on $\omega \times \lambda$. Suppose also that there is an ordinal κ and for each $\sigma \neq \varnothing$ a wellordering W_σ of order type $\rho_\sigma \leq \kappa$ and a map

$$p_\sigma \colon {}^{W_\sigma}[\kappa] \twoheadrightarrow T(\sigma),$$

such that

$$\sigma \subseteq \sigma' \Rightarrow W_\sigma \subseteq W_{\sigma'}$$

and moreover the following two conditions hold:

(i)′ If $\sigma \subseteq \sigma'$, then for $h \in {}^{W_{\sigma'}}[\kappa]$, $p_{\sigma'}(h)\restriction \mathrm{lh}\,\sigma = p_\sigma(h\restriction W_\sigma)$.

(ii)′ If $T(\alpha)$ is not wellfounded, then the union

$$W_\alpha = \bigcup W_{\alpha\restriction n}$$

is a wellordering of order type $\rho_\alpha \leq \kappa$.

Then granting that for each σ

$$\kappa \to (\kappa)^{\omega\cdot\rho_\sigma},$$

we have that T is homogeneous.

Indeed, let ν_σ be the following measure on ${}^{W_\sigma}[\kappa]$:

$$\nu_\sigma(X) = 1 \Leftrightarrow \exists C \subseteq \kappa[C \text{ cub} \wedge {}^{W_\sigma}C\uparrow \subseteq X].$$

Then let

$$\mu_\sigma = (p_\sigma)_*\nu_\sigma.$$

To check property (i) of homogeneity we use (i)′: Indeed let $X \subseteq T(\sigma)$. If $\mu_\sigma(X) = 1$, then there is $C \subseteq \kappa$, C cub with ${}^{W_\sigma}C\uparrow \subseteq p_\sigma^{-1}[X]$. If now $h \in {}^{W_{\sigma'}}C\uparrow$, $h\restriction W_\sigma \in {}^{W_\sigma}C$, thus $p_\sigma(h\restriction W_\sigma) = p_{\sigma'}(h)\restriction \mathrm{lh}\,\sigma \in X$, so $p_{\sigma'}(h) \in \pi_{\sigma'\sigma}^{-1}[X]$. It follows that $\mu_{\sigma'}(\pi_{\sigma'\sigma}^{-1}[X]) = 1$, so that $(\pi_{\sigma'\sigma})_*\mu_{\sigma'}(X) = 1$.

For (ii) we use of course (ii)′: Let $T(\alpha)$ be not wellfounded and let X_n be such that $\mu_{\alpha\restriction n}(X_n) = 1$. Pick C_n cub in κ with ${}^{W_{\alpha\restriction n}}C_n\uparrow \subseteq p_{\alpha\restriction n}^{-1}[X_n]$. Let $C = \bigcap_n C_n$, so that C is cub in κ. Let, since $\rho_\alpha \leq \kappa$, $h \in {}^{W_\alpha}C\uparrow$. If $h_n = h\restriction W_{\alpha\restriction n}$, then $h_n \in {}^{W_{\alpha\restriction n}}C\uparrow$, so that $p_{\alpha\restriction n}(h_n) \in X_n$. Moreover, if $n < m$, then $p_{\alpha\restriction n}(h_n)$ is an initial segment of $p_{\alpha\restriction m}(h_m)$, thus there is $f \in \lambda^\omega$ with $f\restriction n = p_{\alpha\restriction n}(h_n)$ so that $f\restriction n \in X_n$ and we are done.

It is now easy to see that S_1 is an example of such a tree with $\kappa = \omega_1$, $W_\sigma = \langle \mathrm{lh}\,\sigma, <_\sigma\rangle$, $\rho_\sigma = \mathrm{lh}\,\sigma$ and $p_\sigma(u) = u \circ \pi_\sigma$. Moreover, by their construction, S_2, S_3, S_4 are all of that form with $W_\sigma, \rho_\sigma, \pi_\sigma$ as given in §§2–4.

Note. Similar definitions of homogeneity apply to trees on $\omega^\kappa \times \lambda$.

We shall now state and prove a general transfer theorem for homogeneous trees. We say below that $A \subseteq \mathbb{R}$ **admits** the tree T if $A = p[T]$. Similarly for $A \subseteq \mathbb{R} \times \mathbb{R}$, etc.

THEOREM 6.1 (Transfer Theorem for Homogeneous Trees (Kunen, Martin)). Assume $B \subseteq \mathbb{R}^2$ admits the homogeneous tree T (on some $\omega^2 \times \lambda$). Then put

$$\alpha \in A \Leftrightarrow \neg\exists\beta B(\alpha, \beta),$$

and let ρ_σ be the order type of the wellordering $W_\sigma = (T^*(\sigma), <_{\mathrm{BK}})$, where $T^*(\sigma) = \{(\tau_i, u) : (\sigma\restriction\ell_i, \tau_i, u) \in T \wedge 1 \leq i \leq \mathrm{lh}\,\sigma\}$ and let ρ_α be the order type of the wellordering $W_\alpha = \bigcup_n W_{\alpha\restriction n}$, where $\alpha \in A$. Assume there is

$$\kappa \geq \max\{\{\sup \rho_\sigma : \sigma \neq \varnothing \wedge \sigma \in {}^{<\omega}\omega\}, \sup\{\rho_\alpha : \alpha \in A\}\}$$

with

$$\kappa \to (\kappa)^{\omega\cdot\rho_\sigma}$$

for all $\sigma \in {}^{<\omega}\omega$, $\sigma \neq \varnothing$. The A admits a homogeneous tree $\hat{T}$ (on some $\omega \times \hat{\lambda}$).

PROOF. Let $\mu_{\sigma,\tau}$ be the measure on $T(\sigma, \tau)$, where $\mathrm{lh}\,\sigma = \mathrm{lh}\,\tau \neq 0$. Let $\hat{\lambda} = \sup\{i^{\mu_{\sigma,\tau}}(\kappa) : \mathrm{lh}\,\sigma = \mathrm{lh}\,\tau \neq 0\}$. Then define a tree $\hat{T}$ on $\omega \times \hat{\lambda}$ and maps

$$p_\sigma\colon {}^{W_\sigma}[\kappa] \twoheadrightarrow \hat{T}(\sigma)$$

as follows:

Given $h \in {}^{W_\sigma}[\kappa]$, let for $1 \leq i \leq \mathrm{lh}\,\sigma$, $h_i(v) = h(\tau_i, v)$, so that $h_i\colon T(\sigma\restriction\ell_i, \tau_i) \to \kappa$. Abbreviate

$$\mu_{\sigma,i} \equiv \mu_{\sigma\restriction\ell_i,\tau_i}$$

and put

$$p_\sigma(h) = ([h_i]_{\mu_{\sigma,i}})_{1\leq i\leq \mathrm{lh}\,\sigma}$$

and

$$(\sigma, u) \in \hat{T} \Leftrightarrow \exists h \in {}^{W_\sigma}[\kappa](p_\sigma(h) = u) \vee (\sigma = u = \varnothing).$$

First note that if $\sigma \subseteq \sigma'$ and $h \in {}^{W_\sigma}[\kappa]$, then

$$p_{\sigma'}(h)\restriction \mathrm{lh}\,\sigma = ([h_i]_{\mu_{\sigma',i}})_{1\leq i\leq \mathrm{lh}\,\sigma} = p_\sigma(h\restriction W_\sigma),$$

so that $\hat{T}$ is indeed a tree and condition (i)$'$ before is satisfied.

If now $\hat{T}(\alpha)$ is not wellfounded, then as we will see in a moment $\alpha \in A$ so $T(\alpha)$ is wellfounded, therefore $W_\alpha = \bigcup_n W_{\alpha\restriction n}$ is a wellordering and $\rho_\alpha \leq \kappa$, thus (ii)$'$ is also satisfied. Since we have assumed that $\kappa \to (\kappa)^{\omega\cdot\rho_\sigma}$ we have by our preceding discussion that $\hat{T}$ is homogeneous.

So it only remains to show that

$$\alpha \in A \Leftrightarrow \hat{T}(\alpha) \text{ is not wellfounded.}$$

The direction $\Rightarrow$ is clear. To prove the other direction, assume $(\eta_1, \eta_2, \dots) \in [\hat{T}(\alpha)]$. Then for each $n \geq 1$, we can find ${}^n h \in {}^{W_{\alpha\restriction n}}[\kappa]$ such that if we let ${}^n h_i(v) = {}^n h(\tau_i, v)$, for $1 \leq i \leq n$, then $\eta_i = [{}^n h_i]_{\mu_{\alpha\restriction n,i}}$. Since $[{}^n h_i]_{\mu_{\alpha\restriction n,i}} = [{}^{n'} h_i]_{\mu_{\alpha\restriction n',i}}$ for $n, n' \geq i$, let $Z_i \subseteq T(\alpha\restriction \ell_i, \tau_i)$ have $\mu_{\alpha\restriction \ell_i,\tau_i}$-measure 1 and be such that ${}^n h_i(v) = {}^{n'} h_i(v)$, for all $n, n' \geq i$ and all $v \in Z_i$.

If now $\alpha \notin A$, towards a contradiction, find β with $T(\alpha, \beta)$ not wellfounded. Let for $k \geq 1$, $\tau_{i_k} = \beta\restriction k$ so that $i_1 < i_2 < \dots$. Let $X_k = Z_{i_k} \subseteq T(\alpha\restriction k, \beta\restriction k)$. By the homogeneity of T there is $f \in [T(\alpha, \beta)]$ with $f\restriction k \in X_k$ for all $k \geq 1$. Then obviously $(\tau_{i_1}, f\restriction 1) >_{\mathrm{BK}} (\tau_{i_2}, f\restriction 2) >_{\mathrm{BK}} \dots$. But for each $k \geq 1$, if n is large enough, then ${}^n h_{i_k}(f\restriction k) = \theta_k$ is independent of n and also $\theta_k > \theta_{k+1}$, so that $\theta_1 > \theta_2 > \dots$, a contradiction. $\dashv$ (Theorem 6.1)

Let now $\kappa^{\mathbb{R}}$ be the least non-hyperprojective ordinal or equivalently the ordinal of the smallest admissible set containing the reals. Then by Kechris-Kleinberg-Moschovakis-Woodin [KKMW81], there are arbitrarily large $\lambda < \kappa^{\mathbb{R}}$ with $\lambda \to (\lambda)^\lambda$ (and also this holds for $\lambda = \kappa^{\mathbb{R}}$). So it follows that every projective set admits a homogeneous tree or $\omega \times \lambda$ for some $\lambda < \kappa^{\mathbb{R}}$.

Moschovakis has pointed out that in the preceding theorem, it is enough to assume that B admits only a *weakly homogeneous* tree (to conclude again, under the appropriate assumptions, that A carries a *homogeneous* tree). Here a tree T on $\omega \times \lambda$ is called **weakly homogeneous** if for each $\sigma \neq \varnothing$ there is a *partition*

$$T(\sigma) = \bigcup_{i \in I_\sigma} K_{\sigma,i},$$

where each I_σ is a countable set such that the following hold:

(a) If $\sigma \subseteq \sigma'$ and $T(\sigma) = \bigcup_i K_{\sigma,i}$, $T(\sigma') = \bigcup_j K_{\sigma',j}$, then for every j, there is an i such that $K_{\sigma',j}\restriction \mathrm{lh}\,\sigma \equiv \{v\restriction \mathrm{lh}\,\sigma : v \in K_{\sigma',j}\} \subseteq K_{\sigma,i}$.

(b) Each $K_{\sigma,i}$ carries a measure $\mu_{\sigma,i}$ with the following property: If $T(\alpha)$ is not wellfounded and for each $\eta > 0$, $i \in I_{\alpha\restriction n}$, $X_{\alpha\restriction n,i} \subseteq K_{\alpha\restriction n,i}$ and $X_{\alpha\restriction n,i}$ has $\mu_{\alpha\restriction n,i}$-measure 1, then there is $f \in {}^\omega\lambda$ such that for each $n > 0$,

$$f\restriction n \in \bigcup_i X_{\alpha\restriction n,i}.$$

The proof is similar to the one given before and we leave it to the reader. Notice also the simple fact that if $B \subseteq \mathbb{R}^2$ admits a weakly homogeneous tree, then so does

$$C = \{\alpha : \exists\beta B(\alpha, \beta)\}.$$

Note. The concepts and results in this section originate with Kunen [Kun71B] and Martin [Mar77B].

§7. A result of Martin on subsets of $\underset{\sim}{\delta}^1_3$. Let $\mathcal{P} \subseteq \mathbb{R}$ be a universal Π^1_3 set of reals, *i.e.*, assume that $\mathcal{P} \in \Pi^1_3$ and for each $A \subseteq \mathbb{R}$ in Π^1_3, there is a $n \in \omega$ such that $\alpha \in A \Leftrightarrow n^\frown\alpha = (n, \alpha(0), \alpha(1), \dots) \in \mathcal{P}$. Let $\bar{\varphi} = \{\varphi_n\}$ be a regular Π^1_3-scale on $\mathcal{P}$—*i.e.*, each φ_n maps $\mathcal{P}$ onto an initial segment of ordinals (therefore, as is well known, $\text{ran}(\varphi_n) = \underset{\sim}{\boldsymbol{\delta}}^1_3 = \omega_{\omega+1}$). The tree **associated** with this scale is defined by

$$T_3(\bar{\varphi}) = \{(\alpha\restriction n, (\varphi_0(\alpha), \varphi_1(\alpha), \dots, \varphi_{n-1}(\alpha))) : \alpha \in \mathcal{P}, n \in \omega\}.$$

Thus $T_3(\bar{\varphi})$ is a tree on $\omega \times \omega_{\omega+1}$. Also $\mathcal{P} = p[T_3(\bar{\varphi})]$ and for every $\alpha \in \mathcal{P}$, $T_3(\bar{\varphi})(\alpha)$ has an honest leftmost branch, namely $\bar{\varphi}(\alpha)$. Note also that there is a function $K : \omega_{\omega+1} \to \omega_{\omega+1}$ such that

$$(\sigma, (\xi_0, \dots, \xi_{n-1})) \in T_3(\bar{\varphi}) \wedge \xi_0 \leq \xi \Rightarrow \xi_0, \xi_1, \dots, \xi_{n-1} \leq K(\xi).$$

Indeed, if $(\sigma, (\xi_0, \dots, \xi_{n-1})) \in T_3(\bar{\varphi})$, then for some $\alpha \supseteq \sigma$, $\varphi_0(\alpha) = \xi_0, \dots, \varphi_{n-1}(\alpha) = \xi_{n-1}$. Thus $\max\{\xi_0, \xi_1, \dots, \xi_{n-1}\} \leq \sup\{\varphi_n(\alpha) : n \in \omega \wedge \alpha \in \mathcal{P} \wedge \varphi_0(\alpha) \leq \xi\} \stackrel{\text{def}}{=} K(\xi)$. That $K(\xi) < \omega_{\omega+1}$ follows from the fact that for each $\xi < \omega_{\omega+1} = \underset{\sim}{\boldsymbol{\delta}}^1_3$, $\{\alpha : \alpha \in \mathcal{P} \wedge \varphi_0(\alpha) \leq \xi\}$ is $\underset{\sim}{\boldsymbol{\Delta}}^1_3$, so by boundedness, $\{\varphi_n(\alpha) : n \in \omega \wedge \alpha \in \mathcal{P} \wedge \varphi_0(\alpha) \leq \xi\}$ is bounded below $\underset{\sim}{\boldsymbol{\delta}}^1_3$.

If $X \subseteq \omega_{\omega+1}$, then we say that X is Σ^1_n **in the codes** or just Σ^1_n if

$$X^* \stackrel{\text{def}}{=} \{\alpha \in \mathcal{P} : \varphi_0(\alpha) \in X\}$$

is Σ^1_n. One can use the results of Harrington-Kechris [HK81] to show that this is independent of the choice of $\mathcal{P}$, φ_0, where φ_0 is any regular Π^1_3-norm on a universal Π^1_3 set $\mathcal{P}$, provided that $n \geq 4$.

Let also S_3 be the tree associated with $\mathcal{P}$ as in §3.1. Thus again $\mathcal{P} = p[S_3]$. The result below provides an analog of Theorem 1 in Kechris-Moschovakis [KM72].

THEOREM 7.1 (Martin [Mar77B]). *If $X \subseteq \omega_{\omega+1}$ is Σ^1_4, then $X \in \mathbf{L}[S_3, T_3(\bar{\varphi})]$.*

PROOF. Say, putting $T_3(\bar{\varphi}) \equiv T_3$,

$$\begin{aligned} \beta \in X^* &\Leftrightarrow \exists\gamma(n_0{}^\frown\langle\beta, \gamma\rangle \in \mathcal{P}) \\ &\Leftrightarrow \exists\gamma\exists p(n_0{}^\frown\langle\beta, \gamma\rangle, p) \in [T_3], \end{aligned}$$

and

$$\begin{aligned} \alpha \in \mathcal{P} \wedge \beta \in \mathcal{P} \wedge \varphi_0(\alpha) \leq \varphi_0(\beta) &\Leftrightarrow n_1{}^\frown\langle\alpha, \beta\rangle \in \mathcal{P} \\ &\Leftrightarrow \exists f(n_1{}^\frown\langle\alpha, \beta\rangle, f) \in [S_3], \end{aligned}$$

where as usual $\langle\alpha,\beta\rangle = (\alpha(0),\beta(0),\alpha(1),\beta(1),\dots)$. Consider then the following game G_ξ, for $\xi < \omega_{\omega+1}$:

$$\begin{array}{llllll} \text{I} & & \text{II} & & & \\ \alpha(0)\ h(0) & & & & & \\ & \beta(0) & g(0) & \gamma(0) & p(0) & f(0) \\ \alpha(1)\ h(1) & & & & & \\ & \beta(1) & g(1) & \gamma(1) & p(1) & f(1) \\ \vdots & \vdots & & & & \\ \alpha\quad h & \beta & g & \gamma & p & f, \end{array}$$

(where $h, g, p, f \in {}^{\omega}\omega_{\omega+1}$), whose payoff set is defined as follows:

We say that player I has played correctly up to his mth move, for $m \geq 1$, if $(\alpha\restriction m, h\restriction m) \in T_3 \wedge h(0) \leq \xi$. Then note that also $\forall i < m(h(i) \leq K(\xi))$. We say that player II has played correctly up to his mth move, for $m \geq 1$ again, if $(\beta\restriction m, g\restriction m) \in T_3 \wedge g(0) \leq \xi \wedge (n_0{}^\frown\langle\beta,\gamma\rangle\restriction m, p\restriction m) \in T_3 \wedge (n_1{}^\frown\langle\alpha,\beta\rangle\restriction m, f\restriction m) \in S_3$.

Now player II wins iff for all $m \geq 1$: Player I has played correctly up to his mth move $\Rightarrow$ player II has played correctly up to his mth move.

Clearly this is a closed game for player II and it is in $\mathbf{L}[S_3, T_3]$, uniformly on ξ. So it is enough (by the absoluteness of closed games) to show that

$$\xi \in X \Leftrightarrow \text{ player II has a winning strategy in } G_\xi.$$

($\Leftarrow$). Say s is a winning strategy for player II in G_ξ. Let player I play (α, f) where $\alpha \in \mathcal{P}$, $\varphi_0(\alpha) = \xi$ and $h = \bar{\varphi}(\alpha)$. Then player I plays always correctly, so if player II, following his winning strategy s, produces (β, g, γ, p, f) he must have played also always correctly, *i.e.*, $(\beta, g) \in [T_3] \wedge g(0) \leq \xi \wedge (n_0{}^\frown\langle\beta,\gamma\rangle, p) \in [T_3] \wedge (n_1{}^\frown\langle\alpha,\beta\rangle, f) \in [S_3]$, so $\beta \in \mathcal{P}, \varphi_0(\beta) \leq g(0) \leq \xi$ (as $\bar{\varphi}(\beta)$ is the honest leftmost branch of $T_3(\beta)$), $\beta \in X^*$ and $\varphi_0(\alpha) \leq \varphi_0(\beta)$, thus $\varphi_0(\beta) = \varphi_0(\alpha) = \xi$ and $\xi \in X$.

($\Rightarrow$). Assume now player I has a winning strategy t in G_ξ but, towards a contradiction, that $\xi \in X$. Then fix $\beta \in \mathcal{P}$ with $\varphi_0(\beta) = \xi, g = \bar{\varphi}(\beta)$ and γ, p so that $(n_0{}^\frown\langle\beta,\gamma\rangle, p) \in [T_3]$. Let for each $\sigma \in {}^{<\omega}\omega, \sigma \neq \varnothing, \mathcal{V}_\sigma$ be the measure on $S_3(\sigma)$ as in §3.3. Note that since $\omega_{\omega+1} \to (\omega_{\omega+1})^\rho_\nu, \forall\rho < \omega_{\omega+1}, \forall\nu < \omega_{\omega+1}$, we actually have that $\mathcal{V}_\sigma$ is $\omega_{\omega+1}$-additive. Of course these measures satisfy the homogeneity conditions (i), (ii) of §6.

Define now inductively values $\alpha(0), h(0), \alpha(1), h(1), \dots$ and sets $X_1, X_2, \dots$ as follows (recall below that $n_1{}^\frown\langle\alpha,\beta\rangle = (n_1, \alpha(0), \beta(0), \dots)$):

First $\alpha(0), h(0)$ are the values called by t in I's initial move, which is clearly correct. Now for each $(f(0)) \in S_3((n_1))$, if player II plays $\beta(0), g(0), \gamma(0), p(0), f(0)$ in his first move, player I answers by t to play correctly $\alpha(1), h(1)$. In particular, $h(1) < K(\xi)$, so by the $\omega_{\omega+1}$-additivity of $\mathcal{V}_{(n_1)}$, let X_1 be in $\mathcal{V}_{(n_1)}$ and such that $\alpha(1), h(1)$ are always the same for $(f(0)) \in X_1$. This is our

$\alpha(1), h(1)$. Then for each $(f(0), f(1)) \in S_2((n_1, \alpha(0))$ if player II next plays $\beta(1), g(1), \gamma(1), p(1), f(1)$, player I answers following t to play $\alpha(2), h(2)$ which, by an argument exactly as before, is the same for all $(f(0), f(1)) \in X_2$ for some $X_2 \in \mathcal{V}_{(n_1, \alpha(0))}$, etc. Now as $(\alpha, h) \in [T_3]$ and $h(0) \leq \xi$, clearly $\varphi_0(\alpha) \leq \xi$, so $\varphi_0(\alpha) \leq \varphi_0(\beta)$, thus $S_3(n_1{}^\frown\langle\alpha, \beta\rangle)$ is not wellfounded. Since $X_k \in \mathcal{V}_{n_1{}^\frown\langle\alpha,\beta\rangle\restriction k}$ for all $k \geq 1$, we have by condition (ii) of homogeneity that there is f such that $f\restriction k \in X_k$ for each $k \geq 1$. If player II plays now β, g, γ, p, f, he plays always correctly and if player I follows t he plays α, h so that he also plays correctly. But then player II won, a contradiction.

⊣ (Theorem 7.1)

Corollary 7.2. If $X \subseteq \omega_{\omega+1}$, then there is $\alpha \in \mathbb{R}$ such that $X \in \mathbf{L}[X_3, T_3(\vec{\varphi}), \alpha]$.

Proof. By the Moschovakis' Coding Lemma, every $X \subseteq \omega_{\omega+1}$ is $\Sigma^1_4(\alpha)$ for some $\alpha \in \mathbb{R}$. ⊣ (Corollary 7.2)

§8. On the Victoria Delfino Third Problem. Let $\mathcal{P}$ be a universal Π^1_3 set of reals and $\bar{\varphi}$ a regular Π^1_3-scale on it. Let $T_3(\bar{\varphi})$ be its associated tree as in §7. The Victoria Delfino Third Problem (see Kechris-Moschovakis (eds) [Cabal i]) is the question:

$$\text{Is } \mathbf{L}[T_3(\bar{\varphi})] \text{ independent of } \mathcal{P}, \bar{\varphi}?$$

Let also $\tilde{\mathbf{L}}[T_3(\bar{\varphi})] = \bigcup_{\alpha\in\mathbb{R}} \mathbf{L}[T_3(\bar{\varphi}), \alpha]$. Surely the independence of $\tilde{\mathbf{L}}[T_3(\bar{\varphi})]$ from $\mathcal{P}$, $\bar{\varphi}$ would be very strong evidence for an affirmative answer to the above problem. So the following result, despite its dependence on an unproven yet hypothesis is of interest here. Its proof uses methods of Kunen; see Kunen [Kun71D] and Kechris [Kec78].

Theorem 8.1. Assume $\omega_{\omega+1} \to (\omega_{\omega+1})^{\omega_{\omega+1}}$. Then $\wp(\omega_{\omega+1}) \subseteq \tilde{\mathbf{L}}[T_3(\bar{\varphi})]$, so in particular $\tilde{\mathbf{L}}[T_3(\bar{\varphi})]$ is independent of $\mathcal{P}$, $\bar{\varphi}$.

Proof. The heart of the proof is the following:

Lemma 8.2. If $f : \omega_{\omega+1} \to \omega_{\omega+1}$, then there is $g \in \tilde{\mathbf{L}}[T_3(\bar{\varphi})]$ such that $f(\xi) \leq g(\xi), \forall \xi < \omega_{\omega+1}$.

From that it follows that if $C \subseteq \omega_{\omega+1}$ is cub, then there is $\bar{C} \subseteq C$, $\bar{C}$ cub such that $\bar{C} \in \tilde{\mathbf{L}}[T_3(\bar{\varphi})]$. Indeed, let $f : \omega_{\omega+1} \to C$ be the increasing enumeration of C and let g be as in Lemma 8.2. Then if $\bar{C} = \{\xi < \omega_{\omega+1} : \xi \text{ is limit} \wedge \forall \eta < \xi(g(\eta) < \xi)\}$, $\bar{C}$ is cub and if $\xi \in \bar{C}$ then $\forall \eta < \xi(f(\eta) \leq g(\eta) < \xi)$, so $f(\xi) = \xi \in C$, *i.e.*, $\bar{C} \subseteq C$. Now we have

Lemma 8.3. If $\omega_{\omega+1} \to (\omega_{\omega+1})^{\omega_{\omega+1}}$, and for every cub $C \subseteq \omega_{\omega+1}$ there is $\bar{C} \subseteq C$, $\bar{C}$ cub such that $\bar{C} \in \tilde{\mathbf{L}}[T_3(\bar{\varphi})]$, then $\wp(\omega_{\omega+1}) \subseteq \tilde{\mathbf{L}}[T_3(\bar{\varphi})]$.

Proof. Consider the following partition of ${}^{\omega_{\omega+1}}[\omega_{\omega+1}]$:

$$f \in X \Leftrightarrow f \in \tilde{\mathbf{L}}[T_3(\bar{\varphi})].$$

Then let C be cub such that ${}^{\omega_{\omega+1}}C\!\uparrow \subseteq X$ or ${}^{\omega_{\omega+1}}C\!\uparrow \subseteq \bar{X}$. By our hypothesis, C can be assumed to be in $\tilde{\mathbf{L}}[T_3(\bar{\varphi})]$ so that we must have ${}^{\omega_{\omega+1}}C\!\uparrow \subseteq \tilde{\mathbf{L}}[T_3(\bar{\varphi})]$. Let $\{\eta_\theta : \theta < \omega_{\omega+1}\}$ be the increasing enumeration of C and put $H = \{\eta_{\theta+\omega} : \theta < \omega_{\omega+1}\}$. Clearly ${}^{\omega_{\omega+1}}[H] \subseteq \tilde{\mathbf{L}}[T_3(\bar{\varphi})]$. Let now A be an arbitrary unbounded subset of $\omega_{\omega+1}$. Let also f_H be the increasing enumeration of H. Then $f_H[A] \subseteq H$, so if g is the increasing enumeration of $f_H[A]$, $g \in \tilde{\mathbf{L}}[T_3(\bar{\varphi})]$. But $\xi \in A \Leftrightarrow f_H(\xi) \in f_H(A) \Leftrightarrow f_H(\xi) \in \operatorname{ran}(g)$, so $A \in \mathbf{L}[f_H, g] \subseteq \tilde{\mathbf{L}}[T_3(\bar{\varphi})]$ and we are done. ⊣ (Proof of Lemma 8.3)

So we only have to prove Lemma 8.2 above. For that we will play a Solovay-type game which is a variant of a game of Kunen, see Kunen [Kun71D].

Let $\mathcal{S}'$ be a Π^1_2 subset of $\mathbb{R} \times \mathbb{R}$, such that if $\mathcal{S}(\alpha) \Leftrightarrow \exists\beta\mathcal{S}'(\alpha,\beta)$, then $\mathcal{S} \in \Sigma^1_3 \setminus \underset{\sim}{\mathbf{\Pi}}^1_3$. Let

$$\mathcal{S}'(\alpha,\beta) \Leftrightarrow n_0{}^\frown\langle\alpha,\beta\rangle \in \mathcal{P},$$

so that

$$\begin{aligned}\exists\beta\mathcal{S}'(\alpha,\beta) &\Leftrightarrow \exists\beta(n_0{}^\frown\langle\alpha,\beta\rangle \in \mathcal{P})\\ &\Leftrightarrow \exists f\exists\beta\forall k(n_0{}^\frown\langle\alpha,\beta\rangle\restriction k, f\restriction k) \in T_3(\bar{\varphi})\\ &\Leftrightarrow \exists g\forall m(\alpha\restriction m, f\restriction m) \in U,\end{aligned}$$

where

$$\begin{aligned}((a_0,\dots,a_{m-1}),(\xi_0,\dots,\xi_{m-1})) \in U \Leftrightarrow (\xi_0)_0,\dots,(\xi_{m-1})_0 \in \omega \wedge\\ (n_0{}^\frown(a_0,(\xi_0)_0,\dots,a_{m-1},(\xi_{m-1})_0)\restriction m,((\xi_0)_1,\dots,(\xi_{m-1})_1)) \in T_3(\bar{\varphi}),\end{aligned}$$

where $\xi \mapsto ((\xi)_0,(\xi_1))$ is some simple 1-1 correspondence of $\omega_{\omega+1}$ with ${}^2(\omega_{\omega+1})$. Clearly $U \in \mathbf{L}[T_3(\bar{\varphi})]$ and if

$$\mathcal{S}(\alpha) \Leftrightarrow \exists\beta\mathcal{S}'(\alpha,\beta),$$

then $\alpha \in \mathcal{S} \Leftrightarrow U(\alpha)$ is not wellfounded. Now we claim that U is a tree on $\omega \times \lambda$, for some $\lambda < \omega_{\omega+1}$. Indeed in the notation above, if $(\bar{a},\bar{\xi}) \in U$, then there is a γ such that $n_0{}^\frown\gamma \in \mathcal{P}$, $\gamma = \langle\alpha,\beta\rangle$, $\bar{a} \subseteq \alpha$ and $\varphi_i(n_0{}^\frown\gamma) = (\xi_i)_1$ for $i < m = \operatorname{lh}\bar{a}$. Thus $(\alpha,\beta) \in \mathcal{S}'$. But $\mathcal{S}'' = \{n_0{}^\frown\langle\alpha,\beta\rangle : (\alpha,\beta) \in \mathcal{S}'\}$ is a Π^1_2 subset of $\mathcal{P}$, so (by boundedness) there is $\mu < \omega_{\omega+1}$ such that $n_0{}^\frown\gamma \in \mathcal{S}'' \Rightarrow \varphi_i(n_0{}^\frown\gamma) < \mu$ for all i so $(\xi_i)_1 < \mu$, $\forall i < m$, thus there is $\lambda < \omega_{\omega+1}$ such that $\xi_i < \lambda$ and we are done.

Thus if $U(\alpha)$ is wellfounded, $\operatorname{rank}(U(\alpha)) < \omega_{\omega+1}$. Moreover, since $\mathcal{S} \notin \underset{\sim}{\mathbf{\Delta}}^1_3$,

$$\sup\{\operatorname{rank}(U(\alpha)) : U(\alpha) \text{ is wellfounded}\} = \omega_{\omega+1}.$$

Otherwise for some $\rho < \omega_{\omega+1}$

$$\alpha \in \mathcal{S} \Leftrightarrow \neg(\mathrm{rank}(U(\alpha)) < \rho),$$

therefore by Martin's result (see Martin [Mar71B]) that $\utilde{\mathbf{\Delta}}^1_3$ is closed under $< \omega_{\omega+1}$ intersections and unions, $\mathcal{S} \in \utilde{\mathbf{\Delta}}^1_3$, a contradiction.

After these preliminaries consider the following game associated with each $f : \omega_{\omega+1} \to \omega_{\omega+1}$:

$$\begin{array}{cc} \mathrm{I} & \mathrm{II} \\ w & \alpha \end{array}$$

player II wins iff $[w \in \mathcal{P} \Rightarrow U(\alpha)$ is wellfounded $\wedge\, \mathrm{rank}(U(\alpha)) > f(\varphi_0(w))]$.

By a simple boundedness argument and the above remarks, player I cannot have a winning strategy in this game, so player II has a winning strategy s. Define then the following tree $\mathcal{J}$ on $\omega \times \omega_{\omega+1} \times \omega \times \lambda$:

$$\begin{aligned}(\tau, u, a, v) \in \mathcal{J} \Leftrightarrow (\tau, u) \in T_3(\bar{\varphi}) \wedge (a, v) \in U \\ \wedge\, a \text{ is the result of player II playing according} \\ \text{to } s \text{ when player I plays } \tau.\end{aligned}$$

Clearly $\mathcal{J} \in \mathbf{L}[T_3(\bar{\varphi}), s]$. Moreover $\mathcal{J}$ is wellfounded, since if $(w, f, \alpha, g) \in [\mathcal{J}]$ then $(w, f) \in [T_3(\bar{\varphi})]$ so $w \in \mathcal{P}$, and $(\alpha, g) \in U$, thus $U(\alpha)$ is not wellfounded and also α is the result of a run in which player II follows s against player I playing w, a contradiction.

Fix now $\xi < \omega_{\omega+1}$. Let $w \in \mathcal{P}$ be such that $\varphi_0(w) = \xi$. Let α be the result of player II playing according to s while player I plays this w. Finally let $f = \bar{\varphi}(w)$. Then the map

$$v \mapsto (w \restriction \mathrm{lh}\, v, f \restriction \mathrm{lh}\, v, \alpha \restriction \mathrm{lh}\, v, v)$$

is an embedding of $U(\alpha)$ into $\mathcal{J}$. But notice that actually this maps $U(\alpha)$ into

$$\mathcal{J}_{(\xi)} \stackrel{\text{def}}{=} \{(\tau, u, a, v) \in \mathcal{J} : u(0) \leq \xi\}.$$

Thus $\mathrm{rank}(U(\alpha)) \leq \mathrm{rank}(\mathcal{J}_{(\xi)})$. But also $f(\xi) = f(\varphi_0(w)) < \mathrm{rank}(U(\alpha))$, so

$$f(\xi) < \mathrm{rank}(\mathcal{J}_{(\xi)}) \stackrel{\text{def}}{=} g(\xi).$$

As $g \in \mathbf{L}[T_3(\bar{\varphi}), s]$, it will be enough to show that $\mathrm{rank}(\mathcal{J}_{(\xi)}) < \omega_{\omega+1}$. But recall from §7 that for some $K(\xi) < \omega_{\omega+1}$:

$$w \in \mathcal{P} \wedge \varphi_0(w) \leq \xi \Rightarrow \forall i[\varphi_i(w) \leq K(\xi)].$$

Thus if $(\tau, u, a, v) \in \mathcal{J}$, then $u(i) \leq K(u(0))$, since $(\tau, u) \in T_3(\bar{\varphi})$, so there is $w \in \mathcal{P}$ with $\tau \subseteq w$ and $\bar{\varphi}(w) \restriction \mathrm{lh}\, u = u$, so $\varphi_0(w) = u(0)$ and $u(i) = \varphi_i(w) \leq K(u(0))$. So

$$\mathcal{J}_{(\xi)} \subseteq \mathcal{J} \restriction K(\xi),$$

thus

$$f(\xi) \leq g(\xi) = \operatorname{rank}(\mathcal{J}_{(\xi)}) \leq \operatorname{rank}(\mathcal{J}\restriction K(\xi)) < \omega_{\omega+1},$$

where $\mathcal{J}\restriction\theta = \{(\tau, u, a, v) \in \mathcal{J} : u \in {}^{<\omega}\theta\}$. This completes the proof.

⊣ (Lemma 8.2)

⊣ (Theorem 8.1)

REFERENCES

Leo A. Harrington and Alexander S. Kechris

[HK81] *On the determinacy of games on ordinals*, ***Annals of Mathematical Logic***, vol. 20 (1981), pp. 109–154.

Alexander S. Kechris

[Kec78] AD *and projective ordinals*, this volume, originally published in Kechris and Moschovakis [Cabal i], pp. 91–132.

Alexander S. Kechris, Eugene M. Kleinberg, Yiannis N. Moschovakis, and W. Hugh Woodin

[KKMW81] *The axiom of determinacy, strong partition properties, and nonsingular measures*, in Kechris et al. [Cabal ii], pp. 75–99, reprinted in [Cabal I], p. 333–354.

Alexander S. Kechris, Benedikt Löwe, and John R. Steel

[Cabal I] *Games, scales, and Suslin cardinals: the Cabal seminar, volume I*, Lecture Notes in Logic, vol. 31, Cambridge University Press, 2008.

Alexander S. Kechris and Donald A. Martin

[KM78] *On the theory of* Π^1_3 *sets of reals*, ***Bulletin of the American Mathematical Society***, vol. 84 (1978), no. 1, pp. 149–151.

Alexander S. Kechris, Donald A. Martin, and Yiannis N. Moschovakis

[Cabal ii] *Cabal seminar* 77–79, Lecture Notes in Mathematics, no. 839, Berlin, Springer, 1981.

Alexander S. Kechris and Yiannis N. Moschovakis

[KM72] *Two theorems about projective sets*, ***Israel Journal of Mathematics***, vol. 12 (1972), pp. 391–399.

[Cabal i] *Cabal seminar* 76–77, Lecture Notes in Mathematics, no. 689, Berlin, Springer, 1978.

[KM78B] *Notes on the theory of scales*, in *Cabal Seminar* 76–77 [Cabal i], pp. 1–53, reprinted in [Cabal I], p. 28–74.

Kenneth Kunen

[Kun71B] *On* $\underset{\sim}{\delta}^1_5$, circulated note, August 1971.

[Kun71D] *Some singular cardinals*, circulated note, September 1971.

Richard Mansfield

[Man71] *A Souslin operation on* Π^1_2, ***Israel Journal of Mathematics***, vol. 9 (1971), no. 3, pp. 367–379.

Donald A. Martin

[Mar71B] *Projective sets and cardinal numbers: some questions related to the continuum problem*, this volume, originally a preprint, 1971.

[Mar77B] *On subsets of* $\underset{\sim}{\delta}^1_3$, circulated note, January 1977.

Donald A. Martin and Robert M. Solovay
[MS69] *A basis theorem for Σ^1_3 sets of reals*, ***Annals of Mathematics***, vol. 89 (1969), pp. 138–160.

Yiannis N. Moschovakis
[Mos70] *Determinacy and prewellorderings of the continuum*, ***Mathematical logic and foundations of set theory. Proceedings of an international colloquium held under the auspices of the Israel Academy of Sciences and Humanities, Jerusalem, 11–14 November 1968*** (Y. Bar-Hillel, editor), Studies in Logic and the Foundations of Mathematics, North-Holland, Amsterdam-London, 1970, pp. 24–62.

Joseph R. Shoenfield
[Sho61] *The problem of predicativity*, ***Essays on the foundations of mathematics*** (Yehoshua Bar-Hillel, E. I. J. Poznanski, Michael O. Rabin, and Abraham Robinson, editors), Magnes Press, Jerusalem, 1961, pp. 132–139.

Robert M. Solovay
[Sol78A] *A Δ^1_3 coding of the subsets of ω_ω*, this volume, originally published in Kechris and Moschovakis [Cabal i], pp. 133–150.

DEPARTMENT OF MATHEMATICS
CALIFORNIA INSTITUTE OF TECHNOLOGY
PASADENA, CALIFORNIA 91125
UNITED STATES OF AMERICA
E-mail: kechris@caltech.edu

AD AND PROJECTIVE ORDINALS

ALEXANDER S. KECHRIS

This is an unpolished exposition of some work in the theory of projective ordinals under the hypothesis of definable determinacy. This is understood here as the hypothesis that every set of reals in $\mathbf{L}(\mathbb{R})$ is determined. Since the projective ordinals are absolute between the real world and $\mathbf{L}(\mathbb{R})$ we carry this study entirely *within* $\mathbf{L}(\mathbb{R})$. Thus we will use the full Axiom of Determinacy (AD) together with ZF+DC (DC is of course the only choice principle that is preserved under this transition to $\mathbf{L}(\mathbb{R})$).

Starting with the work of Martin, Moschovakis, and Solovay a decade ago, the exciting and unexpected possibility was discovered that one could calculate *precisely* the projective ordinals in terms of the aleph function. Indeed $\underset{\sim}{\boldsymbol{\delta}}^1_1 = \omega_1$ is a classical result and Martin computed that $\underset{\sim}{\boldsymbol{\delta}}^1_2 = \omega_2$, $\underset{\sim}{\boldsymbol{\delta}}^1_3 = \omega_{\omega+1}$ and $\underset{\sim}{\boldsymbol{\delta}}^1_4 = \omega_{\omega+2}$ (the last independently also due to Kunen). The computation of $\underset{\sim}{\boldsymbol{\delta}}^1_5$ and the higher $\underset{\sim}{\boldsymbol{\delta}}^1_n$'s is now the central problem of this theory. Kunen in 1971 has originated a major program towards achieving that goal, developing along the way some very important and powerful techniques. Part of his work is presented in the later sections of this survey and in Solovay's paper [Sol78A]. We are planning to present the rest in a sequel paper, along with other more recent advances in this area.

Work in descriptive set theory over the last ten years has resulted in a basically complete understanding of the analytical sets of the 3rd and 4th level of the analytical hierarchy, fully analogous to that provided by the classical effective theory for the first two. Moreover, recent results in this subject show, in our opinion, that a complete structure theory for all analytical sets at level 5 and beyond is essentially reduced to the problem of the precise calculation of the $\underset{\sim}{\boldsymbol{\delta}}^1_n$'s for $n \geq 5$.

Remark. Since some of the results we state below need actually only weaker forms of AD (like PD, *etc.*) we have put explicitly in the statements of the theorems the set theoretical assumptions which are used to establish them, beyond ZF+DC.

Preparation for this paper was partially supported by NSF Grant MCS 76-17254. The author would like to thank R. M. Solovay for many interesting and helpful discussions on the topics presented in this paper.

Wadge Degrees and Projective Ordinals: The Cabal Seminar, Volume II
Edited by A. S. Kechris, B. Löwe, J. R. Steel
Lecture Notes in Logic, 37

§1. Definitions and the general picture.

DEFINITION 1.1. For all $n \geq 1$, let

$$\underset{\sim}{\delta}^1_n = \sup\{\xi : \xi \text{ is the length of a } \underset{\sim}{\Delta}^1_n \text{ prewellordering of } \mathbb{R} \ (= {}^\omega\omega)\}.$$

The following facts are known, granting AD:

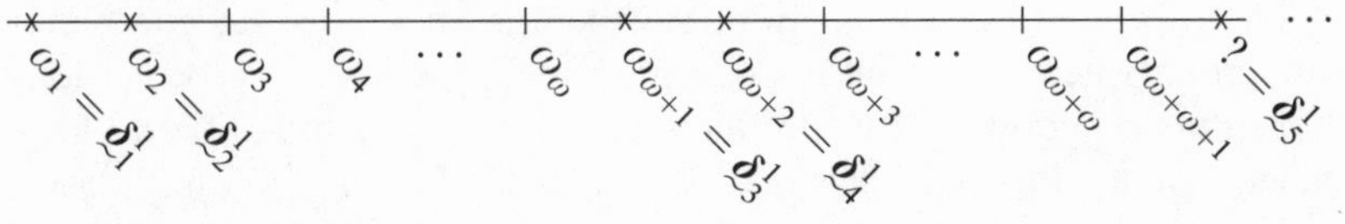

and in general for $n \geq 0$,

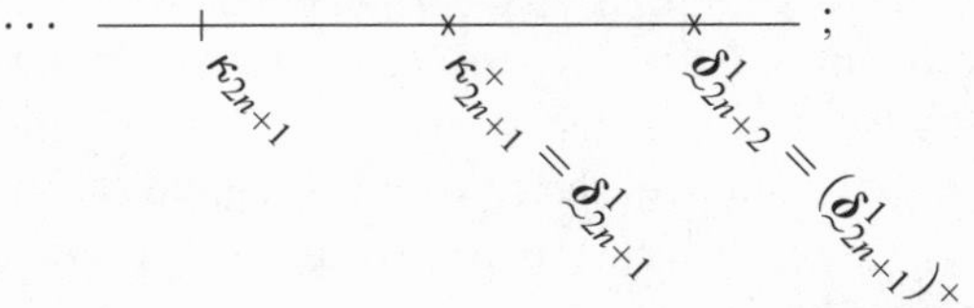

1) All the $\underset{\sim}{\delta}^1_n$ are cardinals.
2) $\underset{\sim}{\delta}^1_{2n+2} = (\underset{\sim}{\delta}^1_{2n+1})^+$.
3) $\underset{\sim}{\delta}^1_{2n+1} = \kappa^+_{2n+1}$, where κ_{2n+1} is a cardinal of cofinality ω.
4) All $\underset{\sim}{\delta}^1_n$ are regular. (Note: without choice this does not follow from the fact that they are successor cardinals.) In fact
5) All $\underset{\sim}{\delta}^1_n$ are measurable.

Proofs of these and other results will be given in the sequel.

§2. For all n, $\underset{\sim}{\delta}^1_n$ is a cardinal.

DEFINITION 2.1. Let $F \subseteq \mathbb{R}$, $\leq$ a prewellordering on F, and let $\varphi\colon F \twoheadrightarrow \xi = \mathrm{lh}(\leq)$, be the canonical norm associated with it (*i.e.*, $\alpha \leq \beta \Leftrightarrow \varphi(\alpha) \leq \varphi(\beta)$). If $f : \xi \to \wp(\mathbb{R})$, where $\wp(\mathbb{R})$ is the power set of $\mathbb{R}$, put

$$\mathrm{Code}(f; \leq) = \{(\alpha, \beta) : \alpha \in F \ \& \ \beta \in f(\varphi(\alpha))\}.$$

If $A \subseteq {}^n\xi$, put

$$\mathrm{Code}(A; \leq) = \{(\alpha_1, \ldots, \alpha_n) \in {}^nF : (\varphi(\alpha_1), \ldots, \varphi(\alpha_n)) \in A\}.$$

DEFINITION 2.2. If $f : \xi \to \wp(\mathbb{R})$, we call $g : \xi \to \wp(\mathbb{R})$ a **choice subfunction of** f if

1) $\forall \eta < \xi, g(\eta) \subseteq f(\eta)$.
2) $\forall \eta < \xi[f(\eta) \neq \varnothing \Rightarrow g(\eta) \neq \varnothing]$.

THEOREM 2.3 (AD) (*The Coding Lemma*; Moschovakis [Mos70]). Let $\leq$ be a $\underset{\sim}{\mathbf{\Delta}}^1_n$ prewellordering of a subset of $\mathbb{R}$ with length ξ. Then every function $f: \xi \to \wp(\mathbb{R})$ has a choice subfunction g such that $\mathrm{Code}(g; \leq)$ is $\underset{\sim}{\mathbf{\Sigma}}^1_n$.

PROOF SKETCH. Let $G \subseteq \mathbb{R} \times \mathbb{R} \times \mathbb{R}$ be universal for the $\underset{\sim}{\mathbf{\Sigma}}^1_n$ subsets of $\mathbb{R} \times \mathbb{R}$; α is a **code** for a $\underset{\sim}{\mathbf{\Sigma}}^1_n$ set $Q \subseteq \mathbb{R} \times \mathbb{R}$ if $Q = G_\alpha = \{(\beta, \gamma) : G(\alpha, \beta, \gamma)\}$. Given $f: \xi \to \wp(\mathbb{R})$, let for all $\eta < \xi$, $f_\eta : \xi \to \wp(\mathbb{R})$, be the restriction of f to η, defined to be $\varnothing$ outside η. Suppose there is an f with no good choice subfunction, where g is **good** if $\mathrm{Code}(g; \leq)$ is $\underset{\sim}{\mathbf{\Sigma}}^1_n$. Let $\eta_0 \leq \xi$ be least such that f_{η_0} has no good choice subfunction. Clearly η_0 is limit. Consider the following game: Player I plays $\alpha \in \mathbb{R}$, player II plays $\beta \in \mathbb{R}$, and player II wins if whenever α codes a good choice subfunction of f_η for some $\eta < \eta_0$, then β codes a good choice subfunction of f_θ, where $\eta < \theta < \eta_0$.

Case I: Player I has a winning strategy. Then for each β there is an $\eta(\beta) < \eta_0$ and a good choice subfunction $g_{\eta(\beta)}$ of $f_{\eta(\beta)}$, with code given by player I's strategy applied to β. Notice that $\sup\{\eta(\beta) : \beta \in \mathbb{R}\}$ must be bounded below η_0 (otherwise the union of the $g_{\eta(\beta)}$ will be a good choice subfunction of f_{η_0}). But since η_0 was chosen least, player II can easily beat player I's winning strategy. Hence Case I never occurs.

Case II: Player II has a winning strategy. Using the recursion theorem, we can find a partial continuous function h such that for all $w \in \mathrm{Field}(\leq)$ with $\varphi(w) < \eta_0, h(w)$ is defined and is a $\underset{\sim}{\mathbf{\Sigma}}^1_n$-code for a good choice subfunction $g_{\theta(w)}$ of $f_{\theta(w)}$ with $\varphi(w) < \theta(w) < \eta_0$. But then if

$$g_{\eta_0}(\eta) = \bigcup_{\substack{w \in \mathrm{Field}(\leq) \\ \varphi(w) < \eta_0}} g_{\theta(w)}(\eta),$$

g_{η_0} is a choice subfunction of f_{η_0} and $(\alpha, \beta) \in \mathrm{Code}(g_{\eta_0}; \leq) \Leftrightarrow \exists w[w \in \mathrm{Field}(\leq) \;\&\; \varphi(w) < \eta_0 \;\&\; G(h(w), \alpha, \beta)]$, hence g_{η_0} is good, contradicting the choice of η_0. ⊣ (Theorem 2.3)

COROLLARY 2.4 (AD). For every $A \subseteq {}^n\xi$, $\mathrm{Code}(A; \leq)$ is $\underset{\sim}{\mathbf{\Delta}}^1_n$.

PROOF. Let us take $n = 1$ for notational simplicity. Let α_0, α_1 be distinct reals, and define $f: \xi \to \wp(\mathbb{R})$ by

$$f(\eta) = \begin{cases} \{\alpha_0\}, & \text{if } \eta \in A \\ \{\alpha_1\}, & \text{if } \eta \notin A. \end{cases}$$

Then the only choice subfunction of f is f itself. Hence $\mathrm{Code}(f; \leq)$ is $\underset{\sim}{\mathbf{\Sigma}}^1_n$ by the Coding Lemma, and

$$\begin{aligned} \alpha \in \mathrm{Code}(A; \leq) &\Leftrightarrow (\alpha, \alpha_0) \in \mathrm{Code}(f; \leq) \\ &\Leftrightarrow \alpha \in \mathrm{Field}(\leq) \;\&\; (\alpha, \alpha_1) \notin \mathrm{Code}(f; \leq), \end{aligned}$$

hence $\mathrm{Code}(A; \leq)$ is $\underset{\sim}{\mathbf{\Delta}}^1_n$. ⊣ (Corollary 2.4)

THEOREM 2.5 (AD) (Moschovakis [Mos70]). For all $n \geq 1$, $\underset{\sim}{\delta}^1_n$ is a cardinal.

PROOF. If not, let $f : \xi \to \underset{\sim}{\delta}^1_n$ be 1-1 and onto, where $\xi < \underset{\sim}{\delta}^1_n$. There is a prewellordering $\leq$ of $\mathbb{R}$ of length ξ which is $\underset{\sim}{\Delta}^1_n$. Let $<^*$ be defined on ξ by

$$\eta <^* \theta \Leftrightarrow f(\eta) < f(\theta).$$

Then by the corollary, Code($<^*; \leq$) is $\underset{\sim}{\Delta}^1_n$. But then Code($<^*; \leq$) is a prewellordering on $\mathbb{R}$ in $\underset{\sim}{\Delta}^1_n$ with length $\underset{\sim}{\delta}^1_n$, contradiction. ⊣ (Theorem 2.5)

§3. The $\underset{\sim}{\delta}^1_n$'s are successor cardinals.

DEFINITION 3.1. Let $A \subseteq \mathbb{R}$. A **norm** on A is a map $\varphi : A \to \text{Ord}$. The **length** of φ is the length of the prewellordering induced by φ on A, *i.e.*, $\alpha \leq^\varphi \beta \Leftrightarrow \varphi(\alpha) \leq \varphi(\beta)$.

If $\underset{\sim}{\Gamma} = \underset{\sim}{\Pi}^1_n$ or $\underset{\sim}{\Sigma}^1_n$ and $A \in \underset{\sim}{\Gamma}$, then φ is a **$\underset{\sim}{\Gamma}$-norm** if the following relations are in $\underset{\sim}{\Gamma}$:

$$\alpha \leq^*_\varphi \beta \Leftrightarrow \alpha \in A \,\&\, [\beta \notin A \vee \varphi(\alpha) \leq \varphi(\beta)],$$
$$\alpha <^*_\varphi \beta \Leftrightarrow \alpha \in A \,\&\, [\beta \notin A \vee \varphi(\alpha) < \varphi(\beta)].$$

If every set in $\underset{\sim}{\Gamma}$ has a $\underset{\sim}{\Gamma}$-norm, we say that $\underset{\sim}{\Gamma}$ has the **prewellordering property**.

THEOREM 3.2 (PD) (*The Prewellordering Theorem*; Martin [Mar68], Moschovakis (see [AM68])). For all $n \geq 0$, $\underset{\sim}{\Pi}^1_{2n+1}$ and $\underset{\sim}{\Sigma}^1_{2n+2}$ have the prewellordering property (and $\underset{\sim}{\Sigma}^1_{2n+1}$, $\underset{\sim}{\Pi}^1_{2n+2}$ do not have the prewellordering property).

DEFINITION 3.3. Let $A, B \subseteq \mathbb{R}$. We say that A is **reducible** to B if there is a total continuous function $f : \mathbb{R} \to \mathbb{R}$ such that $\alpha \in A \Leftrightarrow f(\alpha) \in B$. If $\underset{\sim}{\Gamma} = \underset{\sim}{\Sigma}^1_n$ or $\underset{\sim}{\Pi}^1_n$, a set $A \subseteq \mathbb{R}$ is called **$\underset{\sim}{\Gamma}$-complete** if $A \in \underset{\sim}{\Gamma}$ and every set $B \in \underset{\sim}{\Gamma}$ is reducible to A.

LEMMA 3.4 (AD) (*Wadge's Lemma*). If $A, B \subseteq \mathbb{R}$, then either A is reducible to B or B is reducible to $\mathbb{R} \setminus A$.

PROOF. Consider the game in which player I plays α, player II plays β and player II wins iff $\alpha \in A \Leftrightarrow \beta \in B$. ⊣ (Lemma 3.4)

By this lemma every set in $\underset{\sim}{\Gamma} \setminus \underset{\sim}{\Delta}$ (where $\underset{\sim}{\Gamma} = \underset{\sim}{\Sigma}^1_n$ or $\underset{\sim}{\Pi}^1_n$ and $\underset{\sim}{\Delta} = \underset{\sim}{\Delta}^1_n$) is $\underset{\sim}{\Gamma}$-complete.

THEOREM 3.5 (PD) (Moschovakis [Mos70]). If φ is a $\underset{\sim}{\Pi}^1_{2n+1}$-norm on a $\underset{\sim}{\Pi}^1_{2n+1}$-complete set, then

$$\text{lh}(\varphi) = \underset{\sim}{\delta}^1_{2n+1}.$$

DEFINITION 3.6. A **scale** on a set $A \subseteq \mathbb{R}$ is a sequence of norms $\{\varphi_n\}_{n\in\omega}$ on A such that for every sequence $\{\alpha_i\}_{i\in\omega}$ of members of A, if

1) $\lim_{i\to\infty} \alpha_i = \alpha$, and

2) For each n there is an ordinal λ_n such that $\varphi_n(\alpha_i) = \lambda_n$, for all large enough i, then $\alpha \in A$ and for all n, $\varphi_n(\alpha) \leq \lambda_n$.

The scale $\{\varphi_n\}_{n\in\omega}$ is a **λ-scale** if $\mathrm{lh}(\varphi_n) \leq \lambda$, $\forall n$.

If $\underset{\sim}{\Gamma} = \underset{\sim}{\Sigma}^1_n$ or $\underset{\sim}{\Pi}^1_n$, we call $\{\varphi_n\}_{n\in\omega}$ a **Γ-scale** if the two relations

$$S(n, \alpha, \beta) \Leftrightarrow \alpha \leq^*_{\varphi_n} \beta$$
$$T(n, \alpha, \beta) \Leftrightarrow \alpha <^*_{\varphi_n} \beta$$

are in $\underset{\sim}{\Gamma}$.

THEOREM 3.7 (PD) (*The Scale Theorem*, Moschovakis [Mos71]). For $n \geq 0$, every $\underset{\sim}{\Pi}^1_{2n+1}$ ($\underset{\sim}{\Sigma}^1_{2n+2}$) set admits a $\underset{\sim}{\Pi}^1_{2n+1}$ ($\underset{\sim}{\Sigma}^1_{2n+2}$)-scale.

A **tree** on a set X is a set of finite sequences of members of X closed under initial segments. We will consider many times trees on $\omega \times \lambda$ or $\omega \times \omega \times \lambda$, *etc.*, where λ is an ordinal. Thus if T is a tree on $\omega \times \lambda$, its members are of the form

$$((k_0, \xi_0), (k_1, \xi_1), \ldots, (k_n, \xi_n)), \text{ where } k_i \in \omega \text{ and } \xi_i < \lambda \text{ for all } i \leq n.$$

We will sometimes find it convenient to represent elements of such a T by pairs of tuples of the form

$$((k_0 \ldots k_n), (\xi_0 \ldots \xi_n)).$$

An (infinite) **branch** of tree T on X is a sequence $f \in {}^\omega X$ such that for all n, $f \restriction n \in T$, where $f \restriction n = (f(0), \ldots, f(n-1))$. A branch of a tree on $\omega \times \lambda$ is thus a sequence $g \in {}^\omega(\omega \times \lambda)$, but we will represent it by the unique pair $(\alpha, f) \in {}^\omega\omega \times {}^\omega\lambda$ such that for all n,

$$g(n) = (\alpha(n), f(n)).$$

If J is a tree on $\omega \times \lambda$ put

$$[J] = \{(\alpha, f) : \alpha \in \mathbb{R}, f \in {}^\omega\lambda \text{ and } \forall n(\alpha\restriction n, f\restriction n) \in J\}$$

(*i.e.*, $[J]$ is the set of branches of J), and put $\mathrm{p}[J] = \{\alpha \in \mathbb{R} : \exists f \in {}^\omega\lambda(\alpha, f) \in [J]\}$.

If $\{\varphi_n\}_{n\in\omega}$ is a λ-scale on a set $A \subseteq \mathbb{R}$, the **tree associated with this scale** is the tree on $\omega \times \lambda$ defined by

$$((k_0 \ldots k_n), (\xi_0 \ldots \xi_n)) \in T$$
$$\Leftrightarrow \exists \alpha \in A \text{ such that } \forall i \leq n(\alpha(i) = k_i \text{ and } \varphi_i(\alpha) = \xi_i).$$

CLAIM 3.8. For A, T as above, $\mathrm{p}[T] = A$.

PROOF. $A \subseteq \mathrm{p}[T]$ is obvious. Let $\alpha \in \mathrm{p}[T]$. Find $f \in {}^\omega\lambda$ such that $(\alpha, f) \in [T]$. Then for all n, $(\alpha\restriction n, f\restriction n) \in T$, *i.e.*, there is a sequence of reals $\{\alpha_n\}_{n\in\omega}$ such that $\forall n(\alpha_n \in A)$ and

$$(\alpha\restriction n, f\restriction n) = (\alpha_n\restriction n, (\varphi_0(\alpha_n), \ldots, \varphi_{n-1}(\alpha_n))).$$

Then $\alpha_n \to \alpha$ and $\varphi_n(\alpha_i) = f(n)$ for all $i > n$, hence $\alpha \in A$. ⊣ (Claim 3.8)

DEFINITION 3.9. If T is a tree on $\omega \times \lambda$, put

$$T(\alpha) = \{s \in {}^{<\omega}\lambda : (\alpha \restriction \mathrm{lh}(s), s) \in T\}.$$

Clearly for each α, $T(\alpha)$ is a tree on λ, and if T is the tree associated with a scale on A as above, then by the claim we have

$$\alpha \in A \Leftrightarrow T(\alpha) \text{ has an infinite branch.}$$

DEFINITION 3.10. A set $A \subseteq \mathbb{R}$ is **λ-Suslin** (where λ is an ordinal) if there is a tree T on $\omega \times \lambda$ such that $A = \mathrm{p}[T]$.

The next result is classical for $\utilde{\boldsymbol{\Sigma}}^1_1$, is due to Schoenfield [Sho61] for $\utilde{\boldsymbol{\Sigma}}^1_2$, to Martin-Solovay [MS69] for $\utilde{\boldsymbol{\Sigma}}^1_3$ (see also Mansfield [Man71]) and to Moschovakis [Mos71] in general.

THEOREM 3.11 (PD). (i) For each $n \geq 0$, every $\utilde{\boldsymbol{\Sigma}}^1_{2n+2}$ set is $\utilde{\boldsymbol{\delta}}^1_{2n+1}$-Suslin. (ii) For each $n \geq 0$, every $\utilde{\boldsymbol{\Sigma}}^1_{2n+1}$ set is κ_{2n+1}-Suslin, where κ_{2n+1} is a cardinal $< \utilde{\boldsymbol{\delta}}^1_{2n+1}$.

PROOF. Let $\langle\, ,\rangle \colon \omega \times \kappa \leftrightarrow \kappa$ be a coding of pairs by ordinals less than a cardinal κ, with decoding functions $(\ )_0$ and $(\ )_1$. Then if $B \subseteq \mathbb{R} \times \mathbb{R}$ is κ-Suslin, let T be a tree on $\omega \times \omega \times \kappa$ such that $(\alpha, \beta) \in B \Leftrightarrow \exists f \in {}^{\omega}\kappa \forall n (\alpha \restriction n, \beta \restriction n, f \restriction n) \in T$. Let

$$((k_0, \xi_0), \dots, (k_n, \xi_n)) \in T' \Leftrightarrow$$
$$((k_0, , (\xi_0)_0, (\xi_0)_1), \dots, (k_n, (\xi_n)_0, (\xi_n)_1) \in T.$$

Then clearly $\exists \beta B(\alpha, \beta) \Leftrightarrow \alpha \in \mathrm{p}[T']$, hence $\{\alpha : \exists \beta B(\alpha, \beta)\}$ is also κ-Suslin. So *to prove* (*i*) it is enough to show that every $\utilde{\boldsymbol{\Pi}}^1_{2n+1}$ set is $\utilde{\boldsymbol{\delta}}^1_{2n+1}$-Suslin. But this is obvious by the previous results and the evident fact that a $\utilde{\boldsymbol{\Pi}}^1_{2n+1}$-scale on a $\utilde{\boldsymbol{\Pi}}^1_{2n+1}$ set must be a $\utilde{\boldsymbol{\delta}}^1_{2n+1}$-scale.

We now give the *proof of* (*ii*). By the closure of κ-Suslin sets under real existential quantification, it is enough to show that every $\utilde{\boldsymbol{\Pi}}^1_{2n}$ set is κ-Suslin for some fixed $\kappa < \utilde{\boldsymbol{\delta}}^1_{2n+1}$. The least such κ is the required κ_{2n+1}. Let A be a complete $\utilde{\boldsymbol{\Pi}}^1_{2n}$ set, $\{\varphi_n\}_{n\in\omega}$ a $\utilde{\boldsymbol{\Pi}}^1_{2n+1}$-scale on A. Since A is $\utilde{\boldsymbol{\Delta}}^1_{2n+1}$, $\mathrm{lh}(\varphi_m) < \utilde{\boldsymbol{\delta}}^1_{2n+1}$ for all m. Since any ω-sequence of $\utilde{\boldsymbol{\Delta}}^1_{2n+1}$ prewellorderings can be put together to yield a new $\utilde{\boldsymbol{\Delta}}^1_{2n+1}$ prewellordering of length at least the supremum of the lengths of the original prewellorderings, $\mathrm{cf}(\utilde{\boldsymbol{\delta}}^1_{2n+1}) > \omega$. Hence there is a $\kappa < \utilde{\boldsymbol{\delta}}^1_{2n+1}$ such that $\mathrm{lh}(\varphi_m) < \kappa$, for all m. Hence, by passing from the scale to its associated tree, A is κ-Suslin. ⊣ (Theorem 3.11)

DEFINITION 3.12. A set of reals if **λ-Borel** (λ an ordinal) if it belongs to the smallest class of sets of reals containing the open sets and closed under complements and wellordered unions of length $< \lambda$. This class is denoted by $\boldsymbol{B}_\lambda$.

THEOREM 3.13 (*Separation of κ-Suslin sets*; Luzin for $\kappa = \omega$; see Martin [Mar71B]). If $A, B \subseteq \mathbb{R}$ are κ-Suslin and $A \cap B = \varnothing$ then there is a κ^+-Borel set C which separates them, *i.e.*, $A \subseteq C$ and $C \cap B = \varnothing$.

PROOF. Let $A = \mathrm{p}[T]$, $B = \mathrm{p}[S]$, where T and S are trees on $\omega \times \kappa$. Define a tree U on $\omega \times \kappa \times \kappa$ by

$$(s, u, v) \in U \Leftrightarrow (s, u) \in T \ \& \ (s, v) \in S.$$

Since $A \cap B = \varnothing$, U is well founded, *i.e.*, has no infinite branches. We will define a function on U

$$(s, u, v) \mapsto C_{s,u,v} \subseteq \mathbb{R},$$

by induction on U, such that each $C_{s,u,v} \in \boldsymbol{B}_{\kappa^+}$ and $C_{s,u,v}$ separates $A_{s,u}$ and $B_{s,v}$, where

$$A_{s,u} = \{\alpha \supseteq s : \exists f \supseteq u((\alpha, f) \in [T])\}$$

and

$$B_{s,v} = \{\alpha \supseteq s : \exists f \supseteq v((\alpha, f) \in [S])\}.$$

We can take then $C = C_{\varnothing,\varnothing,\varnothing}$.

Note that $A_{s,u} = \bigcup_{n,\xi} A_{s^\frown n, u^\frown \xi}$ and $B_{s,v} = \bigcup_{m,\eta} B_{s^\frown m, v^\frown \eta}$. So it is enough to define $D_{n,\xi,m,\eta} \in \boldsymbol{B}_{\kappa^+}$ such that $D_{n,\xi,m,\eta}$ separates $A_{s^\frown n, u^\frown \xi}$ from $B_{s^\frown m, v^\frown \eta}$, since we can then take

$$C_{s,u,v} = \bigcup_{n,\xi} \bigcap_{m,\eta} D_{n,\xi,m,\eta}$$

Assume we have defined all $C_{s^\frown n, u^\frown \xi, v^\frown \eta}$, when $(s^\frown n, u^\frown \xi, v^\frown \eta) \in U$.

Case I: $n = m$ and $(s^\frown n, u^\frown \xi, v^\frown \eta) \in U$.
Then take $D_{n,\xi,m,\eta} = C_{s^\frown n, u^\frown \xi, v^\frown \eta}$.

Case II: $n = m$ and $(s^\frown n, u^\frown \xi, v^\frown \eta) \notin U$. Then either $A_{s^\frown n, u^\frown \xi} = \varnothing$ or $B_{s^\frown n, v^\frown \eta} = \varnothing$, so they can be trivially separated.

Case III: $n \neq m$. Then $A_{s^\frown n, u^\frown \xi}$ and $B_{s^\frown n, v^\frown \eta}$ can be separated by disjoint open neighborhoods. ⊣ (Theorem 3.13)

THEOREM 3.14 (*Generalized Suslin Theorem*; see Martin [Mar71B]). If A, $\mathbb{R} \setminus A$ are κ-Suslin, then $A \in \boldsymbol{B}_{\kappa^+}$.

THEOREM 3.15 (AD) ($\subseteq$ Martin [Mar71B]; $\supseteq$ Moschovakis [Mos71]). For all $n \geq 0$, $\boldsymbol{B}_{\boldsymbol{\delta}^1_{2n+1}} = \boldsymbol{\Delta}^1_{2n+1}$.

PROOF. That $\boldsymbol{\Delta}^1_{2n+1} \subseteq \boldsymbol{B}_{\boldsymbol{\delta}^1_{2n+1}}$ follows from the fact that each $\boldsymbol{\Delta}^1_{2n+1}$ set is κ_{2n+1}-Suslin for some $\kappa_{2n+1} < \boldsymbol{\delta}^1_{2n+1}$ (Theorem 3.11 (ii)). For the other direction, it is enough to show that $\boldsymbol{\Delta}^1_{2n+1}$ is closed under unions of length

$< \boldsymbol{\delta}^1_{2n+1}$. If not, let $\theta < \boldsymbol{\delta}^1_{2n+1}$ be least such that for some sequence $\{A_\xi\}_{\xi<\theta}$ of $\boldsymbol{\Delta}^1_{2n+1}$ sets, $\bigcup_{\xi<\theta} A_\xi = A \notin \boldsymbol{\Delta}^1_{2n+1}$. Clearly θ is an uncountable cardinal, and we may assume $A_\xi \subseteq A_\eta$ if $\xi \leq \eta < \theta$ and $A_\lambda = \bigcup_{\xi<\lambda} A_\xi$ if $\lambda = \bigcup \lambda < \theta$. Let $\leq$ be a $\boldsymbol{\Delta}^1_{2n+1}$ prewellordering of $\mathbb{R}$ such that $\mathrm{lh}(\leq) = \theta$. Let $f : \theta \to \wp(\mathbb{R})$ be given by $f(\xi) = \{\varepsilon : \varepsilon$ is a $\boldsymbol{\Delta}^1_{2n+1}$ code of $A_\xi\}$, where ε is a $\boldsymbol{\Delta}^1_{2n+1}$**-code** if $\varepsilon = \langle \varepsilon_0, \varepsilon_1 \rangle$ and the $\boldsymbol{\Pi}^1_{2n+1}$ set coded by ε_0 equals the $\boldsymbol{\Sigma}^1_{2n+1}$ set coded by ε_1. Denote by $\boldsymbol{\Delta}_\varepsilon$ the $\boldsymbol{\Delta}^1_{2n+1}$ set coded by ε, if ε is a $\boldsymbol{\Delta}^1_{2n+1}$-code.

Let g be a choice subfunction of f such that $\mathrm{Code}(g; \leq)$ is $\boldsymbol{\Sigma}^1_{2n+1}$. Then

$$\alpha \in A \Leftrightarrow \exists \beta[(w, \beta) \in \mathrm{Code}(g; \leq) \ \&\ \alpha \in \boldsymbol{\Delta}_\beta].$$

Hence A is $\boldsymbol{\Sigma}^1_{2n+1}$. By Wadge's Lemma, A is $\boldsymbol{\Sigma}^1_{2n+1}$-complete. For $\alpha \in A$, let $\psi(\alpha) =$ the unique $\xi < \theta$ such that $\alpha \in A_{\xi+1} \setminus A_\xi$.

Claim 3.16. ψ is a $\boldsymbol{\Sigma}^1_{2n+1}$-norm (which is a contradiction since it implies that $\boldsymbol{\Sigma}^1_{2n+1}$ has the prewellordering property).

Proof of Claim 3.16. We have $\alpha \leq^*_\psi \beta \Leftrightarrow \exists \xi < \theta[\alpha \in (A_{\xi+1} \setminus A_\xi) \ \&\ \beta \notin A_\xi]$ and $\alpha <^*_\psi \beta \Leftrightarrow \exists \xi < \theta[\alpha \in (A_{\xi+1} \setminus A_\xi) \ \&\ \beta \notin A_{\xi+1}]$. So $\leq^*_\psi$ and $<^*_\psi$ are both unions of $< \boldsymbol{\delta}^1_{2n+1} \boldsymbol{\Delta}^1_{2n+1}$ sets, hence as before they are $\boldsymbol{\Sigma}^1_{2n+1}$.
⊣ (Claim 3.16)

⊣ (Theorem 3.15)

Definition 3.17. If J is a tree on a set X and $u \in {}^{<\omega}X =$ set of finite sequences from X, then $J_u = \{v \in {}^{<\omega}X : u^\frown v \in J\}$.

Notation. $|J| < \xi$ means J is wellfounded and has rank $< \xi$. (Put also $|\varnothing| = -1$.)

Theorem 3.18 (Sierpiński for $\kappa = \omega$). If $A \subseteq \mathbb{R}$ is κ-Suslin, then $A \in \boldsymbol{B}_{\kappa^{++}}$.

Proof. Let $A = \mathrm{p}[T]$, where T is a tree on $\omega \times \kappa$. For each $\xi < \kappa^+$ and $u \in {}^{<\omega}\kappa$ put

$$A^\xi_u = \{\alpha : |T(\alpha)_u| < \xi\}.$$

Then if $\mathrm{lh}(u) = n$,

$$A^0_u = \{\alpha : (\alpha\restriction n, u) \notin T\}$$
$$A^{\xi+1}_u = A^\xi_u \cup \bigcap_{\eta<\lambda} A^\xi_{u^\frown\eta}$$
$$A^\lambda_u = \bigcup_{\xi<\lambda} A^\xi_u, \text{ if } \lambda = \bigcup \lambda > 0.$$

Thus $A^\xi_u \in \boldsymbol{B}_{\kappa^+}$ for all u and ξ. But

$$\begin{aligned}\alpha \notin A &\Leftrightarrow \alpha \notin \mathrm{p}[T]\\ &\Leftrightarrow T(\alpha) \text{ is well founded}\\ &\Leftrightarrow \exists \xi < \kappa^+(|T(\alpha)| < \xi)\\ &\Leftrightarrow \exists \xi < \kappa^+(\alpha \in A^\xi_\varnothing)\end{aligned}$$

⊣ (Theorem 3.18)

THEOREM 3.19 (Martin [Mar71B]). If A is κ-Suslin and $\mathrm{cf}(\kappa) > \omega$, then $A \in \boldsymbol{B}_{\kappa^+}$.

PROOF. let $A = \mathrm{p}[T]$, where T is a tree on $\omega \times \kappa$. Then $\alpha \in A \Leftrightarrow T(\alpha)$ is not well founded $\Leftrightarrow \exists \xi < \kappa(T^\xi(\alpha)$ is not well founded), where $T^\xi = T$ restricted to ordinals $< \xi$. Now apply 3.18. ⊣ (Theorem 3.19)

THEOREM 3.20 (Kechris [Kec74]). For all n, $\utilde{\boldsymbol{\delta}}^1_{2n+1} = (\kappa_{2n+1})^+$, where κ_{2n+1} is a cardinal of cofinality ω.

PROOF. Let (by Theorem 3.11) κ_{2n+1} = least κ such that every $\utilde{\boldsymbol{\Sigma}}^1_{2n+1}$ set is κ-Suslin. If $(\kappa_{2n+1})^{++} \leq \utilde{\boldsymbol{\delta}}^1_{2n+1}$, then every $\utilde{\boldsymbol{\Sigma}}^1_{2n+1}$ set is in $\boldsymbol{B}_{(\kappa_{2n+1})^{++}} \subseteq \boldsymbol{B}_{\utilde{\boldsymbol{\delta}}^1_{2n+1}} = \utilde{\boldsymbol{\Delta}}^1_{2n+1}$, a contradiction. By 3.11, $\kappa_{2n+1} < \utilde{\boldsymbol{\delta}}^1_{2n+1}$, hence $(\kappa_{2n+1})^+ = \utilde{\boldsymbol{\delta}}^1_{2n+1}$. If $\mathrm{cf}(\kappa_{2n+1}) > \omega$, then by 3.19 every $\utilde{\boldsymbol{\Sigma}}^1_{2n+1}$ set is in $\boldsymbol{B}_{(\kappa_{2n+1})^+} = \boldsymbol{B}_{\utilde{\boldsymbol{\delta}}^1_{2n+1}} = \utilde{\boldsymbol{\Delta}}^1_{2n+1}$, contradiction. ⊣ (Theorem 3.20)

THEOREM 3.21 (Kunen [Kun71C], Martin [Mar71B]). If $\prec \subseteq \mathbb{R} \times \mathbb{R}$ is wellfounded and κ-Suslin, then $|\prec| < \kappa^+$.

PROOF. Let $\alpha \prec \beta \Leftrightarrow \exists f \in {}^\omega\kappa((\alpha, \beta, f) \in [T])$, where T is a tree on $\omega \times \omega \times \kappa$. Put

$$T_\prec = \{(\alpha_0, \alpha_1, \dots, \alpha_n) : \alpha_0 \succ \alpha_1 \succ \cdots \succ \alpha_n\}.$$

By induction one easily checks that for each $\alpha \in \mathrm{Field}(\prec)$ and each $\alpha_0, \dots, \alpha_n$ such that $\alpha_0 \succ \alpha_1 \succ \cdots \succ \alpha_n \succ \alpha$ we have

$$|\alpha|_\prec = |(\alpha_0 \dots \alpha_n, \alpha)|_{T_\prec}.$$

Hence $|\prec| \leq |T_\prec|$.

Let S consist of all sequences of the form

$$s = ((s_1, t_1, u_1), \dots, (s_n, t_n, u_n)),$$

where $s_i = t_{i+1}$ for all $i < n$ and $(s_i, t_i, u_i) \in T$, for all $i \leq n$. Thus $s_i, t_i \in {}^{<\omega}\omega$, $u_i \in {}^{<\omega}\kappa$. For s, s' as above, define

$$\begin{aligned}s \succ^* s' \Leftrightarrow\ &\mathrm{lh}(s) < \mathrm{lh}(s') \text{ and for all } i < \mathrm{lh}(s),\\ &(s'_i, t'_i, u'_i \text{ properly extend } s_i, t_i, u_i).\end{aligned}$$

For any α, β such that $\alpha \prec \beta$ let $h_{\alpha,\beta}$ be the leftmost branch of $T(\alpha, \beta)$. Now define $f : T_{\prec} \to S$ by

$$f(\alpha_0 \dots \alpha_n) = ((\alpha_1 \restriction n, \alpha_0 \restriction n, h_{\alpha_1,\alpha_0} \restriction n), \dots, (\alpha_n \restriction n, \alpha_{n-1} \restriction n, h_{\alpha_n,\alpha_{n-1}} \restriction n)).$$

Clearly f embeds in an order preserving way $T_{\prec}$ into $(S, \prec^*)$. It only remains to show $\prec^*$ is wellfounded.

If not, let $s_0 \succ^* s_1 \succ^* s_2 \succ^* \dots$, where

$$s_n = ((s_1^n, t_1^n, u_1^n), \dots, (s_{k_n}^n, t_{k_n}^n, u_{k_n}^n)).$$

Then $k_n \to \infty$, $t_1^n \to \alpha_0$, $s_1^n = t_2^n \to \alpha_1$, $s_2^n = t_3^n \to \alpha_2, \dots$, and $u_1^n \to f_1$, $u_2^n \to f_2, \dots$, where for all n, $(\alpha_{n+1}, \alpha_n, f_{n+1}) \in [T]$. Hence $\alpha_0 \succ \alpha_1 \succ \alpha_2 \succ \dots$, a contradiction. ⊣ (Theorem 3.21)

THEOREM 3.22 (AD) (Kunen [Kun71C], Martin [Mar71B]). For all $n \geq 0$, $(\boldsymbol{\delta}_{2n+1}^1)^+ = \boldsymbol{\delta}_{2n+2}^1$.

PROOF. If φ is a $\boldsymbol{\Pi}_{2n+1}^1$-norm on a $\boldsymbol{\Pi}_{2n+1}^1$-complete set, then $\text{lh}(\varphi) = \boldsymbol{\delta}_{2n+1}^1$. Since its associated prewellordering is $\boldsymbol{\Delta}_{2n+2}^1$, we have $\boldsymbol{\delta}_{2n+1}^1 < \boldsymbol{\delta}_{2n+2}^1$. Hence $\boldsymbol{\delta}_{2n+2}^1 \geq (\boldsymbol{\delta}_{2n+1}^1)^+$. Since every $\boldsymbol{\Sigma}_{2n+2}^1$ relation is $\boldsymbol{\delta}_{2n+1}^1$-Suslin, we have by Theorem 3.21 that $(\boldsymbol{\delta}_{2n+1}^1)^+ \geq \boldsymbol{\delta}_{2n+2}^1$. ⊣ (Theorem 3.22)

THEOREM 3.23 (AD) (Moschovakis [Mos70] for odd n, Kechris [Kec74] for even n). For all n, $\boldsymbol{\delta}_n^1 < \boldsymbol{\delta}_{n+1}^1$.

PROOF. The theorem for n odd follows from 3.22. Suppose $\boldsymbol{\delta}_{2m}^1 = \boldsymbol{\delta}_{2m+1}^1$. Then $\boldsymbol{\delta}_{2m+1}^1 = (\kappa_{2m+1})^+ = \boldsymbol{\delta}_{2m}^1 = (\boldsymbol{\delta}_{2m-1}^1)^+$, hence $\kappa_{2m+1} = \boldsymbol{\delta}_{2m-1}^1$, which is a contradiction since $\text{cf}(\boldsymbol{\delta}_{2m-1}^1) > \omega$. ⊣ (Theorem 3.23)

THEOREM 3.24 (AD) (Moschovakis [Mos70], for odd n; Kunen [Kun71C], Martin [Mar71B], for all n). For all n,

$$\boldsymbol{\delta}_n^1 = \sup\{\xi : \xi \text{ is the length of a } \boldsymbol{\Sigma}_n^1 \text{ wellfounded relation}\}.$$

PROOF. By Theorems 3.11, 3.21 and 3.22. ⊣ (Theorem 3.24)

§4. The $\boldsymbol{\delta}_n^1$'s are regular.

THEOREM 4.1 (AD) (Moschovakis [Mos70] for odd n; Kunen [Kun71C] for all n). For all n, $\boldsymbol{\delta}_n^1$ is regular.

PROOF. Assume not, and let $f : \lambda \to \boldsymbol{\delta}_n^1$ be a cofinal map, with $\lambda < \boldsymbol{\delta}_n^1$. Let $\leq$ be a $\boldsymbol{\Delta}_n^1$ prewellordering of $\mathbb{R}$ of length λ with corresponding norm φ. Let $g(\xi) = \{\alpha : \alpha \text{ is a } \boldsymbol{\Sigma}_n^1 \text{ code of a } \boldsymbol{\Sigma}_n^1 \text{ well founded relation of length } f(\xi)\}$. Let g' be a choice subfunction of g such that $\text{Code}(g'; \leq)$ is $\boldsymbol{\Sigma}_n^1$.

Let $W \subseteq \mathbb{R} \times \mathbb{R} \times \mathbb{R}$ be $\utilde{\Sigma}^1_n$ universal, and put

$$(\alpha, \beta, \gamma) \prec (\alpha', \beta', \gamma') \Leftrightarrow$$
$$[\alpha = \alpha', \beta = \beta', (\alpha, \beta) \in \text{Code}(g'; \leq) \text{ and } (\gamma, \gamma') \in W_\beta].$$

Clearly $\prec$ is $\utilde{\Sigma}^1_n$ and wellfounded. But for any $\xi < \lambda$, if α is such that $\varphi(\alpha) = \xi$, then for any fixed $\beta \in g'(\xi)$ the map

$$\gamma \mapsto (\alpha, \beta, \gamma)$$

embeds W_β into $\prec$. So $|{\prec}| \geq |W_\beta| = f(\xi)$, hence $|{\prec}| = \utilde{\delta}^1_n$, a contradiction.

⊣ (Theorem 4.1)

§5. The $\utilde{\delta}^1_n$'s are measurable.

Theorem 5.1 (AD) (Solovay for $n = 1$ (see [Sol67A]), 2; Martin [Mar71A] for odd n; Kunen [Kun71A] in general). For all n, $\utilde{\delta}^1_n$ is measurable.

Proof. Let $W \subseteq \mathbb{R}^3$ be universal $\utilde{\Sigma}^1_n$ and let

$$S = \{\alpha : W_\alpha \text{ is wellfounded binary relation}\}.$$

For $\alpha \in S$, let $|\alpha| = \text{lh}(W_\alpha)$. Hence

$$\utilde{\delta}^1_n = \sup\{|\alpha| : \alpha \in S\}.$$

For $A \subseteq \utilde{\delta}^1_n$, consider the following game G^A first used (for $n = 1$ and with a different coding) by Solovay in his original proof that ω_1 is measurable. Player I plays α, player II plays β, and player II wins iff

$$[\exists i((\alpha)_i \notin S \text{ or } (\beta)_i \notin S) \text{ and if } i_0 \text{ is the least such } i, \text{ then } (\alpha)_{i_0} \notin S] \text{ or}$$
$$[\forall i((\alpha)_i \in S \text{ and } (\beta)_i \in S) \text{ and}$$
$$\sup\{|(\alpha)_0|, |(\beta)_0|, |(\alpha)_1|, |(\beta)_1|, \dots\} = \sup_i\{|(\alpha)_i|, |(\beta)_i|\} \in A].$$

Here we think of a real α as coding an ω-sequence of reals $\{(\alpha)_i\}_{i\in\omega}$, where $(\alpha)_i(m) = \alpha(\mathrm{p}_i^{m+1})$ and p_i is the ith prime.

Now define $U \subseteq \wp(\utilde{\delta}^1_n)$ by

$$A \in U \Leftrightarrow \text{player II has a winning strategy in } G^A.$$

We show that U is a $\utilde{\delta}^1_n$-additive measure on $\utilde{\delta}^1_n$.

Lemma 5.2. If $A \in U$ and $B \supseteq A$ then $B \in U$.

Proof. Trivial. ⊣ (Lemma 5.2)

Lemma 5.3. If $A, B \in U$ then $A \cap B \in U$.

PROOF. Given reals α, β, let $\alpha \oplus \beta$ be a real such that $(\alpha \oplus \beta)_{2n} = (\alpha)_n$ and $(\alpha \oplus \beta)_{2n+1} = (\beta)_n$ for all n. Suppose player II has a winning strategy τ in G^A, and a winning strategy σ in G^B. To win $G^{A \cap B}$, given a move α of player I, player II simultaneously builds reals β', β'' such that β' is the result of τ against $\alpha \oplus \beta''$, and β'' is the result of σ against $\alpha \oplus \beta'$. Player II's actual play is then $\beta' \oplus \beta''$. If there is i_0 such that $\forall j \leq i_0((\alpha)_j \in S)$ and $(\beta' \oplus \beta'')_{i_0} \notin S$ we are led to a contradiction. If $\forall i((\alpha)_i \in S$ and $(\beta' \oplus \beta'')_i \in S)$, then $\sup_i\{|(\alpha)_i|, |(\beta' \oplus \beta'')_i|\} = \sup\{|(\alpha \oplus \beta')_i|, |(\beta'')_i|\} = \sup\{|(\alpha \oplus \beta'')_i|, |(\beta)_i|\} \in A \cap B$. ⊣ (Lemma 5.3)

LEMMA 5.4. The filter U contains no bounded sets.

PROOF. If A is bounded, then since $\sup\{|\alpha| : \alpha \in S\} = \underset{\sim}{\delta}^1_n$, player I can easily win G^A. ⊣ (Lemma 5.4)

LEMMA 5.5. The filter U is an ultrafilter.

PROOF. We must show that $A \notin U \Rightarrow (\underset{\sim}{\delta}^1_n \setminus A) \in U$. But if player II has no winning strategy in G^A, then player II can essentially follow player I's winning strategy in G^A to win $G^{(\underset{\sim}{\delta}^1_n \setminus A)}$. ⊣ (Lemma 5.5)

LEMMA 5.6. The ultrafilter U is $\underset{\sim}{\delta}^1_n$-additive.

PROOF. Let $\{A_\xi\}_{\xi<\eta<\underset{\sim}{\delta}^1_n}$ be a sequence of $< \underset{\sim}{\delta}^1_n$ members of U. It suffices to show $\bigcap_{\xi<\eta} A_\xi \neq \varnothing$.

Let $\leq$ be a $\underset{\sim}{\Delta}^1_n$ prewellordering of $\mathbb{R}$ of length η with associated norm φ. For $\xi < \eta$ let

$$f(\xi) = \{\tau : \tau \text{ is a winning strategy for player II in } G^{A_\xi}\}.$$

Let g be a choice subfunction of f such that Code$(g; \leq)$ is $\underset{\sim}{\Sigma}^1_n$, say in $\Sigma^1_n(y)$.

CLAIM 5.7. For each $m \geq 0$, there is a function $f_m \colon {}^{m+1}S \to S$ such that for all $\alpha^0, \ldots, \alpha^m \in S$, for all α with $(\alpha)_i = \alpha^i$ if $i \leq m$ and for all $\tau \in \bigcup_{\xi<\eta} g(\xi)$,

$$|f_m(\alpha^0 \ldots \alpha^m)| \geq |(\tau[\alpha])_m|,$$

where $\tau[\alpha]$ = player II's play when player I plays α and player II follows τ.

PROOF. Given $\alpha^0, \ldots, \alpha^m \in S$, consider the following wellfounded relation:

$$\begin{aligned} \langle \alpha, x, \tau, z \rangle \prec_{\alpha^0,\ldots,\alpha^m} \langle \alpha', x', \tau', z' \rangle \Leftrightarrow \alpha = \alpha' \;\&\; x = x' \\ \&\; \tau = \tau' \;\&\; \forall i \leq m((\alpha)_i = \alpha^i) \\ \&\; (x, \tau) \in \text{Code}(g; \leq) \\ \&\; (z, z') \in W_{(\tau[\alpha])_m}. \end{aligned}$$

Then $\prec_{\alpha^0,\ldots,\alpha^m}$ is $\Sigma^1_n(\alpha^0 \ldots \alpha^m, y)$ with length $\geq |(\tau[\alpha])_m|$ for any τ, α as above. Clearly one can find a continuous f_m such that $f_m(\alpha^0, \ldots, \alpha^m)$ is a $\underset{\sim}{\Sigma}^1_n$-code for $\prec_{\alpha^0,\ldots,\alpha^m}$, proving the claim. ⊣ (Claim 5.7)

Now let $\alpha^0 \in S$ and define inductively

$$\alpha^{m+1} = f_m(\alpha^0 \dots \alpha^m).$$

Let $\theta = \sup\{|\alpha^m| : m \in \omega\}$. Given $\xi < \eta$, let player I play α such that $\forall i((\alpha)_i = \alpha^i)$, and let player II play using a strategy τ from $g(\xi)$, producing a real $\beta = \tau[\alpha]$. Then $\forall i((\beta)_i \in S)$ and $\sup\{|(\alpha)_i|, |(\beta)_i|\} = \sup\{|(\alpha)_i|\} = \theta$. Since player II's strategy τ was winning, $\theta \in A_\xi$. Hence $\theta \in \bigcap_{\xi<\eta} A_\xi$, proving the lemma. ⊣ (Lemma 5.6)

⊣ (Theorem 5.1)

§6. Calculating $\underset{\sim}{\delta}^1_n$ for $n \leq 4$.

THEOREM 6.1 (Classical). $\underset{\sim}{\delta}^1_1 = \omega_1$.

PROOF. Every $\underset{\sim}{\Sigma}^1_1$ set is ω-Suslin. So every $\underset{\sim}{\Sigma}^1_1$ wellfounded relation has length $< \omega_1$. ⊣ (Theorem 6.1)

THEOREM 6.2 (AD) (Martin [Mar71B]). $\underset{\sim}{\delta}^1_2 = \omega_2$.

PROOF. Obvious, since $\underset{\sim}{\delta}^1_2 = (\underset{\sim}{\delta}^1_1)^+$. ⊣ (Theorem 6.2)

THEOREM 6.3 ($\forall\alpha$ ($\alpha^\#$ exists)) (Martin-Solovay [MS69]). Every $\underset{\sim}{\Sigma}^1_3$ set is ω_ω-Suslin.

PROOF. We will show that every Π^1_2 set admits a Δ^1_3-scale $\{\varphi_n\}_{n\in\omega}$ such that for all n, $\mathrm{lh}(\varphi_n) < \omega_\omega$. Note that it suffices to prove that the Π^1_2 set $\mathbb{R}^\# = \{\alpha^\# : \alpha \in \mathbb{R}\}$ admits such a scale. Because if $\{\varphi^*_n\}_{n\in\omega}$ is such a scale on $\mathbb{R}^\#$, put $\varphi_n(\alpha) = \langle\varphi^*_0(\alpha^\#), \alpha^\#(0), \varphi^*_1(\alpha), \alpha^\#(1), \dots, \varphi^*_n(\alpha^\#), \alpha^\#(n)\rangle$, where $\langle\,\rangle$ refers to the ordinal of the $2n$-tuple under the lexicographical order. Then $\{\varphi^*_n\}_{n\in\omega}$ is a Δ^1_3-scale when restricted to any Π^1_2 set A: To show this (assuming $\{\varphi^*_n\}_{n\in\omega}$ has the right properties), let $\alpha_i \in A$, $\alpha_i \to \alpha$, and $\varphi_n(\alpha_i) = \lambda_n$, $\forall i \geq n$. This implies that $\alpha_i^\# \to \beta$ for some β. Since each $\varphi^*_n(\alpha_i^\#)$ is eventually constant, $\beta = \bar{\alpha}^\#$ for some $\bar{\alpha}$. Since there is a recursive f such that $f(\beta^\#) = \beta$ for all β, $\bar{\alpha} = \alpha$. To see that $\alpha \in A$, note that for some Π^1_2 formula φ

$$\begin{aligned}\alpha \in A &\Leftrightarrow \mathbf{L}[\alpha] \models \varphi(\alpha)\\ &\Leftrightarrow \alpha^\#(n_0) = 0,\end{aligned}$$

where n_0 is a Gödel number for $\varphi(\dot{\alpha})$ ($\dot{\alpha}$ is the constant symbol denoting α). Hence since for all i, $\alpha_i \in A$, we have $\alpha_i^\#(n_0) = 0$, therefore $\alpha^\#(n_0) = 0$, *i.e.*, $\alpha \in A$.

Finally, to see that $\varphi_n(\alpha) \leq \lambda_n$, pick k large enough so that $\forall i \geq k$, $\forall p \leq n$, $\varphi^\#_p(\alpha_i)$ is constant and $\alpha_i^\# \restriction (n+1) = \alpha^\# \restriction (n+1)$. Then

$$\begin{aligned}\varphi_n(\alpha) &= \langle\varphi^*_0(\alpha^\#),\ \alpha^\#(0), \dots, \varphi^*_n(\alpha^\#),\ \alpha^\#(n)\rangle\\ &\leq \langle\varphi^\#_0(\alpha_k^\#), \alpha_k^\#(0), \dots, \varphi^*_n(\alpha_k^\#), \alpha_k^\#(n)\rangle = \lambda_n.\end{aligned}$$

Since the sharp operation is Δ^1_3, it is clear that $\{\varphi_n\}_{n\in\omega}$ is a Δ^1_3-scale.

So we must produce a Δ^1_3-scale on $\mathbb{R}^\#$ whose norms have length $< \omega_\omega$. Let $\tau'_0, \tau'_1, \tau'_2 \ldots$ be a recursive enumeration of all definable Skolem functions or **terms** (with variables) in the theory $\mathsf{ZF} + \mathbf{V}{=}\mathbf{L}[\dot\alpha] + \dot\alpha \in \mathbb{R}$ (where $\dot\alpha$ is a constant symbol) and put $\tau_n = \text{rank}(\tau'_n)$ so that τ_n takes only ordinal values. Say $\tau_n = \tau_n\ (v_1, \ldots, v_{k_n})$. Then define

$$\varphi^*_n(\alpha^\#) = \tau_n^{\mathbf{L}[\alpha]}(\omega_1, \ldots, \omega_{k_n}).$$

Note that

$$\begin{aligned}\varphi^*_n(\alpha^\#) < \varphi^*_n(\beta^\#) &\Leftrightarrow \tau_n^{\mathbf{L}[\alpha]}(\omega_1, \ldots, \omega_{k_n}) < \tau_n^{\mathbf{L}[\beta]}(\omega_1, \ldots, \omega_{k_n})\\ &\Leftrightarrow \mathbf{L}[\alpha, \beta] \models \psi(\alpha, \beta, \omega_1, \ldots, \omega_{k_n})\\ &\Leftrightarrow \langle \alpha, \beta\rangle^\#(m) = 0,\end{aligned}$$

where m is obtained recursively from n. Hence $\{\varphi^*_n\}_{n\in\omega}$ is a Δ^1_3-scale, if it is a scale.

We also have $\text{lh}(\varphi^*_n) < \omega_\omega$ since in fact $\tau_n^{\mathbf{L}[\alpha]}(\omega_1, \ldots, \omega_{k_n}) < \omega_{k_n+1}$ for all α (because every cardinal is an indiscernible for every $\mathbf{L}[\alpha]$).

To show $\{\varphi^*_n\}_{n\in\omega}$ is a scale, let $\alpha_i^\# \in \mathbb{R}^\#$, $\alpha_i^\# \to \beta$ and $\varphi^*_n(\alpha_i^\#) = \lambda_n$ for $i > n$. Note that

$$\beta \in \mathbb{R}^\# \Leftrightarrow P(\beta)\ \&\ \Gamma(\beta, \omega_1) \text{ is wellfounded},$$

where P is Π^0_1 expressing "β is a set of Gödel numbers of formulas satisfying the syntactical conditions for a remarkable character relative to some real α", and $\Gamma(\beta, \xi)$ is the model of $\mathsf{ZF} + \mathbf{V}{=}\mathbf{L}[\dot\alpha]$ generated by ξ indiscernibles on the basis of β.

Since $P(\alpha_i^\#)$ for all i, we know that $P(\beta)$ holds.

Let $\mathcal{I}^\alpha =$ class of Silver indiscernibles for $\mathbf{L}[\alpha]$. Let $C = \bigcap_{i,j}(\mathcal{I}^{\langle\alpha_i,\alpha_j\rangle} \cap \omega_1)$. Thus C is closed unbounded in ω_1. Let $\{c_\xi\}_{\xi<\omega_1}$ be its increasing enumeration.

Since for any fixed n, $\tau_n^{\mathbf{L}[\alpha_i]}(\omega_1, \ldots, \omega_{k_n})$ becomes eventually constant, the same is true of $\tau_n^{\mathbf{L}[\alpha_i]}(c_{\xi_1}, \ldots, c_{\xi_{k_n}})$ for all $\xi_1 < \cdots < \xi_{k_n} < \omega_1$. So define $f\colon \text{Ord}^{\Gamma(\beta,\omega_1)} \to \text{Ord}$ by $f(\tau_n^{\Gamma(\beta,\omega_1)}(i_{\xi_1}, \ldots, i_{\xi_{k_n}}))$ = eventual value of $\tau_n^{\mathbf{L}[\alpha_i]}(c_{\xi_1}, \ldots, c_{\xi_{k_n}})$, where $I = \{i_\xi : \xi < \omega_1\}$ is a generating set of indiscernibles for $\Gamma(\beta, \omega_1)$.

Claim 6.4. f is well defined and order preserving.

Proof. Suppose $\tau_n^{\Gamma(\beta,\omega_1)}(i_{\xi_1}, \ldots, i_{\xi_{k_n}}) = \tau_m^{\Gamma(\beta,\omega_1)}(i_{\xi'_1}, \ldots i_{\xi'_{k_m}})$, where $\xi_1 < \cdots < \xi_{k_n}$ and $\xi'_1 < \cdots < \xi'_{k_m}$. Then there is a ψ such that $\Gamma(\beta, \omega_1) \models \psi(i_{\eta_1}, \ldots, i_{\eta_\ell}) \Leftrightarrow \tau_n(i_{\xi_1}, \ldots, i_{\xi_{k_n}}) = \tau_m(i_{\xi'_i}, \ldots, i_{\xi'_{k_m}})$, where $\eta_1 < \cdots < \eta_\ell$ is $\{\xi_1 \ldots \xi_{k_n}, \xi'_1 \ldots \xi'_{k_n}\}$ written in increasing order.

Thus $\beta(\ulcorner\psi(v_1,\dots,v_\ell)\urcorner) = 0$. Since $\alpha_i^\# \to \beta_{\mathbf{L}[\alpha_i]}$, the eventual value of $\alpha_i^\#(\ulcorner\psi(v_1,\dots,v_\ell)\urcorner)$ is 0. Hence eventually

$$\tau_n^{\mathbf{L}[\alpha_i]}(c_{\xi_1},\dots,c_{\xi_{k_n}}) = \tau_m^{\mathbf{L}[\alpha_i]}(c_{\xi'_1},\dots,c_{\xi'_{k_m}}),$$

so f is well defined. Similarly f is order preserving. ⊣ (Claim 6.4)

Hence $\Gamma(\beta,\omega_1)$ is well founded, so $\beta \in \mathbb{R}^\#$. So $\beta = \alpha^\#$, where $\alpha_i \to \alpha$. Thus $\Gamma(\beta,\omega_1) = \mathbf{L}_{\omega_1}[\alpha]$. Let $\{i_\xi^\alpha : \xi < \omega_1\}$ be the increasing enumeration of the Silver indiscernibles of $\mathbf{L}_{\omega_1}[\alpha]$. Let $C^* = \{\xi < \omega_1 : c_\xi = i_\xi^\alpha = \xi\}$. Then C^* is closed unbounded, and since f is order preserving, $\forall c_{\xi_1} < \cdots < c_{\xi_{k_n}}$ in C^* we have

$$\begin{aligned}\tau_n^{\mathbf{L}[\alpha]}(c_{\xi_1},\dots,c_{\xi_{k_n}}) = \tau_n^{\mathbf{L}[\alpha]}(i_{\xi_1}^\alpha,\dots,i_{\xi_{k_n}}^\alpha) &\le f(\tau_n^{\mathbf{L}[\alpha]}(i_{\xi_1}^\alpha,\dots,i_{\xi_{k_n}}^n)) \\ &= \text{eventual value of } \tau_n^{\mathbf{L}[\alpha_i]}(c_{\xi_1},\dots,c_{\xi_{k_n}}).\end{aligned}$$

Thus $\tau_n^{\mathbf{L}[\alpha_i]}(\omega_1,\dots,\omega_{k_n}) \le$ eventual value of $\tau_n^{\mathbf{L}[\alpha]}(\omega_1,\dots,\omega_{k_n}) = \lambda_n$, *i.e.*, $\varphi_n^*(\alpha^\#) \le \lambda_n$. Hence $\{\varphi_n^*\}_{n\in\omega}$ is a scale. ⊣ (Theorem 6.3)

Theorem 6.5 (AD) (Martin [Mar71B]). $\underset{\sim}{\delta}_3^1 = \omega_{\omega+1}$.

Proof. We have $\underset{\sim}{\delta}_3^1 = \kappa_3^+$, where $\kappa_3 > \omega$ is a cardinal of cofinality ω and κ_3 is the least cardinal such that every $\underset{\sim}{\Sigma}_3^1$ set is κ_3-Suslin. Hence $\kappa_3 \le \omega_\omega$, so $\kappa_3 = \omega_\omega$, since countable choice implies that ω_ω is the second cardinal of cofinality ω. Hence $\underset{\sim}{\delta}_3^1 = \omega_{\omega+1}$. ⊣ (Theorem 6.5)

Corollary 6.6 (AD) (Kunen [Kun71C], Martin [Mar71B]). $\underset{\sim}{\delta}_4^1 = \omega_{\omega+2}$.

§7. The closed unbounded measure on ω_1.

Theorem 7.1 (AD) (Solovay [Sol67A] for $n = 1$, Moschovakis [Mos70] in general). Let $P \in \underset{\sim}{\Pi}_{2n+1}^1 \setminus \underset{\sim}{\Delta}_{2n+1}^1$, and let φ be a $\underset{\sim}{\Pi}_{2n+1}^1$ norm on P with associated prewellordering $\le$. Then for every $A \subseteq \underset{\sim}{\delta}_{2n+1}^1$, $\mathrm{Code}(A;\le) \in \underset{\sim}{\Pi}_{2n+1}^1$.

Proof. By the Coding Lemma we know that for all $\xi < \underset{\sim}{\delta}_{2n+1}^1$, $\mathrm{Code}(A \cap \xi;\le) \in \underset{\sim}{\Delta}_{2n+1}^1$. Consider the following game: Player I plays w, player II plays α, and player II wins iff $[w \in P \Rightarrow \alpha$ is a $\underset{\sim}{\Delta}_{2n+1}^1$-code of a set $\underset{\sim}{\Delta}_\alpha$ such that

$$\mathrm{Code}(A \cap (\varphi(w)+1);\le) \subseteq \underset{\sim}{\Delta}_\alpha \subseteq \mathrm{Code}(A;\le)].$$

If player I has a winning strategy σ, then $\{\sigma(\alpha) : \alpha \in \mathbb{R}\} = \mathrm{Q}$ is a $\underset{\sim}{\Sigma}_1^1$ subset of P, hence by boundedness, $\xi = \sup\{\varphi(w) : w \in \mathrm{Q}\} < \underset{\sim}{\delta}_{2n+1}^1$. So player II can easily beat this strategy by playing a $\underset{\sim}{\Delta}_{2n+1}^1$-code of $\mathrm{Code}(A \cap (\xi+1);\le)$.

Hence player II has a winning strategy τ, and

$$w \in \mathrm{Code}(A;\le) \Leftrightarrow w \in P \;\&\; w \in \underset{\sim}{\Delta}_{\tau(w)},$$

so $\mathrm{Code}(A;\le)$ is $\underset{\sim}{\Pi}_{2n+1}^1$. ⊣ (Theorem 7.1)

THEOREM 7.2 (AD) (Solovay [Sol67A]). For every $A \subseteq \omega_1$, $\exists \alpha \in \mathbb{R}(A \in \mathbf{L}[\alpha])$.

PROOF. Let WO $= \{\alpha : \alpha$ codes a wellordering of $\omega\}$ and for $\alpha \in$ WO, let $|\alpha| =$ the ordinal coded by α. For $A \subseteq \omega_1$, let $\text{Code}(A) = \{\alpha : |\alpha| \in A\}$. By the above, for every $A \subseteq \omega_1$, $\text{Code}(A) \in \undertilde{\mathbf{\Pi}}^1_1$. We will now show (in ZF+DC) that for any A with $\text{Code}(A) \in \undertilde{\mathbf{\Sigma}}^1_2$, there is an α such that $A \in \mathbf{L}[\alpha]$.

Let P$\in \Sigma^1_2$ be such that $\alpha \in \text{Code}(A) \Leftrightarrow P(\beta_0, \alpha)$ for some β_0. Then $\xi \in A \Leftrightarrow \exists \alpha(P(\beta_0, \alpha) \ \& \ |\alpha| = \xi)$.

Case I: For some γ, $\omega_1^{\mathbf{L}[\gamma]} = \omega_1$.
Then $\xi \in A \Leftrightarrow \mathbf{L}[\gamma, \beta_0] \models \exists \alpha(P(\beta_0, \alpha) \ \& \ |\alpha| = \xi)$, so $A \in \mathbf{L}[\gamma, \beta_0]$.

Case II: For all γ, $\omega_1^{\mathbf{L}[\gamma]} < \omega_1$.
Then if C_ξ is the notion of forcing which collapses ξ to ω (for $\xi < \omega_1$), there are C_ξ-generic over $\mathbf{L}[\beta_0]$ sets (since $(\wp(\xi))^{\mathbf{L}[\beta_0]}$ is countable). Hence

$$\begin{aligned}\xi \in A &\Leftrightarrow \exists \alpha(P(\beta_0, \alpha) \ \& \ |\alpha| = \xi)\\ &\Leftrightarrow \text{For all } C_\xi\text{-generic over } \mathbf{L}[\beta_0]G,\\ &\qquad \mathbf{L}[\beta_0, G] \models \exists \alpha(P(\beta_0, \alpha) \ \& \ |\alpha| = \xi)\\ &\Leftrightarrow \varnothing \Vdash^{\mathbf{L}[\beta_0]}_{C_\xi} \varphi(\hat{\beta}_0, \hat{\xi})\end{aligned}$$

for some formula φ.
Since forcing is definable in $\mathbf{L}[\beta_0]$, this shows that $A \in \mathbf{L}[\beta_0]$. ⊣ (Theorem 7.2)

THEOREM 7.3 (AD) (Solovay [Sol67A]). There is a unique normal measure μ on ω_1, namely

$$\mu(A) = 1 \Leftrightarrow A \text{ contains a closed unbounded set.}$$

PROOF. It suffices to show that for all $A \subseteq \omega_1$, either A or $\omega_1 \setminus A$ contains a cub set. Given $A \subseteq \omega_1$, let α be such that $A \in \mathbf{L}[\alpha]$. Then $A = \tau^{\mathbf{L}[\alpha]}(i^\alpha_{\eta_1}, \dots, i^\alpha_{\eta_k}, i^\alpha_{\xi_1}, \dots, i^\alpha_{\xi_m})$ for some $\eta_1 < \cdots < \eta_k < \omega_1 \leq \xi_1 \cdots < \xi_m$ and some term τ. But then $C = \{i^\alpha_\eta : \eta_k < \eta < \omega_1\}$ is closed unbounded, and either $C \subseteq A$ or $C \subseteq \omega_1 \setminus A$ ⊣ (Theorem 7.3)

§8. Uniform indiscernibles and the ω_n's for $n \leq \omega$.

DEFINITION 8.1. An ordinal u is a uniform indiscernible if $\forall \alpha \in \mathbb{R}(u \in \mathcal{I}^\alpha)$, where $\mathcal{I}^\alpha = \{i^\alpha_\xi\}_{\xi \in \text{Ord}}$ is the class of Silver indiscernibles for $\mathbf{L}[\alpha]$. Let

$$U = \{u_\xi\}_{\xi \in \text{Ord}}$$

by the increasing enumeration of the uniform indiscernibles.

Clearly $u_1 = \omega_1$, U is closed unbounded and every cardinal is in U. Hence $u_\xi \leq \omega_\xi$.

THEOREM 8.2 (AD) (Martin [Mar71B]). $u_\omega = \omega_\omega$.

PROOF. In the proof of Theorem 6.3 we could have used

$$\varphi_n^*(\alpha^\#) = \tau_n^{\mathbf{L}[\alpha]}(u_1, \dots, u_{k_n}),$$

hence $\mathbf{\Sigma}^1_3$ sets are u_ω-Suslin. Hence as in Theorem 6.5

$$(u_\omega)^+ = \boldsymbol{\delta}^1_3 = (\omega_\omega)^+$$

hence $\omega_\omega = u_\omega$. ⊣ (Theorem 8.2)

LEMMA 8.3 ($\forall\alpha(\alpha^\#$ exists)). For every ordinal ξ there is a real α, a term τ, and ordinals $\eta_1 < \dots < \eta_m$ such that $\xi = \tau^{\mathbf{L}[\alpha]}(u_{\eta_1}, \dots, u_{\eta_m})$.

PROOF. By induction on ξ. Clear if $\xi < \aleph_1$. So assume true for all $\xi' < \xi$, and let $\xi \notin U$. Then for some α, $\xi \notin \mathcal{I}^\alpha$. Thus

$$\xi = \sigma^{\mathbf{L}[\alpha]}(i^\alpha_{\theta_1}, \dots, i^\alpha_{\theta_m}, i^\alpha_{\theta_{m+1}}, \dots, i^\alpha_{\theta_k})$$

for some term σ and some

$$i^\alpha_{\theta_1} < \dots < i^\alpha_{\theta_m} < \xi < i^\alpha_{\theta_{m+1}} < \dots < i^\alpha_{\theta_k}.$$

Thus

$$\xi = \sigma^{\mathbf{L}[\alpha]}(i^\alpha_{\theta_1} \dots i^\alpha_{\theta_m}, \vec{\aleph}),$$

where $\vec{\aleph}$ is a sequence of large enough cardinals. Now for all $j \le m$, $i^\alpha_{\theta_j}$ can be defined in some $\mathbf{L}[\beta]$ using uniform indiscernibles (by induction hypothesis) hence

$$\xi = \tau^{\mathbf{L}[\alpha,\beta]}(u_{\rho_1}, \dots, u_{\rho_n}, \vec{\aleph}),$$

for some $\rho_1 < \dots < \rho_n$ and we are done. ⊣ (Lemma 8.3)

DEFINITION 8.4. For λ an ordinal, let

$$\begin{aligned}(\lambda, \alpha)^+ &= \text{ first element of } \mathcal{I}^\alpha > \lambda,\\ (\lambda^+)^\alpha &= \text{ first cardinal in } \mathbf{L}[\alpha] > \lambda.\end{aligned}$$

LEMMA 8.5 ($\forall\alpha(\alpha^\#$ exists)). For all ξ,

$$u_{\xi+1} = \sup_{\alpha\in\mathbb{R}}(u_\xi, \alpha)^+ = \sup_{\alpha\in\mathbb{R}}(u_\xi^+)^\alpha$$

PROOF. Clearly

$$\begin{aligned}u_{\xi+1} &\ge \sup_{\alpha\in\mathbb{R}}(u_\xi^+)^\alpha \ge \sup_{\alpha\in\mathbb{R}}(u_\xi^+)^{\alpha^\#}\\ &\ge \sup_{\alpha\in\mathbb{R}}(u_\xi, \alpha)^+ > u_\xi,\end{aligned}$$

so it suffices to show that $\sup_{\alpha\in\mathbb{R}}(u_\xi, \alpha)^+$ is a uniform indiscernible. Given $\beta \in \mathbb{R}$, $\lambda = \sup_{\alpha\in\mathbb{R}}(u_\xi, \alpha)^+ = \sup_{\alpha\ge_{\mathrm{T}}\beta^\#}(u_\xi, \alpha)^+$ (where $\ge_{\mathrm{T}}$ is Turing reducibility). Hence λ is a sup of members of $\mathcal{I}^\beta$, so $\lambda \in \mathcal{I}^\beta$. Hence $\lambda \in U$. ⊣ (Lemma 8.5)

Lemma 8.6 ($\forall\alpha(\alpha^{\#}$ exists)). If $\xi < u_{\theta+1}$, then there exist $\theta_1 < \cdots < \theta_n \leq \theta$, $\alpha \in \mathbb{R}$ and a term τ such that

$$\xi = \tau^{\mathbf{L}[\alpha]}(u_{\theta_1} \ldots u_{\theta_n}).$$

Proof. We can assume inductively that $u_\theta \leq \xi < u_{\theta+1}$. Then

$$\xi = \sigma^{\mathbf{L}[\beta]}(u_{\theta_1}, \ldots, u_{\theta_{n-1}}, u_{\eta_1}, \ldots, u_{\eta_k}),$$

where

$$u_{\theta_1} < \cdots < u_{\theta_{n-1}} \leq u_\theta < u_{\eta_1} < \cdots < u_{\eta_k},$$

by Lemma 8.3. Find γ such that

$$\xi < (u_\theta, \gamma)^+ < u_{\theta+1}$$

and let $\lambda_1 < \cdots < \lambda_k$ be the first k cardinals above u_θ in $\mathbf{L}[\alpha]$, where $\alpha = \langle \beta^{\#}, \gamma^{\#}\rangle$. Then $\lambda_1, \ldots, \lambda_k \in \mathcal{I}^\beta$ and $(u_\theta, \gamma)^+ < \lambda_1$. Hence $\xi = \sigma^{\mathbf{L}[\beta]}(u_{\theta_1}, \ldots, u_{\theta_{n-1}}, \lambda_1, \ldots, \lambda_k)$ and since $\lambda_1, \ldots, \lambda_k$ are definable in $\mathbf{L}[\alpha]$ from u_θ, we have for some term τ

$$\xi = \tau^{\mathbf{L}[\alpha]}(u_{\theta_1}, \ldots, u_{\theta_{n-1}}, u_\theta). \qquad \dashv \text{(Lemma 8.6)}$$

Theorem 8.7 ($\forall\alpha(\alpha^{\#}$ exists)) (Solovay). For all ξ, $\mathrm{cf}(u_{\xi+1}) = \mathrm{cf}(u_2)$.

Proof. Define $f : u_2 \to u_{\xi+1}$ by $f(\tau^{\mathbf{L}[\alpha]}(u_1)) = \tau^{\mathbf{L}[\alpha]}(u_\xi)$. By indiscernibility, f is well defined and order preserving. Since $\sup_{\alpha\in\mathbb{R}}(u_\xi, \alpha)^+ = u_{\xi+1}$, f is cofinal. ⊣ (Theorem 8.7)

Theorem 8.8 (AD) (Martin [Mar71B]). For all $2 \leq n < \omega$, $\mathrm{cf}(\omega_n) = \omega_2$.

Proof. $u_\omega = \omega_\omega$, hence $\omega_n = u_{k_n+1}$ for some k_n. ⊣ (Theorem 8.8)

Theorem 8.9 (AD) (Kunen, Solovay). For all $1 \leq n \leq \omega$, $u_n = \omega_n$.

Proof. Let μ be the closed unbounded measure on ω_1.
Suppose that for all n, we can prove

$${}^{\omega_1}u_n/\mu \cong u_{n+1}.$$

If for some k, $\mathrm{Card}(u_k) = \mathrm{Card}(u_{k+1})$, then

$$\mathrm{Card}(u_{k+2}) = \mathrm{Card}(({}^{\omega_1}u_{k+1}/\mu)) = \mathrm{Card}(({}^{\omega_1}u_k/\mu)) = \mathrm{Card}(u_{k+1})$$

and similarly $\mathrm{Card}(u_{k+1}) = \mathrm{Card}(u_{k+n})$ for all n. Hence $u_\omega < \omega_\omega$, a contradiction. Hence it is enough to show that $\forall n({}^{\omega_1}u_n/\mu \cong u_{n+1})$.

Lemma 8.10. Let $\tilde{\mathbf{L}} \stackrel{\text{def}}{=} \bigcup_{\alpha\in\mathbb{R}} \mathbf{L}[\alpha]$. Then for all n, ${}^{\omega_1}u_n \subseteq \tilde{\mathbf{L}}$.

Proof. We prove by induction on $\xi < u_\omega$ that ${}^{\omega_1}\xi \subseteq \tilde{\mathbf{L}}$. If for some α, ξ is not a cardinal in $\mathbf{L}[\alpha]$, then this is obvious by induction hypothesis. If ξ is a cardinal in all $\mathbf{L}[\alpha]$'s, then it is a uniform indiscernible, hence has cofinality $=$ $\mathrm{cf}(u_2)$, so it is enough to show that $u_2 = \omega_2$.

To see that $u_2 = \omega_2$: Clearly $u_2 \leq \omega_2$. Suppose $u_2 < \omega_2$. Let $A \subseteq \omega_1 \times \omega_1$ be a well ordering of ω_1 with order type u_2. Then for some $\alpha \in \mathbb{R}$, $A \in \mathbf{L}[\alpha]$. Hence the order type of A is $< (\omega_1^+)^\alpha < u_2$, contradiction. Hence $u_2 = \omega_2$. ⊣ (Lemma 8.10)

Now to complete the proof of Theorem 8.9, let $f \in {}^{\omega_1}u_n$, $n \geq 1$. Then there is a real β such that for some term τ and μ-almost all ξ,

$$f(\xi) = \tau^{\mathbf{L}[\beta]}(\xi; u_1, \dots, u_{n-1}).$$

Indeed, since $f \in \tilde{\mathbf{L}}$ we have by Lemma 8.6 that for some $\alpha \in \mathbb{R}$ and for some term σ

$$f(\xi) = \sigma^{\mathbf{L}[\alpha]}(\xi; u_1, \dots, u_{n-1}, u_n) < u_n,$$

so for all $\xi \in \mathcal{I}^\alpha \cap \omega_1$,

$$\begin{aligned} f(\xi) &= \sigma^{\mathbf{L}[\alpha]}(\xi; u_1, \dots, u_{n-1}, (u_{n-1}, \alpha)^+) \\ &= \tau^{\mathbf{L}[\beta]}(\xi; u_1, \dots, u_{n-1}) \end{aligned}$$

for some term τ and $\beta = \alpha^\#$.

Let $[f]_\mu$ be the equivalence class of f in ${}^{\omega_1}u_n/\mu$ and define

$$j([f]_\mu) = \tau^{\mathbf{L}[\beta]}(u_1; u_2, \dots, u_n) < u_{n+1}.$$

Clearly j is well defined and order preserving by indiscernibility.

Let $\theta < u_{n+1}$. Find τ, α such that $\theta = \tau^{\mathbf{L}[\alpha]}(u_1, \dots, u_n)$. Now define $f \in {}^{\omega_1}u_n$ by

$$f(\xi) = \tau^{\mathbf{L}[\alpha]}(\xi; u_1, \dots, u_{n-1}).$$

Then $j([f]_\mu) = \theta$, so j is onto.

Hence $j\colon {}^{\omega_1}u_n/\mu \cong u_{n+1}$. ⊣ (Theorem 8.9)

To summarize: From AD,

$$\underset{\sim}{\boldsymbol{\delta}}^1_1 = \omega_1 = u_1, \text{ measurable}$$

$$\underset{\sim}{\boldsymbol{\delta}}^1_2 = \omega_2 = u_2, \text{ measurable}$$

For $n \geq 3$, $\omega_n = u_n$ singular, cofinality $= \omega_2$,

$$\underset{\sim}{\boldsymbol{\delta}}^1_3 = \omega_{\omega+1}, \text{ measurable}$$

$$\underset{\sim}{\boldsymbol{\delta}}^1_4 = \omega_{\omega+2}, \text{ measurable.}$$

Proposition 8.11 (AD). For $n \geq 3$, $\underset{\sim}{\boldsymbol{\delta}}^1_n = u_{\underset{\sim}{\boldsymbol{\delta}}^1_n}$.

Proof. If $\underset{\sim}{\boldsymbol{\delta}}^1_n < u_{\underset{\sim}{\boldsymbol{\delta}}^1_n}$, then since $\underset{\sim}{\boldsymbol{\delta}}^1_n$ is a cardinal, $\underset{\sim}{\boldsymbol{\delta}}^1_n = u_\xi$ for some $\xi < \underset{\sim}{\boldsymbol{\delta}}^1_n$. Now ξ cannot be a successor (otherwise $\mathrm{cf}(\underset{\sim}{\boldsymbol{\delta}}^1_n) = \omega_2$). Hence ξ is limit, hence $\underset{\sim}{\boldsymbol{\delta}}^1_n$ is singular, contradiction. ⊣ (Proposition 8.11)

Basic Open Problem 8.12. Compute $\underset{\sim}{\boldsymbol{\delta}}^1_5$.

§9. Back to the real world.

For a moment we interrupt the development of the theory of projective ordinals in the context of ZF+DC+AD, to see what is the picture of these ordinals in a context with Choice and Projective Determinacy only.

THEOREM 9.1.

0) $\underset{\sim}{\boldsymbol{\delta}}^1_1 = \omega_1$.

1) $\underset{\sim}{\boldsymbol{\delta}}^1_2 \leq \omega_2$. (Martin [Mar71B])

$(\forall\alpha(\alpha^\#$ exists$))$ $\underset{\sim}{\boldsymbol{\delta}}^1_3 = u_2$. (Martin [Mar71B])

2) $(\mathsf{AC} + \forall\alpha(\alpha^\#$ exists$))$ $\underset{\sim}{\boldsymbol{\delta}}^1_3 \leq \omega_3$. (Martin [Mar71B])

3) (AC+PD) $\underset{\sim}{\boldsymbol{\delta}}^1_4 \leq \omega_4$. (Kunen [Kun71C], Martin [Mar71B])

4) (PD) $\underset{\sim}{\boldsymbol{\delta}}^1_{2n+2} \leq (\underset{\sim}{\boldsymbol{\delta}}^1_{2n+1})^+$. (Kunen [Kun71C], Martin [Mar71B])

5) (PD) For all n, $\underset{\sim}{\boldsymbol{\delta}}^1_n < \underset{\sim}{\boldsymbol{\delta}}^1_{n+1}$. (Kechris [Kec74], Moschovakis [Mos70])

PROOF. 1) That $\underset{\sim}{\boldsymbol{\delta}}^1_2 \leq \omega_2$ follows from the Kunen-Martin theorem and the fact that every $\underset{\sim}{\boldsymbol{\Sigma}}^1_2$ set is ω_1-Suslin. That $\underset{\sim}{\boldsymbol{\delta}}^1_2 \leq u_2$ follows from the fact that if $A \in \Sigma^1_2(\alpha)$, then $A = \mathrm{p}[T]$, where $T \in \mathbf{L}[\alpha]$ is a tree on $\omega \times \omega_1$. By the proof of the Kunen-Martin theorem the length of a $\Sigma^1_2(\alpha)$ wellfounded relation is $< (\omega_1^+)^\alpha < u_2$, hence $\underset{\sim}{\boldsymbol{\delta}}^1_2 \leq u_2$.

To see that $u_2 \leq \underset{\sim}{\boldsymbol{\delta}}^1_2$: For every $\alpha \in \mathbb{R}$ and for every $\xi < (\omega_1^+)^\alpha$ we can find a term τ such that

$$\xi = \tau^{\mathbf{L}[\alpha]}(i^\alpha_{\xi_1} \dots i^\alpha_{\xi_n}, \omega_1, \vec{\aleph}),$$

for some $\xi_1 < \cdots < \xi_n < \omega_1 < \vec{\aleph}$. Coding the $\xi_1, \dots, \xi_n$ by reals we can easily find a $\Pi^1_1(\alpha^\#)$ prewellordering of reals of length $> (\omega_1^+)^\alpha$, so $u_2 = \sup_\alpha (\omega_1^+)^\alpha \leq \underset{\sim}{\boldsymbol{\delta}}^1_2$.

2) To prove $\underset{\sim}{\boldsymbol{\delta}}^1_3 \leq \omega_3$: We know that every $\underset{\sim}{\boldsymbol{\Sigma}}^1_3$ set is u_ω-Suslin by the Martin-Solovay theorem. Hence $\underset{\sim}{\boldsymbol{\delta}}^1_3 \leq (u_\omega)^+$. Since $\forall n \leq 2$, $\mathrm{cf}(u_n) = \mathrm{cf}(u_2)$, we must have $u_n < \omega_3$. But ω_3 is regular, hence $u_\omega < \omega_3$. Hence $\underset{\sim}{\boldsymbol{\delta}}^1_3 \leq \omega_3$.

3,4) These follow from the fact that every $\underset{\sim}{\boldsymbol{\Sigma}}^1_{2n+2}$ set is $\underset{\sim}{\boldsymbol{\delta}}^1_{2n+1}$-Suslin.

5) To prove $\underset{\sim}{\boldsymbol{\delta}}^1_{2n+1} < \underset{\sim}{\boldsymbol{\delta}}^1_{2n+2}$ use the fact that a $\underset{\sim}{\boldsymbol{\Pi}}^1_{2n+1}$ norm on a complete $\underset{\sim}{\boldsymbol{\Pi}}^1_{2n+1}$ set has length exactly $\underset{\sim}{\boldsymbol{\delta}}^1_{2n+1}$.

To prove that $\underset{\sim}{\boldsymbol{\delta}}^1_{2n} < \underset{\sim}{\boldsymbol{\delta}}^1_{2n+1}$, prove first that

$$\underset{\sim}{\boldsymbol{\delta}}^1_{2n+1} = \sup\{\xi : \xi \text{ is the length of a } \underset{\sim}{\boldsymbol{\Sigma}}^1_{2n+1} \text{ wellfounded relation}\}$$

using the recursion theorem (see Moschovakis [Mos70]). Then let $W \subseteq \mathbb{R} \times \mathbb{R} \times \mathbb{R}$ be universal $\underset{\sim}{\boldsymbol{\Sigma}}^1_{2n}$, and let

$$(\alpha, \beta) \prec (\alpha', \gamma) \Leftrightarrow \alpha = \alpha' \;\&\; W_\alpha \text{ is wellfounded} \;\&\; W(\alpha, \beta, \gamma).$$

Then $\prec$ is $\underset{\sim}{\boldsymbol{\Delta}}^1_{2n+1}$ and dominates every $\underset{\sim}{\boldsymbol{\Sigma}}^1_{2n}$ wellfounded relation. Hence $\underset{\sim}{\boldsymbol{\delta}}^1_{2n} < \underset{\sim}{\boldsymbol{\delta}}^1_{2n+1}$. ⊣ (Theorem 9.1)

THEOREM 9.2 (Sierpiński for $\kappa = \omega$). If A is κ-Suslin, then A is the union of κ^+ sets in $\boldsymbol{B}_{\kappa^+}$.

PROOF. Let $\alpha \in A \Leftrightarrow T(\alpha)$ not wellfounded, T a tree on $\omega \times \kappa$. For $\alpha \in A$ let

$$\psi(\alpha) = \sup\{|T(\alpha)_u| : u \in {}^{<\omega}\kappa \;\&\; T(\alpha)_u \text{ is wellfounded}\} < \kappa^+.$$

For $\xi < \kappa^+$, let $B_\xi = \{\alpha \in A : \psi(\alpha) \leq \xi\}$. Since $A = \bigcup_{\xi<\kappa^+} B_\xi$, it is enough to show each $B_\xi \in \boldsymbol{B}_{\kappa^+}$.

For $\xi < \kappa^+$,

$$\alpha \in B_\xi \Leftrightarrow T(\alpha) \text{ is not wellfounded } \& \quad \forall u[T(\alpha)_u \text{ wellfounded} \Rightarrow |T(\alpha)_u| \leq \xi],$$

therefore

$$\begin{aligned}\alpha \notin B_\xi &\Leftrightarrow T(\alpha) \text{ is wellfounded} \vee \exists u[T(\alpha)_u \text{ is wellfounded } \& \; |T(\alpha)_u| > \xi] \\ &\Leftrightarrow |T(\alpha)| < \xi + 1 \vee \exists u[|T(\alpha)_u| = \xi + 1].\end{aligned}$$

Since by the proof of Theorem 3.18

$$A_u^\xi = \{\alpha : |T(\alpha)_u| < \xi\} \in \boldsymbol{B}_{\kappa^+},$$

we are done. ⊣ (Theorem 9.2)

THEOREM 9.3.
1) (Sierpiński) Every $\underset{\sim}{\boldsymbol{\Sigma}}^1_2$ set is the union of $\aleph_1$ Borel sets.
2) (AC+$\forall\alpha(\alpha^\#$ exists)) (Martin [Mar71B]). Every $\underset{\sim}{\boldsymbol{\Sigma}}^1_3$ set is the union of $\aleph_2$ Borel sets.
3) (AC+PD) (Martin [Mar71B]). Every $\underset{\sim}{\boldsymbol{\Sigma}}^1_4$ set is the union of $\aleph_3$ Borel sets.

PROOF. 1) Every $\underset{\sim}{\boldsymbol{\Pi}}^1_1$ set is the union of $\aleph_1$ Borel sets. Hence every $\underset{\sim}{\boldsymbol{\Sigma}}^1_2$ set is the union of $\aleph_1$ $\underset{\sim}{\boldsymbol{\Sigma}}^1_1$ sets. So it suffices to show that the $\underset{\sim}{\boldsymbol{\Sigma}}^1_1$ sets are unions of $\aleph_1$ Borel sets. This follows from 9.2. (This proof uses AC. This can be avoided by using the uniformization theorem for $\underset{\sim}{\boldsymbol{\Pi}}^1_1$ sets.)

2) Every $\underset{\sim}{\boldsymbol{\Sigma}}^1_3$ set is u_ω-Suslin. Since $\mathrm{cf}(u_n) \leq \aleph_2$ for $n < \omega$, we have $u_\omega < \aleph_3$. So $\underset{\sim}{\boldsymbol{\Sigma}}^1_3$ sets are $\aleph_2$-Suslin. If A is $\underset{\sim}{\boldsymbol{\Sigma}}^1_3$ then for some tree T on $\omega \times \aleph_2$

$$\begin{aligned} a \in A &\Leftrightarrow T(\alpha) \text{ not wellfounded} \\ &\Leftrightarrow \exists \xi < \aleph_2 (T^\xi(\alpha) \text{ not wellfounded}) \text{ (see 3.19)}.\end{aligned}$$

So A is the union of $\aleph_2$ many $\aleph_1$-Suslin sets. By the same argument, each $\aleph_1$-Suslin is the union of $\aleph_1$ many ω-Suslin (*i.e.*, $\underset{\sim}{\boldsymbol{\Sigma}}^1_1$) sets and we are done.

3) Similar, using the fact that every $\underset{\sim}{\boldsymbol{\Sigma}}^1_4$ set is $\underset{\sim}{\boldsymbol{\delta}}^1_3$-Suslin, and the fact that $\underset{\sim}{\boldsymbol{\delta}}^1_3 \leq \aleph_3$. ⊣ (Theorem 9.3)

BASIC OPEN PROBLEM 9.4. Is it true that (from any reasonable hypotheses and AC): $\underset{\sim}{\boldsymbol{\delta}}^1_n \leq \aleph_n$, for $n \geq 5$?

§10. Infinite exponent partition relations and the singular measures μ_λ.

DEFINITION 10.1. If α, β, γ are ordinals with $\gamma \leq \beta \leq \alpha$, we put

$$\alpha \to (\beta)^\gamma$$

iff for every $X \subseteq {}^\gamma[\alpha] = \{f \in {}^\gamma\alpha : f \text{ increasing}\}$ there is an $H \subseteq \alpha$ of order type β such that either ${}^\gamma[H] \subseteq X$ or ${}^\gamma[H] \subseteq \neg X$.

REMARK 10.2. $\mathsf{ZFC} \vdash \neg\exists\kappa(\kappa \to (\omega)^\omega)$.

DEFINITION 10.3. Let κ be a regular cardinal, and let λ be a regular cardinal $< \kappa$. The filter μ_λ is the collection of all subsets of κ which contain a λ-closed unbounded set. ($A \subseteq \kappa$ is **λ-closed** if every increasing λ-sequence from A has its limit in A.)

THEOREM 10.4 (Kleinberg [Kle70]).

1) If κ is a regular uncountable cardinal, $\lambda < \kappa$ a regular cardinal, and $\kappa \to (\kappa)^{\lambda+\lambda}$, then μ_λ is a normal measure.
2) If κ is a regular uncountable cardinal with $< \kappa$ many regular cardinals below κ, and $\forall\xi < \kappa(\kappa \to (\kappa)^\xi)$, then the normal measures on κ are exactly the μ_λ for λ regular $< \kappa$.

PROOF OF 2 FROM 1. By 1) we know each that μ_λ is a normal measure. Let μ be another normal measure. For λ regular $< \kappa$ let

$$E_\lambda = \{\xi < \kappa : \mathrm{cf}(\xi) = \lambda\}.$$

Then the E_λ's are pairwise disjoint, and

$$\bigcup_{\substack{\lambda \text{ regular} \\ \lambda<\kappa}} E_\lambda = \{\xi < \kappa : \xi \text{ limit ordinal}\}.$$

Since there are $< \kappa$ regular cardinals below κ we can find a regular $\lambda_0 < \kappa$ such that

$$E_{\lambda_0} \in \mu.$$

Suppose $\mu \neq \mu_{\lambda_0}$. Then we can find a λ_0-closed unbounded A such that

$$B = \kappa \setminus A \in \mu.$$

For $\xi \in B \cap E_{\lambda_0}$, let

$$g(\xi) = \sup(A \cap \xi).$$

Then $g(\xi) < \xi$ for all $\xi \in B \cap E_{\lambda_0}$, hence g is μ-a.e. constant, which contradicts the unboundedness of A. So $\mu = \mu_{\lambda_0}$. ⊣ (2 from 1)

PROOF OF 1. Assume $\kappa \to (\kappa)^{\lambda+\lambda}$.

To show that μ_λ is a normal measure, let $f : \kappa \to \kappa$ be pressing down. Consider $X \subseteq {}^{\lambda+\lambda}[\kappa]$ given by

$$G \in X \Leftrightarrow f\big(\sup_{\alpha<\lambda} G(\alpha)\big) = f\big(\sup_{\alpha<\lambda} G(\lambda+\alpha)\big]\big).$$

Let $H \subseteq \kappa$ be homogeneous for this partition, with $\mathrm{Card}(H) = \kappa$.

Suppose ${}^{\lambda+\lambda}[H] \subseteq \neg X$. Let C be the set of limits of increasing λ sequences from H. Then C is λ-closed unbounded and for $\xi, \eta \in C$,

$$\xi < \eta \Rightarrow f(\xi) \neq f(\eta),$$

i.e., f is 1-1 on C. Now we inductively define an increasing λ-sequence $\{\gamma_\eta\}_{\eta<\lambda}$ of elements from C as follows:

$$\gamma_0 = \text{least member of } C$$

$$\gamma_\eta = \text{least element } \gamma \text{ of } C \text{ greater than all } \gamma_\theta \text{ for } \theta < \eta$$

which satisfies:

$$\forall\delta(\delta > \gamma \text{ and } \delta \in C \Rightarrow \forall\theta < \eta[f(\delta) > \gamma_\theta].$$

Then if $\gamma = \lim_{\eta<\lambda} \gamma_\eta \in C$ we have $f(\gamma) < \gamma$, hence for some $\eta < \lambda$, $f(\gamma) < \gamma_\eta$, hence $\gamma \leq \gamma_{\eta+1}$, a contradiction.

Hence ${}^{\lambda+\lambda}[H] \subseteq X$. Then if $\xi < \eta$ are both in C, we can find $G \in {}^{\lambda+\lambda}[H]$ such that

$$\xi = \sup_{\alpha<\lambda}(G(\alpha)), \eta = \sup_{\alpha<\lambda} G(\lambda+\alpha)$$

and hence $f(\xi) = f(\eta)$. So f is constant on C, proving normality for μ_λ.

To see now that μ_λ is an ultrafilter, look at the characteristic functions of subsets of κ (which are of course pressing down). To see that μ_λ is κ-additive, let $\{A_\theta\}_{\theta<\rho<\kappa} \subseteq \mu_\lambda$ and suppose $\bigcap_{\theta<\rho} A_\theta = \varnothing$, towards a contradiction. Then $\kappa = \bigcup_{\theta<\rho}(\kappa \setminus A_\theta)$, so consider

$$f(\xi) = \begin{cases} \text{least } \theta < \rho & \text{such that } \xi \notin A_\theta, \text{ if } \xi \geq \rho \\ 0 & \text{otherwise} \end{cases}$$

Then for some $\theta < \rho$, $\{\xi : f(\xi) = \theta\}$ contains a λ-closed unbounded set, *i.e.*, $\mu_\lambda(\kappa \setminus A_\theta) = 1$, a contradiction. ⊣ (1)

§11. Countable exponent partition relations for $\underset{\sim}{\delta}^1_n$, n odd.

We present first in an abstract form Martin's method for proving infinite exponent partition relations from AD. It is a modification of Solovay's technique used in the proof of Theorem 5.1.

LEMMA 11.1 (Martin). Let $\kappa > \omega$ be a regular cardinal, $\lambda \leq \kappa$ an ordinal. Assume:

1) There is $\{C_\xi\}_{\xi<\omega\cdot\lambda}$, with $C_\xi \subseteq \mathbb{R}$, and for each $\xi < \omega\cdot\lambda$ a map $\varepsilon \to f^\xi(\varepsilon)$ from C_ξ into κ such that if $C = \bigcap_{\xi<\omega\cdot\lambda} C_\xi$ and for $\varepsilon \in C$ we let $f_\varepsilon(\xi) = f^\xi(\varepsilon)$ then $\varepsilon \mapsto f_\varepsilon$ maps C onto ${}^{\omega\cdot\lambda}\kappa$.
2) There are $\{C_{\xi,\theta}\}_{\xi<\omega\cdot\lambda,\theta<\kappa}$ such that

$$C_{\xi,\theta} \subseteq \bigcap_{\xi'\leq\xi} C_{\xi'}$$

and if $\sigma\colon \mathbb{R}\to\mathbb{R}$ is continuous and $\sigma[\bigcap_{\xi'\leq\xi} C_{\xi'}] \subseteq C_\xi$ then for all $\xi < \omega\cdot\lambda$ and $\theta < \kappa$ we have

$$G^\sigma(\xi,\theta) \stackrel{\text{def}}{=} \sup\{f^\xi(\sigma(\varepsilon))+1 : \varepsilon\in C_{\xi,\theta}\} < \kappa.$$

3) If $f \in {}^{\omega\cdot\lambda}[\kappa]$ then there is $\varepsilon \in C$ such that $f_\varepsilon = f$ and $\varepsilon \in C_{\xi,f(\xi)}$, $\forall\xi < \omega\cdot\lambda$.

Then:

$$\mathsf{AD}+\mathsf{DC} \Rightarrow \kappa \to (\kappa)^\lambda.$$

Proof. Let $A \subseteq {}^\lambda[\kappa]$ and consider the following game:

I	II
ε^{I}	$\varepsilon^{\mathrm{II}}$

Player II wins iff

(1) $\exists\xi < \omega\cdot\lambda(\varepsilon^{\mathrm{I}} \notin C_\xi \vee \varepsilon^{\mathrm{II}} \notin C_\xi)$ and if ξ_0 is the least such ξ then $\varepsilon^{\mathrm{I}} \notin C_{\xi_0}$, or

(2) $\forall\xi < \omega\cdot\lambda(\varepsilon^{\mathrm{I}} \in C_\xi \wedge \varepsilon^{\mathrm{II}} \in C_\xi)$ and $<\sup_n\{f_{\varepsilon^{\mathrm{I}}}(\omega\cdot\theta+n), f_{\varepsilon^{\mathrm{II}}}(\omega\cdot\theta+n)\}>_{\theta<\lambda} \in A$.

Without loss of generality we can assume that player II has a winning strategy σ. Then by 1) $\sigma[\bigcap_{\xi'\leq\xi} C_{\xi'}] \subseteq C_\xi$, $\forall\xi < \omega\cdot\lambda$. So by 2) above $G^\sigma(\xi,\theta) < \kappa$. By the regularity of κ, let $D \subseteq \kappa$ be closed unbounded such that

$$\rho\in D \Rightarrow \forall\xi<\omega\cdot\lambda\forall\theta<\kappa[\xi<\rho\wedge\theta<\rho \Rightarrow G^\sigma(\xi,\theta)<\rho].$$

Let

$$D\nvdash^\lambda = \{g\in{}^\lambda[D] : \exists f\in{}^{\omega\cdot\lambda}[\kappa]\forall\theta<\lambda(g(\theta) = \sup_n f(\omega\cdot\theta+n)\}.$$

We claim that $D\nvdash^\lambda \subseteq A$, which completes the proof since then ${}^\lambda[H] \subseteq A$, where $H = \{\delta_{\xi+\omega} : \xi < \kappa\}$, where $\{\delta_\nu\}_{\nu<\kappa}$ is the increasing enumeration of D. Let $g \in D\nvdash^\lambda$ and let $f \in {}^{\omega\cdot\lambda}[\kappa]$ be such that $g(\theta) = \sup_n f(\omega\cdot\theta+n)$. Then by 3) above find $\varepsilon \in C$ such that $f_\varepsilon = f$ and $\varepsilon \in C_{\xi,f(\xi)}$. Then for all $\theta < \lambda$ and all $n \in \omega$:

$$f^{\omega\cdot\theta+n}(\sigma(\varepsilon)) < G^\sigma(\omega\cdot\theta+n, f(\omega\cdot\theta+n)) < g(\theta),$$

since

$$\omega\cdot\theta+n \leq f(\omega\cdot\theta+n) < g(\theta) \in D.$$

So for all $\theta < \lambda$

$$\sup_n\{f_\varepsilon(\omega\cdot\theta+n), f_{\sigma(\varepsilon)}(\omega\cdot\theta+n)\} = g(\theta),$$

therefore $g \in A$. ⊣ (Lemma 11.1)

THEOREM 11.2 (AD) (Martin [Mar71A]). For any $n \geq 0$,

$$\underset{\sim}{\boldsymbol{\delta}}^1_{2n+1} \to (\underset{\sim}{\boldsymbol{\delta}}^1_{2n+1})^\lambda, \forall\lambda < \omega_1.$$

PROOF. Fix $t\colon \omega\cdot\lambda \leftrightarrow \omega$. For a real α, set $\alpha_\xi = (\alpha)_i$, where $t(\xi) = i$. Let also W be a complete $\underset{\sim}{\boldsymbol{\Pi}}^1_{2n+1}$ set, φ a $\underset{\sim}{\boldsymbol{\Pi}}^1_{2n+1}$-norm on W with range $\underset{\sim}{\boldsymbol{\delta}}^1_{2n+1}$ and for $\alpha \in W$, write $|\alpha| = \varphi(\alpha)$.

Define now for $\xi < \omega\cdot\lambda$,

$$C_\xi = \{\alpha : \alpha_\xi \in W\}$$

and for $\alpha \in C_\xi$,

$$f^\xi(\alpha) = |\alpha_\xi|.$$

Finally let for $\xi < \omega\cdot\lambda$, $\theta < \underset{\sim}{\boldsymbol{\delta}}^1_{2n+1}$:

$$C_{\xi,\theta} = \{\alpha : \forall\xi' \leq \xi\exists\eta' \leq \theta(\alpha_{\xi'} \in W \wedge |\alpha_{\xi'}| \leq \eta')\}.$$

Now obviously properties 1), 3) of Lemma 11.1 are satisfied so it is enough to verify 2). For that notice that $C_{\xi,\theta} \in \underset{\sim}{\boldsymbol{\Delta}}^1_{2n+1}$, so that if σ is continuous then $\sigma[C_{\xi,\theta}]$ is $\underset{\sim}{\boldsymbol{\Sigma}}^1_{2n+1}$. If also $\sigma[C_{\xi,\theta}] \subseteq C_\xi$, then $\{\alpha_\xi : \alpha \in \sigma[C_{\xi,\theta}]\}$ is a $\underset{\sim}{\boldsymbol{\Sigma}}^1_{2n+1}$ subset of W, so by boundedness

$$G^\sigma(\xi,\theta) = \sup\{|\alpha_\xi| + 1 : \alpha \in \sigma[C_{\xi,\theta}]\} < \underset{\sim}{\boldsymbol{\delta}}^1_{2n+1}$$

and we are done. ⊣ (Theorem 11.2)

§12. $\omega_1 \to (\omega_1)^{\omega_1}$.

DEFINITION 12.1. For $C \subseteq \kappa$, put

$$C^{\not\vdash} = C^{\not\vdash\kappa} = \{f \in {}^\kappa[C] : \exists g \in {}^\kappa[\kappa]\forall\xi, f(\xi) = \sup_n g(\omega\cdot\xi+n)\}.$$

It is not hard to check that

$$\kappa \to (\kappa)^\kappa \Leftrightarrow \forall X \subseteq {}^\kappa[\kappa]\exists C(C \text{ is closed unbounded on } \kappa \text{ and } C^{\not\vdash} \subseteq X \text{ or } C^{\not\vdash} \subseteq \neg X).$$

THEOREM 12.2 (AD) (Martin; see Martin-Paris [MP71]).

$$\omega_1 \to (\omega_1)^{\omega_1}$$

PROOF. We will apply Lemma 11.1 again. Let $\tau_0, \tau_1, \tau_2, \dots$ be a recursive enumeration of terms in the language of ZF + $\mathbf{V}=\mathbf{L}[\dot\alpha]$, which take only ordinal values, as in the proof of Theorem 6.3. For $\xi < \omega_1$ put, using again the notation of 6.3:

$C_\xi = \{\varepsilon : \varepsilon = n^\frown\varepsilon' = \langle n, \varepsilon'(0), \varepsilon'(1), \dots\rangle$ & $P(\varepsilon')$ & the wellfounded part of $\Gamma(\xi + \omega, \varepsilon')$ has an ordinal of order type ξ (denoted also by ξ) & $\tau_n^{\Gamma(\xi+\omega,\varepsilon')}(\xi)$ also belongs to the wellfounded part$\}$.

Then

$$\varepsilon \in C \Leftrightarrow \forall\xi < \omega_1(\varepsilon \in C_\xi) \Leftrightarrow \varepsilon = n^\frown\alpha^\#, \text{ for some } n, \alpha.$$

For $\varepsilon \in C_\xi$, let also

$$f^\xi(\varepsilon) = \tau_n^{\Gamma(\xi+\omega,\varepsilon')}(\xi).$$

Then if $\varepsilon \in C$, say $\varepsilon = n^\frown\alpha^\#$, we have for all $\xi < \omega_1$:

$$f_\varepsilon(\xi) = \tau_n^{\mathbf{L}[\alpha]}(\xi).$$

Finally put for $\xi < \omega_1$, $\theta < \omega_1$,

$$C_{\xi,\theta} = \{\varepsilon : \forall\xi' \leq \xi\exists\eta' \leq \theta(\varepsilon \in C_{\xi'} \wedge \text{ if } \varepsilon = n^\frown\varepsilon', \text{ then } \tau_n^{\Gamma(\xi'+\omega,\varepsilon')}(\xi') \leq \eta')\}.$$

Clearly conditions 1), 3) or Lemma 11.1 are satisfied. To verify also condition 2) note first that each $C_{\xi,\theta}$ is Borel. So if σ is continuous $\sigma[C_{\xi,\theta}]$ is $\utilde{\boldsymbol{\Sigma}}^1_1$. If moreover $\sigma[C_{\xi,\theta}] \subseteq C_\xi$, then an easy boundedness argument shows that $G^\sigma(\xi, \theta) < \omega_1$ and we are done. ⊣ (Theorem 12.2)

§13. The Martin-Paris theorem.

DEFINITION 13.1. Let κ be an uncountable cardinal, μ a normal measure on κ, and assume ${}^\kappa\kappa/\mu \cong \kappa^+$. For each $f \in {}^\kappa\kappa$, let $f(\kappa) = [f]$ (thus $\kappa^+ = \{f(\kappa) : f \in {}^\kappa\kappa\}$). A μ as above is **canonical** if it has the following selection property: If $\pi < \kappa^+$, and $\{\xi_\lambda\}_{\lambda<\pi}$ is a π-sequence of ordinals $< \kappa^+$, then there is a sequence $\{f_\lambda\}_{\lambda<\pi} \subseteq {}^\kappa\kappa$ such that $f_\lambda(\kappa) = \xi_\lambda$.

Note that if such a measure exists, then κ^+ is regular.

THEOREM 13.2 (AD) (Solovay). The measure μ_ω on ω_1 is canonical.

PROOF. If $f \in {}^{\omega_1}\omega_1$, we can find τ, α such that

$$\forall\eta < \omega_1, f(\eta) = \tau^{\mathbf{L}[\alpha]}(\eta).$$

It is easy to check that $f(\omega_1) = \tau^{\mathbf{L}[\alpha]}(\omega_1)$. So in particular

$${}^{\omega_1}\omega_1/\mu_\omega \cong \omega_2.$$

Now let $\{\xi_\lambda\}_{\lambda<\pi}$ be a sequence of ordinals less than ω_2. Without loss of generality $\pi = \omega_1$, and

$$\{\xi_\lambda : \lambda < \omega_1\} = \xi < \omega_2.$$

Define the following prewellordering on ω_1:

$$\lambda \precsim \lambda' \Rightarrow \xi_\lambda \leq \xi_{\lambda'}.$$

By the proof of Solovay's Theorem 7.2, find τ, α such that

$$\precsim = \tau^{\mathbf{L}[\alpha]}(\omega_1).$$

For some term σ then,

$$\xi_\lambda = \sigma^{\mathbf{L}[\alpha]}(\lambda, \omega_1), \forall \lambda < \omega_1.$$

Take $f_\lambda(\eta) = \sigma^{\mathbf{L}[\alpha]}(\lambda, \eta)$. ⊣ (Theorem 13.2)

NOTATION. Let κ be an uncountable cardinal carrying a canonical measure μ. Let $\kappa \leq \pi < \kappa^+$, and fix $h\colon \kappa \leftrightarrow \pi$. For any $\theta < \kappa$, let $\bar\theta$, ρ_θ be such that

$$\rho_\theta\colon \bar\theta \leftrightarrow h[\theta] = \{h(\xi) : \xi < \theta\},$$

with ρ_θ order preserving. Then consider the normal (*i.e.*, increasing and continuous) function $\chi\colon \kappa \to \kappa$ such that $\chi_{\theta+1} - \chi_\theta = \bar\theta$ (where $\chi_\theta = \chi(\theta)$)). For $\lambda \in h[\theta]$ let

$$\chi_{\theta,\lambda} = \chi_\theta + \rho_\theta^{-1}(\lambda).$$

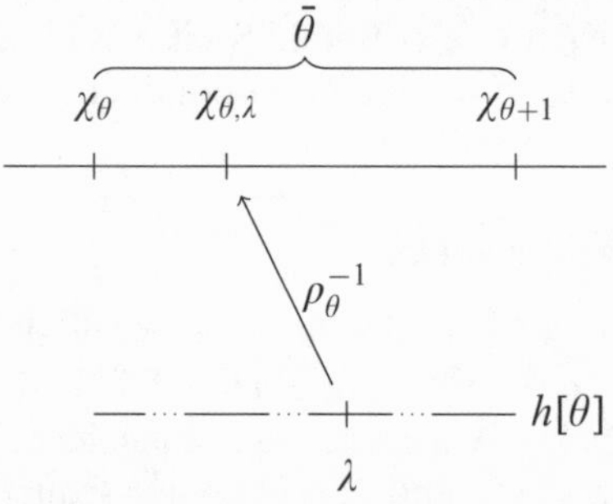

Then $\lambda < \lambda' \in h[\theta] \Rightarrow \chi_{\theta,\lambda} < \chi_{\theta,\lambda'}$.

Thus to each $\lambda < \pi$ we can assign an increasing

$$\psi_\lambda\colon \kappa \to \kappa,$$

(actually ψ_λ is defined from a point on), where

$$\psi_\lambda(\theta) = \chi_{\theta,\lambda}.$$

Then $\lambda < \lambda' \Rightarrow \psi_\lambda(\theta) < \psi_{\lambda'}(\theta)$.

Now let $f \in {}^\kappa[\kappa]$. For $\lambda < \pi$ let

$$f^{[\lambda]}(\theta) = f(\psi_\lambda(\theta)).$$

Thus $f^{[\lambda]} \in {}^{\kappa}[\kappa]$. Now let

$$f^{\tilde{\pi}} \in {}^{\pi}[\kappa^+]$$

be defined by

$$f^{\tilde{\pi}}(\lambda) = f^{[\lambda]}(\kappa).$$

(Recall that for a function $g \in {}^{\kappa}\kappa$, $g(\kappa)$ denotes the image of g in the ultrapower ${}^{\kappa}\kappa/\mu$.) Since $\lambda < \lambda' \Rightarrow$ for all θ from a point on, $\psi_\lambda(\theta) < \psi_{\lambda'}(\theta)$, we see that $\lambda < \lambda' \Rightarrow \{\theta : f^{[\lambda]}(\theta) < f^{[\lambda']}(\theta)\}$ has measure 1, so f^{π} is indeed increasing.

For $A \subseteq \kappa$, let

$$({}^{\kappa}[A])^{\tilde{\pi}} = \{f^{\tilde{\pi}} : f \in {}^{\kappa}[A]\}$$

and let

$$A^* = \{f(\kappa) : f \in {}^{\kappa}A\}.$$

Note that if A is unbounded in κ, then A^* is unbounded in κ^+. (If $f \in {}^{\kappa}\kappa$, define $g \in {}^{\kappa}A$ by $g(\xi) =$ least member of $A > f(\xi)$; then $g(\kappa) < f(\kappa)$).

Theorem 13.3 (AD) (Martin-Paris [MP71]). Let κ be an uncountable cardinal, μ a canonical measure on κ, $\kappa \leq \pi < \kappa^+$, $A \subseteq \kappa$ unbounded. Then

$${}^{\pi}\big(A^* \setminus (\kappa+1)\big) \subseteq ({}^{\kappa}[A])^{\tilde{\pi}}.$$

Corollary 13.4 (AD) (Martin-Paris [MP71]). Let κ be an uncountable cardinal carrying a canonical measure μ. If $\kappa \to (\kappa)^{\kappa}$, then $\forall \pi < \kappa^+$, $\kappa^+ \to (\kappa^+)^{\pi}$. Hence for any regular $\lambda < \kappa^+$, μ_λ is a normal measure on κ^+.

Corollary 13.5 (AD) (Martin-Paris [MP71]).

1) $\forall \pi < \omega_2, \omega_2 \to (\omega_2)^{\pi}$.
2) ω_2 has exactly two normal measures, namely μ_ω, μ_{ω_1}.

Proof of Corollary 13.4. Let $X \subseteq {}^{\pi}[\kappa^+]$, $\kappa \leq \pi < \kappa^+$. Put ${}^{\tilde{\pi}}X = \{f \in {}^{\kappa}[\kappa] : f^{\tilde{\pi}} \in X\}$. Let $H \subseteq \kappa$ have cardinality κ such that, say, ${}^{\kappa}[H] \subseteq {}^{\tilde{\pi}}X$. Then by Theorem 13.3

$${}^{\pi}[H^* \setminus (\kappa+1)] \subseteq ({}^{\kappa}[H])^{\tilde{\pi}} \subseteq ({}^{\tilde{\pi}}X)^{\tilde{\pi}} \subseteq X.$$

⊣ (Corollary 13.4)

Proof of Theorem 13.3. Let $f \in {}^{\pi}[A^* \setminus (\kappa+1)]$. Then find $\{f_\lambda\}_{\lambda<\pi} \subseteq {}^{\kappa}\kappa$ such that

$$f_\lambda(\kappa) = f(\lambda).$$

We want to find $G \in {}^{\kappa}[A]$ such that

$$\forall \lambda < \pi, G^{\tilde{\pi}}(\lambda) = f_\lambda(\kappa),$$

i.e.,

$$\forall \lambda < \pi, G^{[\lambda]}(\kappa) = f_\lambda(\kappa),$$

i.e.,

$$\forall \lambda < \pi, G(\chi_{\theta,\lambda}) = f_\lambda(\theta) \text{ for } \mu\text{-almost all } \theta.$$

For that it is enough to have for μ-almost all θ,

$$\forall \lambda \in h[\theta], G(\chi_{\theta,\lambda}) = f_\lambda(\theta).$$

To prove this we need the following:

LEMMA 13.6. There is a set C of μ-measure 1 such that if $\theta \in C$:

i) $\forall \lambda \in h[\theta](f_\lambda(\theta) \in A)$, and

ii) $\forall \lambda < \lambda' \in h[\theta](f_\lambda(\theta) < f_{\lambda'}(\theta))$.

PROOF OF LEMMA 13.6. For fixed ξ, η, let $C_{\xi,\eta} = \{\theta : f_{h(\xi)}(\theta) \in A$ and $h(\xi) < h(\eta) \Rightarrow f_{h(\xi)}(\theta) < f_{h(\eta)}(\theta)\}$. Then each $C_{\xi,\eta} \in \mu$ (since for all $\lambda < \pi$, $f_\lambda(\kappa) \in A^* \Rightarrow f_\lambda$ is μ-equivalent to some element of ${}^\kappa A$, hence $\{\theta : f_\lambda(\theta) \in A\} \in \mu$).

Now let $C = \{\theta : \forall \xi, \eta < \theta(\theta \in C_{\xi,\eta})\}$. Then C has μ-measure 1 and has the required properties. So the proof of Lemma 13.6 is complete. ⊣ (Lemma 13.6)

To finish the proof of Theorem 13.3: Let θ_0 be large enough so that $0 \in h[\theta_0]$. Now define

$$G(\chi_{\theta,\lambda}) = \begin{cases} f_\lambda(\theta) & \text{if } \theta \in C \setminus \theta_0,\ \lambda \in h[\theta] \text{ and} \\ & f_0(\theta) > \sup_{\zeta < \chi_\theta} G(\xi) \\ \text{least member of } A, \text{ greater} & \text{otherwise.} \\ \text{than all } G(\zeta) \text{ for } \zeta < \chi_{\theta,\lambda} & \end{cases}$$

CLAIM 13.7. For μ-almost all θ, the first definition occurs.

PROOF. Since χ is normal, for μ-almost all θ, $\chi_\theta = \theta = \sup_{\zeta < \chi_\theta} G(\zeta)$. Since $f_0(\kappa) > \kappa$, for μ-almost all θ, $f_0(\theta) > \theta = \sup_{\zeta < \chi_\theta} G(\zeta)$, which proves the claim. ⊣ (Claim 13.7)

By the properties of C given in Lemma 13.6, $G \in {}^\kappa[A]$ and is as desired. ⊣ (Theorem 13.3)

THEOREM 13.8 (AD) (Martin-Paris [MP71]). Let κ be an uncountable cardinal, μ a normal measure on κ, $\kappa_2 \cong {}^\kappa\kappa/\mu$. For each $\pi < \kappa^+$ there is a map $f \mapsto f^{\tilde{\pi}}$ sending ${}^\kappa[\kappa]$ into ${}^\pi[\kappa_2]$, such that if $A \subseteq \kappa$, $F \in {}^\pi[A^* \setminus (\kappa+1)]$, and $\exists G \in {}^\pi({}^\kappa\kappa)$ such that

$$[G(\lambda)]_\mu = F(\lambda),$$

then there is an $f \in {}^\kappa[A]$ with $f^{\tilde{\pi}} = F$.

Corollary 13.9. In the notation of Theorem 13.8, if $\kappa \to (\kappa)^\kappa$, then $\kappa_2 \to (\kappa_2)^\pi, \forall \pi < \omega_1$.

Proof of Theorem 13.8. For $\pi \geq \kappa$ see the proof of Theorem 13.3. Assume $\pi < \kappa$. Define a normal function χ on κ by $\chi_{\theta+1} - \chi_\theta = \pi$. Let $\chi_{\theta,\lambda} = \chi_\theta + \lambda$ for $\lambda < \pi$. For $f \in {}^\kappa\kappa$ let

$$f^{[\lambda]}(\theta) = f(\chi_{\theta,\lambda}),$$

and let $f^{\tilde{\pi}} = \{f^{[\lambda]}\}_{\lambda<\pi}$. Now repeat the proof of Theorem 13.3.

⊣ (Theorem 13.8)

For the corollary just notice that if $A \subseteq \kappa$ has cardinality κ, then $A^* \setminus (\kappa+1)$ has order type κ_2.

§14. The measure μ_ω on $\underset{\sim}{\delta}^1_n$, n odd.

Definition 14.1. Let κ be a cardinal. If W is a wellordering of (a subset of) κ, let for $\xi < \kappa$

$$H_W(\xi) = |W \restriction \xi|,$$

where $W \restriction \xi = W \cap (\xi \times \xi)$. Clearly $H_W : \kappa \to \kappa$.

Theorem 14.2. Let κ be an uncountable cardinal, μ a normal measure on κ. Then for any wellordering W on κ,

$$[H_W]_\mu = |W|.$$

In particular, $\{[H_W]_\mu : W \text{ a wellordering on } \kappa\} = \kappa^+$.

Proof. Notice first that $\{[H_W]_\mu : W \text{ a wellordering on } \kappa\}$ is an initial segment of ordinals. Because if $F : \kappa \to \kappa$ is such that $[F]_\mu < [H_W]_\mu$, then for μ-almost all ξ, $F(\xi) < |W \restriction \xi|$. Hence there is a map $\xi \mapsto \xi^* < \xi$ such that for μ-almost all ξ, $F(\xi) = |W^{\xi^*} \restriction \xi|$ (where for any wellordering W, W^x = initial segment of W determined by x). So by normality, find ξ_0 such that $F(\xi) = |W^{\xi_0} \restriction \xi|$ μ-almost everywhere. Hence $[F]_\mu = [H_{W^{\xi_0}}]_\mu$.

To prove the theorem, it is enough to show that for wellorderings W, V on κ,

$$[H_W]_\mu < [H_V]_\mu \Leftrightarrow |W| < |V|.$$

If $H_W(\xi) < H_V(\xi)$ μ-a.e., let $F_\xi : W \restriction \xi \to V^{\xi^*} \restriction \xi$ be an isomorphism for μ-almost all ξ, with $\xi^* < \xi$. Then by normality, there is ξ_0 such that $F_\xi : W \restriction \xi \to V^{\xi_0} \restriction \xi$ is an isomorphism μ-almost everywhere. For each $\eta \in \text{Field}(W)$, let

$$f_\eta(\xi) = F_\xi(\eta) < \xi,$$

so by normality $f_\eta(\xi) = g(\eta)$ for μ-a.e. ξ (*i.e.*, $g(\eta)$ is the constant value assumed μ-a.e. by f_η). Then $g : W \to V^{\xi_0}$ is an embedding, so $|W| < |V|$. Similarly $H_W(\xi) \leq H_V(\xi)$ μ-a.e. $\Rightarrow |W| \leq |V|$. ⊣ (Theorem 14.2)

COROLLARY 14.3. Let κ be an uncountable cardinal, μ a normal measure on κ. Then the following are equivalent:

1) μ is canonical.
2) ${}^{\kappa}\kappa/\mu \cong \kappa^+$ and κ^+ is regular.
3) Every $F\colon \kappa \to \kappa$ is μ-a.e. equal to some H_W, W a wellordering on κ, and κ^+ is regular.

PROOF. 1) $\Rightarrow$ 2) by definition.

2) $\Leftrightarrow$ 3) by Theorem 14.2.

2) $\Rightarrow$ 1) Enough to show the selection property, and since κ^+ is regular, it is enough to show that if $\rho < \kappa^+$ then we can find $\{f_\lambda\}_{\lambda<\rho}$, where $f_\lambda\colon \kappa \to \kappa$ is such that $[f_\lambda]_\mu = \lambda$. Pick a wellordering W of κ with $|W| = \rho$. For each $\lambda < \rho$ let $\lambda^* < \kappa$ be such that

$$|W^{\lambda^*}| = \lambda.$$

Then let $f_\lambda = H_{W^{\lambda^*}}$. ⊣ (Corollary 14.3)

THEOREM 14.4 (AD) (Kunen [Kun71D]). For each $n \geq 0$, the measure μ_ω on $\underset{\sim}{\delta}^1_{2n+1}$ is canonical.

PROOF. We need some lemmas first.

LEMMA 14.5. There is a $\underset{\sim}{\Pi}^1_{2n+1}$ set $G \subseteq \mathbb{R}$ and a $\underset{\sim}{\delta}^1_{2n+1}$-scale $\{\varphi_m\}_{m\in\omega}$ on G such that if we put $\psi(\alpha) = \sup_m \varphi_m(\alpha)$, then

1) $\varphi_m(\alpha) < \psi(\alpha)$, $\forall \alpha \in G$, $\forall m \in \omega$.
2) $\{\psi(\alpha) : \alpha \in G\} \in \mu_\omega$.
3) If $A \subseteq G$ is $\underset{\sim}{\Sigma}^1_{2n+1}$, then $\sup_{\alpha\in A} \psi(\alpha) < \underset{\sim}{\delta}^1_{2n+1}$.

PROOF. Let W be a $\underset{\sim}{\Sigma}^1_{2n+1}$-complete set of reals, $\{\chi_m\}_{m\in\omega}$ a $\underset{\sim}{\Pi}^1_{2n+1}$-scale on W, where the range of each χ_m is included in $\underset{\sim}{\delta}^1_{2n+1}$. Put for $\alpha \in W$,

$$\bar{\chi}_m(\alpha) = \chi_0(\alpha) + \chi_1(\alpha) + \cdots + \chi_m(\alpha).$$

Then $\{\bar{\chi}_m\}_{m\in\omega}$ is a $\underset{\sim}{\delta}^1_{2n+1}$-scale on W and for all m, $\alpha \in W$, $\bar{\chi}_m(\alpha) < \sup_m \bar{\chi}_m(\alpha)$ (we can clearly assume here without loss of generality that always $\chi_m(\alpha) > 0$). Consider now

$$G = \{\alpha : \forall i, (\alpha)_i \in W\}$$

and for $\alpha \in G$, $m \in \omega$ define

$$\varphi_m(\alpha) = \bar{\chi}_{(m)_0}((\alpha)_{(m)_1}),$$

where $m \mapsto ((m)_0, (m)_1)$ is a 1-1 correspondence between ω and $\omega\times\omega$. Clearly $\{\varphi_m\}_{m\in\omega}$ is a $\underset{\sim}{\delta}^1_{2n+1}$-scale on G and if $\psi(\alpha) = \sup_m \varphi_m(\alpha)$ then properties 1), 3) are satisfied. ⊣ (Lemma 14.5)

CLAIM 14.6. $\{\psi(\alpha) : \alpha \in G\}$ is ω-closed unbounded.

PROOF. Clearly it is unbounded. Let

$$\psi(\alpha^0) < \psi(\alpha^1) < \cdots \to \lambda$$

where $\alpha^0, \alpha^1, \dots \in G$. Let $\alpha \in G$ be such that

$$(\alpha)_i = (\alpha^{(i)_0})_{(i)_1}.$$

Then

$$\varphi_m(\alpha) = \bar{\chi}_{(m)_0}(\alpha_{(m)_1}) = \bar{\chi}_{(n)_0}((\alpha^{(m)_{1,0}})_{(m)_{1,1}}) < \psi(\alpha^{(m)_{1,0}}) < \lambda,$$

where $(m)_{i,j} = ((m)_i)_j$.

If $\theta < \lambda$, find j large enough and κ such that $\varphi_\kappa(\alpha^j) > \theta$. Then if m is such that

$$(m)_0 = (k)_0, (m)_{1,0} = j, (m)_{1,1} = (k)_1,$$

we have $\varphi_m(\alpha) = \bar{\chi}_{(k)_0}((\alpha^j)_{(k)_1}) > \theta$.

Hence $\psi(\alpha) = \lambda$. ⊣ (Claim 14.6)

LEMMA 14.7. There is a tree U on $\omega \times \kappa_{2n+1}$ such that

$$\sup_{\alpha \in \mathbb{R}}\{|U(\alpha)| : U(\alpha) \text{ is wellfounded}\} = \utilde{\boldsymbol{\delta}}^1_{2n+1}.$$

PROOF. Let S be a $\utilde{\boldsymbol{\Sigma}}^1_{2n+1}$-complete set of reals and let U be a tree on $\omega \times \kappa_{2n+1}$ such that $\mathrm{p}[U] = S$. ⊣ (Lemma 14.7)

To prove now the theorem: Let $F : \utilde{\boldsymbol{\delta}}^1_{2n+1} \to \utilde{\boldsymbol{\delta}}^1_{2n+1}$ be given, and consider the following game: Player I plays α, player II plays β, and player II wins iff

$$\alpha \in G \Rightarrow U(\beta) \text{ is wellfounded and } |U(\beta)| > F(\psi(\alpha)).$$

If player I has a winning strategy, then by Lemma 14.5, 3) and Lemma 14.7 we get a contradiction. So assume σ_0 is a winning strategy for player II. Let T be the tree on $\omega \times \utilde{\boldsymbol{\delta}}^1_{2n+1}$ coming from the scale $\{\varphi_n\}_{n\in\omega}$ on G (thus $G = \mathrm{p}[T]$). Then for all α

$$T(\alpha) \text{ not wellfounded}$$
$$\Rightarrow U(\sigma_0[\alpha]) \text{ is wellfounded and } F(\psi(\alpha)) < |U(\sigma_0[\alpha])|. \quad (*)$$

Let $\sigma[\alpha] = \beta \Leftrightarrow \forall n(\sigma{\restriction}n, \alpha{\restriction}n, \beta{\restriction}n) \in S$, S a tree on $\omega \times \omega \times \omega$. Let R be the tree on $\omega \times \omega \times \omega \times \utilde{\boldsymbol{\delta}}^1_{2n+1} \times \kappa_{2n+1}$ defined by

$$(s, a, b, u, v) \in R \Leftrightarrow (s, a, b) \in S \;\&\; (b, v) \in U \;\&\; (a, u) \in T.$$

Then $R(\sigma_0)$ is wellfounded by $(*)$. Suppose $\alpha \in G \;\&\; \psi(\alpha) = \xi > \kappa_{2n+1}$. Let $f \in {}^{\omega}\xi$ be defined by $f(n) = \varphi_n(\alpha)$. Thus if $\beta = \sigma_0[\alpha]$, for any $v \in U(\beta)$ we have

$$(\alpha{\restriction}\mathrm{lh}(v), \beta{\restriction}\mathrm{lh}(v), f{\restriction}\mathrm{lh}(v), v) \in R(\sigma_0){\restriction}\xi$$
$$= \{(a, b, u, v) \in R(\sigma_0) : u \in {}^{<\omega}\xi\}$$

where $\mathrm{lh}(\alpha_0, \ldots, a_{m-1}) = m$. Hence

$$F(\xi) < |U(\alpha_0[\alpha])| \leq |R(\sigma_0)\restriction\xi|.$$

If we let $W(\sigma_0)$ be the Brouwer-Kleene wellordering of $R(\sigma_0)$, viewed as a wellordering of $\boldsymbol{\delta}^1_{2n+1}$ (after identifying ${}^{<\omega}\omega \times {}^{<\omega}\omega \times {}^{<\omega}\boldsymbol{\delta}^1_{2n+1} \times {}^{<\omega}\kappa_{2n+1}$ with $\boldsymbol{\delta}^1_{2n+1}$), then we have by the above

$$F(\xi) < H_{W(\sigma_0)}(\xi)\ \mu_\omega\text{-a.e.}$$

Hence $F(\xi) = H_{W(\sigma_0)^{\xi_0}}(\xi)\ \mu_\omega$-a.e., for some ξ_0. So μ_ω is canonical.

⊣ (Theorem 14.4)

§15. The measures μ_λ, with $\lambda > \omega$, on $\boldsymbol{\delta}^1_n$, n odd.

LEMMA 15.1 (AD). There is a relation $W \subseteq \mathbb{R} \times \boldsymbol{\delta}^1_{2n+1} \times \boldsymbol{\delta}^1_{2n+1}$, with the following properties:

1) If $W_\varepsilon(\xi, \eta) \Leftrightarrow W(\varepsilon, \xi, \eta)$, then for every $F: \boldsymbol{\delta}^1_{2n+1} \to \boldsymbol{\delta}^1_{2n+1}$ there is $\varepsilon_0 \in \mathbb{R}$ and $\xi_0 < \boldsymbol{\delta}^1_{2n+1}$ such that W_{ε_0} is a wellordering and $F(\xi) = |(W_{\varepsilon_0})^{\xi_0}\restriction\xi|$ μ_ω-a.e. In particular, $\sup\{|W_{\varepsilon_0}| : W_{\varepsilon_0} \text{ is a wellordering }\} = \boldsymbol{\delta}^1_{2n+2}$.
2) If $W_{\xi,\eta} = \{\varepsilon : W(\varepsilon, \xi, \eta)\}$ then for each ξ, η, $W_{\xi,\eta}$ is an open set of reals. In particular for each $\xi_0, \xi, \theta < \boldsymbol{\delta}_{2n+1}$, $\{\varepsilon : |(W_\varepsilon)^{\xi_0}\restriction\xi| < \theta\} \in \boldsymbol{\Delta}^1_{2n+1}$.
3) Let $P \subseteq \mathbb{R}$ be a $\boldsymbol{\Pi}^1_{2n+1}$-complete set and φ a $\boldsymbol{\Pi}^1_{2n+1}$-norm on it, $\varphi: P \twoheadrightarrow \boldsymbol{\delta}^1_{2n+1}$. Then there are $\boldsymbol{\Pi}^1_{2n+1}$, $\boldsymbol{\Sigma}^1_{2n+1}$ relations Q, S resp. such that

$$\alpha, \beta \in P \Rightarrow [W(\varepsilon, \varphi(\alpha), \varphi(\beta)) \Leftrightarrow Q(\varepsilon, \alpha, \beta) \Leftrightarrow S(\varepsilon, \alpha, \beta)].$$

PROOF. By the proof of Theorem 14.4 there is a tree T on $\omega \times \boldsymbol{\delta}^1_{2n+1}$ such that if $C: \boldsymbol{\delta}^1_{2n+1} \leftrightarrow {}^{<\omega}\boldsymbol{\delta}^1_{2n+1}$ and we put

$$W(\varepsilon, \xi, \eta) \Leftrightarrow C(\xi), C(\eta) \in T(\varepsilon) \ \&\ C(\xi) <_{\mathrm{BK}} C(\eta),$$

where $<_{\mathrm{BK}}$ is the Brouwer-Kleene ordering, then 1) is satisfied. Also 2) is obvious since if $W(\varepsilon, \xi, \eta)$ holds and $n \geq \mathrm{lh}(C(\xi)), \mathrm{lh}(C(\eta))$ then for all δ with $\delta\restriction n = \varepsilon\restriction n$ we have $W(\delta, \xi, \eta)$. (To compute that $P^\theta_{\xi_0,\xi} = \{\varepsilon : |(W_\varepsilon)^{\xi_0}\restriction\xi| < \theta\}$ is $\boldsymbol{\Delta}^1_{2n+1}$ proceed by induction on θ.)

Finally to check 3): By the proof of Theorem 7.1 we can find $Q^* \subseteq \omega \times \mathbb{R}^2$ and $S^* \subseteq \omega \times \mathbb{R}^2$ in $\boldsymbol{\Pi}^1_{2n+1}$, $\boldsymbol{\Sigma}^1_{2n+1}$ resp., such that (identifying below ${}^{<\omega}\omega$ with ω):

$$\alpha, \beta \in P \Rightarrow [(u, C(\varphi(\alpha))), (u, C(\varphi(\beta))) \in T \wedge C(\varphi(\alpha)) <_{\mathrm{BK}} C(\varphi(\beta)) \\ \Leftrightarrow Q^*(u, \alpha, \beta) \Leftrightarrow S^*(u, \alpha, \beta)].$$

Let then

$$Q(\varepsilon, \alpha, \beta) \Leftrightarrow \exists n Q^*(\varepsilon\restriction n, \alpha, \beta)$$
$$S(\varepsilon, \alpha, \beta) \Leftrightarrow \exists n S^*(\varepsilon\restriction n, \alpha, \beta)$$

Then for $\alpha, \beta \in P$,

$$\begin{aligned} Q(\varepsilon, \alpha, \beta) &\Leftrightarrow \exists n[(\varepsilon \restriction n, C(\varphi(\alpha))), (\varepsilon \restriction n, C(\varphi(\beta)) \in T \\ &\qquad\qquad\qquad \wedge C(\varphi(\alpha)) <_{\mathrm{BK}} C(\varphi(\beta))] \\ &\Leftrightarrow C(\varphi(\alpha)), C(\varphi(\beta)) \in T(\varepsilon) \\ &\qquad\qquad\qquad \wedge C(\varphi(\alpha)) <_{\mathrm{BK}} C(\varphi(\beta)) \\ &\Leftrightarrow W(\varepsilon, \varphi(\alpha), \varphi(\beta)) \end{aligned}$$

and similarly for S. ⊣ (Lemma 15.1)

LEMMA 15.2 (AD). Let $n \geq 0$ be given and assume $\lambda > \omega$ is a regular cardinal $< \underset{\sim}{\delta}^1_{2n+1}$. Then there is a function $F : \underset{\sim}{\delta}^1_{2n+1} \to \underset{\sim}{\delta}^1_{2n+1}$ such that for every wellordering U on $\underset{\sim}{\delta}^1_{2n+1}$ there is a wellordering V on $\underset{\sim}{\delta}^1_{2n+1}$ with $|V| > |U|$ and a cub set $C \subseteq \underset{\sim}{\delta}^1_{2n+1}$ such that $H_V(\xi) < F(\xi), \forall \xi \in E_\lambda \cap C$. In particular if μ_λ is a normal measure, $H_U < F$ μ_λ-a.e. (Recall that $E_\lambda = \{\xi : \mathrm{cf}(\xi) = \lambda\}$.)

PROOF. Let W be as in Lemma 15.1. Put for $\xi \in E_\lambda$

$$F(\xi) = \sup\{|W_\varepsilon \restriction \xi| + 1 : \varepsilon \text{ is such that } \forall \eta < \xi(|W_\varepsilon \restriction \eta| < \xi)\}$$

and let $F(\xi) = 0$ if $\xi \notin E_\lambda$. If now $\xi \in E_\lambda$ *i.e.*, $\mathrm{cf}(\xi) = \lambda > \omega$ then

$$\forall \eta < \xi(W_\varepsilon \restriction \eta \text{ is wellfounded} \Rightarrow W_\varepsilon \restriction \xi \text{ is wellfounded}),$$

so $F(\xi) < \underset{\sim}{\delta}^1_{2n+1}$, by boundedness and Lemma 15.1. Now given a wellordering U find ε_0 such that $|U| < |W_{\varepsilon_0}|$ and then find a closed unbounded C such that

$$\xi \in C \Rightarrow \forall \eta < \xi(|W_{\varepsilon_0} \restriction \eta| < \xi).$$

Then $\xi \in E_\lambda \cap C \Rightarrow F(\xi) > |W_{\varepsilon_0} \restriction \xi|$ and we are done. ⊣ (Lemma 15.2)

THEOREM 15.3 (AD) (Kunen [Kun71D]). If μ_λ, $\lambda > \omega$, is a normal measure on $\underset{\sim}{\delta}^1_{2n+1}$ then ${}^{\underset{\sim}{\delta}^1_{2n+1}}\underset{\sim}{\delta}^1_{2n+1}/\mu_\lambda > (\underset{\sim}{\delta}^1_{2n+1})^+ = \underset{\sim}{\delta}^1_{2n+2}$.

PROOF. If F is as in Lemma 15.2, then by Theorem 14.2 $[F]_{\mu_\lambda} \geq \underset{\sim}{\delta}^1_{2n+2}$. ⊣ (Theorem 15.3)

From Kunen's result (see Solovay [Sol78A]) that $\underset{\sim}{\delta}^1_3 \to (\underset{\sim}{\delta}^1_3)^\lambda$, $\forall \lambda < \underset{\sim}{\delta}^1_3$, it follows that the conclusion of Theorem 15.3 holds for regular $\omega < \lambda < \underset{\sim}{\delta}^1_3$, *i.e.*, for $\lambda = \omega_1, \omega_2$. It also holds for $\lambda = \omega_1$ for any $\underset{\sim}{\delta}^1_{2n+1}$, $n > 0$, by the remarks following 16.1.

§16. Countable exponent partition relations on $\underset{\sim}{\delta}^1_n$, n even.

THEOREM 16.1 (AD) (Kunen [Kun71D]). For all $n \geq 1$, $\underset{\sim}{\delta}^1_{2n} \to (\underset{\sim}{\delta}^1_{2n})^\lambda$, $\forall \lambda < \omega_1$.

Proof. Fix $\lambda < \omega_1$ and a map $t : \omega \cdot \lambda \leftrightarrow \omega$. Let $P \subseteq \mathbb{R}$ be $\underset{\sim}{\Pi}^1_{2n+1}$-complete, φ a $\underset{\sim}{\Pi}^1_{2n+1}$-norm on P with range $\underset{\sim}{\delta}^1_{2n+1}$ and put $|\alpha| = \varphi(\alpha)$. For any real α and any $\xi < \omega \cdot \lambda$ let $\alpha_\xi = (\alpha)_i$, where $t(\xi) = i$.

Fix $X \subseteq {}^\lambda[\delta^1_{2n+1}]$. Then consider the game

$$\begin{array}{cc} \mathrm{I} & \mathrm{II} \\ \varepsilon^{\mathrm{I}}, \alpha^{\mathrm{I}} & \varepsilon^{\mathrm{II}}, \alpha^{\mathrm{II}} \end{array}$$

Player II wins iff

1) $\exists \xi < \omega \cdot \lambda$ such that α^{I}_ξ or $\alpha^{\mathrm{II}}_\xi \notin P$ and for the least such ξ, say ξ_0, $\alpha^{\mathrm{I}}_{\xi_0} \notin P$, or
2) $\forall \xi < \omega \cdot \lambda (\alpha^{\mathrm{I}}_\xi, \alpha^{\mathrm{II}}_\xi \in P)$ and for some $\langle \zeta, \xi \rangle \in \underset{\sim}{\delta}^1_{2n+1} \times (\omega \cdot \lambda)$, $(W_{\varepsilon^{\mathrm{I}}_\xi})^{|\alpha^{\mathrm{I}}_\xi|} \restriction \zeta$ or $(W_{\varepsilon^{\mathrm{II}}_\xi})^{|\alpha^{\mathrm{II}}_\xi|} \restriction \zeta$ is not well ordered and for the lexicographically least such $\langle \zeta_0, \xi_0 \rangle$, $(W_{\varepsilon^{\mathrm{I}}_{\xi_0}})^{|\alpha^{\mathrm{I}}_{\xi_0}|} \restriction \zeta_0$ is not wellordered, or
3) Both 1) and 2) fail and letting $F_{\mathrm{I},\xi}(\zeta) = |(W_{\varepsilon^{\mathrm{I}}_\xi})^{|\alpha^{\mathrm{I}}_\xi|} \restriction \zeta|$ and similarly for player II, we have $\langle \sup_n \{[F_{\mathrm{I},\omega \cdot \vartheta + n}], [F_{\mathrm{II},\omega \cdot \theta + n}]\} \rangle_{\theta < \lambda} \in X$.

Here W is as in Lemma 15.1 and $[F] = [F]_{\mu_\omega}$.

Assume without loss of generality that player II has a winning strategy σ. Put

$$(\varepsilon^{\mathrm{II}}_\sigma, \alpha^{\mathrm{II}}_\sigma) = \sigma[\varepsilon^{\mathrm{I}}, \alpha^{\mathrm{I}}].$$

Let then for $\xi < \omega \cdot \lambda$; $\zeta, \eta, \iota < \underset{\sim}{\delta}^1_{2n+1}$:

$$\begin{aligned} \Theta(\xi, \zeta, \eta, \iota) = \sup\{ &|(W_{(\varepsilon^{\mathrm{II}}_\sigma)_\xi})^{|(\alpha^{\mathrm{II}}_\sigma)_\xi|} \restriction \zeta| + 1 : \\ &\varepsilon^{\mathrm{I}}, \alpha^{\mathrm{I}} \text{ are such that } \forall \bar{\xi} < \omega \cdot \lambda (\alpha^{\mathrm{I}}_{\bar{\xi}} \in P \wedge |\alpha^{\mathrm{I}}_{\bar{\xi}}| < \iota) \\ &\text{and for all } \langle \zeta', \xi' \rangle \leq_{\mathrm{lex}} \langle \zeta, \xi \rangle \\ &|(W_{\varepsilon^{\mathrm{I}}_{\xi'}})^{|\alpha^{\mathrm{I}}_{\xi'}|} \restriction \zeta'| < \eta \}. \end{aligned}$$

By Lemma 15.1, $\Theta(\xi; \zeta, \eta, i) < \underset{\sim}{\delta}^1_{2n+1}$ so we can find a cub C such that

$$\rho \in C \Rightarrow \forall \xi < \omega \cdot \lambda \forall \zeta, \eta, \iota < \rho (\Theta(\xi, \zeta, \eta, \iota) < \rho).$$

Put $H = (C^* \cap E_\omega) \setminus (\underset{\sim}{\delta}^1_{2n+1})$, where C^* = image of C under the embedding generated by the ultrapower relative to μ_ω. We will show that if $f \in H^{\not\nearrow \lambda}$ then $f \in X$.

Fix such an f and then find $g \in {}^{\omega \cdot \lambda}[\underset{\sim}{\delta}^1_{2n+2}]$ such that

$$\lim g(\omega \cdot \theta + n) = f(\theta), \forall \theta < \lambda . n < \omega$$

Then find $\varepsilon^{\mathrm{I}}, \alpha^{\mathrm{I}}$ such that $W_{\varepsilon^{\mathrm{I}}_\xi}$ is wellordered and $\alpha^{\mathrm{I}}_\xi \in$ P for all $\xi < \omega \cdot \lambda$ and such that $[F_{\mathrm{I},\omega \cdot \theta + n}] = g(\omega \cdot \theta + n)$. Put $\varepsilon^{\mathrm{II}} = \varepsilon^{\mathrm{II}}_\sigma$, $\alpha^{\mathrm{II}} = \alpha^{\mathrm{II}}_\sigma$. Then $\alpha^{\mathrm{II}} \in P$ for all $\xi < \omega \cdot \lambda$ and $(W_{\varepsilon^{\mathrm{II}}_\xi})^{|\alpha^{\mathrm{II}}_\xi|}$ is wellordered for all $\xi < \omega \cdot \lambda$, so it is enough to show that

$$[F_{\mathrm{II},\omega\cdot\theta+n}] < f(\theta), \forall\theta < \lambda, \forall n \in \omega.$$

Pick $F \in {}^{\underset{\sim}{\delta}^1_{2n+1}}C$ such that $[F] = f(\theta)$. Then we have to show that

$$F_{\mathrm{II},\omega\cdot\theta+n}(\zeta) < F(\zeta)\ \mu_\omega\text{-a.e.}$$

Let $I \in \mu_\omega$ be such that

$$\begin{aligned} \zeta \in I \Rightarrow\ & 1)\ F(\zeta) > \zeta \\ & 2)\ F(\zeta) > \iota > \sup\{|\alpha^{\mathrm{I}}_\xi| : \xi < \omega\cdot\lambda\} \text{ (for some } \iota) \\ & 3)\ F_{\mathrm{I},\xi'}(\zeta') < \zeta, \forall\xi' < \omega\cdot\lambda, \forall\zeta' < \zeta \\ & 4)\ F_{\mathrm{I},\omega\cdot\theta+n+1}(\zeta) < F(\zeta) \\ & 5)\ F_{\mathrm{I},\xi'}(\zeta) < F_{\mathrm{I},\xi''}(\zeta), \forall\xi' < \xi'' < \omega\cdot\lambda. \end{aligned}$$

Then we claim that for $\zeta \in I$,

$$F_{\mathrm{II},\omega\cdot\theta+n}(\zeta) = |(W_{\varepsilon^{\mathrm{II}}_{\omega\theta+n}})^{|\alpha^{\mathrm{II}}_{\omega\theta+n}|}\restriction\zeta| < F(\zeta),$$

which completes the proof. Since $F(\zeta) \in C$ we have

$$\forall\xi_1 < \omega\cdot\lambda\forall\zeta_1, \eta_1, \iota_1 < F(\zeta)(\Theta(\xi_1, \zeta_1, \eta_1, \iota_1) < F(\zeta)),$$

so since $\zeta < F(\zeta)$, $\iota < F(\zeta)$ we only have to show that for some $\eta < F(\zeta)$ and all $\langle\zeta', \xi'\rangle \leq_{\mathrm{lex}} \langle\zeta, \omega\cdot\theta + n\rangle$ we have $F_{\mathrm{I},\xi}(\zeta') < \eta$. Take

$$\eta = \max\{F_{\mathrm{I},\omega\cdot\theta+n+1}(\zeta), \zeta\} < F(\zeta).$$

Let $\langle\zeta', \xi'\rangle \leq_{\mathrm{lex}} \langle\zeta, \omega\cdot\theta + n\rangle$. Then we have:

Case 1. $\zeta' < \zeta$: Then $F_{\mathrm{I},\xi'}(\zeta') < \zeta \leq \eta$.

Case 2. $\zeta' = \zeta$ and $\xi' \leq \omega\cdot\theta+n$: Then $F_{\mathrm{I},\xi'}(\zeta) < F_{\mathrm{I},\omega\cdot\theta+n+1}(\zeta) \leq \eta < F(\zeta)$ and we are done. ⊣ (Theorem 16.1)

Kechris has recently shown that for all $n \geq 2$, $\underset{\sim}{\delta}^1_n \to (\underset{\sim}{\delta}^1_n)^\lambda$, $\forall\lambda < \omega_2$. Thus μ_{ω_1} is a normal measure for all $\underset{\sim}{\delta}^1_n$, $n \geq 2$.

§17. The measure μ_ω on $\underset{\sim}{\delta}^1_n$, n even.

THEOREM 17.1 (AD) (Kunen [Kun71D]). For all $n \geq 0$, μ_ω is a normal measure on $\underset{\sim}{\delta}^1_{2n+2}$ and is generated by the sets of the form $(C^* \cap E_\omega) \setminus \underset{\sim}{\delta}^1_{2n+1}$, where $C \subseteq \underset{\sim}{\delta}^1_{2n+1}$ is closed unbounded and C^* is the image of C under the embedding generated by the ultrapower relative to μ_ω on $\underset{\sim}{\delta}^1_{2n+1}$.

PROOF. By Theorem 10.4 and §16, μ_ω is a normal measure on $\underset{\sim}{\delta}^1_{2n+1}$. To prove the extra statement we show that if $f : \underset{\sim}{\delta}^1_{2n+2} \to \underset{\sim}{\delta}^1_{2n+2}$ is pressing down there is a set as above on which f is constant. For that consider the partition of ${}^{\omega+\omega}[\underset{\sim}{\delta}^1_{2n+2}]$ as in Theorem 10.4. Then by the proof of Theorem 16.1 there is a closed unbounded $C \subseteq \underset{\sim}{\delta}^1_{2n+1}$ such that if $p \in {}^{\omega+\omega}[C^* \cap E_\omega]$ and $p(\omega) > \lim_{n<\omega} p(n)$ and $p(0) > \underset{\sim}{\delta}^1_{2n+1}$, then

$$f\left(\sup_n(p(n))\right) = f\left(\sup_n(p(\omega+n))\right).$$

Let $D \subseteq C$ be the set of all limit points of C and $E \subseteq D$ the set of all limit points of D. Then both D, E are cub and $E^* \cap E_\omega \subseteq D^* \cap E_\omega \subseteq C^* \cap E_\omega$, while every point of E^* is a limit point of D^*, which in turn is a limit point of C^*. So if $\theta < \eta$ are in $(E^* \cap E_\omega) \setminus \underset{\sim}{\delta}^1_{2n+1}$, find

$$\theta_0 < \theta_1 < \cdots \to \theta < \eta_0 < \eta_1 < \cdots \to \eta$$

$\theta_i, \eta_i \in D^* \setminus \underset{\sim}{\delta}^1_{2n+1}$. Put $p(n) =$ the ωth element of C^* above θ_n and $p(\omega+n) =$ the ωth element of C^* above η_n. Then $p(n) \to \theta$, since $\theta_n \leq p(n) \leq \theta_{n+1}$, and similarly $p(\omega+n) \to \eta$. Since $p \in {}^{\omega+\omega}[C^* \cap E_\omega]$ and $p(\omega) > \lim_n p(n) = \theta$ and $p(0) > \underset{\sim}{\delta}^1_{2n+1}$, we have $f(\sup_n(p(n))) = f(\theta) = f(\sup_n(p(\omega+n))) = f(\eta)$, so f is constant on $(E^* \cap E_\omega) \setminus \underset{\sim}{\delta}^1_{2n+1}$ and we are done. ⊣ (Theorem 17.1)

THEOREM 17.2 (AD) (Kunen [Kun71D]). If $F : \underset{\sim}{\delta}^1_{2n+2} \to \underset{\sim}{\delta}^1_{2n+2}$, there is $J : \underset{\sim}{\delta}^1_{2n+1} \to \underset{\sim}{\delta}^1_{2n+1}$ such that $F(\xi) \leq J^*(\xi)$, for all ξ in $(\underset{\sim}{\delta}^1_{2n+1}, \underset{\sim}{\delta}^1_{2n+2})$, where again $J^* : \underset{\sim}{\delta}^1_{2n+2} \to \underset{\sim}{\delta}^1_{2n+2}$ is the image of J under the embedding generated by μ_ω on $\underset{\sim}{\delta}^1_{2n+1}$.

PROOF. In the notation of Theorem 16.1 consider the game

I	II
$\varepsilon^{\mathrm{I}}, \alpha^{\mathrm{I}}$	$\varepsilon^{\mathrm{II}}$

Player II wins if

1) $\alpha^{\mathrm{I}} \notin P$, or
2) $\alpha^{\mathrm{I}} \in P$ and *either* for some ξ, $(W_{\varepsilon^{\mathrm{I}}})^{|\alpha^{\mathrm{I}}|} \restriction \xi$ or $W_{\varepsilon^{\mathrm{II}}} \restriction \xi$ is not a wellordering and for the least such, say ξ_0, $(W_{\varepsilon^{\mathrm{I}}})^{|\alpha^{\mathrm{I}}|} \restriction \xi_0$ is not a wellordering *or* for all ξ, $(W_{\varepsilon^{\mathrm{I}}})^{|\alpha^{\mathrm{I}}|} \restriction \xi$ and $W_{\varepsilon^{\mathrm{II}}} \restriction \xi$ are wellorderings and if $f_{\mathrm{I}}(\xi) = |(W_{\varepsilon^{\mathrm{I}}})^{|\alpha^{\mathrm{I}}|} \restriction \xi|$ and $f_{\mathrm{II}}(\xi) = |W_{\varepsilon^{\mathrm{II}}} \restriction \xi|$, then $F([f_{\mathrm{I}}]) < [f_{\mathrm{II}}]$, where $[f] = [f]_{\mu_\omega}$.

CLAIM 17.3. Player I does not have a winning strategy.

PROOF. Suppose he had one, τ. Then we will show that there is $K : \underset{\sim}{\delta}^1_{2n+1} \to \underset{\sim}{\delta}^1_{2n+1}$ such that if player II plays correctly so that f_{II} is defined, then if f_{I} is produced following τ, then $[f_{\mathrm{I}}] < [K]$. If player II then "plays f_{II}" such that $[f_{\mathrm{II}}] > \sup\{F(\theta) : \theta < [K]\}$ we immediately have a contradiction. To define K let $(\varepsilon^{\mathrm{I}}_\tau, \alpha^{\mathrm{I}}_\tau) = \tau[\varepsilon^{\mathrm{II}}]$. Then let

$$K(\xi) = \sup\{|(W_{\varepsilon^{\mathrm{I}}_\tau})^{|\alpha^{\mathrm{I}}_\tau|} \restriction \xi| + 1 : \varepsilon^{\mathrm{II}} \text{ is such that} \forall \eta < \xi (|W_{\varepsilon^{\mathrm{II}}} \restriction \eta| < \xi)\}.$$

Then $K(\xi) < \underset{\sim}{\delta}^1_{2n+1}$ by boundedness, since if $W_{\varepsilon^{\mathrm{II}}} \restriction \eta$ is wellordered for all $\eta < \xi$, then $(W_{\varepsilon^{\mathrm{I}}})^{|\alpha^{\mathrm{I}}_\tau|} \restriction \xi$ must be wellordered by the rules of the game and the fact that τ is a winning strategy for player I. Suppose now player II "plays f_{II}" and find C cub in $\boldsymbol{\delta}^1_{2n+1}$ such that

$$\xi \in C \Rightarrow \forall \eta < \xi (f_{\mathrm{II}}(\eta) < \xi).$$

Then if player I produces following his strategy f_{I} we have $f_{\mathrm{I}}(\xi) < K(\xi)$ for all $\xi \in C$, so $[f_{\mathrm{I}}] < [K]$ and the proof of the claim is complete. ⊣ (Claim 17.3)

So player II has a winning strategy σ. Then put for $\xi < \theta < \boldsymbol{\delta}^1_{2n+1}$:

$$J'(\xi, \theta) = \sup\{|W_{\varepsilon^{\mathrm{II}}_\sigma} \restriction \xi| + 1 : \alpha^{\mathrm{I}}, \varepsilon^{\mathrm{I}} \text{ are such that } |\alpha^{\mathrm{I}}| < \xi \text{ and } |(W_{\varepsilon^{\mathrm{I}}})^{|\alpha^{\mathrm{I}}|} \restriction \xi| \leq \theta\},$$

where as usual $\varepsilon^{\mathrm{II}}_\sigma = \sigma[\alpha^{\mathrm{I}}, \varepsilon^{\mathrm{I}}]$. By boundedness again $J'(\xi, \theta) < \boldsymbol{\delta}^1_{2n+1}$. So put for $\theta < \boldsymbol{\delta}^1_{2n+1}$

$$J(\theta) = \sup_{\xi < \theta} J'(\xi, \theta) < \boldsymbol{\delta}^1_{2n+1}.$$

Now we want to prove that

$$F(\xi') < J^*(\xi'), \forall \xi' \in (\boldsymbol{\delta}^1_{2n+1}, \boldsymbol{\delta}^1_{2n+2}).$$

Fix such a ξ' and find $\alpha^{\mathrm{I}}, \varepsilon^{\mathrm{I}}$ such that $[f_{\mathrm{I}}] = \xi'$. Thus we have to show that $F([f_{\mathrm{I}}]) < J^*([f_{\mathrm{I}}])$. Since $F([f_{\mathrm{I}}]) < [f_{\mathrm{II}}]$ it is enough to check that $[f_{\mathrm{II}}] \leq J^*([f_{\mathrm{I}}])$, *i.e.*, $f_{\mathrm{II}}(\xi) \leq J(f_{\mathrm{I}}(\xi))$ μ_ω-a.e. Let $\xi > |\alpha^{\mathrm{I}}|$ and $f_{\mathrm{I}}(\xi) > \xi$ (this happens μ_ω-a.e.). Put $\theta = f_{\mathrm{I}}(\xi) > \xi$. Then

$$J(f_{\mathrm{I}}(\xi)) = J(\theta) \geq J'(\xi, \theta) > f_{\mathrm{II}}(\xi).$$ ⊣ (Theorem 17.2)

Theorem 17.4 (AD) (Kunen [Kun71D]). For all $n \geq 1$, ${}^{\boldsymbol{\delta}^1_{2n+2}}\boldsymbol{\delta}^1_{2n+2}/\mu_\omega = (\boldsymbol{\delta}^1_{2n+2})^+$ and $\mathrm{cf}((\boldsymbol{\delta}^1_{2n+2})^+) = \boldsymbol{\delta}^1_{2n+2}$.

Proof. Let $F : \boldsymbol{\delta}^1_{2n+2} \to \boldsymbol{\delta}^1_{2n+2}$. Find $J : \boldsymbol{\delta}^1_{2n+1} \to \boldsymbol{\delta}^1_{2n+1}$ such that for $\xi > \boldsymbol{\delta}^1_{2n+1}$, $F(\xi) < J^*(\xi)$. Let W be a wellordering on $\boldsymbol{\delta}^1_{2n+1}$ such that $J(\xi) = H_W(\xi)$ μ_ω-a.e., say $J(\xi) = H_W(\xi), \forall \xi \in I \in \mu_\omega$. Then $J^* = (H_W)^* = H_{W^*}$ on I^*, so $F < H_{W^*}$ on $I^* \cap E_\omega$, thus $[F]_{\mu_\omega} < [H_{W^*}]_{\mu_\omega} = |W^*| < (\boldsymbol{\delta}^1_{2n+2})^+$. So ${}^{\boldsymbol{\delta}^1_{2n+2}}\boldsymbol{\delta}^1_{2n+2}/\mu_\omega = (\boldsymbol{\delta}^1_{2n+2})^+$.

Given now $\xi < \boldsymbol{\delta}^1_{2n+2}$ find $f : \boldsymbol{\delta}^1_{2n+1} \to \boldsymbol{\delta}^1_{2n+1}$ such that $[f]_{\mu_\omega} = \xi$. Then $f^* : \boldsymbol{\delta}^1_{2n+2} \to \boldsymbol{\delta}^1_{2n+2}$. Put

$$g(\xi) = [f^*]_{\mu_\omega} < (\boldsymbol{\delta}^1_{2n+2})^+.$$

It is easy to see that g is well defined. Moreover g is cofinal by the preceding fact and Theorem 17.2. ⊣ (Theorem 17.4)

Corollary 17.5 (AD) (Kunen [Kun71E]). For all $n \geq 1$, $\boldsymbol{\delta}^1_{2n} \not\rightarrow (\boldsymbol{\delta}^1_{2n})^{\boldsymbol{\delta}^1_{2n}}$.

Proof. By Theorem 17.4 and Corollary 13.9. ⊣ (Corollary 17.5)

§18. Some singular cardinals.

Let A, B be two transitive classes, $i: A \to B$ a Δ_0-**elementary** embedding, *i.e.*, for every Δ_0 formula φ, $A \models \varphi(a_1, \dots, a_n) \Leftrightarrow B \models \varphi(ia_1, \dots, ia_n)$. Put $i(A) = \bigcup\{ia : a \in A\}$. If A is closed under transitive closure, $i(A)$ is transitive and is therefore equal to the smallest transitive class containing the range of i.

Proof. If $x \in y \in ia$, put $b = \mathrm{TC}(a)$. Then $A \models \forall y \in A \forall x \in y(x \in b)$, so $B \models \forall y \in ia \forall x \in y(x \in ib)$, thus $x \in ib$. ⊣

Put HWO $= \{a : \mathrm{TC}(a)$ is wellorderable$\}$ and for each ordinal κ let $\mathrm{Fn}(\mathrm{HWO}, \kappa) = {}^{\kappa}\mathrm{HWO} \cap \mathrm{HWO} = \{F \in {}^{\kappa}\mathrm{HWO} : \mathrm{ran}(F) \in \mathrm{HWO}\}$. If $\mathcal{U}$ is a countably complete ultrafilter on κ, let $\mathrm{Fn}(\mathrm{HWO}, \kappa)/\mathcal{U}$ be the usual ultrapower. By Łoś' theorem

$$\mathrm{Fn}(\mathrm{HWO}, \kappa)/\mathcal{U} \models \varphi([F_1], \dots, [F_n]) \Rightarrow$$
$$\{\xi \in \kappa : \mathrm{HWO} \models \varphi(F_1(\xi), \dots, F_n(\xi))\} \in \mathcal{U},$$

for $\varphi \in \Delta_0$. So $\mathrm{Fn}(\mathrm{HWO}, \kappa)/\mathcal{U}$ is wellfounded and extensional, thus it can be collapsed to a transitive class $\mathrm{Ult}(\mathrm{HWO}, \mathcal{U})$. Let

$$i^{\mathcal{U}} : \mathrm{HWO} \to \mathrm{Ult}(\mathrm{HWO}, \mathcal{U})$$

be the usual embedding which by the above is Δ_0-elementary. Also

$$\mathrm{Ult}(\mathrm{HWO}, \mathcal{U}) = i^{\mathcal{U}}(\mathrm{HWO}) \subseteq \mathrm{HWO}.$$

Proof. For Ult HWO,$\mathcal{U}) = i^{\mathcal{U}}(\mathrm{HWO})$: If $[F] \in \mathrm{Fn}(\mathrm{HWO}, \kappa)/\mathcal{U}$ and $z = \mathrm{ran}(F)$, then $\{\xi < \kappa : \mathrm{HWO} \models F(\xi) \in z\} \in \mathcal{U}$ so

$$\mathrm{Fn}(\mathrm{HWO}, \kappa)/\mathcal{U} \models [F] \in [\xi \mapsto z],$$

thus $\mathrm{Ult}(\mathrm{HWO}, \mathcal{U}) \models [F] \in i^{\mathcal{U}}(z)$. (We identify here $[F]$ with its collapse.)

For $i^{\mathcal{U}}(\mathrm{HWO}) \subseteq \mathrm{HWO}$: Enough to show that $\{[G] : [G] \in [F]\}$ is wellorderable. Let $<$ be a wellordering of $\mathrm{TC}(\mathrm{ran}(F))$. For $[G], [G'] \in F$ let

$$[G] < [G'] \Rightarrow \{x : G(x) < G(x')\} \in \mathcal{U};$$

then $<$ is a wellordering on $\{[G] : [G] \in [F]\}$. ⊣

Lemma 18.1 (Kunen [Kun71E]). *Let $\kappa \leq \lambda$ be two cardinals such that $\mathrm{cf}(\lambda) = \kappa$. Let $\mathcal{U}$ be a countably complete uniform ultrafilter on κ. Then $i^{\mathcal{U}}(\lambda) \geq \lambda^+$.*

Proof. Fix $\lambda \leq \gamma < \lambda^+$. To show $i^{\mathcal{U}}(\lambda) > \gamma$. Let R be a wellordering of λ of type γ. Let $F : \kappa \xrightarrow{\mathrm{cf}} \lambda$. Then $i^{\mathcal{U}}(\lambda) > [F]_{\mathcal{U}} \geq \sup\{i^{\mathcal{U}}(\xi) : \xi < \lambda\}$. Now $|R{\restriction}\xi| < \lambda$, $\forall \xi < \lambda$, therefore in particular $|R{\restriction}F(\eta)| < \lambda$, $\forall \eta < \kappa$. So $|i^{\mathcal{U}}(R){\restriction}[F]_{\mathcal{U}}| < i^{\mathcal{U}}(\lambda)$. But also $|i^{\mathcal{U}}(R){\restriction}[F]_{\mathcal{U}}| \geq |i^{\mathcal{U}}(R){\restriction}\{i^{\mathcal{U}}(\xi) : \xi < \lambda\}| = |R| = \gamma$, therefore $i^{W}(\lambda) > \gamma$. ⊣ (Lemma 18.1)

Theorem 18.2 (AD) (Kunen [Kun71E]). Let $n \geq 0$, $\underset{\sim}{\delta}^1_{2n+1} = \omega_{p+1}$ and $\underset{\sim}{\delta}^1_{2n+2} = \omega_{p+2}$. Then for each $k \geq 2$,
1) $\mathrm{cf}(\omega_{p+k}) = \omega_{p+2}$.
2) There is Δ_0-elementary $h\colon \mathrm{Ult}(\mathrm{HWO}, \mathcal{U}) \to \mathrm{HWO}$ such that $\omega_{p+k} = \sup\{h(\xi) : \xi < \omega_{p+2}\}$, where $\mathcal{U} = \mu_\omega$ on ω_{p+1}.

Proof. Clearly 2) $\Rightarrow$ 1). We prove now 2) by induction on k. It is trivial for $k = 2$, with $h =$ identity. So assume $k > 2$ and it holds for all $2 \leq k' < k$. If $\mathcal{V} = \mu_\omega$ on ω_{p+2} then by Lemma 18.1, $i^{\mathcal{V}}(\omega_{p+k-1}) \geq \omega_{p+k}$.

Case I. $i^{\mathcal{V}}(\omega_{p+k'}) = \omega_{p+k}$, for some $2 \leq k' < k$.

By induction hypothesis, let h satisfy 2) for $\omega_{p+k'}$, *i.e.*,

$$\omega_{p+k'} = \sup\{h(\xi) : \xi < \omega_{p+2}\}.$$

Define $h^{00}\colon \mathrm{Ult}(\mathrm{HWO}, \mathcal{U}) \to \mathrm{HWO}$ by

$$h^{00}([F]_{\mathcal{U}}) = [h \circ i^{\mathcal{U}} F]_{\mathcal{V}}.$$

First we check that h^{00} is well defined. Assume $F = G$ on $I \in \mathcal{U}$. Then $i^{\mathcal{U}}F = i^{\mathcal{U}}G$ on $i^{\mathcal{U}}I \in \mathcal{V}$, so $h \circ i^{\mathcal{U}}F = h \circ i^{\mathcal{U}}G$ $\mathcal{V}$-a.e. It is equally routine to check that h^{00} is Δ_0-elementary. We shall now prove that

$$\sup\{h^{00}(\xi) : \xi < \omega_{p+2}\} = i^{\mathcal{V}}(\omega_{p+k'}) = \omega_{p+k},$$

which will complete the proof in this case.

First notice that $h^{00}(\xi) < i^{\mathcal{V}}(\omega_{p+k'})$ if $\xi < \omega_{p+2}$, since if $[F]_{\mathcal{U}} = \xi$, where $F\colon \omega_{p+1} \to \omega_{p+1}$, clearly $h \circ i^{\mathcal{U}}F\colon \omega_{p+2} \to \omega_{p+k'}$. Conversely, if $\theta < i^{\mathcal{V}}(\omega_{p+k'})$, then for some $G\colon \omega_{p+2} \to \omega_{p+k'}$, $\theta \leq [G]_{\mathcal{V}}$. Let $F'\colon \omega_{p+2} \to \omega_{p+2}$ be given by

$$F'(\eta) = \text{ least } \zeta < \omega_{p+2} \text{ such that } G(\eta) \leq h(\zeta).$$

Then $G \leq h \circ F'$ everywhere. By Theorem 17.2 let $F\colon \omega_{p+1} \to \omega_{p+1}$ be such that $F' \leq i^{\mathcal{U}}F$ $\mathcal{V}$-a.e. Then $G \leq h \circ i^{\mathcal{U}}F$ $\mathcal{U}$-a.e., so

$$\theta \leq [G]_{\mathcal{V}} \leq [h \circ i^{\mathcal{V}}F]_{\mathcal{V}} = h^{00}([F]_{\mathcal{V}}) = h^{00}(\xi),$$

where $\xi < \omega_{p+2}$ and we are done.

Case II. $i^{\mathcal{V}}(\omega_{p+k'-1}) < \omega_{p+k} < i^{\mathcal{V}}(\omega_{p+k})$, for some $3 \leq k' < k$.

Then find $F\colon \omega_{p+2} \to \omega_{p+k'}$ such that $[F]_{\mathcal{V}} = \omega_{p+k}$. Then F is cofinal in $\omega_{p+k'}$ since otherwise $[F]_{\mathcal{V}} < i^{\mathcal{V}}(\xi)$ for some $\xi < \omega_{p+k'}$. But $\mathrm{Card}(\xi) \leq \omega_{p+k'-1}$ so $\mathrm{Card}(i^{\mathcal{V}}(\xi)) \leq \mathrm{Card}(i^{\mathcal{V}}(\omega_{p+k'-1})) < \omega_{p+k}$, therefore $i^{\mathcal{V}}(\xi) < \omega_{p+k}$, a contradiction. Put

$$h^0 = i^{\mathcal{V}} \circ h,$$

where h comes from our induction hypothesis for $\omega_{p+k'}$. Clearly

$$h^0\colon \mathrm{Ult}(\mathrm{HWO}, \mathcal{U}) \to \mathrm{HWO}$$

is Δ_0-elementary. We will show that h^0 works for $\omega_{\rho+k}$.

Notice first that

$$\sup\{h^0(\alpha) : \alpha < \omega_{\rho+2}\} = [h{\restriction}\omega_{\rho+2}]_{\mathcal{V}}.$$

Indeed, $h^0(\xi) = i^{\mathcal{V}}(h(\xi)) < [h{\restriction}\omega_{\rho+2}]_{\mathcal{V}}$, since $h{\restriction}\omega_{\rho+2}$ is cofinal in $\omega_{\rho+k'}$. On the other hand if $[F]_{\mathcal{V}} < [h{\restriction}\omega_{\rho+2}]_{\mathcal{V}}$, then $F(\xi) < h(\xi)$ $\mathcal{V}$-a.e., so restricting to ξ's of cofinality ω if we let

$$T(\xi) \text{ be the least } \eta < \xi \text{ such that } F(\xi) < h(\eta),$$

then $T(\xi)$ is pressing down $\mathcal{V}$-a.e., because h is continuous at limits of cofinality ω, so $T(\xi) = \theta < \omega_{\rho+2}$ $\mathcal{V}$-a.e. Then $F(\xi) < h(\theta)$ $\mathcal{V}$-a.e., *i.e.*, $[f]_{\mathcal{V}} < i^{\mathcal{V}}(h(\theta))$ and we are done.

So it is enough to show

$$[h{\restriction}\omega_{\rho+2}]_{\mathcal{V}} = \omega_{\rho+k} = [F]_{\mathcal{V}}.$$

Now clearly $[h{\restriction}\omega_{\rho+2}]_{\mathcal{V}} \leq [F]_{\mathcal{V}}$, since F is unbounded in $\omega_{\rho+k'}$. So it suffices to prove that

$$[F]_{\mathcal{V}} = \omega_{\rho+k} \leq [h{\restriction}\omega_{\rho+2}]_{\mathcal{V}}.$$

For that it is again enough to show that

$$([h{\restriction}\omega_{\rho+2}]_{\mathcal{V}})^+ \geq i^{\mathcal{V}}(\omega_{\rho+k'}).$$

Fix $\gamma < i^{\mathcal{V}}(\omega_{\rho+k'})$. Then $\gamma = [H]_{\mathcal{V}}$, where $H\colon \omega_{\rho+2} \to \omega_{\rho+k'}$, so $\gamma \leq [h \circ F]_{\mathcal{V}}$ for some $F\colon \omega_{\rho+2} \to \omega_{\rho+2}$. Find $H_W\colon \omega_{\rho+1} \to \omega_{\rho+1}$ such that $[F]_{\mathcal{V}} \leq [i^{\mathcal{U}} H_W]_{\mathcal{V}} = [H_{i^{\mathcal{U}} W}]_{\mathcal{V}}$. Then $\gamma \leq [h \circ H_{i^{\mathcal{U}} W}] = [h \circ H_V]_{\mathcal{V}}$, where $V = i^{\mathcal{U}} W$ is a wellordering on $\omega_{\rho+2}$. Fix $g \in \mathrm{Ult}(\mathrm{HWO}, \mathcal{U})$ such that for some $I \in \mathcal{V}$, $I \in \mathrm{Ult}(\mathrm{HWO}, \mathcal{U})$, $g(\xi)\colon H_V(\xi) \hookrightarrow \xi$, $\forall \xi \in I$. [To see that these exist, find $P(\xi)\colon |W{\restriction}\xi| \hookrightarrow \xi$, $\forall \xi \in X \in \mathcal{U}$, so $i^{\mathcal{U}} P(\xi)\colon |V{\restriction}\xi| \hookrightarrow \xi$, $\forall \xi \in i^{\mathcal{U}} X \in \mathcal{V}$. Put $g = i^{\mathcal{U}} P$, $I = i^{\mathcal{U}} X$.] Since $g \in \mathrm{Ult}(\mathrm{HWO}, \mathcal{U})$,

$$\xi \in I \Rightarrow h(g(\xi))\colon h \circ H_V(\xi) \hookrightarrow h(\xi),$$

so $[h \circ g]_{\mathcal{V}}\colon [h \circ H_V]_{\mathcal{V}} \hookrightarrow [h{\restriction}\omega_{\rho+2}]_{\mathcal{V}}$, thus $\gamma \leq [h \circ H_V]_{\mathcal{V}} < ([h{\restriction}\omega_{\rho+2}]_{\mathcal{V}})^+$.

⊣ (Theorem 18.2)

REFERENCES

John W. Addison and Yiannis N. Moschovakis

[AM68] *Some consequences of the axiom of definable determinateness*, ***Proceedings of the National Academy of Sciences of the United States of America***, no. 59, 1968, pp. 708–712.

Alexander S. Kechris

[Kec74] *On projective ordinals*, ***The Journal of Symbolic Logic***, vol. 39 (1974), pp. 269–282.

Alexander S. Kechris and Yiannis N. Moschovakis

[Cabal i] *Cabal seminar* 76–77, Lecture Notes in Mathematics, no. 689, Berlin, Springer, 1978.

EUGENE M. KLEINBERG

[Kle70] *Strong partition properties for infinite cardinals*, ***The Journal of Symbolic Logic***, vol. 35 (1970), pp. 410–428.

KENNETH KUNEN

[Kun71A] *Measurability of* $\underset{\sim}{\delta}^1_n$, circulated note, April 1971.

[Kun71C] *A remark on Moschovakis' uniformization theorem*, circulated note, March 1971.

[Kun71D] *Some singular cardinals*, circulated note, September 1971.

[Kun71E] *Some more singular cardinals*, circulated note, September 1971.

RICHARD MANSFIELD

[Man71] *A Souslin operation on* Π^1_2, ***Israel Journal of Mathematics***, vol. 9 (1971), no. 3, pp. 367–379.

DONALD A. MARTIN

[Mar68] *The axiom of determinateness and reduction principles in the analytical hierarchy*, ***Bulletin of the American Mathematical Society***, vol. 74 (1968), pp. 687–689.

[Mar71A] *Determinateness implies many cardinals are measurable*, circulated note, May 1971.

[Mar71B] *Projective sets and cardinal numbers: some questions related to the continuum problem*, this volume, originally a preprint, 1971.

DONALD A. MARTIN AND JEFF B. PARIS

[MP71] AD $\Rightarrow \exists$ *exactly* 2 *normal measures on* ω_2, circulated note, March 1971.

DONALD A. MARTIN AND ROBERT M. SOLOVAY

[MS69] *A basis theorem for* Σ^1_3 *sets of reals*, ***Annals of Mathematics***, vol. 89 (1969), pp. 138–160.

YIANNIS N. MOSCHOVAKIS

[Mos70] *Determinacy and prewellorderings of the continuum*, ***Mathematical logic and foundations of set theory. Proceedings of an international colloquium held under the auspices of the Israel Academy of Sciences and Humanities, Jerusalem, 11–14 November 1968*** (Y. Bar-Hillel, editor), Studies in Logic and the Foundations of Mathematics, North-Holland, Amsterdam-London, 1970, pp. 24–62.

[Mos71] *Uniformization in a playful universe*, ***Bulletin of the American Mathematical Society***, vol. 77 (1971), pp. 731–736.

JOSEPH R. SHOENFIELD

[Sho61] *The problem of predicativity*, ***Essays on the foundations of mathematics*** (Yehoshua Bar-Hillel, E. I. J. Poznanski, Michael O. Rabin, and Abraham Robinson, editors), Magnes Press, Jerusalem, 1961, pp. 132–139.

ROBERT M. SOLOVAY

[Sol67A] *Measurable cardinals and the axiom of determinateness*, lecture notes prepared in connection with the Summer Institute of Axiomatic Set Theory held at UCLA, Summer 1967.

[Sol78A] *A* Δ^1_3 *coding of the subsets of* ω_ω, this volume, originally published in Kechris and Moschovakis [CABAL i], pp. 133–150.

DEPARTMENT OF MATHEMATICS
CALIFORNIA INSTITUTE OF TECHNOLOGY
PASADENA, CALIFORNIA 91125
UNITED STATES OF AMERICA
E-mail: kechris@caltech.edu

A Δ^1_3 CODING OF THE SUBSETS OF ω_ω

ROBERT M. SOLOVAY

§1. Introduction. We present Kunen's [Kun71B] analysis of the measures on ω_ω, from AD+DC, together with some lightface refinements that follow by mixing Kunen's techniques with a theorem of Kechris-Martin. Our purpose in this introduction is (a) to list some applications of the Kunen method (to be presented in Section 3); (b) to give an overview of the more technical results to be proved; (c) to give some idea of the motivation behind the technicalities that follow.

The principle results of type (a) are as follows (AD+DC is assumed throughout):

Let $\lambda < \utilde{\delta}^1_3$. Then $\utilde{\delta}^1_3 \to (\utilde{\delta}^1_3)^\lambda$. This should be compared with the result of Martin that $\utilde{\delta}^1_1 \to (\utilde{\delta}^1_1)^{\utilde{\delta}^1_1}$. It is still open whether or not $\utilde{\delta}^1_3 \to (\utilde{\delta}^1_3)^{\utilde{\delta}^1_3}$. Kunen's proof uses a highly detailed analysis at level 3, and it is not known how to generalize his work to odd n's greater than 3. In particular, it is open whether $\utilde{\delta}^1_5 \to (\utilde{\delta}^1_5)^\lambda$ for all $\lambda < \utilde{\delta}^1_5$, though we have already seen in [Kec78] that $\utilde{\delta}^1_5 \to (\utilde{\delta}^1_5)^\lambda$ for countable λ.

Note that by earlier work in [Kec78], it follows that there are exactly three normal measures on $\utilde{\delta}^1_3$ (concentrating on points of cofinality ω_0, ω_1, and ω_2 respectively).

(b) Our proof will be based on a Δ^1_3 encoding of the subsets of ω_ω (*cf.* Theorem 3.8). To state what this means, recall that the theory of sharps together with the fact that the ω_i's, $1 \leq i < \omega$ are precisely the first ω uniform indiscernibles gives a natural Δ^1_3 encoding of the ordinals $< \omega_\omega$. We shall produce a Δ^1_3 set, C, of codes for subsets of ω_ω so that the relation "the ordinal coded by x lies in the set coded by y" is Δ^1_3.

To get such a coding we need a concrete way of generating all subsets of ω_ω. We prove that every non-empty subset of ω_ω is the countable union of **simple** sets. Here, a simple set is a subset of some ω_m which is the 1-1 image, by some

The author would like to express his thanks to the Sherman Fairchild Distinguished Scholars Program at Caltech for its generous support during the academic year 1976–1977. Thanks also to Greg Ennis for the conscientious painstaking work of transferring a series of lectures to the printed page.

Wadge Degrees and Projective Ordinals: The Cabal Seminar, Volume II
Edited by A. S. Kechris, B. Löwe, J. R. Steel
Lecture Notes in Logic, 37

function constructible from a real, of one of a countable sequence of standard sets, $A^j_{m,k}$. [This is slightly stronger than what we prove below (owing to a different definition of "simple"); it is not hard to prove the stronger claim with Kunen's method.]

By an ingenious reduction, (*cf.* Theorem 3.6) this is reduced to an analysis of measures on ω_ω. Some further simple reductions (which take place in Sections 3.1 and 3.2 below) reduce the problem to the following: We are given integers $m \geq 0$, $k \geq 1$. Each closed unbounded subset C of ω_1 determines in a canonical manner a subset $\tilde{C}$ of ω_{m+1} (the image of C under a suitable elementary embedding of some $\mathbf{L}[z]$ mapping ω_1 into ω_{m+1}). In this way the sets $\tilde{C}^k$ form a filter base for a filter $\mathscr{F}$ on ω^k_{m+1}. Our problem is to characterize the ultrafilters that extend $\mathscr{F}$. The main result proved in Section 2 shows that there are only finitely many such ultrafilters. They live on disjoint sets $A^1_{m,k}, \dots, A^\ell_{m,k}$ (whose union is the set of k-tuples of limit ordinals less than ω_{m+1}) and the decomposition is Δ^1_3 in the codes.

The proof is a refinement of the Martin-Paris analysis of normal measures on ω_2, which was presented in [Kec78]. We replace the study of k-tuples of ordinals less than ω_{m+1} by the study of k-tuples of functions from ω^m_1 into ω_1. We analyse these k-tuples carefully enough so as to be able to imitate the Martin-Paris proof. Of course, in our more general context the technicalities will be considerably greater.

We emphasize that we work in ZF+AD+DC in the following. Any unexplained notation is as in [Kec78].

§2. Classification of tuples of ordinals.

2.1. W is the canonical normal measure on ω_1. W_n is the product measure

$$\underbrace{W \times \cdots \times W}_{n \text{ times}}.$$

If A is a set, A^n is the n-fold Cartesian power. $A^{[n]}$, for A linearly ordered, is the set of strictly increasing n-tuples from A.

2.2. $\tilde{\mathbf{L}} = \bigcup\{\mathbf{L}[x] : x \in \mathbb{R}\}$. For each $x \in \mathbb{R}$, we have an elementary embedding

$$i_{W_n} : \mathbf{L}[x] \to \mathbf{L}[x],$$

given by the transitive realization of the ultrapower. These maps piece together to give a map

$$i_{W_n} : \tilde{\mathbf{L}} \to \tilde{\mathbf{L}}.$$

(Note that this last map is *not* elementary since ω_1 is definable in $\tilde{\mathbf{L}}$ but $i_{W_n}(\omega_1) \neq \omega_1$.)

Note that W_n gives $\omega_1^{[n]}$ measure 1 (and we usually view W_n as a measure on $\omega_1^{[n]}$). If $F\colon \omega_1^{[n]} \to \mathbf{L}[x]$, then F determines (via transitive realization of the ultrapower) an element $[F] \in \mathbf{L}[x]$.

Let $\pi_i\colon \omega_1^{[n]} \to \omega_1$, be given by $\pi_i(\alpha_1, \dots, \alpha_n) = \alpha_i$.

Lemma 2.1. $[\pi_i] = \omega_i$. $i_{W_n}(\omega_1) = \omega_{n+1}$.

Proof. Clearly, $\{\vec{\alpha} \in \omega_1^{[n]} : \mathbf{L}[z^\#] \models \alpha_i \text{ is an indiscernible for } \mathbf{L}[z]\}$ has W_n-measure 1. Thus $[\pi_i]$ is a uniform indiscernible. So clearly is $i_{W_n}(\omega_1)$. Since clearly

$$[\pi_1] < \cdots < [\pi_n] < i_{W_n}(\omega_1),$$

it suffices to show that if $F\colon \omega_1^{[n]} \to \omega_1$ is such that $[F]$ is a uniform indiscernible, then $[F] = [\pi_i]$ for some i.

Now $F \in \mathbf{L}[z]$ for some $z \in \mathbb{R}$ since $F \subseteq \mathbf{L}_{\omega_1}$. By increasing the Turing degree of z, we may suppose F is definable in $\mathbf{L}[z]$ from z, ω_1. Pick i minimal such that $[F] \le [\pi_i]$. (The case when $[F] > [\pi_n]$ is handled similarly.) If $[F] = [\pi_i]$, we are done. Otherwise, for almost all $\vec{\alpha}$, $F(\vec{\alpha})$ is definable in $\mathbf{L}[z]$ from z, and indiscernibles for $\mathbf{L}[z]$ distinct from $F(\vec{\alpha})$. It follows that $F(\vec{\alpha})$ is a.e. not an indiscernible for $\mathbf{L}[z]$. Whence $[F]$ is not a uniform indiscernible. ⊣

Corollary 2.2. Let $\lambda < \omega_{m+1}$. Let $F\colon \omega_1^{[m]} \to \omega_1$ such that $[F] = \lambda$. Let $\tilde{F}\colon \omega_{m+1}^{[m]} \to \omega_{m+1}$ be $i_{W_m}(F)$. Then $\lambda = \tilde{F}(\omega_1, \dots, \omega_m)$.

Proof. By Łoś' theorem, this amounts to

$$F(\vec{\alpha}) = F(\pi_1(\vec{\alpha}), \dots, \pi_m(\vec{\alpha})).$$ ⊣

2.3. We now state the theorem which is the main goal of Section 2.

Theorem 2.3 (Kunen [Kun71B]). Let $m \ge 0$. Let $k \ge 1$. Let

$$X_{m,k} = \{\langle \lambda_1, \dots, \lambda_k \rangle : \text{for } 1 \le i \le k,\ \lambda_i \text{ is a limit ordinal } < \omega_{m+1}\}.$$

We shall construct a decomposition:

$$X_{m,k} = A^1_{m,k} \cup \dots \cup A^\ell_{m,k} \text{ (disjoint union)}$$

with the following properties:

(1) Let C be a cub subset of ω_1. Let $\tilde{C} = i_{W_m}(C)$. Then $\tilde{C}^k \cap A^i_{m,k} \neq \varnothing$.

(2) Sets of the form $\tilde{C}^k \cap A^i_{m,k}$ are the basis for an ultrafilter on $A^i_{m,k}$. (This ultrafilter is clearly countably additive since the intersection of countably many cubs is cub.)

(3) Let $h\colon \omega_1 \to \omega_1$ be normal (*i.e.*, strictly increasing and continuous). Let $\tilde{h}\colon \omega_{m+1} \to \omega_{m+1}$ be $i_{W_m}(h)$. If $\vec{\alpha} \in A^i_{m,k}$, then

$$\langle \tilde{h}(\alpha_i), \dots, \tilde{h}(\alpha_k) \rangle \in A^i_{m,k}.$$

We shall see that properties (1) through (3) characterize the partition of $X_{m,k}$ into the $A^i_{m,k}$'s.

2.4. For certain applications we need that our partition is Δ^1_3 in the codes. A **code** for an ordinal $< \omega_\omega$ is a pair $z = \langle n, x^\# \rangle$, where $x \in \mathbb{R}$. If z is a code as above then we put

$$|z| = \tau_n^{\mathbf{L}[x]}(\omega_1, \ldots, \omega_{k_n})$$

where $\tau_0, \tau_1, \ldots$ is a recursive enumeration of all the terms in the language of $\mathbf{L}[x]$ (x occurring as a constant symbol) taking always ordinal values.

The following result is an effective version (Δ^1_3 replacing $\underset{\sim}{\Delta}^1_3$) of a theorem of Kunen [Kun71B].

THEOREM 2.4.

(1) Let $\ell(m,k)$ be the number of pieces into which $X_{m,k}$ is decomposed. (*cf.* Theorem 2.3) Then the map $\langle m,k \rangle \mapsto \ell(m,k)$ is Δ^1_3.

(2) The following relation $\Phi(k,m,j,x)$ is Δ^1_3: (Here k, m, j are in ω, and x in $\mathbb{R}$) $k \geq 1$; $m \geq 0$; $1 \leq j \leq \ell(m,k)$; for $1 \leq i \leq k$, $(x)_i$ codes a limit ordinal $\lambda_i < \omega_{m+1}$; $\langle \lambda_1, \ldots, \lambda_k \rangle \in A^j_{m,k}$.

2.5. Let HF be the class of hereditarily finite sets (HF $= \mathbf{L}_\omega$, and there is a canonical well-ordering of HF of type ω.)

The decomposition of $X_{m,k}$ will arise in the following way. We will define a map $\Psi_{m,k}\colon X_{m,k} \to$ HF. The range of $\Psi_{m,k}$ will be finite, say $\{x_1, \ldots, x_\ell\}$, where $x_1 < \cdots < x_\ell$ with respect to the canonical well-ordering of HF. Then $A^j_{m,k} = \Psi^{-1}_{m,k}(\{x_j\})$.

The map $\Psi_{m,k}$ will have the following properties:

(α) $\operatorname{ran}(\Psi_{m,k})$ is finite.

(β) $\Psi_{m,k}$ is an invariant: If $h\colon \omega_1 \to \omega_1$ is normal, $\tilde{h} = i_{W_m}(h)$, $\vec{\alpha} \in X_{m,k}$ and $\beta_i = \tilde{h}(\alpha_i)$ for $1 \leq i \leq k$, then

$$\Psi_{m,k}(\vec{\alpha}) = \Psi_{m,k}(\vec{\beta}).$$

(γ) $\Psi_{m,k}$ is Δ^1_3 in the codes: The following relation is Δ^1_3: $m \geq 0$, $k \geq 1$, for $1 \leq i \leq k$ $(x)_i$ codes a limit ordinal $\lambda_i < \omega_{m+1}$ and

$$\Psi_{m,k}(\vec{\lambda}) = s.$$

(Here $s \in$ HF, $x \in \mathbb{R}$, $m, k \in \omega$).

2.6. The case $m = 0$ is trivial but atypical. First we describe the invariant $\Psi_{0,k}$:

$$\Psi_{0,k}(\langle \lambda_1, \ldots, \lambda_k \rangle) = \{\langle i, j \rangle : \lambda_i < \lambda_j\}.$$

It is evident that $\Psi_{0,k}$ has the properties (α), (β), and (γ) of 2.5.

A moment's thought will show that we can effectively tell which $s \in \mathrm{HF}$ lie in $\mathrm{ran}(\Psi_{0,k})$. Thus the portion of 2.4 that relates to $m = 0$ is evident from (γ) holding for $\Psi_{0,k}$ (uniformly in k).

Finally we verify Theorem 2.3 for $X_{0,k}$ with respect to the decomposition induced by $\Psi_{0,k}$. Let $A^j_{0,k}$ be a component of this decomposition. Then there is an integer r, with $1 \leq r \leq k$, integers $s_1, \dots, s_r$ with $1 \leq s_i \leq k$, and integers n_i, $1 \leq i \leq k$, with $1 \leq n_i \leq r$ such that for $\vec{\alpha} \in A^j_{0,k}$, $\alpha_i = \alpha_{s_{n_i}}$; $\alpha_{s_1} < \alpha_{s_2} < \cdots < \alpha_{s_r}$.

By means of the map $A^j_{0,k} \simeq \mathrm{Lim}^{[r]}$: $\langle \alpha_1, \dots, \alpha_n \rangle \mapsto \langle \alpha_{s_1}, \dots, \alpha_{s_r} \rangle$, the claims of Theorem 2.3 for the case $m = 0$ reduce to the following well-known fact: sets of the form $C^{[r]}$, C cub, are a basis for W_r.

2.7. We turn now to the definition of $\Psi_{m,k}$ for $m > 0$. Our strategy will be as follows. We define three simpler invariant functions, $\Psi^1_{m,k}, \Psi^2_{m,k}, \Psi^3_{m,k}$ which satisfy (α), (β), (γ) of 2.5. We then put

$$\Psi_{m,k}(\vec{\alpha}) = \langle \Psi^1_{m,k}(\vec{\alpha}), \dots, \Psi^3_{m,k}(\vec{\alpha}) \rangle.$$

$\Psi_{m,k}$ will then inherit the properties (α), (β), and (γ) from the $\Psi^i_{m,k}$'s.

We now fix $m \geq 1$, $k \geq 1$. Let $\vec{\lambda} \in X_{m,k}$. Pick $F_1, \dots, F_k$ mapping $\omega_1^{[m]}$ into ω_1 such that $[F_i] = \lambda_i$. Pick $z \in \mathbb{R}$ such that $\langle F_i : 1 \leq i \leq k \rangle$ is definable from z, ω_1 in $\mathbf{L}[z]$. Let I_z be the set of canonical indiscernibles for $\mathbf{L}[z]$ which are less than ω_1.

We pick $\vec{\gamma} \in I_z^{[2m]}$. Then

$$\begin{aligned} \Psi^1_{m,k}(\vec{\lambda}) = \{ \langle i, \vec{r}, j, \vec{s} \rangle \; : \; & 1 \leq i, j \leq k; \\ & 1 \leq r_1 < \cdots < r_m \leq 2m; \\ & 1 \leq s_1 < \cdots < s_m \leq 2m; \\ & F_i(\gamma_{r_1}, \dots, \gamma_{r_m}) < F_j(\gamma_{s_1}, \dots, \gamma_{s_m}) \}. \end{aligned}$$

It is necessary to see that $\Psi^1_{m,k}$ depends only on $\vec{\lambda}$ and not on the choices of the F's, z, and γ's. Evidently the choice of the γ's are irrelevant so long as they are chosen in I_z. If $\vec{F}'$, z' are different choices, we can find E cub so that

(a) F_i and F'_i agree on $E^{[m]}$ $(1 \leq i \leq k)$,
(b) $E \subseteq I_z \cap I_{z'}$.

But now if we take $\vec{\gamma} \in E^{[2m]}$, we will get the same value of Ψ^1 from the primed choices as from the unprimed ones.

How about properties (α) through (γ)? Properties (α) and (γ) are evident. As for property (β) note that $[h \circ F_i] = \tilde{h}(\lambda_i)$. Since h is order-preserving, (β) is now clear for Ψ^1.

2.8. Let D be a closed unbounded subset of ω_1. D' is the set of limit points of D. Define a function $H^D_i : D'^{[m]} \to \omega_1$:

$$H^D_i(\vec{\gamma}) = \sup\{F_j(\vec{\delta}) : \vec{\delta} \in D^{[m]} \text{ and } F_j(\vec{\delta}) < F_i(\vec{\gamma})\}.$$

Evidently if E is a cub $\subseteq D$, $[H_i^E] \leq [H_i^D]$, since for $\vec{\gamma} \in E'^{[m]}$, the supremum is over a smaller set for $H_i^E(\vec{\gamma})$ than for $H_i^D(\vec{\gamma})$.

Definition 2.5. i is of type II if for some cub D, $[H_i^D] < [F_i]$.

We put $\Psi^2_{m,k}(\vec{\lambda}) = \{i : i \text{ is of type II}\}$.

As usual we must verify independence of the choice of F's. But if F_j and F'_j agree on $E^{[m]}$, E cub, and for some D, $[H_i^D] < [F_i]$, then by the remark of the paragraph preceding Definition 2.5, $[H_i^{D\cap E}] < [F_i]$. But then $H_i^{D\cap E} = H_i'^{D\cap E}$ (since the F_j's and F'_j's look alike on $E^{[m]}$) so $[H_i'^{D\cap E}] < [F'_i]$. So i is of type II with respect to the F''s (if it is with respect to the F's).

Property (α) is evident. Since if h is normal, it preserves the notion of limit and non-limit, and since $[h \circ F_i] = \tilde{h}(\lambda_i)$, property (β) is clear.

Our proof of property (γ) for Ψ^2 will be preceded by two lemmas.

Lemma 2.6. Let $\vec{\lambda}$, $\vec{F}$, z be as above. Let D, E be cub subsets of ω_1 with $E \subseteq D \subseteq I_z$. Suppose that for some $\vec{\delta} \in E'^{[m]}$, $H_i^E(\vec{\delta}) < F_i(\vec{\delta})$. Then for every $\vec{\gamma} \in D'^{[m]}$, $H_i^D(\vec{\gamma}) < F_i(\vec{\gamma})$.

Proof. Let $\xi = H_i^E(\vec{\delta})$. Let $\vartheta_1, \dots, \vartheta_n$ be ordinals of I_z such that $\xi = \tau^{\mathbf{L}[z]}(\vartheta_1, \dots, \vartheta_n)$. Pick $\vartheta'_1, \dots, \vartheta'_n$ in E so that

(a) $\vartheta_i \leq \vartheta'_i$ $(1 \leq i \leq n)$

(b) $\vec{\vartheta}, \vec{\delta}$ and $\vec{\vartheta}', \vec{\delta}$ are similarly ordered.

(c) If $\alpha_1, \alpha_2 \in \{0, \vartheta'_1, \dots, \vartheta'_n, \delta_1, \dots, \delta_m\}$ with $\alpha_1 < \alpha_2$, then there are at least m members of E strictly between α_1 and α_2.

(There is no difficulty doing this since the δ_i's are limit points of E.)

Let $\xi' = \tau^{\mathbf{L}[z]}(\vartheta'_1, \dots, \vartheta'_n)$. By (a) $\xi \leq \xi'$. By (b) and $E \subseteq I_z$, $\xi' < F_i(\vec{\delta})$.

Next select $\vartheta^*_1, \dots, \vartheta^*_n$ in D with

(d) $\vec{\vartheta}^*, \vec{\gamma}$ similarly ordered to $\vec{\vartheta}', \vec{\delta}$.

(This is possible since the γ_i's are in D' and $\vec{\gamma}$ is similarly ordered to $\vec{\delta}$.) Put $\xi^* = \tau^{\mathbf{L}[z]}(\vec{\vartheta}^*)$. By (d) and the fact that $E \subseteq D \subseteq I_z$, we have $\xi^* < F_i(\vec{\gamma})$. Thus to prove $H_i^D(\vec{\gamma}) < F_i(\vec{\gamma})$ it suffices to show $H_i^D(\vec{\gamma}) \leq \xi^*$.

Deny this towards a contradiction. Then for some $\vec{\eta} \in D^{[m]}$, j, we have $\xi^* < F_j(\vec{\eta}) < F_i(\vec{\gamma})$. By (c), we can choose $\vec{\eta}' \in E^{[m]}$ so that

(e) $\vec{\vartheta}^*, \vec{\eta}, \vec{\gamma}$ are ordered similarly to $\vec{\vartheta}', \vec{\eta}', \vec{\delta}$.

But then by (e) and $E \subseteq D \subseteq I_z$,

(f) $\xi' < F_j(\vec{\eta}') < F_i(\vec{\delta})$.

But this contradicts the fact that $H_i^E(\delta) \leq \xi'$. ⊣

Lemma 2.7. Let $D \subseteq I_z$ be cub. Then if i is not of type II, $H_i^D(\vec{\gamma}) = F_i(\vec{\gamma})$ for all $\vec{\gamma} \in D'^{[m]}$. If i is of type II, $H_i^D(\vec{\gamma}) < F_i(\vec{\gamma})$ for all $\vec{\gamma} \in D'^{[m]}$.

Proof. Suppose that $F_i(\vec{\gamma}) > H_i^D(\vec{\gamma})$ for one $\vec{\gamma} \in D'^{[m]}$. (Note that $F_i(\vec{\gamma}) \geq H_i^D(\vec{\gamma})$ is evident from the definitions.) Apply Lemma 2.6 with $D = E$. Then $F_i(\vec{\gamma}) > H_i^D(\vec{\gamma})$ for every $\vec{\gamma} \in (D')^{[m]}$. Whence $[F_i] > [H_i^D]$, and i is of type II. This proves our first claim.

Next suppose i is of type II. Then for some cub E, $[H_i^E] < [F_i]$. Replace E, if need be, by $E \cap D$. But then Lemma 2.6 guarantees $H_i^D(\vec{\gamma}) < F_i(\vec{\gamma})$ for all $\vec{\gamma} \in D'^{[m]}$ (since the $\vec{\delta}$ needed for Lemma 2.6 is guaranteed by $[H_i^E] < [F_i]$). ⊣

It is now easy to prove that Ψ^2 is "Δ_3^1 in the codes". If $(x)_i$ codes λ_i, we can take x itself for our z. The criterion of Lemma 2.7 can be checked in $\mathbf{L}[z^{\#}]$, hence recursively in $z^{\#\#}$, hence Δ_3^1 in z. This proves that Ψ^2 satisfies (γ).

2.9.

Definition 2.8. $\mathrm{cf}^{\tilde{\mathbf{L}}}(\alpha) = \min\{\mathrm{cf}^{\mathbf{L}[z]}(\alpha) : z \in \mathbb{R}\}$.

Evidently $\mathrm{cf}^{\tilde{\mathbf{L}}}(\alpha) \leq \alpha$. $\mathrm{cf}^{\tilde{\mathbf{L}}}(\alpha)$ is regular in each $\mathbf{L}[z]$. So if α is a limit ordinal $\mathrm{cf}^{\tilde{\mathbf{L}}}$ is either ω or a uniform indiscernible.

It follows that $\mathrm{cf}^{\tilde{\mathbf{L}}}(\lambda_i) = \omega_j$ for some j with $0 \leq j \leq m$. We put

$$\Psi^3_{m,k}(\vec{\lambda}) = \{\langle i, j\rangle : i \text{ is of type II and } \mathrm{cf}(\lambda_i) = \omega_j\}.$$

Evidently Ψ^3 satisfies (α). If $h\colon \omega_1 \to \omega_1$ is normal, then so is $\tilde{h}\colon \omega_{m+1} \to \omega_{m+1}$. If g maps ω_j cofinally into λ_i, $\tilde{h} \circ g$ maps ω_j cofinally into $\tilde{h}(\lambda_i)$. If g is order-preserving, so is $\tilde{h} \circ g$. It follows that $\mathrm{cf}^{\tilde{\mathbf{L}}}(\tilde{h}(\lambda_i)) = \mathrm{cf}^{\tilde{\mathbf{L}}}(\lambda_i)$ and Ψ^3 satisfies (β).

We need the following lemma to show that Ψ^3 satisfies (γ).

Lemma 2.9. Let $z \in \mathbb{R}$. Let λ be an infinite ordinal which is regular in $\mathbf{L}[z]$ but not an indiscernible of $\mathbf{L}[z]$. Then λ is cofinal with ω in $\mathbf{L}[z^{\#}]$.

Proof. We may as well assume $\lambda > \omega$. Let $\gamma = \sup\{\xi < \lambda : \xi = \omega \text{ or } \xi \text{ is an indiscernible for } \mathbf{L}[z]\}$. Then $\omega \leq \gamma \leq \lambda$. If $\gamma = \lambda$, then γ must be an indiscernible for $\mathbf{L}[z]$ (since the class of indiscernibles is closed and $\gamma > \omega$). This contradicts our assumption on λ. So $\gamma < \lambda$.

Let $\vartheta_1, \vartheta_2, \ldots$ be the first ω indiscernibles for $\mathbf{L}[z]$ greater than λ. Then every ordinal $\xi < \lambda$ is definable in $\mathbf{L}[z]$ from ordinals $\leq \gamma$ and some of the ϑ_i's.

We set $S_i = \{\eta < \lambda : \eta \text{ is definable in } \mathbf{L}[z] \text{ from ordinals in } \gamma \cup \{\gamma\} \cup \{\vartheta_1, \ldots, \vartheta_i\}\}$. Then $S_i \in \mathbf{L}[z]$ and in $\mathbf{L}[z]$ S_i has power γ. Since λ is regular in $\mathbf{L}[z]$, $\xi_i = \sup(S_i)$ is less than λ. By the preceding paragraph, the ξ_i's are cofinal in λ. But clearly the sequence ξ_i is definable from λ in $\mathbf{L}[z^{\#}]$. ⊣

Lemma 2.10. Let $\lambda = \tau^{\mathbf{L}[z]}(\omega_1, \ldots, \omega_m)$, λ a limit ordinal. Then $\mathrm{cf}^{\mathbf{L}[z^{\#}]}(\lambda) = \mathrm{cf}^{\tilde{\mathbf{L}}}(\lambda)$.

PROOF. If $\mathrm{cf}^{\mathbf{L}[z]}(\lambda) = \omega_i$, for $0 \le i \le m$, then evidently $\omega_i = \mathrm{cf}^{\mathbf{L}[z^\#]}(\lambda) = \mathrm{cf}^{\tilde{\mathbf{L}}}(\lambda)$. If not, $\mathrm{cf}^{\mathbf{L}[z]}(\lambda)$ is not an indiscernible in $\mathbf{L}[z]$ (since it is definable in $\mathbf{L}[z]$ from $\omega_1, \dots, \omega_m$, but is distinct from each of $\omega_1, \dots, \omega_m$). Whence by Lemma 2.9, $\mathrm{cf}^{\mathbf{L}[z^\#]}(\mathrm{cf}^{\mathbf{L}[z]}(\lambda)) = \omega$. But then clearly $\mathrm{cf}^{\mathbf{L}[z^\#]}(\lambda) = \mathrm{cf}^{\tilde{\mathbf{L}}}(\lambda) = \omega$. ⊣

Lemma 2.10 makes it evident that Ψ^3 is Δ^1_3 in the codes. For again if $(x)_i$ codes λ_i, we can take x as our z and compute $\mathrm{cf}^{\tilde{\mathbf{L}}}(\lambda_i)$ recursively from $x^{\#\#}$.

2.10. Now as promised in 2.5 we put

$$\Psi_{m,k}(\vec{\lambda}) = \langle \Psi^1_{m,k}(\vec{\lambda}), \Psi^2_{m,k}(\vec{\lambda}), \Psi^3_{m,k}(\vec{\lambda}) \rangle$$

for $\vec{\lambda} \in X_{m,k}$. Thus the range of $\Psi_{m,k}$ is a finite subset of HF, say $\{s_1, \dots, s_{\ell(m,k)}\}$, where $s_1 < \cdots < s_{\ell(m,k)}$ in the canonical well-ordering of HF. Let $A^j_{m,k} = \Psi^{-1}_{m,k}(\{s_j\})$.

The proof of part (2) of Theorem 2.4 will follow from part (1) and the fact that Ψ^1, Ψ^2, Ψ^3 all satisfy property (γ) of 2.5. To prove part (1) of Theorem 2.4 we need the Kechris-Martin Theorem.

2.11.

THEOREM 2.11 (Kechris-Martin [KM78]). *Let $A \subseteq \omega_\omega$ be nonempty and Π^1_3 in the codes. Then*

$$\exists x \in \Delta^1_3(|x| \in A).$$

We need the following corollaries:

LEMMA 2.12. *Let $R \subseteq \omega^n_\omega$ be Δ^1_3 in the codes.*
(a) $\neg R$ *is Δ^1_3 in the codes.*
(b) *If $S(\eta_1, \dots, \eta_{n-1}) \Leftrightarrow \exists \xi < \omega_\omega\, R(\xi, \vec{\eta})$, then S is Δ^1_3 in the codes.*
(c) *If $T(\eta_1, \dots, \eta_{n-1}) \Leftrightarrow \forall \xi < \omega_\omega R(\xi, \vec{\eta})$, then T is Δ^1_3 in the codes.*

PROOF. (a) is obvious since the set of codes is Δ^1_3.

(b) Let $R^+(x_1, \dots, x_n) \Leftrightarrow \forall i \le n(x_i$ codes an ordinal $< \omega_\omega$ and $R(|x_1|, \dots, |x_n|))$. Then $S^+(x_2, \dots, x_n) \Leftrightarrow \exists x_1(x_1$ codes an ordinal and $R^+(x_1, x_2, \dots, x_n))$. So S is Σ^1_3 in the codes. Also, by the Kechris-Martin Theorem (2.11), $S^+(x_2, \dots, x_n) \Leftrightarrow \exists x_1 \in \Delta^1_3(x_2, \dots, x_n)\,[x_1$ codes an ordinal and $R^+(x_1, \dots, x_n)]$, so S is Π^1_3 in the codes.

(c) follows from (a) and (b). ⊣

LEMMA 2.13. *Let $R \subseteq \omega^n_\omega$ be Δ^1_3 in the codes, $R \neq \varnothing$. Then $\exists x \in \Delta^1_3\,[(x)_1, \dots, (x)_n$ code ordinals $< \omega_\omega$ & $R(|(x)_1|, \dots, |(x)_n|)]$.*

PROOF. By induction on n. The case $n = 1$ is the Kechris-Martin Theorem (2.11). If $n > 1$, let $S(\eta_1, \dots, \eta_{n-1}) \Leftrightarrow \exists \xi < \omega_\omega\, R(\eta_1, \dots, \eta_{n-1}, \xi)$. By Lemma 2.12 (b), S is Δ^1_3 in the codes, so by the induction hypothesis let $x \in \Delta^1_3$

be such that $S(|(x)_1|, \dots, |(x)_{n-1}|)$. Then $\{\xi : R(|(x)_1|, \dots, |(x)_{n-1}|, \xi)\}$ is $\Delta^1_3(x)$ in the codes, hence Δ^1_3 in the codes, so $\exists y \in \Delta^1_3$ with $R(|(x)_1|, \dots, |(x)_{n-1}|, |y|)$. Let $z \in \Delta^1_3$ be such that $(z)_i = (x)_i$ if $i < n$ and $(z)_n = y$. ⊣

LEMMA 2.14. $\{\langle m, k, s\rangle : s \in \operatorname{ran} \Psi_{m,k}\}$ is Δ^1_3.

PROOF. $s \in \operatorname{ran} \Psi_{m,k} \Leftrightarrow \exists \vec{x}\,[|\vec{x}| \in X_{m,k} \;\&\; \Psi_{m,k}(|\vec{x}|) = s]$. Thus since the components of $\Psi_{m,k}$ all satisfy condition (γ), this is Σ^1_3. But by Lemma 2.13 we can replace the existential quantifiers by $\exists x \in \Delta^1_3$, hence it is Π^1_3. ⊣

PROOF OF THEOREM 2.4. Now Theorem 2.4(1) follows, since

$$\ell(m, k) = \operatorname{Card}(\operatorname{ran} \Psi_{m,k}),$$

so ℓ is clearly a Δ^1_3 function. Theorem 2.4(2) follows easily from this. ⊣ (Theorem 2.4)

2.12. It remains to prove that our partition $X_{m,k} = A^1_{m,k} \cup \dots \cup A^\ell_{m,k}$ has the properties (1) and (2) of Theorem 2.3 (property (3) holds since Ψ^1, Ψ^2, Ψ^3 satisfy condition (β)).

Let Norm $= \{h : \omega_1 \to \omega_1 : h \text{ normal}\}$.

The following is a variant of Martin's partition theorem (12.1 in [Kec78]).

LEMMA 2.15. Let Φ: Norm $\to \{0, 1\}$ be such that $\Phi(h)$ depends only on $h \restriction \mathrm{Lim}$. Then $\exists j \in \{0, 1\}$, $\exists$ cub C such that if $h \in$ Norm, $h : \omega_1 \to C$, then $\Phi(h) = j$.

PROOF. Let $\Psi: \wp(\omega_1) \to \{0, 1\}$ be given by $\Psi(A) = 0$ iff

$$\exists h \in \text{Norm}[\Phi(h) = 0 \text{ and } A = \{h(\omega \cdot \xi + \omega) : \xi < \omega_1\}].$$

By Martin (Theorem 12.1 in [Kec78]) $\exists$ cub C and a $j \in \{0, 1\}$ such that if $A \in C^{\not\,}$, then $\Psi(A) = j$. Say $j = 0$. Given $h \in$ Norm, $h : \omega_1 \to C$, let

$$A = \{h(\omega \cdot \xi + \omega) : \xi < \omega_1\}.$$

Thus $A \in C^{\not\,}$, so $\Psi(A) = 0$. Thus for some $h' \in$ Norm, with $\Phi(h') = 0$ we have

$$A = \{h'(\omega \cdot \xi + \omega) : \xi < \omega_1\}.$$

Since h, h' are both increasing, we must have $\forall \xi < \omega_1 (h(\omega \cdot \xi + \omega) = h'(\omega \cdot \xi + \omega))$. But the closure of $\{\omega \cdot \xi + \omega : \xi < \omega_1\}$ is the set of limit ordinals $< \omega_1$, so since h, h' are continuous they must agree at all limit ordinals. This proves the case $j = 0$, and the case $j = 1$ is easier. ⊣

2.13. Fix $m > 0$, $k \geq 1$, $1 \leq j \leq \ell(m, k)$ such that if $\langle \lambda_1, \dots, \lambda_k\rangle \in A^j_{m,k}$ then at least one $\lambda_i \geq \omega_1$. Note that if $\Psi_{m,k}(A^j_{m,k}) = s$ we can tell using s alone whether $\exists i\, \lambda_i \geq \omega_1$, since if $[F_i] = \lambda_i$ then $\lambda_i < \omega_1$ iff F_i is constant a.e. (W_m), which can be determined from $\Psi^1_{m,k}$. So the property "$\exists i (\lambda_i \geq \omega_1)$" is true of all tuples $\langle \lambda_1, \dots, \lambda_k\rangle$ in $A^j_{m,k}$ if it is true of one tuple. (Note also that if $A^j_{m,k}$ is a subset of ω_1^k then we are done by the $m = 0$ case.)

LEMMA 2.16. $\exists G_1, \dots, G_k : \omega_1^{[m]} \to \omega_1$ such that

(i) $\Psi_{m,k}([G_1], \dots, [G_k]) = s$

(ii) If $\vec{\gamma} \in \omega^k_{m+1}$, $\Psi(\vec{\gamma}) = s$, $C \subseteq \omega_1$ is cub, and $\vec{\gamma} \in \tilde{C}'$, then there is a normal $h : \omega_1 \to C$ such that

$$\langle \tilde{h}([G_1]), \dots, \tilde{h}([G_k]) \rangle = \vec{\gamma}.$$

Granting the lemma, we can prove Theorem 2.3 as follows:

PROOF OF THEOREM 2.3. (1) Given $C \subseteq \omega_1$ cub, let $h : \omega_1 \to C$ be normal. Then $\langle \tilde{h}([G_1]), \dots, \tilde{h}([G_k]) \rangle \in \tilde{C}^k$. By (γ), $\langle \tilde{h}([G_1]), \dots, \tilde{h}([G_k]) \rangle \in A^j_{m,k}$, so $\tilde{C}^k \cap A^j_{m,k} \neq \varnothing$.

(2) Let $B \subseteq A^j_{m,k}$. We want a cub $C \subseteq \omega_1$ such that either

$$\tilde{C}^k \cap A^j_{m,k} \subseteq B$$

or

$$\tilde{C}^k \cap A^j_{m,k} \cap B = \varnothing.$$

Define Φ: Norm $\to \{0, 1\}$ by

$$\Phi(h) = \begin{cases} 0 & \text{if } \langle \tilde{h}([G_1]), \dots, \tilde{h}([G_k]) \rangle \in B \\ 1 & \text{otherwise.} \end{cases}$$

Since $[G_i] \in \text{Lim}$, $G_i(a) \in \text{Lim}$ a.e. Hence since $\langle \tilde{h}([G_1]), \dots, \tilde{h}([G_k]) \rangle = \langle [h \circ G_1], \dots, [h \circ G_k] \rangle$ we have $h \mid \text{Lim} = h' \mid \text{Lim} \Rightarrow \forall i([h \circ G_1] = [h' \circ G_i])$, so $\Phi(h) = \Phi(h')$.

Hence we can apply Lemma 2.15 to get a $j \in \{0, 1\}$ and a cub C such that for all normal $h : \omega_1 \to C$, $\Phi(h) = j$.

If $\vec{\gamma} \in \tilde{C}'^k \cap A^j_{m,k}$ then $\exists h : \omega_1 \to C$ normal with $\vec{\gamma} = \langle \tilde{h}([G_1]), \dots, \tilde{h}([G_k]) \rangle$. So $\vec{\gamma} \in B$ always (never) if $\Phi(h) = 0$ ($\Phi(h) = 1$) for all $h : \omega_1 \to C$. Thus, either $\tilde{C}'^k \cap A^j_{m,k} \subseteq B$ or $\tilde{C}'^k \cap A^j_{m,k} \cap B = \varnothing$.

(3) Holds since Ψ^1, Ψ^2, Ψ^3 satisfy condition (β) from p. 349. ⊣ (Theorem 2.3)

2.14. We now prove Lemma 2.16, thus completing the proof of Kunen's Theorem (2.3).

PROOF OF LEMMA 2.16. We have a fixed invariant $s, \lambda_1, \dots, \lambda_k$ with $\Psi(\vec{\lambda}) = s$, where at least one λ_i is $\geq \omega_1$. Consider all $m + 1$ tuples $\langle i, \alpha_1, \dots, \alpha_m \rangle$ such that $1 \leq i \leq k$, $\alpha_1 < \cdots < \alpha_m < \omega_i$. We define an equivalence relation $\sim$ on these tuples as follows. Given two such tuples $\langle i, \alpha_1, \dots, \alpha_m \rangle$, $\langle j, \beta_1, \dots, \beta_m \rangle$, let $1 \leq r_1 < \cdots < r_m \leq 2m$, $1 \leq s_1 < \cdots < s_m \leq 2m$ be such that $r_1, \dots, r_m, s_1, \dots, s_m$ are ordered similarly to $\alpha_1, \dots, \alpha_m, \beta_1, \dots, \beta_m$. Then put

$$\langle i, \vec{\alpha} \rangle \sim \langle j, \vec{\beta} \rangle \quad \text{iff} \quad \langle i, r_1, \dots, r_m, j, s_1, \dots, s_m \rangle \notin \Psi^1(\vec{\lambda}) \text{ and } \langle i, s_1, \dots, s_m, j, r_1, \dots, r_m \rangle \notin \Psi^1(\vec{\lambda})$$

Let S be the set of $\sim$-equivalence classes. Linearly order S by

$$[i, \alpha_1, \dots, \alpha_m] <_S [j, \beta_1, \dots, \beta_m] \text{ iff } \langle i, r_1, \dots, r_m, j, s_1, \dots, s_m\rangle \in \Psi^1(\vec{\lambda}),$$

where [ ] denotes equivalence class and $r_1, \dots, r_m, s_1, \dots s_m$ are as above. Note that $\langle S, <_S\rangle$ depends only on s not on $\vec{\lambda}$.

Let $[F_i] = \lambda_i$ for $i \leq k$, and let $z \in \mathbb{R}$ be such that $F_1, \dots F_k$ are definable from ω_1 in $\mathbf{L}[z]$ and $\mathrm{cf}^{\mathbf{L}[z]}(\lambda_i) = \mathrm{cf}^{\tilde{\mathbf{L}}}(\lambda_i)$ for $i \leq k$. Let I_z be the indiscernibles for $\mathbf{L}[z] < \omega_1$. Let $h\colon \omega_1 \to I_z$ enumerate I_z in increasing order. Define

$$h^*\colon \{1, \dots, k\} \times \omega_1^{[m]} \to \omega_1$$

by $h^*(i, \alpha_1, \dots, \alpha_m) = F_i(h(\alpha_1), \dots, h(\alpha_m))$.

Claim 2.17. $\sim$ is an equivalence relation.

Proof. By definition of Ψ^1,

$$h^*(i, \alpha_1, \dots, \alpha_m) = h^*(j, \beta_1, \dots, \beta_m) \text{ iff } \langle i, \alpha_1, \dots, \alpha_m\rangle \sim \langle j, \beta_1, \dots, \beta_m\rangle.$$

⊣ (Claim 2.17)

Claim 2.18. $<_S$ is well defined.

Proof. h^* induces a 1-1 order preserving map from S into ω_1 (call this induced map h^* also). ⊣ (Claim 2.18)

Claim 2.19. S is order isomorphic to ω_1.

Proof. Some λ_i is $\geq \omega_1$, hence F_i is not constant a.e. Hence $F_i[I_z^{[m]}]$ has power ω_1, and $h^*[S] \supseteq F_i[I_z^{[m]}]$. ⊣ (Claim 2.19)

For $[i, \alpha_1, \dots, \alpha_m] = x \in S$, say x is of type I (II) if i is of type I (II) (*i.e.*, if $i \notin (\in) \Psi^2(\vec{\lambda})$).

Claim 2.20. The type of $[i, \alpha_1, \dots, \alpha_m]$ does not depend on a choice of representative.

Proof. Let $\gamma_1, \dots, \gamma_m \in \mathrm{Lim}$. Then

$$\begin{aligned}
&h^*([i, \vec{\gamma}]) \text{ is not a limit point of } h^*[S] \\
&\quad\Leftrightarrow \sup\{h^*(j, \vec{\delta}) : h^*(j, \vec{\delta}) < h^*(i, \vec{\gamma})\} < h^*(i, \vec{\gamma}) \\
&\quad\Leftrightarrow \sup\{F_j(h(\delta_1), \dots, h(\delta_m)) : \\
&\qquad\qquad F_j(h(\delta_1), \dots, h(\delta_m)) < F_i(h(\gamma_1), \dots, h(\gamma_m))\} \\
&\qquad < F_i(h(\gamma_1), \dots, h(\gamma_m)) \\
&\quad\Leftrightarrow \forall \vec{\gamma} \in \mathrm{Lim}^{[m]} \sup\{F_j(\overrightarrow{h(\delta)}) : F_j(\overrightarrow{h(\delta)}) < F_i(\overrightarrow{h(\gamma)})\} < F_i(\overrightarrow{h(\gamma)}) \\
&\quad\overset{\text{Lem. 2.7}}{\Leftrightarrow} i \text{ is of type II.}
\end{aligned}$$

⊣ (Claim 2.20)

DEFINITION 2.21. Let $[i, \alpha_1, \dots, \alpha_m] \in S$ be of type II. Then

$$\mathrm{cf}([i, \alpha_1, \dots, \alpha_m]) = \begin{cases} \omega & \text{if } \mathrm{cf}^{\tilde{\mathbf{L}}}(\lambda_i) = \omega \\ \alpha_j & \text{if } \mathrm{cf}^{\tilde{\mathbf{L}}}(\lambda_i) = \omega_j,\ j > 0. \end{cases}$$

CLAIM 2.22. $\mathrm{cf}([i, \alpha_1, \dots, \alpha_m])$ does not depend on choice of representative.

PROOF. Let $[i, \alpha_1, \dots, \alpha_m] = [j, \beta_1, \dots, \beta_m]$. Let $r_1, \dots, r_m, s_1, \dots, s_m$ be integers in $1, \dots, 2m$ such that $\vec{r}$, $\vec{s}$ and $\vec{\alpha}$, $\vec{\beta}$ are similarly ordered. Then by our choice of z, $\mathrm{cf}^{\mathbf{L}[z]}(F_i(\omega_{r_1}, \dots, \omega_{r_m})) = \omega_n$ for some n, $0 \leq n \leq 2m$. Since $F_i(\omega_{r_1}, \dots, \omega_{r_m}) = F_j(\omega_{s_1}, \dots, \omega_{s_m})$, $\mathrm{cf}^{\mathbf{L}[z]}(F_j(\omega_{s_1}, \dots, \omega_{s_m})) = \omega_n$. If $n = 0$, we are done. Otherwise, if $\mathrm{cf}^{\mathbf{L}[z]}(F_i(\omega_{r_1}, \dots, \omega_{r_m})) = \omega_{r_k}$ and $\mathrm{cf}^{\mathbf{L}[z]}(F_j(\omega_{s_1}, \dots, \omega_{s_m})) = \omega_{s_\ell}$, then $\omega_{r_k} = \omega_{s_\ell}$, hence by the similar ordering of $\vec{r}$, $\vec{s}$, and $\vec{\alpha}$, $\vec{\beta}$ we have $\alpha_k = \beta_\ell$. So cf is well defined. ⊣ (Claim 2.22)

Now let $T = S \cup \{\langle x, \alpha \rangle : x \in S,\ x \text{ of type II},\ \exists \vec{\gamma} \in \mathrm{Lim}^{[m]}\, \exists i (x = [i, \vec{\gamma}]) \text{ and } \alpha < \mathrm{cf}(x)\}$. We define $<_T$ on T as follows: for $y_1, y_2 \in T$ $y_1 <_T y_2$ iff

(a) $(y_1 = x_1 \text{ or } y_1 = \langle x_1, \alpha_1 \rangle)$ & $(y_2 = x_2 \text{ or } y_2 = \langle x_2, \alpha_2 \rangle)$ and $x_1 <_S x_2$

or (b) $y_1 = \langle x, \alpha_1 \rangle$ and $[y_2 = x \text{ or } (y_2 = \langle x, \alpha_2 \rangle \text{ and } \alpha_1 < \alpha_2)]$.

Note that $\langle T, <_T \rangle$ depends only on s, not on $\vec{\lambda}$.

CLAIM 2.23. T is order isomorphic with ω_1.

PROOF. T is obtained as follows: Start with S. For certain x in S, we adjoin a countable well-ordered set of elements less than x and greater than any predecessor of x in S. So T is well-ordered, its initial segments are countable, and the power of T is ω_1. ⊣ (Claim 2.23)

So we can identify T with ω_1. Now we can define the functions G_i as stated in the lemma:

$$G_i(\gamma_1, \dots, \gamma_m) = [i, \gamma_1, \dots, \gamma_m]$$

where we view $[i, \gamma_1, \dots, \gamma_m]$ as an element of ω_1 under the isomorphism $T \cong \omega_1$. Thus $G_i : \omega_1^{[m]} \to \omega_1$.

CLAIM 2.24. $\Psi^1([G_1], \dots, [G_k]) = \Psi^1(\vec{\lambda})$.

PROOF. Let $\vec{\gamma} \in I_z^{[2m]}$. Then

$$\begin{aligned}\Psi^1([\vec{G}]) &= \{\langle i, r_1, \dots, r_m, j, s_1, \dots, s_m\rangle : \\ &\qquad 1 \le r_1 < \dots < r_m \le 2m, \\ &\qquad 1 \le s_1 < \dots < s_m \le 2m,\ 1 \le i, j \le k\ \& \\ &\qquad G_i(\gamma_{r_1}, \dots, \gamma_{r_m}) < G_j(\gamma_{s_1}, \dots, \gamma_{s_m})\} \\ &= \{\langle i, r_1, \dots, r_m, j, s_1, \dots, s_m\rangle : \\ &\qquad \dots \text{ and } [i, \gamma_{r_1}, \dots, \gamma_{r_m}] <_S [j, \gamma_{s_1}, \dots, \gamma_{s_m}]\} \\ &= \Psi^1(\vec{\lambda}).\end{aligned}$$

⊣ (Claim 2.24)

CLAIM 2.25. $\Psi^2([G_1], \dots, [G_k]) = \Psi^2(\vec{\lambda})$.

PROOF. We must show i is of type I w.r.t. $\vec{\lambda}$ iff i is of type I w.r.t. $[G_1], \dots [G_k]$. Suppose i is of type I w.r.t. to $\vec{\lambda}$. Then $\forall \vec{\gamma} \in \mathrm{Lim}^{[m]}$,

$$\sup\{F_j(\overrightarrow{h(\delta)}) : F_j(\overrightarrow{h(\delta)}) < F_i(\overrightarrow{h(\gamma)})\} = F_i(\overrightarrow{h(\gamma)}),$$

which by definition of h^* implies that $\forall \vec{\gamma} \in \mathrm{Lim}^{[m]}$,

$$\sup\{h^*([j, \vec{\delta}]) : h^*([j, \vec{\delta}]) < h^*([i, \vec{\gamma}])\} = h^*([i, \vec{\gamma}]).$$

Suppose there is $\vec{\gamma} \in \mathrm{Lim}^{[m]}$ such that

$$\sup\{[j, \vec{\delta}] : [j, \vec{\delta}] < [i, \vec{\gamma}]\} < [i, \vec{\gamma}].$$

Then since i is not of type II, by the way S is embedded in T we must have

$$\sup\{[j, \vec{\delta}] : [j, \vec{\delta}] < [i, \vec{\gamma}]\} \le [k, \vec{\beta}] < [i, \vec{\gamma}]$$

for some $[k, \vec{\beta}] \in S$. This yields a contradiction by applying h^*. Hence for all $\vec{\gamma} \in \mathrm{Lim}^{[m]}$,

$$\sup\{[j, \vec{\delta}] : [j, \vec{\delta}] < [i, \vec{\gamma}]\} = [i, \vec{\gamma}],$$

so i is of type I w.r.t. $[G_1], \dots [G_k]$ by Lemma 2.7.

Clearly (by the definition of T), if i is of type II w.r.t. $\vec{\lambda}$, then $\forall \vec{\gamma} \in \mathrm{Lim}^{[m]}$,

$$\sup\{[j, \vec{\delta}] : [j, \vec{\delta}] < [i, \vec{\gamma}]\} < \langle [i, \vec{\gamma}], 0\rangle < [i, \vec{\gamma}].$$

Hence i is of type II w.r.t. $[G_1], \dots [G_k]$. ⊣ (Claim 2.25)

CLAIM 2.26. $\Psi^3([G_1], \dots [G_k]) = \Psi^3(\vec{\lambda})$.

PROOF. Suppose i is of type II, with $\mathrm{cf}[i, \gamma_1, \dots, \gamma_m] = \gamma_j$. Then by the construction of T, $[i, \gamma_1, \dots, \gamma_m]$ is cofinal with γ_j in T. The construction can be done within $\mathbf{L}$, hence

$$\mathrm{cf}^{\mathbf{L}}\, G_i(\gamma_1, \dots, \gamma_m) = \mathrm{cf}^{\mathbf{L}}(\gamma_j).$$

Hence $\mathrm{cf}^{\mathbf{L}}[G_i] = \mathrm{cf}^{\mathbf{L}}[\pi_j] = \omega_j$. The case $\mathrm{cf} = \omega_0$ is similar. ⊣ (Claim 2.26)

Hence $\Psi([G_1],\dots[G_k]) = s$ as desired. This completes the proof of Lemma 2.16(i).

Now to prove part (ii) of Lemma 2.16, suppose $C \subseteq \omega_1$ is cub with $\lambda_i \in \tilde{C}'$. We will find a normal $h^{***}: \omega_1 \to C$ such that $\tilde{h}^{***}([G_i]) = \lambda_i$.

For the remainder of the proof of the Lemma "definable in $\mathbf{L}[z]$" will mean "definable from z, ω_1 in $\mathbf{L}[z]$."

We have $z \in \mathbb{R}$ with $F_1, \dots, F_k$ definable from z, ω_1 and $\mathrm{cf}^{\mathbf{L}[z]}(\lambda_i) = \mathrm{cf}^{\tilde{\mathbf{L}}}(\lambda_i)$. By increasing the Turing degree of z if necessary, we may assume C is also definable from ω_1 in $\mathbf{L}[z]$. Since $\lambda_i \in \tilde{C}'$, we have $F_i(\gamma_1,\dots,\gamma_m) \in C'$ a.e. Since $C \in \mathbf{L}[z]$, by indescernibility we get

$$\forall \gamma_1 < \cdots < \gamma_m \; F_i(h(\gamma_1),\dots,h(\gamma_m)) \in C'.$$

So $h^*: S \to C'$.

The function h^* is definable in $\mathbf{L}[z^\#]$. Let $x = [i,\gamma_1,\dots,\gamma_m]$ be of type II, all $\gamma_i \in \mathrm{Lim}$. Let $g(x) =$ the least map in $\mathbf{L}[z]$ of $\mathrm{cf}^{\mathbf{L}[z]}(h^*(x))$ into $h^*(x)$ continuously, order preservingly, cofinally, and with range contained in

$$C \cap \{\xi : \xi > \sup(h^*(S) \cap h^*(x))\}.$$

Note that since $h^*(x)$ is not a limit point of $h^*(S)$, such a map exists.

Now define $h^{**}: T \to C$ as follows: if $x \in S$, $h^{**}(x) = h^*(x)$. For $\langle x, \alpha\rangle \in T - S$,

$$h^{**}(\langle x,\alpha\rangle) = \begin{cases} g(x)(\alpha) & \text{if } \mathrm{cf}(h^*(x)) = \omega \\ g(x)(h(\alpha)) & \text{otherwise.} \end{cases}$$

(Note that $\alpha < \mathrm{cf}\, x \Rightarrow h(\alpha) < \mathrm{cf}\, h^*(x)$).

It is an easy exercise to prove that h^{**} is order preserving (using the fact that for all α, $g(x)(\alpha) > \sup(h^*(S) \cap h^*(x))$). Also $h^{**}(G_i(\alpha_1,\dots,\alpha_m)) = h^{**}([i,\alpha_1,\dots,\alpha_m]) = F_i(h(\alpha_1),\dots,h(\alpha_m))$, for all $\vec{\alpha} \in \omega_1^{[m]}$. So since $h(\alpha) = \alpha$ a.e. we have $\tilde{h}^{**}([G_i]) = \lambda_i$. But h^{**} is not normal. To get our desired normal h^{***} we need the following easy result:

FACT 2.27. Let $A \subseteq \omega_1$ have order type ω_1, $B =$ closure of A in ω_1. Let h_A, h_B be the enumerations in order of A, B, respectively. Suppose λ is a limit ordinal with $h_A(\lambda)$ a limit point of A. Then $h_A(\lambda) = h_B(\lambda)$

Thus let h^{***} be the enumeration of the closure of $h^{**}(T)$. Since C is closed, $h^{***}: \omega_1 \to C$. Now if $\gamma_1 < \cdots < \gamma_m$ are limit ordinals, then $[i,\gamma_1,\dots,\gamma_m]$ is a limit point of T, and $h^{**}([i,\gamma_1,\dots,\gamma_m])$ is a limit point of $h^{**}[T]$ (since if $x = [i,\gamma_1,\dots,\gamma_m]$ is of type II and $\mathrm{cf}^{\mathbf{L}[z]}(F_i(\omega_1,\dots,\omega_m)) = \omega_j$, then $\mathrm{cf}^{\mathbf{L}[z]}(h^*(x)) = h(\alpha_j)$). So by the fact, $h^{***}([i,\gamma_1,\dots,\gamma_m]) = h^{**}([i,\gamma_1,\dots,\gamma_m])$), so $\tilde{h}^{***}([\vec{G}]) = \vec{\lambda}$. ⊣ (Lemma 2.16)

§3. Applications.

3.1. Kunen's theorem gives us a partition $(\omega_{m+1} \cap \mathrm{Lim})^k = A^1_{m,k} \cup \cdots \cup A^\ell_{m,k}$ such that each piece carries a canonical measure $V^j_{m,k}$. In this section we prove the following:

THEOREM 3.1 (Kunen [Kun71B]). *Let V be a (countably additive) measure on ω_ω. Then $\exists h \colon \omega_\omega \to \omega^k_{m+1}$ for some m, k such that*

(1) $h \in \tilde{\mathbf{L}}$.
(2) h *is 1-1 a.e.* (V).
(3) $h_*(V) = V^j_{m,k}$ *for some* j.
(4) $\exists g \colon \omega^k_{m+1} \to \omega_\omega$, $g \in \tilde{\mathbf{L}}$, *such that* g *is 1-1 a.e.* $(V^j_{m,k})$ *and* $g = h^{-1}$ *a.e.*

3.2. By countable additivity of V we may assume V is on ω_n for some $1 \leq n < \omega$. We define the **restricted ultrapower** of Ord in the same way as the ordinary ultrapower except we only consider functions $f \colon \omega_n \to \mathrm{Ord}$ which lie in $\tilde{\mathbf{L}}$. Given such an f, $[f]_{\tilde{\mathbf{L}}}$ is its image in the restricted ultrapower.

LEMMA 3.2. *Let $z \in \mathbb{R}$. Then $\mathbf{L}[z] \models \psi([h_1]_{\tilde{\mathbf{L}}}, \dots, [h_m]_{\tilde{\mathbf{L}}})$ iff $\mathbf{L}[z] \models \psi(h_1(\gamma), \dots, h_m(\gamma))$ a.e. (V).*

PROOF. The usual Łoś proof works. ⊣

Recall the following (Lemma 8.6 of [Kec78]).

LEMMA 3.3. *Let $\alpha \in \mathrm{Ord}$. Then there are uniform indiscernibles $\gamma_1, \dots, \gamma_k \leq \alpha$ and a real z such that α is definable in $\mathbf{L}[z]$ from $\gamma_1, \dots, \gamma_k$.*

PROOF OF THEOREM 3.1. Let $z \in \mathbb{R}$ and $[f_1]_{\tilde{\mathbf{L}}} < \cdots < [f_k]_{\tilde{\mathbf{L}}} \leq [\mathrm{id}]_{\tilde{\mathbf{L}}}$ uniform indiscernibles such that $[\mathrm{id}]_{\tilde{\mathbf{L}}} = t^{\mathbf{L}[z]}([f_1]_{\tilde{\mathbf{L}}}, \dots, [f_k]_{\tilde{\mathbf{L}}})$. Let $h \colon \omega_n \to \omega^k_n$ be

$$h(\alpha) = \langle f_1(\alpha), \dots, f_k(\alpha) \rangle.$$

We have a reverse map $g \colon \omega^k_n$ given by $g(\gamma_1, \dots, \gamma_k) = t^{\mathbf{L}[z]}(\gamma_1, \dots, \gamma_k)$.

Clearly $g \circ h = \mathrm{id}$ a.e.

Pick j such that $A^j_{m,k} \in h_*(V)$ (here of course $m = n - 1$). A unique such j exists since there are finitely many $A^j_{m,k}$'s, all disjoint. Let $C \subseteq \omega_1$ be cub. Since $[f_i]$ is a uniform indiscernible for $i = 1, \dots, k$, we have $f_i(\alpha) \in \tilde{C}$ a.e. (V). Thus $\tilde{C}^k \in h_*(V)$. So $h_*(V) = V^j_{m,k}$.

That $h \circ g = \mathrm{id}$ a.e. $(V^j_{m,k})$ follows from the following general fact:

FACT 3.4. *Given (two valued) measure spaces (X, U), (X', U'), $h \colon X \to X'$ such that $h_*(U) = U'$, and $g \colon X' \to X$ such that $g \circ h = \mathrm{id}$ a.e. (U), then $h \circ g = \mathrm{id}$ a.e. (U').*

PROOF OF FACT. If $A \in U$ and $g(h(x)) = x$ for $x \in A$, then let $B = h[A] \in U'$. On B, g is clearly equal to h^{-1}. ⊣ (Fact)

This completes the proof. ⊣ (Theorem 3.1)

3.3.

DEFINITION 3.5. A set $S \subseteq \omega_\omega$ is **simple** if there are m, j, k, C a cub subset of ω_1, and an $F : \omega^k_{m+1} \to \omega_\omega$ such that

(1) F is 1-1 on $A^j_{m,k} \cap \tilde{C}^k$
(2) $S = F[A^j_{m,k} \cap \tilde{C}^k]$
(3) $F \in \tilde{\mathbf{L}}$.

THEOREM 3.6 (Kunen [Kun71B]). If $A \subseteq \omega_\omega$, then A is a countable union of simple sets.

We need the following; where $\Theta = \sup\{\xi : \xi \text{ is a surjective image of } \mathbb{R}\}$.

LEMMA 3.7. If $\mathcal{I}$ is a proper σ-ideal on λ, where $\lambda < \Theta$, then there is a countably additive ultrafilter $\mathcal{U}$ on λ such that $A \in \mathcal{I} \Rightarrow A \notin \mathcal{U}$

PROOF. By Moschovakis [Mos70] let $h : \mathbb{R} \twoheadrightarrow \wp(\lambda)$. So there is a $h_1 : \mathbb{R} \twoheadrightarrow \mathcal{I}$. Let $g : \mathcal{D} \to \lambda$ (where $\mathcal{D}$ = Turing degrees) be given by

$$g(d) = \text{least } \alpha < \lambda \left[\alpha \notin \bigcup\{h(x) : x \leq_{\mathrm{T}} d\}\right].$$

Then if $\mathcal{U}_0$ is the Martin measure on $\mathcal{D}$, $g_*(\mathcal{U}_0) = \mathcal{U}$ is as desired. ⊣

PROOF OF THEOREM 3.6. Fix $A \subseteq \omega_\omega$. Let $\mathcal{I} = \{B \subseteq \bigcup_{i<\omega} A_i : A_i$ is simple & $(A_i \subseteq A$ or $A_i \cap A = \varnothing\}$. $\mathcal{I}$ is a σ-ideal. If $\mathcal{I}$ is not proper, then $A \in \mathcal{I}$ and we are done.

So assume $\mathcal{I}$ is proper, towards a contradiction. By Lemma 3.7, let $\mathcal{U}$ be a countably additive ultrafilter such that $B \in \mathcal{I} \Rightarrow B \notin \mathcal{U}$. By Theorem 3.1 there are functions $H : \omega_\omega \to (\omega_{m+1})^k$ and $G : (\omega_{m+1})^k \to \omega_\omega$, both in $\tilde{\mathbf{L}}$, which demonstrate that $\mathcal{U}$ is equivalent to $V^j_{m,k}$.

Case I: $A \in \mathcal{U}$. Then $H[A] \in V^j_{m,k}$. so $\exists$ cub $C \subseteq \omega_1$ such that $H[A] \supseteq A^j_{m,k} \cap \tilde{C}^k$. Hence $A \supseteq G[A^j_{m,k} \cap \tilde{C}^k]$ and we may assume G is 1-1 on $A^j_{m,k} \cap \tilde{C}^k$. Hence $X = G[A^j_{m,k} \cap \tilde{C}^k]$ is simple, so $X \in \mathcal{I}$. Hence $X \notin \mathcal{U}$, a contradiction to $H_*(V^j_{m,k}) = \mathcal{U}$.

Case II: $A \notin \mathcal{U}$. Proof is similar to Case I. ⊣ (Theorem 3.6)

3.4.

THEOREM 3.8 (Kunen [Kun71B], effectivized). There is a Δ^1_3 coding of subsets of ω_ω, *i.e.*, there is a Δ^1_3 $C_\omega \subseteq \mathbb{R}$ and a map $C_\omega \to \wp(\omega_\omega)$ taking ε to X_ε, such that $\{X_\varepsilon : \varepsilon \in C_\omega\} = \wp(\omega_\omega)$, and the relation "$w$ codes an ordinal $< \omega_\omega$ & $\varepsilon \in C_\varepsilon$ & $|w| \in X_\varepsilon$" is Δ^1_3.

PROOF. Let $\varepsilon \in C^*_\omega \Leftrightarrow \varepsilon = \langle m, k, j, \ulcorner\varphi\urcorner, \alpha, \ulcorner\psi\urcorner\rangle$ & $\{\xi < \omega_1 : \mathbf{L}[\alpha] \models \varphi(\xi)\} = C$ is cub & $\{(\vec{\xi}, \eta) \in (\omega_\omega)^{k+1} : \mathbf{L}[\alpha] \models \psi(\omega_1, \dots, \omega_r, \vec{\xi}, \eta)\} = F$ is a function from ω^k_{m+1} into ω_ω which is 1-1 on $A^j_{m,k} \cap \tilde{C}^k$.

For $\varepsilon \in C^*_\omega$, let $X^*_\varepsilon = F[A^j_{m,k} \cap \tilde{C}^k]$. Put $\varepsilon \in C_\omega \Leftrightarrow \forall i(\varepsilon_i \in C^*_\omega)$. Let

$$X_\varepsilon = \bigcup_i X^*_{\varepsilon_i}$$

for $\varepsilon \in C_\omega$.

Using the Kechris-Martin Theorem and its corollaries (*cf.* 2.11) it can be seen that this coding is Δ^1_3. ⊣

3.5.

Corollary 3.9 (Kunen [Kun71B]). *If $\mathscr{U}$ is an ultrafilter on ω_ω, then $i^{\mathscr{U}}(\underset{\sim}{\boldsymbol{\delta}}^1_3) = \underset{\sim}{\boldsymbol{\delta}}^1_3$.*

Proof. We must show that if $f : \omega_\omega \to \underset{\sim}{\boldsymbol{\delta}}^1_3$, then $[f]_{\mathscr{U}} < \underset{\sim}{\boldsymbol{\delta}}^1_3$. It's enough to consider $f : \omega_\omega \to \omega_\omega$ (since for $\omega_\omega \leq \lambda < \delta^1_3$

$$\operatorname{Card}({}^{\omega_\omega}\lambda/\mathscr{U}) = \operatorname{Card}({}^{\omega_\omega}\omega_\omega/\mathscr{U}).$$

Furthermore, it is enough to show that $i^{V^j_{m,k}}(\omega_\omega) < \underset{\sim}{\boldsymbol{\delta}}^1_3$. To see this: Define

$$\delta \prec \varepsilon \Leftrightarrow \delta, \varepsilon \in C_\omega \;\&\; X_\delta, X_\varepsilon \text{ “are” functions from } A^j_{m,k} \to \omega_\omega \\ \& [X_\delta]_{V^j_{m,k}} < [X_\varepsilon]_{V^j_{m,k}}.$$

This last inequality is expressed by

$$\text{there is a } C \subseteq \omega_1 \text{ cub } [\forall \xi \in \tilde{C}^k \cap A^j_{m,k}(X_\delta(\vec{\xi}) < X_\varepsilon(\vec{\xi}))].$$

Thus $\prec$ is $\underset{\sim}{\boldsymbol{\Sigma}}^1_3$, so it is bounded below δ^1_3. ⊣

3.6.

Theorem 3.10 (Kunen [Kun71B]). *For all $\lambda < \underset{\sim}{\boldsymbol{\delta}}^1_3$, $\underset{\sim}{\boldsymbol{\delta}}^1_3 \to (\underset{\sim}{\boldsymbol{\delta}}^1_3)^\lambda$.*

Proof. Assume $\lambda \geq \omega_\omega$. Fix $t : \omega \cdot \lambda \to \omega_\omega$ 1-1 and onto. For $\rho < \omega_\omega$, define

$$(\varepsilon)_\rho = \begin{cases} \{\langle \eta, \eta' \rangle : \langle \rho, \eta, \eta' \rangle \in X_\varepsilon\} & \text{if } \varepsilon \in C_\omega \\ \varnothing & \text{otherwise.} \end{cases}$$

For $\xi < \omega \cdot \lambda$, let $\varepsilon_\xi = (\varepsilon)_\rho$ where $t(\xi) = \rho$. If ε_ξ “is” a well-ordering of ω_ω, let $|\varepsilon_\xi|$ be its length.

Fix $A \subseteq (\underset{\sim}{\boldsymbol{\delta}}^1_3)^\lambda\uparrow$, and consider the game in which player I plays α, player II plays β, and player II wins if either

(1) For some $\xi < \omega \cdot \lambda$, α_ξ or β_ξ is not a w.o. of ω_ω, and if $\xi_0 =$ the least such ξ, then α_{ξ_0} is not a w.o., or
(2) For all $\xi < \omega \cdot \lambda$, α_ξ and β_ξ are w.o.'s of ω_ω, and

$$\{\sup(\max\{|\alpha_{\omega\cdot\vartheta+n}|, |\beta_{\omega\cdot\vartheta+n}|\})\}_{\vartheta<\lambda} \in A.$$

Assume player II has a winning strategy σ. Then for some $\xi < \lambda$, $\eta < \underset{\sim}{\delta}^1_3$, let

$$\Theta(\xi, \eta) = \sup\{|\sigma[\alpha]_\xi| + 1 : \forall \xi' \leq \xi(\alpha_\xi \text{ is a w.o. of } \omega_\omega \ \& \ \sup\{|\alpha_{\xi'}| : \xi' \leq \xi\} < \eta\}.$$

CLAIM 3.11. $\Theta(\xi, \eta) < \underset{\sim}{\delta}^1_3$.

PROOF OF CLAIM 3.11. It is easy to construct a $\underset{\sim}{\Sigma}^1_3$ wellfounded relation on $\mathbb{R}$ of length $\geq \Theta(\xi, \eta)$. The rest of the proof is as in Lemma 11.1 of [Kec78].

⊣ (Claim 3.11)

⊣ (Theorem 3.10)

REFERENCES

ALEXANDER S. KECHRIS

[Kec78] AD *and projective ordinals*, this volume, originally published in Kechris and Moschovakis [CABAL i], pp. 91–132.

ALEXANDER S. KECHRIS AND DONALD A. MARTIN

[KM78] *On the theory of* Π^1_3 *sets of reals*, ***Bulletin of the American Mathematical Society***, vol. 84 (1978), no. 1, pp. 149–151.

ALEXANDER S. KECHRIS AND YIANNIS N. MOSCHOVAKIS

[CABAL i] *Cabal seminar* 76–77, Lecture Notes in Mathematics, no. 689, Berlin, Springer, 1978.

KENNETH KUNEN

[Kun71B] *On* $\underset{\sim}{\delta}^1_5$, circulated note, August 1971.

YIANNIS N. MOSCHOVAKIS

[Mos70] *Determinacy and prewellorderings of the continuum*, ***Mathematical logic and foundations of set theory. Proceedings of an international colloquium held under the auspices of the Israel Academy of Sciences and Humanities, Jerusalem, 11–14 November 1968*** (Y. Bar-Hillel, editor), Studies in Logic and the Foundations of Mathematics, North-Holland, Amsterdam-London, 1970, pp. 24–62.

DEPARTMENT OF MATHEMATICS
UNIVERSITY OF CALIFORNIA
BERKELEY, CALIFORNIA 94720-3840
UNITED STATES OF AMERICA
E-mail: solovay@gmail.com

AD AND THE PROJECTIVE ORDINALS

STEVE JACKSON

§1. Introduction. The purpose of this paper is to calculate an upper-bound for $\underset{\sim}{\delta}^1_{2n+5}$, $n \geq 0$, assuming certain basic inductive assumptions concerning the lower projective ordinals. In a later paper, we will verify the lower bound for $\underset{\sim}{\delta}^1_{2n+5}$ and establish the inductive assumptions at the next level.

The case $n = 0$ appears in [Jac99], and may be obtained as a special case of the results in this paper.

We assume the reader is familiar with the results in [Kec81A] and [Mar], although the reader may take as "axioms" the results needed. Indeed, the following paper is essentially self-contained, except for a knowledge of the homogeneous tree construction, which is only used indirectly.

For background on the projective ordinals as well as their significance in descriptive set theory, we refer the reader to [Mos80].

We work in AD+DC throughout, although the inductive hypotheses at the lower levels suffice.

§2. Definitions and preliminary results. In this section we define two families of measures; one being essentially the general measures allowed in the homogeneous tree construction, and the other a family of canonical measures. This is the only point in this paper where we use the homogeneous tree construction; the reader not familiar with it may take on faith the fact that our family captures the most general such measure.

We first introduce our main inductive hypotheses:

I_{2n+1}: $\underset{\sim}{\delta}^1_{2n+1}$ has the strong partition relation, $\underset{\sim}{\delta}^1_{2n+3}$ has the weak partition relation, and $\underset{\sim}{\delta}^1_{2n+3} = \aleph_{\tau(2n+1)}$, where $\tau(0) = 1$ and $\tau(k+1) = \omega^{\tau(k)}$ (ordinal exponentiation).

The author wishes to thank Tony Martin for many helpful conversations during the research for this manuscript. This work is an outgrowth of the calculations of $\underset{\sim}{\delta}^1_5$ (the case $n = 0$ of this paper) which the author completed while working with Martin at UCLA. Aside from independently discovering many of the basic techniques and methods of use there, it was a few basic discoveries of Martin (such as [Mar]) which provided the impetus for this research.

Wadge Degrees and Projective Ordinals: The Cabal Seminar, Volume II
Edited by A. S. Kechris, B. Löwe, J. R. Steel
Lecture Notes in Logic, 37

K_{2n+3}:

a) For any measure μ on Ξ_{2n+3} = the predecessor of $\underset{\sim}{\boldsymbol{\delta}}^1_{2n+3}$, and any $\alpha < \underset{\sim}{\boldsymbol{\delta}}^1_{2n+3}$, $j_\mu(\alpha) < \underset{\sim}{\boldsymbol{\delta}}^1_{2n+3}$ where j_μ denotes the embedding from the ultrapower by the measure μ.

b) For any $g\colon \underset{\sim}{\boldsymbol{\delta}}^1_{2n+3} \to \underset{\sim}{\boldsymbol{\delta}}^1_{2n+3}$, and any normal measure V on $\underset{\sim}{\boldsymbol{\delta}}^1_{2n+3}$, there is a tree T on $\underset{\sim}{\boldsymbol{\delta}}^1_{2n+3}$ such that for some measure one set A with respect to V and all $\alpha \in A$, $g(\alpha) \leq |T\restriction(\sup j(\alpha))|$, where the sup ranges over embeddings j corresponding to measures on Ξ_{2n+3} which occur in the homogeneous tree construction on a $\underset{\sim}{\boldsymbol{\Pi}}^1_{2n+2}$ complete set.

We remark that for $n = 0$, I_{2n+1} and $K_{2n+3}(a)$ are well-known theorems of determinacy, and $K_{2n+3}(b)$ is a theorem of Martin's—see [Mar].

We assume I_{2n+1} and K_{2n+3} for the remainder of this paper, and establish the upper bound for $\underset{\sim}{\boldsymbol{\delta}}^1_{2n+5}$, along with some additional auxiliary results.

We introduce the family of canonical measures:

We let W^n_1 = the n-fold product of the ω-cofinal normal measure on ω_1. We identify the domain of W^n_1 with an ordinal by ordering the n-tuples $(\alpha_1, \dots, \alpha_n)$ by α_n first, then α_{n-1}, etc.

We define $S^{1,n}_1$ from the strong partition relation on ω_1 as follows: We let $<^n$ denote the ordering on n-tuples of ordinals $(\alpha_1, \dots, \alpha_n)$, where $\alpha_1 < \cdots < \alpha_n$, defined by

$$(\alpha_1, \dots, \alpha_n) <^n (\beta_1, \dots, \beta_n) \text{ iff } (\alpha_n, \alpha_1, \dots, \alpha_{n-1}) <_{\text{lex}} (\beta_n, \beta_1, \dots, \beta_{n-1}),$$

where $<_{\text{lex}}$ denotes lexicographic ordering. A set A has measure one w.r.t. $S^{1,n}_1$ if there is a c.u.b. subset $C \subseteq \omega_1$ such that for all $F\colon <^n \to C$, order-preserving, everywhere discontinuous, and of uniform cofinality ω, the ordinal $[F]_{W^n_1}$ represented by F w.r.t. W^n_1 is in A. By uniform cofinality ω we mean that there is a function F_2 from the tuples $(\alpha_1, \dots, \alpha_{n-1}, \alpha_n, m)$, where $m \in \omega$, which is order-preserving into ω_1 with the order given by lexicographic ordering on $(\alpha_n, \alpha_1, \dots, \alpha_{n-1}, m)$, and which induces F in the sense that $F(\alpha_1, \dots, \alpha_n) = \sup_{m\in\omega} F_2(\alpha_1, \dots, \alpha_{n-1}, \alpha_n, m)$ for all $\alpha_1, \dots, \alpha_n$.

In general, for a given order-type, we say that functions from that order-type into the ordinals which are order-preserving, everywhere discontinuous, and of uniform cofinality ω (with the obvious meaning) are of the **correct type**.

We assume $W^m_{2n'-1}$ and $S^{\ell,m}_{2n'-1}$ have been defined for $n' \leq n$, $1 \leq \ell \leq 2^{n'} - 1$, and all m, and $W^m_{2n'-1}$ is a measure on $\underset{\sim}{\boldsymbol{\delta}}^1_{2n'-1}$, $S^{\ell,m}_{2n'-1}$ is a measure on $\Xi_{2n'+1}$. We then let W^m_{2n+1} be the measure induced by the weak partition relation on $\underset{\sim}{\boldsymbol{\delta}}^1_{2n+1}$, functions $f\colon \kappa \to \underset{\sim}{\boldsymbol{\delta}}^1_{2n+1}$ of the correct type, and the measure $S^{2^n-1,m}_{2n-1}$. Here $\kappa = \kappa(W^m_{2n+1})$ is the ordinal $\kappa = \operatorname{dom}(S^{2^n-1,m}_{2n-1})$ on which $S^{2^n-1,m}_{2n-1}$ is a measure. That is, A has measure one w.r.t. W^m_{2n+1} if there is a c.u.b. $C \subseteq \underset{\sim}{\boldsymbol{\delta}}^1_{2n+1}$ such that for all $f\colon \operatorname{dom}(S^{2^n-1,m}_{2n-1}) \to C$ of the correct type, $[f]_{S^{2^n-1,m}_{2n-1}} \in A$.

We let $S^{\ell,m}_{2n+1}$ for $2 \leq \ell \leq 2^{n+1}-1$ be the measure induced by the strong partition relation on $\underset{\sim}{\boldsymbol{\delta}}^1_{2n+1}$, functions $F\colon \underset{\sim}{\boldsymbol{\delta}}^1_{2n+1} \to \underset{\sim}{\boldsymbol{\delta}}^1_{2n+1}$ of the correct type, and the measure $\mu^{\vartheta^{\ell,m}}_{2n+1}$ = the measure induced by the weak partition relation on $\underset{\sim}{\boldsymbol{\delta}}^1_{2n+1}$, functions $f\colon \vartheta^{\ell,m} \to \underset{\sim}{\boldsymbol{\delta}}^1_{2n+1}$ of the correct type, and the measure $R^{\ell,m}$ on $\vartheta^{\ell,m}$, where $R^{\ell,m}$ enumerates (w.r.t. ℓ) the measures $W^m_1, S^{1,m}_1, \dots, W^m_{2n-1}, S^{1,m}_{2n-1}, \dots, S^{2^n-1,m}_{2n-1}$. That is, A has measure one w.r.t. $S^{\ell,m}_{2n+1}(\ell > 1)$ if there is a c.u.b. subset $C \subseteq \underset{\sim}{\boldsymbol{\delta}}^1_{2n+1}$ such that for any $F\colon \underset{\sim}{\boldsymbol{\delta}}^1_{2n+1} \to C$ of the correct type $[F]_{\mu^{\vartheta^{\ell,m}}_{2n+1}} \in A$.

We frequently use the abbreviated versions of the above definitions.

For $\ell = 1$, we let $<^m$ denote the ordering on $(\underset{\sim}{\boldsymbol{\delta}}^1_{2n+1})^m$ defined by

$$(\alpha_1, \dots, \alpha_m) <^m (\beta_1, \dots, \beta_m) \text{ iff } (\alpha_m, \alpha_1, \dots, \alpha_{m-1}) <_{\text{lex}} (\beta_m, \beta_1, \dots, \beta_{m-1}),$$

where $<_{\text{lex}}$ again denotes lexicographic ordering. We let $S^{1,m}_{2n+1}$ be the measure (defined analogously to $S^{1,m}_1$) induced by the strong partition relation on $\underset{\sim}{\boldsymbol{\delta}}^1_{2n+1}$, functions $F\colon (\underset{\sim}{\boldsymbol{\delta}}^1_{2n+1})^m \to \underset{\sim}{\boldsymbol{\delta}}^1_{2n+1}$ of the correct type, and the m-fold product of the ω-cofinal normal measure on $\underset{\sim}{\boldsymbol{\delta}}^1_{2n+1}$.

From I_{2n+1}, it follows that $S^{\ell,m}_{2n+1}, W^m_{2n+1}$ are defined.

If v is a measure we let $\vartheta_v = \vartheta(v) = \text{dom}(v)$ be the ordinal on which v is a measure (and bounded subsets have measure zero). When there is no danger of confusion, we use v, ϑ_v interchangeably, and speak of $\alpha \in v$ or $f\colon v \to \text{Ord}$.

We let $R_{2n+1} = \bigcup_m W^m_{2n+1} \cup \bigcup_{\ell,m} S^{\ell,m}_{2n+1}$, and let $R^{\ell,m}_{2n+1}$ denote $S^{\ell,m}_{2n+1}$ if $\ell > 1$ and either $S^{\ell,m}_{2n+1}$ or W^m_{2n+1} if $\ell = 1$.

This completes the definition of the canonical family of measures $R^{\ell,m}_{2n+1}$.

We now define a more general collection of measures $\mathcal{R}_{2n+1} = \mathcal{W}_{2n+1} \cup \mathcal{S}_{2n+1}$, where the measures in $\mathcal{W}_{2n+1}$ are measures on tuples of ordinals $< \underset{\sim}{\boldsymbol{\delta}}^1_{2n+1}$, and the measures in $\mathcal{S}_{2n+1}$ are measures on tuples of ordinals $< \Xi_{2n+3}$. For $v \in \mathcal{R}_{2n+1}$, we let ϑ_v denote the set of tuples of ordinals on which v is a measure. By coding tuples, we may think of v as a measure on an ordinal, which we will also denote by ϑ_v, which should cause no confusion. We proceed by induction on n, and we simultaneously define embeddings $\Pi^{v^j}_{v^i}\colon \vartheta_{v^j} \to \vartheta_{v^i}$, for certain $v^j, v^i \in \mathcal{R}$.

***n* = 0.** $\mathcal{W}_1$ consists of the measures $\mathcal{W}^m_1$ on ω^m_1, where a permutation π^m of $\{1, \dots, m\}$ is associated with each such measure. We identify a measure in $\mathcal{W}_1$ with a measure on an ordinal by identifying $(\alpha_1, \dots, \alpha_m)$ with its order-type in the ordering on these tuples where we order first by α_{π_1}, then α_{π_2}, etc., where $\pi^m = (\pi_1, \pi_2, \dots, \pi_m)$.

If $\mathcal{V}_1 \in \mathcal{W}_1$ with permutation π^m, and $\mathcal{V}_2 \in \mathcal{W}_1$ with permutation π^{m+1} and the first m elements of π^{m+1} are ordered (as integers) as the elements of π^m, then we define $\pi_{v^1}^{v^2} = \pi_1^2$ by $\pi_1^2(\beta_1, \dots, \beta_{m+1}) = (\beta_1, \dots, \hat{\beta}_k, \dots, \beta_{m+1})$ where k is the last element in the permutation π^{m+1}.

$\mathcal{S}_1$ consists of products of basic measures in $\mathcal{S}_1$, where we proceed to define the basic measures in $\mathcal{S}_1$. For fixed n, we consider tuples of the form $S = \langle \alpha_1, i_1, \alpha_2, i_2, \dots, \alpha_m, i_m \rangle$, where $m \leq n$, $\alpha_1, \dots, \alpha_m < \omega_1$, $1 \leq i_1 \leq n_1, \dots, 1 \leq i_m \leq n_m$ for some integers $n_1, \dots, n_m$. We also assume that for each $(i_1, \dots, i_k)$, where $k < n$, we have a permutation $\pi_{(i_1, \dots, i_k)}$ of a size $k+1$ subset of $\{1, \dots, n\}$, beginning with n, such that if $(\vec{\imath})$ extends $(\vec{\jmath})$, then $\pi_{(\vec{\imath})}$ extends $\pi_{(\vec{\jmath})}$. We then require that for $k < n$, and fixed $(i_1, \dots, i_k)$ that the ordinals $(\alpha_1, \dots, \alpha_{k+1})$ are ordered as the integers in $\pi_{(i_1, \dots, i_k)}$. We let $<^s$ denote the Brouwer-Kleene ordering restricted to this set of tuples (*i.e.*, $s_1 <^s s_2$ iff s_1 is less at the least point of disagreement, or if s_1 extends s_2). The measure in $\mathcal{S}_1$ then, is the measure on tuples $(\cdots, \beta^{(i_1, \dots, i_m)}, \cdots)$ indexed by indices $(i_1, \dots, i_m)$, where $m \leq n$ as above, defined using the strong partition relation on ω_1 and functions $F : <^s \to \omega_1$ of the correct type. For fixed F and $(i_1, \dots, i_m)$, the ordinal $\beta^{(i_1, \dots, i_m)}$ is represented with respect to the m-fold product of the ω-cofinal normal measure on ω_1 by the function

$$k(\gamma_1, \dots, \gamma_m) = F(\langle \gamma_{j_1}, i_1, \gamma_{j_2}, i_2, \dots, \gamma_{j_m}, i_m \rangle),$$

where $(\gamma_{j_1}, \dots, \gamma_{j_m})$ is the permutation of $(\gamma_1, \dots, \gamma_m)$ ordered as $\pi_{(i_1, \dots, i_{m-1})}$.

We define Π on basic measures in $\mathcal{S}_1$ as follows: We let $\mathcal{V}_1$, $\mathcal{V}_2$ be basic measures in $\mathcal{S}_1$ as above, with $\mathcal{V}_1$ corresponding to n_1, $\mathcal{V}_2$ corresponding to n_2, and $n_1 < n_2$. We require that if $(i_1, \dots, i_k)$ is an allowed index as in the definition of $\mathcal{V}_1$, then it is allowed in $\mathcal{V}_2$, and for $1 \leq k \leq n_1 - 1$ the integers in $\pi_{(i_1, \dots, i_k)}^{\mathcal{V}^1}$ and $\pi_{(i_1, \dots, i_k)}^{\mathcal{V}^2}$ are ordered similarly. In this case, the tuples S_1 as in the definition of $\mathcal{V}_1$ are tuples S_2 allowed in $\mathcal{V}_2$. Hence, an $F_2 : <^{s_2} \to \omega_1$ of the correct type induces an $F_1 : <^{s_1} \to \omega_1$ of the correct type. This, in turn, induces the map $\Pi_1^2 = \Pi_{\mathcal{V}^1}^{\mathcal{V}^2}$ from $\vartheta_{\mathcal{V}^2}$ into $\vartheta_{\mathcal{V}^1}$. It follows readily that if $A \subseteq \vartheta_{\mathcal{V}^1}$ has measure one w.r.t. $\mathcal{V}^1$ then for almost all $\vartheta < \vartheta_{\mathcal{V}^2}$ w.r.t. $\mathcal{V}^2$, $\Pi_1^2(\vartheta) \in A$.

We extend Π to products of basic measures in $\mathcal{S}_1$ componentwise.

It also follows readily that Π extended to products has the same projection property mentioned just above.

***n* ≥ 1.** We assume $\mathcal{W}_{2m+1}, \mathcal{S}_{2m+1}$ have been defined for $m < n$, and define $\mathcal{W}_{2n+1}, \mathcal{S}_{2n+1}$.

For $\mathcal{V}_1, \mathcal{V}_2$, measures in $\bigcup_{m<n} \mathcal{W}_{2m+1} \cup \bigcup_{m<n} \mathcal{S}_{2m+1}$, we define $\mathcal{V}_2 < \mathcal{V}_1$ if $\Pi_1^2 : \vartheta_{v_2} \to \vartheta_{v_1}$ is defined.

Definition 2.1 ($\mathcal{W}_{2n+1}$). We define an $n - *$ tuple $T = T^n$ to be a function with domain certain tuples of integers $(i_1, \dots, i_m)$, $m \leq n$ and satisfying the following properties:

1) $i_1 \in \operatorname{dom} T$ for $1 \leq i_1 \leq n_1$, for some integer n_1.

2) If $(i_1, \dots, i_\ell) \in \operatorname{dom} T$, then $(i_1, \dots, i_k) \in \operatorname{dom} T$ for $k \leq \ell$.
3) T associates measures to certain of the tuples $(i_1, \dots, i_k)$, where $k < n$, in its domain, which we denote by $v^{(i_1,\dots,i_k)}$, where $v^{(i_1,\dots,i_k)} \in \bigcup_{m<n} \mathcal{R}_{2m+1}$. We require that if $v^{(i_1,\dots,i_{k+1})}$ is defined, then so is $v^{(i_1,\dots,i_k)}$, and $v^{(i_1,\dots,i_{k+1})} < v^{(i_1,\dots,i_k)}$.
4) If $v^{(i_1,\dots,i_k)}$ is defined, then $(i_1, \dots, i_{i+1}) \in \operatorname{dom} T$ for $1 \leq i_{k+1} \leq n_{(i_1,\dots,i_k)}$ for some integer $n_{(i_1,\dots,i_k)}$, and if $(i_1, \dots, i_{k+1}) \in \operatorname{dom} T$, then $v^{(i_1,\dots,i_k)}$ is defined.

For each $n - *$ tuple T, we define a corresponding ordering $<^T$ as follows: We let $\Pi^{(i_1,\dots,i_{k+1})}_{(i_1,\dots,i_k)}$ denote the embedding for $v^{(i_1,\dots,i_{k+1})}, v^{(i_1,\dots,i_k)}$, where defined. The domain of $<^T$ consists of ordinals $\alpha^{(i_1,\dots,i_k)}$ indexed by indices $(i_1, \dots, i_k) \in \operatorname{dom} T$, where $\alpha \in \vartheta^{(i_1,\dots,i_{k-1})}\ (= \vartheta_{v^{(i_1,\dots,i_{k-1})}})$, if $k > 1$. For $k = 1$, we take $\alpha = 1$, and allow $1^{(i_1)} \in \operatorname{dom} <^T$ for some (but not necessarily all—we think of this as being coded in T) $i_1 \in \operatorname{dom} T$. We set $\alpha^{(i_1,\dots,i_k)} <^T \beta^{(j_1,\dots,j_\ell)}$ provided

$$\langle i_1, \Pi^{(i_1,\dots,i_{k-1})}_{(i_1)}(\alpha), i_2, \Pi^{(i_1,\dots,i_{k-1})}_{(i_1,i_2)}(\alpha), \dots, \alpha, i_k\rangle$$
$$<_{\mathrm{BK}} \langle j_1, \Pi^{(j_1,\dots,j_{\ell-1})}_{(j_1)}(\beta), j_2, \Pi^{(j_1,\dots,j_{\ell-1})}_{(j_1,j_2)}(\beta), \dots, \beta, j_k\rangle$$

where $<_{\mathrm{BK}}$ denotes the Brouwer-Kleene ordering of these sequences. We use here the notation (defined inductively by)

$$\Pi^{(i_1,\dots,i_{k-1})}_{(i_1,\dots,i_m)} = \Pi^{(i_1,\dots,i_{m+1})}_{(i_1,\dots,i_m)} \circ \Pi^{(i_1,\dots,i_{k_1})}_{(i_1,\dots,i_{m+1})}, \text{ where}$$
$$\Pi^{(i_1,\dots,i_{k-1})}_{(i_1,\dots,i_{k-1})} = \text{identity.}$$

We let $\mu^{<^T}$ be the measure on tuples $(\cdots, \eta^{(i_1,\dots,i_k)}, \cdots)$ of ordinals $< \underset{\sim}{\delta}^1_{2n+1}$ induced by the weak partition relation on $\underset{\sim}{\delta}^1_{2n+1}$, functions $f : <^T \to \underset{\sim}{\delta}^1_{2n+1}$ of the correct type, and the measures $v^{(i_1,\dots,i_k)}$. That is, A has measure one if there is a c.u.b. $C \subseteq \underset{\sim}{\delta}^1_{2n+1}$ such that for any $f : <^T \to C$ of the correct type, $(\cdots, \eta^{(i_1,\dots,i_k)}, \cdots) \in A$, where $\eta^{(i_1,\dots,i_k)}$ is represented w.r.t. $v^{(i_1,\dots,i_{k-1})}$ by the function $f^{(i_1,\dots,i_k)} : \vartheta^{(i_1,\dots,i_{k-1})} \to \underset{\sim}{\delta}^1_{2n+1}$ defined by $f^{(i_1,\dots,i_k)}(\beta) = f(\beta^{(i_1,\dots,i_k)})$, for $k > 1$, and if $k = 1$ and $i_1 \in \operatorname{dom} <^T$, then $\eta^{(i_1)} = f(1^{i_1})$.

We define $\mathcal{W}_{2n+1}$ to consist of the measures of the form $\mu^{<^T}$ corresponding to all $n - *$ tuples T.

DEFINITION 2.2 (Π on $\mathcal{W}_{2n+1}$). If T_1 is an $n - *$ tuple, and T_2 is an $(n+1) - *$ tuple, we say that T_2 extends T_1 provided:

1) $(i_1, \dots, i_k) \in \operatorname{dom} T_1$, iff $(i_1, \dots, i_k) \in \operatorname{dom} T_2$, for $k \leq n$ and $v^{(i_1,\dots,i_{k-1})}_{T_2} = v^{(i_1,\dots,i_{k-1})}_{T_1}$.
2) If $(i_1, \dots, i_{n+1}) \in \operatorname{dom} T_2$, then $v^{(i_1,\dots,i_n)}_{T_2} < v^{(i_1,\dots,i_{n-1})}_{T_1}$.

3) If $1^{i_1} \in \operatorname{dom} <^{T_1}$ then $1^{i_1} \in \operatorname{dom} <^{T_2}$, and conversely.

We write $T_2 < T_1$ for T_2 extending T_1.

For $T_2 < T_1$, and the corresponding measures $\mu^{<^{T_2}}, \mu^{<^{T_1}}$ on ϑ_2, ϑ_1 (which are equal to $(\underset{\sim}{\boldsymbol{\delta}}^1_{2n+1})^{c_1}, (\underset{\sim}{\boldsymbol{\delta}}^1_{2n+1})^{c_2}$ for some c_1, c_2), we define $\Pi^2_1 \colon \vartheta_2 \to \vartheta_1$ to be the natural projection map defined as follows: If $(\cdots, \eta^{(i_1,\dots,i_k)}, \cdots)$ is a tuple in ϑ_2, where $(i_1, \dots, i_k)$ ranges over $\operatorname{dom} T_2$, we set

$$\Pi^2_1(\dots, \eta^{(i_1,\dots,i_k)}, \dots) = (\dots, \eta^{(i_1,\dots,i_k)}, \dots),$$

where $(i_1, \dots, i_k)$ on the right-hand side now ranges over the indices in $\operatorname{dom} T_1$. We note that if $(\cdots, \eta^{(i_1,\dots,i_k)}, \cdots)$ is represented by $f : <^{T_2} \to \underset{\sim}{\boldsymbol{\delta}}^1_{2n+1}$ (which holds almost everywhere w.r.t. $\mu^{<^{T_2}}$), then f induces a function $\bar{f} : <^{T_1} \to \underset{\sim}{\boldsymbol{\delta}}^1_{2n+1}$, since $<^{T_1}$ is a subordering of $<^{T_2}$, and $\bar{f}$ then represents $\Pi^2_1(\cdots, \eta^{(i_1,\dots,i_k)}, \cdots)$.

We also note that if $A \subseteq \vartheta_1$ has measure one w.r.t. $\mu^{<^{T_1}}$, then for almost all $\vec{\eta}$ w.r.t. $\mu^{<^{T_2}}$, $\Pi^2_1(\vec{\eta}) \in A$.

Definition 2.3 ($\mathcal{S}_{2n+1}$). We fix an $n - *$ tuple T, and associated order in $<^T$. We define, relative to T, a collection $\mathcal{C}$ to be a set $\mathcal{C} = \bigcup_{1 \le m \le n} \mathcal{C}_n$ where:

1) $\mathcal{C}_1$ consists of a_0 sequences for some integer a_0, of the form $C_1 = (i_1^{(1)}, \dots, i_{k(1)}^{(1)}), \dots, C_{a_0} = (i_1^{(a_0)}, \dots, i_{k(a_0)}^{(a_0)})$, of integers between 1 and n_1 (= the number of $(i_1) \in \operatorname{dom} T$).
2) $\mathcal{C}_\ell$ consists of tuples of the form $C_{k_1,\dots,k_\ell}$, where $C_{k_1,\dots,k_{\ell-1}} \in \mathcal{C}_{\ell-1}$. Each $C_{k_1,\dots,k_\ell}$ is a sequence of tuples of the form $C_{k_1,\dots,k_\ell} = (\cdots, (i_1, \dots, i_\ell), \cdots)$, where each tuple $(i_1, \dots, i_\ell) \in C_{k_1,\dots,k_\ell}$, extends $(i_1, \dots, i_{\ell-1}) \in C_{k_1,\dots,k_{\ell-1}}$ and $(i_1, \dots, i_\ell) \in \operatorname{dom} T$.

Corresponding to $T, \mathcal{C}$, we consider sequences S of ordinals $< \underset{\sim}{\boldsymbol{\delta}}^1_{2n+1}$ and integers of the form

$$S = \langle \alpha_0, \mathcal{I}_1, \alpha_1, \dots, \alpha_{j_1}, \mathcal{I}_2, \dots, \alpha_{j_1,j_2}, \dots, \mathcal{I}_n, \dots, \alpha_{j_1,\dots,j_n}, \dots \rangle$$

or an initial segment of such a tuple, where:

1) $\alpha_0 < \underset{\sim}{\boldsymbol{\delta}}^1_{2n+1}$.
2) $1 \le \mathcal{I}_1 \le a_0$. Recall a_0 is the number of elements in $\mathcal{C}_1$.
3) $j_1 =$ the number of elements in $C_{\mathcal{I}_1}$.
4) $\alpha_1, \dots, \alpha_{j_1} < \underset{\sim}{\boldsymbol{\delta}}^1_{2n+1}$.
5) In general, $(\mathcal{I}_1, \dots, \mathcal{I}_k)$ for $k \le n$ is an index corresponding to an element of $\mathcal{C}$.
6) The ordinals $\alpha_{j_1,\dots,j_k}$ following $\mathcal{I}_k$ are indexed by indices $(j_1, \dots, j_k)$ corresponding to elements of $C_{\mathcal{I}_1,\dots,\mathcal{I}_k}$.
7) $\alpha_{j_1,\dots,j_k} < j_{v^{(j_1,\dots,j_{k-1})}}(\alpha_0) =$ the ultrapower of α_0 by the measure $v^{(j_1,\dots,j_{k-1})}$. (If $n_1 \in \operatorname{dom} <^T$, it would be enough to require $\alpha_{j_1}, \dots, \alpha_{j_k} < \alpha_0$ here.)

We let $<^T_{\mathcal{C}}$ denote the Brouwer-Kleene ordering on these tuples. In the definition of S, we assume the ordering on indices $(j_1, \ldots, j_k)$ is the same as their occurrence in $C_{(\mathcal{I}_1,\ldots,\mathcal{I}_k)}$.

We then let $\mathcal{S}_{2n+1}{}^T_{\mathcal{C}}$ be the measure induced by the strong partition relation on $\underset{\sim}{\delta}^1_{2n+1}$, functions $F\colon <^T_{\mathcal{C}} \to \underset{\sim}{\delta}^1_{2n+1}$ of the correct type, and the measure $\mu^{<^T}$. That is, $\mathcal{S}_{2n+1}{}^T_{\mathcal{C}}$ is a measure on tuples of ordinals $(\cdots, \alpha_{(\mathcal{I}_1,\ldots,\mathcal{I}_k)}, \cdots)$, indexed by indices $(\mathcal{I}_1, \ldots, \mathcal{I}_k)$ corresponding to indices in $\mathcal{C}$, where A has measure one w.r.t. $\mathcal{S}_{2n+1}{}^T_{\mathcal{C}}$ if there is a c.u.b. $C \subseteq \underset{\sim}{\delta}^1_{2n+1}$ such that for all $F\colon <^T_{\mathcal{C}} \to C$ of the correct type, $(\cdots, \alpha^F_{(\mathcal{I}_1,\ldots,\mathcal{I}_k)}, \cdots) \in A$, where $\alpha^F_{(\mathcal{I}_1,\ldots,\mathcal{I}_k)}$ is represented w.r.t. $\mu^{<^T}$ by a function $g(\cdots, \xi_{(i_1,\ldots,i_k)}, \cdots)$ defined as follows: if $h\colon <^T \to \underset{\sim}{\delta}^1_{2n+1}$ represents the tuple $(\cdots, \xi_{(i_1,\ldots,i_k)}, \cdots)$ w.r.t. the measures $\upsilon^{(i_1,\ldots,i_k)}$, then we set $g(\cdots, \xi_{(i_1,\ldots,i_k)}, \cdots) = F(S) = F(S^h_{(\mathcal{I}_1,\ldots,\mathcal{I}_k)})$, where

$$S = \langle \alpha_0, \mathcal{I}_1, \alpha_1, \ldots, \alpha_{j_1}, \mathcal{I}_2, \ldots, \alpha_{j_1,j_2}, \ldots, \mathcal{I}_k, \ldots, \alpha_k, \ldots, \alpha_{j_1,\ldots,j_k}, \ldots \rangle$$

and where

1) $\alpha_0 = \sup_{\text{a.e.}}(\operatorname{ran} h)$, where $\sup_{\text{a.e.}}$ denotes the almost everywhere sup w.r.t. the measures $\upsilon^{(i_1,\ldots,i_k)}$. If h is of the correct type everywhere, this is the same as $\sup(\operatorname{ran} h)$.
2) $\alpha_{j_1,\ldots,j_\ell}$ is represented w.r.t. $\upsilon^{(j_1,\ldots,j_{\ell-1})}$ by the sub-function $h^{(j_1,\ldots,j_\ell)}$ of h. That is, $h^{(j_1,\ldots,j_\ell)}\colon \vartheta_{\upsilon^{(j_1,\ldots,j_{\ell-1})}} \to \underset{\sim}{\delta}^1_{2n+1}$ is defined by $h^{(j_1,\ldots,j_\ell)}(\gamma) = h(\gamma^{(j_1,\ldots,j_\ell)})$, for $\gamma \in \vartheta_{\upsilon^{(j_1,\ldots,j_{\ell-1})}}$.

This is well defined. We let $\mathcal{S}_{2n+1}$ consist of the r-fold products of measures of the form $\mathcal{S}_{2n+1}{}^T_{\mathcal{C}}$. We refer to the $\mathcal{S}_{2n+1}{}^T_{\mathcal{C}}$ as the basic measures in $\mathcal{S}_{2n+1}$.

Definition 2.4 (Π on $\mathcal{S}_{2n+1}$). We again define Π component-wise, so it is enough to define it on basic measures of the form $\mathcal{S}_{2n+1}{}^T_{\mathcal{C}}$. We consider measures of the form $\mathcal{S}_{2n+1}{}^{T_1}_{\mathcal{C}_1}, \mathcal{S}_{2n+1}{}^{T_2}_{\mathcal{C}_2}$ which satisfy the following:

1) T_2 is an $(n+1) - *$ tuple which extends the $n - *$ tuple T_1.
2) $\mathcal{C}_1, \mathcal{C}_2$, are collections defined relative to T_1, T_2 respectively.
3) $\mathcal{C}_2$ extends $\mathcal{C}_1$ in the sense that if $C_{(\mathcal{I}_1,\ldots,\mathcal{I}_k)} \in \mathcal{C}_1$, then $C_{(\mathcal{I}_1,\ldots,\mathcal{I}_k)} \in \mathcal{C}_2$.

We define the embedding $\Pi^{T_2,\mathcal{C}_2}_{T_1,\mathcal{C}_1}$ corresponding to $\mathcal{S}_{2n+1}{}^{T_2}_{\mathcal{C}_1}, \mathcal{S}_{2n+1}{}^{T_2}_{\mathcal{C}_2}$ in this case. If a tuple $(\ldots, \alpha_{(\mathcal{I}_1,\ldots,\mathcal{I}_k)}, \ldots)$, $k \leq n+1$ is given in the domain of the measure space $\mathcal{S}_{2n+1}{}^{T_2}_{\mathcal{C}_2}$, and is represented by $F\colon <^{T_2}_{\mathcal{C}_2} \to \underset{\sim}{\delta}^1_{2n+1}$, then F induces by restriction the function $\bar{F}\colon <^{T_1}_{\mathcal{C}_1} \to \underset{\sim}{\delta}^1_{2n+1}$ representing a tuple $(\cdots, \beta_{(\mathcal{I}_1,\ldots,\mathcal{I}_k)}, \cdots)$, $k \leq n$, with respect to $\mu^{<T_1}$. We set

$$\Pi^{T_2,\mathcal{C}_2}_{T_1,\mathcal{C}_1}(\cdots, \alpha_{(\mathcal{I}_1,\ldots,\mathcal{I}_k)}, \cdots) = (\cdots, \beta_{(\mathcal{I}_1,\ldots,\mathcal{I}_k)}, \cdots).$$

It follows readily that this is well-defined on a measure one set w.r.t. $\mathcal{S}_{2n+1}{}^{T_2}_{\mathcal{C}_2}$. We take Π to be zero off this measure one set.

This completes the definitions of the measures $R_{2n+1}, \mathcal{R}_{2n+1}$, and the embeddings Π.

We define an equivalence relation $\sim$ on the measures in R_{2n+1} by induction as follows: We set $\mathcal{V}_1 \sim \mathcal{V}_2$ provided one of the following holds:

1) $\mathcal{V}_1 = S^{1,m_1}_{2n+1}$ and $\mathcal{V}_2 = S^{1,m_2}_{2n+1}$
2) $\mathcal{V}_1 = W^{m_1}_{2n+1}, \mathcal{V}_2 = W^{m_2}_{2n+1}$
3) $\mathcal{V}_1 = S^{\ell_1,m_1}_{2n+1}, \mathcal{V}_2 = S^{\ell_2,m_2}_{2n+1}$, where $\ell_1, \ell_2 > 1$, in which case $\ell_1 = \ell_2$.

So, $\mathcal{V}_1 \sim \mathcal{V}_2$ if $\mathcal{V}_1$ and $\mathcal{V}_2$ are in the same canonical family of measures. We also let $\mathcal{V}_1 < \mathcal{V}_2$ mean that $\mathcal{V}_1 \sim \mathcal{V}_2$ and $m_1 < m_2$ for m_1, m_2 as above.

A useful notation is to let, for $\ell > 1$, $v(S^{\ell,m}_{2n+1})$ denote the measure v such that $S^{\ell,m}_{2n+1}$ is defined from the strong partition relation on $\underset{\sim}{\delta}^1_{2n+1}$, functions $F : \underset{\sim}{\delta}^1_{2n+1} \to \underset{\sim}{\delta}^1_{2n+1}$ of the correct type, and the measure $\mu^v =$ the measure induced by the weak partition on $\underset{\sim}{\delta}^1_{2n+1}$, functions $f : \vartheta_v \to \underset{\sim}{\delta}^1_{2n+1}$ of the correct type and the measure v.

We note that (by induction) 3 above is equivalent to $v(S^{\ell_2,m_2}_{2n+1}) \sim v(S^{\ell_1,m_1}_{2n+1})$.

We prove some basic lemmas which will be used frequently. We recall we are assuming I_{2n+1} throughout.

Lemma 2.5. If $\mathcal{V}_1, \mathcal{V}_2 \in R_{2n+1}$ and $\mathcal{V}_1 \not\sim \mathcal{V}_2$, then $\mathrm{cf}(\vartheta_{\mathcal{V}_1}) \neq \mathrm{cf}(\vartheta_{\mathcal{V}_2})$.

Proof. If not, we fix $\mathcal{V}_1 \not\sim \mathcal{V}_2$, such that $\mathrm{cf}(\vartheta_{\mathcal{V}_1}) = \mathrm{cf}(\vartheta_{\mathcal{V}_2})$.

We consider first the case where $\mathcal{V}_1 = S^{\ell_1,m_1}_{2n+1}, \mathcal{V}_2 = S^{\ell_2,m_2}_{2n+1}$, where $\ell_1, \ell_2 > 1$ and $\ell_1 \neq \ell_2$. We note that $\vartheta_{\mathcal{V}_1}, \vartheta_{\mathcal{V}_2}$ are always limit ordinals. We fix $H : \vartheta_{\mathcal{V}_1} \to \vartheta_{\mathcal{V}_2}$ monotonicially increasing and cofinal. We consider the following partition $\mathcal{P}$: We partition pairs of functions $F_1, F_2 : \underset{\sim}{\delta}^1_{2n+1} \to \underset{\sim}{\delta}^1_{2n+1}$ of the correct type, where we require that for $f_1 : v(\mathcal{V}_1) \to \underset{\sim}{\delta}^1_{2n+1}, f_2 : v(\mathcal{V}_2) \to \underset{\sim}{\delta}^1_{2n+1}$, if $\sup_{\text{a.e.}} f_1 < \sup_{\text{a.e.}} f_2$, then $F_1([f_1]_{v(\mathcal{V}_1)}) < F_2([f_2]_{v(\mathcal{V}_2)})$ and similarly if $\sup_{\text{a.e.}} f_2 < \sup_{\text{a.e.}} f_1$ then $F_2([f_2]_{v(\mathcal{V}_2)}) < F_1([f_1]_{v(\mathcal{V}_1)})$. If $\sup_{\text{a.e.}} f_1 = \sup_{\text{a.e.}} f_2$ (hence one of f_1, f_2 is not of the correct type a.e., since $\mathrm{cf}(\vartheta_{v(\mathcal{V}_1)}) \neq \mathrm{cf}(\vartheta_{v(\mathcal{V}_2)})$ by induction), we require $F_1([f_1]) < F_2([f_2])$. In using the strong partition relation on $\underset{\sim}{\delta}^1_{2n+1}$, we think of the pair F_1, F_2 as being coded by a single function of the correct type. We then partition such F_1, F_2 (or more precisely the single function coding them) according to whether or not $H([F_1]_{\mu^{v_1}}) < [F_2]_{\mu^{v_2}}$, where $v_1 = v_1(\mathcal{V}_1)$, $v_2 = v_2(\mathcal{V}_2)$.

We claim that on the homogeneous side of the partition, the property stated in partition $\mathcal{P}$ holds. We suppose not, and fix a c.u.b. $C \subseteq \underset{\sim}{\delta}^1_{2n+1}$ homogeneous for the contrary side. We assume that C is closed under j_{v_1}, j_{v_2}, the ultrapowers from the measures v_1, v_2 (we use $K_{2n+1}(\mathrm{a})$ here), and C consists of limit ordinals. We fix $F_1 : \underset{\sim}{\delta}^1_{2n+1} \to C$ of the correct type with $F_1(\alpha) > \alpha$ for all α, and fix $F_2 : \underset{\sim}{\delta}^1_{2n+1} \to C$ of the correct type with $[F_2] > H([F_1])$ and $F_2(\alpha) > \alpha$ for all α. We may also assume that $F_2(\alpha)$ is greater than the $j_{v_1}(\alpha)$th element of C after α.

We let C' be a c.u.b. subset of the closure points of C ($= \{\beta \in C :$ for all $\alpha < \beta$, β is the βth element of C after $\alpha\}$) which is closed under F_1, F_2 in the sense that for all $\beta \in C'$, if $\alpha < \beta$ and $f_1 \colon \vartheta_{v_1} \to \alpha$ then $F_1([f_1]_{v_1}) < \beta$, and similarly for F_2.

We then claim that there are functions F_1', F_2' satisfying:

1) $F_1' = F_1$, $F_2' = F_2$, a.e. w.r.t. μ^{v_1} and μ^{v_2} respectively.
2) F_1', F_2' have range a subset of C and are of the correct type.
3) F_1', F_2' are ordered as in the statement of $\mathcal{P}$, everywhere.

We define F_1', F_2' as follows: If $\vartheta \in C'$ and $\operatorname{cf} \vartheta = \operatorname{cf} \vartheta_{v_1}$, then we set $F_1'([f]) = F_1([f])$ for any $f \colon \vartheta_{v_1} \to \underset{\sim}{\boldsymbol{\delta}}^1_{2n+1}$ s.t. $\sup_{\text{a.e.}} f = \vartheta$. If $\vartheta \in C'$ and $\operatorname{cf} \vartheta = \operatorname{cf} \vartheta_{v_2}$, then we set $F_2'([f]) = F_2([f])$ for $f \colon \vartheta_{v_2} \to \underset{\sim}{\boldsymbol{\delta}}^1_{2n+1}$ with $\sup_{\text{a.e.}} f = \vartheta$.

For other values of F_1' and F_2' we proceed inductively. That is, if $F_1'(\vartheta)$ or $F_2'(\vartheta)$ is not defined by the previous paragraph we proceed as follows to define it. Assume $\eta < \underset{\sim}{\boldsymbol{\delta}}^1_{2n+1}$ and $F_1'(\eta_1)$ and $F_2'(\eta_2)$ are defined whenever $\eta_1 = [f_1]_{v_1}$ with $\sup_{\text{a.e.}} f_1 < \eta$ and likewise for η_2. Suppose $\vartheta = [f_1]_{v_1}$, where $\sup_{\text{a.e.}} f_1 = \eta$. If $F_1'(\vartheta)$ is not already defined, set

$$F_1'(\vartheta) = N_C\Big(\max\big(\sup_{\alpha} F_1'(\alpha), \sup_{\beta} F_2'(\beta)\big)\Big),$$

where α ranges over the ordinals less than ϑ and β ranges over ordinals represented by $\bar{f}_2 \colon \vartheta_{v_2} \to \underset{\sim}{\boldsymbol{\delta}}^1_{2n+1}$ with $\sup_{\text{a.e.}} \bar{f}_2 < \eta$. Here $N_C(\gamma)$ denotes the ωth element of C greater than γ. For $\vartheta = [f_2]_{v_2}$ with $\sup_{\text{a.e.}} f_2 = \eta$, if $F_2'(\vartheta)$ is not already defined set

$$F_2'(\vartheta) = N_C\Big(\max\big(\sup_{\alpha} F_1'(\alpha), \sup_{\beta} F_2'(\beta)\big)\Big),$$

where now α ranges over the ordinals represented by $\bar{f}_1 \colon \vartheta_{v_1} \to \underset{\sim}{\boldsymbol{\delta}}^1_{2n+1}$ with $\sup_{\text{a.e.}} \bar{f}_1 \leq \eta$ and β ranges over the ordinals less than ϑ.

We claim that F_1', F_2' satisfy the above properties. Property 1 follows since $F_1' = F_1$ for all $f \colon \vartheta_{v_1} \to C'$ of the correct type, and similarly for F_2'. Property 2 is immediate from the definition of F_1', F_2'. Property 3 follows from the definition of F_1', F_2', the fact that $F_1'(\alpha) > \alpha$, $F_2'(\alpha) > \alpha$, and the definition of C'. For example, for $f_1 \colon \vartheta_{v_1} \to C'$ of the correct type, $F_1'([f_1]) = F_1([f_1]) \geq \sup_{\text{a.e.}} f_1 = \eta \in C'$, and hence $F_1'([f_1]) > \eta > F_2([f_2])$ for all $f_2 \colon \vartheta_{v_2} \to \underset{\sim}{\boldsymbol{\delta}}^1_{2n+1}$ with $\sup_{\text{a.e.}} f_2 < \eta$. It follows easily using the definition of C' that $F_1'([f_1]) > \eta > F_2'([f_2])$ as well. A similar argument shows that if $f_2 \colon \vartheta_{v_2} \to C'$ is of the correct type with $\sup_{\text{a.e.}} f_2 = \eta$, then for all $[f_1]$ with $\sup_{\text{a.e.}} f_1 \leq \eta$ we have $F_2'([f_2]) > F_1'([f_1])$ (here we use the fact that $F_2([f_2])$ is greater than the $j_{v_1}(\eta)$th element of C greater than η).

From property 1, we have that $[F_2'] > H([F_1'])$, which contradicts properties 2, 3 and the definition of C. Hence, on the homogeneous side of the partition, the property stated in partition $\mathcal{P}$ holds.

We fix a c.u.b. $C \subseteq \underset{\sim}{\delta}^1_{2n+1}$ homogeneous for $\mathcal{P}$. We define F_2 as follows. For $\alpha < \underset{\sim}{\delta}^1_{2n+1}$ represented by $f_2 : \vartheta_{v_2} \to \underset{\sim}{\delta}^1_{2n+1}$ with $\sup_{\text{a.e.}} f_2 = \vartheta$, we let $F_2(\alpha)$ be the $\omega \cdot (j_{v_1}(\vartheta) + \alpha + 1)$st element of C greater then ϑ. Note that F_2 is of the correct type with range in C.

We then claim for almost all $[F_1]_{\mu^{v_1}}$ that $H([F_1]) < [F_2]$, contradicting the fact that H is cofinal, and establishing Lemma 2.5.

We fix $F_1 : \underset{\sim}{\delta}^1_{2n+1} \to C$ of the correct type. To prove the claim, it is enough to find F_1', F_2' satisfying:

1) $F_1' = F_1$, $F_2' = F_2$, a.e. w.r.t. μ^{v_1} and μ^{v_2} respectively.
2) F_1', F_2' have range a subset of C and are of the correct type.
3) F_1', F_2' are ordered as in $\mathcal{P}$.

The construction of F_1', F_2' proceeds similarly to the above, starting with a c.u.b. $C' \subseteq \underset{\sim}{\delta}^1_{2n+1}$ contained in the closure points of C, and such that for $\alpha \in C'$ and $\beta < \alpha$, if $\sup_{\text{a.e.}} f_1 \leq \beta$ then $F_1([f_1]) < \alpha$ and likewise if $\sup_{\text{a.e.}} f_2 \leq \beta$ then $F_2([f_2]) < \alpha$. So, if $\sup_{\text{a.e.}} f_1 = \eta \in C'$ with $\operatorname{cf} \eta = \vartheta_{v_1}$, then we set $F_1'([f_1]) = F_1([f_1])$ and likewise if $\sup_{\text{a.e.}} f_2 = \eta \in C'$ with $\operatorname{cf} \eta = \vartheta_{v_2}$, then $F_2'([f_2]) = F_2([f_2])$. For other values of F_1' and F_2' we proceed as before. Again it is easy to check that F_1', F_2' have the required properties. This completes the proof of Lemma 2.5 in the case considered.

If $\ell_1 = 1$, $\ell_2 > 1$ (or vice-versa), the argument is similar.

If $\mathcal{V}_1 = W^m_{2n+1}$, $\mathcal{V}_2 = S^{\ell_2, m_2}_{2n+1}$ (or vice-versa), the argument is again similar. We fix an $H : \underset{\sim}{\delta}^1_{2n+1} \to \vartheta_{v_2}$ cofinal, monotonically increasing, and consider the partition $\mathcal{P}$: we partition pairs of functions f_1, F_2 where $f_1 : \vartheta_{v_1} \to \underset{\sim}{\delta}^1_{2n+1}$ is of the correct type, where $W^n_{2n+1} = \mu^{v_1}_{2n+1}$, and $F_2 : \underset{\sim}{\delta}^1_{2n+1} \to \underset{\sim}{\delta}^1_{2n+1}$ of the correct type, where $\sup f_1 < \inf F_2$. We partition according to whether or not $H([f_1]_{v_1}) < [F_2]_{\mu^{v_2}}$. It follows by a similar argument that on the homogeneous side of the partition, the above property holds. By another similar argument, we contradict the cofinality of H.

If $\mathcal{V}_1 \in \bigcup_{m<n} R_{2m+1}$, $\mathcal{V}_2 = W^m_{2n+1}$, Lemma 2.5 follows from the regularity of $\underset{\sim}{\delta}^1_{2n+1}$, and if $\mathcal{V}_2 = S^{\ell_2, m_2}_{2n+1}$, we contradict the existence of such H by the $\underset{\sim}{\delta}^1_{2n+1}$-additivity of the measure S^{ℓ_2, m_2}_{2n+1}, $\ell_2 \geq 1$. If $\mathcal{V}_1, \mathcal{V}_2 \in \bigcup_m R_{2m+1}$, the lemma follows by induction.

This completes the proof of Lemma 2.5. ⊣ (Lemma 2.5)

We refer the argument in Lemma 2.5, where we "slide" the values in the range of the function on a set of measure zero to produce a desired order type everywhere, as a "sliding argument". This type of argument will be used frequently in the following, and we sometimes abbreviate the details of similar arguments.

Lemma 2.6. For any $\mathcal{V} \in R_{2n+1}$, and any ordinal $\underset{\sim}{\delta}$, and $H: \vartheta_{\mathcal{V}} \to \underset{\sim}{\delta}$, either there is a $\underset{\sim}{\delta}' < \underset{\sim}{\delta}$ such that $H(\alpha) < \underset{\sim}{\delta}'$ for almost all α w.r.t. $\mathcal{V}$, or there is a measure one set A w.r.t. $\mathcal{V}$ such that $H \restriction A$ is dominated by a monotonically increasing function into $\underset{\sim}{\delta}$.

Proof. We consider first the case $\mathcal{V} = S^{\ell,m}_{2n+1}$ with $\ell > 1$. We fix such a $\underset{\sim}{\delta}$ and H. We assume the first clause fails and show the second. In particular, for all $\alpha < \vartheta_{\mathcal{V}}$, we have that for cofinally many $\beta < \vartheta_{\mathcal{V}}$, $H(\alpha) < H(\beta)$. We then consider the partition $\mathcal{P}$: we partition $F_1, F_2 : \underset{\sim}{\delta}^1_{2n+1} \to \underset{\sim}{\delta}^1_{2n+1}$ of the correct type ordered as follows: if α_1, α_2 are represented w.r.t. $v = v(\mathcal{V})$ by $f_1, f_2 : \vartheta_v \to \underset{\sim}{\delta}^1_{2n+1}$, and $\vartheta_1, \vartheta_2 = \sup_{\text{a.e.}} f_1, f_2$ respectively, then if $\vartheta_1 < \vartheta_2$, $F_{1,2}(\alpha_1) < F_{1,2}(\alpha_2)$ (here $F_{1,2}$ denotes either F_1 or F_2). If $\vartheta_1 > \vartheta_2$, then $F_{1,2}(\alpha_1) > F_{1,2}(\alpha_2)$. If $\vartheta_1 = \vartheta_2$, then $F_1(\alpha_1) < F_2(\alpha_2)$. We partition such pairs F_1, F_2 according to whether or not $H([F_1]_{\mu^v}) < H([F_2]_{\mu^v})$. From our above assumption, and a sliding argument as in Lemma 2.5, it follows that on the homogeneous side of the partition, the property stated in $\mathcal{P}$ holds. We fix a c.u.b. subset $C \subseteq \underset{\sim}{\delta}^1_{2n+1}$ homogeneous for $\mathcal{P}$, and let C' be the set of closure points of C. We take A to be the measure one set determined by C'. For $\eta \in A$ represented by $F : \underset{\sim}{\delta}^1_{2n+1} \to C'$ of the correct type, we define $H'(\eta) = \sup(H(\eta'))$, where the sup ranges over η' represented by $F' : \underset{\sim}{\delta}^1_{2n+1} \to C'$ of the correct type, and $\eta' < [F_0]_{\mu^v}$, where $F_0 : \underset{\sim}{\delta}^1_{2n+1} \to \underset{\sim}{\delta}^1_{2n+1}$ is defined by: for α represented by $f : \vartheta_v \to \underset{\sim}{\delta}^1_{2n+1}$ w.r.t. v, $F_0(\alpha) = \sup_{\alpha'} F(\alpha')$, where the sup ranges over α' represented w.r.t. v by f' of the correct type with $\sup_{\text{a.e.}} f' \leq \sup_{\text{a.e.}} f$. This is well-defined in that the equivalence class of F_0 depends only on the equivalence class of F with respect to μ^v.

It is clear that $H' \restriction A$ is monotonic increasing and dominates H. To show that it has the range in $\underset{\sim}{\delta}$, and finish the proof of the claim, it is enough to establish the following claim:

If $F_1, F_2 : \underset{\sim}{\delta}^1_{2n+1} \to C'$ are of the correct type and $(F_1)_0 < (F_2)_0$ a.e. w.r.t. μ^v (where F_0 is defined above for $F : \underset{\sim}{\delta}^1_{2n+1} \to \underset{\sim}{\delta}^1_{2n+1}$), then $H([F_1]) < H([F_2])$.

To establish the claim, it is enough to show, for such F_1, F_2, that there are F_1', F_2' satisfying

1) $F_1' = F_1$, $F_2' = F_2$, a.e. w.r.t. μ^v.
2) F_1', F_2' have range a subset of C and are of the correct type everywhere.
3) F_1', F_2' are ordered as in $\mathcal{P}$.

To establish this we use a sliding argument similar to, but somewhat different from, that of Lemma 2.5. We consider the partition $\mathcal{P}'$: we partition $f : \vartheta_v \to \underset{\sim}{\delta}^1_{2n+1}$ of the correct type according to whether or not $F_2([f]) > (F_1)_0([f])$.

It follows readily that on the homogeneous side of the partition, the property stated in $\mathcal{P}'$ holds. We let $C_2 \subseteq C'$ be a c.u.b. subset of $\underset{\sim}{\delta}^1_{2n+1}$ homogeneous for $\mathcal{P}'$, where C_2 is closed under j_v and $(F_1)_0$, $(F_2)_0$. We let C_2' be the set of closure points of C_2. We define F_1', F_2' as follows:

1) For α represented by $f : \vartheta_v \to C'_2$ of the correct type, we set $F'_1(\alpha) = F_1(\alpha)$ and $F'_2(\alpha) = F_2(\alpha)$.
2) For α not as in 1, if α is represented by $f : \vartheta_v \to \underset{\sim}{\delta}^1_{2n+1}$, and $\vartheta = \sup_{\text{a.e.}} f \in C'_2$, and $f < \vartheta$ a.e. w.r.t. v, we set $F'_1(\alpha) = F_1(\alpha_2)$, $F'_2(\alpha) = F_2(\alpha_2)$, where α_2 is represented by $f_2 \colon \vartheta_v \to \underset{\sim}{\delta}^1_{2n+1}$ defined as follows. We let α' be the least ordinal greater than α which is represented by some $h \colon \vartheta_v \to \vartheta$ which is monotonically increasing (it follows readily from Lemma 2.6 and induction that such α' exists). We fix a monotonic $h \colon \vartheta_v \to \vartheta$ representing α'. For $\beta < \vartheta_v$, we then set $f_2(\beta) =$ the $2\omega \cdot ((\beta - 1) + f(\beta) + 1)$st element of C after $h(\beta)$, where $\beta - 1 = \beta$ if β is a limit ordinal. It follows that α_2 is well-defined.
3) For $\alpha = [f]$ not as in 1, $\vartheta \in C'_2$, and $f = \vartheta$ a.e. w.r.t. v, we set $F'_1(\alpha) = N_C(\sup_{\alpha'} F_1(\alpha'))$, where the sup ranges over α' represented by f' with $\vartheta' = \sup_{\text{a.e.}} f' \le \vartheta$, and $f' < \vartheta'$ almost everywhere. Here $N_C(\beta) =$ the next element in $C > \beta$. We also set $F'_2(\alpha) = N_C(\sup_{\alpha'} F_2(\alpha'))$, with α' ranging over the same set.
4) For $\alpha = [f]$ not as in 1, and $\vartheta = \sup_{\text{a.e.}} f \notin C'_2$, we proceed inductively. We set $F'_1(\alpha) = N_C(\max(\sup_{\alpha' < \alpha} F'_1(\alpha'), \sup_\beta F'_2(\beta)))$, where β ranges over the ordinals represented by $g \colon \vartheta_v \to \underset{\sim}{\delta}^1_{2n+1}$ with $\sup_{\text{a.e.}} g < \vartheta$. Also, $F'_2(\alpha) = N_C(\max(\sup_{\alpha' < \alpha} F'_2(\alpha'), \sup_\beta F'_1(\beta')))$, where now β ranges over the ordinals represented by g with $\sup_{\text{a.e.}} g \le \vartheta$.

It is immediate that $F'_1 = F_1$ and $F'_2 = F_2$ almost everywhere, that F'_1, F'_2 have range a subset of C, and are non-normal of uniform cofinality ω. The fact that F'_1, F'_0 are strictly increasing and ordered as in $\mathcal{P}$ follows from the definitions of C', C'_2, F'_1, and F'_2 upon consideration of several cases. The claim now follows. This establishes Lemma 2.6 in this case. The other cases are similar. ⊣ (Lemma 2.6)

Lemma 2.7. If $\mathcal{V}_1, \mathcal{V}_2 \in \bigcup_{m \le n} R_{2m+1}$, and $\mathcal{V}_1 \not\sim \mathcal{V}_2$ then $\mathcal{V}_1 \times \mathcal{V}_2 = \mathcal{V}_2 \times \mathcal{V}_1$. That is, if $A \subseteq \vartheta_{\mathcal{V}_1} \times \vartheta_{\mathcal{V}_2}$, and A has measure one with respect to $\mathcal{V}_1 \times \mathcal{V}_2$ (that is, for almost all $\alpha \in \vartheta_{\mathcal{V}_1}$, for almost all $\beta \in \vartheta_{\mathcal{V}_2}$, $(\alpha, \beta) \in A$), then A has measure one w.r.t. $\mathcal{V}_2 \times \mathcal{V}_1$ (for almost all $\beta \in \vartheta_{\mathcal{V}_2}$, for almost all $\alpha \in \vartheta_{\mathcal{V}_1}$, $(\alpha, \beta) \in A$)

Proof. We again consider the case $\mathcal{V}_1 = S^{\ell_1, m_1}_{2n+1}$, $\mathcal{V}_2 = S^{\ell_2, m_2}_{2n+1}$, with $\ell_1, \ell_2 > 1$, $\ell_1 \ne \ell_2$, the other cases being similar. We fix A having measure one w.r.t. $\mathcal{V}_1 \times \mathcal{V}_2$, and having measure zero w.r.t. $\mathcal{V}_2 \times \mathcal{V}_1$ towards a contradiction. We consider the partition $\mathcal{P}$: we partition pairs of function $F_1, F_2 \colon \underset{\sim}{\delta}^1_{2n+1} \to \underset{\sim}{\delta}^1_{2n+1}$ of the correct type, and ordered as in the partition of Lemma 2.5, according to whether or not $([F_1], [F_2]) \in A$.

We similarly consider the partition $\mathcal{P}'$: we partition $F_2, F_1 \colon \underset{\sim}{\delta}^1_{2n+1} \to \underset{\sim}{\delta}^1_{2n+1}$ as above, ordered as in the partition of Lemma 2.5 (with roles of F_1, F_2 reversed), according to whether or not $([F_1], [F_2]) \in \neg A$.

It then follows from a sliding argument similar to that of Lemma 2.5 that on the homogeneous side of these partitions, the properties stated in $\mathcal{P}$, $\mathcal{P}'$ hold.

We fix a c.u.b. $C \subseteq \underset{\sim}{\delta}^1_{2n+1}$ homogeneous for $\mathcal{P}, \mathcal{P}'$, and fix $F_1, F_2 : \underset{\sim}{\delta}^1_{2n+1} \to C$ of the correct type, and ordered as in $\mathcal{P}$. Hence $([F_1], [F_2]) \in A$. We then let C' be the set of closure points of a c.u.b. subset of C consisting of limit ordinals, closed under j_{v_1}, j_{v_2} (where $v_1 = v(\mathcal{V}_1)$, $v_2 = v(\mathcal{V}_2)$), and closed under $(F_1)_0, (F_2)_0$ (as in Lemma 2.6). It then follows from a sliding argument as in Lemma 2.5 (using the fact that cf $\vartheta_{v_1} \neq$ cf ϑ_{v_1}) that there are F_2', F_1' such that

1) $F_2' = F_2$, $F_1' = F_1$ almost everywhere.
2) F_2', F_1' have range a subset of C and are of the correct type.
3) F_2', F_1' are ordered as in $\mathcal{P}'$.

Hence it follows that $([F_1'], [F_2']) = ([F_1], [F_2]) \notin A$, a contradiction, which establishes Lemma 2.7. ⊣ (Lemma 2.7)

A simple but frequently used lemma is the following:

Lemma 2.8. For any c.u.b. subset $C \subseteq \underset{\sim}{\delta}^1_{2n+1}$, there is a c.u.b. $C' \subseteq C$ such that for any measure v on Ξ_{2n+1} = the predecessor of $\underset{\sim}{\delta}^1_{2n+1}$, and any $f : \vartheta_v \to C'$, if f is of the correct type almost everywhere with respect to v (*i.e.*, there is a measure one set A w.r.t. v such that $f \restriction A$ is of the correct type), then there is an $f_2 : \vartheta_v \to C$ such that $f_2 = f_1$ almost everywhere w.r.t. v, and such that f_2 is of the correct type everywhere.

Proof. Given C, take C' to be the set of the closure points of a c.u.b. subset of C consisting of limit ordinals. Then for $f : \vartheta_v \to C'$, and $A \subseteq \vartheta_v$ of measure one such that $f \restriction A$ is of the correct type, define f_2 by:

1) for $\alpha \in A$, set $f_2(\alpha) = f(\alpha)$.
2) for $\alpha \notin A$, set $f_2(\alpha) = N_C(\sup_{\alpha' < \alpha} f_2(\alpha'))$.

It follows readily that f_2 has the required properties. ⊣

§3. A Global Embedding Theorem. We let τ denote an $n - *$ tuple with respect to the measures $\bigcup_{m \leq n} \mathcal{S}_{2m+1} \cup \bigcup_{m \leq n} \mathcal{W}_{2m+1}$ and let $<^\tau$ denote the corresponding ordering on tuples of integers and ordinals $< \Xi_{2n+3}$. Since we are assuming I_{2n+1}, $\mu^{<^\tau}_{2n+3}$ is defined.

We state the theorem of the section:

Theorem 3.1. For each $n - *$ tuple τ and corresponding measure $\mu^{<^\tau}_{2n+3}$, there is an integer m such that the ultrapower of $\underset{\sim}{\delta}^1_{2n+3}$ by $\mu^{<^\tau}_{2n+3}$ is less than or equal to the ultrapower of $\underset{\sim}{\delta}^1_{2n+3}$ by W^m_{2n+3}.

In order to prove this, we require an inductive hypothesis:

$\mathcal{B}_{2n+1}$: We let T be an $n - *$ tuple with respect to the measures $\bigcup_{m \leq n-1} \mathcal{S}_{2m+1} \cup \bigcup_{m \leq n-1} \mathcal{W}_{2m+1}$ and let $<^T$ be the corresponding ordering on tuples from Ξ_{2n+1}. Then there is an integer m and a measure $S^{\ell',m}_{2n-1}$, where $\ell' = 2^n - 1$ is maximal, such that $K^{\ell',m}_{2n-1} = \vartheta_{S^{\ell',m}_{2n-1}}$ is greater than the order type of $<^T$, and there is a measure u on Ξ_{2n+1} satisfying the following: there is a family of maps f_α, for $\alpha \in \vartheta_u$, $f_\alpha \colon <^T \to K^{\ell',m}_{2n-1}$ defined almost everywhere with respect to the measures $\upsilon^{(i_1,\dots,i_k)}$ (so actually, f_α is an equivalence class w.r.t. $\mu^{<^T}$), and satisfying

1) f_α is order-preserving almost everywhere w.r.t. the $\upsilon^{(i_1,\dots,i_k)}$.
2) If $A \subseteq K^{\ell',m}_{2n-1}$ has measure one, then for almost all α w.r.t. u, and all indices $(i_1,\dots,i_k) \in \operatorname{dom} T$, for almost all $\beta < \vartheta^{(i_1,\dots,i_{k-1})}$ w.r.t. $\upsilon^{(i_1,\dots,i_{k-1})}$ we have that $f^{(i_1,\dots,i_k)}_\alpha(\beta) = f_\alpha(\beta^{(i_1,\dots,i_k)}) \in A$. This is well defined.
3) For almost all α, $\sup_{\text{a.e.}} f_\alpha < K^{\ell',m}_{2n-1}$. Here, of course $\sup_{\text{a.e.}} f_\alpha$ denotes $\max_{(i_1,\dots,i_k)}(\sup_{\text{a.e. w.r.t. } \upsilon^{(i_1,\dots,i_{k-1})}} f^{(i_1,\dots,i_k)}_\alpha)$.
4) If $S^{\ell',m}_{2n-1}$ satisfies the above, then so does $S^{\ell',m'}_{2n-1}$ for any $m' > m$.

We have defined $\mathcal{B}_{2n+3}$ as well.

We first show that the theorem follows from $\mathcal{B}_{2n+3}$. We produce an embedding M from the ultrapower of $\underset{\sim}{\delta}^1_{2n+3}$ by $\mu^{<^\tau}_{2n+3}$ into the ultrapower by W^m_{2n+3}. We let $(I_1,\dots,I_k,\dots)$ denote the indices in the domain of τ. We let F_1 from $\operatorname{dom}(\mu^{<^\tau}_{2n+3})$ to $\underset{\sim}{\delta}^1_{2n+3}$ be given, representing $[F_1]$ with respect to the measure $\mu^{<^\tau}_{2n+3}$. Recall the elements of $\operatorname{dom}(\mu^{<^\tau}_{2n+3})$ are tuples of the form $(\dots,\alpha^{(I_1,\dots,I_k)},\dots)$ where $\alpha^{(I_1,\dots,I_k)} < \underset{\sim}{\delta}^1_{2n+3}$. We define $M([F_1])$ to be the ordinal represented with respect to W^m_{2n+3} by F_2 defined as follows: given $f \colon K^{\ell,m}_{2n+1} \to \underset{\sim}{\delta}^1_{2n+3}$ (where $\ell = 2^{n+1} - 1$ is maximal), $F_2([f])$ is represented w.r.t. the measure $\mathcal{U}$ (the measure given by $\mathcal{B}_{2n+3}$, this is the analog of u as in $\mathcal{B}_{2n+1}$) by the function which assigns to $\gamma \in \vartheta_{\mathcal{U}}$ the ordinal $F_1(\dots,\beta^{(I_1,\dots,I_k)}_\gamma,\dots)$, where $\beta^{(I_1,\dots,I_k)}_\gamma$ is represented w.r.t. $V^{(I_1,\dots,I_k)}$ (the measures from the $n - *$ tuple τ) by the function $(f \circ \mathcal{F}_\gamma)^{(I_1,\dots,I_k)}$, where $\mathcal{F}_\gamma \colon <^\tau \to K^{\ell,m}_{2n+1}$ is the map corresponding to γ as in $\mathcal{B}_{2n+3}$ and the superscript denotes the subfunction induced by restriction. It follows readily from $\mathcal{B}_{2n+3}$ that the embedding M is well-defined.

We assume $\mathcal{B}_{2n+1}$ throughout the remainder of this section and proceed to establish $\mathcal{B}_{2n+3}$. We fix an $n - *$ tuple τ, and corresponding ordering $<^\tau$ on Ξ_{2n+3}. We let $(I_1,\dots,I_k)$, $V^{(I_1,\dots,I_k)}$ denote indices and measures corresponding to τ, for $k \le n$, $k \le n-1$ respectively. We have that $V^{(I_1,\dots,I_k)} = \prod_r {}^rV^{(I_1,\dots,I_k)}$, where ${}^rV^{(I_1,\dots,I_k)}$ is of the form $\mathcal{S}^{{}^rT(I_1,\dots,I_k)}_{2n+1{}^r\mathcal{C}(I_1,\dots,I_k)}$, for some ${}^rT(I_1,\dots,I_k), {}^r\mathcal{C}(I_1,\dots,I_k)$, or of the form $\mu^{<^{{}^rT(I_1,\dots,I_k)}}_{2n+1}$, or is a measure ${}^r\upsilon^{(I_1,\dots,I_k)} \in \bigcup_{m\le n} R_{2m-1}$. Here each ${}^r\mathcal{C}(I_1,\dots,I_k)$ is a collection defined relative to the $k - *$ tuple ${}^rT(I_1,\dots,I_k)$ (which is $k - *$ tuple relative to the measures in $\bigcup_{m\le n-1} \mathcal{S}_{2m+1} \cup \bigcup_{m\le n-1} \mathcal{W}_{2m+1}$).

From $\mathcal{B}_{2n+1}$, there is an m and measures ${}^r u^{(I_1,\dots,I_k)}$, corresponding to each ${}^r T(I_1,\dots,I_k)$ in the first case above, such that the statement of $\mathcal{B}_{2n+1}$ is satisfied by $<^{{}^r T(I_1,\dots,I_k)}$, $S^{\ell',m}_{2n-1}$ (here $\ell' = 2^n - 1$), and ${}^r u^{(I_1,\dots,I_k)}$. We fix such an m.

We consider $S^{\ell,m}_{2n+1}$, where $\ell = 2^{n+1} - 1$ is maximal, and claim that $\mathcal{B}_{2n+3}$ is satisfied by $<^\tau$ and $S^{\ell,m}_{2n+1}$. We may also take any $m' > m$ in the following.

We define the measure $\mathcal{U}$ as in the statement $\mathcal{B}_{2n+3}$.

We let Ω be the set of tuples of ordinals and integers of the form:

$$\langle \alpha_0, I_1, \dots, {}^r\alpha_1^{(\mathcal{I}_1,\dots,\mathcal{I}_k)}, \dots, I_2, \dots, {}^r\alpha_2^{(\mathcal{I}_1,\dots,\mathcal{I}_k)}, \dots, \\ I_{n-1}, \dots, {}^r\alpha_n^{(\mathcal{I}_1,\dots,\mathcal{I}_k)}, \dots, I_n \rangle,$$

or an initial segment of such a sequence, where:

1) $\alpha_0 < \underset{\sim}{\delta}^1_{2n+1}$
2) The indices $(I_1,\dots,I_k)$ correspond to the indices in the domain of τ.
3) The indices $(\mathcal{I}_1,\dots,\mathcal{I}_k)$ associated with ${}^r\alpha_j$ correspond to the indices $(\mathcal{I}_1,\dots,\mathcal{I}_k)$ in ${}^r\mathcal{C}(I_1,\dots,I_j)$ if ${}^r V^{(I_1,\dots,I_j)} \in \mathcal{S}_{2n+1}$, and to the indices $(i_1,\dots,i_k) \in \operatorname{dom} {}^r T(I_1,\dots,I_j)$ if ${}^r V^{(I_1,\dots,I_j)} \in \mathcal{W}_{2n+1}$. If ${}^r V^{(I_1,\dots,I_j)} \in \bigcup_{m<n} \mathcal{R}_{2m+1}$, we omit the subscripts.
4) a) For fixed r, j, the indices $(\mathcal{I}_1,\dots,\mathcal{I}_k)$ associated with ${}^r\alpha_j$ are ordered as in the fixed (but arbitrary) ordering on them which use to identify the domain of the measure space $\mathcal{S}^{{}^r T(I_1,\dots,I_j)}_{2n+1 {}^r\mathcal{C}(I_1,\dots,I_j)}$ with an ordinal.
   b) For $r \neq r'$ and fixed j, the ordinal ${}^r\alpha_j^{(\mathcal{I}_1,\dots,\mathcal{I}_k)}$ precedes the ordinal ${}^{r'}\alpha_j^{(\mathcal{I}'_1,\dots,\mathcal{I}'_{k'})}$ if in identifying $V^{(I_1,\dots,I_j)} = \prod_r {}^r V^{(I_1,\dots,I_j)}$ with an ordinal, we order by the rth component before the r' component.
5) ${}^r\alpha_j^{(\mathcal{I}_1,\dots,\mathcal{I}_k)} < j_{{}^r u^{(I_1,\dots,I_j)}}(\alpha_0)$ for ${}^r V^{(I_1,\dots,I_j)} \in \mathcal{S}_{2n+1}$, and otherwise ${}^r\alpha_j^{(\mathcal{I}_1,\dots,\mathcal{I}_k)} < \alpha_0$.

We let $<^\Omega$ denote the Brouwer-Kleene ordering on these tuples.

For indices $I_1,\dots,I_k \in \operatorname{dom}\tau$, with $k > 1$, we define a set $M(I_1,\dots,I_k)$ of indexed ordinals and an ordering $<^{M(I_1,\dots,I_k)}$ on them as follows:

$M(I_1,\dots,I_k)$ consists of indexed ordinals of the form ${}^r\vartheta^{(\mathcal{I}_1,\dots,\mathcal{I}_m)}_{(I_1,\dots,I_\ell)}$ where $(I_1,\dots,I_\ell)$ is an initial segment of $(I_1,\dots,I_k)$, $(\mathcal{I}_1,\dots,\mathcal{I}_m)$ is an index in ${}^r\mathcal{C}(I_1,\dots,I_\ell)$ and $\vartheta \in {}^r u^{(I_1,\dots,I_\ell)}$, for ${}^r V^{(I_1,\dots,I_k)} \in \mathcal{S}_{2n+1}$; and of the form ${}^r\vartheta^{(i_1,\dots,i_k)}_{(I_1,\dots,I_\ell)}$, where $(I_1,\dots,I_\ell)$ is an initial segment of $(I_1,\dots,I_k)$, $(i_1,\dots,i_k) \in \operatorname{dom} {}^r T(I_1,\dots,I_\ell)$ and ϑ is in the domain of the measure $v^{(i_1,\dots,i_{k-1})}$ defined by the lower-level $n' - *$ tuple (for some n') ${}^r T(I_1,\dots,I_\ell)$ when ${}^r V^{(I_1,\dots,I_k)} \in \mathcal{W}_{2n+1}$. We also allow in the set $M(I_1,\dots,I_k)$ un-indexed ordinals $\vartheta < \vartheta(K^{\ell',m}_{2n-1})$.

We define ${}^{r_1}\vartheta_{1(I_1,\dots,I_{\ell_1})}^{(\mathcal{I}_1^1,\dots,\mathcal{I}_{m_1}^1)} <^{M(I_1,\dots,I_k)} {}^{r_2}\vartheta_{2(I_1,\dots,I_{\ell_2})}^{(\mathcal{I}_1^2,\dots,\mathcal{I}_{m_2}^2)}$ to hold provided one of the following cases is satisfied:

1) $r_1 \neq r_2$ and ${}^{r_1}V^{(I_1,\dots,I_{\ell_1})}, {}^{r_2}V^{(I_1,\dots,I_{\ell_2})} \in \mathcal{S}_{2n+1}$. In this case, we require that there is a c.u.b. $C \subseteq \underset{\sim}{\delta}_{2n+1}^1$ such that for $\mathcal{F}_1, \mathcal{F}_2 \colon <_{{}^rC(I_1,\dots,I_\ell)}^{{}^rT(I_1,\dots,I_\ell)} \to C$ of the correct type (with $r = r_1$ or r_2 and $\ell = \ell_1$ or ℓ_2 respectively), representing a pair $([{}^{r_1}F],[{}^{r_2}F])$ in the product measure ${}^{r_1}V^{(I_1,\dots,I_{\ell_1})} \times {}^{r_2}V^{(I_1,\dots,I_{\ell_2})}$ (for $r_1 < r_2$, and otherwise representing $([{}^{r_2}F],[{}^{r_1}F])$ w.r.t. ${}^{r_2}V \times {}^{r_1}V$), we have that for almost all $f \colon K_{2n-1}^{\ell',m} \to \underset{\sim}{\delta}_{2n+1}^1$ of the correct type w.r.t. $\mu_{2n+1}^{\ell',m}$ that if $f_1 \colon <^{{}^{r_1}T(I_1,\dots,I_{\ell_1})} \to K_{2n-1}^{\ell',m}$, $f_2 \colon <^{{}^{r_2}T(I_1,\dots,I_{\ell_2})} \to K_{2n-1}^{\ell',m}$, denote the embeddings (determined a.e.) from $\mathcal{B}_{2n+1}$ corresponding to ${}^{r_1}\vartheta^{(I_1,\dots,I_{\ell_1})}$, ${}^{r_2}\vartheta^{(I_1,\dots,I_{\ell_2})}$, and if $g_1 = f \circ f_1$, $g_2 = f \circ f_2$, then ${}^{r_1}F(\langle \mathcal{S}_1\rangle) < {}^{r_2}F(\langle \mathcal{S}_2\rangle)$, where $\mathcal{S}_1$ is the tuple corresponding to $g_1 \colon <^{{}^{r_1}T(I_1,\dots,I_{\ell_1})} \to \underset{\sim}{\delta}_{2n+1}^1$ and $(\mathcal{I}_1^1,\dots,\mathcal{I}_{m_1}^1)$ as in the definition of ${}^{r_1}V$, and similarly for $\mathcal{S}_2$.
2) $r_1 = r_2 = r$ and ${}^rV^{(I_1,\dots,I_k)} \in \mathcal{S}_{2n+1}$ (hence so are ${}^rV^{(I_1,\dots,I_{\ell_1})}$ and ${}^rV^{(I_1,\dots,I_{\ell_2})}$). We then require that there is a c.u.b. $C \subseteq \underset{\sim}{\delta}_{2n+1}^1$ such that for $F \colon <_{{}^rC(I_1,\dots,I_k)}^{{}^rT(I_1,\dots,I_k)} \to C$ of the correct type, with induced functions, $F_1 \colon <_{{}^rC(I_1,\dots,I_{\ell_1})}^{{}^rT(I_1,\dots,I_{\ell_1})} \to C$ and $F_2 \colon <_{{}^rC(I_1,\dots,I_{\ell_2})}^{{}^rT(I_1,\dots,I_{\ell_2})} \to C$, that for almost all $f \colon K_{2n-1}^{\ell',m} \to \underset{\sim}{\delta}_{2n+1}^1$, and f_1, f_2, g_1, g_2 as above $F_1(\langle \mathcal{S}_1\rangle) < F_2(\langle \mathcal{S}_2\rangle)$, where $\mathcal{S}_1$ is the sequence corresponding to g_1 and $(\mathcal{I}_1^1,\dots,\mathcal{I}_{m_1}^1)$, and similarly for $\mathcal{S}_2$.
3) ${}^{r_1}V \in \mathcal{W}_{2n+1}$ and ${}^{r_2}V \in \mathcal{S}_{2n+1}$.
4) $r_1 = r_2 = r$ and ${}^rV^{(I_1,\dots,I_{\ell_1})}, {}^rV^{(I_1,\dots,I_{\ell_2})} \in \mathcal{W}_{2n+1}$. We then require that for almost all $f \colon <^{{}^rT(I_1,\dots,I_k)} \to \underset{\sim}{\delta}_{2n+1}^1$ w.r.t. $\mu_{2n+1}^{<^{{}^rT(I_1,\dots,I_k)}}$ with induced functions $f_1 \colon <^{{}^rT(I_1,\dots,I_{\ell_1})} \to \underset{\sim}{\delta}_{2n+1}^1$, $f_2 \colon <^{{}^rT(I_1,\dots,I_{\ell_2})} \to \underset{\sim}{\delta}_{2n+1}^1$, and subfunctions $f_1^{(i_1^1,\dots,i_{k_1}^1)} \colon \vartheta_{v^{(i_1,\dots,i_{k_1-1})}} \to \underset{\sim}{\delta}_{2n+1}^1$, and $f_2^{(i_1^2,\dots,i_{k_2}^2)}$, that $f_1^{(i_1^1,\dots,i_{k_1}^1)}(\vartheta_1) < f_2^{(i_1^2,\dots,i_{k_2}^2)}(\vartheta_2)$.
5) $r_1 \neq r_2$ and ${}^{r_1}V, {}^{r_2}V \in \mathcal{W}_{2n+1}$. In this case, we require $r_1 < r_2$.
6) ${}^{r_1}V \in \mathcal{W}_{2n+1}$ and $\vartheta_2 \in K_{2n-1}^{\ell',m}$.
7) $\vartheta_1, \vartheta_2 \in K_{2n-1}^{\ell',m}$ and $\vartheta_1 < \vartheta_2$.
8) ${}^{r_1}V \in \mathcal{S}_{2n+1}$ and $\vartheta_2 \in K_{2n-1}^{\ell',m}$. We require that there is a c.u.b. $C \subseteq \underset{\sim}{\delta}_{2n+1}^1$ such that for $F \colon <_{{}^rC(I_1,\dots,I_{\ell_1})}^{{}^rT(I_1,\dots,I_{\ell_1})} \to C$ of the correct type we have that for almost all $f \colon K_{2n-1}^{\ell',m} \to \underset{\sim}{\delta}_{2n+1}^1$ of the correct type, with f_1, g_1 as in case (1), that $F(\langle \mathcal{S}\rangle) < f(\vartheta_2)$, where $\mathcal{S}$ corresponds to g and $(\mathcal{I}_1^1,\dots,\mathcal{I}_{m_1}^1)$ as in the definition of ${}^{r_1}V$.
9) $\vartheta_1 \in K_{2n-1}^{\ell',m}$ and ${}^{r_2}V \in \mathcal{S}_{2n+1}$. Similar to (8) where we now require that $f(\vartheta_1) < F(\langle \mathcal{S}\rangle)$.

We identify $\vartheta_1^{(\dots)}, \vartheta_2^{(\dots)}$ in $<^{M(I_1,\dots,I_k)}$ if neither $\vartheta_1 <^M \vartheta_2$ or $\vartheta_2 <^M \vartheta_1$ holds.

The measure $\mathcal{U}$ is a measure on tuples $(\dots, \eta_{(I_1,\dots,I_k)}, \dots)$, indexed by indices $(I_1, \dots, I_k)$ in τ, which is induced by the strong partition relation on $\underset{\sim}{\boldsymbol{\delta}}^1_{2n+1}$ functions $H\colon <^{\Omega} \to \underset{\sim}{\boldsymbol{\delta}}^1_{2n+1}$ of the correct type, and the measures $\prod_r {}^r u \times \mu_{2n+1}^{<M(I_1,\dots,I_k)}$ for $k > 1$, and $\mu_{2n+1}^{K_{2n-1}^{\ell',m}}$ for $k = 1$. Here $\Pi_r {}^r u$ denotes the subproduct of V consisting of those measures in $\bigcup_{m<n} \mathcal{R}_{2m+1}$.

That is, A has measure one if there is a c.u.b. subset C of $\underset{\sim}{\boldsymbol{\delta}}^1_{2n+1}$ such that for $H\colon <^{\Omega} \to C$ of the correct type, $(\dots, \eta_{(I_1,\dots,I_k)}, \dots) \in A$, where for fixed $(I_1, \dots, I_k)$, $\eta_{(I_1,\dots,I_k)}$ is represented w.r.t. $\prod_r {}^r u \times \mu_{2n+1}^{<M(I_1,\dots,I_k)}$ by the function H_2 which assigns to the pair of tuples

$$(\vartheta_{r_1}, \vartheta_{r_2}, \dots, \vartheta_{r_p}),\ (\gamma_0, \dots, {}^r\gamma_{(I_1,\dots,I_\ell)}^{(\mathcal{I}_1,\dots,\mathcal{I}_m)}, \dots)$$

(where an $h\colon <^{M(I_1,\dots,I_k)} \to \underset{\sim}{\boldsymbol{\delta}}^1_{2n+1}$ represents the γ's w.r.t. the measures ${}^r u^{(I_1,\dots,I_\ell)}$ if ${}^r V^{(I_1,\dots,I_\ell)} \in \mathcal{S}_{2n+1}$, w.r.t. $v^{(i_1,\dots,i_k)}$ if ${}^r V \in \mathcal{W}_{2n+1}$, and w.r.t. $S_{2n-1}^{\ell',m}$ for γ_0) the ordinal $H(\langle T \rangle)$, where T denotes the sequence

$$T = \langle \alpha_0, I_1, \dots, {}^r\alpha_1^{(\mathcal{I}_1,\dots,\mathcal{I}_m)}, \dots, I_j, \dots, {}^r\alpha_j^{(\mathcal{I}_1,\dots,\mathcal{I}_m)}, \dots \rangle,$$

as in the definition of Ω, where:

1) $\alpha_0 = \gamma_0$
2) ${}^r\alpha_j^{(\mathcal{I}_1,\dots,\mathcal{I}_m)} = {}^r\gamma_{(I_1,\dots,I_j)}^{(\mathcal{I}_1,\dots,\mathcal{I}_m)}$ if ${}^r V \in \mathcal{S}_{2n+1}$
3) ${}^r\alpha_j^{(i_1,\dots,i_k)} = {}^r\gamma_{(I_1,\dots,I_j)}^{(i_1,\dots,i_k)}$ if ${}^r V \in \mathcal{W}_{2n+1}$
4) ${}^r\alpha_j = \Pi_{(I_1,\dots,I_j)}^{(I_1,\dots,I_k)}(\vartheta_r)$ if ${}^r V^{(I_1,\dots,I_k)} \in \bigcup_{m<n} \mathcal{R}_{2m+1}$

This completes the definition of the measure $\mathcal{U}$.

We proceed to define the family of embeddings $\mathcal{F}_\gamma$ corresponding to $\mathcal{U}$ as in the statement of $\mathcal{B}_{2n+3}$.

We fix $\gamma \in \vartheta_{\mathcal{U}}$ (the domain of the measure $\mathcal{U}$) which is represented w.r.t. the measures $\prod_r {}^r u \times \mu_{2n+1}^{M(I_1,\dots,I_k)}$ by a function $H\colon <^{\Omega} \to \underset{\sim}{\boldsymbol{\delta}}^1_{2n+1}$ of the correct type (which happens for almost all γ w.r.t. $\mathcal{U}$). We proceed to define $\mathcal{F}_\gamma\colon <^{\tau} \to K_{2n+1}^{\ell,m}$. We fix an element $\alpha^{(I_1,\dots,I_k)}$ in the domain of $<^{\tau}$, where $(I_1, \dots, I_k) \in \operatorname{dom} \tau$, and define $\mathcal{F}_\gamma(\alpha^{(I_1,\dots,I_k)})$ (for almost all α w.r.t. $V^{(I_1,\dots,I_{k-1})}$ for $k > 1$). We see that

$$\alpha = (\dots, {}^r\xi, \dots, {}^r\xi_{(i_1,\dots,i_\ell)}, \dots, {}^r\xi_{(\mathcal{I}_1,\dots,\mathcal{I}_m)}, \dots),$$

where the components ${}^r\xi$ correspond to ${}^r V \in \bigcup_{m<n} \mathcal{R}_{2n+1}$, ${}^r\xi_{(i_1,\dots,i_k)}$ to ${}^r V \in \mathcal{W}_{2n+1}$, and ${}^r\xi_{(I_1,\dots,I_m)}$ to ${}^r V \in \mathcal{S}_{2n+1}$. In the latter two cases, we have functions ${}^r F\colon <^{{}^r T(I_1,\dots,I_k)} \to \underset{\sim}{\boldsymbol{\delta}}^1_{2n+1}$ and ${}^r F\colon <^{{}^r T(I_1,\dots,I_\ell)}_{{}^r C(I_1,\dots,I_k)} \to \underset{\sim}{\boldsymbol{\delta}}^1_{2n+1}$ representing, ${}^r\xi_{(i_1,\dots,i_k)}, {}^r\xi_{(\mathcal{I}_1,\dots,\mathcal{I}_m)}$ w.r.t. the measures $v^{(i_1,\dots,i_k)}$ and the measures $\mu_{2n+1}^{<^{r}T(I_1,\dots,I_k)}$ respectively. We define a function $G\colon \underset{\sim}{\boldsymbol{\delta}}^1_{2n+1} \to \underset{\sim}{\boldsymbol{\delta}}^1_{2n+1}$ representing $\mathcal{F}_\gamma(\alpha^{(I_1,\dots,I_k)})$ w.r.t. $\mu_{2n+1}^{S_{2n-1}^{\ell',m}}$. We let $g\colon K_{2n-1}^{\ell',m} \to \underset{\sim}{\boldsymbol{\delta}}^1_{2n+1}$ of the correct type be given, and define $G([g])$. We set $G([g]) = H(\langle T\rangle)$, where $T = \langle \beta_0, I_1, \dots, {}^r\beta_1^{(\mathcal{I}_1,\dots,\mathcal{I}_m)}, \dots, I_j, \dots, {}^r\beta_j^{(\mathcal{I}_1,\dots,\mathcal{I}_m)}, \dots, I_k\rangle$ is as in the definition of Ω, and the β's are given by:

1) $\beta_0 = [g]_{S^{\ell',m}_{2n-1}}$.

2) If ${}^rV^{(I_1,\dots,I_j)} \in \bigcup_{m<n} \mathcal{R}_{2m+1}$, then ${}^r\beta_j = \Pi^{(I_1,\dots,I_k)}_{(I_1,\dots,I_j)}({}^r\xi)$.

3) If ${}^rV^{(I_1,\dots,I_j)} \in \mathcal{W}_{2n+1}$, then ${}^r\beta_j^{(i_1,\dots,i_k)} = {}^r\xi^{(i_1,\dots,i_k)}$.

4) If ${}^rV^{(I_1,\dots,I_j)} \in \mathcal{S}_{2n+1}$, we represent ${}^r\beta_j^{(\mathcal{I}_1,\dots,\mathcal{I}_m)}$ (where we recall $(\mathcal{I}_1,\dots,\mathcal{I}_m)$ is an index in ${}^r\mathcal{C}(I_1,\dots,I_j)$), w.r.t. ${}^ru^{(I_1,\dots,I_j)}$ by a function ${}^r\mathcal{M}_j^{(\mathcal{I}_1,\dots,\mathcal{I}_m)}$ defined by: for ϑ in the measure space ${}^ru^{(I_1,\dots,I_j)}$, and $f_\vartheta\colon <^{{}^rT(I_1,\dots,I_j)} \to K^{\ell',m}_{2n-1}$ the corresponding embedding from $\mathcal{B}_{2n+1}$ (defined almost everywhere), and $\mathcal{Q} = g \circ f_\vartheta\colon <^{{}^rT(I_1,\dots,I_j)} \to \underset{\sim}{\boldsymbol{\delta}}^1_{2n+1}$, we set ${}^r\mathcal{M}_j^{(\mathcal{I}_1,\dots,\mathcal{I}_m)}(\vartheta) = {}^rF(\langle \mathcal{S}(\mathcal{Q},(I_1,\dots,I_j),(\mathcal{I}_1,\dots,\mathcal{I}_m))\rangle)$, where $\mathcal{S}(\cdots)$ denotes the corresponding sequence as in the definition of ${}^rV$. This is well-defined for almost all g. We use here the fact that the $\vartheta \in K^{\ell',m}_{2n-1}$ are cofinal in the ordering $<^{M(I_1,\dots,I_k)}$.

This completes the definition of $\mathcal{F}_\gamma$. To show that this is well-defined, we verify the following:

1) For fixed H, $({}^{r_1}\xi,\dots,{}^{r_p}\xi)$, ${}^rF$, and g we have that ${}^r\mathcal{M}_j^{(I_1,\dots,I_m)}(\vartheta)$ is well-defined. This follows since f_ϑ is determined almost everywhere w.r.t. the $\upsilon^{(i_1,\dots,i_k)}$, hence so is $\mathcal{Q}$ above, hence the sequence $\mathcal{S}$ above is well-defined.

2) For fixed H, $({}^{r_1}\xi,\dots,{}^{r_p}\xi)$, ${}^rF$, we have that the equivalence class of $\mathcal{F}_\gamma(\alpha^{(I_1,\dots,I_k)})$ is well-defined with respect to $\mu^{S^{\ell',m}_{2n-1}}_{2n+1}$. If $g_1, g_2\colon K^{\ell',m}_{2n-1} \to \underset{\sim}{\boldsymbol{\delta}}^1_{2n+1}$ agree on a measure one set A w.r.t. $S^{\ell',m}_{2n-1}$ then for any $(I_1,\dots,I_j)$ as above, it follows from $\mathcal{B}_{2n+1}$ that for almost all ϑ w.r.t. ${}^ru^{(I_1,\dots,I_j)}$ that $f_\vartheta\colon <^{{}^rT(I_1,\dots,I_j)} \to K^{\ell',m}_{2n-1}$ has range a.e. in A w.r.t. the measures $\upsilon^{(i_1,\dots,i_{k-1})}$, where $(i_1,\dots,i_k) \in \operatorname{dom} {}^rT(I_1,\dots,I_j)$. We are assuming ${}^rV^{(I_1,\dots,I_k)} \in \mathcal{S}_{2n+1}$, the other cases being immediate. Hence, for $\mathcal{Q}_1 = g_1 \circ f_\vartheta$, $\mathcal{Q}_2 = g_2 \circ f_\vartheta$, for each $(\mathcal{I}_1,\dots,\mathcal{I}_m)$ in ${}^r\mathcal{C}(I_1,\dots,I_j)$, it follows that

$$\mathcal{S}(\mathcal{Q}_1,(I_1,\dots,I_j),(\mathcal{I}_1,\dots,\mathcal{I}_m)) = \mathcal{S}(\mathcal{Q}_2,(I_1,\dots,I_j),(\mathcal{I}_1,\dots,\mathcal{I}_m)),$$

and hence ${}^r\mathcal{M}_j^{(\mathcal{I}_1,\dots,\mathcal{I}_m)}(1)(\vartheta) = {}^r\mathcal{M}_j^{(\mathcal{I}_1,\dots,\mathcal{I}_m)}(2)(\vartheta)$ for almost all ϑ w.r.t. ${}^ru^{(I_1,\dots,I_j)}$, and all $(\mathcal{I}_1,\dots,\mathcal{I}_m)$. Hence ${}^r\beta_j^{(\mathcal{I}_1,\dots,\mathcal{I}_m)}(1) = {}^r\beta_j^{(\mathcal{I}_1,\dots,\mathcal{I}_m)}(2)$, and hence the sequences T_1, T_2 corresponding to g_1, g_2 are the same. Hence, $G([g_1]) = G([g_2])$.

3) For fixed $H\colon <^\Omega \to \underset{\sim}{\boldsymbol{\delta}}^1_{2n+1}$, we have that $\mathcal{F}_\gamma$ is well defined w.r.t. the choice of the ${}^rF$ representing $\alpha^{(I_1,\dots,I_k)}$, for ${}^rV^{(I_1,\dots,I_k)} \in \mathcal{S}_{2n+1}$ or $\mathcal{W}_{2n+1}$. For each such r, we let ${}^rF_1, {}^rF_2\colon <^{{}^rT(I_1,\dots,I_k)}_{{}^r\mathcal{C}(I_1,\dots,I_k)} \to \underset{\sim}{\boldsymbol{\delta}}^1_{2n+1}$ or from $<^{{}^rT(I_1,\dots,I_k)} \to \underset{\sim}{\boldsymbol{\delta}}^1_{2n+1}$ be given,

representing ${}^r\alpha^{(I_1,\dots,I_k)}$. We let C be a c.u.b. subset of $\underset{\sim}{\delta}^1_{2n+1}$ such that for r's of the first type, $f\colon <^{{}^rT(I_1,\dots,I_k)} \to C$ of the correct type, $(\mathcal{I}_1,\dots,\mathcal{I}_m)$ an index in ${}^r\mathcal{C}(I_1,\dots,I_k)$, and $\mathcal{S} = \mathcal{S}(f,(I_1,\dots,I_k),(\mathcal{I}_1,\dots,\mathcal{I}_m))$ the corresponding sequence, we have that ${}^rF_1(\langle\mathcal{S}\rangle) = {}^rF_2(\langle\mathcal{S}\rangle)$. For $g\colon K^{\ell',m}_{2n-1} \to C$ of the correct type (where we also assume C is sufficiently closed w.r.t. the ${}^rF$'s so the definition of $\mathcal{F}_\gamma$ makes sense), $j \le k$, $(\mathcal{I}_1,\dots,\mathcal{I}_m)$ an index in ${}^r\mathcal{C}(I_1,\dots,I_j)$ (we consider here the case ${}^rV^{(I_1,\dots,I_j)} \in \mathcal{S}_{2n+1}$) it follows that for almost all ϑ w.r.t. ${}^ru^{(I_1,\dots,I_j)}$ that $Q = g \circ f_\vartheta$ has range almost everywhere in C and is of the correct type almost everywhere, since g is. By Lemma 2.8 we may assume Q has range in C and is of the correct type everywhere. Hence, for almost all ϑ, ${}^r\mathcal{M}_j^{(\mathcal{I}_1,\dots,\mathcal{I}_m)}(1)(\vartheta) = {}^rF_1(\langle\mathcal{S}(Q,(I_1,\dots,I_j),(\mathcal{I}_1,\dots,\mathcal{I}_m))\rangle) = {}^rF_2(\langle\mathcal{S}(Q,(I_1,\dots,I_j),(\mathcal{I}_1,\dots,\mathcal{I}_m))\rangle) = {}^r\mathcal{M}_j^{(\mathcal{I}_1,\dots,\mathcal{I}_m)}(2)(\vartheta)$. Hence, the ${}^r\beta$'s appearing in the sequence T are the same for ${}^rF_1$, ${}^rF_2$ in the case ${}^rV \in \mathcal{S}_{2n+1}$. The other cases are immediate. Hence the value of $G([g])$ is the same when computed using ${}^rF_1$ or ${}^rF_2$.

4) If $H_1, H_2\colon <^\Omega \to \underset{\sim}{\delta}^1_{2n+1}$ of the correct type agree a.e. w.r.t. $\mu_{2n+1}^{<^{M(I_1,\dots,I_k)}}$, then the embeddings $\mathcal{F}_1, \mathcal{F}_2\colon <^\tau \to K^{\ell,m}_{2n+1}$ corresponding to H_1, H_2 agree, for each fixed $(I_1,\dots,I_k) \in \operatorname{dom}\tau$, almost everywhere w.r.t. $V^{(I_1,\dots,I_{k-1})} = \prod_r {}^rV^{(I_1,\dots,I_{k-1})}$, if $k > 1$, and agree if $k = 1$.

We fix a c.u.b. $C \subseteq \underset{\sim}{\delta}^1_{2n+1}$ such that for $(I_1,\dots,I_k) \in \operatorname{dom}\tau$, and the corresponding ordering $<^{M(I_1,\dots,I_k)}$, if $h\colon <^{M(I_1,\dots,I_k)} \to C$ is of the correct type, then for all $({}^{r_1}\xi,\dots,{}^{r_p}\xi) \in A$, a measure one set w.r.t. $\prod_r {}^rv$, we have $H_1(\langle T\rangle) = H_2(\langle T\rangle)$, where $T = \langle \beta_0, I_1,\dots,I_j,\dots,{}^r\beta_j^{(\mathcal{I}_1,\dots,\mathcal{I}_m)},\dots,I_k\rangle$ is the sequence corresponding to h as in the definition of $\mathcal{U}$.

For $k = 1$ we require that for $g\colon K^{\ell',m}_{2n-1} \to C$, of the correct type that $H_1(\langle [g], I_1\rangle) = H_2(\langle [g], I_1\rangle)$.

For each $\vartheta_1^{()}$, $\vartheta_2^{()} \in \operatorname{dom} <^{M(I_1,\dots,I_k)}$, where $\vartheta_1^{()} = {}^r\vartheta^{(\mathcal{I}_1,\dots,\mathcal{I}_{m_1})}_{1(I_1,\dots,I_{\ell_1})}$ or $\vartheta_1^{()} = {}^r\vartheta^{(i_1,\dots,i_k)}_{1(I_1,\dots,I_{\ell_1})}$ or $\vartheta_1^{()} = \vartheta_1 < K^{\ell',m}_{2n-1}$ and similarly for $\vartheta_2^{()}$, it follows that there is a c.u.b. $C_{\vartheta_1,\vartheta_2} \subseteq \underset{\sim}{\delta}^1_{2n+1}$ such that for ${}^rF_1\colon <^{{}^rT(I_1,\dots,I_k)}_{{}^r\mathcal{C}(I_1,\dots,I_k)} \to C_{\vartheta_1,\vartheta_2}$ or ${}^rF_1\colon <^{{}^rT(I_1,\dots,I_k)} \to C_{\vartheta_1,\vartheta_2}$ of the correct type representing components of $\alpha^{(I_1,\dots,I_k)}$ in the measure space $\prod_r {}^rV^{(I_1,\dots,I_k)}$ for ${}^rV \in \mathcal{W}_{2n+1} \cup \mathcal{S}_{2n+1}$, the following are satisfied:

1) a) If ${}^{r_1}V, {}^{r_2}V \in \mathcal{S}_{2n+1}$, and ${}^{r_1}\vartheta^{(I_1^1,\dots,I_{m_1}^1)}_{(I_1^1,\dots,I_{\ell_1}^1)} <^{M(I_1,\dots,I_k)} {}^{r_2}\vartheta^{(I_1^2,\dots,I_{m_2}^2)}_{(I_1^2,\dots,I_{\ell_2}^2)}$ then there is a c.u.b. $D_{\vartheta_1,\vartheta_2} \subseteq \underset{\sim}{\delta}^1_{2n+1}$ (which depends on the ${}^rF$ as well) such that for $g\colon K^{\ell',m}_{2n-1} \to D$ of the correct type,

$$f_{\vartheta_1}\colon <^{{}^{r_1}T(I_1^1,\dots,I_{\ell_1}^1)} \to K^{\ell',m}_{2n-1},\ f_{\vartheta_2}\colon <^{{}^{r_2}T(I_1^2,\dots,I_{\ell_2}^2)} \to K^{\ell',m}_{2n-1},$$

the corresponding embeddings from $\mathcal{B}_{2n+1}$, and $Q_1 = g \circ f_{\vartheta_1}$, $Q_2 = g \circ f_{\vartheta_2}$, then

$$^{r_1}F(\langle S(Q_1, (I_1^1, \ldots, I_{\ell_1}^1), (I_1^1, \ldots, I_{m_1}^1))\rangle) < {}^{r_2}F(\langle S(Q_2, (I_1^2, \ldots, I_{\ell_2}^2), (I_1^2, \ldots, I_{m_2}^2))\rangle).$$

b) similar to a) with $\vartheta_1 >^M \vartheta_2$.

c) if neither $\vartheta_1 <^M \vartheta_2$ nor $\vartheta_2 <^M \vartheta_1$ then we require that ${}^{r_1}F(\langle\cdots\rangle) = {}^{r_2}F(\langle\cdots\rangle)$.

2) a) If ${}^{r_1}V \in \mathcal{S}_{2n+1}, \vartheta_2 \in K_{2n-1}^{\ell',m}$, and ${}^{r_1}\vartheta_{1()}^{()} <^M \vartheta_2$, we proceed as in 1a) where we require that ${}^{r_1}F(\langle\rangle) < g(\vartheta_2)$.

b) and c) similar to b), c) of 1) above.

3) a) similar to 2) a), with $\vartheta_1 \in K_{2n-1}^{\ell',m}$, ${}^{r_2}V \in \mathcal{S}_{2n+1}$ and $\vartheta_1 <^M \vartheta_{2()}^{()}$.

b), c) as above.

4) a) If ${}^{r_1}V, {}^{r_2}V \in \mathcal{W}_{2n+1}$ and ${}^{r_1}\vartheta_{1(I_1^1,\ldots,I_{\ell_1}^1)}^{(i_1^1,\ldots,i_{k_1}^1)} <^M {}^{r_2}\vartheta_{2(I_1^2,\ldots,I_{\ell_2}^2)}^{(i_1^2,\ldots,i_{k_2}^2)}$, then we have ${}^{r_1}F^{(i_1^1,\ldots,i_{k_1}^1)}(\vartheta_1) < {}^{r_2}F^{(i_1^2,\ldots,i_{k_2}^2)}(\vartheta_2)$.

b), c) similar to above.

Since $\vartheta_{r_u(I_1,\ldots,I_k)}$, $\vartheta_{r_v(I_1,\ldots,I_k)}$, $K_{2n-1}^{\ell',m}$ are all less than $\underset{\sim}{\delta}_{2n+1}^1$, it follows readily that there is a c.u.b. $C_2 \subseteq C$ such that for each $\vartheta_1 <^M \vartheta_2$, if the ${}^rF$'s are of the correct type into C_2 representing a tuple in the product space $V^{(I_1,\ldots,I_k)}$, then 1–4 above hold.

For

$$\alpha^{(I_1,\ldots,I_k)} = (\ldots, {}^r\xi, \ldots, {}^r\xi^{(i_1,\ldots,i_k)}, \ldots, {}^r\xi_{(I_1,\ldots,I_l)}^{(\mathcal{I}_1,\ldots,\mathcal{I}_m)}, \ldots)$$

where $({}^{r_1}\xi, \ldots, {}^{r_p}\xi) \in A$, and for r such that ${}^rV \in \mathcal{S}_{2n+1} \cup \mathcal{W}_{2n+1}$ the ${}^r\xi$'s are represented by ${}^rF : <_{{}^rC(I_1,\ldots,I_k)}^{{}^rT(I_1,\ldots,I_k)} \to C_2$, or ${}^rF : <^{{}^rT(I_1,\ldots,I_k)} \to C_2$ of the correct type (and ordered as in the product measure), we claim that $\mathcal{F}_{H_1}(\alpha^{(I_1,\ldots,I_k)}) = \mathcal{F}_{H_2}(\alpha^{(I_1,\ldots,I_k)})$.

For such fixed $({}^{r_1}\xi, \ldots, {}^{r_p}\xi)$ and ${}^rF$, it follows from the $\underset{\sim}{\delta}_{2n+1}^1$-additivity of $\mu_{2n+1}^{K_{2n-1}^{\ell',m}}$ that there is a c.u.b. $D \subseteq \underset{\sim}{\delta}_{2n+1}^1$ such that for $g : K_{2n-1}^{\ell',m} \to D$ of the correct type, 1–4 above hold for all appropriate ϑ_1, ϑ_2. We may also assume $D \subseteq C_2$, and that the least element of $D > \sup {}^rF$ for ${}^rV \in \mathcal{W}_{2n+1}$. It then follows that for $g : K_{2n-1}^{\ell',m} \to D$ of the correct type that the corresponding $h : <^{M(I_1,\ldots,I_k)} \to \underset{\sim}{\delta}_{2n+1}^1$ is order-preserving. Since the ${}^rF$ have range in C_2, and the ${}^rF$ are of the correct type, it also follows that h has range in C_2, and is of the correct type. Hence $\mathcal{F}_{H_1}(\alpha^{(I_1,\ldots,I_k)}) = H_1(\langle T\rangle) = H_2(\langle T\rangle) = \mathcal{F}_{H_2}(\alpha^{(I_1,\ldots,I_k)})$, where T is the sequence corresponding to h and $(I_1, \ldots, I_k)$.

For $k = 1$, the result is immediate.

We now establish that for almost all γ w.r.t. $\mathcal{U}$, that $\mathcal{F}_\gamma$ is order-preserving almost everywhere from $<^\tau$ into $K_{2n+1}^{\ell,m}$. We fix $H : <^\Omega \to \underset{\sim}{\delta}_{2n+1}^1$ of the correct type representing γ (which happens for almost all γ). We show that $\mathcal{F}_\gamma$ is order-preserving almost everywhere. We let $\alpha_1^{(I_1^1,\dots,I_{k_1}^1)}, \alpha_2^{(I_1^2,\dots,I_{k_2}^2)}$ be given, and fix $({}^{r_1}\xi_1,\dots,{}^{r_p}\xi_1)$, and

$$ {}^rF_1\colon <^{{}^rT(I_1^1,\dots,I_{k_1}^1)}_{{}^rC(I_1^1,\dots,I_{k_1}^1)} \to \underset{\sim}{\boldsymbol{\delta}}^1_{2n+1} \text{ or } {}^rF_1\colon <^{{}^rT(I_1^1,\dots,I_{k_1}^1)} \to \underset{\sim}{\boldsymbol{\delta}}^1_{2n+1} $$

of the correct type representing the components of α_1 for ${}^rV \in \mathcal{S}_{2n+1} \cup \mathcal{W}_{2n+1}$, and similarly for α_2, which happens for almost all α_1, α_2. We assume that $\alpha_1^{()} <^\tau \alpha_2^{()}$ and show that $\mathcal{F}_\gamma(\alpha_1^{()}) < \mathcal{F}_\gamma(\alpha_2^{()})$. We let G_1, G_2 represent $\mathcal{F}_\gamma(\alpha_1^{()}), \mathcal{F}_\gamma(\alpha_2^{()})$ w.r.t. $\mu_{2n+1}^{K_{2n-1}^{\ell',m}}$ as in the definition of $\mathcal{F}_\gamma$. We show that for almost all $g\colon K_{2n-1}^{\ell',m} \to \underset{\sim}{\boldsymbol{\delta}}^1_{2n+1}$ of the correct type, $G_1([g]) < G_2([g])$. We consider the following cases:

Case I. $k_1 = 1$ and $k_2 \geq 1$. In this case (from the definition of $<^\tau$) $I_1^1 < I_1^2$. From the definition of $\mathcal{F}$ and the ordering $<^\Omega$ it follows that for almost all g that $G_1([g]) = H(\langle [g], I_1^1\rangle) < H(\langle [g], I_1^2, \dots, {}^r\beta_j^{()}, \dots, \rangle) = G_2([g])$, where the ${}^r\beta_j^{()}$ appear if $k_2 > 1$.

Case II. $k_1 > 1$ and $k_2 = 1$. Hence $I_1^1 \leq I_1^2$. If $I_1^1 < I_1^2$, we proceed as above. If $I_1^1 = I_1^2$, the result follows since for almost all g, $G_1([g]) = H(\langle T_1\rangle) = H(\langle [g], I_1^1, \dots, {}^r\beta_j^{()}\rangle) < H(\langle [g], I_1^2\rangle) = H(\langle T_2\rangle)$, since T_1 extends T_2.

Case III. $k_1, k_2 > 1$. We assume this case for the rest of the argument. We let $\bar{\alpha}_1$ denote the sequence $\bar{\alpha}_1 = \langle I_1^1, \Pi_{(I_1)}^{(I_1,\dots,I_{k_1-1})}(\alpha_1), \dots, I_{k_1-1}^1, \alpha, I_{k_1}^1\rangle$, and similarly for $\bar{\alpha}_2$. We consider the following subcases (a), (b), (c).

Subcase III.a. There is a least point where $\bar{\alpha}_1, \bar{\alpha}_2$ disagree, which of the form I_{q+1}. Hence $I_1^1 = I_1^2 = I_1, \dots, I_q^1 = I_q^2 = I_q$ and $I_{q+1}^1 < I_{q+1}^2$. For fixed $g\colon K_{2n-1}^{l',m} \to \underset{\sim}{\boldsymbol{\delta}}^1_{2n+1}$, we consider the sequences T_1, T_2 as in the definitions of $G_1([g])$, $G_2([g])$ which were used in defining $\mathcal{F}_\gamma(\bar{\alpha}_1)$, $\mathcal{F}_\gamma(\bar{\alpha}_2)$. We claim that for almost all g that T_1 and T_2 agree before the I_{q+1} term. Since $I_{q+1}^1 < I_{q+1}^2$ and H is order-preserving, it then follows that $\mathcal{F}_\gamma(\bar{\alpha}_1) < \mathcal{F}_\gamma(\bar{\alpha}_2)$. We use the following observations, where we use the notation of the definition of the sequence T. For $\bar{q} \leq q$ since $\Pi_{(I_1,\dots,I_{\bar{q}})}^{(I_1,\dots,I_{k_1-1})}(\alpha_1) = \Pi_{(I_1,\dots,I_{\bar{q}})}^{(I_1,\dots,I_{k_2-1})}(\alpha_2)$, it follows that:

1) If $({}^{r_1}\xi(1),\dots,{}^{r_p}\xi(1)), ({}^{r_1}\xi(2),\dots,{}^{r_p}\xi(2))$ enumerate the components of $\Pi_{(I_1,\dots,I_{\bar{q}})}^{(I_1,\dots,I_{k_1-1})}(\alpha_1)$ and $\Pi_{(I_1,\dots,I_{\bar{q}})}^{(I_1,\dots,I_{k_2-1})}(\alpha_2)$ corresponding to measures of the form ${}^rV \in \bigcup_{m<n} \mathcal{R}_{2m+1}$, then ${}^{r_1}\xi(1) = {}^{r_1}\xi(2), \dots, {}^{r_p}\xi(1) = {}^{r_p}\xi(2)$. Hence T_1, T_2 agree on the corresponding terms ${}^r\beta_{\bar{q}}$ for $\bar{q} \leq q$.

2) For r with ${}^rV \in \mathcal{W}_{2n+1}$, the induced functions ${}^rF_1^{(i_1,\dots,i_k)} = {}^rF_2^{(i_1,\dots,i_k)}$ agree almost everywhere w.r.t. $\nu^{(i_1,\dots,i_{k-1})}$ for indices $(i_1,\dots,i_k) \in \operatorname{dom}{}^rT(I_1,\dots,I_{\bar{q}})$. Hence, the ${}^r\beta_{\bar{q}}^{(i_1,\dots,i_k)}$ (as in T_1, T_2) agree for the two sequences T_1, T_2, corresponding to ${}^rF_1, {}^rF_2$ for all $\bar{q} \leq q$.

3) For r with ${}^rV \in \mathcal{S}_{2n+1}$, there is a c.u.b. $C \subseteq \underset{\sim}{\boldsymbol{\delta}}^1_{2n+1}$ such that for $g\colon K_{2n-1}^{\ell',m} \to C$ of the correct type, for all indices $(\mathcal{I}_1,\dots,\mathcal{I}_m)$ in the domain of ${}^r\mathcal{C}(I_1,\dots,I_{\bar{q}})$ we have that for almost all ϑ w.r.t. ${}^ru^{(I_1,\dots,I_{\bar{q}})}$ that ${}^r\mathcal{M}_{\bar{q}}^{\mathcal{I}_1,\dots,\mathcal{I}_m}(1)(\vartheta) = {}^r\mathcal{M}_{\bar{q}}^{\mathcal{I}_1,\dots,\mathcal{I}_m}(2)(\vartheta)$, where $1, 2$ refer to the definitions relative to ${}^rF_1$ and ${}^rF_2$ respectively. Hence it follows that there is a c.u.b. $C \subseteq \underset{\sim}{\boldsymbol{\delta}}^1_{2n+1}$ such that for $g\colon K_{2n-1}^{\ell',m} \to C$, we have that T_1, T_2 agree on the terms ${}^r\beta_{\bar{q}}^{(\mathcal{I}_1,\dots,\mathcal{I}_m)}$ for $\bar{q} \leq q$.

Since H is order-preserving, it follows that $G([g_1]) < G([g_2])$.

Subcase III.b. There is a least position where $\bar{\alpha}_1, \bar{\alpha}_2$ disagree which of the form $\Pi_{(I_1,\dots,I_q)}^{(I_1^1,\dots,I_{k_1-1}^1)}(\alpha_1) < \Pi_{(I_1,\dots,I_q)}^{(I_1^2,\dots,I_{k_2-1}^2)}(\alpha_2)$.

Hence $I_1^1 = I_1^2 \dots, I_q^1 = I_q^2$, and if $\bar{r}$ is the least component with respect to the product $V^{(I_1,\dots,I_q)} = \prod_r {}^rV^{(I_1,\dots,I_q)}$ such that the above tuples disagree, then $\Pi_{(I_1,\dots,I_q)}^{(I_1^1,\dots,I_{k_1-1}^1)}(\alpha_1)(r) < \Pi_{(I_1,\dots,I_q)}^{(I_1^2,\dots,I_{k_2-1}^2)}(\alpha_2)(r)$. It then follows that there is a c.u.b. $C \subseteq \underset{\sim}{\boldsymbol{\delta}}^1_{2n+1}$ such that for $g\colon K_{2n-1}^{\ell',m} \to C$ of the correct type, we have:

1) For $\bar{q} < q$, ${}^r\beta_{\bar{q}}(1) = {}^r\beta_{\bar{q}}(2)$, for all r, where the β's are elements of the sequences T_1, T_2

2) For $r' < \bar{r}$, ${}^{r'}\beta_q(1) = {}^{r'}\beta_q(2)$.

3) If ${}^{\bar{r}}V^{(I_1,\dots,I_q)} \in \bigcup_{m<n} \mathcal{R}_{2m+1}$, then ${}^{\bar{r}}\beta_q(1) = {}^{\bar{r}}\beta_q(2)$, since ${}^{\bar{r}}\beta_q(1) = \Pi_{(I_1,\dots,I_q)}^{(I_1^1,\dots,I_{k_1-1}^1)}(\alpha_1)(r)$ in this case, and similarly for ${}^{\bar{r}}\beta_q(2)$.

4) If ${}^{\bar{r}}V^{(I_1,\dots,I_q)} \in \mathcal{W}_{2m+1}$, then in the fixed ordering we have on the indices $(i_1,\dots,i_k) \in \operatorname{dom}{}^rT(I_1,\dots,I_q)$ which we use to identify $\vartheta_{{}^rV^{(I_1,\dots,I_q)}}$ with an ordinal, if $(\bar{\imath}_1,\dots,\bar{\imath}_k)$ denotes the least index s.t. $\left[{}^{\bar{r}}F_1^{(\bar{\imath}_1,\dots,\bar{\imath}_k)}\right] \neq \left[{}^{\bar{r}}F_2^{(\bar{\imath}_1,\dots,\bar{\imath}_k)}\right]$, then the first is smaller. Hence ${}^{\bar{r}}\beta_q^{(\bar{\imath}_1,\dots,\bar{\imath}_k)}(1) < {}^{\bar{r}}\beta_1^{(\bar{\imath}_1,\dots,\bar{\imath}_k)}(2)$, for the least index $(\bar{\imath}_1,\dots,\bar{\imath}_k)$ where ${}^{\bar{r}}\beta_1^{(\bar{\imath}_1,\dots,\bar{\imath}_k)}(1)$ and ${}^{\bar{r}}\beta_q^{(\bar{\imath}_1,\dots,\bar{\imath}_k)}(2)$ disagree.

5) If ${}^{\bar{r}}V^{(I_1,\dots,I_q)} \in \mathcal{S}_{2n+1}$, then for the least index $(\mathcal{I}_1,\dots,\mathcal{I}_m)$ in the domain of ${}^{\bar{r}}\mathcal{C}(I_1,\dots,I_q)$, such that $\left[{}^{\bar{r}}\mathcal{M}_q^{(\mathcal{I}_1,\dots,\mathcal{I}_m)}(1)\right] \neq \left[{}^{\bar{r}}\mathcal{M}_q^{(\mathcal{I}_1,\dots,\mathcal{I}_m)}(2)\right]$, the first is smaller. This follows since for the least $(\mathcal{I}_1,\dots,\mathcal{I}_m)$ such that $\Pi_{(I_1,\dots,I_q)}^{(I_1^1,\dots,I_{k_1-1}^1)}(\alpha_1)(\mathcal{I}_1,\dots,\mathcal{I}_m) \neq \Pi_{(I_1,\dots,I_q)}^{(I_1^2,\dots,I_{k_2-1}^2)}(\alpha_2)(\mathcal{I}_1,\dots,\mathcal{I}_m)$, the first is

smaller. Hence for the least $(\mathcal{I}_1, \dots, \mathcal{I}_m)$ s.t. ${}^{\bar{r}}\beta_q^{(\mathcal{I}_1,\dots,\mathcal{I}_m)}(1) \neq {}^{\bar{r}}\beta_q^{(\mathcal{I}_1,\dots,\mathcal{I}_m)}(2)$, the first is smaller.

Hence, for such g, it follows that for the least position where T_1, T_2 disagree, the first is smaller. Since H is order-preserving, $G_1([g]) < G_2([g])$.

Subcase III.c. $\bar{\alpha}_1$ is an extension of $\bar{\alpha}_2$. Proceeding as in (a) and (b) above, it follows that for almost all g, the sequence T_1 extends T_2, so $G_1([g]) < G_2([g])$.

Hence, for almost all γ w.r.t. $\mathcal{U}$, we have that $\mathcal{F}_\gamma$ is order-preserving almost everywhere.

We now establish $\mathcal{B}_{2n+3}$. We let $A \subseteq K_{2n+1}^{\ell,m}$ have measure one w.r.t. $S_{2n+1}^{\ell,m}$. We let C be a c.u.b. subset of $\underset{\sim}{\delta}^1_{2n+1}$ such that if $G : \underset{\sim}{\delta}^1_{2n+1} \to C$ is of the correct type, then the ordinal β represented by G w.r.t. $\mu_{2n+1}^{S_{2n-1}^{\ell',m}}$ is in A.

We let B be the measure one set w.r.t. $\mathcal{U}$ determined by C. We fixed $\gamma \in B$, and a function $H : <^{\Omega} \to C$ of the correct type representing γ with respect to the measures $\prod_r {}^r\upsilon^{(I_1,\dots,I_k)} \times \mu_{2n+1}^{<M(I_1,\dots,I_k)}$ for $(I_1, \dots, I_k) \in \operatorname{dom} \tau$. For $\alpha^{(I_1,\dots,I_k)} \in \operatorname{dom} <^{\tau}$ and ${}^rF$ of the correct type representing the components of α for the ${}^rV \in \mathcal{W}_{2n+1} \cup \mathcal{S}_{2n+1}$, (which happens for almost all α), and G representing $\mathcal{F}_\gamma(\alpha^{(I_1,\dots,I_k)})$ w.r.t. $\mu_{2n+1}^{S_{2n-1}^{\ell',m}}$, it follows from the definition of $\mathcal{F}_\gamma$ that for almost all $g : K_{2n-1}^{\ell',m} \to \underset{\sim}{\delta}^1_{2n+1}$, that $G[(g)] = H(\langle T \rangle)$, for the sequence T as in the definition of $\mathcal{F}_\gamma$, and hence $G([g]) \in C$.

Since H is of the correct type, it follows that G is discontinuous and has uniform cofinality ω on a measure one set. Also, since the ordinal β_0 (as in the definition of T) is equal to $[g]$, it follows that G is strictly increasing on a measure one set.

By a sliding argument (as in Lemma 2.7), we may assume G is of the correct type and has range C everywhere. Hence $\mathcal{F}_\gamma(\alpha^{(I_1,\dots,I_k)}) \in B$.

This establishes $\mathcal{B}_{2n+3}$.

§4. A Local Embedding Theorem.

THEOREM 4.1. For any measure $\mathcal{V} \in \bigcup_{m \le n} \mathcal{S}_{2m+1} \cup \bigcup_{m \le n} \mathcal{W}_{2m+1}$, and regular cardinal $\kappa < \underset{\sim}{\delta}^1_{2n+3}$, there is a measure $R \in R_{2n+1}$ (*i.e.*, a canonical measure) and a c.u.b. $C \subseteq \underset{\sim}{\delta}^1_{2n+3}$ such that for $\alpha \in C$ with $\operatorname{cf}(\dot{\alpha}) = \kappa$ we have $j_{\mathcal{V}}(\alpha) \le j_R(\alpha)$. Here $j_{\mathcal{V}}$ and j_R denote the embeddings from the ultrapowers by the measures $\mathcal{V}$ and R respectively.

As in §3, we again use the notation $\mathcal{V} = \prod_r {}^r\mathcal{V}$ where each ${}^r\mathcal{V}$ is basic. Also as in the proof of the global embedding theorem, we require an inductive hypothesis:

H_{2n+1}: For any basic measures ${}^r\mathcal{V}_1 \in \bigcup_{m\leq n} \mathcal{S}_{2m-1} \cup \bigcup_{m\leq n} \mathcal{W}_{2m-1}$, for $1 \leq r \leq r_0$, and corresponding product measure $\mathcal{V}_1 = \prod_{r=1}^{r_0} {}^r\mathcal{V}_1$, there are measures ${}^r\mathcal{V}_2$, for $1 \leq r \leq r_0$, in $\bigcup_{m\leq n} R_{2m-1}$, a measure u on Ξ_{2n-1}, and a map $\alpha \mapsto [f_\alpha]_u$ such that for almost all $\alpha = (\alpha_1, \dots, \alpha_{r_0})$ w.r.t. $\mathcal{V}_2 = \prod_{r=1}^{r_0} {}^r\mathcal{V}_2$ the ordinal $[f_\alpha]_u$ is represented by an $f_\alpha \colon \vartheta_u \to \vartheta_{\mathcal{V}_1}$ defined a.e. w.r.t. u satisfying:

1) For any $A_1 \in \vartheta_{\mathcal{V}_1}$ of measure one w.r.t. $\mathcal{V}_1$, there is an $A_2 \subseteq \vartheta_{\mathcal{V}_2}$ of measure one w.r.t. $\mathcal{V}_2$ such that for all $\alpha \in A_2$, f_α has range u almost everywhere in A_1.
2) There is a measure one set $A_2 \subseteq \vartheta_{\mathcal{V}_2}$ such that for $\alpha \in A_2$, $\sup f_\alpha < \vartheta_{\mathcal{V}_1}$ (that is there are $\beta_1 < \vartheta_{{}^1\mathcal{V}_1}, \dots, \beta_{r_0} < \vartheta_{{}^{r_0}\mathcal{V}_1}$ s.t. for almost all η w.r.t. u, if $f_\alpha(\eta) = (\gamma_1, \dots, \gamma_{r_0})$, then $\gamma_1 < \beta_1, \dots, \gamma_{r_0} < \beta_{r_0}$).
3) If ${}^r\mathcal{V}_1'$, $1 \leq r \leq r_0$ are basic measures in $\bigcup_{m\leq n} \mathcal{S}_{2m-1} \cup \bigcup_{m\leq n} \mathcal{W}_{2m-1}$ such that the embeddings $\Pi_{{}^r\mathcal{V}_1}^{{}^r\mathcal{V}_1'}$ are defined for $1 \leq r \leq r_0$, then there are ${}^r\mathcal{V}_2' \sim {}^r\mathcal{V}_2$ such that 1) and 2) hold for $\mathcal{V}_1'$, $\mathcal{V}_2'$
   We also require that if ${}^r\mathcal{V}_2' > {}^r\mathcal{V}_2$ for $1 \leq r \leq r_0$, then H_{2n+1} holds with $\mathcal{V}_2' = \prod_r {}^r\mathcal{V}_2'$ replacing $\mathcal{V}_2$ (with a possibly different u, etc.).
4) If $\mathcal{V}_1 = \prod_r {}^r\mathcal{V}_1$, where each ${}^r\mathcal{V}_1 \in R_{2n-1}$ and ${}^{r_1}\mathcal{V}_1 \sim {}^{r_2}\mathcal{V}_1$ for $1 \leq r_1 \leq r_2 \leq r_0$, then there is a $\mathcal{V}_2 \in R_{2n-1}$ with $\mathcal{V}_2 \sim {}^r\mathcal{V}_1$, and measure u on Ξ_{2n-1} such that H_{2n+1} 1), 2), 3) above are satisfied (here $\mathcal{V}_2$ is now a single measure rather than a product).

We require the following additional hypothesis:

D_{2n+1}:

1) We let ${}^r\mathcal{V}_1, {}^r\mathcal{V}_2, u$ be as in H_{2n+1} 1), 2), 3). Then for any $1 \leq r \leq r_0$, there is an $r' \leq r_0$ and a measure one set A w.r.t. $\mathcal{V}_2$ such that if $(\beta_1, \dots, \beta_{r'}, \dots, \beta_{r_0})$, $(\bar{\beta}_1, \dots, \bar{\beta}_{r'}, \dots, \bar{\beta}_{r_0})$ are in A and $\beta_{r'} < \bar{\beta}_{r'}$, then for almost all γ w.r.t. u we have that $f_{(\vec{\beta})}(\gamma)(r) < f_{(\vec{\bar{\beta}})}(\gamma)(r)$, these denoting the rth components of these tuples.
   We further require that if ${}^r\mathcal{V}_1' \leq {}^r\mathcal{V}_1$ for $1 \leq r \leq r_0$, are such that H_{2n+1}, 1), 2), 3) are also satisfied by ${}^r\mathcal{V}_1', {}^r\mathcal{V}_2$ (for some u') then the r' for the measures ${}^r\mathcal{V}_1'$ is the same as for the ${}^r\mathcal{V}_1$, and in fact, for $\vec{\beta}, \vec{\bar{\beta}}$ as above, for almost all γ w.r.t. u', $(\Pi_{\mathcal{V}_1}^{\mathcal{V}_1'} f_{(\vec{\beta})}(\gamma))(r) < (\Pi_{\mathcal{V}_1}^{\mathcal{V}_1'} f_{(\vec{\bar{\beta}})}(\gamma))(r)$.
2) If ${}^r\mathcal{V}_1 \in R_{2n+1}$, ${}^{r_1}\mathcal{V}_1 \sim {}^{r_2}\mathcal{V}_1$ for $1 \leq r_1 \leq r_2 \leq r$, $\mathcal{V}_2 \in R_{2n-1}$, and u are as in $H_{2n+1}(4)$, then there is a measure one set A w.r.t. $\mathcal{V}_2$ such that for $\beta_1 < \beta_2$ in A we have that for almost all γ w.r.t. $\mathcal{U}$ that $f_{\beta_1}(\gamma)(r) < f_{\beta_2}(\gamma)(r)$ for all $1 \leq r \leq r_0$.

We first establish Theorem 4.1 from H_{2n+3} (using also our overall inductive hypothesis K_{2n+3}).

We let ${}^r\mathcal{V}_1$, $1 \leq r \leq r_0$, be basic measures in $\bigcup_{m\leq n} \mathcal{S}_{2m+1} \cup \bigcup_{m\leq n} \mathcal{W}_{2n+1}$ and $\mathcal{V}_1 = \prod_r {}^r\mathcal{V}_1$. We let C be a c.u.b. subset of $\underset{\sim}{\delta}^1_{2n+3}$ closed under $j_{\mathcal{V}}$

for all measures $\mathcal{V}$ in $\bigcup_{m\leq n} \mathcal{S}_{2m+1} \cup \bigcup_{m\leq n} \mathcal{W}_{2n+1}$. We first show that there are measures ${}^r\mathcal{V}_2 \in R_{2n+1}$ s.t. for all $\alpha \in C$ we have $j_{\mathcal{V}_1}(\alpha) \leq j_{\mathcal{V}_2}(\alpha)$, where $\mathcal{V}_2 = \prod_r {}^r\mathcal{V}_2$. We let ${}^r\mathcal{V}_2$ be as in H_{2n+3}, and fix $\alpha \in C$. We define an embedding $\mathcal{E}$ from $j_{\mathcal{V}_1}(\alpha)$ into $j_{\mathcal{V}_2}(\alpha)$ as follows. If $F : \vartheta_{\mathcal{V}_1} \to \alpha$, then we represent $\mathcal{E}([F]_{\mathcal{V}_1})$ by $G : \vartheta_{\mathcal{V}_2} \to \alpha$ defined as follows: if $\beta = ({}^1\beta, \dots, {}^{r_0}\beta) \in \vartheta_{\mathcal{V}_2}$, we represent $G(\beta)$ w.r.t. $\mathcal{U}$ (as in H_{2n+3}) by the function $F \circ f_\beta : \vartheta_{\mathcal{U}} \to \alpha$ (here $\beta \mapsto [f_\beta]_{\mathcal{U}}$ is as in H_{2n+3}). It follows form H_{2n+3} and the definition of C that $\mathcal{E}$ is well-defined.

Now assume, by the previous paragraph, that $\mathcal{V}_1 = \prod_r {}^r\mathcal{V}_1$ where ${}^r\mathcal{V}_1 \in \bigcup_{m\leq n} R_{2m+1}$. We fix now a regular cardinal $\kappa < \underset{\sim}{\delta}^1_{2n+3}$ and proceed to show that for some $\mathcal{V}_2 \in \bigcup_{m\leq n} R_{2m+1}$ that $j_{\mathcal{V}_1}(\alpha) \leq j_{\mathcal{V}_2}(\alpha)$ for all $\alpha \in C$ with $\mathrm{cf}(\alpha) = \kappa$. By Lemma 2.7, we may assume that $\mathcal{V}_1$ is of the form $\mathcal{V}_1 = \prod_{r=1}^{q_0} {}^r\mathcal{V}_1 \times \prod_{r=q_0+1}^{r_0} {}^r\mathcal{V}_1$, where ${}^{r_1}\mathcal{V}_1 \sim {}^{r_2}\mathcal{V}_1$ for $1 \leq r_1, r_2 \leq q_0$, and ${}^{r_1}\mathcal{V}_1 \not\sim {}^{r_2}\mathcal{V}_1$ for $r_1 \leq q_0$ and $r_2 > q_0$, and also $\mathrm{cf}\, \vartheta({}^{r_1}\mathcal{V}_1) = \kappa$. It now follows from Lemmas 2.5, 2.6 and the definition of C that if $\mathcal{V}_1' = \prod_{r=q_0+1}^{r_0} {}^r\mathcal{V}_1$ and $\alpha \in C$ then then $j_{\mathcal{V}_1'}(\alpha) = \alpha$. Hence, we may assume without loss of generality that $\mathcal{V}_1 = \prod_r {}^r\mathcal{V}_1$ where ${}^{r_1}\mathcal{V}_1 \sim {}^{r_2}\mathcal{V}_1$ for $1 \leq r_1, r_2 \leq r_0$. We then select $\mathcal{V}_2$ as in $H_{2n+3}(4)$, and proceed as in the previous paragraph to define an embedding from $j_{\mathcal{V}_1}(\alpha)$ into $j_{\mathcal{V}_2}(\alpha)$.

For the rest of this section we assume H_{2n+1} and D_{2n+1}, and proceed to establish H_{2n+3}, D_{2n+3}. We first consider H_{2n+3}.

We let ${}^r\mathcal{V}_1$, for each $1 \leq r \leq r_0$ be basic measures in $\bigcup_{m\leq n} \mathcal{S}_{2m+1} \cup \bigcup_{m\leq n} \mathcal{W}_{2m+1}$. For each r, we will define ${}^r\mathcal{V}_2$ and ${}^r\mathcal{U}_1$ and will take $\mathcal{V}_2 = \prod_r {}^r\mathcal{V}_2$ and $\mathcal{U} = \prod_r {}^r\mathcal{U}$. Hence, in defining $\mathcal{V}_2$ and $\mathcal{U}$ we need only consider a fixed ${}^r\mathcal{V}_1$ and we therefore suppress writing the presuperscript r throughout the definition.

We consider the following cases I, II, III for the definition of $\mathcal{U}$.

Case I. $\mathcal{V}_1$ $(= {}^r\mathcal{V}_1) \in \mathcal{W}_{2n+1}$. We fix an $n - *$ tuple T and corresponding ordering on Ξ_{2n+1} such that $\mathcal{V}_1 = \mu_{2n+1}^{<^T}$. We fix a measure $S_{2n-1}^{\ell',m}$ s.t. the hypothesis B_{2n+1} from the global embedding theorem is satisfied by $<^T$ and $S_{2n-1}^{\ell',m}$, with measure u. We then let $\mathcal{V}_2 = \mu_{2n+1}^{S_{2n-1}^{\ell',m}} = W_{2n+1}^m$, and let the $\mathcal{U}$ in H_{2n+3} be the u from B_{2n+1}.

Case II. $\mathcal{V}_1 \in \mathcal{S}_{2n+1}$ and $1^{(n_1)} \notin \mathrm{dom} <^T$, where n_1 is the largest integer s.t. $n_1 \in \mathrm{dom}\, T$. We are using here the notation of Definitions 2.2, 2.3. Recall that $\mathcal{V}_1$ is of the form $\mathcal{V}_1 = \mathcal{S}_{2n+1}{}^T_{\mathcal{C}}$ for some $n - *$ tuple T and collection $\mathcal{C}$. Recall also that $1^{(n)}$ is in the domain of $<^T$ for some n, $1 \leq n \leq n_1$ (the set of such n is coded into T). We may assume in this case that v^{n_1} (the measure given by T) is defined as otherwise n_1 may be deleted from the domain of T without affecting $<^T$. For all indices $(n_1, i_2, \dots, i_k) \in \mathrm{dom}\, T$ with $v^{(n_1, i_2, \dots, i_{k-1})}$ defined, $v^{(n_1, i_2, \dots, i_{k-1})} < v^{(n_1)}$, that is $\Pi_{(n_1)}^{(n_1, i_2, \dots, i_k)}$ is defined. From $H_{2n+1}(1)$

and (3) it follows that there is a measure $\tilde{v}_2$ s.t. for all $v^{(n_1,i_2,\dots,i_k)}$, H_{2n+1} holds for $v^{(n_1,\dots,i_k)}$, $\tilde{v}_2$ and a measure $u^{(n_1,\dots,i_k)}$, and $\tilde{v}_2 = w_1 \times \cdots \times w_n$ where $w_1, \dots, w_n$ are basic measures in $\bigcup_{m\leq n} R_{2m-1} = \bigcup_{m\leq n} \mathcal{S}_{2m-1} \cup \bigcup_{m\leq n} \mathcal{W}_{2m-1}$. We also let $v^{(n_1)} = \prod_s {}^s v^{(n_1)}$, a product of basic measures. We let $\bar{s}$ be the s such that in identifying $\vartheta_{v^{(n_1)}}$ with $\Pi_s \vartheta_{s_{v^{(n_1)}}}$, we order by $\vartheta_{\bar{s}_{v^{(n_1)}}}$ most significantly. We let $\bar{\bar{s}} \leq n$ be, from D_{2n+1}, the integer s.t. there is a measure one set A w.r.t. $\tilde{v}_2$ such that for $(\beta_1, \dots, \beta_n), (\beta'_1, \dots, \beta'_n) \in A$ and $\beta_{\bar{\bar{s}}} < \beta'_{\bar{\bar{s}}}$ we have that for all $(n_1, i_2, \dots, i_k)$ that for almost all γ w.r.t. $u^{(n_1,\dots,i_k)}$ that $(\Pi^{(n_1,\dots,i_k)}_{(n_1)} f_{\vec{\beta}}(\gamma))(\bar{s}) < (\Pi^{(n_1,\dots,i_k)}_{(n_1)} f_{(\vec{\beta}')}(\gamma))(\bar{s})$.

We let $\tilde{v}_2^2 = \tilde{v}_2 \times \tilde{v}_2 = (w_1 \times \cdots \times w_n) \times (w_1 \times \cdots \times w_n)$, and let $E_1, \dots, E_p$ enumerate the components of $\tilde{v}_2^2$ equivalent (w.r.t. $\sim$) to $w_{\bar{\bar{s}}}$. By $H_{2n+1}(4)$, we let $v_2 \in \bigcup_{m\leq n} R_{2n+1}$ be s.t. H_{2n-1} holds for $E_1 \times \cdots \times E_p$, v_2, and some measure u_2. We then let $\mathcal{V}_2 = S^{\ell,m}_{2n+1}$, where $v(S^{\ell,m}_{2n+1}) = v_2$ (here ℓ is not necessarily maximal). We may also take any $m' \geq m$ in the following.

We proceed to define the measure $\mathcal{U}$.

We define two orderings $<^{\Omega_1}$, and $<^{\Omega_2}$ on tuples of ordinals and integers. The domain of $<^{\Omega_1}$ consists of tuples of the form

$$T = \langle \alpha_0, \mathcal{I}_1, \dots, \alpha^{(i_1)}, \dots, \mathcal{I}_2, \dots, \alpha^{(i_1,i_2)}, \dots, \\ \mathcal{I}_k, \dots, \alpha^{(i_1,\dots,i_k)}, \mathcal{I}_n, \dots, \alpha^{(i_1,\dots,i_n)} \rangle$$

or an initial segment of such, satisfying:

(a) $\alpha_0 \leq \underset{\sim}{\delta}^1_{2n+1}$

(b) $(\mathcal{I}_1, \dots, \mathcal{I}_k)$ is an index in $\mathcal{C}$.

(c) the ordinals $\alpha^{(i_1,\dots,i_k)}$ following $\mathcal{I}_k$ are indexed by indices $(i_1, \dots, i_k)$ in $C_{(\mathcal{I}_1,\dots,\mathcal{I}_k)}$ and occur in the same order as in $C_{(\mathcal{I}_1,\dots,\mathcal{I}_k)}$.

(d) $\alpha^{(i_1,\dots,i_k)} < \alpha_0$.

We let $<^{\Omega_1}$ be the Brouwer-Kleene ordering on the tuples.

We let $<^{\Omega_2}$ denote the usual ordering on $\underset{\sim}{\delta}^1_{2n+1}$.

We define the measure $\mathcal{U}$ as follows: We define A to have $\mathcal{U}$ measure one if there is a c.u.b. $C_2 \subseteq \underset{\sim}{\delta}^1_{2n+1}$ s.t. for all $H_2\colon <^{\Omega_2} \to C_2$ of the correct type, there is a c.u.b. $C_1 \subseteq \underset{\sim}{\delta}^1_{2n+1}$ s.t. for all $H_1\colon <^{\Omega_1} \to C_1$ of the correct type $\vec{\eta} = (\dots, \eta^{(\mathcal{I}_1,\dots,\mathcal{I}_k)}, \dots) \in A$, where for fixed $(\mathcal{I}_1, \dots, \mathcal{I}_k)$, an index in $\mathcal{C}$, $\eta^{(\mathcal{I}_1,\dots,\mathcal{I}_k)}$ is defined through the following sequence of definitions.

(1) $\eta^{(\mathcal{I}_1,\dots,\mathcal{I}_k)}$ is represented w.r.t. $\mu^{<^T}_{2n+1}$ by a function $\bar{F}$,

(2) for fixed $f\colon <^T \to \underset{\sim}{\delta}^1_{2n+1}$ of the correct type, $\bar{F}([f])$ is represented w.r.t. v_2 (as above) by a function g.

(3) for $\delta < \vartheta_{v_2}$ we have $g(\delta) = H_2(\beta)$, where β is represented w.r.t. u_2 (defined above) by a function h_2.

(4) For $\gamma \in \vartheta_{u_2}$, we represent $h_2(\gamma)$ w.r.t. $D_1 \times \cdots \times D_q$ by a function $\bar{g}_1$, where $D_1, \ldots, D_q$ enumerate the components of $\tilde{v}_2^2$ not equivalent to the $E_1, \ldots, E_p$. The function $\bar{g}_1$ is defined as follows. Let $i_1, \ldots, i_q$ enumerate the integers $a \leq 2n$ where $w_a \sim D_i$ for some i, and let $j_1, \ldots, j_p$ enumerate the integers $a \leq 2n$ where $w_a \sim E_1$. Then $\bar{g}_1(\xi_{i_1}, \ldots, \xi_{i_q}) = g_1(\xi_1, \ldots, \xi_n, \xi'_1, \ldots, \xi'_n)$ where $\{\xi_1, \ldots, \xi_n, \xi'_1, \ldots, \xi'_n\} = \{\xi_{i_1}, \ldots, \xi_{i_q}\} \cup \{\xi_{j_1}, \ldots, \xi_{j_p}\}$. Here $(\xi_{j_1}, \ldots, \xi_{j_p}) = f_\delta(\gamma)$, where $\delta \mapsto f_\delta$ is as in H_{2n+1} for $E_1 \times \cdots \times E_p$, v_2, and u_2.

(5) g_1 is defined by $g_1(\xi_1, \ldots, \xi_n, \xi'_1, \ldots, \xi'_n) = H_1(\langle T \rangle)$, where $T = \langle \alpha_0, \mathcal{I}_1, \ldots, \alpha^{i_1}, \ldots, \mathcal{I}_k, \ldots, \alpha^{(i_1, \ldots, i_k)}, \ldots \rangle$ is defined by:

(5a) α_0 is represented w.r.t. $u^{(n_1)}$ by a function h_1^0 defined as follows: We let $f_s : \vartheta_{v^{(n_1)}} \to \underset{\sim}{\delta}^1_{2n+1}$ represent the least equivalence class of an a.e. monotonically increasing function with $\sup_{\text{a.e.}} f_s = \sup_{\text{a.e.}} f^{(n_1)}$. For $\rho \in \vartheta_{u^{(n_1)}}$, we set $h_1^0(\rho) = f_s(f_{(\xi'_1, \ldots, \xi'_n)}(\rho))$. Here, $(\xi'_1, \ldots, \xi'_n) \mapsto f_{(\xi'_1, \ldots \xi'_n)}$ is as in H_{2n+1} for $v^{(n_1)}$, $\tilde{v}_2$, and $u^{(n_1)}$.

(5b) $\alpha^{(n_1, i_2, \ldots, i_k)}$ is represented w.r.t. $u^{(n_1, i_2, \ldots, i_k)}$ by a function $h_1^{(n_1, i_2, \ldots, i_k)}$ defined by: for $\rho \in \vartheta_{u^{(n_1, i_2, \ldots, i_k)}}$, we set $h_1^{(n_1, i_2, \ldots, i_k)}(\rho) = f(\beta^{(n_1, i_2, \ldots, i_k)})$ where $\beta = f_{(\xi_1, \ldots, \xi_n)}(\rho)$, where $(\xi_1, \ldots, \xi_n) \mapsto f_{(\xi_1, \ldots \xi_n)}$ is as in H_{2n+1} for $v^{(n_1, i_2, \ldots, i_k)}$, $\tilde{v}_2$, and $u^{(n_1, i_2, \ldots, i_k)}$.

(5c) $\alpha^{(i_1, \ldots, i_k)}$, for $i_1 < n_1$, is represented w.r.t. $v^{(i_1, \ldots, i_k)}$ by the induced function $f^{(i_1, \ldots, i_k)}$.

We show that $\mathcal{U}$ is well-defined in case II. For fixed H_2, H_1, and $f : <^T \to \underset{\sim}{\delta}^1_{2n+1}$, we first claim that $\bar{F}([f])$ is well defined. By H_{2n+1}, the functions f_δ, $f_{(\xi_1, \ldots, \xi_n)}$, $f_{(\xi'_1, \ldots, \xi'_n)}$ are well-defined almost everywhere. We also note the following facts:

1) For almost all f there is a measure one set A w.r.t. $\tilde{v}_2^2$ s.t. for all tuples $(\xi_1, \ldots, \xi_n, \xi'_1, \ldots, \xi'_n) \in A$ and T the corresponding sequence as above, (for any fixed $(\mathcal{I}_1, \ldots, \mathcal{I}_k)$) we have that $\alpha_0 > \alpha^{(i_1)}, \ldots, \alpha^{(i_1, \ldots, i_k)}$. This follows from H_{2n+1} (i) and (ii).

2) By Lemma 2.7, we have that there is a measure one set E w.r.t. $E_1 \times \cdots \times E_p$ s.t. for $(\xi_{j_1}, \ldots, \xi_{j_p}) \in E$, for almost all $(\xi_{i_1}, \ldots, \xi_{i_q})$ w.r.t. $D_1 \times \cdots \times D_q$, $(\xi_1, \ldots, \xi_n, \xi'_1, \ldots, \xi'_n) \in A$, where, as above, $(\xi_1, \ldots, \xi_n, \xi'_1, \ldots, \xi'_n)$ is the enumeration of $(\xi_{i_1}, \ldots, \xi_{i_q}) \cup (\xi_{j_1}, \ldots, \xi_{j_p})$ by the subscripts.

We next claim that for fixed H_2, H_1 that for almost all $f : <^T \to \underset{\sim}{\delta}^1_{2n+1}$ we have that $\bar{F}([f])$ depends only on the equivalence class of f. We fix

H_2, H_1, $f\colon <^T \to \underset{\sim}{\delta}^1_{2n+1}$ of the correct type, and let f' agree with f a.e. w.r.t. the measures $v^{(i_1,\dots,i_k)}$ (we also assume f is into a sufficiently closed set so that $\bar{F}([f])$ is defined). We also fix f_s, f'_s representing $\sup_{\text{a.e.}} f$ and $\sup_{\text{a.e.}} f'$ w.r.t. $v^{(n_1)}$. We then have that there are measure one sets $A^{(i_1,\dots,i_k)}$ s.t. f_1, f_2 agree for $\vartheta^{(i_1,\dots,i_k)} \in A^{(i_1,\dots,i_k)}$. It also follows readily that there is a measure one set A^0 w.r.t. $v^{(n_1)}$ s.t. for $\alpha \in A^0$, $f_s(\alpha) = f'_s(\alpha)$. (This follows since $\sup_{\text{a.e.}\,\delta} f_1(\langle n_1, \delta\rangle) = \sup_{\text{a.e.}\,\delta} f_2(\langle n_1, \delta\rangle))$. It then follows that for almost all $(\xi_1, \dots, \xi_n, \xi'_1, \dots, \xi'_n)$ w.r.t. $\tilde{v}^2_2$ that (for any fixed $(\mathcal{I}_1, \dots, \mathcal{I}_k)$), $T = T'$, using $H_{2n+1}(1)$, where T, T' denote the sequences as in the definition of $\mathcal{U}$ corresponding to f, f'. It then follows readily from $H_{2n+1}(1)$ and Lemma 2.7 that for almost all δ w.r.t. v_2 that $g(\delta) = g'(\delta)$, where g, g' are as in the definition of $\mathcal{U}$. Hence $\bar{F}(f) = \bar{F}(f')$.

Hence $\mathcal{U}$ is well-defined.

Case III. $\mathcal{V}_1 \in \mathcal{S}_{2n+1}$ and $1^{(n_1)} \in \operatorname{dom} <^T$. We set $\mathcal{V}_2 = S^{1,m}_{2n+1}$ where $m = 2(e+n)$ where e is the total number of indices $(i_1, \dots, i_\ell)$ appearing in all the $C_{(\mathcal{I}_1,\dots,\mathcal{I}_k)}$ for $(\mathcal{I}_1, \dots, \mathcal{I}_k)$ an index in $\mathcal{C}$. Recall that n is the depth of the collection $\mathcal{C}$.

We proceed to define $\mathcal{U}$ ($= {}^r\mathcal{U}$) in this case.

We define measures $\mathcal{U}^{(\mathcal{I}_1,\dots,\mathcal{I}_k)}$ and $\mathcal{U}^{(\mathcal{I}_1,\dots,\mathcal{I}_k)}_{(i_1,\dots,i_\ell)}$ for $(\mathcal{I}_1, \dots, \mathcal{I}_k)$ an index in $\mathcal{C}$, and $(i_1, \dots, i_\ell)$ an index in $\mathcal{C}_{(\mathcal{I}_1,\dots,\mathcal{I}_k)}$.

We let $b =$ the number of indices of the form $(\mathcal{I}_1)$ in $\mathcal{C}$, and we set $\mathcal{U}^{(\mathcal{I}_1)} =$ the b-fold product of the ω-cofinal normal measure on $\underset{\sim}{\delta}^1_{2n+1}$.

For fixed $\mathcal{I}_1$, we let $(i^1), \dots, (i^p)$ denote the integers in $C_{\mathcal{I}_1}$ less than n_1. For $1 \leq q \leq p$ and the index $i_1 = i^q$ in $C_{\mathcal{I}_1}$, we define $\mathcal{U}^{(\mathcal{I}_1)}_{(i_1)}$ as follows. We let $<^{\Omega(\mathcal{I}_1)}_{(i_1)}$ be the lexicographic ordering on tuples $(\alpha_1, \dots, \alpha_q)$ of ordinals $< \underset{\sim}{\delta}^1_{2n+1}$, where $\alpha_2, \dots, \alpha_q < \alpha_1$. We then define A to have measure one if there is a c.u.b. subset $C \subseteq \underset{\sim}{\delta}^1_{2n+1}$ such that for $H\colon <^{\Omega(\mathcal{I}_1)}_{(i_1)} \to C$ of the correct type we have $\alpha \in A$, where α is represented w.r.t. $\mu^{<T}_{2n+1}$ by the function which assigns to $f\colon <^T \to \underset{\sim}{\delta}^1_{2n+1}$ of the correct type the ordinal $H(\langle \alpha_1, \dots, \alpha_q\rangle)$. Here, if f represents the ordinals $\alpha^{(i_1,\dots,i_k)}$, then $\alpha_1 =$ the largest of $\{\alpha^{i^1}, \dots, \alpha^{i^q}\}$, $\alpha_2 =$ the second largest of these, etc.

In general, we let $\mathcal{U}^{(\mathcal{I}_1,\dots,\mathcal{I}_k)}$ be the b-fold product, where $b =$ the number of indices of the form $(\mathcal{I}_1, \dots, \mathcal{I}'_k)$ extending $(\mathcal{I}_1, \dots, \mathcal{I}_{k-1})$ in $\mathcal{C}$, of the measure $\mathcal{U}^{(\mathcal{I}_1,\dots,\mathcal{I}_{k-1})}_{(i_1,\dots,i_\ell)}$, where $(i_1, \dots, i_\ell)$ is the index in $C_{(\mathcal{I}_1,\dots,\mathcal{I}_{k-1})}$ occurring last (note that this really only depends on $(\mathcal{I}_1, \dots, \mathcal{I}_{k-1})$).

For $(i_1, \dots, i_\ell)$ an index in $C_{(\mathcal{I}_1,\dots,\mathcal{I}_k)}$, we define $\mathcal{U}^{(\mathcal{I}_1,\dots,\mathcal{I}_k)}_{(i_1,\dots,i_\ell)}$ as follows. We let $(\vec{i^1}), \dots, (\vec{i^q})$ denote the elements of $C_{(\mathcal{I}_1,\dots,\mathcal{I}_k)}$ preceding (and including)

$(i_1,\dots,i_\ell)$, as well as the elements of $C_{(\mathcal{I}_1)},\dots,C_{(\mathcal{I}_1,\dots,\mathcal{I}_{k-1})}$, except that we exclude (n_1) if it occurs. We let $<^{\Omega(\mathcal{I}_1,\dots,\mathcal{I}_k)}_{(i_1,\dots,i_\ell)}$ be the ordering on tuples $(\alpha_1,\dots,\alpha_q)$ as above. We then define A to have measure one if there is a c.u.b. $C \subseteq \underset{\sim}{\boldsymbol{\delta}}^1_{2n+1}$ such that for all $H : <^{\Omega(\mathcal{I}_1,\dots,\mathcal{I}_k)}_{(i_1,\dots,i_\ell)} \to C$ of the correct type we have $\alpha \in A$, where α is represented w.r.t. $\mu^{<^T}_{2n+1}$ by the function which assigns to $f : <^T \to \underset{\sim}{\boldsymbol{\delta}}^1_{2n+1}$ of the correct type the ordinal $H(\langle \alpha_1,\dots,\alpha_q\rangle)$. Here, α_1 = the largest of $\{\alpha^{(\vec{i^1})},\dots,\alpha^{(\vec{i^q})}\}$, α_2 = the second largest of these, etc., where f represents the ordinals $\alpha^{(i_1,\dots,i_k)}$.

We then let $\mathcal{U}$ be the product:

$$\mathcal{U} = \left[\mathcal{U}^{(\mathcal{I}_1)}\right]^{m+2} \times \prod_{\mathcal{I}_1} \prod_{i_1 \in C_{\mathcal{I}_1}} \left[\mathcal{U}^{(\mathcal{I}_1)}_{(i_1)}\right]^{m+2} \times \cdots \times$$

$$\left[\prod_{(\mathcal{I}_1,\dots,\mathcal{I}_{n-1})\in C} \mathcal{U}^{(\mathcal{I}_1,\dots,\mathcal{I}_n)}\right]^{m+2} \times \prod_{(\mathcal{I}_1,\dots,\mathcal{I}_n)\in C} \;\prod_{(i_1,\dots,i_\ell)\in C_{(\mathcal{I}_1,\dots,\mathcal{I}_n)}} \left[\mathcal{U}^{(\mathcal{I}_1,\dots,\mathcal{I}_n)}_{(i_1,\dots,i_k)}\right]^{m+2},$$

where we order the factors in each product according to the ordering of indices in each $(\mathcal{I}_1,\dots,\mathcal{I}_k)$.

This completes the definition of $\mathcal{U}$ $(= \mathcal{U}^r)$ in this case, and hence completes the definition of $\mathcal{U}$.

We now proceed to define the family of embeddings $\mathcal{F}$ as in H_{2n+3}. We fix $\beta = (\beta_1,\dots,\beta_{r_0}) \in \mathcal{V}_2 = \prod_{r=1}^{r_0} \mathcal{V}_2^r$, and define $\mathcal{F}_\beta$. We fix functions $G_1,\dots,G_{r_0}$ representing $\beta_1,\dots,\beta_{r_0}$ as elements in $\mathcal{V}_2^r$. We fix a tuple $\vartheta = (\vartheta_1,\dots,\vartheta_{r_0}) \in \prod_{r=1}^{r_0} \mathcal{U}^r = \mathcal{U}$. We set

$$\mathcal{F}_\beta(\vartheta) = (\mathcal{F}_1(G_1,\vartheta_1),\dots,\mathcal{F}_{r_0}(G_{r_0},\vartheta_{r_0})) = (\alpha_1,\dots,\alpha_{r_0}),$$

say, where it remains to define $\alpha_r = \mathcal{F}_r(G_r,\vartheta_r)$ for $1 \le r \le r_0$. We again suppress writing the subscript r.

Definition 4.2 (Definition of $\alpha = \mathcal{F}(G,\vartheta)$). We consider the following cases:

Case I. $\mathcal{V}_1 \in \mathcal{W}_{2n+1}$. In this case, $G : S^{\ell',m}_{2n-1} \to \underset{\sim}{\boldsymbol{\delta}}^1_{2n+1}$, and ϑ is in the measure space $\mathcal{U}$ as in B_{2n+1} for T corresponding to $\mathcal{V}_1 = \mu^{<^T}_{2n+1}$. We represent α by $f : <^T \to \underset{\sim}{\boldsymbol{\delta}}^1_{2n+1}$ given by $f = G \circ f_\vartheta$, where $f_\vartheta : <^T \to S^{\ell',m}_{2n-1}$ as in B_{2n+1}. Here α is a tuple $(\cdots,\alpha^{(i_1,\dots,i_k)},\cdots)$. This is well-defined, and f is of the correct type almost everywhere.

Case II. $\mathcal{V}_1 \in \mathcal{S}_{2n+1}$ and $1^{(n_1)} \notin \operatorname{dom} <^T$. We fix functions $\bar{F}^{(\mathcal{I}_1,\dots,\mathcal{I}_k)}$ representing $\vartheta = (\cdots,\vartheta^{(\mathcal{I}_1,\dots,\mathcal{I}_k)},\cdots) \in \operatorname{dom}(\mathcal{U})$ with respect to $\mu^{<^T}_{2n+1}$ (note that there has been a slight change of notation from the definition of $\mathcal{U}$ where there we represented the elements of $\operatorname{dom}(\mathcal{U})$ as $(\cdots,\eta^{(\mathcal{I}_1,\dots,\mathcal{I}_k)},\cdots)$).

We then represent $\alpha = (\cdots, \alpha^{(\mathcal{I}_1,\dots,\mathcal{I}_k)}, \cdots)$ w.r.t. $\mu^{<^T}_{2n+1}$ by the function $[f] \mapsto F^{(\mathcal{I}_1,\dots,\mathcal{I}_k)}([f]) = G(\bar{F}^{(\vec{\mathcal{I}})}([f])$ for $f : <^T \to \underset{\sim}{\delta}^1_{2n+1}$ of the correct type. This is well-defined.

Case III. $\mathcal{V}_1 \in \mathcal{S}_{2n+1}$ and $1^{(n_1)} \in \operatorname{dom} <^T$. We represent $\alpha^{(\mathcal{I}_1,\dots,\mathcal{I}_k)}$, where $\alpha = (\cdots, \alpha^{(\mathcal{I}_1,\dots,\mathcal{I}_k)}, \cdots)$, with respect to $\mu^{<^T}_{2n+1}$ by the function $F^{(\mathcal{I}_1,\dots,\mathcal{I}_k)}$ defined as follows: We fix functions

$$H_0{}^{(\mathcal{I}_1,\dots,\mathcal{I}_k)}_{(i_1,\dots,i_\ell)}, H_1{}^{(\mathcal{I}_1,\dots,\mathcal{I}_k)}_{(i_1,\dots,i_\ell)}, H_p{}^{(\mathcal{I}_1,\dots,\mathcal{I}_k)}_{(i_1,\dots,i_\ell)}, 1 \le p \le m$$

representing the components of ϑ corresponding to the factor $\left[\mathcal{U}^{(\mathcal{I}_1,\dots,\mathcal{I}_k)}_{(i_1,\dots,i_\ell)}\right]^{m+2}$ in $\mathcal{U}$, where the index $(i_1, \dots, i_\ell)$ may not appear. When this index does not appear note that each $H_j^{(\mathcal{I}_1,\dots,\mathcal{I}_k)}$ is actually a b-sequence of functions $H_j^{(\mathcal{I}_1,\dots,\mathcal{I}_k)}(\ell)$, $1 \le \ell \le b$, where b is defined above (when $k = 1$, each $H_j^{(\mathcal{I}_1)}$ is a b-sequence of ordinals less than $\underset{\sim}{\delta}^1_{2n+1}$).

For fixed $f : <^T \to \underset{\sim}{\delta}^1_{2n+1}$, we set $F^{(\mathcal{I}_1,\dots,\mathcal{I}_k)}([f]) = G(\alpha_2, \dots, \alpha_m, \alpha_1)$, where $\alpha_1, \dots, \alpha_m$ are defined as follows:

We set $\alpha_1 = f(\langle n_1 \rangle)$.

We set $\alpha_2 = H_0^{\mathcal{I}_1}(\mathcal{I}_1)$. We recall here that $H_0^{\mathcal{I}_1}$ is a b-sequence of ordinals $< \underset{\sim}{\delta}^1_{2n+1}$ where $\mathcal{I}_1 < b$.

If $k = 1$ and $\mathcal{C}_{\mathcal{I}_1}$ is empty, we set $\alpha_3 = H_2^{\mathcal{I}_1}(\mathcal{I}_1), \dots, \alpha_m = H_{m-1}^{\mathcal{I}_1}(\mathcal{I}_1)$, and otherwise $\alpha_3 = H_1^{\mathcal{I}_1}(\mathcal{I}_1)$. Assume now that $k > 1$.

We let $r_i =$ the number of indices in $C_{(\mathcal{I}_1,\dots,\mathcal{I}_i)}$ for $1 \le i \le k$.

In general, we assume that α_j has been defined for $j \le u = \sum_{q \le i}(2r_q + 2)$, and that α_u is of the form $H_1{}^{(\mathcal{I}_1,\dots,\mathcal{I}_i)}_{(i_1,\dots,i_\ell)}(\dots, \vartheta^{(\vec{\jmath})}, \dots)$ where $(i_1, \dots, i_\ell)$ denotes the last index in $C_{\mathcal{I}_1,\dots,\mathcal{I}_i}$ and $(\vec{\jmath})$ denotes the indices in $C_{\mathcal{I}_1,\dots,\mathcal{I}_i}$ and $C_{\mathcal{I}_1,\dots,\mathcal{I}_{i'}}$ for $i' < i$. Also, the $\vartheta^{(\vec{\jmath})}$ are represented w.r.t. $v^{(\vec{\jmath})}$ by the functions induced from $f : <^T \to \underset{\sim}{\delta}^1_{2n+1}$, and the $\vartheta^{(\vec{\jmath})}$ are ordered as in the definition of $<^{\Omega(\mathcal{I}_1,\dots,\mathcal{I}_i)}_{(i_1,\dots,i_\ell)}$.

We then set

$$\alpha_{u+1} = H_0^{(\mathcal{I}_1,\dots,\mathcal{I}_{i+1})}(\mathcal{I}_{i+1})(\dots, \vartheta^{(\vec{\jmath})}, \dots),$$

which makes sense since the ordering $<^{\Omega(\mathcal{I}_1,\dots,\mathcal{I}_{i+1})}$ used in defining the function $H_0^{(\mathcal{I}_1,\dots,\mathcal{I}_{i+1})}(\mathcal{I}_{i+1})$ is the same as the ordering $<^{\Omega(\mathcal{I}_1,\dots,\mathcal{I}_i)}_{(i_1,\dots,i_\ell)}$ used in defining the $H_j{}^{(\mathcal{I}_1,\dots,\mathcal{I}_i)}_{(i_1,\dots,i_\ell)}$ for $(i_1, \dots, i_\ell)$ the last index in $C_{\mathcal{I}_1,\dots,\mathcal{I}_i}$. We also set $\alpha_{u+2} = H_1^{(\mathcal{I}_1,\dots,\mathcal{I}_{i+1})}(\mathcal{I}_{i+1})(\dots, \vartheta^{(\vec{\jmath})}, \dots)$.

For $2 \le v \le r_{i+1}$ we set

$$\alpha_{u+2\cdot v-1} = H_0{}^{(\mathcal{I}_1,\dots,\mathcal{I}_{i+1})}_{(i_1,\dots,i_\ell)}(\dots, \vartheta^{(\vec{\jmath})}, \dots),$$

where $(i_1, \dots, i_\ell)$ is the $v - 1$st index in $C_{(\mathcal{I}_1,\dots,\mathcal{I}_{i+1})}$ and the $(\vec{\jmath})$'s enumerate the indices in $C_{(\mathcal{I}_1,\dots,\mathcal{I}_{i+1})}$ occurring before $(i_1, \dots, i_\ell)$ in $C_{(\mathcal{I}_1,\dots,\mathcal{I}_{i+1})}$ along with the indices in $C_{(\mathcal{I}_1,\dots,\mathcal{I}_{i'})}$, $i' < i+1$, and occurring in the same order as in $<^{\Omega(\mathcal{I}_1,\dots,\mathcal{I}_{i+1})}_{(i_1,\dots,i_\ell)}$. We also set

$$\alpha_{u+2\cdot v} = H_1{}^{(\mathcal{I}_1,\dots,\mathcal{I}_{i+1})}_{(i_1,\dots,i_\ell)}(\dots,\vartheta^{(\vec{j})},\dots).$$

If $i+1 < k$, we proceed as above for the last index $(i_1,\dots,i_\ell)$ in $C_{(\mathcal{I}_1,\dots,\mathcal{I}_{i+1})}$ as well. Namely, we set

$$\alpha_{u+2\cdot r_{i+1}+1} = H_0{}^{(\mathcal{I}_1,\dots,\mathcal{I}_{i+1})}_{(i_1,\dots,i_\ell)}(\dots,\vartheta^{\vec{j}},\dots),$$

and

$$\alpha_{u+2\cdot r_{i+1}+2} = H_1{}^{(\mathcal{I}_1,\dots,\mathcal{I}_{i+1})}_{(i_1,\dots,i_\ell)}(\dots,\vartheta^{\vec{j}},\dots).$$

If $i+1 = k$, we also set $\alpha_{u+2\cdot r_{i+1}+1} = H_0{}^{(\mathcal{I}_1,\dots,\mathcal{I}_{i+1})}_{(u_1,\dots,i_\ell)}(\dots,\vartheta^{\vec{j}},\dots)$ as above. We then set $\alpha_{u+2\cdot r_{i+1}+c} = H_c{}^{(\mathcal{I}_1,\dots,\mathcal{I}_{i+1})}_{(i_1,\dots,i_\ell)}(\dots,\vartheta^{(\vec{j})},\dots)$, out to $\alpha_m = \alpha_{u+2\cdot r_{i+1}+(m-u-2\cdot r_{i+1})}$.

This completes the definition of $\mathcal{F}(G,\vartheta)$.

We show that $\mathcal{F}(G,\vartheta)$ is well-defined in case (3) (for the fixed function $G\colon <^m \to \underset{\sim}{\delta}^1_{2n+1}$). We let $H_1,\dots,H_p$ and $H'_1,\dots,H'_p$ (for the appropriate integer p) be given representing the same ϑ as in the above definition of $\mathcal{U}$. Hence for all $j \leq p$, if $H_j = H^{(\mathcal{I}_1,\dots,\mathcal{I}_k)}_{(i_1,\dots,i_\ell)}$ then $H_j = H'_j$ almost everywhere with respect to $<^{\Omega(\mathcal{I}_1,\dots,\mathcal{I}_k)}_{(i_1,\dots,i_\ell)}$. More precisely, they agree with respect to the measure $\mu = \mu^{(\mathcal{I}_1,\dots,\mathcal{I}_k)}_{(i_1,\dots,i_\ell)}$ on $(\underset{\sim}{\delta}^1_{2n+1})^q$ implicitly defined in the definition of $\mathcal{U}^{(\mathcal{I}_1,\dots,\mathcal{I}_k)}_{(i_1,\dots,i_\ell)}$. That is, A has μ measure one if for almost all $f\colon <^T \to \underset{\sim}{\delta}^1_{2n+1}$ the corresponding tuple $(\alpha_1,\dots,\alpha_q)$ of ordinals (as in the definition of $\mathcal{U}^{(\mathcal{I}_1,\dots,\mathcal{I}_k)}_{(i_1,\dots,i_\ell)}$) represented by f lies in A. We fix a c.u.b. $C \subseteq \underset{\sim}{\delta}^1_{2n+1}$ such that for $f\colon <^T \to C$ of the correct type, for all $1 \leq j \leq p$ we have $H_j(\alpha_1,\dots,\alpha_q) = H'_j(\alpha_1,\dots,\alpha_q)$, where again $(\alpha_1,\dots,\alpha_q)$ is the sequence represented by f (here q depends on j). It then follows that for $f\colon <^T \to C$ of the correct type that both $\mathcal{F}(G,H)$, $\mathcal{F}(G,H')$ are of the form $G(\beta_2,\dots,\beta_m,\beta_1)$ for the same sequence of ordinals $\beta_1,\dots,\beta_m$. Hence, $\mathcal{F}(G,H) = \mathcal{F}(G,H')$.

We have now shown that $\mathcal{F}(G,\vartheta)$ is well-defined in all cases. We next show that $\mathcal{F}([G],\vartheta)$ is well-defined. That is, $\mathcal{F}_\beta(\vartheta)$ does not depend on the particular choice of the functions $G_1,\dots,G_{r_0}$ but only on their equivalence classes. It is again enough to show this componentwise, and we again suppress the subscript r. We suppose, then, that $[G_1] = [G_2]$, and establish that for almost all ϑ w.r.t. $\mathcal{U}$ that $\mathcal{F}(G_1,\vartheta) = \mathcal{F}(G_2,\vartheta)$. We consider the following cases corresponding to the definition of $\mathcal{F}(G,\vartheta)$.

Case I. $\mathcal{V}_1 \in W_{2n+1}$. If G_1, $G_2\colon S^{\ell',m}_{2n-1} \to \underset{\sim}{\delta}^1_{2n+1}$ agree on a measure one set A w.r.t. $S^{\ell',m}_{2n-1}$, then from $\mathcal{B}_{2n+1}$, we have that for almost all ϑ, that f_ϑ has range in A almost everywhere. Hence, $G_1 \circ f_\vartheta = G_2 \circ f_\vartheta$ agree almost everywhere w.r.t. the $v^{(i_1,\dots,i_k)}$ (and agree if $k = 1$). Since $\mathcal{F}(G_1, \vartheta) = [G_1 \circ f_\vartheta]$ and likewise for $\mathcal{F}(G_2, \vartheta)$, we are done.

Case II. $\mathcal{V}_1 \in \mathcal{S}_{2n+1}$ and $1^{(n_1)} \notin \operatorname{dom} <^T$. We fix $G_1 : \underset{\sim}{\delta}^1_{2n+1} \to \underset{\sim}{\delta}^1_{2n+1}$ of the correct type, and fix G_2 s.t. $G_1 = G_2$ a.e. w.r.t. $\mu_{2n+1}^{S^{\ell',m}_{2n-1}}$. We let C be a c.u.b. subset of $\underset{\sim}{\delta}^1_{2n+1}$ s.t. for $g : S^{\ell',m}_{2n-1} \to C$ of the correct type $G_1([g]) = G_2([g])$. If our claim fails, then we get (from the strong partition relation on $\underset{\sim}{\delta}^1_{2n+1}$) a c.u.b. $C_2 \subseteq C$ s.t. for $H_2 : <^{\Omega_2} \to C_2$ of the correct type, there is a c.u.b. $C_1 (= C_1(H_2))$ s.t. for $H_1 : <^{\Omega_1} \to C_1$ of the correct type we have $\mathcal{F}(G_1; H_2, H_1) \neq \mathcal{F}(G_2; H_2, H_1)$. Here we are using $\mathcal{F}(G; H_2, H_1)$ to mean $\mathcal{F}(G, \vartheta)$ where ϑ is the ordinal derived from H_2 and H_1 as in the definition of $\mathcal{F}(G, \vartheta)$. Fix such H_2, H_1 of the correct type with ranges in C. We fix an index $(\mathcal{I}_1, \dots, \mathcal{I}_k)$. If $\mathcal{F}(G_1; H_2, H_1) = (\dots, \alpha_1^{(\mathcal{I}_1,\dots,\mathcal{I}_k)}, \dots)$ and similarly for $\mathcal{F}(G_2; H_2, H_1)$, and $\alpha_1^{(\mathcal{I}_1,\dots,\mathcal{I}_k)}$, $\alpha_2^{(\mathcal{I}_1,\dots,\mathcal{I}_k)}$ are represented by functions $F_1^{(\vec{\mathcal{I}})}$, $F_2^{(\vec{\mathcal{I}})}$ as in the definition of $\mathcal{F}$, then we show that for almost all $f : <^T \to \underset{\sim}{\delta}^1_{2n+1}$ that $F_1^{(\vec{\mathcal{I}})}([f]) = F_2^{(\vec{\mathcal{I}})}([f])$ to get a contradiction. We have that $F_1^{(\vec{\mathcal{I}})}([f]) = G_1(\bar{F}^{(\vec{\mathcal{I}})}([f]))$, $F_2^{(\vec{\mathcal{I}})}([f]) = G_2(\bar{F}^{(\vec{\mathcal{I}})}([f]))$, where $\bar{F}$ is as in the definition of $\mathcal{U}$ (*i.e.*, it represents ϑ corresponding to H_2, H_1). Recall that $F^{(\vec{\mathcal{I}})}([f])$ is represented with respect to $S^{\ell',m}_{2n-1} (= v(\mathcal{V}_2))$ by a function g, where for $\delta < \vartheta(S^{\ell',m}_{2n-1})$ we have that $g(\delta) = H_2(\gamma)$, for some γ as in the definition of $\mathcal{U}$, and hence $g(\delta) \in C_2 \subseteq C$. Since H_2 is of the correct type, g is of uniform cofinality ω.

It remains to show that g is almost everywhere strictly increasing.

It follows from D_{2n+1} that there is an r' and a measure one set A w.r.t. $\tilde{v}_2^2$ s.t. if $(\xi_1, \dots, \xi_n, \xi'_1, \dots, \xi'_n)$, $(\eta_1, \dots, \eta_n, \eta'_1, \dots, \eta'_n)$ are in A, then:

(1) If $\xi_{r'} < \eta_{r'}$, there then for all indices $(n_1, i_2, \dots, i_k) \in \operatorname{dom} T$ for almost all ρ w.r.t. $u^{(n_1,\dots,i_k)}$ we have

$$\left(\Pi^{(n_1,i_2,\dots,i_k)}_{(n_1)} \circ f_{(\xi_1,\dots,\xi_n)}(\rho)\right)(r) < \left(\Pi^{(n_1,i_2,\dots,i_k)}_{(n_1)} \circ f_{(\eta_1,\dots,\eta_n)}(\rho)\right)(r),$$

where $f_{(\xi_1,\dots,\xi_n)}$, $f_{(\eta_1,\dots,\eta_n)}$ denote embeddings as in H_{2n+1}, and r corresponds to the component of the measure $v^{(n_1)}$ such that in identifying $\vartheta_{v^{(n_1)}}$ with an ordinal, we order first by the rth component of the product measure $v^{(n_1)}$. Hence, we have

$$f((f_{(\xi_1,\dots,\xi_n)}(\rho))^{(n_1,\dots,i_k)}) < f((f_{(\eta_1,\dots,\eta_n)}(\rho))^{(n_1,\dots,i_k)}).$$

(2) If $\xi'_{r'} < \eta'_{r'}$, then for almost all ρ w.r.t. $\mathcal{U}^{(n_1)}$ we have

$$(f_{(\xi'_1,\dots,\xi'_n)}(\rho))(r) < (f_{(\eta'_1,\dots,\eta'_n)}(\rho))(r),$$

and hence $f_s(f_{(\xi'_1,\dots,\xi'_n)}(\rho)) \leq f_s(f_{(\eta'_1,\dots,\eta'_n)}(\rho))$, where f_s is (as in the definition of $\mathcal{U}$) monotonically increasing and represents the minimal equivalence class of a monotonically increasing function with $\sup_{\text{a.e.}} f_s = \sup_{\text{a.e.}} f^{(n_1)}$.

Hence, it follows that if g_1 denotes the function as in the definition of $\mathcal{U}$, then $g_1(\xi_1,\dots,\xi_n,\xi'_1,\dots,\xi'_n) < g_1(\eta_1,\dots,\eta_n,\eta'_1,\dots,\eta'_n)$ for such $\vec{\xi},\vec{\eta}$.

From $H_{2n+1}(i)$ and D_{2n+1} it also follows that there is a measure one set A_2 w.r.t. $S^{\ell',m}_{2n-1}$ such that for $\delta_1 < \delta_2$ in A_2 we have that for almost all ρ w.r.t. u_2 (as in the definition of $\mathcal{U}$, u_2 is the measure from H_{2n+1} for $\tilde{v}^2_2$ and $S^{\ell',m}_{2n-1}$), if $f_{\delta_1}(\rho) = (\xi_{j_1},\dots,\xi_{j_p})$, $f_{\delta_2}(\rho) = (\eta_{j_1},\dots,\eta_{j_p})$, then for almost all $(\gamma_{i_1},\dots,\gamma_{i_q})$ w.r.t. $D_1 \times \cdots \times D_q$ (as in the definition of $\mathcal{U}$), if we consider the corresponding sequences $(\xi_1,\dots,\xi_n,\xi'_1,\dots,\xi'_n)$, $(\eta_1,\dots,\eta_n,\eta'_1,\dots,\eta'_n)$ then $\vec{\xi},\vec{\eta}$ are in A and $\xi_{r'} < \eta_{r'}$, $\xi'_{r'} < \eta'_{r'}$. Hence it follows that for $\delta_1 < \delta_2$ in A_2 we have $g(\delta_1) < g(\delta_2)$, so g is almost everywhere strictly increasing.

Hence $\mathcal{F}([G],\vartheta)$ is well-defined in this case.

Case III. $\mathcal{V}_1 \in \mathcal{S}_{2n+1}$ and $1^{(n_1)} \in \operatorname{dom} <^T$. We let $G_1 \colon <^m \to \underset{\sim}{\boldsymbol{\delta}}^1_{2n+1}$ be of the correct type, and $G_2 \colon <^m \to \underset{\sim}{\boldsymbol{\delta}}^1_{2n+1}$ be such that $G_1 = G_2$ a.e. w.r.t. the m-fold product of the ω-cofinal normal measure on $\underset{\sim}{\boldsymbol{\delta}}^1_{2n+1}$. We let C be a c.u.b. subset of $\underset{\sim}{\boldsymbol{\delta}}^1_{2n+1}$ such that for $\alpha_2 < \cdots < \alpha_m < \alpha_1$ in C of cofinality ω we have $G_1(\alpha_2,\dots,\alpha_m,\alpha_1) = G_2(\alpha_2,\dots,\alpha_m,\alpha_1)$. We fix an index $(\mathcal{I}_1,\dots,\mathcal{I}_k)$ and show $\alpha_1^{(\mathcal{I}_1,\dots,\mathcal{I}_k)} = \alpha_2^{(\mathcal{I}_1,\dots,\mathcal{I}_k)}$ as in the previous case. We fix functions $\vec{H} \colon <^{\Omega(\mathcal{I}_1,\dots,\mathcal{I}_k)}_{(i_1,\dots,i_\ell)} \to C$ of the correct type representing ϑ (which happens for almost all ϑ w.r.t. $\mathcal{U}$). For $F_1^{(\mathcal{I}_1,\dots,\mathcal{I}_k)}$ representing $\alpha_1^{(\mathcal{I}_1,\dots,\mathcal{I}_k)}$ w.r.t. $\mu^{<T}_{2n+1}$ as in the definition of $\mathcal{F}$, we have that for almost all $f \colon <^T \to \underset{\sim}{\boldsymbol{\delta}}^1_{2n+1}$ that $F_1^{(\mathcal{I}_1,\dots,\mathcal{I}_k)}([f]) = G_1(\alpha_2,\dots,\alpha_k,\alpha_1)$ and similarly for F_2, where here $\alpha_1,\dots,\alpha_m$ are as in the definition of $\mathcal{F}$. Since the range of the H's is in C, it follows that the ordinals $\alpha_2,\dots,\alpha_m$ are in C. We may also assume $f \colon <^T \to C$, so that $\alpha_1 \in C$. Also, since the H's are of the correct type, the α_i have cofinality ω. It remains to show that $\alpha_2 < \cdots < \alpha_m < \alpha_1$.

We clearly have that for almost all f, $\alpha_1 = f(\langle n_1 \rangle) > \alpha_i = H_i([f])$, since the $H_i([f])$ do not depend on $f(\langle n_1 \rangle)$. Here we have written $H_i([f])$ to abbreviate H_i evaluated at the sequence of ordinals represented by the appropriate subfunctions of f, as in the definition of $\mathcal{F}$ for this case.

It also follows readily that we may assume (which happens for almost all ϑ) that the $H_1,\dots,H_p$ are chosen s.t. for almost all f, $H_i([f]) < H_{i+1}([f])$. This uses the fact that the indices $\vec{\jmath}$ occurring in the definition of $<^{\Omega(\mathcal{I}_1,\dots,\mathcal{I}_k)}_{(i_1,\dots,i_\ell)}$ corresponding to H_i are a subset of those corresponding to H_{i+1}. Hence it follows that $\alpha_2 < \cdots < \alpha_m$, since if $\alpha_i = H_a([f])$, $a_{i+1} = H_b([f])$, then $a < b$, from the definition of $\mathcal{F}$. Hence $\mathcal{F}([G],\vartheta)$ is well-defined in this case.

This completes the proof that $\mathcal{F}$ is well-defined in all cases. We proceed to establish H_{2n+3}. First we consider $H_{2n+3}(i)$.

We let A have measure one with respect to $\mathcal{V}_1 = \prod_r \mathcal{V}_1^r$. From the definition of the product measure and the strong partition relation on $\underset{\sim}{\boldsymbol{\delta}}^1_{2n+1}$, it follows readily that there is a c.u.b. $C \subseteq \underset{\sim}{\delta}^1_{2n+1}$ such that if $\alpha = (\alpha_1, \dots, \alpha_{r_0})$ is represented by functions $F_1, \dots, F_{r_0}$ (as in the definition of the measures $\mathcal{V}_1^r$) then $\alpha \in A$ provided the following hold:

P1) $F_1, \dots, F_{r_0}$ have range in C and are of the correct type (where $F_i : <^{T_i} \to \underset{\sim}{\delta}^1_{2n+1}$ if $\mathcal{V}_1^i \in \mathcal{W}_{2n+1}$ and $F_i : <^{\Omega_i} \to \underset{\sim}{\delta}^1_{2n+1}$ if $\mathcal{V}_1^i \in \mathcal{S}_{2n+1}$).

P2) a) If $i < j$ and $\mathcal{V}_1^i, \mathcal{V}_1^j \in \mathcal{W}_{2n+1}$, then $\sup_{\text{a.e.}} F_i < \inf F_j$.

b) If $i < j$, $\mathcal{V}_1^i, \mathcal{V}_1^j \in \mathcal{S}_{2n+1}$ and $1^{(n_1)} \notin \operatorname{dom} <^{T_i}$, $1^{(n_1)} \notin \operatorname{dom} <^{T_j}$ (where the n_1 may be different for i, j), and $\operatorname{cf} \vartheta_{v_i^{n_1}} = \operatorname{cf} \vartheta_{v_j^{n_1}}$ (where $v_i^{n_1}$ is the measure v^{n_1} corresponding to $<^{T_i}$ and similarly for j), then there is a c.u.b. $C_2 \subseteq \underset{\sim}{\delta}^1_{2n+1}$ such that if $f_1 : <^{T_i} \to C_2$, $f_2 : <^{T_i} \to C_2$ are of the correct type with $\sup_{\text{a.e.}} f_1 = \sup_{\text{a.e.}} f_2$, then $F_i([f_1]) < F_j([f_2])$.

c) Same as b above where $1^{(n_1)} \in \operatorname{dom} <^{T_i}$ and $1^{(n_1)} \in \operatorname{dom} <^{T_j}$, and we remove the restriction on $\operatorname{cf} \vartheta_{v_i^{n_1}} = \operatorname{cf} \vartheta_{v_j^{n_1}}$. Here of course, $\sup_{\text{a.e.}} f_1$ means $f_1(\langle n_1 \rangle)$ and similarly for f_2.

We fix a c.u.b. set C. We fix $\beta = (\beta_1, \dots, \beta_r, \dots, \beta_{r_0})$ in the measure space $\mathcal{V}_2 = \prod_r \mathcal{V}_2^r$ represented by functions $G_1, \dots, G_r, \dots, G_{r_0}$ satisfying the following (which happens for almost all β w.r.t. $\mathcal{V}_2$):

i) $G_1, \dots, G_{r_0}$ have range in C and are of the correct type.

ii) a) If $i < j$ and $\mathcal{V}_2^i, \mathcal{V}_2^j \in W_{2n+1}$, then $\sup_{\text{a.e.}} G_i < \inf G_j$.

b) If $i < j$ and $\mathcal{V}_2^i, \mathcal{V}_2^j \in S_{2n+1}^{\ell,m}$ with $\ell > 1$ (ℓ, m may depend on i, j) and $\operatorname{cf} \vartheta(S_{2n-1}^{\ell_i,m_i}) = \operatorname{cf} \vartheta(S_{2n-1}^{\ell_j,m_j})$ (this is in fact equivalent to $\ell_i = \ell_j$), then there is a c.u.b. subset $C_2 \subseteq \underset{\sim}{\delta}^1_{2n+1}$ such that for $g_1 : \vartheta(S_{2n-1}^{\ell_i,m_i}) \to C_2$, $g_2 : \vartheta(S_{2n-1}^{\ell_j,m_j}) \to C_2$ of the correct type, if $\sup_{\text{a.e.}} g_1 \leq \sup_{\text{a.e.}} g_2$, then $G_i([g_1]) < G_j([g_2])$.

c) If $i < j$ and $\mathcal{V}_2^i, \mathcal{V}_2^j \in S_{2n+1}^{1,m}$, then there is a c.u.b. $C_2 \subseteq \underset{\sim}{\delta}^1_{2n+1}$ such that for $(\gamma_2, \dots, \gamma_m, \gamma_1)$, $(\delta_2, \dots, \delta_m, \delta_1)$ in C_2, if $\gamma_1 \leq \delta_1$ then $G_1(\gamma_2, \dots, \gamma_m, \gamma_1) < G_2(\delta_2, \dots, \delta_m, \delta_1)$.

We fix $G_1, \dots, G_{r_0}$ satisfying (i) and (ii).

It then follows readily that for almost all $\vartheta = (\vartheta_1, \dots, \vartheta_{r_0})$ w.r.t. $\mathcal{U} = \prod_r \mathcal{U}^r$, where we fix $\vec{H}_1, \dots, \vec{H}_{r_0}$ representing $\vartheta_1, \dots, \vartheta_{r_0}$ (where $\vec{H}_r = (H_{r,1}, H_{r,2})$ if $1^{(n_1)} \notin \operatorname{dom} <^{T_r}$ and $\vec{H}_r = (\dots, H^{(\mathcal{I}_1, \dots, \mathcal{I}_k)}_{(i_1, \dots, i_\ell)}, \dots)$ if $1^{(n_1)} \in \operatorname{dom} <^{T_i}$), that if $\alpha = (\alpha_1, \dots, \alpha_{r_0}) = \mathcal{F}_\beta(\vartheta)$, where α_i is represented by F_i as in the definition of $\mathcal{F}$, then the following are satisfied:

Claim 4.3.

a) If $i < j$ and $\mathcal{V}_1^i, \mathcal{V}_1^j \in \mathcal{W}_{2n+1}$, then $\sup_{\text{a.e.}} F_i < \inf F_j$.

b) If $\mathcal{V}_1^i \in \mathcal{S}_{2n+1}$ and $1^{(n_1)} \notin \operatorname{dom} <^{T_i}$, then for almost all $f_i : <^{T_i} \to \underset{\sim}{\delta}^1_{2n+1}$, if $F_i([f_i]) = G_i([g])$, where $g : \vartheta(S_{2n-1}^{\ell_i,m_i}) \to \underset{\sim}{\delta}^1_{2n+1}$ is as in the definition of $\mathcal{F}$, then $\sup_{\text{a.e.}} g = \sup_{\text{a.e.}} f_i$.

c) If $\mathcal{V}_1^i \in \mathcal{S}_{2n+1}$ and $1^{(n_1)} \in \operatorname{dom} <^{T_i}$, then for almost all $f_i \colon <^{T_i} \to \underset{\sim}{\boldsymbol{\delta}}^1_{2n+1}$, $F_i([f_i]) = G_i(\alpha_2, \dots, \alpha_m, \alpha_1)$, where $\alpha_1 = \sup f_1$

Proof. a) and c) are immediate from the definition of $\mathcal{F}$ and the choice of the G_i. b) follows from H_{2n+1} (i), (ii), the fact that measure one sets are cofinal in v^{n_1}, and the fact that almost all f_i are sufficiently closed with respect to the H_i.

That is, there is a measure on set E w.r.t. $E_1 \times \cdots \times E_p$ (as in the definition of $\mathcal{U}$) s.t. for $(\xi_{i_1}, \dots, \xi_{i_p}) \in E$, for almost all $(\xi_{j_1}, \dots, \xi_{j_q})$ w.r.t. $D_1 \times \cdots \times D_q$ and $(\xi_1, \dots, \xi_n, \xi_1', \dots, \xi_n')$ the enumeration of these ordinals in the order corresponding to the measures in $\tilde{v}_2^2$, we have that $g_1(\xi_1, \dots, \xi_n, \xi_1', \dots, \xi_n') = H_{i,1}(\langle T \rangle)$, where (for a fixed $(\mathcal{I}_1, \dots, \mathcal{I}_k)$), $T = \langle \alpha_0, \mathcal{I}_1, \dots, \mathcal{I}_k, \alpha^{(i_1, \dots, i_k)}, \dots \rangle$ and $\alpha_0, \dots, \alpha^{(i_1, \dots, i_k)} < \sup \operatorname{ran} f_i$. This follows from H_{2n+1} (ii) and the fact that the range of f_i may be taken as closed under $j_{u^{(n_1, i_2, \dots, i_k)}}$.

It further follows from the definition of the measures $E_1, \dots, E_p$ and D_{2n+1} that there is a measure one set E' w.r.t. $E_1 \times \cdots \times E_p$ s.t. if $(\xi_i, \dots, \xi_{i_p}) \in E'$ then $\sup_{\xi_{j_1}, \dots, \xi_{j_q}} g_1(\xi_1, \dots, \xi_n, \xi_1', \dots, \xi_k') < \sup \operatorname{ran} f_i$.

It then follows from another application of H_{2n+1}(ii) that for almost all δ w.r.t. $S^{\ell',m}_{2n-1} (= v(\mathcal{V}_2))$ that $g(\delta)$ is represented w.r.t. u_2 (as in the definition of $\mathcal{U}$, u_2 is the measure from H_{2n+1} corresponding to $E_1 \times \cdots \times E_p$ and $S^{\ell',m}_{2n-1}$) by a function h_2^δ s.t. $\sup_{\text{a.e.}} h_2^\delta(\gamma) < \sup \operatorname{ran} f_i$.

Hence, we may assume that $h_2^\delta(\gamma) < \sup \operatorname{ran} f_i$ almost everywhere. Hence $\sup_{\text{a.e.}} g \le \sup \operatorname{ran} f_i$.

Also $\sup_{\text{a.e.}} g \ge \sup \operatorname{ran} f_i$ follows from H_{2n+1}(i) and the fact that measure on sets in v^{n_1} are cofinal in $\vartheta_{v^{n_1}}$. This establishes (b) of Claim 4.3. ⊣

The second claim P2 above about the F_i now follows from a), b), c) of Claim 4.3 and the choice of the G_i.

Hence, it remains to establish P1 above, and it is enough to establish this componentwise. We again suppress the subscript r. We consider the following cases:

Case I. $\mathcal{V}_1 \in \mathcal{W}_{2n+1}$. In this case, $\mathcal{F}_\beta(\vartheta)$ is represented by $\bar{f} \colon <^T \to \underset{\sim}{\boldsymbol{\delta}}^1_{2n+1}$ given by $\bar{f} = G \circ f_\vartheta$, where $f_\vartheta \colon <^T \to S^{\ell',m}_{2n-1}$ is as $\mathcal{B}_{2n+1}$. Since f_ϑ is of the correct type almost everywhere, and G is of the correct type, it follows that $\bar{f}$ is of the type almost everywhere, and we may assume (by Lemma 2.8) that it is of the correct type everywhere. Since $\operatorname{ran} G \subseteq C$, $\operatorname{ran} \bar{f} \subseteq C$.

Case II. $\mathcal{V}_1 \in \mathcal{S}_{2n+1}$ and $1^{(n_1)} \notin \operatorname{dom} <^T$. We fix $\vec{H}$ of the correct type representing ϑ as in the definition of $\mathcal{U}$ (which happens for almost all ϑ w.r.t. $\mathcal{U}$). Since for almost all $f \colon <^T \to \underset{\sim}{\boldsymbol{\delta}}^1_{2n+1}$ we have that $F([f]) = G([g])$ for some $g : \vartheta(S^{\ell',m}_{2n-1}) \to \underset{\sim}{\boldsymbol{\delta}}^1_{2n+1}$, it follows that for almost all f that $F([f]) \in C$, and F has uniform cofinality ω.

We show that F is order-preserving almost everywhere.

We fix indices $(\mathcal{I}_1, \dots, \mathcal{I}_k)$, $(J_1, \dots, J_\ell)$ in the collection C, and fix functions $f_1, f_2 \colon <^T \to \underset{\sim}{\delta}^1_{2n+1}$ of the correct type, and let

$$S_1 = \langle \alpha_0, \mathcal{I}_1, \dots, \alpha^{(i_1)}, \dots, \mathcal{I}_k, \dots, \alpha^{(i_1, \dots, i_k)}, \dots \rangle,$$
$$S_2 = \langle \beta_0, J_1, \dots, \beta^{(j_1)}, \dots, J_k, \dots, \beta^{(j_1, \dots, j_\ell)}, \dots \rangle$$

denote the corresponding sequences as in the definition of $\mathcal{V}_1$. We assume that $S_1 <_{\mathrm{BK}} S_2$, and proceed to show that $F(\langle S_1 \rangle) < F(\langle S_2 \rangle)$. We consider the following cases:

Subcase II.a. $\alpha_0 < \beta_0$. That is $\sup_{\text{a.e.}} f_1 < \sup_{\text{a.e.}} f_2$. It then follows from H_{2n+1}(i) that for almost all $(\xi_1, \dots, \xi_n, \xi'_1, \dots, \xi'_n)$ that if

$$T_1 = \langle \gamma_0, \mathcal{I}_1, \dots, \gamma^{(i_1)}, \dots, \mathcal{I}_k, \dots, \gamma^{(i_1, \dots, i_k)}, \dots \rangle,$$
$$T_2 = \langle \delta_0, J_1, \dots, \delta^{(j_1)}, \dots, J_\ell, \dots, \delta^{(i_1, \dots, i_\ell)}, \dots \rangle$$

denote the sequences as in the definition of $\mathcal{U}$, then $\gamma_0 < \delta_0$, and hence $T_1 <^\Omega T_2$, and the result follows.

Subcase II.b. $\alpha_0 = \beta_0$ and there is a lexicographically least position in which S_1, S_2 disagree which is of the form $\alpha^{(i_1, \dots, i_m)} < \beta^{(i_1, \dots, i_m)}$, where $\mathcal{I}_1 = J_1, \dots, \mathcal{I}_m = J_m$. It then follows from H_{2n+1}(i) that for almost all sequences $(\xi_1, \dots, \xi_n, \xi'_1, \dots, \xi'_k)$ that T_1, T_2 agree up to $\gamma^{(i_1, \dots, i_m)}$ and that $\gamma^{(i_1, \dots, i_m)} < \delta^{(i_1, \dots, i_m)}$.

Subcase II.c. $\alpha_0 < \beta_0$, and the least position where S_1, S_2 disagree is of the form $\mathcal{I}_m < J_m$. This is similar to the previous case.

Subcase II.d. S_1 extends S_2. As above, by H_{2n+1}(i), we have that for almost all $(\xi_1, \dots, \xi_n, \xi'_1, \dots, \xi'_k)$ that T_1 extends T_2, hence $H_1(\langle T_1 \rangle) < H_1(\langle T_2 \rangle)$, from which the result follows readily.

Case III. $\mathcal{V}_1 \in S_{2n+1}$ and $1^{(n_1)} \in \operatorname{dom} <^T$. We again fix functions $\vec{H}$ representing ϑ as in the definition of $\mathcal{U}$, and fix indices $(\mathcal{I}_1, \dots, \mathcal{I}_k)$, $(J_1, \dots, J_\ell)$, functions f_1, f_2 of the correct type and suppose $S_1 <_{\mathrm{BK}} S_2$, where S_1, S_2 denote the corresponding sequences. We let $(\alpha_2, \dots, \alpha_m, \alpha_1)$, $(\beta_2 \dots, \beta_m, \beta_1)$ denote the sequences of ordinals as in the definition of $\mathcal{F}$. We consider the following cases:

Subcase III.a. $\sup f_1 = f_1(\langle n_1 \rangle) < f_2(\langle n_2 \rangle) = \sup f_2$. In this case, $\alpha_1 < \beta_1$, and $G(\alpha_2, \dots, \alpha_m, \alpha_1) < G(\beta_2, \dots, \beta_m, \beta_1)$ follows from the choice of G.

Subcase III.b. $\alpha_1 = \beta_1$ and there is a lexicographically least position where S_1, S_2 disagree which is the form $\alpha^{(i_1, \dots, i_p)} < \beta^{(i_1, \dots, i_p)}$, where $\mathcal{I}_1 = J_1, \dots, \mathcal{I}_p = J_p$. We then have that for each $H^{(\mathcal{I}_1, \dots, \mathcal{I}_v)}_{(i'_1, \dots, i'_w)}$, where $v < m$ or $v = m$ and $(i'_1, \dots, i'_w)$ precedes $(i_1, \dots, i_p)$, that $H^{(\mathcal{I}_1, \dots, \mathcal{I}_v)}_{(i'_1, \dots, i'_w)}([f_1]) = H^{(\mathcal{I}_1, \dots, \mathcal{I}_v)}_{(i'_1, \dots, i'_w)}([f_2])$, since from the definition of $\mathcal{U}^{(\mathcal{I}_1,\dots,\mathcal{I}_v)}_{(i'_1,\dots,i'_w)}$ we have that $H^{(\mathcal{I}_1,\dots,\mathcal{I}_v)}_{(i'_1,\dots,i'_w)}([f])$ depends only on the ordinals $\gamma^{(i''_1,\dots,i''_{w''})}$ represented by f for indices $(i''_1,\dots,i''_{w''})$ preceding and including $(i'_1,\dots,i'_{w'})$ and hence $\gamma_1^{(i''_1,\dots,i''_{w''})} = \gamma_2^{(i''_1,\dots,i''_{w''})}$ for f_1, f_2. Hence $\alpha_2 = \beta_2,\dots,\alpha_k = \beta_k$ for k corresponding to the H preceding $H^{(\mathcal{I}_1,\dots,\mathcal{I}_p)}_{0(i_1,\dots,i_p)}$. Since $H^{(\mathcal{I}_1,\dots,\mathcal{I}_p)}_{0(i_1,\dots,i_p)}$ is order-preserving and $\gamma_1^{(i_1,\dots,i_p)} < \gamma_2^{(i_1,\dots,i_p)}$, it then follows that $\alpha_{k+1} < \beta_{k+1}$. Since G is order-preserving, it then follows that $G(\alpha_2,\dots,\alpha_m,\alpha_1) < G(\beta_2,\dots,\beta_m,\beta_1)$.

Subcase III.c. $\alpha_1 = \beta_1$, and the least position where S_1, S_2 disagree is of the form $\mathcal{I}_p < J_p$. Proceeding as above, we have that $\alpha_2 = \beta_2,\dots,\alpha_k = \beta_k$ corresponding to the functions H preceding $H^{(\mathcal{I}_1,\dots,\mathcal{I}_p)}(1)$ used in the definition of $\mathcal{F}$. It then follows that $\alpha_{k+1} < \beta_{k+1}$ since $\alpha_{k+1} = H^{(\mathcal{I}_1,\dots,\mathcal{I}_p)}(\mathcal{I}_p)([f]) < H^{(\mathcal{I}_1,\dots,\mathcal{I}_p)}(J_p)([f]) = \beta_{k+1}$, since in the b−fold product measure $\mathcal{U}^{(\mathcal{I}_1,\dots,\mathcal{I}_p)}$, we may assume the functions $H^{(\mathcal{I}_1,\dots,\mathcal{I}_p)}(1),\dots,H^{(\mathcal{I}_1,\dots,\mathcal{I}_p)}(b)$ are such that if $i < j$ then $H^{(\vec{\mathcal{I}})}(i)([f]) < H^{(\vec{\mathcal{I}})}(j)([f])$. Hence, $G(\alpha_2,\dots,\alpha_m,\alpha_1) < G(\beta_2,\dots,\beta_m,\beta_1)$.

Subcase III.d. $\alpha_1 = \beta_1$, and S_1 extends S_2. We again have that $\alpha_2 = \beta_2,\dots,\alpha_k = \beta_k$ corresponding to the H functions preceding $H_0{}^{(\mathcal{I}_1,\dots,\mathcal{I}_\ell)}_{(i_1,\dots,i_p)}$, where $(i_1,\dots,i_p)$ denotes the last index in $C_{(\mathcal{I}_1,\dots,\mathcal{I}_\ell)}$, and $(\mathcal{I}_1,\dots,\mathcal{I}_\ell) = (J_1,\dots,J_\ell)$ here. We then have from the definition of $\mathcal{F}$ that $\alpha_{k+1} = H^{(\mathcal{I}_1,\dots,\mathcal{I}_\ell)}_{0(i_1,\dots,i_p)}([f_1]) = H^{(\mathcal{I}_1,\dots,\mathcal{I}_\ell)}_{0(i_1,\dots,i_p)}([f_2]) = \beta_{k+1}$ as above. Also, $\alpha_{k+2} = H^{(\mathcal{I}_1,\dots,\mathcal{I}_\ell)}_{1(i_1,\dots,i_p)}([f_1]) < H^{(\mathcal{I}_1,\dots,\mathcal{I}_\ell)}_{2(i_1,\dots,i_p)}([f_2]) = \beta_{k+2}$ without loss of generality. Hence $G(\alpha_2,\dots,\alpha_m,\alpha_1) < G(\beta_2,\dots,\beta_m,\beta_1)$.

This establishes P1 above in all cases. This completes the proof of $H_{2n+3}(\mathrm{i})$. We now consider $H_{2n+3}(\mathrm{ii})$. It is enough to establish this componentwise. We consider the following cases:

Case I. $\mathcal{V}_1 \in \mathcal{W}_{2n+1}$. In this case, for all β w.r.t. $\mathcal{V}_2$ represented by a function $G\colon \vartheta(S^{\ell',m}_{2n-1}) \to \underset{\sim}{\delta}^1_{2n+1}$ of the correct type we have from the definition of $\mathcal{F}$ that for almost all ϑ w.r.t. $\mathcal{U}$ that $\mathcal{F}_\beta(\vartheta) = [G \circ f_\vartheta]$, and $\sup_{\text{a.e.}} G \circ f_\vartheta \le \sup_{\text{a.e.}} G$ (in fact strictly $<$), which immediately gives $H_{2n+3}(\mathrm{ii})$.

Case II. $\mathcal{V}_1 \in \mathcal{S}_{2n+1}$ and $1^{(n_1)} \notin \operatorname{dom} T$. We fix β represented by a $G\colon \underset{\sim}{\delta}^1_{2n+1} \to \underset{\sim}{\delta}^1_{2n+1}$ of the correct type. It then follows as in the proof of $H_{2n+3}(\mathrm{i})$ that for almost all ϑ w.r.t. $\mathcal{U}$ that if $\bar{F}$ represents ϑ w.r.t. $\mu^{<^T}_{2n+1}$ that for almost all $f\colon <^T \to \underset{\sim}{\delta}^1_{2n+1}$ that $\bar{F}([f]) = [g]_{S^{\ell',m}_{2n-1}}$, where $\sup_{\text{a.e.}} g = \sup_{\text{a.e.}} f$. Hence, if F represents $\mathcal{F}_\beta(\vartheta)$ w.r.t. $\mu^{<^T}_{2n+1}$ as in the definition of $\mathcal{F}$, then for almost all f we have $F([f]) = G([g]) < G(j_{S^{\ell',m}_{2n-1}}(\sup_{\text{a.e.}} f))$. This establishes $H_{2n+3}(\mathrm{ii})$.

Case III. $\mathcal{V}_1 \in \mathcal{S}_{2n+1}$ and $1^{(n_1)} \in \operatorname{dom} T$. We again fix β in the measure space $\mathcal{V}_2$ represented by a $G\colon <^m \to \underset{\sim}{\boldsymbol{\delta}}^1_{2n+1}$ of the correct type. For any ϑ in the measure space $\mathcal{U}$, it follows the definition of $\mathcal{F}$ that if F represents $\mathcal{F}_\beta(\vartheta)$ w.r.t. $\mu^{<^T}_{2n+1}$, then for almost all $f\colon <^T \to \underset{\sim}{\boldsymbol{\delta}}^1_{2n+1}$ we have that $F([f]) = G(\alpha_2, \dots, \alpha_m, \alpha_1)$, where $\alpha_1 = \sup f = f(\langle n_1 \rangle)$, and $\alpha_2, \dots, \alpha_m < \alpha_1$. Hence $F([f]) < G'(\alpha_1) \equiv \sup_{\alpha_2, \dots, \alpha_{m'} < \alpha_1} G(\alpha_2, \dots, \alpha_m, \alpha_1) < \underset{\sim}{\boldsymbol{\delta}}^1_{2n+1}$. This establishes $H_{2n+3}(\text{ii})$.

We now consider $H_{2n+3}(\text{iii})$. It is enough to establish this componentwise. We again consider the following cases:

Case I. $\mathcal{V}_1 \in \mathcal{W}_{2n+1}$. For $\mathcal{V}_1'$ as in $H_{2n+3}(\text{iii})$ and corresponding measures $\mathcal{V}_2, \mathcal{V}_2'$, we have that $\mathcal{V}_2 = \mu_{2n+1}^{S^{\ell',m}_{2n-1}}$, $\mathcal{V}_2' = \mu_{2n+1}^{S^{\ell',m'}_{2n-1}}$ for some m, m' (here $\ell' = 2^n - 1$ is maximal). From the definition of $\sim$, we have that $\mathcal{V}_2 \sim \mathcal{V}_2'$.

Case II. $\mathcal{V}_1 \in \mathcal{S}_{2n+1}$ and $1^{(n_1)} \notin \operatorname{dom} <^T$. We let $\mathcal{V}_1'$ be as in $H_{2n+3}(\text{iii})$. It follows from the definition of Π that since $\Pi^{\mathcal{V}_1'}_{\mathcal{V}_1}$ is defined, $v^{n_1} = v'^{n_1}$, a measure on $\vartheta(v^{n_1}) = \vartheta_1 \times \cdots \times \vartheta_b$ for some b, where in identifying $\vartheta(v^{n_1})$ with an ordinal, we order first by ϑ_c, say, for some $1 \le c \le b$ (which is the same for $\mathcal{V}_1, \mathcal{V}_1'$). Hence, it follows from $H_{2n+1}(\text{iii})$ that $\tilde{v}^2_2 = (E_1 \times \cdots \times E_p)^2$ where $E_i \sim E_j$ and $E_i \in R_{2n-1}$ for $1 \le i, j \le p$ and also $\tilde{v}'^2_2 = (E_1' \times \cdots \times E_p')^2$, where $E_i' \sim E_i$. Hence, it follows from $H_{2n+1}(\text{iv})$ that $v_2 = v(\mathcal{V}_2) \sim E_i \sim E_i' \sim v_2'$, and hence $\mathcal{V}_2 \sim \mathcal{V}_2'$ from the definition of $\sim$.

Case III. $\mathcal{V}_1 \in \mathcal{S}_{2n+1}$ and $1^{(n_1)} \in \operatorname{dom} <^T$. We let $\mathcal{V}_1'$ be as in $H_{2n+3}(\text{iii})$. Since $\Pi^{\mathcal{V}_1'}_{\mathcal{V}_1}$ is defined, it follows that $n_1 \in \operatorname{dom} <^{T'}$, (since $<^{T'}$ extends $<^T$) and hence $\mathcal{V}_1 = S^{1,m}_{2n+1}$, $\mathcal{V}_1' = S^{\ell,m'}_{2n+1}$ for some m. From the definition of $\sim$ it follows that $\mathcal{V}_1 \sim \mathcal{V}_1'$.

This establishes $H_{2n+3}(\text{iii})$. We now consider $D_{2n+3}(\text{i})$.

For $\mathcal{V}_1 = \prod_r \mathcal{V}_1^r$, $\mathcal{V}_2 = \prod_r \mathcal{V}_2^r$, $\mathcal{U} = \prod_r \mathcal{U}^r$ as in D_{2n+3}, we recall that $\mathcal{F}$ is defined componentwise, that is, for $\beta = (\beta_1, \dots, \beta_{r_0})$ in the measure space $\mathcal{V}_2$, $\vartheta = (\vartheta_1, \dots, \vartheta_{r_0})$ in the measure space $\mathcal{U}$ we have $\mathcal{F}_\beta(\vartheta) = (\mathcal{F}_{\beta_1}(\vartheta_1), \dots, \mathcal{F}_{\beta_{r_0}}(\vartheta_{r_0}))$ in the the measure space $\mathcal{V}_1$. For any fixed $r \le r_0$, we take $r' = r$, and consider $\mathcal{V}_2^r, \mathcal{U}^r$. We consider the following cases.

Case I. $\mathcal{V}_1^r \in \mathcal{W}_{2n+1}$. If $\beta_1^r < \beta_2^r$ are represented by functions $G_1, G_2 : \vartheta(S^{\ell',m}_{2n-1}) \to \underset{\sim}{\boldsymbol{\delta}}^1_{2n+1}$ of the correct type, so $G_1 < G_2$ almost everywhere, then for almost all γ w.r.t. u (as in B_{2n+1}) $[G_1 \circ f_\gamma] < [G_2 \circ f_\gamma]$, using B_{2n+1}. This establishes $D_{2n+3}(\text{i})$ in this case.

Case II. $\mathcal{V}_1^r \in \mathcal{S}_{2n+1}$ and $1^{(n_1)} \notin \operatorname{dom} <^T$. From the definition of $\mathcal{F}$, for $\beta_1^r < \beta_2^r$ represented by $G_1, G_2\colon \underset{\sim}{\boldsymbol{\delta}}^1_{2n+1} \to \underset{\sim}{\boldsymbol{\delta}}^1_{2n+1}$ of the correct type, and $\gamma \in \vartheta(\mathcal{U})$ represented by $\bar{F}$ w.r.t. $\mu^{<^T}_{2n+1}$, if F_1, F_2 represent $\mathcal{F}_{\beta_1^r}(\gamma), \mathcal{F}_{\beta_1^r}(\gamma)$

w.r.t., $\mu_{2n+1}^{<^T}$ then $F_1([f]) = G_1(\bar{F}([f])) < G_2(\bar{F}([f])) = F_2([f])$, since we may assume $\bar{F}$ has the property (which happens for almost all γ w.r.t. $\mathcal{U}$) that for almost all f, $\bar{F}([f])$ is in a measure one set A w.r.t. $\mu_{2n+1}^{S_{2n-1}^{\ell',m}}$ on which $G_1 < G_2$.

Case III. $\mathcal{V}_1^r \in \mathcal{S}_{2n+1}$ and $1^{(n_1)} \in \operatorname{dom} <^T$. If $\beta_1^r < \beta_2^r$ are represented by $G_1, G_2 \colon <^m \to \underset{\sim}{\delta}_{2n+1}^1$ of the correct type, then it follows from the definition of $\mathcal{F}$ that if $\gamma \in \vartheta(\mathcal{U})$ is represented by functions $\vec{H}$ of the correct type (as in the definition of $\mathcal{U}$), and F_1, F_2 represent $\mathcal{F}_{\beta_1^r}(\gamma), \mathcal{F}_{\beta_2^r}(\gamma)$ w.r.t., $\mu_{2n+1}^{<^T}$, then for almost all $f \colon <^T \to \underset{\sim}{\delta}_{2n+1}^1$ we have

$$F_1([f]) = G_1(\alpha_2, \dots, \alpha_m, \alpha_1) < G_2(\alpha_2, \dots, \alpha_m, \alpha_1) = F_2([f]),$$

since for almost all γ we have that for almost all f, the α's above are in a measure one set A on which $G_1 < G_2$.

This establishes $D_{2n+1}(\text{i})$. We now consider $H_{2n+3}(\text{iv})$ and $D_{2n+3}(\text{ii})$. We consider the following cases:

Case I. $\mathcal{V}_1^r \in \mathcal{W}_{2n+1}$, hence is of the form $\mathcal{V}_1^r = \mu_{2n+1}^{S_{2n-1}^{\ell',m_r}}$ ($\ell' = 2^n - 1$ is maximal). We let m be such that B_{2n+1} holds for the ordering corresponding to T defined by $v^r = S_{2n-1}^{\ell',m_r}$ for $1 \le r \le r_0$ and $S_{2n-1}^{\ell',m}$ (so T is the r_0 sum of the S_{2n-1}^{ℓ',m_r}). We let $\mathcal{V}_2^r = \mu_{2n+1}^{S_{2n-1}^{\ell',m}}$, and let $\mathcal{U}$ be the measure as in B_{2n+1}. We construct $\mathcal{F}$ as in $H_{2n+3}(\text{i})$–(iii) in this case. In fact, our previous consideration of $H_{2n+3}(\text{i})$–(iii) for the case $\mathcal{V}_1 \in \mathcal{W}_{2n+1}$ included this case.

$D_{2n+3}(\text{ii})$ follows readily from B_{2n+1}.

Case II. $\mathcal{V}_1^r \in \mathcal{S}_{2n+1}$ and is of the form $S_{2n+1}^{\bar{\ell}_r,m_r}$ where $\bar{\ell}_r > 1$. Recall that $S_{2n+1}^{\bar{\ell}_r,m_r}$ is the measure induced from the strong partition relation on $\underset{\sim}{\delta}_{2n+1}^1$ and a measure $\mu_{2n+1}^{\vartheta^{\bar{\ell}_r,m_r}}$ on $\underset{\sim}{\delta}_{2n+1}^1$. Also, $\mu_{2n+1}^{\vartheta^{\bar{\ell}_r,m_r}}$ is the measure induced from the weak partition relation on $\underset{\sim}{\delta}_{2n+1}^1$ and the measure $R^{\bar{\ell}_r,m_r}$ where the $R^{\bar{\ell},m}$ enumerate the measures $W_1^m, S_1^{1,m}, \dots, W_{2n-1}^m, S_{2n-1}^{1,m}, \dots, S_{2n-1}^{2^n-1,m}$. Since $\mathcal{V}_1^{r_1} \sim \mathcal{V}_1^{r_2}$ it follows that $\bar{\ell}_1 = \bar{\ell}_2 = \dots = \bar{\ell}_{r_0} = \bar{\ell}$, say. We consider the case $R^{\bar{\ell},m_r} = S_{2n-1}^{\ell,m_r}$ the other cases being similar.

Since $S_{2n-1}^{\ell,m_r} \sim S_{2n-1}^{\ell,m_r'}$, it follows from $H_{2n+1}(\text{iv})$ that there is an m s.t. $H_{2n+1}(\text{iv})$ holds for $v_1 = \prod_r S_{2n-1}^{\ell,m_r}$ and $v_2 = S_{2n-1}^{\ell,m}$ for some measure u. We let $\mathcal{V}_2 = S_{2n+1}^{\bar{\ell},m}$.

We let $\mathcal{U}$ be the measure on r_0 tuples of ordinals $(\vartheta_1, \dots, \vartheta_{r_0})$ defined as follows: A has measure one w.r.t. $\mathcal{U}$ if there is a c.u.b. $C \subseteq \underset{\sim}{\delta}_{2n+1}^1$ such that for $H \colon \underset{\sim}{\delta}_{2n+1}^1 \to C$ of the correct type, $(\vartheta_1, \dots, \vartheta_{r_0}) \in A$ where ϑ_r is represented with respect to $\mu_{2n+1}^{S_{2n-1}^{\ell,m_r}}$ by a function $\bar{F}_r$ defined as follows. For

$f\colon \vartheta(S^{\ell,m_r}_{2n-1}) \to \underset{\sim}{\delta}^1_{2n+1}$ of the correct type, $\bar{F}_r([f]) = [g_r]_{S^{\ell,m}_{2n-1}}$, where g_r is defined by: for $\alpha < \vartheta(S^{\ell,m}_{2n-1})$, $g_r(\alpha) = H([h])$, where $h\colon \vartheta(u) \to \underset{\sim}{\delta}^1_{2n+1}$ is given by, for $\beta < \vartheta(u)$, $h(\beta) = f(f_\alpha(\beta)(r))$ and here $\alpha \mapsto f_\alpha$ is as in H_{2n+1}(iv).

This is well defined by H_{2n+1}(iv).

We define $\mathcal{F}$ as follows. If β is represented w.r.t. $\mu_{2n+1}^{S^{\ell,m}_{2n-1}}$ by $G\colon \underset{\sim}{\delta}^1_{2n+1} \to \underset{\sim}{\delta}^1_{2n+1}$ of the correct type, and $\vartheta = (\vartheta_1, \dots, \vartheta_{r_0})$ is in the measure space $\mathcal{U}$ represented by $\bar{F}_1, \dots, \bar{F}_{r_0}$, then we set $\mathcal{F}_\beta(\dots, \vartheta_r, \dots) = (\alpha_1, \dots, \alpha_{r_0})$, where α_r is represented w.r.t. $\mu_{2n+1}^{S^{\ell,m_r}_{2n-1}}$ by F_r, where for $f\colon \vartheta(S^{\ell,m_r}_{2n-1}) \to \underset{\sim}{\delta}^1_{2n+1}$ of the correct type, $F_r([f]) = G(\bar{F}_r([f]))$. This is well defined.

The proofs of H_{2n+3}(iv) and D_{2n+3}(ii) now follows as in H_{2n+3}(i)–(iii) and D_{2n+3}(i) before.

Case III. $\mathcal{V}_1^r \in S_{2n+1}$ and is of the form S^{1,m_r}_{2n+1}. We let $m = (\max_{\ell \le r \le r_o} m_r) + 1$, and we set $\mathcal{V}_2 = S^{1,m}_{2n+1}$. We let $p = r_0 + m$ and $\mathcal{U}$ be the p-fold product of the ω-cofinal normal measure on $\underset{\sim}{\delta}^1_{2n+1}$.

We define $\mathcal{F}$ as follows. For β represented by $G\colon <^m \to \delta_{2n+1}$ of the correct type, and $(\gamma_1, \dots, \gamma_{r_0}, \bar{\gamma}_1, \dots, \bar{\gamma}_m) \in \vartheta(\mathcal{U})$, where $\gamma_i, \bar{\gamma}_i < \underset{\sim}{\delta}^1_{2n+1}$, we set $\mathcal{F}_\beta(\gamma_1, \dots, \bar{\gamma}_m) = (\alpha_1, \dots, \alpha_{r_0})$, where α_r for $1 \le r \le r_0$ is represented by $F_r\colon <^{m_r} \to \underset{\sim}{\delta}^1_{2n+1}$ defined as follows. For $\delta_2 < \cdots < \delta_{m_r} < \delta_1$, we set

$$F_r(\delta_2, \dots, \delta_{m_r}, \delta_1) = G(\gamma_r, \bar{\gamma}_1, \dots, \bar{\gamma}_{m - \bar{m}_r - 1}, \delta_2, \dots, \delta_{m_r}, \delta_1).$$

This is well defined for almost all $\delta_2, \dots, \delta_{m_r}, \delta_1$.

Since $G(\gamma_{r_1}, \vec{\gamma}_2, \delta_1) < G(\gamma_{r_2}, \vec{\gamma}_2, \delta_1)$ for $r_1 < r_2$ (so $\gamma_{r_1} < \gamma_{r_2}$) and any δ_1, $\vec{\gamma}_1, \vec{\gamma}_2$, it follows readily that H_{2n+3}(i) is satisfied. The remaining parts of H_{2n+3} and D_{2n+3} follow readily as before.

This completes the proof of H_{2n+3} and D_{2n+3} and hence of the local embedding theorem.

§5. The Main Lemma. The purpose of this section is to prove a main lemma analyzing functions defined with respect to (in a sense to be made precise) the canonical measures $\bigcup_{m \le n} R_{2m+1} = \bigcup_{m \le n} W^m_{2m+1} \cup \bigcup_{m \le n} S^{\ell,m}_{2n+1}$. We recall that we are still assuming I_{2n+1} and K_{2n+3} as in Section 2.

We outline the methods of this section. We introduce a set $\mathcal{D}$ of "descriptions" which will be finitary objects which "describe" functions $F\colon \underset{\sim}{\delta}^1_{2n+3} \to \underset{\sim}{\delta}^1_{2n+3}$ with respect to the canonical measures, and a lowering operation $\mathcal{L}$ on them. We then introduce a main inductive hypothesis H_{2n+1}, and a main auxiliary lemma $\bar{H}_{2n+1}$ which analyze functions F from $\underset{\sim}{\delta}^1_{2n+3}$ to $\underset{\sim}{\delta}^1_{2n+3}$ in terms of $\mathcal{D}$ and $\mathcal{L}$. We will assume H_{2n+1} and $\bar{H}_{2n+1}$ and establish H_{2n+3}, $\bar{H}_{2n+3}$. We will also require several auxiliary definitions and conditions.

Throughout this section, we will be using the notation $K_1, \dots, K_t$ to denote a sequence of canonical measures in R_{2m+1} where $m < n$ (recall $R_{2m+1} = \bigcup_k W^k_{2m+1} \cup \bigcup_{\ell,k} S^{\ell,k}_{2m+1}$) or of the form W^m_{2n+1} or $S^{\ell,m}_{2n+1}$ for some ℓ, m. Given such a measure K_j, we let $\mathcal{K}_j$ denote the corresponding function space measure. For example, in the case $K_j = S^{\ell,m}_{2n+1}$ and $\ell > 1$, $\mathcal{K}_j$ will be the measure on functions $h_j : \underset{\sim}{\boldsymbol{\delta}}^1_{2n+1} \to \underset{\sim}{\boldsymbol{\delta}}^1_{2n+1}$ of the correct type induced from the strong partition relation on $\underset{\sim}{\boldsymbol{\delta}}^1_{2n+1}$, and if $\ell = 1$, then $\mathcal{K}_j$ will be a measure on functions $h_j : <^m \to \underset{\sim}{\boldsymbol{\delta}}^1_{2n+1}$ of the correct type. We use the notation $h_1, \dots, h_t$ to denote a sequence of functions in the spaces $\mathcal{K}_1, \dots, \mathcal{K}_t$.

Definition 5.1 (Definition of $\mathcal{D}$). We proceed to define the set $\mathcal{D} = \bigcup_n \mathcal{D}_{2n+1} = \bigcup_n \bigcup_{\ell,m} \mathcal{D}^{\ell,m}_{2n+1}$, for $+1 \le \ell \le 2^{n+1} - 1$ or $\ell = -1$. We assume $\mathcal{D}_{2m+1}$ is defined for $m < n$ and define $\mathcal{D}_{2n+1}$. Our definition will be by a simultaneous induction in which we also define an ordering $<$ on $\mathcal{D}_{2n+1}$, two functions h and H associated with descriptions, conditions C, D, A and a numerical function k. We assume these notions are defined for $\mathcal{D}_{2m+1}$, $m < n$.

We define $\mathcal{D}_{2n+1}$ relative to a fixed sequence $K_1, \dots, K_t$ of canonical measures as above. We denote the set of descriptions in $\mathcal{D}_{2n+1}$ which are defined relative to the sequence of measures $K_1, \dots, K_t$ by $\mathcal{D}_{2n+1}(K_1, \dots, K_t)$. So we will have $\mathcal{D}^{\ell,m}_{2n+1} = \bigcup_{\vec{K}} \mathcal{D}^{\ell,m}_{2n+1}(\vec{K})$. When it causes no confusion we frequently just write $\mathcal{D}_{2n+1}$ or $\mathcal{D}^{\ell,m}_{2n+1}$. Fix such a sequence $\vec{K} = K_1, \dots, K_t$ for the rest of the definition, and we proceed to define the set $\mathcal{D}^{\ell,m}_{2n+1} = \mathcal{D}^{\ell,m}_{2n+1}(\vec{K})$.

We define basic and non-basic descriptions and subdivide these into types $-1, 0$, and 1; these types will correspond to $\ell = -1$, $\ell = 1$, and $\ell > 1$ respectively. We will denote a general element of $\mathcal{D}^{\ell,m}_{2n+1}$ by $d^{\ell,m}_{2n+1}$. It will be an indexed tuple of the form $d^{(\mathcal{I}_a)}$, where the index $\mathcal{I}_a$ is one of three forms:

- $\mathcal{I}_a = (f(\bar{K}_1); \bar{K}_2, \dots, \bar{K}_a)$. This will correspond to type 1, that is, $\ell > 1$.
- $\mathcal{I}_a = (f_k; \bar{K}_2, \dots, \bar{K}_a)$. This will correspond to type 0, that is $\ell = 1$.
- $\mathcal{I}_a = (; \bar{K}_2, \dots, \bar{K}_a)$. This will correspond to type -1, that is, $\ell = -1$.

The indices are viewed as formal symbols. The symbols $\bar{K}_2, \dots, \bar{K}_a$ designate measures of the form $\upsilon(K_j)$ for some $1 \le j \le t$, where $K_j \in S_{2n+1}$ (and we recall that $\upsilon(S^{\ell,m}_{2n+1})$ is the measure u such that $S^{\ell,m}_{2n+1}$ induced from the strong partition relation on $\underset{\sim}{\boldsymbol{\delta}}^1_{2n+1}$, functions $F : \underset{\sim}{\boldsymbol{\delta}}^1_{2n+1} \to \underset{\sim}{\boldsymbol{\delta}}^1_{2n+1}$, and the measure μ^u_{2n+1} on $\underset{\sim}{\boldsymbol{\delta}}^1_{2n+1}$). Similarly, f_k and $f(\bar{K}_1)$ are formal symbols, where $\bar{K}_1$ designates the measure $\upsilon(S^{\ell,m}_{2n+1})$ and k is an integer. With a slight abuse of notation, we may also think of the symbol $\bar{K}_j$ (for $j > 1$) as coding a particular integer, which we denote by $r(\bar{K}_j)$, between 1 and t, such that $\bar{K}_j = \upsilon(K_{r(\bar{K}_j)})$. Of course, we may have $\bar{K}_j = \upsilon(K_{r_1}) = \upsilon(K_{r_2})$ for different integers $r_1 \ne r_2$, but will assume that such a particular $r(\bar{K}_j)$ is coded in the symbol $\bar{K}_j$.

Our induction in the following definitions is reverse induction on a function k, which although defined along with $\mathcal{D}_{2n+1}$ below, is actually defined outright.

To define $\mathcal{D}_{2n+1}$, we consider the following cases:

Basic type -1*:*

a) For $n > 0$, we allow descriptions of the form $d_{2n+1}^{-1,m} = d^{(;\bar{K}_2,\dots,\bar{K}_a)}$, where $d = (k;)$, and k is an integer $1 \le k \le t$, such that $K_k \in W_{2n+1}^m$. $k(d) = k$ in this case.

b) For $n > 0$, we allow $d^{(\mathcal{I}_a)} = d^{(;\bar{K}_2,\dots,\bar{K}_a)} = (k; \bar{d}_2)^{s(\mathcal{I}_a)}$, where $\mathcal{I}_a = (\bar{K}_2, \dots, \bar{K}_a)$, s is a formal symbol which may or may not appear, and $\bar{d}_2 \in \bigcup_{m<n} \mathcal{D}_{2m+1}$ is defined relative to the sequence of measures $K_{b_1}, \dots, K_{b_m}, \bar{K}_2, \dots, \bar{K}_a$, where $K_{b_1}, \dots, K_{b_m}$ enumerate the subsequence of $K_{k+1}, \dots, K_t$ consisting of those measures in $\bigcup_{m<n} R_{2m+1}$. We also require that $K_k = W_{2n+1}^m$, and $\bar{d}_2 \in \mathcal{D}_{2n-1}^{\bar{\ell},\bar{m}}$, where $v(K_k) = S_{2n-1}^{\bar{\ell},\bar{m}}$. Also $k(d) = k$ in this case. For $n = 0$, we allow $d = (k; i)$, where $1 \le i \le m$, where $K_k = W_{2n+1}^m$.

Basic type 0*:* We allow $d = d_{2n+1}^{1,m} = (r)$, where r is an integer $1 \le r \le m$. We set $k(d) = \infty$ in this case.

Basic type 1*:* We allow $d_{2n+1}^{\ell,m} = d^{(f(\bar{K}_1);\bar{K}_2,\dots,\bar{K}_a)}$, where $d = (\bar{d}_2)^s$, where the symbol s may or may not appear, $\bar{d}_2$ is defined relative to $\bar{K}_2, \dots, \bar{K}_a$, $\bar{d}_2 \in \mathcal{D}_{2s+1}^{\bar{\ell},\bar{m}}$ where $s < n$, $v(S_{2n+1}^{\ell,m}) = \bar{K}_1$, and $\bar{K}_1 = S_{2s+1}^{\bar{\ell},\bar{m}}$ or $W_{2s+1}^{\bar{m}}$ depending on whether $\bar{\ell} \ge 1$ or $\bar{\ell} = -1$. We set $k(d) = \infty$. We also allow the distinguished description $d = ()$.

Non-Basic Descriptions:

a) If K_k (for some integer $1 \le k \le t$) $= S_{2n+1}^{1,m_k}$, we allow $d = d^{(\mathcal{I}_a)}$ where $d = (k; d_1^{(\mathcal{I}_a)}, \dots, d_r^{(\mathcal{I}_a)})^s$ where s may or may not appear, $\mathcal{I}_a$ is an index of one of the above forms, $r \le m_k$, $r > 1$ if s appears, $d_1^{(\mathcal{I}_a)}, \dots, d_r^{(\mathcal{I}_a)} \in \mathcal{D}_{2n+1}^{\ell,m}$ are defined w.r.t. $K_1, \dots, K_t$ with $k(d_1^{(\mathcal{I}_a)}), \dots, k(d_r^{(\mathcal{I}_a)}) > k$, and $d_1^{(\mathcal{I}_a)} > d_2^{(\mathcal{I}_a)}, \dots, d_r^{(\mathcal{I}_a)}$ w.r.t. the ordering $<$ to be defined below (being defined simultaneously). We set $k(d) = k$.

b) If $K_k = S_{2n+1}^{\ell_k,m_k}$ where $\ell_k > 1$, we allow $d = d^{(\mathcal{I}_a)}$ where $d = (k; d_2^{(\mathcal{I}_a;\bar{K}_{a+1})})^s$, where the symbol s may or may not appear, $\bar{K}_{a+1} = v(K_k)$, d_2 (with the index $\mathcal{I}_{(a+1)} = (\mathcal{I}_a; \bar{K}_{a+1})$) is defined relative to $K_1, \dots, K_t$ and $d_2 \in \mathcal{D}_{2n+1}^{\ell,m}$, with $k(d_2) > k$. We set $k(d) = k$.

This completes the definition of $\mathcal{D}_{2n+1}$.

The functions h, H we will be defining in our induction will have the following properties:

If $d \in \mathcal{D}^{\ell,m}_{2n+1}$ and satisfies condition C relative to $K_1, \dots, K_t$ (where C is to be defined) then for almost all $h_1, \dots, h_t$ w.r.t. $\mathcal{K}_1 \times \cdots \times \mathcal{K}_t$, we have that $h(d; h_1, \dots, h_t)$ will be defined, and be an ordinal.

The function H will have the properties:

a) If d as above with $\ell > 1$ then for almost all $h_1, \dots, h_t$ as above, we have for almost all $f : \vartheta(\bar{K}_1) \to \underset{\sim}{\delta}^1_{2n+1}$ of the correct type (where $\mathcal{I}_a = (f(\bar{K}_1); \bar{K}_2, \dots, \bar{K}_a)$ in this case) that for almost all $\bar{h}_2, \dots, \bar{h}_a$ w.r.t. the product of the function space measures corresponding to $\bar{K}_2, \dots, \bar{K}_a$ that $H(d; h_1, \dots, h_t; f; \bar{h}_2, \dots, \bar{h}_t)$, an ordinal, is defined.
b) Same as above except $\ell = 1$, in which case $\mathcal{I}_a = (f_k; \bar{K}_2, \dots, \bar{K}_a)$, and we require $f : k \to \underset{\sim}{\delta}^1_{2n+1}$ (k an integer).
c) For $\ell = -1$, we have that for almost all $h_1, \dots, h_t, \bar{h}_2, \dots, \bar{h}_a$, that the ordinal

$$H(d; h_1, \dots, h_t, \bar{h}_2, \dots, \bar{h}_a)$$

is defined.

We introduce, in our simultaneous induction, two hypothesis concerning the functions h and H.

F_{2n+1}: If d is defined and satisfies condition C relative to $K_1, \dots, K_t$ then for almost all h_1 if $h_1(1) = h_1(2) = h_1$ almost everywhere (representing elements of K_1), for almost all h_2 if $h_2(1) = h_2(2) = h_2$ almost everywhere, ..., for almost all h_t and $h_t(1) = h_t(2) = h_t$ almost everywhere we have that $h(d; h_1(1), \dots, h_t(1)) = h(d; h_1(2), \dots, h_t(2))$.

F^2_{2n+1}: If $d = d^{(\mathcal{I}_a)}$ is defined and satisfies condition C relative to $K_1, \dots, K_t$, then for almost all h_1 and $h_1(1) = h_1(2) = h_1$ almost everywhere, ..., for almost all h_t and $h_t(1) = h_t(2) = h_t$ almost everywhere we have:

i) If $\mathcal{I}_a = (f(\bar{K}_1); \bar{K}_2, \dots, \bar{K}_a)$, then for almost all $f : \vartheta(\bar{K}_1) \to \underset{\sim}{\delta}^1_{2n+1}$, if $[f_1] = [f_2] = [f]$ (w.r.t. $\bar{K}_1$) then for almost all $\bar{h}_2$ and $\bar{h}_2(1) = \bar{h}_2(2) = \bar{h}_2$, ..., for almost all $\bar{h}_a$ and $\bar{h}_a(1) = \bar{h}_a(2) = \bar{h}_a$ we have

$$H(d; h_1(1), \dots, h_t(1); f_1; \bar{h}_2(1), \dots, \bar{h}_a(1)) \\ = H(d; h_1(2), \dots, h_t(2); f_1; \bar{h}_2(2), \dots, \bar{h}_a(2)).$$

ii) If $\mathcal{I}_a = (f_k; \bar{K}_2, \dots, \bar{K}_a)$, then same as above with $f : k \to \underset{\sim}{\delta}^1_{2n+1}$.
iii) If $\mathcal{I}_a = (\bar{K}_2, \dots, \bar{K}_a)$, then same as above except we omit the f.

In particular, we are assuming F_{2n+1}, F^2_{2n+1} for $m < n$.

We now define conditions C and D. We first consider D. We assume $d \in \mathcal{D}_{2n+1}$ is defined relative to $K_1, \dots, K_t$, and we define when condition D holds for objects of the form d or $(d)^s$ (for $d \in \mathcal{D}_{2n+1}$, $(d)^s$ is not a description, but this does not affect the definitions).

Definition 5.2 (Condition D). Given d or $(d)^s$, where $d = d^{\ell,m}_{2n+1} = \mathcal{D}^{\ell,m}_{2n+1}$ and where d is defined relative to $K_1, \dots, K_t$, we require that d satisfy condition C and further:

a) If $\ell > 1$ (so $\mathcal{I}_a = (f(\bar{K}_1); \bar{K}_2, \dots, \bar{K}_a)$), then if s does not appear we require that for almost all $h_1, \dots, h_t$, if ϑ represents $h(d; h_1, \dots, h_t)$ w.r.t. μ^v_{2n+1} (where $v = v(S^{\ell,m}_{2n+1})$), that there is a measure one set A w.r.t. μ^v_{2n+1} restricted to which ϑ is strictly increasing of uniform cofinality ω. Further, if $C \subseteq \underset{\sim}{\delta}^1_{2n+1}$ is c.u.b. then for almost all $h_1, \dots, h_t$ we have that ϑ has range almost everywhere in C. If s appears, then we require that for almost all $h_1, \dots, h_t$ that $[\vartheta]$ is the supremum of $[\vartheta']$ for ϑ' of the correct type with range a.e. in C.

b) If $\ell = 1$ (so $\mathcal{I}_a = (f_k; \bar{K}_2, \dots, \bar{K}_a)$), then for almost all $h_1, \dots, h_t$, if ϑ represents $h(d; h_1, \dots, h_t)$ w.r.t. the m-fold product of the ω-cofinal normal measure on $\underset{\sim}{\delta}^1_{2n+1}$, then we require that there is a measure one set A restricted to which ϑ is (strictly) order-preserving w.r.t. $<^m$ and of uniform cofinality ω, if s does not appear. In the case where s appears, we require that $[\vartheta]$ is a supremum of ordinals $[\bar{\vartheta}]$ represented by such functions.

c) If $\ell = -1$ (so $\mathcal{I}_a = (\bar{K}_2, \dots, \bar{K}_a)$), then for almost all $h_1, \dots, h_t$, if ϑ represents $h(d; h_1, \dots, h_t)$ w.r.t. $v = v(W^m_{2n+1})$, then there is a measure one set A w.r.t. v restricted to which ϑ is of the correct type, if s does not appear. In the case where s appears, we require $[\vartheta]$ to be the supremum of ordinals $[\bar{\vartheta}]$ represented by functions of the correct type.

We now define condition C.

Definition 5.3 (Condition C). We again let $d^{(\mathcal{I}_a)}$, where $d = d^{\ell,m}_{2n+1} \in \mathcal{D}^{\ell,m}_{2n+1}$ be given and defined relative to $K_1, \dots, K_t$. We let $k = k(d)$. We consider the following cases:

d basic.

1) $\ell = -1$.
   a) $d^{-1,m}_{2n+1} = (k;)$. In this case d satisfies condition C.
   b) $d = (k; \bar{d}_2)^s$, where s may or may not appear. We require $\bar{d}_2$ or $(d_2)^s$ satisfy condition D relative to $K_{b_1}, \dots, K_{b_m}, \bar{K}_2, \dots, \bar{K}_a$ depending on whether s does not or does appear in d.
2) $\ell = 1$ so $d = d^{1,m}_{2n+1} = (r)$ where $1 \leq r \leq m$. In this case d satisfies C.
3) $\ell > 1$, so $d = d^{\ell,m}_{2n+1} = (\bar{d}_2)^s$, where s may or may not appear. We require that $\bar{d}_2$ or $(d_2)^s$ satisfy D w.r.t. $\bar{K}_2, \dots, \bar{K}_a$ depending on whether s does not or does appear in d. If $d = ()$, then we define d to satisfy C.

d non-basic.

1) $K_k = S^{\ell_k, m_k}_{2n+1}$ and $\ell_k > 1$, where $d = d^{\ell,m}_{2n+1} = (k; d_2^{(\mathcal{I}_a, K_{a+1})})^s$, where s may or may not appear. We require d_2 to satisfy C w.r.t. $\bar{K}_2, \dots, \bar{K}_a$,

$\bar{K}_{a+1}$ (defined by induction), and if the symbol s does not appear, we require for almost all $h_1, \dots, h_t, f, \bar{h}_2, \dots, \bar{h}_a$ that the function g defined almost everywhere w.r.t. $\bar{K}_{a+1}$ by $g([h_{a+1}]) = H(d_2; h_1, \dots, h_t; f; \bar{h}_2, \dots, \bar{h}_a, \bar{h}_{a+1})$ is almost everywhere increasing of uniform cofinality ω. This makes sense since H is defined and satisfies F^2_{2n+1}, by induction, for d_2. If s appears, we require that for almost all $h_1, \dots, h_t, f, \bar{h}_2, \dots, \bar{h}_a$ that $[g]$ is the supremum of $[g']$ for g' of the correct type with $\sup_{\text{a.e.}} g = \sup_{\text{a.e.}} g'$.

2) $K_k = S^{1,m_k}_{2n+1}$, where $d = (k; d_1^{(\mathcal{I}_a)}, \dots, d_r^{(\mathcal{I}_a)})^s$, $r \leq m_k$, and s may or not may appear. We then require that $d_2^{(\mathcal{I}_a)} < d_3^{(\mathcal{I}_a)} < \cdots < d_r^{(\mathcal{I}_a)} < d_1^{(\mathcal{I}_a)}$, and that for almost all $h_1, \dots, h_t; f; \bar{h}_2, \dots, \bar{h}_a$ that $H(d_1; h_1, \dots, \bar{h}_a), \dots, H(d^{\mathcal{I}_a}_{r-1}; h_1, \dots, \bar{h}_a)$ have cofinality ω. Also, if s does not appear we further require that $H(d^{\mathcal{I}_a}_{r-1}; h_1, \dots, \bar{h}_a)$ has cofinality ω.

We now define the functions h and H. We first consider H. We assume d is defined and satisfies Condition C relative to $K_1, \dots, K_t$.

We consider the following cases. Recall a description $d = d^{\ell,m}_{2n+1}$ is of type -1 if $\ell = -1$, of type 0 if $\ell = 1$ and of type 1 if $\ell > 1$.

1) d basic of type -1.

a) $d = (k;)$. For fixed $h_1, \dots, h_t, \bar{h}_2, \dots, \bar{h}_a$ where $\mathcal{I}_a = (\bar{K}_2, \dots, \bar{K}_a)$ in this case, we set $H(d; h_1, \dots, h_t, \bar{h}_2, \dots, \bar{h}_a) = \sup_{\text{a.e.}} h_k$, where $K_k = W^{m_k}_{2n+1}$, $\kappa = \kappa(W^{m_k}_{2n+1})$, and $h_k \colon \kappa \to \underset{\sim}{\delta}^1_{2n+1}$ (recall $W^{m_k}_{2n+1}$ is induced by the weak partition property on $\underset{\sim}{\delta}^1_{2n+1}$, functions $h_k \colon \kappa \to \underset{\sim}{\delta}^1_{2n+1}$ and the measure S^{2^n-1,m_k}_{2n-1} on κ).

b) $d = (k; \bar{d}_2)^s$, where s may or may not appear, and $\bar{d}_2$ or $(\bar{d}_2)^s$ satisfies D relative to $K_{b_1}, \dots, K_{b_u}, \bar{K}_2, \dots, \bar{K}_a$. Here $\bar{d}_2 \in \mathcal{D}^{\bar{\ell},\bar{m}}_{2s+1}$ for $s < n$, where $v(W^{m_k}_{2n+1}) = R^{\bar{\ell},\bar{m}}_{2s+1} (= S^{\bar{\ell},\bar{m}}_{2s+1}$ if $\bar{\ell} > 0$, and $= W^{\bar{m}}_{2s+1}$ if $\bar{\ell} = -1)$. Since $\bar{d}_2$ satisfies C, we have by induction that for almost all $h_1, \dots, h_t$, for almost all $\bar{h}_2, \dots, \bar{h}_a$ that $h(\bar{d}_2; h_{b_1}, \dots, h_{b_u}, \bar{h}_2, \dots, \bar{h}_a)$ is defined and is an ordinal in the measure space $v(W^{m_k}_{2n+1})$. We set

$$H(d; h_1, \dots, h_t, \bar{h}_2, \dots, \bar{h}_a) = \begin{cases} h_k(h(\bar{d}_2; \bar{h}_2, \dots, \bar{h}_a, h_{b_u})) & \text{if } d = (k; \bar{d}_2) \\ \sup\limits_{\beta < h(\bar{d}_2; h_{b_1}, \dots, h_{b_u}, \bar{h}_2, \dots, \bar{h}_a)} h_k(\beta) & \text{if } d = (k; \bar{d}_2)^s \end{cases}$$

c) For $n = 0$ and $d = (k; i)$, we set $H(d; h_1, \dots, h_t, \bar{h}_2, \dots, \bar{h}_a) = h_k(i)$. Note in this case that $h_k \colon m_k \to \omega_1$, where $K_k = W^{m_k}_1$.

2) Basic type 0.

Here $d = d^{1,m}_{2n+1} = (r)$, where $r \leq m$. For fixed $h_1, \dots, h_t, f, \bar{h}_2, \dots, \bar{h}_a$ where $f \colon m \to \underset{\sim}{\delta}^1_{2n+1}$, we set $H(d; h_1, \dots, h_t, f, \bar{h}_2, \dots, \bar{h}_a) = f(r)$.

3) Basic type 1.

Here $d^{(\mathcal{I}_a)} = d^{(f(\bar{K}_1);\bar{K}_2,\dots,\bar{K}_a)}$, where $d = (\bar{d_2})^s$, where s may or may not appear and $\bar{d_2}$ is defined and satisfies Condition C relative to $\bar{K}_2, \dots, \bar{K}_a$. Hence by induction, for almost all $h_1, \dots, h_t, f, \bar{h}_2, \dots, \bar{h}_a$, the ordinal $h(\bar{d_2}; \bar{h}_2, \dots, \bar{h}_a)$ is defined. We set

$$H(d; h_1, \dots, h_t, \bar{h}_2, \dots, \bar{h}_a) = \begin{cases} f(h(\bar{d_2}; \bar{h}_2, \dots, \bar{h}_a)) & \text{if } d = (\bar{d_2}) \\ \sup\limits_{\beta < h(\bar{d_2}; \bar{h}_2, \dots, \bar{h}_a)} f(\beta) & \text{if } d = (\bar{d_2})^s \end{cases}$$

This makes sense since $h(\bar{d_2}; \bar{h}_2, \dots, \bar{h}_a)$ is an element of the measure space $\bar{K}_1$.

If $d = ()$, we set $H(d; h_1, \dots, h_t; f, \bar{h}_2, \dots, \bar{h}_a) = \sup_{\text{a.e.}} f$.

4) d non-basic with $K_k = S_{2n+1}^{\ell_k, m_k}$ and $\ell_k > 1$. Here $d = (k; d_2^{(\mathcal{I}_{a+1})})^s$, where s may or may not appear. Since d_2 satisfies Condition C, for almost all $h_1, \dots, h_t, f; \bar{h}_2, \dots, \bar{h}_a, \bar{h}_{a+1}, H(d_2; h_1, \dots, h_t, f, \bar{h}_2, \dots, \bar{h}_a, \bar{h}_{a+1})$ is defined. For fixed $h_1, \dots, h_t, f, \bar{h}_2, \dots, \bar{h}_a$ we let $g\colon \vartheta(\bar{K}_{a+1}) \to \underset{\sim}{\delta}_{2n+1}^1$ be defined by $g([\bar{h}_{a+1}]) = H(d_2; h_1, \dots, h_t, f, \bar{h}_2, \dots, \bar{h}_a, \bar{h}_{a+1})$, which is well defined almost everywhere for almost all $h_1, \dots, h_t, f, \bar{h}_2, \dots, \bar{h}_a$ by F_{2n+1}^2 and induction. We then set

$$H(d_2; h_1, \dots, h_t, f, \bar{h}_2, \dots, \bar{h}_a) = \begin{cases} h_k([g]) & \text{if } d = (k; d_2^{(\mathcal{I}_{a+1})}) \\ \sup\limits_{\beta < [g]} h_k(\beta) & \text{if } d = (k; d_2^{(\mathcal{I}_{a+1})})^s \end{cases}$$

5) d non-basic with $K_k = S_{2n+1}^{1, m_k}$. Here $d = (k; d_1^{\mathcal{I}_a}, \dots, d_r^{\mathcal{I}_a})^s$ with $r \le m_k$, and s may or may not appear. Since $d_1^{\mathcal{I}_a}, \dots, d_r^{\mathcal{I}_a}$ satisfy Condition C, we have by induction that for almost all $h_1, \dots, h_t, f, \bar{h}_2, \dots, \bar{h}_a$ that the ordinals $H(d_i^{\mathcal{I}_a}; h_1, \dots, h_t, f, \bar{h}_2, \dots, \bar{h}_a)$ are defined. Also by induction and the definition of $<$ (given below),

$$H(d_2^{\mathcal{I}_a}; \cdots) < \cdots < H(d_r^{\mathcal{I}_a}; \cdots) < H(d_1^{\mathcal{I}_a}; \cdots).$$

We set

$$H(d; h_1, \dots, h_t, f, \bar{h}_2, \dots, \bar{h}_a) = \\ h_k(H(d_2^{\mathcal{I}_a}; \cdots), \dots, H(d_r^{\mathcal{I}_a}; \cdots), H(d_1^{\mathcal{I}_a}; \cdots)),$$

if s does not appear and $r = m_k$,

$$H(d; h_1, \dots, h_t, f, \bar{h}_2, \dots, \bar{h}_a) = \sup\{h_k(H(d_2^{\mathcal{I}_a}; \cdots), \dots, H(d_r^{\mathcal{I}_a}; \cdots), \\ \beta_{r+1}, \dots, \beta_{m_k}, H(d_1^{\mathcal{I}_a}; \cdots)) : \beta_{r+1}, \dots, \beta_{m_k} < H(d_1^{\mathcal{I}_a}; \cdots)\}$$

if $r < m_k$ and s does not appear, and

$$H(d; h_1, \dots, h_t, f, \bar{h}_2, \dots, \bar{h}_a) = \sup\{h_k(H(d_2^{\mathcal{I}_a}, \dots), \dots, H(d_{r-1}^{\mathcal{I}_a}; \cdots), \beta_r, \dots, \beta_{m_k}, H(d_1^{\mathcal{I}_a}; \cdots)) : \beta_r < H(d_r^{\mathcal{I}_a}; \cdots), \beta_{r+1}, \dots, \beta_{m_k} < H(d_1^{\mathcal{I}_a}; \dots,)\}$$

if s appears.

We now consider h.

If $\mathcal{I}_a = (f(\bar{K}_1); \bar{K}_2, \dots, \bar{K}_a)$, we represent $h(d; h_1, \dots, h_t)$ with respect to $\mu_{2n+1}^{\bar{K}_1}$ by a function $[f] \mapsto h(d; h_1, \dots, h_t, f)$ for $f : \vartheta(\bar{K}_1) \to \underset{\sim}{\delta}_{2n+1}^1$ of the correct type. Similarly, if $\mathcal{I}_a = (f_k; \bar{K}_2, \dots, \bar{K}_a)$, we represent $h(d; h_1, \dots, h_t)$ w.r.t. the k-fold product of the ω-cofinal normal measure on $\underset{\sim}{\delta}_{2n+1}^1$ by $f \mapsto h(d; h_1, \dots, h_t, f)$ where now $f \in (\underset{\sim}{\delta}_{2n+1}^1)^k$. In general. we represent $h(d; h_1, \dots, h_t, f, \bar{h}_2, \dots, \bar{h}_\ell)$ (where f does not appear if $\mathcal{I}_a = (\bar{K}_2, \dots, \bar{K}_a)$) w.r.t. $\bar{K}_{\ell+1}$ by the function

$$[\bar{h}_{\ell+1}] \mapsto h(d; h_1, \dots, h_t, f; \bar{h}_2, \dots, \bar{h}_\ell, \bar{h}_{\ell+1})$$

for $[\bar{h}_{\ell+1}] \in \vartheta(\bar{K}_{\ell+1})$. Finally, we set

$$h(d; h_1, \dots, h_t, f; \bar{h}_2, \dots, \bar{h}_a) = H(d; , h_1, \dots, h_t, f, \bar{h}_2, \dots, \bar{h}_a).$$

By F_{2n+1}^2, this is well defined (where F_{2n+1}^2 is established below).

We now define the ordering $<$ on the set of descriptions of the form $d^{(\mathcal{I}_a)}$, for a fixed index $\mathcal{I}_a$; defined and satisfying Condition C relative to fixed $K_1, \dots, K_t$.

Definition 5.4 (Definition of ordering $<$ on $\mathcal{D}_{2n+1}$). If $d_1 = d_1^{(\mathcal{I}_a)}, d_2 = d_2^{(\mathcal{I}_a)} \in \mathcal{D}_{2n+1}^{\ell,m}(\vec{K})$, we set $d_1 < d_2$ if for almost all $h_1, \dots, h_t, f, \bar{h}_2, \dots, \bar{h}_a$ we have

$$H(d_1; h_1, \dots, h_t, f, \bar{h}_2, \dots, \bar{h}_a) < H(d_2; h_1, \dots, h_t, f, \bar{h}_2, \dots, \bar{h}_a).$$

This is well defined by induction.

Before verifying F_{2n+1} and F_{2n+1}^2, we introduce a further condition.

Definition 5.5 (Condition A). We let $d^{(\mathcal{I}_a)}$ be defined and satisfy Condition C w.r.t. $K_1, \dots, K_t$. If $d = (k; d_2^{(\mathcal{I}_{a+1})})^s$ where $K_k = S_{2n+1}^{\ell_k, m_k}$ with $\ell_k > 1$, and where s may or may not appear, then there is a $\hat{d}^{\mathcal{I}_a}$ satisfying Condition C w.r.t. $K_1, \dots, K_t$, such that for almost all $h_1, \dots, h_t, f, \bar{h}_2, \dots, \bar{h}_a$ we have

$$H(\hat{d}; h_1, \dots, h_t, f, \bar{h}_2, \dots, \bar{h}_k) = \sup_{\text{a.e. } [\bar{h}_{a+1}]} H(d_2; h_1, \dots, h_t, f, \bar{h}_2, \dots, \bar{h}_1, \bar{h}_{a+1}),$$

which makes sense by F_{2n+1}^2. We also require $k(\hat{d}) > k$, and all component tuples of d as well as $\hat{d}$ satisfy A.

We now verify (in our simultaneous induction) F_{2n+1} and F^2_{2n+1}. We note that F^2_{2n+1} implies F_{2n+1}, so we consider F^2_{2n+1}. We consider the following cases:

1) $d^{(\mathcal{I}_a)}$ basic type 1. Here $d = (d_2)^s$, where s may or may not appear. F^2_{2n+1} follows by induction, F_{2s+1} for $s < n$, and the fact that d_2 or $(d_2)^s$ satisfies D w.r.t. $\bar{K}_2, \dots, \bar{K}_a$.
2) $d^{(\mathcal{I}_a)}$ basic type 0. Here $d = (r)$ for some $r \leq m$, where the index $\mathcal{I}_a$ is now of the form $\mathcal{I}_a = (f_m; \bar{K}_2, \dots, \bar{K}_a)$. F^2_{2n+1} follows immediately since $H(d; h_1, \dots, h_t, f_m, \bar{h}_2, \dots, \bar{h}_a) = f_m(r)$ only depends on $f_m : m \to \underset{\sim}{\boldsymbol{\delta}}^1_{2n+1}$.
3) $d^{(\mathcal{I}_a)}$ basic type -1.
   i) $d = (k;)$, where $K_k = W^{m_k}_{2n+1}$. Since (as in the statement of F^2_{2n+1}) $h_k(1) = h_k(2)$ almost everywhere, $\sup_{\text{a.e.}} h_k(1) = \sup_{\text{a.e.}} h_k(2)$, and the results follows.
   ii) $d = (k; \bar{d}_2)^s$, where s may or may not appear. Since d_2 or $(d_2)^s$ satisfies Condition D w.r.t. $K_{b_1}, \dots, K_{b_u}, \bar{K}_2, \dots, \bar{K}_a$, it follows from induction, F_{2s+1} for $s < n$, and the definition of H that F^2_{2n+1} is satisfied.
4) $d^{(\mathcal{I}_a)}$ non-basic where $d = (k; d_2^{(\mathcal{I}_{a+1})})^s$, where s may or may not appear, and $\ell_k > 1$. By induction F^2_{2n+1} holds for d_2 relative to the sequence $h_1, \dots, h_t, f, \bar{h}_2, \dots, \bar{h}_{a+1}$. Hence, for almost all $h_1, \dots, h_t, f, \bar{h}_2, \dots, \bar{h}_a$, the function $g : \vartheta(\bar{K}_{a+1}) \to \underset{\sim}{\boldsymbol{\delta}}^1_{2n+1})$ defined by

$$g([\bar{h}_{a+1}]) = H(d_2; h_1, \dots, h_t, f, \bar{h}_2, \dots, \bar{h}_a, \bar{h}_{a+1})$$

is well defined almost everywhere. Since $k(d_2) > k = k(d)$ it follows from the definition of H that we may assume that g has range almost everywhere in a c.u.b. set $C \subseteq \underset{\sim}{\boldsymbol{\delta}}^1_{2n+1}$ defining a measure one set w.r.t. $\mu^{v_k}_{2n+1}$ (where $v_k = v(K_k)$) where $h_k(1), h_k(2)$ agree. If the symbol s does not appear, then from the definition of Condition C we may assume that g is almost everywhere strictly increasing of uniform cofinality ω. Since we may assume C is sufficiently closed, we may assume g is everywhere increasing of uniform cofinality ω, and has range in C. Hence $h_k(1)([g]) = h_k(2)([g])$, and F^2_{2n+1} follows. If the symbol s appears, $[g]$ is the supremum of $[g']$ where g' is of the correct type. Since C may be taken sufficiently closed, it follows readily that $[g]$ is the sup of $[g']$ where g' is of the correct type having range in C. Hence $\sup_{\beta < [g]} h_k(1)(\beta) = \sup_{\beta < [g]} h_k(2)(\beta)$, and F^2_{2n+1} follows.
5) $d^{(\mathcal{I}_a)}$ non-basic where $d = (k; d^{(\mathcal{I}_a)}, \dots, d_r^{(\mathcal{I}_a)})^{s^*}$, and $K_k = S^{1,m_k}_{2n+1}$, where $r \leq m_k$ and s may or may not appear. We again let $C \subseteq \underset{\sim}{\boldsymbol{\delta}}^1_{2n+1}$ be a c.u.b. set defining a set of measure one w.r.t. the m-fold product of the ω-cofinal normal measure on $\underset{\sim}{\boldsymbol{\delta}}^1_{2n+1}$ where $h_k(1)$ and $h_k(2)$ agree. Since

$k(d_1), \dots, k(d_r) > k$, it follows that for almost all sequences $h_1, \dots, h_t$, f, $\bar{h}_2, \dots, \bar{h}_a$, that the ordinals $\beta_1 = h(d_1; h_1, \dots, h_t, f, \bar{h}_2, \dots, \bar{h}_a), \dots, \beta_r = H(d_r; h_1, \dots, h_t, f, \bar{h}_2, \dots, \bar{h}_a)$ are in C. From the definition of Condition C we also have $\beta_1 < \cdots \beta_r < \beta_1$, and $\beta_1, \dots, \beta_{r-1}$ have cofinality ω, as does β_r if s does not appear. F^2_{2n+1} now follows the definition of H and the fact that C is sufficiently closed, so β_r is a supremum of points in C of cofinality ω.

This completes the simultaneous induction defining $\mathcal{D}, h, H, <$, conditions C, D, and A, and establishing F_{2n+1}, F^2_{2n+1}.

We require a lemma concerning the ordering $<$, which gives a combinatorial reformulation of the ordering.

Lemma 5.6. If $d_1^{(\mathcal{I}_a)}, d_2^{(\mathcal{I}_a)}$ are defined and satisfy conditions C and A relative to $K_1, \dots, K_t$, then $d_1 < d_2$ iff one of the following is satisfied:

I) d_1, d_2 are basic of type -1, so of the form a) $d_1 = (k_1;)$ or b) $d_1 = (k_1; (\bar{d_2})_1)^s$, where s may or may not appear, and similarly for d_2 (where $\bar{d}_1, \bar{d}_2$ are integers for $n = 0$). There we require that one of the following is satisfied:
   1) $k_1 < k_2$.
   2) $k_1 = k_2$ and d_2 is of type (a) and d_1 of type (b).
   3) $k_1 = k_2$, both are of type (b) and $(\bar{d_2})_1 < (\bar{d_2})_2$, where for $n = 0$, by $<$ we mean the usual ordering on the integers.
   4) $k_1 = k_2$, both are of type (b), $(\bar{d_2})_1 = (\bar{d_2})_2$, and d_1 has the symbol s and d_2 does not.

II) d_1, d_2 basic of type 0, so $d_1 = (r_1)$, $d_2 = (r_2)$, and $r_1 < r_2$.

III) d_1, d_2 basic of type 1, so $d_1 = ((\bar{d_2})_1)^s$ or $d_1 = ()$, and similarly for d_2, where s may or may not appear. Then we require $(\bar{d_2})_1 < (\bar{d_2})_2$, or $(\bar{d_2})_1 = (\bar{d_2})_2$ and d_1 involves the symbol s and d_2 does not, or $d_2 = ()$ and d_1 is of the first type.

IV) At least one of d_1, d_2 is non-basic. We then require that one of the following is satisfied:
   1) $k(d_1) > k(d_2)$, in which case we require $\hat{d_2} \geq d_1$ if $K_{k(d_2)} = S^{\ell_k, m_k}_{2n+1}$ with $\ell_k > 1$ (here $\hat{d_2}$ is as in Condition A). If $\ell_k = 1$, then $d_1 \leq (d_2)_1$, where $d = (k; (d_2)_1^{(\mathcal{I}_a)}, (d_2)_2^{(\mathcal{I}_a)}, \dots, (d_2)_r^{(\mathcal{I}_a)})^s$.
   2) $k(d_1) < k(d_2)$, in which case we require $\hat{d_1} < d_2$ if $K_{k(d_1)} = S^{\ell_k, m_k}_{2n+1}$ with $\ell_k > 1$. If $\ell_k = 1$, then we require $(d_1)_1 < d_2$ or $K_{k(d_1)} = W^{m_k}_{2n+1}$.
   3) $k(d_1) = k(d_2) = k$, in which case one of the following is satisfied:
      i) $K_k = S^{\ell_k, m_k}_{2n+1}$ with $\ell_k > 1$, so $d_1 = (k; (d_2)_1)^s, d_2 = (k; (d_2)_2)^s$, where s may or may not appear in d_1, d_2. We require that $(d_2)_1 < (d_2)_2$ or $(d_2)_1 = (d_2)_2$ and d_1 involves s and d_2 does not.

ii) $K_k = S_{2n+1}^{\ell_k,m_k}$ with $\ell_k = 1$, so $d_1 = (k; (d_1)_1^{(\mathcal{I}_a)}, \dots, (d_{r_1})_1^{(\mathcal{I}_a)})^s$, $d_2 = (k; (d_1)_2^{(\mathcal{I}_a)}, \dots, (d_{r_2})_2^{(\mathcal{I}_a)})^s$, where s may or may not appear. We then require that:

a) $(d_1)_1 < (d_1)_2$ or

b) $(d_1)_1 = (d_1)_2$ and there is a $p \leq \min\{r_1, r_2\}$ such that $(d_p)_1 \neq (d_p)_2$, and for the least such p, $(d_p)_1 < (d_p)_2$; or

c) $(d_p)_1 = (d_p)_2$ for $1 \leq p \leq \min\{r_1, r_2\}$ and $r_1 < r_2$ and d_1 involves s; or $r_1 > r_2$ and d_2 does not involve s; or $r_1 = r_2$ and d_1 involves s and d_2 does not.

We also have:

LEMMA 5.7. If $d_1 \neq d_2$, and d_1, d_2 satisfy Conditions C and A, then $d_1 < d_2$ or $d_2 < d_1$.

Lemma 5.7 follows immediately from Lemma 5.6 and the corresponding statement for the ordering $<'$ defined from the statement of Lemma 5.6. This in turn follows readily from a consideration of cases.

Before establishing Lemma 5.6 we introduce the following notation.

DEFINITION 5.8. Suppose $K_k = S_{2n+1}^{\ell_k,m_k}$ and h_k represents the ordinal $[h_k] \in \operatorname{dom}(K_k)$ Recall that if $\ell_k > 1$ then $h_k \colon \underset{\sim}{\boldsymbol{\delta}}_{2n+1}^1 \to \underset{\sim}{\boldsymbol{\delta}}_{2n+1}^1$ of the correct type represents $[h_k]$ with respect to μ_{2n+1}^v where $v = v(S_{2n+1}^{\ell_k,m_k})$ is a measure on $\kappa(S_{2n+1}^{\ell_k,m_k})$. If $\ell_k = 0$ then $h_k \colon <^{m_k}\to \underset{\sim}{\boldsymbol{\delta}}_{2n+1}^1$ is order-preserving. We define $h_k(0)$, the 0*th invariant of* h_k, as follows. If $\ell_k > 1$ we set

$$h_k(0)(\beta) = \sup\{h_k([g]) \colon g \text{ is of the correct type from } \kappa(S_{2n+1}^{\ell_k,m_k}) \text{ to } \beta\},$$

and if $\ell_k = 1$ we set

$$h_k(0)(\beta) = \sup h_k(\beta_2, \dots, \beta_{m_k}, \beta) \colon \beta_2 < \cdots < \beta_{m_k} < \beta\}.$$

To establish Lemma 5.6, it is enough to show that $d_1 <' d_2$ implies $d_1 < d_2$. We assume this for descriptions $d \in \mathcal{D}_{2s+1}$ for $s < n$, and note that for almost all $h_1, \dots, h_t$, we may assume that:

1) If $k_1 < k_2$ and $K_{k_1} = W_{2n+1}^{m_{k_1}}$, $K_{k_2} = W_{2n+1}^{m_{k_2}}$, then $\inf h_{k_2} > \sup h_{k_1}$.

2) If $k_1 < k_2$ and $K_{k_1} = S_{2n+1}^{\ell_{k_1},m_{k_1}}$, $K_{k_2} = S_{2n+1}^{\ell_{k_2},m_{k_2}}$, then the range of h_{k_2} is closed under $h_{k_1}(0)$.

From these two properties and the definition of H, a straightforward induction shows that Lemma 5.6 is satisfied.

We now prove some technical lemmas concerning the family of canonical measures, which we prove in greater detail than required for this paper, and which simplify the description of conditions C and D.

It is convenient to introduce an auxiliary family of canonical measures $\tilde{R}_{2n+1}$ and embeddings $\tilde{\Pi}$.

$\tilde{W}_1^m$: We set $\tilde{W}_1^m = W_1^m$ = the m-fold product of the ω-cofinal normal measure on ω_1, which we identify with an ordinal, as in W_1^m, by ordering by the largest component first, then next largest, etc. We define $\tilde{\Pi}$ on these measures exactly as before on W_1^m, so $\Pi_k^m(\alpha_1, \dots, \alpha_m) = (\alpha_{m-k+1}, \dots, \alpha_m)$.

$\tilde{S}_1^{1,m}$: We define A to have measure on w.r.t. $\tilde{S}_1^{1,m}$ if there is a c.u.b. $C \subseteq \omega_1$ such that for all $h' \colon <^{m+1} \to C$ of the correct type (where $<^{m+1}$ is as in the definition of $S_1^{1,m+1}$), $[h] \in A$, where $h \colon <^m \to C$ is defined by $h(\alpha_2 \dots, \alpha_m, \alpha_1) = \sup_{\alpha_{m+1} < \alpha_1} h'(\alpha_2 \dots, \alpha_{m+1}, \alpha_1)$. This is essentially the same as the definition of $S_1^{1,m}$ except the functions no longer have uniform cofinality ω.

We define $\tilde{\Pi}_{m_1}^{m_2} \colon \vartheta(\tilde{S}_1^{1,m_2}) \to (\tilde{S}_1^{1,m_1})$ (for $m_2 > m_1$) as follows: given α represented by $h \colon <^{m_2} \to \omega_1$, (induced by an $h' \colon <^{m_2+1} \to \omega_1$, as above), we let $\tilde{\Pi}_{m_1}^{m_2}(\alpha)$ be represented by $\bar{h} \colon <^{m_1} \to \omega$ defined by

$$\bar{h}(\alpha_2, \dots, \alpha_{m_1}, \alpha_1) = \sup_{\alpha_{m_1+1}, \dots, \alpha_{m_2}} h(\alpha_2, \dots, \alpha_{m_1}, \alpha_{m_1+1}, \dots, \alpha_{m_2}, \alpha_1).$$

In general, we proceed as follows:

$\tilde{W}_{2n+1}^m$: We define A to have measure one w.r.t. $\tilde{W}_{2n+1}^m$ if there is a c.u.b. $C \subseteq \underset{\sim}{\boldsymbol{\delta}}_{2n+1}^1$ such that for all $f' \colon \vartheta(\tilde{S}_{2n-1}^{\ell,m+1}) \to C$ of the correct type (where, as in the definition of W_{2n+1}^m, $\ell = 2^n - 1$ is maximal), $[f] \in A$, where $f \colon \vartheta(\tilde{S}_{2n-1}^{\ell,m}) \to \underset{\sim}{\boldsymbol{\delta}}_{2n+1}^1$ is defined by $f(\alpha) = \sup_{\beta \colon \tilde{\Pi}_m^{m+1}(\beta) = \alpha} f'(\beta)$, where $\tilde{\Pi}_m^{m+1}$, of course, denotes the embedding from $\vartheta(\tilde{S}_{2n-1}^{\ell,m+1})$ to $\vartheta(\tilde{S}_{2n-1}^{\ell,m})$.

We define $\tilde{\Pi}_{m_1}^{m_2}$ from $\vartheta(W_{2n+1}^{m_2}) \to \vartheta(W_{2n+1}^{m_1})$, for $m_2 > m_1$, as follows: given α represented by $f \colon \vartheta(\tilde{S}_{2n-1}^{\ell,m_2}) \to \underset{\sim}{\boldsymbol{\delta}}_{2n+1}^1$ induced by an f' as above, let $\tilde{\Pi}_{m_1}^{m_2}(\alpha)$ be represented by $\bar{f} \colon \vartheta(\tilde{S}_{2n-1}^{\ell,m_1}) \to \underset{\sim}{\boldsymbol{\delta}}_{2n+1}^1$ defined by $\bar{f}(\alpha) = \sup_{\beta \colon \tilde{\Pi}_{m_1}^{m_2}(\beta) = \alpha} f(\beta)$, where $\tilde{\Pi}_{m_1}^{m_2}$ denotes the embedding from $\tilde{S}_{2n-1}^{\ell,m_2}$ to $\tilde{S}_{2n-1}^{\ell,m_1}$.

$\tilde{S}_{2n-1}^{\ell,m}$: If $\ell = 1$, we proceed as in the definition of $\tilde{S}_1^{1,m}$, using $\underset{\sim}{\boldsymbol{\delta}}_{2n+1}^1$ instead of ω_1. We define $\tilde{\Pi}$ similarly here.

For $\ell > 1$, we let $\tilde{R}_{2s+1}^{\ell',m}$ denote the $(\ell - 1)$st measure in the enumeration of the measures $\tilde{W}_1^m, \tilde{S}_1^{1,m}, \tilde{W}_3^m, \tilde{S}_3^{1,m}, \tilde{S}_3^{2,m}, \dots$, etc., similarly to the definition of $S_{2n+1}^{\ell,m}$. We let then define A to have measure one w.r.t. $\tilde{S}_{2n+1}^{\ell,m}$ if there is a c.u.b. $C \subseteq \underset{\sim}{\boldsymbol{\delta}}_{2n+1}^1$ such that for all $F \colon \underset{\sim}{\boldsymbol{\delta}}_{2n+1}^1 \to C$ of the correct type, $[\bar{F}]_{\mu_{2n+1}^{\tilde{R}_{2s+1}^{\ell',m}}} \in A$. Here $\mu_{2n+1}^{\tilde{R}_{2s+1}^{\ell',m}}$ is the measure on $\underset{\sim}{\boldsymbol{\delta}}_{2n+1}^1$ from the weak partition relation on $\underset{\sim}{\boldsymbol{\delta}}_{2n+1}^1$ and functions $f \colon \vartheta(\tilde{R}_{2s+1}^{\ell',m}) \to \underset{\sim}{\boldsymbol{\delta}}_{2n+1}^1$ which are induced by an $f' \colon \vartheta(\tilde{R}_{2s+1}^{\ell',m+1}) \to \underset{\sim}{\boldsymbol{\delta}}_{2n+1}^1$ of the correct type (via the corresponding embedding $\tilde{\Pi}_m^{m+1}$ for $\tilde{R}_{2s+1}^{\ell',m+1}$ and $\tilde{R}_{2s+1}^{\ell',m}$). Also, $\bar{F}$ is given by $\bar{F}([f]) = \sup\{F([g]) \colon \tilde{\Pi}([g]) = [f]\}$, where $\tilde{\Pi}([g])$, for $g \colon \vartheta(\tilde{R}_{2s+1}^{\ell',m+1}) \to \underset{\sim}{\boldsymbol{\delta}}_{2n+1}^1$, is defined as above.

We define $\tilde{\Pi}^{m_2}_{m_1}$ from $\vartheta(S^{\ell,m_2}_{2s+1})$ to $\vartheta(S^{\ell,m_1}_{2s+1})$ as follows: for α represented w.r.t. $\mu_{2n+1}^{\tilde{R}^{\ell',m_2}_{2s+1}}$ by $\bar{F}$, as above, we represent $\tilde{\Pi}^{m_2}_{m_1}(\alpha)$ w.r.t. $\mu_{2n+1}^{\tilde{R}^{\ell',m_1}_{2s+1}}$ by G defined by: for g induced by an appropriate g', we set $G[(g)] = \sup_f \bar{F}([f])$, where the sup ranges over $f : \vartheta(\tilde{R}^{\ell',m_2}_{2s+1}) \to \underset{\sim}{\boldsymbol{\delta}}^1_{2n+1}$ of the appropriate type (*i.e.*, induced by an f') such that $f(\vartheta) < g(\tilde{\Pi}^{m_2}_{m_1}(\alpha))$ for all ϑ (where $\tilde{\Pi}^{m_2}_{m_1}$ is defined by induction).

We also extend the embeddings slightly to include embeddings Π^{m+1}_m from $R^{\ell,m+1}_{2n+1}$ to $\tilde{R}^{\ell,m}_{2n+1}$ as follows:

If $R^{\ell,m+1}_{2n+1} = W^{m+1}_{2n+1}$, and $f : \vartheta(S^{\ell,m+1}_{2n-1}) \to \underset{\sim}{\boldsymbol{\delta}}^1_{2n+1}$ of the correct type represents α, then we let $\Pi^{m+1}_m(\alpha)$ be represented by $\bar{f} : \vartheta(\tilde{S}^{\ell,m}_{2n-1}) \to \underset{\sim}{\boldsymbol{\delta}}^1_{2n+1}$ defined by $\bar{f}(\beta) = \sup\{f(\beta') : \Pi(\beta') = \beta\}$, where Π denotes the embedding from $S^{\ell,m+1}_{2n-1}$ to $\tilde{S}^{\ell,m}_{2n-1}$, defined by induction. It follows readily that for almost all α (w.r.t. $\mu_{2n+1}^{S^{\ell,m+1}_{2n-1}} = W^{m+1}_{2n+1}$) that $\Pi(\alpha)$ is of the appropriate type, that is, induced by a $f' : \vartheta(\tilde{S}^{\ell,m+1}_{2n-1}) \to \underset{\sim}{\boldsymbol{\delta}}^1_{2n+1}$ of the correct type.

If $R^{\ell,m+1}_{2n+1} = S^{\ell,m+1}_{2n+1}$, and $F : \underset{\sim}{\boldsymbol{\delta}}^1_{2n+1} \to \underset{\sim}{\boldsymbol{\delta}}^1_{2n+1}$ of the correct type represents α w.r.t. $\mu_{2n+1}^{R^{\ell',m+1}_{2s+1}}$, then we represent $\Pi^{m+1}_m(\alpha)$ by a function $\bar{F}$ defined as follows: given $\bar{f} : \vartheta(\tilde{R}^{\ell',m}_{2s+1}) \to \underset{\sim}{\boldsymbol{\delta}}^1_{2n+1}$ induced by an $f' : \vartheta(R^{\ell',m+1}_{2s+1}) \to \underset{\sim}{\boldsymbol{\delta}}^1_{2n+1}$ of the correct type, we set $\bar{F}([\bar{f}]) = \sup\{F([f] : \Pi^{m+1}_m([f]) = [\bar{f}]\}$. Here the Π^{m+1}_m appearing in this formula is as defined in the previous paragraph.

We now state four lemmas which we prove by simultaneous induction on n:

Lemma 5.9. Given any $H : \vartheta(W^m_{2n+1}) = \underset{\sim}{\boldsymbol{\delta}}^1_{2n+1} \to \text{Ord}$ monotonically increasing almost everywhere (*i.e.*, restricted to a measure one set), if $m > 1$ then there is a measure one set A w.r.t. W^m_{2n+1} s.t. $H \restriction A$ is increasing, or $H(\alpha)$ for $\alpha \in A$ depends only on $\Pi^m_{m-1}(\alpha)$. If $m = 1$, then we replace the last part with: $H(\alpha)$ depends only on $\sup_{\text{a.e.}} f$, for $[f] = \alpha$, for $n > 1$, and for $n = 1$, with: H is constant almost everywhere.

Lemma 5.10. Given any $H : \vartheta(S^{\ell,m}_{2n+1}) \to \text{Ord}$ monotonically increasing almost everywhere, there is a measure one set A w.r.t. $S^{\ell,m}_{2n+1}$ s.t. $H \restriction A$ is increasing, or:

1) If $\ell = 1$, $m > 1$, then $H(\alpha)$ depends only on $\Pi^m_{m-1}(\alpha)$ (for $\alpha \in A$).
2) If $\ell = 1$, $m = 1$, then $H \restriction A$ is constant.
3) If $\ell > 1$, $m > 1$, then $H \restriction A$ depends only on $\Pi^m_{m-1}(\alpha)$.
4) If $\ell > 1$, $m = 1$, then $H(\alpha)$, for α represented by $F : \underset{\sim}{\boldsymbol{\delta}}^1_{2n+1} \to \underset{\sim}{\boldsymbol{\delta}}^1_{2n+1}$ of the correct type, depends only on $F(0)$ (recall here Definition 5.8).
5) If $m = 0$, let $S^{\ell,0}_{2n+1}$ be the measure induced from the strong partition relation on $\underset{\sim}{\boldsymbol{\delta}}^1_{2n+1}$, functions $F : \underset{\sim}{\boldsymbol{\delta}}^1_{2n+1} \to \underset{\sim}{\boldsymbol{\delta}}^1_{2n+1}$ of the correct type, and the $\text{cf}(\vartheta(R^{\tilde{\ell},1}_{2s+1}))$ cofinal normal measure on $\underset{\sim}{\boldsymbol{\delta}}^1_{2n+1}$, where $R^{\tilde{\ell},1}_{2s+1} = v(S^{\ell,1}_{2n+1})$.

Then any $H\colon \vartheta(S^{\ell,0}_{2n+1}) \to \mathrm{Ord}$ is either increasing almost everywhere, or constant almost everywhere.

Lemma 5.11. Let $H\colon \vartheta(W^m_{2n+1}) = \underset{\sim}{\boldsymbol{\delta}}^1_{2n+1} \to \mathrm{Ord}$ be pressing down almost everywhere. If $m > 1$, then there is a measure one set A such that for $\alpha \in A$, $H(\alpha)$ depends only on $\Pi^m_{m-1}(\alpha)$. For $m = 1$, $H \restriction A$ is constant.

Lemma 5.12. If $H\colon \vartheta(S^{\ell,m}_{2n+1}) \to \mathrm{Ord}$ is pressing down almost everywhere then there is a measure one set A w.r.t. $S^{\ell,m}_{2n+1}$ such that:

1) If $\ell = 1$, $m > 1$, then $H(\alpha)$ for $\alpha \in A$ depends only on $\Pi^m_{m-1}(\alpha)$.
2) If $\ell = 1$, $m = 1$, then $H \restriction A$ is constant.
3) If $\ell > 1$, $m > 1$, then $H(\alpha)$ for $\alpha \in A$ depends only on $\Pi^m_{m-1}(\alpha)$.
4) If $\ell > 1$, $m = 1$, then $H(\alpha)$ for $\alpha \in A$ depends only on $F(0)$, for F representing α.

We require two additional lemmas for the proofs, which we prove first in a separate induction:

Lemma 5.13. Given any $C \subseteq \underset{\sim}{\boldsymbol{\delta}}^1_{2n+1}$, and $f, g\colon \vartheta(S^{\ell,m}_{2n-1}) \to C$ of the correct type (where $\ell = 2^n - 1$ is maximal) and satisfying $[f] < [g]$ and $\Pi^m_{m-1}([f]) = \Pi^m_{m-1}([g])$, then there are f_2, g_2 satisfying:

1) $[f_2] = [f]$, $[g_2] = [g]$.
2) $f_2(\alpha) < g_2(\alpha) < f_2(\alpha + 1)$ for all α, and f_2, g_2 are of the correct type.
3) f_2, g_2 have range in C.

If $m = 1$, we require f, g satisfy $\sup_{\text{a.e.}} f = \sup_{\text{a.e.}} g$, and have the same conclusion.

Lemma 5.14. The same as Lemma 5.13 except we use $F, G\colon \underset{\sim}{\boldsymbol{\delta}}^1_{2n+1} \to \underset{\sim}{\boldsymbol{\delta}}^1_{2n+1}$ of the correct type with $[F]_{\mu^{R^{\ell,m}_{2s+1}}_{2n+1}} < [G]_{\mu^{R^{\ell,m}_{2s+1}}_{2n+1}}$, and $\Pi^m_{m-1}([F]) = \Pi^m_{m-1}([G])$ if $m > 1$ and if $m = 0$, $[F(0)] = [G(0)]$. We have for $n = 0$, $[F]_{W^m_1} < [G]_{W^m_1}$ in the hypothesis. We have the same conclusion as Lemma 5.13.

We first prove Lemmas 5.13 and 5.14 by induction.

We first consider Lemma 5.13.

We let C be given, and f, g as in Lemma 5.13. We assume $m > 1$, the case $m = 1$ being similar.

We let C_1 be a c.u.b. subset of $\underset{\sim}{\boldsymbol{\delta}}^1_{2n-1}$ such that for α represented by $H\colon \underset{\sim}{\boldsymbol{\delta}}^1_{2n-1} \to C_1$ or $<^m \to C_1$ of the correct type, depending on whether $\ell > 1$ or $\ell = 1$, we have $f(\alpha) < g(\alpha)$ and

$$\sup_{\beta\colon \Pi^m_{m-1}(\beta) = \Pi^m_{m-1}(\alpha)} f(\beta) = \sup_{\beta\colon \Pi^m_{m-1}(\beta) = \Pi^m_{m-1}(\alpha)} g(\beta).$$

We first assume $\ell > 1$. We consider the following partition $\mathcal{P}$: We partition functions $H_1, H_2\colon \underset{\sim}{\boldsymbol{\delta}}^1_{2n-1} \to \underset{\sim}{\boldsymbol{\delta}}^1_{2n-1}$ of the correct type with $H_1(\alpha) < H_2(\alpha) <$

$H_1(\alpha+1)$ for all α, according to whether or not $f([H_2]) > g([H_1])$, where $[H_1], [H_2]$ mean, of course, with respect to the measure $\mu_{2n-1}^{v(S_{2n-1}^{\ell,m})}$.

It follows readily that for H_1, H_2 for the above type that $\Pi_{m-1}^{m}([H_1]) = \Pi_{m-1}^{m}([H_2])$. We claim that on the homogeneous side of the partition the property stated in partition $\mathcal{P}$ holds. We suppose not, and let $C_2 \subseteq \underset{\sim}{\boldsymbol{\delta}}_{2n-1}^1$ be a c.u.b. set homogeneous for the other side. We let $C_3 = C_1 \cap C_2$ and $C_4 \subseteq C_3$ be contained in the closure points of C_3, that is if $\alpha \in C_4$ and $\beta, \gamma < \alpha$, then the $(\omega \cdot \gamma)$th element of C_3 after β is less than α (we use this notation throughout). We fix $H_1 \colon \underset{\sim}{\boldsymbol{\delta}}_{2n-1}^1 \to C_4$ of the correct type. Since $C_4 \subseteq C_1$, it follows that there is an $H_2 \colon \underset{\sim}{\boldsymbol{\delta}}_{2n-1}^1 \to \underset{\sim}{\boldsymbol{\delta}}_{2n-1}^1$ such that $\Pi_{m-1}^{m}([H]) = \Pi_{m-1}^{m}([H_2])$ and $f([H_2]) > g([H_1])$. We may clearly assume that $\operatorname{ran} H_2 \subseteq C_3$, replacing H_2 by a larger function if necessary (*e.g.*, the function $\tilde{H}_2(\alpha) =$ the next ωth element of C_3 after $\max(H_2(\alpha), \sup_{\beta<\alpha} \tilde{H}_2(\beta)))$. By Lemma 5.14 and induction, it now follows that there are functions H_1', H_2' satisfying:

1) $[H_1'] = [H_1], [H_2'] = [H_2]$, and hence $f([H_2']) > g([H_1'])$.
2) H_1', H_2' are of the correct type everywhere and ordered as in $\mathcal{P}$.
3) $\operatorname{ran} H_1', H_2' \subseteq C_3$.

This, however, contradicts $C_3 \subseteq C_2$ and the definition of C_2.

Hence, on the homogeneous side of the partition, the property stated in $\mathcal{P}$ holds. We let $C_2 \subseteq C_1$ be homogeneous for $\mathcal{P}$. We let C_3 be contained in the closure points of C_2 (as above), and C_4 be contained in the closure points of C_3. We then define the functions f_2, g_2 as follows:

For α represented by $H \colon \underset{\sim}{\boldsymbol{\delta}}_{2n-1}^1 \to C_4$ of the correct type, we set $f_2(\alpha) = f(\alpha)$, $g_2(\alpha) = g(\alpha)$. For β not of this form, we let β' be the least ordinal $> \beta$ which is represented by an $H' \colon \underset{\sim}{\boldsymbol{\delta}}_{2n-1}^1 \to C_3$ of the correct type. For $\beta < \alpha$, with α as above, it is easily seen that $\beta' < \alpha$ as well. We fix such an H' representing β', with $H' \subseteq C_3$ everywhere. We then define $H'' \colon \underset{\sim}{\boldsymbol{\delta}}_{2n-1}^1 \to C_2$ by $H''(\gamma) =$ the $(\omega \cdot H_\beta(\gamma))$th element of C_2 after $H'(\gamma)$, where H_β represents β. We let $[H''] = \beta''$. We then set $f_2(\beta) = f(\beta'')$, $g_2(\beta) = g(\beta'')$. This is well-defined. We note that β'' is also of the correct type, since H' was, and $\operatorname{ran} H' \subseteq C_3$. For α as above, that is, represented by H of the correct type with range in C_4, it follows readily that if $\beta < \alpha$ then $\beta'' < \alpha$. From this, and the definitions of C_1, C_2, it now follows that f_2 and g_2 have the desired properties.

The case $\ell = 1$ is similar, using functions $H \colon <^m \to \underset{\sim}{\boldsymbol{\delta}}_{2n-1}^1$ instead.

We now consider Lemma 5.14, We again consider the case $m > 1$, the case $m = 1$ being similar. We fix F, $G \colon \underset{\sim}{\boldsymbol{\delta}}_{2n+1}^1 \to \underset{\sim}{\boldsymbol{\delta}}_{2n+1}^1$ of the correct type as in Lemma 5.14. We consider the following partition $\mathcal{P}$: We partition functions $f, g \colon \vartheta(R_{2s+1}^{\ell,m}) \to \underset{\sim}{\boldsymbol{\delta}}_{2n+1}^1$ of the correct type with $f(\alpha) < g(\alpha) < f(\alpha+1)$ everywhere, according to whether or not $F([g]) > G([f])$.

It follows as in the proof of Lemma 5.13, using induction and Lemma 5.13, that on the homogeneous side of the partition, the property stated in $\mathcal{P}$ holds.

We let $C_2 \subseteq C_1$ be homogeneous for $\mathcal{P}$, where $C_1 \subseteq \underset{\sim}{\delta}^1_{2n+1}$ is a c.u.b. such that for $f : \vartheta(R^{\ell,m}_{2s+1}) \to C_1$ of the correct type, $F([f]) < G([f])$, and

$$\sup_{\beta :\, \Pi^m_{m-1}(\beta)=\Pi^m_{m-1}([f])} F(\beta) = \sup_{\beta :\, \Pi^m_{m-1}(\beta)=\Pi^m_{m-1}([f])} G(\beta).$$

We let C_3 be contained in the closure points of C_2, and C_4 be contained in the closure points of C_3. We then define $F_2(\alpha) = F(\alpha)$ and $G_2(\alpha) = G(\alpha)$ for α represented by $f : \vartheta(R^{\ell,m}_{2s+1}) \to C_4$ of the correct type, and for other β, we defined β', β'' similarly to Lemma 5.13, and set $F_2(\beta) = F(\beta'')$, $G_2(\beta) = G(\beta'')$. It follows that F_1, G_2 have the desired properties.

We now consider Lemma 5.9. We let $H : \underset{\sim}{\delta}^1_{2n+1} \to \mathrm{Ord}$ be monotonically increasing almost everywhere, say restricted to $[f]$ for $f : \vartheta(S^{\ell,m}_{2n-1}) \to C$ of the correct type, where we assume $n \geq 1$ here, the case $n = 0$ being similar. We consider the following partition $\mathcal{P}$: We partition functions f, $g : \vartheta(S^{\ell,m}_{2n-1}) \to \underset{\sim}{\delta}^1_{2n+1}$ of the correct type with $f(\alpha) < g(\alpha) < f(\alpha + 1)$ everywhere, according to whether or not $H([f]) > H([g])$ or not.

We let $C_1 \subseteq \underset{\sim}{\delta}^1_{2n+1}$ be c.u.b. and homogeneous for $\mathcal{P}$, and let C_2 be contained in the closure points on $C \cap C_1$. We then claim that C_2 defines the measure one set as in the statement of Lemma 5.9. This follows from the definition of C_2 and Lemma 5.13. In the event we are on the homogeneous side in which $\mathcal{P}$ holds, we use the trivial fact that if f, g are of the correct type with range in C_2 and $\Pi^m_{m-1}([f]) < \Pi^m_{m-1}([g])$, then there is an f' of the correct type with range in $C \cap C_1$ such that $f < f' < g$ almost everywhere, and $\Pi^m_{m-1}([f']) = \Pi^m_{m-1}([g])$.

The proof of Lemma 5.10 is similar to that of Lemma 5.9, only we partition functions $F, G : \underset{\sim}{\delta}^1_{2n+1} \to \underset{\sim}{\delta}^1_{2n+1}$ instead, and we use Lemma 5.14.

We now consider Lemma 5.11.

We fix $H : \underset{\sim}{\delta}^1_{2n+1} \to \mathrm{Ord}$ pressing down almost everywhere, say on a measure one set w.r.t. $\mu^{S^{\ell,m}_{2n-1}}_{2n+1}$ determined by $C \subseteq \underset{\sim}{\delta}^1_{2n+1}$, where we assume $n \geq 1$, the case $n = 0$ being easier.

We consider the following partition $\mathcal{P}$: We partition functions $f, g : \vartheta(S^{\ell,m}_{2n-1}) \to \underset{\sim}{\delta}^1_{2n+1}$ of the correct type with $f(\alpha) < g(\alpha) < f(\alpha + 1)$ everywhere, according to whether or not $[f] > H([g])$.

It follows readily that on the homogeneous side of the partition, the property stated in $\mathcal{P}$ holds. We let C_1 be homogeneous for $\mathcal{P}$, and let C_2 be contained in the closure points of $C \cap C_1$.

It follows if $f : \vartheta(S^{\ell,m}_{2n-1}) \to C_2$ is of the correct type, and $\bar{f}$ is defined by $\bar{f}(\alpha) =$ the next ωth element of $C \cap C_1$ after $\sup_{\beta > \alpha} f(\beta)$, then $[\bar{f}] > H([f])$.

We now consider the following partition $\mathcal{P}_2$: We partition functions $f, g : \vartheta(S^{\ell,m}_{2n-1}) \to \underset{\sim}{\delta}^1_{2n+1}$ of the correct type with $f(\alpha) < g(\alpha) < f(\alpha + 1)$ everywhere, according to whether or not $H([f]) < H([g])$.

If on the homogeneous side of the partition, the property stated in $\mathcal{P}_2$ fails, then the lemma follows readily from Lemma 5.13.

Otherwise, we let $C_3 \cap C_2$ be homogeneous for $\mathcal{P}_2$, and let C_4 be contained in the closure points of C_3. We fix $f : \vartheta(S^{\ell,m}_{2n-1}) \to C_4$ of the correct type, and let $\bar{f}$ be as above.

We next claim that if f_2 is any function $f_2 \colon \vartheta(S^{\ell,m}_{2n-1}) \to C_3$ of the correct type with $\Pi^m_{m-1}([f_2]) = \Pi^m_{m-1}([f])$, then $[\bar{f}] > H([f_2])$. This will give a contradiction, since it will give an order-preserving map, namely H, from the set of such $[f_2]$ into $[\bar{f}]$, and the former is easily seen to have larger order type.

To prove the claim, we fix $f_2 \colon \vartheta(S^{\ell,m}_{2n-1}) \to \underset{\sim}{\delta}^1_{2n+1}$ of the correct type with $\Pi^m_{m-1}([f]) = \Pi^m_{m-1}([f_2])$. It is enough to establish that $\tilde{f} = \tilde{f}_2$ almost everywhere, where $\tilde{f}(\alpha) = \sup_{\beta > \alpha} \tilde{f}(\beta)$, and similarly for $\tilde{f}_2$. We suppose not, say $\tilde{f}_2 > \tilde{f}$ almost everywhere. We then define the function h on $\vartheta(S^{\ell,m}_{2n-1})$ by $h(\alpha) =$ by least $\beta < \alpha$ such that $f_2(\beta) > \tilde{f}(\alpha)$, defined almost everywhere. Since h is pressing down, it follows from induction and Lemma 5.12 that $h(\alpha)$ depends only on $\Pi^m_{m-1}(\alpha)$ if $m > 1$, and on $F(0)$, for F representing α, if $m = 1$. It follows that $\Pi^m_{m-1}([f]) < \Pi^m_{m-1}([f_2])$, contradicting our assumption. This establishes the claim, and completes the proof of Lemma 5.11.

The proof of Lemma 5.12 is similar, using induction and Lemma 5.11.

Finally, we remark that the above six lemmas are also true with the measure $\tilde{S}^{\ell,m}_{2n+1}$, $\tilde{W}^{\ell,m}_{2n+1}$, the proofs being essentially identical to those give above.

To state the next two lemmas, we require a technical definition:

Definition 5.15. Given two functions $f, g \colon \vartheta(S^{\ell,m}_{2n-1}) \to \mathrm{Ord}$, we say that they are ordered of k-type, where $1 \le k \le m$, if they satisfy the following:

1) f, g are of the correct type.
2) If $\alpha, \beta \in \vartheta(S^{\ell,m}_{2n-1})$ are represented by F, G of the correct type, then $f(\alpha) < g(\beta)$ if $\Pi^m_k(\alpha) \le \Pi^m_k(\beta)$, and $f(\alpha) > g(\beta)$ if $\Pi^m_k(\alpha) > \Pi^m_k(\beta)$.

For other ordinals the ordering is more or less arbitrary; we use the following:

3) If $\alpha_1 < \alpha < \alpha_2$ where α_1, α_2 are of the correct type (*i.e.*, represented by functions of the correct type) with $\Pi^m_k(\alpha_1) = \Pi^m_k(\alpha_2)$, and $\beta \ge$ the sup of all the α' of the correct type with $\Pi^m_k(\alpha') = \Pi^m_k(\alpha_1)$, then $f(\beta) > g(\alpha)$.
4) If α_2 is the least ordinal $> \alpha$ of the correct type, $\alpha < \inf \alpha'$ for all α' of the correct type with $\Pi^m_k(\alpha') = \Pi^m_k(\alpha_2)$, and $\alpha \ge \sup \alpha'$ for all α' of the correct type with $\Pi^m_k(\alpha') < \Pi^m_k(\alpha_2)$, and β also satisfies the above (for the same α_2), then $f(\alpha) < g(\beta)$ if $\alpha \le \beta$, and $f(\alpha) > g(\beta)$ if $\alpha > \beta$.
5) If α_2 is as in above 4, corresponding to α, and $\beta > \inf \alpha'$ for α' of the correct type with $\Pi^m_k(\alpha') = \Pi^m_k(\alpha_2)$, then $f(\beta) > g(\alpha)$.

It is easily seen that this defines a unique way of ordering $\operatorname{ran} f \cup \operatorname{ran} g$.

Definition 5.16. Given $F, G\colon \vartheta(\mu_{2n+1}^{R_{2s+1}^{\ell,m}}) = \underset{\sim}{\delta}_{2n+1}^1 \to \mathrm{Ord}$, we say F, G are ordered of k type, for $1 \le k \le m$, if they satisfy the same 5 conditions above, where α of the correct type means represented by an $f\colon \vartheta(R_{2s+1}^{\ell,m}) \to \underset{\sim}{\delta}_{2n+1}^1$ of the correct type, and we use the corresponding Π_k^m defined for such α. Finally, for $F, G\colon <^m \to \underset{\sim}{\delta}_{2n+1}^1$ of the correct type, F, G are ordered of k-type if for $\vec{\alpha} = (\alpha_2, \dots, \alpha_m, \alpha_1)$, $\vec{\beta} = (\beta_2, \dots, \beta_m, \beta_1)$ then $G(\vec{\alpha}) > F(\vec{\beta})$ if there is an $r < k$ such that $\alpha_r \neq \beta_r$ and for the least such r we have $\alpha_r > \beta_r$, or $\alpha_r = \beta_r$ for all $r < k$ and $\alpha_k \ge \beta_k$. Otherwise we require $G(\vec{\alpha}) < F(\vec{\beta})$.

We now state two lemmas which generalize Lemmas 5.13 and 5.14.

Lemma 5.17. Given any c.u.b. $C \subseteq \underset{\sim}{\delta}_{2n+1}^1$ and $f, g\colon \vartheta(S_{2n-1}^{\ell,m}) \to C$ of the correct type with $\Pi_k^m([f]) = \Pi_k^m([g])$ for some $1 \le k \le m-1$, and $\Pi_{k+1}^m([f]) < \Pi_{k+1}^m([g])$, then there are f_2, g_2 satisfying:
1) $[f_2] = [f]$, $[g_2] = [g]$.
2) f_2, g_2 are ordered of $k+1$-type.
3) f_2, g_2 have range in C.

Also, if $\Pi_1^m([f]) < \Pi_1^m([g])$ then f_2, g_2 are ordered of 1-type if $\sup f = \sup g$, and if $\sup f < \sup g$, then $\inf g_2 > \sup f_2$.

Lemma 5.18. The same as Lemma 5.17 except we use $F, G\colon \vartheta(\mu_{2n+1}^{R_{2s+1}^{\ell,m}}) = \underset{\sim}{\delta}_{2n+1}^1 \to C$, and the corresponding Π_k^m's. If $\Pi_1^m([F]) < \Pi_1^m([G])$ here, we require F_2, G_2 to be ordered of 1-type if $[F(0)] = [G(0)]$, and if $[F(0)] < [G(0)]$, then we require that $F_2(\alpha) < G_2(\beta)$ if $\sup_{\text{a.e.}} f_\alpha \le \sup_{\text{a.e.}} f_\beta$, and if $\sup_{\text{a.e.}} f_\alpha > \sup_{\text{a.e.}} f_\beta$ then $G_2(\beta) < F_2(\alpha)$ for f_α, g_β representing α, β.

Proof of Lemma 5.17. We fix such f, g of the correct type, and suppose $\Pi_k^m([f]) = \Pi_k^m([g])$ and $\Pi_{k+1}^m([f]) < \Pi_{k+1}^m([g])$, the case $\Pi_1^m([f]) < \Pi_1^m([g])$ being similar.

We consider the following partitions:

$\mathcal{P}_1$: We partition functions $H_1, H_2\colon \underset{\sim}{\delta}_{2n-1}^1 \to \underset{\sim}{\delta}_{2n-1}^1$ ordered of $k+1$-type for $\ell > 1$, and $H_1, H_2\colon <^m \to \underset{\sim}{\delta}_{2n-1}^1$ ordered of $k+1$-type if $\ell = 1$, according to whether or not $f([H_2]) > g([H_1])$.

$\mathcal{P}_2$: We partition functions H_1, H_2 as above, ordered of $k+2$-type, according to whether or not $g([H_1]) > f([H_2])$. (If $k = m-1$, we don't consider $\mathcal{P}_2$.)

We then claim that on the homogeneous sides of these partitions that the properties stated in them hold. We consider first $\mathcal{P}_1$

We suppose not, and fix a c.u.b. $C_1 \subseteq \underset{\sim}{\delta}_{2n-1}^1$ homogeneous for the contrary side. We may assume C_1 is contained in a c.u.b. C such that for H having range in C of the correct type,

$$\sup_{\beta\colon \Pi_k^m(\beta) = \Pi_k^m([H])} f(\beta) = \sup_{\beta\colon \Pi_k^m(\beta) = \Pi_k^m([H])} g(\beta).$$

We fix H_1 of the correct type with range in C_1. It follows that there is an H_2 of the correct type with range C_1 such that $f([H_2]) > g([H_1])$ and with $\Pi_k^m([H_2]) = \Pi_k^m([H_1])$, and $\Pi_{k+1}^m([H_2]) > \Pi_{k+1}^m([H_1])$. It then follows from Lemma 5.18 and induction that there are H_1', H_2' ordered of $k+1$-type with range in C_1 and $[H_1'] = [H_1]$, $[H_2'] = [H_2]$. This contradicts the definitions of C_1.

We let C_1 be a c.u.b. subset of $\underset{\sim}{\delta}_{2n-1}^1$ homogeneous for $\mathcal{P}_1$.

It similarly follows that there is a c.u.b. $C_2 \subseteq \underset{\sim}{\delta}_{2n-1}^1$ homogeneous for $\mathcal{P}_2$.

We let $C_3 = C_1 \cap C_2$, and C_4 be contained in the closure points of C_3. Let A_3 and A_4 be the measure one sets (w.r.t. $(S_{2n-1}^{\ell,m})$) they define.

Restricted to A_3, it follows that f, g are ordered of $k+1$-type. For example, if α, β are represented by H_1, H_2 having range in C_3 of the correct type, and $\Pi_{k+1}^m(\beta) > \Pi_{k+1}^m(\alpha)$, then since we may assume without loss of generality that $\Pi_k^m(\beta) = \Pi_k^m(\alpha)$ (since f, g are increasing), we have from Lemma 5.18 and induction that $\alpha = [H_1']$, $\beta = [H_2']$ for some H_1', H_2' ordered of $k+1$-type with range in C_3. So, since $C_3 \subseteq C_1$ we have $f(\beta) > g(\alpha)$. Similarly, if $\Pi_{k+1}^m(\alpha) = \Pi_{k+1}^m(\beta)$ and (without loss of generality), $\Pi_{k+2}^m(\alpha) \leq \Pi_{k+2}^m(\beta)$ for $m < k-1$, then from Lemma 5.18 and induction, and $C_3 \subseteq C_2$ we have $g(\alpha) > f(\beta)$.

We then define f_2, g_2 as follows: For $\alpha \in A_4$, we set $f_2(\alpha) = f(\alpha)$, $g_2(\alpha) = g(\alpha)$. For $\alpha \notin A_4$, we define, by induction, $f_2(\alpha) =$ the ωth element in the range of $f \cup g$ which is greater that $\sup\{f_2(\beta), g_2(\gamma)\}$, where β, γ range over the ordinals such that for any f', g' ordered of $k+1$-type, $f'(\beta) < f'(\alpha)$, $g'(\gamma) < f'(\alpha)$. We similarly define $g_2(\alpha)$, where β, γ now range over ordinals such that $f'(\beta) < g'(\alpha)$, $g'(\gamma) < g'(\alpha)$, for f', g' as above. It then follows readily that f_2, g_2 are ordered of $k+1$-type. This is immediate once it is shown that f_2, g_2 are increasing, and this follows once it is shown that if $\alpha < \beta$, $\alpha \notin A_4$, and $\beta \in A_4$, then $f_2(\alpha) < f_2(\beta)$, and similarly for g_2. This, in turn, follows from the definition of A_4 and $k+1$-type. We use here the facts that if $\gamma_1 < \gamma < \gamma_2$ and γ_1, γ_2 of the correct type with $\Pi_{k+1}^m(\gamma_1) = \Pi_{k+1}^m(\gamma_2)$, then the definition of $k+1$-type requires no value $g(\delta)$ between $f(\gamma_1)$ and $f(\gamma_2)$, and no value $f(\delta)$ between $g(\gamma_1)$ and $g(\gamma_2)$. Also, if $f(\delta)$ is required to be less than $\inf f(\gamma)$ for γ of the correct type with $\Pi_{k+1}^m(\gamma) = \Pi_{k+1}^m(\gamma_0)$ for some fixed γ_0, then δ is less than the least such γ.

This completes the proof of Lemma 5.17. ⊣

The proof of Lemma 5.18 is entirely similar, using induction and Lemma 5.17.

We require two additional technical lemmas.

LEMMA 5.19. Given $H\colon \vartheta(S_{2n-1}^{\ell,m}) \to \mathrm{Ord}$, there is a measure on set A restricted to which H is monotonically increasing.

LEMMA 5.20. Same as Lemma 5.19 using $H : \vartheta(W^m_{2n+1}) = \underset{\sim}{\delta}^1_{2n+1} \to \mathrm{Ord}$, where $n \geq 1$.

PROOF OF LEMMA 5.19. We fix $H : \vartheta(S^{\ell,m}_{2n-1}) \to \mathrm{Ord}$. For each $1 \leq k \leq m$, we consider the following partition $\mathcal{P}_k$: We partition functions $H_1, H_2 : \underset{\sim}{\delta}^1_{2n-1} \to \underset{\sim}{\delta}^1_{2n-1}$ (or if $\ell = 1$ we have $H_1, H_2 : <^m \to \underset{\sim}{\delta}^1_{2n-1}$), ordered of k-type according to whether or not $H([H_1]) \leq H([H_2])$. If $k = m$, then by "ordered of m-type" we mean $H_1(\alpha) < H_2(\alpha) < H_1(\alpha + 1)$ for all α.

For $k = 0$, we also consider the partition $\mathcal{P}_0$, where H_1, H_2 are ordered as in the last clause of Lemma 5.18 for $\ell > 1$. For $\ell = 1$ and $\vec{\alpha} = (\alpha_2, \dots, \alpha_m, \alpha_1)$, $\vec{\beta} = (\beta_2, \dots, \beta_m, \beta_1)$, if $\alpha_1 < \beta_1$ then $H_2(\vec{\alpha}) < H_1(\vec{\beta})$ and if $\alpha_1 \geq \beta_1$, then $H_2(\vec{\alpha}) > H_1(\vec{\beta})$.

It follows from an easy well-foundedness argument that on the homogeneous side of the partition the property stated in $\mathcal{P}_k$ holds. We let C_k, for $0 \leq k \leq m$, be homogeneous for $\mathcal{P}_k$, and $C = \bigcap_k C_k$. If A is the measure one set determined by C, then we claim that $H \restriction A$ is monotonically increasing. For, let $\alpha < \beta$ be in A, say represented by H_1, H_2 of the correct type with range in C. We let $1 \leq k \leq m$ be maximal such that $\Pi^m_k(\alpha) = \Pi^m_k(\beta)$, if such a k exists. In this case, by Lemma 5.18, it follows that there are H_1', H_2' ordered of $k+1$-type with range in $C \subseteq C_k$ with $[H_1'] = [H_1]$, $[H_2'] = [H_2]$. Hence $H([H_1]) = H([H_1']) \leq H([H_2']) = H([H_2])$. If $\Pi^m_1(\alpha) < \Pi^m_1(\beta)$, then H_1', H_2' are ordered as in the last part of Lemma 5.18 (depending on whether $[H_1(0)] < [H_2(0)]$ or not) for $\ell > 1$, and for $\ell = 1$ as in the paragraph above, and $H([H_1]) = H([H_1']) \leq H([H_2']) = H([H_2])$ again follows. ⊣ (Lemma 5.19)

The proof of Lemma 5.20 is similar.

We now state two lemmas which simplify conditions C and D.

LEMMA 5.21. For any c.u.b. $C \subseteq \underset{\sim}{\delta}^1_{2n+1}$, there is a c.u.b. $C' \subseteq C$ such that (for $s \leq n$) for any $f : \vartheta(R^{\ell,m}_{2s-1}) \to C'$ which is monotonically increasing, $[f]$ is the supremum of ordinals $[g]$, for $g : \vartheta(R^{\ell,m}_{2s-1}) \to C$ of the correct type.

LEMMA 5.22. For any c.u.b. $C \subseteq \underset{\sim}{\delta}^1_{2n+1}$, there is a c.u.b. $C' \subseteq C$ such that for any $F : \underset{\sim}{\delta}^1_{2n+1} \to \underset{\sim}{\delta}^1_{2n+1}$ monotonically increasing almost everywhere w.r.t. $\mu^{R^{\ell,m}_{2s+1}}_{2n+1}$, $[F]$ is the supremum of ordinals $[G]$ for $G : \underset{\sim}{\delta}^1_{2n+1} \to C$ of the correct type, provided $[F]$ is not minimal with respect to being non-constant almost everywhere.

We first consider Lemma 5.21.

PROOF OF LEMMA 5.21. We consider the case $R^{\ell,m}_{2s-1} \neq W^m_1$, this case following directly.

For a given C, we let $C' \subseteq C$ be contained in the closure points of C. We fix $f : \vartheta(R^{\ell,m}_{2s-1}) \to C'$ monotonically increasing almost everywhere, say on the measure one set determined by $C_1 \subseteq \underset{\sim}{\delta}^1_{2s-1}$. Applying Lemmas 5.9 and 5.10, we conclude that either f is increasing almost everywhere, or $f(\alpha)$ depends only on $\Pi^m_{m-1}(\alpha)$ almost everywhere. In the latter case, we view f as a function on $f\colon \vartheta(\tilde{R}^{\ell,m-1}_{2s-1})$ (for $m > 1$). That is, there is a $f^{m-1}\colon \vartheta(\tilde{R}^{\ell,m-1}_{2s-1}) \to C'$ with $f(\alpha) = f^{m-1}(\Pi^m_{m-1}(\alpha))$ almost everywhere. Applying Lemmas 5.9 and 5.10 again to f^{m-1}, we get that either f^{m-1} is increasing almost everywhere, or there is an $f^{m-1}\colon \vartheta(\tilde{R}^{\ell,m-2}_{2s-1}) \to C'$ (for $m > 2$) such that $f(\alpha) = f^{m-2}(\Pi^m_{m-2}(\alpha))$ almost everywhere, where $\Pi^m_{m-2}(\alpha) \equiv \tilde{\Pi}^{m-1}_{m-2}(\alpha)(\Pi^m_{m-2}(\alpha))$, etc. Continuing in this manner, we conclude that either for some $0 \leq k \leq m$ there is an $f^k\colon \vartheta(\tilde{R}^{\ell,k}_{2s-1}) \to C'$ (or $f^k\colon \vartheta(R^{\ell,m}_{2s-1}) \to C'$ if $k = m$) with $f(\alpha) = f^k(\Pi^m_k(\alpha))$ almost everywhere (where Π^m_m = identity) and f^k is increasing almost everywhere, or else f is constant almost everywhere. Here, for $k = 0$, by $\tilde{S}^{\ell,0}_{2s-1}$ we mean the measure on $H\colon \underset{\sim}{\delta}^1_{2s-1} \to \underset{\sim}{\delta}^1_{2s-1}$ induced by the measure $S^{\ell,1}_{2n+1}$ on $H'\colon \underset{\sim}{\delta}^1_{2s-1} \to \underset{\sim}{\delta}^1_{2s-1}$ of the correct type and $H = H'(0)$. By W^0_{2s-1} we mean the measure (for $s > 1$) induced by $f\colon \kappa(W^1_{2s-1}) \to \underset{\sim}{\delta}^1_{2s-1}$ of the correct type, the weak partition relation on $\underset{\sim}{\delta}^1_{2s-1}$ and $\sup f$ (this is the κ-cofinal normal measure on $\underset{\sim}{\delta}^1_{2s-1}$). In the case where f is constant almost everywhere the lemma easily follows, so we assume the former.

We fix an $f^k\colon \vartheta(\tilde{R}^{\ell,k}_{2s-1}) \to C'$, where if $k = m$ we use $R^{\ell,m}_{2s-1}$, and let $C_2 \subseteq C_1$ be a c.u.b. subset of $\underset{\sim}{\delta}^1_{2s-1}$ defining a measure on A_2 restricted to which f^k is increasing and $f(\alpha) = f^k(\Pi^m_k(\alpha))$ holds.

We consider the following two cases:

Case 1. There is a c.u.b. $C_3 \subseteq C_2$ defining a measure one set $A_3 \subseteq A_2$ restricted to which f^k is discontinuous, that is, for $\alpha \in A_3$ we have $f^k(\alpha) > \sup_{\beta<\alpha, \beta\in A_2} f^k(\beta)$. We let $f\colon \vartheta(R^{\ell,m}_{2s-1}) \to \underset{\sim}{\delta}^1_{2n+1}$ be given with $[g] < [f]$, and we proceed to show that there is an $\bar{f}$ of the correct type with range in C with $[g] < [\bar{f}] < [f]$. From Lemmas 5.19 and 5.20, there is a $C_4 \subseteq C_3$ defining a measure one set $A_4 \subseteq A_3$ such that $g\restriction A_4$ is monotonic increasing. We then define $\bar{f}$ as follows: for $\alpha \in A_4$, we set $\bar{f}(\alpha)$ = the next ωth point in C after $\max\{g(\alpha), \sup_{\beta<\alpha} \bar{f}(\beta)\}$. For $\alpha \notin A_4$, we define $\bar{f}(\alpha)$ = the next ωth element in C after $\sup_{\beta<\alpha} \bar{f}(\beta)$. It follows readily that for $C_5 \subseteq$ the closure points of C_4, that $\bar{f}\restriction A_5$ is less than $f\restriction A_5$.

Case 2. There is a $C_3 \subseteq C_2$ defining a measure one set $A_3 \subseteq A_2$ restricted to which f^k is continuous, that is, for $\alpha \in A_3$ we have $f^k(\alpha) = \sup_{\beta<\alpha, \beta\in A_2} f^k(\beta)$. Hence, for $[g] < [f]$, for almost all α w.r.t. $R^{\ell,m}_{2s-1}$ we have that $g(\alpha) < f(\alpha) = f^k(\Pi^m_k(\alpha))$, and it follows that for almost all all α that there is a $\beta \in A_3$ with $\beta < \Pi^m_k(\alpha)$ such that $g(\alpha) < f^k(\beta)$. Applying Lemmas 5.11 and 5.12 repeatedly, we conclude that β depends only on

$\Pi^m_{k-1}(\alpha)$, for $k > 1$, and if $k = 1$, then β is constant almost everywhere if $\ell = 1$, and for $\ell > 1$, depends only on $\Pi^m_0(\alpha)$. Here $\Pi^m_0(\alpha)$ denotes $[F(0)]$ for F representing α if $R^{\ell,m}_{2s-1} = S^{\ell,m}_{2s-1}$, and denotes $\sup_{\text{a.e.}} f$ for f representing α if $R^{\ell,m}_{2s-1} = W^m_{2s-1}$. If $k = 0$, then β is constant almost everywhere. Hence, there is an $\bar{f}^{k-1} < f^{k-1}$ (where $f^{k-1}(\alpha) = \sup_{\beta\colon \Pi^k_{k-1}(\beta)=\alpha} f^k(\beta)$) almost everywhere w.r.t. $\tilde{S}^{\ell,k-1}_{2s-1}$ (in the case $k > 1$, or $k \geq 1$ if $\ell > 1$) such that for almost all α w.r.t. $S^{\ell,m}_{2s-1}$ we have $g(\alpha) < \bar{f}^{k-1}\Pi^m_{k-1}(\alpha)$. The result now follows easily from the choice of C^1. If $k = 1$, $\ell = 1$ or $k = 0$, $\ell > 1$, the result similarly follows easily.

This completes the proof of Lemma 5.21. ⊣ (Lemma 5.21)

The proof of Lemma 5.22 is similar.

The last two lemmas allow us to simplify the description of conditions C and D. In condition C, if $d^{(\mathcal{I}_a)}$ is non-basic of the form $d = (k; d_2^{(\mathcal{I}_a;\bar{K}_{a+1})})^s$ where s appears, $\bar{K}_{a+1} = v(K_k)$, and $K_k = S^{\ell_k,m_k}_{2n+1}$, then for almost all functions $h_1, \dots, h_t, f, \bar{h}_2, \dots, \bar{h}_a$, we consider the function $g\colon \vartheta(\bar{K}_{a+1}) \to \underset{\sim}{\delta}^1_{2n+1}$ defined by $g([\bar{h}_{a+1}]) = H(d_2; h_1, \dots, h_t, f, \bar{h}_2, \dots, \bar{h}_{a+1})$, as in the definition of condition C. If $\ell_k > 2$, so $\bar{K}_{a+1}$ is not of the form W^m_1, then from Lemmas 5.21 and 5.22 it follows that d satisfies condition C provided $[g]$ is not minimal amongst all $[g']$ with $\sup_{\text{a.e.}} g' = \sup_{\text{a.e.}} g$. Thus, for $d \in \mathcal{D}_{2n+1}$ with the symbol s appearing on all component tuples, condition C reduces to the following:

1) All basic component tuples of d satisfy condition C; which is just a condition on certain components tuples $\bar{d}_2 \in \mathcal{D}_{2n-1}$ of d.
2) The previous definition in the case d is non-basic with $K_k = S^{\ell_k,m_k}_{2n+1}$ and $\ell_k = 1$ or 2
3) The above non-minimality condition on g, for $\ell_k > 2$.

As for condition D, we note the following. If $\ell > 1$, then $(d^{\ell,m}_{2n+1})^s$ satisfies condition D if $d^{\ell,m}_{2n+1}$ satisfies condition C (relative to $K_1, \dots, K_t$), $\mathcal{I}_a = (f(\bar{K}_1); \cdots)$, and for almost all $h_1, \dots, h_t$ we have that $[h(d; h_1, \dots, h_t)]$ is not minimal subject to being nonconstant almost everywhere w.r.t. $\mu^{\bar{K}_1}_{2n+1}$. This follows from Lemma 5.22 and the fact that for $\mathcal{I}_a = (f(\bar{K}_1); \cdots)$, and $C \subseteq \underset{\sim}{\delta}^1_{2n+1}$ a c.u.b. set, for almost all $h_1, \dots, h_t$ we have that $h(d; h_1, \dots, h_t)$ is represented w.r.t. $\mu^{\bar{K}_1}_{2n+1}$ by a function F having range in C almost everywhere. This follows easily from the definition of h.

These facts will be of use later.

We now proceed to define the lowering operation $\mathcal{L}$ on $\mathcal{D}_{2n+1}$. We first require a preliminary definition.

Definition 5.23. Given $d_1^{(\mathcal{I}_a)}, d_2^{(\mathcal{I}_{a+1})}$ satisfying condition C relative to $K_1, \dots, K_t$ where $\mathcal{I}_{a+1} = (\mathcal{I}_a; \bar{K}_{a+1})$, we say condition $M(d_1^{(\mathcal{I}_a)}, d_2^{(\mathcal{I}_{a+1})})$ is satisfied if for almost all $h_1, \dots, h_t, f, \bar{h}_2, \dots, \bar{h}_a$,

$$H(d_1; h_1, \dots, h_t, f, \bar{h}_2, \dots, \bar{h}_a) = \sup_{\text{a.e. } [\bar{h}_{a+1}]} H(d_2; h_1, \dots, h_t, f, \bar{h}_2, \dots, \bar{h}_{a+1}).$$

We say $M_2(d_1^{(\mathcal{I}_a)}, d_2^{(\mathcal{I}_{a+1})})$ is satisfied if $M(d_1, d_2)$ is satisfied and the function $g : \vartheta(\bar{K}_{a+1}) \to \underset{\sim}{\boldsymbol{\delta}}^1_{2n+1}$ defined by

$$g([\bar{h}_{a+1}]) = H(d_1; h_1, \dots, h_t, f, \bar{h}_2, \dots, \bar{h}_{a+1})$$

is not minimal with respect to the set of $[g']$ for $g : \vartheta(\bar{K}_{a+1}) \to \underset{\sim}{\boldsymbol{\delta}}^1_{2n+1}$ with $\sup_{\text{a.e.}} g' = \sup_{\text{a.e.}} g$.

We will define the operation $\mathcal{L}$ on objects of the form (d) or $(d)^s$, where $d = d^{\ell,m}_{2n+1} \in \mathcal{D}_{2n+1}$, where d or $(d)^s$ satisfies condition D and d satisfies condition A relative to fixed $K_1, \dots, K_t$. To do this, we first define a preliminary operation $\hat{\mathcal{L}}$ on d satisfying conditions C and A. $\hat{\mathcal{L}}(d)$ will also satisfy conditions C and A.

We introduce a notation. For $d^{(\mathcal{I}_a)}$ satisfying condition C relative to $K_1, \dots, K_t$ where $\mathcal{I}_a = (f(\bar{K}_1); \bar{K}_2, \dots, \bar{K}_a)$ (or $f = f_k$ or does not appear) we let $\delta_d(\bar{K}_i)$ for $2 \le i \le a$ be the integer $\delta_d(\bar{K}_i) = j$, where $1 \le j \le t$ such that $v(K_j) = \bar{K}_i$ which was used in the construction of d. We recall that (with a slight abuse of notation), the $\bar{K}_i$ are tagged with integers to such K_j. The function δ_d just recovers this integer.

In defining $\hat{\mathcal{L}}$, we will actually define a more general operation $\hat{\mathcal{L}}^{\restriction k}(d)$, defined for d satisfying conditions C and A relative to $K_1, \dots, K_t$ and satisfying $\delta_d(\bar{K}_a) < k \le k(d)$, and also such that d is not minimal w.r.t. the ordering $<$ restricted to the set of d with $k(d) \ge k$ satisfying conditions C and A.

Also, we will assume inductively that for $m < n$ and any sequence of measures $K_1, \dots, K_t$ in R_{2m+1} such that there is a description $d = d^{(\mathcal{I}_a)}$ or $(d)^s$, for $d \in \mathcal{D}_{2m+1}$, defined satisfying conditions D and A, that there is a canonically defined maximal description $\tilde{d}$ or $(\tilde{d})^s$ satisfying conditions D and A. That is, $\tilde{d} \ge d'$ if d' or $(d')^s$ satisfies conditions D and A. This definition is, of course, relative to fixed $m, \bar{\ell}, \bar{m}$ such that $d \in \mathcal{D}^{\bar{\ell},\bar{m}}_{2m+1}$. We will say explicitly what $\tilde{d}^{\ell,m}_{2n+1}$ is below.

We consider the following cases:

Case I. $k = \infty$, so d basic of type 1 or 0.

Subcase I.a. d basic of type 1, so $d^{(\mathcal{I}_a)} = (\bar{d_2})^s$, where s may or may not appear. If s does not appear, we let $\hat{\mathcal{L}}^{\restriction\infty}(d) = (\bar{d_2})^s$. If s appears and $\bar{d_2}$ is not minimal w.r.t. the operation $\mathcal{L}$ on $\mathcal{D}_{2n-1}$ (which is defined by induction), then we set $\hat{\mathcal{L}}^{\restriction\infty}(d) = (\mathcal{L}(\bar{d_2}))^{(\mathcal{I}_a)}$. If s appears and $\bar{d_2}$ is minimal w.r.t. $\mathcal{L}$ on $\mathcal{D}_{2n-1}$, then d is minimal w.r.t. $\hat{\mathcal{L}}^{\restriction\infty}$, and $\hat{\mathcal{L}}^{\restriction\infty}(d)$ is not defined.

Subcase I.b. d basic of type 0, so $d^{(\mathcal{I}_a)} = d^{(f_m;\bar{K}_2,\dots,\bar{K}_a)}$ where $d = (r)$ and $1 \le r \le m$. If $r > 1$, we set $\hat{\mathcal{L}}^{\restriction\infty}(d) = r - 1$, and if $r = 1$ then d is minimal w.r.t. $\hat{\mathcal{L}}^{\restriction\infty}$ (that is, we do not define $\hat{\mathcal{L}}^{\restriction\infty}(d)$).

Case II. $k < \infty$, $k = k(d)$, and $K_k = W^{m_k}_{2n+1}$. So, d is basic of type -1 of the form $d = (k;\ )$ or $d = (k; \bar{d}_2)^s$, where s may or may not appear, and $\mathcal{I}_a = (\bar{K}_2, \dots, \bar{K}_a)$. (For $n = 0$, $\bar{d}_2$ is an integer and s does not appear).

Subcase II.a. $d = (k;\ )$. We recall that $K_{b_1}, \dots, K_{b_u}$ enumerates the subsequence of $K_1, \dots, K_t$ of measures not of the form $S^{\ell,m}_{2n+1}$ or W^m_{2n+1} (*i.e.*, those of the form $R^{\ell,m}_{2s+1}$ for $s < n$). We let $K_{b(k)}, \dots, K_{b_u}$ enumerate those in $K_{k+1}, \dots, K_t$. We let, in this case $v(K_k) = S^{\bar{\ell}_k,\bar{m}_k}_{2n-1}$ if $n \ge 1$. As above, let $\tilde{d} = \tilde{d}^{\bar{\ell}_k,\bar{m}_k}_{2n-1}$ be the maximal tuple relative to $K_{b(k)}, \dots, K_{b_u}, \bar{K}_1, \dots, \bar{K}_a$ if defined. If $\tilde{d}$ is not defined, we define d to be minimal with respect to $\hat{\mathcal{L}}^{\restriction k}$. If $\tilde{d}$ is defined, we set $\hat{\mathcal{L}}^{\restriction k}(d) = (k; \tilde{d})$ or $(k; \tilde{d})^s$, depending on whether $\tilde{d}$ satisfies condition D or not.

Subcase II.b. $d = (k; \bar{d}_2)^s$, where s may or may not appear.

If $n = 0$, so $d = (k; i)$, we set $\hat{\mathcal{L}}^{\restriction k}(d) = (k; i - 1)$ if $i > 1$, and if $i = 1$, then d is minimal w.r.t. $\hat{\mathcal{L}}^{\restriction k}$. We assume now $n > 0$.

If s does not appear, we set $\hat{\mathcal{L}}^{\restriction k}(d) = (k; \bar{d}_2)^s$.

If s appears and $\bar{d}_2$ is not minimal with respect to $\mathcal{L}$ relative to the sequence of measures $K_{b(k)}, \dots, K_{b_u}, \bar{K}_2, \dots, \bar{K}_a$, we set $\hat{\mathcal{L}}^{\restriction k}(d) = (k; \mathcal{L}'((\tilde{d}_2)^s))$ or $(k; \mathcal{L}'((\tilde{d}_2)^s))^s$ depending on whether $\mathcal{L}((\tilde{d}_2)^s)$ does not or does involve s. Here we use the notation that if $\mathcal{L}((\bar{d}_2)^s) = (d')$ or $(d')^s$, then $\mathcal{L}'((\bar{d}_2)^s) = d'$.

If s appears, and $\bar{d}_2$ is minimal with respect to $\mathcal{L}$ relative to the sequence $K_{b(k)}, \dots, \bar{K}_a$, then d is minimal with respect to $\hat{\mathcal{L}}^{\restriction k}$.

Case III. $k < \infty$, $k = k(d)$, $K_k = S^{\ell_k,m_k}_{2n+1}$ with $\ell_k > 1$. Hence $d = (k; d_2^{(\mathcal{I}_a;\bar{K}_{a+1})})^s$ where s may or may not appear. We let $\hat{d}$ be the tuple from condition A for d.

Subcase III.a. If s does not appear, then $\hat{\mathcal{L}}^{\restriction k}(d) = (k; d_2^{(\mathcal{I}_a;\bar{K}_{a+1})})^s$.

Subcase III.b. $\ell > 2$, s appears, and d_2 is not minimal w.r.t. $\hat{\mathcal{L}}^{\restriction k+1}$. We then set $\hat{\mathcal{L}}^{\restriction k}(d) = (k; \hat{\mathcal{L}}^{\restriction k+1}(d_2))$ if this tuple satisfies condition C and $M(\hat{d}, \hat{\mathcal{L}}^{\restriction k+1}(d_2))$ is satisfied. Otherwise we set $\hat{\mathcal{L}}^{\restriction k}(d) = (k; \hat{\mathcal{L}}^{\restriction k+1}(d_2))^s$ if $M_2(\hat{d}, \hat{\mathcal{L}}^{\restriction k+1}(d_2))$ holds and condition C is satisfied, and if not then we set $\hat{\mathcal{L}}^{\restriction k}(d) = \hat{d}$.

Subcase III.c. $\ell = 2$, s appears, and d_2 is not minimal w.r.t. $\hat{\mathcal{L}}^{\restriction k+1}$. We set $\hat{\mathcal{L}}^{\restriction k}(d) = (k; \hat{\mathcal{L}}^{\restriction k+1(q)}(d_2))^s$, where s appears if $(k; \hat{\mathcal{L}}^{\restriction k+1(q)}(d_2))$ does not satisfy condition C. Here $\hat{\mathcal{L}}^{\restriction k+1(q)}(d_2)$ denotes the least iterate of $\hat{\mathcal{L}}^{\restriction k+1}$ such that $(k; \hat{\mathcal{L}}^{\restriction k+1(q)}(d_2))$ or $(k; \hat{\mathcal{L}}^{\restriction k+1(q)}(d_2))^s$ satisfies condition C. If no such q exists, we set $\hat{\mathcal{L}}^{\restriction k}(d) = \hat{d}$.

Subcase III.d. Otherwise we set $\hat{\mathcal{L}}^{\restriction k}(d) = \hat{d}$.

Case IV. $k < \infty$, $k = k(d)$, $K_k = S^{1,m_k}_{2n+1}$. Hence $d = (k; d_1^{(\mathcal{I}_a)}, \dots, d_r^{(\mathcal{I}_a)})^s$ where s may or may not appear, and $r \leq m_k$.

Subcase IV.a. s appears, $r > 2$, and $\hat{\mathcal{L}}^{\restriction k+1}(d_r)$ is not defined or $\hat{\mathcal{L}}^{\restriction k+1}(d_r) \leq d_{r-1}$. We set $\hat{\mathcal{L}}^{\restriction k}(d) = (k; d_1^{(\mathcal{I}_a)}, \dots, d_{r-1}^{(\mathcal{I}_a)})^s$.

Subcase IV.b. s appears, $r = 2$, and $\hat{\mathcal{L}}^{\restriction k+1}(d_r)$ is not defined. We set $\hat{\mathcal{L}}^{\restriction k}(d) = d_1$.

Subcase IV.c. s appears, $\hat{\mathcal{L}}^{\restriction k+1}(d_r)$ is defined, and if $r > 2$ then $\hat{\mathcal{L}}^{\restriction k+1}(d_r) > d_{r-1}$. Then $\hat{\mathcal{L}}^{\restriction k}(d) = (k; d_1^{(\mathcal{I}_a)}, \dots, d_{r-1}^{(\mathcal{I}_a)}, \hat{\mathcal{L}}^{\restriction k+1}(d_r^{\mathcal{I}_a}))^s$, where s appears if without s this tuple does not satisfy condition C.

Subcase IV.d. s does not appear and $r < m_k$.

Subcase IV.d.i. $\hat{\mathcal{L}}^{\restriction k+1}(d_1)$ is defined, and if $r \geq 2$ then $\hat{\mathcal{L}}^{\restriction k+1}(d_1) > d_r$. We set $\hat{\mathcal{L}}^{\restriction k}(d) = (k; d_1^{(\mathcal{I}_a)}, \dots, d_r^{(\mathcal{I}_a)}, \hat{\mathcal{L}}^{\restriction k+1}(d_1))^s$, where s appears if without s the tuple does not satisfy C.

Subcase IV.d.ii. $\hat{\mathcal{L}}^{\restriction k+1}(d_1)$ is defined, but IV.d.i immediately above fails. We set $\hat{\mathcal{L}}^{\restriction k}(d) = (k; d_1^{(\mathcal{I}_a)}, \dots, d_r^{(\mathcal{I}_a)})^s$.

Subcase IV.d.iii. $\hat{\mathcal{L}}^{\restriction k+1}(d_1)$ not defined. It will follow in this case from our main inductive hypothesis that $r = 1$, since $\hat{\mathcal{L}}^{\restriction k+1}(d_1)$ not defined implies d_1 is minimal w.r.t. the ordering $<$ on the set of d' satisfying conditions C and A and with $k'(d) > k$, which is not the case if $r > 1$. In this case, we set $\hat{\mathcal{L}}^{\restriction k+1}(d) = d_1$.

Subcase IV.e. s does not appear and $r = m_k$. We set $\hat{\mathcal{L}}^{\restriction k}(d) = (k; d_1^{(\mathcal{I}_a)}, \dots, d_r^{(\mathcal{I}_a)})^s$ if $m_k > 1$ and $\hat{\mathcal{L}}^{\restriction k}(d) = d_1$ if $m_k = 1$.

Case V. $k < \infty$, $k < k(d)$, and $K_k = W^{m_k}_{2n+1}$.
If d is not minimal with respect to $\hat{\mathcal{L}}^{\restriction k+1}$, then we set $\hat{\mathcal{L}}^{\restriction k}(d) = \hat{\mathcal{L}}^{\restriction k+1}(d)$.
If d is minimal with respect to $\hat{\mathcal{L}}^{\restriction k+1}$, then we set $\hat{\mathcal{L}}^{\restriction k}(d) = (k;\ )$, a basic -1 description.

Case VI. $k < \infty$, $k < k(d)$, and $K_k = S^{1,m_k}_{2n+1}$.

Subcase VI.a. $\hat{\mathcal{L}}^{\restriction k+1}(d)$ defined and $(k; \hat{\mathcal{L}}^{\restriction k+1}(d))$ satisfies condition C. Then we set $\hat{\mathcal{L}}^{\restriction k}(d) = (k; \hat{\mathcal{L}}^{\restriction k+1}(d))$.

Subcase VI.b. $\hat{\mathcal{L}}^{\restriction k+1}(d)$ defined, but VI.a fails. We set $\hat{\mathcal{L}}^{\restriction k}(d) = \hat{\mathcal{L}}^{\restriction k+1}(d)$.

Subcase VI.c. VI.a and VI.b fail. Then d is minimal with respect to $\hat{\mathcal{L}}^{\restriction k+1}(d)$.

Case VII. $k < \infty$, $k < k(d)$, and $K_k = S^{\ell_k, m_k}_{2n+1}$ with $\ell_k > 1$.

Subcase VII.a. $\hat{\mathcal{L}}^{\restriction k+1}(d)$ defined and for almost all $h_1, \dots, h_t, f, \bar{h}_2, \dots, \bar{h}_a$,

$$\operatorname{cf} H(\hat{\mathcal{L}}^{\restriction k+1}(d); h_1, \dots, h_t, f, \bar{h}_2, \dots, \bar{h}_a) = \operatorname{cf} \kappa(K_k).$$

We set $\hat{\mathcal{L}}^{\restriction k}(d) = (k; \tilde{\hat{\mathcal{L}}}^{\restriction k+1}(d)^{(\mathcal{I}_a; \bar{K}_a+1)})^s$, where $\tilde{\hat{\mathcal{L}}}^{\restriction k+1}(d)$ denotes the tuple obtained by replacing in all component tuples of d the indices $(f(\bar{K}_1); \bar{K}_2, \dots, \bar{K}_a, \cdots)$ by $(f(\bar{K}_1); \bar{K}_2, \dots, \bar{K}_{a+1}, \cdots)$.

Subcase VII.b. $\hat{\mathcal{L}}^{\restriction k+1}(d)$ defined and VII.a fails. We set $\hat{\mathcal{L}}^{\restriction k}(d) = \hat{\mathcal{L}}^{\restriction k+1}(d)$.

Subcase VII.c. $\hat{\mathcal{L}}^{\restriction k+1}(d)$ not defined. Then d is minimal with respect to $\hat{\mathcal{L}}^{\restriction k}$.

Finally for (d) or $(d)^s$ satisfying conditions D and A, we set $\mathcal{L}(d) = (d)^s$, and $\mathcal{L}(d)^s = (\hat{\mathcal{L}}^{\restriction 1(p)}(d))^s$, where s appears if the tuple without s does not satisfy condition D, and $\hat{\mathcal{L}}^{\restriction 1(p)}$ denotes the pth iterate of $\hat{\mathcal{L}}^{\restriction 1}$. Here p is minimal such that $\hat{\mathcal{L}}^{\restriction 1(p)}(d)$ or $(\hat{\mathcal{L}}^{\restriction 1(p)}(d))^s$ satisfies condition D. If no such p exists, then $(d)^s$ is minimal with respect to $\mathcal{L}$.

LEMMA 5.24. If d satisfied conditions C and A relative to $K_1, \dots, K_t$ and $\delta_d(\bar{K}_a) < k \leq k(d)$, then $\hat{\mathcal{L}}^{\restriction k}(d)$ also satisfies conditions C and A. Hence, if $d^{(\mathcal{I}_a)}$ satisfies conditions C and A where $\mathcal{I}_a = (f(\bar{K}_1);\ )$, then $\hat{\mathcal{L}}(d) \equiv \hat{\mathcal{L}}^{\restriction 1}(d)$ satisfies conditions C and A and in this case, if d satisfies condition D, then so does $\mathcal{L}(d)$.

PROOF. The first statement clearly implies the second. The proof of the first statement is routine upon consideration of the cases.

For example, we consider the case $d = (k; d_2^{(\mathcal{I}_a; \bar{K}_{a+1})})^s$, where $K_k = S^{\ell_k, m_k}_{2n+1}$, and $v(K_k) = \bar{K}_{a+1}$. We consider first $\hat{\mathcal{L}}^{\restriction k}(d)$ where $k = k(d)$. If $\hat{\mathcal{L}}^{\restriction k}(d) = (k; \hat{\mathcal{L}}^{\restriction k+1}(d_2))$, then condition C is satisfied by definition. By induction (reverse induction on $k(d)$), $\hat{\mathcal{L}}^{\restriction k+1}(d_2)$ satisfies condition A. Since $M(\hat{d}, \hat{\mathcal{L}}^{\restriction k+1}(d_2))$ is satisfied in this case, condition A is also satisfied by $\hat{\mathcal{L}}^{\restriction k}(d)$, since for $\hat{\mathcal{L}}^{\restriction k}(d)$ we may take $\hat{d}$. Similarly, if $\hat{\mathcal{L}}^{\restriction k}(d) = (k; \hat{\mathcal{L}}^{\restriction k+1}(d_2))^s$, conditions C and A are satisfied. If $\hat{\mathcal{L}}^{\restriction k}(d) = \hat{d}$, conditions C and A are satisfied by definition.

We consider the case with d as above and $k < k(d)$. If $\hat{\mathcal{L}}^{\restriction k}(d) = \hat{\mathcal{L}}^{\restriction k+1}(d)$, we are done by induction. Hence $\hat{\mathcal{L}}^{\restriction k}(d) = (k; \tilde{\hat{\mathcal{L}}}^{\restriction k+1}(d)^{(\mathcal{I}_a; \bar{K}_{a+1})})^s$. By induction, $\hat{\mathcal{L}}^{\restriction k+1}(d)$ satisfies conditions C and A. It follows readily that $\tilde{\hat{\mathcal{L}}}^{\restriction k+1}(d)$ also satisfies conditions C and A (if $M(d_1, d_2)$ is satisfied for component tuples d_1, d_2 of $\hat{\mathcal{L}}^{\restriction k+1}(d)$, then $M(\tilde{d}_1, \tilde{d}_2)$ is also satisfied). Since for almost all

$h_1, \dots, h_t, f, \bar{h}_2, \dots, \bar{h}_a$, the function

$$g([\bar{h}_{a+1}]) = H(\tilde{\mathcal{L}}^{\restriction k+1}(d), h_1, \dots, h_t, f, \bar{h}_2, \dots, \bar{h}_a, \bar{h}_{a+1})$$

is constant almost everywhere in this case, it follows readily that condition C is satisfied, and to satisfy condition A we may take $\hat{d} = \hat{\mathcal{L}}^{\restriction k+1}(d)$. The remaining cases follow similarly. ⊣

Finally, to complete the inductive definition of $\mathcal{L}$, we define the maximal tuple $\tilde{d}^{\ell,m}_{2n+1}$ relative to $K_1, \dots, K_t$.

1) $\ell = -1$. In this case $\tilde{d}^{(\mathcal{I}_a)} = \tilde{d}^{()}$. We let $1 \le w \le t$ denote the largest integer such that $K_w = W^{m_w}_{2n+1}$. We let $a_1, \dots, a_p$ enumerate the integers i from 1 to w with K_i of the form S^{ℓ,m_i}_{2n+1} and $\operatorname{cf} \kappa(S^{\ell,m_i}_{2n+1}) = \operatorname{cf} \kappa(W^{m_w}_{2n+1})$, (or of the form S^{1,m_i}_1 if $n = 0$) or equivalently, $v(S^{\ell,m_i}_{2n+1}) \sim v(W^{m_w}_{2n+1})$. We set $\tilde{d} = (\tilde{d}^{()}_1)^s$ where $\tilde{d}^{()}_1 = (a_1; \tilde{d}^{(\bar{K}_1)}_2)^s$ where $\bar{K}_1 = v(K_{a_1})$, and $\tilde{d}^{(\bar{K}_1)}_2 = (a_2; \tilde{d}^{(\bar{K}_1,\bar{K}_2)}_3)^s$, with $\bar{K}_2 = v(K_{a_2})$. We continue in this manner and finally set $\tilde{d}^{(\bar{K}_1,\dots,\bar{K}_{p-1})}_p = (a_p; \tilde{d}^{(\bar{K}_1,\dots,\bar{K}_p)}_{p+1})^s$ where $\tilde{d}_{p+1} = (w; \,)$ (or $= (w; m_w)$ if $n = 0$).
2) $\ell = 1$. In this case $\tilde{d}^{(\mathcal{I}_a)} = \tilde{d}^{(f_m)}$. We let $b_1, \dots, b_q$ enumerate the measures K_i in $K_1, \dots, K_t$ of the form S^{1,m_i}_{2n+1}. We set $\tilde{d}^{(f_m;)}_\infty = (m)$. We let $\tilde{d}^{(f_m;)}_i = (b_i; \tilde{d}^{(f_m;)}_{i+1})$ for $1 \le i \le q$ (where $\tilde{d}_{p+1} = \tilde{d}_\infty$). We set $\tilde{d} = \tilde{d}_1$ or $(\tilde{d}_1)^s$, depending in whether $\tilde{d}_1$ does or does not satisfy condition D. If b_1 does not exist, $\tilde{d}$ is not defined.
3) $\ell > 1$, $\tilde{d}^{(\mathcal{I}_a)} = \tilde{d}^{(f(\bar{K}_1);)}$. We let $e_1, \dots, e_r$ enumerate the integers i with $K_i = S^{\ell_i,m_i}_{2n+1}$ with $\ell_i > 1$ and $\operatorname{cf} \kappa(S^{\ell_i,m_i}_{2n+1}) = \kappa(S^{\ell,m}_{2n+1})$ (this is equivalent to saying that $v(S^{\ell_i,m_i}_{2n+1}) \sim v(S^{\ell,m}_{2n+1}) = \bar{K}_1$ or equivalently that $\operatorname{cf} \kappa(S^{\ell_i,m_i}_{2n+1}) = \operatorname{cf} \vartheta(\bar{K}_1)$). Set $\tilde{d}^{(f(\bar{K}_1);)}_1 = (e_1; \tilde{d}^{(f(\bar{K}_1);\bar{K}_2)}_2)$, where $\bar{K}_2 = v(K_{e_1})$. Likewise, $\tilde{d}^{(f(\bar{K}_1);\bar{K}_2)}_2 = (e_2; \tilde{d}^{(f(\bar{K}_1);\bar{K}_2,\bar{K}_3)}_3)$ where $\bar{K}_3 = v(K_{e_2})$. We continue in this manner and finally set

$$\tilde{d}^{(f(\bar{K}_1);\bar{K}_2,\dots,\bar{K}_{r-1})}_r = (e_r, \tilde{d}^{(f(\bar{K}_1);\bar{K}_2,\dots,\bar{K}_{r-1},\bar{K}_r)}_\infty)^s$$

where $\tilde{d}^{(f(\bar{K}_1);\bar{K}_2,\dots,\bar{K}_r)}_\infty = ()^{(f(\bar{K}_1);\bar{K}_2,\dots,\bar{K}_r)}$. We set $\tilde{d} = \tilde{d}_1$ or $(\tilde{d}_1)^s$ depending on whether $\tilde{d}_1$ does or does not satisfy condition D. If e_1 does not exist, $\tilde{d}$ is not defined.

We now introduce some additional notation required for the proof.

We fix $K_1, \dots, K_t$ and an index $\mathcal{I}_a$.

We let ϑ be an ordinal. We represent ϑ w.r.t. the measure K_1 by the function g. We write $g(h_1) = g([h_1])$ for $[h_1] \in \vartheta(K_1)$. Thus g is defined almost everywhere with respect to K_1. For a fixed h_1, we represent $g(h_1)$ w.r.t. K_2 by the function $g(h_1, h_2)$. We emphasize that this is defined only for a fixed choice of g, and fixed function representing $g(h_1)$. Continuing, we define $g(h_1, \dots, h_t)$. If $\mathcal{I}_a$ is of the form $\mathcal{I}_a = (f(\bar{K}_1); \bar{K}_2, \dots, \bar{K}_a)$ then we represent $g(h_1, \dots, h_t)$ with respect to $\mu^{\bar{K}_1}_{2n+1}$ by the function $[f] \mapsto g(h_1, \dots, h_t, f)$ for $f : \vartheta(\bar{K}_1) \to \underset{\sim}{\boldsymbol{\delta}}^1_{2n+1}$ of the correct type. If $\mathcal{I}_a = (f_k; \bar{K}_2, \dots, \bar{K}_a)$, then we represent $g(h_1, \dots, h_t)$ with respect to the k-fold product to the ω-cofinal normal measure on $\underset{\sim}{\boldsymbol{\delta}}^1_{2n+1}$ by the function $f \mapsto g(h_1, \dots, h_t, f)$ where now $f : k \to \underset{\sim}{\boldsymbol{\delta}}^1_{2n+1}$. Continuing, we represent $g(h_1, \dots, h_t, f)$ with respect to the measures $\bar{K}_2, \dots, \bar{K}_a$ to get $g(h_1, \dots, h_t, f, \bar{h}_2, \dots, \bar{h}_a)$. Again, $g(h_1, \dots, h_t, f, \bar{h}_2, \dots, \bar{h}_i, \bar{h}_{i+1})$ is only defined after having chosen a specific function representing $g(h_1, \dots, \bar{h}_i)$.

Given d satisfying condition C and A relative to $K_1, \dots, K_t$, we define a sequence of component tuples $d^0, d^1, \dots, d^\ell$ for some $\ell \geq 0$ as follows:

We set $d^0 = d$. We assume $d^i = d^{i(\mathcal{I}_{a_i})}$ has been defined.

a) If $d^{i(\mathcal{I}_{a_i})} = (k; d_2^{(\mathcal{I}_{a_{i+1}})})$ where $K_k = S^{\ell_k, m_k}_{2n+1}$ with $\ell_k > 1$, we set $\ell = i$.
b) If $d^{i(\mathcal{I}_{a_i})} = (k; d_1^{(\mathcal{I}_{a_i})}, \dots, d_r^{(\mathcal{I}_{a_i})})$ where $K_k = S^{1, m_k}_{2n+1}$ and $r = m_k$, we set $\ell = i$.
c) If $d^i = (k; d_2^{(\mathcal{I}_{a_{i+1}})})^s$, we set $d^{i+1} = d_2 = d_2^{(\mathcal{I}_{a_{i+1}})}$.
d) If $d^i = (k; d_1, \dots, d_r)^s$, we set $d^{i+1} = d_r$.
e) If $d^i = (k; d_1, \dots, d_r)$ where $r < m_k$ (here $K_k = S^{1, m_k}_{2n+1}$), we set $d^{i+1} = d_1 = d_1^{(\mathcal{I}_{a_i})}$.
f) If d^i is basic, we set $\ell = i$.

We let $k_i = k(d^i)$, so $1 \leq k_i \leq t$ or $k_i = \infty$, and $k_0 < k_1 < \cdots < k_\ell$.

For a fixed ordinal ϑ or description $\tilde{d}^{i(\mathcal{I}_{a_i})}$ satisfying condition C and A w.r.t. $K_1, \dots, K_t$, where the indices $\mathcal{I}_{a_i}$ are as above, we define the ordinal $h(d; (d^i \to \vartheta))$ or $h(d; (d^i \to \tilde{d}^i))$ as follows. We represent this ordinal with respect to the measures $K_1, \dots, K_t, \mu^{\bar{K}_1}_{2n+1}$ (or the m-fold product of the ω-cofinal normal measure on $\underset{\sim}{\boldsymbol{\delta}}^1_{2n+1}$), and $\bar{K}_2, \dots, \bar{K}_a$ by

$$H(d; h_1, \dots, h_t; f, \bar{h}_2, \dots, \bar{h}_a; (d^i \to \vartheta))$$

or

$$H(d; h_1, \dots, h_t; f, \bar{h}_2, \dots, \bar{h}_a; (d^i \to \tilde{d}^i))$$

where these are defined exactly as $H(d; h_1, \dots, h_t; f, \bar{h}_2, \dots, \bar{h}_a)$ except that in defining the ordinal $H(d^{i-1(\mathcal{I}_{a_{i-1}})}; h_1, \dots, h_t; f, \bar{h}_2, \dots, \bar{h}_{a_{i-1}})$, the ordinal $H(d^i; h_1, \dots, h_t; f, \bar{h}_2, \dots, \bar{h}_{a_i})$ is replaced by

$$g(h_1, \dots, h_t, f, \bar{h}_2, \dots, \bar{h}_{a_i})$$

in the first case and by

$$H(\tilde{d}^i; h_1, \dots, h_t, f, \bar{h}_2, \dots, \bar{h}_{a_i})$$

in the second case.

If C is a c.u.b. subset of $\underset{\sim}{\boldsymbol{\delta}}^1_{2n+1}$, we let $N_C(\alpha)$ be the ωth element of C greater than α. If $g: \underset{\sim}{\boldsymbol{\delta}}^1_{2n+1} \to \underset{\sim}{\boldsymbol{\delta}}^1_{2n+1}$, we let $N_g(\alpha)$ be the ωth element in the range of g greater than α. We abbreviate $N_{h_i}(\alpha)$ by $N_i(\alpha)$.

For $g: \underset{\sim}{\boldsymbol{\delta}}^1_{2n+1} \to \underset{\sim}{\boldsymbol{\delta}}^1_{2n+1}$, we let $h(d;(d^i \to g \circ \tilde{d}^i))$ be defined as above, replacing $H(d^i; h_1, \dots, h_t, f, \bar{h}_2, \dots, \bar{h}_{a_i})$ by

$$g(H(d^i; h_1, \dots, h_t, f, \bar{h}_2, \dots, \bar{h}_{a_i})).$$

If $d = d^{\ell,m}_{2n+1}$ or $(d)^s$ is given and satisfies conditions D and A w.r.t. $K_1, \dots, K_t$, and $g: \underset{\sim}{\boldsymbol{\delta}}^1_{2n+3} \to \underset{\sim}{\boldsymbol{\delta}}^1_{2n+3}$ we define the ordinal $(g; d; K_1, \dots, K_t)$. We represent this with respect to μ^v_{2n+3}, where $v = S^{\ell,m}_{2n+1}$ if $\ell \geq 1$, and $v = W^m_{2n+1}$ if $\ell = -1$. We represent by the function $[f] \mapsto (g; f; d; K_1, \dots, K_t)$, for $f: \vartheta(v) \to \underset{\sim}{\boldsymbol{\delta}}^1_{2n+3}$ of the correct type. We represent this, in turn, w.r.t. K_1 by the function $[h_1] \mapsto (g; f; d; h_1, K_1, \dots, K_t)$, for $[h_1] \in \vartheta(K_1)$. Continuing, we represent in turn with respect to $K_2, \dots, K_t$ by the functions $[h_2] \mapsto (g; f; d; h_1, h_2, K_3, \dots, K_t)$,...,$[h_t] \mapsto (g; f; d; h_1, \dots, h_t)$. Finally, we set $(g; f; d; h_1, \dots, h_t) = g(f(h(d; h_1, \dots, h_t)))$. For $(d)^s$, the procedure is similar, except at the end we use $(g; f; (d)^s; h_1, \dots, h_t) = g(\sup_{\beta < h(d; h_1, \dots, h_t)} f(\beta))$.

We now state the main inductive hypothesis H_{2n+1}.

Definition 5.25 (Main Inductive Hypothesis H_{2n+1}). H_{2n+1} consists of the following assertions.

H_{2n+1}(a): We let $d^{\mathcal{I}_a}$ or $(d)^s$, where $d = d^{\ell,m}_{2n+1} \in \mathcal{D}_{2n+1}$, be defined and satisfying conditions D and A w.r.t. $K_1, \dots, K_t$. We let $v = S^{\ell,m}_{2n+1}$ or $v = W^m_{2n+1}$ depending on whether $\ell \geq 1$ or $\ell = -1$. We assume $\mathcal{I}_a = (f(\bar{K}_1))$ or $(f_m;)$ or $()$, and assume $\mathcal{L}(d)$ is defined. We let $F: \underset{\sim}{\boldsymbol{\delta}}^1_{2n+3} \to \underset{\sim}{\boldsymbol{\delta}}^1_{2n+3}$ be given and satisfy $F < (\text{id}; d; K_1, \dots, K_t)$ almost everywhere w.r.t. μ^v_{2n+3}. Then there is a $g: \underset{\sim}{\boldsymbol{\delta}}^1_{2n+3} \to \underset{\sim}{\boldsymbol{\delta}}^1_{2n+3}$ such that $F < (g; \mathcal{L}(d); K_1, \dots, K_t)$ almost everywhere w.r.t. μ^v_{2n+3}. Similarly for $(d)^s$.

H_{2n+1}(b): As above, except we assume now that $\mathcal{L}(d)$ is not defined. We then have that for some $\alpha < \underset{\sim}{\boldsymbol{\delta}}^1_{2n+3}$, $F(\beta) = \alpha$ for almost all β w.r.t. μ^v_{2n+3}.

H_{2n+1}(c): If for almost all $f: \vartheta(v) \to \underset{\sim}{\boldsymbol{\delta}}^1_{2n+3}$ of the correct type, and almost all $h_2, \dots, h_t$ we have that $F(f, h_1, \dots, h_t) < \sup_{\text{a.e.}} f$, and the maximal tuple $\tilde{d}^{\ell,m}_{2n+1}$ is defined, then for almost all f we have that $F([f]) < (g; f; \tilde{d}; K_1, \dots, K_t)$, for some $g: \underset{\sim}{\boldsymbol{\delta}}^1_{2n+3} \to \underset{\sim}{\boldsymbol{\delta}}^1_{2n+3}$.

H_{2n+1}(d): As above, except $\tilde{d}$ is not defined. We then have that F is constant almost everywhere.

The rest of this section is devoted to a proof of H_{2n+1}. We proceed by induction on n, so we assume H_{2m+1} for $m < n$. We require the following lemma:

LEMMA 5.26 (Cofinality Lemma). We let $d^{(\mathcal{I}_a)}$, where $d = d_{2n+1}^{\ell,m} \in \mathcal{D}_{2n+1}$, be defined and satisfy conditions C and A relative to $K_1, \dots, K_t$. We let $\vartheta \in \mathrm{Ord}$ be represented by $g(h_1, \dots, h_t, f, \bar{h}_2, \dots, \bar{h}_a)$ as defined previously. We assume that for almost all $h_1, \dots, h_t, f, \bar{h}_2, \dots, \bar{h}_a$ that

$$g(h_1, \dots, h_t, f, \bar{h}_2, \dots, \bar{h}_a) < H(d; h_1, \dots, h_t, f, \bar{h}_2, \dots, \bar{h}_a).$$

We then have that there is an ordinal ϑ_ℓ (where ℓ is as in the definition of $d^0, d^1, \dots, d^\ell$) such that if we represent the ordinal ϑ_ℓ as before by $g_\ell(h_1, \dots, h_t, f, \bar{h}_2, \dots, \bar{h}_{a_\ell})$, then

$$g_\ell(h_1, \dots, h_t, f, \bar{h}_2, \dots, \bar{h}_{a_\ell}) < H(d^\ell; h_1, \dots, h_t, f, \bar{h}_2, \dots, \bar{h}_{a_\ell})$$

almost everywhere, and for almost all $h_1, \dots, h_t, f, \bar{h}_2, \dots, \bar{h}_a$ we have

$$g(h_1, \dots, h_t, f, \bar{h}_2, \dots, \bar{h}_a) < H(d; (d^\ell \to \vartheta_\ell); h_1, \dots, h_t, f, \bar{h}_2, \dots, \bar{h}_a).$$

PROOF. By induction on $0 \le i \le \ell$, we establish that there is a ϑ^i such that for almost all $h_1, \dots, h_t, f, \bar{h}_2, \dots, \bar{h}_{a_i}$

$$\vartheta^i(h_1, \dots, h_t, f, \bar{h}_2, \dots, \bar{h}_{a_i}) < H(d^i; h_1, \dots, h_t, f, \bar{h}_2, \dots, \bar{h}_{a_i})$$

and

$$g(h_1, \dots, h_t, f, \bar{h}_2, \dots, \bar{h}_a) < H(d; (d^i \to \vartheta^i); h_1, \dots, h_t, f, \bar{h}_2, \dots, \bar{h}_a).$$

For $i = 0$ this is true by assumption. Assuming true for i, it follows upon consideration of the cases (one of which we consider below) that there is an ordinal ϑ^{i+1} such that

$$\vartheta^{i+1}(h_1, \dots, h_t, f, \bar{h}_2, \dots, \bar{h}_{a_{i+1}}) < H(d^{i+1}; h_1, \dots, h_t, f, \bar{h}_2, \dots, \bar{h}_{a_i})$$

and

$$\vartheta^i(h_1, \dots, h_t, f, \bar{h}_2, \dots, \bar{h}_{a_i}) < H(d^i; (d^{i+1} \to \vartheta^{i+1}); h_1, \dots, h_t, f, \bar{h}_2, \dots, \bar{h}_{a_i})$$

hold almost everywhere. Hence almost everywhere

$$g(h_1, \dots, h_t, f, \bar{h}_2, \dots, \bar{h}_a) < H(d; (d^{i+1} \to \vartheta^{i+1}); h_1, \dots, h_t, f, \bar{h}_2, \dots, \bar{h}_a)$$

follows.

As an example, we consider the case $d^i = (k; d^{i+1})^s$, where $K_k = S_{2n+1}^{\ell_k, m_k}$ with $\ell_k > 1$. So, for almost all $h_1, \dots, h_t, f, \bar{h}_2, \dots, \bar{h}_{a_i}$ we have

$$\begin{aligned}\vartheta^i(h_1, \dots, h_t, f, \bar{h}_2, \dots, \bar{h}_{a_i}) &< H(d^i; h_1, \dots, h_t, f, \bar{h}_2, \dots, \bar{h}_{a_i}) \\ &= \sup_{\beta < [g]} h_k(\beta),\end{aligned}$$

where $g([h_{a_{i+1}}]) = H(d^{i+1}; h_1, \dots, h_t, f, \bar{h}_2, \dots, \bar{h}_{a_{i+1}})$ almost everywhere. Hence, for almost all $h_1, \dots, \bar{h}_{a_i}$, there is a $g' < g$ almost everywhere with

$\vartheta^i(h_1, \ldots, \bar{h}_{a_i}) < h_k([g'])$. Equivalently, for almost all $h_1, \ldots, \bar{h}_{a_{i+1}}$, there is a $\vartheta^{i+1}(h_1, \ldots, \bar{h}_{a_{i+1}}) < H(d^{i+1}; h_1, \ldots, \bar{h}_{a_{i+1}})$ such that

$$\vartheta^i(h_1, \ldots, \bar{h}_{a_{i+1}}) < H(d; (d^{i+1} \to \vartheta^{i+1}); h_1, \ldots, h_t, f, \bar{h}_2, \ldots, \bar{h}_{a_i}).$$

The remaining cases are similar. ⊣

The main part of the proof of the main inductive hypothesis consists of establishing the main inductive lemma, which we now state:

Lemma 5.27 (Main Inductive Lemma). For $1 \le i \le t$ or $i = \infty$ we consider the following statement $\bar{H}(i)$: we let $d^{(\mathcal{I}_a)}$ for $d = d^{\ell,m}_{2n+1}$ be given, where d is defined and satisfies conditions C and A relative to $K_1, \ldots, K_t$, and d is not minimal with respect to $\hat{\mathcal{L}}^{\restriction i}$ where $\delta_d(\bar{K}_a) < i \le k(d)$. Let ϑ be an ordinal. We assume that for almost all $h_1, \ldots, h_t, f, \bar{h}_2, \ldots, \bar{h}_a$ that

$$\vartheta(h_1, \ldots, h_t, f, \bar{h}_2, \ldots, \bar{h}_a) < H(d^{(\mathcal{I}_a)}; h_1, \ldots, h_t, f, \bar{h}_2, \ldots, \bar{h}_a).$$

Then for almost all $h_1, \ldots, h_{i-1}$, there is a c.u.b. $C_i \subseteq \underset{\sim}{\delta}^1_{2n+1}$ such that for almost all $h_i, \ldots, h_t$ we have that

$$\vartheta(h_1, \ldots, h_t) < h(\hat{\mathcal{L}}^{\restriction i}(d); (\hat{\mathcal{L}}^{\restriction i}(d) \to N_{C_i}(\hat{\mathcal{L}}^{\restriction i}(d))); h_1, \ldots, h_t).$$

If $d^{(\mathcal{I}_i)}$ is minimal with respect to $\hat{\mathcal{L}}^{\restriction i}$ (*i.e.*, $\hat{\mathcal{L}}^{\restriction i}(d)$ is not defined), then we require that for almost all $h_1, \ldots, h_{i-1}$, there is an $\alpha_i < \underset{\sim}{\delta}^1_{2n+1}$ such that for almost all $h_i, \ldots, h_t$ we have $\vartheta(h_1, \ldots, h_t) < \alpha_i$.

5.1. Proof of the main inductive lemma. We prove the main inductive lemma by reverse induction on i. We consider the necessary cases.

Case I. $i = \infty$. Hence $d = d^{\ell,m}_{2n+1}$ is basic of type 0 or 1.

Subcase I.a. d basic of type 1, so $d^{(\mathcal{I}_a)} = (d_2)^{s(f(\bar{K}_a);\bar{K}_2,\ldots,\bar{K}_a)}$, where s may or may not appear, $d_2 \in \mathcal{D}^{\bar{\ell},\bar{m}}_{2s+1}$ and $\bar{K}_1 = R^{\bar{\ell},\bar{m}}_{2s+1}$.

Subcase I.a.i. d $(= d^\ell)$ not minimal with respect to $\hat{\mathcal{L}}^{\restriction\infty}$. We let ϑ be given such that for almost all $h_1, \ldots, h_t, f, \bar{h}_2, \ldots, \bar{h}_a$ we have

$$\vartheta(h_1, \ldots, h_t, f, \bar{h}_2, \ldots, \bar{h}_a) < H(d; h_1, \ldots, h_t, f, \bar{h}_2, \ldots, \bar{h}_a).$$

We have in this case that $\hat{\mathcal{L}}^{\restriction\infty} = \mathcal{L}((d_2))$ or $\mathcal{L}((d_2)^s)$ depending on whether s does not or does appear. By induction and $H_{2n-1}(a)$, it follows that for almost all $h_1, \ldots, h_t$, there is a c.u.b. $C \subseteq \underset{\sim}{\delta}^1_{2n+1}$ such that for almost all $f, \bar{h}_2, \ldots, \bar{h}_a$ we have

$$\begin{aligned}
\vartheta(h_1, \ldots, h_t, f, \bar{h}_2, \ldots, \bar{h}_a) &< (N_C; f; \mathcal{L}((d_2)^s); \bar{h}_2, \ldots, \bar{h}_a)\\
&= H(\hat{\mathcal{L}}^{\restriction\infty}(d); (\hat{\mathcal{L}}^{\restriction\infty}(d) \to N_C \circ \hat{\mathcal{L}}^{\restriction\infty}(d)); h_1, \ldots, h_t, f, \bar{h}_2, \ldots, \bar{h}_a),
\end{aligned}$$

and similarly if s does not appear.

Subcase I.a.ii. d is minimal with respect to $\hat{\mathcal{L}}^{\restriction\infty}$. Hence, the symbol s appears, and $(d_2)^s$ is minimal with respect to $\mathcal{L}$. For almost all $h_1, \dots, h_t, f, \bar{h}_2, \dots, \bar{h}_a$,

$$\begin{aligned}\vartheta(h_1, \dots, h_t, f, \bar{h}_2, \dots, \bar{h}_a) &< H(d; h_1, \dots, h_t, f, \bar{h}_2, \dots, \bar{h}_a)\\ &= (f; (d_2)^s; \bar{h}_2, \dots, \bar{h}_a).\end{aligned}$$

Hence it follows by induction and $H_{2n-1}(b)$ that for almost all $h_1, \dots, h_t$, there is an $\alpha < \underset{\sim}{\delta}^1_{2n+1}$ such that for almost all $f_1, \bar{h}_2, \dots, \bar{h}_a$,

$$\vartheta(h_1, \dots, h_t, f, \bar{h}_2, \dots, \bar{h}_a) < \alpha_t.$$

Subcase I.b. d $(= d^\ell)$ basic type 0. Hence $d^{(\mathcal{I}_i)} = (k)^{(f_m; \bar{K}_2, \dots, \bar{K}_a)}$, where $k \le m$.

Subcase I.b.i. d not minimal w.r.t. $\hat{\mathcal{L}}^{\restriction\infty}$, so $k > 1$. Then for almost all $h_1, \dots, h_t$, for almost all $f : m \to \underset{\sim}{\delta}^1_{2n+1}$, for almost all $\bar{h}_2, \dots, \bar{h}_a$, we have

$$\vartheta(h_1, \dots, h_t, f, \bar{h}_2, \dots, \bar{h}_a) < f(k).$$

It follows readily that for almost all $h_1, \dots, h_t$ there is a c.u.b. $C \subseteq \underset{\sim}{\delta}^1_{2n+1}$ such that for almost all $f, \bar{h}_2, \dots, \bar{h}_a$,

$$\begin{aligned}&\vartheta(h_1, \dots, h_t, f, \bar{h}_2, \dots, \bar{h}_a) < N_C(f(k-1))\\ &\quad = H(\hat{\mathcal{L}}^{\restriction\infty}(d); (\hat{\mathcal{L}}^{\restriction\infty}(d) \to N_C \circ \hat{\mathcal{L}}^{\restriction\infty}(d)); h_1, \dots, h_t, f, \bar{h}_2, \dots, \bar{h}_a).\end{aligned}$$

We use here a simple partition (of $f : m \to \underset{\sim}{\delta}^1_{2n+1}$ with an extra value inserted between $f(k-1)$ and $f(k)$) and the countable additivity of the measures $\bar{h}_2, \dots, \bar{h}_a$.

Subcase I.b.ii. d minimal w.r.t. $\hat{\mathcal{L}}^{\restriction\infty}(d)$. Similar to i).

Case II. $i < \infty$ and $i < k(d)$.

Subcase II.a. $K_i = S^{\ell_i, m_i}_{2n+1}$ with $\ell_i > 1$, and d not minimal w.r.t. $\hat{\mathcal{L}}^{\restriction i}$. Hence, d is non-minimal w.r.t. $\hat{\mathcal{L}}^{\restriction i+1}$. We let ϑ be given such that for almost all $h_1, \dots, h_t, f, \bar{h}_2, \dots, \bar{h}_a$ we have

$$\vartheta(h_1, \dots, h_t, f, \bar{h}_2, \dots, \bar{h}_a) < H(d^{(\mathcal{I}_i)}; h_1, \dots, h_t, f, \bar{h}_2, \dots, \bar{h}_a).$$

By induction, for almost all $h_1, \dots, h_i$, there is a c.u.b. $C_{i+1} \subseteq \underset{\sim}{\delta}^1_{2n+1}$ such that for almost all $h_{i+1}, \dots, h_t$,

$$\vartheta(h_1, \dots, h_t) < h(\hat{\mathcal{L}}^{\restriction i+1}(d); \hat{\mathcal{L}}^{\restriction i+1}(d) \to N_{C_{i+1}}(\hat{\mathcal{L}}^{\restriction i+1}(d)); h_1, \dots, h_t).$$

We let κ be such that for almost all $h_1, \dots, h_t, f, \bar{h}_2, \dots, \bar{h}_a$,

$$\kappa = \operatorname{cf} H(\hat{\mathcal{L}}^{\restriction i+1}(d); h_1, \dots, h_t, f, \bar{h}_2, \dots, \bar{h}_a).$$

Subcase II.a.i. $\kappa = \text{cf}\,\kappa(K_i)$.

For fixed $h_1, \dots, h_{i-1}$ such that

$$\vartheta(h_1, \dots, h_{i-1}, K_i, \dots, K_t) < h(d; h_1, \dots, h_{i-1}, K_i, \dots, K_t),$$

we consider the following partition $\mathcal{P}_i(h_1, \dots, h_{i-1})$: we partition functions $h_i \colon \underset{\sim}{\boldsymbol{\delta}}^1_{2n+1} \to \underset{\sim}{\boldsymbol{\delta}}^1_{2n+1}$ of the correct type with the extra value $g(\alpha)$ inserted between

$$h_i(0)(\alpha) = \sup_{f \colon f < \alpha \text{ a.e.}} h_i([f])$$

and the next element in the range of h_i after $h_i(0)(\alpha)$, where g has uniform cofinality ω, according to whether or not for almost all $h_{i+t}, \dots, h_t$,

$$\vartheta(h_1, \dots, h_t) < h(\hat{\mathcal{L}}^{\restriction i+1}(d); \hat{\mathcal{L}}^{\restriction i+1}(d) \to g \circ \hat{\mathcal{L}}^{\restriction i+1}(d); h_1, \dots, h_t).$$

It follows from a sliding argument (as in Lemma 5.18 of this section) that on the homogeneous side of the partition, the property stated in $\mathcal{P}(h_1, \dots, h_{i-1})$ holds. We let C_i be homogeneous for this partition. We therefore have that for almost all $h_1, \dots, h_{i-1}$ there is a C_i such that for almost all h_i (say with range in the closure points of C_i), $h_{i+1}, \dots, h_t$,

$$\begin{aligned}\vartheta(h_1, \dots, h_t) &< h(\hat{\mathcal{L}}^{\restriction i+1}(d); \hat{\mathcal{L}}^{\restriction i+1}(d) \to (N_{C_i} \circ h_i(0)) \circ \hat{\mathcal{L}}^{\restriction i+1}(d); h_1, \dots, h_t) \\ &= h(\hat{\mathcal{L}}^{\restriction i}(d); \hat{\mathcal{L}}^{\restriction i}(d) \to N_{C_i} \circ \hat{\mathcal{L}}^{\restriction i}(d); h_1, \dots, h_t).\end{aligned}$$

The last equality follows from $\hat{\mathcal{L}}^{\restriction i}(d) = (i; \hat{\mathcal{L}}^{\restriction i+1}(d))^s$ and the definition of h in this case.

Subcase II.a.ii. $\kappa \neq \text{cf}(K_i)$.

For fixed $h_1, \dots, h_{i-1}$, we consider $\mathcal{P}(h_1, \dots, h_{i-1})$ as above, and have that on the homogeneous side of the partition, $\mathcal{P}(h_1, \dots, h_{i-1})$ holds. However, for almost all $h_1, \dots, h_t, f, \bar{h}_2, \dots, \bar{h}_a$ we have that

$$H(\hat{\mathcal{L}}^{\restriction i+1}(d); h_1, \dots, h_t, f, \bar{h}_2, \dots, \bar{h}_a) = \sup_{f' \colon f' < H(\hat{\mathcal{L}}^{\restriction i+1}(d); h_1, \dots, h_t, f, \bar{h}_2, \dots, \bar{h}_a) \text{ a.e.}} h_i([f']).$$

since for almost all $\alpha < \underset{\sim}{\boldsymbol{\delta}}^1_{2n+1}$ of cofinality κ we have $\alpha = \sup_{f' \colon f' < \alpha \text{ a.e.}} h_i([f])$ as $\kappa \neq \text{cf}(K_i)$. Hence, for almost all $h_1, \dots, h_t$ we have

$$\begin{aligned}\vartheta(h_1, \dots, h_t) &< h(\hat{\mathcal{L}}^{\restriction i+1}(d); \hat{\mathcal{L}}^{\restriction i+1}(d) \to N_{C_i} \circ \hat{\mathcal{L}}^{\restriction i+1}(d); h_1, \dots, h_t) \\ &= h(\hat{\mathcal{L}}^{\restriction i}(d); \hat{\mathcal{L}}^{\restriction i}(d) \to N_{C_i} \circ \hat{\mathcal{L}}^{\restriction i}(d); h_1, \dots, h_t)\end{aligned}$$

as $\hat{\mathcal{L}}^{\restriction i}(d) = \hat{\mathcal{L}}^{\restriction i+1}(d)$ in this case.

Subcase II.b. $K_i = S^{\ell_i, m_i}_{2n+1}$ with $\ell_i > 1$, and d minimal with respect to $\hat{\mathcal{L}}^{\restriction i}$. Hence, d is minimal with respect to $\hat{\mathcal{L}}^{\restriction i+1}$. Hence, by induction, for almost all $h_1, \dots, h_i$, there is an $\alpha_{i+1} < \underset{\sim}{\boldsymbol{\delta}}^1_{2n+1}$ such that for almost all $h_{i+1}, \dots, h_t$,

$$\vartheta(h_1, \dots, h_t) < \alpha_{i+1}.$$

Hence, by the $\underset{\sim}{\delta}^1_{2n+1}$ additivity of K_i, it follows that for almost all $h_1, \dots, h_{i-1}$, there is an $\alpha_i < \underset{\sim}{\delta}^1_{2n+1}$ s.t. for almost all $h_i, \dots, h_t$, $\vartheta(h_1, \dots, h_t) < \alpha_i$.

Subcase II.c. $K_i = S^{1,m_i}_{2n+1}$ and d not minimal w.r.t. $\hat{\mathcal{L}}^{\restriction i}$. The proof is similar to a, where for fixed $h_1, \dots, h_{i-1}$ as in a, we consider the partition $\mathcal{P}(h_1, \dots, h_{i-1})$: we partition $h_i : <^{m_i} \to \underset{\sim}{\delta}^1_{2n+1}$ of the correct type with the extra value $g(\alpha)$ inserted between

$$h_i(0)(\alpha) = \sup_{\beta_2 < \cdots < \beta_{m_i} < \alpha_i} h_i(\beta_2, \dots \beta_{m_i}, \alpha)$$

and the next element in the range of $h_i(0)(\alpha)$ according to whether or not for almost all $h_{i+1}, \dots, h_t$,

$$\vartheta(h_1, \dots, h_t) < h(\hat{\mathcal{L}}^{\restriction i+1}(d); \hat{\mathcal{L}}^{\restriction i+1}(d) \to g \circ \hat{\mathcal{L}}^{\restriction i+1}(d); h_1, \dots, h_t).$$

It follows that for almost all $h_1, \dots, h_{i-1}$ that on the homogeneous side of the partition, the property stated in $\mathcal{P}(h_1, \dots, h_{i-1})$ holds. Taking cases on whether or not $\mathrm{cf}(H(\hat{\mathcal{L}}^{\restriction i+1}(d); h_1, \dots, h_t, f, \bar{h}_2, \dots, \bar{h}_a)) = \omega$ (and so $\hat{\mathcal{L}}^{\restriction i}(d) = (i; \hat{\mathcal{L}}^{\restriction i+1}(d)^{(\mathcal{I}_a)}))$ or $\hat{\mathcal{L}}^{\restriction i}(d) = \hat{\mathcal{L}}^{\restriction i+1}(d)$, the result follows as in cases II.a.i and II.a.ii above.

Subcase II.d. $K_i = S^{\ell_i, m_i}_{2n+1}$ with $\ell_i = 1$, and d minimal w.r.t. $\hat{\mathcal{L}}^{\restriction i}$. Similar to b) above.

Subcase II.e. $K_i = W^{m_i}_{2n+1}$ and d not minimal w.r.t. $\hat{\mathcal{L}}^{\restriction i+1}$. Once again, by induction for almost all $h_1, \dots, h_i$ there is a C_{i+1} s.t. for almost all $h_{i+1}, \dots, h_t$,

$$\vartheta(h_1, \dots, h_t) < h(\hat{\mathcal{L}}^{\restriction i+1}(d); \hat{\mathcal{L}}^{\restriction i+1}(d) \to N_{C_{i+1}}(\hat{\mathcal{L}}^{\restriction i+1}(d)); h_1, \dots, h_t).$$

By an easy partition and sliding argument (partitioning h_i followed by $g : \underset{\sim}{\delta}^1_{2n+1} \to \underset{\sim}{\delta}^1_{2n+1}$, where $\inf g > \sup h_i$) for almost all $h_1, \dots, h_{i-1}$ there is a C_i such that for almost all $h_i, \dots, h_t$,

$$\vartheta(h_1, \dots, h_t) < h(\hat{\mathcal{L}}^{\restriction i+1}(d); \hat{\mathcal{L}}^{\restriction i+1}(d) \to N_{C_i}(\hat{\mathcal{L}}^{\restriction i+1}(d)); h_1, \dots, h_t).$$

The result follows since $\hat{\mathcal{L}}^{\restriction i}(d) = \hat{\mathcal{L}}^{\restriction i+1}(d)$ in this case.

Subcase II.f. $K_i = W^{m_i}_{2n+1}$ and d minimal w.r.t. $\hat{\mathcal{L}}^{\restriction i+1}$. By induction, for almost all $h_1, \dots, h_i$, there is an $\alpha_{i+1} < \underset{\sim}{\delta}^1_{2n+1}$ s.t. for almost all $h_1, \dots, h_t$,

$$\vartheta(h_1, \dots, h_t) < \alpha_{i+1}.$$

For almost all $h_1, \dots, h_{i-1}$ we consider $\mathcal{P}_i(h_1, \dots, h_{i-1})$ where we partition $h_i : v(K_i) \to \underset{\sim}{\delta}^1_{2n+1}$ of the correct type (or $h_i : m \to \underset{\sim}{\delta}^1_1$ if $n = 0$) with the extra value α after $\sup h_i$ according to whether or not for almost all $h_{i+1}, \dots, h_t$, $\vartheta(h_1, \dots, h_t) < \alpha$. On the homogeneous side, $\mathcal{P}_i$ holds, and, for almost all $h_1, \dots, h_{i-1}$, we let C_i be homogeneous for $\mathcal{P}_i$. We then have that for almost all $h_i, \dots, h_t$ that $\vartheta(h_1, \dots, h_t) < N_{C_i}(\sup_{\text{a.e.}} h_i)$, and the result follows since

$\hat{\mathcal{L}}^{\restriction i}(d) = (i; \,)$ in this case, so

$$H(\hat{\mathcal{L}}^{\restriction i}(d); h_1, \dots, h_t, f, \bar{h}_2, \dots, \bar{h}_a) = \sup_{\text{a.e.}} h_i.$$

The case $n = 0$ is similar.

Subcase II.g. $K_i \in \bigcup_{m<n} R_{2m+1}$. The result follows easily from induction.

Case III. $i < \infty$, $i = k(d) = k$

Subcase III.a. $K_k = S_{2n+1}^{\ell_k, m_k}$ where $\ell_k > 1$.

Subcase III.a.i. $d^{(\mathcal{I}_i)} = (k; d_2^{(\mathcal{I}_a; \bar{K}_{a+1})})^{s(\mathcal{I}_a)}$, where s appears. Recall $\bar{K}_{a+1} = v(K_k)$. We let ϑ be such that for almost all $h_1, \dots, h_t$,

$$\vartheta(h_1, \dots, h_t) < h(d; h_1, \dots, h_t).$$

By the cofinality lemma, there is a ϑ_2 such that for almost all $h_1, \dots, h_t$, f, $\bar{h}_2, \dots, \bar{h}_{a+1}$, we have

$$\vartheta_2(h_1, \dots, h_t, f, \bar{h}_2, \dots, \bar{h}_{a+1}) < H(d_2; h_1, \dots, h_t, f, \bar{h}_2, \dots, \bar{h}_{a+1})$$

and

$$\vartheta(h_1, \dots, h_t) < h(d; (d_2 \to \vartheta_2); h_1, \dots, h_t).$$

We note that d is not minimal w.r.t. to $\hat{\mathcal{L}}^{\restriction i}$ in this case. We assume that d_2 is not minimal with respect to $\hat{\mathcal{L}}^{\restriction i+1}$. We will show in case (iv) below that this is the case. By induction, it then follows that for almost all $h_1, \dots, h_k$ there is a C_{k+1} s.t. for almost all $h_{k+1}, \dots, h_t$,

$$\vartheta_2(h_1, \dots, h_t) < h(\hat{\mathcal{L}}^{\restriction i+1}(d_2); \hat{\mathcal{L}}^{\restriction i+1}(d_2) \to N_{C_{k+1}} \circ \hat{\mathcal{L}}^{\restriction i+1}(d_2); h_1, \dots, h_t),$$

and hence

$$\vartheta(h_1, \dots, h_t) < h(d; d_2 \to N_{C_{k+1}} \circ \hat{\mathcal{L}}^{\restriction i+1}(d_2); h_1, \dots, h_t).$$

We let $\hat{d}$ be the tuple corresponding to d as in condition A.

Subcase III.a.i.1. $M(\hat{d}, \hat{\mathcal{L}}^{\restriction k+1}(d_2))$ is satisfied and $(k; \hat{\mathcal{L}}^{\restriction k+1}(d_2))$ satisfies condition C. That is for almost all $h_1, \dots, h_t, f, \bar{h}_2, \dots, \bar{h}_a$, the function g defined by

$$g([\bar{h}_{a+1}]) = H(\hat{\mathcal{L}}^{\restriction k+1}(d_2); h_1, \dots, h_t, f, \bar{h}_2, \dots, \bar{h}_{a+1})$$

is strictly increasing of uniform cofinality ω. Recall in this case that $\hat{\mathcal{L}}^{\restriction k}(d) = (k; \hat{\mathcal{L}}^{\restriction k+1}(d))$. For almost all $h_1, \dots, h_{k-1}$, we consider the partition $\mathcal{P}_k(h_1, \dots, h_{k-1})$ where we partition functions $h_k : \underset{\sim}{\boldsymbol{\delta}}^1_{2n+1} \to \underset{\sim}{\boldsymbol{\delta}}^1_{2n+1}$ of the correct type, with the extra value $g(\alpha)$ inserted between $h_k(\alpha)$ and $h_k(\alpha+1)$, and g has uniform cofinality ω, according to whether or not for almost all $h_{k+1}, \dots, h_t$,

$$\vartheta(h_1, \dots, h_t) < h(\hat{\mathcal{L}}^{\restriction k}(d); \hat{\mathcal{L}}^{\restriction k}(d) \to N_g \circ \hat{\mathcal{L}}^{\restriction k}(d); h_1, \dots, h_t).$$

We claim that on the homogeneous side of the partition, the property stated in $\mathcal{P}_k(h_1,\dots,h_{k-1})$ holds. We suppose not and let $C_k \subseteq \underset{\sim}{\delta}^1_{2n+1}$ be a c.u.b. set homogeneous for the contrary side. Let C_k^2 be such that for $h_k \colon \underset{\sim}{\delta}^1_{2n+1} \to C_k^2$ of the correct type, there is a C_{k+1} such that for almost all $h_{k+1},\dots,h_t$,

$$\vartheta(h_1,\dots,h_t) < h(d; d_2 \to N_{C_{k+1}} \circ \hat{\mathcal{L}}^{\restriction k+1}(d_2); h_1,\dots,h_t).$$

We let C_k^3 be contained in the closure points of $C_k \cap C_k^2$. We fix now a function $h_k \colon \underset{\sim}{\delta}^1_{2n+1} \to C_k^3$ of the correct type, and fix C_{k+1} as above. We then get h_k^2 and g satisfying:

(1) $h_k^2 = h_k$ almost everywhere w.r.t. $\mu_{2n+1}^{v(K_k)}$.

(2) h_k^2, g are of the correct type and ordered as in $\mathcal{P}_k$.

(3) h_k^2, g have range in C_k^3 (in fact a subset of the range of h_k).

(4) $g([f]) > h_k([N_{C_{k+1}} \circ f])$ for all $f \colon \kappa(K_k) \to \underset{\sim}{\delta}^1_{2n+1}$.

The construction of h_k^2, g follows as in the previous technical lemmas and will be omitted. We then elect $h_{k+1},\dots,h_t$ such that

$$\vartheta(h_1,\dots,h_t) < h(d; d_2 \to N_{C_{k+1}} \circ \hat{\mathcal{L}}^{\restriction k+1}(d_2); h_1,\dots,h_t),$$

and hence

$$\vartheta(h_1,\dots,h_t) < h(\hat{\mathcal{L}}^{\restriction k}(d); \hat{\mathcal{L}}^{\restriction k}(d) \to N_g \circ \hat{\mathcal{L}}^{\restriction k}(d); h_1,\dots,h_t)$$

by (4) above; and also

$$\vartheta(h_1,\dots,h_k^2,\dots,h_t) > h(\hat{\mathcal{L}}^{\restriction k}(d); \hat{\mathcal{L}}^{\restriction k}(d) \to N_g \circ \hat{\mathcal{L}}^{\restriction k}(d); h_1,\dots,h_t)$$

by (2), (3). Also, since $\vartheta(h_1,\dots,h_k) = \vartheta(h_1,\dots,h_k^2)$ by (1), we may assume that $\vartheta(h_1,\dots,h_k,\dots,h_t) = \vartheta(h_1,\dots,h_k^2,\dots,h_t)$. This contradiction establishes that $\mathcal{P}_k(h_1,\dots,h_{k-1})$ holds.

We let C_k be homogeneous for $\mathcal{P}_k$, for fixed $h_1,\dots,h_{k-1}$, and let C_k^2 be contained in the closure points of C_k. It then follows that for almost all h_k, namely $h_k \colon \underset{\sim}{\delta}^1_{2n+1} \to C_k^2$ of the correct type, that for almost all $h_{k+1},\dots,h_t$,

$$\vartheta(h_1,\dots,h_t) < h(\hat{\mathcal{L}}^{\restriction k}(d); \hat{\mathcal{L}}^{\restriction k}(d) \to N_{C_k} \circ \hat{\mathcal{L}}^{\restriction k}(d); h_1,\dots,h_t)$$

and we are done.

Subcase III.a.i.2. $M_2(\hat{d}, \hat{\mathcal{L}}^{\restriction k+1}(d))$ holds, case (i) fails, and $(k; \hat{\mathcal{L}}^{\restriction k+1}(d_2))^s$ satisfies condition C. Recall in this case $\hat{\mathcal{L}}^{\restriction k}(d) = (k; \hat{\mathcal{L}}^{\restriction k+1}(d_2))^s$. For almost all $h_1,\dots,h_{k-1}$, we consider the partition $\mathcal{P}_k(h_1,\dots,h_{k-1})$ where we partition $h_k \colon \underset{\sim}{\delta}^1_{2n+1} \to \underset{\sim}{\delta}^1_{2n+1}$ of the correct type with the extra value $g(\alpha)$ inserted between $\sup_{\beta<\alpha} h_k(\beta)$ and $h_k(\alpha)$ according to whether or not for almost all $h_{k+1},\dots,h_t$,

$$\vartheta(h_1,\dots,h_t) < h(\hat{\mathcal{L}}^{\restriction k}(d); \hat{\mathcal{L}}^{\restriction k}(d) \to N_g \circ \hat{\mathcal{L}}^{\restriction k}(d); h_1,\dots,h_t).$$

If $\mathcal{P}_k$ fails, we fix C_k homogeneous for the contrary side. We define C_k^2, C_k^3 as in (i.1) above, fix $h_k : \underset{\sim}{\delta}^1_{2n+1} \to C_k^3$ of the correct type, and fix C_{k+1} as in (i.1) above. We then get h_k^2, g satisfying:

(1) $h_k^2 = h_k$ almost everywhere w.r.t. $\mu_{2n+1}^{v(K_k)}$.

(2) h_k^2, g are ordered as in $\mathcal{P}_k$ for this case.

(3) h_k^2, g have range in C_k^3.

(4) $g([f]) > h_k([N_{C_{k+1}} \circ f])$ for all $f : \kappa(K_k) \to \underset{\sim}{\delta}^1_{2n+1}$ not of the correct type. We then proceed as in case (i.1) above.

Subcase III.a.i.3. Cases (i.1) and (i.2) fail. In the case, $\ell > 2$, $\hat{\mathcal{L}}^{\restriction k}(d) = \hat{d}$. We first note that $M(\hat{d}, \hat{\mathcal{L}}^{\restriction k+1}(d_2))$ is satisfied. For if not, then for almost all $h_1, \dots, h_t, f, \bar{h}_2, \dots, \bar{h}_a$, we consider the functions $g, \tilde{g}$ defined almost everywhere, where

$$g([h_{a+1}]) = H(d_2; h_1, \dots, h_t, f, \bar{h}_2, \dots, \bar{h}_{a+1})$$

and $\tilde{g} : \vartheta(\bar{K}_{a+1}) \to \underset{\sim}{\delta}^1_{2n+1}$ is minimal subject to $\sup_{\text{a.e.}} \tilde{g} = \sup_{\text{a.e.}} g$. Hence, $[\tilde{g}] < [g]$ for almost all $h_1, \dots, \bar{h}_a$, since $d = (k; d_2)^s$ satisfies condition C. Hence, by induction, for almost all $h_1, \dots, h_k$ there is a C_{k+1} s.t. for almost all $h_{k+1}, \dots, h_t, f, \bar{h}_2, \dots, \bar{h}_a$, $[\tilde{g}] < [g_2]$ where

$$g_2([\bar{h}_{a+1}]) = N_{C_{k+1}}(H(\hat{\mathcal{L}}^{\restriction k+1}(d_2); h_1, \dots, h_t, f, \bar{h}_2, \dots, \bar{h}_{a+1})),$$

a contradiction since $\sup_{\text{a.e.}} \tilde{g} > \sup_{\text{a.e.}} g_2$ for g_2 with a range in a set closed under $N_{C_{k+1}}$, which happens almost everywhere. Hence $M(\hat{d}, \hat{\mathcal{L}}^{\restriction k+1}(d_2))$ holds.

We first consider the case $\ell > 2$. We first claim in this case that $M_2(\hat{d}, \hat{\mathcal{L}}^{\restriction k+1}(d_2))$ is not satisfied. We suppose not. For almost all $h_1, \dots, h_t, f, \bar{h}_2, \dots, \bar{h}_a$, we consider the function $\bar{g}$ where

$$\bar{g}(\bar{h}_{a+1}) = H(\hat{\mathcal{L}}^{\restriction k+1}(d_2); h_1, \dots, h_t, f, \bar{h}_2, \dots, \bar{h}_{a+1}).$$

From our technical lemmas, $[\bar{g}] = \sup_{\text{a.e.}}[g']$ where g' ranges over functions from $\vartheta(\bar{K}_{a+1})$ into $\underset{\sim}{\delta}^1_{2n+1}$ of the correct type. Since $M_2(\hat{d}, \hat{\mathcal{L}}^{\restriction k+1}(d_2))$ is satisfied, $\bar{g} = \sup_{\text{a.e.}}[g'']$, where g'' ranges over functions from $\vartheta(\bar{K}_{a+1})$ into $\underset{\sim}{\delta}^1_{2n+1}$ with $\sup_{\text{a.e.}}[g''] = \sup_{\text{a.e.}} g$. Hence $(k; \hat{\mathcal{L}}^{\restriction k+1}(d_2))^s$ satisfies condition C, contrary to the assumption of this case. Hence $M_2(\hat{d}, \hat{\mathcal{L}}^{\restriction k+1}(d_2))$ fails. By induction, for almost all $h_1, \dots, h_k$ there is a C_{k+1} such that for almost all $h_{k+1}, \dots, h_t$,

$$\vartheta(h_1, \dots, h_t) < h(d; d_2 \to N_{C_{k+1}} \circ \hat{\mathcal{L}}^{\restriction k+1}(d_2); h_1, \dots, h_t).$$

For such fixed $h_1, \dots, h_{k-1}$, we consider the partition $\mathcal{P}_k(h_1, \dots, h_{k-1})$ where we partition $h_k : \underset{\sim}{\delta}^1_{2n+1} \to \underset{\sim}{\delta}^1_{2n+1}$ of the correct type with the extra value $g(\alpha)$ inserted between $\sup_{\beta<\alpha} h_k(0)(\beta)$ (which we recall is equal to $\sup\{h_k([f]) :$

$\sup_{\text{a.e.}} f < \alpha\}$) and the next element in the range of h_k after $\sup_{\beta<\alpha} h_k(0)(\beta)$, where g is of uniform cofinality ω, according to whether or not for almost all $h_{k+1}, \dots, h_t$,

$$\vartheta(h_1, \dots, h_t) < h(\hat{d}; \hat{d} \to N_g \circ \hat{d}; h_1, \dots, h_t).$$

If $\mathcal{P}_k$ fails, we fix C_k homogeneous for the contrary side, let C_k^2 such that for $h_k : \underset{\sim}{\delta}^1_{2n+1} \to C_k^2$ of the correct type there is a C_{k+1} such that the above inequality is satisfied, and let C_k^3 be contained in the closure points of $C_k \cap C_k^2$. We fix $h_k : \underset{\sim}{\delta}^1_{2n+1} \to C_k^3$ of the correct type and C_{k+1}, and get h_k^2, g satisfying:

(1) $h_k^2 = h_k$ a.e. w.r.t. $\mu_{2n+1}^{v(K_k)}$.

(2) h_k^2, g are of the correct type and ordered as in $\mathcal{P}_k$.

(3) h_k^2, g have range in C_k^3.

(4) $g(\alpha) > h_k([N_{C_{k+1}} \circ f])$, where f represents α w.r.t. $v(K_k)$.

The construction of h_k^2, g is similar to that in the proofs of the technical lemmas. We then have that for almost all $h_{k+1}, \dots, h_t, f, \bar{h}_1, \dots, \bar{h}_a$,

$$\vartheta(h_k, \dots, h_t, f, \bar{h}_2, \dots, \bar{h}_a) < h_k([N_{C_{k+1}} \circ \bar{g}])$$

with $\bar{g} : \vartheta(\bar{K}_{a+1}) \to \underset{\sim}{\delta}^1_{2n+1}$ as above, and $[\bar{g}] = \sup_{\text{a.e.}} \bar{g}$. It follows that almost everywhere we have

$$\begin{aligned}\vartheta(h_1, \dots, h_t, f, \bar{h}_1, \dots, \bar{h}_a) &< N_g(\sup_{\text{a.e.}} \bar{g}) \\ &= N_g(h(\hat{d}; h_1, \dots, h_t, f, \bar{h}_2, \dots, \bar{h}_a)).\end{aligned}$$

However, $\vartheta(h_1, \dots, h_k^2, \dots, h_t, f, \bar{h}_2, \dots, \bar{h}_a) = \vartheta(h_1, \dots, h_t, f, \bar{h}_2, \dots, \bar{h}_a)$. This contradiction establishes $\mathcal{P}_k$. The result then follows readily as in previous cases.

We now consider the case $\ell = 2$. Here $v = (K_k) =$ the m-fold product of the normal measure on $\omega_1 = v$ say. We recall $\vartheta(v)$ is identified with an ordinal ($= \omega_1$) by ordering by the largest ordinal first, then the next largest, etc. For almost all $h_1, \dots, h_t, f, \bar{h}_2, \dots, \bar{h}_a$, we again consider the function $g : \vartheta(v) \to \underset{\sim}{\delta}^1_{2n+1}$ defined by

$$g([\bar{h}_{a+1}]) = H(\hat{\mathcal{L}}^{\restriction k+1}(d_2); h_1, \dots, h_t, f, \bar{h}_2, \dots, \bar{h}_{a+1}).$$

Here $\bar{h}_{a+1}$ is a m-tuple of ordinals $< \omega_1$. We first assume $M_2(\hat{d}, \hat{\mathcal{L}}^{\restriction k+1}(d_2))$ is satisfied. For fixed $h_1, \dots, h_{k-1}$, we then consider the partition $\mathcal{P}_k$ where we partition $h_k : \underset{\sim}{\delta}^1_{2n+1} \to \underset{\sim}{\delta}^1_{2n+1}$ of the correct type according to whether or not for almost all $h_{k+1}, \dots, h_t$,

$$\vartheta(h_1, \dots, h_t) < h((k; \hat{\mathcal{L}}^{\restriction k+1}(d_2)); h_1, \dots, h_t).$$

We note that $(k; \hat{\mathcal{L}}^{\restriction k+1}(d_2))$ does not satisfy condition C, but this partition still makes sense. We then claim that on the homogeneous side of the partition the property stated in $\mathcal{P}_k$ holds. We suppose not, and fix $C_k, C_k^2, C_k^3, h_k, C_{k+1}$ as in the previous cases. We then get h_k^2 satisfying:

(1) $h_k^2 = h_k$ a.e. w.r.t. μ^v_{2n+1}.

(2) h_k is of the correct type.

(3) h_k, g has range in C_k^3.

(4) For α represented by $g_\alpha \colon \omega_1^m \to \underset{\sim}{\delta}^1_{2n+1}$ not of the correct type almost everywhere, $h_k^2(\alpha) > h_k([N_{C_{k+1}} \circ g_\alpha])$, where here g_α represents α w.r.t. W_1^m.

We then proceed to a contradiction as in the previous cases, establishing that for almost all $h_1, \dots, h_t$,

$$\vartheta(h_1, \dots, h_t) < h((k; \hat{\mathcal{L}}^{\restriction k+1}(d_2)); h_1, \dots, h_t).$$

We then consider, for fixed $h_1, \dots, h_{k-1}$, the partition $\mathcal{P}_k$ where we partition $h_k \colon \underset{\sim}{\delta}^1_{2n+1} \to \underset{\sim}{\delta}^1_{2n+1}$ of the correct type according to whether or not for almost all $h_{k+1}, \dots, h_t$,

$$\vartheta(h_1, \dots, h_t) < h((k; \hat{\mathcal{L}}^{\restriction k+1}(d_2))^s; h_1, \dots, h_t).$$

If $\mathcal{P}_k$ fails, we again get C_k, C_k^2, C_k^3 and h_k. We then get h_k^2 satisfying (1)–(3) as above and (4): for α represented by $g_\alpha \colon \omega_1^m \to \underset{\sim}{\delta}^1_{2n+1}$ not monotonically increasing a.e. w.r.t. W_1^m, $\sup_{\beta<\alpha} h_k^2(\beta) > h_k(\alpha)$. The construction is similar to that of previous cases, using here the fact that if $g \colon \omega_1^m \to \underset{\sim}{\delta}^1_{2n+1}$ is not monotonically increasing almost everywhere, then $[g]$ is not the supremum of $[g']$ for g' of the correct type. We then proceed to a contradiction establishing that for almost all $h_1, \dots, h_t, \vartheta(h_1, \dots, h_t) < h((k; \hat{\mathcal{L}}^{\restriction k+1}(d_2))^s; h_1, \dots, h_t)$.

If $(k; \hat{\mathcal{L}}^{\restriction k+1(q)}(d_2))$ or $(k; \hat{\mathcal{L}}^{\restriction k+1(q)}(d_2))^s$ satisfies condition C for some q, we then repeat the above argument to establish $\bar{H}_{2n+1}(k)$.

Hence we may assume without loss of generality that $M_2(\hat{d}, \hat{\mathcal{L}}^{\restriction k+1}(d_2))$ is not satisfied. Thus, $\hat{\mathcal{L}}^{\restriction k+1}(d) = \hat{d}$. The result now follows as in the corresponding case for $\ell > 2$.

Subcase III.a.i.4. d_2 minimal with respect to $\hat{\mathcal{L}}^{\restriction k+1}$. We show that this case does not occur. We show that $M_2(\hat{d}, d_2)$ is not satisfied. If it were, then define ϑ_2 such that for almost all $h_1, \dots, h_t, f, \bar{h}_2, \dots, \bar{h}_a$,

$$\vartheta_2(h_1, \dots, h_t, f, \bar{h}_2, \dots, \bar{h}_a) = [\tilde{g}]$$

where $\tilde{g} \colon \vartheta(\bar{K}_{a+1}) \to \underset{\sim}{\delta}^1_{2n+1}$ is minimal s.t.

$$\sup_{\text{a.e.}} \tilde{g} = H(\hat{d}; h_1, \dots, h_t, f, \bar{h}_2, \dots, \bar{h}_a).$$

We then have that for almost all $h_1, \dots, \bar{h}_{a+1}$,

$$\vartheta_2(h_1, \dots, \bar{h}_{a+1}) < H(d_2; h_1, \dots, \bar{h}_{a+1}).$$

However, for almost all $h_1, \dots, h_k$ there is no $\alpha_k < \underset{\sim}{\boldsymbol{\delta}}^1_{2n+1}$ such that for almost all $h_{k+1}, \dots, h_t, f, \bar{h}_2, \dots, \bar{h}_a$ we have $H(\hat{d}; h_1, \dots, \bar{h}_a) < \alpha_k$ (this follows easily from $k(\hat{d}) > k$). This contradicts the minimality of d_2 and induction.

Subcase III.a.ii. $d = (k; d_2^{(\mathcal{I}_a; \bar{K}_{a+1})})$, where s does not appear. In this case $\hat{\mathcal{L}}^{\restriction k}(d) = (k; d_2^{(\mathcal{I}_a; \bar{K}_{a+1})})^s$. Since d satisfies condition C, it follows readily that condition $M_2(\hat{d}, d_2)$ is satisfied. For fixed $h_1, \dots, h_{k-1}$, we consider the partition $\mathcal{P}_k(h_1, \dots, h_{k-1})$ where we partition $h_k : \underset{\sim}{\boldsymbol{\delta}}^1_{2n+1} \to \underset{\sim}{\boldsymbol{\delta}}^1_{2n+1}$ of the correct type with the extra value $g(\alpha)$ inserted between $\sup_{\beta<\alpha} h_k(\beta)$ and $h_k(\alpha)$, where g has uniform cofinality ω, according to whether or not for not almost all $h_{k+1}, \dots, h_t$,

$$\vartheta(h_1, \dots, h_t) < h(\hat{\mathcal{L}}^{\restriction k}(d); \hat{\mathcal{L}}^{\restriction k}(d) \to N_g \circ \hat{\mathcal{L}}^{\restriction k}(d); h_1, \dots, h_t).$$

We claim that on the homogeneous side of the partition $\mathcal{P}_k$ holds. If not, we let C_k be homogeneous for the contrary side, and let C_k^2 be such that for $h_k : \underset{\sim}{\boldsymbol{\delta}}^1_{2n+1} \to C_k^2$ of the correct type, for almost all $h_{k+1}, \dots, h_t$,

$$\vartheta(h_1, \dots, h_t) < H(d; h_1, \dots, h_t).$$

We let C_k^3 be contained in the closure points of $C_k \cap C_k^2$. We fix $h_k : \underset{\sim}{\boldsymbol{\delta}}^1_{2n+1} \to C_k^3$ of the correct type. We let $\tilde{h}_k : \omega \cdot \underset{\sim}{\boldsymbol{\delta}}^1_{2n+1} \to C_k \cap C_k^2$ exhibit that h_k has uniform cofinality ω. Then for almost all $h_{k+1}, \dots, h_t, f, \bar{h}_2, \dots, \bar{h}_a$ there is an $n < \omega$ such that $\vartheta(h_1, \dots, h_t, f, \bar{h}_2, \dots, \bar{h}_a) < \tilde{h}_k(n, [\tilde{g}])$, where $\tilde{g}$ is defined by

$$\tilde{g}([\bar{h}_{a+1}]) = H(d_2; h_1, \dots, h_t, f, \bar{h}_2, \dots, \bar{h}_{a+1}).$$

By countable additivity of the measures, it follows that there is a $g : \underset{\sim}{\boldsymbol{\delta}}^1_{2n+1} \to C_k \cap C_k^2$ of the correct type with h_k, g ordered as in $\mathcal{P}_k$ and such that $g(\alpha) > \tilde{k}_k(n, \alpha)$ for all α, where n is fixed so that the above inequality holds almost everywhere. This contradiction establishes $\mathcal{P}_k$. The result then follows as in the previous cases.

Subcase III.b. $K_k = S_{2n+1}^{\ell_k, m_k}$ where $\ell_k = 1$.

Subcase III.b.i. $d^{(\mathcal{I}_a)} = (k; d_1^{(\mathcal{I}_a)}, \dots, d_r^{(\mathcal{I}_a)})^{s^{(\mathcal{I}_a)}}$, where $2 \leq r \leq m_k$, and s appears. By the cofinality lemma it follows that there is a ϑ_2 such that for almost all $h_1, \dots, h_t, f, \bar{h}_2, \dots, \bar{h}_a$,

$$\vartheta_2(h_1, \dots, h_t, f, \bar{h}_2, \dots, \bar{h}_a) < H(d_r^{(\mathcal{I}_a)}; h_1, \dots, h_t, f, \bar{h}_2, \dots, \bar{h}_a)$$

and such that

$$\vartheta(h_1, \dots, h_t, f, \bar{h}_2, \dots, \bar{h}_a) < H(d; (d_r \to \vartheta_2); h_1, \dots, h_t, f, \bar{h}_2, \dots, \bar{h}_a).$$

Subcase III.b.i.1. $d_r^{(\mathcal{I}_a)}$ non-minimal with respect to $\hat{\mathcal{L}}^{\restriction k+1}$, $\hat{\mathcal{L}}^{\restriction k+1}(d_r) > d_{r-1}$ if $r > 2$ and $\operatorname{cf} H(\hat{\mathcal{L}}^{\restriction k+1}(d_r); h_1, \dots, h_t, f, \bar{h}_2, \dots, \bar{h}_a) = \omega$ almost everywhere. In this case $\hat{\mathcal{L}}^{\restriction k}(d) = (k; d_1^{(\mathcal{I}_a)}, \dots, d_{r-1}^{(\mathcal{I}_a)}, \hat{\mathcal{L}}^{\restriction k+1}(d_r^{(\mathcal{I}_a)}))$. By induction, for almost all $h_1, \dots, h_k$ there is a C_{k+1} such that for almost all $h_{k+1}, \dots, h_t$,

$$\vartheta(h_1, \dots, h_t) < h(d; (d_r \to N_{C_{k+1}} \circ \hat{\mathcal{L}}^{\restriction k+1}(d_r), h_1, \dots, h_t).$$

For such fixed $h_1, \dots, h_{k-1}$, we consider the partition

$\mathcal{P}_k(h_1, \dots, h_{k-1})$: we partition $h_k : <^m \to \underset{\sim}{\delta}^1_{2n+1}$ of the correct type with the extra value $g(\alpha_2, \dots, \alpha_r, \alpha_1)$ inserted between

$$h_k(r)(\alpha_2, \dots, \alpha_r, \alpha_1) := \sup_{\beta_{r+1} < \dots < \beta_{m_k} < \alpha_1} h_k(\alpha_2, \dots, \alpha_r, \beta_{r+1}, \dots, \beta_{m_k}, \alpha_1)$$

and the next element in the range of h_k after $h_k(r)(\alpha_2, \dots, \alpha_r, \alpha_1)$, where g has uniform cofinality ω, according to whether or not for almost all $h_k, \dots, h_t$,

$$\vartheta(h_1, \dots, h_t) < H(\hat{\mathcal{L}}^{\restriction k}(d); \hat{\mathcal{L}}^{\restriction k}(d) \to N_g \circ \hat{\mathcal{L}}^{\restriction k}(d); h_1, \dots, h_t).$$

If $\mathcal{P}_k$ fails, then we let C_k be homogeneous for the contrary side. We let C_k^2 be such that for $h_k : <^m \to C_k^2$ of the correct type there is a C_{k+1} such that for almost all $h_{k+1}, \dots, h_t$ the above inequality (with $N_{C_{k+1}}$) is satisfied. We let C_k^3 be contained in the closure points of $C_k \cup C_k^2$, and fix $h_k : <^m \to C_k^3$ of the correct type, and C_{k+1} as above. We then get h_k^2, g such that:

(1) $h_k^2 = h_k$ a.e. w.r.t. the m-fold product of the ω-cofinal normal measure on $\underset{\sim}{\delta}^1_{2n+1}$.

(2) h_k^2, g are of the correct type and ordered as in $\mathcal{P}_k(h_1, \dots, h_{k-1})$.

(3) h_k^2, g have range in C_k^3.

(4) $g(\alpha_2, \dots, \alpha_r, \alpha_1) > h_k(r)(\alpha_2, \dots, \alpha_{r-1}, N_{C_{k+1}}(\alpha_r), \alpha_1)$, for almost all $\alpha_2, \dots, \alpha_r, \alpha_1$.

The existence of h_k^2 and g follows from an easy sliding argument which we omit. We then elect $h_{k+1}, \dots, h_t$ such that

$$\vartheta(h_1, \dots, h_k, \dots, h_t) = \vartheta(h_1, \dots, h_k^2, \dots, h_t), \vartheta(h_1, \dots, h_t) < h(d; d_r \to N_{C_{k+1}} \circ \hat{\mathcal{L}}^{\restriction k+1}(d_r); h_1, \dots, h_t)$$

and

$$\vartheta(h_1, \dots, h_k^2, \dots, h_t) > h(\hat{\mathcal{L}}^{\restriction k}(d); \hat{\mathcal{L}}^{\restriction k}(d) \to N_g \circ \hat{\mathcal{L}}^{\restriction k}(d); h_1, \dots, h_k^2, \dots, h_t).$$

This contradiction establishes $\mathcal{P}_k$, and the result $\bar{H}_{2n+1}(k)$ then follows as in the previous cases.

Subcase III.b.i.2. $d_r^{(\mathcal{I}_a)}$ non-minimal w.r.t. $\hat{\mathcal{L}}^{\restriction k+1}$, $\hat{\mathcal{L}}^{\restriction k+1}(d_r) > d_{r-1}$ if $r > 2$, and $\operatorname{cf} H(\hat{\mathcal{L}}^{\restriction k+1}(d_r); h_1, \dots, h_t, f, \bar{h}_2, \dots, \bar{h}_a) \neq \omega$ almost everywhere. In this case, $\hat{\mathcal{L}}^{\restriction k}(d) = (k; d_1^{(\mathcal{I}_a)}, \dots, d_{r-1}^{(\mathcal{I}_a)}, \hat{\mathcal{L}}^{\restriction k+1}(d_r^{(\mathcal{I}_a)}))^s$. We proceed as in the previous case, and for fixed $h_1, \dots, h_{k-1}$ consider $\mathcal{P}_k(h_1, \dots, h_{k-1})$ where we partition $h_k \colon <^m \to \underset{\sim}{\boldsymbol{\delta}}^1_{2n+1}$ of the correct type with the extra value $g(\alpha_2, \dots, \alpha_r, \alpha_1)$ inserted between $\sup_{\beta<\alpha_r} h_k(r)(\beta)$ and the next element in the range of h_k after this, where g has uniform cofinality ω, according to whether or not for almost all $h_{k+1}, \dots, h_t$,

$$\vartheta(h_1, \dots, h_t) < H(\hat{\mathcal{L}}^{\restriction k}(d); \hat{\mathcal{L}}^{\restriction k}(d) \to N_g \circ \hat{\mathcal{L}}^{\restriction k}(d); h_1, \dots, h_t).$$

If $\mathcal{P}_k$ fails, we proceed as in the previous case, fixing C_k, C_k^2, C_k^3, h_k, C_{k+1} respectively, and get h_k^2 and g satisfying (1)–(4) as in that case. We the proceed to get a contradiction as in that case, which establishes $\mathcal{P}_k(h_1, \dots, h_{k-1})$, from which $\bar{H}_{2n+1}(k)$ again readily follows.

Subcase III.b.i.3. $d_r^{(\mathcal{I}_a)}$ non-minimal with respect to $\hat{\mathcal{L}}^{\restriction k+1}$, $r > 2$, and either $\hat{\mathcal{L}}^{\restriction k+1}(d_r) \le d_{r-1}$ or d_r is minimal with respect to $\hat{\mathcal{L}}^{\restriction k+1}$. In this case, $\hat{\mathcal{L}}^{\restriction k}(d) = (k; d_1^{(\mathcal{I}_a)}, \dots, d_{r-1}^{(\mathcal{I}_a)})^s$. Actually, the second case cannot arise since d satisfies condition C so $d_{r-1} < d_r$ and (since $k(d_{r-1}) > k$) for almost all $h_1, \dots, h_k$ there is no $\alpha_{k+1} < \underset{\sim}{\boldsymbol{\delta}}^1_{2n+1}$ such that for almost all $h_{k+1}, \dots, h_t$, $\bar{h}_2, \dots, \bar{h}_a$,

$$H(d_{r-1}; h_1, \dots, h_t, f, \bar{h}_2, \dots, \bar{h}_a) < \alpha_{k+1}.$$

For almost all $h_1, \dots, h_{k-1}$, we consider $\mathcal{P}_k(h_1, \dots, h_{k-1})$ where we partition functions $h_k \colon <^m \to \underset{\sim}{\boldsymbol{\delta}}^1_{2n+1}$ of the correct type with the extra value $g(\alpha_2, \dots, \alpha_r, \alpha_1)$ inserted between $\sup_{\beta<\alpha_{r-1}} h_r(r-1)(\alpha_2, \dots, \alpha_{r-2}, \beta, \alpha_1)$ and the next element in the range of h_k after this, where g has uniform cofinality ω, according to whether or not

$$\vartheta(h_1, \dots, h_t) < h(\hat{\mathcal{L}}^{\restriction k}(d); \hat{\mathcal{L}}^{\restriction k}(d) \to N_g \circ \hat{\mathcal{L}}^{\restriction k}(d); h_1, \dots, h_t).$$

If $\mathcal{P}_k$ fails, we fix C_k, C_k^2, C_k^3, h_k, C_{k+1} as in the previous case, and get h_k^2 and g satisfying (1)–(3) as in that case, and (4):

$$g(\alpha_2, \dots, \alpha_r, \alpha_1) > h_k(\alpha_2, \dots, \alpha_{r-2}, \alpha_{r-1}, N_{C_{k+1}}(\alpha_{r-1}), \alpha_1)$$

almost everywhere. We then elect $h_{k+1}, \dots, h_t, f, \bar{h}_2, \dots, \bar{h}_a$ such that $\vartheta(h_1, \dots, h_k, \dots, h_t) = \vartheta(h_1, \dots, h_k^2, \dots, h_t)$,

$$\begin{aligned}
&\vartheta(h_1, \dots, h_k, \dots, \bar{h}_a) < H(d; d_r \to N_{C_{k+1}} \circ \hat{\mathcal{L}}^{\restriction k+1}(d_r); h_1, \dots, h_k, \dots, \bar{h}_a) \\
&= \sup\{h_k(H(d_2; h_1, \dots, \bar{h}_a), \dots, H(d_{r-1}; h_1, \dots, \bar{h}_a), \\
&\qquad N_{C_{k+1}}(H(\hat{\mathcal{L}}^{\restriction k+1}(d_r); h_1, \dots, \bar{h}_a)), \beta_{r+1}, \dots, \beta_{m_k}, H(d_1; h_1, \dots, \bar{h}_a)) \colon \\
&\qquad\qquad \beta_{r+1} < \cdots < \beta_{m_k} < H(d_1; h_1, \dots, \bar{h}_a)\} \\
&\le g(H(d_2; \dots), \dots, H(d_{r-1}; \dots), H(d_1; \dots)) \\
&= N_g(\sup\{h_k(r-1)(H(d_2; \dots), \dots, H(d_{r-2}; \dots), \beta_{r-1}, H(d_1; \dots)) \colon \\
&\qquad \beta_{r-1} < H(d_{r-1}; h_1, \dots, h_k^2, \dots, \bar{h}_a)\}) \\
&= H(\hat{\mathcal{L}}^{\restriction k}(d); \hat{\mathcal{L}}^{\restriction k}(d) \to N_g \circ \hat{\mathcal{L}}^{\restriction k}(d); h_1, \dots, h_k^2, \dots, h_t, f, \bar{h}_2, \dots, \bar{h}_a)
\end{aligned}$$

and also such that

$$\vartheta(h_1,\dots,h_k^2,\dots,h_t) > h(\hat{\mathcal{L}}^{\restriction k}(d);\hat{\mathcal{L}}^{\restriction k}(d)\to N_g\circ\hat{\mathcal{L}}^{\restriction k}(d);\\ h_1,\dots,h_k^2,\dots,h_t).$$

This contradiction establishes $\mathcal{P}_k$ and the result follows as in the previous case.

Subcase III.b.i.4. $d_r^{(\mathcal{I}_a)}$ minimal with respect to $\hat{\mathcal{L}}^{\restriction k+1}$ and $r=2$. In this case we have $\hat{\mathcal{L}}^{\restriction k}(d)=d_1$. By induction, for almost all $h_1,\dots,h_k$ there is an $\alpha_{k+1}<\underset{\sim}{\delta}^1_{2n+1}$ s.t. for almost all $h_{k+1},\dots,h_t$,

$$\vartheta(h_1,\dots,h_t) < h(d;d_r\to\alpha_{k+1},\dots,h_1,\dots,h_t).$$

For almost all $h_1,\dots,h_{k-1}$, we consider the partition $\mathcal{P}_k(h_1,\dots,h_{k-1})$ where we partition functions $h_k\colon <^m\to\underset{\sim}{\delta}^1_{2n+1}$ of the correct type with $g(\alpha)$ inserted between $\sup_{\beta<\alpha}h_k(1)(\beta)$ and the next element in the range of h_k after this, according to whether or not for almost all $h_{k+1},\dots,h_t$,

$$\vartheta(h_1,\dots,h_t) < (\hat{\mathcal{L}}^{\restriction k}(d);\hat{\mathcal{L}}^{\restriction k}(d)\to N_g\circ\hat{\mathcal{L}}^{\restriction k}(d);h_1,\dots,h_t).$$

If $\mathcal{P}_k$ fails, we elect C_k, C_k^2, C_k^3, h_k, α_{k+1} similarly to the previous cases. We then get h_k^2 and g satisfying the usual (1)–(3) and (4):

$$g(\alpha) > \sup_{\beta_3<\dots<\beta_{m_k}<\alpha} h_k(\alpha_{k+1},\beta_3,\dots,\beta_{m_k},\alpha)$$

for all $\alpha>\alpha_{k+1}$. We then elect $h_{k+1},\dots,h_t,f,\bar{h}_2,\dots,\bar{h}_a$ such that

$$\begin{aligned}
&\vartheta(h_1,\dots,h_k^2,\dots,h_t,f,\bar{h}_2,\dots,\bar{h}_a)=\vartheta(h_1,\dots,h_t,f,\bar{h}_2,\dots,\bar{h}_a)\\
&< h(d;d_r\to\alpha_{k+1},h_1,\dots,h_t,f,\bar{h}_2,\dots,\bar{h}_a)\\
&\le N_g(\sup\{h_k^2(\beta_2,\dots,\beta_{m_k},\beta_1)\colon\\
&\qquad \beta_1<H(d_1;h_1,\dots,\bar{h}_a),\beta_2<\dots<\beta_{m_k}<\beta_1\})\\
&= N_g(H(d_1;h_1,\dots,h_k^2,\dots,h_t,f,\bar{h}_2,\dots,\bar{h}_a))\\
&= H(\hat{\mathcal{L}}^{\restriction k}(d);\hat{\mathcal{L}}^{\restriction k}(d)\to N_g\circ\hat{\mathcal{L}}^{\restriction k}(d);h_1,\dots,h_k^2,\dots,h_t,f,\bar{h}_2,\dots,\bar{h}_a).
\end{aligned}$$

This contradiction establishes $\mathcal{P}_k$ from which the result follows.

Subcase III.b.ii. $d^{(\mathcal{I}_a)}=(k;d_1^{(\mathcal{I}_a)},\dots,d_r^{(\mathcal{I}_a)})^{(\mathcal{I}_a)}$, where s does not appear, and $r=m_k$. We consider the case $m_k>1$, the case $m_k=1$ being similar. In this case $\hat{\mathcal{L}}^{\restriction k}(d)=(k;d_1,\dots,d_r)^s$. As in the previous case, we consider $\mathcal{P}_k(h_1,\dots,h_{k-1})$ for almost all $h_1,\dots,h_{k-1}$, where here the extra value $g(\alpha_2,\dots,\alpha_r,\alpha_1)$ is inserted between $\sup_{\beta<m_k}h_k(\alpha_2,\dots,\beta,\alpha_1)$ and the next element in the range of h_k after this. If $\mathcal{P}_k$ fails, we fix C_k homogeneous for the contrary side, let C_k^2 be such that for $h_k\colon <^m\to C_k^2$ of the correct type, for almost all $h_{k+1},\dots,h_t$,

$$\vartheta(h_1,\dots,h_t) < h(d;h_1,\dots,h_t)$$

and let C_k^3 in the closure points of $C_k \cup C_k^2$. We again fix $h_k : \underset{\sim}{\boldsymbol{\delta}}^1_{2n+1} \to C_k^3$ of the correct type, and let $\tilde{h}_k(\alpha_2, \dots, \alpha_{m_k}, n, \alpha_1) \colon \omega \times <^m \to \underset{\sim}{\boldsymbol{\delta}}^1_{2n+1}$ exhibit the uniform cofinality ω of h_k. By countable additivity there is an $n < \omega$ s.t. for almost all $h_{k+1}, \dots, h_t, f, \bar{h}_2, \dots, \bar{h}_a$,

$$\vartheta(h_1, \dots, \bar{h}_a) < \tilde{h}_k(H(d_r, \dots), \dots, n, H(d_1, \dots)).$$

We then get h_k^2, g satisfying the usual (1)–(3) and (4):

$$g(\alpha_2, \dots, \alpha_{m_k}, \alpha_1) > \tilde{h}_k(\alpha_2, \dots, \alpha_{m_k}, n, \alpha_1)$$

for all $\alpha_1, \dots, \alpha_{m_k}$. The result follows as in previous cases.

Subcase III.b.iii. $d^{(\mathcal{I}_a)} = (k; d_1^{(\mathcal{I}_a)}, \dots, d_r^{(\mathcal{I}_a)})^{(\mathcal{I}_a)}$, where s does not appear, and $r < m_k$.

Subcase III.b.iii.1. d_1 is not minimal w.r.t. $\hat{\mathcal{L}}^{\restriction k+1}$, $\hat{\mathcal{L}}^{\restriction k+1}(d_1) > d_r$ if $r \geq 2$, and cf $H(\hat{\mathcal{L}}^{\restriction k+1}(d_1); h_1, \dots, h_t, f, \bar{h}_2, \dots, \bar{h}_a) = \omega$ almost everywhere. In this case $\hat{\mathcal{L}}^{\restriction k}(d) = (k; d_1, \dots, d_r, \hat{\mathcal{L}}^{\restriction k+1}(d_1))$. By the cofinality lemma, there is a ϑ_2 such that for almost all $h_1, \dots, h_t, f, \bar{h}_2, \dots, \bar{h}_a$,

$$\vartheta_2(h_1, \dots, h_t, f, \bar{h}_2, \dots, \bar{h}_a) < H(d_1; h_1, \dots, h_t, f, \bar{h}_2, \dots, \bar{h}_a)$$

and

$$\vartheta(h_1, \dots, \bar{h}_a) < H(\hat{\mathcal{L}}^{\restriction k}(d); \hat{\mathcal{L}}^{\restriction k+1}(d_1) \to \vartheta_2; h_1, \dots, \bar{h}_a).$$

Hence, by induction for almost all $h_1, \dots, h_t$ there is C_{k+1} such that for almost all $h_{k+1}, \dots, h_t$,

$$\vartheta(h_1, \dots, h_t) < H(\hat{\mathcal{L}}^{\restriction k}(d); \hat{\mathcal{L}}^{\restriction k+1}(d_1) \to N_{C_{k+1}} \circ \hat{\mathcal{L}}^{\restriction k+1}(d_1); h_1, \dots, h_t).$$

For fixed $h_1, \dots, h_{k-1}$, we consider $\mathcal{P}_k(h_1, \dots, h_{k-1})$ where we partition h_k of the correct type with $g(\alpha_2, \dots, \alpha_{r+1}, \alpha_1)$ inserted between

$$h_k(r+1)(\alpha_2, \dots, \alpha_{r+1}, \alpha_1)$$

and the next element in the range of h_k according to whether or not for almost all $h_{k+1}, \dots, h_t$,

$$\vartheta(h_1, \dots, h_t) < H(\hat{\mathcal{L}}^{\restriction k}(d); \hat{\mathcal{L}}^{\restriction k}(d) \to N_g \circ \hat{\mathcal{L}}^{\restriction k}(d); h_1, \dots, h_t).$$

If $\mathcal{P}_k$ fails, we fix C_k, C_k^2, C_k^3, h_k, C_{k+1}, as in the previous cases, and get h_k^2 and g satisfying the usual (1)–(3) and (4):

$$g(\alpha_2, \dots, \alpha_{r+1}, \alpha_1) > h_k(r+1)(\alpha_2, \dots, \alpha_r, N_{C_{k+1}}(\alpha_{r+1}), \alpha_1).$$

We then proceed as in the previous cases to establish $\bar{H}_{2n+1}(k)$.

Subcase III.b.iii.2. d_1 not minimal w.r.t. $\hat{\mathcal{L}}^{\restriction k+1}$, $\hat{\mathcal{L}}^{\restriction k+1}(d_1) > d_r$ if $r \geq 2$, and

$$\text{cf } H(\hat{\mathcal{L}}^{\restriction k+1}(d_1); h_1, \dots, h_t, f, \bar{h}_2, \dots, \bar{h}_a) \neq \omega$$

almost everywhere. In this case $\hat{\mathcal{L}}\restriction^k(d) = (k; d_1, \dots, d_r, \hat{\mathcal{L}}\restriction^{k+1}(d_1))^s$. We proceed as in (iii.1) except that in $\mathcal{P}_k(h_1, \dots, h_{k-1})$ we insert the function value $g(\alpha_2, \dots, \alpha_{r+1}, \alpha_1)$ between $\sup_{\beta<\alpha_{r+1}} h_k(r+1)(\alpha_2, \dots, \alpha_r, \beta, \alpha_1)$ and the next element in the range of h_k after this. We proceed as in (iii.1), getting h_k^2 and g which satisfy (1)–(4) of that case, and then proceed to establish $\bar{H}_{2n+1}(k)$ as in that case.

Subcase III.b.iii.3. d_1 not minimal w.r.t. $\hat{\mathcal{L}}\restriction^{k+1}$ and $\hat{\mathcal{L}}\restriction^{k+1}(d_1) \leq d_r$ (where $r \geq 2$). In this case $\hat{\mathcal{L}}\restriction^k(d) = (k; d_1, \dots, d_r)^s$. We proceed as in the above cases, where in $\mathcal{P}_k(h_1, \dots, h_{k-1})$ the function value $g(\alpha_2, \dots, \alpha_r, \alpha_1)$ is inserted between $\sup_{\beta<\alpha_r} h_k(r+1)(\alpha_2, \dots, \alpha_{r-1}, \beta, \alpha_1)$ and the next element in the range of h_k. We proceed as before getting h_k^2 and g which satisfy the usual (1)–(3) and (4):

$$g(\alpha_2, \dots, \alpha_r, \alpha_1) > h_k(r+1)(\alpha_2, \dots, \alpha_r, N_{C_{k+1}}(\alpha_r), \alpha_1).$$

We then finish as in the previous cases.

Subcase III.b.iii.4. d_1 minimal w.r.t. $\hat{\mathcal{L}}\restriction^{k+1}$. It follows readily that $r = 1$. In this case $\hat{\mathcal{L}}\restriction^k(d) = d_1$. By the cofinality lemma and induction, we have that for almost all $h_1, \dots, h_k$ there is $\alpha_{k+1} < \underset{\sim}{\delta}^1_{2n+1}$ such that for almost all $h_{k+1}, \dots, h_t$,

$$\vartheta(h_1, \dots, h_t) < h(\tilde{d}; \tilde{\tilde{d}} \to \alpha_{k+1}; h_1, \dots, h_t),$$

where we define $\tilde{d} = (k; d_1, \tilde{\tilde{d}})$. We assume here that $m_k \geq 2$, the case $m_k = 1$ following similarly. As in the previous cases, we consider $\mathcal{P}_k(h_1, \dots, h_{k-1})$ where the value $g(\alpha)$ is inserted between $\sup_{\beta<\alpha} h_k(1)(\beta)$ and the next element in the range of h_k. If $\mathcal{P}_k$ fails, we fix C_k, C_k^2, C_k^3, h_k, α_{k+1} and get h_k^2 and g satisfying the usual (1)–(3) and (4):

$$g(\alpha) > \sup_{\beta_3<\dots<\beta_{m_k}<\alpha} h_k(\alpha_{k+1}, \beta_3, \dots, \beta_{m_k}, \alpha).$$

We then proceed as in the previous cases to establish $\bar{H}_{2n+1}(k)$.

Subcase III.c. $K_k = W^{m_k}_{2n+1}$. In the following we assume $n > 0$, the case $n = 0$ following easily.

Subcase III.c.i $d^{(\mathcal{I}_a)} = (k;)^{(\mathcal{I}_a)}$ We consider the sequence of measures

$$K_{b(k)}, \dots, K_{b_u}, \bar{K}_2, \dots, \bar{K}_a.$$

where we recall $K_{b(k)}, \dots, K_{b_u}$ enumerates the subsequences of $K_{k+1}, \dots, K_t$ of measures in $\bigcup_{m<n} R_{2m+1}$. We let $S^{\bar{\ell}_k, \bar{m}_k}_{2n-1} = v(K_k)$ (where $\bar{\ell}_k = 2^n - 1$ is maximal here).

Subcase III.c.i.1. the maximal tuple $\tilde{d}^{\bar{\ell}_k, \bar{m}_k}_{2n-1}$ relative to $K_{b(k)}, \dots, K_{b_u}$, $\bar{K}_2$, $\dots, \bar{K}_a$ is defined. Hence for almost all $h_1, \dots, h_{k-1}$, for almost all $h_k, \dots, h_t$,

$f, \bar{h}_2, \dots, \bar{h}_a$, there is an ordinal $\vartheta_2 = \vartheta_2(h_k, \dots, h_t, f, \bar{h}_2, \dots, \bar{h}_a) < \sup_{\text{a.e.}} h_k$ such that $\vartheta(h_1, \dots, h_t, f, \bar{h}_2, \dots, \bar{h}_a) < \vartheta_2$. It follows readily (by $\underset{\sim}{\boldsymbol{\delta}}^1_{2n+1}$-additivity of W^m_{2n+1} and $S^{\ell,m}_{2n+1}$) that for almost all $h_1, \dots, h_{k-1}$ there is an ordinal $\vartheta_2 = \vartheta_2(h_1, \dots, h_{k-1})$ such that for almost all $h_k, h_{b(k)}, \dots, h_{b_u}, \bar{h}_2, \dots, \bar{h}_a$,

$$\vartheta_2(h_k, \dots, \bar{h}_a) < \sup_{\text{a.e.}} h_k$$

and for almost all $h_k, \dots, h_t, f, \bar{h}_2, \dots, \bar{h}_a$,

$$\vartheta(h_1, \dots, h_t, f, \bar{h}_2, \dots, \bar{h}_a) < \vartheta_2(h_1, \dots, h_{k-1})(h_k, h_{b(k)}, \dots, h_{b_u}, \dots, \bar{h}_a).$$

By $\bar{H}_{2n-1}(c)$ it follows that for almost all $h_1, \dots, h_{k-1}$ there is a c.u.b. $C_k \subseteq \underset{\sim}{\boldsymbol{\delta}}^1_{2n+1}$ such that for almost all $h_k, h_{b(k)}, \dots, h_{b_u}, \bar{h}_2, \dots, \bar{h}_a$,

$$\begin{aligned}\vartheta_2(h_1, \dots, h_{k-1})(h_k, h_{b(k)}, \dots, \bar{h}_a) &< (N_{C_k}; h_k; \tilde{d}^{\bar{\ell}_k, \bar{m}_k}_{2n-1}; h_{b(k)}, \dots, \bar{h}_a)\\ &= H(\hat{\mathcal{L}}^{\restriction k}(d); \hat{\mathcal{L}}^{\restriction k}(d) \to N_{C_k} \circ \hat{\mathcal{L}}^{\restriction k}(d); h_1, \dots, h_t, f, \bar{h}_2, \dots, \bar{h}_a).\end{aligned}$$

This establishes $\bar{H}_{2n+1}(k)$ in this case.

Subcase III.c.i.2. the maximal tuple $\tilde{d}^{\bar{\ell}_k, \bar{m}_k}_{2n-1}$ relative to $K_{b(k)}, \dots, K_{b_u}, \bar{K}_2, \dots, \bar{K}_a$ is not defined. In this case d is minimal w.r.t. $\hat{\mathcal{L}}^{\restriction k}$. As above, for almost all $h_1, \dots, h_{k-1}$, there is an ordinal $\vartheta_2 = \vartheta_2(h_1, \dots, h_{k-1})$ such that for almost all $h_k, h_{b(k)}, \dots, \bar{h}_a$,

$$\vartheta_2(h_1, \dots, h_{k-1})(h_k, h_{b(k)}, \dots, \bar{h}_a) < \sup_{\text{a.e.}} h_k$$

and for almost all $h_k, \dots, h_t, f, \bar{h}_2, \dots, \bar{h}_a$,

$$\vartheta(h_1, \dots, h_t, f, \bar{h}_2, \dots, \bar{h}_a) < \vartheta_2(h_1, \dots, h_{k-1})(h_k, h_{b(k)}, \dots, \bar{h}_a).$$

By $\bar{H}_{2n-1}(d)$ it follows that for almost all $h_1, \dots, h_{k-1}$ there is an $\alpha_k < \underset{\sim}{\boldsymbol{\delta}}^1_{2n+1}$ such that for almost all $h_k, \dots, h_t, f, \bar{h}_2, \dots, \bar{h}_a$,

$$\vartheta(h_1, \dots, h_t) < \vartheta_2(h_1, \dots, h_{k-1})(h_k, h_{b(k)}, \dots, \bar{h}_a) < \alpha_k,$$

which establishes $\bar{H}_{2n+1}(k)$ in this case.

Subcase III.c.ii $d = (k; \bar{d}_2)^s$ where s appears. Here $\bar{d}_2$ is defined relative to the sequence $K_{b(k)}, \dots, K_{b_u}, \bar{K}_2, \dots, \bar{K}_a$ and $(\bar{d}_2)^s$ satisfies conditions D and A.

Subcase III.c.ii.1. $(\bar{d}_2)^s$ is non-minimal w.r.t. $\mathcal{L}$ relative to $K_{b(k)}, \dots, K_{b_u}, \bar{K}_2, \dots, \bar{K}_a$. In this case $\hat{\mathcal{L}}^{\restriction k}(d) = (k; \mathcal{L}'((\bar{d}_2)^s))^s$, where s appears if it appears in $\mathcal{L}((d_2)^s)$ (here we use the notation that if $\mathcal{L}((d_2)^s) = (d')^s$, then $\mathcal{L}'((d_2)^s) =$

d'). Proceeding as above, we have that for almost all $h_1, \dots, h_{k-1}$ the ordinal $\vartheta(h_1, \dots, h_{k-1})$ is such that for almost all $h_k, h_{b(k)}, \dots, h_{b_u}, \bar{h}_2, \dots, \bar{h}_a$,

$$\vartheta(h_1, \dots, h_{k-1})(h_k, h_{b(k)}, \dots, \bar{h}_a) < (\mathrm{id}; h_k; (\bar{d_2})^s; h_{b(k)}, \dots, h_{b_u}, \bar{h}_2, \dots, \bar{h}_a).$$

Hence by $H_{2n-1}(a)$ it follows that for almost all $h_1, \dots, h_{k-1}$, there is $C_k \subseteq \underset{\sim}{\boldsymbol{\delta}}^1_{2n+1}$ such that for almost all $h_k, h_{b(k)}, \dots, h_{b_u}, \bar{h}_2, \dots, \bar{h}_a$,

$$\begin{aligned}\vartheta(h_1, \dots, \bar{h}_a) &< (N_{C_k}; h_k; \mathcal{L}((d_2)^s); h_{b(k)}, \dots, \bar{h}_a)\\ &= H(\hat{\mathcal{L}}^{\restriction k}(d); \hat{\mathcal{L}}^{\restriction k}(d) \to N_{C_k} \circ \hat{\mathcal{L}}^{\restriction k}(d); h_1, \dots, h_t, f, \bar{h}_2, \dots, \bar{h}_a),\end{aligned}$$

which establishes $\bar{H}_{2n+1}(k)$ in this case.

Subcase III.c.ii.2. $(\bar{d_2})^s$ is minimal w.r.t. $\mathcal{L}$ relative to $K_{b(k)}, \dots, K_{b_u}, \bar{K}_2, \dots, \bar{K}_a$. In this case d is minimal w.r.t. $\hat{\mathcal{L}}^{\restriction k}$. We proceed as in the previous case, using $H_{2n-1}(b)$ and a simple partition and sliding argument (partitioning h_k with the extra value α_{k+1} inserted before $\inf h_k$).

Subcase III.c.iii $d = (k; \bar{d_2})$ where s does not appear. Here $\hat{\mathcal{L}}^{\restriction k}(d) = (k; \bar{d_2})^s$. We proceed as in case (b)(ii) above.

This establishes $\bar{H}_{2n+1}(k)$ in all cases.

5.2. H_{2n+1}**.** Now we consider H_{2n+1}. We fix the measure $\mathcal{V} = S^{\ell,m}_{2n+1}$ on $\kappa = \vartheta(S^{\ell,m}_{2n+1})$, and consider the measure $\mu^{<\mathcal{V}}_{2n+3}$ on $\underset{\sim}{\boldsymbol{\delta}}^1_{2n+3}$ We assume $F : \underset{\sim}{\boldsymbol{\delta}}^1_{2n+3} \to \underset{\sim}{\boldsymbol{\delta}}^1_{2n+3}$ is given and for almost all $[f]$ w.r.t. $\mu^{<\mathcal{V}}_{2n+3}$ we have $F([f]) < (\mathrm{id}; (d)^s; f; K_1, \dots, K_t)$, where s may or may not appear, and d is defined and satisfies conditions D and A relative to $K_1, \dots, K_t$. Here $d = d^{\ell,m}_{2n+1}$, and $\mathcal{I}_a = (f(\bar{K}_1);)$.

5.3. H_{2n+1}**(a).** We first consider H_{2n+1}(a). We use the notation of the statement of H_{2n+1} and let $\kappa = \vartheta(\mathcal{V})$ be the domain of the measure $\mathcal{V}$. So we have fixed $F : \underset{\sim}{\boldsymbol{\delta}}^1_{2n+3} \to \underset{\sim}{\boldsymbol{\delta}}^1_{2n+3}$ and a description d or $(d)^s$ in $\mathcal{D}_{2n+1}$ and we are assuming $[F] < (\mathrm{id}; d; K_1, \dots, K_t)$ where the equivalence class $[F]$ refers to the measure $\mu^{\mathcal{V}}_{2n+3}$. We consider the following cases.

Case I. s does not appear. In this case $\mathcal{L}((d)) = (d)^s$ We consider the partition $\mathcal{P}$: We partition functions $f : \kappa \to \underset{\sim}{\boldsymbol{\delta}}^1_{2n+3}$ of the correct type with the extra value $g(\alpha)$ inserted between $\sup_{\beta<\alpha} f(\beta)$ and $f(\alpha)$, with g of uniform cofinality ω, according to whether or not $F([f]) < (N_g; \mathcal{L}((d)); f; K_1, \dots, K_t)$. It follows readily by countable additivity of the measures $K_1, \dots, K_t$, that on the homogeneous side of the partition the statement in $\mathcal{P}$ holds. We let C be a c.u.b. subset of $\underset{\sim}{\boldsymbol{\delta}}^1_{2n+3}$ homogeneous for $\mathcal{P}$. It then follows that for almost all F that $F([f]) < (N_C; \mathcal{L}((d)); f; K_1, \dots, K_t)$.

Case II. s appears. We first claim that there is a c.u.b. $C \subseteq \underset{\sim}{\delta}^1_{2n+3}$ such that for almost all f, $F([f]) < (N_C; \hat{\mathcal{L}}(d); f; K_1, \dots, K_t)$. We have that for almost all f, for almost all $h_1, \dots, h_t$, there is a

$$\vartheta(h_1, \dots, h_t) < h(d; h_1, \dots, h_t)$$

such that $F(f; h_1, \dots, h_t) < f(\vartheta(h_1, \dots, h_t))$. Hence, by $\bar{H}_{2n+1}(1)$, for almost all $h_1, \dots, h_t$ there is a c.u.b. $C \subseteq \underset{\sim}{\delta}^1_{2n+1}$ such that for almost all $h_1, \dots, h_t$,

$$\vartheta(h_1, \dots, h_t) < h(\hat{\mathcal{L}}(d); \hat{\mathcal{L}}(d) \to N_C \circ \hat{\mathcal{L}}(d); h_1, \dots, h_t).$$

We consider the partition $\mathcal{P}$: We partition functions $f: \kappa \to \underset{\sim}{\delta}^1_{2n+3}$ of the correct type with the extra value $g(\alpha)$ inserted between the values $f(\alpha)$ and $f(\alpha+1)$, with g of uniform cofinality ω, according to whether or not $F([f]) < (N_g; \mathcal{L}((d)); f; K_1, \dots, K_t)$. We claim that on the homogeneous side of the partition the property stated in $\mathcal{P}$ holds. If not, we fix a c.u.b. C^1 homogeneous for the contrary side, and let $C^2 \subseteq \underset{\sim}{\delta}^1_{2n+3}$ be such that for $f: \kappa \to C^2$ of the correct type there is a $C \subseteq \underset{\sim}{\delta}^1_{2n+1}$ as above. We let C^3 be contained in the closure points of $C^1 \cup C^2$. We fix $f: \kappa \to C^3$ of the correct type, and fix a c.u.b. $C \subseteq \underset{\sim}{\delta}^1_{2n+1}$ as above. We then get f^2, g satisfying:

(1) $f^2 = f$ almost everywhere w.r.t. $\mathcal{V}$.
(2) f^2, g are of the correct type and ordered as in $\mathcal{P}$.
(3) f^2, g have range in C^3 (in fact have range a subset of the range of f)
(4) If $h: \underset{\sim}{\delta}^1_{2n+1} \to \underset{\sim}{\delta}^1_{2n+1}$ (or $h: <^m \to \omega_1$, if $n = 0$) represents $[h]$ w.r.t. $\mu_{2n+1}^{v(S^{\ell,m}_{2n+1})}$, then $g([h]) > f([N_C \circ h])$.

The construction of f^2, g is similar to that given previously and will be omitted. We then fix $h_1, \dots, h_t$ such that

$$F(f; h_1, \dots, h_t) = F(f^2; h_1, \dots, h_t) < f(\vartheta(h_1, \dots, h_t)),$$

and

$$\vartheta(h_1, \dots, h_t) < h(\hat{\mathcal{L}}(d); \hat{\mathcal{L}}(d) \to N_C \circ \hat{\mathcal{L}}; h_1, \dots, h_t),$$

and

$$\begin{aligned} F(f^2; h_1, \dots, h_t) &> (N_g; \hat{\mathcal{L}}(d); f^2; h_1, \dots, h_t) \\ &= N_g(f^2(h(\hat{\mathcal{L}}(d); h_1, \dots, h_t))) \\ &> f(h(\hat{\mathcal{L}}(d); \hat{\mathcal{L}}(d) \to N_C \circ \hat{\mathcal{L}}(d); h_1, \dots, h_t)). \end{aligned}$$

This contradiction establishes $\mathcal{P}$. If $C \subseteq \underset{\sim}{\delta}^1_{2n+3}$ is homogeneous for $\mathcal{P}$, it follows readily that for almost all f, $F([f]) < (N_C; \tilde{\mathcal{L}}(d); f; K_1, \dots, K_t)$ and we are done with the claim.

We next claim that if a description d' satisfies conditions C and A relative to $K_1, \dots, K_t$, but (d') does not satisfy condition D, and if there is a c.u.b. $C \subseteq \underset{\sim}{\delta}^1_{2n+3}$ such that for almost all f,

$$F([f]) < (N_C; d'; f; K_1, \dots, K_t),$$

then for almost all f,

$$F([f]) < (\mathrm{id}; d'; f; K_1, \ldots, K_t).$$

To see this, consider the partition $\mathcal{P}$ where we partition $f : \kappa \to \underset{\sim}{\delta}^1_{2n+3}$ of the correct according to whether or not $F([f]) < (\mathrm{id}; d'; f; K_1, \ldots, K_t)$. If $\mathcal{P}$ fails, we fix a c.u.b. $C^1 \subseteq \underset{\sim}{\delta}^1_{2n+3}$, where $C^1 \subseteq C$ and C^1 is homogeneous for the contrary side, let $C^2 \subseteq \underset{\sim}{\delta}^1_{2n+3}$ be such that for $f : \kappa \to C^2$ of the correct type $F([f]) < (N_C; d'; f; K_1, \ldots, K_t)$ holds, and let C^3 be contained in the closure points of $C^1 \cup C^2$. We fix $f : \kappa \to C^3$ of the correct type. We fix a measure one set A w.r.t. $S^{\ell,m}_{2n+1}$ such that for almost all $h_1, \ldots, h_t$, $h(d'; h_1, \ldots, h_t) \notin A$ (which we can do as (d') does not satisfy condition D). We let C' be a c.u.b. subset of $\underset{\sim}{\delta}^1_{2n+1}$ such that for $h : \underset{\sim}{\delta}^1_{2n+1} \to C'$ of the correct type (or $h : <^m \to C'$ if $n = 0$)we have $[h] \in A$. We let C'' be contained in the closure points of C'. We then get f^2 satisfying:

(1) For h of the correct type having range in C'' we have $f^2([h]) = f([h])$.
(2) f^2 is of the correct type and has range in C^3 (in fact a subset of the range of f).
(3) For $[h]$ not represented by h as in (1), $f^2([h]) > N_f(f([h]))$, hence $f^2([h]) > N_C(f([h]))$.

The existence of f^2 follows from an easy sliding argument. We then fix $h_1, \ldots, h_t$ such that

$$F(f; h_1, \ldots, h_t) = F(f^2; h_1, \ldots, h_t),$$
$$F(f; h_1, \ldots, h_t) < (N_C; d'; f'; h_1, \ldots, h_t) = N_C(f(h(d'; h_1, \ldots, h_t))),$$

and

$$F(f^2; h_1, \ldots, h_t) > (\mathrm{id}; d'; h_1, \ldots, h_t) = f^2(h(d'; h_1, \ldots, h_t)),$$

and also $h(d'; h_1, \ldots, h_t) \notin A$. From this, the last equation, and (3) above it follows that $F(f^2; h_1, \ldots, h_t) > N_C(f(h(d'; h_1, \ldots, h_t)))$. This contradicts the first two equations and establishes that on the homogeneous side the property stated in $\mathcal{P}$ holds. Hence, for almost all f, $F([f]) < (\mathrm{id}; \hat{\mathcal{L}}(d); f; K_1, \ldots, K_t)$.

If $(\hat{\mathcal{L}}(d))$ satisfies condition D, then our first considerations apply, and we are done. If not, then we are in a position to repeat the argument in (i) to get $C \subseteq \underset{\sim}{\delta}^1_{2n+3}$ s.t. for almost all f, $F([f]) < (N_C; (\tilde{\mathcal{L}}(d))^s; f; K_1, \ldots, K_t)$. If $(\hat{\mathcal{L}}(d))^s$ satisfies condition D, we are done. If not an argument similar to the above establishes that for almost all f, $F([f]) < (\mathrm{id}; (\tilde{\mathcal{L}}(d))^s; f; K_1, \ldots, K_t)$. The proof proceeds by considering the partition $\mathcal{P}$ as above, and constructing f^2 satisfying (1), (2) as above and (3): for $[h]$ not represented by a function of the correct type, $f^2([h]) > f(\beta)$ where β is the least ordinal represented by a function of the correct type which is greater than $[h]$. We are now in a position to repeat the previous argument. Repeating this argument, we eventually get a c.u.b. $C \subseteq \underset{\sim}{\delta}^1_{2n+3}$ such that for almost all f, $F([f]) < (N_C; \mathcal{L}(d); f; K_1, \dots, K_t)$, which establishes $H_{2n+1}(\text{a})$.

5.4. H_{2n+1}**(b).** We consider $H_{2n+1}(\text{b})$. We assume that $(d)^s = (d^{\ell,m}_{2n+1})^s$ is minimal w.r.t. $\mathcal{L}$ relative to $K_1, \dots, K_t$.

Case I. d is minimal w.r.t. $\hat{\mathcal{L}}$ relative to $K_1, \dots, K_t$. We have that for almost all $f, h_1, \dots, h_t$ there is a $\vartheta(h_1, \dots, h_t) < h(d; h_1, \dots, h_t)$ such that

$$F(f; h_1, \dots, h_t) < f(\vartheta(h_1, \dots, h_t)).$$

By $\bar{H}_{2n+1}(1)$, for almost all f there is $\alpha < \kappa$ such that for almost all $h_1, \dots, h_t$, $\vartheta(h_1, \dots, h_t) < \alpha$. We consider the partition $\mathcal{P}$ where we partition $f : \kappa \to \underset{\sim}{\delta}^1_{2n+3}$ of the correct with the extra value γ (of cofinality ω) inserted before $\inf f$ according to whether or not $F([f]) < \gamma$. If $\mathcal{P}$ fails, we let C^1 be homogeneous for the contrary side, let C^2 be such that for $f : \kappa \to C^2$ of the correct type $F(f; h_1, \dots, h_t) < f(\alpha)$ almost everywhere (where $\alpha = \alpha(f) < \kappa$ is as above) and let $C^3 = C^1 \cap C^2$. We fix $f : \kappa \to C^3$ of the correct type, let $\alpha < \kappa$ be as above, and get γ, f^2 satisfying:
(1) $f^2 = f$ a.e. w.r.t. $\mathcal{V}$.
(2) $\gamma < \inf f^2$, $\operatorname{cf} \gamma = \omega$, and f^2 is of the correct type.
(3) γ, f^2 have range in C^3.
(4) $\gamma > f(\alpha)$.

Then for almost all $h_1, \dots, h_t$ we have

$$F(f'h_1, \dots, h_t) = F(f^2; h_1, \dots, h_t),$$
$$F(f; h_1, \dots, h_t) < f(\alpha),$$

and

$$F(f^2; h_1, \dots, h_t) > \gamma > f(\alpha).$$

This contradiction establishes $\mathcal{P}$. It follows readily that $H_{2n+1}(\text{b})$ is satisfied.

Case II. d is not minimal w.r.t. $\hat{\mathcal{L}}$ relative to $K_1, \dots, K_t$. It follows as in the proof of $H_{2n+1}(\text{a})$ that for almost all $f, h_1, \dots, h_t$,

$$F(f; h_1, \dots, h_t) < (\text{id}; (\bar{\mathcal{L}}(d))^s; f; h_1, \dots, h_t),$$

and in fact

$$F(f; h_1, \dots, h_t) < (\text{id}; (\bar{\mathcal{L}}^p(d))^s; f; h_1, \dots, h_t),$$

where $(\hat{\mathcal{L}}^p(d))$ denotes the pth iterate of $\hat{\mathcal{L}}$. We elect p minimal such that $\hat{\mathcal{L}}^{p+1}(d)$ is not defined, and then proceed as in (i). This establishes $H_{2n+1}(\text{b})$.

5.5. H_{2n+1}**(c).** We consider $H_{2n+1}(\mathrm{c})$. We assume that F is given s.t. for almost all $f : \kappa \to \underset{\sim}{\delta}^1_{2n+3}$ of the correct type, for almost all $h_1, \dots, h_t$ we have $F([f]) < \sup f$, and the maximal tuple $\tilde{d} = \tilde{d}^{\ell,m}_{2n+1}$ is defined. For almost all $f, h_1, \dots, h_t$, there is an $\alpha < \kappa$ such that $F(f; h_1, \dots, h_t) < f(\alpha)$. It follows readily that for almost all $f, h_1, \dots, h_t$, there is a c.u.b. $C_\infty \subseteq \underset{\sim}{\delta}^1_{2n+1}$ such that

$$\alpha(f; h_1, \dots, h_t) < h(\tilde{d}^\infty; \tilde{d}^\infty \to N_C \circ \tilde{d}^\infty; h_1, \dots, h_t),$$

where $\tilde{d}^\infty$ is the basic type 1 description () with index $\mathcal{I}_a = (f(\bar{K}_1);)$ for $n > 0$ (for $n = 0$, $\tilde{d}^\infty = (m)$, a basic type 0 description). This follows from an easy partition argument (partitioning $f : \kappa \to \underset{\sim}{\delta}^1_{2n+3}$ with the extra value γ after $\sup f$). We recall here that $h(\tilde{d}^\infty; h_1, \dots, h_t)$ is represented by the function $[f] \mapsto H(\tilde{d}^\infty; h_1, \dots, h_t, f) = \sup_{\text{a.e.}} f$. We then proceed as in the proof of $\bar{H}_{2n+1}(k)$ to establish that for $1 \leq k \leq t$ the following holds: for almost all $f, h_1, \dots, h_{k-1}$, there is a c.u.b. $C_k \subseteq \underset{\sim}{\delta}^1_{2n+1}$ such that for almost all $h_1, \dots, h_t$

$$\alpha(h_1, \dots, h_t) < h(\tilde{d}^k; \tilde{d}^k \to N_C \circ \tilde{d}^k; h_1, \dots, h_t),$$

where $\tilde{d}^k$ is as in the definition of the maximal tuple $\tilde{d}$ (which is equal to $\tilde{d}^1$). We then proceed as in $H_{2n+1}(\mathrm{a})$ to establish that there is a c.u.b. $C \subseteq \underset{\sim}{\delta}^1_{2n+3}$ such that for almost all $f, h_1, \dots, h_t$,

$$F(f; h_1, \dots, h_t) < (N_C; (\tilde{d}^{\ell,m}_{2n+1})^s; f; h_1, \dots, h_t),$$

where s appears if $(\tilde{d}^{\ell,m}_{2n+1})$ does not satisfy condition D. This establishes $H_{2n+1}(\mathrm{c})$.

5.6. H_{2n+1}**(d).** We consider $H_{2n+1}(\mathrm{d})$. We again assume that for almost all $f, h_1, \dots, h_t$,

$$F(f; h_1, \dots, h_t) < \sup_{\text{a.e.}} f,$$

but that the maximal description $\tilde{d}^{\ell,m}_{2n+1}$ is not defined. In this case, the integer e_1 as in the definition of $\tilde{d}^{\ell,m}_{2n+1}$ does not exist (or b_1 if $\ell = 1$). As above, we have that for almost all $f, h_1, \dots, h_{k-1}$, there is a c.u.b. $C_\infty \subseteq \underset{\sim}{\delta}^1_{2n+1}$ such that for almost all $h_k, \dots, h_t$ we have

$$\alpha(f, h_1, \dots, h_t) < h(\tilde{d}^\infty \to N_C \circ \tilde{d}^\infty; h_1, \dots, h_t)$$

and $F(f; h_1, \dots, h_t) < f(\alpha)$ (where $\alpha = \alpha(f)$ is as before). Here $\tilde{d}^\infty = ()$, a basic type 1 description (for $n = 0$, $\tilde{d}^\infty = (m)$, a basic type 0 description). We the proceed as in $\bar{H}_{2n+1}(k)$, and since e_1 does not exist we have that for almost all f, there is a c.u.b. $C_1 \subseteq \underset{\sim}{\delta}^1_{2n+1}$ such that for almost all $h_1, \dots, h_t$,

$$\alpha(f; h_1, \dots, h_t) < h(\tilde{d}^\infty; \tilde{d}^\infty \to N_{C_1} \circ \tilde{d}^\infty; h_1, \dots, h_t).$$

We then proceed as in $H_{2n+1}(\mathrm{b})$ to establish $H_{2n+1}(\mathrm{d})$. This establishes H_{2n+1} in all cases.

§6. The Main Theorem. We define an ordering $<_r$ on tuples

$$((d)^s; K_1, \dots, K_t),$$

where s may or may not appear, and $(d)^s$ satisfies conditions D and A relative to $K_1, \dots, K_t$ (or (d) if s does not appear). We let $<_r$ be the transitive relation generated by the following relations:

(1) $((d_1)^s; K_1, \dots, K_t) <_r ((d_2)^s; K_1, \dots, K_t)$ where $(d_1)^s < (d_2)^s$, where s may not appear in d_1 or d_2. Here $(d_1)^s < (d_2)^s$ if $d_2 < d_2$ or $d_1 = d_2$ and d_1 involves s and d_2 does not (*i.e.*, we set $(d_1)^s < (d_1)$).

(2) $(\mathcal{L}((d)^s; K_1, \dots, K_t); K_1, \dots, K_t, K_{t+1}) <_r ((d)^s; K_1, \dots, K_t)$ for all $K_{t+1} \in \bigcup_{m \leq n} R_{2m+1}$

It is easy to see that $<_r$ is well-founded. To be specific, suppose

$$((d_1)^2; K_1, \dots, K_{t_1}) >_r ((d_2)^s; K_1, \dots, K_{t_2}) >_r \dots,$$

where $t_1 \leq t_2 \leq \cdots$, s may or may not not appear in any tuple, and $d_i \in D^{\ell,m}_{2n+1}$ for all i. Then for any p we have that for almost all $f, h_1, \dots, h_{t_p}$ (where $f : \vartheta(S^{\ell,m}_{2n+1}) \to \underset{\sim}{\delta}^1_{2n+3}$) we have that

$$(\text{id}; (d_1)^s; f; h_1, \dots, h_{t_1}) > \cdots > (\text{id}; (d_2)^s; f; h_1, \dots, h_{t_p}).$$

Letting p become arbitrarily large, we define an infinite decreasing sequence of ordinals. Hence $<_r$ is well defined.

Definition 6.1. We let $f((d)^s; K_1, \dots, K_t)$ be the rank of $((d)^s; K_1, \dots, K_t)$ w.r.t. $<_r$ computed according to

$$f((d)^s; K_1, \dots, K_t) = \left(\sup_{\substack{((\bar{d})^s; \bar{K}_1, \dots, \bar{K}_{\bar{t}}) <_r \\ ((d)^s; K_1, \dots, K_t)}} f((\bar{d})^s; \bar{K}_1, \dots, \bar{K}_{\bar{t}}) \right) + 1.$$

Remark 6.2. The rank function of Definition 6.1 is computed in a slightly non-standard manner so that all ranks are successor ordinals.

We introduce the following notation:

Definition 6.3. For any ordinal α we let $E(0, \alpha) = \alpha$ and $E(n+1, \alpha) = \omega^{E(n,\alpha)}$. Also, we let $E(n) = E(n, 1)$.

We now state the main theorem of this section:

Theorem 6.4. Let $(d)^s$, where s may or may not appear, satisfy conditions D and A relative to $K_1, \dots, K_t$, where $d = d^{\ell,m}_{2n+1}$. Let $\mathcal{V} = S^{\ell,m}_{2n+1}$. Let $F : \underset{\sim}{\delta}^1_{2n+3} \to \underset{\sim}{\delta}^1_{2n+3}$ be defined w.r.t. the measure $\mu^{<\mathcal{V}}_{2n+3}$ on $\underset{\sim}{\delta}^1_{2n+3}$ by $F([f]) = (\text{id}; (d)^s; K_1, \dots, K_t)$, for almost all $f : \kappa(S^{\ell,m}_{2n+1}) \to \underset{\sim}{\delta}^1_{2n+3}$. Then in the ultrapower of $\underset{\sim}{\delta}^1_{2n+3}$ by the measure $\mu^{<\mathcal{V}}_{2n+3}$, F represents an ordinal $\leq \aleph_{E(2n+1)+f((d)^s; K_1, \dots, K_t)}$.

Proof. First consider the case $f((d)^s; K_1, \dots, K_t) = 1$. In this case $(d)^s$ is minimal w.r.t. $\mathcal{L}$ relative to $K_1, \dots, K_t$. By H_{2n+1}(b), if $\vartheta < (\text{id}; (d)^s; K_1, \dots, K_t)$ then $\alpha < j_{K_1} \circ \cdots \circ j_{K_t}(\delta)$ for some $\delta < \underset{\sim}{\delta}^1_{2n+3}$. Using our inductive hypothesis I_{2n+1} we have that $\alpha < \underset{\sim}{\delta}^1_{2n+3}$ and so $(\text{id}; (d)^s; K_1, \dots, K_t) \le \underset{\sim}{\delta}^1_{2n+3} = \aleph_{E(2n+1)+1}$.

Next assume $f((d)^s; K_1, \dots, K_t) > 1$. We may assume that $(d)^s$ is non-minimal w.r.t. $\mathcal{L}$. We then have from the main inductive hypothesis (Lemma 5.25) that if

$$F([f]) < (\text{id}; (d)^s; f; K_1, \dots, K_t)$$

for almost all f, then there is a $g \colon \underset{\sim}{\delta}^1_{2n+3} \to \underset{\sim}{\delta}^1_{2n+3}$ such that for almost all f we have

$$F([f]) < (g; \mathcal{L}((d)^s); f; K_1, \dots, K_t).$$

By K_{2n+3}, there is a tree T on $\underset{\sim}{\delta}^1_{2n+3}$ and a c.u.b. $C \subseteq \underset{\sim}{\delta}^1_{2n+3}$ such that for $\alpha \in C$, $g(\alpha) < |T \restriction \sup_j j(\alpha)|$, the supremum ranging over embeddings j from measures in $\bigcup_{m \le n} R_{2m+1}$. We fix such a T and C. Hence for almost all f we have that

$$\begin{aligned} F([f]) &< (|T \restriction \sup_j j|; \mathcal{L}((d)^s); K_1, \dots, K_t) \\ &< (\sup_j j; \mathcal{L}((d)^s); K_1, \dots, K_t)^+. \end{aligned}$$

Here "$\sup_j j$" stands for the function $\alpha \mapsto \sup_j j(\alpha)$ where again the supremum ranges over embeddings j from measures in $\bigcup_{m \le n} R_{2m+1}$. To prove the second inequality, we define an ordering $\lhd$ on the ordinal $(\sup_j j; \mathcal{L}((d)^s); K_1, \dots, K_t)$ as follows: We set $\vartheta_1 \lhd \vartheta_2$ if for almost all $f, h_1, \dots, h_t$,

$$\begin{aligned} &|T \restriction (\sup_j j; \mathcal{L}((d)^s); f, h_1, \dots, h_t)(\vartheta_1(f, h_1, \dots, h_t))| \\ &\qquad < |T \restriction (\sup_j j; \mathcal{L}((d)^s); f, h_1, \dots, h_t)(\vartheta_2(f, h_1, \dots, h_t))|, \end{aligned}$$

where $|T \restriction \alpha(\beta)|$ denotes the rank of β in the tree $T \restriction \alpha$ (we are identifying finite tuples of ordinals with ordinals here). If now

$$\eta < (|T \restriction \sup_j j|; \mathcal{L}((d)^s); K_1, \dots, K_t),$$

then there is a

$$\vartheta = \vartheta(\eta) < (\sup_j j; \mathcal{L}((d)^s); K_1, \dots, K_t)$$

such that for almost all $f, h_1, \dots, h_t$,

$$\eta(f, h_1, \dots, h_t) = |T \restriction (\sup_j j; \mathcal{L}((d)^s); f, h_1, \dots, h_t)(\vartheta(f, h_1, \dots, h_t))|.$$

The map $\eta \mapsto \vartheta(\eta)$ is order-preserving from the ordinal

$$((|T \restriction \sup_j j|; \mathcal{L}((d)^s); K_1, \ldots, K_t)$$

with the usual ordering to the ordinal

$$((\sup_j j; \mathcal{L}((d)^s); K_1, \ldots, K_t)$$

with the ordering $\lhd$. This establishes the above inequalities. It follows immediately that

$$(\mathrm{id}; (d)^s; K_1, \ldots, K_t) \leq (\sup_j j; \mathcal{L}((d)^s); K_1, \ldots, K_t)^+.$$

By countable additivity,

$$(\sup_j j; \mathcal{L}((d)^s); f; K_1, \ldots, K_t) = \sup_j (j; \mathcal{L}((d)^s); K_1, \ldots, K_t),$$

where again the supremum on j ranges over the embeddings corresponding to measures $\bigcup_{m \leq n} R_{2m+1}$. For K_{t+1} in $\bigcup_{m \leq n} R_{2m+1}$, it follows immediately from the definitions that

$$(j_{K_{t+1}}; \mathcal{L}((d)^s); f; K_1, \ldots, K_t) = (\mathrm{id}; \mathcal{L}((d)^s); K_1, \ldots, K_t, K_{t+1}),$$

where $j_{K_{t+1}}$ denotes the embedding from the measure K_{t+1}. Putting this together we have

$$\begin{aligned}(\mathrm{id}; (d)^s; K_1, \ldots, K_t) &\leq [\sup_{K_{t+1}}(\mathrm{id}; \mathcal{L}((d)^s); K_1, \ldots, K_t, K_{t+1})]^+ \\ &\leq \left[\aleph_{E(2n+1)+\sup_{K_{t+1}} f(\mathcal{L}((d)^s); K_1, \ldots, K_t, K_{t+1})}\right]^+ \\ &= \aleph_{E(2n+1)+f((d)^s; K_1, \ldots, K_{t+1})},\end{aligned}$$

using induction for the second inequality and the definition of the rank function for the third. This completes the proof of the theorem. ⊣

§7. A Rank Computation. The main goal of this section is to compute the bound

$$\sup_{d, K_1, \ldots, K_t} f(d; K_1, \ldots, K_t) \leq E(2n+3)$$

We recall (this notation was introduced right after Definition 5.23) that for all $d^{(\mathcal{I}_a)} = d^{(f(\bar{K});)}$ satisfying conditions C and A relative to $K_1, \ldots, K_t$, for all the component tuples $d_2^{f(\bar{K}_1; \bar{K}_2, \ldots, \bar{K}_a)}$ of d we have that the map $\delta_{d_2} : a \to \{1, \ldots, t\}$ is defined. Here $\delta_{d_2}(i)$ is an integer j associated to $\bar{K}_i$ such that $v(K_j) = \bar{K}_i$. It is easy to see that δ_d satisfies the following:

(1) δ_d is strictly increasing.
(2) $v(K_{\delta_d(i)}) = \bar{K}_i$.

(3) for all component tuples $d'^{(\mathcal{I}'_a)}$ of $d^{(\mathcal{I}_a)}$ (where $\mathcal{I}'_a$ extends $\mathcal{I}_a$), $\delta_{d'}$ extends δ_d.
(4) $\delta_d(\bar{K}_a) < k(d)$.

DEFINITION 7.1. We define $<^k_p$ on the set of descriptions $d^{(\mathcal{I}_a)}$ satisfying conditions C and A relative to $K_1, \dots, K_t$ and satisfying $\delta_d(a) < k \leq k(d)$, (where $k \geq 1$ if $\delta_d(a)$ does not exist) to be the ordering generated by the relations:

(1) $(d_1; K_1, \dots, K_t) <^k_p (d_2; K_1, \dots, K_t)$ for $d_1 < d_2$.
(2) $(\hat{\mathcal{L}}^{\restriction k}(d_1; K_1, \dots, K_t); K_1, \dots, K_{t-p}, K_{t+1}, \dots, K_t) \;\; <^k_p \;\; (d_1; K_1, \dots, K_t)$ for all K_{t+1},

$$(\tilde{\mathcal{L}}^{\restriction k+b}_p(d_1; K_1, \dots, K_{t-p}, K_{t+1}, \dots, K_t); K_1, \dots, K_{t-p}, K_{t+1}, K_{t+2}, \dots, K_t) < (d_1; K_1, \dots, K_{t-p}, K_{t+1}, \dots, K_t)$$

for all K_{t+1}, K_{t+2} (where $b = 0$ if $k \leq t - p$ and $b = 1$ if $k > t - p$), etc.

We also define $<^0_p$ to be the ordering generated by the relations:

(1) $(d_1; K_1, \dots, K_t) <^0_p (d_2; K_1, \dots, K_t)$ for $d_1 < d_2$.
(2) $(\tilde{\mathcal{L}}^{\restriction 1}(d_1); K_1, \dots, K_{t-p}, K_{t+1}, \dots, K_t) <^0_p (d_1; K_1, \dots, K_t)$ for all K_{t+1},

$$(\tilde{\mathcal{L}}^{\restriction 1}(d_1); K_1, \dots, K_{t-p}, K_{t+1}, K_{t+2}, \dots, K_t) <^0_p (d_1; K_1, \dots, K_{t-p}, K_{t+1}, \dots, K_t)$$

for all K_{t+1}, K_{t+2}, etc. Here $t+1, t+2$, etc. precede K_1 if $t = p$.

We let $f^k_p(d; K_1, \dots, K_t)$, for $\delta_d(a) < k < k(d)$, be the rank of the tuple $(d; K_1, \dots, K_t)$ with respect to $<^k_p$, computed in the slightly non-standard manner as in the case of $<_r$ (so all ranks are successor ordinals).

We will also define below an auxiliary function $\bar{f}^k_p(d; K_1, \dots, K_t)$ for such d, k, and we will have $E(2n-1) < \bar{f}^k_p(d; K_1, \dots, K_t) < E(2n+3)$. The function $\bar{f}^k_p$ will be an "approximation" to the true rank function f^k_p, but which is more amenable to computation.

We assume (inductively on n) that the corresponding functions, which we denote by $g^k_p, \bar{g}^k_p$ have been defined for $d \in \mathcal{D}_{2n-1}$.

We introduce the following hypotheses concerning $\bar{f}^k_p$:

A^k_p: for $d^{(\mathcal{I}_a)}$ defined and satisfying conditions C and A relative to $K_1, \dots, K_t$ and $\delta_d(a) < k \leq k(d)$ (where $\mathcal{I}_a = (f(\bar{K}_1), \bar{K}_2, \dots, \bar{K}_a)$ is the index for d),

$$\bar{f}^k_p(d; K_1, \dots, K_t) = \bar{f}^{k+b}_p(d; K_1, \dots, K_{t-p}, K_{t+1}, \dots, K_t),$$

for all K_{t+1}, where $b = 0$ if $k \leq t - p$ and $b = 1$ if $k > t - p$. Also, if $p = t$ and $k = 0$, then $\bar{f}^0_t(d_1; K_1, \dots, K_t) = \bar{f}^0_t(d_1; K_{k+1}, K_1, \dots, K_t)$.

B^k_p: let $d^{(\mathcal{I}_a)}$, k be as above, and assume d' is obtained by reindexing d. That is, if $\mathcal{I}_a = (f(\bar{K}_1), \bar{K}_2, \dots, \bar{K}_a)$ is the index of d and $\mathcal{I}'_a = (f(\bar{K}'_1), \bar{K}'_2, \dots, \bar{K}'_{a'})$ is the index of d', where $\delta_{d'}(K'_a) < k(d') = k(d)$, then $\bar{K}_2, \dots, \bar{K}_a$ is a subsequence of $K'_1, \dots, K'_{a'}$, and $d = d'$ except that for each component tuple $d_i^{(\mathcal{I}_{a_i})}$ of d with index

$$\mathcal{I}_{a_i} = (f(\bar{K}_1), \bar{K}_2, \dots, \bar{K}_a, \bar{K}_{a+1}, \dots, \bar{K}_{a_i}),$$

the corresponding component tuple $d_i'^{(\mathcal{I}'_{a_i})}$ of d' has index of the form

$$\mathcal{I}'_{a_i} = (f(\bar{K}_1), \bar{K}'_2, \dots, \bar{K}'_a, \bar{K}_{a+1}, \dots, \bar{K}_{a_i}).$$

We then require that $\bar{f}^k_p(d'; K_1, \dots, K_t) = \bar{f}^k_p(d; K_1, \dots, K_t)$.

R^k_p: suppose $d_1^{(\mathcal{I}_a)}$, $d_2^{(\mathcal{I}_a)}$ are defined and satisfy conditions C and A relative to $K_1, \dots, K_t$, where $\delta_{d_1}(a) < k \le k(d_1)$ and similarly for d_2, and $d_2 < d_1$. Then

$$\bar{f}^{k+b}_p(d_2; K_1, \dots, K_{t-p}, K_{t+1}, \dots, K_t) < \bar{f}^k_p(d_1; K_1, \dots, K_t),$$

where $b = 0$ if $k \le t - p$, and $b = 1$ if $k > t - p$ or $k = 0$.

R_p: if $d \in \mathcal{D}^{\ell,m}_{2n+1}$, then $\bar{f}^k_p(d; K_1, \dots, K_t) < E(2n, m_k)$.

We assume inductively that A^k_p, B^k_p, R^k_p, R_p are satisfied for $d \in \mathcal{D}_{2n-1}$, and the corresponding function $\bar{g}^k_p$ on $\mathcal{D}_{2n-1}$.

Once we have defined the functions $\bar{f}^k_p$, the following definition will define the functions $\bar{f}_p$.

DEFINITION 7.2. Let $d^{(\mathcal{I}_a)}$ satisfy conditions C and A with respect to the sequence $K_1, \dots, K_t$. We set $\bar{f}_p = \bar{f}^1_p$ (and similarly $\bar{g}_p = \bar{g}^1_p$ on $\mathcal{D}_{2n-1}$).

We also introduce the following notation.

DEFINITION 7.3. We let $\mathcal{M}(K_1, \dots, K_t)$ denote the sequence of measures $v(K_i)$ corresponding to those $K_i = S^{\ell_i, m_i}_{2n+1}$ for $\ell_i > 1$.

Assuming $\bar{f}^k_p$ defined we will extend it slightly to a function $\bar{\bar{f}}^k_p$ defined on objects of the form (d) or $(d)^s$, where d satisfies conditions C and A relative to $K_1, \dots, K_t$ (but not necessarily condition D). We do this as follows: For s not appearing, we set

$$\bar{\bar{f}}^k_p((d); K_1, \dots, K_t) = 2 \cdot \bar{f}^k_p(d; K_1, \dots, K_t) + 1,$$

and for s appearing we set

$$\bar{\bar{f}}^k_p((d)^s; K_1, \dots, K_t) = 2 \cdot \bar{f}^k_p(d; K_1, \dots, K_t).$$

If we consider the corresponding versions of A^k_p, B^k_p, R^k_p, for $\bar{\bar{f}}^k_p$, it is immediate that they are satisfied provided they are satisfied for $\bar{f}^k_p$.

It will be convenient for the definition of the $\bar{f}_p^k$ to introduce the following ordinal function.

DEFINITION 7.4. For $m \geq 1$ and $\beta \in \text{Ord}$, the ordinal $\tau_m(\beta)$ is defined (recursively on m) by $\tau_1(\beta) = 2 \cdot \beta$ and

$$\tau_m(\beta) = (m + \tau_{m-1}(\beta) + m) \cdot \beta + (m + \tau_{m-1}(\beta)).$$

We have the following simple lemma.

LEMMA 7.5. For all $m \geq 1$ and $\beta \geq \omega$ we have $\tau_m(\beta) \leq \beta^m \cdot 2$. Furthermore, if $1 \leq k < \ell$ and $\beta \geq \omega$ then

$$\begin{aligned}((\ell+1) + \tau_\ell(\beta) + (\ell+1)) \cdot \beta + (\ell + \tau_{\ell-1}(\beta) + \ell) \cdot \beta + \cdots \\ + ((k+2) + \tau_{k+1}(\beta) + (k+2)) \cdot \beta \\ + ((k+1) + \tau_k(\beta) + (k+1)) \cdot (\beta+1) \\ \leq \tau_{\ell+1}(\beta) + k + 1.\end{aligned}$$

PROOF. Expanding the recursive definition of $\tau_{\ell+1}(\beta)$ and stopping at the $\tau_k(\beta)$ terms immediately gives the result. ⊣

We now proceed to define $\bar{f}_p^k$. We proceed by reverse induction on k.

Case I. $k = \infty$. Hence d basic type 1 or 0.

Subcase I.a. d basic of type 1, so $d^{(\mathcal{I}_a)} = (\bar{d}_2)^{s(f;\bar{K}_2,\dots,\bar{K}_a)}$ where s may or may not appear, $\bar{d}_2 \in \mathcal{D}_{2n-1}$, and $(\bar{d}_2)^s$ satisfies conditions D and A relative to $\bar{K}_2, \dots, \bar{K}_a$ (where s appears here iff s appears in d).

Subcase I.a.i. If s appears we set

$$\bar{f}_p^\infty(d; K_1, \dots, K_t) = E(2n-1) + 2 \cdot \bar{g}_q(\bar{d}_2; \mathcal{M}(K_1, \dots, K_t)) + 1.$$

Subcase I.a.ii. If s does not appear

$$\bar{f}_p^\infty(d; K_1, \dots, K_t) = E(2n-1) + 2 \cdot \bar{g}_q(\bar{d}_2; \mathcal{M}(K_1, \dots, K_t)) + 2,$$

where q is minimal such that $\delta(\tilde{K}_{w-q}) \leq t - p$, where $\tilde{K}_1, \dots, \tilde{K}_w$ enumerates $\mathcal{M}(K_1, \dots, K_t)$ and $\delta(v(K_i)) = i$ for $v(K_i)$ in the sequence $\mathcal{M}(K_1, \dots, K_t)$.

Subcase I.b. d basic type 0, so $d^{(\mathcal{I}_a)} = (k)^{(f_m;\bar{K}_2,\dots,\bar{K}_a)}$, where $k \leq m$. We set

$$\bar{f}_p^\infty(d; K_1, \dots, K_t) = E(2n-1) + m$$

Case II. $k < \infty$, $k < k(d)$, and $k \neq t - p$.

Subcase II.a. $K_k = S_{2n+1}^{\ell_k, m_k}$, with $\ell_k > 1$. We set

$$\bar{f}_p^k(d; K_1, \dots, K_t) = \sum_{\beta < f_p^{k+1}(d; K_1, \dots, K_t)} 2 \cdot (\beta + 1) + 1.$$

Subcase II.b. $K_k = S^{1,m_k}_{2n+1}$.

$$\bar{f}^k_p(d;K_1,\dots,K_t) = \sum_{\beta<\bar{f}^{k+1}_p(d;K_1,\dots,K_t)} (m_k + \tau_{m_k-1}(\beta) + m_k) + 1.$$

Subcase II.c. $K_k = W^{m_k}_{2n+1}$.

$$\bar{f}^k_p(d;K_1,\dots,K_t) = E(2n,m_k) + \bar{f}^{k+1}_p(d;K_1,\dots,K_t).$$

Subcase II.d. K_k is not of these forms.

$$\bar{f}^k_p(d;K_1,\dots,K_t) = \bar{f}^{k+1}_p(d;K_1,\dots,K_t).$$

Subcase II.e. If $k = 0$, we set

$$\bar{f}^k_p(d;K_1,\dots,K_t) = \bar{f}^{k+1}_p(d;K_1,\dots,K_t).$$

Case III. $k < \infty$, $k < k(d)$ and $k = t - p$.

Subcase III.a. $K_k = S^{\ell_k,m_k}_{2n+1}$, with $\ell_k > 1$. We set

$$\bar{f}^k_p(d;K_1,\dots,K_t) = \sum_{\beta<\omega^{\omega^{\bar{f}^{k+1}_p(d;K_1,\dots,K_t)}}} (\beta+1).$$

Subcase III.b. $K_k = S^{1,m_k}_{2n+1}$.

$$\bar{f}^k_p(d;K_1,\dots,K_t) = \sum_{\beta<\omega^{\omega^{\bar{f}^{k+1}_p(d;K_1,\dots,K_t)}}} (m_k + \tau_{m_k-1}(\beta) + m_k).$$

Subcase III.c. $K_k = W^{m_k}_{2n+1}$.

$$\bar{f}^k_p(d;K_1,\dots,K_t) = E(2n,m_k) + \omega^{\omega^{\bar{f}^{k+1}_p(d;K_1,\dots,K_t)}}.$$

Subcase III.d. K_k is not of these forms.

$$\bar{f}^k_p(d;K_1,\dots,K_t) = \omega^{\omega^{\bar{f}^{k+1}_p(d;K_1,\dots,K_t)}}.$$

Subcase III.e. $k = 0$, we set $\bar{f}^k_p(d;K_1,\dots,K_t) = \omega^{\omega^{\bar{f}^{k+1}_p(d;K_1,\dots,K_t)}}$.

Case IV. $k < \infty$, $k = k(d)$, and $k \neq t - p$.

Subcase IV.a. $K_k = S^{\ell_k,m_k}_{2n+1}$, with $\ell_k > 1$. Hence $d^{(\mathcal{I}_a)} = (k; d_2^{(\mathcal{I}_a+1)})^{s(\mathcal{I}_a)}$, where s may or may not appear. We set

$$\bar{f}^k_p(d;K_1,\dots,K_t) = \sum_{\beta<\bar{f}^{k+1}_p(\hat{d};K_1,\dots,K_t)} 2\cdot(\beta+1) + 2\cdot\bar{f}^{k+1}_p(d_2;K_1,\dots,K_t)$$
$$+\ (1 \text{ if } s \text{ appears, and } 2 \text{ if } s \text{ does not appear}).$$

Here $\hat{d}$ is as in condition A.

Subcase IV.b. $K_k = S^{1,m_k}_{2n+1}$. Here $d^{(\mathcal{I}_a)} = (k; d_1^{(\mathcal{I}_a)}, \ldots, d_r^{(\mathcal{I}_a)})^{s(\mathcal{I}_a)}$, where $r \le m_k$ and s may or may not appear.

Subcase IV.b.i. $r = m_k$.

$$\begin{aligned}
\bar{f}^k_p(d; K_1, \ldots, K_t) = &\sum_{\beta < \bar{f}^{k+1}_p(d_1; K_1, \ldots, K_t)} (m_k + \tau_{m_k - 1}(\beta) + m_k) \\
&+ ((m_k - 1) + \tau_{m_k-2}(\bar{f}^{k+1}_p(d_1; K_1, \ldots, K_t)) \\
&\qquad + (m_k - 1)) \cdot \bar{f}^{k+1}_p(d_2; K_1, \ldots, K_t) \\
&+ \cdots \\
&+ (2 + \tau_1(\bar{f}^{k+1}_p(d_1; K_1, \ldots, K_t)) + 2) \cdot \bar{f}^{k+1}_p(d_{r-1}; K_1, \ldots, K_t) \\
&+ \tau_1(\bar{f}^{k+1}_p(d_r; K_1, \ldots, K_t)) + (1 \text{ or } 2)
\end{aligned}$$

depending on whether s appears or not.

Subcase IV.b.ii. $r < m_k$ and s appears.

$$\begin{aligned}
\bar{f}^k_p(d; K_1, \ldots, K_t) = &\sum_{\beta < \bar{f}^{k+1}_p(d_1; K_1, \ldots, K_t)} (m_k + \tau_{m_k - 1}(\beta) + m_k) \\
&+ ((m_k - 1) + \tau_{m_k-2}(\bar{f}^{k+1}_p(d_1; K_1, \ldots, K_t)) \\
&\qquad + (m_k - 1)) \cdot \bar{f}^{k+1}_p(d_2; K_1, \ldots, K_t) \\
&+ \cdots \\
&+ ((m_k - r + 1) + \tau_{m_k-r}(\bar{f}^{k+1}_p(d_1; K_1, \ldots, K_t)) \\
&\qquad + (m_k - r + 1)) \cdot \bar{f}^{k+1}_p(d_r; K_1, \ldots, K_t) + 1
\end{aligned}$$

Subcase IV.b.iii. $r < m_k$ and s does not appear.

$$\begin{aligned}
\bar{f}^k_p(d; K_1, \ldots, K_t) = &\sum_{\beta < \bar{f}^{k+1}_p(d_1; K_1, \ldots, K_t)} (m_k + \tau_{m_k - 1}(\beta) + m_k) \\
&+ ((m_k - 1) + \tau_{m_k-2}(\bar{f}^{k+1}_p(d_1; K_1, \ldots, K_t)) \\
&\qquad + (m_k - 1)) \cdot \bar{f}^{k+1}_p(d_2; K_1, \ldots, K_t) \\
&+ \cdots + ((m_k - r + 1) + \tau_{m_k-r}(\bar{f}^{k+1}_p(d_1; K_1, \ldots, K_t)) \\
&\qquad + (m_k - r + 1)) \cdot \bar{f}^{k+1}_p(d_r; K_1, \ldots, K_t) + 1 \\
&+ \tau_{m_k-r}(\bar{f}^{k+1}_p(d_1; K_1, \ldots, K_t)) + (m_k - r + 1)
\end{aligned}$$

Subcase IV.c. $K_k = W^{m_k}_{2n+1}$

Subcase IV.c.i. $d = (k;)$. We set $\bar{f}^k_p(d; K_1, \ldots, K_t) = E(2n, m_2)$.

Subcase IV.c.ii. $d = (k; \bar{d_2})^s$, where s may or may not appear. If $n > 0$ we set

$$\bar{f}^k_p(d; K_1, \dots, K_t) = E(2n-1) + 2 \cdot (\bar{g}_q(\bar{d_2}; \tilde{K}_1, \dots, \tilde{K}_w) + (1 \text{ or } 2)$$

depending on whether s appears or not. Here $\tilde{K}_1, \dots, \tilde{K}_w$ enumerates the measures in the sequence $K_{b(k)}, \dots, K_{b_u}$ concatenated with the sequence $\mathcal{M}(K_1, \dots, K_{k-1})$. Recall that $K_{b(k)}, \dots, K_{b_u}$ enumerates the subsequence of $K_{k+1}, \dots, K_t$ consisting of the measures in $\bigcup_{m<n} R_{2m+1}$ and $\mathcal{M}(K_1, \dots, K_{k-1})$ denotes the sequence of measures $v(K_i)$ for $i \leq k-1$ with K_i of the form S^{ℓ_i, m_i}_{2n+1} where $\ell_i > 1$. Also, in this formula q is the appropriate integer such that adding a measure to the $K_1, \dots, K_t$ sequences after K_{t-p} corresponds to adding a measure to the $\tilde{K}_1, \dots, \tilde{K}_w$ sequence after $\tilde{K}_{w-q}$. More precisely, extending our previous notation slightly, let the function δ be such that: for K_j in the sequence $\bar{K}_{b(k)}, \dots, \bar{K}_{b_u}$ we have $\delta(K_j) = j$ and for $v(K_j)$ in the sequence $\mathcal{M}(K_1, \dots, K_{k-1})$ we have $\delta(v(K_j)) = j$. Then q is the least integer greater than the length of $\mathcal{M}(K_1, \dots, K_{k-1})$ such that $\delta(\tilde{K}_{w-q}) \leq t-p$. If no such q exists then we set $q = w$. If $n = 0$, so $d = (k; r)$ for $r \leq m_k$, we set $\bar{f}^k_p(d; K_1, \dots, K_t) = r$.

Case V. $k < \infty$, $k = k(d)$, and $k = t - p$.

Subcase V.a. $K_k = S^{\ell_k, m_k}_{2n+1}$, with $\ell_k > 1$. Say $d = (k; d_2)^s$, where s may or may not appear. We set

$$\bar{f}^k_p(d; K_1, \dots, K_t) = \sum_{\beta < \omega^{\omega^{\bar{f}^{k+1}_p(\hat{d}; K_1, \dots, K_t)}}} (\beta + 1) + \omega^{\omega^{\bar{f}^{k+1}_p(d_2; K_1, \dots, K_t)}} + (1 \text{ or } 2)$$

depending on whether s appears or not. Here $\hat{d}$ is as in condition A.

Subcase V.b. $K_k = S^{1, m_k}_{2n+1}$. So, $d^{(\mathcal{I}_a)} = (k; d_1^{(\mathcal{I}_a)}, \dots, d_r^{(\mathcal{I}_a)})^{s(\mathcal{I}_a)}$, where $r \leq m_k$ and s may or may not appear.

Subcase V.b.i. $r = m_k$. We set

$$\begin{aligned}
\bar{f}^k_p(d; K_1, \dots, K_t) = &\sum_{\beta < \omega^{\omega^{\bar{f}^{k+1}_p(d_1; K_1, \dots, K_t)}}} (m_k + \tau_{m_k - 1}(\beta) + m_k) + (m_k - 1) \\
&+ ((m_k - 1) + \tau_{m_k - 2}(\omega^{\omega^{\bar{f}^{k+1}_p(d_1; K_1, \dots, K_t)}}) + (m_k - 1)) \cdot \omega^{\omega^{\bar{f}^{k+1}_p(d_2; K_1, \dots, K_t)}} \\
&+ \cdots \\
&+ (2 + \tau_1(\omega^{\omega^{\bar{f}^{k+1}_p(d_1; K_1, \dots, K_t)}}) + 2) \cdot \omega^{\omega^{\bar{f}^{k+1}_p(d_2; K_1, \dots, K_t)}} \\
&+ \tau_1(\omega^{\omega^{\bar{f}^{k+1}_p(d_r; K_1, \dots, K_t)}}) + (1 \text{ or } 2)
\end{aligned}$$

depending on whether s appears or not.

Subcase V.b.ii. $r < m_k$ and s appears. Similarly to (i) above, $\bar{f}^k_p(d; K_1, \dots, K_t)$ is obtained from the formula in IV.b.ii by replacing terms of the form $\bar{f}^{k+1}_p(\cdots)$ by $\omega^{\omega^{\bar{f}^{k+1}_p(\dots)}}$.

Subcase V.b.iii. $r < m_k$ and s does not appear. As above, using now the formula from IV.b.iii.

Subcase V.c. $K_k = W^{m_k}_{2n+1}$

Subcase V.c.i. $d = (k;)$. We set $\bar{f}^k_p = E(2n, m_k)$.

Subcase V.c.ii. $d = (k; \bar{d}_2)^s$, where s may or may not appear. Here we use the same formula

$$\bar{f}^k_p(d; K_1, \dots, K_t) = E(2n-1) + 2 \cdot (\bar{g}_q(\bar{d}_2; \tilde{K}_1, \dots, \tilde{K}_w) + (1 \text{ or } 2)$$

from IV.c.ii (as there, we use 1 if s appears and 2 if s does not). The measures $\tilde{K}_1, \dots, \tilde{K}_w$ are as in IV.c.ii, and q is also as defined there.

We now proceed to establish A^k_p, B^k_p, and R^k_p for $d \in \mathcal{D}_{2n+1}$. We assume inductively these statements for $d \in \mathcal{D}_{2n-1}$ and the corresponding functions $\bar{g}^k_p$. We establish these simultaneously by reverse induction on k.

We first consider A^k_p. We consider the following cases.

Case I. $k = \infty$

Subcase I.a. d basic of type 1. With notation as in I.a on p. 459 A^k_p follows from A^1_q for $\bar{d}_2 \in \mathcal{D}_{2n-1}$, with q as in I.a on p. 459.

Subcase I.b. d basic of type 0. The result is immediate.

Case II. $k < \infty$, $k < k(d)$, and $k \neq t - p$. The result follows by induction, A^{k+1}_p and the formulas for $\bar{f}^k_p$.

Case III. $k < \infty$, $k < k(d)$, and $k = t - p$. We require the following easy lemma.

Lemma 7.6. If $\alpha = \omega^{\omega^\gamma}$ for some $\gamma \in \mathrm{Ord}$, then $\sum_{\beta<\alpha} \beta^n = \alpha$ for all n.

By induction and A^{k+1}_p, we have

$$\bar{f}^{k+1}_p(d; K_1, \dots, K_t) = \bar{f}^{k+2}_p(d; K_1, \dots, K_{t-p}, K_{t+1}, \dots, K_t).$$

Hence $\omega^{\omega^{\bar{f}^{k+1}_p(d;K_1,\dots,K_t)}} = \omega^{\omega^{\bar{f}^{k+2}_p(d;K_1,\dots,K_{t-p},K_{t+1},\dots,K_t)}}$. From the lemma and the formulas for $\bar{f}^k_p$, it follows that

$$\bar{f}^k_p(d; K_1, \dots, K_t) = \omega^{\omega^{\bar{f}^{k+1}_p(d;K_1,\dots,K_t)}}$$

(or with a $+1$ added) and likewise

$$\bar{f}^{k+1}_p(d; K_1, \dots, K_{t-p}, K_{t+1}, \dots, K_t) = \omega^{\omega^{\bar{f}^{k+2}_p(d;K_1,\dots,K_{t-p},K_{t+1},\dots,K_t)}},$$

hence

$$\bar{f}^k_p(d; K_1, \dots, K_{t-p}, K_{k+1}, \dots, K_t) = \bar{f}^k_p(d; K_1, \dots, K_t).$$

Case IV. $k < \infty$, $k = k(d)$, and $k \neq t - p$. The result follows by induction and A^{k+1}_p, and in case 3 ($K_k = W^{m_k}_{2n+1}$) by A^1_q or A^0_q for $\bar{d}_2 \in \mathcal{D}_{2n-1}$.

Case V. $k < \infty$, $k = k(d)$, and $k = t - p$.

A^k_p then follows from the formulas for $\bar{f}^k_p$, A^{k+1}_p, and Lemma 7.6. For example, in case V.b.i of the definition of $\bar{f}^k_p$ we have

$$\begin{aligned}
\bar{f}^k_p(d; K_1, \dots, K_t) = &\sum_{\beta < \omega^{\omega^{\bar{f}^{k+1}_p(d_1; K_1, \dots, K_t)}}} (m_k + \tau_{m_k - 1}(\beta) + m_k) \\
&+ \tau_{m_k-2}(\omega^{\omega^{\bar{f}^{k+1}_p(d_1; K_1, \dots, K_t)}}) \cdot \omega^{\omega^{f^{k+1}_p(d_2, K_1, \dots, K_t)}} + \cdots \\
&= \omega^{\omega^{\bar{f}^{k+1}_p(d_1; K_1, \dots, K_t)}} + \tau_{m_k-2}(\omega^{\omega^{\bar{f}^{k+1}_p(d_1; K_1, \dots, K_t)}}) \cdot \omega^{\omega^{f^{k+1}_p(d_2, K_1, \dots, K_t)}} \\
&+ \cdots + \tau_1(\omega^{\omega^{f^{k+1}_p(d_r; K_1, \dots, K_t)}}) + (1 \text{ or } 2).
\end{aligned}$$

and

$$\begin{aligned}
&\bar{f}^k_p(d; K_1, \dots, K_{t-p}, K_{t+1}, \dots, K_t) = \\
&\quad \sum_{\beta < \bar{f}^{k+1}_p(d_1; K_1, \dots, K_{t-p}, K_{t+1}, \dots, K_t)} (m_k + \tau_{m_k-1}(\beta) + m_k) \\
&\quad + ((m_k - 1) + \tau_{m_k-2}(\bar{f}^{k+1}_p(d_1; K_1, \dots, K_{t-p}, K_{t+1}, \dots, K_t)) \\
&\qquad\qquad + (m_k - 1)) \cdot f^{k+1}_p(d_2; K_1, \dots, K_t) \\
&\quad + \cdots + \tau_1(f^{k+1}_p(d_r; K_1, \dots, K_{t-p}, K_{t+1}, \dots, K_t)) + (1 \text{ or } 2).
\end{aligned}$$

We further specialize to the case $K_{t+1} = S^{\ell_{t+1}, m_{t+1}}_{2n+1}$ where $\ell_{t+1} > 1$. From III.a we have

$$\bar{f}^{k+1}_p(d_1; K_1, \dots, K_{t-p}, K_{t+1}, \dots, K_t) = \omega^{\omega^{\bar{f}^{k+2}_p(d_1; K_1, \dots, K_{t-p}, K_{t+1}, \dots, K_t)}},$$

and similarly for $d_2, \dots, d_r$. Using Lemma 7.6 it follows that

$$\bar{f}^k_p(d; K_1, \dots, K_t) = \bar{f}^k_p(d; K_1, \dots, K_{t-p}, K_{t+1}, \dots, K_t).$$

In case V.3, we again use A^0_q on $\mathcal{D}_{2n-1}$.

This establishes A^k_p in all cases.

We now consider B^k_p. We again consider the same cases.

Case I. $k = \infty$.

Subcase I.a. $d^{(\mathcal{I}_a)}$ basic of type 1, so $d = (\bar{d}_2)^s$, where s may or may not appear, and $\mathcal{I}_a = (f(\bar{K}_1); \bar{K}_2, \dots, \bar{K}_a)$.

We let d' be a re-indexing of d as in the statement of B_p^k, so $d'^{(\mathcal{I}'_a)} = (\bar{d}_2')^{(f(\bar{K}_1);\bar{K}_2',\ldots,\bar{K}_{a'}')}$, where $\bar{d}_2' = \bar{d}_2$, and $\bar{K}_2, \ldots, \bar{K}_a$ is a subsequence of $\bar{K}_2', \ldots, \bar{K}_{a'}'$. The formula for $\bar{f}_p^k(d)$, however, involves $\bar{g}_q(\bar{d}_2)$ computed relative to the full sequence $\mathcal{M}(K_1, \ldots, K_t)$ and likewise for $f_p^{k+1}(d')$. Hence $\bar{f}_p^k(d; K_1, \ldots, K_t) = \bar{f}_p^k(d'; K_1, \ldots, K_t)$.

Subcase I.b. d basic of type 0. The result is immediate.

Case II. $k < \infty, k < k(d)$, and $k \neq t - p$.
and

Case III. $k < \infty, k < k(d)$, and $k = t - p$.
The result follows immediately from induction and B_p^{k+1}.

Case IV. $k < \infty, k = k(d)$, and $k \neq t - p$
and

Case V. $k < \infty, k = k(d)$, and $k = t - p$.
The result is also immediate from induction and B_p^{k+1}, where in case c.ii, we use the fact that $\mathcal{M}(K_1, \ldots, K_{k-1})$ is the same for both d and d'.

This establishes B_p^k.

We now consider R_p^k. We let $d_1^{(\mathcal{I}_a)}$, $d_2^{(\mathcal{I}_a)}$ satisfying conditions C and A relative to $K_1, \ldots, K_t$ and assume that $d_1 < d_2$. We must establish that

$$\bar{f}_p^{k+b}(d_1; K_1, \ldots, K_{t-p}, K_{k+1}, \ldots, K_t) < \bar{f}_p^k(d_2; K_1, \ldots, K_t),$$

where $b = 0$ if $k \leq t - p$ and $b = 1$ if $k > t - p$ or $k = 0$. We recall that $\delta_{d_1}(\bar{K}_a) = \delta_{d_2}(\bar{K}_a) < k \leq k(d_1)$ or $k(d_2)$. From A_p^k we have that

$$\bar{f}_p^k(d_2; K_1, \ldots, K_t) = \bar{f}_p^{k+b}(d_2; K_1, \ldots, K_{t-p}, K_{t+1}, \ldots, K_t).$$

Note that for $k = 0$, this is part of the statement of A_p^k when $t = p$, and if $p < t$ then $f_p^0(d_2; K_1, \ldots, K_t) = f_p^1(d_2; K_1, \ldots, K_t)$ (by inspection of the formulas) and by A_p^1 this is equal to $f_p^1(d_2; K_1, \ldots, K_{t-p}, K_{t+1}, \ldots, K_t)$. So, in all cases it suffices to establish that

$$\bar{f}_p^{k+b}(d_1; K_1, \ldots, K_{t-p}, K_{k+1}, \ldots, K_t) < \bar{f}_p^{k+b}(d_2; K_1, \ldots, K_{t-p}, K_{k+1}, \ldots, K_t).$$

Changing notation, it suffices to show that if $d_1 < d_2$ both satisfy conditions C and A with respect to $K_1, \ldots, K_t$ then

$$\bar{f}_p^k(d_1; K_1, \ldots, K_t) < \bar{f}_p^k(d_2; K_1, \ldots, K_t).$$

We consider the following cases.

Case I. $k(d_1) > k(d_2)$.

Subcase I.a. $k < k(d_2)$.

Subcase I.a.i. $k \neq t-p$. By induction, $\bar{f}_p^{k+1}(d_1; K_1, \dots, K_t) < \bar{f}_p^{k+1}(d_2; K_1, \dots, K_t)$. From the formulas for $\bar{f}_p^k$ (specifically, those of case II), it follows that $\bar{f}_p^k(d_1; K_1, \dots, K_t) < \bar{f}_p^k(d_2; K_1, \dots, K_t)$. Note that all the terms in the sum in cases II.a and II.b are positive.

Subcase I.a.ii. $k = t-p$. By induction, $\bar{f}_p^{k+1}(d_1; K_1, \dots, K_t) < \bar{f}_p^{k+1}(d_2; K_1, \dots, K_t)$. We have from the formulas for $\bar{f}_p^k$ that $\bar{f}_p^k(d_2; K_1, \dots, K_t) = \omega^{\omega^{\bar{f}_p^{k+1}(d_2;K_1,\dots,K_t)}}$, and $\bar{f}_p^k(d_1; K_1, \dots, K_t) = \omega^{\omega^{\bar{f}_p^{k+1}(d_1;K_1,\dots,K_t)}}$. The result follows immediately.

Subcase I.b. $k = k(d_2)$ and $K_k = S_{2n+1}^{\ell_k, m_k}$ where $\ell_k > 1$. Hence, $d_1 \leq \hat{d}_2$ where $\hat{d}_2$ is as in condition A.

Subcase I.b.i. $k \neq t-p$. By induction, $\bar{f}_p^{k+1}(d_1; K_1, \dots, K_t) \leq \bar{f}_p^{k+1}(\hat{d}_1; K_1, \dots, K_t)$. From formulas II.a and IV.a for $\bar{f}_p^k$, it follows that $\bar{f}_p^k(d_1; K_1, \dots, K_t) < \bar{f}_p^k(d_2; K_1, \dots, K_t)$, since if $d_2 = (k; (d_2)_2)^s$ then $\bar{f}_p^{k+1}((d_2)_2; K_1, \dots, K_t) \geq 1$.

Subcase I.b.ii. $k = t-p$. By induction, $\bar{f}_p^{k+1}(d_1; K_1, \dots, K_t) < \bar{f}_p^{k+1}(\hat{d}_2; K_1, \dots, K_t)$. From formula V.a for $\bar{f}_p^k$ and Lemma 7.6 we have that

$$\bar{f}_p^k(d_2; K_1, \dots, K_t) = \omega^{\omega^{\bar{f}_p^{k+1}(\hat{d}_2;K_1,\dots,K_t)}} + \omega^{\omega^{\bar{f}_p^{k+1}((d_2)_2;K_1,\dots,K_t)}} + (1 \text{ or } 2),$$

where $d_2 = (k; (d_2)_2)^s$ (where s may or may not appear, and we use 1 in the formula if s appears and 2 if it does not). From formula III.a we also have

$$\begin{aligned}\bar{f}_p^k(d_1; K_1, \dots, K_t) &= \omega^{\omega^{\bar{f}_p^{k+1}(d_1;K_1,\dots,K_t)}} \\ &\leq \omega^{\omega^{\bar{f}_p^{k+1}(\hat{d}_2;K_1,\dots,K_t)}} \\ &< \bar{f}_p^k(d_2; K_1, \dots, K_t).\end{aligned}$$

Subcase I.c. $k = k(d_2)$ and $K_k = S_{2n+1}^{1,m_k}$. Hence, $d_1 \leq (d_2)_1$ where we use the notation $d_2 = (k; (d_2)_1, (d_2)_2, \dots, (d_2)_r)^s$ (where s may or may not appear).

Subcase I.c.i. $k \neq t-p$. By induction, $\bar{f}_p^{k+1}(d_1; K_1, \dots, K_t) \leq \bar{f}_p^{k+1}((d_2)_1; K_1, \dots, K_t)$. From formulas II.b and IV.b.i-iii for $\bar{f}_p^k$, it follows that $\bar{f}_p^k(d_1; K_1, \dots, K_t) < \bar{f}_p^k(d_2; K_1, \dots, K_t)$. Note here that the formulas from IV.b start off with the same sum as the formula for II.b, and then contain extra terms which are strictly positive.

Subcase I.c.ii. $k = t-p$. By induction, $\bar{f}_p^{k+1}(d_1; K_1, \dots, K_t) < \bar{f}_p^{k+1}((d_2)_1; K_1, \dots, K_t)$. From formulas V.b.i-iii for $\bar{f}_p^k$ we have that

$$\bar{f}_p^k(d_2; K_1, \dots, K_t) = \omega^{\omega^{\bar{f}_p^{k+1}((d_2)_1;K_1,\dots,K_t)}} + \alpha,$$

where $\alpha \geq 1$. Also, from formula III.b we have $\bar{f}^k_p(d_1; K_1, \dots, K_t) = \omega^{\omega^{\bar{f}^{k+1}_p(d_1; K_1, \dots, K_t)}}$, and the result follows.

Subcase I.d. $k = k(d_2)$ and $K_k = W^{m_k}_{2n+1}$. This case can not arise from the definition of $<$.

Case II. $k(d_1) < k(d_2)$.

Subcase II.a. $k < k(d_1)$. We proceed as in I.a.i or I.a.ii depending on whether $k \neq t - p$ or $k = t - p$.

Subcase II.b. $k = k(d_1)$ and $K_k = S^{\ell_k, m_k}_{2n+1}$ with $\ell_k > 1$. Hence, $\hat{d}_1 < d_2$ (where again $\hat{d}_1$ is as in condition A).

Subcase II.b.i. $k \neq t - p$. By induction, $\bar{f}^{k+1}_p(\hat{d}_1; K_1, \dots, K_t) < \bar{f}^{k+1}_p(d_2; K_1, \dots, K_t)$. From formula IV.a for $\bar{f}^k_p$ we have that

$$\bar{f}^k_p(d_1; K_1, \dots, K_t) = \sum_{\beta < \bar{f}^{k+1}_p(\hat{d}_1; K_1, \dots, K_t)} 2 \cdot (\beta + 1) + 2 \cdot (\bar{f}^{k+1}_p((d_1)_2; K_1, \dots, K_t)) + (1 \text{ or } 2),$$

where $d_1 = (k; (d_1)_2)^s$ and s may or may not appear. From formula II.a we have

$$\bar{f}^k_p(d_2; K_1, \dots, K_t) = \left[\sum_{\beta < \bar{f}^{k+1}_p(d_2; K_1, \dots, K_t)} 2 \cdot (\beta + 1) \right] + 1.$$

Note here that if $d_1 = d_1^{(\mathcal{I}_a)}$ has index $(\mathcal{I}_a)$, then $\hat{d}_1$ also has index $(\mathcal{I}_a)$ while $(d_1)_2$ has index $(\mathcal{I}_{a+1}) = (\mathcal{I}_a, v(K_k))$. By B^{k+1}_p we have that

$$\bar{f}^{k+1}_p(\hat{d}_1; K_1, \dots, K_t) = \bar{f}^{k+1}_p(\hat{d}_1^{(\mathcal{I}_{a+1})}; K_1, \dots, K_t)$$

where $\hat{d}_1^{(\mathcal{I}_{a+1})}$ is obtained from $\hat{d}_1^{(\mathcal{I}_a)}$ by re-indexing as in B. We also have that $(d_1)_2^{(\mathcal{I}_{a+1})} \leq \hat{d}_1^{(\mathcal{I}_{a+1})}$ and hence by induction,

$$\bar{f}^{k+1}_p(\hat{d}_1^{(\mathcal{I}_{a+1})}; K_1, \dots, K_t) \geq \bar{f}^{k+1}_p((d_1)_2^{(\mathcal{I}_{a+1})}; K_1, \dots, K_t).$$

Hence $\bar{f}^{k+1}_p(\hat{d}_1; K_1, \dots, K_t) \geq \bar{f}^{k+1}_p((d_1)_2^{(\mathcal{I}_{a+1})}; K_1, \dots, K_t)$. In fact, we we have strict inequality here unless s appears in d_1 (*i.e.*, $d_1 = (k; (d_1)_2)^s$) So we have

$$\begin{aligned}\bar{f}_p^k(d_1; K_1, \ldots, K_t) &\leq \sum_{\beta \leq \bar{f}_p^{k+1}(\tilde{d}_1; K_1, \ldots, K_t)} 2 \cdot (\beta + 1)\\ &\leq \sum_{\beta < \bar{f}_p^{k+1}(d_2; K_1, \ldots, K_t)} 2 \cdot (\beta + 1)\\ &< \sum_{\beta < \bar{f}_p^{k+1}(d_2; K_1, \ldots, K_t)} 2 \cdot (\beta + 1) + 1\\ &= \bar{f}_p^k(d_2; K_1, \ldots, K_t)\end{aligned}$$

Subcase II.b.ii. $k = t - p$. By induction, $\bar{f}_p^{k+1}(\hat{d}_1; K_1, \ldots, K_t) < \bar{f}_p^{k+1}(d_2; K_1, \ldots, K_t)$. From formula V.a for $\bar{f}_p^k$ we have that

$$\begin{aligned}\bar{f}_p^k(d_1; K_1, \ldots, K_t) &= \sum_{\beta < \omega^{\omega^{\bar{f}_p^{k+1}(\hat{d}_1; K_1, \ldots, K_t)}}} 2 \cdot (\beta + 1)\\ &\quad + \omega^{\omega^{\bar{f}_p^{k+1}((d_1)_2; K_1, \ldots, K_t)}} + (1 \text{ or } 2)\\ &= \omega^{\omega^{\bar{f}_p^{k+1}(\hat{d}_1; K_1, \ldots, K_t)}} + \omega^{\omega^{\bar{f}_p^{k+1}((d_1)_2; K_1, \ldots, K_t)}} + (1 \text{ or } 2).\end{aligned}$$

We also have from III.a that $\bar{f}_p^k(d_2; K_1, \ldots, K_t) = \omega^{\omega^{\bar{f}_p^{k+1}(d_2; K_1, \ldots, K_t)}}$. It also follows from B_p^{k+1}, as in the previous case, that

$$\bar{f}_p^{k+1}((d_1)_2^{(\mathcal{I}_{a+1})}; K_1, \ldots, K_t) \leq \bar{f}_p^{k+1}(\hat{d}_1; K_1, \ldots, K_t).$$

Hence

$$\begin{aligned}\bar{f}_p^k(d_1; K_1, \ldots, K_t) &\leq \omega^{\omega^{\bar{f}_p^{k+1}(\hat{d}_1; K_1, \ldots, K_t)}} \cdot 2 + (1 \text{ or } 2)\\ &< \omega^{\omega^{\bar{f}_p^{k+1}(d_2; K_1, \ldots, K_t)}} = \bar{f}_p^k(d_2; K_1, \ldots, K_t).\end{aligned}$$

Subcase II.c. $k = k(d_1)$ and $K_k = S_{2n+1}^{1,m_k}$. Hence, if $d_1 = (k; (d_1)_1, (d_1)_2, \ldots, (d_1)_r)^s$ (where s may or may not appear) then we have $(d_1)_1 < d_2$.

Subcase II.c.i. $k \neq t - p$. By induction, $\bar{f}_p^{k+1}((d_1)_1; K_1, \ldots, K_t) < \bar{f}_p^{k+1}(d_2; K_1, \ldots, K_t)$. From formulas IV.b.i–iii for $\bar{f}_p^k$ we have that

$$\bar{f}_p^k(d_1; K_1, \ldots, K_t) = \sum_{\beta < \bar{f}_p^{k+1}((d_1)_1; K_1, \ldots, K_t)} (m_k + \tau_{m_k - 1}(\beta) + m_k) + \alpha,$$

where $\alpha \leq \tau_{m_k-1}(\bar{f}_p^{k+1}((d_1)_1; K_1, \ldots, K_t)) + m_k$. This follows from the formulas for $\bar{f}_p^k$, the fact that

$$\bar{f}_p^{k+1}((d_1)_i; K_1, \ldots, K_t) < \bar{f}_p^{k+1}((d_1)_1; K_1, \ldots, K_t)$$

for all $i > 1$, and Lemma 7.5 which gives that

$$\begin{aligned}
&((m_k - 1) + \tau_{m_k-2}(\beta) + (m_k - 1)) \cdot \beta \\
&\quad + ((m_k - 2) + \tau_{m_k-3}(\beta) + (m_k - 2)) \cdot \beta + \cdots \\
&\quad + ((m_k - r + 1) + \tau_{m_k-r}(\beta) + (m_k - r + 1)) \cdot \beta \\
&\quad + \tau_{m_k-r}(\beta) + (m_k - r + 1) \\
&\quad \leq \tau_{m_k-1}(\beta) + (m_k - r + 1) \leq \tau_{m_k-1}(\beta) + m_k
\end{aligned}$$

for all β (where we use here $\beta = \bar{f}_p^{k+1}((d_1)_1; K_1, \ldots, K_t)$). We also have that

$$\bar{f}_p^k(d_2; K_1, \ldots, K_t) = \sum_{\beta < \bar{f}_p^{k+1}(d_2; K_1, \ldots, K_t)} (m_k + \tau_{m_k-1}(\beta) + m_k) + 1,$$

and hence $\bar{f}_p^k(d_1; K_1, \ldots, K_t) < \bar{f}_p^k(d_2; K_1, \ldots, K_t)$.

Subcase II.c.ii. $k = t - p$. As in the previous case we have by induction that

$$\bar{f}_p^{k+1}((d_1)_1; K_1, \ldots, K_t) < \bar{f}_p^{k+1}(d_2; K_1, \ldots, K_t).$$

From formulas V.b.i–iii and Lemma 7.6 we have that

$$\begin{aligned}
\bar{f}_p^k(d_1; K_1, \ldots, K_t) &= \sum_{\beta < \omega^{\omega^{\bar{f}_p^{k+1}((d_1)_1; K_1, \ldots, K_t)}}} (m_k + \tau_{m_k-1}(\beta) + m_k) + \alpha \\
&= \omega^{\omega^{\bar{f}_p^{k+1}((d_1)_1; K_1, \ldots, K_t)}} + \alpha
\end{aligned}$$

where as in the previous case we have

$$\begin{aligned}
\alpha &\leq \tau_{m_k-2}\big(\omega^{\omega^{\bar{f}_p^{k+1}((d_1)_1; K_1, \ldots, K_t)}}\big) \cdot \omega^{\omega^{\bar{f}_p^{k+1}((d_1)_1; K_1, \ldots, K_t)}} \\
&\quad + \tau_{m_k-3}\big(\omega^{\omega^{\bar{f}_p^{k+1}((d_1)_1; K_1, \ldots, K_t)}}\big) \cdot \omega^{\omega^{\bar{f}_p^{k+1}((d_1)_1; K_1, \ldots, K_t)}} \\
&\quad + \cdots + \tau_1\big(\omega^{\omega^{\bar{f}_p^{k+1}((d_1)_1; K_1, \ldots, K_t)}}\big) + m_k \\
&\leq \tau_{m_k-1}\big(\omega^{\omega^{\bar{f}_p^{k+1}((d_1)_1; K_1, \ldots, K_t)}}\big) + m_k.
\end{aligned}$$

We also have from formula III.b and Lemma 7.6 that

$$\begin{aligned}
\bar{f}_p^k(d_2; K_1, \ldots, K_t) &= \sum_{\beta < \omega^{\omega^{\bar{f}_p^{k+1}(d_2; K_1, \ldots, K_t)}}} (m_k + \tau_{m_k-1}(\beta) + m_k) \\
&= \omega^{\omega^{\bar{f}_p^{k+1}(d_2; K_1, \ldots, K_t)}}.
\end{aligned}$$

From Lemma 7.6 we have that

$$\tau_{m_k-1}\big(\omega^{\omega^{\bar{f}_p^{k+1}((d_1)_1; K_1, \ldots, K_t)}}\big) + m_k < \omega^{\omega^{\bar{f}_p^{k+1}(d_2; K_1, \ldots, K_t)}}$$

and it then follows from the above equations that $\bar{f}^k_p(d_1; K_1, \dots, K_t) < \bar{f}^k_p(d_2; K_1, \dots, K_t)$.

Subcase II.d. $k = k(d_1)$ and $K_k = W^{m_k}_{2n+1}$.

Subcase II.d.i. $k \neq t - p$. From formula IV.c for $\bar{f}^k_p$ and $\mathcal{R}_1$ on $\mathcal{D}_{2n-1}$ we have that $\bar{f}^k_p(d_1; K_1, \dots, K_t) \leq E(2n, m_k)$. Also, from formula II.c we have $\bar{f}^k_p(d_2; K_1, \dots, K_t) = E(2n, m_k) + \alpha$, where $\alpha \geq 1$ (in fact $\alpha \geq E(2n - 1)$), and hence the result follows.

Subcase II.d.ii. $k = t - p$. The result follows as in the previous case.

Case III. $k(d_1) = k(d_2)$.

Subcase III.a. $k < k(d_1) = k(d_2)$.

Subcase III.a.i. $k \neq t - p$ and **Subcase III.a.ii.** $k = t - p$. We proceed as in I.a.i and I.a.ii.

Subcase III.b. $k = k(d_1) = k(d_2)$ and $K_k = S^{\ell_k, m_k}_{2n+1}$ with $\ell_k > 1$. Hence $d_1 = (k; \bar{d}^{(\mathcal{I}_{a+i})}_{2,1})^s$, $d_2 = (k; \bar{d}^{(\mathcal{I}_{a+i})}_{2,2})^s$, where s may or may not appear in d_1, d_2.

Subcase III.b.i. $\bar{d}_{2,1} < \bar{d}_{2,2}$.

Subcase III.b.i.1. $k \neq t - p$. We have by induction that $\bar{f}^{k+1}_p(d_{2,1}; K_1, \dots, K_t) < \bar{f}^{k+1}_p(d_{2,2}; K_1, \dots, K_t)$. Since $d_{2,1} < d_{2,2}$, it follows that $\hat{d}_1 \leq \hat{d}_2$, and hence by induction, $\bar{f}^{k+1}_p(\hat{d}_1; K_1, \dots, K_t) < \bar{f}^{k+1}_p(\hat{d}_2; K_1, \dots, K_t)$. From formula IV.a it then follows that $\bar{f}^k_p(d_1; K_1, \dots, K_t) < \bar{f}^k_p(d_2; K_1, \dots, K_t)$. Note here that if we exclude the final term of (1 or 2) from the formulas for these ordinals, then the right-hand side is at least 2 greater than the left-hand side so adding this last term maintains the inequality.

Subcase III.b.i.2. $k = t - p$. Similar to the above case.

Subcase III.b.ii. $d_{2,1} = d_{2,2}$ and the symbol s appears in d_1 and not in d_2. In this case R^k_p is immediate from the formulas for $\bar{f}^k_p$.

Subcase III.c. $k = k(d_1) = k(d_2)$ and $K_k = S^{1,m_k}_{2n+1}$. Here $d_1 = (k; d_{1,1}, d_{2,1}, \dots, d_{r_1,1})^s$ and $d_2 = (k; d_{1,2}, d_{2,2}, \dots, d_{r_2,2})^s$, where s may or may not appear in d_1, d_2.

Subcase III.c.i. $d_{1,1} < d_{1,2}$.

Subcase III.c.i.1. $k \neq t - p$. By induction,

$$\bar{f}^{k+1}_p(d_{1,1}; K_1, \dots, K_t) < \bar{f}^{k+1}_p(d_{1,2}; K_1, \dots, K_t).$$

From the formulas of IV.b for $\bar{f}^k_p$ we have

$$\bar{f}^k_p(d_2; K_1, \dots, K_t) > \sum_{\beta < \bar{f}^{k+1}_p(d_{1,2}; K_1, \dots, K_t)} (m_k + \tau_{m_k - 1}(\beta) + m_k)$$

and

$$\bar{f}_p^k(d_1; K_1, \ldots, K_t) = \left[\sum_{\beta < \bar{f}_p^{k+1}(d_{1,1}; K_1, \ldots, K_t)} (m_k + \tau_{m_k - 1}(\beta) + m_k) \right] + \alpha$$

where $\alpha \leq \tau_{m_k-1}(\bar{f}_p^{k+1}(d_{1,1}; K_1, \ldots, K_t)) + m_k$. Hence it follows that $\bar{f}_p^k(d_1; K_1, \ldots, K_t) < \bar{f}_p^k(d_2; K_1, \ldots, K_t)$.

Subcase III.c.i.2. $k = t - p$. Again, by induction

$$\bar{f}_p^{k+1}(d_{1,1}; K_1, \ldots, K_t) < \bar{f}_p^k(d_{1,2}; K_1, \ldots, K_t).$$

From the formulas of V.b we have $\bar{f}_p^k(d_2; K_1, \ldots, K_t) > \omega^{\omega^{\bar{f}_p^{k+1}(d_{1,2}; K_1, \ldots, K_t)}}$ and $\bar{f}_p^k(d_1; K_1, \ldots, K_t) \leq \omega^{(\omega^{\bar{f}_p^{k+1}(d_{1,1}; K_1, \ldots, K_t)} \cdot m)}$ for some m, which is therefore less than $\bar{f}_p^k(d_2; K_1, \ldots, K_t)$.

Subcase III.c.ii. There is an r, with $2 \leq r \leq \min\{r_1, r_2\}$, such that $d_{i,1} = d_{i,2}$, for $1 \leq i < r$, and $d_{r,1} < d_{r,2}$.

Subcase III.c.ii.1. $k \neq t - p$. By induction, $\bar{f}_p^{k+1}(d_{r,1}; K_1, \ldots, K_t) < \bar{f}_p^{k+1}(d_{r,2}; K_1, \ldots, K_t)$. For $i < r$, let

$$\gamma_i = \bar{f}_p^{k+1}(d_{i,1}; K_1, \ldots, K_t) = \bar{f}_p^{k+1}(d_{i,2}; K_1, \ldots, K_t).$$

From the formulas IV.b for $\bar{f}_p^k$ we have that (note that the smallest of the values occurs in case IV.b.i or IV.b.ii depending on whether $r = m_k$)

$$\begin{aligned}\bar{f}_p^k(d_2; K_1, \ldots, K_t) \geq & \sum_{\beta < \gamma_1} (m_k + \tau_{m_k-1}(\beta) + m_k) + \cdots \\ & + ((m_k - r + 2) + \tau_{m_k - r + 1}(\gamma_1) + (m_k - r + 2)) \cdot \gamma_{r-1} \\ & + ((m_k - r + 1) + \tau_{m_k - r}(\gamma_1) + (m_k - r + 1)) \cdot \bar{f}_p^{k+1}(d_{r,2}; K_1, \ldots, K_t) \\ & + 1.\end{aligned}$$

We also have that (note that the largest of the values occurs in case IV.b.i or IV.b.iii)

$$\begin{aligned}\bar{f}_p^k(d_1; K_1, \ldots, K_t) \leq & \sum_{\beta < \gamma_1} (m_k + \tau_{m_k-1}(\beta) + m_k) + \cdots \\ & + ((m_k - r + 2) + \tau_{m_k - r + 1}(\gamma_1) + (m_k - r + 2)) \cdot \gamma_{r-1} \\ & + ((m_k - r + 1) + \tau_{m_k - r}(\gamma_1) + (m_k - r + 1)) \cdot \bar{f}_p^{k+1}(d_{r,1}; K_1, \ldots, K_t) \\ & + \alpha\end{aligned}$$

where $\alpha \leq \tau_{m_k-r}(\gamma_1) + (m_k - r + 1)$. Hence

$$((m_k - r + 1) + \tau_{m_k-r}(\gamma_1) + (m_k - r + 1)) \cdot \bar{f}_p^{k+1}(d_{r,1}; K_1, \dots, K_t) + \alpha$$
$$\leq ((m_k - r + 1) + \tau_{m_k-r}(\gamma_1) + (m_k - r + 1)) \cdot (\bar{f}_p^{k+1}(d_{r,1}; K_1, \dots, K_t) + 1)$$
$$\leq ((m_k - r + 1) + \tau_{m_k-r}(\gamma_1) + (m_k - r + 1)) \cdot (\bar{f}_p^{k+1}(d_{r,2}; K_1, \dots, K_t)$$

and it follows that $\bar{f}_p^k(d_1; K_1, \dots, K_t) < \bar{f}_p^k(d_2; K_1, \dots, K_t)$.

Subcase III.c.ii.2. $k = t - p$. Similar to the above case.

Subcase III.c.iii. $r_1 < r_2$ and $d_{i,1} = d_{i,2}$ for $1 \leq i \leq r_1$. From the definition of the ordering $<$ on the descriptions we must have that s appears in d_1 (and may or may not appear in d_2).

Subcase III.c.iii.1. $k \neq t - p$. We may in fact also assume that s appears in d_2 as the formula in IV.b.iii results in a strictly larger ordinal than that of formula IV.b.ii. From formula IV.b.ii we see that $\bar{f}_p^k(d_1; K_1, \dots, K_t)$ is of the form $\vartheta + 1$ while $\bar{f}_p^k(d_2; K_1, \dots, K_t)$ is of the form $\vartheta + \alpha + 1$ with $\alpha \geq 1$, and the result follows.

Subcase III.c.iii.2. $k = t - p$. Similar to the above case.

Subcase III.c.iv. $r_1 > r_2$ and $d_{i,1} = d_{i,2}$ for $1 \leq i \leq r_2$. We must have in this case that s does not appear in d_2.

Subcase III.c.iv.1. $k \neq t - p$. For $i \leq r_2$ we again let

$$\gamma_i = \bar{f}_p^{k+1}(d_{i,1}; K_1, \dots, K_t) = \bar{f}_p^{k+1}(d_{i,2}; K_1, \dots, K_t).$$

We may assume that s does not appear in d_1, as this results in the larger value for $\bar{f}_p^{k+1}(d_1; K_1, \dots, K_t)$. From formula IV.b.iii (and IV.b.i if $r_1 = m_k$) we have

$$\bar{f}_p^k(d_1; K_1, \dots, K_t) = \sum_{\beta<\gamma_1}(m_k + \tau_{m_k-1}(\beta) + m_k) + \cdots$$
$$+ ((m_k - r_2 + 1) + \tau_{m_k-r_2}(\gamma_1)) + (m_k - r_2 + 1)) \cdot \gamma_{r_2}$$
$$+ \alpha$$

where from Lemma 7.5

$$\alpha = ((m_k - r_2) + \tau_{m_k-r_2-1}(\gamma_1) + (m_k - r_2)) \cdot \bar{f}_p^{k+1}(d_{r_2+1,1}; K_1, \dots, K_t)$$
$$+ \cdots$$
$$+ ((m_k - r_1 + 1) + \tau_{m_k-r_1}(\gamma_1)$$
$$+ (m_k - r_1 + 1)) \cdot (\bar{f}_p^{k+1}(d_{r_1,1}; K_1, \dots, K_t) + 1)$$
$$\leq \tau_{m_k-r_2}(\gamma_1) + (m_k - r_1 + 1)$$
$$< \tau_{m_k-r_2}(\gamma_1) + (m_k - r_2 + 1).$$

We also have that

$$\bar{f}^k_p(d_2; K_1, \ldots, K_t) = \sum_{\beta<\gamma_1}(m_k + \tau_{m_k-1}(\beta) + m_k) + \cdots$$
$$+ ((m_k - r_2 + 1) + \tau_{m_k-r_2}(\gamma_1) + (m_k - r_2 + 1)) \cdot (\gamma_{r_2} + 1).$$

Thus, $\bar{f}^k_p(d_1; K_1, \ldots, K_t) < \bar{f}^k_p(d_2; K_1, \ldots, K_t)$.

Subcase III.c.iv.2. $k = t - p$. Similar to the above case.

Subcase III.c.v. $r_1 = r_2$ and $d_{i,1} = d_{i,2}$ for $1 \leq i \leq r_1$. We must have that s appears in d_1 and not in d_2. R^k_p then follows immediately from the formulas for $\bar{f}^k_p$.

Subcase III.d. $k = k(d_1) = k(d_2)$ and $K_k = W^{m_k}_{2n+1}$.

Subcase III.d.i. $d_1 = (k; \bar{d}_{2,1})^s$, where s may or may not appear, and $d_2 = (k;)$.

Then $\bar{f}^k_p(d_2; K_1, \ldots, K_t) = E(2n; m_k)$. Also, by R_q and induction, $\bar{g}_q(\bar{d}_{2,1}; K_{b(k)}, \ldots, K_{b(u)}, \mathcal{M}(K_1, \ldots, K_{k-1})) < E(2n; m_k)$, where q, etc. refer to the sequence $K_1, \ldots, K_{t-p}, K_{k+1}, \ldots, K_t$. R^k_p then follows.

Subcase III.d.ii. $d_1 = (k; \bar{d}_{2,1})^s$ and $d_2 = (k; \bar{d}_{2,2})^s$, where $\bar{d}_{2,1} < \bar{d}_{2,2}$, and s may or may not appear in d_1, d_2.

Let q and $\bar{K}_1, \ldots, \bar{K}_w = K_{b(k)}, \ldots, K_{b(u)} {}^\frown \mathcal{M}(K_1, \ldots, K_{k-1})$ be as in IV.c.ii of the definition of $\bar{f}^k_p$ corresponding to $K_1, \ldots, K_t$. Note that q and $\bar{K}_1, \ldots, \bar{K}_w$ only depend on $K_1, \ldots, K_t$, p, and k, and so are the same for both d_1 and d_2. We then have that

$$\bar{f}^k_p(d_1; K_1, \ldots, K_t) = E(2n-1) + 2 \cdot \bar{g}_q(d_{2,1}; \bar{K}_1, \ldots, \bar{K}_w) + (1 \text{ or } 2)$$

and similarly for d_2. By induction,

$$\bar{g}_q(d_{2,1}; \bar{K}_1, \ldots, \bar{K}_w) < \bar{g}_q(d_{2,2}; \bar{K}_1, \ldots, \bar{K}_w)$$

and it follows that $\bar{f}^k_p(d_1; K_1, \ldots, K_t) < \bar{f}^k_p(d_2; K_1, \ldots, K_t)$.

Subcase III.d.iii. As in (ii) immediately above, where $\bar{d}_{2,1} = \bar{d}_{2,2}$. We must have that s appears in d_1 and not in d_2. This case is immediate from the formulas for $\bar{f}^k_p$ (since the last term of the expression for $\bar{f}^k_p(d_1; K_1, \ldots, K_t)$ is $+1$ and for $\bar{f}^k_p(d_2; K_1, \ldots, K_t)$ is $+2$).

R^k_p has now been established in all cases. R_p follows now easily from reverse induction on $k(d)$, R^k_p, the formulas of $\bar{f}^k_p$, and the fact that if $\alpha < E(2n+2, m_k)$ then $\alpha^n < E(2n+2, m_k)$ for all n.

In particular it now follows from R^k_p that for any $(d^{(\mathcal{I}_a)})^s$, where s may not appear, and $\mathcal{I}_a = (f(\bar{K}_1;)$, and d is defined and satisfies conditions C and A relative to $K_1, \ldots, K_t$, that $f((d)^s; K_1, \ldots, K_t) < E(2n+3)$.

§8. The upper bound for $\underset{\sim}{\delta}^1_{2n+5}$. We are now in a position to collect the results of the previous sections and obtain the upper bound for $\underset{\sim}{\delta}^1_{2n+5}$. We recall we are assuming I_{2n+1} and K_{2n+3}.

THEOREM 8.1. $\underset{\sim}{\delta}^1_{2n+5} \leq \aleph_{E(2n+3)+1}$

PROOF. We recall (cf. [Kec81A]) that $\underset{\sim}{\delta}^1_{2n+5} = [\sup j(\underset{\sim}{\delta}^1_{2n+3})]^+$, where the supremum ranges over embeddings j from the ultrapowers by the measures in $\bigcup_{m<n} S_{2m+1} \cup \bigcup_{m<n} W_{2m+3}$, these being, without loss of generality, the most general measures arising from the homogeneous tree construction on a complete $\underset{\sim}{\Pi}^1_{2n+3}$ set. From the global embedding theorem, we may restrict our attention to the measures $\mu_{2n+3}^{S^{k,m}_{2n+1}}$, where $\ell = 2^{n+1} - 1$ is maximal. We fix m, and let $F : \underset{\sim}{\delta}^1_{2n+3} \to \underset{\sim}{\delta}^1_{2n+3}$ be given representing $[F]$ w.r.t. $\mu^{S^{\ell,m}_{2n+1}}$. From the week partition relation on $\underset{\sim}{\delta}^1_{2n+3}$, there is a $g : \underset{\sim}{\delta}^1_{2n+3} \to \underset{\sim}{\delta}^1_{2n+3}$ such that for almost all $f : \vartheta(S^{\ell,m}_{2n+1}) \to \underset{\sim}{\delta}^1_{2n+3}$ we have $F([f]) < g(\sup_{\text{a.e.}} f)$. By the argument of the main theorem of section 6 (Theorem 6.4), it follows that F represents an ordinal $[F] \leq [\sup_{K_1}(\text{id}; f; d; K_1)]^+$, where $d = ()$ is the distinguished description. Repeating the argument gives $(\text{id}; f; d; K_1) \leq [\sup_{K_2}(\text{id}; f; (\tilde{d})^s; K_1, K_2)]^+$, where $(\tilde{d})^{\mathcal{I}_a}$ is the maximal tuple relative to K_1, K_2 (where $\mathcal{I}_a = (f(\bar{K}_1);)$ and $\bar{K}_1 = v(S^{\ell,m}_{2n+1})$). From Theorem 6.4 and R_p (where $p = 0$), it follows that F represents an ordinal less than $\aleph_{E(2n+2,m)}$, and the result follows. ⊣

§9. A Lower Bound for f_p. Our goal in this section is to obtain a lower bound for a certain rank function.

We first define some auxiliary measures. For each regular cardinal $\kappa < \underset{\sim}{\delta}^1_{2n+1}$, we let M_κ be defined using the strong partition relation on $\underset{\sim}{\delta}^1_{2n+1}$, functions $F : \underset{\sim}{\delta}^1_{2n+1} \to \underset{\sim}{\delta}^1_{2n+1}$ of the correct type, and the normal measure on $\underset{\sim}{\delta}^1_{2n+1}$ concentrating in the points of cofinality κ. We let $N = M_{\kappa_1} \times \cdots \times M_{\kappa_p}$, where $\kappa_1, \ldots, \kappa_p$ enumerates the regular cardinals less than $\underset{\sim}{\delta}^1_{2n+1}$.

We next modify slightly the measures $S^{\ell,m}_{2n+1}$ as follows. For $\ell \neq 2$ we let $\tilde{S}^{\ell,m}_{2n+1} = S^{\ell,m}_{2n+1}$. For $\ell = 2$ we define $\tilde{S}^{2,m}_{2n+1}$ as follows. Let $k(m) = 1 + m + m(m-1) + m(m-1)(m-2) + \cdots + m!$ be the number of sequences $\pi = \langle i_1, i_2, \ldots, i_\ell \rangle$, where $0 \leq \ell \leq m$, each $1 \leq i_j \leq m$, and the i_j are all distinct. For such a fixed π, we say a function $f : \omega_1^m \to \underset{\sim}{\delta}^1_{2n+1}$ is of π type if $f(\alpha_1, \ldots, \alpha_m) < f(\beta_1, \ldots, \beta_m)$ whenever $(\alpha_{i_1}, \ldots, \alpha_{i_\ell}) <_{\text{lex}} (\beta_{i_1}, \ldots, \beta_{i_\ell})$, and $f(\vec{\alpha}) = f(\vec{\beta})$ if $(\alpha_{i_1}, \ldots, \alpha_{i_\ell}) = (\beta_{i_1}, \ldots, \beta_{i_\ell})$. It is easy to see (using the weak partition relation on ω_1) that for any $f : \omega_1^m \to \underset{\sim}{\delta}^1_{2n+1}$, there is a π and a measure one set A w.r.t. W_1^m such that $f \restriction A$ is of π type. We define $\tilde{S}^{2,m}_{2n+1}$ to be the measure on $k(m)$ tuples of ordinals defined by: A has measure one if there is a c.u.b. $C \subseteq \underset{\sim}{\delta}^1_{2n+1}$ such that for all $F : \underset{\sim}{\delta}^1_{2n+1} \to \underset{\sim}{\delta}^1_{2n+1}$ of the correct type $(\ldots, \alpha_\pi, \ldots) \in A$, where for a fixed π, α_π is represented with respect to $\mathcal{V} := \mu^{\pi}_{2n+1}$ = the measure on $\underset{\sim}{\delta}^1_{2n+1}$ induced by the weak partition relation on $\underset{\sim}{\delta}^1_{2n+1}$ and functions $f : \omega_1^m \to \underset{\sim}{\delta}^1_{2n+1}$ of π-type which are everywhere discontinuous and of uniform cofinality ω, by $\alpha_\pi([f]) = F([f])$.

We let $B^m_{2n+1} = (S^{1,m}_{2n+1} \times N \times \prod_{\ell=2}^{2^{n+1}-1} \tilde{S}^{\ell,m}_{2n+1})^m$.

We modify some of the previous definitions as follows.

We consider sequences of measures $K'_1, \dots, K'_t$, where each $K'_k = B^{m_k}_{2n+1}$. We let $K^k_1, \dots, K^k_{c_k}$ enumerate the measures in the product measure $B^{m_k}_{2n+1} = K'_k$. We consider $d \in \mathcal{D}_{2n+1}$ to be defined relative to $K'_1, \dots, K'_t$ if d is defined relative to $K^1_1, \dots, K^1_{c_1}, \dots, K^t_1, \dots, K^t_{c_t}$. We let K_k denote the kth element of the sequence $K^1_1, \dots, K^t_{c_t}$.

We allow now $\mathcal{D}_{2m+1}$ to further contain descriptions of the form $d^{(\mathcal{I}_a)} = (k; d_2^{(\mathcal{I}_a)})$, where $K_k = M_\kappa$ for some κ.

We consider d to satisfy condition C relative to $K'_1, \dots, K'_t$ if d satisfies condition C relative to $K^1_1, \dots, K^t_{c_t}$, where we remove the restriction that $r > 1$ if s appears for $K_k = S^{1,m_k}_{2n+1}$ (so now $d = (k; d_1)^s$ will evaluate to the same ordinal as d_1). Also, if $K_k = K_\kappa$ we define $(k; d_2)$ to satisfy condition C, provided

$$\forall^* h^1_1, \dots, h^t_{c_t}, f, \bar{h}_1, \dots, \bar{h}_a \;\; \mathrm{cf}(H(d_2; h^1_1, \dots, h^t_{c_t}, f, \bar{h}_1, \dots, \bar{h}_a)) = \kappa.$$

If $d = (k; d_2)^s$, where $K_k = \tilde{S}^{2,m}_{2n+1}$, then condition condition C imposes no restriction if s appears, and if s does not appear, we define d to satisfy condition C if for almost all $h^1_1, \dots, h^t_{c_t}, f, \bar{h}_1, \dots, \bar{h}_a$, the function $g : \omega_1^m \to \underset{\sim}{\delta}^1_{2n+1}$ given by $g([\bar{h}_{a+1}]) = H(d_2; h^1_1, \dots, h^t_{c_t}, f, \bar{h}_1, \dots, \bar{h}_{a+1})$ is such that for some measure one set A w.r.t. W^m_1, $g \restriction A$ is discontinuous of uniform cofinality ω.

Conditions D and A are as before.

We define modified operations $\mathcal{L}_M, \hat{\mathcal{L}}_M, \hat{\mathcal{L}}^{\restriction k}_M$ on descriptions defined and satisfying condition C relative to $K'_1, \dots, K'_t$. The definition of $\hat{\mathcal{L}}^{\restriction k}_M$ proceeds as in the definition of $\hat{\mathcal{L}}^{\restriction k}$ (pp. 425–428), with the following modifications: (cases here are numbered as in the definition of $\hat{\mathcal{L}}^{\restriction k}$; Cases I, II, and IV–VII are the same).

Case III. If s appears and d_2 not minimal w.r.t. $\hat{\mathcal{L}}^{\restriction k+1}_M$, then we set $\hat{\mathcal{L}}^{\restriction k}_M(d) = (k; \hat{\mathcal{L}}^{\restriction k+1}_M(d_2)^{(\mathcal{I}_{a+1})})$ if this tuple satisfies condition C, and otherwise $= (k; \hat{\mathcal{L}}^{\restriction k+1}_M(d_2)^{(\mathcal{I}_{a+1})})^s$. This case includes now the case $K_k = \tilde{S}^{2,m}_{2n+1}$. If d_2 is minimal with respect to $\hat{\mathcal{L}}^{\restriction k+1}_M$ then d is minimal w.r.t. $\hat{\mathcal{L}}^{\restriction k}_M$.

We also add the following cases to the definition of $\hat{\mathcal{L}}^{\restriction k}_M$ (we continue the numbering of the cases from the definition of $\hat{\mathcal{L}}^{\restriction k}$):

Case VIII. $k < \infty$, $k = k(d)$, and $K_k = M_k$. Hence, $d^{(\mathcal{I}_a)} = (k; d_2^{(\mathcal{I}_a)})$. If d_2 is minimal w.r.t. $\hat{\mathcal{L}}^{\restriction k+1}_M$, then d is minimal w.r.t. $\hat{\mathcal{L}}^{\restriction k}_M$. Otherwise, we set

$\hat{\mathcal{L}}_M^{\restriction k}(d) = (k; \hat{\mathcal{L}}_M^{\restriction k+1}(d_2))$ if this description satisfies condition C, and otherwise $\hat{\mathcal{L}}_M^{\restriction k}(d) = d_2$.

Case IX. $k < \infty$, $k < k(d)$, and $K_k = M_k$. If d is minimal w.r.t. $\hat{\mathcal{L}}_M^{\restriction k+1}$, then d is minimal w.r.t. $\hat{\mathcal{L}}_M^{\restriction k}$. Otherwise, we set $\hat{\mathcal{L}}_M^{\restriction k}(d) = (k; \hat{\mathcal{L}}_M^{\restriction k+1}(d))$ if this description satisfies condition C and if not then $\hat{\mathcal{L}}_M^{\restriction k+1}(d) = \hat{\mathcal{L}}_M^{\restriction k+1}(d)$.

We set $\hat{\mathcal{L}}_M(d) = \hat{\mathcal{L}}_M^{\restriction 1}(d)$ for $d^{(\mathcal{I}_a)}$ with $\mathcal{I}_a = (f(\bar{K}_1);)$.

If d satisfies condition C relative to $K_1', \dots, K_t'$, it is easy to see that $\hat{\mathcal{L}}_M(d)$ also satisfies condition C. Also, property F_{2n+1}^2 is still satisfied for d satisfying condition C.

If d is defined and satisfies condition C relative to $K_1', \dots, K_t'$ and we are given $g \colon \underset{\sim}{\delta}_{2n+3}^1 \to \underset{\sim}{\delta}_{2n+3}^1$, then we define $(\mathrm{id}; f; (d)^s; K_1', \dots, K_t')$ and $(g; f; (d)^s; K_1', \dots, K_t')$ as before, where $f \colon \vartheta(S_{2n+1}^{\ell,m}) \to \underset{\sim}{\delta}_{2n+3}^1$ (here $\bar{K}_1 = v(S_{2n+1}^{\ell,m})$). If s does not appear here, then we require (d) to satisfy condition D. From Lemma 5.21 of section 5, if follows that if s appears, then $(d)^s$ satisfies condition D, so these ordinals are well-defined.

We will establish in part 2 that the ordinals $(\mathrm{id}; d; K_1', \dots, K_t')$, for d satisfying condition C, are all cardinals.

We let $<_M$ denote the ordering on tuples $(d; K_1', \dots, K_t')$, where d is defined and satisfies condition C relative to $K_1', \dots, K_t'$, generated by the relation:

$$(\hat{\mathcal{L}}_M(d; K_1', \dots, K_t'); K_1', \dots, K_t', K_{t+1}') <_M (d; K_1', \dots, K_t')$$

for all K_{t+1}'.

We let $f(d; K_1', \dots, K_t')$ denote the rank of $(d; K_1', \dots, K_t')$ in the ordering $<_M$. It then follows from the above remark (granting the unproven assertion above about the descriptions representing cardinals) that

$$\underset{\sim}{\delta}_{2n+5}^1 \geq \aleph_{\left[\sup_{d, K_1', \dots, K_t'} f(d; K_1', \dots, K_t')\right]} + 1.$$

We now proceed to establish that $\sup_{d, K_1', \dots, K_t'} f(d; K_1', \dots, K_t') \geq E(2n+3)$, yielding the lower bound.

We define two auxiliary orderings $<_q, <_q^k$, where $q \geq 1$ and $1 \leq k \leq c = c_1 + c_2 + \cdots + c_t$, as follows.

We let $<_q$ be generated by the relation:

$$(\hat{\mathcal{L}}_M^{(q)}(d; K_1', \dots, K_t'); K_1', \dots, K_t', K_{t+1}) <_q (d; K_1', \dots, K_t')$$

for all K_{t+1}'.

We let $<_q^k$ be generated by the relation:

$$(\hat{\mathcal{L}}_M^{\restriction k(q)}(d; K_1', \dots, K_t'); K_1', \dots, K_t', K_{t+1}) <_q^k (d; K_1', \dots, K_t').$$

Here $\delta_d(\bar{K}_a) < k \leq k(d)$, d satisfies condition C relative to $K_1', \dots, K_t'$, and $\hat{\mathcal{L}}_M^{(q)}$ denotes the qth iterate of $\hat{\mathcal{L}}_M$.

We let $f_q(d; K'_1, \ldots, K'_t)$ and $f^k_q(d; K'_1, \ldots, K'_t)$ denote the ranks of the tuple $(d; K'_1, \ldots, K'_t)$ in these orderings. We let g_q, g^k_q denote the corresponding rank functions on $\mathcal{D}_{2n-1}$.

We define an auxiliary function $\tilde{f}^k_q(d; K'_1, \ldots, K'_t)$, for $1 \leq k \leq c$, on d satisfying condition C relative to $K'_1, \ldots, K'_t$ and $\delta_d(\bar{K}_a) < k \leq k(d)$. We use the following definition.

DEFINITION 9.1. We define an ordering $\lhd$ on tuples $(\alpha_1, \alpha_j)^s$, where $j = 1$ or 2, $\alpha_2 \leq \alpha_1$, and the symbol s may or may not appear. We define $(\alpha_1, \alpha_j)^s \lhd (\beta_1, \beta_j)^s$ iff one of the following holds:

i) $\alpha_1 < \beta_1$.
ii) $\alpha_1 = \beta_1, j_1 = j_2 = 2$, and $\alpha_2 < \beta_2$.
iii) $\alpha_1 = \beta_1$ and one of the following holds:
  a) $j_1 = 1, j_2 = 2$, $(\vec{\alpha})^s$ involves s, and $\beta_2 > 0$.
  b) $j_1 = 2, j_2 = 1$, and $(\vec{\beta})$ does not involve s.
  c) $j_1 = j_2 = 1$, $(\vec{\alpha})^s$ involves s and $(\vec{\beta})$ does not.

To make $\lhd$ a linear order we further identify (α_1, α_2) with $(\alpha_1, \alpha_2)^s$, and identify $(\alpha_1, 0)$ with $(\alpha_1)^s$. We also identify (α, α) with (α). We let $\tau((\alpha_1, \alpha_j)^s)$ denote the rank of $(\alpha_1, \alpha_j)^s$ in the ordering $\lhd$ with these identifications (so $\tau((\alpha_1, \alpha_2)) = \tau((\alpha_1, \alpha_2)^s)$ and $\tau((\alpha_1, 0)) = r((\alpha_1)^s)$). We also let $\tau(\alpha)$ abbreviate $\tau((\alpha))$.

The reader will note that the ordering $\lhd$ mimics the ordering on descriptions of the form $d = (k; d_1)^s$ or $d = (k; d_1, d_2)^s$ where $K_k = S^{1,m}_{2n+1}$. It can be considered a simplified version of the ordering on these descriptions.

We now define $\bar{f}^k_q$ through the following cases.

Case I. $k(d) = \infty$ and $k = c$.

Subcase I.a. $d = d^{(\mathcal{I}_a)}$ is basic basic type 1, so $d = (\bar{d}_2)^s$, where s may or may not appear. We set $\tilde{f}^k_q(d; K'_1, \ldots, K'_t) = \omega^{\omega^{g_q \cdot 4(d_2; \bar{K}_2, \ldots, \bar{K}_a)}} + 1$, where $\mathcal{I}_a = (f; \bar{K}_2, \ldots, \bar{K}_a)$ as usual, for $g_{q \cdot 4}(d_2; \bar{K}_2, \ldots, \bar{K}_a) > 0$, and otherwise $\tilde{f}^k_q(d; K'_1, \ldots, K'_t) = 0$.

Subcase I.b. d of basic type 0. We set $\tilde{f}^k_q(d; K'_1, \ldots, K'_t) = 0$.

Case II. $k < c$ and $k < k(d)$. We set $\tilde{f}^k_q(d; K'_1, \ldots, K'_t) = \tilde{f}^{k+1}_q(d; K'_1, \ldots, K'_t)$.

Case III. $k < c$ and $k = k(d)$.

Subcase III.a. $K_k = S^{1,m_k}_{2n+1}$. Hence, $d = (k; d_1, \ldots, d_r)^s$, where $r \leq m_k$ and s may or may not appear. We set

$$\tilde{f}^k_q(d; K'_1, \ldots, K'_t) = \tau(((\tilde{f}^{k+1}_q(d_1; K'_1, \ldots, K'_t), \tilde{f}^{k+1}_q(d_2; K'_1, \ldots, K'_t))^s)$$

where τ is the function of Definition 9.1 and s appears here iff it appears in d (if $r = 1$ we don't have the second argument of τ in the formula).

Subcase III.b. $K_k = S_{2n+1}^{\ell_k, m_k}$, with $\ell_k > 1$. Here $d = (k; d_2)^s$, where s may or may not appear. We set $\tilde{f}_q^k(d; K_1', \dots, K_t') = \tilde{f}_q^{k+1}(d_2; K_1', \dots, K_t')$

Subcase III.c. d of basic type -1. We set $\tilde{f}_q^k(d; K_1', \dots, K_t') = 0$

Case IV. $k = c$ and $k = k(d)$. As in case (III) above where we now replace $\tilde{f}_q^{k+1}()$ by $\omega^{\omega^{g_q \cdot 4()}}$ in subcases a and b and retain subcase c. (Actually only subcase b arises due to the definition of B_{2n+1}^m).

We introduce the following hypothesis:

R_q^k: for d satisfying condition C relative to $K_1', \dots, K_t'$, where $\delta_d(\bar{K}_a) < k < k(d)$, we have $f_q^k(d; K_1', \dots, K_t') \geq \tilde{f}_q^k(d; K_1', \dots, K_t')$.

We establish R_q^k by reverse induction on k, and for fixed k by induction on the rank of the tuple with respect to $<$.

We consider the following cases.

Case I. $k(d) = \infty$ and $k = c$. If d is basic of type 0, the result is immediate. Hence d is basic of type 1 of the form $d = (d_2)^s$, where we may assume s appears. We have that

$$\begin{aligned} f_q^c(d; K_1', \dots, K_t') &\geq \sup_{K_{t+1}'}[f_q^c((\mathcal{L}^{(q)}(d_2; \bar{K}_2, \dots, \bar{K}_a))^s; K_1', \dots, K_t', K_{t+1}') + 1] \\ &\geq \sup_{K_{t+1}'}[\sup_{K_{t+2}'} f_q^c((k_1; (k_2; \cdots (k_u; (k_v; (\mathcal{L}^{(q\cdot 2)}(d_2; \bar{K}_2, \dots, \bar{K}_a))^s)) \cdots)); \\ &\qquad\qquad K_1', \dots, K_t', K_{t+1}', K_{t+2}') + 1] \\ &\geq \sup_m \tau^{(m)}(f_q^c((\mathcal{L}^{(q\cdot 2)}(d_2; \bar{K}_2, \dots, \bar{K}_a))^s; K_1', \dots, K_t')). \end{aligned}$$

where $k_1, \dots, k_u$ enumerate the components of K_{t+1}' of the form $S_{2n+1}^{1,m}$ and K_{k_v} is the component of the K_{t+1}' of the form M_κ such that the above description satisfies condition C. Here $\tau^{(m)}$ denotes the mth iterate of τ. The first inequality uses the definition of f_q^k, the second inequality uses also the definition of $\mathcal{L}$ and the form of the measure K_{t+1}', and the third inequality uses the fact that $f_q^k(k; d) \geq \tau(f_q^{k+1}(d))$ for K_k of the form $S_{2n+1}^{1,m}$. We also have that

$$\begin{aligned} &f_q^c((\mathcal{L}^{(q\cdot 2)}(d_2; \bar{K}_2, \dots, \bar{K}_a)^s); K_1', \dots, K_t') \\ &\geq \sup_{K_{t+3}'}[f_q^c((\mathcal{L}^{(q\cdot 3)}(d_2; \bar{K}_2, \dots, \bar{K}_a))^s; K_1', \dots, K_t', K_{t+3}')] \\ &\geq \sup_{K_{t+3}'}[f_q^c((k_1; (k_2; \cdots (k_u; (\mathcal{L}^{(q\cdot 4)}(d_2; \bar{K}_2, \dots, \\ &\qquad \bar{K}_a))^{s(\mathcal{I}_{a+u})})^{s(\mathcal{I}_{a+u-1})} \cdots)^{s(\mathcal{I}_{a+1})})^{s(\mathcal{I}_a)}; K_1', \dots, K_t', K_{t+3}')], \end{aligned}$$

where $k_1, \ldots, k_u$ now enumerate the components of K'_{t+3} of the form S^{ℓ_k, m_k}_{2n+1} with $\ell_k > 1$. By induction this is

$$\geq \sup_{K'_{t+3}} [\tilde{f}^{c'}_q (((\mathcal{L}^{(q\cdot 4)}(d_2; \bar{K}_2, \ldots, \bar{K}_a))^{s(\mathcal{I}_{a+u})}; K'_1, \ldots, K'_t, K'_{t+3})]$$

where c' is the value of c corresponding to the sequence $K'_1, \ldots, K'_t, K'_{t+3}$. This is

$$\begin{aligned} &\geq \sup_{\bar{K}_{a+1}} \omega^{\omega^{[g_{q\cdot 4}(\mathcal{L}^{(q\cdot 4)}(d_2;\bar{K}_2,\ldots,\bar{K}_a);\bar{K}_2,\ldots,\bar{K}_a,\bar{K}_{a+1})]}} \\ &= \omega^{\omega^{\left[\sup_{\bar{K}_{a+1}} (g_{q\cdot 4}(\mathcal{L}^{(q\cdot 4)}(d_2;\bar{K}_2,\ldots,\bar{K}_a);\bar{K}_2,\ldots,\bar{K}_a,\bar{K}_{a+1}))\right]}} \\ &\geq \omega^{\omega^{g_{q\cdot 4}(d_2;\bar{K}_2,\ldots,\bar{K}_a)-1}}. \end{aligned}$$

where $\lambda - 1 = \lambda$ if λ is a limit ordinal. But, for all ordinals β we have that $\sup_m \tau^{(m)}(\omega^{\omega^\beta}) = \omega^{\omega^{\beta+1}}$. Hence,

$$f^c_q(d; K'_1, \ldots, K'_t) \geq \omega^{\omega^{g_{q\cdot 4}(d_2;\bar{K}_2,\ldots,\bar{K}_a)}} = \tilde{f}^c_q(d; K'_1, \ldots, K'_t).$$

Case II. $k < c$ and $k < k(d)$.

Subcase II.a. $k(d) < \infty$. For $\delta_d(\bar{K}_a) < k_1 < k_2 \leq k(d)$, it follows readily that $f^{k_1}_q(d; K'_1, \ldots, K'_t) \geq f^{k_2}_q(d; K'_1, \ldots, K'_t)$. Hence,

$$\begin{aligned} f^k_q(d; K'_1, \ldots, K'_t) &\geq f^{k(d)}_q(d; K'_1, \ldots, K'_t) \\ &\geq \tilde{f}^{k(d)}_q(d; K'_1, \ldots, K'_t) \\ &\geq \tilde{f}^k_q(d; K'_1, \ldots, K'_t). \end{aligned}$$

The second inequality is by induction, and the third is from the definition of $\tilde{f}^k_q$.

Subcase II.b. $k(d) = \infty$.

We have $f^k_q(d; K'_1, \ldots, K'_t) \geq f^c_q(d; K'_1, \ldots, K'_t) \geq \tilde{f}^c_q(d; K'_1, \ldots, K'_t)$, by case I, and $\tilde{f}^c_q(d; K'_1, \ldots, K'_t) = \tilde{f}^k_q(d; K'_1, \ldots, K'_t)$ from the definition of $\tilde{f}^k_q$.

Case III. $k < c$ and $k = k(d)$.

Subcase III.a. $K_k = M_\kappa$. Hence $d = (k; d_2)$. Then,

$$\begin{aligned} f^k_q(d; K'_1, \ldots, K'_t) &\geq f^{k+1}_q(d_2; K'_1, \ldots, K'_t) \\ &\geq \tilde{f}^{k+1}_q(d_2; K'_1, \ldots, K'_t) \\ &= \tilde{f}^k_q(d; K'_1, \ldots, K'_t). \end{aligned}$$

Subcase III.b. $K_k = S^{\ell_k, m_k}_{2n+1}$, where $\ell_k > 1$. Here $d^{(\mathcal{I}_a)} = (k; d_2^{(\mathcal{I}_{a+1})})^s$, where we may assume s appears.

Subcase III.b.i. d_2 is minimal w.r.t. $\hat{\mathcal{L}}_M^{\restriction k+1(q)}$, that is, $\hat{\mathcal{L}}_M^{\restriction k+1(q)}(d_2)$ is not defined. Then by induction, $\tilde{f}_q^{k+1}(d_2; K'_1, \dots, K'_t) = 0$, hence $\tilde{f}_q^k(d; K'_1, \dots, K'_t) = \tilde{f}_q^{k+1}(d_2; K'_1, \dots, K'_t) = 0$, so R_q^k is immediate.

Subcase III.b.ii. $\hat{\mathcal{L}}_M^{\restriction k+1(q)}(d_2)$ is defined.

We first establish that $f_q^k(d; K'_1, \dots, K'_t) \geq f_q^{k+1}(d_2; K'_1, \dots, K'_t)$ for d of this form by induction w.r.t. the ordering $<$. We have that

$$\begin{aligned} f_q^k(d; K'_1, \dots, K'_t) &= \sup_{K'_{t+1}}[f_q^k(\hat{\mathcal{L}}_M^{\restriction k(q)}(d; K'_1, \dots, K'_t); K'_1, \dots, K'_t, K'_{t+1}) + 1] \\ &\geq \sup_{K'_{t+1}}[f_q^k((k; \hat{\mathcal{L}}_M^{\restriction k+1(q)}(d_2; K'_1, \dots, K'_t))^s; K'_1, \dots, K'_{t+1}) + 1] \\ &\geq \sup_{K'_{t+1}}[f_q^{k+1}(\hat{\mathcal{L}}_M^{\restriction k+1(q)}(d_2; K'_1, \dots, K'_t); K'_1, \dots, K'_{t+1}) + 1] \\ &= f_q^{k+1}(d_2; K'_1, \dots, K'_t), \end{aligned}$$

the first inequality using the definition of $\hat{\mathcal{L}}_M^{\restriction k(q)}(d; K'_1, \dots, K'_t)$ and the second inequality being by induction.

We then have that

$$\begin{aligned} f_q^k(d; K'_1, \dots, K'_t) &\geq f_q^{k+1}(d_2; K'_1, \dots, K'_t) \\ &\geq \tilde{f}_q^{k+1}(d_2; K'_1, \dots, K'_t) \\ &= \tilde{f}_q^k(d; K'_1, \dots, K'_t). \end{aligned}$$

Subcase III.c. $K_k = S_{2n+1}^{1,m_k}$. Hence, $d = (k; d_1, \dots, d_r)^s$ where we may assume s appears.

Subcase III.c.i. $r > 2$. We have that $f_q^k(d; K'_1, \dots, K'_t) \geq f_q^k((k; d_1, d_2)^s; K'_1, \dots, K'_t) \geq \tilde{f}_q^k((k; d_1, d_2)^s; K'_1, \dots, K'_t) = \tilde{f}_q^k(d; K'_1, \dots, K'_t)$. We use here the fact that $d' \leq d$ then $f_q^k(d'; K'_1, \dots, K'_t) \leq f_q^k(d; K'_1, \dots, K'_t)$, established by an easy induction.

Subcase III.c.ii. $r = 2$.

Subcase III.c.ii.1. $\hat{\mathcal{L}}_M^{\restriction k+1(q)}(d_2; K'_1, \dots, K'_t)$ is defined. We have that

$$\begin{aligned} &f_q^k(d; K'_1, \dots, K'_t) \\ &\quad \geq \sup_{K'_{t+1}} f_q^k(\hat{\mathcal{L}}_M^{\restriction k(q)}(d; K'_1, \dots, K'_t); K'_1, \dots, K'_{t+1}) + 1 \\ &\quad \geq \sup_{K'_{t+1}} f_q^k((k; d_1, \hat{\mathcal{L}}_M^{\restriction k+1(q)}(d_2; K'_1, \dots, K'_t))^s; K'_1, \dots, K'_{t+1}) + 1, \end{aligned}$$

by cases on whether or not $r = m_k$. By induction, this is

$$\geq \sup_{K'_{t+1}} \tau(f_q^{k+1}(d_1; K'_1, \dots, K'_t), f_q^{k+1}(\hat{\mathcal{L}}_M^{\restriction k+1(q)}(d_2; K'_1, \dots, K'_t); K'_1, \dots, K'_{t+1})) + 1$$

$$\geq \tau(f_q^{k+1}(d_1; K'_1, \dots, K'_t), \sup_{K'_{t+1}} f_q^{k+1}(\hat{\mathcal{L}}_M^{\restriction k+1(q)}(d_2; K'_1, \dots, K'_t); K'_1, \dots, K'_{t+1}) + 1).$$

Hence, $f_q^k(d; K'_1, \dots, K'_t) \geq \tau(f_q^{k+1}(d_1; K'_1, \dots, K'_t), f_q^{k+1}(d_2; K'_1, \dots, K'_t)) = \tilde{f}_q^k(d; K'_1, \dots, K'_t)$.

Subcase III.c.ii.2. $\hat{\mathcal{L}}_M^{\restriction k+1(q)}(d_2; K'_1, \dots, K'_t)$ is not defined.
In this case

$$\begin{aligned} f_q^k(d; K'_1, \dots, K'_t) &\geq f_q^k((k; d_1)^s; K'_1, \dots, K'_t) \\ &\geq \tau((f_q^{k+1}(d_1; K'_1, \dots, K'_t)^s) \\ &= \tau(f_q^{k+1}(d_1; K'_1, \dots, K'_t), 0) \\ &= \tilde{f}_q^k(d; K'_1, \dots, K'_t) \end{aligned}$$

where the second inequality follows by induction, the first equality uses the definition of $\lhd$, and the second equality holds since $f_q^{k+1}(d_2; K'_1, \dots, K'_t) = 0$.

Subcase III.c.iii. $r = 1$ and s appears.

Subcase III.c.iii.1. $\hat{\mathcal{L}}_M^{\restriction k+1(q)}(d_1; K'_1, \dots, K'_t)$ is defined.
We have that

$$\begin{aligned} f_q^k(d; K'_1, \dots, K'_t) &\geq \sup_{K'_{t+1}} [f_q^k(\hat{\mathcal{L}}_M^{\restriction k+1(q)}(d; K'_1, \dots, K'_t); K'_1, \dots, K'_{t+1}) + 1] \\ &\geq \sup_{K'_{t+1}} [f_q^k(k; \hat{\mathcal{L}}_M^{\restriction k+1(q)}(d_1; K'_1, \dots, K'_t); K'_1, \dots, K'_{t+1}) + 1] \\ &\geq \sup_{K'_{t+1}} [\tau(f_q^{k+1}(\hat{\mathcal{L}}_M^{\restriction k+1(q)}(d_1; K'_1, \dots, K'_t); K'_1, \dots, K'_{t+1})) + 1] \\ &= \tau((f_q^{k+1}(d_1; K'_1, \dots, K'_{t+1}))^s) = \tilde{f}_q^k(d; K'_1, \dots, K'_t) \end{aligned}$$

from the definition of τ.

Subcase III.c.iii.2. $\hat{\mathcal{L}}_M^{\restriction k+1(q)}(d_1; K'_1, \dots, K'_t)$ is not defined.
In this case $\tilde{f}_q^{k+1}(d_1; K'_1, \dots, K'_t) = 0$, so R_q^k is immediate.

Subcase III.c.iv. $r = 1$ and s does not appear.

Subcase III.c.iv.1. $\hat{\mathcal{L}}_M^{\restriction k+1(q)}(d_1; K'_1, \dots, K'_t)$ is defined.

Proceeding as above,

$$\begin{aligned}
&f_q^k(d;K'_1,\dots,K'_t)\\
&\qquad \geq \sup_{K'_{t+1}}[f_q^k((k,d_1,\hat{\mathcal{L}}_M^{\restriction k+1(q)}(d_1;K'_1,\dots,K'_t))^s;K'_1,\dots,K'_{t+1})+1]\\
&\qquad \geq \sup_{K'_{t+1}}[\tau(f_q^{k+1}(d_1;K'_1,\dots,K'_t),\\
&\qquad\qquad\qquad f_q^{k+1}(\hat{\mathcal{L}}_M^{\restriction k+1(q)}(d_1;K'_1,\dots,K'_t);K'_1,\dots,K'_{t+1}))+1]\\
&\qquad \geq \tau(f_q^{k+1}(d_1;K'_1,\dots,K'_t),\\
&\qquad\qquad\qquad (\sup_{K'_{t+1}} f_q^{k+1}(\hat{\mathcal{L}}_M^{\restriction k+1(q)}(d_1;K'_1,\dots,K'_t);K'_1,\dots,K'_{t+1})+1))\\
&\qquad = \tilde{f}_q^k(d;K'_1,\dots,K'_t),
\end{aligned}$$

where the last equality uses the definition of $\tilde{f}_q^k$ and the fact that $\tau(\alpha,\alpha) = \tau(\alpha)$.

Subcase III.c.iv.2. $\hat{\mathcal{L}}_M^{\restriction k+1(q)}(d_1;K'_1,\dots,K'_t)$ is not defined.
We have $\tilde{f}_q^{k+1}(d_1;K'_1,\dots,K'_t) = 0$, so R_q^k is immediate.

Case IV. $k = c$ and $k = k(d)$. This case is similar to the previous cases (in fact by a trivial change in the definition of B_{2n+1}^m, so that its last component is in $\bigcup_{m<n} R_{2n+1}$ we may assume that this case does not arise).

This establishes R_q^k in all cases.

If we assume that $\sup_{\bar{d},\bar{K}_1,\dots,\bar{K}_t} g_q(\bar{d};\bar{K}_1,\dots,\bar{K}_t) = E(2n+1)$ for all q, then it follows from R_q and the formulas for $\tilde{f}_q^k$ that $\sup_{d,K'_1,\dots,K'_t} f_q(d,K'_1,\dots,K'_t) = E(2n+3)$ for all q.

REFERENCES

STEPHEN JACKSON

[Jac99] ***A computation of*** $\underset{\sim}{\delta}{}_5^1$, vol. 140, Memoirs of the AMS, no. 670, American Mathematical Society, July 1999.

ALEXANDER S. KECHRIS

[Kec81A] *Homogeneous trees and projective scales*, this volume, originally published in Kechris et al. [CABAL ii], pp. 33–74.

ALEXANDER S. KECHRIS, DONALD A. MARTIN, AND YIANNIS N. MOSCHOVAKIS

[CABAL ii] *Cabal seminar* 77–79, Lecture Notes in Mathematics, no. 839, Berlin, Springer, 1981.

DONALD A. MARTIN

[Mar] AD *and the normal measures on* $\underset{\sim}{\delta}{}_3^1$, unpublished.

YIANNIS N. MOSCHOVAKIS
[Mos80] ***Descriptive set theory***, Studies in Logic and the Foundations of Mathematics, no. 100, North-Holland, Amsterdam, 1980.

DEPARTMENT OF MATHEMATICS
UNIVERSITY OF NORTH TEXAS
P.O. BOX 311430
DENTON, TEXAS 76203-1430
UNITED STATES OF AMERICA
E-mail: jackson@unt.edu

PROJECTIVE SETS AND CARDINAL NUMBERS: SOME QUESTIONS RELATED TO THE CONTINUUM PROBLEM

DONALD A. MARTIN

Editorial Note (2010). This paper was originally written in 1971 and accepted to appear in the *Journal for Symbolic Logic*, but was never published. In 1991, a version incorporating some of the author's corrections from the 1970s was included in a volume of papers for the author's 50th birthday. This retyped paper is based on the 1991 version.

It reflects the state of knowledge from the early 1970s; some of the conjectures made in this paper turned out to be false. For instance at the end of §1, the author conjectures that $\lambda_{2n+1} = \aleph_{\omega \cdot n}$ under AD. While correct for $n = 0, 1$, the conjectured values turned out to be too low, as discussed in [Jac11].

§1. Introduction. A **prewellordering** of a set X is the relation on X induced by a function $f\colon X \to \mathrm{Ord}$, where Ord is the class of all ordinal numbers. In other words, to prewellorder X, divide X into equivalence classes and wellorder the equivalence classes. The **length** of a prewellordering is the order type of the associated wellordering of equivalence classes. There are prewellorderings of the continuum of every length $< (2^{\aleph_0})^+$, where α^+ is the least cardinal (initial ordinal) greater than the ordinal α. For each positive integer n, let $\underset{\sim}{\boldsymbol{\delta}}^1_n$ be the least ordinal other than 0 not the length of a $\underset{\sim}{\boldsymbol{\Delta}}^1_n$ prewellordering of the continuum. (See [Sho67] for information about projective sets.) Note that every $\underset{\sim}{\boldsymbol{\Delta}}^1_n$ prewellordering of the continuum has length $< \underset{\sim}{\boldsymbol{\delta}}^1_n$. It is essentially a classical result that $\underset{\sim}{\boldsymbol{\delta}}^1_1 = \aleph_1$. We shall prove that $\underset{\sim}{\boldsymbol{\delta}}^1_2 \leq \aleph_2$, that, if a measurable cardinal exists, $\underset{\sim}{\boldsymbol{\delta}}^1_3 \leq \aleph_3$, and (as Kunen has independently shown) that, if all projective games are determined, then for all n, we have $\underset{\sim}{\boldsymbol{\delta}}^1_{2n+2} \leq (\underset{\sim}{\boldsymbol{\delta}}^1_{2n+1})^+$), so that, in particular, $\underset{\sim}{\boldsymbol{\delta}}^1_4 \leq \aleph_4$. (See [Mos70] and [Myc64] for information about infinite games and determinacy.) We conjecture that for all n, we have $\underset{\sim}{\boldsymbol{\delta}}^1_n = \aleph_n$. Of course, even $\underset{\sim}{\boldsymbol{\delta}}^1_n \geq \aleph_2$ cannot be proved without refuting the continuum hypothesis, which no one knows how to do at present.

The result that measurable cardinals imply $\underset{\sim}{\boldsymbol{\delta}}^1_3 \leq \aleph_3$ should probably also be counted as independently due to Kunen. We earlier proved that projective determinacy implies $\underset{\sim}{\boldsymbol{\delta}}^1_3 \leq \aleph_3$ and Theorem 3.1 below—independently due to Kunen—allows one to weaken the hypothesis.

Wadge Degrees and Projective Ordinals: The Cabal Seminar, Volume II
Edited by A. S. Kechris, B. Löwe, J. R. Steel
Lecture Notes in Logic, 37

Both Kunen's proof and ours of $\underset{\sim}{\delta}^1_{2n+2} \leq (\underset{\sim}{\delta}^1_{2n+1})^+$ under PD depend on work of Moschovakis [Mos71].

Wellorderings of the continuum might seem a more natural subject than prewellorderings of the continuum. However, a fairly plausible conjecture (and a consequence of projective determinacy) is that no projective wellordering of the continuum exists and that every projective wellordering of a subset of the continuum is countable.

1.1. As is customary in modern descriptive set theory, we work with the space ${}^\omega\omega$ of functions from the natural numbers ω into ω rather than with the reals themselves. ${}^\omega\omega$ is given the product topology, where ω is given the discrete topology. Call a subset of ${}^\omega\omega$ **κ-Suslin** if it has the form

$$\bigcup_{f \in {}^\omega\kappa} \bigcap_{n \in \omega} A_{\overline{f}(n)},$$

where κ is a cardinal (initial von Neumann ordinal), $\overline{f}(n)$ is the sequence $\langle f_0, \ldots, f(n-1)\rangle$, and each $A_{\overline{f}(n)}$ is clopen (closed and open). There are classical results to the effect that every $\underset{\sim}{\Sigma}^1_1$ is $\aleph_0$-Suslin and every $\underset{\sim}{\Sigma}^1_2$ set is $\aleph_1$-Suslin. Every $\aleph_0$-Suslin set is $\underset{\sim}{\Sigma}^1_1$ also, but—at least if a measurable cardinal exists—not every $\aleph_1$-Suslin set is $\underset{\sim}{\Sigma}^1_2$ (although the contrary is consistent with ZFC; see [MS70, pp. 165–166], bearing in mind that being $\aleph_1$-Suslin is equivalent to being the union of $\aleph_1$ Borel sets (3.4 below)). We shall prove, assuming the existence (MC) of a measurable cardinal, that every $\underset{\sim}{\Sigma}^1_3$ set is $\aleph_2$-Suslin. Recently Moschovakis [Mos71] has shown, from projective determinacy (PD), that every $\underset{\sim}{\Sigma}^1_n$ set is λ_n-Suslin for all $n \geq 1$, where $\lambda_{n+1} = \underset{\sim}{\delta}^1_n$ if n is odd. (Hence, in particular, PD implies that every $\underset{\sim}{\Sigma}^1_4$ set is $\aleph_3$-Suslin, which we had already proved by another method.) We conjecture that Moschovakis' result can be improved to show, from PD, that for all n, we have $\lambda_n = \aleph_{n-1}$. This would imply $\underset{\sim}{\delta}^1_n \leq \aleph_n$ for all n, by Theorem 3.1 below. We also conjecture that, for each $n \geq 2$, not every $\underset{\sim}{\Sigma}^1_n$ set is $\aleph_{n-2}$-Suslin. For $n = 2$, this is a classical theorem (since otherwise every $\underset{\sim}{\Sigma}^1_2$ set is $\underset{\sim}{\Sigma}^1_1$). For each n it would follow from $\delta^1_n \geq \aleph_n$ by Theorem 3.1. Note that *every* subset B of ${}^\omega\omega$ is $2^{\aleph_0}$-Suslin. Simply let

$$A_{\overline{f}(n)} = \begin{cases} \text{the empty set} & \text{if } f(0) \notin B; \\ \text{the clopen set determined by } (\overline{f(0)})(n) & \text{if } f(0) \in B, \end{cases}$$

for $f : \omega \to {}^\omega\omega$. So if $2^{\aleph_0} = \aleph_n$ for some n, our conjecture is wrong.

1.2. If A is κ-Suslin and $\aleph_0 < \kappa < \aleph_\omega$, then, as we shall show, A is the union of κ Borel sets.[1] Thus, under suitable assumptions, (as in 1.1), every $\underset{\sim}{\Sigma}^1_{n+1}$ set is the union of $\aleph_n$ Borel sets, for $n = 1, 2, 3$. Our conjectures imply

[1] The converse is true without the restriction $\kappa < \aleph_\omega$.

that this generalizes to all $n \geq 1$ and that not every $\utilde{\Sigma}^1_{n+1}$ set is the union of $< \aleph_n$ Borel sets.

1.3. If α is an ordinal, an algebra of subsets of a set is an **α-algebra** if it is closed under unions and intersections of order type $< \alpha$.[2] If α is an ordinal, $\utilde{B}_\alpha$ is the α-algebra generated by the clopen subsets of ${}^\omega\omega$. We shall mainly be concerned with the algebras $\utilde{B}_{\utilde{\delta}^1_n}$. We have that $\utilde{B}_{\utilde{\delta}^1_1} = \utilde{B}_{\aleph_1}$ is the Borel sets, so the Suslin-Kleene Theorem [Sho67, p. 185] yields that $\utilde{B}_{\utilde{\delta}^1_1} = \utilde{\Delta}^1_1$.

There is no serious hope of generalizing this result, since it is a classical result that $\utilde{B}_{\utilde{\delta}^1_2} \supsetneq \utilde{\Sigma}^1_2$, and we shall prove, assuming MC, that $\utilde{B}_{\utilde{\delta}^1_3} \supsetneq \utilde{\Sigma}^1_3$. Also it follows from PD via Moschovakis [Mos71] that $\utilde{B}_{\utilde{\delta}^1_n} \supsetneq \utilde{\Sigma}^1_n$ for n even.

However, there is a peculiar fact concerning the proofs of these results: The proof for $n = 3$ makes strong use of the axiom of choice, whereas the proof for $n = 2$ uses no choice, and the general proof for even $n > 2$ uses only the axiom of dependent choice (DC). Without the axiom of choice, we are able to prove from MC that $\utilde{\Delta}^1_3 \subseteq \utilde{B}_{\utilde{\delta}^1_3}$, *i.e.*, half the natural generalization of the Suslin-Kleene Theorem for $n = 3$. Using only PD+DC, Moschovakis [Mos71] shows $\utilde{\Delta}^1_n \subseteq \utilde{B}_{\utilde{\delta}^1_n}$, for all odd (and, of course, even) n. The significance of these proofs will be made clear in the next paragraph.

1.4. In §7 we shall prove some consequences of the (false) full axiom of determinacy (AD) (see [Myc64]). Naturally we shall not use the axiom of choice (AC), but we shall use DC. We shall show that AD+DC implies that for all odd n, we have $\utilde{B}_{\delta^1_n} \subseteq \utilde{\Delta}^1_n$.

This result, combined with the Moschovakis theorem mentioned in the last section (which was proved later) gives a Generalized Suslin-Kleene Theorem:

AD+DC implies that for all odd n, we have $\utilde{B}_{\delta^1_n} = \utilde{\Delta}^1_n$.

Unhappily, as we remarked in the last paragraph, this pleasant characterization of $\utilde{\Delta}^1_n$ fails in the real world. Indeed, what remains of the result $\utilde{\Sigma}^1_3 \subseteq \utilde{B}_{\utilde{\delta}^1_3}$, when we extract all use of the axiom of choice from its proof, will be enough—combined with $\utilde{B}_{\utilde{\delta}^1_3} \subseteq \utilde{\Delta}^1_3$, whose proof does not use DC—to prove the following curious consequence of determinacy:

AD implies that $\aleph_n$ is singular for all $n \geq 3$.

Using the "choiceless hull" of a related theorem, we show

AD implies that $\utilde{\delta}^1_3 = \aleph_{\omega+1}$

and

AD+DC implies that $\utilde{\delta}^1_4 = \aleph_{\omega+2}$.

[2] This definition is slightly odd for α not a cardinal. The theorems listed below and in 1.4 explain why we chose this definition.

Since results already mentioned (whose proofs use no more choice than DC)—together with Moschovakis [Mos70]—give that

$$\text{AD+DC implies that } \underset{\sim}{\boldsymbol{\delta}}^1_{2n+2} = (\underset{\sim}{\boldsymbol{\delta}}^1_{2n+1})^+,$$

the $\underset{\sim}{\boldsymbol{\delta}}^1_n$ would all be computed (assuming AD+DC) if we would compute the odd $\underset{\sim}{\boldsymbol{\delta}}^1_n$, $n \geq 5$. Concerning this, Kechris has used Moschovakis' work [Mos71] to generalize partially our results on $\underset{\sim}{\boldsymbol{\delta}}^1_3$:

$$\text{AD+DC implies that } \underset{\sim}{\boldsymbol{\delta}}^1_{2n+2} = (\lambda_{2n+1})^+$$

where λ_{2n+1} is a cardinal cofinal with ω. The natural conjecture would appear to be that $\lambda_{2n+1} = \aleph_{\omega \cdot n}$.

§2. Discussion of the hypotheses.

2.1. We have made several conjectures above. Some of these we hope to be consequences of large cardinal axioms and of projective determinacy. Those which contradict the continuum hypotheses, however, follow neither from known large cardinal axioms nor from, say, determinacy for sets ordinal definable from a real (except possibly through inconsistency). On the other hand it is possible, as far as we know now, that these conjectures can be *refuted* on the basis of the ZFC axioms alone.

Since many of our theorems are of the form PD $\Longrightarrow \varphi$ or MC $\Longrightarrow \varphi$, a few words are in order about the status of these hypotheses.

Note first that MC is essentially a weaker hypothesis. Although PD does not imply MC, PD does imply the existence of inner models with many measurable cardinals (Solovay; see also [Mar77A]). Actually, in this paper we are always able to assume instead of MC only the weaker hypothesis

$$(\forall x)(\exists y)(y = x^\#)$$

which is implied by MC [Sol67B] and which is equivalent [Mar77A] with a certain weak form of Det($\underset{\sim}{\boldsymbol{\Delta}}^1_2$): in the terminology of [Mar77A], Det(α-$\underset{\sim}{\boldsymbol{\Pi}}^1_1$) for all $\alpha < \omega^2$.

We regard both MC and PD as hypotheses for which there is considerable, though nothing remotely like conclusive, evidence. In the case of MC the evidence is mostly *a priori*: analogy with ω and reflection principles (or pseudo-reflection principles) which imply MC. In the case of PD the evidence is mostly *a posteriori*: its consequences (such as the prewellordering theorem [AM68, Mar68]) look *right*—whereas the consequences of, say, $\mathbf{V} = \mathbf{L}$ sometimes do not. (Obviously we are using "*a posteriori*" in a way having nothing to do with sense experience.) One way to increase the evidence for PD would be to prove it from large cardinal axioms—perhaps from the existence of compact or supercompact cardinals.

2.2. The status of full AD is quite different from that of MC and PD: It is known to be false. One may still hope that AD holds in some transitive class satisfying the axioms of set theory other than choice, perhaps even in a class containing all ordinals and all reals. If AD holds in any such class, it also holds in $\mathbf{L}(\mathbb{R})$, the minimal model of set theory containing all reals and all ordinals. $\mathrm{AD}^{\mathbf{L}(\mathbb{R})}$ has been conjectured by several people, including Mycielski, Solovay, and Takeuti. As Solovay perhaps first noted, it is natural to assume DC when one assumes AD, since the axiom of choice implies that $\mathbf{L}(\mathbb{R}) \models \mathrm{DC}$.

If one hopes for such a model M of determinacy, one might also hope that the model is fat enough to look in some respects like the full universe of sets. The author in fact thought at one time that one might have $(\aleph_\alpha)^M = \aleph_\alpha$ for α not too large. Our own results contradict this hypothesis for $\alpha = 3$, since the real $\aleph_3$ is not singular. Nevertheless, it remains possible that, for small enough α the αth regular cardinal in M is the αth regular cardinal.

One reason for proving consequences of AD, is that—assuming $\mathrm{AD}^{\mathbf{L}(\mathbb{R})}$ or PD—such results can sometimes be turned into interesting, though slightly more complicated, theorems about the real world. For example, the assertion $\underset{\sim}{\mathbf{B}}_{\underset{\sim}{\delta}^1_3} = \underset{\sim}{\mathbf{\Delta}}^1_3$ can be modified by considering only special kinds of transfinite unions and intersections, and the modified statement can be proved from, say, PD.

Of course, one could argue in favor of AD+DC that many of its consequences are elegant and appealing, *e.g.*, the Generalized Suslin-Kleene theorem. One might even argue that AD+DC is just as good as the axiom of choice. However, elegance seems less important than truth.

2.3. For obvious reasons, we wish in this paper to keep track of the axioms and hypotheses used in proving our theorems. In particular, we want to keep track of when the axiom of choice (AC) and dependent choice (DC) are used. Since we do not wish to suggest that AC is a hypothesis in the same way that PD is, or merely a formal assumption as AD is, we adopt the following conventions: In the absence of any indication to the contrary, the proof of a theorem uses only the axioms of ZF (not including AC). If AC is used, we indicate its use by writing "THEOREM (AC)". All other hypotheses (including DC) will appear simply as antecedents in conditional theorems, as "AD+DC implies that φ".

§3. κ-Suslin sets.

In this section we prove some properties of κ-Suslin sets. In §§4–6 we shall apply the results of this section to $\underset{\sim}{\mathbf{\Sigma}}^1_n$ sets for various n.

3.1. Finite sequences and trees. As before, if $f : \omega \to X$ then, for each $n \in \omega$, $\overline{f}(n)$ is the sequence $\langle f(0), \ldots, f(n-1)\rangle$. If σ is a finite sequence and $n = \mathrm{lh}(\sigma)$, then $\sigma(0), \ldots, \sigma(n-1)$ are the terms of σ. We shall usually think of a finite sequence of n-tuples as an n-tuple of finite sequences (all of the same length). The collection of all finite sequences of natural numbers we call ${}^{<\omega}\omega$. Let $n \mapsto \sigma_n$ be an effective bijection of ω onto ${}^{<\omega}\omega$ with $\mathrm{lh}(\sigma_n) \leq n$.

By a **tree** on a set X we shall mean a set T of finite sequences of elements of X with the property that if $\sigma \in T$ and σ extends τ, then $\tau \in T$. What we call a "tree" is a special case of what in set theory is usually called a "tree". If T is a tree on $X_1 \times \cdots \times X_n$ and $f_1, \ldots, f_j$ belong to ${}^{\omega}X_1, \ldots, {}^{\omega}X_j$ respectively, then

$$T_{\langle f_1,\ldots,f_j\rangle} = \{\langle \tau_{j+1}, \ldots, \tau_n\rangle : \langle \overline{f_1}(k), \ldots, \overline{f_j}(k), \tau_{j+1}, \ldots, \tau_n\rangle \in T\}.$$

(Recall our convention about sequences of n-tuples.)

If T is a tree on $X_1 \times \cdots \times X_n$ and $Y \subseteq X_j$, then

$$T \restriction Y, j = \{\langle \tau_1, \ldots, \tau_n\rangle \in T : \mathrm{ran}(\tau_j) \subseteq Y\}.$$

We write $T \restriction Y, j$ as $T \restriction Y$ when there is no danger of confusion.

If T is a tree on $X_1 \times \cdots \times X_n$ and $\langle a_1, \ldots, a_n\rangle \in X_1 \times \cdots \times X_n$, then

$$T^{a_1,\ldots,a_n} = \{\langle \tau_1, \ldots, \tau_n\rangle : \langle a_1 * \tau_1, \ldots, a_n * \tau_n\rangle \in T\},$$

where $(a_i * \tau_i)(0) = a_i$ and $(a_i * \tau_i)(k+1) = \tau_i(k)$.

We partially order each tree by setting

$$\tau_1 < \tau_2 \iff \tau_1 \text{ properly extends } \tau_2.$$

We often think of a tree as *being* the partial order of its elements.

Recall that a binary relation R is wellfounded if, for every non-empty subset X of the field of R, there is an $x \in X$ such that no $y \in X$ bears R to x. Using DC, R is wellfounded just in case there are no infinite descending chains with respect to R: *i.e.*, there are no $x_0, x_1, \ldots$ such that $\langle x_1, x_0\rangle \in R$, $\langle x_2, x_1\rangle \in R$, ... Note that if $\langle x, x\rangle \in R$, then $x, x, \ldots$ constitutes an infinite descending chain. (DC is not necessary if the field of R is wellordered.)

Let R be any wellfounded relation. For each element x of the field of R we define an ordinal $|x|^R$:

$$|x|^R = \sup_{\langle x,y\rangle \in R} \{|y|^R + 1\}.$$

The **ordinal** of a wellfounded relation R is $\sup_x\{|x|^R + 1\}$. The **ordinal** of a wellfounded tree T is then $|\Lambda|^T$, where Λ is the empty sequence.

A subset A of $({}^{\omega}\omega)^n$ is **κ-Suslin**, for κ an infinite cardinal (*i.e.*, an infinite initial von Neumann ordinal), if there is a tree T on $\omega^n \times \kappa$ such that

$$\langle x_1, \ldots, x_n\rangle \in A \iff T_{\langle x_1,\ldots,x_n\rangle} \text{ is not wellfounded}.$$

(This definition agrees with that given in §1.1.)

3.2. Theorem 3.1, stated and proved in this section, will yield bounds on many of the $\underset{\sim}{\delta}^1_n$. In 1969, we proved $\underset{\sim}{\delta}^1_2 \leq \aleph_2$ by an argument only slightly different from the appropriate special case of the proof of the theorem given below. The theorem itself was recently proved independently by Kunen and us. Our proof uses forcing, whereas Kunen's is elementary. Indeed, if both proofs are given in detail, Kunen's is almost a sub-proof of ours. We nevertheless feel that our proof may be helpful in that it can be presented without considering combinatorial details.

THEOREM 3.1. If R is a wellfounded relation and R is κ-Suslin, then the ordinal of R is $< \kappa^+$.

PROOF. We first prove the case $\kappa = \aleph_0$. (Kunen's proof does not require considering this case separately.) This case is essentially a classical theorem. Assume R is $\aleph_0$-Suslin ($= \underset{\sim}{\Sigma}^1_1$) and R has ordinal $\geq \aleph_1$. We show, for a contradiction, that the set W of relations r on ω which are wellorderings of ω is $\underset{\sim}{\Sigma}^1_1$:

$$\begin{aligned} r \in W \iff (\exists f)[&f \text{ is a function and } \mathrm{dom}(f) \subseteq {}^\omega\omega \\ &\text{and } \mathrm{ran}(f) = \omega \text{ and } (\forall y)(\forall n)(y \in \mathrm{dom}(f) \\ &\text{and } \langle n, f(y)\rangle \in r \Longrightarrow (\exists z)(z \in \mathrm{dom}(f) \\ &\text{and } \langle z, y\rangle \in R \text{ and } (n = f(z) \text{ or } \langle n, f(z)\rangle \in r)))]. \end{aligned}$$

(If $r \in W$, let $f(y)$ be the number, if any, whose segment in r has order type $|y|^R$.) Furthermore, one easily sees that f may be taken to be countable, so W is $\underset{\sim}{\Sigma}^1_1$. (The choices used in picking a countable f can be made canonically.)

Now we consider the case $\kappa > \aleph_0$. Let T be a tree on $\omega \times \kappa$ witnessing that R is κ-Suslin. To say that there is an infinite descending chain $x_0, x_1, \ldots$ with respect to R is to say that there is an infinite sequence of descending chains in T, witnessing that $\langle x_1, x_0\rangle \in R$, $\langle x_2, x_1\rangle \in R$, *etc.* Shuffling, we produce a tree T^* on $\omega \times \kappa$ which has an infinite descending chain just in case there is an infinite descending chain with respect to R. In addition, R is wellfounded if and only if T^* is wellfounded which in turn is equivalent (without DC) to the statement that T^* has no infinite descending chains.[3]

Let $\mathcal{B}$ be the usual complete Boolean algebra for collapsing κ onto ω: The forcing conditions are the finite functions with domain $\subseteq \omega$ and range $\subseteq \kappa$. In the Boolean-valued universe $V^{(\mathcal{B})}$, we have $(\kappa^+)^{\check{V}} = \omega_1$, where $x \mapsto \check{x}$ is the canonical embedding of V in $V^{(\mathcal{B})}$, extended to proper classes.

We argue in $V^{(\mathcal{B})}$. Let $C := \{\langle x, y\rangle : \check{T}_{\langle x,y\rangle}$ is not wellfounded$\}$. Then $\check{R} \subseteq C$ since $\check{T}_{\langle \check{x},\check{y}\rangle} = (T_{\langle x,y\rangle})\check{}$ and wellfoundedness is absolute. Furthermore, C is wellfounded, since $(\check{T})^* = (T^*)\check{}$ and wellfoundedness is absolute. But $\check{T}$

[3]Kunen completes the proof by verifying that ordinal$(T^*) \geq$ ordinal(R).

is countable, so C is $\aleph_0$-Suslin, whence

$$\text{ordinal}(\check{R}) \leq \text{ordinal}(C) < \omega_1 = (\kappa^+)^{\check{V}}. \qquad \dashv$$

Corollary 3.2 (to proof of Theorem 3.1). If R is a wellfounded relation, and T witnesses that R is κ-Suslin, then the ordinal of R is $<$ the first T-admissible ordinal $> \kappa$.

Proof. The proof is just the lightface analogue of the proof of the theorem, and we omit it. $\dashv$

Using the proof of Theorem 3.1 (either ours or Kunen's) and known results, one can show that $\underset{\sim}{\boldsymbol{\delta}}^1_n$ is the least ordinal not the ordinal of a $\underset{\sim}{\boldsymbol{\Sigma}}^1_n$ wellfounded relation, assuming, say, PD. This fact was noted by both Kunen and us, and Kunen has exploited it in proving that AD+DC implies that for all n, $\underset{\sim}{\boldsymbol{\delta}}^1_n$ is a measurable cardinal. Kunen's proof of the theorem has the advantage over ours that it yields a direct proof of this characterization of $\underset{\sim}{\boldsymbol{\delta}}^1_n$ which furthermore needs to assume only Moschovakis's [Mos71], and so only the determinacy hypotheses of that paper; for $n = 2$ no hypotheses are needed, and for $n = 3$ only "for all x there is a y such that $y = x^{\#}$" is needed, as one can show using §5 below and [MS69]. The case of odd numbers n assuming $\underset{\sim}{\boldsymbol{\Delta}}^1_{n-1}$ determinacy follows from [AM68] and [Mar68] and was already known.

3.3. A subset C of ${}^{\omega}\omega$ **separates** disjoint subsets A and B of ${}^{\omega}\omega$ if $C \supseteq A$ and ${}^{\omega}\omega \setminus C \supseteq B$. The following theorem and its proof are straightforward generalizations of the classical separation theorem (Luzin) for $\underset{\sim}{\boldsymbol{\Sigma}}^1_1$ sets and one of its proofs [Kur58, pp. 393–395].

Theorem 3.3. If A and B are disjoint κ-Suslin sets, then A and B are separated by elements of $\underset{\sim}{\mathbf{B}}_{\kappa^+}$.

Proof. Let T^A and T^B witness that A and B respectively are κ-Suslin. Suppose that for each $m, n \in \omega$ and $\alpha, \beta \in \kappa$, the disjoint sets $\{x : x(0) = m$ and T^A_x has an infinite descending chain beginning with $\langle\alpha\rangle\}$ and $\{x : x(0) = n$ and T^B_x has an infinite descending chain beginning with $\langle\beta\rangle\}$ are separated by elements $C_{m\alpha n\beta}$ of $\underset{\sim}{\mathbf{B}}_{\kappa^+}$. Then

$$C = \bigcup_{m\alpha} \bigcap_{n\beta} C_{m\alpha n\beta}$$

separates A and B. Hence, if A and B are not separable by an element of $\underset{\sim}{\mathbf{B}}_{\kappa^+}$, then we can continue in this way and find functions $x, y \colon \omega \to \omega$ and $f, g \colon \omega \to \kappa$ such that, for each n,

$$A_{\overline{x}(n), \overline{f}(n)} = \{z : \overline{z}(n) = \overline{x}(n) \text{ and } T^A_z \text{ has an infinite descending chain beginning with } \overline{f}(n)\}$$

and the similarly defined $B_{\overline{y}(n),\overline{g}(n)}$ are not separated by any element of $\underset{\sim}{\mathbf{B}}_{\kappa^+}$. Now suppose $\bigcap_n A_{\overline{x}(n),\overline{f}(n)}$ and $\bigcap_n B_{\overline{y}(n),\overline{g}(n)}$ have members x and y respectively. Since $x \in A$ and $y \in B$ and A is disjoint from B, there is an n with $\overline{x}(n) \neq \overline{y}(n)$. But then $\{z : \overline{z}(n) = \overline{x}(n)\}$ separates $A_{\overline{x}(n),\overline{f}(n)}$ and $B_{\overline{y}(n),\overline{g}(n)}$. If either $\bigcap_n A_{\overline{x}(n),\overline{f}(n)}$ or $\bigcap_n B_{\overline{y}(n),\overline{g}(n)}$ is empty, then either the empty set or ${}^{\omega}\omega$ separates some $A_{\overline{x}(n),\overline{f}(n)}$ from $B_{\overline{y}(n),\overline{g}(n)}$. ⊣

The foregoing proof makes a superficial use of the axiom of choice. To avoid AC, observe that the tree

$$T = \{\langle \overline{x}(n), \overline{y}(n), \overline{f}(n), \overline{g}(n)\rangle : \overline{x}(n) = \overline{y}(n) \text{ and } \langle \overline{x}(n), \overline{f}(n)\rangle \in T^A \text{ and } \langle \overline{y}(n), \overline{g}(n)\rangle \in T^B\}$$

is wellfounded, since A and B are disjoint. For non-elements of T, sets separating $A_{\overline{x}(n),\overline{f}(n)}$ and $B_{\overline{y}(n),\overline{g}(n)}$ can be defined trivially, as we have indicated. For elements of T, separating sets are defined by transfinite induction on T.

Corollary 3.4. If A and ${}^{\omega}\omega \setminus A$ are κ-Suslin, then $A \in \underset{\sim}{\mathbf{B}}_{\kappa^+}$.

3.4. Theorem 3.5 in this section implies Theorem 3.3 for the case $\kappa < \aleph_\omega$. However, its proof uses AC, which will be important later. Without AC, we still get a result useful in the context of full AD, and also implying Theorem 3.3 when κ is not cofinal with ω:

Theorem 3.5 (AC). If A is κ-Suslin and $\aleph_0 < \kappa < \aleph_\omega$, then A is a union of κ Borel sets.

Proof. By induction on the cardinal κ. It is a classical result [Kur58, p. 391, Corollary 3] that every $\underset{\sim}{\Sigma}^1_1$ set is a union of $\aleph_1$ Borel sets. The idea of one proof is that, for each countable ordinal α, the set of $x \in A$ such that the Brouwer-Kleene ordering with respect to x (see §4) has wellfounded initial segment of order type $\leq \alpha$ is Borel.[4]

For $n \geq 1$ let T on $\omega \times \aleph_n$ witness that A is $\aleph_n$-Suslin. We have that $A = \{x : T_x \text{ is not wellfounded}\} = \bigcup_{\alpha<\aleph_n} \{x : (T{\restriction}\alpha)_x \text{ is not wellfounded}\}$. By induction, each of the terms of this union is a union of $\leq \kappa$ Borel sets, so we are done. ⊣

Corollary 3.6 (to the proof of Theorem 3.5). If A is κ-Suslin and κ is not cofinal with ω, then $A \in \underset{\sim}{\mathbf{B}}_{\kappa^+}$.

Proof. By the proof of the theorem, it is enough to show that every λ-Suslin set, $\lambda < \kappa$, belongs to $\underset{\sim}{\mathbf{B}}_{\kappa^+}$. For this we show that, for every λ, each λ-Suslin set is an intersection of λ^+ elements of $\underset{\sim}{\mathbf{B}}_{\lambda^+}$ (It can also be shown to be a union

[4]The proof given in [Kur58] is closely related to this one.

of λ^+ elements of B_{λ^+}.) By induction on $\alpha < \lambda^+$, we prove that, for all trees T on $\omega \times \lambda$, the set $\{x : \text{ordinal}(T_x) < \alpha\} \in \underset{\sim}{\mathrm{B}}_{\lambda^+}$. We have that

$$\{x : \text{ordinal}(T_x) < \alpha\} = \bigcup_{\beta<\alpha} \{x : \text{ordinal}(T_x) \leq \beta\}$$
$$= \bigcup_{\beta<\alpha} \bigcap_{\gamma<\beta} \bigcup_{n<\omega} \{x : x(0) = n \text{ and ordinal}(T^{n,\gamma}_{x^*}) < \beta\},$$

where $x^*(m) = x(m+1)$. ($T^{n,\gamma}_x$ is defined in 3.1) To finish the proof, note that $\{x : T_x \text{ not wellfounded}\} = \bigcap_{\alpha<\lambda^+} \{x : T_x \text{ does not have ordinal } < \alpha\}$ for T a tree on $\omega \times \lambda$. ⊣

3.5.

THEOREM 3.7. The class of κ-Suslin sets is closed under projection.

PROOF. Let $C = \{x : (\exists y)(x, y) \in A\}$ with A κ-Suslin. Let T witness that A is κ-Suslin. Then

$$x \in C \iff (\exists y)(T_{\langle x,y\rangle} \text{ is not wellfounded})$$
$$\iff T_x \text{ is not wellfounded.}$$
⊣

§4. **$\underset{\sim}{\Sigma}^1_2$ sets.** That $\underset{\sim}{\Sigma}^1_1$ consists exactly of $\aleph_0$-Suslin sets is just the normal form theorem for $\underset{\sim}{\Sigma}^1_1$ sets. It follows by Theorem 3.1 that $\underset{\sim}{\delta}^1_1 = \aleph_1$, since obviously $\underset{\sim}{\delta}^1_1 \geq \aleph_1$.

To prove (the well-known fact) that every $\underset{\sim}{\Sigma}^1_2$ set is $\aleph_1$-Suslin, it would be enough by Theorem 3.7 to show that every $\underset{\sim}{\Pi}^1_1$ set is $\aleph_1$-Suslin. Nevertheless we shall work directly with $\underset{\sim}{\Sigma}^1_2$ sets so that our work can readily be applied to the study of $\underset{\sim}{\Sigma}^1_2$ sets in 5.3.

Let $A \in \underset{\sim}{\Sigma}^1_2$. Then A is of the form

$$A = \{x : (\exists y \in {}^\omega\omega)(\forall z \in {}^\omega\omega)(\exists n \in \omega)\, R(\overline{x}(n), \overline{y}(n), \overline{z}(n))\}.$$

Choose such an R. Say that $\overline{z}(n)$ is **secured with respect to $\langle x, y\rangle$** if

$$(\exists m \leq n) R(\overline{x}(m), \overline{y}(m), \overline{z}(n))$$

and **unsecured** otherwise. Observe that $\overline{z}(n)$ being secured with respect to $\langle x, y\rangle$ depends only on $\langle \overline{x}(n), \overline{y}(n)\rangle$.

The **Brouwer-Kleene ordering** $<$ of ${}^{<\omega}\omega$ (see 3.1) is defined by

$$\sigma < \tau \iff \sigma \text{ properly extends } \tau$$
$$\text{or } (\exists m)(\sigma(m) < \tau(m) \text{ and } (\forall p < m)(\sigma(p) = \tau(p))).$$

LEMMA 4.1 (Brouwer-Kleene). For A as before, we have $x \in A$ if and only if for some y, the Brouwer-Kleene ordering of the sequences unsecured with respect to $\langle x, y\rangle$ is a wellordering.

Let us define a total ordering $<_{xy}$ of the whole of ${}^{<\omega}\omega$, for each x and $y \in {}^{\omega}\omega$. Set

$$\sigma_n <_{xy} \sigma_m \iff \begin{cases} \sigma_n \text{ secured and } \sigma_m \text{ unsecured, or} \\ \sigma_n \text{ secured and } n < m, \text{ or} \\ \sigma_m \text{ unsecured and } \sigma_n < \sigma_m. \end{cases}$$

In other words, we give the secured sequences the ordering induced by $n \mapsto \sigma_n$ (see 3.1) and place them before the unsecured sequences. Clearly

Lemma 4.2. For A as above, we have $x \in A$ if and only if there is y such that $<_{xy}$ is a wellordering of ${}^{<\omega}\omega$.

Let T be the collection of all $\langle \overline{x}(n), \overline{y}(n), \overline{H}(n) \rangle$, where $x, y \in {}^{\omega}\omega$ and $H \in {}^{\omega}\aleph_1$ such that H^* preserves order $(<_{xy})$, where $H^*(\sigma_n) = H(n)$. Then T is a tree on $\omega \times \omega \times \aleph_1$ (Note that since $\text{lh}(\sigma_n) \leq n$, $\overline{H}(n)$ depends only on $\langle \overline{x}(n), \overline{y}(n) \rangle$.)

Lemma 4.3. For A and T as above, we have $x \in A$ if and only if T_x is not wellfounded.

Proof. We have that $x \in A$ if and only if there is a y such that $<_{xy}$ is a wellordering which in turn is equivalent to the existence of y and H such that H^* preserves $<_{xy}$. This is equivalent to T_x being illfounded. ⊣

Theorem 4.4. Every $\undertilde{\Sigma}^1_2$ set is $\aleph_1$-Suslin.

Corollary 4.5. $\undertilde{\delta}^1_2 \leq \aleph_2$.

Proof. Immediate from Theorem 4.4 and Theorem 3.1. ⊣

Corollary 4.6. Every $\undertilde{\Sigma}^1_2$ set is a union of $\aleph_1$ Borel sets.

Proof. Immediate from the theorem and Theorem 3.5. The proof appears to use AC, since that of Theorem 3.5 does. However, if T is a tree on $\omega \times \alpha$ and $\alpha < \omega_1$, then $\{x : T_x \text{ is not wellfounded}\}$ is *canonically* a union of $\aleph_1$ Borel sets, as can be shown by the same proof as for the case $\alpha = \omega$ (see 3.4). ⊣

Corollary 4.7. $\undertilde{\Sigma}^1_2 \subsetneq \undertilde{\mathrm{B}}_{\undertilde{\delta}^1_2}$.

Proof. We have $\undertilde{\Sigma}^1_2 \subseteq \undertilde{\mathrm{B}}_{\undertilde{\delta}^1_2}$ by Corollary 4.6, since $\undertilde{\delta}^1_2 > \aleph_1$. (To see $\undertilde{\delta}^1_2 > \aleph_1$, let $x < y \iff x$ codes a wellordering of ω and y does not code as short a wellordering. This prewellordering of ${}^{\omega}\omega$ has length $\omega_1 + 1$ and is $\undertilde{\Pi}^1_1$.) $\undertilde{\mathrm{B}}_{\undertilde{\delta}^1_2} \not\subseteq \undertilde{\Sigma}^1_2$ since $\undertilde{\Pi}^1_2 \subseteq \undertilde{\mathrm{B}}_{\undertilde{\delta}^1_2}$. ⊣

Except for Corollary 4.5, the results above are classical. As we mentioned earlier, we proved Corollary 4.5 directly some time before we (and independently Kunen) proved Theorem 3.1.

§5. $\underset{\sim}{\Sigma}^1_3$ sets. As we have remarked, every subset of ${}^\omega\omega$ is $2^{\aleph_0}$-Suslin. In [MS69] it is shown (though only implicitly) that each $\underset{\sim}{\Sigma}^1_3$ set is κ-Suslin for some κ via a tree which can be defined in not too complicated a fashion. Mansfield and the author independently noted this fact and have made use of it. Mansfield [Man71] observes, for example, that the tree in question is ordinal definable from a code for the $\underset{\sim}{\Sigma}^1_3$ set and draws consequences of this. We show below that the tree essentially defined in §7 of [MS69] has cardinality $\leq \aleph_2$, and so that every $\underset{\sim}{\Sigma}^1_3$ set is $\aleph_2$-Suslin.

Our analysis of $\underset{\sim}{\Sigma}^1_3$ sets here more or less follows that of Mansfield [Man71]. This is purely a matter of taste; we could just as well proceed as in [MS69]. We depart from Mansfield in that we use the prewellordering defined in [MS69, §7] rather than using a measurable cardinal directly (since the latter approach does not yield a tree of cardinality $\leq \aleph_2$.) We also depart from [Man71] and [MS69] by defining the tree in a more natural way, so that—in particular—our Lemma 5.12 does not depend upon the axiom of choice for countable sets of sets of reals. We assume familiarity with [Sol67B].

5.1. The orderings $\mathcal{F}^*_n$. In this section we define, for each $n \in \omega$, a well-ordered set $\mathcal{F}^*_n$. In 5.2 we shall use a result of Solovay to prove that $\mathrm{Card}(\mathcal{F}^*_n) \leq \aleph_2$. In 5.3 we prove that every $\underset{\sim}{\Sigma}^1_3$ set is $\mathrm{Card}(\mathcal{F}^*_n)$-Suslin.

5.1.1. If a measurable cardinal exists, then with each $x \in {}^\omega\omega$ is associated a canonical proper class C^x of ordinal numbers with the properties:

a) C^x is closed and unbounded in the order topology.
b) If κ is an uncountable cardinal, then $\mathrm{Card}(C^x \cap \kappa) = \kappa$.
c) If $a_1 < \cdots < a_n$, $b_1 < \cdots < b_n$ are elements of C^x and $\varphi(v_1, \ldots, v_{n+1})$ is a formula of set theory, then

$$\boldsymbol{L}[x] \models \varphi[a_1, \ldots, a_n, x] \iff \boldsymbol{L}[x] \models \varphi[b_1, \ldots, b_n, x].$$

d) $C^x \cup \{x\}$ generates $\boldsymbol{L}[x]$; *i.e.*, every element of $\boldsymbol{L}[x]$ is definable in $\boldsymbol{L}[x]$ from x and elements of C^x via a formula of set theory.

See [Sol67B] for the proof that a unique class C^x satisfying a)–d) exists. We denote the elements of C^x by

$$c^x_1, c^x_2, \ldots, c^x_\alpha, \ldots$$

in order of magnitude.

The set $x^\#$ is the set of Gödel numbers of those sentences of set theory, with constants for x and each c_n, $n \in \omega$, which are true in $\boldsymbol{L}[x]$. The set $x^\#$, if it exists (*i.e.*, if a C^x satisfying a)–d) exists), is the unique element of a certain $\underset{\sim}{\Pi}^1_2$ set [Sol67B], which is empty if $x^\#$ does not exist. The set C^x can be defined from $x^\#$ in a simple fashion.

If $\varphi(v_1, \dots, v_{n+k+2})$ is a formula of set theory and $\alpha_1, \dots, \alpha_k, \beta_1, \dots, \beta_n$ are ordinals with $\alpha_1 < \cdots < \alpha_k$, set

$$\overline{h^x_\varphi}(\beta_1, \dots, \beta_n, c^x_{\alpha_1}, \dots, c^x_{\alpha_k}) = \begin{cases} \mu\gamma(\boldsymbol{L}[x] \models \varphi[\beta_1, \dots, \beta_n, c^x_{\alpha_1}, \dots, c^x_{\alpha_k}, \gamma, x]) \\ \quad \text{if such a } \gamma \text{ exists,} \\ 0 \quad \text{otherwise.} \end{cases}$$

LEMMA 5.1. Assume $x^\#$ exists. If $(\forall i \leq n)(\beta_i < c^x_{\alpha_1} \;\&\; \beta_i < c^x_{\alpha'_1})$ and $\overline{h^x_\varphi}(\beta_1, \dots, \beta_n, c^x_{\alpha_1}, \dots, c^x_{\alpha_k}) < c^x_{\alpha_1}$, then

$$\overline{h^x_\varphi}(\beta_1, \dots, \beta_n, c^x_{\alpha_1}, \dots, c^x_{\alpha_k}) = \overline{h^x_\varphi}(\beta_1, \dots, \beta_n, c^x_{\alpha'_1}, \dots, c^x_{\alpha'_k}).$$

PROOF. A sketch of the proof for the case that each $\beta_i \in C^x$ using essentially our a) and c) is found on page 67 of [Sol67B]. The lemma can be reduced to this special case by replacing the β_i by their definitions as guaranteed by d) and by applying the special case to the definitions. ⊣

By Lemma 5.1, we define

$$\overline{h^x_\varphi}(\beta_1, \dots, \beta_n) = \overline{h^x_\varphi}(\beta_1, \dots, \beta_n, c^x_{\alpha_1}, \dots, c^x_{\alpha_k})$$

for arbitrary sufficiently large $\alpha_1 < \cdots < \alpha_k$.

5.1.2. If $x, y \in {}^\omega\omega$, define

$$[x, y](2n) = x(n)$$
$$[x, y](2n + 1) = y(n).$$

LEMMA 5.2. Assume $[x, y]^\#$ exists. Then $C^{[x,y]} \subseteq C^x \cap C^y$.

PROOF. Let $\alpha \notin C^x$. Then by d) α is defined in $\boldsymbol{L}[x]$ from x and $c^x_{\beta_1} < \cdots < c^x_{\beta_k} < \alpha < c^x_{\beta_{k+1}} < \cdots < c^x_{\beta_n}$. By Lemma 5.1, $c^x_{\beta_{k+1}}, \dots, c^x_{\beta_n}$ may be replaced by any $c^x_{\gamma_1} < \cdots < c^x_{\gamma_{n-k}}$ with $\alpha < c^x_{\gamma_1}$. In particular, by a) they may be replaced by elements of $C^{[x,y]} \cap C^x$. Similarly, each $c^x_{\beta_i}$, $i \leq k$, is defined in $\boldsymbol{L}[x, y]$ from $[x, y]$ and elements of $C^{[x,y]}$ which are $\leq c^x_{\beta_i} < \alpha$, together with elements of $C^{[x,y]}$ which may be chosen larger than α. Hence α is defined in $\boldsymbol{L}[x, y]$ from $[x, y]$ and elements of $C^{[x,y]}$ distinct from α. By c), $\alpha \notin C^{[x,y]}$. ⊣

5.1.3. For each set X and positive integer n, set $[X]^n = \{Y \subseteq X : \text{Card}(Y) = n\}$. If X is an ordered set, we shall think of $[X]^n$ as the collection of all n-tuples $\langle x_1, \dots, x_n \rangle$ with $x_1 < \cdots < x_n$.

For each n, let $\mathcal{F}_n$ be the collection of functions $f : [\aleph_1]^n \to \aleph_1$ such that f is constructible from an element of ${}^\omega\omega$. If $f, g \in \mathcal{F}_n$, we say that $f \sim g$ if there is a closed unbounded $X \subseteq \aleph_1$ such that

$$f(\alpha_1, \dots, \alpha_n) = g(\alpha_1, \dots, \alpha_n) \text{ for all } \langle \alpha_1, \dots, \alpha_n \rangle \in [X]^n,$$

and $f < g$ if there is a closed unbounded $X \subseteq \aleph_1$ such that

$$f(\alpha_1, \dots, \alpha_n) < g(\alpha_1, \dots, \alpha_n) \text{ for } \langle \alpha_1, \dots, \alpha_n \rangle \in [X]^n.$$

For $f \in \mathcal{F}_n$ let $[f]$ be the equivalence class of f with respect to $\sim$. Let $\mathcal{F}_n^* = \{[f] : f \in \mathcal{F}_n\}$. Partially order $\mathcal{F}_n^*$ by the relation induced by $<$.

LEMMA 5.3. *If for all x there is a y such that $y = x^\#$, then $\mathcal{F}_n^*$ is wellordered by $<$.*

PROOF. The ordering is total, since if $f \in \boldsymbol{L}[x]$ and $g \in \boldsymbol{L}[y]$ then one of the following holds for every $\alpha_1 < \cdots < \alpha_n \in C^{[x,y]} \cap \aleph_1$ with α_1 bigger than the countable ordinals used to define f and g in $\boldsymbol{L}[x, y]$ from $[x, y]$ and elements of $C^{[x,y]}$:

$$f(\alpha_1, \dots, \alpha_n) < g(\alpha_1, \dots, \alpha_n)$$
$$f(\alpha_1, \dots, \alpha_n) > g(\alpha_1, \dots, \alpha_n)$$
$$f(\alpha_1, \dots, \alpha_n) = g(\alpha_1, \dots, \alpha_n).$$

If $f_1 > f_2 > \dots$ then, since the intersection of countably many closed unbounded subsets of $\aleph_1$ is closed and unbounded, there is a closed and unbounded $X \subseteq \aleph_1$ such that $\langle \alpha_1, \dots, \alpha_n \rangle \in [X]^n$ implies that $\langle f_i(\alpha_1, \dots, \alpha_n) : i \in \omega \rangle$ is an infinite descending chain. ⊣

We have used DC in showing $\mathcal{F}_n^*$ wellordered. This was not really necessary, since we shall later (Theorem 5.5) define, without any choices, embeddings of the $\mathcal{F}_n^*$ into the ordinals. Hence we have not listed DC as a hypothesis of the lemma.[5]

5.2. The order type of $\mathcal{F}_n^*$.

5.2.1. The class of **uniform indiscernibles** is $C = \bigcap_{x \in {}^\omega\omega} C^x$. We have that C is the class of α that are cardinals in all models $\boldsymbol{L}[x]$, or, equivalently, the class of all α that are x-admissible for all $x \in {}^\omega\omega$, since "α is $x^\#$-admissible" implies that $\alpha \in C^x$, and that implies that α is a cardinal in $\boldsymbol{L}[x]$ which in turn implies that α is x-admissible. We denote the uniform indiscernibles by

$$c_1, c_2, \dots, c_\alpha, \dots$$

in order of magnitude. By a) and b) every uncountable cardinal belongs to C. Clearly $c_1 = \aleph_1$.

Lemma 5.4 is due to Solovay, and we present his proof of it with his permission. Solovay proved the lemma to get a bound on the number of constructible subsets of ${}^\omega\omega$ [Sol67B, p. 51]. The observation that $\mathrm{cf}(d_\alpha) = \mathrm{cf}(d_1)$, which is used in Solovay's proof, is due to us. We originally used it to show that Solovay's proof of "AD implies that $\aleph_2$ is measurable" does not extend to $\aleph_3$.

LEMMA 5.4 (Solovay). *Assume that for all x there is a y such that $y = x^\#$. Then $\mathrm{cf}(c_{\alpha+1}) = \mathrm{cf}(c_2)$ for all $\alpha \geq 1$.*

[5] Jeff Paris pointed out to us that DC is not really used.

Proof. If $c_\alpha = c^x_\beta$ then c_α is definable in $\boldsymbol{L}[x^\#]$ from β and $x^\#$. As in the proof of Lemma 5.2, if $\beta < c_\alpha$ we can derive a contradiction. Thus $c_\alpha = c^x_{c_\alpha}$.[6]

Set $d_\alpha = \sup_{x \in {}^\omega\omega} c^x_{c_\alpha+1}$. We show that $d_\alpha = c_{\alpha+1}$ and $\mathrm{cf}(d_\alpha) = \mathrm{cf}(d_1)$.

Obviously $d_\alpha \le c_{\alpha+1}$. If $d_\alpha < c_{\alpha+1}$ then there is an x such that $d_\alpha \notin C^x$. By a) of 5.1.1 there is a β such that $c^x_\beta < d_\alpha < c^x_{\beta+1}$. By definition of d_α there is a y with $c^x_\beta < c^y_{\alpha+1} \le d_\alpha < c^x_{\beta+1}$. By Lemma 5.2, we get $c^{[x,y]}_{c_\alpha+1} \ge c^x_{c_\beta+1} > d_\alpha$, a contradiction.

To evaluate the cofinality of d_α, note that

$$c^x_{c_1+1} = c^y_{c_1+1} \iff c^x_{c_\alpha+1} = c^y_{c_\alpha+1}$$

by c) of 5.1.1, since

$$c^x_{c^{[x,y]^\#}_\beta+1} = c^y_{c^{[x,y]^\#}_\beta+1}$$

is equivalent to

$$\boldsymbol{L}[[x,y]^\#] \models \varphi\left[c^{[x,y]^\#}_\beta, [x,y]^\#\right]$$

for a certain formula φ, and c_1 and c_α belong to $c^{[x,y]^\#}$. ⊣

5.2.2. Given $f \in \mathcal{F}_n$, we define **the canonical extension of f to $[\mathrm{Ord}]^n$**. The canonical extension of f we also denote by f. If $f \in \mathcal{F}_n$ then, using Lemma 5.1,

$$f(\beta_1,\ldots,\beta_k) = h^x_\varphi(c^x_{\alpha_1},\ldots,c^x_{\alpha_j},\beta_1,\ldots,\beta_k)$$

for some x, φ, and $\alpha_1 < \cdots < \alpha_j < \aleph_1$. We use this same equation to define the canonical extension of f. We must show that the extension is well-defined. If

$$h^x_\varphi(c^x_{\alpha_1},\ldots,c^x_{\alpha_j},\beta_1,\ldots,\beta_k) \ne h^x_\psi(c^y_{\gamma_1},\ldots,c^y_{\gamma_n},\beta_1,\ldots,\beta_k)$$

for some $\beta_1 < \cdots < \beta_k$, then we can define in $\boldsymbol{L}[x,y]$—from $[x,y]$, countable elements of $C^{[x,y]}$ and arbitrary sufficiently large elements of $C^{[x,y]}$—the least β_k such that for some $\beta_1 < \cdots < \beta_{k-1} < \beta_k$ inequality holds. Since $\beta_k < c^{[x,y]}_\eta$ for some η, $\beta_k < c^{[x,y]}_{\aleph_1}$ by c) of 5.1.1.

Theorem 5.5. Assume that for all x there is a y such that $y = x^\#$. Then the order type of $\mathcal{F}^*_n$ is c_{n+1}.

Proof. To show that the order type of $\mathcal{F}^*_n$ is $\le c_{n+1}$, we prove that

$$[f] < [g] \iff f(c_1,\ldots,c_n) < g(c_1,\ldots,c_n)$$

and that $f(c_1,\ldots,c_n) < c_{n+1}$.

---

[6]This fact is not really needed in the proof which follows, but it simplifies the notation.

Suppose for example that $n = 1$ and $f(\beta) = h^x_\varphi(\beta)$ and $g(\beta) = h^y_\psi(\beta)$. Then

$$\begin{aligned} f(c_1) < g(c_1) &\iff h^x_\varphi(c_1) < h^y_\psi(c_1) \\ &\iff (\forall\alpha \geq 1)(h^x_\varphi(c^{[x,y]}_\alpha) < h^y_\psi(c^{[x,y]}_\alpha)) \\ &\iff [f] = [g]. \end{aligned}$$

The argument in the general case is merely more complicated to state, and we omit it.

Since $f(c_1, \ldots, c_n) < c_\alpha$ for some α, we must have $f(c_1, \ldots, c_n) < c_{n+1}$ by c) of 5.1.1.

To prove that the order type of $\mathcal{F}^*_n$ is $\geq c_{n+1}$, we show that every ordinal $< c_{n+1}$ is $f(c_1, \ldots, c_n)$ for some $f \in \mathcal{F}_n$. We proceed by induction on n. If $\gamma < c_{n+1}$, then by the proof of Lemma 5.4, $\gamma < c^x_{c_n+1}$ for some x. By d) of 5.1.1 and Lemma 5.1,

$$\gamma = h^x_\varphi(c^x_{\alpha_1}, \ldots, c^x_{\alpha_k}, c_n)$$

for some φ and $\alpha_1 < \cdots < \alpha_k < c_n$. The desired result follows by induction. ⊣

Corollary 5.6 (AC). Assume that for all x there is a y such that $y = x^\#$. Then the order type of $\mathcal{F}^*_n$ is $< \aleph_3$.

Proof. By Lemma 5.4. ⊣

If the axiom of choice is not assumed, we get only the following result:

Corollary 5.7. Assume that for all x there is a y such that $y = x^\#$. Then the order type of $\mathcal{F}^*_n$ is $\leq \omega_{n+1}$.

Corollary 5.8. Assume that for all x there is a y such that $y = x^\#$. If any ω_k, $k \geq 2$ has cofinality $\neq \mathrm{cf}(c_2)$, then the order type of $\mathcal{F}^*_n$ is $< \omega_k$.

These latter two corollaries will be important in drawing consequences of full determinacy.

5.3. Analysis of $\undertilde{\Sigma}^1_3$ sets. We wish to prove that every $\undertilde{\Sigma}^1_3$ set is $\mathrm{Card}(\bigcup_n \mathcal{F}^*_n)$-Suslin. By Theorem 3.7 it is enough to prove that every $\undertilde{\Pi}^1_2$ set is $\mathrm{Card}(\bigcup_n \mathcal{F}^*_n)$-Suslin.

5.3.1. Let E be $\undertilde{\Pi}^1_2$ and let T be the tree defined from some normal form $\undertilde{\Sigma}^1_2$ representation of the complement A of E as in §4. By varying the relation R (see §4), we may arrange that the empty sequence is unsecured.

We note that T is **homogeneous** in the following sense:

Lemma 5.9. For each $\langle \overline{x}(n), \overline{y}(n)\rangle$, there is a unique permutation π of the sequence $\langle 0, \ldots, n-1\rangle$ such that, for every $\tau : n \to \aleph_1$,

$$\langle \overline{x}(n), \overline{y}(n), \tau\rangle \in T \iff (\forall j, k < n)(\pi(j) < \pi(k) \iff \tau(j) < \tau(k)).$$

Furthermore $\pi(0) = n - 1$.

PROOF. Immediate from the definition of T. We have $\pi(0) = n - 1$ because σ_0 is the empty sequence, which is maximal in the Brouwer-Kleene ordering and unsecured by assumption. ⊣

For convenience, we vary the definition of T (and so of T_x) for the remainder of this section, stipulating that the empty sequence does not belong to T.

If $x \in E$, T_x is wellfounded and so there is a function $\mathcal{H}\colon T_x \to \aleph_1$ which preserves the tree ordering: Set $\mathcal{H}(\langle\sigma, \tau\rangle) = |\langle\sigma, \tau\rangle|^{T_x}$. Then $\mathcal{H}(\langle\sigma, \tau\rangle) < \aleph_1$ by the final clause of Lemma 5.9 and our temporary convention that the empty sequence is not in T. The $\mathcal{H}$ we have defined is, furthermore, constructible from T_x and so from a real (any real coding x and R).

For fixed σ, to define an $\mathcal{H}\colon T_x \to \aleph_1$ which is constructible from a real on all arguments $\langle\sigma, \tau\rangle$ is, by Lemma 5.9, the same thing as choosing an element f_σ of $\mathcal{F}_{\mathrm{lh}(\sigma)}$. Hence any function $\mathcal{H}\colon T_x \to \aleph_1$ constructible from a real determines functions $f_{\sigma_1}, f_{\sigma_2}, \dots$ with $f_{\sigma_n} \in \mathcal{F}_{\mathrm{lh}(\sigma_n)}$ and any such $f_{\sigma_1}, f_{\sigma_2}, \dots$ determine an $\mathcal{H}\colon T_x \to \aleph_1$ (which is constructible from a real, assuming a form of the countable axiom of choice).

Using Lemma 5.9, we may extend T to a tree $\hat{T}$ on $\omega \times \omega \times \mathrm{Ord}$ by setting

$$\langle\sigma, \tau\rangle \in \hat{T}_x \iff (\pi(j) < \pi(k) \iff \tau(j) < \tau(k))$$

for every $\tau\colon n \to \mathrm{Ord}$, with π as in Lemma 5.9.

LEMMA 5.10. Let $Y \subseteq \mathrm{Ord}$ be uncountable. Then $\hat{T}_x \restriction Y$ is wellfounded $\iff$ T_x is wellfounded.

PROOF. If $\hat{T}_x$ is not wellfounded, $\hat{T}_x \restriction Y$ is not wellfounded for some countable Y. By homogeneity, $\hat{T}_x \restriction Z$ is not wellfounded for every Z of order type $\geq$ that of Y. ⊣

Suppose we are given functions $f_{\sigma_1}, f_{\sigma_2}, \dots$ with $f_{\sigma_n} \in \mathcal{F}_{\mathrm{lh}(\sigma_n)}$. Using the canonical extensions of the f_{σ_n}, we can define a **canonical extension**

$$\hat{\mathcal{H}}\colon \hat{T}_x \to \mathrm{Ord}$$

of the $\mathcal{H}\colon T_x \to \aleph_1$ associated with the f_{σ_n}.

LEMMA 5.11. If $\mathcal{H}$ is order preserving, then so is $\hat{\mathcal{H}}$.

PROOF. Similar to the proof in 5.2.2 that the canonical extension of $f \in \mathcal{F}_n$ to $[\mathrm{Ord}]^n$ is well-defined. ⊣

By the proof of Theorem 5.5, an element $[f]$ of $\mathcal{F}_k^*$ determines uniquely $f \restriction [C]^k$. where C is the class of uniform indiscernibles. Therefore, if we pick only $[f_{\sigma_n}]$ and not f_{σ_n} for $n = 1, 2, \dots$, we still define $\mathcal{H}(\sigma_n, \tau)$ for all sequences τ from C.

We are now ready to define a tree $\mathcal{T}$ on $\omega \times \bigcup_n \mathcal{F}_n^*$. Set $\langle\overline{x}(n), \rho\rangle \in \mathcal{T}$ just in case for all $0 < m < n$, we have $\rho(m) \in \mathcal{F}_{\mathrm{lh}(\sigma_m)}^*$ and furthermore $\hat{\mathcal{H}} \restriction (\hat{T} \restriction C)$ is order preserving, where $\mathcal{H}\colon \hat{T} \to \mathrm{Ord}$ is the function determined by setting

$[f_{\sigma_n}] = \rho(n)$, $n = 1, 2, \ldots$. One might note that the latter condition is the same as that $\mathcal{H}\restriction(\hat{T}\restriction\{c_n : n \in \omega\})$ be order preserving.

LEMMA 5.12. $x \in E \iff \mathcal{T}_x$ is not wellfounded.

PROOF. Suppose $x \in E$. Then an order preserving $\mathcal{H}\colon T_x \to \aleph_1$ exists which is constructible from a real. Let $f_{\sigma_1}, f_{\sigma_2}, \ldots$ be the functions associated with $\mathcal{H}$ as above. Lemma 5.11 implies that $[f_{\sigma_1}], [f_{\sigma_2}], \ldots$ yield an infinite descending chain in $\mathcal{T}_x$.

On the other hand, any infinite descending chain in $\mathcal{T}_x$ determines a function

$$\hat{\mathcal{H}}\colon \hat{T}_x\restriction C \to \mathrm{Ord}$$

which preserves order and so witnesses that T_x is wellfounded, by Lemma 5.10. ⊣

THEOREM 5.13. If for all x there is a y such that $y = x^{\#}$, then every $\underset{\sim}{\Sigma}^1_3$ set is $\mathrm{Card}(\bigcup_n \mathcal{F}^*_n)$-Suslin.

COROLLARY 5.14 (AC). If for all x there is a y such that $y = x^{\#}$, then every $\underset{\sim}{\Sigma}^1_3$ set is $\aleph_2$-Suslin.

PROOF. Immediate from Corollary 5.6. ⊣

Without the axiom of choice, we can still use Corollaries 5.7 and 5.8 to get:

COROLLARY 5.15. If for all x there is a y such that $y = x^{\#}$, then every $\underset{\sim}{\Sigma}^1_3$ set is $\aleph_\omega$-Suslin.

COROLLARY 5.16. If for all x there is a y such that $y = x^{\#}$ and if some $\aleph_{n+1}$, $n \geq 1$, has cofinality differing from ω and the cofinality of c_2, then every $\underset{\sim}{\Sigma}^1_3$ set is $\aleph_n$-Suslin.

From Corollary 5.14 and Theorem 3.1, we derive

COROLLARY 5.17 (AC). If for all x there is a y such that $y = x^{\#}$, then $\underset{\sim}{\delta}^1_3 \leq \aleph_3$.

Without the axiom of choice, Corollary 5.15 and Theorem 3.1 yield

COROLLARY 5.18. If for all x there is a y such that $y = x^{\#}$, then $\underset{\sim}{\delta}^1_3 \leq \aleph_{\omega+1}$.

In §7 we show that AD implies $\underset{\sim}{\delta}^1_3 = \aleph_{\omega+1}$.

COROLLARY 5.19 (AC). If for all x there is a y such that $y = x^{\#}$, then every $\underset{\sim}{\Sigma}^1_3$ set is a union of $\aleph_2$ Borel sets.

PROOF. Immediate from Corollary 5.14 and Theorem 3.5. ⊣

Without AC we still get the following result:

COROLLARY 5.20. If for all x there is a y such that $y = x^{\#}$ and if $\mathrm{Card}(c_\omega)$ is not cofinal with ω, then every $\underset{\sim}{\Sigma}^1_3$ set belongs to $\underset{\sim}{\mathrm{B}}_{(c_\omega)^+}$.

PROOF. Immediate from Theorem 5.13 and Corollary 3.6. ⊣

COROLLARY 5.21. If for all x there is a y such that $y = x^{\#}$ and if $\mathrm{Card}(c_\omega)$ is not cofinal with ω, then every $\boldsymbol{\Sigma}^1_3$ set belongs to $\mathbf{B}_{c_2+1}$.

PROOF. By Lemma 5.4, every $\aleph_n$, $n \geq 2$ no larger than c_ω must be cofinal with c_2. The result then follows from Corollary 5.20. ⊣

LEMMA 5.22. $c_\omega < \boldsymbol{\delta}^1_3$.

PROOF. Since $\mathcal{F}^*_n$ is isomorphic to c_{n+1}, we can code ordinals $< c_{n+1}$ by elements of $\mathcal{F}_n$. Elements f of $\mathcal{F}_n$ all have the form

$$f(\beta_1, \dots, \beta_n) = h^x_\varphi(c^x_{\alpha_1}, \dots, c^x_{\alpha_j}, \beta_1, \dots, \beta_n)$$

with the α_i countable. The $c^x_{\alpha_i}$ and x can be coded by some y. We can then code f by y and φ. In §4 of [MS69] it is essentially shown that the resulting prewellordering is $\boldsymbol{\Delta}^1_3$. ⊣

COROLLARY 5.23. If for all x there is a y such that $y = x^{\#}$ and if $\mathrm{Card}(c_\omega)$ is not cofinal with ω, then $\boldsymbol{\Sigma}^1_3 \subseteq \mathbf{B}_{\delta^1_3}$.

PROOF. Immediate from Corollary 5.21 and Lemma 5.22. ⊣

COROLLARY 5.24. If for all x there is a y such that $y = x^{\#}$, then any two disjoint $\boldsymbol{\Sigma}^1_3$ sets are separated by elements of $\mathbf{B}_{\boldsymbol{\delta}^1_3}$.

PROOF. Immediate from Theorem 5.13, Theorem 5.5, Lemma 5.22, and Theorem 3.3. ⊣

COROLLARY 5.25. If for all x there is a y such that $y = x^{\#}$, then $\boldsymbol{\Delta}^1_3 \subseteq \mathbf{B}_{\boldsymbol{\delta}^1_3}$.

5.3.2. Relative to the uniform indiscernibles we can evaluate $\boldsymbol{\delta}^1_2$ exactly:

THEOREM 5.26. If for all x there is a y such that $y = x^{\#}$, then $\boldsymbol{\delta}^1_2 = c_2$.

PROOF. By Corollary 3.2 and by the proof of Theorem 4.4, we see that each wellfounded relation $\boldsymbol{\Sigma}^1_2$ in x has ordinal $<$ the first x-admissible ordinal after $\aleph_1$, which is smaller than c_2.

If $\alpha < c_2$, then $\alpha < c^x_{\aleph_1+1}$ for some x by the proof of Lemma 5.4. Each ordinal $\beta < \alpha$ is coded by $h^x_\varphi(c^x_{\alpha_1}, \dots, c^x_{\alpha_n}, c^x_{\aleph_1})$ for some φ and $\alpha_1, \dots, \alpha_n < \aleph_1$. If we code β by φ and codes for $\alpha_1, \dots, \alpha_n$, the resulting prewellordering is $\boldsymbol{\Pi}^1_1$ in $x^{\#}$. Hence $\alpha < \boldsymbol{\delta}^1_2$. ⊣

§6. Higher levels in the projective hierarchy.

THEOREM 6.1 (Moschovakis [Mos71]). $\mathrm{Det}(\boldsymbol{\Delta}^1_{2n})$+DC implies that every $\boldsymbol{\Sigma}^1_{2n+2}$ set is $\boldsymbol{\delta}^1_{2n}$-Suslin, and every $\boldsymbol{\Sigma}^1_{2n+1}$ set is λ_{2n+1}-Suslin, where $\lambda_{2n+1} < \boldsymbol{\delta}^1_{2n+1}$.

Using results of Moschovakis [Mos70], it can be shown that AD+DC implies that $\boldsymbol{\Sigma}^1_n$ is equal to the set of λ_n-Suslin sets for all $n \geq 1$, where $\lambda_{2m+2} = \boldsymbol{\delta}^1_{2m+1}$.

COROLLARY 6.2 (Independently due to Kunen). $\mathsf{Det}(\underset{\sim}{\mathbf{\Delta}}^1_{2n})+\mathsf{DC}$ implies that $\underset{\sim}{\boldsymbol{\delta}}^1_{2n+2} \leq (\underset{\sim}{\boldsymbol{\delta}}^1_{2n+1})^+$.

PROOF. Immediate from Theorem 3.1. ⊣

COROLLARY 6.3 (AC). $\mathsf{Det}(\underset{\sim}{\mathbf{\Delta}}^1_2)$ implies that $\underset{\sim}{\boldsymbol{\delta}}^1_4 \leq \aleph_4$.

PROOF. From Corollary 5.14 and Corollary 5.17, since $\mathsf{Det}(\underset{\sim}{\mathbf{\Delta}}^1_2)$ implies "for all x there is a y such that $y = x^{\#}$". (See [Fri71A] or [Mar68].) ⊣

COROLLARY 6.4. $\mathsf{Det}(\underset{\sim}{\mathbf{\Delta}}^1_2)+\mathsf{DC}$ implies that $\underset{\sim}{\boldsymbol{\delta}}^1_4 \leq \aleph_{\omega+2}$.

PROOF. From Corollary 5.14 and Corollary 5.18. ⊣

COROLLARY 6.5 (Moschovakis [Mos71]). $\mathsf{Det}(\underset{\sim}{\mathbf{\Delta}}^1_{2n})+\mathsf{DC}$ implies that $\underset{\sim}{\mathbf{\Delta}}^1_{2n+1} \subseteq \underset{\sim}{\mathbf{B}}_{\underset{\sim}{\boldsymbol{\delta}}^1_{2n+1}}$.

PROOF. Immediate from Corollary 3.4.[7] ⊣

COROLLARY 6.6 (AC). If PD holds and if $\underset{\sim}{\boldsymbol{\delta}}^1_n \leq \aleph_n$ for all $n \geq 1$, then every $\underset{\sim}{\mathbf{\Sigma}}^1_{n+1}$ set is a union of $\aleph_n$ Borel sets.

PROOF. Immediate from Theorem 3.5. ⊣

§7. Consequences of the full axiom of determinacy. In this section we show that $\mathsf{AD}+\mathsf{AC}$ implies that $\underset{\sim}{\mathbf{B}}_{\underset{\sim}{\boldsymbol{\delta}}^1_n} \subseteq \underset{\sim}{\mathbf{\Delta}}^1_n$ for all odd n (and so that $\underset{\sim}{\mathbf{B}}_{\underset{\sim}{\boldsymbol{\delta}}^1_n} \subseteq \underset{\sim}{\mathbf{\Delta}}^1_n$ by Moschovakis' Corollary 4 to (his) Theorem 5). This theorem, combined with the results of §5, has surprising consequences concerning the $\underset{\sim}{\boldsymbol{\delta}}^1_n$ and the $\aleph_\alpha$.

We recall that $\mathsf{PWO}(\underset{\sim}{\mathbf{\Sigma}}^1_n)$ is the assertion that there is a map

$$g\colon U \to \mathrm{Ord},$$

where U is a universal $\underset{\sim}{\mathbf{\Sigma}}^1_n$ set, such that there are a $\underset{\sim}{\mathbf{\Sigma}}^1_n$ relation R_1 and a $\underset{\sim}{\mathbf{\Pi}}^1_n$ relation R_2 with

$$g(x) < g(y) \iff \langle x, y\rangle \in R_1$$

and

$$y \in U \Longrightarrow (g(x) < g(y) \iff \langle x, y\rangle \in R_2).$$

$\mathsf{Sep}(\underset{\sim}{\mathbf{\Sigma}}^1_n)$ is the assertion that any two disjoint ($\underset{\sim}{\mathbf{\Sigma}}^1_n$) sets are separated by a $\underset{\sim}{\mathbf{\Delta}}^1_n$ set. $\mathsf{Red}(\underset{\sim}{\mathbf{\Sigma}}^1_n)$ is the assertion that, for any $\underset{\sim}{\mathbf{\Sigma}}^1_n$ sets A and B, there are disjoint $\underset{\sim}{\mathbf{\Sigma}}^1_n$ sets A' and B' with $A' \subseteq A$, $B' \subseteq B$, and $A' \cup B' = A \cup B$.

$\mathsf{PWO}(\underset{\sim}{\mathbf{\Pi}}^1_n)$, $\mathsf{Red}(\underset{\sim}{\mathbf{\Pi}}^1_n)$, and $\mathsf{Sep}(\underset{\sim}{\mathbf{\Pi}}^1_n)$ are similarly defined (putting "$\underset{\sim}{\mathbf{\Pi}}$" for "$\underset{\sim}{\mathbf{\Sigma}}$" everywhere).

Obviously $\mathsf{Red}(\underset{\sim}{\mathbf{\Pi}}^1_n) \Longrightarrow \mathsf{Sep}(\underset{\sim}{\mathbf{\Sigma}}^1_n)$ and $\mathsf{Red}(\underset{\sim}{\mathbf{\Sigma}}^1_n) \Longrightarrow \mathsf{Sep}(\underset{\sim}{\mathbf{\Pi}}^1_n)$. The following lemma is a special case of a well-known general fact. (In particular, it holds with "$\underset{\sim}{\mathbf{\Pi}}$" in place of "$\underset{\sim}{\mathbf{\Sigma}}$".)

---

[7] The case $n = 1$ is, of course, just a weak version of Corollary 5.25.

LEMMA 7.1. $\mathsf{PWO}(\boldsymbol{\Sigma}^1_n) \Longrightarrow \mathsf{Red}(\boldsymbol{\Sigma}^1_n) \Longrightarrow \neg\mathsf{Sep}(\boldsymbol{\Sigma}^1_n)$.

PROOF. Let U, g, R_1 and R_2 be as in the definition of $\mathsf{PWO}(\boldsymbol{\Sigma}^1_n)$. Let $A = \{x : \langle y_0, x\rangle \in U\}$ and $B = \{x : \langle y_1, x\rangle \in U\}$. Let

$$x \in A' \iff x \in A \,\&\, g(\langle y_0, x\rangle) < g(\langle y_1, x\rangle);$$
$$x \in B' \iff x \in B \,\&\, g(\langle y_0, x\rangle) \not< g(\langle y_1, x\rangle).$$

Using the properties of R_1 and R_2, we see that A' and B' are $\boldsymbol{\Sigma}^1_n$ and so that they have the properties as in the definition of $\mathsf{Red}(\boldsymbol{\Sigma}^1_n)$.

Assume that $\mathsf{Red}(\boldsymbol{\Sigma}^1_n)$ and $\mathsf{Sep}(\boldsymbol{\Sigma}^1_n)$ both hold. Consider a $\boldsymbol{\Sigma}^1_n$ enumeration of all pairs of $\boldsymbol{\Sigma}^1_n$ sets. Pull the pairs apart by $\mathsf{Red}(\boldsymbol{\Sigma}^1_n)$ and push them back together by $\mathsf{Sep}(\boldsymbol{\Sigma}^1_n)$. This yields a universal $\boldsymbol{\Delta}^1_n$ set, which an easy diagonal argument shows to be impossible. ⊣

THEOREM 7.2. AD implies that $\mathbf{B}_{\boldsymbol{\delta}^1_n} \subseteq \boldsymbol{\Delta}^1_n$ or $\mathsf{PWO}(\boldsymbol{\Sigma}^1_n)$).

PROOF. We first give the proof of a result of Wadge which shows essentially that any set in $\boldsymbol{\Sigma}^1_n \setminus \boldsymbol{\Pi}^1_n$ is a "universal" $\boldsymbol{\Sigma}^1_n$ set. Say that $A_1 \leq_{\mathrm{W}} A_2$ for $A_1, A_2 \in {}^{\omega}\omega$ if there is a continuous function $f : {}^{\omega}\omega \to {}^{\omega}\omega$ such that $f(A_1) \subseteq A_2$ and $f({}^{\omega}\omega \setminus A_1) \subseteq {}^{\omega}\omega \setminus A_2$.

LEMMA 7.3 (Wadge). AD implies that for any A_1 and A_2, we have either $A_1 \leq_{\mathrm{W}} A_2$ or $A_2 \leq_{\mathrm{W}} {}^{\omega}\omega \setminus A_1$.

PROOF OF LEMMA 7.3. Consider the game which player I wins if I's play belongs to A_2 if and only if player II's play belongs to A_1. A winning strategy for player I gives a continuous function witnessing $A_1 \leq_{\mathrm{W}} A_2$ and a winning strategy for player II gives a function witnessing $A_2 \leq_{\mathrm{W}} {}^{\omega}\omega \setminus A_1$. ⊣ (Lemma 7.3)

Note that we needed determinacy only for the game used in the proof, whose payoff set is closely related to A_1 and A_2. For projective A_1 and A_2, for example, only PD is needed.

Returning to the proof of the theorem, assume AD and that $\mathbf{B}_{\boldsymbol{\delta}^1_n} \not\subseteq \boldsymbol{\Delta}^1_n$ and let η be the least ordinal such that some union of η elements of $\boldsymbol{\Delta}^1_n$ is not $\boldsymbol{\Delta}^1_n$. By our assumption, $\eta < \boldsymbol{\delta}^1_n$. Let A_α, $\alpha < \eta$ be $\boldsymbol{\Delta}^1_n$ sets whose union is not $\boldsymbol{\Delta}^1_n$. Using the minimality of η, we may assume that the family is increasing (i.e., if $\alpha < \beta$, then $A_\alpha \subseteq A_\beta$). Also by the minimality of η, we know that for every $\alpha < \eta$, the set $\widetilde{A}_\alpha = \bigcup_{\beta<\alpha} A_\beta$ is $\boldsymbol{\Delta}^1_n$. Let $x \mapsto |x|$ be a surjection of ${}^{\omega}\omega$ onto η such that the relation $|x| \leq |y|$ is $\boldsymbol{\Delta}^1_n$. This function exists, since $\eta < \boldsymbol{\delta}^1_n$.

By [Mos70, Lemma 6], every union of $< \boldsymbol{\delta}^1_n$ $\boldsymbol{\Sigma}^1_n$ sets is $\boldsymbol{\Sigma}^1_n$, so

$$\{\langle x, y\rangle : x \in A_{|y|}\} \left(= \bigcup_{\alpha<\eta} \{\langle x, y\rangle : x \in A_{|y|} \,\&\, |y| = \alpha\}\right) \in \boldsymbol{\Sigma}^1_n.$$

By the same lemma, the relations $x \notin A_{|y|}$ and $x \notin \widetilde{A}_{|y|}$ are $\mathbf{\Sigma}^1_n$ and we have that $A = \bigcup_{\alpha<\eta} A_\alpha \in \mathbf{\Sigma}^1_n$. Let U be a universal $\mathbf{\Sigma}^1_n$ set, and let $h\colon {}^\omega\omega \times {}^\omega\omega \to {}^\omega\omega$ be the standard homeomorphism. By Wadge's Lemma 7.3,

$$\text{either } h(U) \leq_{\mathrm{W}} A \text{ or } A \leq_{\mathrm{W}} {}^\omega\omega \setminus h(U).$$

The latter alternative implies $A \in \mathbf{\Pi}^1_n$ contrary to hypothesis. Let f then witness $h(U) \leq_{\mathrm{W}} A$. Define

$$g(x) = \mu\alpha[fh(x) \in A_\alpha];$$

$$R_1(x_1, x_2) \iff (\exists y)(fh(x_1) \in A_{|y|} \;\&\; fh(x_2) \notin A_{|y|}) \;\&\; fh(x_2) \in A;$$

$$R_2(x_1, x_2) \iff (\forall y)(fh(x_2) \in A_{|y|} \Longrightarrow fh(x_1) \in \widetilde{A}_{|y|}).$$

U, g, R_1 and R_2 witness $\mathsf{PWO}(\mathbf{\Sigma}^1_n)$. ⊣ (Theorem 7.2)

Note that Theorem 7.2 holds for any reasonable class closed under projection in place of $\mathbf{\Sigma}^1_n$.

Corollary 7.4. If n is odd, then AD+DC implies that $\mathbf{B}_{\boldsymbol{\delta}^1_n} \subseteq \mathbf{\Delta}^1_n$.

Proof. By [AM68] and [Mar68], AD+DC gives $\mathsf{Sep}(\mathbf{\Sigma}^1_n)$, so $\mathsf{PWO}(\mathbf{\Sigma}^1_n)$ is ruled out by Lemma 7.1. ⊣

Corollary 7.5 (Generalized Suslin-Kleene Theorem). If n is odd, then AD+DC implies that $\mathbf{\Delta}^1_n = \mathbf{B}_{\boldsymbol{\delta}^1_n}$.

Proof. Immediate from Corollary 7.4 and Corollary 6.5. (Moschovakis pointed out Corollary 7.5 when he proved Corollary 6.5 [Mos71].) ⊣

In the case $n = 3$, DC was not used in proving "if for all x there is a y such that $y = x^\#$, then $\mathbf{\Delta}^1_3 \subseteq \mathbf{B}_{\boldsymbol{\delta}^1_3}$" (Corollary 5.25). Furthermore, DC is not needed to prove $\mathsf{Red}(\mathbf{\Pi}^1_3)$ and so $\mathsf{Sep}(\mathbf{\Sigma}^1_3)$ from $\mathsf{Det}(\mathbf{\Delta}^1_2)$, as we see as follows: In the proof of Red from PWO, we can get by with the weaker assumption that g maps U into a *linearly* ordered set (instead of a wellordered set such as the ordinals). In the proof of $\mathsf{PWO}(\mathbf{\Pi}^1_3)$ from $\mathsf{Det}(\mathbf{\Delta}^1_2)$ (see, e.g., [Mar68]), DC is used only to prove that g maps into a *well*-ordered set. Hence we have

Corollary 7.6. AD implies that $\mathbf{\Delta}^1_3 = \mathbf{B}_{\boldsymbol{\delta}^1_3}$.

Corollary 7.7. AD implies that $c_\omega = \aleph_\omega$.

Proof. Suppose $c_\omega < \aleph_\omega$. Then $\operatorname{Card} c_\omega = \aleph_n$ for some $n \geq 1$. But AD implies that $\aleph_n$ is not cofinal with ω, since (Friedman; Moschovakis [Mos70]) $2^{\aleph_{n-1}}$ is a surjective image of ${}^\omega\omega$ and the axiom of choice for countable sets of sets of reals follows from AD [Myc64]. The Corollary follows by Corollary 5.23 and Corollary 7.6. ⊣

Kunen and Solovay have independently used Corollary 7.7 to show that $\mathsf{AD} \Longrightarrow c_n = \aleph_n$ for each n.

Parts of Corollaries 7.8 and Corollary 7.11 below depend upon an unpublished result of Solovay.

COROLLARY 7.8. AD implies that for all $n \geq 2$, we have that $\aleph_n$ has cofinality $\aleph_2$.

PROOF. The cardinal $\aleph_2$ is regular, since Solovay (unpublished) has proved $\aleph_2$ measurable. By Lemma 5.4 (Solovay) and the fact that each $\aleph_n$, $n \geq 1$, is some c_m, $m \geq 1$, $\mathrm{cf}(\aleph_n) = \mathrm{cf}(\aleph_2)$. ⊣

COROLLARY 7.9. AD implies that $\underset{\sim}{\boldsymbol{\delta}}^1_3 = \aleph_{\omega+1}$.

PROOF. Assume AD. That $\underset{\sim}{\boldsymbol{\delta}}^1_3 \leq \aleph_{\omega+1}$ follows from Corollary 5.18 and the fact that AD implies that for all x there is a y such that $y = x^{\#}$.

By Lemma 5.12, $\underset{\sim}{\boldsymbol{\delta}}^1_3 > c_\omega = \aleph_\omega$ and, by [Mos70, Theorem 8], $\underset{\sim}{\boldsymbol{\delta}}^1_3$ is a cardinal, so $\underset{\sim}{\boldsymbol{\delta}}^1_3 \geq \aleph_{\omega+1}$. ⊣

COROLLARY 7.10. AD+DC implies that $\underset{\sim}{\boldsymbol{\delta}}^1_4 = \aleph_{\omega+2}$.

PROOF. $\underset{\sim}{\boldsymbol{\delta}}^1_4 > \underset{\sim}{\boldsymbol{\delta}}^1_3$ follows from $\mathsf{PWO}(\underset{\sim}{\boldsymbol{\Pi}}^1_3)$. Also, $\underset{\sim}{\boldsymbol{\delta}}^1_4$ is a cardinal by Moschovakis [Mos70]. The inequality $\underset{\sim}{\boldsymbol{\delta}}^1_4 \leq \aleph_{\omega+2}$ is Corollary 6.3. ⊣

COROLLARY 7.11. AD implies that the nth uncountable regular cardinal $= \underset{\sim}{\boldsymbol{\delta}}^1_n$ for $n = 1, 2, 3$.

PROOF. By Theorem 5.26, $\underset{\sim}{\boldsymbol{\delta}}^1_2 = c_2 = \aleph_2$, since $\underset{\sim}{\boldsymbol{\delta}}^1_2$ is a cardinal [Mos70, Theorem 8]. The cardinal $\aleph_2$ is regular by Solovay's result mentioned in the proof of Corollary 7.7.[8]

The case $n = 3$ follows from the case $n = 2$, Corollaries 7.8 and 7.9, and the fact that $\underset{\sim}{\boldsymbol{\delta}}^1_3$ is a regular cardinal. (If $\underset{\sim}{\boldsymbol{\delta}}^1_3$ were singular, we would have $\underset{\sim}{\boldsymbol{\Sigma}}^1_3 \subseteq \mathbf{B}_{(\underset{\sim}{\boldsymbol{\delta}}^1_3)^+} = \mathbf{B}_{\underset{\sim}{\boldsymbol{\delta}}^1_3}$ by the proof of Corollary 3.6. [Mos70] shows $\underset{\sim}{\boldsymbol{\delta}}^1_3$ regular from AD, but he needs DC.) ⊣

Kunen has recently shown that $\underset{\sim}{\boldsymbol{\delta}}^1_n$ is measurable for all n (we earlier showed this for all odd n), and so Corollary 7.11 can be extended (assuming DC) to cover the case $n = 4$. Furthermore "regular" can be replaced by "measurable" in the statement of that corollary.

Given our results and the result of Kunen just alluded to, we know that $\aleph_1$, $\aleph_2$, $\aleph_{\omega+1}$ and $\aleph_{\omega+2}$ are measurable ($\aleph_1$ and $\aleph_2$ due to Solovay) while $\aleph_3, \aleph_4, \ldots$ are singular, assuming AD+DC. One would expect $\aleph_{\omega+3}, \ldots$ to be singular, $\aleph_{\omega\cdot 2+1}$ and $\aleph_{\omega\cdot 2+2}$ to be measurable, etc. This expectation is reinforced by Kechris' recent result that $\mathsf{AD}+\mathsf{DC} \Longrightarrow$ each odd $\underset{\sim}{\boldsymbol{\delta}}^1_n$ is the successor of a

[8] Solovay's proof in fact shows that c_2 is a measurable cardinal.

cardinal cofinal with ω. Recall also that 1 and 2 are regular, while $3, 4, \ldots$ are singular.

REFERENCES

JOHN W. ADDISON AND YIANNIS N. MOSCHOVAKIS

[AM68] *Some consequences of the axiom of definable determinateness*, ***Proceedings of the National Academy of Sciences of the United States of America***, no. 59, 1968, pp. 708–712.

HARVEY FRIEDMAN

[Fri71A] *Determinateness in the low projective hierarchy*, ***Fundamenta Mathematicae***, vol. 72 (1971), pp. 79–95.

STEPHEN JACKSON

[Jac11] *Projective ordinals. Introduction to Part IV*, 2011, this volume.

CASIMIR KURATOWSKI

[Kur58] ***Topologie. Vol. I***, 4ème ed., Monografie Matematyczne, vol. 20, Państwowe Wydawnictwo Naukowe, Warsaw, 1958.

RICHARD MANSFIELD

[Man71] *A Souslin operation on* Π^1_2, ***Israel Journal of Mathematics***, vol. 9 (1971), no. 3, pp. 367–379.

DONALD A. MARTIN

[Mar68] *The axiom of determinateness and reduction principles in the analytical hierarchy*, ***Bulletin of the American Mathematical Society***, vol. 74 (1968), pp. 687–689.

[Mar77A] α-Π^1_1 *games*, planned but unfinished paper, 1977.

DONALD A. MARTIN AND ROBERT M. SOLOVAY

[MS69] *A basis theorem for* Σ^1_3 *sets of reals*, ***Annals of Mathematics***, vol. 89 (1969), pp. 138–160.

[MS70] *Internal Cohen extensions*, ***Annals of Mathematical Logic***, vol. 2 (1970), no. 2, pp. 143–178.

YIANNIS N. MOSCHOVAKIS

[Mos70] *Determinacy and prewellorderings of the continuum*, ***Mathematical logic and foundations of set theory. Proceedings of an international colloquium held under the auspices of the Israel Academy of Sciences and Humanities, Jerusalem, 11–14 November 1968*** (Y. Bar-Hillel, editor), Studies in Logic and the Foundations of Mathematics, North-Holland, Amsterdam-London, 1970, pp. 24–62.

[Mos71] *Uniformization in a playful universe*, ***Bulletin of the American Mathematical Society***, vol. 77 (1971), pp. 731–736.

JAN MYCIELSKI

[Myc64] *On the axiom of determinateness*, ***Fundamenta Mathematicae***, vol. 53 (1964), pp. 205–224.

JOSEPH R. SHOENFIELD

[Sho67] ***Mathematical logic***, Addison-Wesley, 1967.

ROBERT M. SOLOVAY

[Sol67B] *A nonconstructible* Δ^1_3 *set of integers*, ***Transactions of the American Mathematical Society***, vol. 127 (1967), no. 1, pp. 50–75.

DEPARTMENT OF MATHEMATICS
UNIVERSITY OF CALIFORNIA
LOS ANGELES, CALIFORNIA 90024
UNITED STATES OF AMERICA
E-mail: dam@math.ucla.edu

REGULAR CARDINALS WITHOUT THE WEAK PARTITION PROPERTY

STEVE JACKSON

§1. **Introduction.** We work throughout in the theory ZF+DC, though our main results will require AD as well. Recall that $\kappa \rightarrow (\kappa)^{\vartheta}$ if for all partitions $\mathcal{P}\colon (\kappa)^{\vartheta} \rightarrow \{0,1\}$ of the increasing functions from ϑ to κ into two pieces, there is a **homogeneous** set $H \subseteq \kappa$ of size κ. That is, there is an $i \in \{0,1\}$ such that $\mathcal{P}$ restricted to $(H)^{\vartheta}$ has constant value i. We say κ has the **weak partition property**, $\kappa \rightarrow \kappa^{<\kappa}$, if $\kappa \rightarrow \kappa^{\vartheta}$ for all $\vartheta < \kappa$, and κ has the **strong partition property** if $\kappa \rightarrow \kappa^{\kappa}$.

Our purpose is to prove a result which shows, assuming AD, that there are many regular cardinals without the weak partition property. In contrast, a result of Steel [Ste95] shows, assuming AD + **V**=**L**($\mathbb{R}$), that every regular cardinal below Θ is measurable. It is probably true (but not proven) that every regular Suslin cardinal has the strong partition relation, so the problem hinges on the regular cardinals between Suslin cardinals. The first two of these are the even projective ordinals $\underset{\sim}{\delta}^1_2 = \aleph_2$, $\underset{\sim}{\delta}^1_4 = \aleph_{\omega+2}$, and a theorem of Martin and Paris (see corollary 13.3 of [Kec78]) gives that they have the weak partition relation. However, between $\underset{\sim}{\delta}^1_4$ and $\underset{\sim}{\delta}^1_5$ there are two additional regular cardinals, $\aleph_{\omega\cdot 2+1}$ and $\aleph_{\omega^{\omega}+1}$. In general, there are $2^n - 2$ regular cardinals strictly between $\underset{\sim}{\delta}^1_{2n}$ and $\underset{\sim}{\delta}^1_{2n+1}$. Our result here implies that these cardinals do not have the weak partition property. In fact, we show exactly what exponent partition relations they satisfy.

By a **measure** on a set (usually an ordinal) we mean a countably additive ultrafilter. Recall from AD that every ultrafilter on a set is countably additive, and so is a measure. If ν is a measure and $f\colon \mathrm{dom}(\nu) \rightarrow \mathrm{Ord}$, we write $[f]_{\nu}$ for the ordinal represented by the equivalence class of f in the ultrapower by the measure ν.

§2. **Negative Partition Results.** We say a function $f\colon \vartheta \rightarrow \mathrm{Ord}$ has **uniform cofinality** ω if there is a $f'\colon \omega\cdot\vartheta \rightarrow \mathrm{Ord}$ such that for all $\alpha < \vartheta$ $f(\alpha) = \sup_{\beta<\omega\cdot(\alpha+1)} f'(\beta)$. We say f is of the **correct type** if f is increasing, everywhere discontinuous, and of uniform cofinality ω. We similarly define f

Wadge Degrees and Projective Ordinals: The Cabal Seminar, Volume II
Edited by A. S. Kechris, B. Löwe, J. R. Steel
Lecture Notes in Logic, 37

having uniform cofinality ω_1, etc. If $\prec$ is a well-ordering, we say a function $f\colon \mathrm{dom}(\prec) \to \mathrm{Ord}$ has the correct type if f is order-preserving with respect to $\prec$, everywhere discontinuous, and of uniform cofinality ω.

We employ a useful notational convention throughout. If $\alpha \in \mathrm{Ord}$, and ν is a measure (on an ordinal), we write $\forall^*_\nu \beta\ P(\alpha(\beta))$ to mean: if $[f]_\nu = \alpha$, then $\forall^*_\nu \beta\ P(f(\beta))$. Similarly, if ν_1, ν_2 are measures and $\alpha \in \mathrm{Ord}$, $\forall^*_{\nu_1} \beta\, \forall^*_{\nu_2} \gamma\ P(\alpha(\beta, \gamma))$ abbreviates: if $[f]_{\nu_1} = \alpha$, then $\forall^*_{\nu_1} \beta$ if $[g]_{\nu_2} = f(\beta)$, then $\forall^*_{\nu_2} \gamma\ P(g(\gamma))$.

The following fact is well-known. It says that the partition property can be reformulated using c.u.b. homogeneous sets provided we restrict the "type" of the function.

Fact. For ϑ with $\vartheta = \omega \cdot \vartheta$, $\kappa \to (\kappa)^\vartheta$ iff for all partitions $\mathcal{P}\colon (\kappa)^\vartheta \to \{0, 1\}$ there is an $i \in \{0, 1\}$ and a c.u.b. $C \subseteq \kappa$ such that $\forall f\colon \vartheta \to C$ of the correct type, $\mathcal{P}(f) = i$.

Proof. Assume $\kappa \to (\kappa)^\vartheta$, and let $\mathcal{P}\colon (\kappa)^\vartheta \to \{0, 1\}$. Define $\mathcal{P}'\colon (\kappa)^{\omega\cdot\vartheta} = (\kappa)^\vartheta \to \{0, 1\}$ by: $\mathcal{P}'(f') = \mathcal{P}(f)$, where $f(\alpha) := \sup_{\beta<\omega\cdot(\alpha+1)} f'(\beta)$. If H is homogeneous for $\mathcal{P}'$, and C is the set of limit points of H, then C is as required. Conversely, given $\mathcal{P}\colon (\kappa)^\vartheta \to \{0, 1\}$, and a c.u.b. set C as in the fact, the function $h(\alpha) :=$ the $\omega \cdot (\alpha + 1)$st element of C enumerates a set H homogeneous for $\mathcal{P}$. ⊣

We will henceforth officially use the c.u.b. version of the partition property (equivalent to the original definition for $\vartheta = \omega \cdot \vartheta$).

We fix some notation we will use for the rest of this section. We assume κ is a cardinal with the strong partition property, and $\omega < \bar{\kappa} < \kappa$ is a regular cardinal. The c.u.b. filter restricted to points of cofinality $\bar{\kappa}$ defines a normal measure on κ which we denote by ν. Similarly, ν_ω, etc., denotes the ω-cofinal normal measure on κ. We will assume that the c.u.b. filter restricted to points of cofinality ω of $\bar{\kappa}$ also defines a normal measure on $\bar{\kappa}$, which we denote by $\bar{\nu}$. Let $\mathcal{M}$ denote the transitive collapse of the ultrapower of V by ν. For μ a measure, let j_μ denote the embedding into the (transitive collapse of) the ultrapower of V by μ.

We state now our main result.

Theorem 2.1. Suppose $\kappa \to (\kappa)^\kappa$, $\omega < \bar{\kappa} < \kappa$ is regular, let $\nu, \bar{\nu}$ be as above, and assume κ is closed under $j_{\bar{\nu}}$. Then $\lambda := j_\nu(\kappa)$ does not have the weak partition property, in fact, $\lambda \not\to (\lambda)^\vartheta$ for $\vartheta = [h]_\nu$ where $h(\alpha) = j_{\bar{\nu}}(\alpha)$.

The next lemma is similar to the arguments of the Martin-Paris theorem.

Lemma 2.2. Let $\lambda = j_\nu(\kappa)$, and $\vartheta < \lambda$. Fix $h\colon \kappa \to \kappa$ with $[h]_\nu = \vartheta$. Let $\prec$ be lexicographic ordering on pairs (α, β) with $\alpha < \kappa$, $\beta < h(\alpha)$. Then $\lambda \to (\lambda)^\vartheta$ iff $(*)$ for every $g\colon \vartheta \to (\lambda - \kappa)$ of the correct type, there is a $G\colon \mathrm{dom}(\prec) \to \lambda$ of the correct type with $[G]_\nu = g$.

REMARK 2.3. $(*)$ implies that $\mathcal{P}(\vartheta) \subseteq \mathcal{M}$.

PROOF. Assume first $(*)$, and let $\mathcal{P}$ be a given partition of the functions $g : \vartheta \to \lambda$ of the correct type. Let $\mathcal{P}'$ be the partition of G: $\mathrm{dom}(\prec) \to \kappa$ of the correct type defined by: $\mathcal{P}'(G) = \mathcal{P}([G]_\nu)$. Note that if G: $\mathrm{dom}(\prec) \to \kappa$ is of the correct type, then $[G]_\nu : \vartheta \to \lambda$ is of the correct type. Let $C \subseteq \kappa$ be c.u.b. and homogeneous for $\mathcal{P}'$, say for the 1 side. Let $D = j_\nu(C) \subseteq \lambda$. Easily, D is c.u.b. in λ. Suppose $g : \vartheta \to D$ is of the correct type. By $(*)$, let G: $\mathrm{dom}(\prec) \to \kappa$ be of the correct type with $[G]_\nu = g$. Since g has range in D, we easily have that $\forall^*_\nu \alpha\ \forall \beta < h(\alpha)\ G(\alpha, \beta) \in C$. By changing the value of $G(\alpha, \beta)$ for α off a ν measure one set, we may assume G has range in C. Then, $\mathcal{P}(g) = \mathcal{P}'(G) = 1$, and we are done.

Suppose now $\lambda \to (\lambda)^\vartheta$. First consider the partition $\mathcal{P}$ of functions $g : \vartheta \to \lambda$ of the correct type, with $\mathcal{P}(g) = 1$ iff $g \in \mathcal{M}$ and there is a function of the correct type representing g. Let $D \subseteq \lambda$ be c.u.b. and homogeneous for $\mathcal{P}$. Suppose D were homogeneous for the 0 side of $\mathcal{P}$. There is a c.u.b. $C \subseteq \kappa$ such that $j_\nu(C) - \kappa \subseteq D$. To see this, partition functions $f_1, f_2 \colon \kappa \to \kappa$ of the correct type with $f_1(\alpha) < f_2(\alpha) < f_1(\alpha + 1)$ according to whether $[f_2]_\nu > N_D([f_1]_\nu)$, where $N_D(\alpha) :=$ the least element of D greater than α. Using the fact that changing f_1, f_2 off a c.u.b. set does not change $[f_1]_\nu, [f_2]_\nu$, an easy argument shows that on the homogeneous side the stated property holds. If $\overline{C}$ is homogeneous for this partition, and $C = \overline{C}'$ is the set of limit points of $\overline{C}$, then $j_\nu(C) - \kappa \subseteq D$. Fix G: $\mathrm{dom}(\prec) \to C$ of the correct type. Then $g := [G]_\nu : \vartheta \to D$ is of the correct type, a contradiction. Thus, D is homogeneous for the 1 side of $\mathcal{P}$.

As above, let $C \subseteq \kappa$ be c.u.b. such that $j_\nu(C) - \kappa \subseteq D$. Suppose $g : \vartheta \to \lambda$ is of the correct type. Let $g_2 \colon \vartheta \to \lambda$ be defined by $g_2(\alpha) = g(\alpha)$th element of $j_\nu(C)$. Thus, g_2 is of the correct type with range in D. So we may let G_2: $\mathrm{dom}(\prec) \to \kappa$ be of the correct type with $[G_2]_\nu = g_2$. We may assume G_2 has range everywhere in C. Since each $g_2(\alpha)$ is an ω limit of points in $j_\nu(C)$, we may also assume that each $G_2(\alpha, \beta)$ is a limit point of C. Let G: $\mathrm{dom}(\prec) \to \kappa$ be such that $G_2(\alpha, \beta) = G(\alpha, \beta)$th element of C. It follows that G is of the correct type. Also, $[G]_\nu = g$, and we are done. ⊣

LEMMA 2.4. Assume κ is closed under $j_{\overline{\nu}}$, and let $\rho = [h]_\nu$, where $h(\alpha) = j_{\overline{\nu}}(\alpha)$. Then $\mathcal{P}(\rho) \not\subseteq \mathcal{M}$.

PROOF. Let $A = \{\alpha : \kappa < \alpha < \rho \wedge \mathrm{cf}(\alpha) = \overline{\kappa}\}$. Suppose $A \in \mathcal{M}$. Let $A = [F]_\nu$, and we may assume $F(\alpha) \subseteq h(\alpha) - \alpha$ for all $\alpha < \kappa$. Consider the partition $\mathcal{P}$: we partition $u \colon \overline{\kappa} \to \kappa$ which are increasing, everywhere discontinuous, and with $u(\beta)$ of uniform cofinality β, according to whether $[u]_{\overline{\nu}} \in F(\sup(u))$.

Suppose first that on the homogeneous side of $\mathcal{P}$ the stated property fails. Let $\overline{C} \subseteq \kappa$ be homogeneous for the contrary side, and $C = (\overline{C})'$. For $\alpha \in C$

of cofinality $\overline{\kappa}$, define $f(\alpha)$ by: $\forall^*_{\overline{\nu}}\beta\ f(\alpha)(\beta) =$ the βth element of $\overline{C} > \alpha(\beta)$. For such α and $\gamma < \overline{\kappa}$, define $f^\gamma(\alpha)$ by: $\forall^*_{\overline{\nu}}\beta\ f^\gamma(\alpha)(\beta) =$ the γth element of $\overline{C} > \alpha(\beta)$. By the normality of $\overline{\nu}$, $\sup_{\gamma<\overline{\kappa}} f^\gamma(\alpha) = f(\alpha)$. Thus, $[f]_\nu$ has cofinality $\overline{\kappa}$, and hence $\forall^*_\nu\alpha\ f(\alpha) \in F(\alpha)$. However, $\forall^*_\nu\alpha$, $f(\alpha) = [u]_{\overline{\nu}}$ where $u\colon \overline{\kappa} \to \overline{C}$ is increasing, discontinuous, and $u(\beta)$ has uniform cofinality β. Thus, $f(\alpha) = [u]_{\overline{\nu}} \notin F(\alpha)$, a contradiction.

Next, suppose that on the homogeneous side of $\mathcal{P}$ the stated property holds. Let $\overline{C}$ be homogeneous, and $C = (\overline{C})'$. Fix $k\colon \kappa \to C$ increasing, discontinuous, with $k(\alpha)$ of uniform cofinality α. Let ℓ be defined on pairs (α, β) with $\beta < \alpha$ such that $k(\alpha) = \sup_{\beta<\alpha} \ell(\alpha, \beta)$. For $\alpha \in C$ of cofinality $\overline{\kappa}$ closed under k, define $f(\alpha)$ by: $\forall^*_{\overline{\nu}}\beta\ f(\alpha)(\beta) = k(\alpha(\beta))$. We claim that $f(\alpha)$ has uniform cofinality α. To see this, define for α as above, and $\gamma < \alpha$, $f_2(\alpha, \gamma)$ by: $\forall^*_{\overline{\nu}}\beta\ f_2(\alpha, \gamma)(\beta) = \ell(\alpha(\beta), \gamma)$. If $\eta < f(\alpha)$, then $\forall^*_{\overline{\nu}}\beta\ \exists\gamma < \alpha(\beta)\ \eta(\beta) < \ell(\alpha(\beta), \gamma)$. Thus, we have that $\exists\gamma < \alpha\ \forall^*_{\overline{\nu}}\beta\ \eta(\beta) < \ell(\alpha(\beta), \gamma)$. So, $\exists\gamma < \alpha\ \eta < f_2(\alpha, \gamma)$. Thus, $f(\alpha) = \sup_{\gamma<\alpha} f_2(\alpha, \gamma)$, and proves the claim. It follows that $\mathrm{cf}([f]_\nu) = \kappa$, so $[f]_\nu \notin A$. Thus, $\forall^*_\nu\alpha\ f(\alpha) \notin F(\alpha)$. Fix for the moment $\alpha < \kappa$ of cofinality $\overline{\kappa}$ closed under k. Let $v\colon \overline{\kappa} \to \alpha$ be increasing, continuous, and cofinal, so that $[v]_{\overline{\nu}} = \alpha$. Then $f(\alpha) = [u]_{\overline{\nu}}$, where $u(\beta) = k(v(\beta))$. Clearly u is increasing, discontinuous, cofinal in α, and has range in C. Also, $u(\beta)$ has uniform cofinality β, since $u(\beta) = k(v(\beta)) = \sup_{\gamma<v(\beta)} \ell(v(\beta), \gamma) = \sup_{\delta<\beta} \ell(v(\beta), v(\delta))$. Hence, $f(\alpha) = [u]_{\overline{\nu}} \in F(\alpha)$, a contradiction. ⊣

From Lemmas 2.2 and 2.4, Theorem 2.1 is now immediate.

Corollary 2.5 (ZF+DC+AD). The cardinals $\aleph_{\omega\cdot 2+1}$, $\aleph_{\omega^\omega+1}$ do not have the weak partition property. In fact, for $\lambda = \aleph_{\omega\cdot 2+1}$ or $\lambda = \aleph_{\omega^\omega+1}$, we have $\lambda \not\to (\lambda)^{\underset{\sim}{\delta}^1_4}$.

Proof. Let $\kappa = \underset{\sim}{\delta}^1_3$, and $\overline{\kappa} = \omega_1$ or $\overline{\kappa} = \omega_2$. From [Jac99], κ has the strong partition property, and $j_\nu(\kappa) = \aleph_{\omega\cdot 2+1}$ or $\aleph_{\omega^\omega+1}$ respectively. Also (see [Kec78]) $j_{\nu_\omega}(\kappa) = \kappa^+$, and the ω-cofinal c.u.b. filter is a normal measure $\overline{\nu}$ on $\overline{\kappa}$. Let $h(\alpha) := j_{\overline{\nu}}(\alpha)$. To finish, we show that $\kappa^+ = j_{\nu_\omega}(\kappa) = [h]_\nu$.

Define an embedding $\pi\colon j_{\nu_\omega}(\kappa) \to [h]_\nu$ as follows. If $f\colon \kappa \to \kappa$, let $\pi([f]_{\nu_\omega}) = [g]_\nu$, where for $\alpha < \kappa$ of cofinality $\overline{\kappa}$, $\forall^*_{\overline{\nu}}\beta\ g(\alpha)(\beta) = f(\alpha(\beta))$. This is easily well-defined and an embedding, so $\kappa^+ \le [h]_\nu$.

Suppose now $f\colon \kappa \to \kappa$ and $[f]_\nu < [h]_\nu$. Partition $u\colon \overline{\kappa} \to \kappa$ of the correct type according to whether $[u]_{\overline{\nu}} > f(\sup(u))$. Easily, on the homogeneous side this holds. Let $C \subseteq \kappa$ be homogeneous, and $p(\alpha) =$ the ωth element of C greater than α. Then, $\forall^*_\nu\alpha\ \forall^*_{\overline{\nu}}\beta\ f(\alpha)(\beta) < p(\alpha(\beta))$. From the Kunen analysis (see [Kec78]), there is a wellordering W on κ such that $\forall^*_{\nu_\omega}\beta < \kappa\ p(\beta) < |W{\restriction}\beta|$. Let $W{\restriction}\beta(\gamma)$ denote $W{\restriction}\beta$ restricted to ordinals which are W less than γ. Thus,

$$\forall^*_\nu \alpha < \kappa\, \forall^*_{\overline{\nu}} \beta < \overline{\kappa}\, \exists \gamma < \alpha(\beta)\ f(\alpha)(\beta) < |(W \restriction \alpha(\beta))(\gamma)|.$$

By normality of $\overline{\nu}$ and ν we have:

$$\exists \gamma < \kappa\, \forall^*_\nu \alpha < \kappa\, \forall^*_{\overline{\nu}} \beta < \overline{\kappa}\ f(\alpha)(\beta) < |W \restriction \alpha(\beta)(\gamma)|.$$

Thus, if we define a well-order $\prec$ of κ by $\gamma_1 \prec \gamma_2$ iff $\forall^*_\nu \alpha\, \forall^*_{\overline{\nu}} \beta\, |W \restriction \alpha(\beta)(\gamma_1)| < |W \restriction \alpha(\beta)(\gamma_2)|$, we have $|\prec| \geq [f]_\nu$. ⊣

Corollary 2.6 (ZF+DC+AD). There are measurable cardinals without the weak partition property.

§3. Positive Partition Results. In this section we specialize to λ within the projective hierarchy. If λ is a regular cardinal, $\underset{\sim}{\delta}^1_{2n} < \lambda < \underset{\sim}{\delta}^1_{2n+1}$, then according to Theorem 2.1 $\lambda \not\rightarrow (\lambda)^{\underset{\sim}{\delta}^1_{2n}}$. We are assuming here $\underset{\sim}{\delta}^1_{2n-1} \rightarrow (\underset{\sim}{\delta}^1_{2n-1})^{\underset{\sim}{\delta}^1_{2n-1}}$ and $\underset{\sim}{\delta}^1_{2n-1}$ is closed under $j_{\overline{\nu}}$ for all regular $\overline{\kappa} < \underset{\sim}{\delta}^1_{2n-1}$. These are facts from the projective hierarchy analysis (the reader may consult [Jac99] for the complete details below $\underset{\sim}{\delta}^1_5$). Thus, the best one could expect is $\lambda \rightarrow (\lambda)^{<\underset{\sim}{\delta}^1_{2n}}$. We show in this section that this is the case. The proof of the next lemma uses some of the general theory of descriptions, as developed in [Jac99], or [Jac88] for the general case. Since it is not feasible to review the general theory here, we merely sketch the proof and illustrate with a specific example for $n = 2$, and $\lambda = \aleph_{\omega^\omega+1}$.

We briefly recall some terminology used in the theory of descriptions. Let $<_r$ be the order on $(\omega_1)^r$ defined by:

$$(\alpha_1, \dots, \alpha_r) <_r (\beta_1, \dots, \beta_r) \leftrightarrow (\alpha_r, \alpha_1, \dots, \alpha_{r-1}) <_{\text{lex}} (\beta_r, \beta_1, \dots, \beta_{r-1}).$$

Let W^r_1 be the r-fold product of $W^1_1 :=$ the normal measure on ω_1. Let S^r_1 be the measure on $\aleph_{r+1}$ induced by the strong partition relation on ω_1, functions $h\colon \operatorname{dom}(<_r) \rightarrow \omega_1$ of the correct type, and the measure W^r_1 on $(\omega_1)^r$. Thus, S^1_1 is the ω-cofinal normal measure on ω_2. If $h\colon \operatorname{dom}(<_r) \rightarrow \omega_1$ is of the correct type, we define the **invariants** of h as follows. For $1 \leq j \leq r-1$, we define

$$h(j)(\alpha_1, \dots, \alpha_j) = \sup\{h(\alpha_1, \dots, \alpha_{j-1}, \beta_{j+1}, \dots, \beta_{r-1}, \alpha_j) : \\ \alpha_{j-1} < \beta_{j+1} < \cdots < \beta_{r-1} < \alpha_j\}.$$

[If $j = 1$ we regard α_0 as being 0 in this formula.] We also define $h(r) = h$. Similarly, for $1 \leq j \leq r-1$ we define

$$h^s(j)(\alpha_1, \dots, \alpha_j) = \sup\{h(\alpha_1, \dots, \alpha_{j-2}, \beta_{j-1}, \beta_j, \dots, \beta_{r-1}, \alpha_j) : \\ \beta_{j-1} < \alpha_{j-1}, \beta_{j-1} < \beta_j < \cdots < \beta_{r-1} < \alpha_j\}$$

The description analysis shows that every regular cardinal λ with $\underset{\sim}{\boldsymbol{\delta}}^1_{2n-1} < \lambda < \underset{\sim}{\boldsymbol{\delta}}^1_{2n+1}$ is of the form $\lambda = j_\nu(\underset{\sim}{\boldsymbol{\delta}}^1_{2n-1})$, where ν is the $\overline{\kappa}$-cofinal normal measure on $\underset{\sim}{\boldsymbol{\delta}}^1_{2n-1}$, for some regular $\overline{\kappa} < \underset{\sim}{\boldsymbol{\delta}}^1_{2n-1}$.

Lemma 3.1 (ZF+DC+AD). *Let $\lambda = j_\nu(\underset{\sim}{\boldsymbol{\delta}}^1_{2n-1})$ be a regular cardinal, $\underset{\sim}{\boldsymbol{\delta}}^1_{2n-1} < \lambda < \underset{\sim}{\boldsymbol{\delta}}^1_{2n+1}$. Then for every successor cardinal $\underset{\sim}{\boldsymbol{\delta}}^1_{2n-1} < \tau^+ \leq \lambda$, $\mathrm{cf}(\tau^+) > \underset{\sim}{\boldsymbol{\delta}}^1_{2n-1}$. Furthermore, if $\tau \leq \beta < \tau^+$, then there is a well-ordering of τ of length β in $\mathcal{M}$ (the ultrapower of V by ν).*

Remark 3.2. $\mathrm{cf}(\tau^+) > \underset{\sim}{\boldsymbol{\delta}}^1_{2n-1}$ holds for all $\underset{\sim}{\boldsymbol{\delta}}^1_{2n-1} < \tau^+ < \underset{\sim}{\boldsymbol{\delta}}^1_{2n+1}$, and can be shown without the description theory, a result of Kechris and Woodin [KW80].

Proof. We consider the case $n = 2$, $\overline{\kappa} = \omega_2$, so $\lambda = \aleph_{\omega^\omega+1}$. If $\underset{\sim}{\boldsymbol{\delta}}^1_3 < \tau^+ < \aleph_{\omega^\omega+1}$, then τ^+ is represented by a description. The reader can consult [Jac99] for the complete definition of description and associated terminology. We briefly recall here what the definitions amount to in the particular case we are considering. In this case one can see that $\tau^+ = (\mathrm{id}; d; S^r_1; W^\ell_1)$ for some integers r, ℓ, and description d. Here $\mathrm{id}\colon \underset{\sim}{\boldsymbol{\delta}}^1_3 \to \underset{\sim}{\boldsymbol{\delta}}^1_3$ is the identity function. In general, for $g\colon \underset{\sim}{\boldsymbol{\delta}}^1_3 \to \underset{\sim}{\boldsymbol{\delta}}^1_3$, $(g; d; S^r_1, W^\ell_1)$ is defined to be the ordinal represented with respect to ν (the ω_2-cofinal normal measure on $\underset{\sim}{\boldsymbol{\delta}}^1_3$) by the function which assigns to $\alpha = [f]_{S^1_1}$ (here f will be increasing, continuous, with $\sup(f) = \alpha$) the value $(g; f; d; S^r_1, W^\ell_1)$. This, in turn, is represented with respect to S^r_1 by $[h]_{W^1_r} \to (g; f; d; h; W^\ell_1)$, which is represented with respect to W^ℓ_1 by $(\gamma_1, \dots, \gamma_\ell) \to (g; f; d; h; \gamma_1, \dots, \gamma_\ell) := g(f(d; h; \gamma_1, \dots, \gamma_\ell))$. Finally, $(d; h; \gamma_1, \dots, \gamma_\ell) < \omega_2$ is represented with respect to W^1_1 by $\beta \to (d; h; \gamma_1, \dots, \gamma_\ell)(\beta)$. This last ordinal will be of the form $h(j)(\gamma_{a_1}, \dots, \gamma_{a_j}, \beta)$, or $h^s(j)(\gamma_{a_1}, \dots, \gamma_{a_j}, \beta)$ (depending on d). It is not difficult to check that this definition is well-defined.

To illustrate, if $\tau^+ = \aleph_{\omega^5\cdot 2+\omega^3\cdot 2+1}$, then $\tau^+ = (\mathrm{id}; d; S^7_1, W^8_1)$, where d is such that $(d; h, \gamma_1, \dots, \gamma_8)(\beta) = h^s_1(4)(\gamma_3, \gamma_4, \gamma_7, \beta)$.

There is a description, denoted $\mathcal{L}(d)$ in the notation of [Jac99], such that if $\delta < \tau^+$, then $\exists g\colon \underset{\sim}{\boldsymbol{\delta}}^1_3 \to \underset{\sim}{\boldsymbol{\delta}}^1_3\ \delta < (g; \mathcal{L}(d); S^r_1, W^\ell_1)$. The ordinal $(g; \mathcal{L}(d); S^r_1; W^\ell_1)$ depends only on $[g]_{\nu_\omega}$ or $[g]_{\nu_{\omega_1}}$, depending on whether $\forall^*_{S^r_1}[h]\ \forall^*_{W^\ell_1}\gamma_1, \dots, \gamma_\ell\ \mathrm{cf}\,(d; h; \gamma_1, \dots, \gamma_\ell) = \omega$ (case 1) or ω_1 (case 2). Letting g vary, we obtain a cofinal embedding from either $\aleph_{\omega+2} = j_{\nu_\omega}(\underset{\sim}{\boldsymbol{\delta}}^1_3)$ or $\aleph_{\omega\cdot 2+1} = j_{\nu_{\omega_1}}(\underset{\sim}{\boldsymbol{\delta}}^1_3)$ into τ^+.

For example, for τ^+ as above, $\mathcal{L}(d)$ is such that

$$(\mathcal{L}(d); h; \gamma_1, \dots, \gamma_8)(\beta) = h(4)(\gamma_3, \gamma_4, \gamma_6, \beta).$$

In this case, $\forall^* h\ \forall^*\gamma_1, \dots, \gamma_8\ (\mathcal{L}(d); h; \gamma_1, \dots, \gamma_8)$ has cofinality ω_1. The map $[g]_{\nu_{\omega_1}} \to (g; \mathcal{L}(d); S^r_1, W^\ell_1)$ defines a cofinal map from $\aleph_{\omega\cdot 2+1}$ to τ^+.

Suppose now $\delta < \tau^+$, and let $g\colon \underset{\sim}{\delta}^1_3 \to \underset{\sim}{\delta}^1_3$ be such that $\delta < (g; \mathcal{L}(d); S^r_1, W^\ell_1)$. There is a wellfounded tree T on $\underset{\sim}{\delta}^1_3$ such that $\forall^*_{\nu_\omega} \alpha < \underset{\sim}{\delta}^1_3\ g(\alpha) < |T\restriction\alpha|$, and $\forall^*_{\nu_{\omega_1}} \alpha\ g(\alpha) < |T\restriction \sup_m(j_{W^m_1}(\alpha))|$. The first part is due to Kunen, and the second to Martin (who proved also a similar result for ν_{ω_2}, replacing W^m_1 by S^m_1; see [Jac99]). Fix such a tree T. T defines, in $\mathcal{M}$, a wellordering of $\tau = (\mathrm{id}; \mathcal{L}(d); S^r_1, W^\ell_1)$ in case 1, or of $\tau = (\sup_m j_{W^m_1}; \mathcal{L}(d); S^r_1, W^\ell_1)$, in case 2, of length $\geq \delta$. Here $\sup_m j_{W^m_1}$ denotes the function $\alpha \mapsto \sup_m j_{W^m_1}(\alpha)$.

For example, in case 2, for $\alpha = [f]_{S^1_1} < \underset{\sim}{\delta}^1_3$ of cofinality ω_2 and sufficiently closed, we define a wellordering $<_\alpha$ of $\Omega_\alpha \doteq (\sup_m j_{W^m_1}; f; \mathcal{L}(d); S^r_1, W^\ell_1)(\alpha)$ as follows (note that $\Omega_\alpha < \underset{\sim}{\delta}^1_3$ and depends only on α, not on the choice of f, and we may assume f is increasing and continuous with $\sup(f) = \alpha$). For $\eta_1, \eta_2 < \Omega_\alpha$, define:

$$\eta_1 <_\alpha \eta_2 \leftrightarrow \forall^*[h]_{W^r_1}\ \forall^*\gamma_1, \dots, \gamma_\ell\ |(T\restriction\Omega([h], \vec{\gamma}))(\eta_1([h], \vec{\gamma}))| < |(T\restriction\Omega([h], \vec{\gamma}))(\eta_2([h], \vec{\gamma}))|.$$

The key point is that this computation can be done entirely locally, that is, just using $T\restriction\alpha$. Then $[\alpha \mapsto <_\alpha]_\nu$ is the required well-ordering. ⊣

THEOREM 3.3 (ZF+DC+AD). Let $\underset{\sim}{\delta}^1_{2n} \leq \lambda < \underset{\sim}{\delta}^1_{2n+1}$ be a regular cardinal. Then $\lambda \to (\lambda)^{<\underset{\sim}{\delta}^1_{2n}}$.

PROOF. Fix $\vartheta < \underset{\sim}{\delta}^1_{2n}$. We verify $(*)$ of Lemma 2.2. We first show that any $g\colon \vartheta \to \lambda$ is in $\mathcal{M}$. Let $\tau \leq \lambda$ be the least cardinal such that $\exists g\colon \vartheta \to \tau$ with $g \notin \mathcal{M}$. Easily, $\tau \geq \underset{\sim}{\delta}^1_{2n}$ (using Lemma 3.1 and the fact that every subset of $\underset{\sim}{\delta}^1_3$ is in $\mathcal{M}$). If τ is a successor, then From Lemma 3.1 we may assume $\mathrm{ran}(g)$ is bounded in τ. By Lemma 3.1 again, there is a bijection between $\sup(g)$ and τ^- in $\mathcal{M}$. This produces a $g'\colon \vartheta \to \tau^-$ not in $\mathcal{M}$, a contradiction. Since $\mathcal{M}$ is closed under ω sequences, τ is not a limit cardinal (all the projective ordinals $\underset{\sim}{\delta}^1_n$ are below $\aleph_{\omega_1}$), and we are done.

Let $[h]_\nu = \vartheta$, and $\prec$ be lexicographic ordering on pairs (α, β) with $\alpha < \underset{\sim}{\delta}^1_{2n-1}$, $\beta < h(\alpha)$. We next show that if $g\colon \vartheta \to \tau \leq \lambda$ has uniform cofinality ω, then there is a $G\colon \mathrm{dom}(\prec) \to \underset{\sim}{\delta}^1_{2n-1}$ of uniform cofinality ω with $[G]_\nu = g$. Let τ again be a minimal counterexample, and assume $g\colon \vartheta \to \tau$. Easily $\tau \geq \underset{\sim}{\delta}^1_{2n}$, and τ is a successor cardinal. Fix $G\colon \mathrm{dom}(\prec) \to \underset{\sim}{\delta}^1_{2n-1}$ with $[G]_\nu = g$. Let $[t]_\nu = \tau^-$, and $[W]_\nu$ be a well-ordering of τ^- of length at least $\sup(g)$. We may assume that for all $\alpha < \underset{\sim}{\delta}^1_{2n-1}$ that $W(\alpha)$ is a wellordering of $t(\alpha)$ of length at least $\sup_{\beta<h(\alpha)} G(\alpha, \beta)$. Define $F\colon \mathrm{dom}(\prec) \to \underset{\sim}{\delta}^1_{2n-1}$ by:

$$F(\alpha, \beta) = \text{ the least } \delta < t(\alpha) \text{ such that } \{|\delta'|_{W(\alpha)}\colon \delta' < \delta\} \text{ is cofinal in } G(\alpha, \beta).$$

If F has uniform cofinality ω, then so does G. To see this, suppose that $F'\colon \operatorname{dom}(\prec) \times \omega \to \underset{\sim}{\delta}^1_{2n-1}$ is such that $F(\alpha, \beta) = \sup_n F'(\alpha, \beta, n)$ for all α, β. Then define $G'(\alpha, \beta, n) = \sup\{|\delta|_{W(\alpha)} : \delta < G'(\alpha, \beta, n)\}$. Easily, $G(\alpha, \beta) = \sup_n G'(\alpha, \beta, n)$ for all α, β. To see F has uniform cofinality ω, it suffices, by minimality of τ to show that $f := [F]_\nu\colon \vartheta \to \tau^-$ has uniform cofinality ω. Let $g'\colon \vartheta \times \omega \to \tau$ induce g, that is, $g(\gamma) = \sup_n g'(\gamma, n)$. Define then $f'\colon \vartheta \times \omega \to \tau^-$ as follows. If $g'(\gamma, n) = [u]_\nu$, then set $f'(\gamma, n) = [v]_\nu$, where $v(\alpha) :=$ least δ such that $\sup\{|\delta'|_{W(\alpha)}\colon \delta' < \delta\} \geq u(\alpha)$, if one exists, and otherwise $v(\alpha) = 0$. Easily, $\sup_n f'(\alpha, n) = f(\alpha)$ for all α.

Finally, if $g\colon \vartheta \to (\lambda - \underset{\sim}{\delta}^1_{2n-1})$ is of the correct type, let $G\colon \operatorname{dom}(\prec) \to \underset{\sim}{\delta}^1_{2n-1}$ have uniform cofinality ω with $[G]_\nu = g$. Then for ν almost all α, $G(\alpha)$ is increasing and discontinuous from $h(\alpha)$ to $\underset{\sim}{\delta}^1_{2n-1}$. Changing G off a ν measure one set, we may assume G is of the correct type. ⊣

REFERENCES

Stephen Jackson

[Jac88] AD *and the projective ordinals*, this volume, originally published in Kechris et al. [Cabal iv], pp. 117–220.

[Jac99] ***A computation of*** $\underset{\sim}{\delta}^1_5$, vol. 140, Memoirs of the AMS, no. 670, American Mathematical Society, July 1999.

Alexander S. Kechris

[Kec78] AD *and projective ordinals*, this volume, originally published in Kechris and Moschovakis [Cabal i], pp. 91–132.

Alexander S. Kechris, Benedikt Löwe, and John R. Steel

[Cabal I] *Games, scales, and Suslin cardinals: the Cabal seminar, volume I*, Lecture Notes in Logic, vol. 31, Cambridge University Press, 2008.

Alexander S. Kechris, Donald A. Martin, and John R. Steel

[Cabal iv] *Cabal seminar* 81–85, Lecture Notes in Mathematics, no. 1333, Berlin, Springer, 1988.

Alexander S. Kechris and Yiannis N. Moschovakis

[Cabal i] *Cabal seminar* 76–77, Lecture Notes in Mathematics, no. 689, Berlin, Springer, 1978.

Alexander S. Kechris and W. Hugh Woodin

[KW80] *Generic codes for uncountable ordinals, partition properties, and elementary embeddings*, circulated manuscript, 1980, reprinted in [Cabal I], p. 379–397.

Yiannis N. Moschovakis

[Mos80] ***Descriptive set theory***, Studies in Logic and the Foundations of Mathematics, no. 100, North-Holland, Amsterdam, 1980.

John R. Steel

[Ste95] ***HOD***$^{\mathbf{L}(\mathbb{R})}$ *is a core model below* Θ, ***The Bulletin of Symbolic Logic***, vol. 1 (1995), pp. 75–84.

DEPARTMENT OF MATHEMATICS
UNIVERSITY OF NORTH TEXAS
P.O. BOX 311430
DENTON, TEXAS 76203-1430
UNITED STATES OF AMERICA
E-mail: jackson@unt.edu

BIBLIOGRAPHY

JOHN W. ADDISON

[Add54] *On Certain Points of the Theory of Recursive Functions*, ***Ph.D. thesis***, University of Wisconsin–Madison, 1954.

[Add04] *Tarski's theory of definability: common themes in descriptive set theory, recursive function theory, classical pure logic, and finite-universe logic*, ***Annals of Pure and Applied Logic***, vol. 126 (2004), no. 1-3, pp. 77–92.

JOHN W. ADDISON AND YIANNIS N. MOSCHOVAKIS

[AM68] *Some consequences of the axiom of definable determinateness*, ***Proceedings of the National Academy of Sciences of the United States of America***, no. 59, 1968, pp. 708–712.

ALESSANDRO ANDRETTA

[And03] *Equivalence between Wadge and Lipschitz determinacy*, ***Annals of Pure and Applied Logic***, vol. 123 (2003), no. 1–3, pp. 163–192.

[And06] *More on Wadge determinacy*, ***Annals of Pure and Applied Logic***, vol. 144 (2006), no. 1–3, pp. 2–32.

ALESSANDRO ANDRETTA, GREGORY HJORTH, AND ITAY NEEMAN

[AHN07] *Effective cardinals of boldface pointclasses*, ***Journal of Mathematical Logic***, vol. 7 (2007), no. 1, pp. 35–92.

ALESSANDRO ANDRETTA AND DONALD A. MARTIN

[AM03] *Borel-Wadge degrees*, ***Fundamenta Mathematicae***, vol. 177 (2003), no. 2, pp. 175–192.

ROBERT J. AUMANN AND LLOYD S. SHAPLEY

[AS74] ***Values of non-atomic games***, Princeton University Press, 1974.

JOHN F. BARNES

[Bar65] *The classification of the closed-open and the recursive sets of number theoretic functions*, ***Ph.D. thesis***, UC Berkeley, 1965.

HOWARD S. BECKER

[Bec85] *A property equivalent to the existence of scales*, ***Transactions of the American Mathematical Society***, vol. 287 (1985), pp. 591–612.

HOWARD S. BECKER AND ALEXANDER S. KECHRIS

[BK96] ***The descriptive set theory of Polish group actions***, London Mathematical Society Lecture Note Series, vol. 232, Cambridge University Press, Cambridge, 1996.

J. BURGESS AND D. MILLER

[BM75] *Remarks on invariant descriptive set theory*, ***Fundamenta Mathematicae***, vol. 90 (1975), pp. 53–75.

Wadge Degrees and Projective Ordinals: The Cabal Seminar, Volume II
Edited by A. S. Kechris, B. Löwe, J. R. Steel
Lecture Notes in Logic, 37

CHEN-LIAN CHUANG

[Chu82] *The propagation of scales by game quantifiers*, ***Ph.D. thesis***, UCLA, 1982.

MORTON DAVIS

[Dav64] *Infinite games of perfect information*, ***Advances in game theory*** (Melvin Dresher, Lloyd S. Shapley, and Alan W. Tucker, editors), Annals of Mathematical Studies, vol. 52, 1964, pp. 85–101.

ACHIM DITZEN

[Dit92] *Definable equivalence relations on Polish spaces*, ***Ph.D. thesis***, California Institute of Technology, 1992.

JACQUES DUPARC

[Dup01] *Wadge hierarchy and Veblen hierarchy. I. Borel sets of finite rank*, ***The Journal of Symbolic Logic***, vol. 66 (2001), no. 1, pp. 56–86.

[Dup03] *A hierarchy of deterministic context-free ω-languages*, ***Theoretical Computer Science***, vol. 290 (2003), no. 3, pp. 1253–1300.

JACQUES DUPARC, OLIVIER FINKEL, AND JEAN-PIERRE RESSAYRE

[DFR01] *Computer science and the fine structure of Borel sets*, ***Theoretical Computer Science***, vol. 257 (2001), no. 1–2, pp. 85–105.

HARVEY FRIEDMAN

[Fri71A] *Determinateness in the low projective hierarchy*, ***Fundamenta Mathematicae***, vol. 72 (1971), pp. 79–95.

[Fri71B] *Higher set theory and mathematical practice*, ***Annals of Mathematical Logic***, vol. 2 (1971), no. 3, pp. 325–357.

HARVEY FRIEDMAN AND LEE STANLEY

[FS89] *A Borel reducibility theory for classes of countable structures*, ***The Journal of Symbolic Logic***, vol. 54 (1989), no. 3, pp. 894–914.

LEO A. HARRINGTON

[Har78] *Analytic determinacy and $0^\#$*, ***The Journal of Symbolic Logic***, vol. 43 (1978), pp. 685–693.

LEO A. HARRINGTON AND ALEXANDER S. KECHRIS

[HK81] *On the determinacy of games on ordinals*, ***Annals of Mathematical Logic***, vol. 20 (1981), pp. 109–154.

LEO A. HARRINGTON, ALEXANDER S. KECHRIS, AND ALAIN LOUVEAU

[HKL90] *A Glimm–Effros dichotomy for Borel equivalence relations*, ***Journal of the American Mathematical Society***, vol. 3 (1990), pp. 902–928.

LEO A. HARRINGTON AND RAMEZ-LABIB SAMI

[HS79] *Equivalence relations, projective and beyond*, ***Logic Colloquium '78. Proceedings of the Colloquium held in Mons, August 24–September 1, 1978*** (Maurice Boffa, Dirk van Dalen, and Kenneth McAloon, editors), Studies in Logic and the Foundations of Mathematics, vol. 97, North-Holland, Amsterdam, 1979, pp. 247–264.

FELIX HAUSDORFF

[Hau57] ***Set theory***, Chelsea, New York, 1957, translated by J. R. Aumann.

GREGORY HJORTH

[Hjo95] *A dichotomy for the definable universe*, ***The Journal of Symbolic Logic***, vol. 60 (1995), no. 4, pp. 1199–1207.

[Hjo96] Π^1_2 *Wadge degrees*, ***Annals of Pure and Applied Logic***, vol. 77 (1996), no. 1, pp. 53–74.

[Hjo98] *An absoluteness principle for Borel sets*, ***The Journal of Symbolic Logic***, vol. 63 (1998), no. 2, pp. 663–693.

[Hjo01] *A boundedness lemma for iterations*, ***The Journal of Symbolic Logic***, vol. 66 (2001), no. 3, pp. 1058–1072.

[Hjo02] *Cardinalities in the projective hierarchy*, ***The Journal of Symbolic Logic***, vol. 67 (2002), no. 4, pp. 1351–1372.

Gregory Hjorth and Alexander S. Kechris

[HK01] *Recent developments in the theory of Borel reducibility*, ***Fundamenta Mathematicae***, vol. 170 (2001), no. 1–2, pp. 21–52.

Stephen Jackson

[Jac88] AD *and the projective ordinals*, this volume, originally published in Kechris et al. [Cabal iv], pp. 117–220.

[Jac90A] *A new proof of the strong partition relation on* ω_1, ***Transactions of the American Mathematical Society***, vol. 320 (1990), no. 2, pp. 737–745.

[Jac90B] *Partition properties and well-ordered sequences*, ***Annals of Pure and Applied Logic***, vol. 48 (1990), no. 1, pp. 81–101.

[Jac91] *Admissible Suslin cardinals in* $\mathrm{L}(\mathbb{R})$, ***The Journal of Symbolic Logic***, vol. 56 (1991), no. 1, pp. 260–275.

[Jac99] ***A computation of*** $\boldsymbol{\utilde{\delta}^1_5}$, vol. 140, Memoirs of the AMS, no. 670, American Mathematical Society, July 1999.

[Jac08] *Suslin cardinals, partition properties, homogeneity. Introduction to Part II*, in Kechris et al. [Cabal I], pp. 273–313.

[Jac10] *Structural consequences of* AD, in Kanamori and Foreman [KF10], pp. 1753–1876.

[Jac11] *Projective ordinals. Introduction to Part IV*, 2011, this volume.

Stephen Jackson and Farid Khafizov

[JK] *Descriptions and cardinals below* $\utilde{\delta}^1_5$, in submission.

Stephen Jackson and Benedikt Löwe

[JL] *Canonical measure assignments*, in submission.

Thomas J. Jech

[Jec71] ***Lectures in set theory, with particular emphasis on the method of forcing***, Lecture Notes in Mathematics, Vol. 217, Springer-Verlag, Berlin, 1971.

Thomas John

[Joh86] *Recursion in Kolmogorov's R-operator and the ordinal* σ_3, ***The Journal of Symbolic Logic***, vol. 51 (1986), no. 1, pp. 1–11.

Akihiro Kanamori and Matthew Foreman

[KF10] ***Handbook of set theory***, Springer, 2010.

L. Kantorovich and E. Livenson

[KL32] *Memoir on the analytical operations and projective sets I*, ***Fundamenta Mathematicae***, vol. 18 (1932), pp. 214–279.

Alexander S. Kechris

[Kec73] *Measure and category in effective descriptive set theory*, ***Annals of Mathematical Logic***, vol. 5 (1973), no. 4, pp. 337–384.

[Kec74] *On projective ordinals*, ***The Journal of Symbolic Logic***, vol. 39 (1974), pp. 269–282.

[Kec75] *The theory of countable analytical sets*, ***Transactions of the American Mathematical Society***, vol. 202 (1975), pp. 259–297.

[Kec77A] AD *and infinite exponent partition relations*, circulated manuscript, 1977.

[Kec77B] *Classifying projective-like hierarchies*, ***Bulletin of the Greek Mathematical Society***, vol. 18 (1977), pp. 254–275.

[Kec78] AD *and projective ordinals*, this volume, originally published in Kechris and Moschovakis [CABAL i], pp. 91–132.

[Kec81A] *Homogeneous trees and projective scales*, this volume, originally published in Kechris et al. [CABAL ii], pp. 33–74.

[Kec81B] *Suslin cardinals, κ-Suslin sets, and the scale property in the hyperprojective hierarchy*, in Kechris et al. [CABAL ii], pp. 127–146, reprinted in [CABAL I], p. 314–332.

[Kec92] *The structure of Borel equivalence relations in Polish spaces*, ***Set theory of the continuum. Papers from the workshop held in Berkeley, California, October 16–20, 1989*** (H. Judah, W. Just, and H. Woodin, editors), Mathematical Sciences Research Institute Publications, vol. 26, Springer, New York, 1992, pp. 89–102.

[Kec94] ***Classical descriptive set theory***, Graduate Texts in Mathematics, vol. 156, Springer, 1994.

ALEXANDER S. KECHRIS, EUGENE M. KLEINBERG, YIANNIS N. MOSCHOVAKIS, AND W. HUGH WOODIN

[KKMW81] *The axiom of determinacy, strong partition properties, and nonsingular measures*, in Kechris et al. [CABAL ii], pp. 75–99, reprinted in [CABAL I], p. 333–354.

ALEXANDER S. KECHRIS, BENEDIKT LÖWE, AND JOHN R. STEEL

[CABAL I] *Games, scales, and Suslin cardinals: the Cabal seminar, volume I*, Lecture Notes in Logic, vol. 31, Cambridge University Press, 2008.

ALEXANDER S. KECHRIS AND DONALD A. MARTIN

[KM78] *On the theory of* Π^1_3 *sets of reals*, ***Bulletin of the American Mathematical Society***, vol. 84 (1978), no. 1, pp. 149–151.

ALEXANDER S. KECHRIS, DONALD A. MARTIN, AND YIANNIS N. MOSCHOVAKIS

[CABAL ii] *Cabal seminar* 77–79, Lecture Notes in Mathematics, no. 839, Berlin, Springer, 1981.

[CABAL iii] *Cabal seminar* 79–81, Lecture Notes in Mathematics, no. 1019, Berlin, Springer, 1983.

ALEXANDER S. KECHRIS, DONALD A. MARTIN, AND JOHN R. STEEL

[CABAL iv] *Cabal seminar* 81–85, Lecture Notes in Mathematics, no. 1333, Berlin, Springer, 1988.

ALEXANDER S. KECHRIS AND YIANNIS N. MOSCHOVAKIS

[KM72] *Two theorems about projective sets*, ***Israel Journal of Mathematics***, vol. 12 (1972), pp. 391–399.

[CABAL i] *Cabal seminar* 76–77, Lecture Notes in Mathematics, no. 689, Berlin, Springer, 1978.

[KM78B] *Notes on the theory of scales*, in *Cabal Seminar* 76–77 [CABAL i], pp. 1–53, reprinted in [CABAL I], p. 28–74.

ALEXANDER S. KECHRIS, ROBERT M. SOLOVAY, AND JOHN R. STEEL

[KSS81] *The axiom of determinacy and the prewellordering property*, this volume, originally published in Kechris et al. [CABAL ii], pp. 101–125.

ALEXANDER S. KECHRIS AND W. HUGH WOODIN

[KW80] *Generic codes for uncountable ordinals, partition properties, and elementary embeddings*, circulated manuscript, 1980, reprinted in [CABAL I], p. 379–397.

STEPHEN C. KLEENE

[Kle50] *A symmetric form of Gödel's theorem*, ***Indagationes Mathematicae***, vol. 12 (1950), pp. 244–246.

EUGENE M. KLEINBERG

[Kle70] *Strong partition properties for infinite cardinals*, ***The Journal of Symbolic Logic***, vol. 35 (1970), pp. 410–428.

KENNETH KUNEN

[Kun71A] *Measurability of* $\underset{\sim}{\delta}^1_n$, circulated note, April 1971.

[Kun71B] *On* $\underset{\sim}{\delta}^1_5$, circulated note, August 1971.

[Kun71C] *A remark on Moschovakis' uniformization theorem*, circulated note, March 1971.

[Kun71D] *Some singular cardinals*, circulated note, September 1971.

[Kun71E] *Some more singular cardinals*, circulated note, September 1971.

CASIMIR KURATOWSKI

[Kur58] ***Topologie. Vol. I***, 4ème ed., Monografie Matematyczne, vol. 20, Państwowe Wydawnictwo Naukowe, Warsaw, 1958.

KAZIMIERZ KURATOWSKI

[Kur66] ***Topology***, vol. 1, Academic Press, New York and London, 1966.

RICHARD LAVER

[Lav71] *On Fraïssé's order type conjecture*, ***Annals of Mathematics***, vol. 93 (1971), pp. 89–111.

ALAIN LOUVEAU

[Lou80] *A separation theorem for* Σ^1_1 *sets*, ***Transactions of the American Mathematical Society***, vol. 260 (1980), no. 2, pp. 363–378.

[Lou83] *Some results in the Wadge hierarchy of Borel sets*, this volume, originally published in Kechris et al. [CABAL iii], pp. 28–55.

[Lou92] *Classifying Borel structures*, ***Set Theory of the Continuum. Papers from the workshop held in Berkeley, California, October 16–20, 1989*** (H. Judah, W. Just, and H. Woodin, editors), Mathematical Sciences Research Institute Publications, vol. 26, Springer, New York, 1992, pp. 103–112.

[Lou94] *On the reducibility order between Borel equivalence relations*, ***Logic, Methodology and Philosophy of Science, IX. Proceedings of the Ninth International Congress held in Uppsala, August 7–14, 1991*** (Dag Prawitz, Brian Skyrms, and Dag Westerståhl, editors), Studies in Logic and the Foundations of Mathematics, vol. 134, North-Holland, Amsterdam, 1994.

ALAIN LOUVEAU AND JEAN SAINT-RAYMOND

[LSR87] *Borel classes and closed games: Wadge-type and Hurewicz-type results*, ***Transactions of the American Mathematical Society***, vol. 304 (1987), no. 2, pp. 431–467.

[LSR88A] *Les propriétés de réduction et de norme pour les classes de Boréliens*, ***Fundamenta Mathematicae***, vol. 131 (1988), no. 3, pp. 223–243.

[LSR88B] *The strength of Borel Wadge determinacy*, this volume, originally published in Kechris et al. [CABAL iv], pp. 1–30.

[LSR90] *On the quasi-ordering of Borel linear orders under embeddability*, ***The Journal of Symbolic Logic***, vol. 55 (1990), no. 2, pp. 537–560.

NIKOLAI LUZIN

[Luz30] ***Leçons sur les ensembles analytiques et leurs applications***, Collection de monographies sur la théorie des fonctions, Gauthier-Villars, Paris, 1930.

NIKOLAI LUZIN AND WACLAW SIERPIŃSKI

[LS29] *Sur les classes des constituantes d'un complémentaire analytique*, ***Comptes rendus hebdomadaires des séances de l'Académie des Sciences***, vol. 189 (1929), pp. 794–796.

RICHARD MANSFIELD

[Man71] *A Souslin operation on* Π^1_2, ***Israel Journal of Mathematics***, vol. 9 (1971), no. 3, pp. 367–379.

RICHARD MANSFIELD AND GALEN WEITKAMP

[MW85] ***Recursive aspects of descriptive set theory***, Oxford Logic Guides, vol. 11, The Clarendon Press Oxford University Press, New York, 1985, With a chapter by Stephen Simpson.

DONALD A. MARTIN

[Mar] AD *and the normal measures on* $\underset{\sim}{\delta}^1_3$, unpublished.

[Mar68] *The axiom of determinateness and reduction principles in the analytical hierarchy*, ***Bulletin of the American Mathematical Society***, vol. 74 (1968), pp. 687–689.

[Mar70] *Measurable cardinals and analytic games*, ***Fundamenta Mathematicae***, vol. 66 (1970), pp. 287–291.

[Mar71A] *Determinateness implies many cardinals are measurable*, circulated note, May 1971.

[Mar71B] *Projective sets and cardinal numbers: some questions related to the continuum problem*, this volume, originally a preprint, 1971.

[Mar73] *The Wadge degrees are wellordered*, unpublished, 1973.

[Mar75] *Borel determinacy*, ***Annals of Mathematics***, vol. 102 (1975), no. 2, pp. 363–371.

[Mar77A] α-Π^1_1 *games*, planned but unfinished paper, 1977.

[Mar77B] *On subsets of* $\underset{\sim}{\delta}^1_3$, circulated note, January 1977.

DONALD A. MARTIN AND JEFF B. PARIS

[MP71] AD $\Rightarrow \exists$ *exactly* 2 *normal measures on* ω_2, circulated note, March 1971.

DONALD A. MARTIN AND ROBERT M. SOLOVAY

[MS69] *A basis theorem for* Σ^1_3 *sets of reals*, ***Annals of Mathematics***, vol. 89 (1969), pp. 138–160.

[MS70] *Internal Cohen extensions*, ***Annals of Mathematical Logic***, vol. 2 (1970), no. 2, pp. 143–178.

DONALD A. MARTIN AND JOHN R. STEEL

[MS83] *The extent of scales in* $\mathbf{L}(\mathbb{R})$, in Kechris et al. [CABAL iii], pp. 86–96, reprinted in [CABAL I], p. 110–120.

YIANNIS N. MOSCHOVAKIS

[Mos67] *Hyperanalytic predicates*, ***Transactions of the American Mathematical Society***, vol. 129 (1967), pp. 249–282.

[Mos70] *Determinacy and prewellorderings of the continuum*, ***Mathematical logic and foundations of set theory. Proceedings of an international colloquium held under the auspices of the Israel Academy of Sciences and Humanities, Jerusalem, 11–14 November 1968*** (Y. Bar-Hillel, editor), Studies in Logic and the Foundations of Mathematics, North-Holland, Amsterdam-London, 1970, pp. 24–62.

[Mos71] *Uniformization in a playful universe*, ***Bulletin of the American Mathematical Society***, vol. 77 (1971), pp. 731–736.

[Mos74] ***Elementary induction on abstract structures***, North-Holland, 1974.

[Mos80] ***Descriptive set theory***, Studies in Logic and the Foundations of Mathematics, no. 100, North-Holland, Amsterdam, 1980.

LUCA MOTTO ROS

[MR07] *General reducibilities for sets of reals*, ***Ph.D. thesis***, Politecnico di Torino, 2007.

JAN MYCIELSKI

[Myc64] *On the axiom of determinateness*, ***Fundamenta Mathematicae***, vol. 53 (1964), pp. 205–224.

JAN MYCIELSKI AND HUGO STEINHAUS

[MS62] *A mathematical axiom contradicting the axiom of choice*, ***Bulletin de l'Académie Polonaise des Sciences***, vol. 10 (1962), pp. 1–3.

JOHN C. OXTOBY

[Oxt71] ***Measure and category***, Springer, 1971.

NEIL ROBERTSON AND P. D. SEYMOUR

[RS04] *Graph minors. XX. Wagner's conjecture*, ***Journal of Combinatorial Theory. Series B***, vol. 92 (2004), no. 2, pp. 325–357.

HARTLEY ROGERS

[Rog59] *Computing degrees of unsolvability*, ***Mathematische Annalen***, vol. 138 (1959), pp. 125–140.

JEAN SAINT-RAYMOND

[SR76] *Fonctions boréliennes sur un quotient*, ***Bulletin des Sciences Mathématiques***, vol. 100 (1976), pp. 141–147.

JOSEPH R. SHOENFIELD

[Sho61] *The problem of predicativity*, ***Essays on the foundations of mathematics*** (Yehoshua Bar-Hillel, E. I. J. Poznanski, Michael O. Rabin, and Abraham Robinson, editors), Magnes Press, Jerusalem, 1961, pp. 132–139.

[Sho67] ***Mathematical logic***, Addison-Wesley, 1967.

WACLAW SIERPIŃSKI

[Sie52] ***General topology***, Mathematical Expositions, No. 7, University of Toronto Press, Toronto, 1952, Translated by C. Cecilia Krieger.

ROMAN SIKORSKI

[Sik58] *Some examples of Borel sets*, ***Colloquium Mathematicum***, vol. 5 (1958), pp. 170–171.

JACK SILVER

[Sil80] *Counting the number of equivalence classes of Borel and coanalytic equivalence relations*, ***Annals of Mathematical Logic***, vol. 18 (1980), no. 1, pp. 1–28.

JOHN SIMMS

[Sim79] *Semihypermeasurables and* $\Pi^0_1(\Pi^1_1)$ *games*, ***Ph.D. thesis***, Rockefeller University, 1979.

ROBERT M. SOLOVAY

[Sol67A] *Measurable cardinals and the axiom of determinateness*, lecture notes prepared in connection with the Summer Institute of Axiomatic Set Theory held at UCLA, Summer 1967.

[Sol67B] *A nonconstructible* Δ^1_3 *set of integers*, ***Transactions of the American Mathematical Society***, vol. 127 (1967), no. 1, pp. 50–75.

[Sol78A] *A* Δ^1_3 *coding of the subsets of* ω_ω, this volume, originally published in Kechris and Moschovakis [CABAL i], pp. 133–150.

[Sol78B] *The independence of* DC *from* AD, in Kechris and Moschovakis [CABAL i], pp. 171–184.

JOHN R. STEEL

[Ste77] *Determinateness and subsystems of analysis*, ***Ph.D. thesis***, Berkeley, 1977.

[Ste80] *Analytic sets and Borel isomorphisms*, ***Fundamenta Mathematicae***, vol. 108 (1980), no. 2, pp. 83–88.

[Ste81A] *Closure properties of pointclasses*, this volume, originally published in Kechris et al. [CABAL ii], pp. 147–163.

[Ste81B] *Determinateness and the separation property*, ***The Journal of Symbolic Logic***, vol. 46 (1981), no. 1, pp. 41–44.

[Ste82] *A classification of jump operators*, ***The Journal of Symbolic Logic***, vol. 47 (1982), no. 2, pp. 347–358.

[Ste83] *Scales in* $\mathbf{L}(\mathbb{R})$, in Kechris et al. [CABAL iii], pp. 107–156, reprinted in [CABAL I], p. 130–175.

[Ste95] ***HOD***$^{\mathbf{L}(\mathbb{R})}$ *is a core model below* Θ, ***The Bulletin of Symbolic Logic***, vol. 1 (1995), pp. 75–84.

JOHN R. STEEL AND ROBERT VAN WESEP

[SVW82] *Two consequences of determinacy consistent with choice*, ***Transactions of the American Mathematical Society***, (1982), no. 272, pp. 67–85.

MIKHAIL YA. SUSLIN

[Sus17] *Sur une définition des ensembles mesurables* B *sans nombres transfinis*, ***Comptes Rendus Hebdomadaires des Séances de l'Académie des Sciences***, vol. 164 (1917), pp. 88–91.

ALFRED TARSKI

[Tar00] *Address at the Princeton University Bicentennial Conference on Problems of Mathematics* (*December* 17–19, 1946), ***The Bulletin of Symbolic Logic***, vol. 6 (2000), no. 1, pp. 1–44.

FONS VAN ENGELEN, ARNOLD W. MILLER, AND JOHN R. STEEL

[vEMS87] *Rigid Borel sets and better quasi-order theory*, ***Proceedings of the AMS-IMS-SIAM joint summer research conference on applications of mathematical logic to finite combinatorics held at Humboldt State University, Arcata, Calif., August 4–10, 1985*** (Stephen G. Simpson, editor), Contemporary Mathematics, vol. 65, American Mathematical Society, Providence, RI, 1987, pp. 199–222.

ROBERT VAN WESEP

[Van77] *Subsystems of second-order arithmetric, and descriptive set theory under the axiom of determinateness*, ***Ph.D. thesis***, University of California, Berkeley, 1977.

[Van78A] *Separation principles and the axiom of determinateness*, ***The Journal of Symbolic Logic***, vol. 43 (1978), no. 1, pp. 77–81.

[Van78B] *Wadge degrees and descriptive set theory*, this volume, originally published in Kechris and Moschovakis [CABAL i], pp. 151–170.

OSWALD VEBLEN

[Veb08] *Continuous increasing functions of finite and transfinite ordinals*, ***Transactions of the American Mathematical Society***, vol. 9 (1908), no. 3, pp. 280–292.

WILLIAM W. WADGE

[Wad84] *Reducibility and determinateness on the Baire space*, ***Ph.D. thesis***, University of California, Berkeley, 1984.

[Wad11] *Early investigations of the degrees of Borel sets*, 2011, this volume.

W. HUGH WOODIN

[Woo99] ***The axiom of determinacy, forcing axioms, and the nonstationary ideal***, De Gruyter Series in Logic and its Applications, Walter de Gruyter, Berlin, 1999.

ERNST ZERMELO

[Zer13] *Über eine Anwendung der Mengenlehre auf die Theorie des Schachspiels*, ***Proceedings of the Fifth International Congress of Mathematicians*** (E. W. Hobson and A. E. H. Love, editors), vol. 2, 1913, pp. 501–504.